SPECIAL TRIANGLES

NAME	CHARACTERISTIC	EXAMPLES
Right Triangle	Triangle has a right angle.	90°
Isosceles Triangle	Triangle has two equal sides.	$AB = BC$
Equilateral Triangle	Triangle has three equal sides.	$AB = BC = CA$
Similar Triangles	Corresponding angles are equal; corresponding sides are proportional.	$A = D, B = E, C = F$ $\frac{AB}{DE} = \frac{AC}{DF} = \frac{BC}{EF}$

ANNOTATED
INSTRUCTOR'S EDITION

INTERMEDIATE
ALGEBRA

SEVENTH EDITION

ANNOTATED INSTRUCTOR'S EDITION

INTERMEDIATE ALGEBRA

SEVENTH EDITION

MARGARET L. LIAL
AMERICAN RIVER COLLEGE

E. JOHN HORNSBY, JR.
UNIVERSITY OF NEW ORLEANS

CHARLES D. MILLER

HarperCollinsCollegePublishers

Sponsoring Editor: Karin E. Wagner
Developmental Editor: Sandi Goldstein
Project Editor: Cathy Wacaser
Design Administrator: Jess Schaal
Text and Cover Design: Lesiak/Crampton Design Inc: Lucy Lesiak
Photo Researcher: Rosemary Hunter
Production Administrator: Randee Wire
Compositor: Interactive Composition Corporation
Printer and Binder: R.R. Donnelley & Sons Company
Cover Printer: Phoenix Color Corporation

Intermediate Algebra, Seventh Edition

HarperCollins® and ⬛® are registered trademarks of HarperCollins Publishers Inc.

Library of Congress Cataloging-in-Publication Data
Lial, Margaret L.
   Intermediate algebra / Margaret L. Lial, E. John Hornsby, Jr.,
Charles D. Miller. -- 7th ed.
      p.   cm.
   Includes index.
   ISBN 0-673-99059-1 (Student Edition)
   ISBN 0-673-99540-2 (Annotated Instructor's Edition)
   1. Algebra.   I. Hornsby, E. John. II. Miller, Charles David.
   III. Title.
QA152.2.L52   1996

   512.9--dc20   95–10263
                                         CIP

95 96 97 98 9 8 7 6 5 4 3 2 1

CONTENTS

CHAPTER 6 QUADRATIC EQUATIONS AND INEQUALITIES 321

CHAPTER 7 THE STRAIGHT LINE 373

CHAPTER 8 SYSTEMS OF LINEAR EQUATIONS 451

CHAPTER 9   QUADRATIC FUNCTIONS AND THE CONIC SECTIONS   513

CHAPTER 10   INVERSE, EXPONENTIAL, AND LOGARITHMIC FUNCTIONS   587

CHAPTER 11 SEQUENCES AND SERIES 649

PREFACE

This seventh edition of *Intermediate Algebra* is designed to give a thorough treatment of those topics in algebra necessary for success in later courses. Although we assume that most students using this book will have had a previous course in algebra, all necessary ideas are introduced or reviewed as needed.

The text retains the successful features of previous editions: learning objectives for each section; careful exposition; fully developed examples; cautions and notes; and boxes that set off important definitions, formulas, rules, and procedures. In this new edition, we have made several content changes to follow the guidelines set forth in the *Curriculum and Evaluation Standards for School Mathematics,* published by the National Council of Teachers of Mathematics.

CHANGES IN CONTENT

- More than 80 percent of the exercises are new to this edition and many of these exercises now incorporate real data and graphics.
- Beginning with Chapter 2, cumulative reviews appear after each chapter, covering material learned up to that point.
- In this text we view calculators as a means of allowing students to spend more time on the conceptual nature of mathematics and less time on the mechanics of computation with paper and pencil. We have included an introduction to graphics calculators, and the use of both the scientific and graphics calculator is discussed throughout the book wherever appropriate. Some sections include exercises that require a calculator.
- Graphics calculator text and exercises are included as appropriate throughout the book. This material is designed so that it can easily be incorporated into the course, treated separately, or omitted, as the instructor chooses.

FEATURES

The following pages illustrate important features. These features are designed to assist students in the learning process and deepen their understanding of the underlying principles and interrelations between topics.

Connections boxes that provide connections to the real world or to other mathematical concepts or other disciplines open each chapter and appear throughout the text—almost one for every section. Most of these include thought-provoking questions for writing or class discussion. The Connections provide motivation for the topic under discussion, show how mathematics is used in many different aspects of life, give some historical background, and provide a larger context for the current material.

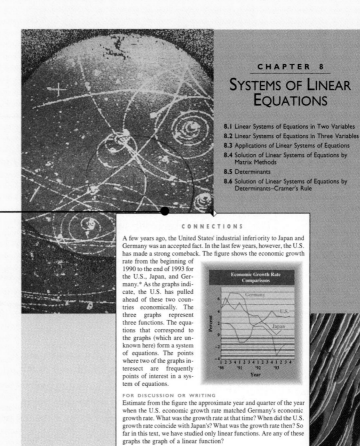

CONNECTIONS

A few years ago, the United States' industrial inferiority to Japan and Germany was an accepted fact. In the last few years, however, the U.S. has made a strong comeback. The figure shows the economic growth rate from the beginning of 1990 to the end of 1993 for the U.S., Japan, and Germany.* As the graphs indicate, the U.S. has pulled ahead of these two countries economically. The three graphs represent three functions. The equations that correspond to the graphs (which are unknown here) form a system of equations. The points where two of the graphs intersect are frequently points of interest in a system of equations.

FOR DISCUSSION OR WRITING

Estimate from the figure the approximate year and quarter of the year when the U.S. economic growth rate matched Germany's economic growth rate. What was the growth rate at that time? When did the U.S. growth rate coincide with Japan's? What was the growth rate then? So far in this text, we have studied only linear functions. Are any of these graphs the graph of a linear function?

* Graph, "The U.S. Economy Is Growing," from *The New York Times*, February 27, 1994. Copyright © 1994 by The New York Times Company. Reprinted by permission.

451

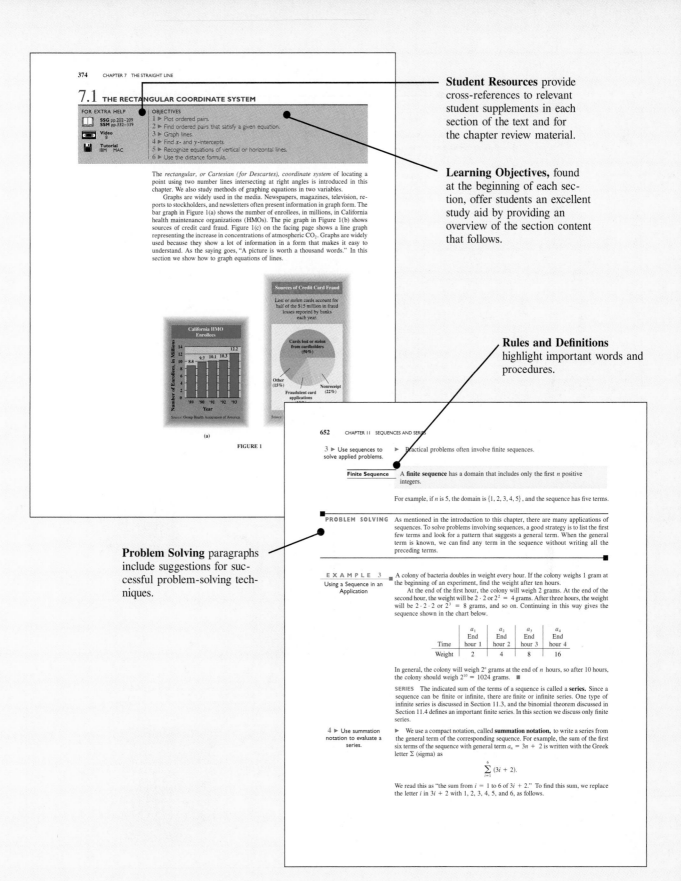

Student Resources provide cross-references to relevant student supplements in each section of the text and for the chapter review material.

Learning Objectives, found at the beginning of each section, offer students an excellent study aid by providing an overview of the section content that follows.

Rules and Definitions highlight important words and procedures.

Problem Solving paragraphs include suggestions for successful problem-solving techniques.

The content of the two textbook pages shown:

7.1 THE RECTANGULAR COORDINATE SYSTEM

FOR EXTRA HELP

SSG pp.202–209
SSM pp.332–339

Video 8

Tutorial IBM MAC

OBJECTIVES

1 ▶ Plot ordered pairs.
2 ▶ Find ordered pairs that satisfy a given equation.
3 ▶ Graph lines.
4 ▶ Find x- and y-intercepts.
5 ▶ Recognize equations of vertical or horizontal lines.
6 ▶ Use the distance formula.

The *rectangular, or Cartesian (for Descartes), coordinate system* of locating a point using two number lines intersecting at right angles is introduced in this chapter. We also study methods of graphing equations in two variables.

Graphs are widely used in the media. Newspapers, magazines, television, reports to stockholders, and newsletters often present information in graph form. The bar graph in Figure 1(a) shows the number of enrollees, in millions, in California health maintenance organizations (HMOs). The pie graph in Figure 1(b) shows sources of credit card fraud. Figure 1(c) on the facing page shows a line graph representing the increase in concentrations of atmospheric CO_2. Graphs are widely used because they show a lot of information in a form that makes it easy to understand. As the saying goes, "A picture is worth a thousand words." In this section we show how to graph equations of lines.

(a)

FIGURE 1

---

3 ▶ Use sequences to solve applied problems.

▶ Practical problems often involve finite sequences.

Finite Sequence   A **finite sequence** has a domain that includes only the first n positive integers.

For example, if n is 5, the domain is $\{1, 2, 3, 4, 5\}$, and the sequence has five terms.

PROBLEM SOLVING   As mentioned in the introduction to this chapter, there are many applications of sequences. To solve problems involving sequences, a good strategy is to list the first few terms and look for a pattern that suggests a general term. When the general term is known, we can find any term in the sequence without writing all the preceding terms.

EXAMPLE 3
Using a Sequence in an Application

A colony of bacteria doubles in weight every hour. If the colony weighs 1 gram at the beginning of an experiment, find the weight after ten hours.

At the end of the first hour, the colony will weigh 2 grams. At the end of the second hour, the weight will be $2 \cdot 2$ or $2^2 = 4$ grams. After three hours, the weight will be $2 \cdot 2 \cdot 2$ or $2^3 = 8$ grams, and so on. Continuing in this way gives the sequence shown in the chart below.

Time	a_1 End hour 1	a_2 End hour 2	a_3 End hour 3	a_4 End hour 4
Weight	2	4	8	16

In general, the colony will weigh 2^n grams at the end of n hours, so after 10 hours, the colony should weigh $2^{10} = 1024$ grams. ■

SERIES   The indicated sum of the terms of a sequence is called a **series.** Since a sequence can be finite or infinite, there are finite or infinite series. One type of infinite series is discussed in Section 11.3, and the binomial theorem discussed in Section 11.4 defines an important finite series. In this section we discuss only finite series.

4 ▶ Use summation notation to evaluate a series.

▶ We use a compact notation, called **summation notation,** to write a series from the general term of the corresponding sequence. For example, the sum of the first six terms of the sequence with general term $a_n = 3n + 2$ is written with the Greek letter Σ (sigma) as

$$\sum_{i=1}^{6} (3i + 2).$$

We read this as "the sum from $i = 1$ to 6 of $3i + 2$." To find this sum, we replace the letter i in $3i + 2$ with 1, 2, 3, 4, 5, and 6, as follows.

xiii

EXAMPLE 5
Factoring by Grouping

Factor $3x - 3y - ax + ay$.

Grouping terms as above gives

$$(3x - 3y) + (-ax + ay),$$

or

$$3(x - y) + a(-x + y).$$

There is no simple common factor here. However, if we factor out $-a$, we get

$$3(x - y) - a(x - y),$$

which equals

$$(x - y)(3 - a). \blacksquare$$

NOTE   In Example 5, different grouping would lead to the product

$$(a - 3)(y - x).$$

As we can verify by multiplication, this is another correct factorization.

In some cases, factoring by grouping requires that the terms be rearranged before the groupings are made.

EXAMPLE 6
Factoring by Grouping

Factor $p^2q^2 - 10 - 2q^2 + 5p^2$.

The first two terms have no common factor except 1 (nor do the last two terms). We can group the terms as follows:

$$(p^2q^2 - 2q^2) + (5p^2 - 10) \qquad \text{Group terms.}$$
$$q^2(p^2 - 2) + 5(p^2 - 2) \qquad \text{Factor out the common factors.}$$
$$(p^2 - 2)(q^2 + 5) \qquad \text{Factor out } p^2 - 2. \blacksquare$$

CAUTION   It is a common error to stop at the step

$$q^3(p^2 - 2) + 5(p^2 - 2).$$

This expression is *not in factored form* because it is a *sum* of two terms: $q^3(p^2 - 2)$ and $5(p^2 - 2)$.

EXAMPLE 7
Factoring by Grouping

Factor each of the following polynomials by grouping.

(a) $6x^2 - 4x - 15x + 10$

Work as above. Note that we must factor -5 rather than 5 from the second group in order to get a common factor of $3x - 2$.

$$(6x^2 - 4x) + (-15x + 10) = 2x(3x - 2) - 5(3x - 2) \qquad \text{Factor out } 2x \text{ and } -5.$$
$$= (3x - 2)(2x - 5) \qquad \text{Factor out } (3x - 2).$$

(b) $6ax + 12bx + a + 2b$

Group the first two terms and the last two terms.

$$6ax + 12bx + a + 2b = (6ax + 12bx) + (a + 2b)$$

Titled Examples include detailed, step-by-step solutions and descriptive side comments.

Notes draw attention to important comments to help students solve problems and understand concepts.

Cautions highlight common student errors and difficulties.

Graph Reading Activities A variety of graphs from both popular and professional media sources are used throughout the text to help students visualize mathematics and interpret real data.

HT LINE

...hange to solve the problem. See Example 8.

...e monthly rates
...from 1980 to
...roximated by a

...ided for 1980 and
...erage rate of

...ain how a *positive*
rate of change affects the consumer in a situation such as the one illustrated by this graph.

Cable TV Rates
Average Monthly Rates for Basic Service

1980: $7.85
1992: $18.85

Year
Sources: Census Bureau, Paul Kagan Associates.

56. Assuming a linear relationship, what is the average rate of change for cable industry revenues over the period from 1990 to 1992?

Cable Industry Revenues
Annual Operating Revenue

'90 $21.7 billion
'91 $22.7 billion
'92 $24.9 billion

Sources: Census Bureau, Paul Kagan Associates.

57. The 1993 Annual Report of AT&T cites the following figures concerning the global growth of international telephone calls: From 47.5 billion minutes in 1993, traffic is expected to rise to 60 billion minutes in 1995. Assuming a linear relationship, what is the average rate of change for this time period? (*Source:* TeleGeography, 1993, Washington, D.C.)

58. The market for international phone calls during the ten-year period from 1986 to 1995 is depicted in the accompanying bar graph. The tops of the bars approximate a straight line. Assuming that the traffic volume at the beginning of this period was 18 billion minutes and at the end was 60 billion minutes, what was the average rate of change for the ten-year period?

The Market for International Calls
Traffic Volume (1986–1995) in Billions of Minutes

Year

◈ MATHEMATICAL CONNECTIONS (Exercises 51–54) ◈

Work Exercises 51–54 in order to see the connections between systems of linear equations and the graphs of linear functions.

51. Solve the system

$$3x + y = 6$$
$$-2x + 3y = 7.$$

Use elimination or substitution.

52. For the first equation in the system of Exercise 51, solve for y and rename it $f(x)$. What special kind of function is f?

53. For the second equation in the system of Exercise 51, solve for y and rename it $g(x)$. What special kind of function is g?

⊚ **54.** Use the result of Exercise 51 to fill in the blanks with the appropriate responses:

Because the graphs of f and g are straight lines that are neither parallel nor coincide, they intersect in exactly _____ point. The coordinates of the point are (_____ , _____). Using functional notation, this is given by $f($ _____ $) =$ _____ and $g($ _____ $) =$ _____.

◈

55. The table shown was generated by a graphics calculator. The functions defined by y_1 and y_2 are linear. Based on the table, what are the coordinates of the point of intersection of the graphs?

56. The table shown was generated by a graphics calculator. The functions defined by y_1 and y_2 are linear.
(a) Use the methods of Chapter 7 to find the equation for y_1.
(b) Use the methods of Chapter 7 to find the equation for y_2.
(c) Solve the system of equations formed by y_1 and y_2.

57. The solution set of the system

$$y_1 = 3x - 5$$
$$y_2 = -4x + 2$$

is $\{(1, -2)\}$. Using slopes and y-intercepts, determine which one of the two calculator-generated graphs is the appropriate one for this system.

(a)

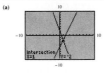

(b)

Mathematical Connections exercises tie together different topics and highlight the relationships among various concepts and skills.

Exercises are carefully paired (evens with odds) and are graded with regard to increasing difficulty. Exercises now include conceptual, writing, challenging, and calculator (both scientific and graphics calculators) types.

▲ **40.** $\dfrac{9x - 8}{4x^2 + 25} < 0$

▲ **42.** $\dfrac{(5x - 3)^2}{2x + 1} \le 0$

◈ MATHEMATICAL CONNECTIONS (Exercises 43–46) ◈

A rock is projected vertically upward from the ground. Its distance s in feet above the ground after t seconds is given by the equation $s = -16t^2 + 256t$. Work Exercises 43–46 in order to see how quadratic equations and inequalities are connected.

43. At what times will the rock be 624 feet above the ground? (*Hint*: Set $s = 624$ and solve the quadratic *equation*.)

44. At what times will the rock be more than 624 feet above the ground? (*Hint*: Set $s > 624$ and solve the quadratic *inequality*.)

45. At what times will the rock be at ground level? (*Hint*: Set $s = 0$ and solve the quadratic *equation*.)

46. At what times will the rock be less than 624 feet above the ground? (*Hint*: Set $s < 624$, solve the quadratic *inequality*, and observe the solutions in Exercise 45 to determine the smallest and largest possible values of t.)

◈

REVIEW EXERCISES

For the equation $3x + 2y = 24$, find y for the given value of x. See Section 1.2.

47. $x = 0$ **48.** $x = -2$ **49.** $x = 8$ **50.** $x = 1.5$

CHAPTER 6 SUMMARY

KEY TERMS

6.2 quadratic formula discriminant	6.3 quadratic in form	6.5 quadratic inequality rational inequality

QUICK REVIEW

CONCEPTS	EXAMPLES
6.1 COMPLETING THE SQUARE	
Square Root Property (a) If $x^2 = b$, $b \ge 0$, then $x = \sqrt{b}$ or $x = -\sqrt{b}$.	Solve. (a) $(x - 1)^2 = 8$ $\qquad x - 1 = \pm\sqrt{8} = \pm 2\sqrt{2}$ $\qquad x = 1 \pm 2\sqrt{2}$
(b) If $x^2 = -b$, $b > 0$, then $x = i\sqrt{b}$ or $x = -i\sqrt{b}$.	(b) $x^2 = -5$ $\qquad x = i\sqrt{5}$ or $x = -i\sqrt{5}$

Chapter Summaries present students with a helpful, section-referenced overview of each chapter. The Chapter Summary is followed by the Chapter Review Exercises, the Chapter Test, and the Cumulative Review Exercises.

Connections

Connections boxes that provide connections to the real world or to other mathematical concepts or other disciplines open each chapter and appear throughout the text—almost one for every section. Most of these include thought-provoking questions for writing or class discussion. The Connections provide motivation for the topic under discussion, show how mathematics is used in many different aspects of life, give some historical background, and provide a larger context for the current material.

Calculator Coverage

Connections boxes that focus on scientific and graphics calculators address the growing interest in the use of calculator technology, as well as provide extrinsic motivation. These Connections boxes are included in addition to the calculator passages that are part of the regular text and are optional in nature. Each such Connections box is specially marked with a scientific or graphics calculator symbol. Corresponding exercises are included in the exercise sets.

Graph Reading Activities

A variety of graphs from both popular and professional media sources are used throughout the text to help students visualize mathematics and interpret real data.

EXERCISES

More than 80 percent of the exercises are new to this edition. Care has been taken to pair exercises (evens with odds) and to grade the exercises with regard to increasing difficulty. Exercises now include a number of special types:

Mathematical Connections exercises tie together different topics and highlight the relationships among various concepts and skills. These multiple-skill and multiple-concept exercises sharpen students' problem-solving techniques and improve students' critical thinking abilities. Mathematical Connections exercises are included in most exercise sets and are grouped under a special heading.

Conceptual and writing exercises are designed to require a deeper understanding of concepts. 250 of the more than 650 exercises require the student to respond by writing a few sentences. Answers are not given for the writing exercises because they are open-ended, and instructors may use them in different ways.

Challenging exercises require the student to go beyond the examples in the text.

Cumulative Reviews end each chapter. These begin with Chapter 2 and test the topics covered from the beginning of the text up to that point.

Calculator exercises are included in some sections and require the student to use a scientific calculator. Other sections include passages on the use of graphics calculators accompanied by appropriately designed exercises. These are grouped and identified with a special symbol so that they can be easily omitted if the instructor prefers.

Applications have been updated and rewritten to include interesting and realistic information. In many cases they use actual data from current events, sports, and other sources.

SUPPLEMENTS

Our extensive supplemental package includes an Annotated Instructor's Edition, testing materials, solutions, software, and videotapes.

For the Instructor

Annotated Instructor's Edition   This edition provides instructors with immediate access to the answers to every exercise in the text, with the exception of writing exercises. Each answer is printed in color next to the corresponding text exercise.

Symbols are used to identify the conceptual ⊙, writing 🖹, and challenging ▲ exercises to assist in making homework assignments. Scientific 🖩 and graphics 🖩 calculator exercises are marked in both the student and instructor texts. Additional exercises, called Chalkboard Exercises, parallel almost every example, and Teaching Tips corresponding to the text discussions are also included in the Annotated Instructor's Edition.

Instructor's Test Manual   The Instructor's Test Manual includes short-answer and multiple-choice versions of a placement test; six forms of chapter tests for each chapter, including four open response and two multiple-choice forms; short-answer and multiple-choice forms of a final examination; and an extensive set of additional exercises (including more "mixed" exercises) providing 10 to 20 extra exercises for each textbook objective that instructors may use as an additional source of questions for tests, quizzes, or student review of difficult topics. Finally, this manual also includes a list of all conceptual, writing, connection, challenging, and calculator exercises.

Instructor's Solution Manual   This book includes detailed, worked-out solutions to each section exercise in the book, including conceptual *and* writing exercises. This manual also includes a list of all conceptual, writing, connection, challenging, and calculator exercises.

Instructor's Answer Manual   This manual includes answers to all exercises and a list of conceptual, writing, connection, challenging, and calculator exercises.

HarperCollins Test Generator/Editor for Mathematics with QuizMaster   Available in IBM (both DOS and Windows applications) and Macintosh versions, the Test Generator is fully networkable. The Test Generator enables instructors to select questions by objective, section, or chapter, or to use a ready-made test for each chapter. The Editor allows instructors to edit any preexisting data or to easily create their own questions. The software is algorithm driven, so the instructor may regenerate constants while maintaining problem type, providing a very large number of test or quiz items in multiple-choice and/or open response formats for one or more test forms. The system features printed graphics and accurate mathematics symbols. **QuizMaster** enables instructors to create tests and quizzes using the Test Generator/Editor and save them to disk so students can take the test or quiz on a stand-alone computer or network. **QuizMaster** then grades the test or quiz and allows the instructor to create reports on individual students or entire classes. CLAST and TASP versions of this package are also available for IBM and Mac machines.

For the Student

Student's Solution Manual   This book contains solutions to every odd-numbered section exercise (including conceptual and writing exercises) as well as solutions to all Connections, chapter review exercises, chapter tests, and cumulative review exercises. (ISBN 0-673-99541-0)

Student's Study Guide   This book provides additional practice problems and reinforcement for each learning objective in the textbook. Self-tests are included at the end of every chapter. (ISBN 0-673-99539-9)

Interactive Mathematics Tutorial Software with Management System   This innovative package is also available in DOS, Windows, and Macintosh versions and is fully networkable. As with the Test Generator/Editor, this software is algorithm driven, which automatically regenerates constants so that the numbers rarely repeat in a problem type when students revisit any particular section. The tutorial is objective-based, self-paced, and provides unlimited opportunities to review lessons and to practice problem solving. If students give a wrong answer, they can ask to see the problem worked out and get a textbook page reference. Many problems include hints for first incorrect responses. Tools such as an on-line glossary and Quick Reviews provide definitions and examples, and an on-line calculator aids students in computation. The program is menu-driven for ease of use and on-screen help can be obtained at any time with a single keystroke. Students' scores are calculated at the end of each lesson and can be printed for a permanent record. The optional **Management System** lets instructors record student scores on disk and print diagnostic reports for individual students or classes. CLAST and TASP versions of this tutorial are also available for both IBM and Mac machines. This software may also be purchased by students for home use. Student versions include record-keeping and practice tests.

Videotapes   A new videotape series has been developed to accompany *Intermediate Algebra,* Seventh Edition. In a separate lesson for each section in the book, the series covers all objectives, topics, and problem-solving techniques discussed in the text.

Overcoming Math Anxiety   This book, written by Randy Davidson and Ellen Levitov, includes step-by-step guides to problem solving, note taking, and applied problems. Students can discover the reasons behind math anxiety and ways to overcome those obstacles. The book also will help them learn relaxation techniques, build better math skills, and improve study habits. (ISBN 0-06-501651-3)

OTHER BOOKS IN THIS SERIES

Other textbooks in this series include: *Beginning Algebra,* Seventh Edition, *Intermediate Algebra with Early Graphs and Functions,* Seventh Edition, *Algebra for College Students,* Third Edition, and *Beginning and Intermediate Algebra.*

ACKNOWLEDGMENTS

We appreciate the many contributions of users of the sixth edition of the book. We also wish to thank our reviewers for their insightful comments and suggestions:

Emily Anne Battle, *University of Montevallo*
Chuck Beals, *Hartnell College*
Alan A. Bishop, *Western Illinois University*
Randall Brian, *Vincennes University*
Linda Britton, *Kellogg Community College*
Irene Brown, *Mercyhurst College*
Sister Marguerite Chandler, *St. Leo College*

Elaine Cheng, *Brookhaven College*
John W. Coburn, *St. Louis Community College at Florissant Valley*
Beverly Conner, *Polk Community College*
Linda F. Crabtree, *Longview Community College*
John DeCoursey, *Vincennes University*
Guy DiJulio, *Ohlone College*

Susan M. Dimick, *Spokane Community College*

Paul Dirks, *Miami-Dade Community College—Downtown Campus*

Glenn DiStefano, *Louisana State University at Alexandria*

Warrene C. Ferry, *Jones County Junior College*

Jeanne Fitzgerald, *Phoenix College*

Grace P. Foster, *Beaufort County Community College*

Bill Fried, *Napa Valley College*

Larry Friesen, *Butler County Community College*

Herb Gamage, *Alpena Community College*

Mark E. Gann, *Southwest Missouri State University*

Susan A. Garstka, *Moraine Valley Community College*

Frank Gentry, *Texas State Technical College at Waco*

Mary Lou Hammond, *Spokane Community College*

Janell Golden Hawkins, *Tennessee Technological University*

Sandra A. Gordon, *Aiken Technical College*

E. L. Hutton, *Glendale Community College*

David Jackson, *Parks College of St. Louis University*

Andrew W. Jonca, *Pacific University*

Robert Kaiden, *Lorain County Community College*

Robert C. Knapp, *University of Wisconsin at Whitewater*

Rachel Lamp, *North Iowa Area Community College*

Inessa Levi, *University of Louisville*

William R. Livingston, *Missouri Southern State College*

Wilma R. Lott, *South Georgia College*

Steven E. Martin, *Richard Bland College*

Maria Maspons, *Miami-Dade Community College—Downtown Campus*

Saralyn E. Mathis, *West Virginia Institute of Technology*

Robert McCown, *Florida Community College at Jacksonville*

Gael Mericle, *Mankato State University*

Philip Y. Meyer, *Skyline College*

John Michels, *Chemeketa Community College*

Lamar Middleton, *Polk Community College*

Peggy Miller, *University of Nebraska at Kearney*

Ann Moskol, *Rhode Island College*

Barbara Muenster, *St. Ambrose University*

Ebrahim K. Nikfarjam, *Guilford Technical Community College*

Thomas J. Peneski, *University of Wisconsin Center—Sheboygan*

Susan Poston, *Chemeketa Community College*

Glenn Powers, *Western Kentucky University*

Clay Prall, *Northeastern Junior College*

Thea Prettyman, *Essex Community College*

Beverly Ridenhour, *Utah State University*

Sharon Rigley, *American River College*

Wiley H. Russell, *Lake City Community College*

Philip Savoye, *Mansfield University of Pennsylvania*

Ladan Scott, *Brookhaven College*

Marsha Self, *El Paso Community College*

James Sells, *San Jacinto College—Central Campus*

Ellen Shatto, *Harrisburg Area Community College*

Bruce Sisko, *Belleville Area College*

Courtney Small, *Robert Morris College*

William H. Snyder, *Bainbridge College*

John W. Sullivan, *College of DuPage*

Gwen H. Terwilliger, *University of Toledo*

Shelley Tingey, *Central Arizona College*

Ralph Tippins, *Citrus College*

Stephanie Unruh, *Hutchinson Community College*

Anita L. Walker, *Blue Ridge Community College*

Jane Watkins, *Calhoun Community College*

Donald Wemmie, *Southeastern Community College*

Mary Whalen, *University of Wisconsin at Stevens Point*

Jacquelyn E. White, *Community College of Denver*

Gail G. Wiltse, *St. John's River Community College*

Cheryll Wingard, *Community College of Aurora*

Walter Wooden, *Broward Community College—North Campus*

Kelly Wyatt, *Umpqua Community College*

Michelle A. Wyatt, *Community College of Southern Nevada*

As always, Paul Eldersveld, *College of DuPage,* has done an outstanding job of coordinating all the print supplements for us.

We also wish to thank those who did an excellent job checking all the answers:

K. S. Asal, *Broward Community College—North Campus*

Douglas E. Cameron, *University of Akron*

Udema Childress, *East Tennessee State University*

Susan Costa, *Broward Community College—Central Campus*

Don Davis, *Lakeland Community College*

John DeCoursey, *Vincennes University*

Eugenia E. Fitzgerald, *Phoenix College*

Lyenatte S. Goff, *Glendale Community College*

William Livingston, *Missouri Southern State College*

Robert McCown, *Florida Community College—Downtown Campus*

Gael Mericle, *Mankato State Universtiy*

Kathleen L. Pellissier

Donna M. Szott, *Community College of Allegheny County—South Campus*

Our appreciation also goes to Tommy Thompson, *Cedar Valley College,* for his suggestions for the feature "To the Student: Success in Algebra."

Special thanks go to the dedicated staff at HarperCollins who have worked so long and hard to make this book a success: Karin Wagner, Sandi Goldstein, Cathy Wacaser, Anne Kelly, George Duda, Ed Moura, and Linda Youngman.

Margaret L. Lial
E. John Hornsby, Jr.

TO THE STUDENT: SUCCESS IN ALGEBRA

The main reason students have difficulty with mathematics is that they don't know how to study it. Studying mathematics *is* different from studying subjects like English or history. The key to success is regular practice.

This should not be surprising. After all, can you learn to play the piano or to ski well without a lot of regular practice? The same thing is true for learning mathematics. Working problems nearly every day is the key to becoming successful. Here is a list of things you can do to help you succeed in studying algebra.

1. *Attend class regularly.* Pay attention in class to what your teacher says and does, and make careful notes. In particular, note the problems the teacher works on the board and copy the complete solutions. Keep these notes separate from your homework to avoid confusion when you read them over later.

2. Don't hesitate to ask questions in class. It is not a sign of weakness, but of strength. There are always other students with the same question who are too shy to ask.

3. *Read your text carefully.* Many students read only enough to get by, usually only the examples. Reading the complete section will help you to be successful with the homework problems. Most exercises are keyed to specific examples or objectives that will explain the procedures for working them.

4. Before you start on your homework assignment, rework the problems the teacher worked in class. This will reinforce what you have learned. Many students say, "I understand it perfectly when you do it, but I get stuck when I try to work the problem myself."

5. Do your homework assignment only *after* reading the text and reviewing your notes from class. Check your work with the answers in the back of the book. If you get a problem wrong and are unable to see why, mark that problem and ask your instructor about it. Then practice working additional problems of the same type to reinforce what you have learned.

6. Work as neatly as you can. Write your symbols neatly, and make sure the problems are clearly separated from each other. Working neatly will help you to think clearly and also make it easier to review the homework before a test.

7. After you have completed a homework assignment, look over the text again. Try to decide what the main ideas are in the lesson. Often they are clearly highlighted or boxed in the text.

8. Use the chapter test at the end of each chapter as a practice test. Work through the problems under test conditions, without referring to the text or the answers until you are finished. You may want to time yourself to see

how long it takes you. When you have finished, check your answers against those in the back of the book and study those problems that you missed. Answers are referenced to the appropriate sections of the text.

9. Keep any quizzes and tests that are returned to you and use them when you study for future tests and the final exam. These quizzes and tests indicate what your instructor considers most important. Be sure to correct any problems on these tests that you missed, so you will have the corrected work to study.

10. Don't worry if you do not understand a new topic right away. As you read more about it and work through the problems, you will gain understanding. Each time you look back at a topic you will understand it a little better. No one understands each topic completely right from the start.

AN INTRODUCTION TO GRAPHICS CALCULATORS*

CAPABILITIES OF GRAPHICS CALCULATORS

Graphics calculators are a result of the amazingly rapid evolution in computer technology toward packaging more power into smaller "boxes." These machines have powerful graphing capabilities in addition to the full range of features found on programmable scientific calculators. Instead of a one-line display, graphics calculators typically can show up to eight lines of text. This makes it much easier to keep track of the steps of your work, whether you are doing routine computations, entering a long mathematical function, or writing a program. Like programmable scientific calculators, graphics calculators are capable of doing many things we have previously come to expect only from computers. Programs can be written relatively easily, and after they are stored in memory, the programs are always available. Like computers, graphics calculators can be programmed to include graphic displays as part of the program.

It takes some study to learn how to use graphics calculators, but they are much easier to master than most computers—and they can go wherever you do! New models with added features, more memory, greater ease of use, and other improvements are frequently being introduced. The most popular brand names at this time are Casio, Sharp, Texas Instruments, and Hewlett-Packard.

GRAPHICS CALCULATOR FEATURES

Every graphics calculator has keys for the usual operations of arithmetic and all commonly used functions (square root, x^2, log, ln, and so on). All of them can graph functions of the form $y = f(x)$ and have programming capabilities. Except for the cheapest ones, graphics calculators may have a variety of additional features. Before buying one, you should consider which features you are likely to need.

Many of the features of the typical graphics calculator do not play a role in the topics covered in this book, and further mathematics courses will be needed in order to fully appreciate their power. The more advanced models have the capability to perform some symbolic manipulations (such as factoring, adding polynomials, and performing operations with rational algebraic expressions). However, these calculators are more expensive and more difficult to learn to use because of their increased complexity and the fact that they do not use standard algebraic order of operations.

ADVICE ON USING A GRAPHICS CALCULATOR

1. **BASICS** Graphics calculators have forty-nine or more keys. Most modern desk-top computers have 101 keys on their keyboards. With fewer keys, each key must be used for more actions, so you will find special mode-changing keys such as **"2nd," "shift," "alpha,"** and **"mode"**. Become familiar with

* Prepared by Jim Eckerman of *American River College.*

xxiii

the capabilities of the machine, the layout of the keyboard, how to adjust the screen contrast, and so on. Remember that a graphics calculator is composed of two parts: the machinery *and* the owner's manual!

2. **EDITING** When keying in expressions, you can pause at any time and use the arrow keys, located at the upper right of the keyboard, to move the cursor to any point in the text. You can then make changes by using the **"DEL"** key to delete and the **"INS"** key to insert material. ["INS" is the "second function" of "DEL."] After an expression has been entered or a calculation made, it can still be edited by using the **edit/replay** feature, available on almost every calculator model. On TI calculators (except TI-81), use "2nd, ENTER" to return to the previously entered expression. On Casio calculators, use the left or right arrow key, and on Sharp models use "2nd, up arrow."

3. **SCIENTIFIC NOTATION** Learn how to enter and read data in **scientific notation** form. This form is used when the numbers become too large or too small (too many zeros between the decimal point and the first significant digit) for the machine's display.

4. **FUNCTION GRAPHING**

   A. **Setting the Range or Window** Learn to set **"RANGE"** or **"WINDOW"** values to delineate a window that is appropriate for the function you are graphing before using the **"Graph"** key. This involves keying in the minimum and maximum values of x and y that will be displayed on the screen, along with the distance to be used between tick marks along the axes. If you do not do this, you will often find your graph screen blank! Usually one can quickly find a point on the graph by substitution of either zero or one for x. Use the **"RANGE"** or **"WINDOW"** command to set the x-values to the left and right of the x-coordinate of this point and set the y-values above and below the y-coordinate of this point. Then the **"zoom out"** feature can be used to see more of the graph.

   B. **Using the Function Memory** Learn how to redraw graphs without reentering the function. Often you will need to change the **"RANGE"** settings several times before you get the "window" that is most appropriate for your function. All graphics calculators allow you to do this without reentering the function. If you plan to graph a particular function often, then you should either store it in the **function memory** (which is labeled "$y = $" on TI and "EQTN" on Sharp models) or store it in **program memory.** Of course, you have to write a program with your function as part of the program in order to make use of the program memory. Once entered into the machine's memory, this function can be used at any time.

   C. **Using the Trace Feature** With the **"Trace"** feature, the left/right arrows can be used to move the cursor along the last curve plotted, and the values of x and y will be displayed for each point plotted on the screen. If more than one graph was plotted, one can move the cursor vertically between the different graphs by using the up/down arrows.

   D. **Using the Zoom Feature** The **"zoom"** feature allows a quick redrawing of your graph using smaller ranges of values for x and y (**"zoom in"**) or larger ranges of values (**"zoom out"**). Thus one can easily examine the behavior of the graph of a function within the close vicinity of a particular point or the general behavior as seen from farther away. Using **"zoom box,"** a box can be drawn for a particular region for closer inspection of the graph within that region.

SOLVING EQUATIONS AND SYSTEMS GRAPHICALLY

Some mathematical procedures can be quite difficult or even impossible to do algebraically but can be done easily and to a very high degree of accuracy using a graphics calculator. Listed below are some examples.

1. **SOLVING EQUATIONS OF THE FORM** $f(x) = k$ The quickest way to solve this type of equation is to form the new equation $y = f(x) - k$ and then use the built-in Equation Solver or Root Finder feature. If your calculator does not have this capability or if you prefer to see the solution graphically, you can find the roots of $y = f(x) - k$ by locating the points where the graph crosses the x-axis.
2. **SOLVING SYSTEMS OF TWO EQUATIONS IN TWO UNKNOWNS** Some graphics calculators have built-in programs for solving systems of linear equations, but all of them can be used to solve systems of two equations (linear or not) by finding intersection points. If both equations can be solved explicitly for one variable in terms of the other, then the two equations can be put into the forms $y = f(x)$ and $y = g(x)$. Then we can graph both in the same window and zoom in on the point or points where the two curves intersect.

PROGRAMMING

Many formulas are used often enough to justify automating the process of evaluating them. The distance formula and the quadratic formula are two examples of this. Graphics calculators are able to store programs that will perform such tasks. The realm of *programming* is much different from that of *using* a graphics calculator. Many programs are available from the manufacturers and the literature that they support. Some students like to experiment and program their own calculators. The old adage "The sky's the limit" is certainly applicable to programming, and it is up to the user's imagination as to how far he or she wishes to take the programming capability of the calculator.

SOME SUGGESTIONS FOR REDUCING FRUSTRATION

We all find ways to make even the simplest machines do the wrong things without even trying. One of the more common problems with graphics calculators is getting a blank screen when a graph was expected. This usually results from not setting the **"WINDOW"** values appropriately before graphing the function, although it could also easily result from incorrectly entering the function.

A common problem that is particularly annoying is interpreting cryptic error messages such as **"Syn ERROR"** and **"Ma ERROR."** (Keep that manual handy!) The most common mistakes are made entering formulas and using special functions. For example, on the Casio machines **"Syn ERROR"** means that a mistake was made when entering the function or operation, such as entering "Graph $Y = \log x^y$" instead of "Graph $Y = \log x^2$".

Another common error is having more right parentheses than left parentheses. To confuse us further, these same machines think it is perfectly OK to have more left parentheses than right parentheses. For instance, the expression $5(3 - 4(2 + 7)$ has two left parentheses and one right parenthesis, but it will be evaluated as $5(3 - 4(2 + 7))$. The message **"Ma ERROR"** appears when a number is too large or when a number is not allowed. If you try to find the 1000th power of ten or divide a number by zero you will most certainly see some kind of error message. By pressing one of the **"cursor"** keys on the Casio you will see the cursor blinking at the location of the error in your expression. When the Texas Instruments models detect an error, they display a special menu that lists a code number and a name for the type of error. For certain types of errors the choice **"Go to error"** is offered. The TI-85 will display the number 9.99999999 E999 but shows "ERROR 01 OVERFLOW" for 10 E999 (which means 10 times the 999th power of ten). The latter and other more advanced machines display "(0, 2)" when asked to find the square root of -4. The "(0, 2)" represents the complex number $0 + 2i$.

SOME FINAL COMMENTS

While studying mathematics it is important to learn the mathematical concepts well enough to make intelligent decisions about when to use and when not to use "high tech" aids such as computers and graphics calculators. These machines make it easy to experiment with graphs of mathematical relations. One can learn much about the behavior of different types of functions by playing "what if" games with the formulas. However, in a timed test situation you may find yourself spending too much time working with the graphics calculator when a quick algebraic solution and a rough sketch with pencil and paper are more appropriate.

To get the most return on your investment, learn to use as many features of your machine as possible. Of course, some of the features may not be of use to you, so feel free to ignore them. A first session of two or three hours with your graphics calculator and your user's manual is essential. Be sure to keep your manual handy, referring to it when needed.

A final word of caution: These machines are fun to use, but they can be addictive. So set time limits for yourself, or you may find that your graphics calculator has been more of a detriment than a help!

The following features are special to the *Annotated Instructor's Edition*.

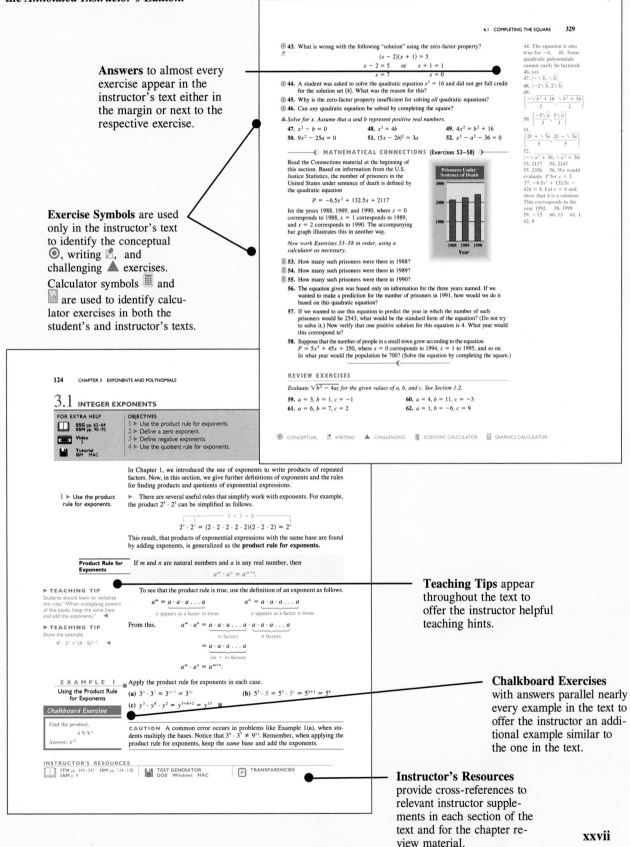

Answers to almost every exercise appear in the instructor's text either in the margin or next to the respective exercise.

Exercise Symbols are used only in the instructor's text to identify the conceptual ◉, writing ✐, and challenging ▲ exercises. Calculator symbols ▦ and ▨ are used to identify calculator exercises in both the student's and instructor's texts.

Teaching Tips appear throughout the text to offer the instructor helpful teaching hints.

Chalkboard Exercises with answers parallel nearly every example in the text to offer the instructor an additional example similar to the one in the text.

Instructor's Resources provide cross-references to relevant instructor supplements in each section of the text and for the chapter review material.

THE REAL NUMBERS

CONNECTIONS

Mathematics is not a finished product. It is still evolving and much important work has occurred recently. In 1993 Andrew Weil rocked the mathematics world by claiming to prove Fermat's last theorem, stated by the famous French mathematician Pierre de Fermat in the 17th century. The problem is a simple one: If n represents any whole number larger than 2, there is no solution in positive integers to the equation "x to the nth power plus y to the nth power equals z to the nth power." With a little knowledge of algebra, the theorem can be stated more briefly as "For $n > 2$, $x^n + y^n \neq z^n$." Mathematicians have tried to prove the theorem ever since Fermat left a note in the margin of a book claiming he had found an "admirable proof of this theorem, but the margin is too narrow to contain it." Wiles had been working on the problem for five years. His proof builds on the work of scores of mathematicians over the centuries. Mathematicians are currently looking closely at his work to see if it is, in fact, correct.

One of the most important reasons for studying algebra is that it has proved so useful in solving problems that arise in the real world. We can think of algebra as a language with symbols, rules, and patterns that must be learned in order to become fluent in it. This chapter reviews some of the basic symbols and rules of algebra that are studied in elementary algebra.

1.1 BASIC TERMS

FOR EXTRA HELP	OBJECTIVES
📖 **SSG** pp. 1–7 **SSM** pp. 1–4	1 ▶ Write sets.
📼 **Video** 1	2 ▶ Know the common sets of numbers.
	3 ▶ Use number lines.
💾 **Tutorial** IBM MAC	4 ▶ Find additive inverses.
	5 ▶ Use absolute value.
	6 ▶ Use inequality symbols.
	7 ▶ Graph sets of real numbers.

1 ▶ Write sets.

▶ A basic term used in algebra is **set,** a collection of objects, numbers, or ideas called the **elements** or **members** of the set. In algebra, the elements in a set are usually numbers. **Set braces,** { }, are used to enclose the elements. For example, 2 is an element of the set {1, 2, 3}.

A set can be defined either by listing or by describing its elements. For example,

$$S = \{\text{Oregon, Ohio, Oklahoma}\}$$

defines the set S by *listing* its elements. The same set might be *described* by saying that set S is the set of all states in the United States whose names begin with the letter "O."

Set S above has a finite number of elements. Some sets contain an infinite number of elements, such as

$$N = \{1, 2, 3, 4, 5, 6, \ldots\},$$

where the three dots show that the list continues in the same pattern. Set N is called the set of **natural numbers,** or **counting numbers.** A set containing no elements, such as the set of natural numbers less than 1, is called the **empty set,** or **null set,** usually written $\emptyset$. The empty set may also be written as { }.

CAUTION It is not correct to write $\{\emptyset\}$ for the empty set. $\{\emptyset\}$ represents a set with one element (the empty set) and thus is not an empty set. Also, the number 0 is not the same as the empty set; do not use the symbol $\emptyset$ for zero. Always use only the notations $\emptyset$ or { } for the empty set.

To write the fact that 2 is an element of the set {0, 1, 2, 3}, we use the symbol $\in$ (read "is an element of"):

$$2 \in \{0, 1, 2, 3\}.$$

The number 2 is also an element of the set of natural numbers N above, so we may write

$$2 \in N.$$

To show that 0 is *not* an element of set N, we draw a slash through the symbol $\in$:

$$0 \notin N.$$

Two sets are **equal** if they contain exactly the same elements. For example,

$$\{1, 2\} = \{2, 1\},$$

because the sets contain the same elements. (The order doesn't matter.) On the other hand, $\{1, 2\} \neq \{0, 1, 2\}$ ($\neq$ means "is not equal to") since one set contains the element 0 while the other does not.

In algebra, letters called **variables** are often used to represent numbers or to define sets of numbers. For example,

$$\{x \mid x \text{ is a natural number between 3 and 15}\}$$

(read "the set of all elements x such that x is a natural number between 3 and 15") defines the set

$$\{4, 5, 6, 7, \ldots, 14\}.$$

▶ **TEACHING TIP**
Set-builder notation will be used to specify solutions of dependent equations in the chapter on systems of equations. ◄

The notation $\{x \mid x$ is a natural number between 3 and 15$\}$ is an example of **set-builder notation.**

$$\{x \mid x \text{ has \textbf{property} } P\}$$

the set of   all elements x such that x has a given property P

EXAMPLE 1

Listing the Elements in a Set

List the elements in the set.

(a) $\{x \mid x$ is a natural number less than 4$\}$
The natural numbers less than 4 are 1, 2, and 3. The given set is

$$\{1, 2, 3\}.$$

(b) $\{y \mid y$ is one of the first five even natural numbers$\} = \{2, 4, 6, 8, 10\}$

(c) $\{z \mid z$ is a natural number at least 7$\}$
The set of natural numbers at least 7 is an infinite set; write it with three dots as

$$\{7, 8, 9, 10, \ldots\}. \blacksquare$$

Chalkboard Exercise

List the elements in the set $\{x \mid x$ is a natural number greater than 12$\}$.

Answer: $\{13, 14, 15, \ldots\}$

EXAMPLE 2

Using Set-Builder Notation to Describe a Set

Use set-builder notation to describe the set.

(a) $\{1, 3, 5, 7, 9\}$
There are several ways to describe a set with set-builder notation. One way to describe this set is

$$\{y \mid y \text{ is one of the first five odd natural numbers}\}.$$

(b) $\{5, 10, 15, \ldots\}$
This set can be described as $\{d \mid d$ is a multiple of 5 greater than 0$\}$. $\blacksquare$

Chalkboard Exercise

Use set-builder notation to describe the set $\{-2, -1, 0, 1, 2, \ldots\}$.

Answer: $\{x \mid x$ is an integer greater than $-3\}$

2 ▶ Know the common sets of numbers.

▶ The following sets of numbers will be used throughout the book.

Sets of Numbers

Real Numbers	$\{x \mid x \text{ is represented by a point on a number line}\}$*
Natural Numbers or Counting Numbers	$\{1, 2, 3, 4, 5, 6, 7, 8, \ldots\}$
Whole Numbers	$\{0, 1, 2, 3, 4, 5, 6, \ldots\}$
Integers	$\{\ldots, -3, -2, -1, 0, 1, 2, 3, \ldots\}$
Rational Numbers	$\left\{\dfrac{p}{q} \middle\vert p \text{ and } q \text{ are integers, with } q \neq 0\right\}$ or all terminating or repeating decimals
Irrational Numbers	$\{x \mid x \text{ is a real number that is not rational}\}$ or all nonterminating and nonrepeating decimals

Examples of irrational numbers include most square roots, such as $\sqrt{7}$, $\sqrt{11}$, $\sqrt{2}$, and $-\sqrt{5}$. (Some square roots *are* rational: $\sqrt{16} = 4$, $\sqrt{100} = 10$, and so on.) Another irrational number is π, the ratio of the circumference of a circle to its diameter. All irrational numbers are real numbers.

Real numbers can also be defined in terms of decimals. By repeated subdivisions, any decimal can be located (at least in theory) as a point on a number line. Because of this, the set of real numbers can be defined as the set of all decimals. Also, the set of rational numbers can be shown to be the set of all repeating or terminating decimals. A bar over the series of numerals that repeat is used to indicate a repeating decimal. For example, some decimal forms of rational numbers are $.\overline{6} = 2/3$ (the 6 repeats), $.25 = 1/4$, $.2 = 1/5$, $.\overline{142857} = 1/7$ (the block of digits 142857 repeats), $.4\overline{3} = 13/30$ (the 3 repeats), and so on. The set of irrational numbers is the set of decimals that do not repeat and do not terminate.

The relationships among these various sets of numbers are shown in Figure 1; in particular, the figure shows that the set of real numbers includes both the rational and the irrational numbers.

▶ **TEACHING TIP**
Learning the relationship among the sets of real numbers is difficult for students. Use Figure I and relate each set to a department within a department, as in a large retail store. ◀

EXAMPLE 3
Identifying Examples of Number Sets

(a) 0, $2/3$, $-9/64$, $28/7$ (or 4), 2.45, and $1.\overline{37}$ are rational numbers.

(b) $\sqrt{3}$, π, $-\sqrt{2}$, and $\sqrt{7} + \sqrt{3}$ are irrational numbers.

(c) -8, $12/2$, $-3/1$, and $75/5$ are integers.

(d) All the numbers in parts (a), (b), and (c) above are real numbers.

(e) $4/0$ is undefined, since the definition requires the denominator of a rational number to be nonzero. (However $0/4 = 0$, which is a real number.) ■

Chalkboard Exercise

Select all the words from the following list that apply to the number $\sqrt{4}$.
 Whole number
 Rational number
 Irrational number
 Real number
 Undefined

Answer: $(\sqrt{4} = 2)$ whole number, rational number, real number

---

*An example of a number that is not represented by a point on a number line is $\sqrt{-1}$. This number, which is not a real number, is discussed in Chapter 5.

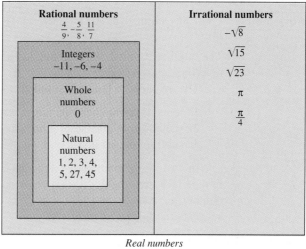

Real numbers

(a)

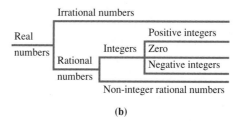

(b)

FIGURE 1

3 ▶ Use number lines.

▶ A good way to picture a set of numbers is to use a **number line.** To construct a number line, choose any point on a horizontal line and label it 0. Then choose a point to the right of 0 and label it 1. The distance from 0 to 1 establishes a scale that can be used to locate more points, with positive numbers to the right of 0 and negative numbers to the left of 0. A number line is shown in Figure 2.

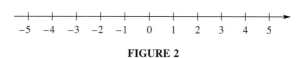

FIGURE 2

Each number is called the **coordinate** of the point that it labels, while the point is the **graph of the number.** A number line with several selected points graphed on it is shown in Figure 3.

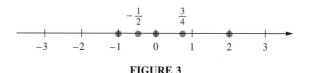

FIGURE 3

4 ▶ Find additive inverses.

▶ Two numbers that are the same distance from 0 on the number line but on opposite sides of 0 are called **additive inverses, negatives,** or **opposites,** of each other. For example, 5 is the additive inverse of -5, and -5 is the additive inverse of 5.

| Additive Inverse | For any number a, the number $-a$ is the **additive inverse** of a. |

The number 0 is its own additive inverse. The sum of a number and its additive inverse is always zero.

The symbol "$-$" can be used to indicate any of the following three things:

1. a negative number, such as -9 or -15;
2. the additive inverse of a number, as in "-4 is the additive inverse of 4";
3. subtraction, as in $12 - 3$.

In writing the number $-(-5)$, the symbol "$-$" is being used in two ways: the first $-$ indicates the additive inverse of -5, and the second indicates a negative number, -5. Since the additive inverse of -5 is 5, then $-(-5) = 5$. This example suggests the following property.

| Double Negative | For any number a, $-(-a) = a$. |

Numbers written with a positive or negative sign, such as $+4$, $+8$, -9, -5, and so on, are called **signed numbers.** Positive numbers can be called signed numbers even if the positive sign is left off.

E X A M P L E   4
Finding Additive Inverses

The following list shows several signed numbers and the additive inverse of each.

Number	Additive Inverse
6	-6
-4	$-(-4)$, or 4
$\frac{2}{3}$	$-\frac{2}{3}$
0	-0, or 0

5 ▶ Use absolute value.

▶   The **absolute value** of a number a is the distance on the number line from 0 to a and is written $|a|$. For example, the absolute value of 5 is the same as the absolute value of -5, since each number lies five units from 0. See Figure 4. That is, both

$$|5| = 5 \quad \text{and} \quad |-5| = 5.$$

Distance is 5, so $|-5| = 5$.   Distance is 5, so $|5| = 5$.

FIGURE 4

NOTE   Since absolute value represents distance, and since distance is never negative, **the absolute value of a number is never negative.**

The formal definition of absolute value is as follows.

Absolute Value

$$|a| = \begin{cases} a & \text{if } a \text{ is positive or } 0 \\ -a & \text{if } a \text{ is negative} \end{cases}$$

NOTE The second part of this definition, $|a| = -a$ if a is negative, is tricky. If a is a *negative* number, then $-a$, the additive inverse or opposite of a, is a positive number, so that $|a|$ is positive.

EXAMPLE 5

Evaluating Absolute Value Expressions

Find the indicated value.

(a) $|13| = 13$ **(b)** $|-2| = -(-2) = 2$ **(c)** $|0| = 0$

(d) $-|8|$

Evaluate the absolute value first. Then find the additive inverse.

$$-|8| = -(8) = -8$$

(e) $-|-8|$

Work as in (d): $|-8| = 8$, so

$$-|-8| = -(8) = -8.$$

(f) $|-2| + |5|$

Evaluate each absolute value first, then add.

$$|-2| + |5| = 2 + 5 = 7 \quad \blacksquare$$

Chalkboard Exercise

Find the indicated value.

$$|8| + |-1|$$

Answer: 9

▶ **TEACHING TIP**

Compare these examples in class: $-|-5|$ and $-(-5)$. ◄

6 ▶ Use inequality symbols.

▶   A statement that two numbers are *not* equal is called an **inequality.** For example, the numbers 4/2 and 3 are not equal. Write this inequality as

$$\frac{4}{2} \neq 3,$$

where the slash through the equals sign is read "does not equal."

When two numbers are not equal, one must be smaller than the other. The symbol $<$ means "is smaller than" or "is less than." For example, $8 < 9$, $-6 < 15$, and $0 < 4/3$. Similarly, "is greater than" is written with the symbol $>$. For example, $12 > 5$, $9 > -2$, and $6/5 > 0$. The number line in Figure 5 shows the graphs of the numbers 4 and 9. The graphs show that $4 < 9$. Starting at 4, add the positive number 5 to get 9. As this example suggests, $a < b$ means that there exists a positive number c such that $a + c = b$.

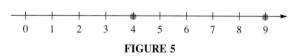

FIGURE 5

On the number line, the smaller of two given numbers is always located to the left of the other. Also, if a is less than b, then b is greater than a. The geometric definitions of $<$ and $>$ are as follows.

Definitions of $<$ and $>$

On the number line,

$a < b$ if a is to the left of b; $b > a$ if b is to the right of a.

EXAMPLE 6
Using a Number Line to Determine Order

Chalkboard Exercise

Insert < or > in the blank.

−4 ______ −8

Answer: >

(a) As shown on the number line in Figure 6, −6 is located to the left of 1. For this reason, $-6 < 1$. Also, $1 > -6$.

(b) From the same number line, $-5 < -2$, or $-2 > -5$. ■

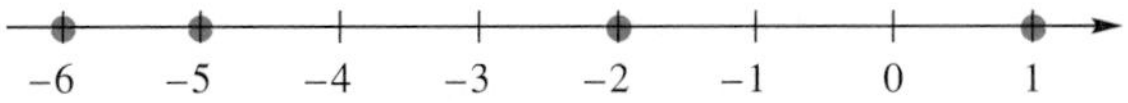

FIGURE 6

The following box summarizes results about positive and negative numbers. The same statement is given in both words and symbols.

Words	*Symbols*
Every negative number is smaller than 0.	If a is negative, then $a < 0$.
Every positive number is larger than 0.	If a is positive, then $a > 0$.

0 is neither positive nor negative.

In addition to the symbols < and >, the following symbols often are used. (To complete the list, the symbols < and > are also included.)

Symbols of Inequality

Meaning	*Example*
< is less than	$-4 < -1$
> is greater than	$3 > -2$
≤ is less than or equal to	$6 \leq 6$
≥ is greater than or equal to	$-8 \geq -10$

EXAMPLE 7
Interpreting Inequality Symbols

Chalkboard Exercise

True or False.

3(4) > 2(6)

Answer: false

The following table shows several statements and the reason that each is true.

Statement	*Reason*
$6 \leq 8$	$6 < 8$
$-2 \leq -2$	$-2 = -2$
$-9 \geq -12$	$-9 > -12$
$-3 \geq -3$	$-3 = -3$
$6 \cdot 4 \leq 5(5)$	$24 < 25$

In the last line, recall that the dot in $6 \cdot 4$ indicates the product 6×4, or 24. Also, 5(5) means 5×5, or 25. The statement is thus $24 \leq 25$, which is true. ■

7 ▶ Graph sets of real numbers.

▶ Inequality symbols and variables can be used to write sets of real numbers. For example, the set $\{x \mid x > -2\}$ is made up of all the real numbers greater than −2. On a number line, we show the elements of this set (the set of all real numbers to the right of −2) by drawing a line from −2 to the right. We use a parenthesis at −2 since −2 is not an element of the given set. The result, shown in Figure 7, is called the **graph of the set** $\{x \mid x > -2\}$.

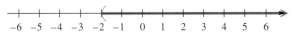

FIGURE 7

The set of numbers greater than -2 is an example of an **interval** on the number line. A simplified notation, called **interval notation,** is used for writing intervals. For example, using this notation, the interval of all numbers greater than -2 is written as $(-2, \infty)$. The infinity symbol ∞ does not indicate a number; it is used to show that the interval includes all real numbers greater than -2. The left parenthesis indicates that -2 is not included. A parenthesis is always used next to the infinity symbol in interval notation. The set of all real numbers is written in interval notation as $(-\infty, \infty)$.

E X A M P L E   8

Graphing an Inequality Written in Interval Notation

Write $\{x \mid x < 4\}$ in interval notation and graph the interval.

The interval is written as $(-\infty, 4)$. The graph is shown in Figure 8. Since the elements of the set are all the real numbers *less* than 4, the graph extends to the left. ■

-6 -5 -4 -3 -2 -1 0 1 2 3 4 5 6

FIGURE 8

Chalkboard Exercise

Write in interval notation and graph.

$$\{x \mid x \le 5\}$$

Answer: $(-\infty, 5]$

2 3 4 5 6

The set $\{x \mid x \le -6\}$ contains all the real numbers less than or equal to -6. To show that -6 itself is part of the set, a *square bracket* is used at -6, as shown in Figure 9. In interval notation, this set is written as $(-\infty, -6]$.

-8 -6 -4 -2 0 2

FIGURE 9

E X A M P L E   9

Graphing an Inequality Written in Interval Notation

Write $\{x \mid x \ge -4\}$ in interval notation and graph the interval.

This set is written in interval notation as $[-4, \infty)$. The graph is shown in Figure 10. A square bracket is used at -4 since -4 is part of the set. ■

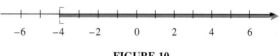

-6 -4 -2 0 2 4 6

FIGURE 10

Chalkboard Exercise

Write in interval notation and graph.

$$\{x \mid x > 0\}$$

Answer: $(0, \infty)$

-1 0 1 2 3

NOTE   In a previous course you may have graphed $\{x \mid x > -2\}$ using an open circle instead of a parenthesis at -2. Also, you may have graphed $\{x \mid x \ge -4\}$ using a solid dot instead of a bracket at -4. The interval notation we use in this book is preferred in more advanced courses.

▶ **TEACHING TIP**

Encourage oral participation when presenting the sets of the real numbers using exercises from this section. ◀

It is common to graph sets of numbers that are *between* two given numbers. For example, the set $\{x \mid -2 < x < 4\}$ is made up of all those real numbers between -2 and 4, but not the numbers -2 and 4 themselves. This set is written in interval notation as $(-2, 4)$. The graph has a heavy line between -2 and 4 with parentheses at -2 and 4. See Figure 11. The inequality $-2 < x < 4$ is read "x is greater than -2 and less than 4," or "x is between -2 and 4."

-4 -2 0 2 4 6

FIGURE 11

EXAMPLE 10

Graphing a Three-Part Inequality

Write in interval notation and graph $\{x \mid 3 < x \le 10\}$.

Use a parenthesis at 3 and a square bracket at 10 to get $(3, 10]$ in interval notation. The graph is shown in Figure 12. Read the inequality $3 < x \le 10$ as "3 is less than x and x is less than or equal to 10," or "x is between 3 and 10, excluding 3 and including 10." ∎

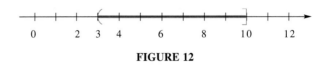

FIGURE 12

Chalkboard Exercise

Write in interval notation and graph.

$\{x \mid -4 \le x < 2\}$

Answer: $[-4, 2)$

1.1 EXERCISES

1. false   2. true
3. true   4. false
5. true   6. true
7. $\{1, 2, 3, 4, 5\}$
8. $\{1, 2, 3, 4, 5, 6, 7, 8\}$
9. $\{5, 6, 7, 8, \ldots\}$
10. $\{9, 10, 11, 12, \ldots\}$
11. $\{10, 12, 14, 16, \ldots\}$
12. $\{\ldots, -7, -5, -3, -1\}$
13. $\emptyset$ 14. $\emptyset$
15. $\{-4, 4\}$ 16. $\{-7, 7\}$
17. $\{0, 3, 6, 9, \ldots\}$
18. $\{7, 14, 21, 28, \ldots\}$
19. yes   20. no
21. $\{x \mid x$ is a multiple of 4 greater than 0$\}$
22. $\{x \mid x$ is a multiple of 3$\}$
23. $\{x \mid x$ is an even natural number less than or equal to 8$\}$ 24. $\{x \mid x$ is an integer between 10 and 15$\}$

Decide whether the statement concerning real numbers is true or false.

1. There is a number that has a negative absolute value.

2. The absolute value of a negative number is equal to the additive inverse of that number.

3. Every number has an additive inverse.

4. The absolute value of any number must be positive.

5. $.25, \dfrac{3}{8}$, and $.1\overline{6}$ are all rational numbers.

6. Division of a number by zero is undefined.

Write the set by listing its elements. See Example 1.

7. $\{x \mid x$ is a natural number less than 6$\}$

8. $\{m \mid m$ is a natural number less than 9$\}$

9. $\{z \mid z$ is an integer greater than 4$\}$

10. $\{y \mid y$ is an integer greater than 8$\}$

11. $\{a \mid a$ is an even integer greater than 8$\}$

12. $\{k \mid k$ is an odd integer less than 1$\}$

13. $\{x \mid x$ is an irrational number that is also rational$\}$

14. $\{r \mid r$ is a number that is both positive and negative$\}$

15. $\{p \mid p$ is a number whose absolute value is 4$\}$

16. $\{w \mid w$ is a number whose absolute value is 7$\}$

17. $\{z \mid z$ is a whole number and a multiple of 3$\}$

18. $\{n \mid n$ is a counting number and a multiple of 7$\}$

◉ 19. A student claimed that $\{x \mid x$ is a natural number greater than 3$\}$ and $\{y \mid y$ is a natural number greater than 3$\}$ actually name the same set, even though different variables are used. Was this student correct?

◉ 20. A student claimed that $\{\emptyset\}$ and $\emptyset$ name the same set. Was this student correct?

Write the set using set-builder notation. (More than one description is possible.) See Example 2.

21. $\{4, 8, 12, 16, \ldots\}$

22. $\{\ldots, -6, -3, 0, 3, 6, \ldots\}$

23. $\{2, 4, 6, 8\}$

24. $\{11, 12, 13, 14\}$

 CONCEPTUAL   WRITING   ▲ CHALLENGING    SCIENTIFIC CALCULATOR    GRAPHICS CALCULATOR

*Which elements of the given set are **(a)** natural numbers, **(b)** whole numbers,*
*(c) integers, **(d)** rational numbers, **(e)** irrational numbers, **(f)** real numbers,*
(g) undefined? See Example 3.

25. $\left\{ -8, -\sqrt{5}, -.6, 0, \dfrac{1}{0}, \dfrac{3}{4}, \sqrt{3}, 4, 5, \dfrac{13}{2}, 17, \dfrac{40}{2} \right\}$

26. $\left\{ -9, -\sqrt{6}, -.7, 0, \dfrac{2}{0}, \dfrac{6}{7}, \sqrt{7}, 3, 8, \dfrac{21}{2}, 13, \dfrac{75}{5} \right\}$

Graph the elements of the set on a number line.

27. $\{-3, -1, 0, 4, 6\}$

28. $\{-4, -2, 0, 3, 5\}$

29. $\left\{ -\dfrac{2}{3}, 0, \dfrac{4}{5}, \dfrac{12}{5}, \dfrac{9}{2}, 4.8 \right\}$

30. $\left\{ -\dfrac{6}{5}, -\dfrac{1}{4}, 0, \dfrac{5}{6}, \dfrac{13}{4}, \dfrac{11}{2}, 5.2 \right\}$

◎ **31.** Explain the difference between the graph of a number and the coordinate of a point.

◎ **32.** Explain why the real numbers .36 and $.\overline{36}$ have different points as graphs on a number line.

Find the value of the expression. See Example 5.

33. $|-8|$ **34.** $|-11|$ **35.** $-|5|$ **36.** $-|17|$

37. $-|-2|$ **38.** $-|-8|$ **39.** $-|4.5|$ **40.** $-|12.6|$

41. $|-2| + |3|$ **42.** $|-16| + |12|$ **43.** $|-9| - |-3|$

44. $|-10| - |-5|$ **45.** $|-9| + |-13|$ **46.** $|-13| + |-21|$

47. $|-1| + |-2| - |-3|$ **48.** $|-6| + |-4| - |-10|$

Refer to a number line to answer true or false to the statement. See Example 6.

49. $-6 < -2$ **50.** $-4 < -3$ **51.** $-4 > -3$ **52.** $-2 > -1$

53. $3 > -2$ **54.** $5 > -3$ **55.** $-3 \geq -3$ **56.** $-4 \leq -4$

◎ **57.** An inequality of the form "$a < b$" may also be written "$b > a$." Write $-3 < 2$ using this alternate form, and explain why both inequalities are true.

◎ **58.** If $x > 0$ is a false statement for a given value of x, then is $x < 0$ necessarily a true statement? If not, explain why.

Use an inequality symbol to write the statement. See Example 7.

59. 6 is less than 11.   **60.** -4 is less than 12.

61. 4 is greater than x.   **62.** 7 is greater than y.

63. $3t - 4$ is less than or equal to 10.   **64.** $5x + 4$ is greater than or equal to 19.

65. 5 is greater than or equal to 5.   **66.** -3 is less than or equal to -3.

67. t is between -3 and 5.   **68.** r is between -4 and 12.

69. $3x$ is between -3 and 4, including -3 and excluding 4.

70. $5y$ is between -2 and 6, excluding -2 and including 6.

71. $5x + 3$ is not equal to 0.

72. $6x + 7$ is not equal to -3.

Answers (margin):

25. (a) 4, 5, 17, $\dfrac{40}{2}$ (or 20)

(b) 0, 4, 5, 17, $\dfrac{40}{2}$

(c) -8, 0, 4, 5, 17, $\dfrac{40}{2}$

(d) -8, $-.6$, 0, $\dfrac{3}{4}$, 4, 5, $\dfrac{13}{2}$, 17, $\dfrac{40}{2}$ (e) $-\sqrt{5}$, $\sqrt{3}$

(f) All are real numbers except $\dfrac{1}{0}$.   (g) $\dfrac{1}{0}$

26. (a) 3, 8, 13, $\dfrac{75}{5}$ (or 15)

(b) 0, 3, 8, 13, $\dfrac{75}{5}$

(c) -9, 0, 3, 8, 13, $\dfrac{75}{5}$

(d) -9, $-.7$, 0, $\dfrac{6}{7}$, 3, 8, $\dfrac{21}{2}$, 13, $\dfrac{75}{5}$ (e) $-\sqrt{6}$, $\sqrt{7}$

(f) All are real numbers except $\dfrac{2}{0}$.   (g) $\dfrac{2}{0}$

27. (number line, points at $-4, -2, 0, 2, 4, 6, 8$)

28. (number line, points at $-6, -4, -2, 0, 2, 4, 6$)

29. $-\dfrac{2}{3}$, $\dfrac{4}{5}$, $\dfrac{12}{5}$, $\dfrac{9}{2}$, 4.8 (number line -1 to 5)

30. (number line $-\dfrac{6}{5}$, $-\dfrac{1}{4}$, $\dfrac{5}{6}$, $\dfrac{13}{4}$, 5.2, $\dfrac{11}{2}$; -1 to 5)

33. 8   34. 11   35. -5

36. -17 37. -2 38. -8

39. -4.5 40. -12.6

41. 5   42. 28   43. 6

44. 5   45. 22

46. 34   47. 0   48. 0

49. true   50. true

51. false   52. false

53. true   54. true

55. true   56. true

59. $6 < 11$ 60. $-4 < 12$

61. $4 > x$ 62. $7 > y$

63. $3t - 4 \leq 10$

64. $5x + 4 \geq 19$

65. $5 \geq 5$ 66. $-3 \leq -3$

67. $-3 < t < 5$

68. $-4 < r < 12$

69. $-3 \leq 3x < 4$

70. $-2 < 5y \leq 6$

71. $5x + 3 \neq 0$

72. $6x + 7 \neq -3$

◎ CONCEPTUAL   ✎ WRITING   ▲ CHALLENGING   ▦ SCIENTIFIC CALCULATOR   ▦ GRAPHICS CALCULATOR

73. $3 \geq 2$ 74. $4 \leq 5$
75. $-3 \leq -3$
76. $-6 \geq -6$
77. $5 \not< 3$ 78. $6 \not> 7$
79. Pacific Ocean, Indian Ocean, Caribbean Sea, South China Sea, Gulf of California   80. Point Success, Ranier, Matlalcueyetl, Steele, McKinley   81. true
82. false
83. $(-2, \infty)$
84. $(-\infty, 5)$
85. $(-\infty, 6]$
86. $[-3, \infty)$
87. $(0, 3.5)$
88. $(-4, 6.1)$
89. $[2, 7]$
90. $[-3, -2]$
91. $(-4, 3]$
92. $[3, 6)$
93. $(0, 3]$
94. $[-1, 6)$
95. October 1991, November 1991, February 1992, and March 1992
96. January 1991 and February 1991

The slash symbol, /, is used to obtain the *negation* of the meaning of a symbol. We know that if $a = b$ is true, then $a \neq b$ is false, for example. The slash symbol is also used to negate inequality: "$a \not< b$" is read "a is not less than b" and "$a \not> b$" is read "a is not greater than b." The chart shows how these symbols are equivalent to $\geq$ and $\leq$, respectively.

Symbolism with Slash	Equivalent Statement
$a \not< b$	$a \geq b$
$a \not> b$	$a \leq b$

▲*Write an equivalent statement based on the explanation above.*

73. $3 \not< 2$ **74.** $4 \not> 5$ **75.** $-3 \not> -3$

76. $-6 \not< -6$ **77.** $5 \geq 3$ **78.** $6 \leq 7$

◉ Sea level refers to the surface of the ocean. The depth of a body of water such as an ocean or sea can be expressed as a negative number, representing average depth in feet below sea level. On the other hand, the altitude of a mountain can be expressed as a positive number, indicating its height in feet above sea level. The following chart gives selected depths and heights.

Bodies of Water	Average Depth in Feet (as a negative number)	Mountain	Altitude in Feet (as a positive number)
Pacific Ocean	−12,925	McKinley	20,320
South China Sea	−4802	Point Success	14,150
Gulf of California	−2375	Matlalcueyetl	14,636
Caribbean Sea	−8448	Ranier	14,410
Indian Ocean	−12,598	Steele	16,644

79. List the bodies of water in order, starting with the deepest and ending with the shallowest.

80. List the mountains in order, starting with the shortest and ending with the tallest.

81. True or false: The absolute value of the depth of the Pacific Ocean is greater than the absolute value of the depth of the Indian Ocean.

82. True or false: The absolute value of the depth of the Gulf of California is greater than the absolute value of the depth of the Caribbean Sea.

Write the set using interval notation and graph the set on a number line. See Examples 8–10.

83. $\{x \mid x > -2\}$ **84.** $\{x \mid x < 5\}$ **85.** $\{x \mid x \leq 6\}$
86. $\{x \mid x \geq -3\}$ **87.** $\{x \mid 0 < x < 3.5\}$ **88.** $\{x \mid -4 < x < 6.1\}$
89. $\{x \mid 2 \leq x \leq 7\}$ **90.** $\{x \mid -3 \leq x \leq -2\}$ **91.** $\{x \mid -4 < x \leq 3\}$
92. $\{x \mid 3 \leq x < 6\}$ **93.** $\{x \mid 0 < x \leq 3\}$ **94.** $\{x \mid -1 \leq x < 6\}$

◉ *The graphs show the number of initial public stock offerings (IPOs) and the amounts generated from IPOs from January 1991 through May 1992. Use the graphs to answer the following questions.*

95. During which months was the number of IPOs greater than 50 billion?

96. During which months was the number of IPOs less than or equal to 10 billion?

97. During which months was the amount raised greater than 3.5 billion dollars?

98. If x represents the amount raised in July 1991, and y represents the amount raised in September 1991, which one of the following is true: $x > y$ or $x < y$?

97. December 1991, February 1992, and March 1992   98. $x < y$

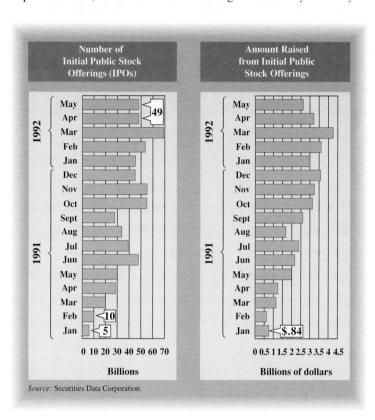

Source: Securities Data Corporation.

1.2 OPERATIONS ON REAL NUMBERS

FOR EXTRA HELP	OBJECTIVES
SSG pp. 7–16 SSM pp. 4–9	1 ▶ Add and subtract real numbers.
Video 1	2 ▶ Find the distance between two points.
	3 ▶ Multiply and divide real numbers.
Tutorial IBM MAC	4 ▶ Use exponents and roots.
	5 ▶ Learn the order of operations.
	6 ▶ Evaluate expressions for given values of variables.

In this section we review the rules for the four operations with signed numbers: addition, subtraction, multiplication, and division.

1 ▶ Add and subtract real numbers.

▶ The answer to an addition problem is called the **sum.** The rules for addition of real numbers follow.

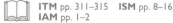

Adding Real Numbers

Like Signs

Add two numbers with the *same* sign by adding their absolute values. The sign of the answer (either $+$ or $-$) is the same as the sign of the two numbers.

Unlike Signs

Add two numbers with *different* signs by subtracting the absolute values of the numbers. The answer is positive if the positive number has the larger absolute value. The answer is negative if the negative number has the larger absolute value.

For example, to add -12 and -8, first find their absolute values:

$$|-12| = 12 \quad \text{and} \quad |-8| = 8.$$

Since these numbers have the *same* sign, add their absolute values: $12 + 8 = 20$. Give the sum the sign of the two numbers. Since both numbers are negative, the sign is negative and

$$-12 + (-8) = -20.$$

Find $-17 + 11$ by subtracting the absolute values, since these numbers have different signs.

$$|-17| = 17 \quad \text{and} \quad |11| = 11$$
$$17 - 11 = 6$$

Give the result the sign of the number with the larger absolute value.

$$-17 + 11 = -6$$

└── Negative since $|-17| > |11|$

E X A M P L E   1

Adding Real Numbers

━━━━━━━━━━━━━━━━━
Chalkboard Exercise

Add $-\dfrac{3}{8} + \dfrac{1}{4}$.

Answer: $-\dfrac{1}{8}$
━━━━━━━━━━━━━━━━━

■ Add.

(a) $(-6) + (-3) = -(6 + 3) = -9$

(b) $(-12) + (-4) = -(12 + 4) = -16$

(c) $4 + (-1) = 3$ **(d)** $-9 + 16 = 7$

(e) $-\dfrac{1}{4} + \dfrac{2}{3} = -\dfrac{3}{12} + \dfrac{8}{12} = \dfrac{5}{12}$ **(f)** $-16 + 12 = -4$

(g) $-\dfrac{7}{8} + \dfrac{1}{3} = -\dfrac{21}{24} + \dfrac{8}{24} = -\dfrac{13}{24}$ ■

We now turn our attention to subtraction of real numbers. The result of subtraction is called the **difference.** Thus, the difference between 7 and 5 is 2. To see how subtraction should be defined, compare the two statements below.

$$7 - 5 = 2$$
$$7 + (-5) = 2$$

In a similar way,

$$9 - 3 = 9 + (-3).$$

That is, to subtract 3 from 9, add the additive inverse of 3 to 9. These examples suggest the following rule for subtraction.

Definition of Subtraction	For all real numbers a and b, $$a - b = a + (-b).$$ (Change the sign of the second number and add.)

This method of observing patterns and similarities and generalizing from them as we did above is used often in mathematics. Looking at many examples strengthens our confidence in such generalizations. If possible, though, mathematicians prefer to prove the results using previously established facts.

EXAMPLE 2
Subtracting Real Numbers

Subtract.

Change to addition.
Change sign of second number.

(a) $6 - 8 = 6 + (-8) = -2$

Changed to addition
Sign changed.

(b) $-12 - 4 = -12 + (-4) = -16$

(c) $-10 - (-7) = -10 + [-(-7)]$ This step is often omitted.
$$= -10 + 7$$
$$= -3$$

(d) $8.43 - (-5.27) = 8.43 + 5.27 = 13.70$ ∎

When a problem with both addition and subtraction is being worked, perform the additions and subtractions in order from the left, as in the following example. Do not forget to work inside the parentheses or brackets first.

EXAMPLE 3
Adding and Subtracting
Real Numbers

Perform the operations.

(a) $15 - (-3) - 5 - 12 = (15 + 3) - 5 - 12$
$$= 18 - 5 - 12$$
$$= 13 - 12$$
$$= 1$$

(b) $-9 - [8 - (-4)] + 6 = -9 - [8 + 4] + 6$
$$= -9 - [4] + 6$$
$$= -9 + 4 + 6$$
$$= -5 + 6$$
$$= 1$$ ∎

2 ▶ Find the distance between two points.

▶   One application of subtraction is finding the distance between two points on a number line. The number line in Figure 13 shows several points. Find the distance between the points 4 and 7 by subtracting the numbers: $7 - 4 = 3$. Since distance is never negative, we must be careful to subtract in such a way that the answer is not negative. Or, to avoid this problem altogether, take the absolute value of the difference. Then the distance between 4 and 7 is either

$$|7 - 4| = 3 \qquad \text{or} \qquad |4 - 7| = 3.$$

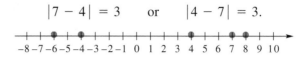

FIGURE 13

Distance

The **distance** between two points on a number line is the absolute value of the difference between the numbers.

E X A M P L E   4

Finding Distance Between Points on the Number Line

Find the distance between the following pairs of points from Figure 13.

(a) 8 and -4

Find the absolute value of the difference of the numbers, taken in either order.

$$|8 - (-4)| = 12 \qquad \text{or} \qquad |-4 - 8| = 12$$

(b) -4 and -6

$$|-4 - (-6)| = 2 \qquad \text{or} \qquad |-6 - (-4)| = 2 \qquad ■$$

3 ▶ Multiply and divide real numbers.

▶   A **product** is the answer to a multiplication problem. For example, 24 is the product of 8 and 3. The rules for products of real numbers are given below.

Multiplying Real Numbers

Like Signs

The product of two numbers with the *same* sign is positive.

Unlike Signs

The product of two numbers with *different* signs is negative.

E X A M P L E   5

Multiplying Real Numbers

Multiply.

(a) $-3(-9) = 27$

(b) $-\dfrac{3}{4}\left(-\dfrac{5}{3} \right) = \dfrac{5}{4}$

(c) $7 \cdot 9 = 63$

(d) $-.05(.3) = -.015$

(e) $\dfrac{2}{3}(-3) = -2$

(f) $-\dfrac{5}{8}\left(\dfrac{12}{13} \right) = -\dfrac{15}{26} \qquad ■$

▶ **TEACHING TIP**

Restate the rules for addition and subtraction of real numbers presented earlier alongside the rules for multiplication and division. ◀

Addition and multiplication are the basic operations on real numbers. Subtraction was defined in terms of addition, and similarly, division is defined in terms of multiplication. The result of dividing two numbers is called the **quotient.** The quotient of the real numbers a and b ($b \neq 0$) is the real number q such that $a = bq$. That is,

$$\frac{a}{b} = q \qquad \text{only if} \qquad a = bq.$$

For example,

$$\frac{36}{9} = 4 \qquad \text{since} \qquad 36 = 9 \cdot 4.$$

Also,

$$\frac{-12}{-2} = 6 \qquad \text{since} \qquad -12 = -2(6).$$

This definition of division is the reason division by 0 is undefined. To see why, suppose that the quotient of a nonzero real number a and 0 is the real number q, or

$$\frac{a}{0} = q.$$

By the definition of division, this means that

$$a = 0 \cdot q.$$

However, $0 \cdot q = 0$ for *every* number q, so no such quotient q is possible. On the other hand, if $a = 0$, we have

$$\frac{0}{0} = q, \qquad \text{or} \qquad 0 = q \cdot 0.$$

This statement is true for all values of q, so there is no single quotient $0/0$.

CAUTION Remember that division by 0 is always undefined.

The division of two numbers can be restated as multiplication.

Dividing Real Numbers

If a and b are real numbers and $b \neq 0$, then

$$\frac{a}{b} = a \cdot \frac{1}{b}.$$

If $b \neq 0$, $1/b$ is the **reciprocal** of b. This rule for division is the reason we "multiply by the reciprocal of the denominator" or "invert the denominator and multiply" when a division problem involves fractions.

EXAMPLE 6
Dividing Real Numbers

Find the quotient.

(a) $\dfrac{24}{-6} = 24\left(-\dfrac{1}{6}\right) = -4$

(b) $\dfrac{-\dfrac{2}{3}}{-\dfrac{1}{2}} = -\dfrac{2}{3}\left(-\dfrac{2}{1}\right) = \dfrac{4}{3}$ ■

Chalkboard Exercise

Divide $\dfrac{-15}{-3}$.

Answer: 5

NOTE Since division is equivalent to multiplication by the reciprocal, the rules for the sign of the quotient are the same as for the sign of the product.

The rules for multiplication and division suggest the results given below.

The fractions $\dfrac{-x}{y}$, $-\dfrac{x}{y}$, and $\dfrac{x}{-y}$ are equal.

Also, the fractions $\dfrac{x}{y}$ and $\dfrac{-x}{-y}$ are equal. (Assume $y \neq 0$).

CAUTION Every fraction has three signs: the sign of the numerator, the sign of the denominator, and the sign of the fraction itself. As shown above, changing any two of these three signs does not change the value of the fraction. Changing only one sign, or changing all three, *does* change the value.

---

4 ▶ Use exponents and roots.

▶ A **factor** of a given number is any number that divides evenly (without remainder) into the given number. For example, 2 and 6 are factors of 12 since $2 \cdot 6 = 12$. Other factors of 12 include 4 and 3, 12 and 1, -4 and -3, -12 and -1, and -6 and -2. A number is in **factored form** if it is expressed as a product of two or more numbers.

▶ **TEACHING TIP**
Introduce the power (x^y) key and the root key on a scientific calculator here. These keys will be discussed in more detail in Chapter 5. ◀

In algebra, exponents are used as a way of writing the products of repeated factors. For example, the product $2 \cdot 2 \cdot 2 \cdot 2 \cdot 2$ is written

$$2 \cdot 2 \cdot 2 \cdot 2 \cdot 2 = 2^5.$$

The number 5 shows that 2 appears as a factor five times. The number 5 is the **exponent,** 2 is the **base,** and 2^5 is an **exponential** or **power.** Multiplying out the five 2s gives

$$2^5 = 2 \cdot 2 \cdot 2 \cdot 2 \cdot 2 = 32.$$

Definition of Exponent

If a is a real number and n is a natural number,

$$a^n = \underbrace{a \cdot a \cdot a \ldots a}_{n \text{ factors of } a}.$$

EXAMPLE 7
Evaluating an Exponential

Evaluate the exponential.

(a) $5^2 = 5 \cdot 5 = 25$
Read 5^2 as "5 squared."

(b) $9^3 = 9 \cdot 9 \cdot 9 = 729$
Read 9^3 as "9 cubed."

(c) $2^6 = 2 \cdot 2 \cdot 2 \cdot 2 \cdot 2 \cdot 2 = 64$
Read 2^6 as "2 to the sixth power" or just "2 to the sixth." ■

Chalkboard Exercise

Evaluate the exponential 3^4.

Answer: 81

We need to be careful when evaluating an exponential with a negative sign. Compare the results in the next example.

EXAMPLE 8
Evaluating Exponentials with Negative Signs

Evaluate the exponential.

(a) $(-3)^5 = (-3)(-3)(-3)(-3)(-3) = -243$

(b) $(-2)^6 = (-2)(-2)(-2)(-2)(-2)(-2) = 64$

(c) $-2^6 = -(2 \cdot 2 \cdot 2 \cdot 2 \cdot 2 \cdot 2) = -64$ ■

Chalkboard Exercise

Evaluate the exponential -3^2.

Answer: -9

Example 8 suggests the following generalizations.

> The product of an odd number of negative factors is negative.
> The product of an even number of negative factors is positive.

CAUTION As shown by Examples 8(b) and (c), it is important to be careful to distinguish between $-a^n$ and $(-a)^n$.

$$-a^n = -1\underbrace{(a \cdot a \cdot a \cdots a)}_{n \text{ factors of } a}$$

$$(-a)^n = \underbrace{(-a)(-a) \cdots (-a)}_{n \text{ factors of } -a}$$

◆ Answers will vary.

◆ **C O N N E C T I O N S** ◆

In Example 7, we used the terms "squared" and "cubed" to refer to powers of 2 and 3, respectively. The term "squared" comes from the figure of a square, which has the same measure for both length and width, as shown in the figure. Similarly, the term "cubed" comes from the figure of a cube. As shown in the figure, the length, width, and height of a cube have the same measure.

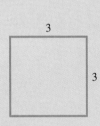

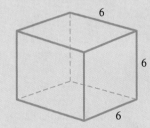

(a) $3 \cdot 3 = 3$ squared, or 3^2 (b) $6 \cdot 6 \cdot 6 = 6$ cubed, or 6^3

FOR DISCUSSION OR WRITING
Why do you suppose there is no special terminology similar to the words squared and cubed for powers that are higher than three?

As we saw in Example 7, $5^2 = 5 \cdot 5 = 25$, so that 5 squared is 25. The opposite of squaring a number is called taking its **square root.** For example, a square root of 25 is 5. Another square root of 25 is -5, since $(-5)^2 = 25$. Thus, 25 has two square roots, 5 and -5. The positive square root of a number is written with the symbol $\sqrt{}$. For example, the positive square root of 25 is written $\sqrt{25} = 5$. The negative square root of 25 is written $-\sqrt{25}$. Since the square of any nonzero real number is positive, a number like $\sqrt{-4}$ is not a real number.

E X A M P L E   9

Finding Square Roots

Find the root.

(a) $\sqrt{36} = 6$ since 6 is positive and $6^2 = 36$.

(b) $\sqrt{144} = 12$ since $12^2 = 144$.

(c) $\sqrt{0} = 0$ since $0^2 = 0$.

Chalkboard Exercise

Find the root.

$$-\sqrt{\dfrac{121}{81}}$$

Answer: $-\dfrac{11}{9}$

(d) $\sqrt{\dfrac{9}{16}} = \dfrac{3}{4}$

(e) $-\sqrt{100} = -10$

(f) $\sqrt{-16}$ is not a real number. ∎

CAUTION   The symbol $\sqrt{}$ is used only for the *positive* square root, except that $\sqrt{0} = 0$.

Since 6 cubed is $6^3 = 6 \cdot 6 \cdot 6 = 216$, the **cube root** of 216 is 6. We write this as

$$\sqrt[3]{216} = 6.$$

In the same way, the **fourth root** of 81 is 3, written

$$\sqrt[4]{81} = 3.$$

The number -3 is also a fourth root of 81, but the symbol $\sqrt[4]{}$ is reserved for roots that are not negative. Negative roots are discussed in Chapter 5.

EXAMPLE   10

Finding Higher Roots

Find the root.

(a) $\sqrt[3]{27} = 3$ since $3^3 = 27$.

(b) $\sqrt[3]{125} = 5$ since $5^3 = 125$.

(c) $\sqrt[4]{16} = 2$ since $2^4 = 16$.

(d) $\sqrt[5]{32} = 2$ since $2^5 = 32$.

(e) $\sqrt[7]{128} = 2$ since $2^7 = 128$. ∎

Chalkboard Exercise

Find the root.

$$\sqrt[4]{\dfrac{16}{625}}$$

Answer: $\dfrac{2}{5}$

◈ A calculator will show an error message for $\sqrt{-25}$. Some calculators will also show an error message for $\sqrt[3]{-125}$, even though $\sqrt[3]{-125} = -5$.

◇ **CONNECTIONS** ◇

Scientific calculators have a key labeled y^x (or x^y) to compute powers of numbers. For instance, 2^5 is found by entering [2] [y^x] [5] [=]. The result is 32. This key can also be used to find roots by using it together with the key labeled 2nd or INV, which produces the inverse operation—that is, the operation that "undoes" exponentiation. For example, to find the fifth root of 32, enter [32] [INV] [y^x] [5] [=] to get 2. Most scientific calculators also have a key labeled $\sqrt{x}$ for finding square roots. With this key, to find $\sqrt{25}$, enter [25] [$\sqrt{x}$] [=] to get 5. (Graphics calculators use different keystrokes.)

FOR DISCUSSION OR WRITING

Experiment with finding powers and roots using your calculator. Then use the calculator to find $\sqrt{-25}$ and $\sqrt[3]{-125}$. Use both methods to find the square root. Are the results what you expected? Explain why or why not. What does this tell you about using a calculator?

5 ▶ Learn the order of operations.

▶   Given a problem such as $5 + 2 \cdot 3$, should 5 and 2 be added first or should 2 and 3 be multiplied first? When a problem involves more than one operation, we use the following order of operations. (This is the order used by computers and many calculators.)

Order of Operations

If parentheses, square brackets, or fraction bars are present:

Step 1 Work separately above and below any fraction bar.
Step 2 Use the rules below within each set of parentheses or square brackets. Start with the innermost and work outward.

If no parentheses or brackets are present:

Step 1 Evaluate all powers and roots.
Step 2 Do any multiplications or divisions in the order in which they occur, working from left to right.
Step 3 Do any additions or subtractions in the order in which they occur, working from left to right.

▶ **TEACHING TIP**
Show how $5 + 2 \cdot 3$ suggests two possible answers: 21 or 11; however, only 11 is correct according to the order of operations. ◀

E X A M P L E   11
Using Order of Operations

■ Simplify $5 + 2 \cdot 3$.
To do this, first multiply, and then add.

$$5 + 2 \cdot 3 = 5 + 6 \quad \text{Multiply.}$$
$$= 11 \quad \text{Add.} \ ■$$

E X A M P L E   12
Using Order of Operations

■ Simplify $4 \cdot 3^2 + 7 - (2 + 8)$.
Work inside the parentheses first.

$$4 \cdot 3^2 + 7 - (2 + 8) = 4 \cdot 3^2 + 7 - 10$$

For Example 11

Simplify powers and roots. Since $3^2 = 3 \cdot 3 = 9$,

$$4 \cdot 3^2 + 7 - 10 = 4 \cdot 9 + 7 - 10.$$

Chalkboard Exercise
Simplify $5 \cdot 9 + 2 \cdot 4$.   Answer: 53

Do all multiplications or divisions, working from left to right.

$$4 \cdot 9 + 7 - 10 = 36 + 7 - 10$$

Chalkboard Exercise
Simplify $(4 + 2) - 3^2 - (8 - 3)$.   Answer: -8

Finally, do all additions or subtractions, working from left to right.

$$36 + 7 - 10 = 43 - 10$$
$$= 33 \ ■$$

E X A M P L E   13
Using Order of Operations

■ Simplify $\dfrac{1}{2} \cdot 4 + (6 \div 3 \cdot 7)$.
Work inside the parentheses first, doing the division before the multiplication.

Chalkboard Exercise
Simplify $-\dfrac{4}{7}(-14) - 6\left(-\dfrac{2}{3}\right) + \dfrac{1}{2}(-12)$.   Answer: 6

$$\frac{1}{2} \cdot 4 + (6 \div 3 \cdot 7) = \frac{1}{2} \cdot 4 + (2 \cdot 7) \quad \text{Divide.}$$
$$= 2 + (14) \quad \text{Multiply.}$$
$$= 16 \quad \text{Add.} \ ■$$

E X A M P L E   14
Using Order of Operations

■ Simplify $\dfrac{5 + (-2^3)(2)}{6 \cdot \sqrt{9} - 9 \cdot 2}$.
The division bar is also a grouping symbol. The numerator and denominator must always be calculated separately before performing the division.

$$\frac{5 + (-2^3)(2)}{6 \cdot \sqrt{9} - 9 \cdot 2} = \frac{5 + (-8)(2)}{6 \cdot 3 - 9 \cdot 2}$$ Evaluate powers and roots.

$$= \frac{5 - 16}{18 - 18}$$ Multiply.

$$= \frac{-11}{0}$$ Subtract.

Since division by zero is not possible, the given expression is undefined. ■

6 ▶ Evaluate expressions for given values of variables.

▶ In many problems, an expression is given, along with the values of the variables in the expression. The expression can be *evaluated* by substituting the numerical values for the variables.

E X A M P L E   15
Evaluating Expressions

Evaluate the expression when $m = -4$, $n = 5$, and $p = -6$.

(a) $5m - 9n$
Replace m with -4 and n with 5.

$$5m - 9n = 5(-4) - 9(5) = -20 - 45 = -65$$

(b) $\dfrac{m + 2n}{4p} = \dfrac{-4 + 2(5)}{4(-6)} = \dfrac{-4 + 10}{-24} = \dfrac{6}{-24} = -\dfrac{1}{4}$

(c) $-3m^2 + n^3$
Replace m with -4 and n with 5.

$$-3m^2 + n^3 = -3(-4)^2 + 5^3$$
$$= -3(16) + 125$$
$$= -48 + 125 = 77 \quad ■$$

CAUTION When evaluating expressions, it is a good idea to use parentheses around any negative numbers that are substituted for variables. Notice the placement of the parentheses in Example 15(c) assures that -4 is squared, giving a positive result. Writing (-4^2) would lead to -16, which would be incorrect.

1.2 EXERCISES

◉ *Decide whether the statement is always true, sometimes true, or never true. If it is sometimes true, give an example where it is true and where it is false.*

1. The sum of two negative numbers is negative.

2. The sum of two positive numbers is positive.

3. The sum of a negative number and a positive number is positive.

4. The sum of a negative number and a positive number is negative.

5. The product of two numbers with like signs is positive.

6. The product of two numbers with different signs is positive.

7. The difference between two positive numbers is negative.

8. The difference between two negative numbers is positive.

9. The sum of a positive number and a negative number is 0.

10. The difference between a positive number and a negative number is 0.

Add or subtract as indicated. See Examples 1–3.

11. $13 + (-4)$

12. $19 + (-13)$

13. $-6 + (-13)$

14. $-8 + (-15)$

15. $-\dfrac{7}{3} + \dfrac{3}{4}$

16. $-\dfrac{5}{6} + \dfrac{3}{8}$

17. $-.125 + .312$

18. $-.235 + .455$

19. $-8 - (-12) - (2 - 6)$

20. $-3 + (-14) + (-5 + 3)$

21. $\left(-\dfrac{5}{4} - \dfrac{2}{3}\right) + \dfrac{1}{6}$

22. $\left(-\dfrac{5}{8} + \dfrac{1}{4}\right) - \left(-\dfrac{1}{4}\right)$

23. $(-.382) + (4 - .6)$

24. $(3 - 2.94) - (-.63)$

The sketch shows a number line with several points labeled. Find the distance between the pair of points. See Example 4.

25. *A and B*

26. *A and C*

27. *D and F*

28. *E and C*

 **MATHEMATICAL CONNECTIONS** (Exercises 29–32)

In Section 1.1 we discussed the meanings of $a < b$, $a = b$, and $a > b$. Choose two numbers a and b such that $a < b$. Work Exercises 29–32 in order.

29. Find the difference $a - b$.

30. How does the answer in Exercise 29 compare to 0? (Is it greater than, less than, or equal to 0?)

31. Repeat Exercise 29 with different values for a and b.

32. How does the answer in Exercise 31 compare to 0? Based on your observations in these exercises, complete the following statement: If $a < b$, then $a - b$ _____ 0.

⊚ **33.** Give an example of a difference between two negative numbers that is equal to 5.

⊚ **34.** Give an example of a sum of a positive number and a negative number that is equal to 4.

⊚ **35.** A statement that is often heard is "Two negatives give a positive." When is this true? When is it not true? Give a more precise statement that conveys this message.

⊚ **36.** Explain why the reciprocal of a nonzero number must have the same sign as the number.

Multiply or divide. See Examples 5 and 6.

37. $(-15)(-3)$

38. $(-12)(-4)$

39. $\dfrac{3}{4}(-20)(-12)$

40. $-\dfrac{2}{5}(-15)(-3)$

41. $-3.45(-2.14)$

42. $-2.4(-2.45)$

43. $\dfrac{-100}{-25}$

44. $\dfrac{-300}{-60}$

45. $\dfrac{\frac{12}{13}}{-\frac{4}{3}}$

46. $\dfrac{\frac{5}{6}}{-\frac{1}{30}}$

47. $\dfrac{5}{0}$

48. $\dfrac{-1}{0}$

---

*Exercise groups marked MATHEMATICAL CONNECTIONS are designed to show interrelationships among concepts currently being studied and concepts that have appeared in previous sections. They should be worked in the order in which they appear. Both even and odd answers for these exercises are given in the answer section.

⊚ CONCEPTUAL   ✎ WRITING   ▲ CHALLENGING   ▦ SCIENTIFIC CALCULATOR   ▦ GRAPHICS CALCULATOR

51. true   52. true
53. true   54. false
55. false   56. true
57. true   58. true
59. false   60. true
61. 11   62. 19
63. .021952
64. .753571   65. $\dfrac{49}{100}$
66. $-\dfrac{49}{100}$ 67. -30
68. -20 69. -7
70. -3 71. 10   72. 4
The number of digits
displayed will vary in
Exercises 75–80.
75. 136.011029
76. 20.8038458
77. 4.534873083
78. 7.57094167
79. 1.659870127
80. 1.549424047   81. not a
real number   82. positive
83. 29   84. 27
85. -79 86. -45
87. 39   88. -2
89. -2 90. 4   91. 2
92. undefined
93. undefined
94. $-\dfrac{4}{5}$

◉ **49.** In your own words, explain why division by 0 is undefined.

◉ **50.** In your own words, explain why dividing 0 by a nonzero number will always give a quotient of 0.

◉ *Decide whether the statement is true or false.*

51. $(-2)^7$ is a negative number.

52. $(-2)^8$ is a positive number.

53. The product of 8 positive factors and 8 negative factors is positive.

54. The product of 3 positive factors and 3 negative factors is positive.

55. $-4^6 = (-4)^6$

56. $-4^7 = (-4)^7$

57. $\sqrt{16}$ is a positive number.

58. $3 + 5 \cdot 6 = 3 + (5 \cdot 6)$

59. In the exponential -3^5, -3 is the base.

60. $\sqrt[3]{a}$ has the same sign as a for all nonzero real numbers a.

Evaluate. See Examples 7–10.

61. $\sqrt{121}$ **62.** $\sqrt{361}$ **63.** $.28^3$ **64.** $.91^3$

65. $\left(-\dfrac{7}{10}\right)^2$ **66.** $-\dfrac{7}{10}^2$ **67.** $-\sqrt{900}$ **68.** $-\sqrt{400}$

69. $-\sqrt[3]{343}$ **70.** $-\sqrt[3]{27}$ **71.** $\sqrt[4]{10,000}$ **72.** $\sqrt[5]{1024}$

◉ **73.** Why is it incorrect to say that $\sqrt{16}$ is equal to 4 or -4?

◉ **74.** Explain why $\sqrt[3]{-1000}$ is equal to $-\sqrt[3]{1000}$.

▦ *Find the following roots on a calculator. Show as many digits as your calculator displays.*

75. $\sqrt{18,499}$ **76.** $\sqrt{432.8}$ **77.** $\sqrt[3]{93.26}$
78. $\sqrt[3]{433.96}$ **79.** $\sqrt[5]{12.6}$ **80.** $\sqrt[5]{8.93}$

◉ **81.** If a is a positive number, is $-\sqrt{-a}$ positive, negative, or not a real number?

◉ **82.** If a is a positive number, is $-\sqrt[3]{-a}$ positive, negative, or not a real number?

Perform the operation where possible, using the order of operations. See Examples 11–14.

83. $-7(-3) - (-2^3)$ **84.** $-4 - 3(-2) + 5^2$

85. $|-6 - 5|(-8) + 3^2$ **86.** $(-6 - 3)|-2 - 3|$

87. $(-8 - 5)(-2 - 1)$ **88.** $\dfrac{(-10 + 4) \cdot (-3)}{-7 - 2}$

89. $\dfrac{(-6 + 3) \cdot (-4)}{-5 - 1}$ **90.** $\dfrac{2(-5 + 3)}{-2^2} - \dfrac{(-3^2 + 2)3}{3 - (-4)}$

91. $\dfrac{3(-5 - 2^2)}{3^2} + \dfrac{(3^2 \cdot 2 - 7)(5)}{5 \cdot 2 - (-1)}$ **92.** $\dfrac{2(-5) + (-3)(-2^2)}{-6 + 5 + 1}$

93. $\dfrac{3(-4) + (-5)(-2)}{2^3 - 2 + (-6)}$ **94.** $-\dfrac{4}{5}[6(-4) + (-5)(-5)]$

95. $-\dfrac{1}{4}[3(-5) + 7(-5) + 1(-2)]$

▲ **96.** $\dfrac{5 - 3\left(\dfrac{-5 - 9}{-7}\right) - 6}{-9 - 11 + 3 \cdot 7}$

▲ **97.** $\dfrac{-4\left(\dfrac{12 - (-8)}{3 \cdot 2 + 4}\right) - 5(-1 - 7)}{-9 - (-7) - [-5 - (-8)]}$

◎ **98.** Write a paragraph explaining how you would evaluate the expression
✐ $(a + 2b)(-3b^2 + \sqrt{c})$, if $a = 4$, $b = 5$, and $c = 16$.

Evaluate the expression if $a = -3$, $b = 64$, and $c = 6$. See Example 15.

99. $3a + \sqrt{b}$

100. $-2a - \sqrt{b}$

101. $\sqrt[3]{b} + c - a$

102. $\sqrt[3]{b} - c + a$

103. $4a^3 + 2c$

104. $-3a^4 - 3c$

105. $\dfrac{2c + a^3}{4b + 6a}$

106. $\dfrac{3c + a^2}{2b - 6c}$

Solve the problem.

107. The highest temperature ever recorded in Juneau, Alaska, was 90° Fahrenheit. The lowest temperature ever recorded there was −22° Fahrenheit. What is the difference between these two temperatures?

108. On August 10, 1936, a temperature of 120° Fahrenheit was recorded in Arkansas. On February 13, 1905, Arkansas recorded a temperature of −29° Fahrenheit. What is the difference between these two temperatures?

109. Telescope Peak, altitude 11,049 feet, is next to Death Valley, 282 feet below sea level. Find the difference between these altitudes.

110. The surface of the Dead Sea has altitude 1299 feet below sea level. A stunt pilot is flying 80 feet above that surface. How much altitude must she gain to clear a 3852-foot pass by 225 feet?

◎ *Use the graph to answer the questions in Exercises 111 and 112.**

111. What is the difference between the net incomes of Southwest Airlines and Delta Airlines?

112. What is the sum of the net incomes of all the airlines listed?

Answers (margin):

95. 13

96. −7   97. $-\dfrac{32}{5}$

99. −1   100. −2

101. 13   102. −5

103. −96   104. −261

105. $-\dfrac{15}{238}$ 106. $\dfrac{27}{92}$

107. 112° Fahrenheit

108. 149° Fahrenheit

109. 11,331 feet

110. 5296 feet

111. $266 million

112. −$1503 million

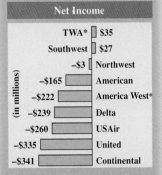

AIRLINE INCOME

The current air fare war may hurt airlines already in financial trouble. U.S. airlines lost almost $2 billion in a recent year.

Net Income

TWA* — $35

Southwest — $27

−$3 — Northwest

−$165 — American

−$222 — America West*

−$239 — Delta

−$260 — USAir

−$335 — United

−$341 — Continental

(in millions)

*Reorganized under bankruptcy law

Source: Air Transport Association of America.

*Graph for Exercises 111 and 112, "Airline Income," from *The Times-Picayune,* May 30, 1992. Copyright © 1992 by The Times-Picayune. Reprinted by permission.

◎ CONCEPTUAL   WRITING   ▲ CHALLENGING   SCIENTIFIC CALCULATOR   ▦ GRAPHICS CALCULATOR

1.3 PROPERTIES OF REAL NUMBERS

FOR EXTRA HELP	OBJECTIVES
📖 **SSG** pp. 16–19 **SSM** pp. 9–11	1 ▶ Use the distributive property.
📼 **Video** 1	2 ▶ Use the inverse properties.
	3 ▶ Use the identity properties.
💾 **Tutorial** IBM MAC	4 ▶ Use the commutative and associative properties.
	5 ▶ Use the multiplication property of zero.

◊ Answers will vary.

◊ **C O N N E C T I O N S** ◊

The study of any subject is simplified when we know the properties of the subject. The property of water that it boils at 100°C and freezes at 0°C, changing its form at each of these temperatures, helps us to predict its behavior. Learning the general properties that govern how numbers behave is an important part of our study of algebra.

FOR DISCUSSION OR WRITING

List some basic properties of numbers and the operations of addition, subtraction, multiplication, and division that you have observed.

In this section we discuss many of the basic properties of the real numbers. These properties are results that have been observed to occur consistently in work with numbers, so they have been generalized to apply to expressions with variables as well.

1 ▶ Use the distributive property.

▶ The properties we discuss in this section are used in simplifying algebraic expressions. For example, notice that

$$2(3 + 5) = 2 \cdot 8 = 16$$

and

$$2 \cdot 3 + 2 \cdot 5 = 6 + 10 = 16$$

so that

$$2(3 + 5) = 2 \cdot 3 + 2 \cdot 5.$$

This idea is illustrated by the divided rectangle in Figure 14. Similarly,

$$-4[5 + (-3)] = -4(2) = -8$$

and

$$-4(5) + (-4)(-3) = -20 + 12 = -8$$

so

$$-4[5 + (-3)] = -4(5) + (-4)(-3).$$

▶ **TEACHING TIP**

Mention to students that the distributive property will be used in factoring. Students should observe that when using this property, parentheses used on one side are removed on the other side.

Also, note that there can be more than two terms inside parentheses separated by + or − signs. ◄

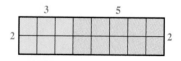

Area of left part is $2 \cdot 3 = 6$.
Area of right part is $2 \cdot 5 = 10$.
Area of total is $2(3 + 5) = 16$.

FIGURE 14

These examples suggest the **distributive property.**

| Distributive Property | For any real numbers a, b, and c, |

$$a(b + c) = ab + ac \quad \text{and} \quad (b + c)a = ba + ca.$$

The distributive property can also be written

$$ab + ac = a(b + c)$$

This property is important because it provides a way to change a *product* $(a(b + c))$ to a *sum* $(ab + ac)$ or a sum to a product. When the form $a(b + c) = ab + ac$ is used, we sometimes refer to it as "removing parentheses."

E X A M P L E   1

Using the Distributive Property

Use the distributive property to rewrite the expression.

(a) $3(x + y)$

In the statement of the property, let $a = 3$, $b = x$ and $c = y$. Then

$$3(x + y) = 3x + 3y.$$

(b) $-2(5 + k) = -2(5) + (-2)(k)$
$$= -10 - 2k$$

(c) $4x + 8x$

Use the second form of the property.

$$4x + 8x = (4 + 8)x = 12x$$

(d) $3r - 7r = 3r + (-7)r$ Definition of subtraction
$$= [3 + (-7)]r \quad \text{Distributive property}$$
$$= -4r$$

(e) $5p + 7q$

Since there is no common factor here, we cannot use the distributive property to simplify the expression. ∎

As illustrated in Example 1(d), the distributive property can also be used for subtraction, so that

$$a(b - c) = ab - ac.$$

2 ▶ Use the inverse properties.

▶ In Section 1.1 we saw that the additive inverse of a number a is $-a$. For example, 3 and -3 are additive inverses, as are -8 and 8. The number 0 is its own additive inverse. In Section 1.2, we saw that two numbers with a product of 1 are reciprocals. Another name for a reciprocal is **multiplicative inverse.** This is similar to the idea of an additive inverse. Thus, 4 and $1/4$ are multiplicative inverses, and so are $-2/3$ and $-3/2$. (Note that a pair of reciprocals has the same sign.) These properties are called the **inverse properties** of addition and multiplication.

| Inverse Properties | For any real number a, there is a single real number $-a$, such that |

$$a + (-a) = 0 \quad \text{and} \quad -a + a = 0.$$

For any nonzero real number a, there is a single real number $1/a$ such that

$$a \cdot \frac{1}{a} = 1 \quad \text{and} \quad \frac{1}{a} \cdot a = 1.$$

Examples showing how these properties are used are given later in this section.

3 ▶ Use the identity properties.

▶   The numbers 0 and 1 each have a special property. Zero is the only number that can be added to any number to get that number. That is, adding 0 leaves the identity of a number unchanged. For this reason, 0 is called the **identity element for addition.** In a similar way, multiplying by 1 leaves the identity of any number unchanged, so 1 is the **identity element for multiplication.** The following **identity properties** summarize this discussion.

Identity Properties

For any real number a,

$$a + 0 = 0 + a = a$$

and

$$a \cdot 1 = 1 \cdot a = a.$$

The identity property for 1 is especially useful in simplifying algebraic expressions.

E X A M P L E   2
Using the Identity Property $1 \cdot a = a$

Use the identity property for 1 to rewrite the expression so the distributive property can be used.

(a) $12m + m$

$$
\begin{aligned}
12m + m &= 12m + 1m &&\text{Identity property} \\
&= (12 + 1)m &&\text{Distributive property} \\
&= 13m
\end{aligned}
$$

(b)
$$
\begin{aligned}
y + y &= 1y + 1y &&\text{Identity property} \\
&= (1 + 1)y &&\text{Distributive property} \\
&= 2y
\end{aligned}
$$

(c)
$$
\begin{aligned}
-(m - 5n) &= -1(m - 5n) &&\text{Identity property} \\
&= -1 \cdot m + (-1)(-5n) &&\text{Distributive property} \\
&= -m + 5n \quad ■
\end{aligned}
$$

Expressions such as $12m$ and $5n$ from Example 2 are examples of *terms*. A **term** is a number or the product of a number and one or more variables raised to powers. Terms with exactly the same variables raised to exactly the same powers are called **like terms.** The number in the product is called the **numerical coefficient** or just the **coefficient.** For example, in the term $5p$, the coefficient is 5.

4 ▶ Use the commutative and associative properties.

▶   Simplifying expressions as in Examples 2(a) and (b) is called **combining like terms.** Only like terms may be combined. To combine like terms in an expression such as

$$-2m + 5m + 3 - 6m + 8$$

we need two more properties. We are familiar with the fact that

$$3 + 9 = 12 \quad \text{and} \quad 9 + 3 = 12.$$

Also,

$$3 \cdot 9 = 27 \quad \text{and} \quad 9 \cdot 3 = 27.$$

Furthermore, notice that

$$(5 + 7) + (-2) = 12 + (-2) = 10$$

and

$$5 + [7 + (-2)] = 5 + 5 = 10.$$

Also,

$$(5 \cdot 7)(-2) = 35(-2) = -70$$

and

$$(5)[7(-2)] = 5(-14) = -70.$$

These observations suggest the following properties.

Commutative Properties	For any real numbers a and b, $$a + b = b + a$$ $$ab = ba.$$

Associative Properties	For any real numbers a, b, and c, $$a + (b + c) = (a + b) + c$$ $$a(bc) = (ab)c.$$

NOTE The associative properties are used to *regroup* the terms of an expression. The commutative properties are used to change the *order* of the terms in an expression.

EXAMPLE 3

Using the Commutative and Associative Properties

Use the properties to combine like terms: $-2m + 5m + 3 - 6m + 8.$

$$-2m + 5m + 3 - 6m + 8$$
$$= (-2m + 5m) + 3 - 6m + 8 \qquad \text{Order of operations}$$
$$= 3m + 3 - 6m + 8 \qquad \text{Distributive property}$$

By the order of operations, the next step would be to add $3m$ and 3, but they are unlike terms. To get $3m$ and $-6m$ together, use the associative and commutative properties. Begin by putting in parentheses and brackets according to the order of operations.

$$[(3m + 3) - 6m] + 8$$
$$= [3m + (3 - 6m)] + 8 \qquad \text{Associative property}$$
$$= [3m + (-6m + 3)] + 8 \qquad \text{Commutative property}$$
$$= [(3m + [-6m]) + 3] + 8 \qquad \text{Associative property}$$
$$= (-3m + 3) + 8 \qquad \text{Combine like terms.}$$
$$= -3m + (3 + 8) \qquad \text{Associative property}$$
$$= -3m + 11 \qquad \text{Add.}$$

In practice, many of the steps are not written down, but you should realize that the commutative and associative properties are used whenever the terms in an expression are rearranged in order to combine like terms. ■

EXAMPLE 4

Using the Properties of Real Numbers

Use the properties to simplify the expression.

(a) $5y^2 - 8y^2 - 6y^2 + 11y^2$

$$5y^2 - 8y^2 - 6y^2 + 11y^2 = (5 - 8 - 6 + 11)y^2 = 2y^2$$

(b) $-2(m - 3)$

$$-2(m - 3) = -2(m) - (-2)(3) = -2m + 6$$

(c) $3x^3 + 4 - 5(x^3 + 1) - 8$

First use the distributive property to eliminate the parentheses.

$$3x^3 + 4 - 5(x^3 + 1) - 8 = 3x^3 + 4 - 5x^3 - 5 - 8$$

Next use the commutative and associative properties to rearrange the terms; then combine like terms.

$$= 3x^3 - 5x^3 + 4 - 5 - 8$$
$$= -2x^3 - 9$$

(d) $8 - (3m + 2)$

Think of $8 - (3m + 2)$ as $8 - 1 \cdot (3m + 2)$.

$$8 - 1 \cdot (3m + 2) = 8 - 3m - 2 = 6 - 3m$$

(e) $(3x)(5)(y) = [(3x)(5)]y$ Order of operations

$$= [3(x \cdot 5)]y \qquad \text{Associative property}$$
$$= [3(5x)]y \qquad \text{Commutative property}$$
$$= [(3 \cdot 5)x]y \qquad \text{Associative property}$$
$$= (15x)y$$
$$= 15(xy) \qquad \text{Associative property}$$
$$= 15xy$$

As mentioned earlier, many of these steps usually are not written out. ■

5 ▶ Use the multiplication property of zero.

▶   The additive identity property gives a special property of zero, namely that $a + 0 = a$ for any real number a. The **multiplication property of zero** gives a special property of zero that involves multiplication: The product of any real number and zero is zero.

Multiplication Property of Zero

For all real numbers a,

$$a \cdot 0 = 0 \qquad \text{and} \qquad 0 \cdot a = 0.$$

Chalkboard Exercise

Simplify the expression $6 - (2x + 7) - 3$.

Answer: $-2x - 4$

▶ **TEACHING TIP**
Encourage oral participation by choosing exercises from this section which require students to identify the property being used. ◀

1.3 EXERCISES

Answer the question or respond to the statement in a complete sentence.

1. What is the identity element for addition?
2. What is the identity element for multiplication?
3. What is meant by *like terms*?
4. What is the coefficient in the term $-6x^2z$?
5. What is the distinction between the commutative and associative properties?
6. Explain the multiplication property of zero.

Use the properties of real numbers to simplify the expression. See Examples 1 and 2.

7. $5k + 3k$
8. $6a + 5a$
9. $-9r + 7r$
10. $-4n + 6n$
11. $-8z + 4w$
12. $-12k + 3r$
13. $-a + 7a$
14. $-s + 9s$
15. $2(m + p)$
16. $3(a + b)$
17. $-12(x - y)$
18. $-10(p - q)$
19. $-5(2d + f)$
20. $-2(3m + n)$

Simplify the expression by removing parentheses and combining terms. See Examples 1–4.

21. $4x + 3x + 7 + 19$
22. $5m + 9m + 8 + 14$
23. $-12y + 4y + 3 + 2y$
24. $-5r - 9r + 8r - 5$
25. $-6p + 11p - 4p + 6 + 5$
26. $-8x - 5x + 3x - 12 + 9$
27. $3(k + 2) - 5k + 6 + 3$
28. $5(r - 3) + 6r - 2r + 4$
29. $-2(m + 1) + 3(m - 4)$
30. $6(a - 5) - 4(a + 6)$
31. $.25(8 + 4p) - .5(6 + 2p)$
32. $.4(10 - 5x) - .8(5 + 10x)$
33. $-(2p + 5) + 3(2p + 4) - 2p$
34. $-(7m - 12) - 2(4m + 7) - 8m$
35. $2 + 3(2z - 5) - 3(4z + 6) - 8$
36. $-4 + 4(4k - 3) - 6(2k + 8) + 7$

Complete the statement so that the indicated property is illustrated. Simplify the answer, if possible.

37. $5x + 8x =$ _____
   (distributive property)

38. $9y - 6y =$ _____
   (distributive property)

39. $5(9r) =$ _____
   (associative property)

40. $-4 + (12 + 8) =$ _____
   (associative property)

41. $5x + 9y =$ _____
   (commutative property)

42. $-5 \cdot 7 =$ _____
   (commutative property)

43. $1 \cdot 7 =$ _____
   (identity property)

44. $-12x + 0 =$ _____
   (identity property)

1. The identity element for addition is 0.   2. The identity element for multiplication is 1.
3. Like terms are terms with exactly the same variables raised to exactly the same powers.
4. The coefficient in the term $-6x^2z$ is -6.
5. The commutative properties state that the *order* in which the terms are operated on does not affect the answer, while the associative properties state that the *grouping* of the terms does not affect the answer.
6. The multiplication property of zero states that the product of any real number and zero is zero.   7. $8k$ 8. $11a$
9. $-2r$ 10. $2n$
11. $-8z + 4w$ (cannot be simplified)
12. $-12k + 3r$ (cannot be simplified)   13. $6a$
14. $8s$ 15. $2m + 2p$
16. $3a + 3b$
17. $-12x + 12y$
18. $-10p + 10q$
19. $-10d - 5f$
20. $-6m - 2n$
21. $7x + 26$
22. $14m + 22$
23. $-6y + 3$
24. $-6r - 5$
25. $p + 11$ 26. $-10x - 3$
27. $-2k + 15$
28. $9r - 11$
29. $m - 14$ 30. $2a - 54$
31. -1 32. $-10x$
33. $2p + 7$ 34. $-23m - 2$
35. $-6z - 39$
36. $4k - 57$
37. $(5 + 8)x = 13x$
38. $(9 - 6)y = 3y$
39. $(5 \cdot 9)r = 45r$
40. $(-4 + 12) + 8 = 16$
41. $9y + 5x$
42. $7 \cdot (-5) = -35$
43. 7 44. $-12x$

 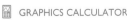

45. 0   46. 1
47. $8(-4) + 8x =$
$-32 + 8x$
48. $3x - 3y + 3z$
49. 0   50. 0
51. associative property of
addition   52. associative
property of addition
53. commutative property
of addition
54. associative property of
addition   55. distributive
property   56. arithmetic
facts   57. Answers will
vary. One example is
washing your face and
brushing your teeth.
58. Answers will vary. One
example is waking up and
going to sleep.
59. $2 + 6 \cdot 5 =$
$2 + 30 = 32$, which does
not equal $8 \cdot 5 = 40$.
60. $4 \cdot 5 - 5 = 15$, which
does not equal 4.
61. 1900   62. 2700
63. 75   64. 87.5
65. 431   66. 48
67. Yes. Any nonzero
numbers a and b that have
the same absolute value
satisfy $\dfrac{a}{b} = \dfrac{b}{a}$.
68. no   69. No. One
example is $7 + (5 \cdot 3) =$
$(7 + 5)(7 + 3)$, which is
false.

45. $-\dfrac{1}{4}ty + \dfrac{1}{4}ty = $ _____
(inverse property)

46. $-\dfrac{9}{8}\left(-\dfrac{8}{9}\right) = $ _____
(inverse property)

47. $8(-4 + x) = $ _____
(distributive property)

48. $3(x - y + z) = $ _____
(distributive property)

49. $0(.875x + 9y - 88z) = $ _____
(multiplication property of 0)

50. $0(35t^2 - 8t + 12) = $ _____
(multiplication property of 0)

◆ **MATHEMATICAL CONNECTIONS** (Exercises 51–56) ◆

While it may seem that simplifying the expression $3x + 4 + 2x + 7$ to $5x + 11$ is fairly easy, there are several important steps that require mathematical justification. These steps are usually done mentally. For now, provide the property that justifies the statement in the simplification. (These steps could be done in other orders.)

51. $3x + 4 + 2x + 7 = (3x + 4) + (2x + 7)$

52. $= 3x + (4 + 2x) + 7$

53. $= 3x + (2x + 4) + 7$

54. $= (3x + 2x) + (4 + 7)$

55. $= (3 + 2)x + (4 + 7)$

56. $= 5x + 11$

◉ **57.** Give an "everyday" example of a commutative operation.
◉ **58.** Give an "everyday" example of inverse operations.
◉ **59.** Replace x with 5 to show that $2 + 6x \neq 8x$.
◉ **60.** Replace x with 5 to show that $4x - x \neq 4$.

◉ *Use the distributive property to calculate the following values in your head.*

61. $96 \cdot 19 + 4 \cdot 19$ **62.** $27 \cdot 60 + 27 \cdot 40$ **63.** $58 \cdot \dfrac{3}{2} - 8 \cdot \dfrac{3}{2}$

64. $8.75(15) - 8.75(5)$ **65.** $4.31(69) + 4.31(31)$ **66.** $\dfrac{8}{5}(17) + \dfrac{8}{5}(13)$

▲ **67.** Are there any two different numbers a and b for which $a/b = b/a$? Give an example if
◉   your answer is yes.

▲ **68.** Do *any* different numbers satisfy the statement $a - b = b - a$? Give an example if
◉   your answer is yes.

▲ **69.** By the distributive property, $a(b + c) = ab + ac$. This property is more completely
◉   named the **distributive property of multiplication over addition.** Is there a
         distributive property of addition over multiplication? That is, does

$$a + (b \cdot c) = (a + b)(a + c)$$

for all real numbers, a, b, and c? To find out, try various sample values of a, b, and c.

◉ **70.** Explain how the distributive property is used in combining like terms.

CHAPTER 1 SUMMARY

KEY TERMS

1.1 set
   elements (members)
   empty set (null set)
   variable
   set-builder notation
   number line
   coordinate
   graph
   additive inverse
      (negative,
      opposite)
   signed numbers
   absolute value
   inequality
   interval
   interval notation

1.2 sum
   difference
   product
   quotient
   reciprocal

factor
factored form
exponent
base
exponential (power)
square root
cube root

1.3 multiplicative
      inverse
   identity element for
      addition
   identity element for
      multiplication
   term
   like terms
   coefficient
      (numerical
      coefficient)
   combining like
      terms

NEW SYMBOLS

$\{a, b\}$	set containing the elements a and b
$\emptyset$	the empty set
$\in$	is an element of (a set)
$\notin$	is not an element of
$\neq$	is not equal to
$\{x \mid x \text{ has property } P\}$	the set of all x, such that x has property P
$\lvert x \rvert$	the absolute value of x
$<$	is less than
$\leq$	is less than or equal to
$>$	is greater than
$\geq$	is greater than or equal to
(a, ∞)	the interval $\{x \mid x > a\}$
$(-\infty, a)$	the interval $\{x \mid x < a\}$
$(a, b]$	the interval $\{x \mid a < x \leq b\}$
a^m	m factors of a
$\sqrt{a}$	the square root of a
$\sqrt[n]{a}$	the nth root of a

QUICK REVIEW

CONCEPTS	EXAMPLES
1.1 BASIC TERMS	
Sets of Numbers	
Real Numbers $\{x \mid x \text{ is represented by a point on a number line}\}$	$-3, .7, \pi, -\dfrac{2}{3}$
Natural Numbers $\{1, 2, 3, 4, \ldots\}$	$10, 25, 143$
Whole Numbers $\{0, 1, 2, 3, 4, \ldots\}$	$0, 8, 47$
Integers $\{\ldots, -2, -1, 0, 1, 2, \ldots\}$	$-22, -7, 0, 4, 9$
Rational Numbers $\left\{\dfrac{p}{q} \,\middle|\, p \text{ and } q \text{ are integers, } q \neq 0\right\}$, or all terminating or repeating decimals	$-\dfrac{2}{3}, -.14, 0, 6, \dfrac{5}{8}, .\overline{3}$
Irrational Numbers $\{x \mid x \text{ is a real number that is not rational}\}$ or all nonterminating, nonrepeating decimals	$\pi, .125469 \ldots, \sqrt{3}, -\sqrt{22}$

CONCEPTS	EXAMPLES

1.2 OPERATIONS ON REAL NUMBERS

Addition

Same sign: Add the absolute values. The sum has the same sign as the numbers.

Add: $-2 + (-7)$.
$$-2 + (-7) = -(2 + 7)$$
$$= -9$$

Different signs: Subtract the absolute values. The answer has the sign of the number with the larger absolute value.

$$-5 + 8 = 8 - 5 = 3$$
$$-12 + 4 = -(12 - 4) = -8$$

Subtraction

Change the sign of the second number and add.

Subtract: $-5 - (-3)$.
$$-5 - (-3) = -5 + 3 = -2$$

Multiplication

Same sign: The product is positive.
Different signs: The product is negative.

Multiply: $(-3)(-8) = 24$.
Multiply: $(-7)(5) = -35$.

Division

Same sign: The quotient is positive.

Divide: $\dfrac{-15}{-5} = 3$.

Different signs: The quotient is negative.

Divide: $\dfrac{-24}{12} = -2$.

The product of an even number of negative factors is positive.

$(-5)^6$ is positive.

The product of an odd number of negative factors is negative.

$(-5)^7$ is negative.

Order of Operations

1. Work separately above and below any fraction bar.

2. Use the rules below within each set of parentheses or square brackets. Start with the innermost set and work outward.

3. Evaluate all powers and roots.

4. Do any multiplications or divisions in the order in which they occur, working from left to right.

5. Do any additions or subtractions in the order in which they occur, working from left to right.

Perform the indicated operations.

$$\frac{12 + 3}{5 \cdot 2} = \frac{15}{10} = \frac{3}{2}$$

$$(-6)[2^2 - (3 + 4)] + 3 = (-6)[2^2 - 7] + 3$$
$$= (-6)[4 - 7] + 3$$
$$= (-6)[-3] + 3$$
$$= 18 + 3$$
$$= 21$$

1.3 PROPERTIES OF REAL NUMBERS

For any real numbers, a, b, and c:

Distributive Property
$$a(b + c) = ab + ac$$

$$12(4 + 2) = 12 \cdot 4 + 12 \cdot 2$$

CONCEPTS	EXAMPLES
Inverse Properties $a + (-a) = 0$ and $-a + a = 0$ $a \cdot \dfrac{1}{a} = 1$ and $\dfrac{1}{a} \cdot a = 1$ $(a \neq 0)$	$5 + (-5) = 0$ $-\dfrac{1}{3}(-3) = 1$
Identity Properties $a + 0 = a$ and $0 + a = a$ $a \cdot 1 = a$ and $1 \cdot a = a$	$-32 + 0 = -32$ $(17.5)(1) = 17.5$
Associative Properties $a + (b + c) = (a + b) + c$ $\quad a(bc) = (ab)c$	$7 + (5 + 3) = (7 + 5) + 3$ $-4(6 \cdot 3) = (-4 \cdot 6)3$
Commutative Properties $a + b = b + a$ $\quad ab = ba$	$9 + (-3) = -3 + 9$ $6(-4) = (-4)6$
Multiplication Property of Zero $a \cdot 0 = 0$ and $0 \cdot a = 0$	$47 \cdot 0 = 0 \qquad 0 \cdot -18 = 0$

CHAPTER 1 REVIEW EXERCISES

For help with any of these exercises, look in the section given in brackets.

[1.1] *Graph the set on a number line.*

1. $\left\{ -4, -1, 2, \dfrac{9}{4}, 4 \right\}$

2. $\left\{ -5, -\dfrac{11}{4}, -.5, 0, 3, \dfrac{13}{3} \right\}$

Find the value of the expression.

3. $|-16|$

4. $|23|$

5. $-|-4|$

6. $-|-8| + |-3|$

Let set $S = \left\{ -9, -\dfrac{4}{3}, -\sqrt{10}, 0, \dfrac{5}{3}, \sqrt{7}, \dfrac{12}{3} \right\}$. *Simplify the elements of S as necessary and then list the elements that belong to the specified set.*

7. whole numbers

8. integers

9. rational numbers

10. real numbers

Write the set by listing its elements.

11. $\{x \mid x$ is a natural number between 3 and 9$\}$

12. $\{y \mid y$ is a whole number less than 4$\}$

1.

2.

3. 16   4. 23   5. −4
6. −5   7. 0, 4
8. −9, 0, 4
9. $-9, -\dfrac{4}{3}, 0, \dfrac{5}{3}, 4$
10. All are real numbers.
11. {4, 5, 6, 7, 8}
12. {0, 1, 2, 3}

RESOURCES FOR REVIEW

📖 **ITM** pp. 307–318   **ISM** pp. 21–27
IAM pp. 2–3

📖 **SSG** pp. 1–19
SSM pp. 12–19

💾 **TEST GENERATOR**
DOS   Windows   MAC

◎ CONCEPTUAL   📝 WRITING   ▲ CHALLENGING   🔢 SCIENTIFIC CALCULATOR   📱 GRAPHICS CALCULATOR

13. true   14. false
15. true
16. $(-\infty, -5)$

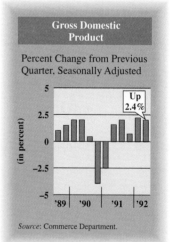

17. $(-2, 3]$

18. $\dfrac{41}{24}$ 19. $-\dfrac{1}{2}$

20. -3 21. -17.09

22. -39 23. -1 24. $\dfrac{23}{20}$

25. $-\dfrac{5}{18}$ 26. -35

28. To subtract $a - b$, write as an addition problem, $a + (-b)$, and add.   29. -90 30. $\dfrac{2}{3}$

31. -11.408 32. 2

33. -15 34. 3.21

35. true   36. false

37. $\dfrac{5}{7 - 7}$ is undefined.

Write true *or* false *for the inequality.*

13. $4 \cdot 2 \le |12 - 4|$

14. $2 + |-2| > 4$

15. $4(3 + 7) > -|40|$

Write in interval notation and graph.

16. $\{x \mid x < -5\}$

17. $\{x \mid -2 < x \le 3\}$

[1.2] *Add or subtract, as indicated.*

18. $-\dfrac{5}{8} - \left(-\dfrac{7}{3}\right)$

19. $-\dfrac{4}{5} - \left(-\dfrac{3}{10}\right)$

20. $-5 + (-11) + 20 - 7$

21. $-9.42 + 1.83 - 7.6 - 1.9$

22. $-15 + (-13) + (-11)$

23. $-1 - 3 - (-10) + (-7)$

24. $\dfrac{3}{4} - \left(\dfrac{1}{2} - \dfrac{9}{10}\right)$

25. $-\dfrac{2}{3} - \left(\dfrac{1}{6} - \dfrac{5}{9}\right)$

26. $-|-12| - |-9| + (-4) - |10|$

27. State in your own words how to determine the sign of the sum of two numbers.

28. How is subtraction related to addition?

Find the product.

29. $2(-5)(-3)(-3)$

30. $-\dfrac{3}{7}\left(-\dfrac{14}{9}\right)$

31. $-4.6(2.48)$

Find the quotient.

32. $\dfrac{-38}{-19}$

33. $\dfrac{75}{-5}$

34. $\dfrac{-2.3754}{-.74}$

Use the graph to answer true *or* false *to the statement.* *

35. The percent change from the previous quarter during the first quarter of 1991 was greater than that of the fourth quarter of 1990.

36. The sum of the percent change from the previous quarter in the fourth quarter of 1990 and that of the first quarter of 1992 is a positive number.

37. Which one of the following is undefined? $\dfrac{5}{7 - 7}$ or $\dfrac{7 - 7}{5}$

---

* Graph for Exercises 35 and 36, "Gross Domestic Product," from *The Times-Picayune,* May 30, 1992. Copyright © 1992 by The Times-Picayune. Reprinted by permission.

Evaluate the expression.

38. $\left(\dfrac{3}{7}\right)^3$ **39.** $(-5)^3$ **40.** -5^3 **41.** $(1.7)^2$

Find the root. If it is not a real number, say so.

42. $\sqrt{400}$ **43.** $\sqrt[3]{27}$ **44.** $\sqrt[3]{-343}$ **45.** $\sqrt[4]{81}$ **46.** $\sqrt[6]{-64}$

Use the order of operations to simplify the expression.

47. $-14\left(\dfrac{3}{7}\right) + 6 \div 3$

48. $-\dfrac{2}{3}[5(-2) + 8 - 4^3]$

49. $\dfrac{-4(\sqrt{25}) - (-3)(-5)}{3 + (-6)(\sqrt{9})}$

50. $\dfrac{-5(3^2) + 9(\sqrt{4}) - 5}{6 - 5(\sqrt[3]{-8})}$

Let $k = -4$, $m = 2$, and $n = 16$, and evaluate the expression.

51. $4k - 7m$

52. $-3(\sqrt{16}) + m + 5k$

53. $-2(3k^2 + 5m)$

54. $\dfrac{4m^3 - 3n}{7k^2 - 10}$

◉ **55.** In order to evaluate $(3 + 2)^2$, should you work within the parentheses first, or should you square 3 and square 2 and then add?

56. By replacing a with 4 and b with 6, show that $(a + b)^2 \neq a^2 + b^2$.

[1.3] *Use the properties of real numbers to simplify the expression.*

57. $2q + 19q$ **58.** $13z - 17z$ **59.** $-m + 6m$

60. $5p - p$ **61.** $-2(k + 3)$ **62.** $6(r + 3)$

63. $9(2m + 3n)$ **64.** $-(3k - 4h)$

65. $-(-p + 6q) - (2p - 3q)$ **66.** $-2x + 5 - 4x + 1$

67. $-3y + 6 - 5 + 4y$ **68.** $2a + 3 - a - 1 - a - 2$

69. $-2(k - 1) + 3k - k$ **70.** $-3(4m - 2) + 2(3m - 1) - 4(3m + 1)$

Complete the statement so that the indicated property is illustrated. Simplify the answer, if possible.

71. $2x + 3x = $ _____
     (distributive property)

72. $-4 \cdot 1 = $ _____
     (identity property)

73. $2(4x) = $ _____
     (associative property)

74. $-3 + 13 = $ _____
     (commutative property)

75. $-3 + 3 = $ _____
     (inverse property)

76. $5(x + z) = $ _____
     (distributive property)

77. $0 + 7 = $ _____
     (identity property)

---

38. $\dfrac{27}{343}$ 39. -125
40. -125 41. 2.89
42. 20   43. 3
44. -7 45. 3
46. not a real number
47. -4 48. 44   49. $\dfrac{7}{3}$
50. -2 51. -30
52. -30 53. -116
54. $-\dfrac{8}{51}$ 55. Work
within the parentheses
first.   56. $(4 + 6)^2 =$
$10^2 = 100$; $4^2 + 6^2 = 52$;
$100 \neq 52$ 57. $21q$
58. $-4z$ 59. $5m$
60. $4p$ 61. $-2k - 6$
62. $6r + 18$
63. $18m + 27n$
64. $-3k + 4h$
65. $-p - 3q$
66. $-6x + 6$
67. $y + 1$ 68. 0
69. 2   70. $-18m$
71. $(2 + 3)x = 5x$
72. -4
73. $(2 \cdot 4)x = 8x$
74. $13 + (-3) = 10$
75. 0   76. $5x + 5z$
77. 7

---

78. 1
79. $(3 + 5 + 6)a = 14a$
80. 0   81. $\dfrac{256}{625}$ 82. 25
83. 31   84. 9   85. 0
86. -2 87. $\dfrac{4}{3}$ 88. -6.16
89. -9 90. -29
91. $-\dfrac{47}{3}$ 92. It is greater than -16.
93. (a) no   (b) yes
94. $\dfrac{47}{3}$ 95. no   96. yes
97. $\dfrac{47}{3}$ 98. $-\dfrac{3}{47}$
99. yes   100. No, the new answer is $-\dfrac{29}{3}$.

78. $8 \cdot \dfrac{1}{8} =$ _____
(inverse property)

79. $3a + 5a + 6a =$ _____
(distributive property)

80. $\dfrac{9}{28} \cdot 0 =$ _____
(multiplication property of 0)

MIXED REVIEW EXERCISES*

Perform the indicated operations.

81. $\left(-\dfrac{4}{5}\right)^4$

82. $-\dfrac{5}{8}(-40)$

83. $-25\left(-\dfrac{4}{5}\right) + 3^3 - 32 \div \sqrt{4}$

84. $-8 + |-14| + |-3|$

85. $\dfrac{6 \cdot \sqrt{4} - 3 \cdot \sqrt{16}}{-2 \cdot 5 + 7(-3) - 10}$

86. $-\sqrt[5]{32}$

87. $-\dfrac{10}{21} \div -\dfrac{5}{14}$

88. $.8 - 4.9 - 3.2 + 1.14$

89. -3^2

90. $-\dfrac{4.64}{.16}$

─────◆ **MATHEMATICAL CONNECTIONS** (Exercises 91–100)† ◆─────

Evaluate the expression $\dfrac{2}{3}x - y^2 - 3z$ *for* $x = 5$, $y = -4$, *and* $z = 1$. *Then respond to the questions or statements in Exercises 91–100, working them in order.*

91. What is the value of the expression for these particular values of x, y, and z?

92. Is the value of the expression greater than -16 or less than -16?

93. Is the value of the expression **(a)** an integer   **(b)** a rational number?

94. What is the absolute value of the expression?

95. Is the square root of the value of the expression a real number?

96. Is the cube root of the value of the expression a real number?

97. Give the additive inverse of the value of the expression.

98. Give the multiplicative inverse of the value of the expression.

99. If parentheses are placed around the first two terms of the expression, will you obtain the same answer? If not, what is the new answer?

100. If parentheses are placed around the last two terms of the expression, will you obtain the same answer? If not, what is the new answer?

─────────◆─────────

*The order of exercises in this final group does not correspond to the order in which topics occur in the chapter. This random ordering should help you prepare for the chapter test in yet another way.

† Each Chapter Review includes a group of exercises designed to show interrelationships among the concepts studied in the chapter, as well as previous chapters (when applicable).

CHAPTER 1 TEST

1. Graph $\left\{-3, .75, \dfrac{5}{3}, 5, 6.3\right\}$ on a number line.

Let $A = \left\{-\sqrt{6}, -1, -.5, 0, 3, 7.5, \dfrac{24}{2}\right\}$. *First simplify each element as needed and then list the elements from A that belong to the set.*

2. whole numbers

3. integers

4. rational numbers

5. real numbers

Write the set in interval notation, and graph it.

6. $\{x \mid x < -3\}$

7. $\{y \mid -4 < y \le 2\}$

Peform the indicated operations.

8. $-6 + 14 + (-11) - (-3)$

9. $10 - 4 \cdot 3 + 6(-4)$

10. $7 - 4^2 + 2(6) + (-4)^2$

11. $\dfrac{10 - 24 + (-6)}{\sqrt{16}(-5)}$

12. $\dfrac{-2[3 - (-1 - 2) + 2]}{\sqrt{9}(-3) - (-2)}$

13. $\dfrac{8 \cdot 4 - 3^2 \cdot 5 - 2(-1)}{-3 \cdot 2^3 + 1}$

Find the indicated root. If the number is not real, say so.

14. $\sqrt{196}$

15. $-\sqrt{225}$

16. $\sqrt[3]{-27}$

17. $\sqrt[4]{-16}$

18. For the expression $\sqrt[n]{a}$, under what conditions will its value be **(a)** positive **(b)** negative **(c)** zero?

Evaluate the expression if $k = -3$, $m = -3$, *and* $r = 25$.

19. $\sqrt{r} + 2k - m$

20. $\dfrac{8k + 2m^2}{r - 2}$

21. Use the properties of real numbers to simplify $-3(2k - 4) + 4(3k - 5) - 2 + 4k$.

22. How does the subtraction sign affect the terms $-4r$ and 6 when simplifying $(3r + 8) - (-4r + 6)$? What is the simplified form?

Match the statement with the appropriate property. Answers may be used more than once.

23. $6 + (-6) = 0$

24. $4 + 5 = 5 + 4$

25. $-2 + (3 + 6) = (-2 + 3) + 6$

26. $5x + 15x = (5 + 15)x$

27. $13 \cdot 0 = 0$

28. $-9 + 0 = -9$

29. $4 \cdot 1 = 4$

30. $(a + b) + c = (b + a) + c$

A. Distributive property

B. Inverse Property

C. Identity Property

D. Associative Property

E. Commutative Property

F. Multiplication Property of Zero

[1.1] 1.

2. 0, 3, 12
3. −1, 0, 3, 12
4. −1, −.5, 0, 3, 7.5, 12
5. All are real numbers.
6. $(-\infty, -3)$

7. $(-4, 2]$

[1.2] 8. 0   9. −26

10. 19   11. 1   12. $\dfrac{16}{7}$

13. $\dfrac{11}{23}$ 14. 14   15. −15

16. −3   17. not a real number   18. (a) *a* must be positive.   (b) *n* must be odd and *a* must be negative.   (c) *a* must be zero.   19. 2   20. $-\dfrac{6}{23}$

[1.3] 21. $10k - 10$
22. It changes the sign of each term. The simplified form is $7r + 2$.
23. B   24. E   25. D
26. A   27. F   28. C
29. C   30. E

RESOURCES FOR TEST

 ITM pp. 15–36   **ISM** pp. 27–30
IAM p. 1

 **SSG** pp. 20–21
SSM pp. 19–21

 **TEST GENERATOR**
DOS   Windows   MAC

 CONCEPTUAL   WRITING    CHALLENGING   SCIENTIFIC CALCULATOR    GRAPHICS CALCULATOR

LINEAR EQUATIONS AND INEQUALITIES

CONNECTIONS

To use mathematics to solve a real-world problem, we must first set up a **mathematical model,** in other words, a mathematical description of the situation. Constructing such a model requires a good understanding of the situation to be modeled and a familiarity with relevant mathematical techniques. A great deal of mathematical theory is available for building models. Yet, the very richness and diversity of contemporary mathematics often deters people in other fields from finding the mathematical tools they need to create a model. It is helpful to understand the basic mathematics that is available for constructing these models. In this chapter we look at the mathematics of *linear models,* which are used for data whose graphs can be approximated by a straight line.

2.1 LINEAR EQUATIONS IN ONE VARIABLE

FOR EXTRA HELP

SSG pp. 22–28
SSM pp. 22–27

Video
I

Tutorial
IBM   MAC

OBJECTIVES

1 ▶ Define linear equations.
2 ▶ Solve linear equations using the addition and multiplication
    properties of equality.
3 ▶ Solve linear equations using the distributive property.
4 ▶ Solve linear equations with fractions and decimals.
5 ▶ Identify conditional equations, contradictions, and identities.

An **algebraic expression** is the result of performing the basic operations of addition, subtraction, multiplication, and division (except by 0), or extraction of roots on any collection of variables and numbers. Some examples of algebraic expressions include

$$8x + 9, \qquad \sqrt{y} + 4, \qquad \text{and} \qquad \frac{x^3 y^8}{z}.$$

1 ▶ Define linear equations.

▶ Applications of mathematics often lead to **equations,** statements that two algebraic expressions are equal.

Linear Equation

An equation is **linear** if it can be written in the form

$$ax = b,$$

where a and b are real numbers, with $a \neq 0$.

Examples of linear equations include

$$x + 1 = -2, \qquad y - 3 = 5, \qquad \text{and} \qquad 2k + 5 = 10.$$

A linear equation is also called a **first-degree** equation, since the highest power on the variable is one.

If the variable in an equation is replaced by a real number that makes the statement true, then that number is a **solution** of the equation. For example, 8 is a solution of the equation $y - 3 = 5$, since replacing y with 8 gives a true statement. An equation is **solved** by finding its **solution set,** the set of all solutions. The solution set of the equation $y - 3 = 5$ is $\{8\}$.

2 ▶ Solve linear equations using the addition and multiplication properties of equality.

▶ **Equivalent equations** are equations with the same solution set. Equations are generally solved by starting with a given equation and producing a series of simpler equivalent equations. For example,

$$8x + 1 = 17, \qquad 8x = 16, \qquad \text{and} \qquad x = 2$$

are all equivalent equations since each has the same solution set, $\{2\}$. We use the addition and multiplication properties of equality to produce equivalent equations.

INSTRUCTOR'S RESOURCES

   **ITM** pp. 319–321   **ISM** pp. 31–41
             IAM p. 3

TEST GENERATOR
DOS   Windows   MAC

 TRANSPARENCIES

Addition and Multiplication Properties of Equality

Addition Property of Equality

For all real numbers a, b, and c, the equations

$$a = b \quad \text{and} \quad a + c = b + c$$

are equivalent. (The same number may be added to both sides of an equation without changing the solution set.)

Multiplication Property of Equality

For all real numbers, a, b, and c, where $c \neq 0$, the equations

$$a = b \quad \text{and} \quad ac = bc$$

are equivalent. (Both sides of an equation may be multiplied by the same nonzero number without changing the solution set.)

By the addition property, the same number may be added to both sides of an equation without affecting the solution set. By the multiplication property, both sides of an equation may be multiplied by the same nonzero number to produce an equivalent equation. Because subtraction and division are defined in terms of addition and multiplication, respectively, these properties can be extended: The same number may be subtracted from both sides of an equation, and both sides may be divided by the same nonzero number.

EXAMPLE 1
Solving a Linear Equation

Chalkboard Exercise

Solve $-7 + 3y - 9y = 12y - 5$.

Answer: $\left\{ -\dfrac{1}{9} \right\}$

Solve $4y - 2y - 5 = 4 + 6y + 3$.

$$2y - 5 = 7 + 6y \qquad \text{Combine terms.}$$
$$2y - 5 + 5 = 7 + 6y + 5 \qquad \text{Add 5.}$$
$$2y = 12 + 6y$$
$$2y - 6y = 12 + 6y - 6y \qquad \text{Subtract } 6y.$$
$$-4y = 12$$
$$\frac{-4y}{-4} = \frac{12}{-4} \qquad \text{Divide by } -4.$$
$$y = -3$$

▶ **TEACHING TIP**
Remind students that only one equals sign should be used per line in the solution of an equation. Some bad habits may have developed in beginning algebra. ◀

To be sure that -3 is the solution, check by substituting back into the *original* equation (not an intermediate one).

$$4y - 2y - 5 = 4 + 6y + 3 \qquad \text{Given equation}$$
$$4(-3) - 2(-3) - 5 = 4 + 6(-3) + 3 \quad ? \quad \text{Let } y = -3.$$
$$-12 + 6 - 5 = 4 - 18 + 3 \quad ? \quad \text{Multiply.}$$
$$-11 = -11 \qquad \text{True}$$

Since a true statement is obtained, -3 is the solution. The solution set is $\{-3\}$. ■

◆ 1. (a) 1940 (b) 1964
2. men: 37.22 seconds;
women: 39.06 seconds

◇ **CONNECTIONS** ◇

A mathematical model that can be used to predict the value of one variable given the value of another can be found by using pairs of data. This process is called *curve fitting*. In a later chapter, we will see how linear equations like those discussed below are developed from the data.

By studying the winning times in the 500-meter speed-skating event at Olympic games back to 1900, it was found that the winning times for men were closely approximated by the equation

$$y_m = 46.338 - .097x,$$

where y_m is the time in seconds needed to win for men, with x the Olympic year and $x = 0$ corresponding to 1900. (For example, 1994 winning times would be estimated by replacing x with $1994 - 1900 = 94$.) The corresponding equation for women is

$$y_w = 57.484 - .196x.$$

FOR DISCUSSION OR WRITING

1. Find the year in which **(a)** $y_m = 42.458$; **(b)** $y_w = 44.940$.
2. Use the equations to predict the winning times for men and women in 1994. How do they compare with the actual results?

We use the following steps to solve a linear equation in one variable. (Not all equations require all these steps.)

Solving a Linear Equation in One Variable

Step 1 **Clear fractions.** Eliminate any fractions by multiplying both sides by a common denominator.

Step 2 **Simplify each side separately.** Simplify each side of the equation as much as possible by using the distributive property to clear parentheses and by combining like terms as needed.

Step 3 **Isolate the variable terms on one side.** Use the addition property to get all terms with variables on one side of the equation and all numbers on the other.

Step 4 **Isolate the variable.** Use the multiplication property to get an equation with just the variable (with coefficient 1) on one side.

Step 5 **Check.** Check by substituting back into the original equation.

3 ▶ Solve linear equations using the distributive property.

▶ In Example 1 we did not use Step 1 and the distributive property in Step 2 as given above. Many other equations, however, will require one or both of these steps, as shown in the next examples.

EXAMPLE 2

Using the Distributive Property to Solve a Linear Equation

Solve $2(k - 5) + 3k = k + 6$.

Since there are no fractions in this equation, Step 1 does not apply. Begin by using the distributive property to simplify and combine terms on the left side of the equation (Step 2).

$$2(k - 5) + 3k = k + 6$$
$$2k - 10 + 3k = k + 6 \qquad \text{Distributive property}$$
$$5k - 10 = k + 6 \qquad \text{Combine like terms.}$$

Next, use the addition property of equality (Step 3).

$$5k - 10 + 10 = k + 6 + 10 \qquad \text{Add } 10.$$

$$5k = k + 16$$

$$5k - k = k + 16 - k \qquad \text{Subtract } k.$$

$$4k = 16 \qquad \text{Combine like terms.}$$

The multiplication property of equality is used to get just k on the left. (Step 4).

$$\frac{4k}{4} = \frac{16}{4} \qquad \text{Divide by 4.}$$

$$k = 4$$

Check that the solution set is $\{4\}$ by substituting 4 for k in the original equation (Step 5). ■

In the rest of the examples in this section, we do not identify the steps by number.

4 ▶ Solve linear equations with fractions and decimals.

▶ When fractions or decimals appear as coefficients in equations, our work can be made easier if we multiply both sides of the equation by the least common denominator of all the fractions. This is an application of the multiplication property of equality, and it produces an equivalent equation with integer coefficients. The next examples illustrate this idea.

E X A M P L E 3

Solving a Linear Equation with Fractions

Solve $\dfrac{x + 7}{6} + \dfrac{2x - 8}{2} = -4$.

Start by eliminating the fractions. Multiply both sides by the least common denominator, 6.

$$6\left[\frac{x + 7}{6} + \frac{2x - 8}{2}\right] = 6 \cdot (-4)$$

$$6\left(\frac{x + 7}{6}\right) + 6\left(\frac{2x - 8}{2}\right) = 6(-4)$$

$$x + 7 + 3(2x - 8) = -24$$

$$x + 7 + 6x - 24 = -24 \qquad \text{Distributive property}$$

$$7x - 17 = -24 \qquad \text{Combine terms.}$$

$$7x - 17 + 17 = -24 + 17 \qquad \text{Add 17.}$$

$$7x = -7$$

$$\frac{7x}{7} = \frac{-7}{7} \qquad \text{Divide by 7.}$$

$$x = -1$$

▶ **TEACHING TIP**

Emphasize that each term of the equation must be multiplied by the LCD. It is best to identify how many terms are in the equation before multiplying. ◀

Check by substituting -1 for x in the original equation.

$$\frac{x + 7}{6} + \frac{2x - 8}{2} = -4$$

$$\frac{-1 + 7}{6} + \frac{2(-1) - 8}{2} = -4$$

$$\frac{6}{6} + \frac{-10}{2} = -4$$

$$1 - 5 = -4 \qquad \text{True}$$

The solution checks, so the solution set is $\{-1\}$. ∎

In later sections we solve problems involving interest rates and concentrations of solutions. These problems involve percents that are converted to decimal numbers. The equations that are used to solve such problems involve decimal coefficients. We can clear these decimals by multiplying by the largest power of 10 necessary to obtain integer coefficients. The next example shows how this is done.

EXAMPLE 4
Solving a Linear Equation
with Decimals

Solve $.06x + .09(15 - x) = .07(15)$.

Since each decimal number is given in hundredths, multiply both sides of the equation by 100. (This is done by moving the decimal points two places to the right.)

$$.06x + .09(15 - x) = .07(15)$$
$$6x + 9(15 - x) = 7(15) \qquad \text{Multiply by } 100.$$
$$6x + 9(15) - 9x = 105 \qquad \text{Distributive property}$$
$$-3x + 135 = 105 \qquad \text{Combine like terms.}$$
$$-3x + 135 - 135 = 105 - 135 \qquad \text{Subtract } 135.$$
$$-3x = -30$$
$$\frac{-3x}{-3} = \frac{-30}{-3} \qquad \text{Divide by } -3.$$
$$x = 10$$

Check to verify that the solution set is $\{10\}$. ∎

CAUTION When multiplying the term $.09(15 - x)$ by 100 in Example 4, do not multiply both .09 and $15 - x$ by 100. This step is not an application of the distributive property, but of the associative property. The correct procedure is

$$100[.09(15 - x)] = [100(.09)](15 - x) \qquad \text{Associative property}$$
$$= 9(15 - x). \qquad \text{Multiply.}$$

---

5 ▶ Identify conditional
equations, contradictions,
and identities.

▶ All the equations above had a solution set containing one element; for example, $2(k - 5) + 3k = k + 6$ has solution set $\{4\}$. Some linear equations, however, have no solutions, while others have an infinite number of solutions. The chart below gives the names of these types of equations.

Type of Linear Equation	Number of Solutions
Conditional	One
Contradiction	None, solution set $\emptyset$
Identity	Infinite, solution set {all real numbers}

The next example shows how to recognize these types of equations.

EXAMPLE 5
Identifying Conditional
Equations, Identities, and
Contradictions

Solve each equation. Decide whether it is a conditional equation, an identity, or a contradiction.

(a) $5x - 9 = 4(x - 3)$

Work as in the previous examples.

$$5x - 9 = 4(x - 3)$$
$$5x - 9 = 4x - 12 \qquad \text{Distributive property}$$
$$5x - 9 - 4x = 4x - 12 - 4x. \qquad \text{Subtract } 4x.$$
$$x - 9 = -12 \qquad \text{Combine like terms.}$$
$$x - 9 + 9 = -12 + 9 \qquad \text{Add 9.}$$
$$x = -3$$

The solution set, $\{-3\}$, has one element, so $5x - 9 = 4(x - 3)$ is a *conditional equation*.

(b) $5x - 15 = 5(x - 3)$

Use the distributive property on the right side.

$$5x - 15 = 5x - 15$$

Both sides of the equation are *exactly the same*, so any real number would make the equation true. For this reason, the solution set is the set of all real numbers, and the equation $5x - 15 = 5(x - 3)$ is an *identity*.

(c) $5x - 15 = 5(x - 4)$

Use the distributive property.

$$5x - 15 = 5x - 20 \qquad \text{Distributive property}$$
$$5x - 15 - 5x = 5x - 20 - 5x \qquad \text{Subtract } 5x.$$
$$-15 = -20 \qquad \text{False}$$

Since the result, $-15 = -20$, is *false*, the equation has no solution. The solution set is ∅. The equation $5x - 15 = 5(x - 4)$ is a *contradiction*. ■

2.1 EXERCISES

Decide whether the given number is a solution of the equation.

1. $-6x = -24;\quad 4$

2. $8r = 56;\quad 7$

3. $5x + 2 = 3;\quad \dfrac{2}{5}$

4. $6y - 4 = 4;\quad \dfrac{1}{2}$

5. $9x + 2x = 6x;\quad 0$

6. $-2p + 10p = 7p;\quad 0$

⊙ **7.** Which one of the following equations is not a linear equation in x?

    (a) $3x + x - 1 = 0$ **(b)** $8 = x^2$ **(c)** $6x + 2 = 9$ **(d)** $\dfrac{1}{2}x - \dfrac{1}{4}x = 0$

⊙ **8.** If two equations are equivalent, they have the same _____ _____.

Solve the equation. See Examples 1 and 2.

9. $7k + 8 = 1$ **10.** $5m - 4 = 21$ **11.** $8 - 8x = -16$ **12.** $9 - 2r = 15$

13. $7y - 5y + 15 = y + 8$ **14.** $2x + 4 - x = 4x - 5$

1. yes   2. yes   3. no
4. no   5. yes   6. yes
7. (b)   8. solution set
9. $\{-1\}$ 10. $\{5\}$ 11. $\{3\}$
12. $\{-3\}$ 13. $\{-7\}$
14. $\{3\}$

⊙ CONCEPTUAL   ✎ WRITING   ▲ CHALLENGING   ▦ SCIENTIFIC CALCULATOR   ▥ GRAPHICS CALCULATOR

15. $12w + 15w - 9 + 5 = -3w + 5 - 9$

16. $-4t + 5t - 8 + 4 = 6t - 4$

17. $2(x + 3) = -4(x + 1)$

18. $4(y - 9) = 8(y + 3)$

19. $3(2w + 1) - 2(w - 2) = 5$

20. $4(x - 2) + 2(x + 3) = 6$

21. $2x + 3(x - 4) = 2(x - 3)$

22. $6y - 3(5y + 2) = 4(1 - y)$

23. $6p - 4(3 - 2p) = 5(p - 4) - 10$

24. $-2k - 3(4 - 2k) = 2(k - 3) + 2$

25. $-[2z - (5z + 2)] = 2 + (2z + 7)$

26. $-[6x - (4x + 8)] = 9 + (6x + 3)$

27. $-(9 - 3a) - (4 + 2a) - 3 = -(2 - 5a) + (-a) + 1$

28. $-(-2 + 4x) - (3 - 4x) + 5 = -(-3 + 6x) + x + 1$

29. Explain in your own words the steps used to solve a linear equation.

30. In order to solve the linear equation $\dfrac{8y}{3} - \dfrac{2y}{4} = -13$, we may multiply both sides by the least common denominator of all the fractions in the equation. What is this least common denominator?

31. In order to solve the linear equation $.05y + .12(y + 5000) = 940$, we may multiply both sides by a power of 10 so that all coefficients are integers. What is the smallest power of 10 that will accomplish this goal?

32. Suppose that in solving the equation

$$\frac{1}{3}y + \frac{1}{2}y = \frac{1}{6}y,$$

you begin by multiplying both sides by 12, rather than the *least* common denominator, 6. Should you get the correct solution anyway? Explain.

Solve the equation. See Examples 3 and 4.

33. $\dfrac{3x}{4} + \dfrac{5x}{2} = 13$

34. $\dfrac{8y}{3} - \dfrac{2y}{4} = -13$

35. $\dfrac{x - 8}{5} + \dfrac{8}{5} = -\dfrac{x}{3}$

36. $\dfrac{2r - 3}{7} + \dfrac{3}{7} = -\dfrac{r}{3}$

37. $\dfrac{4t + 1}{3} = \dfrac{t + 5}{6} + \dfrac{t - 3}{6}$

38. $\dfrac{2x + 5}{5} = \dfrac{3x + 1}{2} + \dfrac{-x + 7}{2}$

39. $.05y + .12(y + 5000) = 940$

40. $.09k + .13(k + 300) = 61$

41. $.02(50) + .08r = .04(50 + r)$

42. $.20(14,000) + .14t = .18(14,000 + t)$

43. $.05x + .10(200 - x) = .45x$

44. $.08x + .12(260 - x) = .48x$

45. The equation $x^2 + 2 = x^2 + 2$ is called a(n) _____ , because its solution set is {all real numbers}. The equation $x + 1 = x + 2$ is called a(n) _____ , because it has no solutions.

46. Which one of the following is a conditional equation?
   (a) $2x + 1 = 3$ **(b)** $x = 3x - 2x$ **(c)** $2(x + 2) = 2x + 2$
   (d) $5x - 3 = 4x + x - 5 + 2$

Decide whether the equation is conditional, an identity, or a contradiction. Give the solution set. See Example 5.

47. $-2p + 5p - 9 = 3(p - 4) - 5$

48. $-6k + 2k - 11 = -2(2k - 3) + 4$

49. $-11m + 4(m - 3) + 6m = 4m - 12$

50. $3p - 5(p + 4) + 9 = -11 + 15p$

51. $7[2 - (3 + 4r)] - 2r = -9 + 2(1 - 15r)$

52. $4[6 - (1 + 2m)] + 10m = 2(10 - 3m) + 8m$

 **MATHEMATICAL CONNECTIONS** (Exercises 53–58)

Work Exercises 53–58 in order.

53. Use the methods of this section to solve the equation $2[3x + (x - 2)] = 9x + 4$.

54. If an equation is *equivalent* to the one in Exercise 53, what must its solution set be?

55. Let us now consider the following linear equation in x:

$$-4(x + 2) - 3(x + 5) = k.$$

Assume that k is some real number constant. Evaluate the left side of the equation for the value of x you found in Exercise 53. What is your result?

56. What must be the value of k in the equation in Exercise 55 for that equation to be equivalent to the one in Exercise 53?

57. Solve the equation in Exercise 55 with the value of k you found in Exercise 56.

58. Use the concepts presented in Exercises 53–57 to find the value of k that will make the equations $3x + k = 11$ and $5x - 8 = 22$ equivalent.

53. $\{-8\}$ 54. $\{-8\}$
55. 33 56. 33 57. $\{-8\}$
58. -7 59. equivalent
60. equivalent
61. not equivalent
62. not equivalent
63. not equivalent
64. not equivalent
65. 2820 million dollars;
1996 (when $x = 7$)
66. 570 million dollars;
1998 (when $x = 7$)
67. (a) 230 billion dollars
(b) 275 billion dollars
(c) 300 billion dollars
(d) 335 billion dollars

 Decide if the pair of equations is equivalent.

59. $5x = 10$ and $\dfrac{5x}{x + 2} = \dfrac{10}{x + 2}$

60. $x + 1 = 9$ and $\dfrac{x + 1}{8} = \dfrac{9}{8}$

61. $y = -3$ and $\dfrac{y}{y + 3} = \dfrac{-3}{y + 3}$

62. $m = 1$ and $\dfrac{m + 1}{m - 1} = \dfrac{2}{m - 1}$

63. $k = 4$ and $k^2 = 16$

64. $p^2 = 36$ and $p = 6$

▲ 65. The mathematical model $y = 420x + 720$ approximates the worldwide credit card fraud losses between the years 1989 and 1993, where $x = 0$ corresponds to 1989, $x = 1$ corresponds to 1990, and so on, and y is in millions of dollars. Based on this model, what would be the approximate amount of credit card fraud losses in 1994? In what year would losses reach 3660 million dollars (that is, \$3,660,000,000).

▲ 66. According to research done by the Beverage Marketing Corporation, ready-to-drink iced tea sales have boomed during the past few years. The model $y = 310x + 260$ approximates the revenue generated, where $x = 0$ corresponds to 1991, and y is in millions of dollars. Based on this model, what would be the revenue generated in 1992? In what year would revenue reach 2430 million dollars?

▲ 67. The accompanying bar graph gives a pictorial representation of bank credit card charges in each year from 1989 to 1992. Use the graph to estimate the charges in each of the following years:
(a) 1989   (b) 1990
(c) 1991   (d) 1992.

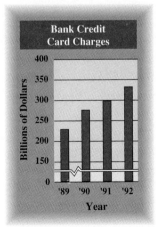

68. (a) 378 billion dollars
69. 36   70. .45   71. 72
72. 960   73. 50
74. 140   75. $\frac{39}{2}$ or $19\frac{1}{2}$
76. $\frac{147}{2}$ or $73\frac{1}{2}$

▲ **68.** Based on information from RAM Research, the amount of bank credit card charges in each year between 1989 and 1992 can be approximated by the linear model $y = 37x + 230$, where $x = 0$ corresponds to 1989, and y is in billions of dollars.

    (a) According to this model, what would be the charges in 1993?

    ▨ **(b)** In 1993, the actual charges were approximately 422 billion dollars. How might you explain the discrepancy between the predicted amount from part (a) and the actual amount charged?

REVIEW EXERCISES

Most of the exercise sets in the rest of this book end with a few review exercises. These are designed to help you review ideas that were previously studied in order to prepare you for the section or sections to follow. In each case, we cite the section or sections in which the concepts were first presented. Refer to the appropriate section(s) if you need help.

Evaluate the expression using the given value. See Section 1.2.

69. $2L + 2W$; $L = 10, W = 8$

70. rt; $r = .15, t = 3$

71. $\frac{1}{3}Bh$; $B = 27, h = 8$

72. prt; $p = 8000, r = .06, t = 2$

73. $\frac{5}{9}(F - 32)$; $F = 122$

74. $\frac{9}{5}C + 32$; $C = 60$

75. $\frac{1}{2}h(b + B)$; $B = 9, b = 4, h = 3$

76. $\frac{1}{2}bh$; $b = 21, h = 7$

2.2 FORMULAS AND TOPICS FROM GEOMETRY

FOR EXTRA HELP	OBJECTIVES
SSG pp. 28–33 **SSM** pp. 27–32	1 ▶ Solve a formula for a specified variable.
Video I	2 ▶ Solve applied problems using formulas.
Tutorial IBM MAC	3 ▶ Solve problems about angle measures.

◈ .10 or 10%; .0676 or 6.76%; .1125 or 11.25%

◈▷ **C O N N E C T I O N S** ◁◈

Formulas are an important part of applied mathematics, since they provide relationships among several quantities in a particular setting. Formulas are found in geometry, the mathematics of finance, branches of science, and many other fields.

    The formula

$$A = \frac{24f}{b(p + 1)}$$

gives the approximate annual interest rate for a consumer loan paid off with monthly payments. Here f is the finance charge on the loan, p is the number of payments, and b is the original amount of the loan.

INSTRUCTOR'S RESOURCES

📖 ITM pp. 321–325 ISM pp. 41–50 IAM pp. 3–4	💻 TEST GENERATOR DOS Windows MAC	⊞ TRANSPARENCIES

Models for many applied problems already exist; they are called *formulas*. A formula is a mathematical expression in which letters are used to describe a relationship. You have probably used formulas in other courses or in your daily life. Some formulas that we will be using are

$$d = rt, \qquad I = prt, \qquad \text{and} \qquad P = 2L + 2W.$$

A list of the formulas used in this book is given inside the covers of this book. Metric units are used in some of the exercises, although a working knowledge of the metric system is not needed to work the problems.

1 ▶ Solve a formula for a specified variable.

▶ In some applications, the necessary formula is solved for one of its variables, which may not be the unknown number that must be found. The following examples show how to solve a formula for any one of its variables. This process is called **solving for a specified variable.** Notice how the steps used in these examples are very similar to those used in solving a linear equation. Keep in mind that when you are solving for a specified variable, treat that variable as if it were the only one, and treat all other variables as if they were numbers.

EXAMPLE 1
Solving for a Specified Variable

Solve the formula $I = prt$ for t.

This formula gives the amount of simple interest, I, in terms of the principal (the amount of money deposited), p, the yearly rate of interest, r, and time in years, t. Solve this formula for t by assuming that I, p, and r are constants (having a fixed value) and that t is the variable. Then use the properties of the previous section as follows.

$$I = prt$$
$$I = (pr)t \qquad \text{Associative property}$$
$$\frac{I}{pr} = \frac{(pr)t}{pr} \qquad \text{Divide by } pr.$$
$$\frac{I}{pr} = t$$

Chalkboard Exercise

Solve $I = prt$ for p.

Answer: $p = \dfrac{I}{rt}$

The result is a formula for t, time in years. ∎

▶ TEACHING TIP
In order to solve for a specified variable, say for t in $d = rt$, first solve a similar equation with numerals, such as $8 = 4t$, to show the steps. Then show the parallel steps with d and r back in the equation. You may wish to do them side by side. ◀

While the process of solving for a specified variable uses the same steps as solving a linear equation from Section 2.1, the following additional suggestions may be helpful.

Solving for a Specified Variable

Step 1 Use the addition or multiplication properties as necessary to get all terms containing the specified variable on one side of the equation.

Step 2 All terms not containing the specified variable should be on the other side of the equation.

Step 3 If necessary, use the distributive property to write the side with the specified variable as the product of that variable and a sum of terms.

In general, follow the steps used for solving linear equations given in Section 2.1.

EXAMPLE 2

Solving for a Specified Variable

Solve the formula $P = 2L + 2W$ for W.

This formula gives the relationship between the perimeter of a rectangle, P, the length of the rectangle, L, and the width of the rectangle, W. See Figure 1. Solve the formula for W by getting W alone on one side of the equals sign. To begin, subtract $2L$ from both sides.

$$P = 2L + 2W$$
$$P - 2L = 2L + 2W - 2L \qquad \text{Subtract } 2L.$$
$$P - 2L = 2W$$
$$\frac{P - 2L}{2} = \frac{2W}{2} \qquad \text{Divide both sides by 2.}$$
$$\frac{P - 2L}{2} = W \quad \blacksquare$$

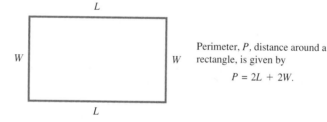

Chalkboard Exercise

Solve the formula $m = 2k + 3b$ for k.

Answer: $k = \dfrac{m - 3b}{2}$

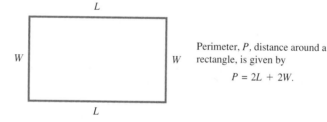

Perimeter, P, distance around a rectangle, is given by
$$P = 2L + 2W.$$

FIGURE 1

CAUTION Do not try to express the result in Example 2 by eliminating the 2 in the numerator and denominator. The 2 in the numerator is not a factor of the entire numerator. Expressing fractions in lowest terms will be explained in a later chapter.

EXAMPLE 3

Using the Distributive Property to Solve for a Specified Variable

Given the surface area, height, and width of a rectangular solid, write a formula for the length.

The formula for the surface area of a rectangular solid is found in the inside covers of this book. It is

$$A = 2HW + 2LW + 2LH,$$

where A represents the surface area, H represents the height, W represents the width, and L represents the length. See Figure 2. To solve for L, write the equation so that only the terms involving L appear on the right side.

$$A = 2HW + 2LW + 2LH$$
$$A - 2HW = 2LW + 2LH \qquad \text{Subtract } 2HW.$$

Next, use the distributive property on the right to write $2LW + 2LH$ so that L, the variable for which we are solving, is a factor.

$$A - 2HW = L(2W + 2H) \qquad \text{Distributive property}$$

Finally, divide both sides by the coefficient of L, which is $2W + 2H$.

$$\frac{A - 2HW}{2W + 2H} = L \qquad \text{or} \qquad L = \frac{A - 2HW}{2W + 2H} \quad \blacksquare$$

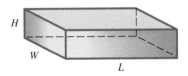

FIGURE 2

CAUTION The most common error in working a problem like Example 3 is not using the distributive property correctly. We must write the expression so that the variable for which we are solving is a *factor,* so that we can divide by its coefficient in the final step.

2 ▶ Solve applied problems using formulas.

▶ The next examples show how applications can be solved using formulas.

PROBLEM SOLVING As seen in the examples, it may be convenient to first solve for the particular unknown variable before substituting the given values. This is particularly useful when we wish to substitute several different values for the same variable.

EXAMPLE 4
Finding Average Speed

Janet Branson found that on the average it took her 3/4 hour each day to drive a distance of 15 miles to work. What was her average speed?

The formula needed for this problem is the distance formula, $d = rt$, where d represents distance traveled, r the rate, and t the time. Find the speed r, by solving $d = rt$ for r.

$$d = rt$$
$$\frac{d}{t} = \frac{rt}{t} \qquad \text{Divide by } t.$$
$$\frac{d}{t} = r$$

Now find r by substituting the given values of d and t into this formula.

$$r = \frac{d}{t}$$

$$r = \frac{15}{\frac{3}{4}} \qquad \text{Let } d = 15, t = \tfrac{3}{4}.$$

$$r = 15 \div \frac{3}{4} = 15 \cdot \frac{4}{3}$$

$$r = 20$$

Her average speed was 20 miles per hour. ■

When a consumer loan is paid off ahead of schedule, the finance charge is smaller than if the loan were paid off over its scheduled life. By one method, called *the rule of 78,* the amount of *unearned interest* (finance charge that need not be paid) is given by

$$u = f \cdot \frac{k(k+1)}{n(n+1)},$$

where u is the amount of unearned interest (money saved) when a loan scheduled to run n payments is paid off k payments ahead of schedule. The total scheduled finance charge is f.

Actually, if uniform monthly payments are made, the rule results in more interest in the early months of the loan than is strictly proper. Thus any refund for early repayment would be less by this rule than it really should be. However, the method is still often used, and its accuracy is acceptable for fairly short-term consumer loans. (With tables and calculators readily available, very accurate interest and principal allocations of loan payments are easily determined.)

The next example illustrates the use of the rule of 78.

EXAMPLE 5
Using the Rule of 78

Julie is scheduled to pay off a loan in 60 monthly payments. If the total finance charge is $500 and the loan is paid off after 50 payments, find the unearned interest, u, and the new total finance charge.

Answer: $u = \$15.03$; total finance charge = $\$484.97$

Juan Ortega is scheduled to pay off a loan in 36 monthly payments. If the total finance charge is $360, and the loan is paid off after 24 payments (that is, 12 payments ahead of schedule), find the unearned interest, u.

A calculator is helpful in this problem. Here $f = 360$, $n = 36$, and $k = 12$. Use the formula above.

$$u = f \cdot \frac{k(k+1)}{n(n+1)} = 360 \cdot \frac{12(13)}{36(37)} = 42.16$$

A total of $42.16 of the $360 finance charge need not be paid. The total finance charge for this loan is $360 − $42.16 = $317.84. ■

3 ▶ Solve problems about angle measures.

▶   In this section and the next, we will look at some properties of angles and their measures that lend themselves to algebraic interpretation. Recall that a basic unit of measure of angles is the **degree;** an angle that measures one degree (1°) is 1/360 of a complete revolution. See Figure 3.

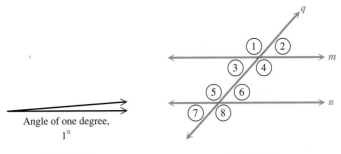

Angle of one degree,
1°

FIGURE 3   **FIGURE 4**

Parallel lines are lines that lie in the same plane and do not intersect. Figure 4 shows parallel lines *m* and *n*. When a line *q* intersects the two parallel lines, *q* is called a **transversal.** In Figure 4, the transversal intersecting the parallel lines forms eight angles, indicated by the circled numbers.

Pairs of angles like ① and ④, ② and ③, ⑤ and ⑧, and ⑥ and ⑦, are called **vertical angles.** They lie "opposite" each other. It is shown in geometry that *vertical angles have the same measure.*

Angles ① through ⑧ possess some special properties regarding their degree measures. The following chart gives their names with respect to each other, and the rules regarding their measures.

Name	Sketch	Rule
Alternate interior angles	④ ⑤ (also ③ and ⑥)	Angle measures are equal.
Alternate exterior angles	① ⑧ (also ② and ⑦)	Angle measures are equal.
Interior angles on same side of transversal	④ ⑥ (also ③ and ⑤)	Angle measures add to 180°.
Corresponding angles	② ⑥ (also ① and ⑤, ③ and ⑦, ④ and ⑧)	Angle measures are equal.

Use the information in the chart to solve the next problem.

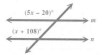
Find x, given that lines m and n are parallel. Then find the measure of each marked angle.

In Figure 5, the marked angles are alternate exterior angles, which are equal. This gives

$$3x + 2 = 5x - 40$$
$$42 = 2x$$
$$21 = x.$$

One angle has a measure of $3x + 2 = 3 \cdot 21 + 2 = 65$ degrees, and the other has a measure of $5x - 40 = 5 \cdot 21 - 40 = 65$ degrees. ■

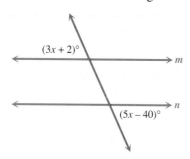

FIGURE 5

2.2 EXERCISES

1. Write an explanation, in step-by-step form, of the procedure you would use to solve the formula $P = 2L + 2W$ for L.

2. One source for geometric formulas gives the formula for the perimeter of a rectangle as $P = 2L + 2W$, while another gives it as $P = 2(L + W)$. Are these equivalent? If so, what property justifies their equivalence?

3. Suppose the formula $A = 2HW + 2LW + 2LH$ is "solved for L" as follows:

$$A = 2HW + 2LW + 2LH$$
$$A - 2LW - 2HW = 2LH$$
$$\frac{A - 2LW - 2HW}{2H} = L.$$

While there are no algebraic errors here, what is wrong with the final line, if we are interested in solving for L?

4. When a formula is solved for a particular variable, several different equivalent forms may be possible. If we solve $A = \frac{1}{2} bh$ for h, one possible correct answer is

$$h = \frac{2A}{b}.$$

Which one of the following is *not* equivalent to this?

(a) $h = 2\left(\dfrac{A}{b}\right)$ **(b)** $h = 2A\left(\dfrac{1}{b}\right)$ **(c)** $h = \dfrac{A}{\frac{1}{2}b}$ **(d)** $h = \dfrac{\frac{1}{2}A}{b}$

Solve the formula for the specified variable. See Examples 1–3.

5. $d = rt$;   for r (distance)

6. $I = prt$;   for r (simple interest)

7. $A = bh$;   for b (area of a parallelogram)

8. $P = 2L + 2W$;   for L (perimeter of a rectangle)

9. $P = a + b + c$;   for a (perimeter of a triangle)

10. $V = LWH$;   for W (volume of a rectangular solid)

11. $A = \dfrac{1}{2}bh$;   for h (area of a triangle)

12. $C = 2\pi r$;   for r (circumference of a circle)

13. $S = 2\pi rh + 2\pi r^2$;   for h (surface area of a right circular cylinder)

14. $A = \dfrac{1}{2}(B + b)h$;   for B (area of a trapezoid)

15. $C = \dfrac{5}{9}(F - 32)$;   for F   (Fahrenheit to Celsius)

16. $F = \dfrac{9}{5}C + 32$;   for C   (Celsius to Fahrenheit)

17. $A = 2HW + 2LW + 2LH$;   for H (surface area of a rectangular solid)

18. $V = \dfrac{1}{3}Bh$;   for h (volume of a right pyramid)

———————◆ **MATHEMATICAL CONNECTIONS** (Exercises 19–24) ◆———————

Consider the following equations:

First Equation   *Second Equation*

$$x = \frac{5x + 8}{3} \qquad t = \frac{bt + k}{c} \quad (c \neq 0).$$

Solving the second equation for t follows the same logic as solving the first equation for x. When solving for t, we treat all other variables as though they were constants.

The following group of exercises illustrates the "parallel logic" of solving for x and solving for t. All must be worked sequentially.

19. (a) Clear the first equation of fractions by multiplying through by 3.
  (b) Clear the second equation of fractions by multiplying through by c.

20. (a) Get the terms involving x on the left side of the first equation by subtracting $5x$ from both sides.
  (b) Get the terms involving t on the left side of the second equation by subtracting bt from both sides.

21. (a) Combine like terms on the left side of the first equation. What property allows us to write $3x - 5x$ as $(3 - 5)x = -2x$?
  (b) Write the expression on the left side of the second equation so that t is a factor. What property allows us to do this?

22. (a) Divide both sides of the first equation by the coefficient of x.
  (b) Divide both sides of the second equation by the coefficient of t.

◉ **23.** Look at your answer for the second equation. What restriction must be placed on the variables? Why is this necessary?

▱ **24.** Write a short paragraph summarizing what you have learned in this group of exercises.

*Solve the problem. Refer to the inside covers of this book if you need to look up a formula.
See Example 4.*

25. In 1992 Al Unser, Jr. won the Indianapolis 500 (mile) with a speed of 134.5 miles per hour. Find his time.*

26. In 1975, rain shortened the Indianapolis 500 to 435 miles. It was won by Bobby Unser who averaged 149.2 miles per hour. What was his time?*

27. The lowest temperature ever recorded in Arizona was −40 degrees Celsius on January 7, 1971. Find the corresponding Fahrenheit temperature.

28. The melting point of brass is 900 degrees Celsius. Find the corresponding Fahrenheit temperature.

29. The base of the Great Pyramid of Cheops is a square whose perimeter is 920 meters. What is the length of each side of this square?

30. The Peachtree Plaza Hotel in Atlanta is in the shape of a cylinder with radius 46 meters and height 220 meters. Find its volume to the nearest hundred thousand.

31. Faye Korn traveled from Kansas City to Louisville, a distance of 520 miles, in 10 hours. Find her rate in miles per hour.

32. The distance from Melbourne to London is 10,500 miles. If a jet averages 500 miles per hour between the two cities, what is its travel time in hours?

33. A circle has a circumference of 30π meters. Find the radius of the circle.

34. The circumference of a circle is 120π meters. Find the radius of the circle.

35. The surface area of a can is 32π square inches. The radius of the can is 2 inches. Find the height of the can.

36. A can has a surface area of 12π square inches. The radius of the can is 1 inch. Find the height of the can.

37. A cord of wood contains 128 cubic feet of wood. If a stack of wood is 4 feet wide and 4 feet high, how long must it be if it contains exactly 1 cord?

38. Give two sets of possible dimensions for a stack of wood that contains 1.5 cords. (See Exercise 37.)

39. In order to purchase fencing to go around a yard, would you need to use perimeter or area to decide how much to buy?

40. In order to purchase sod for a lawn, would you need to use perimeter or area to decide how much to buy?

Solve the problem.

41. A mixture of alcohol and water contains a total of 36 ounces of liquid. There are 9 ounces of pure alcohol in the mixture. What percent of the mixture is water? What percent is alcohol?

42. A mixture of acid and water is 35% acid. If the mixture contains a total of 40 liters, how many liters of pure acid are in the mixutre? How many liters of pure water are in the mixture?

43. In a survey of 263,000 consumers aged 14 years or older, 24% did not read the Sunday paper. How many of these consumers did not read the Sunday paper? (*Source:* Impact Resources)

44. A study showed that 17% of 18–24 year olds snore. In a group of 2500 people in this age bracket, how many could we expect to snore? (*Source:* The Better Sleep Council)

---

*Data from *The Universal Almanac,* 1993. John W. Wright, General Editor. Andrews and McMeel, Kansas City and New York, p. 673.

Refer to the graph to answer the questions in Exercises 45 and 46. (Source: Research Partnership)

45. In a typical group of 3500 automatic teller card users, how many would we expect to use their cards 6–9 times per month?

46. In a typical group of 4000 automatic teller card users, how many of them would we expect to use their cards at least once?

Banking on Automatic Tellers

60% of consumers in the USA have automatic-teller cards. How many times per month they use them:

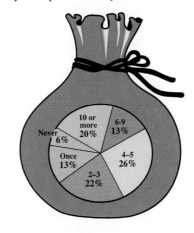

47. A real estate agent earned $6300 commission on a property sale of $210,000. What is her rate of commission?

48. A certificate of deposit for one year pays $85 simple interest on a principal of $3400. What is the interest rate being paid on this deposit?

Refer to the formula for the rule of 78 to solve each of the following. See Example 5.

49. Rhonda Alessi bought a new Ford and agreed to pay it off in 36 monthly payments. The total finance charge is $700. Find the unearned interest if she pays the loan off 4 payments ahead of schedule.

50. Paul Lorio bought a car and agreed to pay in 36 monthly payments. The total finance charge on the loan was $600. With 12 payments remaining, Paul decided to pay the loan in full. Find the amount of unearned interest.

51. The finance charge on a loan taken out by Vic Denicola is $380.50. If there were 24 equal monthly installments needed to repay the loan, and the loan is paid in full with 8 months remaining, find the amount of unearned interest.

52. Adrian Ortega is scheduled to repay a loan in 24 equal monthly installments. The total finance charge on the loan is $450. With 9 payments remaining he decides to repay the loan in full. Find the amount of unearned interest.

The formula

$$A = \frac{24f}{b(p + 1)}$$

gives the approximate annual interest rate for a consumer loan paid off with monthly payments. Here f is the finance charge on the loan, p is the number of payments, and b is the original amount of the loan.

Solve the following problems using the formula above.

53. Find the approximate annual interest rate for an installment loan to be repaid in 24 monthly installments. The finance charge on the loan is $200, and the original loan balance is $1920.

54. Find the approximate annual interest rate for an automobile loan to be repaid in 36 monthly installments. The finance charge on the loan is $740 and the amount financed is $3600.

 CONCEPTUAL   WRITING   ▲ CHALLENGING    SCIENTIFIC CALCULATOR    GRAPHICS CALCULATOR

55. $x = -8$; 49°; 49°
56. $x = 15$; 122°; 122°
57. $x = 27$; 49°; 49°
58. $x = 28$; 117°; 117°
59. $x = 47$; 48°; 132°
60. $x = 13$; 141°; 141°
61. $\{12\}$ 62. $\{55\}$
63. $\left\{\dfrac{1}{2}\right\}$ 64. $\{3000\}$
65. -3 66. -4

In the given figure, find the value of x, and then find the measure of each marked angle. In Exercises 57–60, assume that lines m and n are parallel. See Example 6.

55.

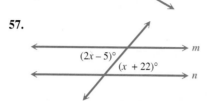

56.

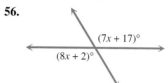

57.

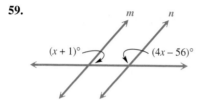

58.

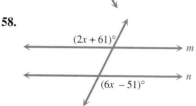

59.

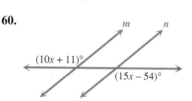

60.

REVIEW EXERCISES

Solve the equation. See Section 2.1.

61. $4x + 4(x + 7) = 124$

62. $x + .20x = 66$

63. $\dfrac{5}{3} + \dfrac{2}{3}x = 2$

64. $.07x + .05(9000 - x) = 510$

Use the order of operations to evaluate. See Section 1.2.

65. the product of -3 and 5, divided by 1 less than 6

66. half of -18 added to the reciprocal of $\dfrac{1}{5}$

2.3 APPLICATIONS OF LINEAR EQUATIONS

FOR EXTRA HELP

📖 **SSG** pp. 33–39
SSM pp. 32–38

📼 **Video** 2

💾 **Tutorial**
IBM MAC

OBJECTIVES

1 ▶ Translate from word expressions to mathematical expressions.
2 ▶ Write equations from given information.
3 ▶ Solve problems about unknown numerical quantities.
4 ▶ Solve problems about percents, simple interest, and mixture.
5 ▶ Solve problems about angles: supplementary, complementary, and sums of angles in triangles.

1 ▶ Translate from word expressions to mathematical expressions.

▶ In this section, we discuss ways to produce a mathematical model of a real situation. This often involves translating verbal statements into mathematical statements. Although the problems we will be working with are simple ones, the methods we use also apply to more difficult problems later.

INSTRUCTOR'S RESOURCES

📖 **ITM** pp. 326–331 **ISM** pp. 50–61
IAM pp. 4

💾 **TEST GENERATOR**
DOS Windows MAC

✳ **TRANSPARENCIES**

PROBLEM SOLVING

Usually there are key words and phrases in the verbal problem that translate into mathematical expressions involving the operations of addition, subtraction, multiplication, and division. Translations of some commonly used expressions follow.

► **TEACHING TIP**
As an oral exercise, have the class compile a list of key words and expressions categorized as to arithmetic operations. ◄

Translation from Words to Mathematical Expressions

Verbal Expression	Mathematical Expression
Addition	
The sum of a number and 7	$x + 7$
6 more than a number	$x + 6$
3 plus 8	$3 + 8$
24 added to a number	$x + 24$
A number increased by 5	$x + 5$
The sum of two numbers	$x + y$
Subtraction	
2 less than a number	$x - 2$
12 minus a number	$12 - x$
A number decreased by 12	$x - 12$
The difference between two numbers	$x - y$
A number subtracted from 10	$10 - x$
Multiplication	
16 times a number	$16x$
Some number multiplied by 6	$6x$
$\frac{2}{3}$ of some number (used only with fractions and percent)	$\frac{2}{3}x$
Twice (2 times) a number	$2x$
The product of two numbers	xy
Division	
The quotient of 8 and some number	$\frac{8}{x}$
A number divided by 13	$\frac{x}{13}$
The ratio of two numbers or the quotient of two numbers	$\frac{x}{y}$

CAUTION Because subtraction and division are not commutative operations, it is important to correctly translate expressions involving them. For example, "2 less than a number" is translated as $x - 2$, *not* $2 - x$. "A number subtracted from 10" is expressed as $10 - x$, not $x - 10$. For division, it is understood that the number doing the dividing is the denominator, and the number that is divided is the numerator. For example, "a number divided by 13" and "13 divided into x" both translate as $x/13$. Similarly, "the quotient of x and y" is translated as x/y.

2 ▶ Write equations from given information.

▶ The symbol for equality, =, is often indicated by the word "is." In fact, since equal mathematical expressions represent different names for the same number, any words that indicate the idea of "sameness" indicate translation to =.

EXAMPLE 1

Translating Sentences into Equations

Translate the following verbal sentences into equations.

Verbal Sentence	Equation
Twice a number, decreased by 3, is 42.	$2x - 3 = 42$
If the product of a number and 12 is decreased by 7, the result is 105.	$12x - 7 = 105$
A number divided by the sum of 4 and the number is 28.	$\dfrac{x}{4 + x} = 28$
The quotient of a number and 4, plus the number, is 10.	$\dfrac{x}{4} + x = 10.$ ■

▶ **TEACHING TIP**
Include as an example: "Subtract 15 from a number." Rewrite this expression as: "From a number, subtract 15." ◀

■ **PROBLEM SOLVING**   Throughout this book we will be examining different types of applications. While there is no one method that will allow us to solve all types of applied problems, there are some guidelines that are helpful. They are listed below.

Solving an Applied Problem

For Example 1

Step 1 **Determine what you are asked to find.** Read the problem carefully. Decide what is given and what must be found. Choose a variable and write down exactly what it represents.

Step 2 **Write down any other pertinent information.** If there are other unknown quantities, express them using the variable. Draw figures or diagrams if they apply.

Step 3 **Write an equation.** Write an equation expressing the relationships among the quantities given in the problem.

Step 4 **Solve the equation.** Use the methods of the earlier sections to solve the equation.

Step 5 **Answer the question(s) of the problem.** Reread the problem and make sure that you answer the question or questions posed. In some cases, you will need to give more than just the solution of the equation.

Step 6 **Check.** Check your solution by using the original words of the problem. Be sure that your answer makes sense.

◇ Our steps 1, 3, 4, and 6 correspond to Polya's steps.

◆ **CONNECTIONS** ◆

Probably the most famous study of problem-solving techniques was developed by George Polya (1888–1985). Among his many publications was the modern classic *How to Solve It.* In this book, Polya proposed a four-step process for problem solving.

POLYA'S FOUR-STEP PROCESS FOR PROBLEM SOLVING

1. **Understand the problem.** You must first decide what you are to find.
2. **Devise a plan.** Here are some strategies that may prove useful.

Problem-Solving Strategies
If a formula applies, use it.
Write an equation and solve it.
Draw a sketch.

Make a table or a chart.
Look for a pattern.
Use trial and error.
Work backward.

We used the first of these strategies in the previous section. In this section we will use the next three strategies.

3. **Carry out the plan.** This is where the algebraic techniques you are learning in this book can be helpful.
4. **Look back and check.** Is your answer reasonable? Does it answer the question that was asked?

FOR DISCUSSION OR WRITING
Compare Polya's four steps with the six steps for problem solving given above. Which of our steps correspond with each of Polya's steps?

3 ► Solve problems about unknown numerical quantities.

► The next example illustrates the use of the six steps outlined for problem solving.

E X A M P L E   2
Finding Unknown Numerical Quantities

The Perry brothers, Jim and Gaylord, were two outstanding pitchers in the major leagues during the past few decades. Together, they won 529 games. Gaylord won 99 games more than Jim. How many games did each brother win?

Step 1 We are asked to find the number of games each brother won. We must choose a variable to represent the number of wins of one of the men.

Let j = the number of wins for Jim.

Step 2 We must also find the number of wins for Gaylord. Since he won 99 games more than Jim,

let $j + 99$ = Gaylord's number of wins.

Step 3 The sum of the numbers of wins is 529, so we can now write an equation.

$$\underbrace{\text{Jim's wins}}_{j} + \underbrace{\text{Gaylord's wins}}_{(j + 99)} = \underbrace{529}_{529}$$

Step 4 Solve the equation.

$$j + (j + 99) = 529$$
$$2j + 99 = 529 \qquad \text{Combine like terms.}$$
$$2j = 430 \qquad \text{Subtract 99.}$$
$$j = 215 \qquad \text{Divide by 2.}$$

Step 5 Since j represents the number of Jim's wins, Jim won 215 games. Gaylord won $j + 99 = 215 + 99 = 314$ games.

Step 6 314 is 99 more than 215, and the sum of 314 and 215 is 529.

The words of the problem are satisfied, and our solution checks. ■

Chalkboard Exercise

Cindy is 5 years older than three times the age of her son, Cody. The sum of their ages is 45. Find their ages.

Answer: Cindy is 35 years old; Cody is 10 years old.

CAUTION A common error in solving applied problems is forgetting to answer all the questions asked in the problem. In Example 2, we were asked for the number of wins for *each* brother, so there was an extra step at the end in order to find Gaylord's number.

4 ▶ Solve problems about percents, simple interest, and mixture.

▶   The next few examples in this section involve percent, which means per one hundred. For example, 17% means 17/100, or .17, and 109% means 1.09.

  (We will not number the steps in the following examples; see if you can identify them.)

E X A M P L E 3

Solving a Percent Problem

In 1993 there were 154 long distance area codes in the United States. This was an increase of 79% over the number when the area code plan originated in 1947. How many area codes were there in 1947?

> Find: the number of area codes in 1947
> Given: the number in 1947, increased by 79%, was 154.
> Let x = the number of area codes in 1947;
> .79x = the increase.

From the given information, we get the equation:

the number in 1947 + the increase = 154.
$$x + .79x = 154$$

Solve the equation.

$$1x + .79x = 154 \qquad \text{Identity property}$$
$$1.79x = 154 \qquad \text{Combine terms.}$$
$$x = 86 \qquad \text{Divide by 1.79. (Use a calculator if you wish.)}$$

There were 86 area codes in 1947. Check that the increase, $154 - 86 = 68$, is 79% of 86. ■

CAUTION Watch out for two common errors that occur in solving problems like the one in Example 3. First, do not try to find 79% of the number of calls in 1993 and then subtract from that amount. Second, avoid writing the equation incorrectly. It would be *wrong* to write the equation as

$$x + .79 = 154.$$

The decimal .79 must be multiplied by x, as shown in the example.

The next example shows how to solve a mixture problem involving different concentrations. Notice how a sketch and a table are used.

EXAMPLE 4

Solving a Mixture Problem

A chemist must mix 8 liters of a 40% solution of potassium chloride with some 70% solution to get a mixture that is a 50% solution. How much of the 70% solution should be used?

The information in the problem is illustrated in Figure 6.

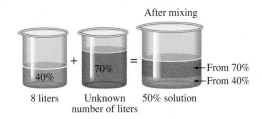

FIGURE 6

Let x = the number of liters of the 70% solution that should be used. The information in the problem is used to get the following table.

Strength	Liters of Solution	Liters of Pure Potassium Chloride
40%	8	.40(8) = 3.2
70%	x	.70x
50%	8 + x	.50(8 + x)

Sum must equal

The numbers in the right-hand column were found by multiplying the strengths and the numbers of liters. The number of liters of pure potassium chloride in the 40% solution plus the number of liters in the 70% solution must equal the number of liters in the 50% solution.

$$3.2 + .70x = .50(8 + x)$$
$$3.2 + .70x = 4 + .50x \qquad \text{Distributive property}$$
$$.20x = .8 \qquad \text{Subtract 3.2 and .50}x.$$
$$x = 4 \qquad \text{Divide by .20.}$$

The chemist must use 4 liters of the 70% solution. ■

PROBLEM SOLVING A chart was used in Example 4 to organize the information. Some students like to use "box diagrams" instead. Either method works well if you are careful while filling in the information. The next example uses box diagrams.

EXAMPLE 5

Solving an Investment Problem

After winning the state lottery, Michael Chin has $40,000 to invest. He will put part of the money in an account paying 4% simple interest, and the remainder into stocks paying 6% simple interest. His accountant tells him the total annual income from these investments should be $2040. How much should he invest at each rate?

Let
$$x = \text{the amount invested at 4\%}$$
$$40,000 - x = \text{the amount invested at 6\%.}$$

Chalkboard Exercise

A man has \$34,000 to invest. He invests some at 17% and the balance at 20%. His total annual interest income is \$6245. Find the amount invested at each rate.

Answer: \$18,500 at 17%; \$15,500 at 20%

In the box diagram in Figure 7, fill in the important information. The formula for interest is $I = prt$. The interest earned in one year at 4% is

$$x(.04)(1) = .04x.$$

Amount invested	x	+	$40{,}000 - x$	=	2040	Total interest
Rate	.04		.06			

FIGURE 7

The interest at 6% is

$$(40{,}000 - x)(.06)(1) = .06(40{,}000 - x).$$

Since the total interest is \$2040,

interest at 4% + interest at 6% = total interest

$$.04x + .06(40{,}000 - x) = 2040.$$

Solve the equation.

$$.04x + 2400 - .06x = 2040 \quad \text{Distributive property}$$
$$2400 - .02x = 2040 \quad \text{Combine terms.}$$
$$-.02x = -360 \quad \text{Subtract 2400.}$$
$$x = 18{,}000 \quad \text{Divide by } -.02.$$

Chin should invest \$18,000 at 4% and \$40,000 − \$18,000 = \$22,000 at 6%. Check by finding the annual interest at each rate; they should add up to \$2040. ■

NOTE Refer to Example 5. We chose to let the variable represent the amount invested at 4%. Students often ask "Can I let the variable represent the other unknown?" The answer is yes. The equation will be different, but in the end the two answers will be the same. You might wish to work Example 5, letting the variable equal the amount invested at 6%.

You may notice that the investment problems in this chapter deal with simple interest. In most "real-world" applications, compound interest is used. However, more advanced methods (covered in a later chapter) are needed for compound interest problems, so we will deal only with simple interest until then.

5 ▶ Solve problems about angles: supplementary, complementary, and sums of angles in triangles.

▶ In the last section we saw some angle relationships with regard to vertical angles and angles formed by transversals intersecting parallel lines. Here, we examine some other concepts involving angle measures.

Figure 8 shows some special angles, with their names. A **right angle** measures 90°; a **straight angle** measures 180°. If the sum of the measures of two angles is 90°, the angles are **complementary angles,** and they are called **complements** of each other. If the sum of the measures of two angles is 180°, the angles are **supplementary angles,** and they are called **supplements** of each other.

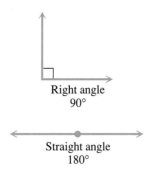

Right angle
90°

Straight angle
180°

FIGURE 8

One of the important results of Euclidean geometry (the geometry of the Greek mathematician Euclid) is that the sum of the angle measures of any triangle is 180°. This property, along with the others mentioned above, is used in the next example.

E X A M P L E 6
Solving Problems About
Angle Measures

(a) Find the value of x, and determine the measure of each angle in Figure 9.

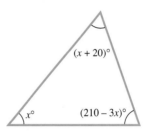

FIGURE 9

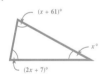
Since the three marked angles are angles of a triangle, their sum must be 180°. Write the equation indicating this, and then solve.

$$x + (x + 20) + (210 - 3x) = 180$$
$$-x + 230 = 180 \qquad \text{Combine like terms.}$$
$$-x = -50 \qquad \text{Subtract } 230.$$
$$x = 50 \qquad \text{Divide by } -1.$$

One angle measures 50°, another measures $x + 20 = 50 + 20 = 70°$, and the third measures $210 - 3x = 210 - 3(50) = 60°$. Since $50° + 70° + 60° = 180°$, the answers are correct.

(b) The supplement of an angle measures 10° more than three times its complement. Find the measure of the angle.

Let $\qquad x =$ the degree measure of the angle. Then,

$180 - x =$ the degree measure of its supplement, and

$90 - x =$ the degree measure of its complement.

Now use the words of the problem to write the equation.

three times
supplement   measures   10 more than   its complement

$$180 - x \quad = \quad 10 + \quad 3(90 - x)$$

Solve the equation.

$$180 - x = 10 + 270 - 3x \qquad \text{Distributive property}$$
$$180 - x = 280 - 3x$$
$$2x = 100 \qquad \text{Add } 3x; \text{ subtract } 180.$$
$$x = 50 \qquad \text{Divide by 2.}$$

The angle measures 50°. Since its supplement (130°) is 10° more than three times its complement (40°), that is

$$130 = 10 + 3(40)$$

is true, the answer checks.   ■

2.3 EXERCISES

2. (a) $x - 9$ (b) $9 < x$;
(a) is an expression
3. $x - 13$ 4. $x - 12$
5. $7 + x$ 6. $x + 12$
7. $8(x + 12)$
8. $(x - 9)(x + 6)$ 9. $\dfrac{x}{6}$
10. $\dfrac{6}{x}$ $(x \neq 0)$ 11. $\dfrac{x}{12}$
12. $\dfrac{12}{x}$ $(x \neq 0)$ 13. $\dfrac{6}{7}x$
14. $.19x$ or $\dfrac{19}{100}x$
15. (d)

1. In your own words, list the six steps suggested in the box marked "Solving an Applied Problem" in this section.

2. If x represents the number, express the following using algebraic symbols:
(a) 9 less than a number
(b) 9 is less than a number
Which one of these is an *expression*?

Translate the verbal phrase into a mathematical expression. Use x to represent the unknown number. See Example 1.

3. a number decreased by 13

4. a number decreased by 12

5. 7 increased by a number

6. 12 more than a number

7. the product of 8 and 12 more than a number

8. the product of 9 less than a number and 6 more than the number

9. the quotient of a number and 6

10. the quotient of 6 and a nonzero number

11. the ratio of a number and 12

12. the ratio of 12 and a nonzero number

13. $\dfrac{6}{7}$ of a number

14. 19% of a number

15. Which one of the following is *not* a valid translation of "30% of a number"?
(a) $.30x$ (b) $.3x$ (c) $\dfrac{3x}{10}$ (d) $30x$

16. Explain the difference between the answers to Exercises 11 and 12.

Solve the problem. See Example 2.

17. Labrador retrievers and Rottweilers rank as the top two dog breeds using figures provided by the American Kennel Club. In 1992, there were 25,434 more Labrador retrievers registered than Rottweilers, and together there were 216,324 dogs of these two breeds registered. How many of each breed were registered?

18. According to figures provided by the Air Transport Association of America, the Boeing B747–400 and the MacDonald Douglas L1011–100/200 are among the air carriers with maximum passenger seating. The Boeing seats 110 more passengers than the MacDonald Douglas, and together the two models seat 696 passengers. What is the seating capacity of each model?

19. Two of the highest paid business executives in a recent year were Mike Eisner, chairman of Disney, and Ed Horrigan, vice chairman of RJR Nabisco. Together their salaries totaled 61.8 million dollars. Eisner earned 18.4 million dollars more than Horrigan. What was the salary for each executive?

20. In a recent year, the two U.S. industrial corporations with the highest sales were General Motors and Ford Motor. Their sales together totaled 213.5 billion dollars. Ford Motor sales were 28.7 billion dollars less than General Motors. What were the sales for each corporation?

21. Babe Ruth and Rogers Hornsby were two great hitters. Together they got 5803 base hits in their careers. Hornsby got 57 more hits than Ruth. How many base hits did each get?

22. In the 1992 presidential election, Bill Clinton and George Bush together received 538 electoral votes. Clinton received 202 more votes than Bush. How many votes did each candidate receive?

Refer to the accompanying graph to help you answer the questions in Exercises 23–26.

23. According to figures provided by the Tobacco Institute, Department of Agriculture, the top two states in the tobacco industry together employ 167,281 people. The leading state employs 17,663 fewer than twice as many as the second leading state. What are these states and how many people does each employ?

24. New York employs 1446 fewer people than California, and California employs 2511 fewer people than Virginia. The three states together employ 127,143 people. How many people work in the tobacco industry in each of these states?

◉ 25. After working Exercises 23 and 24, look at your answers and see if they correspond to the graph. Remember that a graph such as this provides a visual estimate, and not exact values.

▦ 26. Suppose that the five states listed accounted for 60% of all tobacco-related jobs in the United States. If this were the case, what would be the total number of such jobs in the country? Round your answer to the nearest hundred.

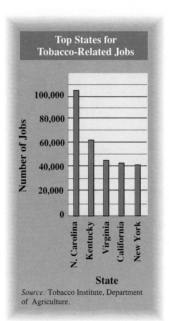

Top States for Tobacco-Related Jobs

Number of Jobs / State

Source: Tobacco Institute, Department of Agriculture.

17. Labarador retrievers: 120,879; Rottweilers: 95,445   18. Boeing: 403; MacDonald Douglas: 293
19. Eisner: 40.1 million dollars; Horrigan: 21.7 million dollars
20. General Motors: 121.1 billion dollars;   Ford Motor: 92.4 billion dollars
21. Ruth: 2873 hits; Hornsby: 2930 hits
22. Clinton: 370 votes; Bush: 168 votes   23. The leading state, North Carolina, employs 105,633 people, and the second leading state, Kentucky, employs 61,648 people.
24. New York: 40,580 people;   California: 42,026 people;   Virginia: 44,537 people   26. 490,700

 CONCEPTUAL   WRITING   CHALLENGING   SCIENTIFIC CALCULATOR    GRAPHICS CALCULATOR

27. 7.0%   **28.** 11.9%
29. $2008   **30.** 129.9
31. $122.28   **32.** $149,460
33. (a) $800 - x$
(b) $800 - y$
34. (a) $.05x$; $.10(800 - x)$
(b) $.05y$; $.10(800 - y)$
35. (a) $.05x +$
$.10(800 - x) = 800(.0875)$
(b) $.05y + .10(800 - y) =$
$800(.0875)$ **36.** (a) $200
at 5%; $600 at 10%
(b) 200 liters of 5% acid;
600 liters of 10% acid

▦ *Solve the problem. See Example 3.*

27. In a recent 10-year period, composite scores on the ACT exam dropped from 19.9 to 18.5. What percent decrease was this? (*Source:* The American College Testing Program)

28. In one recent year, the number of participitants in the ACT exam was 730,000. Later, in 1990, a total of 817,000 people took the exam. What percent increase was this? (*Source:* The American College Testing Program)

29. In 1980, the average tuition for public 4-year universities in the United States was $840 for full-time students. By 1990, it had risen approximately 139%. What was the approximate cost in 1990? (*Source:* National Center for Education Statistics, U.S. Dept. of Education)

30. The consumer price index (CPI) in June, 1991 was 136.0. This represented a 4.7% increase from June, 1990. What was the CPI in June, 1990?

31. At the end of a day, Jeff Hornsby found that the total cash register receipts at the motel where he works amounted to $1650.78. This included the 8% sales tax charged. Find the amount of the tax.

32. Dongming Wei sold his house for $159,000. He got this amount knowing that he would have to pay a 6% comission to his agent. What amount did he come away with after the agent was paid?

──────── ◆ **MATHEMATICAL CONNECTIONS** (Exercises 33–38) ◆ ────────

Consider the following two problems.

Problem A
Jack has $800 invested in two accounts. One pays 5% interest per year and the other pays 10% interest per year. The amount of yearly interest is the same as he would get if the entire $800 was invested at 8.75%. How much does he have invested at each rate?

Problem B
Jill has 800 liters of acid solution. She obtained it by mixing some 5% acid with some 10% acid. Her final mixture of 800 liters is 8.75% acid. How much of each of the 5% and 10% mixtures did she use to get her final mixture?

In Problem A , let x represent the amount invested at 5% interest, and in Problem B, let y represent the amount of 5% acid used. Work the following problems in sequence.

33. (a) Write an expression in x that represents Jack's amount of money invested at 10% in Problem A.
(b) Write an expression in y that represents Jill's amount of 10% acid mixture used in Problem B.

34. (a) Write expressions that represent the amount of interest Jack earns per year at 5% and at 10%.
(b) Write expressions that represent the amount of pure acid in Jill's 5% and 10% acid mixtures.

35. (a) The sum of the two expressions in part (a) of Exercise 34 must equal the total amount of interest earned in one year. Write an equation representing this fact.
(b) The sum of the two expression in part (b) of Exercise 34 must equal the amount of pure acid in the final mixture. Write an equation representing this fact.

36. (a) Solve Problem A.
(b) Solve Problem B.

◉ **37.** Write a paragraph explaining the similarities between the solution processes used in solving Problems A and B.

◉ **38.** Meredith Many explained these two processes as "stuff plus stuff equals stuff." What did she mean by this?

──────── ◆ ────────

Solve the problem. See Example 4.

39. Ten liters of a 4% acid solution must be mixed with a 10% solution to get a 6% solution. How many liters of the 10% solution are needed?

40. How many liters of a 14% alcohol solution must be mixed with 20 liters of a 50% solution to get a 30% solution?

41. In a chemistry class, 12 liters of a 12% alcohol solution must be mixed with a 20% solution to get a 14% solution. How many liters of the 20% solution are needed?

42. How many liters of a 10% alcohol solution must be mixed with 40 liters of a 50% solution to get a 40% solution?

43. A medicated first aid spray on the market is 78% alcohol by volume. If the manufacturer has 50 liters of the spray containing 70% alcohol, how much pure alcohol should be added so that the final mixture is the required 78% alcohol? (*Hint:* Pure alcohol is 100% alcohol.)

44. How much water must be added to 3 gallons of a 4% insecticide solution to reduce the concentration to 3%? (*Hint:* Water is 0% insecticide.)

45. It is necessary to have a 40% antifreeze solution in the radiator of a certain car. The radiator now holds 20 liters of 20% solution. How many liters of this should be drained and replaced with 100% antifreeze to get the desired strength? (*Hint:* The number of liters drained is equal to the number of liters replaced.)

46. A tank holds 80 liters of a chemical solution. Currently, the solution has a strength of 30%. How much of this should be drained and replaced with a 70% solution to get a final strength of 40%?

Solve the problem. See Example 5.

47. George Duda earned $12,000 last year by giving tennis lessons. He invested part at 3% simple interest and the rest at 4%. He earned a total of $440 in interest. How much did he invest at each rate?

48. Lakeisha Holiday won $60,000 on a slot machine in Las Vegas. She invested part at 2% simple interest and the rest at 3%. She earned a total of $1600 in interest. How much was invested at each rate?

49. Julie Tarr invested some money at 4.5% simple interest and $1000 less than twice this amount at 3%. Her total annual income from the interest was $1020. How much was invested at each rate?

50. Mary Johnson invested some money at 3.5% simple interest, and $5000 more than 3 times this amount at 4%. She earned $1440 in interest. How much did she invest at each rate?

51. Ed Moura has $29,000 invested in stocks paying 5%. How much additional money should he invest in certificates of deposit paying 2% so that the average return on the two investments is 3%?

52. Margo Deal placed $15,000 in an account paying 6%. How much additional money should she deposit at 4% so that the average return on the two investments is 5.5%?

⊚ *Fill in the blank or blanks with the correct response.*

53. The sum of the measures of the angles of any triangle is _____ degrees.

54. If two angles are complementary, the sum of their measures is _____ degrees.

55. If two angles are supplementary, the sum of their measures is _____ degrees.

56. The measure of a straight angle is _____ degrees, while the measure of a right angle is _____ degrees.

---

39. 5 liters   **40.** 25 liters
41. 4 liters
42. $13\frac{1}{3}$ liters
43. $18\frac{2}{11}$ liters
44. 1 gallon
45. 5 liters   **46.** 20 liters
47. $4000 at 3%; $8000 at 4%   **48.** $20,000 at 2%; $40,000 at 3%
49. $10,000 at 4.5%; $19,000 at 3%   **50.** $8000 at 3.5%; $29,000 at 4%
51. $58,000   **52.** $5000
53. 180   **54.** 90   **55.** 180
56. 180;   90

 CONCEPTUAL   ✍ WRITING   ▲ CHALLENGING   🖩 SCIENTIFIC CALCULATOR   🖳 GRAPHICS CALCULATOR

57. 20°;   30°;   130°
58. 60°;   60°;   60°
59. 65°;   115°   60. 64°;
26°   61. 180°   62. We
list the measures in the
order in which the angles
are numbered: 55°, 65°,
60°, 65°, 60°, 120°, 60°, 60°,
55°, 55°   63. 200   64. 32
65. 6   66. 19   67. 12
68. 5

Solve the problem. See Example 6.

57. Find the measure of each angle in the triangle.

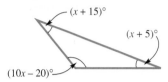

58. Find the measure of each angle in the triangle.

59. Find the measure of each marked angle.

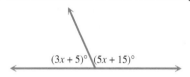

60. Together, the two marked angles form a right angle. Find the measure of each angle.

$(5x - 1)°$
$(2x)°$

▲ *Use the diagram to answer Exercises 61 and*
◉ *62. Assume that m and n are parallel, and use the ideas about angle measures covered in Section 2.2 and this section.*

61. What must be the sum of the measures of angles 2, 8, and 9?

62. Find the measures of the numbered angles.

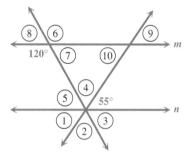

REVIEW EXERCISES

Solve the problem. See Section 2.2.

63. Use $d = rt$ to find d if $r = 50$ and $t = 4$.

64. Use $P = 2L + 2W$ to find P if $L = 10$ and $W = 6$.

65. Use $P = 2L + 2W$ to find W if $P = 80$ and $L = 34$.

66. Use $P = a + b + c$ to find a if $b = 13$, $c = 14$, and $P = 46$.

67. Use $A = \frac{1}{2}(B + b)h$ to find h if $A = 156$, $b = 12$, and $B = 14$.

68. Use $d = rt$ to find r if $d = 75$ and $t = 15$.

2.4 MORE APPLICATIONS OF LINEAR EQUATIONS

FOR EXTRA HELP	OBJECTIVES
📖 **SSG** pp. 40–43 **SSM** pp. 38–43	**1** ▶ Solve problems about different denominations of money.
📼 **Video** 2	**2** ▶ Solve problems about uniform motion.
💾 **Tutorial** IBM MAC	**3** ▶ Solve problems about geometric figures. **4** ▶ Solve problems about consecutive integers.

INSTRUCTOR'S RESOURCES

📖 **ITM** pp. 331–336 **ISM** pp. 61–71   💾 **TEST GENERATOR**   ✳️ **TRANSPARENCIES**
IAM p. 4   DOS   Windows   MAC

There are four common applications of linear equations that we did not discuss in Section 2.3: problems about different denominations of money, uniform motion problems, problems about geometric figures, and consecutive integer problems. The first type is very similar to the simple interest type problems seen in Section 2.3. Uniform motion problems use the formula $d = rt$. Problems about geometric figures usually require the use of the formulas studied in Section 2.2.

1 ▶ Solve problems about different denominations of money.

▶ The first example shows how to solve problems about different denominations of money.

PROBLEM SOLVING

In problems involving money, use the basic fact that

$$\begin{bmatrix} \text{Number of monetary} \\ \text{units of the same kind} \end{bmatrix} \times [\text{Denomination}] = \begin{bmatrix} \text{Total monetary} \\ \text{value} \end{bmatrix}.$$

For example, 30 dimes have a monetary value of $30(.10) = 3.00$ dollars. Fifteen five-dollar bills have a value of $15(5) = 75$ dollars.

EXAMPLE 1

Solving a Problem About Denominations of Money

▶ **TEACHING TIP**

Mention that one must be consistent when working with problems involving money. Either work the problem entirely without decimal points (as numbers of cents) or entirely with decimal points. ◀

For a bill totaling $5.65, a cashier received 25 coins consisting of nickels and quarters. How many of each type of coin did the cashier receive?

Let $\qquad\qquad x = $ the number of nickels received;

$\qquad 25 - x = $ the number of quarters received.

The information for this problem may be arranged as shown in Figure 10. Multiply the numbers of coins by the denominations, and add the results to get 5.65.

$$.05x + .25(25 - x) = 5.65$$
$$5x + 25(25 - x) = 565 \qquad \text{Multiply by } 100.$$
$$5x + 625 - 25x = 565 \qquad \text{Distributive property}$$
$$-20x = -60 \qquad \text{Combine terms; subtract } 625.$$
$$x = 3 \qquad \text{Divide by } -20.$$

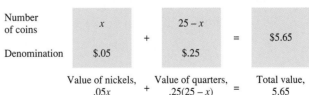

FIGURE 10

The cashier has 3 nickels and $25 - 3 = 22$ quarters. Check to see that the total value of these coins is $5.65. ■

CAUTION Be sure that your answer is reasonable when working problems like Example 1. Since you are dealing with a number of coins, an answer can neither be negative nor a fraction.

2 ▶ Solve problems about uniform motion.

▶ The next examples involve uniform motion.

PROBLEM SOLVING

Uniform motion problems use the distance formula, $d = rt$. In this formula, when rate (or speed) is given in miles per hour, time must be given in hours. When solving such problems, draw a sketch to illustrate what is happening in the problem, and make a chart to summarize the information of the problem. (*Note:* Difficulties often arise in uniform motion problems because the step of drawing a sketch is neglected. The sketch will determine how to set up the equation.)

EXAMPLE 2

Solving a Motion Problem

Two cars leave the same place at the same time, one going east and the other west. The eastbound car averages 40 miles per hour, while the westbound car averages 50 miles per hour. In how many hours will they be 300 miles apart?

A sketch shows what is happening in the problem: The cars are going in *opposite* directions. See Figure 11.

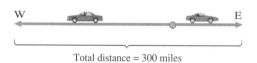

Total distance = 300 miles

FIGURE 11

Chalkboard Exercise

Two cars leave the same town at the same time. One travels north at 60 miles per hour and the other south at 45 miles per hour. In how many hours will they be 420 miles apart?

Answer: 4 hours

Let x represent the time traveled by each car. Summarize the information of the problem in a chart.

	Rate	Time	Distance
Eastbound car	40	x	**40x**
Westbound car	50	x	**50x**

We find the distances traveled by the cars, $40x$ and $50x$, from the formula $d = rt$. When the expressions for rate and time are entered in the chart, *fill in the distance expression by multiplying rate by time.*

From the sketch in Figure 11, the sum of the two distances is 300.

$$40x + 50x = 300$$
$$90x = 300 \qquad \text{Combine terms.}$$
$$x = \frac{300}{90} \qquad \text{Divide by 90.}$$
$$x = \frac{10}{3} \qquad \text{Lowest terms}$$

The cars travel $\frac{10}{3} = 3\frac{1}{3}$ hours, or 3 hours and 20 minutes. ■

CAUTION   It is a common error to write 300 as the distance for each car in Example 2. Three hundred miles is the *total* distance traveled.

Example 2 involved motion in opposite directions. The next example deals with motion in the same direction.

EXAMPLE 3
Solving a Motion Problem

Jeff Bezzone can bike to work in 3/4 hour. When he takes the bus, the trip takes 1/4 hour. If the bus travels 20 miles per hour faster than Jeff rides his bike, how far is it to his workplace?

Although the problem asks for a distance, it is easier here to let x be Jeff's speed when he rides his bike to work. Then the speed of the bus is $x + 20$. By the distance formula, for the trip by bike,

$$d = rt = x \cdot \frac{3}{4} = \frac{3}{4}x,$$

and by bus,

$$d = rt = (x + 20) \cdot \frac{1}{4} = \frac{1}{4}(x + 20).$$

Summarize the information of the problem in a chart.

	Rate	Time	Distance
Bike	x	$\frac{3}{4}$	$\frac{3}{4}x$
Bus	$x + 20$	$\frac{1}{4}$	$\frac{1}{4}(x + 20)$

The key to setting up the correct equation is to understand that the distance in each case is the same. See Figure 12.

Home   Workplace

FIGURE 12

Since the distance is the same in both cases,

$$\frac{3}{4}x = \frac{1}{4}(x + 20).$$

Solve the equation. First multiply on both sides by 4.

$$4\left(\frac{3}{4}x\right) = 4\left(\frac{1}{4}\right)(x + 20)$$

$$3x = x + 20 \qquad \text{Multiply; distributive property}$$

$$2x = 20 \qquad \text{Subtract } x.$$

$$x = 10 \qquad \text{Divide by 2.}$$

Now answer the question in the problem. The required distance is given by

$$d = \frac{3}{4}x = \frac{3}{4}(10) = \frac{30}{4} = 7.5.$$

Check by finding the distance using

$$d = \frac{1}{4}(x + 20) = \frac{1}{4}(10 + 20) = \frac{30}{4} = 7.5,$$

the same result. The required distance is 7.5 miles. ■

NOTE As mentioned in Example 3, it was easier to let the variable represent a quantity other than the one that we were asked to find. This is the case in some problems. It takes practice to learn when this approach is the best, and practice means working lots of problems!

---

3 ▶ Solve problems about geometric figures.

▶ In Section 2.2 we saw how formulas can be used to find areas and perimeters of geometric figures.

PROBLEM SOLVING When applied problems deal with geometric figures, the appropriate formula often provides the equation for the problem. The next example illustrates how a figure and a formula are used in an application involving geometry.

E X A M P L E 4

Solving a Problem About a Geometric Figure

A label is in the shape of a rectangle. The length of the rectangle is 1 centimeter more than twice the width. The perimeter is 110 centimeters. Find the length and the width. See Figure 13.

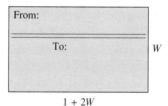

$1 + 2W$

FIGURE 13

Let W = the width

$1 + 2W$ = the length (one more than twice the width).

The perimeter of a rectangle is given by the formula $P = 2L + 2W$. Replace P in this formula with 110 and L with $1 + 2W$, giving

$$P = 2L + 2W$$
$$110 = 2(1 + 2W) + 2W$$
$$110 = 2 + 4W + 2W \qquad \text{Distributive property}$$
$$110 = 2 + 6W \qquad \text{Combine terms.}$$
$$108 = 6W \qquad \text{Subtract 2.}$$
$$18 = W. \qquad \text{Divide by 6.}$$

The width of the label is 18 centimeters, and the length is $1 + 2W = 1 + 2(18) = 1 + 36 = 37$ centimeters. ■

4 ▶ Solve problems about consecutive integers.

▶ Consecutive integers are integers that follow each other in counting order, such as 8, 9, and 10, or 33, 34, and 35. The final example involves such integers.

E X A M P L E   5

Solving a Problem Involving Consecutive Integers

Find three consecutive integers such that the sum of the first and third, increased by 3, is 50 more than the second.

Let x represent the first of the unknown integers. Then $x + 1$ will be the second, and $x + 2$ will be the third. The equation we need can be found by going back to the words of the original problem.

Sum of the first and third	increased by 3	is	50 more than the second.
↓	↓	↓	↓
$x + (x + 2)$	$+ 3$	$=$	$(x + 1) + 50$

Now solve the equation.

$$x + x + 2 + 3 = x + 1 + 50$$
$$2x + 5 = x + 51$$
$$x + 5 = 51 \qquad \text{Subtract } x.$$
$$x = 46 \qquad \text{Subtract 5.}$$

The first integer is $x = 46$, the second is $x + 1 = 47$, and the third is $x + 2 = 48$. The three integers are 46, 47, and 48. Check by substituting these numbers back into the words of the original problem. ■

Chalkboard Exercise

Find four consecutive integers such that the sum of the first three is 54 more than the fourth.

Answer: 27, 28, 29, 30

2.4 EXERCISES

◉ *Provide a short answer to the question.*

1. What amount of money is found in a coin purse containing 21 nickels and 14 dimes?

2. The distance from Melbourne to London is 10,500 miles. If a jet averages 500 miles per hour between the two cities, what is its travel time in hours?

3. Anh Nguyen traveled from Louisville to Kansas City, a distance of 520 miles, in 10 hours. What was his rate in miles per hour?

4. A square has perimeter 20 inches. What would be the perimeter of an equilateral triangle whose sides each measure the same length as the side of the square?

◉ *Use the graph to answer the questions in Exercises 5 and 6.*

5. What is the best estimate of the number of new cars sold for personal use that were leased in 1993, if a *total* of 4 million cars were sold?
   (a) 2 million   (b) 2.5 million
   (c) 1 million   (d) 1.5 million

6. What is the best estimate of the percent of new cars sold in 1987 that were *not* leased?
   (a) 15%   (b) 75%   (c) 85%
   (d) 25%

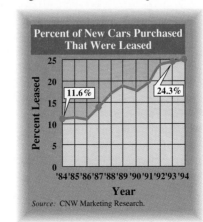

Percent of New Cars Purchased That Were Leased

11.6%   24.3%

Percent Leased

'84 '85 '86 '87 '88 '89 '90 '91 '92 '93 '94

Year

Source: CNW Marketing Research.

1. $2.45   2. 21 hours
3. 52 miles per hour
4. 15 inches
5. (c)   6. (c)

Ⓓ CONCEPTUAL   ✐ WRITING   ▲ CHALLENGING   ▦ SCIENTIFIC CALCULATOR   ▦ GRAPHICS CALCULATOR

7. Leslie Cobar's piggy bank has 36 coins. Some are quarters and the rest are half dollars. If the total value of the coins is $14.75, how many of each denomination does Leslie have?

8. Liz Harold has a jar in her office that contains 47 coins. Some are pennies and the rest are dimes. If the total value of the coins is $2.18 how many of each denomination does she have?

9. Sam Abo-zahrah has a box of coins that he uses when playing poker with his friends. The box currently contains 44 coins, consisting of pennies, dimes, and quarters. The number of pennies is equal to the number of dimes, and the total value is $4.37. How many of each denomination of coin does he have in the box?

10. Roma Sherry found some coins while looking under her sofa pillows. There were equal numbers of nickels and quarters, and twice as many half dollars as quarters. If she found $2.60 in all, how many of each denomination of coin did she find?

11. In the nineteenth century, the United States minted two-cent and three-cent pieces. Toni Tusa has three times as many three-cent pieces as two-cent pieces, and the face value of these coins is $1.21. How many of each denomination does she have?

12. Mary Catherine Dooley collects U.S. gold coins. She has a collection of 80 coins. Some are $10 coins and the rest are $20 coins. If the face value of the coins is $1060, how many of each denomination does she have?

13. The school production of *Our Town* was a big success. For opening night, 410 tickets were sold. Students paid $1.50 each, while nonstudents paid $3.50 each. If a total of $825 was collected, how many students and how many nonstudents attended?

14. A total of 550 people attended a Boston Pops concert. Floor tickets cost $20 each, while balcony tickets cost $14 each. If a total of $10,400 was collected, how many of each type of ticket were sold?

15. In your own words, explain how the problems in Exercises 7–14 are similar to the investment problems studied in Section 2.3.

16. In the nineteenth century, the United States minted half-cent coins. If an applied problem involved half-cent coins, what decimal number would represent this denomination?

Solve the problem. See Examples 2 and 3.

17. A train leaves Little Rock, Arkansas, and travels north at 85 kilometers per hour. Another train leaves at the same time and travels south at 95 kilometers per hour. How long will it take before they are 315 kilometers apart?

18. Two steamers leave a port on a river at the same time, traveling in opposite directions. Each is traveling 22 miles per hour. How long will it take for them to be 110 miles apart?

19. Joey and Mark commute to work, traveling in opposite directions. Joey leaves the house at 8:00 A.M. and averages 35 miles per hour. Mark leaves at 8:15 A.M. and averages 40 miles per hour. At what time will they be 140 miles apart?

20. Jeff leaves his house on his bicycle at 8:30 A.M. and averages 5 miles per hour. His wife, Joan, leaves at 9:00 A.M., following the same path and averaging 8 miles per hour. At what time will Joan catch up with Jeff?

21. When Tri drives his car to work, the trip takes 30 minutes. When he rides the bus, it takes 45 minutes. The average speed of the bus is 12 miles per hour less than his speed when driving. Find the distance he travels to work.

22. Latoya can get to school in 15 minutes if she rides her bike. It takes her 45 minutes if she walks. Her speed when walking is 10 miles per hour slower than her speed when riding. What is her speed when she rides?

23. A pleasure boat on the Mississippi River traveled from Baton Rouge to New Orleans with a stop at White Castle. On the first part of the trip, the boat traveled at an average speed of 10 miles per hour. From White Castle to New Orleans the average speed was 15 miles per hour. The entire trip covered 100 miles. How long did the entire trip take if the two parts each took the same number of hours?

24. Steve leaves Nashville to visit his cousin David in Napa, 80 miles away. He travels at an average speed of 50 miles per hour. One-half hour later David leaves to visit Steve, traveling at an average speed of 60 miles per hour. How long after David leaves will it be before they meet?

25. In a run for charity Janet runs at a speed of 5 miles per hour. Paula leaves 10 minutes after Janet and runs at 6 miles per hour. How long will it take for Paula to catch up with Janet? (*Hint:* Change minutes to hours.)

⦿**26.** Read over Example 3 in this section. The solution of the equation is 10. Why is 10 miles per hour not the answer to the problem?

Solve the problem. See Example 4.

27. The John Hancock Center in Chicago has a rectangular base. The length of the base measures 65 feet less than twice the width. The perimeter of this base is 860 feet. What are the dimensions of the base?

28. The John Hancock Center (Exercise 27) tapers as it rises. The top floor is rectangular and has perimeter 520 feet. The width of the top floor measures 20 feet more than one-half its length. What are the dimensions of the top floor?

29. The Bermuda Triangle supposedly causes trouble for aircraft pilots. It has a perimeter of 3075 miles. The shortest side measures 75 miles less than the middle side, and the longest side measures 375 miles more than the middle side. Find the lengths of the three sides.

30. The Vietnam War Memorial in Washington, D.C., is in the shape of an unenclosed isosceles triangle. If the two walls of equal length were joined by a straight line of 438 feet, the perimeter of the resulting triangle would be 931.5 feet. Find the lengths of the two walls.

31. A farmer wishes to enclose a rectangular region with 210 meters of fencing in such a way that the length is twice the width and the region is divided into two equal parts, as shown in the figure. What length and width should be used?

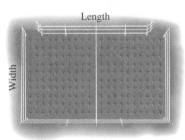
Length
Width

32. Joshua Rogers has a sheet of tin 12 centimeters by 16 centimeters. He plans to make a box by cutting equal squares out of each of the four corners and folding up the remaining edges. How large a square should he cut so that the finished box will have a length that is 5 centimeters less than twice the width?

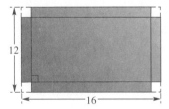
12
16

Solve each problem. See Example 5.

33. Two pages facing each other in this book have 153 as the sum of their page numbers. What are the two page numbers?

34. If I add my current age to the age I will be next year on this date, the sum will be 87 years. How old will I be ten years from today?

23. 8 hours   24. $\frac{1}{2}$ hour
25. $\frac{5}{6}$ hour   26. The problem asks for the distance Jeff travels to the workplace, so we must multiply the rate, 10 miles per hour, by the time $\frac{3}{4}$ hour, to get the distance, 7.5 miles.
27. width: 165 feet; length: 265 feet   28. width: 100 feet;   length: 160 feet
29. 850 miles, 925 miles, 1300 miles   30. 246.75 feet
31. length: 60 meters; width: 30 meters
32. $1\frac{1}{2}$ centimeters
33. 76 and 77
34. 53 years old

⦿ CONCEPTUAL   ✍ WRITING   ▲ CHALLENGING   ▦ SCIENTIFIC CALCULATOR   ▦ GRAPHICS CALCULATOR

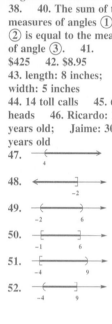

35. Find three consecutive integers such that the sum of the first and twice the second is 17 more than twice the third.

36. Find four consecutive integers such that the sum of the first three is 60 more than the fourth.

◇ **MATHEMATICAL CONNECTIONS** (Exercises 37–40) ◇

Consider the following two figures. Work Exercises 37–40 in order.

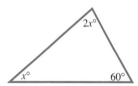
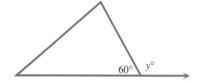

37. Solve for the measures of the unknown angles in Figure A.

38. Solve for the measure of the unknown angle marked $y°$ in Figure B.

39. Add the measures of the two angles you found in Exercise 37. How does the sum compare to the measure of the angle you found in Exercise 38?

40. From Exercises 37–39, make a conjecture (an educated guess) about the relationship among the angles marked ①, ②, and ③ in the figure shown here.

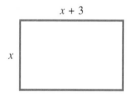

▲ *The following problems are of various types, and all may be solved using the strategies presented in this section and the previous one. Solve the problem.*

41. An electronics store offered a videodisc player for $255. This price was the sale price, after the regular price had been discounted 40%. What was the regular price?

42. After a discount of 30%, the sale price of *The Parents' Guide to Kids' Sports* was $6.27. What was the regular price of the book? (Give your answer to the nearest 5¢.)

43. The length of a rectangle is 3 inches more than its width. If the length were decreased by 2 inches and the width were increased by 1 inch, the perimeter of the resulting rectangle would be 24 inches. Find the dimensions of the original rectangle.

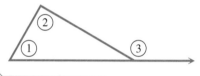

44. The monthly phone bill includes a monthly charge of $10 for local calls plus an additional charge of 50¢ for each toll call in a certain area. A federal tax of 5% is added to the total bill. If all calls were local or within the 50¢ area and the total bill was $17.85, find the number of toll calls made.

45. A grocer buys lettuce for $10.40 a crate. Of the lettuce he buys, 10% cannot be sold. If he charges 40¢ for each head he sells and makes a profit of 20¢ on each head he buys, how many heads of lettuce are in the crate?

46. Jaime is twice as old as his brother, Ricardo. In five years, Ricardo will be the same age as his brother was ten years ago. What are their present ages?

REVIEW EXERCISES

Graph the interval. See Section 1.1.

47. $(4, \infty)$

48. $(-\infty, -2]$

49. $(-2, 6)$

50. $[-1, 6]$

51. $[-4, 9)$

52. $(-4, 9]$

2.5 LINEAR INEQUALITIES IN ONE VARIABLE

FOR EXTRA HELP	OBJECTIVES
SSG pp. 43–48 **SSM** pp. 44–50 **Video** 2 **Tutorial** IBM MAC	1 ▶ Solve linear inequalities using the addition property. 2 ▶ Solve linear inequalities using the multiplication property. 3 ▶ Solve linear inequalities having three parts. 4 ▶ Solve applied problems using linear inequalities.

We have seen how to write and solve linear equations to model certain types of problems. Now we will learn to write and solve linear *inequalities* and see how they apply to practical problems.

◈ Answers will vary.

◆ C O N N E C T I O N S ◆

In recent years, many states have increased their maximum speed limit on rural highways to 65 miles per hour. Inequalities can help determine whether driving at this faster speed is more cost effective. The cost of shipping can be expressed as the cost of gas plus the cost of paying a driver to deliver the goods. Writing expressions in terms of the driver's hourly wage for the cost of shipping at 55 miles per hour and at 65 miles per hour, and setting the cost at 65 miles per hour greater than or equal to the cost at 55 miles per hour produces an inequality. The solution of the inequality gives the minimum hourly wage at which it is more cost effective to travel at 55 miles per hour.

FOR DISCUSSION OR WRITING
What other factors might affect the shipping cost? What assumptions are implicit in the discussion of this problem?

An **inequality** says that two expressions are *not* equal. A **linear inequality in one variable** is an inequality such as

$$x + 5 < 2, \qquad y - 3 \geq 5, \qquad \text{or} \qquad 2k + 5 \leq 10.$$

Linear Inequality

A linear inequality in one variable can be written in the form

$$ax < b,$$

where a and b are real numbers, with $a \neq 0$.

(Throughout this section we give the definitions and rules only for $<$, but they are also valid for $>$, $\leq$, and $\geq$.)

1 ▶ Solve linear inequalities using the addition property.

▶ An inequality is solved by finding all numbers that make the inequality true. Usually, an inequality has an infinite number of solutions. These solutions, like the solutions of equations, are found by producing a series of simpler equivalent inequalities. **Equivalent inequalities** are inequalities with the same solution set. The inequalities in this chain of equivalent inequalities are found with the addition and multiplication properties of inequality.

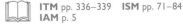

Addition Property of Inequality	For all real numbers a, b, and c, the inequalities

$$a < b \quad \text{and} \quad a + c < b + c$$

are equivalent. (The same number may be added to both sides of an inequality without changing the solution set.)

As with equations, the addition property can be used to *subtract* the same number from both sides of an inequality.

EXAMPLE 1

Using the Addition Property of Inequality

Solve $x - 7 < -12$.

Add 7 to both sides.

$$x - 7 + 7 < -12 + 7$$
$$x < -5$$

The solution set, $(-\infty, -5)$, is graphed in Figure 14. ■

FIGURE 14

Interval notation was discussed briefly in Section 1.2. Because interval notation is very convenient in writing solution sets of inequalities, we now summarize the various types of intervals. The symbolism $(-\infty, \infty)$ is often used to represent the set of all real numbers.

▶ **TEACHING TIP**
Again, remind students that parentheses and not brackets are used with infinity symbols when using interval notation. ◀

Type of Interval	Set	Interval Notation	Graph
Open interval	$\{x \mid a < x\}$	(a, ∞)	
	$\{x \mid a < x < b\}$	(a, b)	
	$\{x \mid x < b\}$	$(-\infty, b)$	
Half-open interval	$\{x \mid a \le x\}$	$[a, \infty)$	
	$\{x \mid a < x \le b\}$	$(a, b]$	
	$\{x \mid a \le x < b\}$	$[a, b)$	
	$\{x \mid x \le b\}$	$(-\infty, b]$	
Closed interval	$\{x \mid a \le x \le b\}$	$[a, b]$	

EXAMPLE 2

Using the Addition
Property of Inequality

Solve the inequality $14 + 2m \leq 3m$, and graph the solution set.

First, subtract $2m$ from both sides.

$$14 + 2m - 2m \leq 3m - 2m \qquad \text{Subtract } 2m.$$
$$14 \leq m \qquad \text{Combine like terms.}$$

The inequality $14 \leq m$ (14 is less than or equal to m) can also be written $m \geq 14$ (m is greater than or equal to 14). Notice that in each case, the inequality symbol points to the smaller expression, 14. The solution set in interval notation is $[14, \infty)$. The graph is shown in Figure 15. ■

FIGURE 15

Chalkboard Exercise

Solve the inequality
$5x + 3 \geq 4x - 1$ and graph
the solution set.

Answer: $[-4, \infty)$

▶ **TEACHING TIP**
Sometimes the solution process will produce the variable on the right side. Remind students that $5 > x$ is equivalent to $x < 5$. Students have an easier time graphing $x < 5$. ◀

CAUTION Errors often occur in graphing inequalities where the variable term is on the right side. (This is probably due to the fact that we read from left to right.) To guard against such errors, it is a good idea to rewrite these inequalities so that the variable is on the left, as discussed in Example 2.

---

2 ▶ Solve linear inequalities using the multiplication property.

▶ An inequality such as $3x \leq 15$ can be solved by dividing both sides by 3. This is done with the multiplication property of inequality, which is a little more involved than the corresponding property for equations. To see how this property works, start with the true statement

$$-2 < 5.$$

Multiply both sides by, say, 8.

$$-2(8) < 5(8)$$
$$-16 < 40 \qquad \text{True}$$

This gives a true statement. Start again with $-2 < 5$, and this time multiply both sides by -8.

$$-2(-8) < 5(-8)$$
$$16 < -40 \qquad \text{False}$$

▶ **TEACHING TIP**
Demonstrate the rule for reversing the inequality symbol by using the examples

$$-2x < 10 \quad \text{and} \quad 2x < -10,$$

emphasizing that in the first one, the symbol must be reversed, but in the second, it remains unchanged. ◀

The result, $16 < -40$, is false. To make it true, change the direction of the inequality symbol to get

$$16 > -40.$$

As these examples suggest, multiplying both sides of an inequality by a *negative* number forces the direction of the inequality symbol to be reversed. The same is true for dividing by a negative number, since division is defined in terms of multiplication.

Multiplication Property of Inequality

For all real numbers a, b, and c, with $c \neq 0$,

(a) the inequalities

$$a < b \qquad \text{and} \qquad ac < bc$$

are equivalent if $c > 0$;

(b) the inequalities

$$a < b \qquad \text{and} \qquad ac > bc$$

are equivalent if $c < 0$. (Both sides of an inequality may be multiplied by a *positive* number without changing the direction of the inequality symbol. Multiplying or dividing by a *negative* number requires that the inequality symbol be reversed.)

CAUTION It is a common error to forget to reverse the direction of the inequality symbol when multiplying or dividing by a negative number.

EXAMPLE 3

Using the Multiplication Property of Inequality

Solve each inequality, then graph each solution.

(a) $5m \leq -30$

Use the multiplication property to divide both sides by 5. Since $5 > 0$, do *not* reverse the inequality symbol.

$$5m \leq -30$$
$$\frac{5m}{5} \leq \frac{-30}{5} \qquad \text{Divide by 5.}$$
$$m \leq -6$$

The solution set, graphed in Figure 16, is the interval $(-\infty, -6]$.

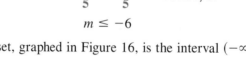

FIGURE 16

(b) $-4k \leq 32$

Divide both sides by -4. Since $-4 < 0$, the inequality symbol must be reversed.

$$-4k \leq 32$$
$$\frac{-4k}{-4} \geq \frac{32}{-4} \qquad \text{Divide by } -4 \text{ and reverse symbol.}$$
$$k \geq -8$$

Figure 17 shows the graph of the solution set, $[-8, \infty)$. ■

FIGURE 17

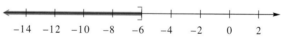

The steps used in solving a linear inequality are given below.

Solving a Linear Inequality

Step 1 Simplify each side of the inequality as much as possible by using the distributive property to clear parentheses and by combining like terms as needed.

Step 2 Use the addition property of inequality to change the inequality so that all terms with variables are on one side and all terms without variables are on the other side.

Step 3 Use the multiplication property to change the inequality to the form $x < k$ or $x > k$.

Remember: Reverse the direction of the inequality symbol **only** when multiplying or dividing both sides of an inequality by a **negative** number.

EXAMPLE 4

Solving a Linear Inequality Using the Distributive Property

Solve $2 - 4(r - 3) < 3(5 - r) + 5$, and graph the solution set.

We begin by using the distributive property to clear parentheses.

Step 1
$$2 - 4(r - 3) < 3(5 - r) + 5$$
$$2 - 4r + 12 < 15 - 3r + 5 \qquad \text{Distributive property}$$
$$14 - 4r < 20 - 3r$$

Step 2
$$14 - 4r + 3r < 20 - 3r + 3r \qquad \text{Add } 3r \text{ to both sides.}$$
$$14 - r < 20$$
$$14 - r - 14 < 20 - 14 \qquad \text{Subtract } 14.$$
$$-r < 6$$

We solve for r by multiplying both sides of the inequality by -1. Since -1 is negative, change the direction of the inequality symbol.

Step 3
$$(-1)(-r) > (-1)(6) \qquad \text{Multiply by } -1, \text{ change } < \text{ to } >.$$
$$r > -6$$

The solution set, $(-6, \infty)$, is graphed in Figure 18. ∎

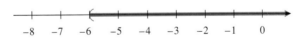

FIGURE 18

Chalkboard Exercise

Solve $6(x - 1) + 3x \geq -x - 3(x + 2)$ and graph the solution set.

Answer: $[0, \infty)$

-2 -1 0 1 2 3 4

3 ▶ Solve linear inequalties having three parts.

▶ For some applications, it is necessary to work with an inequality such as

$$3 < x + 2 < 8,$$

where $x + 2$ is *between* 3 and 8. To solve this inequality, we subtract 2 from each of the three parts of the inequality, giving

$$3 - 2 < x + 2 - 2 < 8 - 2$$
$$1 < x < 6.$$

The solution set, $(1, 6)$, is graphed in Figure 19.

FIGURE 19

CAUTION When using inequalities with three parts like the one above, it is important to have the numbers in the correct positions. It would be *wrong* to write the inequality as $8 < x + 2 < 3$, since this would imply that $8 < 3$, a false statement. In general, three-part inequalities are written so that the symbols point in the same direction, and they both point toward the smaller number.

EXAMPLE 5

Solving a Three-Part Inequality

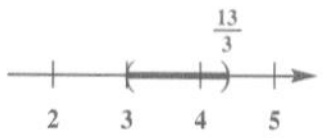

Solve the inequality $-2 \le 3k - 1 \le 5$ and graph the solution set.

To begin, we add 1 to each of the three parts.

$$-2 + 1 \le 3k - 1 + 1 \le 5 + 1 \qquad \text{Add 1.}$$

$$-1 \le 3k \le 6$$

$$\frac{-1}{3} \le \frac{3k}{3} \le \frac{6}{3} \qquad \text{Divide by 3.}$$

$$-\frac{1}{3} \le k \le 2$$

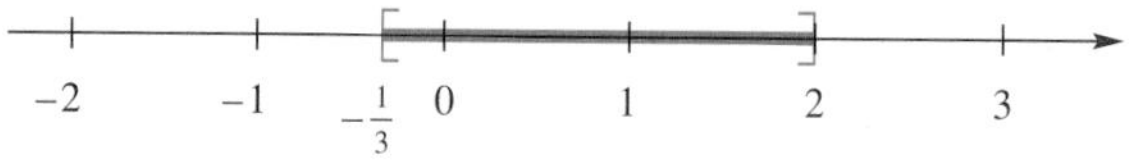

FIGURE 20

A graph of the solution set, $\left[-\frac{1}{3}, 2\right]$, is shown in Figure 20. ■

The types of solutions to linear equations or linear inequalities are shown in the box that follows.

Solution Sets of Linear Equations and Inequalities	Equation or Inequality	Typical Solution Set	Graph of Solution Set
	Linear equation $ax = b$	$\{p\}$	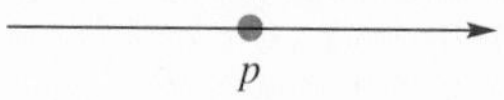 p
	Linear inequality $ax < b$	$(-\infty, p)$ or (p, ∞)	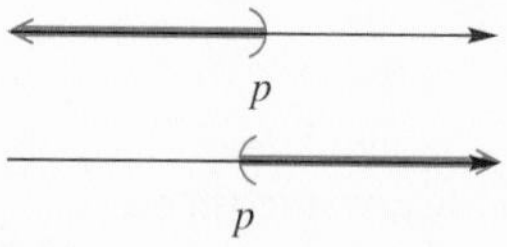 p
	Three-part inequality $b < ax < c$	(p, q)	p q

4 ▶ Solve applied problems using linear inequalities.

▶ There are several phrases that denote inequality. Some of them were discussed in Chapter 1. In addition to the familiar "is less than" and "is greater than" which are examples of **strict** inequalities, the expressions "is no more than," "is at least," and others also denote inequalities. These are called **nonstrict.**

Chalkboard Exercise

Solve the inequality $5 < 3x - 4 < 9$ and graph the solution set.

Answer: $\left(3, \dfrac{13}{3}\right)$

PROBLEM SOLVING   Expressions for nonstrict inequalities sometimes appear in applied problems that are solved using inequalities. The chart below shows how these expressions are interpreted.

Word Expression	Interpretation	Word Expression	Interpretation
a is at least b	$a \geq b$	a is at most b	$a \leq b$
a is no less than b	$a \geq b$	a is no more than b	$a \leq b$

The final example shows how an applied problem can be solved using a linear inequality.

EXAMPLE 6

Solving an Application Using a Linear Inequality

A rental company charges $15.00 to rent a chain saw, plus $2.00 per hour. Rusty Brauner can spend no more than $35.00 to clear some logs from his yard. What is the maximum amount of time he can keep the rented saw?

Let h = the number of hours he can rent the saw. He must pay $15.00, plus $2.00h$, to rent the saw for h hours, and this amount must be *no more than* $35.00.

Cost of renting	is no more than	35 dollars.	
$15 + 2h$	$\leq$	35	
$15 + 2h - 15$	$\leq$	$35 - 15$	Subtract 15.
$2h$	$\leq$	20	
h	$\leq$	10	Divide by 2.

He can keep the saw for a maximum of 10 hours. (Of course, he may keep it for less time, as indicated by the inequality $h \leq 10$.)   ∎

Chalkboard Exercise

Teresa has been saving dimes and nickels. She has three times as many nickels as dimes and has at least 48 coins. What is the smallest number of nickels she might have?

Answer: 36 nickels

2.5 EXERCISES

1. Explain how to determine whether to use parentheses or brackets when graphing the solution set of an inequality.

2. When is it necessary to reverse the direction of the inequality sign when solving an inequality?
   2. **Reverse the sign when multiplying or dividing by a negative number.**

Solve the inequality. Give the solution set in both interval and graph forms. See Examples 1–4.

3. $x + 3 > 5$
3. $(2, \infty)$

4. $y - 9 > -6$
4. $(3, \infty)$

5. $5r \leq -15$
5. $(-\infty, -3]$

6. $12m \leq -36$
6. $(-\infty, -3]$

7. $4x + 1 \geq 21$
7. $[5, \infty)$

8. $5t + 2 \geq 52$
8. $[10, \infty)$

9. $\dfrac{3k - 1}{4} > 5$
9. $(7, \infty)$

10. $\dfrac{5z - 6}{8} < 8$
10. $(-\infty, 14)$

11. $-4x < 16$
11. $(-4, \infty)$

⊙ CONCEPTUAL   ✏ WRITING   ▲ CHALLENGING   🖩 SCIENTIFIC CALCULATOR   🖥 GRAPHICS CALCULATOR

12. $-2m > 10$

12. $(-\infty, -5)$

13. $-\dfrac{3}{4}r \geq 30$

13. $(-\infty, -40]$

14. $-\dfrac{2}{3}y \leq 12$

14. $[-18, \infty)$

15. $-1.5y \leq -\dfrac{9}{2}$

15. $[3, \infty)$

16. $-.4x \geq -\dfrac{4}{25}$

16. $\left(-\infty, \dfrac{2}{5}\right]$

17. $-1.3m \geq -5.2$

17. $(-\infty, 4]$

18. $-2.5y \leq -1.25$

18. $[.5, \infty)$

19. $\dfrac{2k - 5}{-4} > 5$

19. $\left(-\infty, -\dfrac{15}{2}\right)$

20. $\dfrac{3z - 2}{-5} < 6$

20. $\left(-\dfrac{28}{3}, \infty\right)$

21. $y + 4(2y - 1) \geq y$

21. $\left[\dfrac{1}{2}, \infty\right)$

22. $m - 2(m - 4) \leq 3m$

22. $[2, \infty)$

23. $-(4 + r) + 2 - 3r < -14$

23. $(3, \infty)$

24. $-(9 + k) - 5 + 4k \geq 4$

24. $[6, \infty)$

25. $-3(z - 6) > 2z - 2$

25. $(-\infty, 4)$

26. $-2(y + 4) \leq 6y + 16$

26. $[-3, \infty)$

27. $\dfrac{2}{3}(3k - 1) \geq \dfrac{3}{2}(2k - 3)$

27. $\left(-\infty, \dfrac{23}{6}\right]$

28. $\dfrac{7}{5}(10m - 1) < \dfrac{2}{3}(6m + 5)$

28. $\left(-\infty, \dfrac{71}{150}\right)$

29. $-\dfrac{1}{4}(p + 6) + \dfrac{3}{2}(2p - 5) < 10$

29. $\left(-\infty, \dfrac{76}{11}\right)$

30. $\dfrac{3}{5}(k - 2) - \dfrac{1}{4}(2k - 7) \leq 3$

30. $\left(-\infty, \dfrac{49}{2}\right]$

▲ **31.** $3(2x - 4) - 4x < 2x + 3$

31. $(-\infty, \infty)$

▲ **32.** $7(4 - x) + 5x < 2(16 - x)$

32. $(-\infty, \infty)$

▲ **33.** $8\left(\dfrac{1}{2}x + 3\right) < 8\left(\dfrac{1}{2}x - 1\right)$

33. $\emptyset$

▲ **34.** $10x + 2(x - 4) < 12x - 10$

34. $\emptyset$

◎ **35.** A student solved the inequality $5x < -20$ by dividing both sides by 5 and reversing the direction of the inequality symbol. His reasoning was that since -20 is a negative number, reversing the direction of the symbol was required. This is incorrect. Explain why.

◎ **36.** Explain the difference between a strict inequality and a nonstrict inequality.

─────◆ **MATHEMATICAL CONNECTIONS** (Exercises 37–42) ◆─────

Work Exercises 37–42 in order.

37. Solve the linear equation $5(x + 3) - 2(x - 4) = 2(x + 7)$, and graph the solution on a number line.

37. $\{-9\}$

38. Solve the linear inequality $5(x + 3) - 2(x - 4) > 2(x + 7)$, and graph the solutions on a number line.

38. $(-9, \infty)$

39. Solve the linear inequality $5(x + 3) - 2(x - 4) < 2(x + 7)$, and graph the solutions on a number line.

39. $(-\infty, -9)$

40. Graph all the solution sets of the equations and inequalities in Exercises 37–39 on the same number line. What set do you obtain?

40. We obtain the set of all real numbers.

-9

◉ **41.** Based on the results of Exercises 37–39, complete the following using a conjecture (educated guess): The solution set of $-3(x + 2) = 3x + 12$ is $\{-3\}$, and the solution set of $-3(x + 2) < 3x + 12$ is $(-3, \infty)$. Therefore the solution set of $-3(x + 2) > 3x + 12$ is _____ .

41. $(-\infty, -3)$

◉ **42.** Write a paragraph commenting on the following statement: Equality is the boundary between less than and greater than.

Solve the inequality. Give the solution set in both interval and graph forms. See Example 5.

43. $-4 < x - 5 < 6$

43. $(1, 11)$

1   11

44. $-1 < x + 1 < 8$

44. $(-2, 7)$

-2 7

45. $-9 \leq k + 5 \leq 15$

45. $[-14, 10]$

-14 10

46. $-4 \leq m + 3 \leq 10$

46. $[-7, 7]$

-7 7

47. $-6 \leq 2z + 4 \leq 16$

47. $[-5, 6]$

-5 6

48. $-15 < 3p + 6 < -12$

48. $(-7, -6)$

-7 -6

49. $-19 \leq 3x - 5 \leq 1$

49. $\left[-\dfrac{14}{3}, 2\right]$

$-\frac{14}{3}$ 2

50. $-16 < 3t + 2 < -10$

50. $(-6, -4)$

-6 -4

51. $-1 \leq \dfrac{2x - 5}{6} \leq 5$

51. $\left[-\dfrac{1}{2}, \dfrac{35}{2}\right]$

$-\frac{1}{2}$ $\frac{35}{2}$

52. $-3 \leq \dfrac{3m + 1}{4} \leq 3$

52. $\left[-\dfrac{13}{3}, \dfrac{11}{3}\right]$

$-\frac{13}{3}$ $\frac{11}{3}$

53. $4 \leq 5 - 9x < 8$

53. $\left(-\dfrac{1}{3}, \dfrac{1}{9}\right]$

$-\frac{1}{3}$ $\frac{1}{9}$

54. $4 \leq 3 - 2x < 8$

54. $\left(-\dfrac{5}{2}, -\dfrac{1}{2}\right]$

$-\frac{5}{2}$ $-\frac{1}{2}$

Answer the questions in Exercises 55–58 based on the given graph.

55. In which months did the percent of tornadoes exceed 7.7%?

55. April, May, June, July

56. In which months was the percent of tornadoes at least 12.9%?

56. April, May, June

57. The data used to determine the graph was based on the number of tornadoes sighted in the United States during the last twenty years. A total of 17,252 tornadoes were reported. In which months were fewer than 1500 reported?

57. January, February, March, August, September, October, November, December

58. How many more tornadoes occurred during March than October? (Use the total given in Exercise 57.)

58. 690

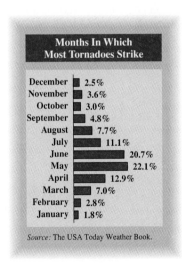

Months In Which Most Tornadoes Strike

Month	Percent
December	2.5%
November	3.6%
October	3.0%
September	4.8%
August	7.7%
July	11.1%
June	20.7%
May	22.1%
April	12.9%
March	7.0%
February	2.8%
January	1.8%

Source: The USA Today Weather Book.

Solve the problem. See Example 6.

59. According to information provided by the Taxi and Limousine Commission, in January 1990, taxicabs in New York charged $1.50 for the first 1/5 mile and $.25 for each additional 1/5 mile. Amos has only $3.75 in his pocket. What is the maximum distance he can travel (not including a tip for the cabbie)?

59. 2 miles

60. In July 1979, New York taxicab fares were $.90 for the first 1/7 mile and $.10 for each additional 1/7 mile. Based on the information given in Exercise 59 and the answer you found, how much farther could Amos have traveled in 1979 than in 1990?

60. approximately 2 miles farther

61. Margaret Boyle earned scores of 90 and 82 on her first two tests in English Literature. What score must she make on her third test to keep an average of 84 or greater?

61. at least 80

62. Shannon Mulkey scored 92 and 96 on her first two tests in Methods in Teaching Mathematics. What score must she make on her third test to keep an average of 90 or greater?

62. at least 82

63. A couple wishes to rent a car for one day while on vacation. Ford Automobile Rental wants $15.00 per day and 14¢ per mile, while Chevrolet-For-A-Day wants $14.00 per day and 16¢ per mile. After how many miles would the price to rent the Chevrolet exceed the price to rent a Ford?

63. 50 miles

64. Jane and Terry Brandsma went to Long Island for a week. They needed to rent a car, so they checked out two rental firms. Avis wanted $28 per day, with no mileage fee. Downtown Toyota wanted $108 per week and 14¢ per mile. How many miles would they have to drive before the Avis price is less than the Toyota price?

64. 628.6 miles

A product will produce a profit only when the revenue R from selling the product exceeds the cost C of producing it. In Exercises 65 and 66 find the smallest whole number of units x that must be sold for the business to show a profit for the item described.

65. Peripheral Visions, Inc. finds that the cost to produce x studio quality videotapes is $C = 20x + 100$, while the revenue produced from them is $R = 24x$ (C and R in dollars).

65. 26 tapes

66. Speedy Delivery finds that the cost to make x deliveries is $C = 3x + 2300$, while the revenue produced from them is $R = 5.50x$ (C and R in dollars).

66. 921 deliveries

▲ **67.** Suppose that $-1 < x < 5$. Complete this inequality: _____ $< -x <$ _____ .

◉ **67.** $-5 < -x < 1$

▲ **68.** Suppose that $4 < y < 1$. What can you say about y?

◉ **68. There is no such number y, since $4 \not< 1$.**

▲ **69.** Write an explanation of what is wrong with the following argument.

◉
🖉   Let a and b be numbers, with $a > b$. Certainly, $2 > 1$. Multiply both sides of the inequality by $b - a$.

$$2(b - a) > 1(b - a)$$
$$2b - 2a > b - a$$
$$2b - b > 2a - a$$
$$b > a$$

But the final inequality is impossible, since we know that $a > b$ from the given information.

▲ 70. Assume that $0 < a < b$, and go through the following steps.

$$a < b$$
$$ab < b^2$$
$$ab - b^2 < 0$$
$$b(a - b) < 0$$

Divide both sides by $a - b$ to get $b < 0$. This implies that b is negative. We originally assumed that b is positive. What is wrong with this argument?

REVIEW EXERCISES

Each exercise requires the graph of two inequalities. Graph them and respond to the statement that follows. See Section 1.1.

71. (a) Graph $x > 4$.   **(b)** Graph $x < 5$.
(c) Describe in your own words the set of numbers belonging to *both* of these sets at the same time.
71. (a)    **(b)**

72. (a) Graph $y < 7$.   **(b)** Graph $y < 9$.
(c) Describe in your own words the set of numbers belonging to *both* of these sets at the same time.
72. (a)    **(b)**

73. (a) Graph $t < 5$.   **(b)** Graph $t > 4$.
(c) Describe in your own words the set of numbers belonging to *either one or both* of these sets.
73. (a) _____   **(b)** _____

74. (a) Graph $s < -3$.   **(b)** Graph $s > -1$.
(c) Describe in your own words the set of numbers belonging to *neither* of these sets.
74. (a) _____   **(b)** _____

2.6 SET OPERATIONS AND COMPOUND INEQUALITIES

FOR EXTRA HELP	OBJECTIVES
SSG pp. 48–51 **SSM** pp. 50–55	**1 ▶** Find the intersection of two sets.
Video 2	**2 ▶** Solve compound inequalities with the word *and*.
Tutorial IBM MAC	**3 ▶** Find the union of two sets. **4 ▶** Solve compound inequalities with the word *or*.

The words *and* and *or* are very important in interpreting certain kinds of equations and inequalities in algebra. They are also used when studying sets. In this section we discuss the use of these two words as they relate to sets and inequalities.

A **compound inequality** consists of two inequalities linked by a connective word such as *and* or *or*. Examples of compound inequalities are

$$x + 1 \leq 9 \quad \textbf{and} \quad x - 2 \geq 3$$

and

$$2x > 4 \quad \textbf{or} \quad 3x - 6 < 5.$$

1 ▶ Find the intersection of two sets.

▶ We start by looking at the use of the word "and" with sets. The intersection of sets is defined below.

___

Intersection of Sets

For any two sets A and B, the **intersection** of A and B, symbolized $A \cap B$, is defined as follows:

$$A \cap B = \{x \mid x \text{ is an element of } A \textbf{ and } x \text{ is an element of } B\}.$$

___

EXAMPLE 1

Finding the Intersection of Two Sets

Let $A = \{1, 2, 3, 4\}$ and $B = \{2, 4, 6\}$. Find $A \cap B$.

The set $A \cap B$ contains those elements that belong to both A *and* B at the same time: the numbers 2 and 4. Therefore,

$$A \cap B = \{1, 2, 3, 4\} \cap \{2, 4, 6\}$$
$$= \{2, 4\}. \quad \blacksquare$$

2 ▶ Solve compound inequalities with the word *and*.

▶ To solve a compound inequality with the word *and*, we use the following steps.

___

Solving Inequalities with *and*

Step 1 Solve each inequality in the compound inequality individually.

Step 2 Since the inequalities are joined with *and*, the solution will include all numbers that satisfy both solutions in Step 1 at the same time (the intersection of the solution sets).

___

The next example shows how a compound inequality with *and* is solved.

EXAMPLE 2

Solving a Compound Inequality with *and*

Solve the compound inequality

$$x + 1 \leq 9 \quad \text{and} \quad x - 2 \geq 3.$$

Step 1 directs that we solve each inequality in the compound inequality individually.

$$x + 1 \leq 9 \quad \text{and} \quad x - 2 \geq 3$$
$$x \leq 8 \quad \text{and} \quad x \geq 5$$

Now we apply Step 2. Since the inequalities are joined with the word *and*, the solution set will include all numbers that satisfy both solutions in Step 1 at the same time. Thus, the compound inequality is true whenever $x \leq 8$ and $x \geq 5$ are both true. The top graph in Figure 21 shows $x \leq 8$ and the bottom graph shows $x \geq 5$.

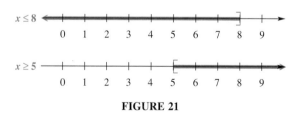

FIGURE 21

▶ **TEACHING TIP**

It is helpful to draw one number line, draw the graphs of the inequalities above the number line, and then graph the intersection on the original number line. The intersection represents the set of points where the two original graphs "overlap." Pay attention to the graphs of the endpoint(s). ◀

Next, we find the intersection of the two graphs in Figure 21 to get the solution set of the compound inequality. The solution set consists of all numbers between 5 and 8, including both 5 and 8. This is the intersection of the two graphs, written in interval notation as [5, 8]. See Figure 22. ■

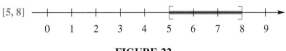

FIGURE 22

EXAMPLE 3

Solving a Compound Inequality with *and*

■ Solve the compound inequality

$$-3x - 2 > 4 \qquad \text{and} \qquad 5x - 1 \leq -21.$$

We begin by solving $-3x - 2 > 4$ and $5x - 1 \leq -21$ separately.

$$-3x - 2 > 4 \qquad \text{and} \qquad 5x - 1 \leq -21$$
$$-3x > 6 \qquad \text{and} \qquad 5x \leq -20$$
$$x < -2 \qquad \text{and} \qquad x \leq -4$$

The graphs of $x < -2$ and $x \leq -4$ are shown in Figure 23.

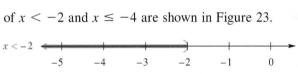

FIGURE 23

Now we find all values of x that satisfy both conditions; that is, the real numbers that are less than -2 and also less than or equal to -4. As shown by the graph in Figure 24, the solution set is $(-\infty, -4]$. ■

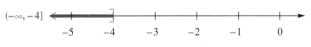

FIGURE 24

EXAMPLE 4

Solving a Compound Inequality with *and*

■ Solve $x + 2 < 5$ and $x - 10 > 2$.

First solve each inequality separately.

$$x + 2 < 5 \qquad \text{and} \qquad x - 10 > 2$$
$$x < 3 \qquad \text{and} \qquad x > 12$$

The graphs of $x < 3$ and $x > 12$ are shown in Figure 25.

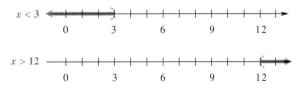

FIGURE 25

There is no number that is both less than 3 *and* greater than 12, so the given compound inequality has no solution. The solution set is ∅. See Figure 26. ∎

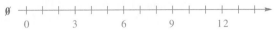

FIGURE 26

3 ▶ Find the union of two sets.

▶ We now discuss the union of two sets, which involves the use of the word "or."

Union of Sets For any two sets A and B, the **union** of A and B, symbolized $A \cup B$, is defined as follows:

$$A \cup B = \{x \mid x \text{ is an element of } A \textbf{ or } x \text{ is an element of } B\}.$$

E X A M P L E   5 ∎
Finding the Union of Two Sets

Find the union of the sets $A = \{1, 2, 3, 4\}$ and $B = \{2, 4, 6\}$.

We begin by listing all the elements of set A: 1, 2, 3, 4. Then we list any additional elements from set B. In this case the elements 2 and 4 are already listed, so the only additional element is 6. Therefore,

$$A \cup B = \{1, 2, 3, 4\} \cup \{2, 4, 6\} = \{1, 2, 3, 4, 6\}.$$

The union consists of all elements in either A *or* B (or both). ∎

Chalkboard Exercise

Let $A = \{3, 4, 5, 6\}$ and $B = \{5, 6, 7\}$. Find $A \cup B$.

Answer: $\{3, 4, 5, 6, 7\}$

Notice in Example 5, that even though the elements 2 and 4 appeared in both sets A and B, they are only written once in $A \cup B$. It is not necessary to write them more than once in the union.

4 ▶ Solve compound inequalities with the word *or*.

▶ To solve compound inequalities with the word *or*, we use the following steps.

Solving Inequalities with *or*

Step 1   Solve each inequality in the compound inequality individually.
Step 2   Since the inequalities are joined with *or*, the solution will include all numbers that satisfy either one or both of the solutions in Step 1 (the union of the solution sets).

The next examples show how to solve a compound inequality with *or*.

E X A M P L E   6 ∎
Solving a Compound Inequality with *or*

Solve $6x - 4 < 2x$ or $-3x \le -9$.
Solve each inequality separately (Step 1).

$$6x - 4 < 2x \qquad \text{or} \qquad -3x \le -9$$
$$4x < 4$$
$$x < 1 \qquad \text{or} \qquad x \ge 3$$

The graphs of these two inequalities are shown in Figure 27.

Chalkboard Exercise

Solve and graph.

$$y - 1 > 2 \ \text{ or }$$
$$3y + 5 < 2y + 6$$

Answer: $(-\infty, 1) \cup (3, \infty)$

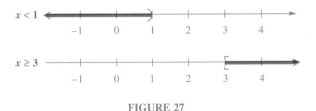

FIGURE 27

Since the inequalities are joined with *or*, find the union of the two sets (Step 2). The union is shown in Figure 28, and is written

$$(-\infty, 1) \cup [3, \infty). \quad \blacksquare$$

FIGURE 28

CAUTION   When interval notation is used to write the solution of Example 6, it *must* be written as

$$(-\infty, 1) \cup [3, \infty).$$

There is no short-cut way to write the solution of a union.

EXAMPLE 7

Solving a Compound Inequality with *or*

Solve $-4x + 1 \geq 9$ or $5x + 3 \geq -12$.

Solve each inequality separately.

$$
\begin{array}{rclcrcl}
-4x + 1 &\geq& 9 & \quad\text{or}\quad & 5x + 3 &\geq& -12 \\
-4x &\geq& 8 & \quad\text{or}\quad & 5x &\geq& -15 \\
x &\leq& -2 & \quad\text{or}\quad & x &\geq& -3
\end{array}
$$

The graphs of these two inequalities are shown in Figure 29.

FIGURE 29

By taking the union, we obtain every real number as a solution, since every real number satisfies at least one of the two inequalities. The set of all real numbers is written in interval notation as

$$(-\infty, \infty).$$

The graph of the solution set is shown in Figure 30. ■

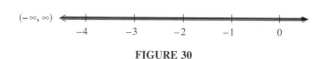

FIGURE 30

2.6 EXERCISES

Answer column (left):

1. true   2. false;   The intersection is Ø.   3. true
4. true   5. false;   6 is not included in the union.
6. false;   The intersection is {6}.   7. {1, 3, 5} or B
8. {1, 3, 5} or B   9. {4} or D   10. {1}   11. Ø
12. Ø
13. {1, 2, 3, 4, 5, 6} or A
14. {1, 3, 4, 5}
15. {1, 3, 5, 6}
16. {1, 3, 5, 6}
17. {1, 4, 6}   18. {1, 4, 6}
19. Each is equal to {1}. This illustrates the associative property of set intersection.   20. Each is equal to {1, 2, 3, 4, 5, 6} or A. This illustrates the associative property of set union.

23.
      −3   2

24.   0   5

25.   2

26.   6

27. [5, 9]
    5   9

28. [2, 6]
    2   6

29. (−3, −1)
    −3   −1

30. (−1, 4)
    −1   4

31. (−∞, 4]
       4

32. (−∞, 6]
       6

33.   2   4

34.   −5   6

Exercises (right):

⊙ *Decide whether the statement is true or false. If it is false, explain why.*

1. The union of the set of rational numbers and the set of irrational numbers is the set of real numbers.

2. The intersection of the set of rational numbers and the set of irrational numbers is the set {0}.

3. The union of the solution sets of $2x + 1 = 3$, $2x + 1 > 3$, and $2x + 1 < 3$ is the set of real numbers.

4. The intersection of the sets $\{x \mid x \geq 5\}$ and $\{x \mid x \leq 5\}$ is {5}.

5. The union of the sets $(-\infty, 6)$ and $(6, \infty)$ is $(-\infty, \infty)$.

6. The intersection of the sets $[6, \infty)$ and $(-\infty, 6]$ is Ø.

Let $A = \{1, 2, 3, 4, 5, 6\}$, $B = \{1, 3, 5\}$, $C = \{1, 6\}$, and $D = \{4\}$. Specify each of the following sets. See Examples 1 and 5.

7. $A \cap B$ **8.** $B \cap A$ **9.** $A \cap D$ **10.** $B \cap C$

11. $B \cap \emptyset$ **12.** $A \cap \emptyset$ **13.** $A \cup B$ **14.** $B \cup D$

15. $B \cup C$ **16.** $C \cup B$ **17.** $C \cup D$ **18.** $D \cup C$

⊙ **19.** Use the sets A, B, and C for Exercises 7–18 to show that $A \cap (B \cap C)$ is equal to $(A \cap B) \cap C$. This is true for any choices of sets. What property does this illustrate? (*Hint:* See Section 1.3.)

⊙ **20.** Repeat Exercise 19, showing that $A \cup (B \cup C)$ is equal to $(A \cup B) \cup C$.

⊙ **21.** Write a few sentences showing how the concept of intersection can be applied to a real-life situation.

⊙ **22.** A compound inequality uses one of the words *and* or *or*. Explain how you will determine whether to use *intersection* or *union* when graphing the solution set.

Two sets are specified by graphs. Graph the intersection of the two sets.

23.

24.

25.

26.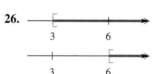

For the compound inequality, give the solution set in both interval and graph forms. See Examples 2–4.

27. $x - 3 \leq 6$ and $x + 2 \geq 7$ **28.** $x + 5 \leq 11$ and $x - 3 \geq -1$

29. $3x < -3$ and $x + 3 > 0$ **30.** $3x > -3$ and $x + 2 < 6$

31. $3x - 4 \leq 8$ and $4x - 1 \leq 15$ **32.** $7x + 6 \leq 48$ and $-4x \geq -24$

Two sets are specified by graphs. Graph the union of the two sets.

33.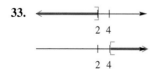
      2 4

      2 4

34.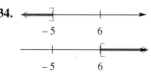
      −5   6

      −5   6

⊙ CONCEPTUAL   ✐ WRITING   ▲ CHALLENGING   ▦ SCIENTIFIC CALCULATOR   ▦ GRAPHICS CALCULATOR

35.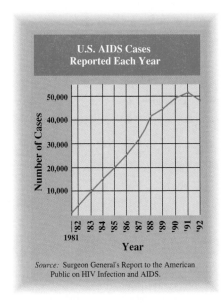

36.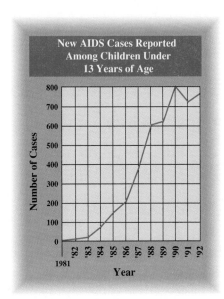

For the compound inequality, give the solution set in both interval and graph forms. See Examples 6 and 7.

37. $x + 2 > 7$ or $x - 1 < -6$

38. $x + 1 > 3$ or $x + 4 < 2$

39. $x + 1 > 3$ or $4x - 1 < -5$

40. $3x < x + 12$ or $x + 1 > 10$

Express the set in the simplest interval form.

41. $(-\infty, -1] \cap [-4, \infty)$

42. $[-1, \infty) \cap (-\infty, 9]$

43. $(-\infty, -6] \cap [-9, \infty)$

44. $(-\infty, 3) \cup (-\infty, -2)$

45. $(5, 11] \cap [6, \infty)$

46. $[-9, 1] \cap (-\infty, -3)$

47. $[3, 6] \cap (4, 9)$

48. $[-1, 2] \cup (0, 5)$

For the compound inequality, decide whether intersection or union should be used. Then give the solution set in both interval and graph forms. See Examples 2, 3, 4, 6, and 7.

49. $x < -1$ and $x > -5$

50. $x > -1$ and $x < 7$

51. $x < 4$ or $x < -2$

52. $x < 5$ or $x < -3$

53. $x + 1 \geq 5$ and $x - 2 \leq 10$

54. $2x - 6 \leq -18$ and $2x \geq -18$

55. $-3x \leq -6$ or $-3x \geq 0$

56. $-8x \leq -24$ or $-5x \geq 15$

Use the graphs to answer the questions in Exercises 57 and 58.

U.S. AIDS Cases Reported Each Year

Number of Cases

Year

Source: Surgeon General's Report to the American Public on HIV Infection and AIDS.

New AIDS Cases Reported Among Children Under 13 Years of Age

Number of Cases

Year

57. In which years did the number of U.S. AIDS cases exceed 30,000 *and* the new AIDS cases among children under thirteen years of age exceed 400?

58. In which years was the number of U.S. AIDS cases greater than 40,000 *or* the new AIDS cases among children under thirteen years of age less than 200?

35.

36.

37. $(-\infty, -5) \cup (5, \infty)$

38. $(-\infty, -2) \cup (2, \infty)$

39. $(-\infty, -1) \cup (2, \infty)$

40. $(-\infty, 6) \cup (9, \infty)$

41. $[-4, -1]$

42. $[-1, 9]$

43. $[-9, -6]$

44. $(-\infty, 3)$

45. $[6, 11]$ **46.** $[-9, -3]$

47. $(4, 6]$ **48.** $[-1, 5)$

49. intersection; $(-5, -1)$

50. intersection; $(-1, 7)$

51. union; $(-\infty, 4)$

52. union; $(-\infty, 5)$

53. intersection; $[4, 12]$

54. intersection; $[-9, -6]$

55. union; $(-\infty, 0] \cup [2, \infty)$

56. union; $(-\infty, -3] \cup [3, \infty)$

57. 1988, 1989, 1990, 1991, 1992

58. 1981, 1982, 1983, 1984, 1985, 1988, 1989, 1990, 1991, 1992

59. Mario, Joe   60. none
of them   61. none of them
62. Luigi, Than
63. $[-6, \infty)$ 64. $(-1, \infty)$
65. $(-3, 2)$ 66. $\left[-\dfrac{5}{3}, 3\right)$
67. -21 68. -13

◆ **MATHEMATICAL CONNECTIONS** (Exercises 59–62) ◆

The figures represent the backyards of neighbors Luigi, Mario, Than, and Joe. Find the area and the perimeter of each yard. Suppose that each resident has 150 feet of fencing, and enough sod to cover 1400 square feet of lawn. Give the name or names of the residents whose yards satisfy the following descriptions.

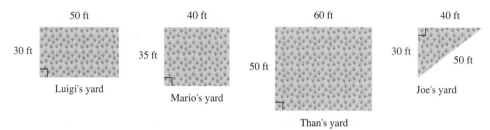

59. the yard can be fenced *and* the yard can be sodded

60. the yard can be fenced *and* the yard cannot be sodded

61. the yard cannot be fenced *and* the yard can be sodded

62. the yard cannot be fenced *and* the yard cannot be sodded

◆

REVIEW EXERCISES

Solve the inequality. See Section 2.5.

63. $2y - 4 \le 3y + 2$

64. $5t - 8 < 6t - 7$

65. $-5 < 2r + 1 < 5$

66. $-7 \le 3w - 2 < 7$

Evaluate. See Sections 1.1 and 1.2.

67. $-|6| - |-11| + (-4)$

68. $(-5) - |-9| + |5 - 4|$

2.7 ABSOLUTE VALUE EQUATIONS AND INEQUALITIES

FOR EXTRA HELP

 SSG pp. 51–58
SSM pp. 55–60

 Video 2

 Tutorial
IBM   MAC

OBJECTIVES

1 ▶ Use the distance definition of absolute value.
2 ▶ Solve equations of the form $|ax + b| = k$, for $k > 0$.
3 ▶ Solve inequalities of the form $|ax + b| < k$ and of the form $|ax + b| > k$, for $k > 0$.
4 ▶ Solve absolute value equations that involve rewriting.
5 ▶ Solve absolute value equations of the form $|ax + b| = |cx + d|$.
6 ▶ Solve absolute value equations and inequalities that have only nonpositive constants on one side.

Sometimes a model requires that a linear expression must be positive. We can use absolute value to ensure that this will happen. Such situations are expressed as absolute value equations or inequalities.

INSTRUCTOR'S RESOURCES

 ITM pp. 342–344   **ISM** pp. 92–102
IAM pp. 6–7

 TEST GENERATOR
DOS   Windows   MAC

 TRANSPARENCIES

◆ Answers will vary.

▶ **TEACHING TIP**

Students usually have difficulty with this section. Handle absolute value equations as three cases. The two-solution case must have the absolute value expression alone on one side and equal to a positive number. The one-solution case must have the absolute value expression equal to zero. The no-solution case must have the absolute value expression equal to a negative number.   ◀

◆ **CONNECTIONS** ◆

Absolute value is used to find the relative error of a measurement in science, engineering, and other fields. If x_t represents the true value of a measurement and x represents the measured value, then the

$$\text{relative error in } x = \left| \frac{x_t - x}{x_t} \right|.$$

In many situations in the work world, the relative error must be less than some predetermined amount. For example, suppose a machine filling quart milk cartons is set for a relative error no greater than .05. With this tolerance level, how many ounces may a carton contain?

Here $x_t = 32$ ounces, the relative error $= .05$ ounces, and we must find x, given

$$\left| \frac{32 - x}{32} \right| = \left| 1 - \frac{x}{32} \right| \leq .05.$$

FOR DISCUSSION OR WRITING

What are some other situations where relative error is important? Consider the areas of science, manufacturing, and engineering, for example.

1 ▶ Use the distance definition of absolute value.

▶ **TEACHING TIP**

Here is one strategy for teaching the type of inequality in Example 2 on the next page (assuming that the absolute value expression is on the left).

Indicate that the $>$ symbol points "outward." Show an "outward" graph:

⟵—— ——⟶

Place the individual inequalities with each portion of the graph, then solve and label the graph. The word "or" is placed in the blank portion between the two portions of the graph. Finally, the solution set in interval notation can be extracted from the graph.   ◀

▶   In Chapter 1 we saw that the absolute value of a number x, written $|x|$, represents the distance from x to 0 on the number line. For example, the solutions of $|x| = 4$ are 4 and -4, as shown in Figure 31. Since absolute value represents distance from 0, it is reasonable to interpret the solutions of $|x| > 4$ to be all numbers that are *more* than 4 units from 0. The set $(-\infty, -4) \cup (4, \infty)$ fits this description. Figure 32 shows the graph of the solution set of $|x| > 4$. The solution set of $|x| < 4$ consists of all numbers that are *less* than 4 units from 0 on the number line. Another way of thinking of this is to think of all numbers between -4 and 4. This set of numbers is given by $(-4, 4)$, as shown in Figure 33.

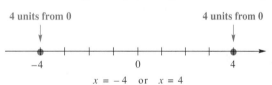

FIGURE 31

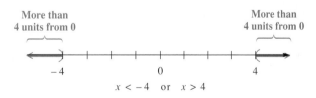

FIGURE 32

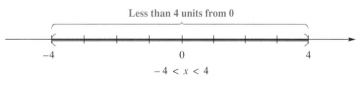

FIGURE 33

The equation and inequalities just described are examples of **absolute value equations and inequalities.** These are equations and inequalities that involve the absolute value of a variable expression. These generally take the form

$$|ax + b| = k, \qquad |ax + b| > k, \qquad \text{or} \qquad |ax + b| < k$$

where k is a positive number. We may solve them by rewriting them as compound equations or inequalities. The various methods of solving them are explained in the following examples.

2 ▶ Solve equations of the form $|ax + b| = k$, for $k > 0$.

▶ The first example shows how to solve a typical absolute value equation. Remember that since absolute value refers to distance from the origin, an absolute value equation will have two cases.

EXAMPLE 1

Solving an Absolute Value Equation

Solve $|2x + 1| = 7$.

For $|2x + 1|$ to equal 7, $2x + 1$ must be 7 units from 0 on the number line. This can happen only when $2x + 1 = 7$ or $2x + 1 = -7$. We solve this compound equation as follows.

$$
\begin{array}{ccc}
2x + 1 = 7 & \text{or} & 2x + 1 = -7 \\
2x = 6 & \text{or} & 2x = -8 \\
x = 3 & \text{or} & x = -4
\end{array}
$$

The solution set is $\{-4, 3\}$. Its graph is shown in Figure 34. ■

FIGURE 34

3 ▶ Solve inequalities of the form $|ax + b| < k$ and of the form $|ax + b| > k$, for $k > 0$.

▶ We now discuss how to solve absolute value inequalities.

EXAMPLE 2

Solving an Absolute Value Inequality with $>$

Solve $|2x + 1| > 7$.

This absolute value inequality must be rewritten as

$$2x + 1 > 7 \qquad \text{or} \qquad 2x + 1 < -7,$$

because $2x + 1$ must represent a number that is *more* than 7 units from 0 on either side of the number line. Now we solve the compound inequality.

$$
\begin{array}{ccc}
2x + 1 > 7 & \text{or} & 2x + 1 < -7 \\
2x > 6 & \text{or} & 2x < -8 \\
x > 3 & \text{or} & x < -4
\end{array}
$$

The solution set, $(-\infty, -4) \cup (3, \infty)$, is graphed in Figure 35. Notice that the graph consists of two intervals. ■

FIGURE 35

EXAMPLE 3

Solving an Absolute Value Inequality with $<$

Solve $|2x + 1| < 7$.

Here the expression $2x + 1$ must represent a number that is less than 7 units from 0 on the number line. Another way of thinking of this is to realize that $2x + 1$ must be between -7 and 7. This is written as the three-part inequality

$$-7 < 2x + 1 < 7.$$

We solved such inequalities in Section 2.5 by working with all three parts at the same time.

$$-7 < 2x + 1 < 7$$
$$-8 < 2x < 6 \qquad \text{Subtract 1 from each part.}$$
$$-4 < x < 3 \qquad \text{Divide each part by 2.}$$

The solution set is $(-4, 3)$, and the graph consists of a single interval, as shown in Figure 36. ■

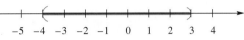

FIGURE 36

Look back at Figures 34, 35, and 36. These are the graphs of $|2x + 1| = 7$, $|2x + 1| > 7$, and $|2x + 1| < 7$. If we find the union of the three sets, we get the set of all real numbers. This is because for any value of x, $|2x + 1|$ will satisfy one and only one of the following: it is equal to 7, greater than 7, or less than 7.

CAUTION When solving absolute value equations and inequalities of the types in Examples 1, 2, and 3, be sure to remember the following.

1. The methods described apply when the constant is alone on one side of the equation or inequality and is *positive*.
2. Absolute value inequalities written in the form $|ax + b| > k$ and absolute value equations translate into "or" compound statements.
3. Absolute value inequalities in the form $|ax + b| < k$ translate into "and" compound statements, which may be written as three-part inequalities.
4. An "or" statement *cannot* be written in three parts. It would be *incorrect* to use

$$-7 > 2x + 1 > 7$$

in Example 2, since this would imply that $-7 > 7$, which is false.

4 ▶ Solve absolute value equations that involve rewriting.

▶ Sometimes an absolute value equation or inequality is given in a form that requires some rewriting before it can be set up as a compound statement. The next example illustrates this process for an absolute value equation.

EXAMPLE 4

Solving an Absolute Value Equation Requiring Rewriting

Solve the equation $|x + 3| + 5 = 12$.

First get the absolute value alone on one side of the equals sign. Do this by subtracting 5 on each side.

$$|x + 3| + 5 - 5 = 12 - 5 \qquad \text{Subtract 5.}$$
$$|x + 3| = 7$$

Then use the method shown in Example 1.

$$x + 3 = 7 \qquad \text{or} \qquad x + 3 = -7$$
$$x = 4 \qquad \text{or} \qquad x = -10$$

Check that the solution set is $\{4, -10\}$ by substituting in the original equation. ■

Solving an absolute value *inequality* requiring rewriting is done in a similar manner.

The methods used in Examples 1, 2, and 3 are summarized here.

Solving Absolute Value Equations and Inequalities

Let k be a positive number, and p and q be two numbers.

1. To solve $|ax + b| = k$, solve the compound equation

$$ax + b = k \quad \text{or} \quad ax + b = -k.$$

The solution set is of the form $\{p, q\}$, with two numbers.

2. To solve $|ax + b| > k$, solve the compound inequality

$$ax + b > k \quad \text{or} \quad ax + b < -k.$$

The solution set is of the form $(-\infty, p) \cup (q, \infty)$, which consists of two separate intervals.

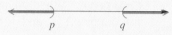

3. To solve $|ax + b| < k$, solve the compound inequality

$$-k < ax + b < k.$$

The solution set is of the form (p, q), a single interval.

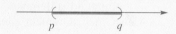

▶ **TEACHING TIP**
Students will want to use the methods shown earlier for the special cases in Example 6. They should be reminded to pay attention to the fact that the absolute value of a quantity cannot be negative. Ask them to check the "sense" of their answers. If an answer does not make sense, they should then check to see if the inequality is one of these special cases. ◀

NOTE   Some people prefer to write the compound statements in parts 1 and 2 of the summary as

$$ax + b = k \quad \text{or} \quad -(ax + b) = k$$

and

$$ax + b > k \quad \text{or} \quad -(ax + b) > k.$$

These forms are equivalent to those we give in the summary and produce the same results.

5 ▶ Solve absolute value equations of the form $|ax + b| = |cx + d|$.

▶   The next example shows how to solve an absolute value equation with two absolute value expressions. For two expressions to have the same absolute value, they must either be equal or be negatives of each other.

Solving $|ax + b| = |cx + d|$

To solve an absolute value equation of the form

$$|ax + b| = |cx + d|,$$

solve the compound equation

$$ax + b = cx + d \quad \textbf{or} \quad ax + b = -(cx + d).$$

EXAMPLE 5

Solving an Equation with Two Absolute Values

Solve the equation $|z + 6| = |2z - 3|$.

This equation is satisfied either if $z + 6$ and $2z - 3$ are equal to each other, or if $z + 6$ and $2z - 3$ are negatives of each other. Thus, we have

$$z + 6 = 2z - 3 \quad \text{or} \quad z + 6 = -(2z - 3).$$

Solve each equation.

$$6 + 3 = 2z - z \quad \text{or} \quad z + 6 = -2z + 3$$
$$3z = -3$$
$$9 = z \quad \text{or} \quad z = -1$$

The solution set is $\{9, -1\}$. ■

6 ▶ Solve absolute value equations and inequalities that have only nonpositive constants on one side.

▶ When a typical absolute value equation or inequality involves a *negative* constant or *zero* alone on one side, simply use the properties of absolute value to solve. Keep in mind the following.

1. The absolute value of an expression can never be negative: $|a| \geq 0$ for all real numbers a.
2. The absolute value of an expression equals 0 only when the expression is equal to 0.

The next two examples illustrate these special cases.

EXAMPLE 6

Solving Special Cases of Absolute Value Equations

Solve each equation.

(a) $|5r - 3| = -4$
Since the absolute value of an expression can never be negative, there are no solutions for this equation. The solution set is ∅.

(b) $|7x - 3| = 0$
The expresson $7x - 3$ will equal 0 *only* for the solution of the equation

$$7x - 3 = 0.$$

The solution of this equation is $3/7$. The solution set is $\{3/7\}$. It consists of only one element. ■

EXAMPLE 7

Solving Special Cases of Absolute Value Inequalities

Solve each of the following inequalities.

(a) $|x| \geq -4$
The absolute value of a number is never negative. For this reason, $|x| \geq -4$ is true for *all* real numbers. The solution set is $(-\infty, \infty)$.

(b) $|k + 6| < -2$
There is no number whose absolute value is less than -2, so this inequality has no solution. The solution set is ∅.

(c) $|m - 7| \leq 0$
The value of $|m - 7|$ will never be less than 0. However, $|m - 7|$ will equal 0 when $m = 7$. Therefore, the solution set is $\{7\}$. ■

2.7 EXERCISES

Keeping in mind that the absolute value of a number can be interpreted as the distance between the graph of the number and 0 on the number line, match the absolute value equation or inequality with the graph of its solution set.

Choices

◉ **1.** $|x| = 5$ A.

$|x| < 5$ B.

$|x| > 5$ C.

$|x| \le 5$ D.

$|x| \ge 5$ E.

1. E;   C;   D;   B;   A

Choices

◉ **2.** $|x| = 9$ A.

$|x| > 9$ B.

$|x| \ge 9$ C.

$|x| < 9$ D.

$|x| \le 9$ E.

2. E;   D;   A;   C;   B

✎ **3.** Explain when to use *and* and when to use *or* if you are solving an absolute value equation or inequality of the form $|ax + b| = k$, $|ax + b| < k$, or $|ax + b| > k$, where k is a positive number.

◉ **4.** How many solutions will $|ax + b| = k$, have if **(a)** $k = 0$;   **(b)** $k > 0$;   **(c)** $k < 0$?

4. (a) one   (b) two   (c) none

Solve each equation. See Example 1.

5. $|x| = 12$

5. $\{-12, 12\}$

6. $|k| = 14$

6. $\{-14, 14\}$

7. $|4x| = 20$

7. $\{-5, 5\}$

8. $|5x| = 30$

8. $\{-6, 6\}$

9. $|y - 3| = 9$

9. $\{-6, 12\}$

10. $|p - 5| = 13$

10. $\{-8, 18\}$

11. $|2x + 1| = 7$

11. $\{-4, 3\}$

12. $|2y + 3| = 19$

12. $\{-11, 8\}$

13. $|4r - 5| = 17$

13. $\left\{-3, \dfrac{11}{2}\right\}$

14. $|5t - 1| = 21$

14. $\left\{-4, \dfrac{22}{5}\right\}$

15. $|2y + 5| = 14$

15. $\left\{-\dfrac{19}{2}, \dfrac{9}{2}\right\}$

16. $|2x - 9| = 18$

16. $\left\{-\dfrac{9}{2}, \dfrac{27}{2}\right\}$

17. $\left|\dfrac{1}{2}x + 3\right| = 2$

17. $\{-10, -2\}$

18. $\left|\dfrac{2}{3}q - 1\right| = 5$

18. $\{-6, 9\}$

19. $\left|1 - \dfrac{3}{4}k\right| = 7$

19. $\left\{-8, \dfrac{32}{3}\right\}$

20. $\left|2 - \dfrac{5}{2}m\right| = 14$

20. $\left\{-\dfrac{24}{5}, \dfrac{32}{5}\right\}$

◉ CONCEPTUAL   ✎ WRITING   ▲ CHALLENGING   ▦ SCIENTIFIC CALCULATOR   ▦ GRAPHICS CALCULATOR

⊙ 21. The graph of the solution set of $|2x + 1| = 9$ is given here.

$$-5 \qquad 0 \qquad 4$$

Without actually doing the algebraic work, graph the solution set of each inequality, referring to the graph above.

(a) $|2x + 1| < 9$ **(b)** $|2x + 1| > 9$

21. (a) ⊢⟨++++++++⟩+⊢→   (b) ⟨+⟩+++++++⟨+⟩→
       −5 0 4   −5 0 4

⊙ 22. The graph of the solution set of $|3y - 4| < 5$ is given here.

$$-\frac{1}{3} \qquad\qquad 3$$

Without actually doing the algebraic work, graph the solution set of each of the following, referring to the graph above.

(a) $|3y - 4| = 5$ **(b)** $|3y - 4| > 5$

22. (a) 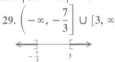   (b)

Solve the inequality and graph the solution set. See Example 2.

23. $|x| > 3$
23. $(-\infty, -3) \cup (3, \infty)$

⟵——⟶   ⟵——⟶
−3   3

24. $|y| > 5$
24. $(-\infty, -5) \cup (5, \infty)$

⟵——⟶   ⟵——⟶
−5   5

25. $|k| \geq 4$
25. $(-\infty, -4] \cup [4, \infty)$

⟵——⟶   ⟵——⟶
−4   4

26. $|r| \geq 6$
26. $(-\infty, -6] \cup [6, \infty)$

⟵——⟶   ⟵——⟶
−6   6

27. $|t + 2| > 10$
27. $(-\infty, -12) \cup (8, \infty)$

⟵——⟶   ⟵——⟶
−12   8

28. $|r + 5| > 20$
28. $(-\infty, -25) \cup (15, \infty)$

⟵——⟶   ⟵——⟶
−25   15

29. $|3x - 1| \geq 8$
29. $\left(-\infty, -\frac{7}{3}\right] \cup [3, \infty)$

⟵——⟶   ⟵——⟶
$-\frac{7}{3}$ 3

30. $|4x + 1| \geq 21$
30. $\left(-\infty, -\frac{11}{2}\right] \cup [5, \infty)$

⟵——⟶   ⟵——⟶
$-\frac{11}{2}$ 5

31. $|3 - x| > 5$
31. $(-\infty, -2) \cup (8, \infty)$

⟵——⟶   ⟵——⟶
−2   8

32. $|5 - x| > 3$
32. $(-\infty, 2) \cup (8, \infty)$

⟵——⟶   ⟵——⟶
2   8

Solve the inequality and graph the solution set. See Example 3. (Hint: Compare your answers to those in Exercises 23–32.)

33. $|x| \leq 3$
33. $[-3, 3]$

⊢——⊣→
−3   3

34. $|y| \leq 5$
34. $[-5, 5]$

⊢——⊣→
−5   5

35. $|k| < 4$
35. $(-4, 4)$

⟨——⟩→
−4   4

36. $|r| < 6$
36. $(-6, 6)$

⟨——⟩→
−6   6

37. $|t + 2| \leq 10$
37. $[-12, 8]$

⊢——⊣→
−12   8

38. $|r + 5| \leq 20$
38. $[-25, 15]$

⊢——⊣→
−25   15

39. $|3x - 1| < 8$
39. $\left(-\frac{7}{3}, 3\right)$

⟨——⟩→
$-\frac{7}{3}$ 3

40. $|4x + 1| < 21$
40. $\left(-\frac{11}{2}, 5\right)$

⟨——⟩→
$-\frac{11}{2}$ 5

41. $|3 - x| \leq 5$
41. $[-2, 8]$

⊢——⊣→
−2   8

42. $|5 - x| \leq 3$
42. $[2, 8]$

⊢——⊣→
2   8

Decide on which method you should use to solve the given absolute value equation or inequality. Find the solution set and graph. See Examples 1–3.

43. $|-4 + k| > 9$

43. $(-\infty, -5) \cup (13, \infty)$

44. $|-3 + t| > 8$

44. $(-\infty, -5) \cup (11, \infty)$

45. $|7 + 2z| = 5$

45. $\{-6, -1\}$

46. $|9 - 3p| = 3$

46. $\{2, 4\}$

47. $|3r - 1| \le 11$

47. $\left[-\dfrac{10}{3}, 4\right]$

48. $|2s - 6| \le 6$

48. $[0, 6]$

49. $|-6x - 6| \le 1$

49. $\left[-\dfrac{7}{6}, -\dfrac{5}{6}\right]$

50. $|-2x - 6| \le 5$

50. $\left[-\dfrac{11}{2}, -\dfrac{1}{2}\right]$

▲**51.** Write an absolute value equation in the variable x that states that the distance between x and 4 is equal to 9.

51. $|x - 4| = 9$ (or $|4 - x| = 9$)

▲**52.** Write an absolute value inequality in the variable x that states that the distance between $2x$ and -3 is **(a)** greater than 4   **(b)** less than 4.

52. (a) $|2x + 3| > 4$ (or $|-3 - 2x| > 4$)   (b) $|2x + 3| < 4$ (or $|-3 - 2x| < 4$)

Solve the equation or inequality. Give the solution set in set notation for equations and in interval notation for inequalities. See Example 4.

53. $|x + 4| + 1 = 2$

53. $\{-5, -3\}$

54. $|y + 5| - 2 = 12$

54. $\{-19, 9\}$

55. $|2x + 1| + 3 > 8$

55. $(-\infty, -3) \cup (2, \infty)$

56. $|6x - 1| - 2 > 6$

56. $\left(-\infty, -\dfrac{7}{6}\right) \cup \left(\dfrac{3}{2}, \infty\right)$

57. $|x + 5| - 6 \le -1$

57. $[-10, 0]$

58. $|r - 2| - 3 \le 4$

58. $[-5, 9]$

Solve the equation. See Example 5.

59. $|3x + 1| = |2x + 4|$

59. $\{-1, 3\}$

60. $|7x + 12| = |x - 8|$

60. $\left\{-\dfrac{10}{3}, -\dfrac{1}{2}\right\}$

61. $\left|m - \dfrac{1}{2}\right| = \left|\dfrac{1}{2}m - 2\right|$

61. $\left\{-3, \dfrac{5}{3}\right\}$

62. $\left|\dfrac{2}{3}r - 2\right| = \left|\dfrac{1}{3}r + 3\right|$

62. $\{-1, 15\}$

63. $|6x| = |9x + 1|$

63. $\left\{-\dfrac{1}{3}, -\dfrac{1}{15}\right\}$

64. $|13y| = |2y + 1|$

64. $\left\{-\dfrac{1}{15}, \dfrac{1}{11}\right\}$

65. $|2p - 6| = |2p + 11|$

65. $\left\{-\dfrac{5}{4}\right\}$

66. $|3x - 1| = |3x + 9|$

66. $\left\{-\dfrac{4}{3}\right\}$

Solve the equation or inequality. See Examples 6 and 7.

67. $|12t - 3| = -8$

67. ∅

68. $|13w + 1| = -3$

68. ∅

69. $|4x + 1| = 0$

69. $\left\{-\dfrac{1}{4}\right\}$

70. $|6r - 2| = 0$

70. $\left\{\dfrac{1}{3}\right\}$

71. $|2q - 1| < -6$

71. ∅

72. $|8n + 4| < -4$

72. ∅

73. $|x + 5| > -9$

73. $(-\infty, \infty)$

74. $|x + 9| > -3$

74. $(-\infty, \infty)$

75. $|7x + 3| \le 0$

75. $\left\{-\dfrac{3}{7}\right\}$

76. $|4x - 1| \le 0$

76. $\left\{\dfrac{1}{4}\right\}$

77. $|5x - 2| \ge 0$

77. $(-\infty, \infty)$

78. $|4 + 7x| \ge 0$

78. $(-\infty, \infty)$

79. $|10z + 7| > 0$

79. $\left(-\infty, -\dfrac{7}{10}\right) \cup \left(-\dfrac{7}{10}, \infty\right)$

80. $|4x + 1| > 0$

80. $\left(-\infty, -\dfrac{1}{4}\right) \cup \left(-\dfrac{1}{4}, \infty\right)$

———————— ◆ **MATHEMATICAL CONNECTIONS** (Exercises 81–84) ◆ ————————

The ten tallest buildings in Kansas City, Missouri, are listed along with their heights.

Building	Height (in feet)
One Kansas City Place	626
AT&T Town Pavillion	590
Hyatt Regency	504
Kansas City Power and Light	476
City Hall	443
Federal Office Building	413
Commerce Tower	402
City Center Square	402
Southwest Bell Telephone	394
Pershing Road Associates	352

Use this information to work through Exercises 81–84 in order.

81. To find the average of a group of numbers, we add the numbers and then divide by the number of items added. Use a calculator to find the average of the heights.

81. 460.2 feet

82. Let k represent the average height of these buildings. If a height x satisfies the inequality

$$|x - k| < t,$$

then the height is said to be within t feet of the average. Using your result from Exercise 81, list the buildings that are within 50 feet of the average.

82. Federal Office Building, City Hall, Kansas City Power and Light, Hyatt Regency

83. Repeat Exercise 82, but find the buildings that are within 75 feet of the average.

83. Southwest Bell Telephone, City Center Square, Commerce Tower, Federal Office Building, City Hall, Kansas City Power and Light, Hyatt Regency

84. (a) Write an absolute value inequality that describes the height of a building that is *not* within 75 feet of the average.
  (b) Solve the inequality you wrote in part (a).
  ⊙ **(c)** Use the result of part (b) to find the buildings that are not within 75 feet of the average.
  (d) Confirm that your answer to part (c) makes sense by comparing it with your answer to Exercise 83.

84. (a) $|x - 460.2| \geq 75$ (b) $x \geq 535.2$ or $x \leq 385.2$ (c) Pershing Road Associates, AT&T Town Pavillion, One Kansas City Place   (d) It makes sense because it includes all buildings *not* listed earlier.

———————— ◆ ————————

Solve the problem.

85. Solve the inequality found in the Connections at the beginning of this section.

85. [30.4, 33.6]

86. A study showed that 90% of the adult population gets an average of x hours of sleep per night, where

$$|x - 6.5| \leq 1.0.$$

What is the interval of hours of sleep corresponding to this inequality?

86. [5.5, 7.5]

REVIEW EXERCISES

Evaluate the exponential expression. See Section 1.2.

87. 2^6 **88.** $(-4)^5$ **89.** $(-5)^4$ **90.** -5^4 **91.** $\left(\dfrac{3}{4}\right)^4$ **92.** $\left(-\dfrac{3}{2}\right)^5$

87. 64   88. -1024 89. 625   90. -625 91. $\dfrac{81}{256}$ 92. $-\dfrac{243}{32}$

SUMMARY: SOLVING LINEAR AND ABSOLUTE VALUE EQUATIONS AND INEQUALITIES

Students often have difficulty distinguishing between the various types of equations and inequalities introduced in this chapter. This section of miscellaneous equations and inequalities provides practice in solving all such types. You might wish to refer to the boxes in this chapter that summarize the various methods of solution. Solve the equation or inequality.

1. $4z + 1 = 49$
1. $\{12\}$

2. $|m - 1| = 6$
2. $\{-5, 7\}$

3. $6q - 9 = 12 + 3q$
3. $\{7\}$

4. $3p + 7 = 9 + 8p$
4. $\left\{-\dfrac{2}{5}\right\}$

5. $|a + 3| = -4$
5. $\emptyset$

6. $2m + 1 \leq m$
6. $(-\infty, -1]$

7. $8r + 2 \geq 5r$
7. $\left[-\dfrac{2}{3}, \infty\right)$

8. $4(a - 11) + 3a = 20a - 31$
8. $\{-1\}$

9. $2q - 1 = -7$
9. $\{-3\}$

10. $|3q - 7| - 4 = 0$
10. $\left\{1, \dfrac{11}{3}\right\}$

11. $6z - 5 \leq 3z + 10$
11. $(-\infty, 5]$

12. $|5z - 8| + 9 \geq 7$
12. $(-\infty, \infty)$

13. $9y - 3(y + 1) = 8y - 7$
13. $\{2\}$

14. $|y| \geq 8$
14. $(-\infty, -8] \cup [8, \infty)$

15. $9y - 5 \geq 9y + 3$
15. $\emptyset$

16. $13p - 5 > 13p - 8$
16. $(-\infty, \infty)$

17. $|q| < 5.5$
17. $(-5.5, 5.5)$

18. $4z - 1 = 12 + z$
18. $\left\{\dfrac{13}{3}\right\}$

19. $\dfrac{2}{3}y + 8 = \dfrac{1}{4}y$
19. $\left\{-\dfrac{96}{5}\right\}$

20. $-\dfrac{5}{8}y \geq -20$
20. $(-\infty, 32]$

21. $\dfrac{1}{4}p < -6$
21. $(-\infty, -24)$

22. $7z - 3 + 2z = 9z - 8z$
22. $\left\{\dfrac{3}{8}\right\}$

23. $\dfrac{3}{5}q - \dfrac{1}{10} = 2$
23. $\left\{\dfrac{7}{2}\right\}$

24. $|r - 1| < 7$
24. $(-6, 8)$

25. $r + 9 + 7r = 4(3 + 2r) - 3$
25. $(-\infty, \infty)$

26. $6 - 3(2 - p) < 2(1 + p) + 3$
26. $(-\infty, 5)$

27. $|2p - 3| > 11$
27. $(-\infty, -4) \cup (7, \infty)$

28. $\dfrac{x}{4} - \dfrac{2x}{3} = -10$
28. $\{24\}$

29. $|5a + 1| \leq 0$
29. $\left\{-\dfrac{1}{5}\right\}$

30. $5z - (3 + z) \geq 2(3z + 1)$
30. $\left(-\infty, -\dfrac{5}{2}\right]$

31. $-2 \leq 3x - 1 \leq 8$
31. $\left[-\dfrac{1}{3}, 3\right]$

32. $-1 \leq 6 - x \leq 5$
32. $[1, 7]$

33. $|7z - 1| = |5z + 3|$
33. $\left\{-\dfrac{1}{6}, 2\right\}$

34. $|p + 2| = |p + 4|$
34. $\{-3\}$

35. $|1 - 3x| \geq 4$

35. $(-\infty, -1] \cup \left[\dfrac{5}{3}, \infty\right)$

37. $-(m + 4) + 2 = 3m + 8$

37. $\left\{-\dfrac{5}{2}\right\}$

39. $-6 \leq \dfrac{3}{2} - x \leq 6$

39. $\left[-\dfrac{9}{2}, \dfrac{15}{2}\right]$

41. $|y - 1| \geq -6$

41. $(-\infty, \infty)$

43. $8q - (1 - q) = 3(1 + 3q) - 4$

43. $(-\infty, \infty)$

45. $|r - 5| = |r + 9|$

45. $\{-2\}$

47. $2x + 1 > 5 \quad \text{or} \quad 3x + 4 < 1$

47. $(-\infty, -1) \cup (2, \infty)$

36. $\dfrac{1}{2} \leq \dfrac{2}{3}r \leq \dfrac{5}{4}$

36. $\left[\dfrac{3}{4}, \dfrac{15}{8}\right]$

38. $\dfrac{p}{6} - \dfrac{3p}{5} = p - 86$

38. $\{60\}$

40. $|5 - y| < 4$

40. $(1, 9)$

42. $|2r - 5| = |r + 4|$

42. $\left\{\dfrac{1}{3}, 9\right\}$

44. $8y - (y + 3) = -(2y + 1) - 12$

44. $\left\{-\dfrac{10}{9}\right\}$

46. $|r + 2| < -3$

46. $\emptyset$

48. $1 - 2x \geq 5 \quad \text{and} \quad 7 + 3x \geq -2$

48. $[-3, -2]$

CHAPTER 2 SUMMARY

KEY TERMS

2.1 mathematical model
algebraic expression
equation
linear equation
(first-degree
equation in one
variable)
solution
solution set
equivalent equations
addition and
multiplication
properties of
equality
conditional equation
contradiction
identity

2.2 formula
degree
parallel lines
transversal
vertical angles

2.3 right angle
straight angle
complementary
angles
(complements)

supplementary
angles
(supplements)

2.5 inequality
linear inequality in
one variable
equivalent
inequalities
addition and
multiplication
properties of
inequality
interval
strict inequality
nonstrict inequality

2.6 intersection
compound
inequality
union

2.7 absolute value
equation
absolute value
inequality

NEW SYMBOLS

$1°$ one degree
$\cap$ set intersection
$\cup$ set union

⊙ CONCEPTUAL   ✎ WRITING   ▲ CHALLENGING   ▦ SCIENTIFIC CALCULATOR   ▦ GRAPHICS CALCULATOR

QUICK REVIEW

CONCEPTS	EXAMPLES

2.1   LINEAR EQUATIONS IN ONE VARIABLE

Solving a Linear Equation
If necessary, eliminate fractions by multiplying both sides by the LCD. Simplify each side, and then use the addition property of equality to get the variables on one side and the numbers on the other. Combine terms if possible, and then use the multiplication property of equality to make the coefficient of the variable equal to 1. Check by substituting into the original equation.

Solve the equation.

$$4(8 - 3t) = 32 - 8(t + 2)$$
$$32 - 12t = 32 - 8t - 16$$
$$32 - 12t = 16 - 8t$$
$$32 - 12t + 12t = 16 - 8t + 12t$$
$$32 = 16 + 4t$$
$$32 - 16 = 16 + 4t - 16$$
$$16 = 4t$$
$$\frac{16}{4} = \frac{4t}{4}$$
$$4 = t$$

The solution set is {4}. This can be checked by substituting 4 for t in the original equation.

2.2   FORMULAS AND TOPICS FROM GEOMETRY

Solving for a Specified Variable
Use the addition or multiplication properties as necessary to get all terms with the specified variable on one side of the equals sign, and all other terms on the other side. If necessary, use the distributive property to write the terms with the specified variable as the product of that variable and a sum of terms. Complete the solution.

Solve for h: $A = \frac{1}{2}bh$.

$$A = \frac{1}{2}bh$$
$$2A = 2\left(\frac{1}{2}bh\right)$$
$$2A = bh$$
$$\frac{2A}{b} = h$$

2.3   APPLICATIONS OF LINEAR EQUATIONS

Solving an Applied Problem
Step 1   Determine what you are asked to find.

Step 2   Write down any other pertinent information.

How many liters of 30% alcohol solution and 80% alcohol solution must be mixed to obtain 100 liters of 50% alcohol solution?

Let x = number of liters of 30% solution needed;

$100 - x$ = number of liters of 80% solution needed.

The information of the problem is summarized in the chart that follows.

Liters	Concentration	Liters of Pure Alcohol
x	.30	.30x
$100 - x$	.80	.80$(100 - x)$
100	.50	.50(100)

CONCEPTS	EXAMPLES
Step 3 Write an equation.	The equation is $$.30x + .80(100 - x) = .50(100).$$
Step 4 Solve the equation. *Step 5* Answer the question(s) of the problem. *Step 6* Check.	Solving this equation gives $x = 60$. 60 liters of 30% alcohol and $100 - 60 = 40$ liters of 80% alcohol should be used. Check this result.

2.4   MORE APPLICATIONS OF LINEAR EQUATIONS

To solve a uniform motion problem, draw a sketch and make a chart. Use the formula $d = rt$.	Two cars start from towns 400 miles apart and travel toward each other. They meet after 4 hours. Find the speed of each car if one travels 20 miles per hour faster than the other. Let $\quad\quad x =$ speed of the slower car in miles per hour; $\quad x + 20 =$ speed of the faster car. Use the information in the problem, and $d = rt$ to complete the chart.

	r	t	d
Slower car	x	4	$4x$
Faster car	$x + 20$	4	$4(x + 20)$

A sketch shows that the sum of the distances, $4x$ and $4(x + 20)$, must be 400.

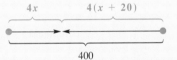

The equation is

$$4x + 4(x + 20) = 400.$$

Solving this equation gives $x = 40$. The slower car travels 40 miles per hour and the faster car travels $40 + 20 = 60$ miles per hour.

2.5   LINEAR INEQUALITIES IN ONE VARIABLE

Solving a Linear Inequality Simplify each side separately, combining like terms and removing parentheses. Use the addition property of inequality to get the variables on one side of the inequality sign and the numbers on the other. Combine like terms, and then use the multiplication property to change the inequality to the form $x < k$ or $x > k$. If an inequality is multiplied or divided by a *negative* number, the inequality symbol *must be reversed.*	Solve $3(x + 2) - 5x \le 12$. $$3x + 6 - 5x \le 12$$ $$-2x + 6 \le 12$$ $$-2x \le 6$$ $$\frac{-2x}{-2} \ge \frac{6}{-2}$$ $$x \ge -3$$ The solution set is $[-3, \infty)$ and is graphed below. 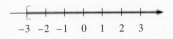

CONCEPTS	EXAMPLES

2.6 SET OPERATIONS AND COMPOUND INEQUALITIES

Solving a Compound Inequality Solve each inequality in the compound inequality individually. If the inequalities are joined with *and,* the solution is the intersection of the two individual solutions. If the inequalities are joined with *or,* the solution is the union of the two individual solutions.	Solve $x + 1 > 2$ and $2x < 6$. $\qquad x + 1 > 2 \qquad$ and $\qquad 2x < 6$ $\qquad\quad x > 1 \qquad$ and $\qquad\ x < 3$ 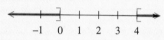 The solution set is $(1, 3)$. Solve $x \geq 4 \qquad$ or $\qquad x \leq 0$. The solution set is $(-\infty, 0] \cup [4, \infty)$. 

2.7 ABSOLUTE VALUE EQUATIONS AND INEQUALITIES

Let k be a positive number. To solve $\lvert ax + b \rvert = k$, solve the compound equation $\qquad ax + b = k \quad$ or $\quad ax + b = -k$.	Solve $\lvert x - 7 \rvert = 3$. $\qquad x - 7 = 3 \qquad$ or $\qquad x - 7 = -3$ $\qquad\quad x = 10 \qquad\qquad\qquad x = 4$ The solution set is $\{4, 10\}$. 
To solve $\lvert ax + b \rvert > k$, solve the compound inequality $\qquad ax + b > k \qquad$ or $\qquad ax + b < -k$.	Solve $\lvert x - 7 \rvert > 3$. $\qquad x - 7 > 3 \qquad$ or $\qquad x - 7 < -3$ $\qquad\quad x > 10 \qquad$ or $\qquad\quad x < 4$ The solution set is $(-\infty, 4) \cup (10, \infty)$. 
To solve $\lvert ax + b \rvert < k$, solve the compound inequality $\qquad -k < ax + b < k$.	Solve $\lvert x - 7 \rvert < 3$. $\qquad -3 < x - 7 < 3$ $\qquad\quad 4 < x < 10 \qquad$ Add 7. The solution set is $(4, 10)$. 
To solve an absolute value equation of the form $\qquad \lvert ax + b \rvert = \lvert cx + d \rvert$ solve the compound equation $\qquad ax + b = cx + d \qquad$ or $\qquad ax + b = -(cx + d)$.	Solve $\lvert x + 2 \rvert = \lvert 2x - 6 \rvert$. $\quad x + 2 = 2x - 6 \qquad$ or $\qquad x + 2 = -(2x - 6)$ $\qquad\quad x = 8 \qquad\qquad$ or $\qquad\qquad x = \dfrac{4}{3}$ The solution set is $\left\{ \dfrac{4}{3}, 8 \right\}$.

CHAPTER 2 REVIEW EXERCISES

[2.1] *Solve the equation.*

1. $-(8 + 3y) + 5 = 2y + 6$

2. $-(r + 5) - (2 + 7r) + 8r = 3r - 8$

3. $\dfrac{m - 2}{4} + \dfrac{m + 2}{2} = 8$

4. $\dfrac{2q + 1}{3} - \dfrac{q - 1}{4} = 0$

5. $5(2x - 3) = 6(x - 1) + 4x$

6. $-3x + 2(4x + 5) = 10$

7. $-\dfrac{3}{4}x = -12$

8. $.05x + .03(1200 - x) = 42$

Decide whether the given equation is conditional, an identity, or a contradiction. Give the solution set.

9. $7r - 3(2r - 5) + 5 + 3r = 4r + 20$

10. $8p - 4p - (p - 7) + 9p + 6 = 12p - 7$

11. $-2r + 6(r - 1) + 3r - (4 - r) = -(r + 5) - 5$

12. $-(2y - 5) - y + 7(y + 2) = 6y + 14 - (2y - 5)$

[2.2] *Solve the formula for the specified variable.*

13. $V = LWH$;   for H

14. $A = \dfrac{1}{2}(B + b)h$;   for b

15. $C = \pi d$;   for d

Solve the equation for x.

16. $Q = 2x + 3y - 4z$

17. $M = -\dfrac{1}{4}(x + 3y)$

18. $P = \dfrac{3}{4}x - 12$

Find the unknown value.

19. Two sides of a triangle measure 4.6 and 12.2 centimeters. The perimeter is 28.4 centimeters. Find the measure of the third side.

20. A newspaper recycling collection bin is in the shape of a box, 1.5 feet wide and 5 feet long. If the volume of the bin is 75 cubic feet, find the height.

21. The area of a mural that is in the shape of a rectangle is 132 square feet. Its length is 16.5 feet. What is its width?

22. If Kyung Ho deposits $1200 at 3% simple annual interest, how long will it take for the deposit to earn $126?

23. Find the simple interest rate that Francesco Castellucio is getting, if a principal of $30,000 earns $7800 interest in 4 years.

24. On March 12, 1990, the high temperature in Morganza, Louisiana, was 41 degrees Fahrenheit. Find the corresponding Celsius temperature.

25. Water boils at 212 degrees Fahrenheit. Find the corresponding Celsius temperature.

26. If a child has a fever of 40 degrees Celsius, what is the child's temperature in Fahrenheit?

27. The lowest temperature ever recorded in Arizona was -40 degrees Celsius, on January 7, 1971. What was the corresponding Fahrenheit temperature?

1. $\left\{-\dfrac{9}{5}\right\}$ **2.** $\left\{\dfrac{1}{3}\right\}$

3. $\{10\}$ **4.** $\left\{-\dfrac{7}{5}\right\}$ **5.** 0

6. $\{0\}$ **7.** $\{16\}$
8. $\{300\}$ **9.** identity; $(-\infty, \infty)$
10. contradiction;   0
11. conditional; $\{0\}$
12. identity; $(-\infty, \infty)$

13. $H = \dfrac{V}{LW}$

14. $b = \dfrac{2A - Bh}{h}$ or

$b = \dfrac{2A}{h} - B$ **15.** $d = \dfrac{C}{\pi}$

16. $x = \dfrac{Q - 3y + 4z}{2}$

17. $x = -4M - 3y$

18. $x = \dfrac{4}{3}(P + 12)$ or

$x = \dfrac{4}{3}P + 16$ **19.** 11.6

centimeters   **20.** 10 feet
21. 8 feet   **22.** 3.5 years
23. 6.5%   **24.** 5°C
25. 100°C   **26.** 104°F
27. −40°F

RESOURCES FOR REVIEW

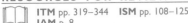 **ITM** pp. 319–344   **ISM** pp. 108–125
IAM p. 8

 **SSG** pp. 22–58
SSM pp. 63–80

 **TEST GENERATOR**
DOS   Windows   MAC

 ◎ CONCEPTUAL   ☑ WRITING   ▲ CHALLENGING   ▦ SCIENTIFIC CALCULATOR   ▦ GRAPHICS CALCULATOR

28. 7.3%   29. 66.6%
30. $22.50
31. 105°; 105°
32. 120°; 60°
33. 37.4%

28. For a 10-day period in early June 1993, projected sales of U.S.-built cars and trucks were 11.7 million, up from 10.9 million a year earlier. What percent increase does this represent?

29. In 1975 there were 161,927 federally licensed firearm dealers in the United States. In 1993 there were 269,712 such dealers. What percent increase does this represent? (*Source:* Bureau of Alcohol, Tobacco, and Firearms)

30. A loan has a finance charge of $450. The loan was scheduled to run for 24 months. Find the unearned interest if the loan is paid off with 5 payments left.

31. Find the measure of each marked angle.

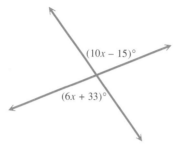

$(10x - 15)°$

$(6x + 33)°$

32. Find the measure of each marked angle, given that lines m and n are parallel.

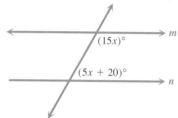

$(15x)°$ m

$(5x + 20)°$ n

33. The economic reform plan proposed by President Clinton in 1993 was projected to save $108 billion by the end of 1999. The accompanying chart shows the sources of savings and how much would be saved. What percent of the total savings is represented by staff cuts?*

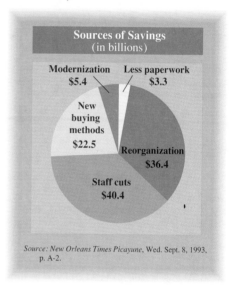

Sources of Savings
(in billions)

Modernization
$5.4

Less paperwork
$3.3

New buying methods
$22.5

Reorganization
$36.4

Staff cuts
$40.4

Source: New Orleans Times Picayune, Wed. Sept. 8, 1993, p. A-2.

34. The formula for the area of a trapezoid is $A = \frac{1}{2}h(B + b)$. Suppose that you know the values of B, b, and h. Write *in words* the procedure you would use to find A. It may begin as follows: "To find the area of this trapezoid, I would first add . . . ."

---

* Graph for Exercise 33, "Sources of Savings," from *The Times-Picayune*, September 8, 1993, p. A-2. Copyright © 1993 by The Times-Picayune. Reprinted by permission.

[2.3] *Write the word phrase as a mathematical expression, using x as the variable.*

35. three times a number

36. the difference between 9 and twice a number

37. half a number, added to 5

38. the product of 4 and a number, subtracted from 8

Solve the problem.

39. When 9 is added to three times a number, the result is 15. Find the number.

40. The product of 6, and a number decreased by 2, is 18. Find the number.

41. After a number is increased by 20%, the result is 36. Find the number.

42. A number is decreased by 35%, giving 260. Find the number.

43. A candy clerk has three times as many kilograms of chocolate creams as peanut clusters. The clerk has 48 kilograms of the two candies altogether. How many kilograms of peanut clusters does the clerk have?

44. Kevin Connors invested some money at 6% and $4000 less than this amount at 4%. Find the amount invested at each rate if his total annual interest income is $840.

45. How many liters of a 20% solution of a chemical should be mixed with 15 liters of a 50% solution to get a mixture that is 30% chemical?

46. According to the Wilderness Society, in the early 1990s there were 1631 breeding pairs of northern spotted owls in California and Washington. California had 289 more pairs than Washington. How many pairs were there in each state?

47. The supplement of an angle measures 25° more than twice its complement. Find the measure of the angle.

48. The complement of an angle measures 10° less than one-fifth of its supplement. Find the measure of the angle.

49. Find the measure of each angle in the triangle.

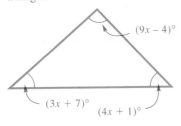

50. Find the measure of each marked angle.

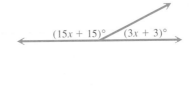

[2.4] *Solve the problem.*

51. A total of 1096 people attended a Garth Brooks concert. Reserved seats cost $25 each and general admission seats cost $20 each. If $26,170 was collected, how many of each type of seat were sold?

52. There were 311 tickets sold for a soccer game, some for students and some for nonstudents. Student tickets cost 25¢ each and nonstudent tickets cost 75¢ each. The total receipts were $108.75. How many of each type of ticket were sold?

53. Suzanne Gainey wishes to mix 30 pounds of candy worth $6 per pound with candy worth $3 per pound to get a mixture worth $5 per pound. How much of the $3 candy should be used?

35. $3x$ 36. $9 - 2x$
37. $5 + \frac{1}{2}x$ 38. $8 - 4x$
39. 2   40. 5   41. 30
42. 400   43. 12 kilograms
44. $10,000 at 6%;
$6000 at 4%   45. 30 liters
46. California: 960 pairs;
Washington: 671 pairs
47. 25°   48. 80°
49. 40°; 45°; 95°
50. 150°; 30°   51. 850
reserved;   246 general
admission   52. 249
students;   62 nonstudents
53. 15 pounds

54. 50 kilometers per hour;
65 kilometers per hour
55. 6 inches; 12 inches;
16 inches 56. length: 13
meters; width: 8 meters
57. 6 inches
58. approximately 70,710
square feet 59. 52, 53
60. 46, 47, 48
61. $(-9, \infty)$
62. $(-\infty, -3]$
63. $\left(\dfrac{3}{2}, \infty\right)$
64. $\left(-\infty, -\dfrac{14}{9}\right)$
65. $[-3, \infty)$ 66. $[-3, 12]$
67. $[3, 5)$ 68. $\left(-3, \dfrac{7}{2}\right)$
69. $\left(\dfrac{59}{31}, \infty\right)$
70. $\left(-\infty, \dfrac{14}{17}\right)$
71. $(-\infty, 1]$
72. any grade greater than
or equal to 61% 73. 87
points
75. {a, c} 76. {a}
77. {a, c, e, f, g}
78. {a, b, c, d, e, f, g}
79. $(-\infty, 3)$

80. (6, 8)

81. $\emptyset$
82. $(-\infty, -1] \cup (5, \infty)$

83. $(-\infty, \infty)$

84. $(-\infty, -2] \cup [7, \infty)$

85. $(-3, 4)$ 86. $(-\infty, 2)$
87. $(4, \infty)$ 88. $(1, \infty)$

---

54. Two cars leave towns 230 kilometers apart at the same time, traveling directly toward one another. One car travels 15 kilometers per hour slower than the other. They pass one another 2 hours later. What are their speeds?

55. The perimeter of a triangle is 34 inches. The middle side is twice as long as the shortest side. The longest side is 2 inches less than three times the shortest side. Find the lengths of the three sides.

56. The length of a rectangle is 3 meters less than twice the width. The perimeter of the rectangle is 42 meters. Find the length and the width of the rectangle.

57. A square is such that if each side were increased by 4 inches, the perimeter would be 8 inches less than twice the perimeter of the original square. Find the length of a side of the original square.

58. The world's largest painting shows over 800 species of the animal kingdom. It is 1.86 miles long and 7.2 feet wide. What is the area of the painting in square feet? (*Hint:* 1 mile = 5280 feet)

59. The sum of two consecutive integers is 105. What are the integers?

60. The sum of the smallest and largest of three consecutive integers is 47 more than the middle integer. What are the integers?

[2.5] *Solve the inequality. Express the solution set in interval form.*

61. $-\dfrac{2}{3}k < 6$

62. $-5x - 4 \geq 11$

63. $\dfrac{6a + 3}{-4} < -3$

64. $\dfrac{9y + 5}{-3} > 3$

65. $5 - (6 - 4k) \geq 2k - 7$

66. $-6 \leq 2k \leq 24$

67. $8 \leq 3y - 1 < 14$

68. $-4 < 3 - 2k < 9$

69. $\dfrac{5}{3}(m - 2) + \dfrac{2}{5}(m + 1) > 1$

70. $\dfrac{3}{4}(a - 2) - \dfrac{1}{3}(5 - 2a) < -2$

71. $-.3x + 2.1(x - 4) \leq -6.6$

72. To pass algebra, a student must have an average of at least 70% on five tests. On the first four tests, a student has grades of 75%, 79%, 64%, and 71%. What possible grades on the fifth test would guarantee a passing grade in the class?

73. Samantha has a total of 815 points so far in her algebra class. At the end of the course she must have 82% of the 1100 points possible in order to get a B. What is the lowest score she can earn on the 100-point final to get a B in the class?

74. While solving the inequality $10x + 2(x - 4) < 12x - 13$, a student did all the work correctly and obtained the statement $-8 < -13$. The student did not know what to do at this point, because the variable "disappeared." How would you explain to the student the interpretation of this result?

[2.6] *Let* $A = \{a, b, c, d\}$, $B = \{a, c, e, f\}$, *and* $C = \{a, e, f, g\}$. *Find the set.*

75. $A \cap B$ **76.** $A \cap C$ **77.** $B \cup C$ **78.** $A \cup C$

Solve the compound inequality. Graph the solution.

79. $x \leq 4$ and $x < 3$

80. $x > 6$ and $x < 8$

81. $x + 4 > 12$ and $x - 2 < 1$

82. $x > 5$ or $x \leq -1$

83. $x - 4 > 6$ or $x + 3 \leq 18$

84. $-5x + 1 \geq 11$ or $3x + 5 \geq 26$

Express the union or intersection in the simplest interval form.

85. $(-3, \infty) \cap (-\infty, 4)$

86. $(-\infty, 6) \cap (-\infty, 2)$

87. $(4, \infty) \cup (9, \infty)$

88. $(1, 2) \cup (1, \infty)$

According to figures provided by the Equal Employment Opportunity Commission, Bureau of Labor Statistics, the following are the median weekly earnings of full-time workers by occupation for men and women.

Occupation	Men	Women
Managerial and professional specialty	$753	$527
Mathematical and computer scientists	$923	$707
Waiters and waitresses	$281	$205
Bus drivers	$411	$321

⊙ *Give the occupation that satisfies the description.*

89. The median earnings for men are less than $900 *and* for women are greater than $500.

90. The median earnings for men are greater than $900 *or* for women are greater than $600.

[2.7] *Solve the absolute value equation.*

91. $|x| = 7$

92. $|y + 2| = 9$

93. $|3k - 7| = 8$

94. $|z - 4| = -12$

95. $|2k - 7| + 4 = 11$

96. $|4a + 2| - 7 = -3$

97. $|3p + 1| = |p + 2|$

98. $|2m - 1| = |2m + 3|$

Solve the absolute value inequality. Give the solution set in interval form.

99. $|p| < 14$

100. $|-y + 6| \leq 7$

101. $|2p + 5| \leq 1$

102. $|x + 1| \geq -3$

103. $|5r - 1| > 9$

104. $|3k + 6| \geq 0$

105. $|11x - 3| \leq -2$

106. $|11x - 3| \leq 0$

⊙ **107.** Write an inequality that states that the distance between x and 14 is greater than 12.

⊙ **108.** The solution set of $|3x + 4| = 7$ is shown on the number line.
   (a) What is the solution set of $|3x + 4| \geq 7$?
   (b) What is the solution set of $|3x + 4| \leq 7$?

MIXED REVIEW EXERCISES*

Solve.

109. $5 - (6 - 4k) > 2k - 5$

110. $S = 2HW + 2LW + 2LH$;   for L

111. $x < 3$ and $x \geq -2$

112. $-4(3 + 2m) - m = -3m$

113. A rectangle has a perimeter of 46 centimeters. The width is 8 centimeters. Find the length.

114. $-6z \leq 72$

---

* The order of exercises in this final group does not correspond to the order in which topics occur in the chapter. This random ordering should help you prepare for the chapter test.

Answers (right column):

89. managerial and professional specialty
90. mathematical and computer scientists
91. $\{-7, 7\}$ 92. $\{-11, 7\}$
93. $\left\{-\dfrac{1}{3}, 5\right\}$ 94. ∅
95. $\{0, 7\}$ 96. $\left\{-\dfrac{3}{2}, \dfrac{1}{2}\right\}$
97. $\left\{-\dfrac{3}{4}, \dfrac{1}{2}\right\}$ 98. $\left\{-\dfrac{1}{2}\right\}$
99. $(-14, 14)$
100. $[-1, 13]$
101. $[-3, -2]$
102. $(-\infty, \infty)$
103. $\left(-\infty, -\dfrac{8}{5}\right) \cup (2, \infty)$
104. $(-\infty, \infty)$
105. ∅   106. $\left\{\dfrac{3}{11}\right\}$
107. $|x - 14| > 12$ (or $|14 - x| > 12$)
108. (a) $\left(-\infty, -\dfrac{11}{3}\right] \cup [1, \infty)$ (b) $\left[-\dfrac{11}{3}, 1\right]$
109. $(-2, \infty)$
110. $L = \dfrac{S - 2HW}{2W + 2H}$
111. $[-2, 3)$ 112. $\{-2\}$
113. 15 centimeters
114. $[-12, \infty)$

 CONCEPTUAL   WRITING   CHALLENGING   SCIENTIFIC CALCULATOR    GRAPHICS CALCULATOR

115. $\left(-\infty, -\dfrac{13}{5}\right) \cup (3, \infty)$

116. $(-\infty, \infty)$

117. 5 liters

118. $\left\{-4, -\dfrac{2}{3}\right\}$

119. {30}   **120.** $[-4, -2]$

121. $\left\{1, \dfrac{11}{3}\right\}$

122. Reagan: 525 votes;   Mondale: 13 votes

123. ∅   **124.** 80°; 80°

125. $\dfrac{21}{2} = 10.5$;

$0.5x + .25(50 - x) = 10.5$

126. 10^2 or 100

127. {10}   ++++•++++→
                      10

128. $(-\infty, 10)$

←+++++)+++→
        10

129. $(10, \infty)$

++++++(++++→
        10

130. $(-\infty, \infty)$ **131.** ∅

132. The union will be $(-\infty, \infty)$ and the intersection will be ∅.

---

115. $|5r - 1| > 14$

116. $x \geq -2$ or $x < 4$

117. How many liters of a 20% solution of a chemical should be mixed with 10 liters of a 50% solution to get a 40% mixture?

118. $|m - 1| = |2m + 3|$

119. $\dfrac{3y}{5} - \dfrac{y}{2} = 3$

120. $|m + 3| \leq 1$

121. $|3k - 7| = 4$

122. In the 1984 presidential election, Ronald Reagan and Walter Mondale together received 538 electoral votes. Reagan received 512 more votes than Mondale in the landslide. How many votes did each man receive?

123. $5(2x - 7) = 2(5x + 3)$

124. Find the measure of each marked angle.

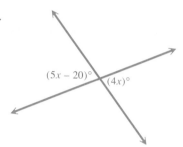

$(5x - 20)°$ $(4x)°$

◆ **MATHEMATICAL CONNECTIONS** (Exercises 125–132) ◆

Work these exercises in order.

Consider the following equation.

$$.05x + .25(50 - x) = \frac{21}{2}$$

125. Convert the fraction $\dfrac{21}{2}$ to a decimal numeral, and write the equation with its decimal equivalent.

126. By what power of 10 should you multiply both sides so that all coefficients are whole numbers?

127. Solve the equation, and graph the solution on a number line.

128. Solve the inequality $.05x + .25(50 - x) > \dfrac{21}{2}$, and graph its solution set on a number line.

129. Solve the inequality $.05x + .25(50 - x) < \dfrac{21}{2}$, and graph its solution set on a number line.

130. What is the *union* of the solution sets found in Exercises 127, 128, and 129?

131. Choose any two of the three solution sets found in Exercises 127, 128, and 129. What is the intersection of these two sets?

◉ **132.** Suppose that you start with any linear equation that has a single solution. Then you solve the two inequalities associated with that equation, one with < and the other with >. Based on your observations so far, what might you conclude about the solution sets?

CHAPTER 2 TEST

Solve the equation.

1. $3(2y - 2) - 4(y + 6) = 3y + 8 + y$

2. $.08x + .06(x + 9) = 1.24$

3. $\dfrac{x + 6}{10} + \dfrac{x - 4}{15} = \dfrac{x + 2}{6}$

4. Decide whether the equation

$$3x - (2 - x) + 4x + 2 = 8x + 3$$

is *conditional,* an *identity,* or a *contradiction.* Give its solution set.

5. Solve for v: $-16t^2 + vt - S = 0$.

Solve the problem.

6. The Daytona 500 (mile) race was shortened to 450 miles in 1974 due to the energy crisis. In that year it was won by Richard Petty, who averaged 140.9 miles per hour. What was Petty's time?

7. A certificate of deposit pays $862.50 in simple interest for one year on a principal of $23,000. What is the rate of interest?

8. In a certain South Dakota county, 6118 residents live in poverty. This represents 63.1% of the population of the county. What is the population of the county?

9. Charles Dawkins invested some money at 3% simple interest and some at 5% simple interest. The total amount of his investments was $28,000, and the interest he earned during the first year was $1240. How much did he invest at each rate?

10. Find the measure of each angle.

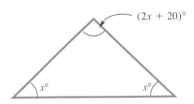

11. The accompanying graph shows the trend in the catch level of haddock during the years 1985 to 1992. If the value of haddock is $1.65 per pound, how much less was the 1992 value of haddock catch than in 1985? (Assume the same price for both years.)

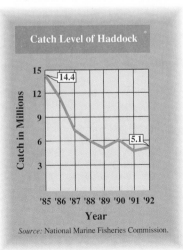

Catch Level of Haddock

Source: National Marine Fisheries Commission.

[2.1] 1. $\{-19\}$ 2. $\{5\}$
3. $(-\infty, \infty)$
4. contradiction; $\emptyset$
[2.2] 5. $v = \dfrac{S + 16t^2}{t}$
[2.3, 2.4] 6. 3.2 hours
7. 3.75%   8. 9696
residents   9. $8000 at
3%; $20,000 at 5%
10. 40°; 40°; 100°
11. $15,345,000

ITM pp. 37–62 **ISM** pp. 125–128
IAM 8–9

SSG pp. 59–61
SSM pp. 81–84

TEST GENERATOR
DOS Windows MAC

 CONCEPTUAL   WRITING   ▲ CHALLENGING   SCIENTIFIC CALCULATOR   GRAPHICS CALCULATOR

[2.5] 12. We must reverse the direction of the inequality symbol.

13. $[1, \infty)$

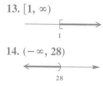

14. $(-\infty, 28)$

15. $[-3, 3]$

16. (c)   17. 82%

[2.6] 18. (a) $\{1, 5\}$
(b) $\{1, 2, 5, 7, 9, 12\}$
19. (a) $[2, 9)$
(b) $(-\infty, 3) \cup [6, \infty)$

[2.7] 20. $\left\{-1, \dfrac{5}{2}\right\}$

21. $(-\infty, -1) \cup \left(\dfrac{5}{2}, \infty\right)$

22. $\left(-1, \dfrac{5}{2}\right)$

23. $\left\{-\dfrac{5}{7}, \dfrac{11}{3}\right\}$

24. $\left(\dfrac{1}{3}, \dfrac{7}{3}\right)$ 25. (a) $\emptyset$
(b) $(-\infty, \infty)$ (c) $\emptyset$

⊙ **12.** What is the special rule that must be remembered when multiplying or dividing both sides of an inequality by a negative number?

Solve the inequality. Give the solution set in both interval and graph forms.

13. $4 - 6(x + 3) \le -2 - 3(x + 6) + 3x$

14. $-\dfrac{4}{7}x > -16$

15. $-6 \le \dfrac{4}{3}x - 2 \le 2$

⊙ **16.** Which one of the following inequalities is equivalent to $x < -3$?
(a) $-3x < 9$ (b) $-3x > -9$ (c) $-3x > 9$ (d) $-3x < -9$

17. A student must have an average grade of at least 80% on the four tests in a course to get a grade of B. The student had grades of 83%, 76%, and 79% on the first three tests. What minimum grade on the fourth test would guarantee a B in the course?

18. Let $A = \{1, 2, 5, 7\}$ and $B = \{1, 5, 9, 12\}$. Find
(a) $A \cap B$ (b) $A \cup B$.

19. Solve the compound inequality.
(a) $3k \ge 6$ and $k - 4 < 5$ (b) $-4x \le -24$ or $4x - 2 < 10$

Solve the absolute value equation or inequality.

20. $|4x - 3| = 7$ **21.** $|4x - 3| > 7$

22. $|4x - 3| < 7$ **23.** $|3 - 5x| = |2x + 8|$

24. $|-3x + 4| - 4 < -1$

⊙ **25.** If $k < 0$, what is the solution set of
(a) $|5x + 3| < k$ (b) $|5x + 3| > k$ (c) $|5x + 3| = k$?

CUMULATIVE REVIEW (Chapters 1–2)

1. 9, 6   2. 0, 9, 6   3. $-8,$
0, 9, 6   4. $-8, -\dfrac{2}{3}, 0, \dfrac{4}{5},$
9, 6   5. $-\sqrt{6}$ 6. All are
real numbers.   7. $-\dfrac{22}{21}$
8. 8   9. 8   10. 0
11. -243 12. $\dfrac{216}{343}$
13. $-\dfrac{8}{27}$ 14. -4096
15. $\sqrt{-36}$ is not a real
number.   16. $\dfrac{4 + 4}{4 - 4}$ is
undefined.

Beginning with this chapter each chapter in the text will conclude with a set of cumulative review exercises designed to cover the major topics from the beginning of the course. This feature allows the student to contantly review topics that have been introduced up to that point.

Let $A = \left\{-8, -\dfrac{2}{3}, -\sqrt{6}, 0, \dfrac{4}{5}, 9, \sqrt{36}\right\}$. *Simplify the elements of A as necessary and then list the elements that belong to the set.*

1. natural numbers   **2.** whole numbers   **3.** integers

4. rational numbers   **5.** irrational numbers   **6.** real numbers

Add or subtract, as indicated.

7. $-\dfrac{4}{3} - \left(-\dfrac{2}{7}\right)$ **8.** $|-4| - |2| + |-6|$

9. $(-2)^4 + (-2)^3$ **10.** $\sqrt{25} - \sqrt[3]{125}$

Evaluate the expression.

11. $(-3)^5$ **12.** $\left(\dfrac{6}{7}\right)^3$ **13.** $\left(-\dfrac{2}{3}\right)^3$ **14.** -4^6

⊙**15.** Which one of the following is not a real number: $-\sqrt{36}$ or $\sqrt{-36}$?

⊙**16.** Which one of the following is undefined: $\dfrac{4 - 4}{4 + 4}$ or $\dfrac{4 + 4}{4 - 4}$?

Evaluate if a = 2, b = −3, and c = 4.

17. $-3a + 2b - c$ **18.** $-2b^2 - 4c$ **19.** $-8(a^2 + b^3)$ **20.** $\dfrac{3a^3 - b}{4 + 3c}$

Use the properties of real numbers to simplify the expression.

21. $-7r + 5 - 13r + 12$ **22.** $-(3k + 8) - 2(4k - 7) + 3(8k + 12)$

Identify the property of real numbers illustrated by the equation.

23. $(a + b) + 4 = 4 + (a + b)$ **24.** $4x + 12x = (4 + 12)x$

25. $-9 + 9 = 0$

26. What is the reciprocal, or multiplicative inverse, of $-\dfrac{2}{3}$?

Solve the equation.

27. $-4x + 7(2x + 3) = 7x + 36$ **28.** $-\dfrac{3}{5}x + \dfrac{2}{3}x = 2$

29. $.06x + .03(100 + x) = 4.35$ **30.** $P = a + b + c;$ for b

Solve the inequality. Give the solution set in both interval and graph forms.

31. $3 - 2(x + 7) \le -x + 3$ **32.** $-4 < 5 - 3x \le 0$

33. $2x + 1 > 5$ or $2 - x > 2$ **34.** $|-7k + 3| \ge 4$

Solve the problem.

35. Rita Lieux invested some money at 7% interest and the same amount at 10%. Her total interest for the year was $150 less than one-tenth of the total amount she invested. How much did she invest at each rate?

36. A dietician must use three foods, A, B, and C, in a diet. He must include twice as many grams of food A as food C, and 5 grams of food B. The three foods must total at most 24 grams. What is the largest amount of food C that the dietician can use?

37. Louise James got scores of 88 and 78 on her first two tests. What score must she make on her third test to keep an average of 80 or greater?

38. Jack and Jill are running in the Fresh Water Fun Run. Jack runs at 7 miles per hour and Jill runs at 5 miles per hour. If they start at the same time, how long will it be before Jack is $\dfrac{1}{4}$ mile ahead of Jill?

39. How much pure alcohol should be added to 7 liters of 10% alcohol to increase the concentration to 30% alcohol?

40. A coin collection contains 29 coins. It consists of cents, nickels, and quarters. The number of quarters is 4 less than the number of nickels, and the face value of the collection is $2.69. How many of each denomination are there in the collection?

Clark's rule is a formula used in reducing drug dosage according to weight from the recommended adult dosage to a child dosage. It is as follows.

$$\frac{\text{Weight of child in pounds}}{150} \times \text{adult dose} = \text{child's dose}$$

41. Find a child's dosage if the child weighs 55 pounds and the recommended adult dosage is 120 milligrams.

42. Find a child's dosage if the child weighs 75 pounds and the recommended adult dosage is 40 drops.

Answers:

17. -16
18. -34 19. 184
20. $\dfrac{27}{16}$ 21. $-20r + 17$
22. $13k + 42$
23. commutative property
24. distributive property
25. inverse property
26. $-\dfrac{3}{2}$ 27. {5}
28. {30}   29. {15}
30. $b = P - a - c$
31. $[-14, \infty)$

32. $\left[\dfrac{5}{3}, 3\right)$

33. $(-\infty, 0) \cup (2, \infty)$

34. $\left(-\infty, -\dfrac{1}{7}\right] \cup [1, \infty)$

35. $5000   36. $6\dfrac{1}{3}$ grams

37. 74 or greater

38. $\dfrac{1}{8}$ hour   39. 2 liters

40. 9 cents, 12 nickels, 8 quarters

41. 44 milligrams

42. 20 drops

---

43. (a) $317,000
(b) 55%   44. 25.7

43. According to the accompanying graph,
  (a) by how much did the average payment of catastrophic claims go up between 1988 and 1991?
  (b) by what *percent* did the amount of average payment of the claims increase from 1988 to 1991?

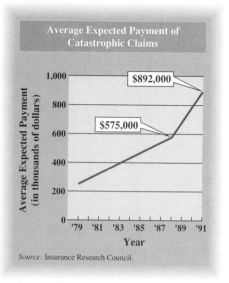

Average Expected Payment of Catastrophic Claims

$892,000

$575,000

Average Expected Payment (in thousands of dollars)

Year

Source: Insurance Research Council.

44. The body-mass index, or BMI, of a person is given by the formula

$$\text{BMI} = \frac{704 \text{ (weight in pounds)}}{(\text{height in inches})^2}.$$

Ken Griffey, Jr. is listed as being 6 feet, 3 inches tall and weighing 205 pounds. What is his BMI?

EXPONENTS AND POLYNOMIALS

CONNECTIONS

The expression $-.006x^4 + .14x^3 - .05x^2 + 2x$ is used to determine the concentration of dye injected into a cardiac patient's bloodstream x seconds after injection. This algebraic expression is called a *polynomial*. Polynomials are used as models in medicine, biology, physics, business, and many other fields. In this chapter we study properties of exponents and operations on polynomials. The ability to factor polynomials is an essential skill for learning and for applying later topics in this book.

3.1 INTEGER EXPONENTS

FOR EXTRA HELP

 SSG pp. 62–64
SSM pp. 90–92

 Video
3

 Tutorial
IBM   MAC

OBJECTIVES

1 ▶ Use the product rule for exponents.
2 ▶ Define a zero exponent.
3 ▶ Define negative exponents.
4 ▶ Use the quotient rule for exponents.

In Chapter 1, we introduced the use of exponents to write products of repeated factors. Now, in this section, we give further definitions of exponents and the rules for finding products and quotients of exponential expressions.

1 ▶ Use the product rule for exponents.

▶ There are several useful rules that simplify work with exponents. For example, the product $2^5 \cdot 2^3$ can be simplified as follows.

$$2^5 \cdot 2^3 = (2 \cdot 2 \cdot 2 \cdot 2 \cdot 2)(2 \cdot 2 \cdot 2) = 2^8$$
$$5 + 3 = 8$$

This result, that products of exponential expressions with the same base are found by adding exponents, is generalized as the **product rule for exponents.**

Product Rule for Exponents

If m and n are natural numbers and a is any real number, then
$$a^m \cdot a^n = a^{m+n}.$$

▶ **TEACHING TIP**
Students should learn to verbalize this rule: "When multiplying powers of like bases, keep the same base and add the exponents." ◀

▶ **TEACHING TIP**
Show the example
$4^2 \cdot 5^3 \neq (4 \cdot 5)^{2+3}$. ◀

To see that the product rule is true, use the definition of an exponent as follows.

$$a^m = \underbrace{a \cdot a \cdot a \dots a}_{a \text{ appears as a factor } m \text{ times}} \qquad a^n = \underbrace{a \cdot a \cdot a \dots a}_{a \text{ appears as a factor } n \text{ times}}$$

From this,
$$a^m \cdot a^n = \underbrace{a \cdot a \cdot a \dots a}_{m \text{ factors}} \cdot \underbrace{a \cdot a \cdot a \dots a}_{n \text{ factors}}$$
$$= \underbrace{a \cdot a \cdot a \dots a}_{(m+n) \text{ factors}}$$
$$a^m \cdot a^n = a^{m+n}.$$

EXAMPLE 1

Using the Product Rule for Exponents

Apply the product rule for exponents in each case.

(a) $3^4 \cdot 3^7 = 3^{4+7} = 3^{11}$ **(b)** $5^3 \cdot 5 = 5^3 \cdot 5^1 = 5^{3+1} = 5^4$

(c) $y^3 \cdot y^8 \cdot y^2 = y^{3+8+2} = y^{13}$ ■

Chalkboard Exercise

Find the product.
$$k^4 k^3 k^6$$
Answer: k^{13}

CAUTION A common error occurs in problems like Example 1(a), when students multiply the bases. Notice that $3^4 \cdot 3^7 \neq 9^{11}$. Remember, when applying the product rule for exponents, keep the *same* base and add the exponents.

INSTRUCTOR'S RESOURCES

 ITM pp. 345–347 **ISM** pp. 134–138
IAM p. 9

 TEST GENERATOR
DOS   Windows   MAC

 TRANSPARENCIES

E X A M P L E 2

Using the Product Rule for Exponents

Find each product.

(a) $(5y^2)(-3y^4)$

Use the associative and commutative properties as necessary to multiply the numbers and multiply the variables.

$$
\begin{aligned}
(5y^2)(-3y^4) &= 5(-3)y^2y^4 \\
&= -15y^{2+4} \\
&= -15y^6
\end{aligned}
$$

(b) $(7p^3q)(2p^5q^2) = 7(2)p^3p^5qq^2 = 14p^8q^3$ ■

2 ▶ Define a zero exponent.

▶ So far we have discussed only positive exponents. Let us consider how we might define a zero exponent. Suppose we multiply 4^2 by 4^0. By the product rule,

$$4^2 \cdot 4^0 = 4^{2+0} = 4^2.$$

For the product rule to hold true, 4^0 must equal 1, and so a^0 is defined as follows for any nonzero real number.

Zero Exponent

If a is any nonzero real number, then

$$a^0 = 1.$$

The symbol 0^0 is undefined.

E X A M P L E 3

Using Zero as an Exponent

Evaluate each expression.

(a) $12^0 = 1$ **(b)** $(-6)^0 = 1$

(c) $-6^0 = -(6^0) = -1$ **(d)** $5^0 + 12^0 = 1 + 1 = 2$

(e) $(8k)^0 = 1$ $(k \neq 0)$ ■

3 ▶ Define negative exponents.

For Example 3

▶ How should we define a negative exponent? Using the product rule again,

$$8^2 \cdot 8^{-2} = 8^{2+(-2)} = 8^0 = 1.$$

This indicates that 8^{-2} is the reciprocal of 8^2. But $\dfrac{1}{8^2}$ is the reciprocal of 8^2, and a number can have only one reciprocal. Thus, we must define 8^{-2} to equal $\dfrac{1}{8^2}$, so negative exponents are defined as follows.

Negative Exponent

For any natural number n and any nonzero real number a,

$$a^{-n} = \frac{1}{a^n}.$$

With this definition, the expression a^n is meaningful for any integer exponent n and any nonzero real number a.

EXAMPLE 4

Using Negative Exponents

■ Write the following expressions with only positive exponents.

(a) $2^{-3} = \dfrac{1}{2^3} = \dfrac{1}{8}$

(b) $3^{-2} = \dfrac{1}{3^2} = \dfrac{1}{9}$

(c) $6^{-1} = \dfrac{1}{6^1} = \dfrac{1}{6}$

(d) $(5z)^{-3} = \dfrac{1}{(5z)^3}, \quad z \neq 0$

(e) $5z^{-3} = 5\left(\dfrac{1}{z^3}\right) = \dfrac{5}{z^3}, \quad z \neq 0$

(f) $(5z^2)^{-3} = \dfrac{1}{(5z^2)^3}, \quad z \neq 0$

(g) $-m^{-2} = -\dfrac{1}{m^2}, \quad m \neq 0$

(h) $(-m)^{-4} = \dfrac{1}{(-m)^4}, \quad m \neq 0$ ■

CAUTION A negative exponent does not necessarily lead to a negative number; negative exponents lead to reciprocals, as shown below.

Expression	Example	
a^{-m}	$3^{-2} = \dfrac{1}{3^2} = \dfrac{1}{9}$	Not negative
$-a^{-m}$	$-3^{-2} = -\dfrac{1}{3^2} = -\dfrac{1}{9}$	Negative

---

EXAMPLE 5

Using Negative Exponents

■ Evaluate each of the following expressions.

(a) $3^{-1} + 4^{-1}$

Since $3^{-1} = \dfrac{1}{3}$ and $4^{-1} = \dfrac{1}{4}$,

$$3^{-1} + 4^{-1} = \dfrac{1}{3} + \dfrac{1}{4} = \dfrac{4}{12} + \dfrac{3}{12} = \dfrac{7}{12}.$$

(b) $5^{-1} - 2^{-1} = \dfrac{1}{5} - \dfrac{1}{2} = \dfrac{2}{10} - \dfrac{5}{10} = -\dfrac{3}{10}$ ■

CAUTION In Example 5, note that

$$3^{-1} + 4^{-1} \neq (3 + 4)^{-1}$$

because

$$3^{-1} + 4^{-1} = \dfrac{7}{12}$$

but

$$(3 + 4)^{-1} = 7^{-1} = \dfrac{1}{7}.$$

Also,

$$5^{-1} - 2^{-1} \neq (5 - 2)^{-1}$$

because

$$5^{-1} - 2^{-1} = -\dfrac{3}{10}$$

but

$$(5 - 2)^{-1} = 3^{-1} = \dfrac{1}{3}.$$

We will see this use of negative exponents again in the next chapter.

EXAMPLE 6 ■
Using Negative Exponents

Evaluate each expression.

(a) $\dfrac{1}{2^{-3}} = \dfrac{1}{\dfrac{1}{2^3}} = 1 \div \dfrac{1}{2^3} = 1 \cdot \dfrac{2^3}{1} = 2^3 = 8$

(b) $\dfrac{1}{3^{-2}} = \dfrac{1}{\dfrac{1}{3^2}} = 3^2 = 9$

(c) $\dfrac{2^{-3}}{3^{-2}} = \dfrac{\dfrac{1}{2^3}}{\dfrac{1}{3^2}} = \dfrac{1}{2^3} \cdot \dfrac{3^2}{1} = \dfrac{3^2}{2^3} = \dfrac{9}{8}$ ■

Example 6 suggests the following generalizations.

Special Rules for Negative Exponents

If $a \neq 0$ and $b \neq 0$, $\qquad \dfrac{1}{a^{-n}} = a^n \qquad$ and $\qquad \dfrac{a^{-n}}{b^{-m}} = \dfrac{b^m}{a^n}$.

4 ▶ Use the quotient rule for exponents.

▶ A quotient, such as $\dfrac{a^8}{a^3}$, can be simplified in much the same way as a product. (In all quotients of this type, assume that the denominator is not zero.) Using the definition of an exponent,

$$\dfrac{a^8}{a^3} = \dfrac{a \cdot a \cdot a \cdot a \cdot a \cdot a \cdot a \cdot a}{a \cdot a \cdot a}$$

$$= a \cdot a \cdot a \cdot a \cdot a$$

$$= a^5.$$

Notice that $8 - 3 = 5$. In the same way,

$$\dfrac{a^3}{a^8} = \dfrac{a \cdot a \cdot a}{a \cdot a \cdot a \cdot a \cdot a \cdot a \cdot a \cdot a}$$

$$\dfrac{a^3}{a^8} = \dfrac{1}{a^5}.$$

Here again, $8 - 3 = 5$. These examples suggest the following quotient rule for exponents.

Quotient Rule for Exponents

If a is any nonzero real number and m and n are integers, then

$$\dfrac{a^m}{a^n} = a^{m-n}.$$

EXAMPLE 7
Using the Quotient Rule for Exponents

Apply the quotient rule for exponents in each case.

Numerator exponent

Denominator exponent

(a) $\dfrac{3^7}{3^2} = 3^{7-2} = 3^5$

Minus sign

(b) $\dfrac{p^6}{p^2} = p^{6-2} = p^4 \quad (p \neq 0)$ **(c)** $\dfrac{12^{10}}{12^9} = 12^{10-9} = 12^1 = 12$

(d) $\dfrac{7^4}{7^6} = 7^{4-6} = 7^{-2} = \dfrac{1}{7^2}$ **(e)** $\dfrac{k^7}{k^{12}} = k^{7-12} = k^{-5} = \dfrac{1}{k^5} \quad (k \neq 0)$ ■

EXAMPLE 8
Using the Quotient Rule for Exponents

Write each quotient using only positive exponents.

(a) $\dfrac{2^7}{2^{-3}} = 2^{7-(-3)}$

Since $7 - (-3) = 10$, $\dfrac{2^7}{2^{-3}} = 2^{10}$.

(b) $\dfrac{8^{-2}}{8^5} = 8^{-2-5} = 8^{-7} = \dfrac{1}{8^7}$

(c) $\dfrac{6^{-5}}{6^{-2}} = 6^{-5-(-2)} = 6^{-3} = \dfrac{1}{6^3}$

(d) $\dfrac{6}{6^{-1}} = \dfrac{6^1}{6^{-1}} = 6^{1-(-1)} = 6^2$

(e) $\dfrac{z^{-5}}{z^{-8}} = z^{-5-(-8)} = z^3 \quad (z \neq 0)$

(f) $\dfrac{q^{5r}}{q^{3r}} = q^{5r-3r} = q^{2r}$,   if r is an integer and $q \neq 0$ ■

▶ **TEACHING TIP**
Students have difficulty using the quotient rule with negative exponents. You may wish to show that Example 8(e) can also be worked by first converting to positive exponents and then using the quotient rule.

Be sure that an exponent of 1 is included when no exponent is shown, as in Example 8(d). ◀

CAUTION As seen in Example 8, we must be very careful when working with quotients that involve negative exponents in the denominator. Always be sure to write the numerator exponent, then a minus sign, and then the denominator exponent.

As suggested by Example 8(f), the rules for exponents also apply with variable exponents, as long as the variables used as exponents represent integers. The product and quotient rules for exponents are summarized below.

Summary of Product and Quotient Rules

When multiplying expressions such as a^m and a^n where the base is the same, keep the same base and add the exponents. When dividing such expressions, keep the same base and subtract the exponent of the denominator from the exponent of the numerator.

3.1 EXERCISES

Use the product and/or quotient rules as needed to simplify the expression. Write the answer with only positive exponents. Assume that all variables represent nonzero real numbers. See Examples 1, 2, 4, 7, and 8.

1. $a^5 \cdot a^3$

2. $x^{15} \cdot x^4$

3. $y^5 \cdot y^4 \cdot y^{-3}$

4. $k^3 \cdot k^9 \cdot k^{-8}$

5. $(9x^2y^3)(-2x^3y^5)$

6. $(-3x^5y^4)(-5xy^2)$

7. $\dfrac{p^{19}}{p^5}$

8. $\dfrac{q^{13}}{q^7}$

9. $\dfrac{z^{-6}}{z^{-12}}$

10. $\dfrac{r^{-4}}{r^{-9}}$

11. $\dfrac{r^{13}r^{-4}}{r^{-2}r^{-5}}$

12. $\dfrac{z^{-4}z^{-2}}{z^3z^{-1}}$

13. $7k^2(-2k)(4k^{-5})$

14. $3a^2(-5a^{-6})(-2a)$

15. $-4(2x^3)(3x)$

16. $6(5z^3)(2zw^2)$

17. $\dfrac{(3pq)q^2}{6p^2q^4}$

18. $\dfrac{(-8xy)(y^3)}{4x^5y^4}$

19. $\dfrac{6x^{-5}y^{-2}}{(3x^{-3})(2x^{-2}y^{-2})}$

20. $\dfrac{-8(x^2y^{-4})}{(4xy^{-5})(-2xy)}$

◉ **21.** Your friend evaluated $4^5 \cdot 4^2$ as 16^7. Explain to him why his answer is incorrect.

◉ **22.** Consider the expressions $-a^n$ and $(-a)^n$. In some cases they are equal and in some cases they are not. Using $n = 2, 3, 4, 5,$ and 6 and $a = 2$, draw a conclusion as to when they are equal and when they are opposites.

Evaluate. See Examples 3–8.

23. $\left(\dfrac{2}{3}\right)^2$

24. $\left(\dfrac{4}{3}\right)^3$

25. 4^{-3}

26. 5^{-2}

27. -4^{-3}

28. -5^{-2}

29. $(-4)^{-3}$

30. $(-5)^{-2}$

31. $\dfrac{1}{3^{-2}}$

32. $\dfrac{1}{6^{-1}}$

33. $\dfrac{-3^{-1}}{4^{-2}}$

34. $\dfrac{2^{-3}}{-3^{-1}}$

35. $\left(\dfrac{2}{3}\right)^{-3}$

36. $\left(\dfrac{5}{4}\right)^{-2}$

37. $3^{-1} + 2^{-1}$

38. $4^{-1} + 5^{-1}$

39. $6^{-1} - 4^{-1}$

40. $8^{-1} - 16^{-1}$

41. $(6 - 4)^{-1}$

42. $(8 - 16)^{-1}$

43. $\dfrac{3^{-5}}{3^{-2}}$

44. $\dfrac{2^{-4}}{2^{-3}}$

45. $\dfrac{9^{-1}}{-9}$

46. $\dfrac{8^{-1}}{-8}$

◆ **MATHEMATICAL CONNECTIONS** (Exercises 47–52) ◆

Answer these exercises in order.

47. The expression $\dfrac{25}{25}$ simplifies to _____ .

48. Using exponential notation, $25 = 5^{\underline{\quad}}$. (Fill in the blank with an exponent.)

49. Write $\dfrac{25}{25}$ using the expression you found in Exercise 48, so that it is expressed as a quotient of powers of 5.

Answers column:

1. a^8 2. x^{19} 3. y^6
4. k^4 5. $-18x^5y^8$
6. $15x^6y^6$ 7. p^{14} 8. q^6
9. z^6 10. r^5 11. r^{16}
12. $\dfrac{1}{z^8}$ 13. $-\dfrac{56}{k^2}$ 14. $\dfrac{30}{a^3}$
15. $-24x^4$ 16. $60z^4w^2$
17. $\dfrac{1}{2pq}$ 18. $-\dfrac{2}{x^4}$ 19. 1
20. 1   22. When n is even the expressions are opposites. When n is odd they are equal.
23. $\dfrac{4}{9}$ 24. $\dfrac{64}{27}$ 25. $\dfrac{1}{64}$
26. $\dfrac{1}{25}$ 27. $-\dfrac{1}{64}$
28. $-\dfrac{1}{25}$ 29. $-\dfrac{1}{64}$
30. $\dfrac{1}{25}$ 31. 9   32. 6
33. $-\dfrac{16}{3}$ 34. $-\dfrac{3}{8}$
35. $\dfrac{27}{8}$ 36. $\dfrac{16}{25}$ 37. $\dfrac{5}{6}$
38. $\dfrac{9}{20}$ 39. $-\dfrac{1}{12}$ 40. $\dfrac{1}{16}$
41. $\dfrac{1}{2}$ 42. $-\dfrac{1}{8}$ 43. $\dfrac{1}{27}$
44. $\dfrac{1}{2}$ 45. $-\dfrac{1}{81}$ 46. $-\dfrac{1}{64}$
47. 1   48. 2   49. $\dfrac{5^2}{5^2}$

 CONCEPTUAL   WRITING   CHALLENGING    SCIENTIFIC CALCULATOR   GRAPHICS CALCULATOR

50. 5^0 **51.** $1 = 5^0$ because

they are both equal to $\dfrac{25}{25}$.

52. If $a \ne 0$, $a^0 = 1$.
53. 1   **54.** 1   **55.** -1
56. -1 **57.** 1   **58.** 1
59. 1   **60.** 1   **61.** -2
62. -2 **63.** (c)   **64.** (d)
65. 15   **66.** 12   **67.** $-\dfrac{1}{k^3}$

68. $-\dfrac{1}{r^4}$ **69.** $\dfrac{1}{(3x^4)^2}$

70. $\dfrac{1}{(2y^3)^3}$ **71.** $-\dfrac{12}{r^8}$

72. $-\dfrac{6}{m^7}$ **73.** $\dfrac{1}{5}$ **74.** $\dfrac{1}{8}$

75. -1 **76.** 6   **79.** p^{8q}
80. z^{2m+4} **81.** a^{3r+3}
82. k^{3+7y} **83.** m^a
84. b^{6q-1}
85. $\dfrac{1}{5}$; $\dfrac{5}{6}$; $\dfrac{1}{5} \ne \dfrac{5}{6}$

86. $\dfrac{6}{5}$; 5; $\dfrac{6}{5} \ne 5$

87. 25; 13; $25 \ne 13$
88. 125; 35; $125 \ne 35$

50. Use the quotient rule to simplify the expression you wrote in Exercise 49. Write your answer as a power of 5.

⊙ 51. Compare your answers in Exercises 47 and 50. Explain why they must be equal.

52. What definition from this section does your result support?

Evaluate the expression. Assume that all variables represent nonzero real numbers. See Example 3.

53. 25^0 **54.** 14^0 **55.** -7^0 **56.** -10^0

57. $(-7)^0$ **58.** $(-10)^0$ **59.** $\left(\dfrac{1}{\pi} + \sqrt{2}\right)^0$ **60.** $(-\sqrt[3]{7} - \pi^2)^0$

61. $-4^0 - m^0$ **62.** $-8^0 - k^0$

63. Which one of the following is equal to 1 $(a \ne 0)$?
   (a) $3a^0$ **(b)** $-3a^0$ **(c)** $(3a)^0$ **(d)** $3(-a)^0$

64. Which one of the following does not represent the reciprocal of x $(x \ne 0)$?

   (a) x^{-1} **(b)** $\dfrac{1}{x}$ **(c)** $\left(\dfrac{1}{x^{-1}}\right)^{-1}$ **(d)** $-x$

Use any or all of the ideas of this section to simplify the expression. Do not use negative exponents in your answers. Assume that all variables represent nonzero real numbers. See Examples 1–8.

65. $15^{-16} \cdot 15^{17}$ **66.** $12^{-8} \cdot 12^9$ **67.** $-k^{-3}$ **68.** $-r^{-4}$

69. $(3x^4)^{-2}$ **70.** $(2y^3)^{-3}$ **71.** $-12r^{-8}$ **72.** $-6m^{-7}$

73. $\dfrac{5^{-3}}{5^{-2}}$ **74.** $\dfrac{8^{-4}}{8^{-3}}$ **75.** $\left(\dfrac{1}{2}\right)^{-1} - \left(\dfrac{1}{3}\right)^{-1}$ **76.** $\left(\dfrac{1}{2} - \dfrac{1}{3}\right)^{-1}$

⊙ 77. Consider the expression $\dfrac{1}{25}$. Use the facts that $1 = 5^0$ and $25 = 5^2$, and apply the quotient rule. Now look at your result and write a short paragraph to explain how this exercise supports the definition of negative exponent.

⊙ 78. Your friend thinks that $(-3)^{-2}$ is a negative number. Explain to her why this is incorrect.

▲ *Use the product rule or the quotient rule to simplify the expression. Write all answers without denominators. Assume that all variables used as exponents represent integers and that all other variables represent nonzero real numbers.*

79. $p^q \cdot p^{7q}$ **80.** $z^{m+1} \cdot z^{m+3}$ **81.** $a^r \cdot a^{r+1} \cdot a^{r+2}$

82. $k^{3+y} \cdot k^y \cdot k^{5y}$ **83.** $\dfrac{m^{2a} \cdot m^{3a}}{m^{4a}}$ **84.** $\dfrac{b^{5q} \cdot b^q}{b}$

▲ *Many students believe that both expressions given below represent the same quantity. This* ⊙ *is incorrect. Show that the expressions represent different quantities by replacing x with 2 and y with 3.*

85. $(x + y)^{-1}$; $x^{-1} + y^{-1}$ **86.** $(x^{-1} + y^{-1})^{-1}$; $x + y$
87. $(x + y)^2$; $x^2 + y^2$ **88.** $(x + y)^3$; $x^3 + y^3$

REVIEW EXERCISES

Simplify the expression, writing answers in exponential form. See Section 1.3.

89. $\left(\dfrac{1}{2}\right)^2 \cdot \left(\dfrac{1}{2}\right)^2 \cdot \left(\dfrac{1}{2}\right)^3$

90. $(-2)^3 \cdot (-2)^3$

91. $-(2^3) \cdot (2^3)$

92. $\left(\dfrac{3}{4}\right)^2 \cdot \left(\dfrac{3}{4}\right)^2$

89. $\left(\dfrac{1}{2}\right)^7$
90. $(-2)^6$ or 2^6
91. -2^6
92. $\left(\dfrac{3}{4}\right)^4$ or $\dfrac{3^4}{4^4}$

3.2 FURTHER PROPERTIES OF EXPONENTS

FOR EXTRA HELP	OBJECTIVES
SSG pp. 65–70 **SSM** pp. 92–98	**1** ▶ Use the power rules for exponents.
Video 3	**2** ▶ Simplify exponential expressions.
Tutorial IBM MAC	**3** ▶ Use the rules for exponents with scientific notation.

In the previous section we learned some important rules for working with exponents. In this section we complete the rules for exponents by introducing the power rules.

1 ▶ Use the power rules for exponents.

▶ The expression $(3^4)^2$ can be simplified as $(3^4)^2 = 3^4 \cdot 3^4 = 3^{4+4} = 3^8$, where $4 \cdot 2 = 8$. This example suggests the first of the **power rules for exponents;** the other two parts can be demonstrated with similar examples.

Power Rules for Exponents

If a and b are real numbers and m and n are integers, then

$$(a^m)^n = a^{mn}$$
$$(ab)^m = a^m b^m$$
$$\left(\frac{a}{b}\right)^m = \frac{a^m}{b^m} \quad (b \neq 0).$$

▶ **TEACHING TIP**

Verbalize these rules as follows.

1. To raise a power to a power, multiply exponents.
2. To raise a product to a power, raise each factor to that power.
3. To raise a quotient to a power, raise the numerator and the denominator to that power. ◀

The parts of this rule can be illustrated in the same way as the product rule. In the statements of rules for exponents, we always assume that zero never appears to a negative power.

E X A M P L E   1

Using the Power Rules for Exponents

■ Use a power rule in each case.

(a) $(p^8)^3 = p^{8 \cdot 3} = p^{24}$

(b) $\left(\dfrac{2}{3}\right)^4 = \dfrac{2^4}{3^4} = \dfrac{16}{81}$

(c) $(3y)^4 = 3^4 y^4 = 81 y^4$

(d) $(6p^7)^2 = 6^2 p^{7 \cdot 2} = 6^2 p^{14} = 36 p^{14}$

(e) $\left(\dfrac{-2m^5}{z}\right)^3 = \dfrac{(-2)^3 m^{5 \cdot 3}}{z^3} = \dfrac{(-2)^3 m^{15}}{z^3} = \dfrac{-8m^{15}}{z^3} \quad (z \neq 0)$ ■

Chalkboard Exercise

Simplify using the power rule.
$(-3y^5)^2$

Answer: $9y^{10}$

INSTRUCTOR'S RESOURCES

 ITM pp. 347–349   **ISM** pp. 138–149
IAM p. 10

 TEST GENERATOR DOS Windows MAC

 **TRANSPARENCIES**

 CONCEPTUAL   WRITING   ▲ CHALLENGING    SCIENTIFIC CALCULATOR    GRAPHICS CALCULATOR

▶ **TEACHING TIP**
When showing examples of the power rules, include: $(x + y)^2 \neq x^2 + y^2$. Also, be sure that each factor in the base has an exponent. That is, have students write in "1" where required. ◀

The reciprocal of a^n is $\dfrac{1}{a^n} = \left(\dfrac{1}{a}\right)^n$. Also, by definition, a^n and a^{-n} are reciprocals since

$$a^n \cdot a^{-n} = a^n \cdot \frac{1}{a^n} = 1.$$

Thus, since both are reciprocals of a^n,

$$a^{-n} = \left(\frac{1}{a}\right)^n.$$

Some examples of this result are

$$6^{-3} = \left(\frac{1}{6}\right)^3 \qquad \text{and} \qquad \left(\frac{1}{3}\right)^{-2} = 3^2.$$

The discussion above can be generalized as follows.

Special Rules for Negative Exponents

Any nonzero number raised to the negative nth power is equal to the reciprocal of that number raised to the nth power. That is, if $a \neq 0$ and $b \neq 0$,

$$a^{-n} = \left(\frac{1}{a}\right)^n \qquad \text{and} \qquad \left(\frac{a}{b}\right)^{-n} = \left(\frac{b}{a}\right)^n.$$

EXAMPLE 2
Using Negative Exponents with Fractions

Write the following expressions with only positive exponents and then evaluate.

(a) $\left(\dfrac{3}{7}\right)^{-2} = \left(\dfrac{7}{3}\right)^2 = \dfrac{49}{9}$
(b) $\left(\dfrac{4}{5}\right)^{-3} = \left(\dfrac{5}{4}\right)^3 = \dfrac{125}{64}$ ■

The definitions and rules of this section and the previous one are summarized below.

Definitions and Rules for Exponents

For all integers m and n and all real numbers a and b:

Product Rule $\qquad a^m \cdot a^n = a^{m+n}$

Quotient Rule $\qquad \dfrac{a^m}{a^n} = a^{m-n} \quad (a \neq 0)$

Zero Exponent $\qquad a^0 = 1 \quad (a \neq 0)$

Negative Exponent $\qquad a^{-n} = \dfrac{1}{a^n} \quad (a \neq 0)$

Power Rules $\qquad (a^m)^n = a^{mn}$

$\qquad\qquad\qquad\quad (ab)^m = a^m b^m$

$\qquad\qquad\qquad\quad \left(\dfrac{a}{b}\right)^m = \dfrac{a^m}{b^m} \quad (b \neq 0)$

Special Rules $\qquad a^{-n} = \left(\dfrac{1}{a}\right)^n \quad (a \neq 0)$

$\qquad\qquad\qquad\quad \left(\dfrac{a}{b}\right)^{-n} = \left(\dfrac{b}{a}\right)^n \quad (a, b \neq 0)$

For Example 2
Chalkboard Exercise

Write the expression with only positive exponents and evaluate.

$$\left(\frac{2}{3}\right)^{-4}$$

Answer: $\dfrac{81}{16}$

▶ **TEACHING TIP**
Alternatively, in Example 2, show that a negative exponent can be first applied to both the numerator and the denominator, which are then converted to positive exponents in the final answer. ◀

2 ▶ Simplify exponential expressions.

▶ In the next three examples, we show how the definitions and rules for exponents are used to write expressions in equivalent forms.

EXAMPLE 3

Using the Definitions and Rules for Exponents

Simplify each expression so that no negative exponents appear in the final result.

(a) $3^2 \cdot 3^{-5} = 3^{2+(-5)} = 3^{-3} = \dfrac{1}{3^3}$ or $\dfrac{1}{27}$

(b) $x^{-3} \cdot x^{-4} \cdot x^2 = x^{-3+(-4)+2} = x^{-5} = \dfrac{1}{x^5}$ $(x \ne 0)$

(c) $(2^5)^{-3} = 2^{5(-3)} = 2^{-15} = \dfrac{1}{2^{15}}$

(d) $(4^{-2})^{-5} = 4^{(-2)(-5)} = 4^{10}$

(e) $(x^{-4})^6 = x^{(-4)6} = x^{-24} = \dfrac{1}{x^{24}}$ $(x \ne 0)$

(f) $(xy)^3 = x^3 y^3$ ■

CAUTION As shown in Example 3(f), $(xy)^3 = x^3 y^3$, so that $(xy)^3 \ne xy^3$. Remember that ab^m is *not* the same as $(ab)^m$.

EXAMPLE 4

Using Combined Definitions and Rules for Exponents

Simplify. Assume that all variables represent nonzero real numbers.

(a) $\dfrac{x^{-4}y^2}{x^2 y^{-5}} = \dfrac{x^{-4}}{x^2} \cdot \dfrac{y^2}{y^{-5}}$

$\qquad = x^{-4-2} \cdot y^{2-(-5)}$

$\qquad = x^{-6} y^7$

$\qquad = \dfrac{y^7}{x^6}$

(b) $(2^3 x^{-2})^{-2} = (2^3)^{-2} \cdot (x^{-2})^{-2}$

$\qquad\qquad = 2^{-6} x^4$

$\qquad\qquad = \dfrac{x^4}{2^6}$ or $\dfrac{x^4}{64}$ ■

NOTE Expressions like those in Example 4 can be correctly simplified in more than one way. For instance, we could simplify Example 4(a) as follows.

$$\dfrac{x^{-4}y^2}{x^2 y^{-5}} = \dfrac{y^5 y^2}{x^4 x^2} \qquad \text{Use } \dfrac{a^{-n}}{b^{-m}} = \dfrac{b^m}{a^n}.$$

$$= \dfrac{y^7}{x^6} \qquad \text{Product rule}$$

Recall that the rules for exponents can also be used with variable exponents, as we show in the next example. Assume that all variables used as exponents represent integers.

EXAMPLE 5

Working with Variable Exponents

Apply the rules for exponents in each case.

(a) $m^k \cdot m^{5-k} = m^{k+(5-k)}$

$\qquad\qquad = m^5$ $(m \ne 0)$

(b) $\dfrac{r^{2k}}{r^{5k}} = r^{2k-5k}$

$\qquad = r^{-3k} = \dfrac{1}{r^{3k}}$ $(r \ne 0)$ ■

With the work of the first two sections of this chapter, a^n is meaningful for all integer exponents n. In a later chapter this will be extended to include all *rational* values of n.

◇ Answers will vary.

◇ **C O N N E C T I O N S** ◇

Scientists often need to use numbers that are very large or very small. For example, the number of one-celled organisms that will sustain a whale for a few hours is 400,000,000,000,000, and the shortest wavelength of visible light is approximately .0000004 meter. It is simpler to write these numbers using *scientific notation*. Scientific notation is an important use of exponents. We discuss scientific notation in the rest of this section.

FOR DISCUSSION OR WRITING
Try to find examples of scientific notation used in science classes at your school. (*Hint:* One example is given in Section 10.6 of this book.)

3 ▶ Use the rules for exponents with scientific notation.

▶ In *scientific notation,* a number is written with the decimal point after the first nonzero digit and multiplied by a power of 10, as indicated in the following definition.

Scientific Notation

A number is written in **scientific notation** when it is expressed in the form

$$a \times 10^n$$

where $1 \leq |a| < 10$, and n is an integer.

For example, in scientific notation,

$$8000 = 8 \times 1000 = 8 \times 10^3.$$

The following numbers are not in scientific notation.

$$.230 \times 10^4 \qquad\qquad 46.5 \times 10^{-3}$$

.230 is less than 1   46.5 is greater than 10

The steps involved in writing a number in scientific notation are given below. (If the number is negative, ignore the minus sign, go through these steps, and then attach a minus sign to the result.)

Converting to Scientific Notation

Step 1   Place a caret, ∧, to the right of the first nonzero digit, where the decimal point will be placed.

Step 2   Count the number of digits from the caret to the decimal point. This number gives the absolute value of the exponent on ten.

Step 3   Decide whether multiplying by 10^n should make the result of Step 1 larger or smaller. The exponent should be positive to make the result larger; it should be negative to make the result smaller.

It is helpful to remember that

$$\text{for } n \geq 1, 10^{-n} < 1 \qquad \text{and} \qquad 10^n \geq 10.$$

Write each number in scientific notation.

(a) 820,000

Place a caret to the right of the 8 (the first nonzero digit) to mark the new location of the decimal point.

$$8\!\wedge\!20,000$$

Count from the caret to the decimal point, which is understood to be after the last 0.

8 20,000. ← Decimal point

↑
Count 5 places

Since the number 8.2 is to be made larger, the exponent on 10 is positive.

$$820,000 = 8.2 \times 10^5$$

(b) .000072

Count from right to left.

$$.00007\,2$$

5 places

Since the number 7.2 is to be made smaller, the exponent on 10 is negative.

$$.000072 = 7.2 \times 10^{-5}\quad\blacksquare$$

To convert a number written in scientific notation to standard notation, just work in reverse.

**Converting from
Scientific Notation**

Multiplying a number by a positive power of 10 makes the number larger, so move the decimal point to the right if n is positive in 10^n. Multiplying by a negative power of 10 makes a number smaller, so move the decimal point to the left if n is negative. If n is zero, leave the decimal point where it is.

Write the following numbers without scientific notation.

(a) 6.93×10^5

$$6.93000$$

5 places

The decimal point was moved 5 places to the right. (It was necessary to attach 3 zeros.)

$$6.93 \times 10^5 = 693,000$$

(b) $3.52 \times 10^7 = 35,200,000$

(c) 4.7×10^{-6}

$$000004.7$$

6 places

The decimal point was moved 6 places to the left.

$$4.7 \times 10^{-6} = .0000047$$

(d) $1.083 \times 10^0 = 1.083$ ∎

When problems require operations with numbers that are very large and/or very small, it is often advantageous to write the numbers in scientific notation first, and then perform the calculations using the rules for exponents. The next example illustrates this.

EXAMPLE 8
Using Scientific Notation
in Computation

■ Find $\dfrac{1,920,000 \times .0015}{.000032 \times 45,000}$.

First, express all numbers in scientific notation.

$$\frac{1,920,000 \times .0015}{.000032 \times 45,000} = \frac{1.92 \times 10^6 \times 1.5 \times 10^{-3}}{3.2 \times 10^{-5} \times 4.5 \times 10^4}$$

Next, use the commutative and associative properties and the rules for exponents to simplify the expression.

$$\frac{1,920,000 \times .0015}{.000032 \times 45,000} = \frac{1.92 \times 1.5 \times 10^6 \times 10^{-3}}{3.2 \times 4.5 \times 10^{-5} \times 10^4}$$

$$= \frac{1.92 \times 1.5}{3.2 \times 4.5} \times 10^4$$

$$= .2 \times 10^4$$

$$= (2 \times 10^{-1}) \times 10^4$$

$$= 2 \times 10^3$$

The expression is equal to 2×10^3, or 2000. ∎

Chalkboard Exercise

Find $\dfrac{200,000 \times .0003}{.06}$.

Answer: 1000

◇ Answers will vary.

▶ **TEACHING TIP**
It is helpful to show how the EXP or EE keys work on scientific calculators. Most calculators have an eight-digit display, so they cannot handle extremely large or extremely small numbers found in applications unless the numbers are entered in scientific notation. ◀

◇ **CONNECTIONS** ◇

Many scientific calculators accept and print numbers in scientific notation using a key marked EE or EXP. Using a calculator with an EE key to perform the calculations in Example 8, enter the following:

| 1.92 | EE | 6 | × | 1.5 | EE | 3 | ± |

-3

| ÷ | (| 3.2 | EE | 5 | ± | × | 4.5 | EE | 4 |) | = |

The EXP key is used in the same way. The result should be 2. 03, which means $2 \times 10^3 = 2000$. Notice that the negative exponent -3 is entered by pressing 3 and then ±. The exponent of -5 is handled the same way. Also, be careful to enclose the denominator in parentheses, or 4.5×10^4 will be multiplied by the rest of the expression, not divided into it.

EXAMPLE 9

Using Scientific Notation to Solve a Problem

A certain computer can perform an algorithm in .00000000036 second. How long would it take the computer to perform 5.2 billion of these algorithms? (One billion = 1,000,000,000)

In order to solve this problem, we must multiply the time per algorithm by the number of algorithms.

$$.00000000036 \times 5{,}200{,}000{,}000 = \text{total time}$$

Write each number in scientific notation, and then use rules for exponents.

$$.00000000036 \times 5{,}200{,}000{,}000$$
$$= (3.6 \times 10^{-10}) \times (5.2 \times 10^{9})$$
$$= (3.6 \times 5.2)(10^{-10} \times 10^{9}) \qquad \text{Commutative and associative properties}$$
$$= 18.72 \times 10^{-1} \qquad \text{Product rule}$$
$$= 1.872 \qquad \text{Convert to standard notation.}$$

It would take the computer 1.872 seconds. ■

Chalkboard Exercise

The distance to the sun is 9.3×10^7 miles. How long would it take a rocket, traveling at 3.2×10^3 miles per hour, to reach the sun?

Answer: about 2.9×10^4 hours

▶ **TEACHING TIP**

Explain that the exercises in this section can often be worked in several different manners in order to obtain the correct answer. Emphasize that there is only one correct final answer, no matter how the exercise is worked. ◀

3.2 EXERCISES

◉ **1.** Which one of the following is correct?

(a) $-\dfrac{3}{4} = \left(\dfrac{3}{4}\right)^{-1}$ 　(b) $\dfrac{3^{-1}}{4^{-1}} = \left(\dfrac{4}{3}\right)^{-1}$ 　(c) $\dfrac{3^{-1}}{4} = \dfrac{3}{4^{-1}}$ 　(d) $\dfrac{3^{-1}}{4^{-1}} = \left(\dfrac{3}{4}\right)^{-1}$

◉ **2.** Which one of the following is incorrect?

(a) $(3r)^{-2} = 3^{-2}r^{-2}$ 　(b) $3r^{-2} = (3r)^{-2}$ 　(c) $(3r)^{-2} = \dfrac{1}{(3r)^2}$ 　(d) $(3r)^{-2} = \dfrac{r^{-2}}{9}$

◉ *Decide whether the expression has been simplified correctly. If not, correct it.*

3. $(ab)^2 = ab^2$ 　**4.** $(xy)^4 = xy^4$ 　**5.** $(5x)^3 = 5^3x^3$ 　**6.** $(2k)^5 = 2^5k^5$

7. $\left(\dfrac{4}{a}\right)^5 = \dfrac{4^5}{a}$ 　**8.** $\left(\dfrac{7}{y}\right)^3 = \dfrac{7^3}{y}$ 　**9.** $(z^4)^5 = z^9$ 　**10.** $(m^3)^4 = m^7$

Write the expression with a positive exponent. See Example 2.

11. $\left(\dfrac{3}{4}\right)^{-2}$ 　**12.** $\left(\dfrac{2}{5}\right)^{-3}$ 　**13.** $\left(\dfrac{6}{5}\right)^{-1}$ 　**14.** $\left(\dfrac{8}{3}\right)^{-2}$

Simplify the expression using only positive exponents. Assume that variables represent nonzero real numbers. See Examples 1–4.

15. $(2^{-3} \cdot 5^{-1})^3$ 　**16.** $(5^{-4} \cdot 6^{-2})^3$ 　**17.** $(5^{-4} \cdot 6^{-2})^{-3}$ 　**18.** $(2^{-5} \cdot 3^{-4})^{-1}$

19. $(k^2)^{-3}k^4$ 　**20.** $(x^3)^{-4}x^5$ 　**21.** $-4r^{-2}(r^4)^2$ 　**22.** $-2m^{-1}(m^3)^2$

23. $(5a^{-1})^4(a^2)^{-3}$ 　**24.** $(3p^{-4})^2(p^3)^{-1}$ 　**25.** $(z^{-4}x^3)^{-1}$ 　**26.** $(y^{-2}z^4)^{-3}$

27. $\dfrac{(p^{-2})^3}{5p^4}$ 　**28.** $\dfrac{(m^4)^{-1}}{9m^3}$ 　**29.** $\dfrac{4a^5(a^{-1})^3}{(a^{-2})^{-2}}$ 　**30.** $\dfrac{12k^{-2}(k^{-3})^{-4}}{6k^5}$

31. $\dfrac{(-y^{-4})^2}{6(y^{-5})^{-1}}$ 　**32.** $\dfrac{2(-m^{-1})^{-4}}{9(m^{-3})^2}$ 　**33.** $\dfrac{(2k)^2m^{-5}}{(km)^{-3}}$ 　**34.** $\dfrac{(3rs)^{-2}}{3^2r^2s^{-4}}$

35. $\dfrac{(2k)^2k^3}{k^{-1}k^{-5}}(5k^{-2})^{-3}$ 　**36.** $\dfrac{(3r^2)^2r^{-5}}{r^{-2}r^3}(2r^{-6})^2$ 　**37.** $\left(\dfrac{3k^{-2}}{k^4}\right)^{-1} \cdot \dfrac{2}{k}$ 　**38.** $\left(\dfrac{7m^{-2}}{m^{-3}}\right)^{-2} \cdot \dfrac{m^3}{4}$

Answers (right column):

1. (d) 　2. (b)
3. incorrect; a^2b^2
4. incorrect; x^4y^4
5. correct 　6. correct
7. incorrect; $\dfrac{4^5}{a^5}$
8. incorrect; $\dfrac{7^3}{y^3}$
9. incorrect; z^{20}
10. incorrect; m^{12}
15. $\dfrac{1}{2^9 \cdot 5^3}$ 　16. $\dfrac{1}{5^{12} \cdot 6^6}$
17. $5^{12} \cdot 6^6$
18. $2^5 \cdot 3^4$ 　19. $\dfrac{1}{k^2}$ 　20. $\dfrac{1}{x^7}$
21. $-4r^6$ 　22. $-2m^5$
23. $\dfrac{5^4}{a^{10}}$ 　24. $\dfrac{3^2}{p^{11}}$ 　25. $\dfrac{z^4}{x^3}$
26. $\dfrac{y^6}{z^{12}}$ 　27. $\dfrac{1}{5p^{10}}$
28. $\dfrac{1}{9m^7}$ 　29. $\dfrac{4}{a^2}$ 　30. $2k^5$
31. $\dfrac{1}{6y^{13}}$ 　32. $\dfrac{2m^{10}}{9}$
33. $\dfrac{2^2k^5}{m^2}$ 　34. $\dfrac{s^2}{3^4r^4}$
35. $\dfrac{2^2k^{17}}{5^3}$ 　36. $\dfrac{3^2 \cdot 2^2}{r^{14}}$
37. $\dfrac{2k^5}{3}$ 　38. $\dfrac{m}{7^2 \cdot 4}$

◉ CONCEPTUAL 　✎ WRITING 　▲ CHALLENGING 　▦ SCIENTIFIC CALCULATOR 　▨ GRAPHICS CALCULATOR

39. $\left(\dfrac{b^6}{a^3}\right)^{-2}$; $\dfrac{b^{-12}}{a^{-6}} = \dfrac{a^6}{b^{12}}$

40. $\dfrac{a^{16}b^{-4}}{a^{10}b^8}$; $\dfrac{a^{16-10}}{b^{8+4}} = \dfrac{a^6}{b^{12}}$

41. They are the same.

42. Both methods are correct. 43. 5.3×10^2

44. 1.6×10^3

45. 8.3×10^{-1}

46. 7.2×10^{-3}

47. 6.92×10^{-6}

48. 8.75×10^{-1}

49. -3.85×10^4

50. -9.76×10^8

51. 72,000   52. 891

53. .00254   54. .000542

55. $-60,000$ 56. -9000

57. .000012   58. .0000027

59. .06   60. .004

61. .0000025   62. .000002

63. 200,000

64. 1,000,000   65. 3000

66. 200,000   67. $3838.38

68. approximately 9.474×10^{-7} parsec

69. 63,360 inches in a mile

70. 3×10^2 seconds

—————— ◆ **MATHEMATICAL CONNECTIONS** (Exercises 39–42) ◇——

Consider the expression

$$\left(\frac{a^{-8}b^2}{a^{-5}b^{-4}}\right)^{-2} \quad (a, b \neq 0).$$

If we wish to simplify the expression so that only positive exponents appear in the answer, there are several ways we can approach the problem. However, all methods should yield the same answer.

Work Exercises 39–42 in order.

39. Simplify the expression by first writing the fraction in parentheses using only positive exponents. Then apply the exponent -2 using the power rule $\left(\dfrac{a}{b}\right)^m = \dfrac{a^m}{b^m}$. Finally, simplify this result.

40. Simplify the expression by first applying the exponent -2 to each factor within the parentheses. Then simplify this result.

41. How do your answers in Exercises 39 and 40 compare?

⊚ **42.** After seeing results from Exercises 39 and 40, how would you answer the question "Which one of these methods is the correct way of simplifying the expression?"

Write each number in scientific notation. See Example 6.

43. 530   **44.** 1600   **45.** .830   **46.** .0072

47. .00000692   **48.** .875   **49.** $-38,500$ **50.** $-976,000,000$

Write each of the following in standard notation. See Example 7.

51. 7.2×10^4 **52.** 8.91×10^2 **53.** 2.54×10^{-3}

54. 5.42×10^{-4} **55.** -6×10^4 **56.** -9×10^3

57. 1.2×10^{-5} **58.** 2.7×10^{-6}

Use the rules for exponents to find each value. See Example 8.

59. $\dfrac{12 \times 10^4}{2 \times 10^6}$ **60.** $\dfrac{16 \times 10^5}{4 \times 10^8}$ **61.** $\dfrac{3 \times 10^{-2}}{12 \times 10^3}$

62. $\dfrac{5 \times 10^{-3}}{25 \times 10^2}$ **63.** $\dfrac{.05 \times 1600}{.0004}$ **64.** $\dfrac{.003 \times 40,000}{.00012}$

65. $\dfrac{20,000 \times .018}{300 \times .0004}$ **66.** $\dfrac{840,000 \times .03}{.00021 \times 600}$

Use scientific notation to complete the exercise. See Example 9.

67. In a state lottery, a player must choose six numbers from 1 through 40. It can be shown that there are 3.83838×10^6 different ways to do this. Suppose that a group of 1000 persons decides to purchase tickets for all these numbers and each ticket costs $1.00. How much should each person expect to pay?

▦ **68.** A parsec, a unit of length used in astronomy, is 19×10^{12} miles. The mean distance of Uranus from the sun is 1.8×10^7 miles. How many parsecs is Uranus from the sun?

▦ **69.** An inch is approximately 1.57828×10^{-5} mile. Find the reciprocal of this number to determine the number of inches in a mile.

70. The speed of light is approximately 3×10^{10} centimeters per second. How long will it take light to travel 9×10^{12} centimeters?

71. The distance to the sun is 9.3×10^7 miles. How long would it take a rocket, traveling at 2.9×10^3 miles per hour, to reach the sun?

72. A *light-year* is the distance that light travels in one year. Find the number of miles in a light-year if light travels 1.86×10^5 miles per second.

73. Use the information given in the previous two exercises to find the number of minutes necessary for light from the sun to reach the Earth.

74. A computer can do one addition in 1.4×10^{-7} seconds. How long would it take the computer to do a trillion (10^{12}) calculations? Give the answer in seconds and then in hours.

 **75.** The planet Mercury has a mean distance from the sun of 3.6×10^7 miles, while the mean distance of Venus to the sun is 6.7×10^7 miles. How long would it take a spacecraft traveling at 1.55×10^3 miles per hour to travel from Venus to Mercury? (Give your answer in hours, without scientific notation.)

76. Use the information from the previous exercise to find the number of days it would take the spacecraft to travel from Venus to Mercury. Round your answer to the nearest whole number of days.

77. According to Bode's law, the distance d of the nth planet from the sun is

$$d = \frac{3(2^{n-2}) + 4}{10},$$

in astronomical units. Find the distance of each of the following planets from the sun.
(a) Venus $(n = 2)$ **(b)** Earth $(n = 3)$ **(c)** Mars $(n = 4)$

78. When the distance between the centers of the Moon and the Earth is 4.60×10^8 meters, an object on the line joining the centers of the Moon and the Earth exerts the same gravitational force on each when it is 4.14×10^8 meters from the center of the Earth. How far is the object from the center of the Moon at that point?

79. Assume that the volume of the Earth is 5×10^{14} cubic meters and that the volume of a bacterium is 2.5×10^{-16} cubic meters. If the Earth could be packed full of bacteria, how many would it contain?

80. Our galaxy is approximately 1.2×10^{17} kilometers across. Suppose a spaceship could travel at 1.5×10^5 kilometers per second (half the speed of light). Find the approximate number of years needed for the spaceship to cross the galaxy.

The graph shows the number of "800" phone calls made in the AT&T system (in dark blue) plus others made (estimated, in light blue), in billions of calls (1 billion = 1,000,000,000).

81. In which year was the number of calls made in the AT&T system approximately 7×10^9?

82. In which year was the total number of calls approximately 1.7×10^{10}?

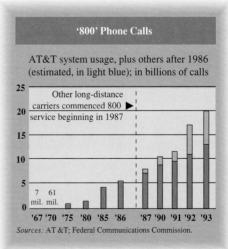

71. approximately $3.2 \times 10^4 = 32,000$ hours (about 3.7 years)
72. approximately 5.87×10^{12} miles
73. 500 seconds, which is approximately 8.3 minutes
74. 1.4×10^5 seconds, or approximately 38.9 hours
75. 20,000 hours
76. 833 days
77. (a) .7 astronomical unit (b) 1 astronomical unit (c) 1.6 astronomical units
78. 4.6×10^7 meters or 46,000,000 meters
79. 2×10^{30} bacteria
80. 2.54×10^4 years, or 25,400 years
81. 1987 **82.** 1992

 ⊙ CONCEPTUAL   ✎ WRITING   ▲ CHALLENGING   ▦ SCIENTIFIC CALCULATOR   ▦ GRAPHICS CALCULATOR

83. p^{2y+2} 84. $3y^{3m-2}$
85. a^{-4-2k} 86. m^{4+6p}
87. r^{p^2+3p-6} 88. z^{2q^2-7q}
89. $-\dfrac{3}{32m^8p^4}$ 90. $\dfrac{15}{8y^5z^3}$
91. $\dfrac{2}{3y^4}$ 92. $\dfrac{1}{54m^{11}x^9}$
93. $\dfrac{3p^8}{16q^{14}}$ 94. $\dfrac{108b^9}{25a^{14}}$
95. $9x$ 96. $-16p$
97. $4-11y$ 98. $5+5q$
99. $2x-2$
100. $20q-10$

Simplify the expression. Assume that all variables used as exponents represent integers and that all other variables represent nonzero real numbers. See Example 5.

83. $\dfrac{p^y p^{4y+2}}{(p^y)^3}$

84. $\dfrac{(y^2)^m(3y^{m+1})}{(y^{-1})^{-3}}$

85. $\dfrac{(a^{4+k}a^{3k})^{-1}}{(a^2)^{-k}}$

86. $\dfrac{(m^{2-p}\cdot m^{3p})^2}{(m^{-2})^p}$

87. $\dfrac{r^{-p}(r^{p+2})^p}{(r^{p-3})^{-2}}$

88. $\dfrac{z^{-3q}(z^{q-5})^q}{(z^{q+1})^{-q}}$

Simplify the expression. Write answers with only positive exponents. Assume that all variables represent nonzero real numbers.

89. $\dfrac{(2m^2p^3)^2(4m^2p)^{-2}}{(-3mp^4)^{-1}(2m^3p^4)^3}$

90. $\dfrac{(-5y^3z^4)^2(2yz^5)^{-2}}{10(y^4z)^3(3y^3z^2)^{-1}}$

91. $\dfrac{(-3y^3x^3)(-4y^4x^2)(x^2)^{-4}}{18x^3y^2(y^3)^3(x^3)^{-2}}$

92. $\dfrac{(2m^3x^2)^{-1}(3m^4x)^{-3}}{(m^2x^3)^3(m^2x)^{-5}}$

93. $\left(\dfrac{p^2q^{-1}}{2p^{-2}}\right)^2\cdot\left(\dfrac{p^3\cdot 4q^{-2}}{3q^{-5}}\right)^{-1}\cdot\left(\dfrac{pq^{-5}}{q^{-2}}\right)^3$

94. $\left(\dfrac{a^6b^{-2}}{2a^{-2}}\right)^{-1}\cdot\left(\dfrac{6a^{-2}}{5b^{-4}}\right)^2\cdot\left(\dfrac{2b^{-1}a^2}{3b^{-2}}\right)^{-1}$

REVIEW EXERCISES

Combine like terms. Remove all parentheses first, if necessary. See Section 1.3.

95. $9x+5x-x+8x-12x$

96. $3p+2p-8p-13p$

97. $6-4(3-y)+5(2-3y)$

98. $2+3(5+q)-2(6-q)$

99. $7x-(5+5x)+3$

100. $12q-(6-8q)-4$

3.3 ADDITION AND SUBTRACTION OF POLYNOMIALS

FOR EXTRA HELP

 SSG pp. 70–74
SSM pp. 98–102

 Video 3

 Tutorial IBM MAC

OBJECTIVES

1 ▶ Know the basic definitions for polynomials.
2 ▶ Identify monomials, binomials, and trinomials.
3 ▶ Find the degree of a polynomial.
4 ▶ Add and subtract polynomials.
5 ▶ Use $P(x)$ notation.

Just as whole numbers are the basis of arithmetic, *polynomials* are fundamental in algebra. A **term of a polynomial** is a number or the product of a number and one or more variables raised to a nonnegative power. Examples of terms include

$$4x,\qquad \frac{1}{2}m^5,\qquad -7z^9,\qquad 6x^2z,\qquad \text{and}\qquad 9.$$

The number in the product is called the **numerical coefficient,** or just the **coefficient.** In the term $8x^3$, the coefficient is 8. In the term $-4p^5$, it is -4. The coefficient of the term k is understood to be 1. The coefficient of $-r$ is -1. More generally, any factor in a term is the coefficient of the product of the remaining factors. For example, $3x^2$ is the coefficient of y in the term $3x^2y$, and $3y$ is the coefficient of x^2 in $3x^2y$.

INSTRUCTOR'S RESOURCES

ITM pp. 350–352 **ISM** pp. 149–157
IAM p. 10

TEST GENERATOR
DOS Windows MAC

TRANSPARENCIES

1 ▶ Know the basic definitions for polynomials.

▶ Any combination of variables or constants joined by the basic operations of addition, subtraction, multiplication, division (except by zero), or extraction of roots is called an **algebraic expression.** The simplest kind of algebraic expression is a **polynomial.**

| Polynomial | A **polynomial** is a term or a finite sum of terms in which all variables have whole number exponents and no variables appear in denominators. |

Examples of polynomials include

$$3x - 5, \qquad 4m^3 - 5m^2p + 8, \qquad \text{and} \qquad -5t^2s^3.$$

Even though the expression $3x - 5$ involves subtraction, it is called a sum of terms, since it could be written as $3x + (-5)$. Also, $-5t^2s^3$ can be thought of as a sum of terms by writing it as $0 + (-5t^2s^3)$.

Most of the polynomials used in this book contain only one variable. A polynomial containing only the variable x is called a **polynomial in x.** A polynomial in one variable is written in **descending powers** of the variable if the exponents on the terms of the polynomial decrease from left to right. For example,

$$x^5 - 6x^2 + 12x - 5$$

is a polynomial in descending powers of x.

NOTE The term -5 in the polynomial above can be thought of as $-5x^0$, since $-5x^0 = -5(1) = -5$.

For Example 1

Chalkboard Exercise

Write the polynomial in descending powers.

$-3z^4 + 2z^3 + z^5 - 6z$

Answer: $z^5 - 3z^4 + 2z^3 - 6z$

EXAMPLE 1

Writing Polynomials in Descending Powers

Write each of the following in descending powers of the variable.

(a) $y - 6y^3 + 8y^5 - 9y^4 + 12$
Write the polynomial as

$$8y^5 - 9y^4 - 6y^3 + y + 12.$$

(b) $-2 + m + 6m^2 - 4m^3$ would be written as $-4m^3 + 6m^2 + m - 2.$ ■

2 ▶ Identify monomials, binomials, and trinomials.

▶ Polynomials with a certain number of terms are so common that they are given special names. A polynomial of exactly three terms is a **trinomial,** and a polynomial with exactly two terms is a **binomial.** A single-term polynomial is a **monomial.**

EXAMPLE 2

Identifying Types of Polynomials

In the list below, we give examples of monomials, binomials, and trinomials, as well as polynomials that are none of these.

Monomials	$5x, \quad 7m^9, \quad -8$
Binomials	$3x^2 - 6, \quad 11y + 8, \quad 5k + 15$
Trinomials	$y^2 + 11y + 6, \quad 8p^3 - 7p + 2m, \quad -3 + 2k^5 + 9z^4$
None of these	$p^3 - 5p^2 + 2p - 5, \quad -9z^3 + 5c^3 + 2m^5 + 11r^2 - 7r$ ■

Chalkboard Exercise

Identify the polynomial as a trinomial, binomial, monomial, or none of these.

$12m^4 - 6m^2$

Answer: binomial

3 ▶ Find the degree of a polynomial.

▶ The **degree of a term** with one variable is the exponent on the variable. For example, the degree of $2x^3$ is 3, the degree of $-x^4$ is 4, and the degree of $17x$ is 1. The degree of a term in more than one variable is defined to be the sum of the exponents of the variables. For example, the degree of $5x^3y^7$ is 10, because $3 + 7 = 10$.

The largest degree of any of the terms in a polynomial is called the **degree of the polynomial.** In most cases, we will be interested in finding the degree of a polynomial in one variable. For example, the degree of $4x^3 - 2x^2 - 3x + 7$ is 3, because the largest degree of any term is 3 (the degree of $4x^3$).

E X A M P L E   3

Finding the Degree of a Polynomial

Find the degree of each polynomial.

— Largest exponent is 2.

(a) $9x^2 - 5x + 8$

The largest exponent is **2**, so the polynomial is of degree **2**.

(b) $17m^9 + 8m^{14} - 9m^3$

This polynomial is of degree **14**.

(c) $5x$

The degree is 1, since $5x = 5x^1$.

(d) -2

A constant term, other than zero, is of degree zero.

(e) $5a^2b^5$

The degree is the sum of the exponents, $2 + 5 = 7$.

(f) $x^3y^9 + 12xy^4 + 7xy$

The degrees of the terms are 12, 5, and 2. Therefore, the degree of the polynomial is 12, which is the largest degree of any term in the polynomial. ■

NOTE The number 0 has no degree, since 0 times a variable to any power is 0.

---

4 ▶ Add and subtract polynomials.

▶ We use the distributive property to simplify polynomials by combining terms. For example, to simplify $x^3 + 4x^2 + 5x^2 - 1$, we use the distributive property as follows.

$$x^3 + 4x^2 + 5x^2 - 1 = x^3 + (4 + 5)x^2 - 1$$
$$= x^3 + 9x^2 - 1$$

On the other hand, it is not possible to combine the terms in the polynomial $4x + 5x^2$. As these examples suggest, only terms containing exactly the same variables to the same powers may be combined. As mentioned in Chapter 1, such terms are called *like terms.*

E X A M P L E   4

Combining Like Terms

Combine terms.

(a) $-5y^3 + 8y^3 - y^3$

These like terms may be combined by the distributive property.

$$-5y^3 + 8y^3 - y^3 = (-5 + 8 - 1)y^3 = 2y^3$$

Chalkboard Exercise

Combine terms.

$2z^4 + 3x^4 + 5z^4 - 9x^4$

Answer: $7z^4 - 6x^4$

(b) $6x + 5y - 9x + 2y$

Use the associative and commutative properties to rewrite the expression with all the x's together and all the y's together.

$$6x + 5y - 9x + 2y = 6x - 9x + 5y + 2y$$

Now combine like terms.

$$= -3x + 7y$$

Since $-3x$ and $7y$ are unlike terms, no further simplification is possible.

(c) $5x^2y - 6xy^2 + 9x^2y + 13xy^2 = 5x^2y + 9x^2y - 6xy^2 + 13xy^2$
$$= 14x^2y + 7xy^2 \quad ■$$

CAUTION Remember that only like terms can be combined.

We use the following rule to add two polynomials.

Adding Polynomials	**Add** two polynomials by combining like terms.

EXAMPLE 5

Adding Polynomials Horizontally

Chalkboard Exercise

Add
$(-5p^3 + 6p^2) + (8p^3 - 12p^2)$.

Answer: $3p^3 - 6p^2$

Add $4k^2 - 5k + 2$ and $-9k^2 + 3k - 7$.

Use the commutative and associative properties to rearrange the polynomials so that like terms are together. Then use the distributive property to combine like terms.

$$(4k^2 - 5k + 2) + (-9k^2 + 3k - 7) = 4k^2 - 9k^2 - 5k + 3k + 2 - 7$$
$$= -5k^2 - 2k - 5 \quad ■$$

EXAMPLE 6

Adding Polynomials Vertically

The two polynomials in Example 5 can also be added vertically by lining up like terms in columns. Then add by columns.

$$\begin{array}{r} 4k^2 - 5k + 2 \\ -9k^2 + 3k - 7 \\ \hline -5k^2 - 2k - 5 \end{array} \quad ■$$

EXAMPLE 7

Adding Polynomials
For Example 6

Chalkboard Exercise

Add. $\begin{array}{r} -6r^5 + 2r^3 - r^2 \\ 8r^5 - 2r^3 + 5r^2 \\ \hline \end{array}$

Answer: $2r^5 + 4r^2$

Add $3a^5 - 9a^3 + 4a^2$ and $-8a^5 + 8a^3 + 2$.

$$(3a^5 - 9a^3 + 4a^2) + (-8a^5 + 8a^3 + 2)$$
$$= 3a^5 - 8a^5 - 9a^3 + 8a^3 + 4a^2 + 2$$
$$= -5a^5 - a^3 + 4a^2 + 2 \qquad \text{Combine like terms.}$$

These same two polynomials can be added by placing them in columns, with like terms in the same columns.

$$\begin{array}{r} 3a^5 - 9a^3 + 4a^2 \\ -8a^5 + 8a^3 \qquad + 2 \\ \hline -5a^5 - a^3 + 4a^2 + 2 \end{array}$$

For many people, there is less chance of error with vertical addition. ■

Chalkboard Exercise

Add $(12y^2 - 7y + 9) + (-4y^2 - 11y + 5)$.

Answer: $8y^2 - 18y + 14$

In Chapter 1, subtraction of real numbers was defined as

$$a - b = a + (-b).$$

That is, add the first number and the negative (or opposite) of the second. We can give a similar definition for subtraction of polynomials by defining the **negative of a polynomial** as that polynomial with every sign changed.

Subtracting Polynomials	**Subtract** two polynomials by adding the first polynomial and the negative of the *second* polynomial.

EXAMPLE 8

Subtracting Polynomials Horizontally

Subtract: $(-6m^2 - 8m + 5) - (-5m^2 + 7m - 8).$

Change every sign in the second polynomial and add.

$$(-6m^2 - 8m + 5) - (-5m^2 + 7m - 8)$$
$$= -6m^2 - 8m + 5 + 5m^2 - 7m + 8$$

Now add by combining like terms.

$$= -6m^2 + 5m^2 - 8m - 7m + 5 + 8$$
$$= -m^2 - 15m + 13$$

Check by adding the sum, $-m^2 - 15m + 13$, to the second polynomial. The result should be the first polynomial. ■

EXAMPLE 9

Subtracting Polynomials Vertically

Use the same polynomials as in Example 8 and subtract in columns.

Write the first polynomial above the second, lining up like terms in columns.

$$\begin{array}{r} -6m^2 - 8m + 5 \\ -5m^2 + 7m - 8 \end{array}$$

Change all the signs in the second polynomial, and add.

$$\begin{array}{r} -6m^2 - 8m + 5 \\ +5m^2 - 7m + 8 \\ \hline -m^2 - 15m + 13 \end{array}$$

All signs changed

Add in columns. ■

5 ▶ Use $P(x)$ notation.

▶ Sometimes one problem will involve several polynomials. To keep track of these polynomials, capital letters can be used to name the polynomials. For example, $P(x)$, read "P of x," or, "the value of P at x," can be used to represent the polynomial $3x^2 - 5x + 7$, so that

$$P(x) = 3x^2 - 5x + 7.$$

The x in $P(x)$ is used to show that x is the variable in the polynomial.

If $x = -2$, then $P(x) = 3x^2 - 5x + 7$ takes on the value

$$P(-2) = 3(-2)^2 - 5(-2) + 7 \qquad \text{Let } x = -2.$$
$$= 3 \cdot 4 + 10 + 7$$
$$= 29.$$

We will see a more general use of this notation in a later chapter.

CAUTION Note that $P(x)$ does *not* mean P times x. It is a special notation to name a polynomial in x.

EXAMPLE 10

Using $P(x)$ Notation

Let $P(x) = 4x^3 - x^2 + 5$. Find each of the following.

(a) $P(3)$

First, substitute 3 for x.

$$P(x) = 4x^3 - x^2 + 5$$
$$P(3) = 4 \cdot 3^3 - 3^2 + 5$$

Now use the order of operations from Chapter 1.

$$= 4 \cdot 27 - 9 + 5$$
$$= 108 - 9 + 5$$
$$= 104$$

(b) $P(-4) = 4 \cdot (-4)^3 - (-4)^2 + 5 \qquad$ Let $x = -4$.
$$= 4 \cdot (-64) - 16 + 5$$
$$= -267 \quad \blacksquare$$

◆ 1. $P(1) = 1$, $P(2) = 2$
2. 4, 8, 16   3. $P(3) = 4$,
$P(4) = 8$, $P(5) = 16$
4. Because the pattern 1, 2, 4, 8, 16 emerges, most people will predict 32, since the terms are doubling each time. However, $P(6) = 31$ (not 32). See the figure.

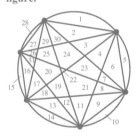

◇ CONNECTIONS ◇

The polynomial

$$P(x) = \frac{1}{24}x^4 - \frac{1}{4}x^3 + \frac{23}{24}x^2 - \frac{3}{4}x + 1$$

will give the number of interior regions formed in a circle if x points on the circumference are joined by all possible chords. See the figure.

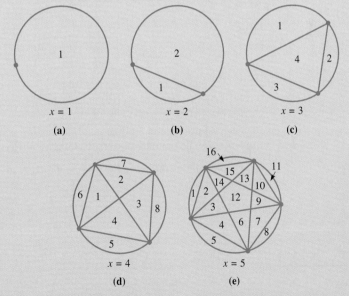

For example, in (a) we have 1 point and since no chords can be drawn, we have only 1 interior region. In (b) there are 2 points and 2 interior regions are formed.

FOR DISCUSSION OR WRITING
1. Verify that $P(1) = 1$ and $P(2) = 2$.
2. Based on the appropriate figure alone, what should be the value of $P(3)$? $P(4)$? $P(5)$?
3. Verify your answers in Item 2 by evaluating $P(3)$, $P(4)$, and $P(5)$.
4. Observe a pattern in the results of the first three items. Use the pattern to predict $P(6)$. Does your prediction equal $P(6)$?

3.3 EXERCISES

1. 7; 1   2. 3; 1
3. −15; 2   4. −27; 3
5. 1; 4   6. 1; 6
7. −1; 6   8. −1; 6
11. neither
12. descending
13. ascending
14. neither
15. descending
16. ascending
17. monomial; 0
18. monomial; 0
19. binomial; 1
20. binomial; 5
21. trinomial; 3
22. trinomial; 2
23. none of these; 5
24. none of these; 4
25. (a)   26. Answers will vary. One example is $7x^5 + 2x^3 - 6x^2 + 9x$.
27. $12p - 4$ 28. $19m + 3$
29. $-9p^2 + 11p - 9$
30. $3y^3 + 12y - 1$
31. $14m^2 - 13m + 6$
32. $-7z^2 + 7z - 3$
33. $8q^3 - 7q + 11$
34. $2y^3 - 11y^2 - 5y + 8$

Give the coefficient and the degree of the term. See Example 3.

1. $7z$
2. $3r$
3. $-15p^2$
4. $-27k^3$
5. x^4
6. y^6
7. $-mn^5$
8. $-a^5b$

9. The exponent in the expression 4^5 is 5. Explain why the degree of 4^5 is not 5. What is its degree?

10. A polynomial with *exactly* two terms is called a binomial. Why is the word *exactly* important in this definition?

We defined a polynomial written in descending powers in the text. See Example 1. Sometimes we write a polynomial in ascending powers, with the degree of the terms increasing from left to right. Decide whether the polynomial is written in descending powers, ascending powers, or neither.

11. $2x^3 + x - 3x^2$
12. $3y^5 + y^4 - 2y^3 + y$
13. $4p^3 - 8p^5 + p^7$
14. $q^2 + 3q^4 - 2q + 1$
15. $-m^3 + 5m^2 + 3m + 10$
16. $4 - x + 3x^2$

Identify the polynomial as a monomial, binomial, trinomial, or none of these. Also give the degree. See Examples 2 and 3.

17. 25
18. 5
19. $7m - 22$
20. $-x^2 + 6x^5$
21. $2r^3 + 4r^2 + 5r$
22. $5z^2 - 6z + 7$
23. $-6p^4q - 3p^3q^2 + 2pq^3 - q^4$
24. $8s^3t - 4s^2t^2 + 2st^3 + 9$

25. Which one of the following is a trinomial in descending powers, having degree 6?
(a) $5x^6 - 4x^5 + 12$ (b) $6x^5 - x^6 + 4$ (c) $2x + 4x^2 - x^6$
(d) $4x^6 - 6x^4 + 9x^2 - 8$

26. Give an example of a polynomial of four terms in the variable x, having degree 5, written in descending powers, lacking a fourth degree term.

Add or subtract as indicated. See Examples 6, 7, and 9.

27. Add.
$21p - 8$
$-9p + 4$

28. Add.
$15m - 9$
$4m + 12$

29. Add.
$-12p^2 + 4p - 1$
$3p^2 + 7p - 8$

30. Add.
$-6y^3 + 8y + 5$
$9y^3 + 4y - 6$

31. Subtract.
$6m^2 - 11m + 5$
$-8m^2 + 2m - 1$

32. Subtract.
$-4z^2 + 2z - 1$
$3z^2 - 5z + 2$

33. Subtract.
$5q^3 - 5q + 2$
$-3q^3 + 2q - 9$

34. Subtract.
$6y^3 - 9y^2 + 8$
$4y^3 + 2y^2 + 5y$

 CONCEPTUAL    WRITING   CHALLENGING    SCIENTIFIC CALCULATOR    GRAPHICS CALCULATOR

———◆ **MATHEMATICAL CONNECTIONS** (Exercises 35–40) ◇———

Our modern system of numeration, the Hindu-Arabic system, is a place value system that uses ten as its base. For example, the numeral 4353 means 4 thousands, 3 hundreds, 5 tens, and 3 ones. The two 3s in the numeral have different meanings because they appear in different place values.

A whole number in the Hindu-Arabic system can be written in expanded form using powers of ten. In the case of 4353, we have

$$(4 \times 10^3) + (3 \times 10^2) + (5 \times 10^1) + (3 \times 10^0).$$

If we consider the polynomial $4x^3 + 3x^2 + 5x + 3$, by letting $x = 10$ we get 4353. Thus we see that our usual method of writing whole numbers is just a specific case of writing a polynomial in descending powers of the variable.

Work Exercises 35–40 in order.

35. Write 241 in expanded form.

⊙ **36.** If you were to add 4353 to 241 by hand, using a vertical format, why would the following *not* lead to the correct answer?

$$\begin{array}{r} 4353 \\ + \ 241 \\ \hline \end{array}$$

📝 **37.** Explain the rule you learned years ago about how to add numbers in a vertical column.

⊙ **38.** If you were to add $4x^3 + 3x^2 + 5x + 3$ to $2x^2 + 4x + 1$ using a vertical format, why would the following *not* lead to the correct answer?

$$\begin{array}{r} 4x^3 + 3x^2 + 5x + 3 \\ + \ 2x^2 + 4x \ + 1 \\ \hline \end{array}$$

📝 **39.** Explain the rule you learned in this section how to add polynomials in a vertical column.

⊙ **40.** While we recognize that $4x^2 + 7x + 3$ and $7x + 3 + 4x^2$ represent the same
📝   expression, we also recognize that 473 and 734 represent two different numbers. Comment on why this is so.

———◆———

Combine terms. See Examples 4, 5, 7, and 8.

41. $5z^4 + 3z^4$ **42.** $8r^5 - 2r^5$ **43.** $-m^3 + 2m^3 + 6m^3$

44. $3p^4 + 5p^4 - 2p^4$ **45.** $x + x + x + x + x$ **46.** $z - z - z + z$

47. $y^2 + 7y - 4y^2$ **48.** $2c^2 - 4 + 8 - c^2$ **49.** $2k + 3k^2 + 5k^2 - 7$

50. $4x^2 + 2x - 6x^2 - 6$ **51.** $n^4 - 2n^3 + n^2 - 3n^4 + n^3$

52. $2q^3 + 3q^2 - 4q - q^3 + 5q^2$ **53.** $[4 - (2 + 3m)] + (6m + 9)$

54. $[8a - (3a + 4)] - (5a - 3)$ **55.** $[(6 + 3p) - (2p + 1)] - (2p + 9)$

56 $[(4x - 8) - (-1 + x)] - (11x + 5)$ **57.** $(3p^2 + 2p - 5) + (7p^2 - 4p^3 + 3p)$

58. $(y^3 + 3y + 2) + (4y^3 - 3y^2 + 2y - 1)$ **59.** $(2x^5 - 2x^4 + x^3 - 1) + (x^4 - 3x^3 + 2)$

60. $(y^2 + 3y) + [2y - (5y^2 + 3y + 4)]$ **61.** $(9a - 5a) - [2a - (4a + 3)]$

62. $(6r - 6r^2) - [(2r - 5r^2) - (3r + r^2)]$

For each polynomial, find **(a)** $P(-1)$ *and* **(b)** $P(2)$. *See Example 10.*

63. $P(x) = 6x - 4$ **64.** $P(x) = -2x + 5$

65. $P(x) = x^2 - 3x + 4$ **66.** $P(x) = 3x^2 + x - 5$

67. $P(x) = 5x^4 - 3x^2 + 6$ **68.** $P(x) = -4x^4 + 2x^2 - 1$

35. $(2 \times 10^2) + (4 \times 10^1) + (1 \times 10^0)$
36. Corresponding place values are not aligned in columns.
38. Corresponding powers of the variable are not aligned in columns.
41. $8z^4$ 42. $6r^5$ 43. $7m^3$
44. $6p^4$ 45. $5x$ 46. 0
47. $7y - 3y^2$ 48. $c^2 + 4$
49. $8k^2 + 2k - 7$
50. $-2x^2 + 2x - 6$
51. $-2n^4 - n^3 + n^2$
52. $q^3 + 8q^2 - 4q$
53. $3m + 11$ 54. -1
55. $-p - 4$ 56. $-8x - 12$
57. $-4p^3 + 10p^2 + 5p - 5$
58. $5y^3 - 3y^2 + 5y + 1$
59. $2x^5 - x^4 - 2x^3 + 1$
60. $-4y^2 + 2y - 4$
61. $6a + 3$ 62. $7r$
63. (a) -10 (b) 8
64. (a) 7   (b) 1   65. (a) 8
(b) 2   66. (a) -3 (b) 9
67. (a) 8   (b) 74
68. (a) -3 (b) -57

⊙ CONCEPTUAL   📝 WRITING   ▲ CHALLENGING   ▦ SCIENTIFIC CALCULATOR   ▦ GRAPHICS CALCULATOR

69. 22.5   70. 95   71. 187.5
72. 180   73. −137.5
75. $1.05 billion
76. $1.25 billion
77. $1.48 billion
78. $1.76 billion
79. $3y^2 − 4y + 2$
80. $5z^2 − 7x + 2m$
81. $−4m^2 + 4n^2 − 7n$
82. $−9m^3 − 6m^2 + 6m$
83. $y^4 − 4y^2 − 4$
84. $13rt + 7r − 5t$
85. $−3xy − 11z^2$
86. $−19p + 14$
87. $10z^2 − 16z$ 88. 20k
89. $5a^{3x} + 2a^{2x} + 3a^x + 2$
90. $−2m^{3y} − 12m^{2y} − m^y + 4m$
91. $−8k^{4p} − 9k^{2p} + 2$
92. $−r^{3x} − 2r^{2x} + r^x − 2$
93. $3p^{2k} + 8p^k − 7$
94. $−5z^{3m} − 4z^{2m} − 8z^m$
95. 0   96. $−2y^2 + 8y − 7$
97. $8z^2 + 4z + 2$
98. $−3k^3 + 3k^2 + 11k$
99. $12m^5$ 100. $35z^6$
101. $−6a^3b^9$ 102. $12k^7t^3$
103. $60x^3y^4$ 104. $18m^5n^4$

🖩 *In the Connections box at the beginning of this chapter, we saw that the polynomial* $−.006x^4 + .14x^3 − .05x^2 + 2x$ *models the concentration of dye in a cardiac patient's bloodstream x seconds after injection. Use a calculator to find the concentration of dye after the following numbers of seconds.*

69. 5   **70.** 10   **71.** 15   **72.** 20   **73.** 25

◎ **74.** What appears to be happening to the concentration of dye as time increases? What
📝   happens to the model between $x = 20$ and $x = 25$? What does it tell us about the
      concentration of dye in the bloodstream?

🖩 *The polynomial* $P(x) = .0033x^3 + .005x^2 + .1917x + 1.05$ *gives the approximate total revenues of home shopping channels, in billions of dollars, where* $x = 0$ *corresponds to 1988,* $x = 1$ *corresponds to 1989, and so on. This is an example of a mathematical model, and is valid for the years from 1988 through 1991. Use a calculator and determine these revenues for the year given. (Source: Jupiter Communications)*

75. 1988   **76.** 1989   **77.** 1990   **78.** 1991

▲ *Perform the operations.*

79. Subtract $4y^2 − 2y + 3$ from $7y^2 − 6y + 5$.

80. Subtract $−(−4x + 2z^2 + 3m)$ from $[(2z^2 − 3x + m) + (z^2 − 2m)]$.

81. Subtract $(3m^2 − 5n^2 + 2n) + [−(3m^2 − 4n^2)]$ from $−4m^2 + 3n^2 − 5n$.

82. $[−(4m^2 − 8m + 4m^3) − (3m^2 + 2m + 5m^3)] + m^2$

83. $[−(y^4 − y^2 + 1) − (y^4 + 2y^2 + 1)] + (3y^4 − 3y^2 − 2)$

84. $(8rt + 6r − 4t) + (3rt − 4r + t) − (−2rt − 5r + 2t)$

85. $[(6xy − 10z^2) + (3z^2 − 2xy)] − [(5xy + 3z^2) − (−2xy − z^2)]$

86. $(2p − [3p − 6]) − [(5p − (8 − 9p)) + 4p]$

87. $−(3z^2 + 5z − [2z^2 − 6z]) + [(8z^2 − [5z − z^2]) + 2z^2]$

88. $5k − (5k − [2k − (4k − 8k)]) + 11k − (9k − 12k)$

89. $(3a^{2x} + a^x − 4) + (5a^{3x} − a^{2x} + 2a^x + 6)$

90. $(2m^{3y} − 4m^{2y} + m) − (4m^{3y} + 8m^{2y} + m^y − 3m)$

91. $(−3k^{4p} − 6k^{2p} + 10) − (5k^{4p} + 3k^{2p} + 8)$

92. $(7r^{3x} − 5r^{2x} + 6r^x + 1) + (−8r^{3x} + 3r^{2x} − 5r^x − 3)$

93. $(5p^{2k} + 3p^k − 4) + [(6p^k − 3) − (2p^{2k} + p^k)]$

94. $[(−z^{3m} − 2z^{2m} + z^m) − 4z^{3m}] − (2z^{2m} + 9z^m)$

▲**95.** Let $P(x) = 3x^2 − 5x$ and $Q(x) = 5x − 3x^2$. Find $P(x) + Q(x)$.

▲**96.** Let $P(y) = 5y^2 − 2y + 3$ and $Q(y) = 3y^2 + 6y − 4$. Find $Q(y) − P(y)$.

▲**97.** Let $P(z) = 2z − 3z^2$, $Q(z) = 3z^2 + 2$, and $R(z) = 6z + 2z^2$. Find
      $Q(z) + R(z) − P(z)$.

▲**98.** Let $P(k) = −5k + 3k^2$, $Q(k) = 2k − 3k^3$, and $R(k) = 4k + 6k^2$. Find
      $Q(k) − P(k) + R(k)$.

REVIEW EXERCISES

Find the product. See Section 3.1.

99. $3m^3(4m^2)$ **100.** $5z^2(7z^4)$ **101.** $−3b^5(2a^3b^4)$

102. $−4k^2(−3k^5t^3)$ **103.** $12x^2y(5xy^3)$ **104.** $6mn^3(3m^4n)$

3.4 MULTIPLICATION OF POLYNOMIALS

OBJECTIVES

1 ▶ Multiply terms.
2 ▶ Multiply any two polynomials.
3 ▶ Multiply binomials.
4 ▶ Find the product of the sum and difference of two terms.
5 ▶ Find the square of a binomial.

The previous section showed how polynomials are added and subtracted. In this section we discuss polynomial multiplication.

1 ▶ **Multiply terms.**

▶ **TEACHING TIP**

Warn students that while the mechanics of multiplication presented in this section are simple, errors often occur due to carelessness with signs. ◀

▶ Recall that the product of the two terms $3x^4$ and $5x^3$ is found by using the commutative and associative properties, along with the rules for exponents.

$$(3x^4)(5x^3) = 3 \cdot 5 \cdot x^4 \cdot x^3$$
$$= 15x^{4+3}$$
$$= 15x^7$$

EXAMPLE 1

Multiplying Monomials

Find the following products.

(a) $(-4a^3)(3a^5) = (-4)(3)a^3 \cdot a^5 = -12a^8$

(b) $(2m^2z^4)(8m^3z^2) = (2)(8)m^2 \cdot m^3 \cdot z^4 \cdot z^2 = 16m^5z^6$ ■

2 ▶ **Multiply any two polynomials.**

▶ The distributive property can be used to extend this process to find the product of any two polynomials.

EXAMPLE 2

Multiplying a Monomial and a Polynomial

(a) Find the product of -2 and $8x^3 - 9x^2$.
Use the distributive property.

$$-2(8x^3 - 9x^2) = -2(8x^3) - 2(-9x^2)$$
$$= -16x^3 + 18x^2$$

For Example 1

Chalkboard Exercise

Find the product.

$$(8k^3y)(9ky^3)$$

Answer: $72k^4y^4$

(b) Find the product of $5x^2$ and $-4x^2 + 3x - 2$.

$$5x^2(-4x^2 + 3x - 2) = 5x^2(-4x^2) + 5x^2(3x) + 5x^2(-2)$$
$$= -20x^4 + 15x^3 - 10x^2$$ ■

EXAMPLE 3

Multiplying Two Polynomials

Find the product of $3x - 4$ and $2x^2 + x$.
Use the distributive property to multiply each term of $2x^2 + x$ by $3x - 4$.

$$(3x - 4)(2x^2 + x) = (3x - 4)(2x^2) + (3x - 4)(x)$$

For Example 2

Chalkboard Exercise

Find the product.

$$-2r(9r - 5)$$

Answer: $-18r^2 + 10r$

Here $3x - 4$ has been treated as a single expression so that the distributive property could be used. Now use the distributive property twice again.

$$= 3x(2x^2) + (-4)(2x^2) + (3x)(x) + (-4)(x)$$
$$= 6x^3 - 8x^2 + 3x^2 - 4x$$
$$= 6x^3 - 5x^2 - 4x$$ ■

It is often easier to multiply polynomials by writing them vertically. To find the product from Example 3, $(3x - 4)(2x^2 + x)$, vertically, proceed as follows. (Notice how this process is similar to that of finding the product of two numbers, such as 24×78.)

1. Multiply x and $3x - 4$.

$$\begin{array}{r} 3x - 4 \\ 2x^2 + x \\ \hline \end{array}$$
$$x(3x - 4) \rightarrow 3x^2 - 4x$$

2. Multiply $2x^2$ and $3x - 4$. Line up like terms of the products in columns.

$$\begin{array}{r} 3x - 4 \\ 2x^2 + x \\ \hline 3x^2 - 4x \end{array}$$
$$2x^2(3x - 4) \rightarrow 6x^3 - 8x^2$$

3. Combine like terms.

$$\overline{6x^3 - 5x^2 - 4x}$$

E X A M P L E 4

Multiplying Polynomials
Vertically

Find the product of $5a - 2b$ and $3a + b$.

$$\begin{array}{r} 5a - 2b \\ 3a + b \\ \hline 5ab - 2b^2 \qquad \leftarrow b(5a - 2b) \\ 15a^2 - 6ab \qquad \leftarrow 3a(5a - 2b) \\ \hline 15a^2 - ab - 2b^2 \end{array}$$

E X A M P L E 5

Multiplying Polynomials
Vertically

Find the product of $3m^3 - 2m^2 + 4$ and $3m - 5$.

$$\begin{array}{r} 3m^3 - 2m^2 + 4 \\ 3m - 5 \\ \hline -15m^3 + 10m^2 \qquad\quad - 20 \qquad \text{-5 times $3m^3 - 2m^2 + 4$} \\ 9m^4 - 6m^3 \qquad\quad + 12m \qquad\qquad \text{$3m$ times $3m^3 - 2m^2 + 4$} \\ \hline 9m^4 - 21m^3 + 10m^2 + 12m - 20 \qquad \text{Combine like terms.} \end{array}$$

3 ▶ Multiply binomials.

▶ In working with polynomials, the product of two binomials occurs repeatedly. In the rest of this section, we discuss a shortcut method for finding these products. You will recall that a binomial is a polynomial with just two terms, such as $3x - 4$ or $2x + 3$. We can find the product of the binomials $3x - 4$ and $2x + 3$ using the distributive property as follows.

$$\begin{aligned} (3x - 4)(2x + 3) &= 3x(2x + 3) - 4(2x + 3) \\ &= 3x(2x) + 3x(3) - 4(2x) - 4(3) \\ &= 6x^2 + 9x - 8x - 12 \end{aligned}$$

Before combining like terms to find the simplest form of the answer, let us check the origin of each of the four terms in the sum. First, $6x^2$ is the product of the two *first* terms.

$$(3x - 4)(2x + 3) \qquad (3x)(2x) = 6x^2 \qquad \text{First terms}$$

To get $9x$, the *outside* terms are multiplied.

$$(3x - 4)(2x + 3) \qquad 3x(3) = 9x \qquad \text{Outside terms}$$

The term $-8x$ comes from the *inside* terms

$$(3x - 4)(2x + 3) \qquad -4(2x) = -8x \qquad \text{Inside terms}$$

Finally, -12 comes from the *last* terms.

$$(3x - 4)(2x + 3) \qquad -4(3) = -12 \qquad \text{Last terms}$$

The product is found by combining these four results.

$$(3x - 4)(2x + 3) = 6x^2 + 9x - 8x - 12$$
$$= 6x^2 + x - 12$$

To keep track of the order of multiplying these terms, we use the initials FOIL (first, outside, inside, last). All the steps of the FOIL method can be done as follows. Try to do as many of these steps as possible in your head.

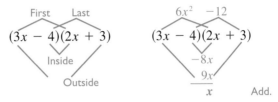

The FOIL method will be very helpful in factoring, which is discussed in the rest of this chapter.

Use the FOIL method to find $(4m - 5)(3m + 1)$.
Find the product of the first terms.

$$(4m - 5)(3m + 1) \qquad (4m)(3m) = 12m^2$$

Multiply the outside terms.

$$(4m - 5)(3m + 1) \qquad (4m)(1) = 4m$$

Find the product of the inside terms.

$$(4m - 5)(3m + 1) \qquad (-5)(3m) = -15m$$

Multiply the last terms.

$$(4m - 5)(3m + 1) \qquad (-5)(1) = -5$$

Simplify by combining the four terms obtained above.

$$(4m - 5)(3m + 1) = 12m^2 + 4m - 15m - 5$$
$$= 12m^2 - 11m - 5$$

The procedure can be written in compact form as follows.

$$
\begin{array}{c}
\overset{12m^2 \qquad -5}{\overgroup{(4m - 5)(3m + 1)}} \\
-15m \\
4m \\
\hline
-11m \qquad \text{Add.}
\end{array}
$$

Combine these four results to get $12m^2 - 11m - 5$. ■

EXAMPLE 7

Using the FOIL Method

(a) $(6a - 5b)(3a + 4b) = 18a^2 + 24ab - 15ab - 20b^2$

First   Outside   Inside   Last

$= 18a^2 + 9ab - 20b^2$

(b) $(2k + 3z)(5k - 3z) = 10k^2 - 6kz + 15kz - 9z^2$

$= 10k^2 + 9kz - 9z^2$ ∎

4 ▶ Find the product of the sum and difference of two terms.

▶ A special type of binomial product that occurs frequently is the product of the sum and difference of the same two terms. By the FOIL method, the product $(x + y)(x - y)$ is

$$(x + y)(x - y) = x^2 - xy + xy - y^2$$
$$= x^2 - y^2.$$

Product of the Sum and Difference of Two Terms

The **product of the sum and difference of two terms** is the difference of the squares of the terms, or

$$(x + y)(x - y) = x^2 - y^2.$$

EXAMPLE 8

Multiplying the Sum and Difference of Two Terms

Find the following products.

(a) $(p + 7)(p - 7) = p^2 - 7^2 = p^2 - 49$

(b) $(2r + 5)(2r - 5) = (2r)^2 - 5^2$

$= 2^2 r^2 - 25$

$= 4r^2 - 25$

(c) $(6m + 5n)(6m - 5n) = (6m)^2 - (5n)^2$

$= 36m^2 - 25n^2$ ∎

◆ Answers will vary.

◆**CONNECTIONS**◆

The special product $(a + b)(a - b) = a^2 - b^2$ can be used to perform some multiplication problems. For example,

$$51 \times 49 = (50 + 1)(50 - 1)$$
$$= 50^2 - 1^2 = 2500 - 1$$
$$= 2499$$
$$102 \times 98 = (100 + 2)(100 - 2)$$
$$= 100^2 - 2^2$$
$$= 10{,}000 - 4$$
$$= 9996.$$

Once these patterns are recognized, multiplications of this type can be done mentally.

FOR DISCUSSION OR WRITING

With a friend, make up some problems of this type and challenge one another.

▶ Another special binomial product is the *square of a binomial*. To find the square of $x + y$, or $(x + y)^2$, multiply $x + y$ and $x + y$.

$$(x + y)^2 = (x + y)(x + y) = x^2 + xy + xy + y^2$$
$$= x^2 + 2xy + y^2$$

A similar result is true for the square of a difference.

Square of a Binomial

The **square of a binomial** is the sum of the square of the first term, twice the product of the two terms, and the square of the last term.

$$(x + y)^2 = x^2 + 2xy + y^2$$
$$(x - y)^2 = x^2 - 2xy + y^2$$

EXAMPLE 9

Squaring a Binomial

Find the following products.

(a) $(m + 7)^2 = m^2 + 2 \cdot m \cdot 7 + 7^2$
$$= m^2 + 14m + 49$$

(b) $(p - 5)^2 = p^2 - 2 \cdot p \cdot 5 + 5^2$
$$= p^2 - 10p + 25$$

(c) $(2p + 3v)^2 = (2p)^2 + 2(2p)(3v) + (3v)^2$
$$= 4p^2 + 12pv + 9v^2$$

(d) $(3r - 5s)^2 = (3r)^2 - 2(3r)(5s) + (5s)^2$
$$= 9r^2 - 30rs + 25s^2 \quad ■$$

CAUTION As the products in the definition of the square of a binomial show,

$$(x + y)^2 \neq x^2 + y^2.$$

Also, more generally,

$$(x + y)^n \neq x^n + y^n.$$

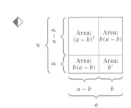

◆ CONNECTIONS ◆

The special product

$$(a + b)^2 = a^2 + 2ab + b^2$$

can be illustrated geometrically using the diagram shown here. The side of the large square has length $a + b$, so the area of the square is

$$(a + b)^2.$$

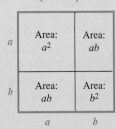

The large square is made up of two smaller squares and two congruent rectangles. The sum of the areas of these figures is

$$a^2 + 2ab + b^2.$$

Since these expressions represent the same quantity, they must be equal, thus giving us the pattern for squaring a binomial.

FOR DISCUSSION OR WRITING
Draw a figure and give a similar proof for $(a - b)^2 = a^2 - 2ab + b^2$.

The special products of this section are now summarized.

Special Products

Product of the Sum and Difference of Two Terms

$$(x + y)(x - y) = x^2 - y^2$$

Square of a Binomial

$$(x + y)^2 = x^2 + 2xy + y^2$$
$$(x - y)^2 = x^2 - 2xy + y^2$$

We can use the patterns for the product of the sum and difference of two terms and the square of a binomial with more complicated products, as the following example shows.

EXAMPLE 10
Multiplying More Complicated Binomials

Use the special products to multiply the following polynomials.

(a) $[(3p - 2) + 5q][(3p - 2) - 5q]$

$= (3p - 2)^2 - (5q)^2$ Product of sum and difference of terms

$= 9p^2 - 12p + 4 - 25q^2$ Square both quantities.

(b) $[(2z + r) + 1]^2 = (2z + r)^2 + 2(2z + r)(1) + 1^2$ Square of a binomial

$= 4z^2 + 4zr + r^2 + 4z + 2r + 1$ Square again; distributive property. ∎

Chalkboard Exercise

Find the product.

$(x - y + z)(x - y - z)$

Answer: $x^2 - 2xy + y^2 - z^2$

3.4 EXERCISES

3. $-24m^5$ 4. $-20p^6$
5. $-6x^2 + 15x$
6. $-30y^2 - 5y$
7. $-2q^3 - 3q^4$
8. $-12a^4 + 3a^5$
9. $18k^4 + 12k^3 + 6k^2$
10. $10r^5 - 15r^4 - 20r^3$
11. $6m^3 + m^2 - 14m - 3$
12. $4z^3 + 10z^2 + 14z - 10$
13. $-d^4 + 6d^3 + 2d^2 - 13d + 6$
14. $-q^4 + 3q^3 - 6q^2 + 16q + 6$

 1. Explain in your own words how to find the product of two monomials with numerical coefficients.

 2. Explain in your own words how to find the product of two polynomials.

Find each product. See Examples 1–7.

3. $-8m^3(3m^2)$ **4.** $4p^2(-5p^4)$ **5.** $3x(-2x + 5)$

6. $5y(-6y - 1)$ **7.** $-q^3(2 + 3q)$ **8.** $-3a^4(4 - a)$

9. $6k^2(3k^2 + 2k + 1)$ **10.** $5r^3(2r^2 - 3r - 4)$

11. $(2m + 3)(3m^2 - 4m - 1)$ **12.** $(4z - 2)(z^2 + 3z + 5)$

13. $(-d + 6)(d^3 - 2d + 1)$ **14.** $(-q + 3)(q^3 + 6q + 2)$

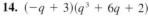

◉ CONCEPTUAL   ✐ WRITING   ▲ CHALLENGING   ▦ SCIENTIFIC CALCULATOR   ▦ GRAPHICS CALCULATOR

15. $2y + 3$
  $3y - 4$

16. $5m - 3$
  $2m + 6$

17. $-b^2 + 3b + 3$
  $2b + 4$

18. $-r^2 - 4r + 8$
  $3r - 2$

19. $5m - 3n$
  $5m + 3n$

20. $2k + 6q$
  $2k - 6q$

21. $(m + 5)(m - 8)$

22. $(p - 6)(p + 4)$

23. $(2z + 1)(3z - 4)$

24. $(8y - 3)(2y + 1)$

25. $3k - 2$
  $5k + 1$

26. $2m - 5$
  $m + 7$

27. $6m^2 + 2m - 1$
  $2m + 3$

28. $-y^2 + 2y + 1$
  $3y - 5$

29. $2z^3 - 5z^2 + 8z - 1$
  $4z + 3$

30. $3z^4 - 2z^3 + \ z - 5$
  $2z - 5$

31. $-x^2 + 8x - 3$
  $2x^3 + 5x$

32. $2k^2 + 6k + 5$
  $-k^2 + 2k$

33. $2p^2 + 3p + 6$
  $3p^2 - 4p - 1$

34. $5y^2 - 2y + 4$
  $2y^2 + \ y + 3$

⊙ **35.** What type of polynomials can be multiplied by the FOIL method? Describe the method in your own words.

⊙ **36.** Explain why the product of the sum and difference of two terms is not a trinomial.

Find the product. See Example 8.

37. $(2p - 3)(2p + 3)$

38. $(3x - 8)(3x + 8)$

39. $(5m - 1)(5m + 1)$

40. $(6y + 3)(6y - 3)$

41. $(3a + 2c)(3a - 2c)$

42. $(5r - 4s)(5r + 4s)$

43. $\left(4x - \dfrac{2}{3}\right)\left(4x + \dfrac{2}{3}\right)$

44. $\left(3t + \dfrac{5}{4}\right)\left(3t - \dfrac{5}{4}\right)$

45. $(4m + 7n^2)(4m - 7n^2)$

46. $(2k^2 + 6h)(2k^2 - 6h)$

47. $(5y^3 + 2)(5y^3 - 2)$

48. $(3x^3 + 4)(3x^3 - 4)$

Find the square. See Example 9.

49. $(y - 5)^2$

50. $(a - 3)^2$

51. $(2p + 7)^2$

52. $(3z + 8)^2$

53. $(4n - 3m)^2$

54. $(5r - 7s)^2$

55. $\left(k - \dfrac{5}{7}p\right)^2$

56. $\left(q - \dfrac{3}{4}r\right)^2$

⊙ **57.** Explain how the expressions $(x + y)^2$ and $x^2 + y^2$ differ.

⊙ **58.** Explain how you can find the product $101 \cdot 99$ using the special product $(a + b)(a - b) = a^2 - b^2$.

Find the product. See Example 10.

59. $[(5x + 1) + 6y]^2$

60. $[(3m - 2) + p]^2$

61. $[(2a + b) - 3]^2$

62. $[(4k + h) - 4]^2$

63. $[(2a + b) - 3][(2a + b) + 3]$

64. $[(m + p) + 5][(m + p) - 5]$

▲ *Find the product.*

65. $(2a + b)(3a^2 + 2ab + b^2)$

66. $(m - 5p)(m^2 - 2mp + 3p^2)$

67. $(4z - x)(z^3 - 4z^2x + 2zx^2 - x^3)$

68. $(3r + 2s)(r^3 + 2r^2s - rs^2 + 2s^3)$

69. $(m^2 - 2mp + p^2)(m^2 + 2mp - p^2)$

70. $(3 + x + y)(-3 + x - y)$

⊙ CONCEPTUAL   ✐ WRITING   ▲ CHALLENGING   ▦ SCIENTIFIC CALCULATOR   ▦ GRAPHICS CALCULATOR

15. $6y^2 + y - 12$
16. $10m^2 + 24m - 18$
17. $-2b^3 + 2b^2 + 18b + 12$
18. $-3r^3 - 10r^2 + 32r - 16$
19. $25m^2 - 9n^2$
20. $4k^2 - 36q^2$
21. $m^2 - 3m - 40$
22. $p^2 - 2p - 24$
23. $6z^2 - 5z - 4$
24. $16y^2 + 2y - 3$
25. $15k^2 - 7k - 2$
26. $2m^2 + 9m - 35$
27. $12m^3 + 22m^2 + 4m - 3$
28. $-3y^3 + 11y^2 - 7y - 5$
29. $8z^4 - 14z^3 + 17z^2 + 20z - 3$ 30. $6z^5 - 19z^4 + 10z^3 + 2z^2 - 15z + 25$ 31. $-2x^5 + 16x^4 - 11x^3 + 40x^2 - 15x$
32. $-2k^4 - 2k^3 + 7k^2 + 10k$ 33. $6p^4 + p^3 + 4p^2 - 27p - 6$
34. $10y^4 + y^3 + 21y^2 - 2y + 12$
37. $4p^2 - 9$ 38. $9x^2 - 64$
39. $25m^2 - 1$
40. $36y^2 - 9$
41. $9a^2 - 4c^2$
42. $25r^2 - 16s^2$
43. $16x^2 - \dfrac{4}{9}$
44. $9t^2 - \dfrac{25}{16}$
45. $16m^2 - 49n^4$
46. $4k^4 - 36h^2$
47. $25y^6 - 4$
48. $9x^6 - 16$
49. $y^2 - 10y + 25$
50. $a^2 - 6a + 9$
51. $4p^2 + 28p + 49$
52. $9z^2 + 48z + 64$
53. $16n^2 - 24nm + 9m^2$
54. $25r^2 - 70rs + 49s^2$
55. $k^2 - \dfrac{10}{7}kp + \dfrac{25}{49}p^2$
56. $q^2 - \dfrac{3}{2}qr + \dfrac{9}{16}r^2$
59. $25x^2 + 10x + 1 + 60xy + 12y + 36y^2$
60. $9m^2 - 12m + 4 + 6mp - 4p + p^2$
61. $4a^2 + 4ab + b^2 - 12a - 6b + 9$
62. $16k^2 + 8kh + h^2 - 32k - 8h + 16$

63. $4a^2 + 4ab + b^2 - 9$
64. $m^2 + 2mp + p^2 - 25$
65. $6a^3 + 7a^2b + 4ab^2 + b^3$ 66. $m^3 - 7m^2p + 13mp^2 - 15p^3$
67. $4z^4 - 17z^3x + 12z^2x^2 - 6zx^3 + x^4$
68. $3r^4 + 8r^3s + r^2s^2 + 4rs^3 + 4s^4$ 69. $m^4 - 4m^2p^2 + 4mp^3 - p^4$
70. $-9 - 6y + x^2 - y^2$
71. $a^3 - 7ab^2 - 6b^3$
72. $2m^3 - 5m^2p + mp^2 + 2p^3$ 73. $y^3 + 6y^2 + 12y + 8$ 74. $z^3 - 9z^2 + 27z - 27$ 75. $q^4 - 8q^3 + 24q^2 - 32q + 16$
76. $r^4 + 12r^3 + 54r^2 + 108r + 81$
77. 49;   25;   49 $\neq$ 25
78. 343;   91;   343 $\neq$ 91
79. 2401;   337; 2401 $\neq$ 337
80. 16,807; 1267;   16,807 $\neq$ 1267   81. Although they are equal for this *particular* case, they are not equal *in general.*
82. (d)   83. $\dfrac{9}{2}x^2 - 2y^2$
84. $x^4 + 16x^2 + 64$
85. $15x^2 - 2x - 24$
86. $x^3 + \dfrac{7}{2}x^2 + 7x + 6$
87. $a - b$
88. $A = s^2$; $(a - b)^2$
89. $(a - b)b$ or $ab - b^2$; $2ab - 2b^2$ 90. b^2

71. $(a + b)(a + 2b)(a - 3b)$
72. $(m - p)(m - 2p)(2m + p)$
73. $(y + 2)^3$
74. $(z - 3)^3$
75. $(q - 2)^4$
76. $(r + 3)^4$

In Exercises 77–80, two expressions are given. Replace x with 3 and y with 4 to show that, in general, the two expressions are not equal to each other.

77. $(x + y)^2$; $x^2 + y^2$
78. $(x + y)^3$; $x^3 + y^3$
79. $(x + y)^4$; $x^4 + y^4$
80. $(x + y)^5$; $x^5 + y^5$

⊙ 81. A student claims that the two expressions in Exercise 77 *must* be equal to each other, because if we let $x = 0$ and $y = 1$, each expression simplifies to 1. How would you respond to the student's claim?

82. Which one of the following is equivalent to $(x + y)^{-2}$?

(a) $\dfrac{1}{(x + y)^{-2}}$ (b) $\dfrac{1}{x^{-2} + y^{-2}}$ (c) $x^{-2} + y^{-2}$ (d) $\dfrac{1}{x^2 + 2xy + y^2}$

Write a mathematical expression for the area of the figure. Express it as a polynomial in descending powers of the variable x.

83.

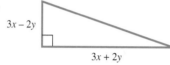

$3x - 2y$

$3x + 2y$

84.

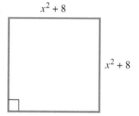

$x^2 + 8$

$x^2 + 8$

85.

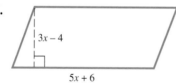

$3x - 4$

$5x + 6$

86.

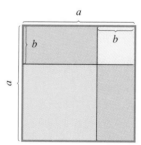

$x^2 + 2x + 4$

$2x + 3$

◆ **MATHEMATICAL CONNECTIONS** (Exercises 87–94) ◆

Consider the figure. Work Exercises 87–94 in order.

87. What is the length of each side of the blue square in terms of a and b?

88. What is the formula for the area of a square? Use the formula to write an expression, in the form of a product, for the area of the blue square.

89. Each green rectangle has an area of _____ . Therefore, the total area in green is represented by the polynomial _____ .

90. The yellow square has an area of _____ .

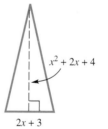

a

b

b

a

91. The area of the entire colored region is represented by _____ , because each side of the entire colored region has length _____ .

92. The area of the blue square is equal to the area of the entire colored region minus the total area of the green squares minus the area of the yellow square. Write this as a simplified polynomial in a and b.

93. What must be true about the expressions for the area of the blue square you found in Exercises 88 and 92?

94. Write a statement of equality based on your answer in Exercise 93. How does this reinforce one of the main ideas of this section?

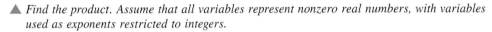

▲ *Find the product. Assume that all variables represent nonzero real numbers, with variables used as exponents restricted to integers.*

95. $6b^p(2b^{p-2} + b^{2p})$

96. $-5r^a(3r^5 - 2r^{-a})$

97. $z^{m-4}(3z^{m+2} - 4z^{-m})$

98. $y^{k-5}(2y^{3-k} - 7y^{5k})$

99. $(6k^n + 1)(2k^n - 3)$

100. $(9y^z - 2)(y^z + 5)$

101. $(8k^m - 3y^{3m})(2k^{2m} + 7y^m)$

102. $(4q^{2z} - p^{5z})(3q^{5z} - p^{2z})$

103. $(y^{2n} + 1)(y^{2n} - 1)$

104. $(3z^{5p} + 2)(3z^{5p} - 2)$

REVIEW EXERCISES

Use the distributive property to rewrite the expression. See Section 1.3.

105. $9 \cdot 6 + 9 \cdot r^2$

106. $8 \cdot y - 8 \cdot 5t$

107. $7(2x) - 7(3z)$

108. $4(8p) + 4(9y)$

109. $3x(x + 1) + 4(x + 1)$

110. $4p(2p - 3) + 5(2p - 3)$

Simplify. Write the answer with positive exponents. Assume all variables represent positive numbers. See Section 3.2.

111. $18z^3w(zw^2)^4$

112. $8z^4(a^2b)^5$

113. $12p^4q^{-2}(5pq)^{-1}$

114. $(5m^2n)^3(-mn^{-1})^2$

Answers (right column):

91. a^2; a
92. $a^2 - (2ab - 2b^2) - b^2 = a^2 - 2ab + b^2$
93. They must be equal to each other.
94. $(a - b)^2 = a^2 - 2ab + b^2$; This reinforces the special product for the square of a binomial difference.
95. $12b^{2p-2} + 6b^{3p}$
96. $-15r^{a+5} + 10$
97. $3z^{2m-2} - 4z^{-4}$
98. $2y^{-2} - 7y^{6k-5}$
99. $12k^{2n} - 16k^n - 3$
100. $9y^{2z} + 43y^z - 10$
101. $16k^{3m} + 56k^my^m - 6y^{3m}k^{2m} - 21y^{4m}$
102. $12q^{7z} - 4q^{2z}p^{2z} - 3p^{5z}q^{5z} + p^{7z}$
103. $y^{4n} - 1$
104. $9z^{10p} - 4$
105. $9(6 + r^2)$
106. $8(y - 5t)$
107. $7(2x - 3z)$
108. $4(8p + 9y)$
109. $(3x + 4)(x + 1)$
110. $(4p + 5)(2p - 3)$
111. $18z^7w^9$
112. $8a^{10}b^5z^4$
113. $\dfrac{12p^3}{5q^3}$ 114. $125m^8n$

3.5 GREATEST COMMON FACTORS; FACTORING BY GROUPING

FOR EXTRA HELP	OBJECTIVES
📖 **SSG** pp. 79–82 **SSM** pp. 107–110	1 ▶ Factor out the greatest common factor.
📼 **Video** 3	2 ▶ Factor by grouping.
💾 **Tutorial** IBM MAC	

Writing a polynomial as the product of two or more simpler polynomials is called **factoring** the polynomial. For example, the product of $3x$ and $5x - 2$ is $15x^2 - 6x$, and $15x^2 - 6x$ can be factored as the product $3x(5x - 2)$.

$$3x(5x - 2) = 15x^2 - 6x \qquad \text{Multiplication}$$

$$15x^2 - 6x = 3x(5x - 2) \qquad \text{Factoring}$$

INSTRUCTOR'S RESOURCES

 **ITM** pp. 355–357   **ISM** pp. 165–171   **IAM** p. 11   **TEST GENERATOR** DOS Windows MAC   **TRANSPARENCIES**

 CONCEPTUAL   WRITING   ▲ CHALLENGING   SCIENTIFIC CALCULATOR    GRAPHICS CALCULATOR

Notice that both multiplication and factoring are examples of the distributive property, used in opposite directions. Factoring is the reverse of multiplying.

1 ▶ Factor out the greatest common factor.

▶ **TEACHING TIP**
Mention that the greatest common factor is never larger than the smallest coefficient (although it can be smaller). ◀

▶ The first step in factoring a polynomial is to find the *greatest common factor* for the terms of the polynomial. The **greatest common factor** is the largest term that is a factor of all terms in the polynomial. For example, the greatest common factor for $8x + 12$ is 4, since 4 is the largest number that is a factor of both $8x$ and 12. Using the distributive property.

$$8x + 12 = 4(2x) + 4(3) = 4(2x + 3).$$

As a check, multiply 4 and $2x + 3$. The result should be $8x + 12$. This process is called **factoring out the greatest common factor.**

E X A M P L E   1

Factoring Out the Greatest Common Factor

Chalkboard Exercise

Factor out the greatest common factor.

$$5z + 5$$

Answer: $5(z + 1)$

Factor out the greatest common factor.

(a) $9z - 18$
Since 9 is the greatest common factor,

$$9z - 18 = 9 \cdot z - 9 \cdot 2 = 9(z - 2).$$

(b) $56m + 35p = 7(8m + 5p)$

(c) $2y + 5$ There is no common factor other than 1.

(d) $12 + 24z$

$$12 + 24z = 12 \cdot 1 + 12 \cdot 2z$$
$$= 12(1 + 2z) \qquad \text{12 is the greatest common factor.}$$

(e) $r(3x + 2) - 8s(3x + 2)$
In this polynomial, the greatest common factor is the binomial $(3x + 2)$.

$$r(3x + 2) - 8s(3x + 2) = (3x + 2)(r - 8s) \quad ■$$

CAUTION   Refer to Example 1(d). It is very common to forget to write the 1. Be careful to include it as needed. Remember to check answers by multiplication.

E X A M P L E   2

Factoring Out the Greatest Common Factor

Chalkboard Exercise

Factor out the greatest common factor.

$$100m^5 - 50m^4 + 25m^3$$

Answer: $25m^3(4m^2 - 2m + 1)$

▶ **TEACHING TIP**
Include with your examples one that cannot be factored and one with a common numerical factor but no common variable factor in all terms. ◀

Factor out the greatest common factor.

(a) $9x^2 + 12x^3$
The numerical part of the greatest common factor is 3, the largest number that divides into both 9 and 12. For the variable parts, x^2 and x^3, use the least exponent that appears on x; here the least exponent is 2. The greatest common factor is $3x^2$.

$$9x^2 + 12x^3 = 3x^2(3) + 3x^2(4x)$$
$$= 3x^2(3 + 4x)$$

(b) $32p^4 - 24p^3 + 40p^5$
The greatest common numerical factor is 8. Since the least exponent on p is 3, the greatest common factor is $8p^3$.

$$32p^4 - 24p^3 + 40p^5 = 8p^3(4p) + 8p^3(-3) + 8p^3(5p^2)$$
$$= 8p^3(4p - 3 + 5p^2)$$

(c) $3k^4 - 15k^7 + 24k^9 = 3k^4(1 - 5k^3 + 8k^5) \quad ■$

Factor out the greatest common factor.

(a) $24m^3n^2 - 18m^2n + 6m^4n^3$

The numerical part of the greatest common factor is 6. Find the variable part by writing each variable with its least exponent. Here 2 is the least exponent on m that appears, while 1 is the least exponent on n. Finally, $6m^2n$ is the greatest common factor.

$$24m^3n^2 - 18m^2n + 6m^4n^3$$
$$= (6m^2n)(4mn) + (6m^2n)(-3) + (6m^2n)(m^2n^2)$$
$$= 6m^2n(4mn - 3 + m^2n^2)$$

(b) $25x^2y^3 + 30y^5 - 15x^4y^7 = 5y^3(5x^2 + 6y^2 - 3x^4y^4)$ ∎

Factor out the greatest common factor from

$$-a^3 + 3a^2 - 5a.$$

There are two ways to factor this polynomial, both of which are correct. First, a could be used as the common factor, giving

$$-a^3 + 3a^2 - 5a = a(-a^2) + a(3a) + a(-5)$$
$$= a(-a^2 + 3a - 5).$$

Alternatively, $-a$ could be used as the common factor.

$$-a^3 + 3a^2 - 5a = -a(a^2) + (-a)(-3a) + (-a)(5)$$
$$= -a(a^2 - 3a + 5) \quad ∎$$

NOTE In Example 4 we showed two ways of factoring a polynomial. Sometimes, in a particular problem there will be a reason to prefer one of these forms over the other, but both are correct. The answer section in this book will usually give only one of these forms, the one where the common factor has a positive coefficient, but either is correct.

2 ▶ Factor by grouping.

▶ Sometimes a polynomial has a greatest common factor of 1, but it still may be possible to factor the polynomial by a suitable rearrangement of the terms. This process is called **factoring by grouping.** It is usually used when a polynomial has more than three terms.

For example, to factor the polynomial

$$ax - ay + bx - by,$$

group the terms as follows.

Terms with common factors
$$\downarrow \qquad \downarrow$$
$$(ax - ay) + (bx - by)$$

Then factor $ax - ay$ as $a(x - y)$ and factor $bx - by$ as $b(x - y)$ to get

$$ax - ay + bx - by = a(x - y) + b(x - y).$$

On the right, the common factor is $x - y$. The final factored form is

$$ax - ay + bx - by = (x - y)(a + b).$$

EXAMPLE 5

Factoring by Grouping

Factor $3x - 3y - ax + ay$.

Grouping terms as above gives

$$(3x - 3y) + (-ax + ay),$$

or

$$3(x - y) + a(-x + y).$$

There is no simple common factor here. However, if we factor out $-a$, we get

$$3(x - y) - a(x - y),$$

which equals

$$(x - y)(3 - a). \quad \blacksquare$$

NOTE In Example 5, different grouping would lead to the product

$$(a - 3)(y - x).$$

As we can verify by multiplication, this is another correct factorization.

In some cases, factoring by grouping requires that the terms be rearranged before the groupings are made.

EXAMPLE 6

Factoring by Grouping

Factor $p^2q^2 - 10 - 2q^2 + 5p^2$.

The first two terms have no common factor except 1 (nor do the last two terms). We can group the terms as follows.

$$(p^2q^2 - 2q^2) + (5p^2 - 10) \qquad \text{Group terms.}$$
$$q^2(p^2 - 2) + 5(p^2 - 2) \qquad \text{Factor out the common factors.}$$
$$(p^2 - 2)(q^2 + 5) \qquad \text{Factor out } p^2 - 2. \quad \blacksquare$$

▶ **TEACHING TIP**

After showing several examples of factoring by grouping, mention that the binomial factors after the first step must be identical. This fact helps to determine the coefficient of the second term before completing the factorization. ◀

CAUTION It is a common error to stop at the step

$$q^2(p^2 - 2) + 5(p^2 - 2).$$

This expression is *not in factored form* because it is a *sum* of two terms: $q^2(p^2 - 2)$ and $5(p^2 - 2)$.

EXAMPLE 7

Factoring by Grouping

Factor each of the following polynomials by grouping.

(a) $6x^2 - 4x - 15x + 10$

Work as above. Note that we must factor -5 rather than 5 from the second group in order to get a common factor of $3x - 2$.

$$(6x^2 - 4x) + (-15x + 10) = 2x(3x - 2) - 5(3x - 2) \qquad \text{Factor out } 2x \text{ and } -5.$$
$$= (3x - 2)(2x - 5) \qquad \text{Factor out } (3x - 2).$$

(b) $6ax + 12bx + a + 2b$

Group the first two terms and the last two terms.

$$6ax + 12bx + a + 2b = (6ax + 12bx) + (a + 2b)$$

Now factor $6x$ from the first group, and use the identity property of multiplication to introduce the factor 1 in the second group.

$$(6ax + 12bx) + (a + 2b) = 6x(a + 2b) + 1(a + 2b)$$
$$= (a + 2b)(6x + 1) \qquad \text{Factor out } (a + 2b).$$

Again, as in Example 1(d), be careful not to forget the 1. ■

The steps used in factoring by grouping are listed below.

Factoring by Grouping	*Step 1*	**Group terms.** Collect the terms into groups so that each group has a common factor.
	Step 2	**Factor within the groups.** Factor out the common factor in each group.
	Step 3	**Factor the entire polynomial.** If each group now has a common factor, factor it out. If not, try a different grouping.

3.5 EXERCISES

⊙ **1.** Explain in your own words what it means to factor a polynomial.

✎

⊙ **2.** What is the first step in attempting to factor a polynomial?

3. What is the greatest common factor of the following terms? $7z^2(m + n)^4, 9z^3(m + n)^5$

⊙ **4.** Which one of the following is an example of a polynomial in factored form?
  (a) $3x^2y^3 + 6x^2(2x + y)$ **(b)** $5(x + y)^2 - 10(x + y)^3$
  (c) $(-2 + 3x)[5y^2 + 4y + 3]$ **(d)** $(3x + 4)(5x - y) - (3x + 4)(2x - 1)$

Factor out the greatest common factor. Simplify the factors, if possible. See Examples 1–4.

5. $12m + 60$ **6.** $15r - 27$ **7.** $8k^3 + 24k$
5. $12(m + 5)$ 6. $3(5r - 9)$ 7. $8k(k^2 + 3)$

8. $9z^4 + 72z$ **9.** $xy - 5xy^2$ **10.** $5h^2j + hj$
8. $9z(z^3 + 8)$ 9. $xy(1 - 5y)$ 10. $hj(5h + 1)$

11. $-4p^3q^4 - 2p^2q^5$ **12.** $-3z^5w^2 - 18z^3w^4$
11. $-2p^2q^4(2p + q)$ 12. $-3z^3w^2(z^2 + 6w^2)$

13. $21x^5 + 35x^4 - 14x^3$ **14.** $6k^3 - 36k^4 + 48k^5$
13. $7x^3(3x^2 + 5x - 2)$ 14. $6k^3(1 - 6k + 8k^2)$

15. $15a^2c^3 - 25ac^2 + 5ac$ **16.** $15y^3z^3 + 27y^2z^4 - 36yz^5$
15. $5ac(3ac^2 - 5c + 1)$ 16. $3yz^3(5y^2 + 9yz - 12z^2)$

17. $-27m^3p^5 + 36m^4p^3 - 72m^5p^4$ **18.** $-50r^4t^2 + 80r^3t^3 - 90r^2t^4$
17. $-9m^3p^3(3p^2 - 4m + 8m^2p)$ 18. $-10r^2t^2(5r^2 - 8rt + 9t^2)$

19. $(m - 4)(m + 2) + (m - 4)(m + 3)$ **20.** $(z - 5)(z + 7) + (z - 5)(z + 9)$
19. $(m - 4)(2m + 5)$ 20. $2(z - 5)(z + 8)$

21. $(2z - 1)(z + 6) - (2z - 1)(z - 5)$ **22.** $(3x + 2)(x - 4) - (3x + 2)(x + 8)$
21. $11(2z - 1)$ 22. $-12(3x + 2)$

23. $-y^5(r + w) - y^6(z + k)$ **24.** $-r^6(m + n) - r^7(p + q)$
23. $-y^5(r + w + yz + yk)$ 24. $-r^6(m + n + rp + rq)$

25. $5(2 - x)^2 - (2 - x)^3 + 4(2 - x)$ **26.** $3(5 - x)^4 + 2(5 - x)^3 - (5 - x)^2$
25. $(2 - x)(10 - x - x^2)$ 26. $(5 - x)^2(3x^2 - 32x + 84)$ \

▲ **27.** $4(3 - x)^2 - (3 - x)^3 + 3(3 - x)$ ▲ **28.** $2(t - s) + 4(t - s)^2 - (t - s)^3$
27. $(3 - x)(6 + 2x - x^2)$ 28. $(t - s)(2 + 4t - 4s - t^2 + 2ts - s^2)$

⊙ CONCEPTUAL   ✎ WRITING   ▲ CHALLENGING   ▦ SCIENTIFIC CALCULATOR   GRAPHICS CALCULATOR

2. Factor out the greatest common factor.
3. $z^2(m + n)^4$
4. (c)

33. (a)

35. $2x^2(-x^3 + 3x + 2)$ or $-2x^2(x^3 - 3x - 2)$

36. $5a^3(-1 + 2a - 3a^2)$ or $-5a^3(1 - 2a + 3a^2)$

37. $16a^2m^3 \cdot (-2a^2m^2 - 1 - 4a^3m^3)$ or $-16a^2m^3 \cdot (2a^2m^2 + 1 + 4a^3m^3)$

38. $16z^3n^5 \cdot (-9z^8 + n^6 - 2zn^2)$ or $-16z^3n^5 \cdot (9z^8 - n^6 + 2zn^2)$

▲ **29.** $15(2z + 1)^3 + 10(2z + 1)^2 - 25(2z + 1)$

29. $20z(2z + 1)(3z + 4)$

▲ **30.** $6(a + 2b)^2 - 4(a + 2b)^3 + 12(a + 2b)^4$

30. $2(a + 2b)^2(3 - 2a - 4b + 6a^2 + 24ab + 24b^2)$

▲ **31.** $5(m + p)^3 - 10(m + p)^2 - 15(m + p)^4$

31. $5(m + p)^2(m + p - 2 - 3m^2 - 6mp - 3p^2)$

▲ **32.** $-9a^2(p + q) - 3a^3(p + q)^2 + 6a(p + q)^3$

32. $-3a(p + q)(3a + a^2p + a^2q - 2p^2 - 4pq - 2q^2)$

◉ **33.** Which one of the following is the factorization of

$$6x^3y^4 - 12x^5y^2 + 24x^4y^8$$

that has the greatest common factor as one of the factors?

(a) $6x^3y^2(y^2 - 2x^2 + 4xy^6)$ **(b)** $6xy(x^2y^3 - 2x^4y + 4x^3y^7)$

(c) $2x^3y^2(3y^2 - 6x^2 + 12xy^6)$ **(d)** $6x^2y^2(xy^2 - 2x^3 + 4x^2y^6)$

◉ **34.** When directed to factor the polynomial $4x^2y^5 - 8xy^3$ completely, a student responded with $2xy^3(2xy^2 - 4)$. When the teacher did not give him full credit, he complained because when his answer is multiplied out, the result is the original polynomial. Was the teacher justified in her grading? Why or why not?

Factor the polynomial twice. First, use a common factor with a positive coefficient, and then use a common factor with a negative coefficient. See Example 4.

35. $-2x^5 + 6x^3 + 4x^2$

36. $-5a^3 + 10a^4 - 15a^5$

37. $-32a^4m^5 - 16a^2m^3 - 64a^5m^6$

38. $-144z^{11}n^5 + 16z^3n^{11} - 32z^4n^7$

————◆ **MATHEMATICAL CONNECTIONS** (Exercises 39–46) ◆————

A natural number greater than 1 whose only natural number factors are 1 and itself is called a **prime number.** For example, the first few prime numbers are 2, 3, 5, 7, 11, and 13. Any other natural number greater than 1 is called a **composite number.** A composite number is composed of prime number factors in one and only one way (if order of the factors is ignored). Some of the first few composite numbers and their prime factorizations are

$$4 = 2^2, \quad 6 = 2 \cdot 3, \quad 8 = 2^3, \quad 9 = 3^2, \quad 10 = 2 \cdot 5, \quad 12 = 2^2 \cdot 3.$$

One way of finding the prime factorization of a composite number is a *factor tree,* as shown below. It does not matter how we start if we have a choice—the prime factors will be the same.

Here is the prime factorization of 360 by a factor tree:

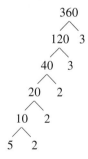

Therefore, the prime factorization of 360 is

$$2^3 \cdot 3^2 \cdot 5.$$

Now use this idea to answer Exercises 39–46. Work them in order.

39. Find the prime factorization of 60.

40. Find the prime factorization of 420.

41. Since $420 = 360 + 60$, the number 420 can be written as the sum of the prime factored forms of 360 (see the example) and 60 (see Exercise 39). Write 420 in this manner.

42. What is the greatest common factor of 360 and 60? Give it in factored form.

43. Write your expression from Exercise 41 in factored form by factoring out the greatest common factor you found in Exercise 42.

44. One of the factors obtained in Exercise 42 is a sum. Add the terms to find the prime number that it represents. (The sum will not always be a prime.)

45. Based on your answers in Exercises 43–44, what is the prime factored form of $360 + 60$?

46. Compare your answers in Exercises 40 and 45. Are they the same?

Factor by grouping. See Examples 5–7.

47. $mx + 3qx + my + 3qy$

48. $2k + 2h + jk + jh$

49. $10m + 2n + 5mk + nk$

50. $3ma + 3mb + 2ab + 2b^2$

51. $m^2 - 3m - 15 + 5m$

52. $z^2 - 6z - 54 + 9z$

53. $p^2 - 4zq + pq - 4pz$

54. $r^2 - 9tw + 3rw - 3rt$

55. $3a^2 + 15a - 10 - 2a$

56. $7k + 2k^2 - 6k - 21$

57. $-15p^2 + 5pq - 6pq + 2q^2$

58. $-6r^2 + 9rs + 8rs - 12s^2$

59. $-3a^3 - 3ab^2 + 2a^2b + 2b^3$

60. $-16m^3 + 4m^2p^2 - 4mp + p^3$

61. $4 + xy - 2y - 2x$

62. $2ab^2 - 4 - 8b^2 + a$

63. $8 + 9y^4 - 6y^3 - 12y$

64. $x^3y^2 - 3 - 3y^2 + x^3$

65. $1 - a + ab - b$

66. $2ab^2 - 8b^2 + a - 4$

67. Refer to Exercise 65 above. The factored form as given in the answer section in the back of the text is $(1 - a)(1 - b)$. As mentioned in the text, sometimes other acceptable factored forms can be given. Which one of the following is *not* a factored form of $1 - a + ab - b$?
(a) $(a - 1)(b - 1)$ (b) $(-a + 1)(-b + 1)$ (c) $(-1 + a)(-1 + b)$
(d) $(1 - a)(b + 1)$

68. Refer to Exercise 66 above. One form of the answer is $(2b^2 + 1)(a - 4)$. Give two other acceptable factored forms of $2ab^2 - 8b^2 + a - 4$.

▲ *Factor the polynomial. Assume that all variables used as exponents represent positive integers.*

69. $p^{6m} - 2p^{4m}$

70. $4r^{3z} + 8r^{5z}$

71. $q^{3k} + 2q^{2k} + 3q^k$

72. $z^{2x} - z^x + z^{3x}$

73. $y^{r+5} + y^{r+4} + y^{r+2}$

74. $8k^{2z+3} + 2k^{2z+1} + 12k^{2z}$

75. $r^pm^p + q^pm^p - r^pz^p - q^pz^p$

76. $6a^zb^z - 10b^z - 3a^zc^z + 5c^z$

▲ *Factor out the greatest common factor. Assume all variables are nonzero.*

77. $3m^{-5} + m^{-3}$

78. $k^{-2} + 2k^{-4}$

79. $3p^{-3} + 2p^{-2} - 4p^{-1}$

80. $-5y^{-3} + 8y^{-2} + y^{-1}$

39. $2^2 \cdot 3 \cdot 5$
40. $2^2 \cdot 3 \cdot 5 \cdot 7$
41. $2^3 \cdot 3^2 \cdot 5 + 2^2 \cdot 3 \cdot 5$
42. $2^2 \cdot 3 \cdot 5$
43. $2^2 \cdot 3 \cdot 5(2 \cdot 3 + 1)$
44. It represents 7.
45. $2^2 \cdot 3 \cdot 5 \cdot 7$
46. Yes, the answers are the same.
47. $(m + 3q)(x + y)$
48. $(k + h)(2 + j)$
49. $(5m + n)(2 + k)$
50. $(a + b)(3m + 2b)$
51. $(m - 3)(m + 5)$
52. $(z - 6)(z + 9)$
53. $(p + q)(p - 4z)$
54. $(r + 3w)(r - 3t)$
55. $(a + 5)(3a - 2)$
56. $(2k + 7)(k - 3)$
57. $(-3p + q)(5p + 2q)$
58. $(-3r + 4s)(2r - 3s)$
59. $(a^2 + b^2)(-3a + 2b)$
60. $(4m^2 + p)(-4m + p^2)$
61. $(y - 2)(x - 2)$
62. $(a - 4)(2b^2 + 1)$
63. $(3y - 2)(3y^3 - 4)$
64. $(x^3 - 3)(y^2 + 1)$
65. $(1 - a)(1 - b)$
66. $(a - 4)(2b^2 + 1)$
67. (d)
68. $(-2b^2 - 1)(4 - a)$; $-(2b^2 + 1)(-a + 4)$
69. $p^{4m}(p^{2m} - 2)$
70. $4r^{3z}(1 + 2r^{2z})$
71. $q^k(q^{2k} + 2q^k + 3)$
72. $z^x(z^x - 1 + z^{2x})$
73. $y^{r+2}(y^3 + y^2 + 1)$
74. $2k^{2z}(4k^3 + k + 6)$
75. $(r^p + q^p)(m^p - z^p)$
76. $(3a^z - 5)(2b^z - c^z)$
77. $m^{-5}(3 + m^2)$ or $\dfrac{m^2 + 3}{m^5}$
78. $k^{-4}(k^2 + 2)$ or $\dfrac{k^2 + 2}{k^4}$
79. $p^{-3}(3 + 2p - 4p^2)$ or $\dfrac{-4p^2 + 2p + 3}{p^3}$
80. $y^{-3}(-5 + 8y + y^2)$ or $\dfrac{y^2 + 8y - 5}{y^3}$

81. $m^2 - m - 6$
82. $k^2 + 6k - 7$
83. $10m^2 - 3m - 27$
84. $20y^2 + 14y - 12$
85. $64z^2 - 48az + 9a^2$
86. $25x^2 - 4t^2$
87. $2x^4 - 17x^2 + 30$
88. $6y^6 + y^3 - 12$
89. $2x^3 - 11x^2 - 21x$
90. $15t^3 - 110t^2 - 80t$

REVIEW EXERCISES

Use the FOIL method to find the product. See Section 3.4.

81. $(m - 3)(m + 2)$ **82.** $(k + 7)(k - 1)$ **83.** $(2m + 3)(5m - 9)$

84. $(4y - 2)(5y + 6)$ **85.** $(8z - 3a)(8z - 3a)$ **86.** $(5x - 2t)(5x + 2t)$

87. $(2x^2 - 5)(x^2 - 6)$ **88.** $(3y^3 - 4)(2y^3 + 3)$

89. $x(2x + 3)(x - 7)$ **90.** $5t(3t + 2)(t - 8)$

3.6 FACTORING TRINOMIALS

FOR EXTRA HELP	OBJECTIVES

SSG pp. 83–88
 SSM pp. 111–116
 	
  Video 4
 	
  Tutorial IBM   MAC
	1 ▶ Factor trinomials when the coefficient of the squared term is **1**.
	 2 ▶ Factor trinomials when the coefficient of the squared term is not **1**.
	 3 ▶ Use an alternative method for factoring trinomials.
	 4 ▶ Factor by substitution.

The product of $x + 3$ and $x - 5$ is

$$(x + 3)(x - 5) = x^2 - 5x + 3x - 15$$
$$= x^2 - 2x - 15.$$

Also, by this result, the *factored form* of $x^2 - 2x - 15$ is $(x + 3)(x - 5)$.

$$\text{Factored form} \longrightarrow (x + 3)(x - 5) = x^2 - 2x - 15 \longleftarrow \text{Product}$$

with *Multiplication* indicated above and *Factoring* indicated below.

1 ▶ Factor trinomials when the coefficient of the squared term is 1.

▶ We show how to factor trinomials in this section, beginning with those having 1 as the coefficient of the squared term. Let us start by analyzing the example above, $x^2 - 2x - 15$. Since multiplying and factoring are inverses, factoring trinomials involves using FOIL backwards. As shown below, the x^2 term came from multiplying x and x, and -15 came from multiplying 3 and -5.

▶ **TEACHING TIP**
Emphasize the importance of this section as it relates to subsequent sections and the next chapter. A great deal of practice is required to become proficient in factoring. ◀

Product of x and x is x^2.

$$(x + 3)(x - 5) = x^2 - 2x - 15$$

Product of 3 and -5 is -15.

We found the $-2x$ in $x^2 - 2x - 15$ by multiplying the outside terms, and then the inside terms, and adding.

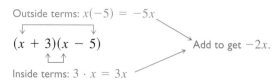

Outside terms: $x(-5) = -5x$

$$(x + 3)(x - 5)$$

Add to get $-2x$.

Inside terms: $3 \cdot x = 3x$

Based on this example, we factor a trinomial $x^2 + bx + c$, with 1 as the coefficient of the squared term, by following these steps.

Factoring $x^2 + bx + c$

Step 1 **Find pairs whose product is c.** Find all pairs of numbers whose product is the third term of the trinomial (c).

Step 2 **Find pairs whose sum is b.** Choose the pair whose sum is the coefficient of the middle term (b).

If there are no such numbers, the polynomial cannot be factored.

A polynomial that cannot be factored is called **prime.**

EXAMPLE 1
Factoring a Trinomial in $x^2 + bx + c$ Form

Factor $r^2 + 8r + 12$.

Look for two numbers with a product of 12 and a sum of 8. Of all pairs of numbers having a product of 12, only the pair 6 and 2 has a sum of 8. Therefore,

$$r^2 + 8r + 12 = (r + 6)(r + 2).$$

Because of the commutative property, it would be equally correct to write $(r + 2)(r + 6)$. ∎

EXAMPLE 2
Factoring a Trinomial in $x^2 + bx + c$ Form

For Example 3

Factor $y^2 + 2y - 35$.

Step 1 Find pairs of numbers whose product is -35.

$$-35(1)$$
$$35(-1)$$
$$7(-5)$$
$$5(-7)$$

Step 2 Write sums of those numbers.

$$-35 + 1 = -34$$
$$35 + (-1) = 34$$
$$7 + (-5) = 2 \leftarrow \text{Coefficient of the middle term}$$
$$5 + (-7) = -2$$

The required numbers are 7 and -5.

$$y^2 + 2y - 35 = (y + 7)(y - 5)$$

Check by finding the product of $y + 7$ and $y - 5$. ∎

EXAMPLE 3
Recognizing a Prime Polynomial

Factor $m^2 + 6m + 7$.

Look for two numbers whose product is 7 and whose sum is 6. Only two pairs of integers, 7 and 1 and -7 and -1, give a product of 7. Neither of these pairs has a sum of 6, so $m^2 + 6m + 7$ cannot be factored, and is prime. ∎

EXAMPLE 4
Factoring a Trinomial in Two Variables

Factor $p^2 + 6ap - 16a^2$.

Look for two expressions whose sum is $6a$ and whose product is $-16a^2$. The quantities $8a$ and $-2a$ have the necessary sum and product, so

$$p^2 + 6ap - 16a^2 = (p + 8a)(p - 2a). ∎$$

EXAMPLE .5

Factoring a Trinomial with a Common Factor

Factor $16y^3 - 32y^2 - 48y$.

Start by factoring out the greatest common factor, $16y$.

$$16y^3 - 32y^2 - 48y = 16y(y^2 - 2y - 3)$$

To factor $y^2 - 2y - 3$, look for two integers whose sum is -2 and whose product is -3. The necessary integers are -3 and 1, with

$$16y^3 - 32y^2 - 48y = 16y(y - 3)(y + 1). \quad \blacksquare$$

CAUTION In factoring, always look for a common factor first. Do not forget to write the common factor as part of the answer.

2 ▶ Factor trinomials when the coefficient of the squared term is not 1.

▶ We can use a generalization of the method shown above to factor a trinomial of the form $ax^2 + bx + c$, where the coefficient of the squared term is not equal to 1. We will factor $3x^2 + 7x + 2$ to see how this method works. First, we identify the values of a, b, and c.

$$\begin{array}{ccc} ax^2 & + \ bx & + \ c \\ \downarrow & \downarrow & \downarrow \\ 3x^2 & + \ 7x & + \ 2 \end{array}$$

$$a = 3, \quad b = 7, \quad c = 2$$

The product ac is $3 \cdot 2 = 6$, so we must find integers having a product of 6 and a sum of 7 (since the middle term is 7). The necessary integers are 1 and 6, so we write $7x$ as $1x + 6x$, or $x + 6x$, giving

$$3x^2 + 7x + 2 = 3x^2 + \underbrace{x + 6x} + 2.$$
$$x + 6x = 7x$$

Now we factor by grouping.

$$= x(3x + 1) + 2(3x + 1)$$
$$3x^2 + 7x + 2 = (3x + 1)(x + 2)$$

EXAMPLE 6

Factoring a Trinomial in $ax^2 + bx + c$ Form

Factor $12r^2 - 5r - 2$.

Since $a = 12$, $b = -5$, and $c = -2$, the product ac is -24. The two integers whose product is -24 and whose sum is -5 are -8 and 3.

$$\begin{array}{ll} 12r^2 - 5r - 2 = 12r^2 + 3r - 8r - 2 & \text{Write } -5r \text{ as } 3r - 8r. \\ \qquad\qquad\quad = 3r(4r + 1) - 2(4r + 1) & \text{Factor by grouping.} \\ 12r^2 - 5r - 2 = (4r + 1)(3r - 2) & \text{Factor out the common factor.} \quad \blacksquare \end{array}$$

3 ▶ Use an alternative method for factoring trinomials.

▶ An alternative approach, the method of trying repeated combinations and using FOIL, is especially helpful when the product ac is large. This method is shown using the same polynomials as above.

EXAMPLE 7

Factoring a Trinomial in $ax^2 + bx + c$ Form

Factor each of the following.

(a) $3x^2 + 7x + 2$

To factor this polynomial, find the correct numbers to put in the blanks.

$$3x^2 + 7x + 2 = (\underline{\quad} x + \underline{\quad})(\underline{\quad} x + \underline{\quad})$$

Plus signs are used since all the signs in the polynomial are plus. The first two expressions have a product of $3x^2$, so they must be $3x$ and x.

$$3x^2 + 7x + 2 = (3x + \underline{})(x + \underline{})$$

The product of the two last terms must be 2, so the numbers must be 2 and 1. There is a choice. The 2 could be used with the $3x$ or with the x. Only one of these choices can give the correct middle term. Use FOIL to try each one.

$$(3x + 2)(x + 1) \qquad (3x + 1)(x + 2)$$

$3x$ / $2x$ $\qquad$ $6x$ / x

$3x + 2x = 5x$ $\qquad$ $6x + x = 7x$

Wrong middle term $\qquad$ Correct middle term

Therefore, $3x^2 + 7x + 2 = (3x + 1)(x + 2)$.

(b) $12r^2 - 5r - 2$

To reduce the number of trials, we note that the trinomial has no common factor. This means that neither of its factors can have a common factor. We should keep this in mind as we choose factors. Let us try 4 and 3 for the two first terms. If these do not work, we will make another choice.

$$12r^2 - 5r - 2 = (4r\underline{})(3r\underline{})$$

We do not know what signs to use yet. The factors of -2 are -2 and 1 or 2 and -1. Try both possibilities.

$$(4r - 2)(3r + 1) \qquad (4r - 1)(3r + 2)$$

Wrong: $4r - 2$ has a $\qquad$ $8r$ / $-3r$
common factor. $\qquad$ $8r - 3r = 5r$

Wrong middle term

The middle term on the right is $5r$, instead of the $-5r$ that is needed. We get $-5r$ by exchanging the middle signs.

$$(4r + 1)(3r - 2)$$

$-8r$ / $3r$

$-8r + 3r = -5r$

Correct middle term

Thus, $12r^2 - 5r - 2 = (4r + 1)(3r - 2)$. ∎

NOTE As shown in Example 7(b), if the terms of a polynomial have no common factor (except 1), then none of the terms of its factors have a common factor. Remembering this will eliminate some potential factors.

The alternative method of factoring a trinomial $ax^2 + bx + c$ not having 1 as coefficient of the squared term is summarized on the next page.

Factoring $ax^2 + bx + c$	*Step 1*	**Find pairs whose product is *a*.** Write all pairs of factors of the coefficient of the squared term (*a*).
	Step 2	**Find pairs whose product is *c*.** Write all pairs of factors of the last term (*c*).
	Step 3	**Choose inner and outer terms.** Use FOIL and various combinations of the factors from Steps 1 and 2 until the necessary middle term is found.

If no such combinations exist, the polynomial is prime.

EXAMPLE 8

Factoring a Trinomial in Two Variables

Factor $18m^2 - 19mx - 12x^2$.

There is no common factor (except 1). Go through the steps to factor the trinomial. There are many possible factors of both 18 and -12. As a general rule, the middle-sized factors should be tried first. Let's try 6 and 3 for 18 and -3 and 4 for -12.

$$(6m - 3x)(3m + 4x) \qquad (6m + 4x)(3m - 3x)$$
$\cdot$ Wrong: common factor $\qquad\qquad$ Wrong: common factors

Since 6 and 3 do not work as factors of 18, try 9 and 2 instead, with -4 and 3 as factors of -12.

$$(9m + 3x)(2m - 4x) \qquad (9m - 4x)(2m + 3x)$$

Wrong: common factors

$$27mx$$
$$-8mx$$
$$27mx + (-8mx) = 19mx$$

The result on the right differs from the correct middle term only in the sign, so exchange the middle signs.

$$18m^2 - 19mx - 12x^2 = (9m + 4x)(2m - 3x) \quad \blacksquare$$

EXAMPLE 9

Factoring a Trinomial with a Common Factor

Factor $16y^3 + 24y^2 - 16y$.

This polynomial has a greatest common factor of $8y$. Factor this out first.

$$16y^3 + 24y^2 - 16y = 8y(2y^2 + 3y - 2)$$

Factor $2y^2 + 3y - 2$ by either method given above.

$$16y^3 + 24y^2 - 16y = 8y(2y - 1)(y + 2)$$

Do not forget to write the common factor in front. $\quad \blacksquare$

When factoring a trinomial of the form $ax^2 + bx + c$ with a negative, it is helpful to begin by factoring out -1, as shown in the next example.

EXAMPLE 10

Factoring $ax^2 + bx + c$, $a < 0$

Factor $-3x^2 + 16x + 12$.

While it is possible to factor this polynomial directly, it is helpful to start by factoring out -1. Then proceed as in the earlier examples.

$$-3x^2 + 16x + 12 = -1(3x^2 - 16x - 12)$$
$$= -1(3x + 2)(x - 6) \quad \blacksquare$$

NOTE The factored form given in Example 10 can be written in other ways. Two of them are

$$(-3x - 2)(x - 6) \quad \text{and} \quad (3x + 2)(-x + 6).$$

Verify that these both give the original polynomial when multiplied.

---

4 ▶ Factor by substitution.

▶ Sometimes we can factor a more complicated polynomial by making a substitution of one variable for an expression. The next examples show this **method of substitution.** The method of substitution is used when a particular polynomial appears to various powers in a more involved polynomial.

EXAMPLE 11

Factoring a Trinomial Using Substitution

Factor $2(x + 3)^2 + 5(x + 3) - 12$.

Since the binomial $x + 3$ appears to the powers of 2 and 1, let the substitution variable represent $x + 3$. We may choose any letter we wish except x. Let us choose y to equal $x + 3$.

$$2(x + 3)^2 + 5(x + 3) - 12 = 2y^2 + 5y - 12 \qquad \text{Let } y = x + 3.$$
$$= (2y - 3)(y + 4) \qquad \text{Factor.}$$

Now replace y with $x + 3$ to get

$$2(x + 3)^2 + 5(x + 3) - 12 = [2(x + 3) - 3][(x + 3) + 4]$$
$$= (2x + 6 - 3)(x + 7)$$
$$= (2x + 3)(x + 7). \quad ■$$

Chalkboard Exercise

Factor
$8(z + 5)^2 - 2(z + 5) - 3$.

Answer: $(4z + 17)(2z + 11)$

CAUTION Do not forget the final substitution.

EXAMPLE 12

Factoring a Trinomial in $ax^4 + bx^2 + c$ Form

Factor $6y^4 + 7y^2 - 20$.

The variable y appears to powers in which the larger exponent is twice the smaller exponent. In a case such as this, let the substitution variable equal the smaller power. Here, let $m = y^2$. Since $y^4 = (y^2)^2 = m^2$, the given trinomial becomes

$$6m^2 + 7m - 20,$$

which is factored as follows:

$$6m^2 + 7m - 20 = (3m - 4)(2m + 5).$$

Since $m = y^2$,

$$6y^4 + 7y^2 - 20 = (3y^2 - 4)(2y^2 + 5). \quad ■$$

Chalkboard Exercise

Factor $6r^4 - 13r^2 + 5$.

Answer: $(3r^2 - 5)(2r^2 - 1)$

▶ TEACHING TIP

Students may discover that some polynomials factored using substitution can also be done by first simplifying the expression and then using the FOIL method. Point out that this is a good time to learn to use the substitution method, since it will appear later in equation solving techniques. ◀

NOTE Some students feel comfortable enough about factoring to factor polynomials like the one in Example 12 directly, without using the substitution method.

3.6 EXERCISES

2. Yes, they were both correct, because each product is equal to the original polynomial.
3. $(a - 5)(a + 3)$
4. $(m + 10)(m - 3)$
5. $(p - 8)(p + 7)$
6. $(k - 5)(k - 6)$
7. $(-m + 6)(m - 10)$
8. $(-p - 3)(p - 9)$
9. $(a + 5b)(a - 7b)$
10. $(z + 3w)(z + 5w)$
11. prime
12. $(k - 4h)(k - 7h)$
13. $(xy + 9)(xy + 2)$
14. prime
15. $(-6m + 5)(m + 3)$
16. $(-3y - 2)(5y - 9)$
17. $(5x - 6)(2x + 3)$
18. $(4k + 7)(2k + 5)$
19. $(4k + 3)(5k + 8)$
20. $(9z - 1)(3z + 5)$
21. $(3a - 2b)(5a - 4b)$
22. prime   23. $(6m - 5)^2$
24. $(5r - 9)^2$ 25. prime
26. $(7c + 2d)(2c - 3d)$
27. $(2xz - 1)(3xz + 4)$
28. $(4mn - 3)(2mn - 1)$
29. $3(4x + 5)(2x + 1)$
30. $2(6x - 1)(3x + 2)$
31. $5(a + 6)(3a - 4)$
32. $2(3a + 7)(2a - 3)$
33. $4m(m + 5)(m - 2)$
34. $5z(z + 5)(z + 4)$
35. $11x(x - 6)(x - 4)$
36. $9k(k + 7)(k - 3)$
37. $2xy^3(x - 12y)(x - 12y)$
38. $6mn^2(m - 5n)(m + n)$
39. $3(3a + 2)(2a - 3)$
40. $10(5r - 2)(2r - 1)$
41. $6a(a - 3)(a + 5)$
42. $3m^2(m + 6)(m - 4)$
43. $13y(y + 4)(y - 1)$
44. $4p(p + 8)(p - 2)$
45. $2xy^3(x - 12y)^2$
46. $6mn^2(m - 5n)(m + n)$
47. $-(x - 9)(x + 2)$
46. $6mn^2(m - 5n)(m + n)$
47. $-(x - 9)(x + 2)$
48. $-(p + 3)(p + 4)$
49. $-(9a + 5)(2a - 3)$
50. $-(6x + 1)(x - 4)$
51. $-r(7r + 1)(2r - 3)$
52. $-h(10h + 1)(h - 3)$
55. $-18a^3 + 27a^2b + 315ab^2$
56. $120z^4 - 36xz^3 - 12z^2x^2$
57. no
58. 1, 3, 5, 9, 15, 45;   no

1. Multiply the polynomials in each pair.
   (a) $(x + 4)(x - 3)$ **(b)** $(2x + 7)(x - 5)$ **(c)** $(3x + 8)(4x - 3)$
   $(x - 4)(x + 3)$ $(2x - 7)(x + 5)$ $(3x - 8)(4x + 3)$

1. (a) $x^2 + x - 12;$ (b) $2x^2 - 3x - 35;$ (c) $12x^2 + 23x - 24;$
   $x^2 - x - 12$ $2x^2 + 3x - 35$ $12x^2 - 23x - 24$

⊙ **(d)** In your answers, how does the sign of the term involving x compare in the pair of products in each case?   (d) The signs are opposites.

⊙ **(e)** Based on your results in parts (a)–(d), complete the following statement: When factoring a trinomial into a product of binomials, if when we multiply our proposed factors, the middle term is the opposite of that term in the trinomial, then we need only _____ in order to find the correct factors.
   (e) reverse the signs of the second terms of the binomials

⊙ **2.** When factoring the polynomial $-4x^2 - 29x + 24$, Margo obtained $(-4x + 3)(x + 8)$, while Steve got $(4x - 3)(-x - 8)$. Were they both correct? Why or why not?

Factor the trinomial. See Examples 1–10.

3. $a^2 - 2a - 15$ **4.** $m^2 + 7m - 30$ **5.** $p^2 - p - 56$
6. $k^2 - 11k + 30$ **7.** $-m^2 + 16m - 60$ **8.** $-p^2 + 6p + 27$
9. $a^2 - 2ab - 35b^2$ **10.** $z^2 + 8zw + 15w^2$ **11.** $y^2 - 3yq - 15q^2$
12. $k^2 - 11hk + 28h^2$ **13.** $x^2y^2 + 11xy + 18$ **14.** $p^2q^2 - 5pq - 18$
15. $-6m^2 - 13m + 15$ **16.** $-15y^2 + 17y + 18$ **17.** $10x^2 + 3x - 18$
18. $8k^2 + 34k + 35$ **19.** $20k^2 + 47k + 24$ **20.** $27z^2 + 42z - 5$
21. $15a^2 - 22ab + 8b^2$ **22.** $15p^2 + 24pq + 8q^2$ **23.** $36m^2 - 60m + 25$
24. $25r^2 - 90r + 81$ **25.** $40x^2 + xy + 6y^2$ **26.** $14c^2 - 17cd - 6d^2$
27. $6x^2z^2 + 5xz - 4$ **28.** $8m^2n^2 - 10mn + 3$ **29.** $24x^2 + 42x + 15$
30. $36x^2 + 18x - 4$ **31.** $15a^2 + 70a - 120$ **32.** $12a^2 + 10a - 42$
33. $4m^3 + 12m^2 - 40m$ **34.** $5z^3 + 45z^2 + 100z$
35. $11x^3 - 110x^2 + 264x$ **36.** $9k^3 + 36k^2 - 189k$
37. $2x^3y^3 - 48x^2y^4 + 288xy^5$ **38.** $6m^3n^2 - 24m^2n^3 - 30mn^4$
39. $18a^2 - 15a - 18$ **40.** $100r^2 - 90r + 20$
41. $6a^3 + 12a^2 - 90a$ **42.** $3m^4 + 6m^3 - 72m^2$
43. $13y^3 + 39y^2 - 52y$ **44.** $4p^3 + 24p^2 - 64p$
45. $2x^3y^3 - 48x^2y^4 + 288xy^5$ **46.** $6m^3n^2 - 24m^2n^3 - 30mn^4$
47. $-x^2 + 7x + 18$ **48.** $-p^2 - 7p - 12$
49. $-18a^2 + 17a + 15$ **50.** $-6x^2 + 23x + 4$
51. $-14r^3 + 19r^2 + 3r$ **52.** $-10h^3 + 29h^2 + 3h$

⊙ **53.** When a student was given the polynomial $4x^2 + 2x - 20$ to factor completely on a test, the student lost some credit when her answer was $(4x + 10)(x - 2)$. She complained to her teacher that when we multiply $(4x + 10)(x - 2)$, we get the original polynomial. Write a short explanation of why she lost some credit for her answer, even though the product is indeed $4x^2 + 2x - 20$.

⊙ **54.** Write an explanation as to why most people would find it more difficult to factor $36x^2 - 44x - 15$ than $37x^2 - 183x - 10$.

⊙ **55.** Find a polynomial that can be factored as $-9a(a - 5b)(2a + 7b)$.

⊙ **56.** Find a polynomial that can be factored as $12z^2(5z + x)(2z - x)$.

⊙ CONCEPTUAL   ✐ WRITING   ▲ CHALLENGING   ▦ SCIENTIFIC CALCULATOR   ▦ GRAPHICS CALCULATOR

——◆ **MATHEMATICAL CONNECTIONS** (Exercises 57–64) ◆——

Refer to the note following Example 7 in this section. Then work Exercises 57–64 in order.

57. Is 2 a factor of the composite number 45?

58. List all positive integer factors of 45. Is 2 a factor of any of these factors?

59. Is 3 a factor of 20?

60. List all positive integer factors of 20. Is 3 a factor of any of these factors?

61. Is 5 a factor of $10x^2 + 29x + 10$?

62. Factor $10x^2 + 29x + 10$. Is 5 a factor of either of its factors?

◎ **63.** Suppose that k is an odd integer and you are asked to factor $2x^2 + kx + 8$. Why is $2x + 4$ not a possible choice in factoring this polynomial?

 64. The polynomial $12y^2 - 11y - 15$ can be factored using the methods of this section. Explain why $3y + 15$ cannot be one of its factors.

——————◇——————

Factor the polynomial completely. Assume that all variables used as exponents represent positive integers. See Examples 11 and 12.

65. $p^4 - 10p^2 + 16$ **66.** $k^4 + 10k^2 + 9$ **67.** $2x^4 - 9x^2 - 18$

68. $6z^4 + z^2 - 1$ **69.** $16x^4 + 16x^2 + 3$ **70.** $9r^4 + 9r^2 + 2$

71. $12p^6 - 32p^3r + 5r^2$ **72.** $2y^6 + 7xy^3 + 6x^2$

73. $10(k + 1)^2 - 7(k + 1) + 1$ **74.** $4(m - 5)^2 - 4(m - 5) - 15$

75. $3(m + p)^2 - 7(m + p) - 20$ **76.** $4(x - y)^2 - 23(x - y) - 6$

▲ **77.** $a^2(a + b)^2 - ab(a + b)^2 - 6b^2(a + b)^2$ ▲ **78.** $m^2(m - p) + mp(m - p) - 2p^2(m - p)$

▲ **79.** $p^2(p + q) + 4pq(p + q) + 3q^2(p + q)$ ▲ **80.** $2k^2(5 - y) - 7k(5 - y) + 5(5 - y)$

▲ **81.** $z^2(z - x) - zx(x - z) - 2x^2(z - x)$ ▲ **82.** $r^2(r - s) - 5rs(s - r) - 6s^2(r - s)$

▲ **83.** $p^{2n} - p^n - 6$ ▲ **84.** $k^{2y} + 4k^y - 5$ ▲ **85.** $6z^{4r} - 5z^{2r} - 4$

▲ **86.** $12a^{4p} + 11a^{2p} + 2$ ▲ **87.** $36k^{3r} + 30k^{2r} + 4k^r$ ▲ **88.** $30y^{7a} - 26y^{6a} - 40y^{5a}$

REVIEW EXERCISES

Find the product. See Section 3.4.

89. $(3x - 5)(3x + 5)$ **90.** $(8m + 3)(8m - 3)$ **91.** $(p + 3q)^2$

92. $(2z - 7)^2$ **93.** $(y + 3)(y^2 - 3y + 9)$ **94.** $(3m - 1)(9m^2 + 3m + 1)$

3.7 SPECIAL FACTORING

FOR EXTRA HELP

📖 **SSG** pp. 88–91
  SSM pp. 116–121

📼 **Video** 4

💾 **Tutorial** IBM MAC

OBJECTIVES

1 ▶ Factor the difference of two squares.
2 ▶ Factor a perfect square trinomial.
3 ▶ Factor the difference of two cubes.
4 ▶ Factor the sum of two cubes.

Certain types of factoring occur so often that they deserve special study in this section.

59. no   **60.** 1, 2, 4, 5, 10, 20; no   **61.** no
62. $(5x + 2)(2x + 5)$; no
63. Since k is odd, 2 is not a factor of $2x^2 + kx + 8$, and because 2 is a factor of $2x + 4$, the binomial $2x + 4$ cannot be a factor.
65. $(p^2 - 8)(p^2 - 2)$
66. $(k^2 + 1)(k^2 + 9)$
67. $(2x^2 + 3)(x^2 - 6)$
68. $(3z^2 - 1)(2z^2 + 1)$
69. $(4x^2 + 3)(4x^2 + 1)$
70. $(3r^2 + 1)(3r^2 + 2)$
71. $(6p^3 - r)(2p^3 - 5r)$
72. $(2y^3 + 3x)(y^3 + 2x)$
73. $(5k + 4)(2k + 1)$
74. $(2m - 15)(2m - 7)$
75. $(3m + 3p + 5) \cdot (m + p - 4)$
76. $(4x - 4y + 1) \cdot (x - y - 6)$
77. $(a + b)^2(a - 3b) \cdot (a + 2b)$
78. $(m - p)^2(m + 2p)$
79. $(p + q)^2(p + 3q)$
80. $(5 - y)(2k - 5)(k - 1)$
81. $(z - x)^2(z + 2x)$
82. $(r - s)^2(r + 6s)$
83. $(p^n - 3)(p^n + 2)$
84. $(k^y + 5)(k^y - 1)$
85. $(2z^{2r} + 1)(3z^{2r} - 4)$
86. $(3a^{2p} + 2)(4a^{2p} + 1)$
87. $2k^r(6k^r + 1)(3k^r + 2)$
88. $2y^{5a}(5y^a + 4)(3y^a - 5)$
89. $9x^2 - 25$
90. $64m^2 - 9$
91. $p^2 + 6pq + 9q^2$
92. $4z^2 - 28z + 49$
93. $y^3 + 27$
94. $27m^3 - 1$

INSTRUCTOR'S RESOURCES

📖 **ITM** pp. 359–360 **ISM** pp. 182–190   💾 **TEST GENERATOR** DOS Windows MAC   **TRANSPARENCIES**
  IAM p. 12

◎ CONCEPTUAL   ✎ WRITING   ▲ CHALLENGING   ▦ SCIENTIFIC CALCULATOR   ▦ GRAPHICS CALCULATOR

1 ▶ Factor the difference of two squares.

▶ As discussed earlier in this chapter, the product of the sum and difference of two terms leads to the **difference of two squares,** a pattern that is useful in factoring.

Difference of Two Squares	$x^2 - y^2 = (x + y)(x - y)$

E X A M P L E   1

Factoring the Difference of Squares

Factor each difference of squares.

(a) $16m^2 - 49p^2 = (4m)^2 - (7p)^2$
$$= (4m + 7p)(4m - 7p)$$

(b) $81k^2 - 121a^2 = (9k)^2 - (11a)^2$
$$= (9k + 11a)(9k - 11a)$$

(c) $(m - 2p)^2 - 16 = (m - 2p)^2 - 4^2$
$$= [(m - 2p) + 4][(m - 2p) - 4]$$
$$= (m - 2p + 4)(m - 2p - 4) \quad \blacksquare$$

▶ **TEACHING TIP**
Mention that perfect square variable terms will have exponents that are even (divisible by two). ◀

CAUTION Assuming no greatest common factor (except 1), it is not possible to factor (with real numbers) a *sum* of two squares such as $x^2 + 25$. In particular, $x^2 + y^2 \neq (x + y)^2$, as shown in Objective 2.

◆ Answers will vary.

◆ **C O N N E C T I O N S** ◆

The difference of squares property can be shown geometrically. In the figure below,

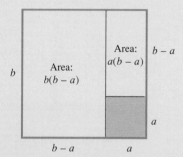

area of the large square − area of the small square

= area of the large rectangle + area of the small rectangle.

$$b^2 - a^2 = b(b - a) + a(b - a)$$
$$= (b - a)(b + a) \qquad \text{Factor out } b - a.$$

FOR DISCUSSION OR WRITING
Explain how a similar geometric proof applies to $b^3 - a^3 = (b - a) \cdot (b^2 + ba + a^2)$, using a cube with sides of length b.

2 ▶ Factor a perfect square trinomial.

▶ Two other special products from Section 3.4 lead to the following rules for factoring.

Perfect Square Trinomial

$$x^2 + 2xy + y^2 = (x + y)^2$$
$$x^2 - 2xy + y^2 = (x - y)^2$$

▶ **TEACHING TIP**

Work several examples of the form $(x + y)^2$ and $(x - y)^2$ to establish the pattern. State that factoring is a reversal of this process. Some trinomials, like $x^2 + 5x + 4$, may appear to be perfect square trinomials but require the methods of Section 3.6 instead.

The trinomial $x^2 + 2xy + y^2$ is the square of $x + y$. For this reason, the trinomial $x^2 + 2xy + y^2$ is called a **perfect square trinomial.** In this pattern, both the first and the last terms of the trinomial must be perfect squares. In the factored form, twice the product of the first and the last terms must give the middle term of the trinomial. It is important to understand these patterns in terms of words, since they occur with many different symbols (other than x and y).

$$4m^2 + 20m + 25 \qquad\qquad p^2 - 8p + 64$$

Square trinomial | Not a square trinomial; middle term should be $16p$.

EXAMPLE 2

Factoring Perfect Square Trinomials

Chalkboard Exercise

Factor $x^2 - 2x + 1 - y^2$.

Answer:
$(x - 1 + y)(x - 1 - y)$

Factor each of the following perfect square trinomials.

(a) $144p^2 - 120p + 25$

Here $144p^2 = (12p)^2$ and $25 = 5^2$. The sign on the middle term is $-$, so if $144p^2 - 120p + 25$ is a perfect square trinomial, it will have to be

$$(12p - 5)^2.$$

Take twice the product of the two terms to see if this is correct:

$$2(12p)(-5) = -120p,$$

which is the middle term of the given trinomial. Thus,

$$144p^2 - 120p + 25 = (12p - 5)^2.$$

(b) $4m^2 + 20mn + 49n^2$

If this is a square trinomial, it will equal $(2m + 7n)^2$. By the pattern in the box, if multiplied out, this squared binomial has a middle term of $2(2m)(7n) = 28mn$, which *does not equal* $20mn$. Verify that this trinomial cannot be factored by the methods of the previous section either. It is prime.

(c) $(r + 5)^2 + 6(r + 5) + 9 = [(r + 5) + 3]^2$
$$= (r + 8)^2,$$

since $2(r + 5)(3) = 6(r + 5)$, the middle term.

(d) $m^2 - 8m + 16 - p^2$

The first three terms here are the square of a binomial; group them together, and factor as follows.

$$(m^2 - 8m + 16) - p^2 = (m - 4)^2 - p^2$$

The result is the difference of two squares. Factor again to get

$$(m - 4)^2 - p^2 = (m - 4 + p)(m - 4 - p). \quad\blacksquare$$

Perfect square trinomials, of course, can be factored using the general methods shown earlier for other trinomials. The patterns given here provide a "shortcut."

3 ▶ Factor the difference of two cubes.

▶ The **difference of two cubes,** $x^3 - y^3$, can be factored as follows.

Difference of Two Cubes

$$x^3 - y^3 = (x - y)(x^2 + xy + y^2)$$

We could check this pattern by finding the product of $x - y$ and $x^2 + xy + y^2$.

E X A M P L E   3

Factoring the Difference of Cubes

Factor each difference of cubes.

(a) $m^3 - 8 = m^3 - 2^3 = (m - 2)(m^2 + 2m + 2^2)$

Check:

$$\overset{m^3}{\overbrace{(m - 2)}} \overset{-8}{} (m^2 + 2m + 4)$$

$-2m$

Opposite of the product of the roots gives the middle term.

(b) $27x^3 - 8y^3 = (3x)^3 - (2y)^3$
$= (3x - 2y)[(3x)^2 + (3x)(2y) + (2y)^2]$
$= (3x - 2y)(9x^2 + 6xy + 4y^2)$

(c) $1000k^3 - 27n^3 = (10k)^3 - (3n)^3$
$= (10k - 3n)[(10k)^2 + (10k)(3n) + (3n)^2]$
$= (10k - 3n)(100k^2 + 30kn + 9n^2)$ ∎

4 ▶ Factor the sum of two cubes.

▶ While an expression of the form $x^2 + y^2$ (a sum of two squares) cannot be factored with real numbers, the **sum of two cubes** is factored as follows.

Sum of Two Cubes

$$x^3 + y^3 = (x + y)(x^2 - xy + y^2)$$

To verify this result, find the product of $x + y$ and $x^2 - xy + y^2$. Compare this pattern with the pattern for the difference of two cubes.

NOTE The sign of the second term in the binomial factor of the sum or difference of cubes rule is *always the same* as the sign in the original problem. In the trinomial factor, the first and last terms are *always positive*; the sign of the middle term is *the opposite of* the sign of the second term in the binomial factor.

E X A M P L E   4

Factoring the Sum of Cubes

Factor each sum of cubes.

(a) $r^3 + 27 = r^3 + 3^3 = (r + 3)(r^2 - 3r + 3^2)$
$= (r + 3)(r^2 - 3r + 9)$

(b) $27z^3 + 125 = (3z)^3 + 5^3 = (3z + 5)[(3z)^2 - (3z)(5) + 5^2]$
$= (3z + 5)(9z^2 - 15z + 25)$

(c) $125t^3 + 216s^6 = (5t)^3 + (6s^2)^3$

$$= (5t + 6s^2)[(5t)^2 - (5t)(6s^2) + (6s^2)^2]$$

$$= (5t + 6s^2)(25t^2 - 30ts^2 + 36s^4) \quad \blacksquare$$

CAUTION A common error is to think that there is a 2 as coefficient of xy when factoring the sum or difference of two cubes. There is no 2 there, so in general, expressions of the form $x^2 + xy + y^2$ and $x^2 - xy + y^2$ cannot be factored further.

The special types of factoring in this section are summarized here. *These should be memorized.*

Special Types of Factoring		
Difference of two squares	$x^2 - y^2 = (x + y)(x - y)$
Perfect square trinomial	$x^2 + 2xy + y^2 = (x + y)^2$
	$x^2 - 2xy + y^2 = (x - y)^2$
Difference of two cubes	$x^3 - y^3 = (x - y)(x^2 + xy + y^2)$
Sum of two cubes	$x^3 + y^3 = (x + y)(x^2 - xy + y^2)$

3.7 EXERCISES

⊙ **1.** Which of the following binomials are differences of squares?
   (a) $64 - m^2$ **(b)** $2x^2 - 25$ **(c)** $k^2 + 9$ **(d)** $4z^4 - 49$

⊙ **2.** Which of the following binomials are sums or differences of cubes?
   (a) $64 + y^3$ **(b)** $125 - p^6$ **(c)** $9x^3 + 125$ **(d)** $(x + y)^3 - 1$

⊙ **3.** Which of the following trinomials are perfect squares?
   (a) $x^2 - 8x - 16$ **(b)** $4m^2 + 20m + 25$ **(c)** $9z^4 + 30z^2 + 25$
   (d) $25a^2 - 45a + 81$

⊙ **4.** Of the twelve polynomials listed in Exercises 1–3, which ones can be factored using the methods of this section?

Factor the polynomial. See Examples 1–4.

5. $p^2 - 16$ **6.** $k^2 - 9$ **7.** $25x^2 - 4$

8. $36m^2 - 25$ **9.** $9a^2 - 49b^2$ **10.** $16c^2 - 49d^2$

11. $64m^4 - 4y^4$ **12.** $243x^4 - 3t^4$ **13.** $(y + z)^2 - 81$

14. $(h + k)^2 - 9$ **15.** $16 - (x + 3y)^2$ **16.** $64 - (r + 2t)^2$

17. $(p + q)^2 - (p - q)^2$ **18.** $(a + b)^2 - (a - b)^2$ **19.** $k^2 - 6k + 9$

20. $x^2 + 10x + 25$ **21.** $4z^2 + 4zw + w^2$ **22.** $9y^2 + 6yz + z^2$

23. $16m^2 - 8m + 1 - n^2$ **24.** $25c^2 - 20c + 4 - d^2$ **25.** $4r^2 - 12r + 9 - s^2$

26. $9a^2 - 24a + 16 - b^2$ **27.** $x^2 - y^2 + 2y - 1$ **28.** $-k^2 - h^2 + 2kh + 4$

29. $98m^2 + 84mn + 18n^2$ **30.** $80z^2 - 40zw + 5w^2$ **31.** $(p + q)^2 + 2(p + q) + 1$

1. (a), (d)   2. (a), (b), (d)
3. (b), (c)   4. 1(a), 1(d), 2(a), 2(b), 2(d), 3(b), 3(c)
5. $(p + 4)(p - 4)$
6. $(k + 3)(k - 3)$
7. $(5x + 2)(5x - 2)$
8. $(6m + 5)(6m - 5)$
9. $(3a + 7b)(3a - 7b)$
10. $(4c + 7d)(4c - 7d)$
11. $4(4m^2 + y^2)(2m + y) \cdot (2m - y)$
12. $3(9x^2 + t^2) \cdot (3x + t)(3x - t)$
13. $(y + z + 9)(y + z - 9)$
14. $(h + k + 3)(h + k - 3)$
15. $(4 + x + 3y) \cdot (4 - x - 3y)$
16. $(8 + r + 2t)(8 - r - 2t)$
17. $4pq$ 18. $4ab$
19. $(k - 3)^2$ 20. $(x + 5)^2$
21. $(2z + w)^2$ 22. $(3y + z)^2$
23. $(4m - 1 + n) \cdot (4m - 1 - n)$
24. $(5c - 2 + d) \cdot (5c - 2 - d)$
25. $(2r - 3 + s) \cdot (2r - 3 - s)$

26. $(3a - 4 + b)(3a - 4 - b)$
27. $(x + y - 1)(x - y + 1)$
28. $(2 + k - h)(2 - k + h)$
29. $2(7m + 3n)^2$
30. $5(4z - w)^2$
31. $(p + q + 1)^2$
32. $(x + y + 3)^2$
33. $(a - b + 4)^2$
34. $(m - n + 2)^2$
35. $(2x - y)(4x^2 + 2xy + y^2)$
36. $(z + 5p)(z^2 - 5zp + 25p^2)$
37. $(4g + 3h) \cdot$
$(16g^2 - 12gh + 9h^2)$
38. $(3a - 2b) \cdot$
$(9a^2 + 6ab + 4b^2)$
39. $3(2n + 3p) \cdot$
$(4n^2 - 6np + 9p^2)$
40. $2(5x - 2y) \cdot$
$(25x^2 + 10xy + 4y^2)$
46. $x^4 + x^2y^2 + y^4$
47. The product must equal
$x^4 + x^2y^2 + y^4$. Multiply
$(x^2 + xy + y^2)(x^2 - xy + y^2)$
to verify this.
48. Start by factoring as the
difference of squares.
49. 16   50. 64   51. 25
52. 9   53. −56 or 56
54. −12 or 12

32. $(x + y)^2 + 6(x + y) + 9$
34. $(m - n)^2 + 4(m - n) + 4$
36. $z^3 + 125p^3$
38. $27a^3 - 8b^3$
40. $250x^3 - 16y^3$
42. $(p - q)^3 + 125$
42. $(p - q + 5)(p^2 - 2pq + q^2 - 5p + 5q + 25)$

33. $(a - b)^2 + 8(a - b) + 16$
35. $8x^3 - y^3$
37. $64g^3 + 27h^3$
39. $24n^3 + 81p^3$
41. $(y + z)^3 - 64$
41. $(y + z - 4)(y^2 + 2yz + z^2 + 4y + 4z + 16)$

◆ **MATHEMATICAL CONNECTIONS** (Exercises 43–48) ◆

The binomial $x^6 - y^6$ may be considered either as a difference of squares or a difference of cubes. Keep this in mind as you work Exercises 43–48 in order.

43. Factor $x^6 - y^6$ by first factoring as a difference of squares. Then factor further by considering one of the factors as the sum of cubes and the other factor as the difference of cubes.

43. $(x^3 - y^3)(x^3 + y^3);$ $(x - y)(x^2 + xy + y^2)(x + y)(x^2 - xy + y^2)$

44. Based on your answer in Exercise 43, fill in the blank with the correct factors so that $x^6 - y^6$ is factored completely:

$$x^6 - y^6 = (x - y)(x + y) \underline{\hspace{4cm}}.$$

44. $(x^2 + xy + y^2)(x^2 - xy + y^2)$

45. Factor $x^6 - y^6$ by first factoring as a difference of cubes. Then factor further by considering one of the factors as the difference of squares.

45. $(x^2 - y^2)(x^4 + x^2y^2 + y^4);$ $(x - y)(x + y)(x^4 + x^2y^2 + y^4)$

46. Based on your answer in Exercise 45, fill in the blank with the correct factor so that $x^6 - y^6$ is factored:

$$x^6 - y^6 = (x - y)(x + y) \underline{\hspace{4cm}}.$$

◉ 47. Notice that the factor you wrote in the blank in Exercise 46 is a fourth degree polynomial, while the two factors you wrote in the blank in Exercise 44 are both second degree polynomials. What must be true about the product of the two factors you wrote in the blank in Exercise 44? Verify this.

◉ 48. If you have a choice of factoring as the difference of squares or the difference of cubes, which method should you start with so that you may more easily obtain the complete factorization of the polynomial? Base the answer on your results in Exercises 43–47 and the methods of factoring explained in this section.

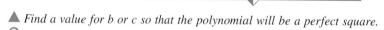

▲ *Find a value for b or c so that the polynomial will be a perfect square.*
◉
49. $p^2 + 8p + c$ 50. $y^2 - 16y + c$ 51. $9z^2 + 30z + c$
52. $16r^2 - 24r + c$ 53. $16x^2 + bx + 49$ 54. $36y^2 + by + 1$

In some cases, the method of factoring by grouping can be combined with the methods of special factoring dicussed in this section. For example, to factor $8x^3 + 4x^2 + 27y^3 - 9y^2$, we proceed as follows.

$$8x^3 + 4x^2 + 27y^3 - 9y^2 = (8x^3 + 27y^3) + (4x^2 - 9y^2)$$ Associative and commutative properties

$$= (2x + 3y)(4x^2 - 6xy + 9y^2) + (2x + 3y)(2x - 3y)$$ Factor within groups.
$$= (2x + 3y)[(4x^2 - 6xy + 9y^2) + (2x - 3y)]$$ Factor out the greatest common factor, $2x + 3y$.

$$= (2x + 3y)(4x^2 - 6xy + 9y^2 + 2x - 3y)$$ Combine terms.

It is important to note in problems such as this, how we choose to group in the first step is essential to factoring correctly. If we reach a "dead end," then we should group differently and try again.

 Use the method described above to factor the polynomial.

55. $27x^3 + 9x^2 + y^3 - y^2$
   55. $(3x + y)(9x^2 - 3xy + y^2 + 3x - y)$

56. $125p^3 + 25p^2 + 8q^3 - 4q^2$
   56. $(5p + 2q)(25p^2 - 10pq + 4q^2 + 5p - 2q)$

57. $1000k^3 + 20k - m^3 - 2m$
   57. $(10k - m)(100k^2 + 10km + m^2 + 2)$

58. $27a^3 + 15a - 64b^3 - 20b$
   58. $(3a - 4b)(9a^2 + 12ab + 16b^2 + 5)$

59. $y^4 + y^3 + y + 1$
   59. $(y + 1)^2(y^2 - y + 1)$

60. $8t^4 - 24t^3 + t - 3$
   60. $(t - 3)(2t + 1)(4t^2 - 2t + 1)$

61. $10x^2 + 5x^3 - 10y^2 + 5y^3$
   61. $5(x + y)(2x - 2y + x^2 - xy + y^2)$

62. $64m^2 - 512m^3 - 81n^2 + 729n^3$
   62. $(8m - 9n)(8m + 9n - 64m^2 - 72mn - 81n^2)$

 Factor the polynomial. Assume that all variables used as exponents represent positive integers.

63. $16m^{4x} - 9$

64. $100m^{2q} - 81$

65. $64r^{8z} - 1$

66. $4 - 49x^{4y}$

67. $100m^{2z} - 9p^{8z}$

68. $16k^{8b} - 25m^{4b}$

69. $9a^{4z} - 30a^{2z} + 25$

70. $121p^{8k} + 44p^{4k} + 4$

71. $x^{3n} - 8$

72. $216 + b^{3k}$

63. $(4m^{2x} + 3)(4m^{2x} - 3)$
64. $(10m^q + 9)(10m^q - 9)$
65. $(8r^{4z} + 1)(8r^{4z} - 1)$
66. $(2 + 7x^{2y})(2 - 7x^{2y})$
67. $(10m^z + 3p^{4z}) \cdot$
   $(10m^z - 3p^{4z})$
68. $(4k^{4b} + 5m^{2b}) \cdot$
   $(4k^{4b} - 5m^{2b})$
69. $(3a^{2z} - 5)^2$
70. $(11p^{4k} + 2)^2$
71. $(x^n - 2)(x^{2n} + 2x^n + 4)$
72. $(6 + b^k) \cdot$
   $(36 - 6b^2 + b^{2k})$
73. $8y^3(3 - 2y^2 + 8y^4)$
74. $s^3t(st^4 - 1 + s^6t^2)$
75. $(2x + y)(a - b)$
76. $(y^2 + 5)(x^2 + 3)$
77. $(y - 2)(y + 1)$
78. $(p + 7)(p - 3)$
79. $(3t - s)(2t + 7s)$
80. $(5k + 2m)(2k + 3m)$

REVIEW EXERCISES

Factor completely. See Sections 3.5 and 3.6.

73. $24y^3 - 16y^5 + 64y^7$

74. $s^4t^5 - s^3t + s^9t^3$

75. $2ax + ay - 2bx - by$

76. $x^2y^2 + 3y^2 + 5x^2 + 15$

77. $y^2 - y - 2$

78. $p^2 + 4p - 21$

79. $6t^2 + 19ts - 7s^2$

80. $10k^2 + 19km + 6m^2$

3.8 GENERAL METHODS OF FACTORING

FOR EXTRA HELP	OBJECTIVES
SSG pp. 91–94 **SSM** pp. 121–125	1 ▶ Know the first step in trying to factor a polynomial.
Video 4	2 ▶ Know the rules for factoring binomials.
Tutorial IBM MAC	3 ▶ Know the rules for factoring trinomials. 4 ▶ Know the rules for factoring polynomials of more than three terms.

Polynomials are factored by using the methods discussed earlier in this chapter. This section shows how to decide which method to use in factoring a particular polynomial. The factoring process is complete when the polynomial is in the form described on the next page.

INSTRUCTOR'S RESOURCES

 ITM pp. 361–362 **ISM** pp. 190–199
IAM p. 13

 TEST GENERATOR
DOS Windows MAC

 TRANSPARENCIES

 CONCEPTUAL   WRITING   CHALLENGING   SCIENTIFIC CALCULATOR   GRAPHICS CALCULATOR

Definition of Factored Form

A polynomial is in **factored form** when the following are satisfied.

1. The polynomial is written as a product of prime polynomials with integer coefficients.
2. All the polynomial factors are prime, except that a monomial factor need not be factored completely.

The order of the factors does not matter.

For example, $9x^2(x + 2)$ is the factored form of $9x^3 + 18x^2$. Because of the second rule above, it is not necessary to factor $9x^2$ as $3 \cdot 3 \cdot x \cdot x$.

1 ▶ Know the first step in trying to factor a polynomial.

▶ The first step in trying to factor a polynomial is always the same, regardless of the number of terms in the polynomial.

The first step in factoring any polynomial is to factor out any common factor.

E X A M P L E   1

Factoring Out a Common Factor

Factor the polynomial.

(a) $9p + 45 = 9(p + 5)$

(b) $5z^2 + 11z^3 + 9z^4 = z^2(5 + 11z + 9z^2)$

(c) $8m^2p^2 + 4mp = 4mp(2mp + 1)$ ■

2 ▶ Know the rules for factoring binomials.

▶ If the polynomial to be factored is a binomial, it will be necessary to use one of the following rules.

For a **binomial** (two terms), check for the following.

Difference of two squares	$x^2 - y^2 = (x + y)(x - y)$
Difference of two cubes	$x^3 - y^3 = (x - y)(x^2 + xy + y^2)$
Sum of two cubes	$x^3 + y^3 = (x + y)(x^2 - xy + y^2)$

CAUTION The sum of two squares usually cannot be factored with real numbers.

E X A M P L E   2

Factoring Binomials

Factor the polynomial, if possible.

(a) $64m^2 - 9n^2 = (8m)^2 - (3n)^2$ Difference of two squares

$= (8m + 3n)(8m - 3n)$

(b) $8p^3 - 27 = (2p)^3 - 3^3$ Difference of two cubes

$= (2p - 3)[(2p)^2 + (2p)(3) + 3^2]$

$= (2p - 3)(4p^2 + 6p + 9)$

(c) $100m^3 + 1 = (10m)^3 + 1^3$ Sum of two cubes

$$= (10m + 1)[(10m)^2 - (10m)(1) + 1^2]$$

$$= (10m + 1)(100m^2 - 10m + 1)$$

(d) $25m^2 + 121$ is prime. It is the sum of two squares. ■

3 ▶ Know the rules for factoring trinomials.

▶ If the polynomial to be factored is a trinomial, proceed as follows.

For a **trinomial** (three terms), first see if it is a perfect square trinomial of the form

$$x^2 + 2xy + y^2 = (x + y)^2,$$

or $x^2 - 2xy + y^2 = (x - y)^2.$

If it is not, use the methods of Section 3.6.

E X A M P L E   3

Factoring Trinomials

Factor the polynomial.

(a) $p^2 + 10p + 25 = (p + 5)^2$ Perfect square trinomial.

(b) $49z^2 - 42z + 9 = (7z - 3)^2$ Perfect square trinomial.

(c) $y^2 - 5y - 6 = (y - 6)(y + 1)$

     The numbers -6 and 1 have a product of -6 and a sum of -5.

(d) $r^2 + 18r + 72 = (r + 6)(r + 12)$

(e) $2k^2 - k - 6 = (2k + 3)(k - 2)$

     Use either method from Section 3.6.

(f) $28z^2 + 6z - 10 = 2(14z^2 + 3z - 5)$ Factor out the common factor.

$$= 2(7z + 5)(2z - 1) ■$$

4 ▶ Know the rules for factoring polynomials of more than three terms.

▶ If the polynomial to be factored has more than three terms, use the following guideline.

If the polynomial has more than three terms, try factoring by grouping.

E X A M P L E   4

Factoring Polynomials with More than Three Terms

Factor the polynomial.

(a) $xy^2 - y^3 + x^3 - x^2y = y^2(x - y) + x^2(x - y)$

$$= (x - y)(y^2 + x^2)$$

(b) $20k^3 + 4k^2 - 45k - 9 = (20k^3 + 4k^2) - (45k + 9)$ Be careful with signs.

$$= 4k^2(5k + 1) - 9(5k + 1)$$

$$= (5k + 1)(4k^2 - 9)$$ $5k + 1$ is a common factor.

$$= (5k + 1)(2k + 3)(2k - 3)$$ Difference of two squares

(c) $4a^2 + 4a + 1 - b^2 = (4a^2 + 4a + 1) - b^2$ Associative property

$$= (2a + 1)^2 - b^2$$ Perfect square trinomial

$$= (2a + 1 + b)(2a + 1 - b)$$ Difference of two squares

▶ **TEACHING TIP**
Include the example: Factor
$x^2 + 2xy + y^2 - 1$. Students often
have difficulty with the proper
grouping in factoring such a
polynomial. ◀

(d) $8m^3 + 4m^2 - n^3 - n^2$

First, notice that the terms must be rearranged before grouping because

$$(8m^3 + 4m^2) - (n^3 + n^2) = 4m^2(2m + 1) - n^2(n + 1),$$

which cannot be factored further. Write the polynomial as follows.

$8m^3 + 4m^2 - n^3 - n^2$

$\quad = (8m^3 - n^3) + (4m^2 - n^2)$ Group the cubes and squares.

$\quad = (2m - n)(4m^2 + 2mn + n^2)$
$\qquad + (2m - n)(2m + n)$ Factor each group.

$\quad = (2m - n)(4m^2 + 2mn + n^2 + 2m + n)$ Factor out the common factor. ■

The steps used in factoring a polynomial are summarized here.

Factoring a Polynomial		
	Step 1	Factor out any common factor.
	Step 2a	If the polynomial is a binomial, check to see if it is the difference of two squares, the difference of two cubes, or the sum of two cubes.
	Step 2b	If the polynomial is a trinomial, check to see if it is a perfect square trinomial. If it is not, factor as in Section 3.6.
	Step 2c	If the polynomial has more than three terms, try to factor by grouping.

Remember that the factored form can always be checked by multiplication.

3.8 EXERCISES

Factor the polynomial. See Examples 1–4.

1. $100a^2 - 9b^2$ **2.** $10r^2 + 13r - 3$ **3.** $3p^4 - 3p^3 - 90p^2$

4. $k^4 - 16$ **5.** $3a^2pq + 3abpq - 90b^2pq$ **6.** $49z^2 - 16$

7. $225p^2 + 256$ **8.** $x^3 - 1000$ **9.** $6b^2 - 17b - 3$

10. $k^2 - 6k - 16$ **11.** $18m^3n + 3m^2n^2 - 6mn^3$ **12.** $6t^2 + 19tu - 77u^2$

13. $2p^2 + 11pq + 15q^2$ **14.** $40p - 32r$ **15.** $9m^2 - 45m + 18m^3$

16. $4k^2 + 28kr + 49r^2$ **17.** $54m^3 - 2000$ **18.** $mn - 2n + 5m - 10$

19. $2a^2 - 7a - 4$ **20.** $9m^2 - 30mn + 25n^2$ **21.** $kq - 9q + kr - 9r$

22. $56k^3 - 875$ **23.** $9r^2 + 100$ **24.** $16z^3x^2 - 32z^2x$

◉ **25.** The polynomial in Exercise 23, $9r^2 + 100$, is an example of a sum of two squares. In general, a sum of two squares cannot be factored. Anne Kelly, a perceptive algebra student, commented that $9x^2 + 36y^2$ is a sum of two squares that *can* be factored. Factor this sum of two squares. Under what conditions can a sum of two squares be factored?

◉ **26.** The prime factorization of 35 is $5 \cdot 7$. Write 35 as the sum of two cubes, and factor using the method described in Section 3.7. Do the factors equal 5 and 7?

Answers (left column):

1. $(10a + 3b)(10a - 3b)$
2. $(5r - 1)(2r + 3)$
3. $3p^2(p - 6)(p + 5)$
4. $(k^2 + 4)(k + 2)(k - 2)$
5. $3pq(a + 6b)(a - 5b)$
6. $(7z + 4)(7z - 4)$ 7. prime
8. $(x - 10)(x^2 + 10x + 100)$
9. $(6b + 1)(b - 3)$
10. $(k - 8)(k + 2)$
11. $3mn(3m + 2n)(2m - n)$
12. $(3t - 7u)(2t + 11u)$
13. $(2p + 5q)(p + 3q)$
14. $8(5p - 4r)$
15. $9m(m - 5 + 2m^2)$
16. $(2k + 7r)^2$
17. $2(3m - 10) \cdot$ $(9m^2 + 30m + 100)$
18. $(n + 5)(m - 2)$
19. $(2a + 1)(a - 4)$
20. $(3m - 5n)^2$
21. $(k - 9)(q + r)$
22. $7(2k - 5)(4k^2 + 10k + 25)$

◉ CONCEPTUAL   ✐ WRITING   ▲ CHALLENGING   ▦ SCIENTIFIC CALCULATOR   ▦ GRAPHICS CALCULATOR

Factor the polynomial. See Examples 1–4.

27. $x^4 - 625$

28. $2m^2 - mn - 15n^2$

29. $p^3 + 64$

30. $48y^2z^3 - 28y^3z^4$

31. $64m^2 - 625$

32. $14z^2 - 3zk - 2k^2$

33. $12z^3 - 6z^2 + 18z$

34. $225k^2 - 36r^2$

35. $256b^2 - 400c^2$

36. $z^2 - zp - 20p^2$

37. $1000z^3 + 512$

38. $64m^2 - 25n^2$

39. $10r^2 + 23rs - 5s^2$

40. $12k^2 - 17kq - 5q^2$

41. $24p^3q + 52p^2q^2 + 20pq^3$

42. $32x^2 + 16x^3 - 24x^5$

43. $48k^4 - 243$

44. $14x^2 - 25xq - 25q^2$

45. $m^3 + m^2 - n^3 - n^2$

46. $64x^3 + y^3 - 16x^2 + y^2$

47. $x^2 - 4m^2 - 4mn - n^2$

48. $4r^2 - s^2 - 2st - t^2$

49. The area of the rectangle shown is given by the polynomial $2W^2 + 9W$, where W represents the width. Express the length of the rectangle in terms of W.

W

The area is $2W^2 + 9W$.

50. The area of the square shown is given by the polynomial $4x^2 + 12xy + 9y^2$. Find the length of a side of the square in terms of x and y.

The area is $4x^2 + 12xy + 9y^2$.

Factor the polynomial. See Examples 1–4.

51. $2x^2 - 2x - 40$
51. $2(x + 4)(x - 5)$

52. $27x^3 - 3y^3$
52. $3(9x^3 - y^3)$

53. $(2m + n)^2 - (2m - n)^2$
53. $8mn$

54. $(3k + 5)^2 - 4(3k + 5) + 4$
54. $9(k + 1)^2$

55. $50p^2 - 162$
55. $2(5p + 9)(5p - 9)$

56. $y^2 + 3y - 10$
56. $(y + 5)(y - 2)$

57. $12m^2rx + 4mnrx + 40n^2rx$
57. $4rx(3m^2 + mn + 10n^2)$

58. $18p^2 + 53pr - 35r^2$
58. $(9p - 5r)(2p + 7r)$

59. $21a^2 - 5ab - 4b^2$
59. $(7a - 4b)(3a + b)$

60. $x^2 - 2xy + y^2 - 4$
60. $(x - y + 2)(x - y - 2)$

61. $x^2 - y^2 - 4$
61. prime

62. $(5r + 2s)^2 - 6(5r + 2s) + 9$
62. $(5r + 2s - 3)^2$

63. $(p + 8q)^2 - 10(p + 8q) + 25$
63. $(p + 8q - 5)^2$

64. $z^4 - 9z^2 + 20$
64. $(z^2 - 5)(z + 2)(z - 2)$

65. $21m^4 - 32m^2 - 5$
65. $(7m^2 + 1)(3m^2 - 5)$

66. $(x - y)^3 - (27 - y)^3$
66. $(x - 27)(x^2 - 3xy + 3y^2 + 27x - 81y + 729)$

67. $(r + 2t)^3 + (r - 3t)^3$
67. $(2r - t)(r^2 - rt + 19t^2)$

68. $16x^3 + 32x^2 - 9x - 18$
68. $(4x + 3)(4x - 3)(x + 2)$

69. $x^5 + 3x^4 - x - 3$
69. $(x + 3)(x^2 + 1)(x + 1)(x - 1)$

70. $x^{16} - 1$
70. $(x^8 + 1)(x^4 + 1)(x^2 + 1)(x + 1)(x - 1)$

71. $m^2 - 4m + 4 - n^2 + 6n - 9$
71. $(m + n - 5)(m - n + 1)$

72. $x^2 + 4 + x^2y + 4y$
72. $(x^2 + 4)(1 + y)$

23. prime
24. $16z^2x(zx - 2)$
25. $9(x^2 + 4y^2)$; A sum of two squares can be factored when the greatest common factor is not 1.
26. $35 = 2^3 + 3^3$; $35 = (2 + 3)(2^2 - 2 \cdot 3 + 3^2)$; yes
27. $(x - 5)(x + 5)(x^2 + 25)$
28. $(2m + 5n)(m - 3n)$
29. $(p + 4)(p^2 - 4p + 16)$
30. $4y^2z^3(12 - 7yz)$
31. $(8m + 25)(8m - 25)$
32. $(7z + 2k)(2z - k)$
33. $6z(2z^2 - z + 3)$
34. $9(5k + 2r)(5k - 2r)$
35. $16(4b + 5c)(4b - 5c)$
36. $(z - 5p)(z + 4p)$
37. $8(5z + 4) \cdot (25z^2 - 20z + 16)$
38. $(8m + 5n)(8m - 5n)$
39. $(5r - s)(2r + 5s)$
40. $(4k + q)(3k - 5q)$
41. $4pq(2p + q)(3p + 5q)$
42. $8x^2(4 + 2x - 3x^3)$
43. $3(4k^2 + 9) \cdot (2k + 3)(2k - 3)$
44. $(7x + 5q)(2x - 5q)$
45. $(m - n) \cdot (m^2 + mn + n^2 + m + n)$
46. $(4x + y) \cdot (16x^2 - 4xy + y^2 - 4x + y)$
47. $(x - 2m - n) \cdot (x + 2m + n)$
48. $(2r - s - t) \cdot (2r + s + t)$
49. $2w + 9$
50. $2x + 3y$

73. $a(1 + y)$
75. $(x^2 + 4)(1 + y)$
77. $(3m^k + 2)(m^k - 3)$
78. $2a^x(1 + 2a^x + 4a^{2x})$
79. $3c^y(5c^{3y} + c^y - 2)$
80. $2(p^a - 2)(p^a + 7)$
81. $(z^{2x} + w^x)(z^{2x} - w^x)$
82. $(x^p + y^p)(x^p - y^p)$
83. $\left\{-\dfrac{2}{3}\right\}$ 84. $\left\{\dfrac{7}{2}\right\}$
85. $\{0\}$ 86. $\{0\}$
87. $\{-10\}$ 88. $\{-8\}$

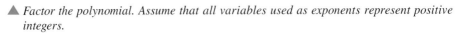

MATHEMATICAL CONNECTIONS (Exercises 73–76)

The polynomial in Exercise 72 factors as $(x^2 + 4)(1 + y)$. Students often forget to include the first term in the second binomial (that is, 1). Work Exercises 73–76 in order, so that you can see a connection between this problem and a simpler one.

73. Factor $a + ay$ by taking out the greatest common factor.

74. If you worked Exercise 73 correctly, you should have obtained $a(1 + y)$. Explain why the 1 must appear in the binomial.

75. Begin to factor $x^2 + 4 + x^2y + 4y$ by grouping the first two terms and the last two terms. Then complete the factorization.

76. If you worked Exercise 75 correctly, you should have obtained $(x^2 + 4)(1 + y)$. Remembering that $x^2 + 4$ is taking the place of a in Exercises 73 and 74, explain why 1 must appear in the second binomial.

▲ *Factor the polynomial. Assume that all variables used as exponents represent positive integers.*

77. $3m^{2k} - 7m^k - 6$ **78.** $2a^x + 4a^{2x} + 8a^{3x}$ **79.** $15c^{4y} + 3c^{2y} - 6c^y$

80. $2p^{2a} + 10p^a - 28$ **81.** $z^{4x} - w^{2x}$ **82.** $x^{2p} - y^{2p}$

REVIEW EXERCISES

Solve the equation. See Section 2.1.

83. $3x + 2 = 0$ **84.** $-2x + 7 = 0$ **85.** $5x = 0$

86. $-8x = 0$ **87.** $\dfrac{1}{2}t + 5 = 0$ **88.** $-\dfrac{3}{4}x - 6 = 0$

3.9 SOLVING EQUATIONS BY FACTORING

FOR EXTRA HELP	OBJECTIVES
SSG pp. 95–98 **SSM** pp. 126–132	1 ▶ Learn the zero-factor property.
Video 4	2 ▶ Use the zero-factor property to solve equations.
Tutorial IBM MAC	3 ▶ Solve applied problems that require the zero-factor property.

The equations that we have solved so far in this book have been linear equations. Recall that in a linear equation, the largest power of the variable is 1. In order to solve equations of degree greater than 1, other methods must be developed. In this section we learn one of these methods, which involves factoring.

1 ▶ Learn the zero-factor property.

▶ Some equations that cannot be solved by other methods can be solved by factoring. This process depends on a special property of the number 0, called the **zero-factor property.**

| Zero-Factor Property | If two numbers have a product of 0, then at least one of the numbers must be 0. That is, if $ab = 0$, then either $a = 0$ or $b = 0$. |

To prove the zero-factor property, we first assume $a \neq 0$. (If a does equal 0, then the property is proved already.) If $a \neq 0$, then $1/a$ exists, and both sides of $ab = 0$ can be multiplied by $1/a$ to get

$$\frac{1}{a} \cdot ab = \frac{1}{a} \cdot 0$$

$$b = 0.$$

Thus, if $a \neq 0$, then $b = 0$, and the property is proved.

2 ▶ Use the zero-factor property to solve equations.

▶ The next examples show how to use the zero-factor property to solve equations.

E X A M P L E   1

Using the Zero-Factor Property to Solve an Equation

Solve the equation $(x + 6)(2x - 3) = 0$.

Here the product of $x + 6$ and $2x - 3$ is 0. By the zero-factor property, this can be true only if $x + 6$ equals 0 or if $2x - 3$ equals 0. That is,

$$x + 6 = 0 \quad \text{or} \quad 2x - 3 = 0.$$

Solve these two equations.

$$x + 6 = 0 \quad \text{or} \quad 2x - 3 = 0$$
$$x = -6 \quad \text{or} \quad 2x = 3$$
$$x = \frac{3}{2}$$

To check these solutions, first replace x with -6 in the original equation. Then go back and replace x with $3/2$. This check shows that the solution set is $\{-6, 3/2\}$. ∎

Since the product $(x + 6)(2x - 3)$ equals $2x^2 + 9x - 18$, the equation of Example 1 has a squared term and is an example of a *quadratic equation*.

| Quadratic Equation | An equation that can be written in the form

$$ax^2 + bx + c = 0,$$

where $a \neq 0$, is a **quadratic equation.** The form given is called **standard form.** |

Quadratic equations are discussed in more detail in a later chapter. Many quadratic equations can be solved by factoring.

E X A M P L E   2

Solving a Quadratic Equation by Factoring

Solve the equation $2p^2 + 3p = 2$.

The zero-factor property requires a product of two factors equal to 0. We get 0 on one side of the equals sign in this equation by subtracting 2 from both sides.

Chalkboard Exercise

Solve $(8a + 3)(2a + 1) = 0$.

Answer: $\left\{ -\dfrac{3}{8}, -\dfrac{1}{2} \right\}$

Chalkboard Exercise

Solve $15m^2 + 7m = 2$.

Answer: $\left\{\dfrac{1}{5}, -\dfrac{2}{3}\right\}$

$$2p^2 + 3p = 2$$

$$
\begin{array}{ll}
2p^2 + 3p - 2 = 0 & \text{Standard form} \\
(2p - 1)(p + 2) = 0 & \text{Factor on the left.}
\end{array}
$$

$$
\begin{array}{lll}
2p - 1 = 0 & \text{or} \quad p + 2 = 0 & \text{Zero-factor property} \\
2p = 1 & \text{or} \quad\quad p = -2 & \text{Solve each equation.} \\
p = \dfrac{1}{2}
\end{array}
$$

The solution set is $\{1/2, -2\}$. Check by substitution in the original equation. ∎

EXAMPLE 3

Solving a Quadratic Equation with a Missing Term

Solve $5z^2 - 25z = 0$.

   This quadratic equation has a missing term. Comparing it with the general form $ax^2 + bx + c = 0$ shows that $c = 0$. The zero-factor property still can be used, however, since $5z^2 - 25z = 5z(z - 5)$.

$$
\begin{array}{ll}
5z^2 - 25z = 0 & \text{Given equation} \\
5z(z - 5) = 0 & \text{Factor.}
\end{array}
$$

$$
\begin{array}{lll}
5z = 0 & \text{or} \quad z - 5 = 0 & \text{Zero-factor property} \\
z = 0 & \text{or} \quad\quad z = 5 &
\end{array}
$$

Chalkboard Exercise

Solve $p^2 = -12p$.

Answer: $\{0, -12\}$

The solutions are 0 and 5, as can be verified by substituting in the original equation. The solution set is $\{0, 5\}$. ∎

EXAMPLE 4

Solving an Equation Requiring Rewriting

Solve $(2q + 1)(q + 1) = 2(1 - q) + 6$.

   Put the equation in the standard form $ax^2 + bx + c = 0$ by first multiplying on each side.

$$
\begin{array}{ll}
(2q + 1)(q + 1) = 2(1 - q) + 6 & \\
2q^2 + 3q + 1 = 2 - 2q + 6 & \\
2q^2 + 5q - 7 = 0 & \text{Combine terms; get 0 on the right side.} \\
(2q + 7)(q - 1) = 0 & \text{Factor.}
\end{array}
$$

Chalkboard Exercise

Solve
$(a + 6)(a - 2) = 2 + a - 10$.

Answer: $\{1, -4\}$

$$
\begin{array}{lll}
2q + 7 = 0 & \text{or} \quad q - 1 = 0 & \text{Zero-factor property} \\
2q = -7 & \text{or} \quad\quad q = 1 & \\
q = -\dfrac{7}{2}
\end{array}
$$

Check that the solution set is $\{-7/2, 1\}$. ∎

   In summary, use the following steps to solve an equation by factoring.

Solving an Equation by Factoring

Step 1 **Write in standard form.** Rewrite the equation if necessary so that one side is 0.

Step 2 **Factor.** Factor the polynomial.

Step 3 **Use the zero-factor property.** Place each variable factor equal to zero, using the zero-factor property.

Step 4 **Find the solution(s).** Solve each equation formed in Step 3.

Step 5 **Check.** Check each solution in the original equation.

▶ **TEACHING TIP**
Emphasize that 0 must be alone on one side before the zero-factor property can be used. ◀

CAUTION If $ab = 0$, then $a = 0$ or $b = 0$. However, if $ab = 6$, for example, it is not necessarily true that $a = 6$ or $b = 6$; it is very likely that *neither* $a = 6$ *nor* $b = 6$. It is important to remember that the zero-factor property works only for a product equal to *zero*.

◆ 8 seconds; 4 seconds

◆ **C O N N E C T I O N S** ◆

Quadratic equations are used to describe the distance a falling object or a propelled object travels in a specific time. For example, if a toy rocket is launched vertically upward from ground level with an initial velocity of 128 feet per second, then its height h after t seconds is given by the equation $h = -16t^2 + 128t$ (if air resistance is neglected). By substituting values for t or h into the equation, corresponding values of the other value can be found. In particular, if h is zero, the rocket is at ground level, and the positive value of t gives the time required for it to go up and return to the ground.

FOR DISCUSSION OR WRITING
How many seconds will it take for the rocket to return to the ground? By experimenting with different values of t, try to find the time it will take for the rocket to reach its maximum height.

The equations we have solved so far in this section have all been quadratic equations. The zero-factor property can be extended to solve certain equations of degree 3 or higher, as shown in the next example.

E X A M P L E   5

Solving an Equation of Degree 3

Solve $-x^3 + x^2 = -6x$.

Start by adding $6x$ to both sides to get 0 on the right side.

$$-x^3 + x^2 + 6x = 0.$$

To make the factoring step easier, multiply both sides by -1.

$$x^3 - x^2 - 6x = 0$$
$$x(x^2 - x - 6) = 0 \qquad \text{Factor.}$$
$$x(x - 3)(x + 2) = 0 \qquad \text{Factor the trinomial.}$$

Use the zero-factor property, extended to include the three variable factors.

$$x = 0 \qquad \text{or} \qquad x - 3 = 0 \qquad \text{or} \qquad x + 2 = 0$$
$$x = 3 \qquad\qquad\qquad x = -2$$

The solution set is $\{0, 3, -2\}$. ■

Chalkboard Exercise

Solve $3n^3 + n^2 = 4n$.

Answer: $\left\{ 0, -\dfrac{4}{3}, 1 \right\}$

3 ▶ Solve applied problems that require the zero-factor property.

▶ In the next example, we show an application that leads to a quadratic equation.

E X A M P L E   6

Using a Quadratic Equation in an Application

Some surveyors are surveying a lot that is in the shape of a parallelogram. They find that the longer sides of the parallelogram are each 8 meters longer than the distance between them. The area of the lot is 48 square meters. Find the length of the longer sides and the distance between them.

Chalkboard Exercise

The length of a room is 2 meters less than three times the width. The area of the room is 96 square meters. Find the width of the room.

Answer: 6 meters

Let x represent the distance between the longer sides as shown in Figure 1. Then $x + 8$ is the length of each longer side. The area of a parallelogram is given by $A = bh$ where b is the length of the longer side and h is the distance between the longer sides. Here $b = x + 8$ and $h = x$.

$$A = bh$$
$$48 = (x + 8)x \qquad \text{Let } A = 48, b = x + 8, h = x.$$
$$48 = x^2 + 8x \qquad \text{Distributive property}$$
$$0 = x^2 + 8x - 48 \qquad \text{Subtract 48.}$$
$$0 = (x + 12)(x - 4) \qquad \text{Factor.}$$
$$x + 12 = 0 \qquad \text{or} \qquad x - 4 = 0 \qquad \text{Zero-factor property}$$
$$x = -12 \qquad \text{or} \qquad x = 4$$

A parallelogram cannot have a side with a negative length, so reject -12 as a solution. The only possible solution is 4, so the distance between the longer sides is 4 meters. The length of the longer sides is $4 + 8 = 12$ meters. ■

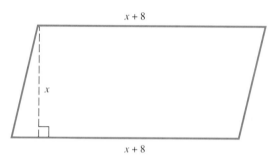

FIGURE 1

When applications lead to quadratic equations, it often happens that a solution of the equation does not satisfy the physical requirements of the problem, as in Example 6. In such cases, we must reject this solution of the equation as an answer to the problem.

3.9 EXERCISES

2. (d); The polynomial is not in factored form.
3. $\{5, -10\}$ 4. $\{-3, -7\}$
5. $\left\{\dfrac{5}{2}, -\dfrac{8}{3}\right\}$ 6. $\left\{\dfrac{4}{3}, -\dfrac{5}{2}\right\}$
7. $\left\{-6, \dfrac{3}{4}, 1\right\}$
8. $\left\{2, 7, -\dfrac{9}{2}\right\}$

◉ 1. Explain in your own words how the zero-factor property is used in solving a quadratic equation.

◉ 2. One of the following equations is *not* in proper form for using the zero-factor property. Which one is it, and why is it not in proper form?
(a) $(x + 2)(x - 6) = 0$ (b) $x(3x - 7) = 0$ (c) $3t(t + 8)(t - 9) = 0$
(d) $y(y - 3) + 6(y - 3) = 0$

Solve the equation by using the zero-factor property. See Example 1.

3. $(x - 5)(x + 10) = 0$
4. $(y + 3)(y + 7) = 0$
5. $(2k - 5)(3k + 8) = 0$
6. $(3q - 4)(2q + 5) = 0$
7. $(m + 6)(4m - 3)(m - 1) = 0$
8. $(z - 2)(z - 7)(2z + 9) = 0$

◉ CONCEPTUAL   📝 WRITING   ▲ CHALLENGING   ▦ SCIENTIFIC CALCULATOR   ▦ GRAPHICS CALCULATOR

◉ **9.** In trying to solve $(x + 4)(x - 1) = 1$, a student reasons that since $1 \cdot 1 = 1$, the
📝   equation is solved by solving

$$x + 4 = 1 \quad \text{or} \quad x - 1 = 1.$$

Explain the error in this reasoning. What is the correct way to solve this equation?

◉ **10.** In solving the equation $4(x - 3)(x + 7) = 0$, a student writes
📝

$$4 = 0 \quad \text{or} \quad x - 3 = 0 \quad \text{or} \quad x + 7 = 0.$$

Then the student becomes confused, not knowing how to handle the equation $4 = 0$.
Explain how to handle this, and give the solutions of the equation.

Solve the equation by using the zero-factor property. See Examples 2 and 3.

11. $m^2 - 3m - 10 = 0$ **12.** $x^2 + x - 12 = 0$ **13.** $z^2 + 9z + 18 = 0$

14. $x^2 - 18x + 80 = 0$ **15.** $2x^2 = 7x + 4$ **16.** $2x^2 = 3 - x$

17. $15k^2 - 7k = 4$

18. $3c^2 + 3 = -10c$

19. $2y^2 - 12 - 4y = y^2 - 3y$

20. $3p^2 + 9p + 30 = 2p^2 - 2p$

21. $8m^2 - 72 = 0$

22. $6m^2 - 54 = 0$

23. $5k^2 + 3k = 0$

24. $9t^2 - 5t = 0$

25. $16x^2 + 24x + 9 = 0$

26. $9y^2 + 6y + 1 = 0$

27. $4x^2 = 9$

28. $16y^2 = 25$

———————◆ **MATHEMATICAL CONNECTIONS** (Exercises 29–36) ◆———————

While the zero-factor property can be used to solve the quadratic equations presented in
this section, it can also be used "in reverse" to write quadratic equations with particular
solutions. Work Exercises 29–33 in order to write a quadratic equation with solution set
$\{-4, 8\}$. Then work Exercises 34–36.

29. Write a binomial in the form $x + k$ such that when $x = -4$, the value of the binomial
is 0.

30. Write a binomial in the form $x + k$ such that when $x = 8$, the value of the binomial
is 0.

31. Write your answers to Exercises 29 and 30 as a product.

32. Multiply the result in Exercise 31 to obtain a quadratic trinomial.

◉ **33.** Set the quadratic trinomial from Exercise 32 equal to 0. Without doing any other work,
what must be its two solutions?

◉ **34.** Use the concepts of Exercises 29–33 to write a quadratic equation with solution set
$\{-5, 9\}$. Express it in the form $ax^2 + bx + c = 0$.

◉ **35.** Use the concepts of Exercises 29–33 to write a quadratic equation with solution set
$\left\{-\dfrac{2}{3}, \dfrac{5}{6}\right\}$. Express it in the form $ax^2 + bx + c = 0$, where a, b, and c are integers.

▲ **36.** Use the concepts of Exercises 29–33 to write a quadratic equation with solution set
$\{-9\}$. Express it in the form $ax^2 + bx + c = 0$, where a, b, and c are integers.

————————◆————————

Solve the equation by using the zero-factor property. See Example 4.

37. $(x - 3)(x + 5) = -7$

38. $(x + 8)(x - 2) = -21$

39. $(2x + 1)(x - 3) = 6x + 3$

40. $(3x + 2)(x - 3) = 7x - 1$

41. $(x + 3)(x - 6) = (2x + 2)(x - 6)$

42. $(2x + 1)(x + 5) = (x + 11)(x + 3)$

11. $\{-2, 5\}$ **12.** $\{-4, 3\}$
13. $\{-6, -3\}$ **14.** $\{8, 10\}$

15. $\left\{-\dfrac{1}{2}, 4\right\}$

16. $\left\{-\dfrac{3}{2}, 1\right\}$

17. $\left\{-\dfrac{1}{3}, \dfrac{4}{5}\right\}$

18. $\left\{-\dfrac{1}{3}, -3\right\}$

19. $\{-3, 4\}$ **20.** $\{-6, -5\}$

21. $\{-3, 3\}$ **22.** $\{-3, 3\}$

23. $\left\{-\dfrac{3}{5}, 0\right\}$ **24.** $\left\{0, \dfrac{5}{9}\right\}$

25. $\left\{-\dfrac{3}{4}\right\}$ **26.** $\left\{-\dfrac{1}{3}\right\}$

27. $\left\{-\dfrac{3}{2}, \dfrac{3}{2}\right\}$

28. $\left\{-\dfrac{5}{4}, \dfrac{5}{4}\right\}$

29. $x + 4$ **30.** $x - 8$
31. $(x + 4)(x - 8)$
32. $x^2 - 4x - 32$
33. $x^2 - 4x - 32 = 0$;
The solutions are -4 and
8.   **34.** $x^2 - 4x - 45 = 0$
35. $18x^2 - 3x - 10 = 0$
36. $x^2 + 18x + 81 = 0$
37. $\{-4, 2\}$ **38.** $\{-5, -1\}$

39. $\left\{-\dfrac{1}{2}, 6\right\}$

40. $\left\{-\dfrac{1}{3}, 5\right\}$

41. $\{1, 6\}$ **42.** $\{-4, 7\}$

◉ CONCEPTUAL   📝 WRITING   ▲ CHALLENGING   ▥ SCIENTIFIC CALCULATOR   ▦ GRAPHICS CALCULATOR

45. $\left\{-\dfrac{1}{2}, 0, 5\right\}$

46. $\left\{-\dfrac{1}{3}, 0, \dfrac{5}{2}\right\}$

47. $\left\{-\dfrac{4}{3}, 0, \dfrac{4}{3}\right\}$

48. $\left\{-\dfrac{8}{5}, 0, \dfrac{8}{5}\right\}$

49. $\left\{-\dfrac{5}{2}, -1, 1\right\}$

50. $\left\{-7, -\dfrac{1}{2}, 7\right\}$

51. By dividing both sides by a variable expression, she "lost" the solution 0.

52. (d)   53. $\left\{-\dfrac{1}{2}, 6\right\}$

54. $\left\{-\dfrac{15}{8}, -1\right\}$

55. $\left\{-\dfrac{2}{3}, \dfrac{4}{15}\right\}$

56. $\left\{-\dfrac{5}{2}, -1\right\}$

57. $\{2, 4\}$ 58. $\left\{-\dfrac{3}{2}, \dfrac{1}{4}\right\}$

59. $\left\{-\dfrac{3}{2}, \dfrac{1}{2}\right\}$

60. $\left\{-1, -\dfrac{1}{4}\right\}$

61. width: 16 feet; length: 20 feet
62. smaller square: 4 inches;   larger square: 13 inches   63. base: 8 meters;   height: 11 meters

43. Explain why a quadratic equation of the form $ax^2 + bx = 0$ must have 0 in its solution set.

44. Explain why a third degree equation of the form $ax^3 + bx^2 + cx = 0$ must have 0 in its solution set.

Solve the equation by using the zero-factor property. See Example 5.

45. $2x^3 - 9x^2 - 5x = 0$

46. $6x^3 - 13x^2 - 5x = 0$

47. $9t^3 = 16t$

48. $25y^3 = 64y$

49. $2r^3 + 5r^2 - 2r - 5 = 0$

50. $2p^3 + p^2 - 98p - 49 = 0$

51. A student tried to solve the equation in Exercise 47 by first dividing both sides by t, obtaining $9t^2 = 16$. She then solved the resulting equation by the zero-factor property to get the solution set $\left\{-\dfrac{4}{3}, \dfrac{4}{3}\right\}$. What was incorrect about her procedure?

52. Without actually solving each equation, determine which one of the following has 0 in its solution set.
(a) $4x^2 - 25 = 0$ **(b)** $x^2 + 2x - 3 = 0$ **(c)** $6x^2 + 9x + 1 = 0$
(d) $x^3 + 4x^2 = 3x$

▲ *Solve the equation by using the zero-factor property.*

53. $2(x - 1)^2 - 7(x - 1) - 15 = 0$

54. $4(2k + 3)^2 - (2k + 3) - 3 = 0$

55. $5(3a - 1)^2 + 3 = -16(3a - 1)$

56. $2(m + 3)^2 = 5(m + 3) - 2$

57. $(x - 1)^2 - (2x - 5)^2 = 0$

58. $(3y + 1)^2 - (y - 2)^2 = 0$

59. $(2k - 3)^2 = 16k^2$

60. $9p^2 = (5p + 2)^2$

Solve the problem by writing a quadratic equation and then solving it using the zero-factor property. See Example 6.

61. A garden has an area of 320 square feet. Its length is 4 feet more than its width. What are the dimensions of the garden?

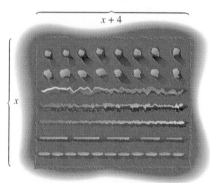

$x + 4$

x

62. There are two squares such that the side of the larger square is 1 inch longer than 3 times the length of the side of the smaller square. If the difference between their areas is 153 square inches, find the lengths of the sides of the squares.

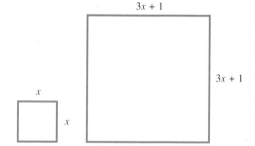

$3x + 1$

$3x + 1$

x

x

63. A sign has the shape of a triangle. The length of the base is 3 meters less than the height. What are the measures of the base and the height, if the area is 44 square meters?

64. The base of a parallelogram is 7 feet more than the height. If the area of the parallelogram is 60 square feet, what are the measures of the base and the height?

65. As mentioned in the Connections box in this section, if a toy rocket is launched vertically upward from ground level with an initial velocity of 128 feet per second, then its height h after t seconds is given by the equation $h = -16t^2 + 128t$ (if air resistance is neglected). When will the rocket be 240 feet above the ground?

66. After how many seconds will the rocket in Exercise 65 be 112 feet above the ground?

67. If an object is thrown upward with an initial velocity of 64 feet per second from a height of 80 feet, then its height t seconds after it is thrown is defined by the expression $-16t^2 + 64t + 80$. How long after it is thrown will it hit the ground? (*Hint:* When it hits the ground, its height is 0 feet.)

▲**68.** A box with no top is to be constructed from a piece of cardboard whose length measures 6 inches more than its width. The box is to be formed by cutting squares that measure 2 inches on each side from the four corners, and then folding up the sides. If the volume of the box will be 110 cubic inches, what are the dimensions of the piece of cardboard?

▲**69.** The surface area of the box with open top shown in the figure is 161 square inches. Find the dimensions of the base. (*Hint:* The surface area S is given by the formula $S = x^2 + 4xh$.)

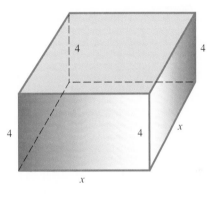

64. base: 12 feet; height: 5 feet
65. after 3 seconds and 5 seconds
66. 1 second and 7 seconds
67. 5 seconds   **68.** length: 15 inches;   width: 9 inches
69. Each side measures 7 inches.   **70.** −6 and −5 or 5 and 6   **71.** −9 and −8 or 8 and 9   **72.** 50 feet by 100 feet   **73.** 100 feet by 300 feet
74. $6\frac{1}{4}$ seconds   **75.** $4p$
76. $-\dfrac{b}{3a^2}$
77. $-\dfrac{3}{4m^4n^3}$
78. $\dfrac{15}{24}$ **79.** $\dfrac{36}{75}$ **80.** $\dfrac{40}{15}$

70. Find two consecutive integers such that the sum of their squares is 61.

71. Find two consecutive integers such that their product is 72.

▲**72.** A farmer has 300 feet of fencing and wants to enclose a rectangular area of 5000 square feet. What dimensions should she use?

▲**73.** A rectangular landfill has an area of 30,000 square feet. Its length is 200 feet more than its width. What are the dimensions of the landfill?

74. If a baseball is dropped from a helicopter 625 feet above the ground, then its distance in feet from the ground t seconds later is given by the expression $-16t^2 + 625$. How long after it is dropped will it hit the ground?

REVIEW EXERCISES

Simplify. See Section 3.1.

75. $\dfrac{12p^2}{3p}$

76. $\dfrac{-50a^4b^5}{150a^6b^4}$

77. $\dfrac{-27m^2n^5}{36m^6n^8}$

Write the fraction with the indicated denominator.

78. $\dfrac{5}{8} = \dfrac{?}{24}$

79. $\dfrac{12}{25} = \dfrac{?}{75}$

80. $\dfrac{8}{3} = \dfrac{?}{15}$

◉ CONCEPTUAL   ☑ WRITING   ▲ CHALLENGING   ▦ SCIENTIFIC CALCULATOR   ▨ GRAPHICS CALCULATOR

CHAPTER 3 SUMMARY

NEW SYMBOLS

$P(x)$ polynomial in x (read "P of x")

QUICK REVIEW

CONCEPTS	EXAMPLES
3.1 INTEGER EXPONENTS	
Definitions and Rules for Exponents	
Product Rule: $a^m \cdot a^n = a^{m+n}$	$3^4 \cdot 3^2 = 3^6$
Quotient Rule: $\dfrac{a^m}{a^n} = a^{m-n}$	$\dfrac{2^5}{2^3} = 2^2$
Negative Exponent: $a^{-n} = \dfrac{1}{a^n}$	$5^{-2} = \dfrac{1}{5^2}$
$\dfrac{a^{-n}}{b^{-m}} = \dfrac{b^m}{a^n}$	$\dfrac{5^{-3}}{4^{-6}} = \dfrac{4^6}{5^3}$
Zero Exponent: $a^0 = 1$	$27^0 = 1, \quad (-5)^0 = 1$
3.2 FURTHER PROPERTIES OF EXPONENTS	
Power Rules	
$\qquad (a^m)^n = a^{mn}$	$(6^3)^4 = 6^{12}$
$\qquad (ab)^m = a^m b^m$	$(5p)^4 = 5^4 p^4$
$\qquad \left(\dfrac{a}{b}\right)^n = \dfrac{a^n}{b^n}$	$\left(\dfrac{2}{3}\right)^5 = \dfrac{2^5}{3^5}$
$\qquad a^{-n} = \left(\dfrac{1}{a}\right)^n$	$4^{-3} = \left(\dfrac{1}{4}\right)^3$
$\qquad \left(\dfrac{a}{b}\right)^{-n} = \left(\dfrac{b}{a}\right)^n$	$\left(\dfrac{4}{7}\right)^{-2} = \left(\dfrac{7}{4}\right)^2$

CONCEPTS	EXAMPLES
3.3 ADDITION AND SUBTRACTION OF POLYNOMIALS	
Add or subtract polynomials by combining like terms.	Add: $(x^2 - 2x + 3) + (2x^2 - 8)$. $$= 3x^2 - 2x - 5$$ Subtract: $(5x^4 + 3x^2) - (7x^4 + x^2 - x)$. $$= -2x^4 + 2x^2 + x$$
3.4 MULTIPLICATION OF POLYNOMIALS	
Use the commutative and associative properties and the rules for exponents to multiply two polynomials.	Multiply: $(x^3 + 3x)(4x^2 - 5x + 2)$. $$= 4x^5 + 12x^3 - 5x^4 - 15x^2 + 2x^3 + 6x$$ $$= 4x^5 - 5x^4 + 14x^3 - 15x^2 + 6x$$
Special Products $$(x + y)(x - y) = x^2 - y^2$$ $$(x + y)^2 = x^2 + 2xy + y^2$$ $$(x - y)^2 = x^2 - 2xy + y^2$$	$$(3m + 8)(3m - 8) = 9m^2 - 64$$ $$(5a + 3b)^2 = 25a^2 + 30ab + 9b^2$$ $$(2k - 1)^2 = 4k^2 - 4k + 1$$
To multiply two binomials in general, use the FOIL method. Multiply the First terms, the Outside terms, the Inside terms, and the Last terms.	$(2x + 3)(x - 7)$ $$= 2x(x) + 2x(-7) + 3x + 3(-7)$$ $$= 2x^2 - 14x + 3x - 21$$ $$= 2x^2 - 11x - 21$$
3.5 GREATEST COMMON FACTORS; FACTORING BY GROUPING	
Factoring Out the Greatest Common Factor The product of the largest common numerical factor and the variable of lowest degree common to every term in a polynomial is the greatest common factor of the terms of the polynomial.	Factor: $4x^2y - 50xy^2 = 2^2x^2y - 2 \cdot 5^2xy^2$. The greatest common factor is $2xy$. $$4x^2y - 50xy^2 = 2xy(2x - 25y)$$
Factoring by Grouping Group the terms so that each group has a common factor. Factor out the common factor in each group. If the groups now have a common factor, factor it out. If not, try a different grouping.	$5a - 5b - ax + bx = (5a - 5b) + (-ax + bx)$ $$= 5(a - b) - x(a - b)$$ $$= (a - b)(5 - x)$$
3.6 FACTORING TRINOMIALS	
Choose factors of the coefficient of the first term and factors of the coefficient of the last term. Use combinations of these factors to find the correct middle term of the trinomial.	Factor: $15x^2 + 14x - 8$. Factors of $15x^2$ are $5x$ and $3x$, $1x$ and $15x$. Factors of -8 are -4 and 2, 4 and -2, -1 and 8, -8 and 1. $15x^2 + 14x - 8 = (5x - 2)(3x + 4)$
3.7 SPECIAL FACTORING	
Difference of Two Squares $$x^2 - y^2 = (x + y)(x - y)$$	Factor: $$4m^2 - 25n^2 = (2m)^2 - (5n)^2$$ $$= (2m + 5n)(2m - 5n)$$
Perfect Square Trinomials $$x^2 + 2xy + y^2 = (x + y)^2$$ $$x^2 - 2xy + y^2 = (x - y)^2$$	$$9y^2 + 6y + 1 = (3y + 1)^2$$ $$16p^2 - 56p + 49 = (4p - 7)^2$$

CONCEPTS	EXAMPLES
Difference of Two Cubes $\quad x^3 - y^3 = (x - y)(x^2 + xy + y^2)$	$8 - 27a^3 = (2 - 3a)(4 + 6a + 9a^2)$
Sum of Two Cubes $\quad x^3 + y^3 = (x + y)(x^2 - xy + y^2)$	$64z^3 + 1 = (4z + 1)(16z^2 - 4z + 1)$

3.8 GENERAL METHODS OF FACTORING

Factor out any common factors. For a binomial, check for the difference of two squares, the difference of two cubes, or the sum of two cubes. For a trinomial, see if it is a perfect square. If not, factor as in Section 3.6. For more than three terms, try factoring by grouping.	Factor: $ak^3 + 2ak^2 - 9ak - 18a$. $\quad = a(k^3 + 2k^2 - 9k - 18)$ $\quad = a[(k^3 + 2k^2) - (9k + 18)]$ $\quad = a[k^2(k + 2) - 9(k + 2)]$ $\quad = a[(k + 2)(k^2 - 9)]$ $\quad = a(k + 2)(k - 3)(k + 3)$

3.9 SOLVING EQUATIONS BY FACTORING

Step 1	Rewrite the equation if necessary so that one side is 0.	Solve: $\qquad 2x^2 + 5x = 3.$ $\qquad 2x^2 + 5x - 3 = 0$
Step 2	Factor the polynomial.	$\qquad (2x - 1)(x + 3) = 0$
Step 3	Set each factor equal to 0.	$2x - 1 = 0 \quad$ or $\quad x + 3 = 0$
Step 4	Solve each equation.	$2x = 1 \qquad\qquad x = -3$ $x = \dfrac{1}{2}$
Step 5	Check each solution.	A check verifies that the solution set is $\{1/2, -3\}$.

CHAPTER 3 REVIEW EXERCISES

1. $-12x^2y^8$ 2. $\dfrac{-2n}{m^5}$

3. $\dfrac{10p^8}{q^7}$ 5. 64 6. $\dfrac{1}{81}$

7. -125 8. 18 9. $\dfrac{81}{16}$

10. $\dfrac{16}{25}$ 11. $\dfrac{11}{30}$ 12. $\dfrac{1}{11}$

13. 0 14. For example, let $x = 2$ and $y = 3$. Then $(x + y)^{-1} = (2 + 3)^{-1} = \dfrac{1}{5}$, and $x^{-1} + y^{-1} = \dfrac{1}{2} + \dfrac{1}{3} = \dfrac{5}{6} \neq \dfrac{1}{5}$.

[3.1] *Use the product rule and/or the quotient rule to simplify. Write the answer with only positive exponents. Assume that all variables represent nonzero real numbers.*

1. $(-3x^4y^3)(4x^{-2}y^5)$ **2.** $\dfrac{6m^{-4}n^3}{-3mn^2}$ **3.** $\dfrac{(5p^{-2}q)(4p^5q^{-3})}{2p^{-5}q^5}$

4. Explain the difference between the expressions $(-6)^0$ and -6^0.

Evaluate.

5. 4^3 **6.** $\left(\dfrac{1}{3}\right)^4$ **7.** $(-5)^3$

8. $\dfrac{2}{(-3)^{-2}}$ **9.** $\left(\dfrac{2}{3}\right)^{-4}$ **10.** $\left(\dfrac{5}{4}\right)^{-2}$

11. $5^{-1} + 6^{-1}$ **12.** $(5 + 6)^{-1}$ **13.** $-3^0 + 3^0$

14. Give an example to show that $(x + y)^{-1} \neq x^{-1} + y^{-1}$ by choosing specific values for x and y.

RESOURCES FOR REVIEW

 ITM pp. 345–365 ISM pp. 211–224
 IAM p. 14

 SSG pp. 62–98
 SSM pp.132–146

 TEST GENERATOR
 DOS Windows MAC

⊙ CONCEPTUAL ✍ WRITING ▲ CHALLENGING ▦ SCIENTIFIC CALCULATOR ▨ GRAPHICS CALCULATOR

[3.2] *Simplify. Write answers with only positive exponents. Assume that all variables represent positive real numbers.*

15. $(3^{-4})^2$

16. $(x^{-4})^{-2}$

17. $(xy^{-3})^{-2}$

18. $(z^{-3})^3 z^{-6}$

19. $(5m^{-3})^2 (m^4)^{-3}$

20. $\dfrac{(3r)^2 r^4}{r^{-2} r^{-3}} (9r^{-3})^{-2}$

21. $\left(\dfrac{5z^{-3}}{z^{-1}}\right)\dfrac{5}{z^2}$

22. $\left(\dfrac{6m^{-4}}{m^{-9}}\right)^{-1}\left(\dfrac{m^{-2}}{16}\right)$

23. $\left(\dfrac{3r^5}{5r^{-3}}\right)^{-2}\left(\dfrac{9r^{-1}}{2r^{-5}}\right)^3$

▲ **24.** $\left(\dfrac{a^{-2}b^{-1}}{3a^2}\right)^{-2}\left(\dfrac{b^{-2} \cdot 3a^4}{2b^{-3}}\right)^{-2}\left(\dfrac{a^{-4}b^5}{a^3}\right)^{-2}$

Simplify. Assume that all variables used as exponents represent integers and that all other variables represent nonzero real numbers.

▲ **25.** $\dfrac{m^q \cdot m^{5q-3}}{m^{-2q}}$

▲ **26.** $\dfrac{(z^{-1+a} \cdot z^{3a})^{-1}}{(z^{-a})^{-2}}$

◉ **27.** Give an example to show that $(x^2 + y^2)^2 \neq x^4 + y^4$ by choosing specific values for x and y.

Write in scientific notation.

28. 13,450

29. .0000000765

30. .138

Write without scientific notation.

31. 1.21×10^6

32. 5.8×10^{-3}

33. 6.3×10^{-1}

Use scientific notation to compute.

34. $\dfrac{16 \times 10^4}{8 \times 10^8}$

35. $\dfrac{6 \times 10^{-2}}{4 \times 10^{-5}}$

36. $\dfrac{.0000000164}{.0004}$

37. $\dfrac{.0009 \times 12,000,000}{400,000}$

◉ **38.** Explain why multiplying by 10^{-a} is the same as dividing by 10^a.

The population of Fresno, California is approximately 3.45×10^5. According to the 1994 World Almanac, the population density is 5,449 per square mile.

39. Write the population density in scientific notation.

▦ **40.** To the nearest whole number, what is the area of Fresno?

[3.3] *Give the coefficient of the term.*

41. $14p^5$

42. $-z$

43. $504p^3 r^5$

For the polynomial, (a) write in descending powers, (b) identify as monomial, binomial, trinomial, or none of these, and (c) give the degree.

44. $9k + 11k^3 - 3k^2$

45. $14m^6 + 9m^7$

46. $-5y^4 + 3y^3 + 7y^2 - 2y$

47. $-7q^5 r^3$

◉ **48.** Give an example of a polynomial in the variable x such that it has degree 5, is lacking a third degree term, and is in descending powers of the variable.

For the polynomial, find (a) $P(-2)$ and (b) $P(3)$.

49. $P(x) = -6x + 15$

50. $P(x) = -2x^2 + 5x + 7$

---

15. $\dfrac{1}{3^8}$ **16.** x^8 **17.** $\dfrac{y^6}{x^2}$

18. $\dfrac{1}{z^{15}}$ **19.** $\dfrac{25}{m^{18}}$ **20.** $\dfrac{r^{17}}{9}$

21. $\dfrac{25}{z^4}$ **22.** $\dfrac{1}{96m^7}$

23. $\dfrac{2025}{8r^4}$ **24.** $\dfrac{4a^{14}}{b^{10}}$

25. m^{8q-3} **26.** z^{1-6a}

27. For example, let $x = 2$ and $y = 3$. Then $(x^2 + y^2)^2 = (2^2 + 3^2)^2 = 169$. $x^4 + y^4 = 2^4 + 3^4 = 97 \neq 169$.

28. 1.345×10^4

29. 7.65×10^{-8}

30. 1.38×10^{-1}

31. $1,210,000$

32. $.0058$ **33.** $.63$

34. 2×10^{-4} or $.0002$

35. 1.5×10^3 or 1500

36. 4.1×10^{-5} or $.000041$

37. 2.7×10^{-2} or $.027$

39. 5.449×10^3

40. 63 square miles

41. 14   **42.** -1 **43.** 504

44. (a) $11k^3 - 3k^2 + 9k$
(b) trinomial   (c) 3

45. (a) $9m^7 + 14m^6$
(b) binomial   (c) 7

46. (a) $-5y^4 + 3y^3 + 7y^2 - 2y$
(b) none of these   (c) 4

47. (a) $-7q^5 r^3$
(b) monomial   (c) 8

48. For example, $x^5 + 2x^4 - x^2 + x + 2$

49. (a) 27   (b) -3

50. (a) -11 (b) 4

◉ CONCEPTUAL   WRITING   ▲ CHALLENGING   ▦ SCIENTIFIC CALCULATOR   ▦ GRAPHICS CALCULATOR

51. $-x^2 - 3x + 1$
52. $18m - 10$
53. $-5y^3 - 4y^2 + 6y - 12$
54. $6a^3 - 4a^2 - 16a + 15$
55. $8y^2 - 9y + 5$
56. $12x^2 + 8x + 5$
57. $-15b^3 - 50b$
58. $-12k^3 - 42k$
59. $15m^2 - 7m - 2$
60. $14y^2 + 5y - 24$
61. $6w^2 - 13wt + 6t^2$
62. $10p^4 + 30p^3 -$
$8p^2 - 24p$ 63. $3q^3 -$
$13q^2 - 14q + 20$
64. $9z^4 - 12z^3 + 16z^2 -$
$11z + 2$ 65. $36r^4 - 1$
66. $z^2 - \dfrac{9}{25}$

67. $16m^2 + 24m + 9$
68. $4n^2 - 40n + 100$
69. $18y^{2q-1} - 45$
70. $3m^3 + 4m^{2p-3}$
71. $6z^{2r} - 11x^{2r}z^r - 10x^{4r}$
72. $k^{2m} - 4k^m q^m + 4q^{2m}$
73. $6p(2p - 1)$
74. $7y(3y + 5)$
75. $4qb(3q + 2b - 5q^2b)$
76. $6rt(r^2 - 5rt + 3t^2)$
77. $(x + 3)(x - 3)$
78. $(z + 1)(3z - 1)$
79. $(m + q)(4 + n)$
80. $(x + y)(x + 5)$
81. $(m + 3)(2 - a)$
82. $(a - b)(2m - p)$
83. $(3p - 4)(p + 1)$
84. $(3k - 2)(2k + 5)$
85. $(3r + 1)(4r - 3)$
86. $(2m + 5)(5m + 6)$
87. $(2k - h)(5k - 3h)$
88. prime
89. $2x(4 + x)(3 - x)$
90. $3b(2b - 5)(b + 1)$
91. $(y^2 + 4)(y^2 - 2)$
92. $(2k^2 + 1)(k^2 - 3)$
93. $(p + 2)^2(p + 3)(p - 2)$
94. $(3r + 16)(r + 1)$
95. $(5m^{2q} - 3)(2m^{2q} + 1)$
96. $(6p^x + 1)(2p^x - 5)$
98. $p + 1$

Add or subtract as indicated.

51. Add.
$$3x^2 - 5x + 6$$
$$\underline{-4x^2 + 2x - 5}$$

52. Subtract.
$$10m - 4$$
$$\underline{-8m + 6}$$

53. Subtract.
$$-5y^3 \quad\ + 8y - 3$$
$$\underline{4y^2 + 2y + 9}$$

54. $(4a^3 - 9a + 15) - (-2a^3 + 4a^2 + 7a)$ **55.** $(3y^2 + 2y - 1) + (5y^2 - 11y + 6)$

56. Find the perimeter of the triangle.

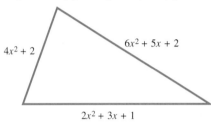

$4x^2 + 2$ *$6x^2 + 5x + 2$*

$2x^2 + 3x + 1$

[3.4] *Find the product.*

57. $-5b(3b^2 + 10)$ **58.** $-6k(2k^2 + 7)$ **59.** $(3m - 2)(5m + 1)$

60. $(7y - 8)(2y + 3)$ **61.** $(3w - 2t)(2w - 3t)$ **62.** $(2p^2 + 6p)(5p^2 - 4)$

63. $(3q^2 + 2q - 4)(q - 5)$ **64.** $(3z^3 - 2z^2 + 4z - 1)(3z - 2)$

65. $(6r^2 - 1)(6r^2 + 1)$ **66.** $\left(z + \dfrac{3}{5}\right)\left(z - \dfrac{3}{5}\right)$

67. $(4m + 3)^2$ **68.** $(2n - 10)^2$

▲ *Find the product. Assume that all variables represent nonzero real numbers, with variables used as exponents restricted to integers.*

69. $9y^q(2y^{q-1} - 5y^{-q})$ **70.** $m^{p-2}(3m^{5-p} + 4m^{-1+p})$

71. $(2z^r - 5x^{2r})(3z^r + 2x^{2r})$ **72.** $(k^m - 2q^m)^2$

[3.5] *Factor out the greatest common factor.*

73. $12p^2 - 6p$ **74.** $21y^2 + 35y$

75. $12q^2b + 8qb^2 - 20q^3b^2$ **76.** $6r^3t - 30r^2t^2 + 18rt^3$

77. $(x + 3)(4x - 1) - (x + 3)(3x + 2)$ **78.** $(z + 1)(z - 4) + (z + 1)(2z + 3)$

Factor by grouping.

79. $4m + nq + mn + 4q$ **80.** $x^2 + 5y + 5x + xy$

81. $2m + 6 - am - 3a$ **82.** $2am - 2bm - ap + bp$

[3.6] *Factor completely.*

83. $3p^2 - p - 4$ **84.** $6k^2 + 11k - 10$ **85.** $12r^2 - 5r - 3$

86. $10m^2 + 37m + 30$ **87.** $10k^2 - 11kh + 3h^2$ **88.** $9x^2 + 4xy - 2y^2$

89. $24x - 2x^2 - 2x^3$ **90.** $6b^3 - 9b^2 - 15b$

91. $y^4 + 2y^2 - 8$ **92.** $2k^4 - 5k^2 - 3$

93. $p^2(p + 2)^2 + p(p + 2)^2 - 6(p + 2)^2$ **94.** $3(r + 5)^2 - 11(r + 5) - 4$

▲ **95.** $10m^{4q} - m^{2q} - 3$ ▲ **96.** $12p^{2x} - 28p^x - 5$

◉ **97.** When asked to factor $x^2y^2 - 6x^2 + 5y^2 - 30$, a student gave the following as his answer: $x^2(y^2 - 6) + 5(y^2 - 6)$. Why is this answer incorrect? What is the correct answer?

◉ **98.** If the area of a rectangle is represented by $4p^2 + 3p - 1$ and the length is $4p - 1$, what is the width in terms of p?

[3.7] *Factor completely.*

99. $16x^2 - 25$

100. $9t^2 - 49$

101. $x^2 + 14x + 49$

102. $9k^2 - 12k + 4$

103. $r^3 + 27$

104. $125x^3 - 1$

105. $m^6 - 1$

106. $x^8 - 1$

107. $x^2 + 6x + 9 - 25y^2$

108. $(a + b)^3 - (a - b)^3$

109. $x^5 - x^3 - 8x^2 + 8$

▲ **110.** $25k^{4b} - 30k^{2b} + 9$

▲ **111.** $x^{4y} + 6x^{2y} + 9$

▲ **112.** $121a^{10m} - 144b^{6m}$

▲ **113.** $125r^{3k} + 1$

[3.9]

114. Which one of the following equations is not in a form that allows solving directly by the zero-factor property?
(a) $(2x + 9)(x - 3) + (4x + 7)(x - 3) = 0$ **(b)** $3x(x - 7) = 0$
(c) $(x - 5)(x + 3)(9x + 3) = 0$ **(d)** $(x - 4)^2 = 0$

115. For the equation in Exercise 114 that cannot be solved directly by the zero-factor property **(a)** put it in the form that does allow this procedure and **(b)** solve the equation.

Use the zero-factor property to solve the equation.

116. $(5x + 2)(x + 1) = 0$

117. $p^2 - 5p + 6 = 0$

118. $q^2 + 2q = 8$

119. $6z^2 = 5z + 50$

120. $6r^2 + 7r = 3$

121. $8k^2 + 14k + 3 = 0$

122. $-4m^2 + 36 = 0$

123. $6y^2 + 9y = 0$

124. $(2x + 1)(x - 2) = -3$

125. $(r + 2)(r - 2) = (r - 2)(r + 3) - 2$

126. $2x^3 - x^2 - 28x = 0$

127. $-t^3 - 3t^2 + 4t + 12 = 0$

128. $(r + 2)(5r^2 - 9r - 18) = 0$

Solve the problem.

129. A triangular wall brace has the shape of a right triangle. One of the perpendicular sides is 1 foot longer than twice the other. The area enclosed by the triangle is 10.5 square feet. Find the shorter of the perpendicular sides.

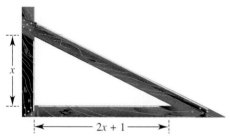

The area is 10.5 square feet.

130. A rectangular parking lot has a length 20 feet more than its width. Its area is 2400 square feet. What are the dimensions of the lot?

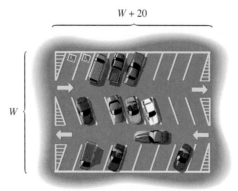

The area is 2400 square feet.

99. $(4x + 5)(4x - 5)$
100. $(3t + 7)(3t - 7)$
101. $(x + 7)^2$
102. $(3k - 2)^2$
103. $(r + 3)(r^2 - 3r + 9)$
104. $(5x - 1) \cdot$
$(25x^2 + 5x + 1)$
105. $(m + 1) \cdot$
$(m^2 - m + 1) \cdot$
$(m - 1)(m^2 + m + 1)$
106. $(x^4 + 1) \cdot$
$(x^2 + 1)(x + 1)(x - 1)$
107. $(x + 3 + 5y) \cdot$
$(x + 3 - 5y)$
108. $2b(3a^2 + b^2)$
109. $(x + 1)(x - 1)(x - 2) \cdot$
$(x^2 + 2x + 4)$
110. $(5k^{2b} - 3)^2$
111. $(x^{2y} + 3)^2$
112. $(11a^{5m} + 12b^{3m}) \cdot$
$(11a^{5m} - 12b^{3m})$
113. $(5r^k + 1) \cdot$
$(25r^{2k} - 5r^k + 1)$
114. (a)
115. (a) $(x - 3) \cdot$
$(6x + 16) = 0$
(b) $\left\{ 3, -\dfrac{8}{3} \right\}$
116. $\left\{ -1, -\dfrac{2}{5} \right\}$
117. $\{2, 3\}$ **118.** $\{-4, 2\}$
119. $\left\{ -\dfrac{5}{2}, \dfrac{10}{3} \right\}$
120. $\left\{ -\dfrac{3}{2}, \dfrac{1}{3} \right\}$
121. $\left\{ -\dfrac{3}{2}, -\dfrac{1}{4} \right\}$
122. $\{-3, 3\}$
123. $\left\{ -\dfrac{3}{2}, 0 \right\}$
124. $\left\{ \dfrac{1}{2}, 1 \right\}$ **125.** $\{4\}$
126. $\left\{ -\dfrac{7}{2}, 0, 4 \right\}$
127. $\{-3, -2, 2\}$
128. $\left\{ -2, -\dfrac{6}{5}, 3 \right\}$
129. 3 feet   **130.** length: 60 feet;   width: 40 feet

 CONCEPTUAL   WRITING   ▲ CHALLENGING   SCIENTIFIC CALCULATOR   GRAPHICS CALCULATOR

131. after 16 seconds
132. after 1 second and after 15 seconds
134. $8x^2 - 10x - 3$
135. $\dfrac{y^4}{36}$ 136. $\dfrac{1}{125}$
137. $\dfrac{1}{16y^{18}}$
138. $-14 + 16w - 8w^2$
139. $21p^9 + 7p^8 + 14p^7$
140. $-\dfrac{1}{5z^9}$ 141. -9
142. $\dfrac{1250z^7x^6}{9}$
143. $-3k^2 + 4k - 7$
144. $a(6 - m)(5 + m)$
145. $k(11 + 12k)$
146. $(2 - a)(4 + 2a + a^2)$
147. prime
148. $5y^2(3y + 4)$
149. $(5z - 3m)^2$
150. $\left\{ -\dfrac{3}{5}, 4 \right\}$
151. $\{-1, 0, 1\}$
152. 6 inches

A rock is thrown directly upward. After t seconds, its height in feet is given by the expression $-16t^2 + 256t$ *(if air resistance is neglected).*

131. When will the rock return to the ground?

132. After how many seconds will it be 240 feet above the ground?

◉ **133.** Explain why the question in Exercise 132 has two answers.

MIXED REVIEW EXERCISES

Perform the indicated operations, then simplify. Write answers with only positive exponents. Assume all variables represent nonzero real numbers.

134. $(4x + 1)(2x - 3)$

135. $\dfrac{6^{-1}y^3(y^2)^{-2}}{6y^{-4}(y^{-1})}$

136. 5^{-3}

137. $(y^6)^{-5}(2y^{-3})^{-4}$

138. $(-5 + 11w) + (6 + 5w) + (-15 - 8w^2)$

139. $7p^5(3p^4 + p^3 + 2p^2)$

140. $\dfrac{(-z^{-2})^3}{5(z^{-3})^{-1}}$

141. $-(-3)^2$

142. $\dfrac{(5z^2x^3)^2(2zx^2)^{-1}}{(-10zx^{-3})^{-2}(3z^{-1}x^{-4})^2}$

143. $(2k - 1) - (3k^2 - 2k + 6)$

Factor completely.

144. $30a + am - am^2$

145. $11k + 12k^2$

146. $8 - a^3$

147. $9x^2 + 13xy - 3y^2$

148. $15y^3 + 20y^2$

149. $25z^2 - 30zm + 9m^2$

Solve.

150. $5x^2 - 17x - 12 = 0$

151. $x^3 - x = 0$

152. The length of a rectangular picture frame is 2 inches longer than its width. The area enclosed by the frame is 48 square inches. What is the width?

───────◆ **MATHEMATICAL CONNECTIONS** (Exercises 153–158) ◇───────

In this chapter we studied a variety of types of factoring. To show how several of these types may be used in factoring a polynomial, work Exercises 153–158 in order. We will factor the polynomial

$$x^{14} - x^2 - 4x^{13} + 4x + 4x^{12} - 4.$$

153. Start by grouping the first two terms, the next two terms, and the last two terms. (Be careful with signs when grouping the middle two terms.)

153. $(x^{14} - x^2) - (4x^{13} - 4x) + (4x^{12} - 4)$

154. Factor out the greatest common factor within each group.

154. $x^2(x^{12} - 1) - 4x(x^{12} - 1) + 4(x^{12} - 1)$

155. The three terms of the polynomial now have $x^{12} - 1$ as their greatest common factor, so factor it out.

155. $(x^{12} - 1)(x^2 - 4x + 4)$

156. The polynomial is now factored, but not completely. Factor the trinomial as a perfect square trinomial, and factor the binomial as the difference of two squares.

156. $(x^6 - 1)(x^6 + 1)(x - 2)^2$

157. One of the factors is now the sum of two cubes, and another is the difference of two squares. Perform these factorizations.

157. $(x^3 - 1)(x^3 + 1)(x^2 + 1)(x^4 - x^2 + 1)(x - 2)^2$

158. In your result from Exercise 157, you should have obtained a sum of two cubes and a difference of two cubes. Perform these factorizations to obtain the complete factorization of the polynomial.

158. $(x - 1)(x^2 + x + 1)(x + 1)(x^2 - x + 1)(x^2 + 1)(x^4 - x^2 + 1) (x - 2)^2$

CHAPTER 3 TEST

◎ **1.** Give an example to show that $(2a)^{-3}$ is not equal to $\dfrac{2}{a^3}$ in general by choosing a specific value for a.

Simplify. Write answers with only positive exponents. Assume that all variables represent nonzero real numbers.

2. $\left(\dfrac{3}{2}\right)^{-2}\left(\dfrac{3}{2}\right)^{5}\left(\dfrac{3}{2}\right)^{-6}$ **3.** $(-4m^{-2}n^4)^{-1}$ **4.** $\dfrac{5^{-3}a^{-2}}{2(a^3)^{-3}}$ **5.** $(-12)^0$

6. $(3a^{-2}b^{-2})^{-3} \cdot (2^{-2}ab^{-3})^2$ $(a \neq 0, b \neq 0)$

7. Write 3.7×10^{-6} without using scientific notation.

8. Use scientific notation to evaluate $\dfrac{(2,500,000)(.00003)}{(.05)(5,000,000)}$.

▦ **9.** In 1992, the estimated population of Luxembourg was 3.92×10^5. The population density was 400 people per square mile. Base on this information, what is the area of Luxembourg to the nearest square mile?

Perform the indicated operations.

10. $(4x^3 - 3x^2 + 2x - 5) - (3x^3 + 11x + 8) + (x^2 - x)$

11. $(5x - 3)(2x + 1)$ **12.** $(2m - 5)(3m^2 + 4m - 5)$

13. $(6x + y)(6x - y)$ **14.** $(3k + q)^2$

15. $[2y + (3z - x)][2y - (3z - x)]$ **16.** If $P(x) = -3x^2 - 5x + 2$, find $P(4)$.

In Exercises 17–24, factor the polynomial.

17. $11z^2 - 44z$ **18.** $(h - 1)(3h + 4) - (h - 1)(h + 2)$

19. $3x + by + bx + 3y$ **20.** $4p^2 + 3pq - q^2$

21. $16a^2 + 40ab + 25b^2$ **22.** $y^3 - 216$

23. $9k^2 - 121j^2$ **24.** $6k^4 - k^2 - 35$

Answers (right column):

[3.1–3.2] 1. For example, if $a = 4$, $(2 \cdot 4)^{-3} = 8^{-3} = \dfrac{1}{512}$, and $\dfrac{2}{a^3} = \dfrac{2}{4^3} = \dfrac{1}{32}$. $\dfrac{1}{512} \neq \dfrac{1}{32}$. 2. $\dfrac{2^3}{3^3}$

3. $-\dfrac{m^2}{4n^4}$ 4. $\dfrac{a^7}{5^3 \cdot 2}$

5. 1 6. $\dfrac{a^8}{3^3 \cdot 2^4}$

7. .0000037

8. 3×10^{-4} or .0003

9. 980 square miles

[3.3] 10. $x^3 - 2x^2 - 10x - 13$

[3.4] 11. $10x^2 - x - 3$

12. $6m^3 - 7m^2 - 30m + 25$ 13. $36x^2 - y^2$

14. $9k^2 + 6kq + q^2$

15. $4y^2 - 9z^2 + 6zx - x^2$

[3.3] 16. -66

[3.5] 17. $11z(z - 4)$

18. $2(h - 1)(h + 1)$

19. $(x + y)(3 + b)$

[3.6–3.7]

20. $(4p - q)(p + q)$

21. $(4a + 5b)^2$

22. $(y - 6)(y^2 + 6y + 36)$

23. $(3k + 11j)(3k - 11j)$

24. $(2k^2 - 5)(3k^2 + 7)$

RESOURCES FOR TEST

| 📖 **ITM** pp. 63–78 **ISM** pp. 224–227 **IAM** pp. 14–15 | 📖 **SSG** pp. 99–101 **SSM** pp. 146–149 | 💾 **TEST GENERATOR** DOS Windows MAC |

◎ CONCEPTUAL   ✎ WRITING   ▲ CHALLENGING   ▦ SCIENTIFIC CALCULATOR   ▦ GRAPHICS CALCULATOR

25. (d)

[3.9] 26. $\left\{-2, -\dfrac{2}{3}\right\}$

27. $\left\{\dfrac{1}{5}, \dfrac{3}{2}\right\}$ 28. $\left\{-\dfrac{2}{5}, 1\right\}$

22. length: 8 inches; width: 5 inches

30. 2 seconds and 4 seconds

◉ **25.** Which one of the following is not a factored form of $-x^2 - x + 12$?
   (a) $(3 - x)(x + 4)$ **(b)** $-(x - 3)(x + 4)$ **(c)** $(-x + 3)(x + 4)$
   (d) $(x - 3)(-x + 4)$

In Exercises 26–28, solve the equation by using the zero-factor property.

26. $3x^2 + 8x + 4 = 0$ **27.** $10x^2 = 17x - 3$ **28.** $5m(m - 1) = 2(1 - m)$

Solve the problem.

29. The area of the rectangle shown is 40 square inches. Find the length and the width of the rectangle.

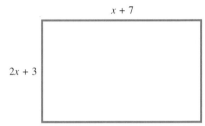

The area is 40 square inches.

30. A ball is thrown straight upward from ground level. After t seconds, its height (in feet) is given by the polynomial $-16t^2 + 96t$. Determine the number of seconds that must elapse before the ball is 128 feet high.

CUMULATIVE REVIEW (Chapters 1–3)

1. $-2m + 6$ 2. $4m - 3$
3. $2x^2 + 5x + 4$
4. $(2, \infty)$ 5. $(-\infty, 1]$
6. $(-3, 5]$ 7. no   8. -24
9. 204   10. 56
11. undefined   12. 10
13. $\left\{\dfrac{7}{6}\right\}$ 14. $\{-1\}$
15. $\left(-\infty, \dfrac{15}{4}\right]$
16. $\left(-\dfrac{1}{2}, \infty\right)$ 17. $(2, 3)$
18. $(-\infty, 2) \cup (3, \infty)$
19. $\left\{-\dfrac{16}{5}, 2\right\}$
20. $(-11, 7)$
21. $(-\infty, -2] \cup [7, \infty)$
22. 2 hours

Use the properties of real numbers to simplify.

1. $-2(m - 3)$ **2.** $-(-4m + 3)$ **3.** $3x^2 - 4x + 4 + 9x - x^2$

Write in interval notation.

4. $\{x \mid x > 2\}$ **5.** $\{x \mid x \le 1\}$ **6.** $\{x \mid -3 < x \le 5\}$

7. Is $\sqrt{\dfrac{-2 + 4}{-5}}$ a real number?

Evaluate if $p = -4$, $q = -2$, and $r = 5$.

8. $-3(2q - 3p)$ **9.** $8r^2 + q^2$ **10.** $|p|^3 - |q^3|$

11. $\dfrac{\sqrt{r}}{-p + 2q}$ **12.** $\dfrac{5p + 6r^2}{p^2 + q - 1}$

Solve.

13. $2z - 5 + 3z = 4 - (z + 2)$ **14.** $\dfrac{3a - 1}{5} + \dfrac{a + 2}{2} = -\dfrac{3}{10}$

15. $-\dfrac{4}{3}d \ge -5$ **16.** $3 - 2(m + 3) < 4m$

17. $2k + 4 < 10$ and $3k - 1 > 5$ **18.** $2k + 4 > 10$ or $3k - 1 < 5$
19. $|5x + 3| - 10 = 3$ **20.** $|x + 2| < 9$
21. $|2y - 5| \ge 9$

22. Two planes leave the Dallas-Fort Worth airport at the same time. One travels east at 550 miles per hour, and the other travels west at 500 miles per hour. Assuming no wind, how long will it take for the planes to be 2100 miles apart?

23. A small pottery shop makes two kinds of planters, glazed and unglazed. The glazed type requires 1/2 hour to throw on the wheel. The unglazed type takes one hour. The wheel is available no more than 10 hours per day. To satisfy customer demand, the shop must make twice as many unglazed pots as glazed. Find the smallest number of glazed pots they can make.

Evaluate.

24. 7^3

25. -3^0

26. $\left(\dfrac{1}{4}\right)^{-2}$

Perform the indicated operations. Assume no denominators are zero.

27. $(3x^2y^{-1})^{-2}(2x^{-3}y)^{-1}$

28. $\dfrac{5m^{-2}y^3}{3m^{-3}y^{-1}}$

29. $(2p + 3)(5p^2 - 4p - 8)$

30. $(3x^3 + 4x^2 - 7) - (2x^3 - 8x^2 + 3x)$

Factor.

31. $16w^2 + 50wz - 21z^2$

32. $4x^2 - 4x + 1 - y^2$

33. $4y^2 - 36y + 81$

34. $100x^4 - 81$

35. $8p^3 + 27$

Solve.

36. $(p - 1)(2p + 3)(p + 4) = 0$

37. $9q^2 = 6q - 1$

38. $5x^2 - 45 = 0$

39. A game board has the shape of a rectangle. The longer sides are each 2 inches longer than the distance between them. The area of the board is 288 square inches. Find the length of the longer sides and the distance between them.

40. A sign is to have the shape of a triangle with a height 3 meters greater than the length of the base. How long should the base be if the area is to be 14 square meters?

23. 4 glazed pots
24. 343   25. -1 26. 16
27. $\dfrac{y}{18x}$ 28. $\dfrac{5my^4}{3}$
29. $10p^3 + 7p^2 - 28p - 24$
30. $x^3 + 12x^2 - 3x - 7$
31. $(2w + 7z)(8w - 3z)$
32. $(2x - 1 + y)\cdot$
$(2x - 1 - y)$
33. $(2y - 9)^2$
34. $(10x^2 + 9)(10x^2 - 9)$
35. $(2p + 3)(4p^2 - 6p + 9)$
36. $\left\{-4, -\dfrac{3}{2}, 1\right\}$
37. $\left\{\dfrac{1}{3}\right\}$ 38. $\{-3, 3\}$
39. longer sides: 18 inches;   distance between: 16 inches
40. 4 meters

 CONCEPTUAL   WRITING   CHALLENGING    SCIENTIFIC CALCULATOR    GRAPHICS CALCULATOR

RATIONAL EXPRESSIONS

CONNECTIONS

Rational expressions play the same role in algebra as rational numbers do in arithmetic. (Note the word "ratio" in the name.) Many useful models require rational expressions. For example, the amount of heating oil produced (in gallons per day) by an oil refinery is modeled by the rational expression

$$\frac{125{,}000 - 25x}{125 + 2x},$$

where x is the amount of gasoline produced per day (in hundreds of gallons). Formulas from physics, chemistry, engineering, optics, and economics involve rational expressions. In this chapter we discuss the basic operations with rational expressions, as well as solving equations with rational expressions.

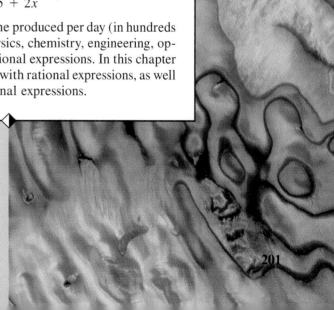

As we shall see, work with rational expressions depends heavily on the techniques of factoring presented in the previous chapter. We will also be using the rules for operations on fractions from arithmetic.

4.1 MULTIPLICATION AND DIVISION OF RATIONAL EXPRESSIONS

FOR EXTRA HELP	OBJECTIVES
SSG pp. 102–108 **SSM** pp. 154–158	**1 ▶** Define rational expressions.
Video 4	**2 ▶** Find the numbers that make a rational expression undefined. **3 ▶** Write rational expressions in lowest terms.
Tutorial IBM MAC	**4 ▶** Multiply rational expressions. **5 ▶** Find reciprocals for rational expressions. **6 ▶** Divide rational expressions.

1 ▶ Define rational expressions.

▶   In arithmetic, a rational number is the quotient of two integers, with the denominator not 0. In algebra this idea is generalized: A **rational expression** or *algebraic fraction* is the quotient of two polynomials, again with the denominator not 0. For example,

$$\frac{m+4}{m-2}, \qquad \frac{8x^2 - 2x + 5}{4x^2 + 5x}, \qquad \text{and} \qquad x^5 \left(\text{or } \frac{x^5}{1}\right)$$

are all rational expressions. In other words, rational expressions are the elements of the set

$$\left\{ \frac{P}{Q} \;\middle|\; P, Q \text{ polynomials, with } Q \neq 0 \right\}.$$

2 ▶ Find the numbers that make a rational expression undefined.

▶   Any number can be used as a replacement for the variable in a rational expression, except for values that make the denominator 0. With a zero denominator, the rational expression is undefined. For example, the number 5 cannot be used as a replacement for x in

$$\frac{2}{x-5}$$

since 5 would make the denominator equal 0.

EXAMPLE 1

Finding Numbers that Make Rational Expressions Undefined

Find all numbers that make the rational expression undefined.

(a) $\dfrac{3}{7k - 14}$

The only values that cannot be used are those that make the denominator 0. We find these values by setting the denominator equal to 0 and solving the resulting equation.

$$7k - 14 = 0$$
$$7k = 14 \qquad \text{Add 14.}$$
$$k = 2 \qquad \text{Divide by 7.}$$

The number 2 cannot be used as a replacement for k; 2 makes the rational expression undefined.

INSTRUCTOR'S RESOURCES

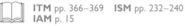 ITM pp. 366–369   ISM pp. 232–240    TEST GENERATOR   TRANSPARENCIES
IAM p. 15   DOS   Windows   MAC

Chalkboard Exercise

Find all numbers that make the rational expression undefined.

$$\frac{k + 2}{k^2 + 1}$$

Answer: Any real number can replace k.

▶ **TEACHING TIP**

Use the example $x/(x^2 + 4)$ to demonstrate a situation in which there is no value of x that makes the denominator equal to zero. ◀

3 ▶ Write rational expressions in lowest terms.

(b) $\dfrac{3 + p}{p^2 - 4p + 3}$

Set the denominator equal to 0.

$$p^2 - 4p + 3 = 0$$
$$(p - 3)(p - 1) = 0 \qquad \text{Factor.}$$
$$p - 3 = 0 \quad \text{or} \quad p - 1 = 0 \qquad \text{Zero-factor property}$$
$$p = 3 \quad \text{or} \qquad p = 1$$

Both 3 and 1 make the rational expression undefined.

(c) $\dfrac{8x + 2}{3}$

The denominator can never be 0, so any real number can replace x in the rational expression. ■

▶ In arithmetic, the fraction $15/20$ is written in lowest terms by dividing the numerator and denominator by 5, to get $3/4$. In a similar manner, we use the **fundamental principle of rational numbers** to write rational expressions in lowest terms.

Fundamental Principle of Rational Numbers

If a/b is a rational number and if c is any nonzero real number, then

$$\frac{a}{b} = \frac{ac}{bc}.$$

(The numerator and denominator of a rational number may either be multiplied or divided by the same nonzero number without changing the value of the rational number.)

▶ **TEACHING TIP**

Good factoring skill is required for this and subsequent sections. Show examples using most of the factoring techniques introduced in Chapter 3.

Show an example factoring out -1: $\dfrac{a - b}{2} \cdot \dfrac{1}{b - a}$. ◀

Since a rational expression is a quotient of two polynomials, and since the value of a polynomial is a real number for all values of the variables for which it is defined, any statement that applies to rational numbers will also apply to rational expressions. In particular, the fundamental principle may be used to write rational expressions in lowest terms. We do this as follows.

Writing in Lowest Terms

Step 1 **Factor.** Factor both numerator and denominator to find their greatest common factor.

Step 2 **Reduce.** Apply the fundamental principle.

EXAMPLE 2

Writing Rational Expressions in Lowest Terms

Write each rational expression in lowest terms.

(a) $\dfrac{8k}{16} = \dfrac{k \cdot 8}{2 \cdot 8} = \dfrac{k}{2}$

Here we divided the numerator and denominator by 8, the greatest common factor of $8k$ and 16.

▶ **TEACHING TIP**

Students will insist on "cancelling" like factors. Be sure that they understand that they are dividing by the greatest common factor, and that all completely cancelled terms should be replaced by 1. (Although we never use the terminology "cancel" in the text, we have used it in these teaching tips, which are intended only for teachers, not students.) ◀

▶ **TEACHING TIP**

Before beginning each section in this chapter, repeat the command: "Factor first, cancel second." ◀

(b) $\dfrac{12x^3y^2}{6x^4y} = \dfrac{2y \cdot 6x^3y}{x \cdot 6x^3y} = \dfrac{2y}{x}$

Divide by the greatest common factor, $6x^3y$. (Here 3 is the least exponent on x, and 1 the least exponent on y.)

(c) $\dfrac{a^2 - a - 6}{a^2 + 5a + 6}$

Start by factoring the numerator and denominator.

$$\frac{a^2 - a - 6}{a^2 + 5a + 6} = \frac{(a - 3)(a + 2)}{(a + 3)(a + 2)}$$

Divide the numerator and denominator by $a + 2$ to get

$$\frac{a^2 - a - 6}{a^2 + 5a + 6} = \frac{a - 3}{a + 3}.$$

(d) $\dfrac{y^2 - 4}{2y + 4} = \dfrac{(y + 2)(y - 2)}{2(y + 2)} = \dfrac{y - 2}{2}$ ■

CAUTION   One of the most common errors in algebra involves incorrect use of the fundamental principle of rational numbers. Only common *factors* may be divided. For example,

$$\frac{y - 2}{2} \neq y \qquad \text{or} \qquad y - 1$$

because the 2 in $y - 2$ is not a *factor* of the numerator. It is essential to *factor* before writing a fraction in lowest terms.

In the rational expression

$$\frac{a^2 - a - 6}{a^2 + 5a + 6}, \qquad \text{or} \qquad \frac{(a - 3)(a + 2)}{(a + 3)(a + 2)},$$

a can take any value except -3 or -2. In the rational expression

$$\frac{a - 3}{a + 3}$$

a cannot equal -3. Because of this,

$$\frac{a^2 - a - 6}{a^2 + 5a + 6} = \frac{a - 3}{a + 3}$$

for all values of a except -3 or -2. From now on such statements of equality will be made with the understanding that they apply only for those real numbers that make neither denominator equal 0, and we will no longer state these restrictions.

E X A M P L E   3

Writing Rational Expressions in Lowest Terms

Write each rational expression in lowest terms.

(a) $\dfrac{m - 3}{3 - m}$

In this rational expression, the numerator and denominator are opposites. The given expression can be written in lowest terms by writing the denominator as $3 - m = -1(m - 3)$, giving

$$\frac{m - 3}{3 - m} = \frac{m - 3}{-1(m - 3)} = \frac{1}{-1} = -1.$$

(b) $\dfrac{r^2 - 16}{4 - r} = \dfrac{(r + 4)(r - 4)}{4 - r}$

$$= \frac{(r + 4)(r - 4)}{-1(r - 4)} \qquad \text{Write } 4 - r \text{ as } -1(r - 4).$$

$$= \frac{r + 4}{-1} \qquad \text{Lowest terms}$$

$$= -(r + 4) \qquad \text{or} \qquad -r - 4 \quad \blacksquare$$

Working as in Examples 3(a) and (b), the quotient

$$\frac{a}{-a} \quad (a \ne 0)$$

can be simplified as

$$\frac{a}{-a} = \frac{a}{-1(a)} = \frac{1}{-1} = -1.$$

The following generalization applies.

> The quotient of two quantities that differ only in sign is -1.

Based on this result,

$$\frac{q - 7}{7 - q} = -1, \qquad \frac{-5a + 2b}{5a - 2b} = -1,$$

but

$$\frac{r - 2}{r + 2}$$

cannot be simplified further.

4 ▶ Multiply rational expressions.

▶ To multiply two rational expressions, we multiply the numerators and multiply the denominators. The product should be simplified by writing it in lowest terms. In practice, we usually simplify before performing any multiplication.

Multiplying Rational Expressions	
Step 1	**Factor.** Factor all numerators and denominators as completely as possible.
Step 2	**Reduce.** Apply the fundamental principle.
Step 3	**Multiply.** Multiply remaining factors in the numerator and remaining factors in the denominator.
Step 4	**Check.** Check to be sure the product is in lowest terms.

▶ **TEACHING TIP**
Rewriting $b - a$ as $-1(a - b)$ can be justified as "factoring out a factor of -1." Multiply out $-1(a - b)$ to justify this last result. ◀

EXAMPLE 4

Multiplying Rational Expressions

Multiply.

(a) $\dfrac{3x^2}{5} \cdot \dfrac{10}{x^3} = \dfrac{3x^2 \cdot 10}{5 \cdot x^3} = \dfrac{30x^2}{5x^3} = \dfrac{6 \cdot 5x^2}{x \cdot 5x^2} = \dfrac{6}{x}$

Apply the fundamental principle using $5x^2$ to write the product in lowest terms. Notice that common factors in the numerator and denominator can be eliminated *before* multiplying the numerator factors and the denominator factors as follows.

$$\dfrac{3x^2}{5} \cdot \dfrac{10}{x^3} = \dfrac{3x^2}{5} \cdot \dfrac{2 \cdot 5}{x \cdot x^2} = \dfrac{6}{x}$$

This is the most efficient way to multiply fractions.

(b) $\dfrac{5p - 5}{p} \cdot \dfrac{3p^2}{10p - 10}$

Factor where possible.

$\dfrac{5p - 5}{p} \cdot \dfrac{3p^2}{10p - 10} = \dfrac{5(p - 1)}{p} \cdot \dfrac{3p \cdot p}{2 \cdot 5(p - 1)}$ Factor.

$= \dfrac{1}{1} \cdot \dfrac{3p}{2}$ Lowest terms

$= \dfrac{3p}{2}$ Multiply.

(c) $\dfrac{k^2 + 2k - 15}{k^2 - 4k + 3} \cdot \dfrac{k^2 - k}{k^2 + k - 20} = \dfrac{(k + 5)(k - 3)}{(k - 3)(k - 1)} \cdot \dfrac{k(k - 1)}{(k + 5)(k - 4)}$

$= \dfrac{k}{k - 4}$

(d) $(p - 4) \cdot \dfrac{3}{5p - 20}$

$(p - 4) \cdot \dfrac{3}{5p - 20} = \dfrac{p - 4}{1} \cdot \dfrac{3}{5p - 20}$ Write $p - 4$ as $\frac{p-4}{1}$.

$= \dfrac{p - 4}{1} \cdot \dfrac{3}{5(p - 4)}$ Factor.

$= \dfrac{3}{5}$

(e) $\dfrac{a^2b^3c^4}{(ab^2)^2c} \cdot \dfrac{(a^2b)^3c^2}{(abc^3)^2}$

$\dfrac{a^2b^3c^4}{(ab^2)^2c} \cdot \dfrac{(a^2b)^3c^2}{(abc^3)^2} = \dfrac{a^2b^3c^4}{a^2b^4c} \cdot \dfrac{a^6b^3c^2}{a^2b^2c^6}$ Power rules for exponents

Use the definition of multiplication and then the product and quotient rules for exponents.

$= \dfrac{a^8b^6c^6}{a^4b^6c^7}$ Product rule for exponents

$= a^{8-4}b^{6-6}c^{6-7}$ Quotient rule for exponents

$= a^4b^0c^{-1}$

$= \dfrac{a^4}{c}$ $b^0 = 1; \ c^{-1} = \frac{1}{c}$

▶ **TEACHING TIP**
Repeat the statement "Factor first, cancel second." ◀

▶ **TEACHING TIP**
Again, students will use cancellation techniques. Insist that complete cancellations are replaced with 1 and other results placed above and below cancellations where they can be found. ◀

(f) $\dfrac{x^2 + 2x}{x + 1} \cdot \dfrac{x^2 - 1}{x^3 + x^2}$

$\dfrac{x^2 + 2x}{x + 1} \cdot \dfrac{x^2 - 1}{x^3 + x^2} = \dfrac{x(x + 2)}{x + 1} \cdot \dfrac{(x + 1)(x - 1)}{x^2(x + 1)}$ Factor where possible.

$= \dfrac{x(x + 2)(x + 1)(x - 1)}{x^2(x + 1)(x + 1)}$ Multiply.

$\dfrac{x^2 + 2x}{x + 1} \cdot \dfrac{x^2 - 1}{x^3 + x^2} = \dfrac{(x + 2)(x - 1)}{x(x + 1)}$ Lowest terms ∎

5 ▶ Find reciprocals for rational expressions.

▶ Recall that rational numbers a/b and c/d are reciprocals of each other if the numbers have a product of 1. The **reciprocal** of a rational expression can be defined in the same way: Two rational expressions are reciprocals of each other if they have a product of 1. Recall that 0 has no reciprocal.

E X A M P L E 5

Finding the Reciprocal of a Rational Expression

The following list shows several rational expressions and the reciprocals of each. In each case, check that the product of the rational expression and its reciprocal is 1.

Rational Expression	Reciprocal
$\dfrac{5}{k}$	$\dfrac{k}{5}$
$\dfrac{m^2 - 9m}{2}$	$\dfrac{2}{m^2 - 9m}$
$\dfrac{0}{4}$	undefined ∎

This example suggests the following procedure.

Reciprocal   To find the reciprocal of a nonzero rational expression, invert the rational expression.

6 ▶ Divide rational expressions.

▶ Division of rational expressions follows the rule for division of rational numbers.

Dividing Rational Expressions   To divide two rational expressions, *multiply* the first by the reciprocal of the second.

E X A M P L E 6

Dividing Rational Expressions

Divide.

(a) $\dfrac{2z}{9} \div \dfrac{5z^2}{18} = \dfrac{2z}{9} \cdot \dfrac{18}{5z^2}$ Multiply by the reciprocal of the divisor.

$= \dfrac{4}{5z}$

(b) $\dfrac{8k - 16}{3k} \div \dfrac{3k - 6}{4k^2} = \dfrac{8k - 16}{3k} \cdot \dfrac{4k^2}{3k - 6}$ Multiply by the reciprocal.

$= \dfrac{8(k - 2)}{3k} \cdot \dfrac{4k^2}{3(k - 2)}$ Factor.

$= \dfrac{32k}{9}$

(c) $\dfrac{5m^2 + 17m - 12}{3m^2 + 7m - 20} \div \dfrac{5m^2 + 2m - 3}{15m^2 - 34m + 15}$

$= \dfrac{(5m - 3)(m + 4)}{(m + 4)(3m - 5)} \div \dfrac{(5m - 3)(m + 1)}{(3m - 5)(5m - 3)}$

$= \dfrac{(5m - 3)(m + 4)}{(m + 4)(3m - 5)} \cdot \dfrac{(3m - 5)(5m - 3)}{(5m - 3)(m + 1)}$

$= \dfrac{5m - 3}{m + 1}$ ■

EXAMPLE 7
Dividing Rational Expressions

■ Divide $\dfrac{m^2pq^3}{mp^4}$ by $\dfrac{m^5p^2q}{mpq^2}$.

Use the definitions of division and multiplication and the properties of exponents.

$\dfrac{m^2pq^3}{mp^4} \div \dfrac{m^5p^2q}{mpq^2} = \dfrac{m^2pq^3}{mp^4} \cdot \dfrac{mpq^2}{m^5p^2q}$

$= \dfrac{m^3p^2q^5}{m^6p^6q}$

$= \dfrac{q^4}{m^3p^4}$ ■

Chalkboard Exercise

Divide $\dfrac{x^2y^2z^2}{yz^3} \div \dfrac{xy^4}{xy^2z}$.

Answer: $\dfrac{x^2}{y}$

4.1 EXERCISES

3. 7   4. −3   5. $-\dfrac{1}{7}$

6. $-\dfrac{7}{2}$ 7. 0   8. 0

9. $-2, \dfrac{3}{2}$ 10. $-6, \dfrac{7}{3}$

11. none   12. none
13. none   14. none

◉ **1.** In Example 1(a), we show that the rational expression $\dfrac{3}{7k - 14}$ is undefined for $k = 2$. Explain in your own words why this is so. In general, how do we find the value or values for which a rational expression is undefined?

◉ **2.** The rational expression $\dfrac{x + 1}{x^2 + 3}$ is never undefined. Explain why this is so.

Find all real numbers that make the rational expression undefined. See Example 1.

3. $\dfrac{z}{z - 7}$ **4.** $\dfrac{r}{r + 3}$ **5.** $\dfrac{6p - 5}{7p + 1}$

6. $\dfrac{8x - 3}{2x + 7}$ **7.** $\dfrac{12y + 3}{y}$ **8.** $\dfrac{9y + 8}{y}$

9. $\dfrac{3x + 1}{2x^2 + x - 6}$ **10.** $\dfrac{2x + 4}{3x^2 + 11x - 42}$ **11.** $\dfrac{x + 2}{14}$

12. $\dfrac{y - 9}{26}$ **13.** $\dfrac{2x^2 - 3x + 4}{3x^2 + 8}$ **14.** $\dfrac{9x^2 - 8x + 3}{4x^2 + 1}$

◉ CONCEPTUAL   ✎ WRITING   ▲ CHALLENGING   ▦ SCIENTIFIC CALCULATOR   ▦ GRAPHICS CALCULATOR

15. (a) Identify the two *terms* in the numerator and the two *terms* in the denominator of the rational expression $\dfrac{x^2 + 4x}{x + 4}$.

⊙ **(b)** Describe the steps you would use to rewrite this rational expression in lowest terms. (*Hint:* It simplifies to x.)

⊙ **16.** Only one of the following rational expressions can be expressed in lower terms. Which one is it?

(a) $\dfrac{x^2 + 2}{x^2}$ **(b)** $\dfrac{x^2 + 2}{2}$ **(c)** $\dfrac{x^2 + y^2}{y^2}$ **(d)** $\dfrac{x^2 - 5x}{x}$

A ratio is a quotient of two quantities. Ratios provide a way to compare two quantities. Use the graphs to write the ratios in Exercises 17–20. Then express the quotient in decimal form to the nearest hundredth using a calculator.

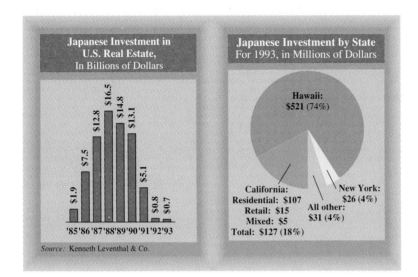

17. The greatest number of dollars (in billions) invested in the U.S. by the Japanese in one year compared to the total investment in the years shown.

18. The least amount invested in one year compared to the greatest amount invested in one year.

19. The 1993 investment in New York compared to the total investment in 1993.

20. The 1993 investment in California compared to the amount invested in that year in Hawaii.

Write the rational expression in lowest terms. See Example 2.

21. $\dfrac{24x^2y^4}{18xy^5}$

22. $\dfrac{36m^4n^3}{24m^2n^5}$

23. $\dfrac{(x + 4)(x - 3)}{(x + 5)(x + 4)}$

24. $\dfrac{(2x + 7)(x - 1)}{(2x + 3)(2x + 7)}$

25. $\dfrac{4x(x + 3)}{8x^2(x - 3)}$

26. $\dfrac{5y^2(y + 8)}{15y(y - 8)}$

27. $\dfrac{3x + 7}{3}$

28. $\dfrac{4x - 9}{4}$

29. $\dfrac{6m + 18}{7m + 21}$

30. $\dfrac{5r - 20}{3r - 12}$

31. $\dfrac{3z^2 + z}{18z + 6}$

32. $\dfrac{2x^2 - 5x}{16x - 40}$

15. (a) numerator: x^2, $4x$; denominator: x, 4
16. (d)
17. $\dfrac{16.5}{73.2}$, .23
18. $\dfrac{.7}{16.5}$, .04
19. $\dfrac{26}{705}$, .04 **20.** $\dfrac{127}{521}$, .24
21. $\dfrac{4x}{3y}$ **22.** $\dfrac{3m^2}{2n^2}$
23. $\dfrac{x - 3}{x + 5}$
24. $\dfrac{x - 1}{2x + 3}$ **25.** $\dfrac{x + 3}{2x(x - 3)}$
26. $\dfrac{y(y + 8)}{3(y - 8)}$
27. already in lowest terms
28. already in lowest terms
29. $\dfrac{6}{7}$ **30.** $\dfrac{5}{3}$ **31.** $\dfrac{z}{6}$
32. $\dfrac{x}{8}$

33. $\dfrac{2}{t-3}$ 34. $\dfrac{5}{s+5}$

35. $\dfrac{x-3}{x+1}$ 36. $\dfrac{y-7}{y-1}$

37. $\dfrac{4x+1}{4x+3}$ 38. $\dfrac{6x-5}{4x-5}$

39. $a^2 - ab + b^2$

40. $r^2 + rs + s^2$

41. $\dfrac{c+6d}{c-d}$ 42. $\dfrac{s+3t}{s+4t}$

43. $\dfrac{a+b}{a-b}$ 44. $\dfrac{y+w}{y-w}$

45. (b) 46. (b) and (d)

47. -1 48. -1 In

Exercises 49–54, there are
several other acceptable
ways to express the answer.

49. $-(x+y)$

50. $-(m+n)$

51. $-\dfrac{x+y}{x-y}$

52. $-\dfrac{x+2}{x-2}$ 53. $-\dfrac{1}{2}$

54. $-\dfrac{1}{3}$

55. already in lowest terms

56. already in lowest terms

59. $\dfrac{3y}{x^2}$ 60. $5ab^2$

61. $\dfrac{3a^3b^2}{4}$ 62. $\dfrac{t^2}{16s^3}$

63. $\dfrac{7x}{6}$ 64. $\dfrac{8}{5x}$

65. $-\dfrac{p+5}{2p}$ (There are

other ways.)

66. $-\dfrac{a+1}{2a}$ (There are

other ways.)

67. $\dfrac{-m(m+7)}{m+1}$ (There are

other ways.)

68. $\dfrac{-11(k+2)}{3k}$ (There are

other ways.)

69. -2 70. $-\dfrac{9}{2}$

71. $\dfrac{x+4}{x-4}$ 72. $\dfrac{t+3}{t-3}$

---

33. $\dfrac{2t+6}{t^2-9}$

34. $\dfrac{5s-25}{s^2-25}$

35. $\dfrac{x^2+2x-15}{x^2+6x+5}$

36. $\dfrac{y^2-5y-14}{y^2+y-2}$

37. $\dfrac{8x^2-10x-3}{8x^2-6x-9}$

38. $\dfrac{12x^2-4x-5}{8x^2-6x-5}$

39. $\dfrac{a^3+b^3}{a+b}$

40. $\dfrac{r^3-s^3}{r-s}$

41. $\dfrac{2c^2+2cd-60d^2}{2c^2-12cd+10d^2}$

42. $\dfrac{3s^2-9st-54t^2}{3s^2-6st-72t^2}$

43. $\dfrac{ac-ad+bc-bd}{ac-ad-bc+bd}$

44. $\dfrac{2xy+2xw+y+w}{2xy+y-2xw-w}$

⊙ **45.** Only one of the following rational expressions is *not* equivalent to

$$\dfrac{x-3}{4-x}.$$

Which one is it?

(a) $\dfrac{3-x}{x-4}$ (b) $\dfrac{x+3}{4+x}$ (c) $-\dfrac{3-x}{4-x}$ (d) $-\dfrac{x-3}{x-4}$

⊙ **46.** Which of the following rational expressions equals -1?

(a) $\dfrac{2x+3}{2x-3}$ (b) $\dfrac{2x-3}{3-2x}$ (c) $\dfrac{2x+3}{3+2x}$ (d) $\dfrac{2x+3}{-2x-3}$

Write the rational expression in lowest terms. See Example 3.

47. $\dfrac{7-b}{b-7}$ **48.** $\dfrac{r-13}{13-r}$ **49.** $\dfrac{x^2-y^2}{y-x}$ **50.** $\dfrac{m^2-n^2}{n-m}$

51. $\dfrac{(a-3)(x+y)}{(3-a)(x-y)}$ **52.** $\dfrac{(8-p)(x+2)}{(p-8)(x-2)}$ **53.** $\dfrac{5k-10}{20-10k}$

54. $\dfrac{7x-21}{63-21x}$ **55.** $\dfrac{a^2-b^2}{a^2+b^2}$ **56.** $\dfrac{p^2+q^2}{p^2-q^2}$

📝 **57.** Explain in a few words how to multiply rational expressions.

📝 **58.** Explain in a few words how to divide rational expressions.

Multiply or divide as indicated. See Examples 4 and 6.

59. $\dfrac{x^3}{3y}\cdot\dfrac{9y^2}{x^5}$

60. $\dfrac{a^4}{5b^2}\cdot\dfrac{25b^4}{a^3}$

61. $\dfrac{5a^4b^2}{16a^2b}\div\dfrac{25a^2b}{60a^3b^2}$

62. $\dfrac{s^3t^2}{10s^2t^4}\div\dfrac{8s^4t^2}{5t^6}$

63. $\dfrac{4x}{8x+4}\cdot\dfrac{14x+7}{6}$

64. $\dfrac{12x-20}{5x}\cdot\dfrac{6}{9x-15}$

65. $\dfrac{p^2-25}{4p}\cdot\dfrac{2}{5-p}$

66. $\dfrac{a^2-1}{4a}\cdot\dfrac{2}{1-a}$

67. $\dfrac{m^2-49}{m+1}\div\dfrac{7-m}{m}$

68. $\dfrac{k^2-4}{3k^2}\div\dfrac{2-k}{11k}$

69. $\dfrac{12x-10y}{3x+2y}\cdot\dfrac{6x+4y}{10y-12x}$

70. $\dfrac{9s-12t}{2s+2t}\cdot\dfrac{3s+3t}{4t-3s}$

71. $\dfrac{x^2-25}{x^2+x-20}\cdot\dfrac{x^2+7x+12}{x^2-2x-15}$

72. $\dfrac{t^2-49}{t^2+4t-21}\cdot\dfrac{t^2+8t+15}{t^2-2t-35}$

73. $\dfrac{6x^2 + 5xy - 6y^2}{12x^2 - 11xy + 2y^2} \div \dfrac{4x^2 - 12xy + 9y^2}{8x^2 - 14xy + 3y^2}$

74. $\dfrac{8a^2 - 6ab - 9b^2}{6a^2 - 5ab - 6b^2} \div \dfrac{4a^2 + 11ab + 6b^2}{9a^2 + 12ab + 4b^2}$

75. $\dfrac{3k^2 + 17kp + 10p^2}{6k^2 + 13kp - 5p^2} \div \dfrac{6k^2 + kp - 2p^2}{6k^2 - 5kp + p^2}$

76. $\dfrac{16c^2 + 24cd + 9d^2}{16c^2 - 16cd + 3d^2} \div \dfrac{16c^2 - 9d^2}{16c^2 - 24cd + 9d^2}$

▲ **77.** $\left(\dfrac{6k^2 - 13k - 5}{k^2 + 7k} \div \dfrac{2k - 5}{k^3 + 6k^2 - 7k}\right) \cdot \dfrac{k^2 - 5k + 6}{3k^2 - 8k - 3}$

▲ **78.** $\left(\dfrac{2x^3 + 3x^2 - 2x}{3x - 15} \div \dfrac{2x^3 - x^2}{x^2 - 3x - 10}\right) \cdot \dfrac{5x^2 - 10x}{3x^2 + 12x + 12}$

▲ **79.** $\dfrac{a^2(2a + b) + 6a(2a + b) + 5(2a + b)}{3a^2(2a + b) - 2a(2a + b) + (2a + b)} \div \dfrac{a + 1}{a - 1}$

▲ **80.** $\dfrac{2x^2(x - 3z) - 5x(x - 3z) + 2(x - 3z)}{4x^2(x - 3z) - 11x(x - 3z) + 6(x - 3z)} \div \dfrac{4x - 3}{4x + 1}$

◆ **MATHEMATICAL CONNECTIONS** (Exercises 81–86) ◆

We prefer that rational expressions be written in lowest terms, as are numerical fractions. This first requires factoring both the numerator and denominator of the rational expression. So, work with rational expressions requires a good understanding of factoring and recognizing algebraic factors.

Work the following exercises in order.

81. Factor $3x^2 + 2x - 5$.

82. Factor $9x^2 - 25$.

83. What rational expression in lowest terms is equivalent to

$$\dfrac{3x^2 + 2x - 5}{9x^2 - 25}?$$

84. Write another rational expression that is equivalent to the expression in Exercise 83.

85. (a) Multiply $\dfrac{3x^2 + 2x - 5}{9x^2 - 25}$ by $\dfrac{3x - 5}{x - 1}$.

   (b) Divide $\dfrac{3x^2 + 2x - 5}{9x^2 - 25}$ by $\dfrac{x - 1}{3x - 5}$.

◉ **86.** How do your answers in Exercise 85 compare? Complete the following sentence.

For rational expressions $\dfrac{P}{Q}$ and $\dfrac{R}{S}$, $\dfrac{P}{Q} \cdot$ _____ is equivalent to $\dfrac{P}{Q} \div$ _____ .

◆

REVIEW EXERCISES

Add or subtract as indicated. See Section 1.2.

87. $\dfrac{4}{7} + \dfrac{1}{3} - \dfrac{1}{2}$

88. $\dfrac{9}{10} - \left(-\dfrac{1}{3}\right)$

89. $-\dfrac{3}{4} + \dfrac{1}{12}$

90. $-\dfrac{2}{3} + \left(-\dfrac{4}{7}\right) - \dfrac{1}{8}$

73. $\dfrac{2x + 3y}{2x - 3y}$ **74.** $\dfrac{3a + 2b}{a + 2b}$

75. $\dfrac{k + 5p}{2k + 5p}$

76. $\dfrac{4c + 3d}{4c - d}$

77. $(k - 1)(k - 2)$

78. $\dfrac{5(x - 2)}{9}$

79. $\dfrac{(a + 5)(a - 1)}{3a^2 - 2a + 1}$

80. $\dfrac{(2x - 1)(4x + 1)}{(4x - 3)^2}$

81. $(3x + 5)(x - 1)$

82. $(3x + 5)(3x - 5)$

83. $\dfrac{x - 1}{3x - 5}$

84. Many answers are possible. One example is $\dfrac{(3x^2 + 2x - 5)(4x)}{(9x^2 - 25)(4x)} = \dfrac{12x^3 + 8x^2 - 20x}{36x^3 - 100x}$.

85. (a) 1   (b) 1

86. They are the same. $\dfrac{R}{S}, \dfrac{S}{R}$

87. $\dfrac{17}{42}$ **88.** $\dfrac{37}{30}$ **89.** $-\dfrac{2}{3}$

90. $-\dfrac{229}{168}$

◉ CONCEPTUAL   ✎ WRITING   ▲ CHALLENGING   ▦ SCIENTIFIC CALCULATOR   ▦ GRAPHICS CALCULATOR

4.2 ADDITION AND SUBTRACTION OF RATIONAL EXPRESSIONS

FOR EXTRA HELP

 SSG pp. 108–111
SSM pp. 159–165

 Video
5

 Tutorial
IBM   MAC

OBJECTIVES

1 ▶ Add and subtract rational expressions with the same denominator.
2 ▶ Find a least common denominator.
3 ▶ Add and subtract rational expressions with different denominators.

1 ▶ Add and subtract
rational expressions with
the same denominator.

▶ The sum or difference of two rational expressions is found by following the definitions of addition and subtraction of rational numbers.

The following steps are used to add or subtract rational expressions.

| **Adding or Subtracting**
Rational Expressions | *Step 1(a)* | If the denominators are the same, add or subtract the numerators. Place the result over the common denominator. |
| | *Step 1(b)* | If the denominators are different, first find the least common denominator. Write all rational expressions with this least common denominator, and then add or subtract the numerators. Place the result over the common denominator. |
| | *Step 2* | Write all answers in lowest terms. |

E X A M P L E   1

Adding and Subtracting
Rational Expressions with
the Same Denominators

Add or subtract as indicated.

(a) $\dfrac{3y}{5} + \dfrac{x}{5} = \dfrac{3y + x}{5}$

The denominators of these rational expressions are the same, so just add the numerators, and place the sum over the common denominator.

(b) $\dfrac{7}{2r^2} - \dfrac{11}{2r^2} = \dfrac{7 - 11}{2r^2} = \dfrac{-4}{2r^2} = -\dfrac{2}{r^2}$ Lowest terms

Subtract the numerators since the denominators are the same, and keep the common denominator.

Chalkboard Exercise

Subtract
$\dfrac{2y - 1}{y^2 + y - 2} - \dfrac{y}{y^2 + y - 2}$.

Answer: $\dfrac{1}{y + 2}$

(c) $\dfrac{m}{m^2 - p^2} + \dfrac{p}{m^2 - p^2} = \dfrac{m + p}{m^2 - p^2}$ Add numerators.

$= \dfrac{m + p}{(m + p)(m - p)}$ Factor.

$= \dfrac{1}{m - p}$ Lowest terms

▶ **TEACHING TIP**

Begin this section by noting that when we add or subtract fractions, our answer is generally a fraction (that is, we keep the common denominator).

Start with a numerical example, $\dfrac{5}{12} + \dfrac{3}{12}$, then lead into algebraic examples. ◀

(d) $\dfrac{4}{x^2 + 2x - 8} + \dfrac{x}{x^2 + 2x - 8} = \dfrac{4 + x}{x^2 + 2x - 8}$

$= \dfrac{4 + x}{(x - 2)(x + 4)}$

$= \dfrac{1}{x - 2}$ ∎

INSTRUCTOR'S RESOURCES

 ITM pp. 369–372   **ISM** pp. 240–251
IAM p. 16

 TEST GENERATOR
DOS   Windows   MAC

 TRANSPARENCIES

2 ▶ Find a least common denominator.

▶ TEACHING TIP

It is worth insisting that students write answers in *factored* form where possible in the event that the simplified result can be reduced. ◀

Finding the Least Common Denominator

▶ The rational expressions in each part of Example 1 had the same denominators. As we mentioned above, if the rational expressions to be added or subtracted have different denominators, it is necessary to first find their **least common denominator,** an expression divisible by the denominator of each of the rational expressions. We find the least common denominator for a group of rational expressions as shown below.

Step 1 Factor each denominator.
Step 2 The least common denominator is the product of all different factors from each denominator, with each factor raised to the *greatest* power that occurs in any denominator.

EXAMPLE 2

Finding Least Common Denominators

Find the least common denominator for each pair of denominators.

(a) $5xy^2$, $2x^3y$
Each denominator is already factored.

$$5xy^2 = 5 \cdot x \cdot y^2$$
$$2x^3y = 2 \cdot x^3 \cdot y$$

Highest exponent on x is 3.

$$\text{LCD} = 5 \cdot 2 \cdot x^3 \cdot y^2 \quad \longleftarrow \text{Highest exponent on } y \text{ is 2.}$$
$$= 10x^3y^2$$

(b) $k - 3$, k
The least common denominator, an expression divisible by both $k - 3$ and k, is

$$k(k - 3).$$

It is often best to leave a least common denominator in factored form.

(c) $y^2 - 2y - 8$, $y^2 + 3y + 2$
Factor the denominators to get

$$y^2 - 2y - 8 = (y - 4)(y + 2)$$
$$y^2 + 3y + 2 = (y + 2)(y + 1).$$

The least common denominator, divisible by both polynomials, is

$$(y - 4)(y + 2)(y + 1).$$

(d) $8z - 24$, $5z^2 - 15z$

$$8z - 24 = 8(z - 3) \quad \text{and} \quad 5z^2 - 15z = 5z(z - 3) \quad \text{Factor.}$$

The least common denominator is

$$40z(z - 3). \quad \blacksquare$$

▶ TEACHING TIP

A skill needs to be developed for extracting the least common denominator (LCD) before proceeding to the next step: adding and subtracting fractions. Include the following example: Find the LCD for the denominators $3(x + 2)^2$, $4(x + 2)$, $15(x^2 - 4)$. ◀

3 ▶ Add and subtract rational expressions with different denominators.

▶ As we mentioned earlier, if the rational expressions to be added or subtracted have different denominators, it is necessary to rewrite the rational expressions with their least common denominator. This is done by multiplying the numerator and the denominator of each rational expression by the factors required to get the least common denominator. This procedure is valid because each rational expression is being multiplied by a form of 1, the identity element for multiplication.

EXAMPLE 3

Adding and Subtracting Rational Expressions with Different Denominators

Add or subtract as indicated.

(a) $\dfrac{5}{2p} + \dfrac{3}{8p}$

The least common denominator for $2p$ and $8p$ is $8p$. To write the first rational expression with a denominator of $8p$, multiply by 4/4.

$$\dfrac{5}{2p} + \dfrac{3}{8p} = \dfrac{5 \cdot 4}{2p \cdot 4} + \dfrac{3}{8p} \qquad \text{Fundamental principle}$$

$$= \dfrac{20}{8p} + \dfrac{3}{8p}$$

$$= \dfrac{20 + 3}{8p} \qquad \text{Add numerators.}$$

$$= \dfrac{23}{8p}$$

▶ **TEACHING TIP**
The difficulty with these problems is keeping track of the steps involved. Try these basic steps.

1. Write down the *original problem*.
2. Multiply the numerators and denominators in the original problem by the appropriate expression to get the LCD (*fraction building*).
3. Combine the numerators, then put over the LCD (*the new problem*).
4. Factor the numerator and *reduce* if possible. ◀

(b) $\dfrac{6}{r} - \dfrac{5}{r-3}$

The least common denominator is $r(r-3)$. Rewrite each rational expression with this denominator.

$$\dfrac{6}{r} - \dfrac{5}{r-3} = \dfrac{6(r-3)}{r(r-3)} - \dfrac{r \cdot 5}{r(r-3)} \qquad \text{Fundamental principle}$$

$$= \dfrac{6(r-3) - 5r}{r(r-3)} \qquad \text{Subtract numerators.}$$

$$= \dfrac{6r - 18 - 5r}{r(r-3)} \qquad \text{Distributive property}$$

$$= \dfrac{r - 18}{r(r-3)} \qquad \text{Combine terms in numerator.} \blacksquare$$

▶ **TEACHING TIP**
It is beneficial, after having worked a variety of examples from this section, to have students work out, either individually or in a group, one challenging problem on their own. ◀

CAUTION One of the most common sign errors in algebra occurs when a rational expression with two or more terms in the numerator is being subtracted. Remember that in this situation, the subtraction sign must be distributed to *every* term in the numerator of the fraction that follows it. Read Example 4 carefully to see how this is done.

EXAMPLE 4

Subtracting Rational Expressions

Subtract.

(a) $\dfrac{3}{k+1} - \dfrac{k-2}{k+3}$

The least common denominator is $(k+1)(k+3)$.

$$\dfrac{3}{k+1} - \dfrac{k-2}{k+3} = \dfrac{3(k+3)}{(k+1)(k+3)} - \dfrac{(k-2)(k+1)}{(k+3)(k+1)}$$

$$= \dfrac{3(k+3) - (k-2)(k+1)}{(k+1)(k+3)} \qquad \text{Subtract.}$$

$$= \dfrac{3k + 9 - (k^2 - k - 2)}{(k+1)(k+3)} \qquad \text{Multiply in numerator.}$$

$$= \frac{3k + 9 - k^2 + k + 2}{(k + 1)(k + 3)} \qquad \text{Distributive property}$$

$$= \frac{-k^2 + 4k + 11}{(k + 1)(k + 3)} \qquad \text{Combine terms.}$$

(b) $\dfrac{1}{q - 1} - \dfrac{1}{q + 1} = \dfrac{1(q + 1)}{(q - 1)(q + 1)} - \dfrac{1(q - 1)}{(q + 1)(q - 1)}$ Get a common denominator.

$$= \frac{(q + 1) - (q - 1)}{(q - 1)(q + 1)} \qquad \text{Subtract.}$$

$$= \frac{q + 1 - q + 1}{(q - 1)(q + 1)}$$

$$= \frac{2}{(q - 1)(q + 1)} \qquad \text{Combine terms.}$$

(c) $\dfrac{m + 4}{m^2 - 2m - 3} - \dfrac{2m - 3}{m^2 - 5m + 6}$

$$= \frac{m + 4}{(m - 3)(m + 1)} - \frac{2m - 3}{(m - 3)(m - 2)} \qquad \text{Factor each denominator.}$$

The least common denominator is $(m - 3)(m + 1)(m - 2)$.

$$= \frac{(m + 4)(m - 2)}{(m - 3)(m + 1)(m - 2)} - \frac{(2m - 3)(m + 1)}{(m - 3)(m - 2)(m + 1)}$$

$$= \frac{(m + 4)(m - 2) - (2m - 3)(m + 1)}{(m - 3)(m + 1)(m - 2)} \qquad \text{Subtract.}$$

$$= \frac{m^2 + 2m - 8 - (2m^2 - m - 3)}{(m - 3)(m + 1)(m - 2)} \qquad \text{Multiply.}$$

$$= \frac{m^2 + 2m - 8 - 2m^2 + m + 3}{(m - 3)(m + 1)(m - 2)} \qquad \text{Distributive property}$$

$$= \frac{-m^2 + 3m - 5}{(m - 3)(m + 1)(m - 2)} \qquad \text{Combine terms.} \ \blacksquare$$

EXAMPLE 5

Adding Rational
Expressions with
Opposites in
Denominators

■ Add.

$$\frac{a}{(a - 1)^2} + \frac{2a}{1 - a^2} = \frac{a}{(a - 1)^2} + \frac{2a}{(1 - a)(1 + a)} \qquad \text{Factor denominators.}$$

$$= \frac{a}{(a - 1)^2} + \frac{-1 \cdot 2a}{-1(1 - a)(1 + a)} \qquad a - 1 \text{ and } 1 - a \text{ are opposites.}$$

$$= \frac{a}{(a - 1)^2} + \frac{-2a}{(a - 1)(a + 1)}$$

$$= \frac{a(a + 1)}{(a - 1)^2(a + 1)} + \frac{-2a(a - 1)}{(a - 1)(a + 1)(a - 1)}$$

$$= \frac{a(a + 1) - 2a(a - 1)}{(a - 1)^2(a + 1)} \qquad \text{Add numerators.}$$

$$= \frac{a^2 + a - 2a^2 + 2a}{(a - 1)^2(a + 1)}$$

$$= \frac{-a^2 + 3a}{(a - 1)^2(a + 1)} \qquad \text{Combine terms.} \ \blacksquare$$

Chalkboard Exercise

Add $\dfrac{2y}{y - 3} + \dfrac{y + 1}{9 - y^2}$.

Answer: $\dfrac{-2y^2 - 5y + 1}{(3 + y)(3 - y)}$

4.2 EXERCISES

Answers (left column):

3. $\dfrac{9}{t}$ 4. $\dfrac{14}{r}$ 5. $\dfrac{2}{x}$ 6. $\dfrac{1}{y}$

7. 1   8. 2   9. $x - 5$

10. $y - 6$ 11. $\dfrac{5}{p + 3}$

12. $\dfrac{2}{x - 4}$ 13. $a - b$

14. $p + q$ 15. $72x^4y^5$

16. $72a^5b^4$ 17. $z(z - 2)$

18. $k(k + 3)$

19. $2(y + 4)$ 20. $3(r - 7)$

21. $30(x + 3)$

22. $12(c + 3)$

23. $(m + n)(m - n)$

24. $(r + s)(r - s)$

25. $x(x - 4)(x + 1)$

26. $y(y - 2)(y - 6)$

27. $(t + 5)(t - 2)(2t - 3)$

28. $(s + 1)(s - 4)(3s - 2)$

29. $2y(y + 3)(y - 3)$

30. $9x(x + 2)(x - 2)$

31. $6x^2(x + 1)$

32. $11y^5(y - 1)$

35. $\dfrac{31}{3t}$ 36. $\dfrac{29}{4x}$

37. $\dfrac{5 - 22x}{12x^2y}$ 38. $\dfrac{7 - 4a^2b}{18a^3b^2}$

39. $\dfrac{1}{x(x - 1)}$

40. $\dfrac{2x + 3}{x(x - 3)}$

41. $\dfrac{5a^2 - 7a}{(a + 1)(a - 3)}$

42. $\dfrac{5x^2 - 2x}{(x + 4)(x - 7)}$

43. $\dfrac{-6x + 3}{x - 4}$ or $\dfrac{6x - 3}{4 - x}$

44. $\dfrac{-10t - 2}{t - 2}$ or $\dfrac{10t + 2}{2 - t}$

---

1. Write a step-by-step method of adding or subtracting rational expressions that have a common denominator.

2. Write a step-by-step method of adding or subtracting rational expressions that have different denominators.

Add or subtract as indicated. Write the answer in lowest terms. See Example 1.

3. $\dfrac{7}{t} + \dfrac{2}{t}$

4. $\dfrac{5}{r} + \dfrac{9}{r}$

5. $\dfrac{11}{5x} - \dfrac{1}{5x}$

6. $\dfrac{7}{4y} - \dfrac{3}{4y}$

7. $\dfrac{5x + 4}{6x + 5} + \dfrac{x + 1}{6x + 5}$

8. $\dfrac{6y + 12}{4y + 3} + \dfrac{2y - 6}{4y + 3}$

9. $\dfrac{x^2}{x + 5} - \dfrac{25}{x + 5}$

10. $\dfrac{y^2}{y + 6} - \dfrac{36}{y + 6}$

11. $\dfrac{-3p + 7}{p^2 + 7p + 12} + \dfrac{8p + 13}{p^2 + 7p + 12}$

12. $\dfrac{5x + 6}{x^2 + x - 20} + \dfrac{4 - 3x}{x^2 + x - 20}$

13. $\dfrac{a^3}{a^2 + ab + b^2} - \dfrac{b^3}{a^2 + ab + b^2}$

14. $\dfrac{p^3}{p^2 - pq + q^2} + \dfrac{q^3}{p^2 - pq + q^2}$

Find the least common denominator for the group of denominators. See Example 2.

15. $18x^2y^3, 24x^4y^5$

16. $24a^3b^4, 18a^5b^2$

17. $z - 2, z$

18. $k + 3, k$

19. $2y + 8, y + 4$

20. $3r - 21, r - 7$

21. $6x + 18, 5x + 15$

22. $4c + 12, 6c + 18$

23. $m + n, m - n$

24. $r + s, r - s$

25. $\dfrac{x + 8}{x^2 - 3x - 4}, \dfrac{-9}{x + x^2}$

26. $\dfrac{y + 1}{y^2 - 8y + 12}, \dfrac{-3}{y^2 - 6y}$

27. $\dfrac{t}{2t^2 + 7t - 15}, \dfrac{t}{t^2 + 3t - 10}$

28. $\dfrac{s}{s^2 - 3s - 4}, \dfrac{s}{3s^2 + s - 2}$

29. $\dfrac{y}{2y + 6}, \dfrac{3}{y^2 - 9}, \dfrac{6}{y}$

30. $\dfrac{x}{9x + 18}, \dfrac{-1}{x^2 - 4}, \dfrac{5}{x}$

31. $\dfrac{5}{6x}, \dfrac{3}{x^2}, \dfrac{7}{x + 1}$

32. $\dfrac{12}{11y}, \dfrac{5}{y^5}, \dfrac{9}{y - 1}$

33. One student added two rational expressions and obtained the answer $\dfrac{3}{5 - y}$. Another student obtained the answer $\dfrac{-3}{y - 5}$ for the same problem. Is it possible that both answers are correct? Explain.

34. What is *wrong* with the following work?

$$\dfrac{x}{x + 2} - \dfrac{4x - 1}{x + 2} = \dfrac{x - 4x - 1}{x + 2} = \dfrac{-3x - 1}{x + 2}$$

Add or subtract as indicated. Write the answer in lowest terms. See Examples 3–5.

35. $\dfrac{8}{t} + \dfrac{7}{3t}$

36. $\dfrac{5}{x} + \dfrac{9}{4x}$

37. $\dfrac{5}{12x^2y} - \dfrac{11}{6xy}$

◉ CONCEPTUAL   ✎ WRITING   ▲ CHALLENGING   ▦ SCIENTIFIC CALCULATOR   ▦ GRAPHICS CALCULATOR

38. $\dfrac{7}{18a^3b^2} - \dfrac{2}{9ab}$

39. $\dfrac{1}{x-1} - \dfrac{1}{x}$

40. $\dfrac{3}{x-3} - \dfrac{1}{x}$

41. $\dfrac{3a}{a+1} + \dfrac{2a}{a-3}$

42. $\dfrac{2x}{x+4} + \dfrac{3x}{x-7}$

43. $\dfrac{3x+2}{4-x} + \dfrac{5-3x}{x-4}$

44. $\dfrac{5t+3}{2-t} + \dfrac{1-5t}{t-2}$

45. $\dfrac{-3w+2z}{w-z} - \dfrac{4w-z}{z-w}$

46. $\dfrac{2b-5a}{a-b} - \dfrac{6a-b}{b-a}$

47. $\dfrac{4x}{x-1} - \dfrac{2}{x+1} - \dfrac{4}{x^2-1}$

48. $\dfrac{4}{x+3} - \dfrac{x}{x-3} - \dfrac{18}{x^2-9}$

49. $\dfrac{5}{x-2} + \dfrac{1}{x} + \dfrac{2}{x^2-2x}$

50. $\dfrac{5x}{x-3} + \dfrac{2}{x} + \dfrac{6}{x^2-3x}$

51. $\dfrac{3x}{x+1} + \dfrac{4}{x-1} - \dfrac{6}{x^2-1}$

52. $\dfrac{5x}{x+3} + \dfrac{x+2}{x} - \dfrac{6}{x^2+3x}$

53. $\dfrac{4}{x+1} + \dfrac{1}{x^2-x+1} - \dfrac{12}{x^3+1}$

54. $\dfrac{5}{x+2} + \dfrac{2}{x^2-2x+4} - \dfrac{60}{x^3+8}$

▲ **55.** $\dfrac{2x+4}{x+3} + \dfrac{3}{x} - \dfrac{6}{x^2+3x}$

▲ **56.** $\dfrac{4x+1}{x+5} - \dfrac{2}{x} + \dfrac{10}{x^2+5x}$

▲ **57.** $\dfrac{5x-y}{x^2+xy-2y^2} - \dfrac{3x+2y}{x^2+5xy-6y^2}$

▲ **58.** $\dfrac{6x+5y}{6x^2+5xy-4y^2} - \dfrac{x+2y}{9x^2-16y^2}$

▲ **59.** $\dfrac{r+s}{3r^2+2rs-s^2} - \dfrac{s-r}{6r^2-5rs+s^2}$

▲ **60.** $\dfrac{3y}{y^2+yz-2z^2} + \dfrac{4y-1}{y^2-z^2}$

◉ **61.** If $x = 4$ and $y = 2$, show that $\dfrac{1}{x} + \dfrac{1}{y} \neq \dfrac{1}{x+y}$.

◉ **62.** Use $x = 3$ and $y = 5$ to show that $\dfrac{1}{x} - \dfrac{1}{y} \neq \dfrac{1}{x-y}$.

◉ **63.** Add $\dfrac{2}{m+1} + \dfrac{5}{m}$ using each of the common denominators given below. Be sure all answers are given in lowest terms.

    (a) $m^2(m+1)$ **(b)** $m(m+1)$ **(c)** $m(m+1)^2$ **(d)** $m^2(m+1)^2$

◉ **64.** Is it necessary to find the *least* common denominator when adding or subtracting fractions, or would any common denominator work? What advantage is there in using the *least* common denominator? See Exercise 63.

—————◆ **MATHEMATICAL CONNECTIONS** (Exercises 65–68) ◇—————

Students often confuse the steps used to multiply (or divide) rational expressions with the steps for adding (or subtracting) rational expressions. Consider the following problems.

    Problem A   *Addition*   Problem B   *Multiplication*

$$\dfrac{-x}{4xy+3y^2} + \dfrac{8x+6y}{16x^2-9y^2} \qquad \dfrac{-x}{4xy+3y^2} \cdot \dfrac{8x+6y}{16x^2-9y^2}$$

The following exercises compare the steps for performing these operations on the expressions. Work them in order.

65. (a) Factor the denominators in Problem A.
    (b) Factor all numerators and denominators in Problem B.

Answer column (right side):

45. $\dfrac{w+z}{w-z}$ or $\dfrac{-w-z}{z-w}$

46. $\dfrac{a+b}{a-b}$ or $\dfrac{-a-b}{b-a}$

47. $\dfrac{2(2x-1)}{x-1}$

48. $\dfrac{-x^2+x-30}{(x-3)(x+3)}$

49. $\dfrac{6}{x-2}$ **50.** $\dfrac{5x+2}{x-3}$

51. $\dfrac{3x-2}{x-1}$ **52.** $\dfrac{6x+5}{x+3}$

53. $\dfrac{4x-7}{x^2-x+1}$

54. $\dfrac{5x-18}{x^2-2x+4}$

55. $\dfrac{2x+1}{x}$ **56.** $\dfrac{4x-1}{x+5}$

57. $\dfrac{2x^2+21xy-10y^2}{(x+2y)(x-y)(x+6y)}$

58. $\dfrac{16x^2-12xy-18y^2}{(3x+4y)(2x-y)(3x-4y)}$

59. $\dfrac{3r-2s}{(2r-s)(3r-s)}$

60. $\dfrac{7y^2+11yz-y-2z}{(y+z)(y-z)(y+2z)}$

61. If $x = 4$ and $y = 2$, then
$\dfrac{1}{x} + \dfrac{1}{y} = \dfrac{1}{4} + \dfrac{1}{2} = \dfrac{3}{4}$
which does not equal
$\dfrac{1}{x+y} = \dfrac{1}{4+2} = \dfrac{1}{6}$.

62. If $x = 3$ and $y = 5$,
then $\dfrac{1}{x} - \dfrac{1}{y} = \dfrac{1}{3} - \dfrac{1}{5} = \dfrac{2}{15}$,
which does not equal
$\dfrac{1}{x-y} = \dfrac{1}{3-5} = -\dfrac{1}{2}$.

63. (a)–(d) All equal $\dfrac{7m+5}{m(m+1)}$.

65. (a) $\dfrac{-x}{y(4x+3y)} + \dfrac{8x+6y}{(4x+3y)(4x-3y)}$

(b) $\dfrac{-x}{y(4x+3y)} \cdot \dfrac{2(4x+3y)}{(4x+3y)(4x-3y)}$

66. (a) $\dfrac{-x(4x-3y)}{y(4x+3y)(4x-3y)}+$

$\dfrac{y(8x+6y)}{y(4x+3y)(4x-3y)}$

(b) $\dfrac{-2x(4x+3y)}{y(4x+3y)^2(4x-3y)}$

67. (a) $\dfrac{-4x^2+11xy+6y^2}{y(4x+3y)(4x-3y)}$

(b) $\dfrac{-2x}{y(4x+3y)(4x-3y)}$

69. $\dfrac{4}{15}$ **70.** 1 **71.** $\dfrac{7}{17}$

72. $-\dfrac{16}{15}$

66. (a) In Problem A, find the least common denominator, and write each fraction with the least common denominator.
(b) In Problem B, write a single fraction with the product of the numerator factors in the numerator and the product of the denominator factors in the denominator.

67. (a) Add the numerators in Problem A and write the sum over the least common denominator. Write in lowest terms, if necessary.
(b) Use the fundamental property to write Problem B in lowest terms.

 68. Discuss the similarities and differences in these two processes.

REVIEW EXERCISES

Simplify. See Section 1.2.

69. $\dfrac{\dfrac{5}{9}-\dfrac{1}{3}}{\dfrac{2}{3}+\dfrac{1}{6}}$

70. $\dfrac{\dfrac{7}{8}-\dfrac{3}{2}}{-\dfrac{1}{4}-\dfrac{3}{8}}$

71. $\dfrac{2-\dfrac{1}{4}}{\dfrac{5}{4}+3}$

72. $\dfrac{\dfrac{4}{3}-2}{1-\dfrac{3}{8}}$

4.3 COMPLEX FRACTIONS

FOR EXTRA HELP

SSG pp. 111–115
SSM pp. 165–169

Video 5

Tutorial
IBM   MAC

OBJECTIVES
1 ▶ Simplify complex fractions by simplifying numerator and denominator.
2 ▶ Simplify complex fractions by multiplying by a common denominator.
3 ▶ Simplify rational expressions with negative exponents.

A **complex fraction** is an expression having a fraction in the numerator, denominator, or both. Examples of complex fractions include

$$\frac{1+\dfrac{1}{x}}{2},\qquad \frac{\dfrac{4}{y}}{6-\dfrac{3}{y}},\qquad \text{and}\qquad \frac{\dfrac{m^2-9}{m+1}}{\dfrac{m+3}{m^2-1}}.$$

1 ▶ Simplify complex fractions by simplifying numerator and denominator.

▶ There are two different methods for simplifying complex fractions.

Simplifying Complex Fractions

Method 1 Simplify the numerator and denominator separately, as much as possible. Then multiply the numerator by the reciprocal of the denominator. Simplify the resulting fraction, if possible.

INSTRUCTOR'S RESOURCES

ITM pp. 372–375 **ISM** pp. 251–257
IAM p. 16

TEST GENERATOR
DOS   Windows   MAC

TRANSPARENCIES

In the second step, we are writing the complex fraction as a quotient of two rational expressions and dividing. Before performing this step, be sure that both the numerator and denominator are single fractions.

EXAMPLE 1
Simplifying Complex Fractions by Method **1**

Use Method 1 to simplify each complex fraction.

(a) $\dfrac{\dfrac{x + 1}{x}}{\dfrac{x - 1}{2x}}$

Both the numerator and the denominator are already simplified, so multiply the numerator by the reciprocal of the denominator.

$$\frac{\dfrac{x + 1}{x}}{\dfrac{x - 1}{2x}} = \frac{x + 1}{x} \div \frac{x - 1}{2x} \qquad \text{Write as a division problem.}$$

$$= \frac{x + 1}{x} \cdot \frac{2x}{x - 1} \qquad \text{Reciprocal of } \tfrac{x - 1}{2x}$$

$$= \frac{2(x + 1)}{x - 1}$$

(b) $\dfrac{2 + \dfrac{1}{y}}{3 - \dfrac{2}{y}} = \dfrac{\dfrac{2y}{y} + \dfrac{1}{y}}{\dfrac{3y}{y} - \dfrac{2}{y}} = \dfrac{\dfrac{2y + 1}{y}}{\dfrac{3y - 2}{y}}$ Simplify numerator and denominator.

$$= \frac{2y + 1}{y} \cdot \frac{y}{3y - 2} \qquad \text{Reciprocal of } \tfrac{3y - 2}{y}$$

$$= \frac{2y + 1}{3y - 2} \qquad \blacksquare$$

▶ **TEACHING TIP**
It is certainly a matter of choice as to which method you choose. Show a simple example using both methods.

Again, remind students to write answers in factored form in the event that an answer can be reduced.   ◀

2 ▶ Simplify complex fractions by multiplying by a common denominator.

▶   The second method for simplifying complex fractions uses the identity property for multiplication.

Simplifying Complex Fractions

Method 2   Multiply the numerator and denominator of the complex fraction by the least common denominator of the fractions in the numerator and the fractions in the denominator of the complex fraction. Then simplify the resulting fraction, if possible.

EXAMPLE 2
Simplifying Complex Fractions by Method **2**

Use Method 2 to simplify each complex fraction.

(a) $\dfrac{2 + \dfrac{1}{y}}{3 - \dfrac{2}{y}}$

Multiply the numerator and denominator by the least common denominator of all the fractions in the numerator and the denominator of the complex fraction. Here the least common denominator is y.

$$\frac{2+\dfrac{1}{y}}{3-\dfrac{2}{y}}=\frac{\left(2+\dfrac{1}{y}\right)\cdot y}{\left(3-\dfrac{2}{y}\right)\cdot y}$$ 　Multiply numerator and denominator by y.

$$=\frac{2\cdot y+\dfrac{1}{y}\cdot y}{3\cdot y-\dfrac{2}{y}\cdot y}$$ 　Use the distributive property.

$$=\frac{2y+1}{3y-2}$$

Compare this method of solution with that used in Example 1(b) above.

(b) $\dfrac{2p+\dfrac{5}{p-1}}{3p-\dfrac{2}{p}}$

The least common denominator is $p(p-1)$.

$$\frac{2p+\dfrac{5}{p-1}}{3p-\dfrac{2}{p}}=\frac{2p[p(p-1)]+\dfrac{5}{p-1}\cdot p(p-1)}{3p[p(p-1)]-\dfrac{2}{p}\cdot p(p-1)}$$

$$=\frac{2p[p(p-1)]+5p}{3p[p(p-1)]-2(p-1)}$$

$$=\frac{2p^3-2p^2+5p}{3p^3-3p^2-2p+2}$$ ∎

◆ 1. Answers will vary. 2. $\dfrac{34}{21}$; 1, 2, 1.5, 1.$\overline{6}$, 1.6, 1.625, 1.615384615, 1.619047619; They seem to approach a number close to 1.62.

> ◇ **CONNECTIONS** ◇
>
> An infinite complex fraction, such as
>
> $$1+\cfrac{1}{1+\cfrac{1}{1+\cdots}}$$
>
> is called a *continued fraction*. Its value can be approximated by adding more and more terms as follows.
>
> $$1,\ 1+\frac{1}{1},\ 1+\cfrac{1}{1+\dfrac{1}{1}},\ 1+\cfrac{1}{1+\cfrac{1}{1+1}},\ 1+\cfrac{1}{1+\cfrac{1}{1+\cfrac{1}{1+1}}},\ \ldots$$
>
> These approximations form the sequence
>
> $$\frac{1}{1},\frac{2}{1},\frac{3}{2},\frac{5}{3},\frac{8}{5},\frac{13}{8},\frac{21}{13},\ \ldots$$

Both the numerators and the denominators form another sequence, called the *Fibonacci sequence,* which is found in nature, such as in the number of spirals in many pine cones and pineapples, as well as in the number of bees in succeeding generations. The sequence of fractions approaches a number called the *golden ratio.* This ratio has been widely used in art and architecture as producing the most pleasing ratio of length to width in a figure.

FOR DISCUSSION OR WRITING

1. Look up the golden rectangle in a reference book and list examples of its use in art and architecture.
2. Find the next number in the sequence of fractions given above. Use a calculator to write the fractions in the sequence, including the ones you find, in decimal form. To the nearest hundredth, what decimal number do they seem to approach?

3 ▶ Simplify rational expressions with negative exponents.

▶ Rational expressions and complex fractions often involve negative exponents. To simplify such expressions, we begin by rewriting the expressions with only positive exponents.

EXAMPLE 3

Simplifying Expressions with Negative Exponents

Simplify each of the following.

(a) $\dfrac{m^{-1} + p^{-2}}{2m^{-2} - p^{-1}}$

First write the expression with only positive exponents using the definition of a negative exponent.

$$\frac{m^{-1} + p^{-2}}{2m^{-2} - p^{-1}} = \frac{\dfrac{1}{m} + \dfrac{1}{p^2}}{\dfrac{2}{m^2} - \dfrac{1}{p}}$$

Note that the 2 in $2m^{-2}$ is not raised to the -2 power, so $2m^{-2} = \dfrac{2}{m^2}$. Simplify the complex fraction by multiplying numerator and denominator by the least common denominator, m^2p^2.

$$\frac{\dfrac{1}{m} + \dfrac{1}{p^2}}{\dfrac{2}{m^2} - \dfrac{1}{p}} = \frac{m^2p^2 \cdot \dfrac{1}{m} + m^2p^2 \cdot \dfrac{1}{p^2}}{m^2p^2 \cdot \dfrac{2}{m^2} - m^2p^2 \cdot \dfrac{1}{p}} = \frac{mp^2 + m^2}{2p^2 - m^2p}$$

(b) $\dfrac{k^{-1}}{k^{-1} + 1} = \dfrac{\dfrac{1}{k}}{\dfrac{1}{k} + 1}$ Write with positive exponents.

$$= \frac{k \cdot \dfrac{1}{k}}{k \cdot \dfrac{1}{k} + k \cdot 1}$$ Use Method 2.

$$= \frac{1}{1 + k}$$

Chalkboard Exercise

Simplify $\dfrac{a^{-2} + b^{-1}}{a^{-1} - 5b^{-3}}$.

Answer: $\dfrac{b^3 + a^2b^2}{ab^3 - 5a^2}$

▶ **TEACHING TIP**

A common mistake is for students to write $a^{-1} + b^{-1}$ as $1/(a + b)$. Point out that $a^{-1} + b^{-1} \neq (a + b)^{-1}$. ◀

4.3 EXERCISES

2. identity property for multiplication

3. $\dfrac{2x}{x-1}$ **4.** $\dfrac{4t}{t+4}$

5. $\dfrac{2(k+1)}{3k-1}$

6. $\dfrac{2(1-r)}{-1-r}$ **7.** $\dfrac{5x^2}{9z^3}$

8. $\dfrac{2}{3yx}$ **9.** $\dfrac{1+x}{-1+x}$

10. $\dfrac{2-k}{2+k}$ **11.** $\dfrac{y+x}{y-x}$

12. $\dfrac{s-t}{s+t}$ **13.** $4x$

14. $\dfrac{5y}{2}$ **15.** $x+4y$

16. $\dfrac{1}{2t+3s}$ **17.** $\dfrac{3y}{2}$

18. $\dfrac{3p^2-2p}{3p-10}$

19. $\dfrac{x^2+5x+4}{x^2+5x+10}$

20. $\dfrac{y^2-2y-3}{y^2+y-1}$

21. $\dfrac{m^2+6m-4}{m(m-1)}$

22. $\dfrac{m^2-m-2}{m(m-1)}$

23. $\dfrac{m^2+6m-4}{m^2-m-2}$

◉ **1.** Explain in your own words Method 1 of simplifying complex fractions.

◉ **2.** Method 2 of simplifying complex fractions says that we can multiply both the numerator and the denominator of the complex fraction by the same nonzero expression. What property of real numbers from Section 1.3 justifies this method?

Use either method to simplify the complex fraction. See Examples 1 and 2.

3. $\dfrac{\dfrac{12}{x-1}}{\dfrac{6}{x}}$ **4.** $\dfrac{\dfrac{24}{t+4}}{\dfrac{6}{t}}$ **5.** $\dfrac{\dfrac{k+1}{2k}}{\dfrac{3k-1}{4k}}$

6. $\dfrac{\dfrac{1-r}{4r}}{\dfrac{-1-r}{8r}}$ **7.** $\dfrac{\dfrac{4z^2x^4}{9}}{\dfrac{12x^2z^5}{15}}$ **8.** $\dfrac{\dfrac{3y^2x^3}{8}}{\dfrac{9y^3x^4}{16}}$

9. $\dfrac{\dfrac{1}{x}+1}{-\dfrac{1}{x}+1}$ **10.** $\dfrac{\dfrac{2}{k}-1}{\dfrac{2}{k}+1}$ **11.** $\dfrac{\dfrac{3}{x}+\dfrac{3}{y}}{\dfrac{3}{x}-\dfrac{3}{y}}$

12. $\dfrac{\dfrac{4}{t}-\dfrac{4}{s}}{\dfrac{4}{t}+\dfrac{4}{s}}$ **13.** $\dfrac{\dfrac{8x-24y}{10}}{\dfrac{x-3y}{5x}}$ **14.** $\dfrac{\dfrac{10x-5y}{12}}{\dfrac{2x-y}{6y}}$

15. $\dfrac{\dfrac{x^2-16y^2}{xy}}{\dfrac{1}{y}-\dfrac{4}{x}}$ **16.** $\dfrac{\dfrac{2}{s}-\dfrac{3}{t}}{\dfrac{4t^2-9s^2}{st}}$ **17.** $\dfrac{y-\dfrac{y-3}{3}}{\dfrac{4}{9}+\dfrac{2}{3y}}$

18. $\dfrac{p-\dfrac{p+2}{4}}{\dfrac{3}{4}-\dfrac{5}{2p}}$ **19.** $\dfrac{\dfrac{x+2}{x}+\dfrac{1}{x+2}}{\dfrac{5}{x}+\dfrac{x}{x+2}}$ **20.** $\dfrac{\dfrac{y+3}{y}-\dfrac{4}{y-1}}{\dfrac{y}{y-1}+\dfrac{1}{y}}$

◆ **MATHEMATICAL CONNECTIONS** (Exercises 21–26) ◆

Simplifying a complex fraction by Method 1 is a good way to review the methods of adding, subtracting, multiplying, and dividing rational expressions. Method 2 gives a good review of the fundamental theorem of rational expressions. Refer to the complex fraction below and work the following exercises in order.

$$\dfrac{\dfrac{4}{m}+\dfrac{m+2}{m-1}}{\dfrac{m+2}{m}-\dfrac{2}{m-1}}$$

21. Add the fractions in the numerator.

22. Subtract as indicated in the denominator.

23. Divide your answer from Exercise 21 by your answer from Exercise 22.

◉ CONCEPTUAL   ✐ WRITING   ▲ CHALLENGING   ▦ SCIENTIFIC CALCULATOR   ▦ GRAPHICS CALCULATOR

24. Go back to the original complex fraction and find the least common denominator of all denominators.

25. Multiply the numerator and denominator of the complex fraction by your answer from Exercise 24.

26. Your answers for Exercises 23 and 25 should be the same. Write a paragraph comparing the two methods. Which method do you prefer? Explain why.

◆

Simplify the expression, using only positive exponents in your answer. See Example 3.

27. $\dfrac{1}{x^{-2} + y^{-2}}$

28. $\dfrac{1}{p^{-2} - q^{-2}}$

29. $\dfrac{x^{-2} + y^{-2}}{x^{-1} + y^{-1}}$

30. $\dfrac{x^{-1} - y^{-1}}{x^{-2} - y^{-2}}$

31. $(r^{-1} + s^{-1})^{-1}$

32. $((2k)^{-1} + (4s)^{-1})^{-1}$

33. (a) Start with the complex fraction $\dfrac{\dfrac{3}{mp} - \dfrac{4}{p} + \dfrac{8}{m}}{2m^{-1} - 3p^{-1}}$ and write it so that there are no negative exponents in your expression.

(b) Explain why $\dfrac{\dfrac{3}{mp} - \dfrac{4}{p} + \dfrac{8}{m}}{\dfrac{1}{2m} - \dfrac{1}{3p}}$ would *not* be a correct response in part (a).

(c) Simplify the complex fraction in part (a).

34. Is $\dfrac{m^{-1} + n^{-1}}{m^{-2} + n^{-2}} = \dfrac{m^2 + n^2}{m + n}$ a true statement? If not, explain why.

▲ *Simplify.*

35. $1 - \dfrac{3}{3 - \dfrac{1}{2y}}$

36. $2 - \dfrac{1}{2 + \dfrac{1}{x}}$

37. $\dfrac{1}{p + \dfrac{1}{p + \dfrac{1}{1 + p}}}$

38. $1 - \dfrac{1}{1 - \dfrac{1}{1 - \dfrac{1}{x - 1}}}$

REVIEW EXERCISES

Perform the indicated operations. See Sections 3.1 and 3.3.

39. $\dfrac{12p^7}{6p^3}$

40. $\dfrac{-9y^{11}}{3y}$

41. $\dfrac{-8a^3b^7}{6a^5b}$

42. $\dfrac{20r^3s^5}{15rs^9}$

43. Subtract.

$\quad -3a^2 + 4a - 5$
$\quad \underline{5a^2 + 3a - 9}$

44. Subtract.

$\quad -4p^2 - 8p + 5$
$\quad \underline{3p^2 + 2p + 9}$

45. Subtract.

$\quad 10x^3 - 8x^2 + 4x$
$\quad \underline{12x^3 + 5x^2 - 7x}$

46. Subtract.

$\quad 2z^3 - 5z^2 + 9z$
$\quad \underline{8z^3 + 2z^2 - 11z}$

---

24. $m(m - 1)$

25. $\dfrac{m^2 + 6m - 4}{m^2 - m - 2}$

27. $\dfrac{x^2y^2}{y^2 + x^2}$

28. $\dfrac{p^2q^2}{q^2 - p^2}$ or $\dfrac{p^2q^2}{(q - p)(q + p)}$

29. $\dfrac{y^2 + x^2}{xy^2 + x^2y}$ or $\dfrac{y^2 + x^2}{xy(y + x)}$

30. $\dfrac{xy}{y + x}$

31. $\dfrac{rs}{s + r}$

32. $\dfrac{4ks}{2s + k}$

33. (a) $\dfrac{\dfrac{3}{mp} - \dfrac{4}{p} + \dfrac{8}{m}}{\dfrac{2}{m} - \dfrac{3}{p}}$

(c) $\dfrac{3 - 4m + 8p}{2p - 3m}$

35. $\dfrac{-1}{6y - 1}$ or $\dfrac{1}{1 - 6y}$

36. $\dfrac{3x + 2}{2x + 1}$

37. $\dfrac{p + p^2 + 1}{p^3 + p^2 + 2p + 1}$

38. $x - 1$ **39.** $2p^4$

40. $-3y^{10}$ **41.** $\dfrac{-4b^6}{3a^2}$

42. $\dfrac{4r^2}{3s^4}$

43. $-8a^2 + a + 4$

44. $-7p^2 - 10p - 4$

45. $-2x^3 - 13x^2 + 11x$

46. $-6z^3 - 7z^2 + 20z$

 CONCEPTUAL WRITING ▲ CHALLENGING SCIENTIFIC CALCULATOR GRAPHICS CALCULATOR

4.4 DIVISION OF POLYNOMIALS

FOR EXTRA HELP

SSG pp. 115–119
SSM pp. 169–172

Video
5

Tutorial
IBM   MAC

OBJECTIVES

1 ▶ Divide a polynomial by a monomial.
2 ▶ Divide a polynomial by a polynomial of two or more terms.

---

1 ▶ Divide a polynomial by a monomial.

▶ In the previous chapter, we showed how to add, subtract, and multiply polynomials. We discuss polynomial division in this section, beginning with the division of a polynomial by a monomial. (Recall that a monomial is a single term, such as $8x$, $-9m^4$, or $11y^2$.)

Dividing by a Monomial

To divide a polynomial by a monomial, divide each term in the polynomial by the monomial, and then write each quotient in lowest terms.

EXAMPLE 1
Dividing a Polynomial by a Monomial

Divide $15x^2 - 12x + 6$ by 3.

Divide each term of the polynomial by 3. Then write the result in lowest terms.

$$\frac{15x^2 - 12x + 6}{3} = \frac{15x^2}{3} - \frac{12x}{3} + \frac{6}{3}$$

$$= 5x^2 - 4x + 2$$

Check this answer by multiplying it by the divisor, 3. You should get $15x^2 - 12x + 6$ as the result.

$$3(5x^2 - 4x + 2) = 15x^2 - 12x + 6$$

Divisor Quotient   Original polynomial ▪

> **Chalkboard Exercise**
>
> Divide $\dfrac{6z^2 + 12z + 18}{2z}$.
>
> Answer: $3z + 6 + \dfrac{9}{z}$

EXAMPLE 2
Dividing a Polynomial by a Monomial

Find each quotient.

(a) $\dfrac{5m^3 - 9m^2 + 10m}{5m^2} = \dfrac{5m^3}{5m^2} - \dfrac{9m^2}{5m^2} + \dfrac{10m}{5m^2}$ Divide each term by $5m^2$.

$$= m - \frac{9}{5} + \frac{2}{m}$$ Write in lowest terms.

This result is not a polynomial; the quotient of two polynomials need not be a polynomial.

(b) $\dfrac{8xy^2 - 9x^2y + 6x^2y^2}{x^2y^2} = \dfrac{8xy^2}{x^2y^2} - \dfrac{9x^2y}{x^2y^2} + \dfrac{6x^2y^2}{x^2y^2}$

$$= \frac{8}{x} - \frac{9}{y} + 6 \quad ▪$$

> **Chalkboard Exercise**
>
> Divide $\dfrac{8a^2b^2 - 20ab^3}{4a^3b}$.
>
> Answer: $\dfrac{2b}{a} - \dfrac{5b^2}{a^2}$

INSTRUCTOR'S RESOURCES

ITM pp. 375–378   **ISM** pp. 258–263
IAM p. 17

TEST GENERATOR
DOS Windows MAC

TRANSPARENCIES

2 ▶ Divide a polynomial by a polynomial of two or more terms.

▶ **TEACHING TIP**
Try an arithmetic example to show the basic steps in long division.

$$25\overline{)5860}$$

Identify the divisor, dividend, and quotient in the above example, and review how to check a result. ◀

E X A M P L E   3

Dividing a Polynomial by a Polynomial

Chalkboard Exercise

Divide $\dfrac{2k^2 + 17k + 30}{k + 6}$.

Answer: $2k + 5$

▶ Earlier, we saw that the quotient of two polynomials can be found by factoring and then dividing out any common factors. For instance,

$$\frac{2x^2 + 5x - 3}{4x - 2} = \frac{(2x - 1)(x + 3)}{2(2x - 1)} = \frac{x + 3}{2}.$$

When the polynomials in a quotient of two polynomials have no common factors or cannot be factored, they can be divided by a process very similar to that for dividing one whole number by another. The following examples show how this is done.

Divide $2m^2 + m - 10$ by $m - 2$.

Write the problem, making sure that both polynomials are written in descending powers of the variables.

$$m - 2\overline{)2m^2 + m - 10}$$

Divide the first term of $2m^2 + m - 10$ by the first term of $m - 2$. Since $2m^2/m = 2m$, place this result above the division line.

$$\begin{array}{r} 2m \quad\longleftarrow \text{ Result of } \frac{2m^2}{m} \\ m - 2\overline{)2m^2 + \ m - 10} \end{array}$$

Multiply $m - 2$ and $2m$, and write the result below $2m^2 + m - 10$.

▶ **TEACHING TIP**
Note that the number of terms on the second line (the line being subtracted) is the same as the number of terms in the divisor. ◀

$$\begin{array}{r} 2m \\ m - 2\overline{)2m^2 + \ m - 10} \\ \underline{2m^2 - 4m} \quad\longleftarrow 2m(m - 2) = 2m^2 - 4m \end{array}$$

Now subtract $2m^2 - 4m$ from $2m^2 + m$. Do this by changing the signs on $2m^2 - 4m$ and *adding*.

▶ **TEACHING TIP**
It helps to have students show the sign change on the second line, then circle this change. ◀

$$\begin{array}{r} 2m \\ m - 2\overline{)2m^2 + \ m - 10} \\ \underline{2m^2 - 4m} \\ 5m \quad\longleftarrow \text{ Subtract.} \end{array}$$

Bring down -10 and continue by dividing $5m$ by m.

$$\begin{array}{r} 2m + \ 5 \longleftarrow \frac{5m}{m} = 5 \\ m - 2\overline{)2m^2 + \ m - 10} \\ \underline{2m^2 - 4m} \\ 5m - 10 \longleftarrow \text{Bring down } -10. \\ \underline{5m - 10} \longleftarrow 5(m - 2) = 5m - 10 \\ 0 \longleftarrow \text{Subtract.} \end{array}$$

Finally, $\dfrac{2m^2 + m - 10}{m - 2} = 2m + 5$. Check by multiplying $m - 2$ and $2m + 5$. The result should be $2m^2 + m - 10$. Since there is no remainder, this quotient could have been found by factoring and writing in lowest terms. ∎

E X A M P L E   4

Dividing a Polynomial with a Missing Term

Divide $3x^3 - 2x + 5$ by $x - 3$.

Make sure that $3x^3 - 2x + 5$ is in descending powers of the variable. Add a term with a 0 coefficient for the missing x^2 term.

$$\underset{\hphantom{x - 3}}{\overset{\displaystyle\text{Missing term}}{\ }}$$

$$x - 3\overline{)3x^3 + 0x^2 - 2x + 5}$$

Start with $\dfrac{3x^3}{x} = 3x^2$.

$$
\begin{array}{r}
3x^2 \\
x - 3\overline{)3x^3 + 0x^2 - 2x + 5} \\
3x^3 - 9x^2
\end{array}
\quad
\begin{array}{l}
\leftarrow \frac{3x^3}{x} = 3x^2 \\[4pt]
\leftarrow 3x^2(x - 3)
\end{array}
$$

Subtract by changing the signs on $3x^3 - 9x^2$ and adding.

$$
\begin{array}{r}
3x^2 \\
x - 3\overline{)3x^3 + 0x^2 - 2x + 5} \\
\underline{3x^3 - 9x^2} \\
9x^2 \qquad \leftarrow \text{Subtract.}
\end{array}
$$

Bring down the next term.

$$
\begin{array}{r}
3x^2 \\
x - 3\overline{)3x^3 + 0x^2 - 2x + 5} \\
\underline{3x^3 - 9x^2} \\
9x^2 - 2x \qquad \leftarrow \text{Bring down } -2x.
\end{array}
$$

In the next step, $\dfrac{9x^2}{x} = 9x$.

$$
\begin{array}{r}
3x^2 + 9x \qquad\quad \leftarrow \frac{9x^2}{x} = 9x \\
x - 3\overline{)3x^3 + 0x^2 - \ 2x + 5} \\
\underline{3x^3 - 9x^2} \\
9x^2 - \ 2x \\
\underline{9x^2 - 27x} \qquad \leftarrow 9x(x - 3) \\
25x + 5 \leftarrow \text{Subtract and bring down 5.}
\end{array}
$$

Finally, $\dfrac{25x}{x} = 25$.

$$
\begin{array}{r}
3x^2 + \ 9x + 25 \qquad \leftarrow \frac{25x}{x} = 25 \\
x - 3\overline{)3x^3 + 0x^2 - \ 2x + \ 5} \\
\underline{3x^3 - 9x^2} \\
9x^2 - \ 2x \\
\underline{9x^2 - 27x} \\
25x + \ 5 \\
\underline{25x - 75} \quad \leftarrow 25(x - 3) \\
80 \quad \leftarrow \text{Subtract.}
\end{array}
$$

We write the remainder of 80 as the numerator of the fraction $\dfrac{80}{x - 3}$. In summary,

$$
\frac{3x^3 - 2x + 5}{x - 3} = 3x^2 + 9x + 25 + \frac{80}{x - 3}.
$$

Check by multiplying $x - 3$ and $3x^2 + 9x + 25$ and adding 80 to the result. You should get $3x^3 - 2x + 5$. ∎

CAUTION Don't forget the $+$ sign when adding the remainder to a quotient.

Sometimes the division process requires a fractional coefficient in the quotient.

E X A M P L E 5

Getting a Fractional
Coefficient in the
Quotient

Divide $2p^3 + 5p^2 + p - 2$ by $2p + 2$.

$$\frac{3p^2}{2p} = \frac{3}{2}p$$

$$
\begin{array}{r}
p^2 + \dfrac{3}{2}p - 1 \\
2p + 2\overline{)2p^3 + 5p^2 + p - 2} \\
\underline{2p^3 + 2p^2} \\
3p^2 + p \\
\underline{3p^2 + 3p} \\
-2p - 2 \\
\underline{-2p - 2} \\
0
\end{array}
$$

Since the remainder is 0, the quotient is $p^2 + \dfrac{3}{2}p - 1$. ∎

E X A M P L E 6

Dividing by a Polynomial
with a Missing Term

Divide $6r^4 + 9r^3 + 2r^2 - 8r + 7$ by $3r^2 - 2$.

The polynomial $3r^2 - 2$ has a missing term. Write it as $3r^2 + 0r - 2$ and divide as usual.

$$
\begin{array}{r}
2r^2 + 3r + 2 \\
3r^2 + 0r - 2\overline{)6r^4 + 9r^3 + 2r^2 - 8r + 7} \\
\underline{6r^4 + 0r^3 - 4r^2} \\
9r^3 + 6r^2 - 8r \\
\underline{9r^3 + 0r^2 - 6r} \\
6r^2 - 2r + 7 \\
\underline{6r^2 + 0r - 4} \\
-2r + 11
\end{array}
$$

Since the degree of the remainder, $-2r + 11$, is less than the degree of $3r^2 - 2$, the division process is now finished. The result is written

$$2r^2 + 3r + 2 + \frac{-2r + 11}{3r^2 - 2}.\quad ∎$$

▶ **TEACHING TIP**

Have students work, either
individually or in a group, an
example that has a missing dividend
term and a remainder in the
answer. ◀

4.4 EXERCISES

◉ **1.** When dividing a monomial by a monomial, we use the rules for
_____.

◉ **2.** Before dividing a polynomial by a polynomial (not monomials), what form should both polynomials have?

Complete the division.

3.
$$
\begin{array}{r}
r^2 \\
3r - 1\overline{)3r^3 - 22r^2 + 25r - 6} \\
\underline{3r^3 - r^2} \\
-21r^2
\end{array}
$$

4.
$$
\begin{array}{r}
3b^2 \\
2b - 5\overline{)6b^3 - 7b^2 - 4b - 40} \\
\underline{6b^3 - 15b^2} \\
8b^2
\end{array}
$$

Divide. See Examples 1–6.

5. $\dfrac{9y^2 + 12y - 15}{3y}$

6. $\dfrac{80r^2 - 40r + 10}{10r}$

1. exponents
2. descending powers
3. $r^2 - 7r + 6$
4. $3b^2 + 4b + 8$
5. $3y + 4 - \dfrac{5}{y}$
6. $8r - 4 + \dfrac{1}{r}$

 CONCEPTUAL   WRITING    CHALLENGING   SCIENTIFIC CALCULATOR    GRAPHICS CALCULATOR

7. $3m + 5 + \dfrac{6}{m}$

8. $8 - \dfrac{9}{x} + \dfrac{3}{2x^2}$

9. $n - \dfrac{3n^2}{2m} + 2$

10. $\dfrac{3}{2k} + \dfrac{7}{2h} - \dfrac{7}{4hk}$

11. $\dfrac{2y}{x} + \dfrac{3}{4} + \dfrac{3w}{x}$

12. $\dfrac{2b}{a} + \dfrac{5}{3} + \dfrac{3c}{a}$

13. $y - 3$ 14. $q + 8$

15. $t + 5$ 16. $k - 4$

17. $z^2 + 3$ 18. $p^2 + 6$

19. $x^2 + 2x - 3 + \dfrac{6}{4x + 1}$

20. $2z^2 - 4z + 1 + \dfrac{-10}{5z - 3}$

21. $2x - 5 + \dfrac{-4x + 5}{3x^2 - 2x + 4}$

22.

$4m - 3 + \dfrac{4m + 5}{2m^2 - 3m + 6}$

23. $2k^2 + 3k - 1$

24. $2y^2 + 2$

25. $9z^2 - 4z + 1 +$

$\dfrac{-z + 6}{z^2 - z + 2}$

26. $q^2 + 3q - 5 +$

$\dfrac{-10}{2q^2 - q + 2}$

27. $t^3 + 6t^2 + 5t + 4$

28. $t + 4$

29. $t^2 + 2t - 3 + \dfrac{16}{t + 4}$

30. $118\dfrac{1}{7}$ 31. $118\dfrac{1}{7}$

32. They are the same.

33. The correct quotient is $a^2 + 1$.

34. Both methods can be used to divide polynomials if the quotient has no remainder.

35. $\dfrac{2}{3}x - 1$ 36. $\dfrac{1}{2}m + 1$

37. $\dfrac{3}{4}a - 2 + \dfrac{1}{4a + 3}$

38. $\dfrac{3}{5}q + 1 + \dfrac{-1}{5q - 2}$

7. $\dfrac{15m^3 + 25m^2 + 30m}{5m^2}$

9. $\dfrac{14m^2n^2 - 21mn^3 + 28m^2n}{14m^2n}$

11. $\dfrac{8wxy^2 + 3wx^2y + 12w^2xy}{4wx^2y}$

13. $\dfrac{y^2 + 3y - 18}{y + 6}$

15. $\dfrac{3t^2 + 17t + 10}{3t + 2}$

17. $(2z^3 - 5z^2 + 6z - 15) \div (2z - 5)$

19. $(4x^3 + 9x^2 - 10x + 3) \div (4x + 1)$

21. $\dfrac{6x^3 - 19x^2 + 14x - 15}{3x^2 - 2x + 4}$

23. $\dfrac{4k^4 + 6k^3 + 3k - 1}{2k^2 + 1}$

25. $(9z^4 - 13z^3 + 23z^2 - 10z + 8) \div (z^2 - z + 2)$

26. $(2q^4 + 5q^3 - 11q^2 + 11q - 20) \div (2q^2 - q + 2)$

8. $\dfrac{64x^3 - 72x^2 + 12x}{8x^3}$

10. $\dfrac{24h^2k + 56hk^2 - 28hk}{16h^2k^2}$

12. $\dfrac{12ab^2c + 10a^2bc + 18abc^2}{6a^2bc}$

14. $\dfrac{q^2 + 4q - 32}{q - 4}$

16. $\dfrac{2k^2 - 3k - 20}{2k + 5}$

18. $(3p^3 + p^2 + 18p + 6) \div (3p + 1)$

20. $(10z^3 - 26z^2 + 17z - 13) \div (5z - 3)$

22. $\dfrac{8m^3 - 18m^2 + 37m - 13}{2m^2 - 3m + 6}$

24. $\dfrac{6y^4 + 4y^3 + 4y - 6}{3y^2 + 2y - 3}$

◆ **MATHEMATICAL CONNECTIONS** (Exercises 27–32) ◇

Let $10 = t$. *Then* $100 = t^2$, $1000 = t^3$, *and so on. We can write integers using* t *as follows:* $23 = 20 + 3 = 2t + 3$, $547 = 500 + 40 + 7 = 5t^2 + 4t + 7$, $1080 = 1000 + 80 = t^3 + 8t$. *To see the close connection between division of polynomials and division of integers, work the following exercises in order.*

27. Write 1654 using t.

28. Write 14 using t.

29. Divide your answer to Exercise 27 by your answer to Exercise 28. What is the quotient?

30. Divide 1654 by 14. (Use a calculator if you like, but write any decimal as a common fraction.)

31. Write your answer to Exercise 29 without t, substituting $t = 10$, $t^2 = 100$, and so on.

32. What do you observe about the quotients in Exercises 29 and 30?

◆

◉ 33. We can check a division problem by multiplying the quotient by the divisor. If the division is correct, the product is the dividend.

$$\text{Divisor} \to a\overline{)b} \quad \overset{c \; \leftarrow \text{ Quotient}}{\underset{\leftarrow \text{ Dividend}}{}}$$

Use this method to check the following division problem: $(4a^3 + 5a^2 + 4a + 5) \div (4a + 5) = a^2 + a + 1$. If the quotient is incorrect, find the correct quotient, and check it.

◉ 34. Look at Exercise 15. Factor the numerator and write the fraction in lowest terms. Compare your result with the quotient found in Exercise 15. What can you conclude?

▲ *Divide.*

35. $\left(2x^2 - \dfrac{7}{3}x - 1\right) \div (3x + 1)$

36. $\left(m^2 + \dfrac{7}{2}m + 3\right) \div (2m + 3)$

37. $\left(3a^2 - \dfrac{23}{4}a - 5\right) \div (4a + 3)$

38. $\left(3q^2 + \dfrac{19}{5}q - 3\right) \div (5q - 2)$

Solve the applied problem.

39. $2p + 7$
40. $m^2 + 3m - 7$
kilometers per hour
42. $Q(x) = 2x^2 - 3x + 5$;
$R(x) = 3$ 43. -6 44. 9
45. -21 46. -11
47. -6 48. -110

 39. The volume of a box is $2p^3 + 15p^2 + 28p$. The height is p and the length is $p + 4$; find the width.

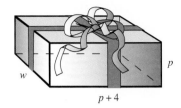

 40. Suppose a car goes $2m^3 + 15m^2 + 13m - 63$ kilometers in $2m + 9$ hours. Find its rate of speed.

 41. Evaluate the polynomial $p^3 - 4p^2 + 3p - 5$ when $p = -1$. Then divide the
 polynomial by $p + 1$. Compare the remainder in the division to the first answer. What do these results suggest?

▲ **42.** Let $P(x) = 4x^3 - 8x^2 + 13x - 2$, and $D(x) = 2x - 1$. Use division to find polynomials $Q(x)$ and $R(x)$ so that $P(x) = Q(x) \cdot D(x) + R(x)$.

REVIEW EXERCISES

Evaluate the polynomial as indicated. See Section 3.2.

43. $P(x) = x^2 + 2x - 9$; $P(1)$

44. $P(x) = 3x^2 - 5x + 7$; $P(2)$

45. $P(x) = -x^2 + 8x + 12$; $P(-3)$

46. $P(x) = -x^2 + 2x - 8$; $P(-1)$

47. $P(x) = 2x^3 + x^2 - 5x - 4$; $P(-2)$

48. $P(x) = 3x^3 - x^2 + 4x - 8$; $P(-3)$

4.5 SYNTHETIC DIVISION

FOR EXTRA HELP	OBJECTIVES
📖 **SSG** pp. 119–122 **SSM** pp. 172--175	**1** ▶ Use synthetic division to divide by a polynomial of the form $x - k$.
📼 **Video** 5	**2** ▶ Use the remainder theorem to evaluate a polynomial.
💾 **Tutorial** IBM MAC	**3** ▶ Decide whether a given number is a solution of an equation.

1 ▶ Use synthetic division to divide by a polynomial of the form $x - k$.

▶ Many times when one polynomial is divided by a second, the second polynomial is of the form $x - k$, where the coefficient of the x term is 1. There is a shortcut way for doing these divisions. To see how this shortcut works, look first at the left below, where the division of $3x^3 - 2x + 5$ by $x - 3$ is shown. (Notice that a 0 was inserted for the missing x^2 term.)

$$
\begin{array}{r}
3x^2 + 9x + 25 \\
x - 3{\overline{\smash{\big)}\,3x^3 + 0x^2 - 2x + 5}} \\
\underline{3x^3 - 9x^2} \\
9x^2 - 2x \\
\underline{9x^2 - 27x} \\
25x + 5 \\
\underline{25x - 75} \\
80
\end{array}
\qquad
\begin{array}{r}
3 \quad\; 9 \quad\; 25 \\
1 - 3{\overline{\smash{\big)}\,3 \quad\; 0 \quad -2 \quad\; 5}} \\
\underline{3 \; -9} \\
9 \quad -2 \\
\underline{9 \quad -27} \\
25 \quad\; 5 \\
\underline{25 \; -75} \\
80
\end{array}
$$

On the right, exactly the same division is shown written without the variables. All the numbers in color on the right are repetitions of the numbers directly above them, so they may be omitted, as shown on the left below.

$$
\begin{array}{r}
3 \quad\ 9 \quad\ 25 \\
1-3\overline{)3 \quad\ 0 \ -2 \quad\ 5} \\
\underline{-9} \\
9 \ -2 \\
\underline{-27} \\
25 \quad\ 5 \\
\underline{-75} \\
80
\end{array}
\qquad
\begin{array}{r}
3 \quad\ 9 \quad\ 25 \\
1-3\overline{)3 \quad\ 0 \ -2 \quad\ 5} \\
\underline{-9} \\
9 \\
\underline{-27} \\
25 \\
\underline{-75} \\
80
\end{array}
$$

The numbers in color on the left are again repetitions of the numbers directly above them; they too may be omitted, as shown on the right above.

Now the problem can be condensed. If the 3 in the dividend is brought down to the beginning of the bottom row, the top row can be omitted, since it duplicates the bottom row.

$$
\begin{array}{r}
1-3\overline{)3 \quad\ \ 0 \ \ -2 \quad\ \ 5} \\
-9 \ -27 \ -75 \\
\hline
3 \quad\ 9 \quad\ 25 \quad\ 80
\end{array}
$$

Finally, the 1 at the upper left can be omitted. Also, to simplify the arithmetic, subtraction in the second row is replaced by addition. We compensate for this by changing the -3 at upper left to its additive inverse, 3. The result of doing all this is shown below.

Additive inverse

$$
\begin{array}{r}
3\overline{)3 \quad\ \ 0 \ \ -2 \quad\ \ 5} \\
9 \quad\ 27 \quad\ 75 \leftarrow \text{Signs changed} \\
\hline
3 \quad\ 9 \quad\ 25 \quad\ 80
\end{array}
$$

The quotient is read from the bottom row.

$$3x^2 + 9x + 25 + \frac{80}{x-3}$$

The first three numbers in the bottom row are used to obtain a polynomial with degree 1 less than the degree of the dividend. The last number gives the remainder.

Synthetic Division

This shortcut procedure is called **synthetic division.** It is used only when dividing a polynomial by a binomial of the form $x - k$.

EXAMPLE 1

Using Synthetic Division

Use synthetic division to divide $5x^2 + 16x + 15$ by $x + 2$.

As mentioned above, synthetic division can be used only when dividing by a polynomial of the form $x - k$. Get $x + 2$ in this form by writing it as

$$x + 2 = x - (-2),$$

where $k = -2$. Now write the coefficients of $5x^2 + 16x + 15$, placing -2 to the left.

$x + 2$ leads to -2

$$-2\overline{)5 \quad 16 \quad 15} \leftarrow \text{Coefficients}$$

Bring down the 5, and multiply: $-2 \cdot 5 = -10$.

$$
\begin{array}{r|rrr}
-2) & 5 & 16 & 15 \\
& \downarrow & -10 & \\
\hline
& 5 & &
\end{array}
$$

Add 16 and -10, getting 6. Multiply 6 and -2 to get -12.

$$
\begin{array}{r|rrr}
-2) & 5 & 16 & 15 \\
& & -10 & -12 \\
\hline
& 5 & 6 &
\end{array}
$$

Add 15 and -12, getting 3.

$$
\begin{array}{r|rrr}
-2) & 5 & 16 & 15 \\
& & -10 & -12 \\
\hline
& 5 & 6 & 3
\end{array}
$$

The result is read from the bottom row.

$$\frac{5x^2 + 16x + 15}{x + 2} = 5x + 6 + \frac{3}{x + 2} \qquad ■$$

EXAMPLE 2

Using Synthetic Division with Missing Terms

Use synthetic division to find

$$(-4x^5 + x^4 + 6x^3 + 2x^2 + 50) \div (x - 2).$$

Use the steps given above, inserting a 0 for the missing x term.

$$
\begin{array}{r|rrrrrr}
2) & -4 & 1 & 6 & 2 & 0 & 50 \\
& & -8 & -14 & -16 & -28 & -56 \\
\hline
& -4 & -7 & -8 & -14 & -28 & -6
\end{array}
$$

Read the result from the bottom row.

$$\frac{-4x^5 + x^4 + 6x^3 + 2x^2 + 50}{x - 2} = -4x^4 - 7x^3 - 8x^2 - 14x - 28 + \frac{-6}{x - 2} \qquad ■$$

2 ► Use the remainder theorem to evaluate a polynomial.

► We can use synthetic division to evaluate polynomials. For example, in the synthetic division of Example 2, where the polynomial was divided by $x - 2$, the remainder was -6.

Replacing x in the polynomial with 2 gives

$$
\begin{aligned}
-4x^5 + x^4 + 6x^3 + 2x^2 + 50 &= -4 \cdot 2^5 + 2^4 + 6 \cdot 2^3 + 2 \cdot 2^2 + 50 \\
&= -4 \cdot 32 + 16 + 6 \cdot 8 + 2 \cdot 4 + 50 \\
&= -128 + 16 + 48 + 8 + 50 \\
&= -6,
\end{aligned}
$$

the same number as the remainder; that is, dividing by $x - 2$ produced a remainder equal to the result when x is replaced with 2. This always happens, as the following remainder theorem states.

Remainder Theorem

If the polynomial $P(x)$ is divided by $x - k$, then the remainder is equal to $P(k)$.

This result is proved in more advanced courses.

EXAMPLE 3

Using the Remainder
Theorem

▶ **TEACHING TIP**

Find $P(-2)$ using methods from
Chapter 3 for comparison. ◀

Let $P(x) = 2x^3 - 5x^2 - 3x + 11$. Find $P(-2)$.
Use the remainder theorem; divide $P(x)$ by $x - (-2)$.

$$
\begin{array}{r}
-2)\overline{2 \quad -5 \quad -3 \quad 11} \\
-4 \quad 18 \quad -30 \\
\overline{2 \quad -9 \quad 15 \quad -19} \leftarrow \text{Remainder}
\end{array}
$$

By this result, $P(-2) = -19$. ∎

3 ▶ Decide whether a
given number is a
solution of an equation.

▶ The remainder theorem also can be used to show that a given number is a solution of an equation.

EXAMPLE 4

Using the Remainder
Theorem

Show that -5 is a solution of the equation

$$2x^4 + 12x^3 + 6x^2 - 5x + 75 = 0.$$

One way to show that -5 is a solution is by substituting -5 for x in the equation. However, an easier way is to use synthetic division and the remainder theorem given above.

Proposed solution →
$$
\begin{array}{r}
-5)\overline{2 \quad 12 \quad 6 \quad -5 \quad 75} \\
-10 \quad -10 \quad 20 \quad -75 \\
\overline{2 \quad 2 \quad -4 \quad 15 \quad 0}
\end{array}
$$

Chalkboard Exercise

Use synthetic division to decide
whether or not 2 is a solution
for the following equation.

$3r^3 - 11r^2 + 17r - 14 = 0$

Answer: yes

Since the remainder is 0, the polynomial has a value of 0 when $k = -5$, and so -5 is a solution of the given equation. ∎

The synthetic division above also shows that $x - (-5)$ divides the polynomial with 0 remainder. Thus $x - (-5) = x + 5$ is a *factor* of the polynomial and

$$2x^4 + 12x^3 + 6x^2 - 5x + 75 = (x + 5)(2x^3 + 2x^2 - 4x + 15).$$

The second factor is the quotient polynomial found in the last row of the synthetic division.

4.5 EXERCISES

2. binomials of the form
$x - k$ 3. $x - 5$
4. $x - 7$ 5. $4m - 1$
6. $3k + 4$

7. $2a + 4 + \dfrac{5}{a + 2}$

8. $4y + 11 + \dfrac{24}{y - 4}$

9. $p - 4 + \dfrac{9}{p + 1}$

10. $z + 9 + \dfrac{39}{z - 5}$

11. $4a^2 + a + 3$
12. $5p^2 - 11p + 14$
13. $x^4 + 2x^3 + 2x^2 +$

$7x + 10 + \dfrac{18}{x - 2}$

◎ **1.** What is the purpose of synthetic division?

◎ **2.** What type of polynomial divisors may be used with synthetic division?

Use synthetic division to perform the division. See Examples 1 and 2.

3. $\dfrac{x^2 - 6x + 5}{x - 1}$

4. $\dfrac{x^2 - 4x - 21}{x + 3}$

5. $\dfrac{4m^2 + 19m - 5}{m + 5}$

6. $\dfrac{3k^2 - 5k - 12}{k - 3}$

7. $\dfrac{2a^2 + 8a + 13}{a + 2}$

8. $\dfrac{4y^2 - 5y - 20}{y - 4}$

9. $(p^2 - 3p + 5) \div (p + 1)$

10. $(z^2 + 4z - 6) \div (z - 5)$

11. $\dfrac{4a^3 - 3a^2 + 2a - 3}{a - 1}$

12. $\dfrac{5p^3 - 6p^2 + 3p + 14}{p + 1}$

13. $(x^5 - 2x^3 + 3x^2 - 4x - 2) \div (x - 2)$

◎ CONCEPTUAL   ✎ WRITING   ▲ CHALLENGING   ▦ SCIENTIFIC CALCULATOR   ▧ GRAPHICS CALCULATOR

14. $(2y^5 - 5y^4 - 3y^2 - 6y - 23) \div (y - 3)$

15. $(-4r^6 - 3r^5 - 3r^4 + 5r^3 - 6r^2 + 3r + 3) \div (r - 1)$

16. $(2t^6 - 3t^5 + 2t^4 - 5t^3 + 6t^2 - 3t - 2) \div (t - 2)$

17. $(-3y^5 + 2y^4 - 5y^3 - 6y^2 - 1) \div (y + 2)$ **18.** $(m^6 + 2m^4 - 5m + 11) \div (m - 2)$

19. $\dfrac{y^3 + 1}{y - 1}$ **20.** $\dfrac{z^4 + 81}{z - 3}$

Use the remainder theorem to find P(k). See Example 3.

21. $P(x) = 2x^3 - 4x^2 + 5x - 3; \quad k = 2$ **22.** $P(y) = y^3 + 3y^2 - y + 5; \quad k = -1$

23. $P(r) = -r^3 - 5r^2 - 4r - 2; \quad k = -4$ **24.** $P(z) = -z^3 + 5z^2 - 3z + 4; \quad k = 3$

25. $P(y) = 2y^3 - 4y^2 + 5y - 33; \quad k = 3$ **26.** $P(x) = x^3 - 3x^2 + 4x - 4; \quad k = 2$

27. Explain why a zero remainder in synthetic division of $P(x)$ by k indicates that k is a solution of the equation $P(x) = 0$.

──────◆ **MATHEMATICAL CONNECTIONS** (Exercises 28–32) ◆──────

In Section 4.4 we saw the close connection between polynomial division and writing a quotient of polynomials in lowest terms after factoring the numerator. Now we can show a connection between dividing one polynomial by another and factoring the first polynomial. Let $P(x) = 2x^2 + 5x - 12$. Work the following exercises in order.

28. Factor $P(x)$.

29. Solve $P(x) = 0$.

30. Find $P(-4)$ and $P\left(\dfrac{3}{2}\right)$.

31. Complete the following sentence. If $P(a) = 0$, then $x - $ _____ is a factor of $P(x)$.

32. Use the conclusion in Exercise 31 to decide whether $x - 3$ is a factor of $Q(x) = 3x^3 - 4x^2 - 17x + 6$. Factor $Q(x)$ completely.

──────────────◆──────────────

Use synthetic division to decide whether the given number is a solution of the equation. See Example 4.

33. $x^3 - 2x^2 - 3x + 10 = 0; \quad x = -2$

34. $x^3 - 3x^2 - x + 10 = 0; \quad x = -2$

35. $m^4 + 2m^3 - 3m^2 + 8m - 8 = 0; \quad m = -2$

36. $r^4 - r^3 - 6r^2 + 5r + 10 = 0; \quad r = -2$

37. $3a^3 + 2a^2 - 2a + 11 = 0; \quad a = -2$

38. $3z^3 + 10z^2 + 3z - 9 = 0; \quad z = -2$

39. $2x^3 - x^2 - 13x + 24 = 0; \quad x = -3$

40. $5p^3 + 22p^2 + p - 28 = 0; \quad p = -4$

REVIEW EXERCISES

Use the distributive property to rewrite the expression. See Section 1.3.

41. $5 \cdot 8 + 5x$ **42.** $12z - 12 \cdot 4$ **43.** $6(5m) - 6(3n)$

44. $17(4j) + 17(9k)$ **45.** $x(x + 1) + y(x + 1)$ **46.** $p(2p - 3) + 5(2p - 3)$

---

14. $2y^4 + y^3 + 3y^2 + 6y + 12 + \dfrac{13}{y - 3}$

15. $-4r^5 - 7r^4 - 10r^3 - 5r^2 - 11r - 8 + \dfrac{-5}{r - 1}$

16. $2t^5 + t^4 + 4t^3 + 3t^2 + 12t + 21 + \dfrac{40}{t - 2}$

17. $-3y^4 + 8y^3 - 21y^2 + 36y - 72 + \dfrac{143}{y + 2}$

18. $m^5 + 2m^4 + 6m^3 + 12m^2 + 24m + 43 + \dfrac{97}{m - 2}$

19. $y^2 + y + 1 + \dfrac{2}{y - 1}$

20. $z^3 + 3z^2 + 9z + 27 + \dfrac{162}{z - 3}$ **21.** 7   **22.** 8

23. -2 **24.** 13   **25.** 0

26. 0

28. $(2x - 3)(x + 4)$

29. $\left\{\dfrac{3}{2}, -4\right\}$

30. $P(-4) = 0, \; P\left(\dfrac{3}{2}\right) = 0$

31. a **32.** Yes, $x - 3$ is a factor. $Q(x) = (x - 3)(3x - 1)(x + 2)$

33. yes   **34.** no   **35.** no

36. yes   **37.** no   **38.** no

39. yes   **40.** yes

41. $5(8 + x)$

42. $12(z - 4)$

43. $6(5m - 3n)$

44. $17(4j + 9k)$

45. $(x + 1)(x + y)$

46. $(2p - 3)(p + 5)$

 CONCEPTUAL   WRITING    CHALLENGING   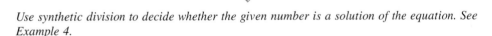 SCIENTIFIC CALCULATOR    GRAPHICS CALCULATOR

4.6 EQUATIONS WITH RATIONAL EXPRESSIONS

FOR EXTRA HELP	OBJECTIVES
SSG pp. 123–126 **SSM** pp. 175–179	**1** ▶ Solve equations with rational expressions. **2** ▶ Know when potential solutions must be checked.
Video 5	
💾 **Tutorial** IBM MAC	

1 ▶ Solve equations with rational expressions.

▶ The easiest way to solve most equations inolving rational expressions is to multiply all the terms in the equation by the least common denominator. This step will clear the equation of all denominators as the next examples show. We can do this only with equations.

EXAMPLE 1

Solving an Equation with Rational Expressions

■ Solve $\dfrac{2x}{5} - \dfrac{x}{3} = 2$.

The least common denominator for $2x/5$ and $x/3$ is 15, so we multiply both sides of the equation by 15.

$$15\left(\frac{2x}{5} - \frac{x}{3}\right) = 15(2)$$

$$15\left(\frac{2x}{5}\right) - 15\left(\frac{x}{3}\right) = 15(2) \qquad \text{Distributive property}$$

$$6x - 5x = 30 \qquad \text{Multiply.}$$

$$x = 30 \qquad \text{Combine terms.}$$

By substituting 30 for x in the original equation, check that the solution set for the given equation is $\{30\}$. ■

EXAMPLE 2

Solving an Equation with Rational Expressions

■ Solve $\dfrac{2}{y} - \dfrac{3}{2} = \dfrac{7}{2y}$.

Multiply both sides by the least common denominator, $2y$.

$$2y\left(\frac{2}{y} - \frac{3}{2}\right) = 2y\left(\frac{7}{2y}\right)$$

$$2y\left(\frac{2}{y}\right) - 2y\left(\frac{3}{2}\right) = 2y\left(\frac{7}{2y}\right) \qquad \text{Distributive property}$$

$$4 - 3y = 7 \qquad \text{Multiply.}$$

$$-3y = 3 \qquad \text{Subtract 4.}$$

$$y = -1 \qquad \text{Divide by } -3.$$

To see if -1 is a solution of the equation, replace y with -1 in the original equation.

▶ **TEACHING TIP**

When adding or subtracting fractions, we keep the common denominator in our answer. However, when solving an equation, we eliminate the denominators by multiplying each term by the LCD. Show examples contrasting the two methods. ◀

INSTRUCTOR'S RESOURCES

 ITM pp. 381–383   **ISM** pp. 268–276
 IAM p. 17

 **TEST GENERATOR**
 DOS Windows MAC

 TRANSPARENCIES

$$\frac{2}{-1} - \frac{3}{2} = \frac{7}{2(-1)} \quad ?$$

$$-\frac{4}{2} - \frac{3}{2} = -\frac{7}{2} \qquad \text{True}$$

Since -1 checks, the solution set is $\{-1\}$. ■

2 ▶ Know when potential solutions must be checked.

▶ Because the first step in solving a rational equation is to multiply both sides of the equation by a common denominator, in many cases it is *necessary* to check the solutions.

▶ TEACHING TIP

Be aware that students will discard a good solution because of an arithmetic error in checking their answer. A solution will not check if a possible solution produces zero in any denominator of the original equation. ◀

CAUTION When both sides of an equation are multiplied by a *variable* expression, it is possible that the resulting "solutions" are not actually solutions of the given equation.

EXAMPLE 3

Solving an Equation with No Solution

■ Solve $\dfrac{2}{m-3} - \dfrac{3}{m+3} = \dfrac{12}{m^2-9}$.

The least common denominator is $(m+3)(m-3)$, which is used to multiply both sides. Since $(m+3)(m-3) = 0$ when $m = 3$ or $m = -3$, the solution cannot be 3 or -3.

$$(m+3)(m-3) \cdot \frac{2}{m-3} - (m+3)(m-3) \cdot \frac{3}{m+3}$$
$$= (m+3)(m-3) \cdot \frac{12}{m^2-9}$$

$$2(m+3) - 3(m-3) = 12 \qquad \text{Multiply.}$$
$$2m + 6 - 3m + 9 = 12 \qquad \text{Distributive property}$$
$$-m + 15 = 12 \qquad \text{Combine terms.}$$
$$-m = -3$$
$$m = 3$$

Chalkboard Exercise

Solve
$\dfrac{3}{a+1} = \dfrac{1}{a-1} - \dfrac{2}{a^2-1}$.

Answer: ∅

Since both sides were multiplied by a variable expression, we must check the potential solution, 3.

$$\frac{2}{3-3} - \frac{3}{3+3} = \frac{12}{3^2-9} \quad ?$$

$$\frac{2}{0} - \frac{3}{6} = \frac{12}{0} \quad ?$$

Division by 0 is undefined. The given equation has no solution, and the solution set is ∅. We predicted this outcome at the beginning of the solution. ■

EXAMPLE 4

Solving an Equation with Rational Expressions

■ Solve $\dfrac{3}{p^2+p-2} - \dfrac{1}{p^2-1} = \dfrac{7}{2(p^2+3p+2)}$.

Factor to find the least common denominator $2(p-1)(p+2)(p+1)$, so that $p \neq 1, p \neq -2, p \neq -1$. Multiply both sides by the least common denominator.

$$2(p-1)(p+2)(p+1)\left[\frac{3}{(p+2)(p-1)}\right]$$

$$-2(p-1)(p+2)(p+1)\left[\frac{1}{(p+1)(p-1)}\right]$$

$$=2(p-1)(p+2)(p+1)\left[\frac{7}{2(p+2)(p+1)}\right]$$

Now simplify.

$$2\cdot 3(p+1)-2(p+2)=7(p-1) \qquad \text{Multiply.}$$
$$6p+6-2p-4=7p-7 \qquad \text{Distributive property}$$
$$4p+2=7p-7 \qquad \text{Combine terms.}$$
$$9=3p$$
$$3=p$$

Substitute 3 for p in the original equation to check that the solution set is $\{3\}$.   ■

EXAMPLE 5

Solving an Equation that Leads to a Quadratic Equation

■ Solve $\dfrac{2}{3x+1}=\dfrac{1}{x}-\dfrac{6x}{3x+1}$.

Multiply both sides by $x(3x+1)$. The resulting equation is

$$2x=(3x+1)-6x^2.$$

Now solve. Since the equation is quadratic, get 0 on the right side.

$$6x^2-3x+2x-1=0 \qquad \text{Standard form}$$
$$6x^2-x-1=0 \qquad \text{Combine terms.}$$
$$(3x+1)(2x-1)=0 \qquad \text{Factor.}$$
$$3x+1=0 \qquad \text{or} \qquad 2x-1=0 \qquad \text{Zero-factor property}$$
$$x=-\frac{1}{3} \qquad \text{or} \qquad x=\frac{1}{2}$$

Using $-\dfrac{1}{3}$ in the original equation causes the denominator $3x+1$ to equal 0, so it is not a solution. The solution set is $\left\{\dfrac{1}{2}\right\}$.   ■

◇ 1. 14,000 gallons
2. It decreases.

◇ **CONNECTIONS** ◇

At the beginning of this chapter, the rational expression that modeled the amount of heating oil per day (in gallons) produced by an oil refinery was given as

$$\frac{125,000-25x}{125+2x},$$

where x is the amount of gasoline produced per day (in hundreds of gallons). If the refinery has a specific amount of heating oil it must produce, say y gallons per day, then the amount of gasoline produced is determined by solving the equation

$$\frac{125,000-25x}{125+2x}=y$$

for x.

FOR DISCUSSION OR WRITING

Suppose the refinery must produce 300,000 gallons of heating oil per day to meet the needs of its customers.

1. How much gasoline can be produced per day?

2. What happens to the amount of gasoline that can be produced as the amount of heating oil produced increases?

4.6 EXERCISES

From the explanation in this section, we know that any values that would cause a denominator to equal zero must be excluded from possible solutions of equations that have variable expressions appearing in denominators. Without actually solving the equation, list all possible numbers that would have to be rejected if they appeared as potential solutions.

1. $\dfrac{1}{x + 1} - \dfrac{1}{x - 2} = 0$

2. $\dfrac{3}{x + 4} - \dfrac{2}{x - 9} = 0$

3. $\dfrac{5}{3x + 5} - \dfrac{1}{x} = \dfrac{1}{2x + 3}$

4. $\dfrac{6}{4x + 7} - \dfrac{3}{x} = \dfrac{5}{6x - 13}$

5. $\dfrac{3x + 1}{x - 4} = \dfrac{6x + 5}{2x - 7}$

6. $\dfrac{4x - 1}{2x + 3} = \dfrac{12x - 25}{6x - 2}$

7. $\dfrac{x + 5}{10} - \dfrac{2x + 3}{5} = \dfrac{x}{20}$

 **8.** Is it possible that any potential solutions to the equation

$$\dfrac{x + 7}{4} - \dfrac{x + 3}{3} = \dfrac{x}{12}$$

would have to be rejected? Explain why or why not.

Solve the equation. See Examples 1–5.

9. $\dfrac{x}{4} - \dfrac{x}{6} = \dfrac{2}{3}$

10. $\dfrac{y}{10} + \dfrac{3y}{5} = -\dfrac{7}{2}$

11. $\dfrac{x + 8}{5} = \dfrac{6 + x}{3}$

12. $\dfrac{r + 1}{4} = \dfrac{1 + 2r}{5}$

13. $\dfrac{x - 4}{x + 6} = \dfrac{2x + 3}{2x - 1}$

14. $\dfrac{5x - 8}{x + 2} = \dfrac{5x - 1}{x + 3}$

15. $\dfrac{3x + 1}{x - 4} = \dfrac{6x + 5}{2x - 7}$

16. $\dfrac{4x - 1}{2x + 3} = \dfrac{12x - 25}{6x - 2}$

17. $\dfrac{-5}{2x} + \dfrac{3}{4x} = \dfrac{-7}{4}$

18. $\dfrac{6}{5x} - \dfrac{2}{3x} = \dfrac{-8}{45}$

19. $x - \dfrac{24}{x} = -2$

20. $p + \dfrac{15}{p} = -8$

21. $\dfrac{1}{y - 1} + \dfrac{5}{12} = \dfrac{-4}{3y - 3}$

22. $\dfrac{4}{m + 2} - \dfrac{11}{9} = \dfrac{5}{3m + 6}$

23. $\dfrac{3}{4m + 2} = \dfrac{17}{2} - \dfrac{7}{2m + 1}$

24. $\dfrac{3}{k + 2} - \dfrac{2}{k^2 - 4} = \dfrac{1}{k - 2}$

25. $\dfrac{3}{x - 2} + \dfrac{21}{x^2 - 4} = \dfrac{14}{x + 2}$

26. $\dfrac{1}{y + 2} + \dfrac{3}{y + 7} = \dfrac{5}{y^2 + 9y + 14}$

1. $-1, 2$ 2. $-4, 9$

3. $-\dfrac{5}{3}, 0, -\dfrac{3}{2}$

4. $-\dfrac{7}{4}, 0, \dfrac{13}{6}$

5. $4, \dfrac{7}{2}$ 6. $-\dfrac{3}{2}, \dfrac{1}{3}$

7. There are no numbers that would have to be rejected.

9. $\{8\}$ 10. $\{-5\}$ 11. $\{-3\}$

12. $\left\{\dfrac{1}{3}\right\}$ 13. $\left\{-\dfrac{7}{12}\right\}$

14. $\{-11\}$ 15. $\emptyset$ 16. $\emptyset$

17. $\{1\}$ 18. $\{-3\}$

19. $\{-6, 4\}$ 20. $\{-5, -3\}$

21. $\left\{-\dfrac{23}{5}\right\}$ 22. $\left\{-\dfrac{1}{11}\right\}$

23. $\{0\}$ 24. $\{5\}$ 25. $\{5\}$

26. $\emptyset$

 CONCEPTUAL   WRITING   ▲ CHALLENGING   ▦ SCIENTIFIC CALCULATOR   ▤ GRAPHICS CALCULATOR

27. 0   28. 0

29. {−2}   30. {0}   31. 0

32. {x | x ≠ −3, x ≠ 3}

33. $\left\{ x \mid x \neq -\dfrac{3}{2}, x \neq \dfrac{3}{2} \right\}$

35. 6

36. (a) 3x + 2x = −30

(b) $\dfrac{3x}{6} + \dfrac{2x}{6}$

37. (a) {−6}   (b) $\dfrac{5x}{6}$

27. $\dfrac{1}{t+3} + \dfrac{4}{t+5} = \dfrac{2}{t^2+8t+15}$

28. $\dfrac{6}{w+3} + \dfrac{-7}{w-5} = \dfrac{-48}{w^2-2w-15}$

29. $\dfrac{2x}{x-3} + \dfrac{4}{x+3} = \dfrac{-24}{x^2-9}$

30. $\dfrac{2}{4x+7} + \dfrac{x}{3} = \dfrac{6}{12x+21}$

31. $\dfrac{7}{x-4} + \dfrac{3}{x} = \dfrac{-12}{x^2-4x}$

▲ **32.** $\dfrac{5x+14}{x^2-9} = \dfrac{-2x^2-5x+2}{x^2-9} + \dfrac{2x+4}{x-3}$

▲ **33.** $\dfrac{4x-7}{4x^2-9} = \dfrac{-2x^2+5x-4}{4x^2-9} + \dfrac{x+1}{2x+3}$

◉ **34.** Make up an equation similar to the one in Exercise 9, and then solve it. (*Hint:* Start with the answer, and work backward.)

 **MATHEMATICAL CONNECTIONS** (Exercises 35–40)

A common student error is to confuse an *equation,* such as

$$\frac{x}{2} + \frac{x}{3} = -5,$$

with an *addition problem,* such as

$$\frac{x}{2} + \frac{x}{3}.$$

Look for the equals sign to distinguish between them. Equations are solved to get a numerical answer, while addition (or subtraction) problems result in simplified expressions. Consider the following equation and expression.

$$\frac{x}{2} + \frac{x}{3} = -5 \qquad \frac{x}{2} + \frac{x}{3}$$

Although they look very much alike, the steps we use to solve the equation or to simplify the expression are different. We begin the same way in each case by finding the least common denominator. The following exercises will lead you through the steps for each problem.

35. Find the least common denominator for each problem.

36. (a) Multiply both sides of the equation by the least common denominator, and simplify both sides.
(b) Use the fundamental principle to rewrite both terms of the expression with the common denominator.

37. (a) Combine terms on the left side of the equation. Solve for x.
(b) Combine terms in the expression by adding numerators and keeping the common denominator.

◉ **38.** How do the answers differ in parts (a) and (b) of Exercise 37 ?

◉ **39.** Explain the difference between *simplifying the expression* $\dfrac{2}{x+1} + \dfrac{3}{x-2} - \dfrac{6}{x^2-x-2}$ and *solving the equation* $\dfrac{2}{x+1} + \dfrac{3}{x-2} = \dfrac{6}{x^2-x-2}.$

◉ **40.** What is wrong with the following problem? "Solve $\dfrac{2x+1}{3x-4} + \dfrac{1}{2x+3}.$"

REVIEW EXERCISES

Solve the formula for the specified variable. See Section 2.2.

41. $d = rt$;   for t

42. $I = prt$;   for r

43. $P = a + b + c$;   for c

44. $A = \dfrac{1}{2}h(b + B)$;   for B

These exercises are a combination of the work with applications from Section 2.3 and the work with solving equations involving rational expressions from Section 4.6. Write an equation and then solve it to find the unknown number.

45. If the reciprocal of a number is added to half the number, the result is $\dfrac{9}{4}$. Find the number.

46. The reciprocal of a number added to the reciprocal of 1 more than the number gives a sum of $\dfrac{11}{30}$. Find the number.

47. If half a number is added to two-thirds of the same number, the result is 10. Find the number.

48. If twice a number is subtracted from one-third of the number, the result is 5. Find the number.

41. $t = \dfrac{d}{r}$ **42.** $r = \dfrac{I}{pt}$

43. $c = P - a - b$

44. $B = \dfrac{2A}{h} - b$ or

$B = \dfrac{2A - bh}{h}$ **45.** $\dfrac{1}{2}$ or 4

46. $-\dfrac{6}{11}$ or 5   **47.** $\dfrac{60}{7}$

48. -3

SUMMARY OF RATIONAL EXPRESSIONS: EQUATIONS VERSUS EXPRESSIONS

The Mathematical Connections in the exercises for Section 4.6 demonstrated the distinction between rational equations and rational expressions. In the following exercises, either perform the indicated operation or solve the given equation, as appropriate.

1. $\dfrac{x}{2} - \dfrac{x}{4} = 5$

2. $\dfrac{8x^4 z}{12x^3 z^2} \cdot \dfrac{7x}{3x^5}$

3. $\dfrac{4x - 20}{x^2 - 25} \cdot \dfrac{(x + 5)^2}{10}$

4. $\dfrac{6}{7x} - \dfrac{4}{x}$

5. $\dfrac{\dfrac{1}{x} + \dfrac{1}{y}}{\dfrac{1}{x} - \dfrac{1}{y}}$

6. $\dfrac{5}{7t} = \dfrac{52}{7} - \dfrac{3}{t}$

7. $\dfrac{x - 5}{3} + \dfrac{1}{3} = \dfrac{x - 2}{5}$

8. $\dfrac{7}{6x} + \dfrac{5}{8x}$

9. $\dfrac{4}{x} - \dfrac{8}{x + 1} = 0$

10. $\dfrac{\dfrac{6}{x + 1} - \dfrac{1}{x}}{\dfrac{2}{x} - \dfrac{4}{x + 1}}$

11. $\dfrac{8}{r + 2} - \dfrac{7}{4r + 8}$

12. $\dfrac{x}{x + y} + \dfrac{2y}{x - y}$

13. $\dfrac{3p^2 - 6p}{p + 5} \div \dfrac{p^2 - 4}{8p + 40}$

14. $\dfrac{x - 2}{9} \cdot \dfrac{5}{8 - 4x}$

15. $\dfrac{a - 4}{3} + \dfrac{11}{6} = \dfrac{a + 1}{2}$

16. $\dfrac{b^2 + b - 6}{b^2 + 2b - 8} \cdot \dfrac{b^2 + 8b + 16}{3b + 12}$

1. $\{20\}$ **2.** $\dfrac{14}{9x^3 z}$

3. $\dfrac{2(x + 5)}{5}$ **4.** $\dfrac{-22}{7x}$

5. $\dfrac{y + x}{y - x}$ **6.** $\left\{\dfrac{1}{2}\right\}$

7. $\{7\}$ **8.** $\dfrac{43}{24x}$ **9.** $\{1\}$

10. $\dfrac{5x - 1}{-2x + 2}$ or

$\dfrac{5x - 1}{-2(x - 1)}$

11. $\dfrac{25}{4(r + 2)}$

12. $\dfrac{x^2 + xy + 2y^2}{(x + y)(x - y)}$

13. $\dfrac{24p}{p + 2}$ **14.** $-\dfrac{5}{36}$

15. $\{0\}$ **16.** $\dfrac{b + 3}{3}$

 CONCEPTUAL   WRITING   CHALLENGING   SCIENTIFIC CALCULATOR   GRAPHICS CALCULATOR

17. $\dfrac{5}{3z}$

18. $\dfrac{2x + 10}{x(x - 2)(x + 2)}$

19. $\{2\}$ **20.** $\dfrac{y + x}{x - y}$

21. $\dfrac{-x}{3x + 5y}$ **22.** $\{-13\}$

23. $\dfrac{3}{2s - 5r}$ **24.** $\dfrac{3y + 2}{y + 3}$

25. $\left\{\dfrac{5}{4}\right\}$ **26.** 0

27. $\dfrac{2z - 3}{2z + 3}$ **28.** $\dfrac{-1}{x - 3}$ or

$\dfrac{1}{3 - x}$ **29.** $\dfrac{t - 2}{8}$

30. $\{-10\}$

31. $\dfrac{13x + 28}{2x(x + 4)(x - 4)}$

32. 0

33. $\dfrac{k(2k^2 - 2k + 5)}{(k - 1)(3k^2 - 2)}$

34. $\{-3, 2\}$

---

17. $\dfrac{10z^2 - 5z}{3z^3 - 6z^2} \div \dfrac{2z^2 + 5z - 3}{z^2 + z - 6}$

19. $\dfrac{6}{t + 1} + \dfrac{4}{5t + 5} = \dfrac{34}{15}$

21. $\dfrac{\dfrac{5}{x} - \dfrac{3}{y}}{\dfrac{9x^2 - 25y^2}{x^2y}}$

23. $\dfrac{2r^{-1} + 5s^{-1}}{\dfrac{4s^2 - 25r^2}{3rs}}$

25. $\dfrac{8}{3k + 9} - \dfrac{8}{15} = \dfrac{2}{5k + 15}$

27. $\dfrac{6z^2 - 5z - 6}{6z^2 + 5z - 6} \cdot \dfrac{12z^2 - 17z + 6}{12z^2 - z - 6}$

29. $\dfrac{\dfrac{t}{4} - \dfrac{1}{t}}{1 + \dfrac{t + 4}{t}}$

31. $\dfrac{7}{2x^2 - 8x} + \dfrac{3}{x^2 - 16}$

▲ **33.** $\dfrac{2k + \dfrac{5}{k - 1}}{3k - \dfrac{2}{k}}$

---

18. $\dfrac{5}{x^2 - 2x} - \dfrac{3}{x^2 - 4}$

20. $\dfrac{x^{-1} + y^{-1}}{y^{-1} - x^{-1}}$

22. $\dfrac{-2}{a^2 + 2a - 3} - \dfrac{5}{3 - 3a} = \dfrac{4}{3a + 9}$

24. $\dfrac{4y^2 - 13y + 3}{2y^2 - 9y + 9} \div \dfrac{4y^2 + 11y - 3}{6y^2 - 5y - 6}$

26. $\dfrac{3r}{r - 2} = 1 + \dfrac{6}{r - 2}$

28. $\dfrac{-1}{3 - x} - \dfrac{2}{x - 3}$

30. $\dfrac{2}{y + 1} - \dfrac{3}{y^2 - y - 2} = \dfrac{3}{y - 2}$

32. $\dfrac{3}{y - 3} - \dfrac{3}{y^2 - 5y + 6} = \dfrac{2}{y - 2}$

34. $1 + \dfrac{1}{x} = \dfrac{6}{x^2}$

4.7 APPLICATIONS

FOR EXTRA HELP	OBJECTIVES
SSG pp. 126–134 **SSM** pp. 182–189	**1** ▶ Find the value of an unknown variable in a formula.
Video 6	**2** ▶ Solve a formula for a specified variable.
Tutorial IBM MAC	**3** ▶ Solve applications using proportions. **4** ▶ Solve applications about distance, rate, and time. **5** ▶ Solve applications about work rates.

As we have seen, applications may require models that are rational expressions. In this section, we show additional examples of such models. Many formulas involve rational expressions. Problems about proportions, rates of work, and rates of travel also may require rational expressions.

1 ▶ Find the value of an unknown variable in a formula.

▶ The first example shows how to find the value of an unknown variable in a formula.

INSTRUCTOR'S RESOURCES

ITM pp. 383–389   **ISM** pp. 282–295   TEST GENERATOR DOS Windows MAC   TRANSPARENCIES
IAM p. 18

EXAMPLE 1

Finding the Value of a Variable in a Formula

In physics, the focal length, f, of a lens is given by the formula

$$\frac{1}{f} = \frac{1}{p} + \frac{1}{q},$$

where p is the distance from the object to the lens and q is the distance from the lens to the image. Find q if $p = 20$ centimeters and $f = 10$ centimeters.

Replace f with 10 and p with 20.

$$\frac{1}{f} = \frac{1}{p} + \frac{1}{q}$$

$$\frac{1}{10} = \frac{1}{20} + \frac{1}{q} \qquad \text{Let } f = 10, p = 20.$$

Multiply both sides by the least common denominator, $20q$.

$$20q \cdot \frac{1}{10} = 20q \cdot \frac{1}{20} + 20q \cdot \frac{1}{q}$$

$$2q = q + 20$$

$$q = 20$$

The distance from the lens to the image is 20 centimeters. ∎

Chalkboard Exercise

Use the formula in Example 1 to find p if $f = 15$ and $q = 25$.

Answer: $\dfrac{75}{2}$

2 ▶ Solve a formula for a specified variable.

▶ In the next example we show how to solve a formula for a specified variable.

EXAMPLE 2

Solving a Formula for a Specified Variable

Solve $\dfrac{1}{f} = \dfrac{1}{p} + \dfrac{1}{q}$ for p.

Begin by multiplying both sides by the common denominator fpq.

$$fpq \cdot \frac{1}{f} = fpq\left(\frac{1}{p} + \frac{1}{q}\right)$$

$$pq = fq + fp \qquad \text{Distributive property}$$

Get the terms with p (the specified variable) on the same side of the equation. To do this, subtract fp on both sides.

$$pq - fp = fq \qquad \text{Subtract } fp.$$

$$p(q - f) = fq \qquad \text{Distributive property; factor out } p.$$

$$p = \frac{fq}{q - f} \qquad \text{Divide by } q - f. \ ∎$$

Chalkboard Exercise

Solve $\dfrac{3}{p} + \dfrac{3}{q} = \dfrac{5}{r}$ for q.

Answer: $q = \dfrac{3rp}{5p - 3r}$

EXAMPLE 3

Solving a Formula for a Specified Variable

Solve $I = \dfrac{nE}{R + nr}$ for n.

First, multiply both sides by $R + nr$.

$$(R + nr)I = (R + nr)\frac{nE}{R + nr}$$

$$RI + nrI = nE$$

$$RI = nE - nrI \qquad \text{Subtract } nrI.$$

$$RI = n(E - rI) \qquad \text{Distributive property; factor out } n.$$

$$\frac{RI}{E - rI} = n \qquad \text{Divide by } E - rI. \ ∎$$

Chalkboard Exercise

Solve for r.

$$I = \frac{nE}{R + nr}$$

Answer: $r = \dfrac{nE - IR}{In}$

CAUTION Refer to the steps in Examples 2 and 3 that use the distributive property. This is a step that often gives students difficulty. Remember that the variable for which you are solving *must* be a factor on one side of the equation so that in the last step, both sides are divided by the remaining factor there. The *distributive property* allows us to perform this factorization.

PROBLEM SOLVING

We are now able to solve problems that translate into equations with rational expressions. The strategy for solving these problems is the same as we have used in earlier chapters. Notice how we continue to use the six steps for problem solving from Chapter 2.

3 ▶ Solve applications using proportions.

▶   A **ratio** is a comparison of two quantities with the same units. The ratio of a to b may be written in any of the following ways.

$$a \text{ to } b, \qquad a:b, \qquad \text{or} \qquad \frac{a}{b}$$

Ratios are usually written as quotients in algebra. A **proportion** is a statement that two ratios are equal. Proportions are a useful and important type of rational equation.

EXAMPLE 4
Solving a Proportion

In 1990, 15 of every 100 Americans had no health insurance coverage. The population at that time was about 246 million. How many million had no health insurance?

Let x = the number (in millions) who had no health insurance. To get an equation, we set up a proportion. The ratio x to 246 should equal the ratio 15 to 100. Write the proportion and solve the equation.

$$\frac{15}{100} = \frac{x}{246}$$

$$24{,}600\left(\frac{15}{100}\right) = 24{,}600\left(\frac{x}{246}\right) \qquad \text{Multiply by a common denominator.}$$

$$246(15) = 100x \qquad \text{Simplify.}$$

$$3690 = 100x$$

$$x = 36.9$$

There were 36.9 million Americans with no health insurance. Check that this number compared to 246 million is equivalent to 15/100. ■

A comparison of two quantities with different units is called a **rate.** Two equal rates can be expressed as a proportion. It is important to be sure the two rates are expressed with the units in the same order.

EXAMPLE 5
Solving a Proportion
Involving Rates

Marissa's car uses 10 gallons of gas to travel 210 miles. She has 5 gallons of gas in the car and she wants to know how much more gas she will need to drive 640 miles. If we assume the car continues to use gas at the same rate, how many more gallons will she need?

We can set up a proportion.

Let x = the additional amount of gas needed.

$$\underset{\text{miles}}{\overset{\text{gallons}}{}} \quad \frac{10}{210} = \frac{5 + x}{640} \quad \underset{\text{miles}}{\overset{\text{gallons}}{}}$$

The LCD is $10 \cdot 21 \cdot 64$.

$$10 \cdot 21 \cdot 64 \left(\frac{10}{210}\right) = 10 \cdot 21 \cdot 64 \left(\frac{5 + x}{640}\right)$$

$$64 \cdot 10 = 21(5 + x)$$

$$30.5 = 5 + x \qquad \text{Divide 640 by 21; round to the nearest tenth.}$$

$$25.5 = x$$

Marissa will need 25.5 more gallons of gas. Check the answer in the words of the problem. The 25.5 gallons plus the 5 gallons equals 30.5 gallons.

$$\frac{30.5}{640} = .0476$$

$$\frac{10}{210} = .0476$$

Since the rates are equal, the solution is correct. ■

4 ▶ Solve applications about distance, rate, and time.

▶ A familiar example of a rate is speed, which is the ratio of distance to time. The next examples use the distance formula $d = rt$ introduced in Chapter 2.

E X A M P L E 6

Solving a Problem About Distance, Rate, and Time

At the airport, Cheryl and Bill are walking to the gate (at the same speed) to catch their flight to Akron, Ohio. Since Bill wants a window seat, he steps onto the moving sidewalk and continues to walk while Cheryl uses the stationary sidewalk. If the sidewalk moves at 1 meter per second and Bill saves 50 seconds covering the 300-meter distance, what is their walking speed?

Let x represent their walking speed. Then Cheryl travels at x meters per second and Bill travels at $x + 1$ meters per second. Since Bill's time is 50 seconds less than Cheryl's time, express their times in terms of the known distances and the variable rates. Start with $d = rt$ and divide both sides by r to get

$$t = \frac{d}{r}.$$

For Cheryl, distance is 300 meters and the rate is x. Cheryl's time is

$$t = \frac{d}{r} = \frac{300}{x}.$$

Bill goes 300 meters at a rate of $x + 1$, so his time is

$$t = \frac{d}{r} = \frac{300}{x + 1}.$$

▶ **TEACHING TIP**
In a problem like the one in Example 6, the equation for a distance-rate-time problem will come from the *time* portion of the table. Quite simply, the equation will involve addition, subtraction, or setting the two time quantities equal to each other.

Try to form the equation from a literal translation of the time portion of the problem. ◀

This information is summarized in the following chart.

	d	r	t
Cheryl	300	x	$\dfrac{300}{x}$
Bill	300	$x + 1$	$\dfrac{300}{x + 1}$

Now use the information given in the problem about the times to write an equation.

$$\underset{\text{time}}{\text{Bill's}} \quad \underset{}{\text{is}} \quad \underset{\text{time}}{\text{Cheryl's}} \quad \underset{\text{seconds.}}{\text{less 50}}$$

$$\frac{300}{x + 1} = \frac{300}{x} - 50$$

The common denominator is $x(x + 1)$. Multiply both sides by $x(x + 1)$.

$$x(x + 1)\left(\frac{300}{x + 1}\right) = x(x + 1)\left(\frac{300}{x} - 50\right)$$

$$300x = 300(x + 1) - 50x(x + 1)$$

$$300x = 300x + 300 - 50x^2 - 50x$$

$$0 = 50x^2 + 50x - 300 \qquad \text{Subtract } 300x; \text{ multiply by } -1.$$

$$0 = x^2 + x - 6 \qquad \text{Divide both sides by } 50.$$

$$0 = (x + 3)(x - 2) \qquad \text{Factor.}$$

$$x + 3 = 0 \qquad \text{or} \qquad x - 2 = 0$$

$$x = -3 \qquad \text{or} \qquad x = 2$$

Discard the negative answer. Their walking speed is 2 meters per second. Check the solution in the words of the original problem. ∎

5 ▶ Solve applications about work rates.

▶ Problems about work are closely related to the distance problems we discussed in Section 2.4.

PROBLEM SOLVING

People work at different rates. Let the letters r, t, and A represent the rate at which the work is done, the time required, and the amount of work accomplished, respectively. Then $A = rt$. Notice the similarity to the distance formula, $d = rt$. The amount of work is often measured in terms of jobs accomplished. Thus, if 1 job is completed, $A = 1$, and the formula gives

$$1 = rt$$

$$r = \frac{1}{t}$$

as the rate.

Rate of Work

If a job can be accomplished in t units of time, then the rate of work is

$$\frac{1}{t} \text{ job per unit of time.}$$

In solving a work problem, we begin by using this fact to express all rates of work.

EXAMPLE 7

Solving a Problem About Work

Chalkboard Exercise

Stan needs 45 minutes to do the dishes, while Bobbie can do them in 30 minutes. How long will it take them if they work together?

Answer: 18 minutes

▶ **TEACHING TIP**

Be sure that students know that the variable is in the numerator of the equation for an example like this. ◀

◇ **48 miles per hour**

Lindsay and Kareem are working on a neighborhood cleanup. Kareem can clean up all the trash in the area in 7 hours, while Lindsay can do the same job in 5 hours. How long will it take them if they work together?

Let x = the number of hours it will take the two people working together. Just as we made a chart for the distance formula, $d = rt$, we can make a chart here for $A = rt$, with $A = 1$. Since $A = 1$, the rate for each person will be $1/t$, where t is the time it takes each person to complete the job alone. For example, since Kareem can clean up all the trash in 7 hours, his rate is $1/7$ of the job per hour. Similarly, Lindsay's rate is $1/5$ of the job per hour. Fill in the chart as shown.

Worker	Rate	Time Working Together	Fractional Part of the Job Done
Kareem	$\dfrac{1}{7}$	x	$\dfrac{1}{7}x$
Lindsay	$\dfrac{1}{5}$	x	$\dfrac{1}{5}x$

Since together they complete 1 job, the sum of the fractional parts accomplished by each of them should equal 1.

$$\underset{\text{by Kareem}}{\text{part done}} + \underset{\text{by Lindsay}}{\text{part done}} = \underset{\text{job}}{1 \text{ whole}}$$

$$\frac{1}{7}x + \frac{1}{5}x = 1$$

Solve this equation. The least common denominator is 35.

$$35\left(\frac{1}{7}x + \frac{1}{5}x\right) = 35 \cdot 1$$

$$5x + 7x = 35$$

$$12x = 35$$

$$x = \frac{35}{12}$$

Working together, Kareem and Lindsay can do the entire job in 35/12 hours, or 2 hours and 55 minutes. Check this result in the original problem. ■

◇ **CONNECTIONS** ◇

It is very common for people to average two rates by adding and then dividing by two, as we find the average for two numbers. However, a rate is a different kind of average, and we cannot average rates as we do numbers. If a car travels from A to B at 40 miles per hour and returns at 60 miles per hour, what is its rate for the entire trip? The correct answer is not 50 miles per hour! Using the distance-rate-time relationship and letting x = the distance between A and B, we can simplify a complex fraction to find the correct answer.

$$\underset{\text{entire trip}}{\text{Rate for}} = \frac{\text{Total distance}}{\text{Total time}} = \frac{x + x}{\dfrac{x}{40} + \dfrac{x}{60}}$$

FOR DISCUSSION OR WRITING

Simplify the complex fraction to find the rate for the entire trip. Notice that x (the distance) in the problem was eliminated. Why do you suppose the distance does not matter?

4.7 EXERCISES

1. (a)   2. (c)   3. (a)
4. (d)   5. 1.349
6. 65.625
7. 24   8. $\dfrac{25}{4}$

9. $G = \dfrac{Fd^2}{Mm}$

10. $M = \dfrac{Fd^2}{Gm}$

11. $a = \dfrac{bc}{c + b}$

12. $b = \dfrac{ac}{c - a}$ or
$b = \dfrac{-ac}{a - c}$

◉ *In Exercises 1–4, a familiar formula is given. Give the letter of the choice that is an equivalent form of the given formula (involving a rational expression).*

1. $d = rt$ (motion)

   (a) $r = \dfrac{d}{t}$ **(b)** $t = \dfrac{r}{d}$

   (c) $r = \dfrac{t}{d}$ **(d)** $d = \dfrac{t}{r}$

2. $I = prt$ (simple interest)

   (a) $p = \dfrac{r}{It}$ **(b)** $r = \dfrac{It}{p}$

   (c) $p = \dfrac{I}{rt}$ **(d)** $t = \dfrac{pr}{I}$

3. $A = \dfrac{1}{2}bh$ (mathematics)

   (a) $h = \dfrac{2A}{b}$ **(b)** $h = \dfrac{b}{2A}$

   (c) $b = \dfrac{2}{Ah}$ **(d)** $b = \dfrac{2h}{A}$

4. $PVT = pvt$ (chemistry)

   (a) $t = \dfrac{pV}{PvT}$ **(b)** $p = \dfrac{vt}{PVT}$

   (c) $v = \dfrac{pt}{PVT}$ **(d)** $p = \dfrac{PVT}{vt}$

Solve the problem. See Example 1.

5. The gravitational force between two masses is given by

$$F = \dfrac{GMm}{d^2}.$$

Find M if $F = 10$, $G = 6.67 \times 10^{-11}$, $m = 1$, and $d = 3 \times 10^{-6}$.

6. A gas law in chemistry says that

$$\dfrac{PV}{T} = \dfrac{pv}{t}.$$

Suppose that $T = 300$, $t = 350$, $V = 9$, $P = 50$, and $v = 8$. Find p.

7. In work with electric circuits, the formula

$$\dfrac{1}{a} = \dfrac{1}{b} + \dfrac{1}{c}$$

occurs. Find b if $a = 8$ and $c = 12$.

8. A formula from anthropology says that

$$c = \dfrac{100b}{L}.$$

Find L if $c = 80$ and $b = 5$.

Solve the formula for the specified variable. See Examples 2 and 3.

9. $F = \dfrac{GMm}{d^2}$ for G (physics)

10. $F = \dfrac{GMm}{d^2}$ for M (physics)

11. $\dfrac{1}{a} = \dfrac{1}{b} + \dfrac{1}{c}$ for a (electricity)

12. $\dfrac{1}{a} = \dfrac{1}{b} + \dfrac{1}{c}$ for b (electricity)

◉ CONCEPTUAL   ✎ WRITING   ▲ CHALLENGING   ▦ SCIENTIFIC CALCULATOR   ▦ GRAPHICS CALCULATOR

13. $\dfrac{PV}{T} = \dfrac{pv}{t}$ for v (chemistry)

14. $\dfrac{PV}{T} = \dfrac{pv}{t}$ for T (chemistry)

15. $\dfrac{E}{e} = \dfrac{R + r}{r}$ for e (engineering)

16. $a = \dfrac{V - v}{t}$ for V (physics)

17. $A = \dfrac{1}{2}h(B + b)$ for b (mathematics)

18. $S = \dfrac{n}{2}(a + \ell)d$ for n (mathematics)

19. $A = \dfrac{Rr}{R + r}$ for R (engineering)

20. $\dfrac{E}{e} = \dfrac{R + r}{r}$ for r (engineering)

⊙ **21.** To solve the equation $m = \dfrac{ab}{a - b}$ for a, what is the first step?

⊙ **22.** Suppose you get the equation

$$rp - rq = p + q$$

to be solved for r. What is the next step?

Use proportions to solve the following problems in your head.

23. In a mathematics class, 3 of every 4 students is a girl. If there are 20 students in the class, how many are girls? How many are boys?

24. In a certain southern state, sales tax on a purchase of $1.50 is $.12. What is the sales tax on a purchase of $6.00?

25. If Marin can mow her yard in 2 hours, what is her rate (in job per hour)?

26. A bus traveling from Atlanta to Detroit averages 50 miles per hour and takes 14 hours to make the trip. How far is it from Atlanta to Detroit?

Use the figure to answer the following questions in Exercises 27–30. Round numbers from the graph to the nearest 1/2 billion.

27. In 1986, what was the ratio of the box office revenue to home video revenue?

28. What was the home video revenue in 1992? in 1993?

29. What was the ratio of box office revenue to home video revenue in 1993? in 1992?

30. What home video revenue in 1993 would have been in the same proportion to box office revenue as was the case in 1992?

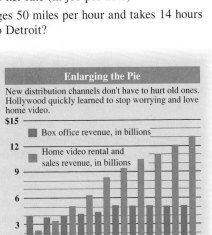

13. $v = \dfrac{PVt}{pT}$ **14.** $T = \dfrac{PVt}{pv}$

15. $e = \dfrac{Er}{R + r}$

16. $V = at + v$

17. $b = \dfrac{2A}{h} - B$

or $b = \dfrac{2A - Bh}{h}$

18. $n = \dfrac{2S}{(a + \ell)d}$

19. $R = \dfrac{Ar}{r - A}$

20. $r = \dfrac{eR}{E - e}$

21. Multiply both sides by $a - b$.

22. Factor out r on the left.

23. 15 girls, 5 boys

24. $.48

25. $\dfrac{1}{2}$ job per hour

26. 700 miles   **27.** $\dfrac{4}{5}$

28. $12 billion; $13 billion

29. $\dfrac{5}{13}$; $\dfrac{5}{12}$

30. 12 billion

 CONCEPTUAL   WRITING   ▲ CHALLENGING   SCIENTIFIC CALCULATOR   GRAPHICS CALCULATOR

31. $300   32. $70,880
33. 25,000   34. 20,000
There are two ways to set
up this problem.
35. Find the distance from
Dr. Dawson's office to his
home.
Method 1: Let x represent
his time riding his bike to
the office.
Method 2: Let x represent
the distance.
36. *Method 1:*

	d	r	t
bike	$12x$	12	x
car	$36\left(x - \dfrac{1}{4}\right)$	36	$x - \dfrac{1}{4}$

Method 2:

	d	r	t
bike	x	12	$\dfrac{x}{12}$
car	x	36	$\dfrac{x}{36}$

37. *Method 1:*
$$12x = 36\left(x - \frac{1}{4}\right);$$

Method 2: $\dfrac{x}{36} = \dfrac{x}{12} - \dfrac{1}{4}$

38. *Method 1:* $x = \dfrac{3}{8}$ hour;

Method 2: $x = \dfrac{9}{2}$ miles

39. The distance is $\dfrac{9}{2}$ or 4.5

miles.   40. $\dfrac{9}{2}$ miles at 12

miles per hour takes

$\dfrac{\frac{9}{2}}{12} = \dfrac{3}{8}$ hour; $\dfrac{9}{2}$ miles

at 36 miles per hour takes

$\dfrac{\frac{9}{2}}{36} = \dfrac{1}{8}$ hour. $\dfrac{3}{8} - \dfrac{1}{4} = \dfrac{1}{8}$

as required.
41. *Step 1:* Find the
distance from Tulsa to
Detroit. Let x represent
that distance.
Step 2:

d	r	t
x	50	$\dfrac{x}{50}$
x	60	$\dfrac{x}{60}$

Use a proportion to solve the problem. See Examples 4 and 5.

31. In 1991, a single person could expect to pay $4,500 in income taxes on an adjusted gross income of $30,000. How much more tax should the person expect to pay if her adjusted gross income increased $2000, knowing that this would not move her into a higher tax bracket?

32. In 1990, the top-performing mutual fund would have produced income of $44,300 on an investment of $100,000. If this investment had been increased to $260,000, how much more income would have been produced?

33. Biologists tagged 500 fish in a lake on January 1. On February 1 they returned and collected a random sample of 400 fish, 8 of which had been previously tagged. Approximately how many fish does the lake have based on this experiment?

34. Suppose that in the experiment of Exercise 33, 10 of the previously tagged fish were collected on February 1. What would be the estimate of the fish population?

◆ **MATHEMATICAL CONNECTIONS** (Exercises 35–40) ◆

The six steps for solving an applied problem given in Chapter 2 will help you to solve any applied problem. In these exercises, you are asked to apply the six steps to the following problem.

If Dr. Dawson rides his bike to his office, he averages 12 miles per hour. If he drives his car, he averages 36 miles per hour. His time driving is $\frac{1}{4}$ hour less than his time riding his bike. How far is his office from home?

35. *Step 1*   Determine what you are asked to find. Choose a variable and write down what it represents.

36. *Step 2*   Write down any other pertinent information. Express it using the variable, if appropriate. Use a chart or a diagram, if appropriate.

37. *Step 3*   Write an equation.

38. *Step 4*   Solve the equation.

39. *Step 5*   Answer the question(s) of the problem.

40. *Step 6*   Check your answer in the words of the original problem.

◆

Go through the six steps in Exercises 35–40 to solve the problem. See Example 6. Give the answers for each step in Exercise 41.

41. Driving from Tulsa to Detroit, Jeff averaged 50 miles per hour. He figured that if he had averaged 60 miles per hour, his driving time would have decreased 3 hours. How far is it from Tulsa to Detroit?

42. A private plane traveled from San Francisco to a secret rendezvous. It averaged 200 miles per hour. On the return trip, the average speed was 300 miles per hour. If the total traveling time was 4 hours, how far from San Francisco was the secret rendezvous?

43. Johnny averages 30 miles per hour when he drives on the old highway to his favorite fishing hole, while he averages 50 miles per hour when most of his route is on the interstate. If both routes are the same length, and he saves 2 hours by traveling on the interstate, how far away is the fishing hole?

44. On the first part of a trip to Carmel traveling on the freeway, Marge averaged 60 miles per hour. On the rest of the trip, which was 10 miles longer than the first part, she averaged 50 miles per hour. Find the total distance to Carmel if the second part of the trip took 30 minutes more than the first part.

45. While on vacation, Jim and Annie decided to drive all day. During the first part of their trip on the highway they averaged 60 miles per hour. When they got to Houston, traffic caused them to average only 30 miles per hour. The distance they drove in Houston was 100 miles less than their distance on the highway. What was their total driving distance, if they spent 50 minutes more on the highway than they did in Houston?

46. Kellen's boat goes 12 miles per hour. Find the speed of the current of the river if she can go 6 miles upstream in the same amount of time she can go 10 miles downstream.

	Distance	Rate	Time
Downstream		$12 + x$	
Upstream		$12 - x$	

⊙ 47. Explain the similarities between the methods of solving problems about distance, rate, and time and problems about work.

⊙ 48. If one person takes 3 hours to do a job and another takes 4 hours to do the same job, why is "$3\frac{1}{2}$ hours" *not* a valid answer to the problem "How long will it take for them to do the job working together?"?

Solve the problem. See Example 7.

49. Butch can paint a room in 8 hours, but it takes his wife, Peggy, only 5 hours to paint the same room. How long will it take them to paint the room if they work together?

50. Bill and Julie want to pick up the mess that their grandson, J.W., has made in his playroom. Bill could do it in 15 minutes working alone. Julie, working alone, could clean it in 12 minutes. How long will it take them if they work together?

51. Bernard and Carolyn Goldstein are refinishing a table. Working alone, Bernard could do the job in 7 hours. If the two work together, the job takes 5 hours. How long will it take Carolyn to refinish the table if she works alone?

52. Mike can paint a room in 6 hours working alone. If Dee helps him, the job takes 4 hours. How long would it take Dee to do the job if she worked alone?

53. A winery has a barrel to hold chardonnay. An inlet pipe can fill the barrel in 9 hours, while an outlet pipe can empty it in 12 hours. How long will it take to fill the barrel if both the outlet and the inlet pipes are open?

54. If a vat of acid can be filled by an inlet pipe in 10 hours, and emptied by an outlet pipe in 20 hours, how long will it take to fill the vat if both pipes are open?

55. An inlet pipe can fill a barrel of ink in 6 hours, while an outlet pipe can empty it in 8 hours. Through an error both pipes are left on. How long will it take for the barrel to fill?

56. Suppose that Hortense and Mort can clean their entire house in 7 hours, while their toddler, Mimi, just by being around, can completely mess it up in only 2 hours. If Hortense and Mort clean the house while Mimi is at her grandma's, and then start cleaning up after Mimi the minute she gets home, how long does it take from the time Mimi gets home until the whole place is a shambles?

Step 3: $\dfrac{x}{60} = \dfrac{x}{50} - 3$

Step 4: $x = 900$
Step 5: The distance is 900 miles.
Step 6: Check: 900 miles at 50 miles per hour takes 18 hours; 900 miles at 60 miles per hour takes 15 hours; $15 = 18 - 3$ as required. **42.** 480 miles
43. 150 miles
44. 190 miles
45. 200 miles
46. 3 miles per hour
48. The answer *must* be less than 3 hours.
49. $\dfrac{40}{13}$ or $3\dfrac{1}{13}$ hours
50. $\dfrac{20}{3}$ or $6\dfrac{2}{3}$ minutes
51. $17\dfrac{1}{2}$ hours
52. 12 hours
53. 36 hours
54. 20 hours
55. 24 hours
56. $\dfrac{14}{5}$ or $2\dfrac{4}{5}$ hours

57. $x = \dfrac{7}{2}$; $AC = 8$;
$DF = 12$

58. $y = 2$;   side $NP = 8$,
side $MP = \dfrac{16}{3}$;   side
$RT = 6$, side $QT = 4$

59. 9   60. 25   61. 4
62. −10   63. 3   64. .5

In geometry, it is shown that two triangles with corresponding angles equal, called similar triangles, *have corresponding sides proportional. For example, in the figure, angle A = angle D, angle B = angle E, and angle C = angle F, so the triangles are similar. Then the following ratios of corresponding sides are equal.*

$$\frac{4}{6} = \frac{6}{9} = \frac{2x + 1}{2x + 5}$$

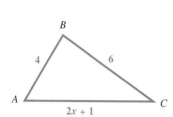
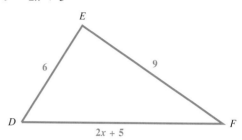

57. Solve for x using the proportion given above to find the lengths of the third sides of the triangles.

58. Suppose the triangles shown below are similar. Find y and the lengths of the two longest sides of each triangle.

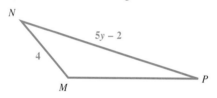
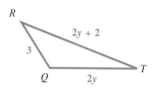

REVIEW EXERCISES

Evaluate the root. See Section 1.2.

59. $\sqrt{81}$ 60. $\sqrt{625}$ 61. $\sqrt[3]{64}$ 62. $-\sqrt[3]{1000}$ 63. $\sqrt[4]{81}$ 64. $\sqrt{.25}$

CHAPTER 4 SUMMARY

KEY TERMS

4.1 rational expression
reciprocals

4.2 least common
denominator

4.3 complex fraction

4.5 synthetic division

4.7 ratio
proportion
rate

QUICK REVIEW

CONCEPTS	EXAMPLES
4.1 MULTIPLICATION AND DIVISION OF RATIONAL EXPRESSIONS	
Writing Rational Expressions in Lowest Terms	Write in lowest terms.
Step 1 Factor the numerator and the denominator completely.	$$\frac{2x + 8}{x^2 - 16} = \frac{2(x + 4)}{(x - 4)(x + 4)}$$
Step 2 Apply the fundamental principle.	$$= \frac{2}{x - 4}$$

CONCEPTS	EXAMPLES
Multiplying Rational Expressions Factor numerators and denominators. Replace all pairs of common factors in numerators and denominators by 1. Multiply remaining factors in the numerator and in the denominator.	Multiply. $\dfrac{x^2 + 2x + 1}{x^2 - 1} \cdot \dfrac{5}{3x + 3}$ $= \dfrac{(x + 1)^2}{(x - 1)(x + 1)} \cdot \dfrac{5}{3(x + 1)}$ $= \dfrac{5}{3(x - 1)}$
Dividing Rational Expressions Multiply the first by the reciprocal of the second.	Divide. $\dfrac{2x + 5}{x - 3} \div \dfrac{2x^2 + 3x - 5}{x^2 - 9}$ $= \dfrac{2x + 5}{x - 3} \cdot \dfrac{(x + 3)(x - 3)}{(2x + 5)(x - 1)}$ $= \dfrac{x + 3}{x - 1}$

4.2 ADDITION AND SUBTRACTION OF RATIONAL EXPRESSIONS

Adding or Subtracting Rational Expressions If the denominators are the same, add or subtract the numerators. Place the result over the common denominator. If the denominators are different, write all rational expressions with the LCD. Then add or subtract the like fractions. Be sure answer is in lowest terms.	Subtract. $\dfrac{1}{x + 6} - \dfrac{3}{x + 2}$ $= \dfrac{x + 2}{(x + 6)(x + 2)} - \dfrac{3(x + 6)}{(x + 6)(x + 2)}$ $= \dfrac{x + 2 - 3(x + 6)}{(x + 6)(x + 2)}$ $= \dfrac{x + 2 - 3x - 18}{(x + 6)(x + 2)}$ $= \dfrac{-2x - 16}{(x + 6)(x + 2)}$

4.3 COMPLEX FRACTIONS

Simplifying Complex Fractions *Method 1* Simplify the numerator and denominator separately, as much as possible. Then multiply the numerator by the reciprocal of the denominator. Write the answer in lowest terms.	Simplify the complex fraction. *Method 1* $\dfrac{\dfrac{1}{2} + \dfrac{1}{3}}{\dfrac{1}{4} - \dfrac{1}{2}} = \dfrac{\dfrac{3}{6} + \dfrac{2}{6}}{\dfrac{1}{4} - \dfrac{2}{4}}$ $= \dfrac{\dfrac{5}{6}}{\dfrac{-1}{4}} = \dfrac{5}{6} \cdot \dfrac{4}{-1}$ $= \dfrac{20}{-6} = -\dfrac{10}{3}$
Method 2 Multiply the numerator and denominator of the complex fraction by the least common denominator of all fractions appearing in the complex fraction. Then simplify the results.	*Method 2* $\dfrac{\dfrac{1}{2} + \dfrac{1}{3}}{\dfrac{1}{4} - \dfrac{1}{2}} = \dfrac{12\left(\dfrac{1}{2}\right) + 12\left(\dfrac{1}{3}\right)}{12\left(\dfrac{1}{4}\right) - 12\left(\dfrac{1}{2}\right)}$ $= \dfrac{6 + 4}{3 - 6} = \dfrac{10}{-3} = -\dfrac{10}{3}$

CONCEPTS	EXAMPLES

4.4 DIVISION OF POLYNOMIALS

Dividing by a Monomial
To divide a polynomial by a monomial, divide each term in the polynomial by the monomial, and then write each fraction in lowest terms.

Divide. $\dfrac{2x^3 - 4x^2 + 6x - 8}{2x}$

$$= \dfrac{2x^3}{2x} - \dfrac{4x^2}{2x} + \dfrac{6x}{2x} - \dfrac{8}{2x}$$

$$= x^2 - 2x + 3 - \dfrac{4}{x}$$

Dividing by a Polynomial
Use the "long division" process.

Divide. $\dfrac{m^3 - m^2 + 2m + 5}{m + 1}$

$$
\begin{array}{r}
m^2 - 2m + 4 \\
m + 1\overline{)m^3 - m^2 + 2m + 5} \\
\underline{m^3 + m^2} \\
-2m^2 + 2m \\
\underline{-2m^2 - 2m} \\
4m + 5 \\
\underline{4m + 4} \\
1
\end{array}
$$

The quotient is

$$m^2 - 2m + 4 + \dfrac{1}{m + 1}.$$

4.6 EQUATIONS WITH RATIONAL EXPRESSIONS

To solve an equation involving rational expressions, multiply all the terms in the equation by the least common denominator. Then solve the resulting equation. Each potential solution must be checked to make sure that no denominator in the original equation is 0.

Solve for x.

$$\frac{1}{x} + x = \frac{26}{5}$$

$$5 + 5x^2 = 26x \qquad \text{Multiply by } 5x.$$

$$5x^2 - 26x + 5 = 0$$

$$(5x - 1)(x - 5) = 0$$

$$x = \frac{1}{5} \qquad \text{or} \qquad x = 5$$

Both check. The solution set is $\left\{\dfrac{1}{5}, 5\right\}$.

4.7 APPLICATIONS

To solve a formula for a particular variable, get that variable alone on one side by following the method described in Section 4.6.

Solve for L.

$$c = \frac{100b}{L}$$

$$cL = 100b \qquad \text{Multiply by } L.$$

$$L = \frac{100b}{c} \qquad \text{Divide by } c.$$

CONCEPTS	EXAMPLES
If an applied problem translates into an equation with rational expressions, solve the equation using the method described in Section 4.6.	If the 6.4 ounce tube of Crest toothpaste costs $1.89, what should the 8.2 ounce tube cost? Let x represent the unknown cost. Use a proportion. $$\frac{6.4}{1.89} = \frac{8.2}{x}$$ $$1.89x\left(\frac{6.4}{1.89}\right) = 1.89x\left(\frac{8.2}{x}\right)$$ $$6.4x = 15.498$$ $$x = 2.42$$ The 8.2 ounce tube should cost $2.42. Check by comparing the two ratios.

CHAPTER 4 REVIEW EXERCISES

◉ **1.** Write a few sentences defining the following terms and distinguishing among them: rational expression, fraction, complex fraction.

[4.1] *Find all real numbers that make the rational expression undefined.*

2. $\dfrac{1}{p + 3}$

3. $\dfrac{-7}{3t + 18}$

4. $\dfrac{5r + 17}{r^2 - 7r + 10}$

✎ **5.** List the steps you would use to write $\dfrac{x^2 + 3x}{5x + 15}$ in lowest terms.

Write in lowest terms.

6. $\dfrac{55m^4n^3}{10m^5n}$

7. $\dfrac{12x^2 + 6x}{24x + 12}$

8. $\dfrac{y^2 + 3y - 10}{y^2 - 5y + 6}$

9. $\dfrac{25m^2 - n^2}{25m^2 - 10mn + n^2}$

10. $\dfrac{r - 2}{4 - r^2}$

◉ **11.** What is meant by the reciprocal of a rational expression?
✎

Multiply or divide. Write the answer in lowest terms.

12. $\dfrac{25p^3q^2}{8p^4q} \div \dfrac{15pq^2}{16p^5}$

13. $\dfrac{(2y + 3)^2}{5y} \cdot \dfrac{15y^3}{4y^2 - 9}$

14. $\dfrac{w^2 - 16}{w} \cdot \dfrac{3}{4 - w}$

15. $\dfrac{y^2 + 2y}{y^2 + y - 2} \div \dfrac{y - 5}{y^2 + 4y - 5}$

16. $\dfrac{z^2 - z - 6}{z - 6} \cdot \dfrac{z^2 - 6z}{z^2 + 2z - 15}$

17. $\dfrac{m^3 - n^3}{m^2 - n^2} \div \dfrac{m^2 + mn + n^2}{m + n}$

Answers (margin):

2. -3 3. -6
4. $2, 5$ 6. $\dfrac{11n^2}{2m}$
7. $\dfrac{x}{2}$ 8. $\dfrac{y + 5}{y - 3}$
9. $\dfrac{5m + n}{5m - n}$ 10. $\dfrac{-1}{2 + r}$
(There are other ways.)
12. $\dfrac{10p^3}{3q}$
13. $\dfrac{3y^2(2y + 3)}{2y - 3}$
14. $\dfrac{-3(w + 4)}{w}$
15. $\dfrac{y(y + 5)}{y - 5}$
16. $\dfrac{z(z + 2)}{z + 5}$ 17. 1

RESOURCES FOR REVIEW

 ITM pp. 366–389 ISM pp. 295–308
IAM p. 19

 SSG pp. 102–134
SSM pp. 189–202

 TEST GENERATOR
DOS Windows MAC

◉ CONCEPTUAL ✎ WRITING ▲ CHALLENGING ▦ SCIENTIFIC CALCULATOR ▦ GRAPHICS CALCULATOR

18. The terms x and 5 in the denominator cannot be separated into denominators of two fractions. The correct simplified form is x.

20. $60x$ 21. $96b^5$

22. $9r^2(3r + 1)$

23. $45(2k + 1)$

24. $(3x - 1)(2x + 5)$ · $(3x + 4)$

25. $\dfrac{16z - 3}{2z^2}$

26. $\dfrac{8}{t - 2}$ or $\dfrac{-8}{2 - t}$

27. 12   28. $\dfrac{71}{30(a + 2)}$

29. $\dfrac{3x + 2}{x - 5}$

30. $\dfrac{13r^2 + 5rs}{(5r + s)(2r - s)(r + s)}$

31. $\dfrac{17}{12}$ 32. $\dfrac{5}{7}$

33. no; denominator; numerators; least common denominator

34. .67, .71, .75; $\dfrac{2}{3} < \dfrac{5}{7} < \dfrac{3}{4}$

It lies between the original two fractions.

35. numerators; denominators

36. $\dfrac{2}{5}$

◉ **18.** What is *wrong* with the following work?

$$\frac{x^2 + 5x}{x + 5} = \frac{x^2}{x} + \frac{5x}{5} = x + 5$$

What is the *correct* simplified form?

◉ **19.** How would you explain finding the least common denominator to a classmate?

[4.2] *Find the least common denominator for the group of rational expressions.*

20. $\dfrac{7x}{12}, \ \dfrac{8}{15x}$

21. $\dfrac{5a}{32b^3}, \ \dfrac{31}{24b^5}$

22. $\dfrac{17}{9r^2}, \ \dfrac{5r - 3}{3r + 1}$

23. $\dfrac{14}{6k + 3}, \ \dfrac{7k^2 + 2k + 1}{10k + 5}, \ \dfrac{-11k}{18k + 9}$

24. $\dfrac{4x - 9}{6x^2 + 13x - 5}, \ \dfrac{x + 15}{9x^2 + 9x - 4}$

Add or subtract as indicated.

25. $\dfrac{8}{z} - \dfrac{3}{2z^2}$

26. $\dfrac{3}{t - 2} - \dfrac{5}{2 - t}$

27. $\dfrac{5y + 13}{y + 1} - \dfrac{1 - 7y}{y + 1}$

28. $\dfrac{6}{5a + 10} + \dfrac{7}{6a + 12}$

29. $\dfrac{3x}{x - 5} + \dfrac{20}{x^2 - 25} + \dfrac{2}{x + 5}$

30. $\dfrac{3r}{10r^2 - 3rs - s^2} + \dfrac{2r}{2r^2 + rs - s^2}$

◆ **MATHEMATICAL CONNECTIONS** (Exercises 31–36) ◆

A simple method for finding a rational number between any two given rational numbers was discovered by an eighth grade student, Robert McKay.* His teacher, Laurence Sherzer, after verifying that the method is indeed correct in general, named it McKay's theorem. Work the following exercises in order to develop McKay's theorem for the rational numbers $\frac{3}{4}$ and $\frac{2}{3}$ and distinguish it from addition of rational numbers.

31. Add the two rational numbers in the usual way.

32. Form a new fraction by adding the numerators and adding the denominators.

33. Are the answers to Exercises 31 and 32 the same? Complete this sentence: The correct way to add two rational numbers is to write them with the same _____ , then add their _____ and use the _____ as the denominator of the sum.

34. Use a calculator to get a decimal equivalent (to the nearest hundredth) for the answer to Exercise 32 and the original two rational numbers. Write the three rational numbers in increasing order. What is the position of the new rational number? This is an example of McKay's theorem.

35. Complete the following statement of McKay's theorem: To find a number between two positive rational numbers, add the _____ and add the _____ .

36. Use McKay's theorem to find a rational number between $\dfrac{1}{3}$ and $\dfrac{1}{2}$.

---

* From *The Mathematics Teacher*, March 1973.

[4.3] *Simplify the complex fraction.*

37. $\dfrac{\dfrac{3}{t} + 2}{\dfrac{4}{t} - 7}$

38. $\dfrac{\dfrac{4m^5n^6}{mn}}{\dfrac{8m^7n^3}{m^4n^2}}$

39. $\dfrac{\dfrac{r + 2s}{20}}{\dfrac{8r + 16s}{5}}$

40. $\dfrac{3x^{-1} - 5}{6 + x^{-1}}$

41. $\dfrac{\dfrac{3}{p} - \dfrac{2}{q}}{\dfrac{9q^2 - 4p^2}{qp}}$

[4.4] *Divide.*

42. $\dfrac{18x^2 - 32x + 12}{12x}$

43. $\dfrac{4y^3 - 12y^2 + 5y}{4y}$

44. $\dfrac{15k^2 + 11k - 17}{3k - 2}$

45. $\dfrac{5p^4 + 15p^3 - 33p^2 - 9p + 18}{5p^2 - 3}$

46. $\dfrac{2p^3 + 9p^2 + 27}{2p - 3}$

47. $\dfrac{12y^4 + 7y^2 - 2y + 1}{3y^2 + 1}$

[4.5] *Use synthetic division to perform the division.*

48. $\dfrac{3p^2 - p - 2}{p - 1}$

49. $\dfrac{10k^2 - 3k - 15}{k + 2}$

50. $(2k^3 - 5k^2 + 12) \div (k - 3)$

51. $(-a^4 + 19a^2 + 18a + 15) \div (a + 4)$

Use synthetic division to decide whether or not -5 is a solution of the equation.

52. $2w^3 + 8w^2 - 14w - 20 = 0$

53. $-3q^4 + 2q^3 + 5q^2 - 9q + 1 = 0$

Use synthetic division to evaluate $P(k)$ for the given value of k.

54. $P(x) = 3x^3 - 5x^2 + 4x - 1; \quad k = -1$

55. $P(z) = z^4 - 2z^3 - 9z - 5; \quad k = 3$

[4.6] *Solve the equation.*

56. $\dfrac{4}{3} - \dfrac{1}{x} = \dfrac{1}{3x}$

57. $\dfrac{1}{t + 4} + \dfrac{1}{2} = \dfrac{3}{2t + 8}$

58. $\dfrac{-5m}{m + 1} + \dfrac{m}{3m + 3} = \dfrac{56}{6m + 6}$

59. $\dfrac{2}{k - 1} - \dfrac{4k + 1}{k^2 - 1} = \dfrac{-1}{k + 1}$

60. $\dfrac{x + 3}{x^2 - 5x + 4} - \dfrac{1}{x} = \dfrac{2}{x^2 - 4x}$

61. $\dfrac{5}{x + 2} + \dfrac{3}{x + 3} = \dfrac{x}{x^2 + 5x + 6}$

 62. After solving the equation

$$\frac{3}{x - 3} - \frac{2}{x - 2} = \frac{3}{x^2 - 5x + 6},$$

a student got $x = 3$ as her final step. She could not understand why the answer in the back of the book was "0", because she checked her algebra several times and was sure that all her algebraic work was correct. Was she wrong or was the answer in the back of the book wrong? Explain.

Answers (right column):

37. $\dfrac{3 + 2t}{4 - 7t}$

38. $\dfrac{mn^4}{2}$ 39. $\dfrac{1}{32}$

40. $\dfrac{3 - 5x}{6x + 1}$ 41. $\dfrac{1}{3q + 2p}$

42. $\dfrac{3x}{2} - \dfrac{8}{3} + \dfrac{1}{x}$

43. $y^2 - 3y + \dfrac{5}{4}$

44. $5k + 7 + \dfrac{-3}{3k - 2}$

45. $p^2 + 3p - 6$

46. $p^2 + 6p + $

$9 + \dfrac{54}{2p - 3}$

47. $4y^2 + 1 + \dfrac{-2y}{3y^2 + 1}$

48. $3p + 2$

49. $10k - 23 + \dfrac{31}{k + 2}$

50. $2k^2 + k + 3 + \dfrac{21}{k - 3}$

51. $-a^3 + 4a^2 + 3a + 6 + $

$\dfrac{-9}{a + 4}$

52. yes   53. no

54. -13 55. -5 56. $\{1\}$

57. $\{-3\}$ 58. $\{-2\}$

59. $\{0\}$ 60. $\left\{\dfrac{1}{3}\right\}$ 61. $\emptyset$

---

64. $\dfrac{15}{2}$ **65.** $m = \dfrac{Fd^2}{GM}$

66. $\ell = \dfrac{2S}{n} - a$ or

$\ell = \dfrac{2S - na}{n}$

67. $m = \dfrac{Mv - \mu M}{\mu}$ or

$m = \dfrac{Mv}{\mu} - M$

68. $R = \dfrac{nE - Inr}{I}$ or

$R = \dfrac{nE}{I} - nr$

69. \$21.06   **70.** 10,725
71. bus: 50 miles per hour;
train: 60 miles per hour
72. 16 kilometers per hour
73. $\dfrac{24}{5}$ or $4\dfrac{4}{5}$ minutes

74. $\dfrac{18}{5}$ or $3\dfrac{3}{5}$ hours

75. $\dfrac{6m + 5}{3m^2}$

76. $\dfrac{k - 3}{36k^2 + 6k + 1}$

77. $\dfrac{x^2 - 6}{2(2x + 1)}$

78. $\dfrac{x(9x + 1)}{3x + 1}$

79. $k^2 - 7k + 6$
80. $\dfrac{11}{3 - x}$ or $\dfrac{-11}{x - 3}$

81. $\dfrac{1}{3}$ **82.** $\dfrac{s^2 + t^2}{st(s - t)}$

83. $\dfrac{5a^2 + 4ab + 12b^2}{(a + 3b)(a - 2b)(a + b)}$

84. no

⊙ 63. Explain the difference between simplifying the expression $\dfrac{4}{x} + \dfrac{1}{2} - \dfrac{1}{3}$ and solving the

equation $\dfrac{4}{x} + \dfrac{1}{2} = \dfrac{1}{3}$.

[4.7] *Work the problem.*

64. According to a law from physics, $\dfrac{1}{A} = \dfrac{1}{B} + \dfrac{1}{C}$. Find A if $B = 30$ and $C = 10$.

Solve the formula for the specified variable.

65. $F = \dfrac{GMm}{d^2}$ for m (physics)   **66.** $S = \dfrac{n}{2}(a + \ell)$ for ℓ (mathematics)

67. $\mu = \dfrac{Mv}{M + m}$ for m (electronics)   **68.** $I = \dfrac{nE}{R + nr}$ for R (electricity)

Solve the problem.

69. At a certain gasoline station, 3 gallons of unleaded gasoline cost \$4.86. How much would 13 gallons of the same gasoline cost?

70. In a sample of 2000 registered voters, 1430 responded that they were in favor of increasing funding to the local animal shelter. If it is predicted that 15,000 people will vote in the election, how many would vote for for this funding if the survey proves to be accurate?

71. A bus can travel 80 miles in the same time that a train goes 96 miles. The speed of the train is 10 miles per hour faster than the speed of the bus. Find both speeds.

72. A river has a current of 4 kilometers per hour. Find the speed of Lynn McTernan's boat in still water if it goes 40 kilometers downstream in the same time that it takes to go 24 kilometers upstream.

73. A sink can be filled by a cold-water tap in 8 minutes, and filled by the hot-water tap in 12 minutes. How long would it take to fill the sink with both taps open?

74. Jane Brandsma and Mark Cairnes need to sort a pile of bottles at the recycling center. Working alone, Jane could do the entire job in 9 hours, while Mark could do the entire job in 6 hours. How long will it take them if they work together?

MIXED REVIEW EXERCISES

Perform the indicated operations.

75. $\dfrac{2}{m} + \dfrac{5}{3m^2}$

76. $\dfrac{k^2 - 6k + 9}{1 - 216k^3} \cdot \dfrac{6k^2 + 17k - 3}{9 - k^2}$

77. $\dfrac{\dfrac{-3}{x} + \dfrac{x}{2}}{1 + \dfrac{x + 1}{x}}$

78. $\dfrac{9x^2 + 46x + 5}{3x^2 - 2x - 1} \div \dfrac{x^2 + 11x + 30}{x^3 + 5x^2 - 6x}$

79. $(k^3 - 4k^2 - 15k + 18) \div (k + 3)$

80. $\dfrac{9}{3 - x} - \dfrac{2}{x - 3}$

81. $\dfrac{4y + 16}{30} \div \dfrac{2y + 8}{5}$

82. $\dfrac{t^{-2} + s^{-2}}{t^{-1} - s^{-1}}$

83. $\dfrac{4a}{a^2 - ab - 2b^2} - \dfrac{6b - a}{a^2 + 4ab + 3b^2}$

84. Use synthetic division to decide whether 3 is a solution of $7z^3 - z^2 + 5z - 3 = 0$.

Solve.

85. $A = \dfrac{Rr}{R + r}$ for r

86. $1 - \dfrac{5}{r} = \dfrac{-4}{r^2}$

87. $\dfrac{3x}{x - 4} + \dfrac{2}{x} = \dfrac{48}{x^2 - 4x}$

88. The hot-water tap can fill a tub in 20 minutes. The cold-water tap takes 15 minutes to fill the tub. How long would it take to fill the tub with both taps open?

85. $r = \dfrac{AR}{R - A}$

or $r = \dfrac{-AR}{A - R}$

86. $\{1, 4\}$ **87.** $\left\{ -\dfrac{14}{3} \right\}$

88. $\dfrac{60}{7}$ or $8\dfrac{4}{7}$ minutes

CHAPTER 4 TEST

1. Find all real numbers that make the following expression undefined: $\dfrac{2k - 1}{3k^2 + 2k - 8}$.

2. Write $\dfrac{6x^2 - 13x - 5}{9x^3 - x}$ in lowest terms.

Multiply or divide.

3. $\dfrac{4x^2y^5}{7xy^8} \div \dfrac{8xy^6}{21xy}$

4. $\dfrac{y^2 - 16}{y^2 - 25} \cdot \dfrac{y^2 + 2y - 15}{y^2 - 7y + 12}$

5. $\dfrac{x^2 - 9}{x^3 + 3x^2} \div \dfrac{x^2 + x - 12}{x^3 + 9x^2 + 20x}$

6. Find the least common denominator for the following group of denominators: $t^2 + t - 6, t^2 + 3t, t^2$

Add or subtract as indicated.

7. $\dfrac{7}{6t^2} - \dfrac{1}{3t}$

8. $\dfrac{9}{x - 7} + \dfrac{4}{x + 7}$

9. $\dfrac{6}{x + 4} + \dfrac{1}{x + 2} - \dfrac{3x}{x^2 + 6x + 8}$

Simplify the complex fraction.

10. $\dfrac{\dfrac{12}{r + 4}}{\dfrac{11}{6r + 24}}$

11. $\dfrac{\dfrac{1}{a} - \dfrac{1}{b}}{\dfrac{a}{b} - \dfrac{b}{a}}$

Divide.

12. $\dfrac{16p^3 - 32p^2 + 24p}{4p^2}$

13. $\dfrac{9q^4 - 18q^3 + 11q^2 + 10q - 10}{3q - 2}$

14. $\dfrac{6y^4 - 4y^3 + 5y^2 + 6y - 9}{2y^2 + 3}$

15. Use synthetic division to decide whether 4 is a solution of $x^4 - 8x^3 + 21x^2 - 14x - 24 = 0$.

16. Use synthetic division to divide $9x^5 + 40x^4 - 23x^3 + 8x^2 - 6x + 22$ by $x + 5$.

[4.1] **1.** $-2, \dfrac{4}{3}$

2. $\dfrac{2x - 5}{x(3x - 1)}$

3. $\dfrac{3x}{2y^8}$ **4.** $\dfrac{y + 4}{y - 5}$

5. $\dfrac{x + 5}{x}$

[4.2] **6.** $t^2(t + 3)(t - 2)$

7. $\dfrac{7 - 2t}{6t^2}$

8. $\dfrac{13x + 35}{(x - 7)(x + 7)}$

9. $\dfrac{4}{x + 2}$

[4.3] **10.** $\dfrac{72}{11}$ **11.** $\dfrac{-1}{a + b}$

[4.4] **12.** $4p - 8 + \dfrac{6}{p}$

13. $3q^3 - 4q^2 + q + 4 + \dfrac{-2}{3q - 2}$

14. $3y^2 - 2y - 2 + \dfrac{12y - 3}{2y^2 + 3}$

[4.5] **15.** yes

16. $9x^4 - 5x^3 + 2x^2 - 2x + 4 + \dfrac{2}{x + 5}$

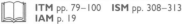

[4.2, 4.6] 17. (a) $\dfrac{11(x - 6)}{12}$

(b) {6}

[4.6] 18. $\left\{\dfrac{1}{2}\right\}$ 19. {5}

20. A solution cannot make a denominator zero.

[4.7] 21. $r = \dfrac{En - IRn}{I}$

or $r = \dfrac{En}{I} - Rn$

22. $\dfrac{48}{5}$ 23. $\dfrac{45}{14}$ or

$3\dfrac{3}{14}$ hours

24. 15 miles per hour

25. 48,000

17. One of the following is an expression to be simplified by algebraic operations, and the other is an equation to be solved. Simplify the one that is an expression, and solve the one that is an equation.

(a) $\dfrac{2x}{3} + \dfrac{x}{4} - \dfrac{11}{2}$ (b) $\dfrac{2x}{3} + \dfrac{x}{4} = \dfrac{11}{2}$

Solve the equation.

18. $\dfrac{1}{x} - \dfrac{4}{3x} = \dfrac{1}{x - 2}$

19. $\dfrac{y}{y + 2} - \dfrac{1}{y - 2} = \dfrac{8}{y^2 - 4}$

20. Checking solutions of an equation in Chapters 1–3 verified that the algebraic steps were performed correctly. When an equation includes a term with a variable denominator, what additional reason *requires* that the solutions be checked?

21. Solve for the variable r in this formula from the field of electronics: $I = \dfrac{En}{Rn + r}$.

22. A formula involving pulleys is $P = \dfrac{W(R - r)}{2R}$. Find R if $W = 120$, $r = 8$, and $P = 10$.

Solve the problem.

23. Wayne can do a job in 9 hours, while Susan can do the same job in 5 hours. How long would it take them to do the job if they worked together?

24. The current of a river runs at 3 miles per hour. Nana's boat can go 36 miles downstream in the same time that it takes to go 24 miles upstream. Find the speed of the boat in still water.

25. Biologists collected a sample of 600 fish from Lake Linda on May 1 and tagged each of them. When they returned on June 1, a new sample of 800 fish was collected, and 10 of these had been previously tagged. Use this experiment to determine the approximate fish population of Lake Linda.

CUMULATIVE REVIEW (Chapters 1–4)

1. −199   2. 455   3. 14

4. $\left\{-\dfrac{15}{4}\right\}$ 5. $\left\{\dfrac{2}{3}, 2\right\}$

6. $x = \dfrac{d - by}{a - c}$ or

$x = \dfrac{by - d}{c - a}$

7. {11}   8. $\left(-\infty, \dfrac{240}{13}\right]$

9. $\left[-2, \dfrac{2}{3}\right]$ 10. $(-\infty, \infty)$

11. 180 votes, 225 votes

Evaluate if $x = -4$, $y = 3$, and $z = 6$.

1. $|2x| + 3y - z^3$

2. $-5(x^3 - y^3)$

3. $\dfrac{2x^2 - x + z}{3y - z}$

Solve the equation.

4. $7(2x + 3) - 4(2x + 1) = 2(x + 1)$

5. $|6x - 8| - 4 = 0$

6. $ax + by = cx + d$ for x

7. $.04x + .06(x - 1) = 1.04$

Solve the inequality.

8. $\dfrac{2}{3}y + \dfrac{5}{12}y \le 20$

9. $|3x + 2| \le 4$

10. $|12t + 7| \ge 0$

Solve the problem.

11. In an election, one candidate received 45 more votes than the other. The total number of votes cast in the election was 405. Find the number of votes received by each candidate.

12. A triangle has an area of 42 square meters. The base is 14 meters long. Find the height of the triangle.

13. A jar contains only cents, nickels, and dimes. The number of dimes is 1 more than the number of nickels, and the number of cents is 6 more than the number of nickels. How many of each denomination can be found in the jar, if the total value is $4.80?

14. Two angles of a triangle have the same measure. The measure of the third angle is 4° less than twice the measure of each of the equal angles. Find the measures of the three angles.

Write in scientific notation.

15. 35,000

16. .0000000076

Write without scientific notation.

17. 5.6×10^9

18. 4.89×10^{-6}

Simplify. Write the answer with only positive exponents. Assume that all variables represent nonzero real numbers.

19. $\dfrac{3x^{-4}y^3}{7^{-1}x^5y^{-4}}$

20. $\left(\dfrac{a^{-3}b^4}{a^2b^{-1}}\right)^{-2}$

21. $\left(\dfrac{m^{-4}n^2}{m^2n^{-3}}\right) \cdot \left(\dfrac{m^5n^{-1}}{m^{-2}n^5}\right)$

Perform the indicated operations.

22. $(3y^2 - 2y + 6) - (-y^2 + 5y + 12)$

23. $9(-3x^3 - 4x + 12) + 2(x^3 - x^2 + 3)$

24. $-6x^4(x^2 - 3x + 2)$

25. $(4f + 3)(3f - 1)$

26. $(x + y)(x^2 - xy + y^2)$

27. $(7t^3 + 8)(7t^3 - 8)$

28. $\left(\dfrac{1}{4}x + 5\right)^2$

29. $\dfrac{24x^5 - 18x^3 + 12x^2 + 6x}{6x}$

30. $(3x^3 + 13x^2 - 17x - 7) \div (3x + 1)$

31. If $P(x) = -4x^3 + 2x - 8$, find $P(-2)$.

32. Use synthetic division to divide $(2x^4 + 3x^3 - 8x^2 + x + 2)$ by $(x - 1)$.

Factor the polynomial completely.

33. $2x^2 - 13x - 45$

34. $100t^4 - 25$

35. $8p^3 + 125$

36. Solve the equation $3x^2 + 4x = 7$.

Write the rational expression in lowest terms.

37. $\dfrac{y^2 - 16}{y^2 - 8y + 16}$

38. $\dfrac{8x^2 - 18}{8x^2 + 4x - 12}$

Perform the indicated operations. Express the answer in lowest terms.

39. $\dfrac{x + 4}{x - 2} + \dfrac{2x - 10}{x - 2}$

40. $\dfrac{2}{a + b} - \dfrac{3}{a - b}$

41. $\dfrac{2x}{2x - 1} + \dfrac{4}{2x + 1} + \dfrac{8}{4x^2 - 1}$

42. $\dfrac{5}{x^3 - y^3} + \dfrac{3}{x^2 + xy + y^2}$

12. 6 meters
13. 35 cents, 29 nickels, 30 dimes
14. 46°, 46°, 88°
15. 3.5×10^4
16. 7.6×10^{-9}
17. 5,600,000,000
18. .00000489
19. $\dfrac{21y^7}{x^9}$ 20. $\dfrac{a^{10}}{b^{10}}$ 21. $\dfrac{m}{n}$
22. $4y^2 - 7y - 6$
23. $-25x^3 - 2x^2 - 36x + 114$
24. $-6x^6 + 18x^5 - 12x^4$
25. $12f^2 + 5f - 3$
26. $x^3 + y^3$
27. $49t^6 - 64$
28. $\dfrac{1}{16}x^2 + \dfrac{5}{2}x + 25$
29. $4x^4 - 3x^2 + 2x + 1$
30. $x^2 + 4x - 7$ 31. 20
32. $2x^3 + 5x^2 - 3x - 2$
33. $(2x + 5)(x - 9)$
34. $25(2t^2 + 1)(2t^2 - 1)$
35. $(2p + 5) \cdot (4p^2 - 10p + 25)$
36. $\left\{-\dfrac{7}{3}, 1\right\}$
37. $\dfrac{y + 4}{y - 4}$ 38. $\dfrac{2x - 3}{2(x - 1)}$
39. 3   40. $\dfrac{-a - 5b}{(a + b)(a - b)}$
41. $\dfrac{2(x + 2)}{2x - 1}$
42. $\dfrac{5 + 3x - 3y}{(x - y)(x^2 + xy + y^2)}$

43. $\{-4\}$

44. $q = \dfrac{fp}{p - f}$ or

$q = \dfrac{-fp}{f - p}$

45. 150 miles per hour

46. $\dfrac{6}{5}$ or $1\dfrac{1}{5}$ hours

43. Solve the equation $\dfrac{-3x}{x + 1} + \dfrac{4x + 1}{x} = \dfrac{-3}{x^2 + x}$.

44. Solve the formula for q: $\dfrac{1}{f} = \dfrac{1}{p} + \dfrac{1}{q}$.

Solve the problem.

45. Lucinda can fly her plane 200 miles against the wind in the same time it takes her to fly 300 miles with the wind. The wind blows at 30 miles per hour. Find the speed of her plane in still air.

46. Machine A can complete a certain job in 2 hours. To speed up the work, Machine B, which could complete the job alone in 3 hours, is brought in to help. How long will it take the two machines to complete the job working together?

ROOTS AND RADICALS

CONNECTIONS

In the previous chapter, we introduced rational expressions, the algebraic extensions of rational numbers. Recall that non-rational real numbers are called *irrational numbers*. The most familiar irrational numbers are square roots of numbers that are not squares, such as $\sqrt{2}$, and $\sqrt{5}$. In fact, most nth roots are irrational. Another familiar irrational number is π. The number π has fascinated humankind since it was first realized that this number could not be expressed as the ratio of two integers. By 2000 B.C. its significance was understood and a rough approximation had been found for it. The computation of π has engrossed both mathematicians and laypersons ever since. In mid-1991, Gregory and David Chudnovsky had calculated the decimal digits of π to 2,260,321,336 digits. In this chapter, we will study roots, their notation, and their properties.

5.1 RATIONAL EXPONENTS AND RADICALS

FOR EXTRA HELP

 SSG pp. 138–142
SSM pp. 213–216

 Video
6

Tutorial
IBM MAC

OBJECTIVES

1 ▶ Define $a^{1/n}$.
2 ▶ Use radical notation for nth roots.
3 ▶ Define $a^{m/n}$.
4 ▶ Convert between rational exponents and radicals.

In this section, we extend the definitions and rules for exponents to include rational exponents as well as integer exponents. In Chapter 1, nth roots were written with radicals; in this section we shall see that nth roots can also be written with exponents.

1 ▶ Define $a^{1/n}$.

▶ If the rules for exponents are still to be valid, then the product $(3^{1/2})^2 = 3^{1/2} \cdot 3^{1/2}$ should be found by adding exponents.

$$(3^{1/2})^2 = 3^{1/2} \cdot 3^{1/2}$$
$$= 3^{1/2+1/2}$$
$$= 3^1 = 3$$

However, by definition, $(\sqrt{3})^2 = 3$. Since both $(3^{1/2})^2$ and $(\sqrt{3})^2$ are equal to 3, we must have

$$3^{1/2} = \sqrt{3}.$$

This suggests the following generalization.

nth Root, n Even

If n is an *even* positive integer and if a is positive, then $a^{1/n}$ is the positive real number whose nth power is a, and $a^{1/n}$ is the **principal nth root of a.**

EXAMPLE 1

Evaluating Exponentials of the Form $a^{1/n}$, n Even

Evaluate each exponential expression.

(a) $64^{1/2} = (8^2)^{1/2} = 8^1 = 8$

(b) $10,000^{1/4} = (10^4)^{1/4} = 10$

(c) $-144^{1/2} = -(144^{1/2}) = -12$

(d) $(-144)^{1/2}$ is not a real number; only nonnegative numbers have real square roots. ◼

Chalkboard Exercise

Simplify $-16^{1/4}$.

Answer: -2

With *odd* values of n, the base may be negative.

nth Root, n Odd

If n is an *odd* positive integer, for any real number a, $a^{1/n}$ is the nth root of a.

▶ **TEACHING TIP**

Take some time to show students how to use the square root, exponential, and reciprocal keys on their scientific calculators. ◀

Because only even powers have two roots, only even powers require a principal nth root, which is the positive root. Odd powers of a have the same sign as a itself.

INSTRUCTOR'S RESOURCES

 ITM pp. 390–392 **ISM** pp. 319–325
IAM pp. 20

 **TEST GENERATOR**
DOS Windows MAC

 **TRANSPARENCIES**

EXAMPLE 2

Evaluating Exponentials of
the Form $a^{1/n}$, n Odd

Chalkboard Exercise

Simplify $32^{1/5}$.

Answer: 2

◆ yes; No, because T
depends on h^3, not h.

Evaluate each exponential expression.

(a) $64^{1/3} = (4^3)^{1/3} = 4^1 = 4$

(b) $(-64)^{1/3} = [(-4)^3]^{1/3} = (-4)^1 = -4$

(c) $(-32)^{1/5} = [(-2)^5]^{1/5} = -2$

(d) $\left(\dfrac{1}{8}\right)^{1/3} = \left[\left(\dfrac{1}{2}\right)^3\right]^{1/3} = \dfrac{1}{2}$ ∎

◆ **CONNECTIONS** ◆

Many formulas and other models used in applications involve roots and may be expressed using fractional exponents. For example, the threshold weight T for a person is the weight above which the risk of death increases greatly. The threshold weight in pounds for men aged 40–49 is related to height in inches by the formula

$$h = 12.3T^{1/3}.$$

Several other applications that involve fractional exponents or roots are given in the exercises.

FOR DISCUSSION OR WRITING
Would the threshold weight for men in the age group described above be greater for taller men than for shorter? Is this relationship direct—that is, if one man is 1.5 times as tall as another, is his corresponding threshold weight 1.5 times as large? Why or why not?

2 ▶ Use radical notation
for *n*th roots.

▶ The discussion of *n*th roots leads naturally to radical notation. It is common for *n*th roots to be written with a radical sign, first introduced in Chapter 1.

Radical Notation If *n* is a positive integer greater than 1, and if $a^{1/n}$ is a real number, then

$$a^{1/n} = \sqrt[n]{a}.$$

The number *a* is the **radicand**, *n* is the **index** or **order**, and the expression $\sqrt[n]{a}$ is a **radical.** The second root of a number, $\sqrt[2]{a}$, is its principal *square root*. It is customary to omit the index 2 and write this square root of *a* as $\sqrt{a}$.

EXAMPLE 3

Simplifying Roots Using
Radical Notation

Chalkboard Exercise

Simplify $\sqrt[6]{64}$.

Answer: 2

Find each root that is a real number.

(a) $-\sqrt{25} = -5$

(b) $\sqrt{\dfrac{9}{16}} = \dfrac{3}{4}$

(c) $\sqrt[3]{125} = 5$, since $5^3 = 125$.

(d) $\sqrt[3]{-216} = -6$, since $(-6)^3 = -216$.

(e) $\sqrt[4]{-16}$ is not a real number. ∎

As mentioned earlier, a positive number has *two* even *n*th roots, one positive and one negative. The symbol $\sqrt{a}$ is used only for the *nonnegative* root, the principal nth root. The negative nth root is written $-\sqrt[n]{a}$. (Sometimes the two roots are written $\pm\sqrt[n]{a}$, read "positive or negative nth root of a.")

A square root of a^2 (where $a \neq 0$) is a number that can be squared to give a^2. This number is either a or $-a$. Since the symbol $\sqrt{a^2}$ represents the *nonnegative* square root, we must write $\sqrt{a^2}$ with absolute value bars, as $|a|$, because a may be a negative number.

$\overline{\sqrt{a^2}}$ For any real number a, $\qquad \sqrt{a^2} = |a|$.

This result can be generalized to any nth root.

$\overline{\sqrt[n]{a^n}}$ If n is an **even** positive integer, $\qquad \sqrt[n]{a^n} = |a|$,

and if n is an **odd** positive integer, $\qquad \sqrt[n]{a^n} = a$.

EXAMPLE 4
Simplifying Roots with Absolute Value

Use absolute value as necessary to evaluate each radical expression.

(a) $\sqrt{5^2} = |5| = 5$

(b) $\sqrt{(-5)^2} = |-5| = 5$

(c) $\sqrt[6]{(-3)^6} = |-3| = 3$

(d) $\sqrt[5]{(-4)^5} = -4$ $\qquad$ 5 is odd; no absolute value is necessary.

(e) $-\sqrt[4]{(-9)^4} = -|-9| = -9$ ∎

Chalkboard Exercise

Simplify $\sqrt[3]{(-7)^3}$.

Answer: -7

3 ▶ Define $a^{m/n}$.

▶ **TEACHING TIP**
Try writing $a^{m/n}$ as $a^{\text{power } m/\text{index } n}$ for a while to reinforce the meaning of a rational exponent. ◀

▶ We can now define the more general expression $a^{m/n}$, where m/n is any rational number, starting with a rational number m/n where both m and n are positive integers, with m/n written in lowest terms. If the earlier rules of exponents are to be valid for rational exponents, then we should have

$$a^{m/n} = a^{(1/n)m}$$
$$= (a^{1/n})^m,$$

provided that $a^{1/n}$ is a real number. This result leads to the definition of $a^{m/n}$.

$\overline{a^{m/n}}$ If m and n are positive integers with m/n in lowest terms, then

$$a^{m/n} = (a^{1/n})^m,$$

provided that $a^{1/n}$ is a real number. If $a^{1/n}$ is not a real number, then $a^{m/n}$ is not a real number.

The expression $a^{-m/n}$ is defined as follows.

$\overline{a^{-m/n}}$

$$a^{-m/n} = \frac{1}{a^{m/n}} \qquad (a \neq 0),$$

provided that $a^{m/n}$ is a real number.

EXAMPLE 5

Evaluating Exponentials of the Form $a^{m/n}$

Evaluate each exponential expression that is a real number.

(a) $36^{3/2} = (36^{1/2})^3 = 6^3 = 216$

(b) $125^{2/3} = (125^{1/3})^2 = 5^2 = 25$

(c) $-4^{5/2} = -(4^{5/2}) = -(4^{1/2})^5 = -(2)^5 = -32$

(d) $(-27)^{2/3} = [(-27)^{1/3}]^2 = (-3)^2 = 9$

(e) $(-100)^{3/2}$ is not a real number, since $(-100)^{1/2}$ is not a real number.

(f) $16^{-3/4} = \dfrac{1}{16^{3/4}} = \dfrac{1}{(16^{1/4})^3} = \dfrac{1}{2^3} = \dfrac{1}{8}$ ∎

Chalkboard Exercise

Simplify $100^{3/2}$.

Answer: 1000

▶ **TEACHING TIP**

Emphasize that a negative exponent does not always lead to a negative result. For example,

$$8^{-1/3} = 1/2,$$

not -2, as many students will be tempted to say. ◀

We get an alternative definition of m/n by using the power rule for exponents a little differently than in the definition above. If all indicated roots are real numbers,

$$a^{m/n} = a^{m(1/n)}$$
$$= (a^m)^{1/n},$$

so that

$$a^{m/n} = (a^m)^{1/n}.$$

With this result, $a^{m/n}$ can be defined in either of two ways.

$\underline{a^{m/n}}$ If all indicated roots are real numbers, then $a^{m/n} = (a^{1/n})^m = (a^m)^{1/n}.$

▶ **TEACHING TIP**

Explain how, in most cases, it is easier to evaluate the root first, and then the power. ◀

An expression such as $27^{2/3}$ can now be evaluated in two ways:

$$27^{2/3} = (27^{1/3})^2 = 3^2 = 9$$
$$27^{2/3} = (27^2)^{1/3} = 729^{1/3} = 9.$$

In most cases, it is easier to use $(a^{1/n})^m$.

---

4 ▶ Convert between rational exponents and radicals.

▶ Until now, expressions of the form $a^{m/n}$ have been evaluated without introducing radical notation. However, we have seen that for appropriate values of a and n, $a^{1/n} = \sqrt[n]{a}$, and so it is not difficult to extend our discussion to using radical notation for $a^{m/n}$.

Radical Form of $a^{m/n}$ If all indicated roots are real numbers, $a^{m/n} = (\sqrt[n]{a})^m$ or $a^{m/n} = \sqrt[n]{a^m}.$

EXAMPLE 6

Converting Between Rational Exponents and Radicals

Write each of the following with radicals. Assume that all variables represent positive real numbers. Use the first definition in the box above.

(a) $13^{1/2} = \sqrt{13}$

(b) $6^{3/4} = (\sqrt[4]{6})^3$

(c) $9m^{5/8} = 9(\sqrt[8]{m})^5$

(d) $6x^{2/3} - (4x)^{3/5} = 6(\sqrt[3]{x})^2 - (\sqrt[5]{4x})^3$

(e) $r^{-2/3} = \dfrac{1}{r^{2/3}} = \dfrac{1}{(\sqrt[3]{r})^2}$

(f) $(a^2 + b^2)^{1/2} = \sqrt{a^2 + b^2}$ ∎

CAUTION Refer to Example 6(f). The expression $\sqrt{a^2 + b^2}$ is *not equal to* $a + b$. In general,

$$\sqrt[n]{a^n + b^n} \neq a + b.$$

EXAMPLE 7

Converting Between Radicals and Rational Exponents

Replace all radicals with rational exponents. Simplify. Assume that all variables represent positive real numbers.

(a) $\sqrt{10} = 10^{1/2}$

(b) $\sqrt[4]{3^8} = 3^{8/4} = 3^2 = 9$

(c) $\sqrt[6]{z^6} = z$ ∎

NOTE Refer to Example 7(c). It was not necessary to use absolute value bars, since the directions specifically stated that the variable represents a positive real number. Since the absolute value of the positive real number z is z itself, the answer is simply z. When working exercises involving radicals, we will often make the assumption that variables represent positive real numbers, which will eliminate the need for absolute value.

Scientific calculators have keys for powers and roots, and they are helpful when working problems with nth roots. These operations are discussed in detail in the next section.

5.1 EXERCISES

1. cube (or third); 8; 2;
second; 4; 4   2. (a)
3. (a) m must be even
(b) m must be odd
4. It is not a real number.
7. 13   8. 11   9. 9   10. 8
11. 2   12. 5   13. $\dfrac{8}{9}$
14. $\dfrac{2}{3}$

◉ **1.** Fill in the blanks with the correct responses: One way to evaluate $8^{2/3}$ is to first find the _____ root of _____ , which is _____ . Then raise that result to the _____ power, to get an answer of _____ . Therefore, $8^{2/3} =$ _____ .

◉ **2.** Which one of the following is a positive number?
   (a) $(-27)^{2/3}$ **(b)** $(-64)^{5/3}$ **(c)** $(-100)^{1/2}$ **(d)** $(-32)^{1/5}$

◉ **3.** If a is a negative number and n is odd, what must be true about m for $a^{m/n}$ to be
   (a) positive   **(b)** negative?

◉ **4.** If a is negative and n is even, what can be said about $a^{1/n}$?

◉ **5.** Explain why $(-64)^{1/2}$ is not a real number, while $-64^{1/2}$ is a real number.

◉ 🖉 **6.** Explain why $a^{1/n}$ is defined to be equal to $\sqrt[n]{a}$ when $\sqrt[n]{a}$ is real.

Simplify the expression involving rational exponents. See Examples 1, 2, and 5.

7. $169^{1/2}$ **8.** $121^{1/2}$ **9.** $729^{1/3}$ **10.** $512^{1/3}$

11. $16^{1/4}$ **12.** $625^{1/4}$ **13.** $\left(\dfrac{64}{81}\right)^{1/2}$ **14.** $\left(\dfrac{8}{27}\right)^{1/3}$

◉ CONCEPTUAL   🖉 WRITING   ▲ CHALLENGING   SCIENTIFIC CALCULATOR   ▦ GRAPHICS CALCULATOR

15. $(-27)^{1/3}$ **16.** $(-32)^{1/5}$ **17.** $100^{3/2}$ **18.** $64^{3/2}$

19. $-4^{5/2}$ **20.** $-32^{3/5}$ **21.** $(-144)^{1/2}$ **22.** $(-36)^{1/2}$

23. $64^{-3/2}$ **24.** $81^{-3/2}$ **25.** $\left(-\dfrac{8}{27}\right)^{-2/3}$ **26.** $\left(-\dfrac{64}{125}\right)^{-2/3}$

Find the root if it is a real number. Use a calculator as needed. See Examples 3 and 4.

27. $\sqrt{36}$ **28.** $\sqrt{100}$ **29.** $\sqrt{\dfrac{64}{81}}$ **30.** $\sqrt{\dfrac{100}{9}}$

31. $-\sqrt{-169}$ **32.** $-\sqrt{-400}$ **33.** $\sqrt[3]{216}$ **34.** $\sqrt[3]{343}$

35. $\sqrt[3]{-64}$ **36.** $\sqrt[3]{-125}$ **37.** $-\sqrt[3]{512}$ **38.** $-\sqrt[3]{1000}$

39. $-\sqrt[4]{81}$ **40.** $-\sqrt[4]{256}$ **41.** $\sqrt[4]{-16}$ **42.** $\sqrt[4]{-81}$

43. $\sqrt[6]{(-2)^6}$ **44.** $\sqrt[6]{(-4)^6}$ **45.** $\sqrt[5]{(-9)^5}$ **46.** $\sqrt[5]{(-8)^5}$

47. $\sqrt{x^2}$ **48.** $-\sqrt{x^2}$ **49.** $\sqrt[3]{x^3}$ **50.** $-\sqrt[3]{x^3}$

51. $\sqrt[3]{x^{15}}$ **52.** $\sqrt[4]{k^{20}}$

◆ MATHEMATICAL CONNECTIONS (Exercises 53–62) **◆**

Every positive number a has two even nth roots, the principal root $\sqrt[n]{a}$, which is positive, and a negative root $-\sqrt[n]{a}$. The following exercises, which should be worked in order, explore connections between these roots.

53. Find the square roots of 16.

54. Find the principal square root of 16.

55. Find $16^{1/2}$ and $-\sqrt{16}$. Which of these is the principal square root?

56. What is the solution set of $x^2 = 16$?

57. Find the fourth roots of 81.

58. Find the principal fourth root of 81.

59. Find $\sqrt[4]{81}$ and $-81^{1/4}$. Which one is the principal fourth root?

60. Give the solution set of $x^4 = 81$.

61. Explain what is meant by $\pm 25^{1/2}$.

62. Explain why $\sqrt[4]{x^4}$ is simplified as $|x|$, but $\sqrt[3]{x^3}$ is simply x.

◆

63. Consider the expression $-\sqrt{-a}$. Decide whether it is positive, negative, zero, or not a real number if
(a) $a > 0$;   **(b)** $a < 0$;   **(c)** $a = 0$.

64. If n is odd, under what conditions is $\sqrt[n]{a}$
(a) positive;   **(b)** negative;   **(c)** zero?

65. Observe the following simplification:
$$\sqrt[3]{5^{12}} = 5^{12/3} = 5^4 = 625.$$

Now fill in the blanks with the correct responses:

One way to evaluate $\sqrt[3]{5^{12}}$ is to raise _____ to the power obtained when _____ is divided by _____ . The resulting exponent is _____ , which gives a final answer of _____ .

66. Which one of the following is *not* equal to $\sqrt[4]{9^2}$?
(a) 3   **(b)** $\sqrt{9}$ **(c)** $9^{1/2}$ **(d)** -3

15. -3 **16.** -2
17. 1000   **18.** 512
19. -32 **20.** -8
21. not a real number
22. not a real number
23. $\dfrac{1}{512}$ **24.** $\dfrac{1}{729}$
25. $\dfrac{9}{4}$ **26.** $\dfrac{25}{16}$ **27.** 6
28. 10   **29.** $\dfrac{8}{9}$ **30.** $\dfrac{10}{3}$
31. not a real number
32. not a real number
33. 6   **34.** 7   **35.** -4
36. -5 **37.** -8 **38.** -10
39. -3 **40.** -4
41. not a real number
42. not a real number
43. 2   **44.** 4   **45.** -9
46. -8 **47.** $|x|$
48. $-|x|$ **49.** x **50.** $-x$
51. x^5 **52.** $|k|^5$ **53.** 4 and -4 **54.** 4   **55.** 4, -4; 4   **56.** $\{4, -4\}$
57. 3 and -3 **58.** 3
59. 3, -3; 3   **60.** $\{3, -3\}$
63. (a) not a real number
(b) negative   (c) zero
64. (a) a must be positive $(a > 0)$ (b) a must be negative $(a < 0)$ (c) a must be zero $(a = 0)$
65. 5; 12; 3; 4; 625
66. (d)

◉ CONCEPTUAL   **✎** WRITING   **▲** CHALLENGING   **▦** SCIENTIFIC CALCULATOR   **▤** GRAPHICS CALCULATOR

67. $\sqrt{12}$
68. $\sqrt{3}$ 69. $(\sqrt[4]{8})^3$
70. $(\sqrt[3]{7})^2$ 71. $(\sqrt[8]{9q})^5$
72. $(\sqrt[4]{3p})^3$ 73. $\dfrac{1}{(\sqrt{2m})^3}$
74. $\dfrac{1}{(\sqrt[5]{5y})^3}$
75. $(\sqrt[3]{2y + x})^2$
76. $(\sqrt{r + 2z})^3$
77. $\dfrac{1}{(\sqrt[3]{3m^4 + 2k^2})^2}$
78. $\dfrac{1}{(\sqrt[6]{5x^2 + 3z^3})^5}$
79. $\sqrt{a^2 + b^2} =$
$\sqrt{3^2 + 4^2} = 5$; $a + b =$
$3 + 4 = 7$; $5 \neq 7$
81. 64   82. 3125   83. 64
84. 36   85. x^{10} 86. r^{25}
87. $\sqrt[6]{x^5}$ 88. $\sqrt[20]{y^{13}}$
89. $y\sqrt{7y}$ 90. $y^2\sqrt{10y}$
91. $\sqrt[15]{t^8}$ 92. $\sqrt[12]{w^7}$
93. 10 miles
94. 74 inches, 6.2 feet
95. 1.6 seconds
96. 4.5 hours

Write with radicals. Assume that all variables represent positive real numbers. See Example 6.

67. $12^{1/2}$ **68.** $3^{1/2}$ **69.** $8^{3/4}$

70. $7^{2/3}$ **71.** $(9q)^{5/8}$ **72.** $(3p)^{3/4}$

73. $(2m)^{-3/2}$ **74.** $(5y)^{-3/5}$ **75.** $(2y + x)^{2/3}$

76. $(r + 2z)^{3/2}$ **77.** $(3m^4 + 2k^2)^{-2/3}$ **78.** $(5x^2 + 3z^3)^{-5/6}$

◉ **79.** Show that in general, $\sqrt{a^2 + b^2} \neq a + b$ by replacing a with 3 and b with 4.

◉ **80.** Suppose that someone claims that $\sqrt[n]{a^n + b^n}$ must equal $a + b$, since when we let $a = 1$ and $b = 0$, a true statement results:

$$\sqrt[n]{a^n + b^n} = \sqrt[n]{1^n + 0^n} = \sqrt[n]{1^n} = 1 = 1 + 0$$
$$= a + b.$$

Write an explanation of why this person has given faulty reasoning.

Simplify by first converting to rational exponents. Assume that all variables represent positive real numbers. See Example 7.

81. $\sqrt{2^{12}}$ **82.** $\sqrt{5^{10}}$ **83.** $\sqrt[3]{4^9}$ **84.** $\sqrt[4]{6^8}$

85. $\sqrt{x^{20}}$ **86.** $\sqrt{r^{50}}$ **87.** $\sqrt[3]{x} \cdot \sqrt{x}$ **88.** $\sqrt[4]{y} \cdot \sqrt[5]{y^2}$

89. $\sqrt[4]{49y^6}$ **90.** $\sqrt[4]{100y^{10}}$ **91.** $\dfrac{\sqrt[3]{t^4}}{\sqrt[5]{t^4}}$ **92.** $\dfrac{\sqrt[4]{w^3}}{\sqrt[6]{w}}$

93. According to an article in *The World Scanner Report* (August, 1991), the distance D, in miles, to the horizon from an observer's point of view over water or "flat" earth is given by

$$D = (2H)^{1/2}$$

where H is the height of the point of view, in feet. If a person whose eyes are 6 feet above ground level is standing at the top of a hill 44 feet above the "flat" earth, approximately how far to the horizon will she be able to see?

▦ **94.** As mentioned in the Connections box, the threshold weight T, in pounds, for a person is the weight above which the risk of death increases greatly. The threshold weight in pounds for men aged 40–49 is related to height in inches by the equation

$$h = 12.3T^{1/3}.$$

What height corresponds to a threshold weight of 216 pounds for a 43-year-old man? Round your answer to the nearest inch, and then to the nearest tenth of a foot.

95. The time for one complete swing of a simple pendulum is

$$t = 2\pi\sqrt{\dfrac{L}{g}},$$

where t is time in seconds, L is the length of the pendulum in feet, and g, the force due to gravity, is about 32 feet per second per second. Find the time of a complete swing of a 2-foot pendulum to the nearest tenth of a second.

96. Meteorologists can determine the duration of a storm by using the formula $T = .07D^{3/2}$, where D is the diameter of the storm in miles and T is the time in hours. Find the duration of a storm with a diameter of 16 miles. Round your answer to the nearest tenth of an hour.

REVIEW EXERCISES

Use laws of exponents to simplify the expression. Use only positive exponents in your answers. All variables represent positive numbers. See Sections 3.1 and 3.2.

97. $x^{-3}(x^2)^{-4}$

98. $\dfrac{y^{-1}x^{-2}}{(x^2)^{-3}}$

99. $(r^{-2}s^{-3})^{-4}$

100. $\dfrac{(s^{-6}t^4)^{-2}}{(s^2t^{-3})^{-1}}$

101. $(3x^2y^{-7})(-2xy^{-6})$

102. $(5p^{-3}q^2)(4pq^{-1})$

97. $\dfrac{1}{x^{11}}$ **98.** $\dfrac{x^4}{y}$

99. r^8s^{12} **100.** $\dfrac{s^{14}}{t^{11}}$

101. $\dfrac{-6x^3}{y^{13}}$ **102.** $\dfrac{20q}{p^2}$

5.2 MORE ABOUT RATIONAL EXPONENTS AND RADICALS

FOR EXTRA HELP

 SSG pp. 142–145
SSM pp. 216–219

Video
6

 Tutorial
IBM   MAC

OBJECTIVES

1 ▶ Use the rules of exponents with rational exponents.
2 ▶ Use a calculator to find roots.

1 ▶ Use the rules of exponents with rational exponents.

▶   Earlier we introduced rules of exponents for integer exponents. In the previous section, rational exponents were defined in such a way that the power rule for exponents is still valid. With this definition of rational exponents, it can be shown that all the familiar properties of exponents are also valid for rational exponents.

Rules for Rational Exponents

Let r and s be rational numbers. For all real numbers a and b for which the indicated expressions exist:

$$a^r \cdot a^s = a^{r+s} \qquad a^{-r} = \frac{1}{a^r} \qquad \frac{a^r}{a^s} = a^{r-s} \qquad \left(\frac{a}{b}\right)^{-r} = \frac{b^r}{a^r}$$

$$(a^r)^s = a^{rs} \qquad (ab)^r = a^rb^r \qquad \left(\frac{a}{b}\right)^r = \frac{a^r}{b^r} \qquad a^{-r} = \left(\frac{1}{a}\right)^r.$$

EXAMPLE 1

Applying Rules for Rational Exponents

Write with only positive exponents. Assume that all variables represent positive real numbers.

(a) $2^{1/2} \cdot 2^{1/4} = 2^{1/2+1/4} = 2^{3/4}$

(b) $\dfrac{5^{2/3}}{5^{7/3}} = 5^{2/3-7/3} = 5^{-5/3} = \dfrac{1}{5^{5/3}}$

(c) $\dfrac{(x^{1/2}y^{2/3})^4}{y} = \dfrac{(x^{1/2})^4(y^{2/3})^4}{y}$ Power rule

$\qquad = \dfrac{x^2y^{8/3}}{y^1}$ Power rule

$\qquad = x^2y^{8/3-1}$ Quotient rule

$\qquad = x^2y^{5/3}$

Chalkboard Exercise

Simplify. Write with only positive exponents. Assume that all variables represent positive real numbers.

$$\dfrac{(x^2y)^{1/2}}{x^{3/4}y^{-1/4}}$$

Answer: $x^{1/4}y^{3/4}$

INSTRUCTOR'S RESOURCES

 ITM pp. 392–394   **ISM** pp. 325–331
IAM pp. 20–21

 **TEST GENERATOR**
DOS   Windows   MAC

 **TRANSPARENCIES**

 CONCEPTUAL   WRITING   CHALLENGING   SCIENTIFIC CALCULATOR    GRAPHICS CALCULATOR

(d) $m^{3/4}(m^{5/4} - m^{1/4}) = m^{3/4} \cdot m^{5/4} - m^{3/4} \cdot m^{1/4}$ Distributive property

$$= m^{3/4+5/4} - m^{3/4+1/4}$$ Product rule

$$= m^{8/4} - m^{4/4}$$

$$= m^2 - m$$

Do not make the common mistake of multiplying exponents in the first step.

(e) $\dfrac{(p^q)^{5/2}}{p^{q-1}} = \dfrac{p^{5q/2}}{p^{q-1}}$

$$= p^{(5q/2)-(q-1)} = p^{(5q/2)-q+1} = p^{(3q/2)+1}$$ Assume q is a rational number. ■

CAUTION Errors often occur in exercises like the ones in Example 1 because students try to convert the expressions to radical form. Remember that the rules of exponents apply here.

EXAMPLE 2

Applying Rules for Rational Exponents

Replace all radicals with rational exponents, and then apply the rules for rational exponents. Leave answers in exponential form. Assume that all variables represent positive real numbers.

(a) $\sqrt[3]{x^2} \cdot \sqrt[4]{x} = x^{2/3} \cdot x^{1/4}$ Convert to rational exponents.

$$= x^{2/3+1/4}$$ Use product rule for exponents.

$$= x^{8/12+3/12}$$ Get a common denominator.

$$= x^{11/12}$$

(b) $\dfrac{\sqrt{x^3}}{\sqrt[3]{x^2}} = \dfrac{x^{3/2}}{x^{2/3}} = x^{3/2-2/3} = x^{5/6}$

(c) $\sqrt{\sqrt[4]{z}} = \sqrt{z^{1/4}} = (z^{1/4})^{1/2} = z^{1/8}$ ■

2 ▶ Use a calculator to find roots.

▶ So far in this chapter we have discussed definitions, properties, and rules for rational exponents and radicals. An important application of radicals involves finding approximations of roots that are not rational numbers. Many expressions involving radicals and rational exponents represent *rational* numbers. For example,

$$9^{3/2} = (9^{1/2})^3 = (\sqrt{9})^3 = 3^3 = 27$$

is a rational number. However, there is no rational number equal to $\sqrt{15}$ or $\sqrt[3]{10}$, for example. These numbers are irrational. We can find decimal approximations of such irrational numbers with scientific calculators.

To use a calculator to approximate $\sqrt{15}$, enter the number 15 and press the key marked $\boxed{\sqrt{}}$. The display will then read 3.8729833. (There may be fewer or more decimal places, depending on the model of calculator used.) In this book we will usually show approximations rounded to three decimal places. Therefore,

$$\sqrt{15} \approx 3.873$$

where $\approx$ means "is approximately equal to."

EXAMPLE 3

Finding Approximations for Square Roots

Find a decimal approximation for each of the following, using a calculator.

(a) $\sqrt{39} \approx 6.245$

(b) $\sqrt{83} \approx 9.110$

(c) $-\sqrt{72} \approx -8.485$

To find the negative square root, first find the positive square root and then take its negative, using the sign changing key $\boxed{\pm}$.

(d) $\sqrt{770} \approx 27.749$

(e) $-\sqrt{420} \approx -20.494$ ∎

◇ **C O N N E C T I O N S** ◇

Scientific calculators have the capability to find roots other than square roots. The key marked $\boxed{y^x}$ is the exponential key. For many scientific calculators, using the inverse key $\left(\boxed{\text{INV}}\right)$ in conjunction with the exponential key will allow us to take roots. For example, to find an approximation for $\sqrt[5]{467}$ (the fifth root of 467), the typical sequence of keystrokes is as follows.

$\boxed{4}\ \boxed{6}\ \boxed{7}\ \boxed{\text{INV}}\ \boxed{y^x}\ \boxed{5}\ \boxed{=}$

The display should read 3.4187188. (More or fewer decimal places may show up depending on the model of calculator.) This decimal is an approximation of $\sqrt[5]{467}$, and rounding to the nearest thousandth, we get

$$\sqrt[5]{467} \approx 3.419.$$

If your calculator does not follow the same sequence of keystrokes just described, consult the owner's manual.

There is a simple way to check that a calculator approximation is in the "ballpark". For instance, since 16 is a little larger than 15, $\sqrt{16} = 4$ should be a little larger than $\sqrt{15}$. Thus 3.873 is a reasonable approximation of $\sqrt{15}$.

E X A M P L E   4

Finding Approximations for Higher Roots

Find a decimal approximation for each of the following, using a calculator.

(a) $\sqrt[3]{943} \approx 9.806$ **(b)** $\sqrt[4]{992} \approx 5.612$ **(c)** $\sqrt[5]{10,847} \approx 6.413$

(d) $629^{1/7}$

This can be found either by evaluating $1/7$ in the calculator and using the $\boxed{y^x}$ key directly, with $y = 629$ and $x = 1/7$, or by writing it as $\sqrt[7]{629}$ and then using the method described above. In either case, the approximation is 2.511.

(e) $1.97^{5/2}$

Again, we can find the approximation by raising 1.97 to the $5/2$ power, or by taking the square root of 1.97 and raising this result to the fifth power, since

$$1.97^{5/2} = (\sqrt{1.97})^5.$$

The approximation is 5.447. ∎

5.2 EXERCISES

 1. State the product rule for exponents in your own words.

 2. State the quotient rule for exponents in your own words.

◉ **3.** By the product rule for exponents, we know that $2^{1/4} \cdot 2^{1/5} = 2^{9/20}$. However, there is no exponent rule to simplify $3^{1/4} \cdot 2^{1/5}$. Why?

◉ **4.** What is wrong with the statement $(4^{2/3})^{1/3} = 4^{(2/3+1/3)} = 4$?

3. The product rule for exponents applies only if the bases are the same.

◉ CONCEPTUAL   ☑ WRITING   ▲ CHALLENGING   ▦ SCIENTIFIC CALCULATOR   ▨ GRAPHICS CALCULATOR

4. When one base is raised to a power and then raised to another power, we multiply the exponents, so $(4^{2/3})^{1/3} = 4^{(2/3)(1/3)} = 4^{2/9}$.

5. 9   6. 36   7. 4   8. 25
9. y 10. r 11. $k^{2/3}$
12. $z^{3/2}$ 13. $9x^8y^{10}$
14. $512p^6q^9$
15. $\dfrac{1}{x^{10/3}}$ 16. $\dfrac{1}{p^{7/4}}$
17. $\dfrac{1}{m^{1/4}n^{3/4}}$ 18. $\dfrac{1}{b^{19/12}}$
19. $\dfrac{c^{11/3}}{b^{11/4}}$ 20. $\dfrac{a^{5/2}}{m^{23/12}}$
21. $p + 2p^2$
22. $3z^{5/4} + 5z^2$
23. $k^{7/4} - k^{3/4}$
24. $r^{11/10} + r^{27/20}$
25. $6 + 18a$ 26. $4m - 16$
27. $x^{8/5} - 4x^{2/5}$
28. $y^{3/7} - 3y^{10/7}$ 29. $y^{5a/6}$
30. $x^{23b/4}$ 31. $x^{3a/2}$
32. $\dfrac{1}{x^{3b}}$ 33. $x^{(a-b)/(ab)} \cdot y$
34. $\dfrac{1}{x^{a/2}}$ 35. $x^{-1/2}$
36. $m^{5/2}$
37. $k^{-3/4}$ 38. $x^{-1/2}(3 - 4x)$
39. $m^{5/2}(m^{1/2} - 3)$
40. $k^{-3/4}(9 + 2k^{1/2})$
41. $t^{-1/2}(4 + 7t^2)$
42. $x^{-1/3}(8x - 5)$

Use the rules of exponents to simplify the expression. Write all answers with positive exponents. Assume that all variables represent positive real numbers. See Example 1.

5. $3^{1/2} \cdot 3^{3/2}$

6. $6^{4/3} \cdot 6^{2/3}$

7. $\dfrac{64^{5/3}}{64^{4/3}}$

8. $\dfrac{125^{7/3}}{125^{5/3}}$

9. $y^{7/3} \cdot y^{-4/3}$

10. $r^{-8/9} \cdot r^{17/9}$

11. $\dfrac{k^{1/3}}{k^{2/3} \cdot k^{-1}}$

12. $\dfrac{z^{3/4}}{z^{5/4} \cdot z^{-2}}$

13. $(27x^{12}y^{15})^{2/3}$

14. $(64p^4q^6)^{3/2}$

15. $\dfrac{(x^{2/3})^2}{(x^2)^{7/3}}$

16. $\dfrac{(p^3)^{1/4}}{(p^{5/4})^2}$

17. $\dfrac{m^{3/4}n^{-1/4}}{(m^2n)^{1/2}}$

18. $\dfrac{(a^2b^5)^{-1/4}}{(a^{-3}b^2)^{1/6}}$

▲ **19.** $\left(\dfrac{b^{-3/2}}{c^{-5/3}}\right)^2 (b^{-1/4}c^{-1/3})^{-1}$

▲ **20.** $\left(\dfrac{m^{-2/3}}{a^{-3/4}}\right)^4 (m^{-3/8}a^{1/4})^{-2}$

Multiply. Assume that all variables represent positive real numbers. See Example 1(d).

21. $p^{2/3}(p^{1/3} + 2p^{4/3})$

22. $z^{5/8}(3z^{5/8} + 5z^{11/8})$

23. $k^{1/4}(k^{3/2} - k^{1/2})$

24. $r^{3/5}(r^{1/2} + r^{3/4})$

25. $6a^{7/4}(a^{-7/4} + 3a^{-3/4})$

26. $4m^{5/3}(m^{-2/3} - 4m^{-5/3})$

27. $x^{-3/5}(x^{11/5} - 4x)$

28. $y^{-11/7}(y^2 - 3y^3)$

▲ *Apply the rules for exponents to simplify. Assume that x and y are positive real numbers, that a and b are rational numbers, and all indicated roots are real numbers. See Example 1(e).*

29. $y^{a/2}y^{a/3}$

30. $(x^{3/4})^b(x^5)^b$

31. $\dfrac{x^{5a/3}}{x^{a/6}}$

32. $\left(\dfrac{x^b}{x^{3b}}\right)^{1/2}\left(\dfrac{x^{-2b}}{x^{-4b}}\right)^{-1}$

33. $\left(\dfrac{x^{a/b}y^a}{x}\right)^{1/a}$

34. $\dfrac{x^a x^{a/2}}{x^{2a}}$

◆ **MATHEMATICAL CONNECTIONS** (Exercises 35–42) ◆

Earlier, we factored expressions like $x^4 - x^5$ by factoring out the greatest common factor, so that $x^4 - x^5 = x^4(1 - x)$. We can adapt this approach to factor expressions with rational exponents. When one or more of the exponents is negative or a fraction, we use the ordering on the number line discussed in Chapter 1 to decide on the common factor. In this type of factoring, we want the binomial factor to have only positive exponents, so we always factor out the variable with the *smallest* exponent. A positive exponent is greater than a negative exponent, so in $7z^{5/8} + z^{-3/4}$, we factor out $z^{-3/4}$, because $-3/4$ is smaller than $5/8$.

Find the appropriate common factor in Exercises 35–37.

35. $3x^{-1/2} - 4x^{1/2}$

36. $m^3 - 3m^{5/2}$

37. $9k^{-3/4} + 2k^{-1/4}$

38. Factor $3x^{-1/2} - 4x^{1/2}$.

39. Factor $m^3 - 3m^{5/2}$.

40. Factor $9k^{-3/4} + 2k^{-1/4}$.

▲ *Use the method discussed above to factor the following.*

41. $\dfrac{4}{\sqrt{t}} + 7\sqrt{t^3}$

42. $8\sqrt[3]{x^2} - \dfrac{5}{\sqrt[3]{x}}$

43. Find the error in the following. $-1 = (-1)^1 = (-1)^{2/2} = [(-1)^2]^{1/2} = 1^{1/2} = 1$

Choose values of x and y to show the following.

44. $(x^{1/2} + y^{1/2})^2 \neq x + y$

45. $(x^2 + y^2)^{1/2} \neq x + y$

46. Suppose someone claims that $a^{1/2}(a^{1/4} - a^{1/2})$ simplifies to $a^{1/8} - a^{1/4}$. Explain to the person how the error occurred, and then give the correct answer.

Write with rational exponents, and then apply the properties of exponents. Assume that all radicands represent positive real numbers. Give answers in exponential form. See Example 2.

47. $\sqrt[5]{x^3} \cdot \sqrt[4]{x}$

48. $\sqrt[6]{y^5} \cdot \sqrt[3]{y^2}$

49. $\dfrac{\sqrt{x^5}}{x^4}$

50. $\dfrac{\sqrt[3]{k^5}}{\sqrt[3]{k^7}}$

51. $\sqrt{y} \cdot \sqrt[3]{yz}$

52. $\sqrt[3]{xz}\sqrt{z}$

53. $\sqrt[4]{\sqrt[3]{m}}$

54. $\sqrt[3]{\sqrt{k}}$

Use estimation skills to choose the letter of the closest approximation of the square root shown. Do not use a calculator and do not use a table of square roots.

55. $\sqrt{83}$
   (a) 9   **(b)** 7   **(c)** 8   **(d)** 10

56. $\sqrt{47}$
   (a) 5   **(b)** 6   **(c)** 7   **(d)** 8

57. $\sqrt{123.5}$
   (a) 9   **(b)** 10   **(c)** 11   **(d)** 12

58. $\sqrt{67.8}$
   (a) 7   **(b)** 8   **(c)** 9   **(d)** 10

Suppose that a rectangle has length $\sqrt{98}$ and width $\sqrt{26}$.

$\sqrt{98}$

$\sqrt{26}$

59. Which one of the following is the best estimate of its area?
   (a) 2500   **(b)** 250
   (c) 50   **(d)** 100

60. Which one of the following is the best estimate of its perimeter?
   (a) 15   **(b)** 250
   (c) 100   **(d)** 30

61. If a whole number is a four-digit perfect square and its units digit is 9, what are the possibilities for the units digit of its real square roots?

62. Explain why a perfect square cannot have 7 as its units digit.

Find a decimal approximation for the radical expression. Round the answer to three decimal places. See Example 3.

63. $\sqrt{9483}$
64. $\sqrt{6825}$
65. $\sqrt{284.361}$
66. $\sqrt{846.104}$
67. $\sqrt{7}$
68. $\sqrt{11}$
69. $-\sqrt{19}$
70. $-\sqrt{56}$
71. $-\sqrt{82}$
72. $-\sqrt{91}$

Use a calculator to find a decimal approximation for the radical expression. Round the answer to three decimal places. See Example 4.

73. $\sqrt[3]{423}$
74. $\sqrt[3]{555}$
75. $\sqrt[4]{100}$
76. $\sqrt[4]{250}$
77. $\sqrt[5]{23.8}$
78. $\sqrt[5]{98.4}$
79. $59^{2/3}$
80. $86^{7/8}$
81. $26^{-2/5}$
82. $104^{-5/4}$

43. In the definition of $a^{1/n}$, if n is even we must have $a \geq 0$. Thus $(-1)^{2/2} = [(-1)^2]^{1/2}$ is not a valid step.   **44.** For example, let $x = 9$ and $y = 16$. $(x^{1/2} + y^{1/2})^2 = (9^{1/2} + 16^{1/2})^2 = (3 + 4)^2 = 49;$ $x + y = 9 + 16 = 25;$ $49 \neq 25$
45. For example, let $x = 3$ and $y = 4$. $(x^2 + y^2)^{1/2} = (3^2 + 4^2)^{1/2} = 25^{1/2} = 5;$ $x + y = 3 + 4 = 7;$ $5 \neq 7$ **47.** $x^{17/20}$ **48.** $y^{3/2}$
49. $\dfrac{1}{x^{3/2}}$ **50.** $\dfrac{1}{k^{2/3}}$
51. $y^{5/6}z^{1/3}$ **52.** $x^{1/3}z^{5/6}$
53. $m^{1/12}$ **54.** $k^{1/6}$
55. (a)   **56.** (c)   **57.** (c)
58. (b)   **59.** (c)   **60.** (d)
61. 3, 7
63. 97.381   **64.** 82.614
65. 16.863   **66.** 29.088
67. 2.646   **68.** 3.317
69. −4.359   **70.** −7.483
71. −9.055   **72.** −9.539
73. 7.507   **74.** 8.218
75. 3.162   **76.** 3.976
77. 1.885   **78.** 2.504
79. 15.155   **80.** 49.282
81. .272   **82.** .003

 CONCEPTUAL   WRITING   CHALLENGING   SCIENTIFIC CALCULATOR    GRAPHICS CALCULATOR

84. 3.141592653
86. 1; yes
87. 392,000 square miles
88. 411,000 cycles per
second 89. r^3
90. $\dfrac{x^5}{16}$ 91. $\dfrac{q^8}{112}$ 92. 1

83. Many scientific calculators are designed so that negative bases are not allowed when using the exponential key. For example, attempting to evaluate $(-4)^2$ on a popular calculator gives an error message. Write an explanation of how one can use the key to evaluate powers of negative bases by considering the sign of the final answer first.

84. Use a calculator to find an approximation of $\sqrt[4]{\dfrac{2143}{22}}$, giving as many decimal places as the calculator shows. This number is the same as the decimal approximation of the irrational number π, correct to eight places.

85. Write an explanation of why the fourth root of a number can be found on a calculator by touching the square root key twice.

86. Use your calculator to evaluate 5^0. Is your answer the same as you would expect to find when using the rules of exponents?

87. *Heron's formula* gives a method of finding the area of a triangle if the lengths of its sides are known. Suppose that a, b, and c are the lengths of the sides. Let s denote one-half of the perimeter of the triangle (called the *semiperimeter*); that is,

$$s = \frac{1}{2}(a + b + c).$$

Then the area of the triangle is

$$A = \sqrt{s(s - a)(s - b)(s - c)}.$$

Find the area of the Bermuda Triangle, if the "sides" of this triangle measure approximately 850 miles, 925 miles, and 1300 miles. Give your answer to the nearest thousand.

88. In electronics, the resonant frequency f of a circuit may be found by the formula

$$f = \frac{1}{2\pi \sqrt{LC}},$$

where f is in cycles per second, L is in henrys, and C is in farads. Find the resonant frequency f if $L = 5 \times 10^{-4}$ henrys and $C = 3 \times 10^{-10}$ farads. Give your answer to the nearest thousand. (Use $\pi \approx 3.14159$.)

REVIEW EXERCISES

Multiply. Write answers with only positive exponents. Assume that all variables represent positive real numbers. See Sections 3.1 and 3.2.

89. $r^{-4} \cdot r^{13} \cdot (r^{-2})^3$

90. $(4x^3)^{-1}(2x^{-4})^{-2}$

91. $(4p^{-2}q^{-3})^{-2}(7p^4q^{-2})^{-1}$

92. $(-13x^{15}y^{-2})^{-3} \cdot (-13x^{15}y^{-2})^3$

5.3 SIMPLIFYING RADICALS

FOR EXTRA HELP	OBJECTIVES
📖 **SSG** pp. 145–149 **SSM** pp. 219–222	1 ▶ Use the product rule for radicals.
📼 Video 6	2 ▶ Use the quotient rule for radicals.
	3 ▶ Simplify radicals.
💾 Tutorial IBM MAC	4 ▶ Simplify radicals by using different indexes.
	5 ▶ Use the Pythagorean formula to find the length of a side of a right triangle.

INSTRUCTOR'S RESOURCES

📖 **ITM** pp. 395–399 **ISM** pp. 332–337 **IAM** p. 21	💾 **TEST GENERATOR** DOS Windows MAC	✳ **TRANSPARENCIES**

We have seen that the rules for exponents can be used to simplify products and quotients of radicals by writing the radicands with rational exponents. However, sometimes we prefer to work directly with roots written with radicals. In this section, the rules for exponents are used to develop product and quotient rules for radicals.

1 ▸ Use the product rule for radicals.

▸ The product of $\sqrt{36}$ and $\sqrt{4}$ is

$$\sqrt{36} \cdot \sqrt{4} = 6 \cdot 2 = 12.$$

Multiplying 36 and 4 and then taking the square root gives

$$\sqrt{36 \cdot 4} = \sqrt{144} = 12,$$

the same answer. This result is an example of the **product rule for radicals.**

Product Rule for Radicals

If $\sqrt[n]{a}$ and $\sqrt[n]{b}$ are real numbers and n is a natural number, then

$$\sqrt[n]{a} \cdot \sqrt[n]{b} = \sqrt[n]{ab}.$$

(The product of two radicals is the radical of the product.)

We can justify the product rule using rational exponents. Since $\sqrt[n]{a} = a^{1/n}$ and $\sqrt[n]{b} = b^{1/n}$,

$$\sqrt[n]{a} \cdot \sqrt[n]{b} = a^{1/n} \cdot b^{1/n} = (ab)^{1/n} = \sqrt[n]{ab}.$$

CAUTION The product rule is applied only when the radicals have the same indexes.

EXAMPLE 1
Using the Product Rule

Multiply. Assume that all variables represent positive real numbers.

(a) $\sqrt{5} \cdot \sqrt{7} = \sqrt{5 \cdot 7} = \sqrt{35}$

(b) $\sqrt{11} \cdot \sqrt{p} = \sqrt{11p}$

(c) $\sqrt{7} \cdot \sqrt{11xyz} = \sqrt{77xyz}$

(d) $\sqrt{\dfrac{7}{y}} \cdot \sqrt{\dfrac{3}{p}} = \sqrt{\dfrac{21}{yp}}$ ∎

Chalkboard Exercise

Multiply. Assume that all variables represent positive real numbers.

$$\sqrt{10y} \cdot \sqrt{3k}$$

Answer: $\sqrt{30yk}$

EXAMPLE 2
Using the Product Rule

Multiply. Assume that all variables represent positive real numbers.

(a) $\sqrt[3]{3} \cdot \sqrt[3]{12} = \sqrt[3]{3 \cdot 12} = \sqrt[3]{36}$

(b) $\sqrt[5]{9} \cdot \sqrt[5]{7} = \sqrt[5]{63}$

(c) $\sqrt[4]{8y} \cdot \sqrt[4]{3r^2} = \sqrt[4]{24yr^2}$

(d) $\sqrt[6]{10m^4} \cdot \sqrt[6]{5m} = \sqrt[6]{50m^5}$

(e) $\sqrt[4]{2} \cdot \sqrt[5]{2}$ cannot be simplified by the product rule, because the two indexes (4 and 5) are different. ∎

Chalkboard Exercise

Multiply. Assume all variables represent positive real numbers.

$$\sqrt[5]{9y^2x} \cdot \sqrt[5]{8xy^2}$$

Answer: $\sqrt[5]{72y^4x^2}$

2 ▶ Use the quotient rule for radicals.

▶ The quotient rule for radicals is similar to the product rule.

Quotient Rule for Radicals

If $\sqrt[n]{a}$ and $\sqrt[n]{b}$ are real numbers, $b \neq 0$, and n is a natural number, then

$$\sqrt[n]{\frac{a}{b}} = \frac{\sqrt[n]{a}}{\sqrt[n]{b}}.$$

(The radical of a quotient is the quotient of the radicals.)

The quotient rule can be justified, like the product rule, using the properties of exponents. It, too, is used only when the radicals have the same indexes.

EXAMPLE 3
Using the Quotient Rule

Simplify. Assume that all variables represent positive real numbers.

(a) $\sqrt{\frac{16}{25}} = \frac{\sqrt{16}}{\sqrt{25}} = \frac{4}{5}$

(b) $\sqrt{\frac{7}{36}} = \frac{\sqrt{7}}{\sqrt{36}} = \frac{\sqrt{7}}{6}$

(c) $\sqrt[3]{-\frac{8}{125}} = \sqrt[3]{\frac{-8}{125}} = \frac{\sqrt[3]{-8}}{\sqrt[3]{125}} = \frac{-2}{5} = -\frac{2}{5}$

(d) $\sqrt[3]{\frac{7}{216}} = \frac{\sqrt[3]{7}}{\sqrt[3]{216}} = \frac{\sqrt[3]{7}}{6}$

(e) $\sqrt{\frac{x}{9}} = \frac{\sqrt{x}}{\sqrt{9}} = \frac{\sqrt{x}}{3}$

(f) $\sqrt{\frac{m^4}{25}} = \frac{\sqrt{m^4}}{\sqrt{25}} = \frac{m^2}{5}$ ▪

Chalkboard Exercise

Simplify. Assume that all variables represent positive real numbers.

$$\sqrt{\frac{y^8}{16}}$$

Answer: $\dfrac{y^4}{4}$

▶ **TEACHING TIP**
While most students will know perfect squares through 100, many of them do not know the larger squares, nor do they know perfect cubes and fourth powers. Encourage them to learn them so that simplifying radical expressions will be made easier. ◀

---

3 ▶ Simplify radicals.

▶ One of the main uses of the product and quotient rules is in simplifying radicals. A radical is **simplified** if the following four conditions are met.

Simplified Radical

1. The radicand has no factor raised to a power greater than or equal to the index.
2. The radicand has no fractions.
3. No denominator contains a radical.
4. Exponents in the radicand and the index of the radical have no common factor (except 1).

EXAMPLE 4
Simplifying Roots of Numbers

Simplify.

(a) $\sqrt{24}$

Check to see if 24 is divisible by a perfect square (square of a natural number), such as 4, 9, ... Choose the largest perfect square that divides into 24. The largest such number is 4. Write 24 as the product of 4 and 6, and then use the product rule.

$$\sqrt{24} = \sqrt{4 \cdot 6} = \sqrt{4} \cdot \sqrt{6} = 2\sqrt{6}$$

▶ **TEACHING TIP**

Display a partial list of squares, cubes, or fourth powers from which the largest perfect power that divides into the given number is chosen.

You might wish to show that while $\sqrt{108}$ can be simplified by first factoring it as $\sqrt{9 \cdot 12}$, another step will be required, since 12 has the perfect square factor 4. Then explain that it is advantageous to choose the *largest* perfect square factor to start; otherwise, there will be more work to do. ◀

(b) $\sqrt{108}$

The number 108 is divisible by the perfect square 36. If this is not obvious, try factoring 108 into prime factors.

$$\begin{aligned} \sqrt{108} &= \sqrt{2^2 \cdot 3^3} \\ &= \sqrt{2^2 \cdot 3^2 \cdot 3} \\ &= 2 \cdot 3 \cdot \sqrt{3} \qquad \text{Product rule} \\ &= 6\sqrt{3} \end{aligned}$$

(c) $\sqrt{10}$

No perfect square (other than 1) divides into 10, so $\sqrt{10}$ cannot be simplified further.

(d) $\sqrt[3]{16}$

Look for the largest perfect *cube* that divides into 16. The number 8 satisfies this condition, so write 16 as $8 \cdot 2$.

$$\sqrt[3]{16} = \sqrt[3]{8 \cdot 2} = \sqrt[3]{8} \cdot \sqrt[3]{2} = 2\sqrt[3]{2}$$

(e) $\sqrt[4]{162} = \sqrt[4]{81 \cdot 2}$ 81 is a perfect 4th power.

$\qquad\quad = \sqrt[4]{81} \cdot \sqrt[4]{2}$

$\qquad\quad = 3\sqrt[4]{2}$ ■

EXAMPLE 5

Simplifying Roots of Variable Expressions

Simplify. Assume that all variables represent positive real numbers.

(a) $\sqrt{16m^3} = \sqrt{16m^2 \cdot m}$ $16m^2$ is the largest perfect square that divides $16m^3$.

$\qquad\qquad = \sqrt{16m^2} \cdot \sqrt{m}$

$\qquad\qquad = 4m\sqrt{m}$

No absolute value bars are needed around m because of the assumption that all variables represent *positive* real numbers.

(b) $\sqrt{200k^7q^8} = \sqrt{10^2 \cdot 2 \cdot (k^3)^2 \cdot k \cdot (q^4)^2}$

$\qquad\qquad\quad = \sqrt{10^2 \cdot (k^3)^2 \cdot (q^4)^2 \cdot 2 \cdot k}$

$\qquad\qquad\quad = 10k^3q^4\sqrt{2k}$

▶ **TEACHING TIP**

Warn students that they should not forget to write the root index after simplifying a radical with an index greater than 2. ◀

(c) $\sqrt[3]{8x^4y^5} = \sqrt[3]{(8x^3y^3)(xy^2)}$ $8x^3y^3$ is the largest perfect cube that divides $8x^4y^5$.

$\qquad\qquad\quad = \sqrt[3]{8x^3y^3} \cdot \sqrt[3]{xy^2}$

$\qquad\qquad\quad = 2xy\sqrt[3]{xy^2}$

(d) $\sqrt[4]{32y^9} = \sqrt[4]{(16y^8)(2y)}$ $16y^8$ is the largest 4th power that divides $32y^9$.

$\qquad\qquad\; = \sqrt[4]{16y^8} \cdot \sqrt[4]{2y}$

$\qquad\qquad\; = 2y^2\sqrt[4]{2y}$ ■

NOTE From Example 5 we see that if a variable is raised to a power with an exponent divisible by 2, it is a perfect square. If it is raised to a power with an exponent divisible by 3, it is a perfect cube. In general, if it is raised to a power with an exponent divisible by n, it is a perfect nth power.

---

4 ▶ Simplify radicals by using different indexes.

▶ The conditions for a simplified radical given above state that an exponent in the radicand and the index of the radical should have no common factor. The next example shows how to simplify radicals with such common factors.

EXAMPLE 6

Simplify Radicals by Using Smaller Indexes

Simplify. Assume that all variables represent positive real numbers.

(a) $\sqrt[9]{5^6}$

Write this radical using rational exponents and then write the exponent in lowest terms. Express the answer as a radical.

$$\sqrt[9]{5^6} = 5^{6/9} = 5^{2/3} = \sqrt[3]{5^2} \quad \text{or} \quad \sqrt[3]{25}$$

(b) $\sqrt[4]{p^2} = p^{2/4} = p^{1/2} = \sqrt{p}$

Recall the assumption $p > 0$. ∎

Chalkboard Exercise

Simplify. Assume all variables represent positive real numbers.

$\sqrt[6]{t^2}$

Answer: $\sqrt[3]{t}$

These examples suggest the following rule.

> If m is an integer, n and k are natural numbers, and all indicated roots exist,
> $$\sqrt[kn]{a^{km}} = \sqrt[n]{a^m}.$$

The next example shows how we simplify the product of two radicals having different indexes.

EXAMPLE 7

Multiplying Radicals with Different Indexes

Simplify $\sqrt{7} \cdot \sqrt[3]{2}$.

Since the indexes, 2 and 3, have a least common index of 6, use rational exponents to write each radical as a sixth root.

$$\sqrt{7} = 7^{1/2} = 7^{3/6} = \sqrt[6]{7^3} = \sqrt[6]{343}$$
$$\sqrt[3]{2} = 2^{1/3} = 2^{2/6} = \sqrt[6]{2^2} = \sqrt[6]{4}$$

Therefore,

$$\sqrt{7} \cdot \sqrt[3]{2} = \sqrt[6]{343} \cdot \sqrt[6]{4} = \sqrt[6]{1372}. \quad \text{Product rule} \quad ∎$$

Chalkboard Exercise

Simplify $\sqrt{5} \cdot \sqrt[3]{4}$.

Answer: $\sqrt[6]{2000}$

5 ▶ Use the Pythagorean formula to find the length of a side of a right triangle.

▶ One useful application of radicals occurs when using the **Pythagorean formula** from geometry to find the length of one side of a right triangle when the lengths of the other sides are known.

Pythagorean Formula

If c is the length of the longest side of a right triangle and a and b are the lengths of the shorter sides, then

$$c^2 = a^2 + b^2.$$

▶ **TEACHING TIP**

Demonstrate this formula with some familiar Pythagorean triples: 3-4-5, 5-12-13, and 7-24-25, for example. ◀

The longest side is the **hypotenuse** and the two shorter sides are the **legs** of the triangle. The hypotenuse is the side opposite the right angle.

From the formula $c^2 = a^2 + b^2$, the length of the hypotenuse is given by $c = \sqrt{a^2 + b^2}$.

EXAMPLE 8
Using the Pythagorean Formula

Use the Pythagorean formula to find the length of the hypotenuse in the triangle in Figure 1.

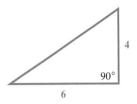

90°

4

6

FIGURE 1

By the formula, the length of the hypotenuse is

$$c = \sqrt{a^2 + b^2}$$
$$= \sqrt{4^2 + 6^2} \qquad \text{Let } a = 4 \text{ and } b = 6.$$
$$= \sqrt{16 + 36}$$
$$= \sqrt{52} = \sqrt{4 \cdot 13} \qquad \text{Factor.}$$
$$= \sqrt{4} \cdot \sqrt{13} \qquad \text{Product rule}$$
$$= 2\sqrt{13}. \quad \blacksquare$$

CAUTION When using the equation $c^2 = a^2 + b^2$, be sure that the length of the hypotenuse is substituted for c, and that the lengths of the legs are substituted for a and b. Errors often occur because values are substituted incorrectly.

◇ no; no

◇ **CONNECTIONS** ◇

The Pythagorean formula is undoubtedly one of the most widely used and oldest formulas we have. It is very important in trigonometry, which is used in surveying, drafting, engineering, navigation, and many other fields. There is evidence that the Babylonians knew the concept quite well. Although attributed to Pythagoras, it was known to every surveyor from Egypt to China for a thousand years before Pythagoras. In the 1939 movie *The Wizard of Oz,* the Scarecrow asks the Wizard for a brain. When the Wizard presents him with a diploma, granting him a Th.D (Doctor of Thinkology), the Scarecrow recites the following:

> The sum of the square roots of any two sides of an isosceles triangle is equal to the square root of the remaining side . . . .
> Oh joy! Rapture! I've got a brain.

FOR DISCUSSION OR WRITING
Did the Scarecrow recite the Pythagorean formula? (An *isosceles triangle* is a triangle with two equal sides.) Is his statement true? Explain.

5.3 EXERCISES

Left margin answers:

1. true; Both are equal to $4\sqrt{3}$ and approximately 6.92820323.   2. true; Both are equal to $6\sqrt{2}$ and approximately 8.485281374.
3. true; Both are equal to $6\sqrt{2}$ and approximately 8.485281374.   4. true; Both are equal to $30\sqrt{2}$ and approximately 42.42640687.   5. (d)
6. No, because they are only approximations that agree in the first few decimal places.   7. $\sqrt{30}$
8. $\sqrt{30}$ 9. $\sqrt[3]{14xy}$
10. $\sqrt[3]{36xy}$ 11. $\sqrt[4]{33}$
12. $\sqrt[4]{54}$ 13. $\dfrac{8}{11}$ 14. $\dfrac{4}{7}$
15. $\dfrac{\sqrt{3}}{5}$ 16. $\dfrac{\sqrt{13}}{7}$
17. $\dfrac{\sqrt{x}}{5}$ 18. $\dfrac{\sqrt{k}}{10}$ 19. $\dfrac{p^3}{9}$
20. $\dfrac{w^5}{6}$ 21. $\dfrac{3}{4}$ 22. $\dfrac{6}{5}$
23. $-\dfrac{\sqrt[3]{r^2}}{2}$ 24. $-\dfrac{\sqrt[3]{t}}{5}$
25. $2\sqrt{7}$ 26. $6\sqrt{2}$
27. $-4\sqrt{2}$ 28. $-4\sqrt{3}$
29. $10\sqrt{3}$ 30. $5\sqrt{6}$
31. $4\sqrt[3]{2}$ 32. $2\sqrt[3]{3}$
33. $-2\sqrt[3]{2}$ 34. $-5\sqrt[3]{2}$
35. $2\sqrt[3]{5}$ 36. $5\sqrt[3]{3}$
37. $-4\sqrt[4]{2}$ 38. $-5\sqrt[4]{2}$
39. $2\sqrt[5]{2}$ 40. $2\sqrt[5]{4}$
41. His reasoning was incorrect. Here 8 is a term and not a factor.
43. $6k\sqrt{2}$ 44. $3m\sqrt{2}$
45. $\dfrac{3\sqrt[3]{3}}{4}$ 46. $\dfrac{\sqrt[3]{4}}{3}$
47. $11x^3$ 48. $16z^6$

Decide whether each statement is true or false by using the simplification procedures explained in this section. Then support your answer by finding calculator approximations for each expression.

1. $2\sqrt{12} = \sqrt{48}$ **2.** $\sqrt{72} = 2\sqrt{18}$ **3.** $3\sqrt{8} = 2\sqrt{18}$ **4.** $5\sqrt{72} = 6\sqrt{50}$

5. Which one of the following is *not* equal to $\sqrt{\dfrac{1}{2}}$? (Do not use calculator approximations.)

(a) $\sqrt{.5}$ (b) $\sqrt{\dfrac{2}{4}}$ (c) $\sqrt{\dfrac{3}{6}}$ (d) $\dfrac{\sqrt{4}}{\sqrt{16}}$

6. Use the π key on your calculator to get a value for π. In Exercise 84, Section 5.2, we found that $\sqrt[4]{\dfrac{2143}{22}} \approx 3.141592653$. Does this mean that $\pi = \sqrt[4]{\dfrac{2143}{22}}$? Why or why not?

Multiply. See Examples 1 and 2.

7. $\sqrt{5} \cdot \sqrt{6}$ **8.** $\sqrt{10} \cdot \sqrt{3}$ **9.** $\sqrt[3]{7x} \cdot \sqrt[3]{2y}$
10. $\sqrt[3]{9x} \cdot \sqrt[3]{4y}$ **11.** $\sqrt[4]{11} \cdot \sqrt[4]{3}$ **12.** $\sqrt[4]{6} \cdot \sqrt[4]{9}$

Simplify each radical. Assume that all variables represent nonnegative real numbers. See Example 3.

13. $\sqrt{\dfrac{64}{121}}$ **14.** $\sqrt{\dfrac{16}{49}}$ **15.** $\sqrt{\dfrac{3}{25}}$ **16.** $\sqrt{\dfrac{13}{49}}$

17. $\sqrt{\dfrac{x}{25}}$ **18.** $\sqrt{\dfrac{k}{100}}$ **19.** $\sqrt{\dfrac{p^6}{81}}$ **20.** $\sqrt{\dfrac{w^{10}}{36}}$

21. $\sqrt[3]{\dfrac{27}{64}}$ **22.** $\sqrt[3]{\dfrac{216}{125}}$ **23.** $\sqrt[3]{-\dfrac{r^2}{8}}$ **24.** $\sqrt[3]{-\dfrac{t}{125}}$

Express each of the following in simplified form. See Example 4.

25. $\sqrt{28}$ **26.** $\sqrt{72}$ **27.** $-\sqrt{32}$ **28.** $-\sqrt{48}$
29. $\sqrt{300}$ **30.** $\sqrt{150}$ **31.** $\sqrt[3]{128}$ **32.** $\sqrt[3]{24}$
33. $\sqrt[3]{-16}$ **34.** $\sqrt[3]{-250}$ **35.** $\sqrt[3]{40}$ **36.** $\sqrt[3]{375}$
37. $-\sqrt[4]{512}$ **38.** $-\sqrt[4]{1250}$ **39.** $\sqrt[5]{64}$ **40.** $\sqrt[5]{128}$

41. A student claimed that $\sqrt[3]{14}$ is not in simplified form, since $14 = 8 + 6$, and 8 is a perfect cube. Was his reasoning correct? Why or why not?

42. Explain in your own words why $\sqrt[3]{k^4}$ is not a simplified radical.

Express each of the following in simplified form. Assume that all variables represent positive real numbers. See Example 5.

43. $\sqrt{72k^2}$ **44.** $\sqrt{18m^2}$ **45.** $\sqrt{\dfrac{81}{64}}$

46. $\sqrt[3]{\dfrac{32}{216}}$ **47.** $\sqrt{121x^6}$ **48.** $\sqrt{256z^{12}}$

○ CONCEPTUAL   ✎ WRITING   ▲ CHALLENGING   ▦ SCIENTIFIC CALCULATOR   ▦ GRAPHICS CALCULATOR

49. $-\sqrt[3]{27t^{12}}$

50. $-\sqrt[3]{64y^{18}}$

51. $-\sqrt{100m^8z^4}$

52. $-\sqrt{25t^6s^{20}}$

53. $-\sqrt[3]{-125a^6b^9c^{12}}$

54. $-\sqrt[3]{-216y^{15}x^6z^3}$

55. $\sqrt{\dfrac{1}{16}r^8t^{20}}$

56. $\sqrt[4]{\dfrac{81}{256}t^{12}u^8}$

57. $-\sqrt{13x^7y^8}$

58. $-\sqrt{23k^9p^{14}}$

59. $\sqrt[3]{8z^6w^9}$

60. $\sqrt[3]{64a^{15}b^{12}}$

61. $\sqrt[3]{-16z^5t^7}$

62. $\sqrt[3]{-81m^4n^{10}}$

63. $\sqrt[4]{81x^{12}y^{16}}$

64. $\sqrt[4]{81t^8u^{28}}$

65. $-\sqrt[4]{162r^{15}s^9}$

66. $-\sqrt[4]{32k^5m^9}$

67. $\sqrt{\dfrac{y^{11}}{36}}$

68. $\sqrt{\dfrac{v^{13}}{49}}$

69. $\sqrt[3]{\dfrac{x^{16}}{27}}$

70. $\sqrt[3]{\dfrac{y^{17}}{125}}$

Simplify the radical. Assume that all variables represent positive real numbers. See Example 6.

71. $\sqrt[4]{48^2}$

72. $\sqrt[4]{50^2}$

73. $\sqrt[4]{25}$

74. $\sqrt[6]{8}$

75. $\sqrt[10]{x^{25}}$

76. $\sqrt[12]{x^{44}}$

Simplify by first writing the radicals as radicals with the same index. Then multiply. Assume that all variables represent positive real numbers. See Example 7.

77. $\sqrt[3]{4} \cdot \sqrt{3}$

78. $\sqrt[3]{5} \cdot \sqrt{6}$

79. $\sqrt[3]{3} \cdot \sqrt[4]{4}$

80. $\sqrt[5]{7} \cdot \sqrt[3]{5}$

81. $\sqrt{x} \cdot \sqrt[3]{x}$

82. $\sqrt[3]{y} \cdot \sqrt[4]{y}$

Find the missing length in the right triangle. Simplify the answer if necessary. (Hint: If one leg is unknown, write the formula as $a = \sqrt{c^2 - b^2}$ or $b = \sqrt{c^2 - a^2}$.) See Example 8.

83.

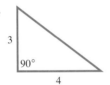

84.

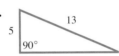

85.

86.

In Exercises 87 and 88 give the answer first as a simplified radical, and then as an approximation correct to the place value specified.

87. The illumination I, in footcandles, produced by a light source is related to the distance d, in feet, from the light source by the equation

$$d = \sqrt{\dfrac{k}{I}}$$

where k is a constant. If $k = 640$ how far from the light source will the illumination be 2 footcandles? Give the exact value, and then round to the nearest tenth of a foot.

88. The length of the diagonal of a box is given by

$$D = \sqrt{L^2 + W^2 + H^2}$$

where L, W, and H are the length, the width, and the height of the box. Find the length of the diagonal, D, of a box that is 4 feet long, 3 feet high, and 2 feet wide. Give the exact value, and then round to the nearest tenth of a foot.

49. $-3t^4$ **50.** $-4y^6$

51. $-10m^4z^2$ **52.** $-5t^3s^{10}$

53. $5a^2b^3c^4$ **54.** $6y^5x^2z$

55. $\dfrac{1}{2}r^2t^5$ **56.** $\dfrac{3}{4}t^3u^2$

57. $-x^3y^4\sqrt{13x}$

58. $-k^4p^7\sqrt{23k}$

59. $2z^2w^3$ **60.** $4a^5b^4$

61. $-2zt^2\sqrt[3]{2z^2t}$

62. $-3mn^3\sqrt[3]{3mn}$

63. $3x^3y^4$ **64.** $3t^2u^7$

65. $-3r^3s^2\sqrt[4]{2r^3s}$

66. $-2km^2\sqrt[4]{2km}$

67. $\dfrac{y^5\sqrt{y}}{6}$ **68.** $\dfrac{v^6\sqrt{v}}{7}$

69. $\dfrac{x^5\sqrt[3]{x}}{3}$ **70.** $\dfrac{y^5\sqrt[3]{y^2}}{5}$

71. $4\sqrt{3}$ **72.** $5\sqrt{2}$

73. $\sqrt{5}$ **74.** $\sqrt{2}$

75. $x^2\sqrt{x}$ **76.** $x^3\sqrt[3]{x^2}$

77. $\sqrt[6]{432}$ **78.** $\sqrt[6]{5400}$

79. $\sqrt[12]{6912}$

80. $\sqrt[15]{2,573,571,875}$

81. $\sqrt[6]{x^5}$ **82.** $\sqrt[12]{y^7}$

83. 5   **84.** 12   **85.** $8\sqrt{2}$

86. $3\sqrt{5}$

87. $8\sqrt{5}$ feet;   17.9 feet

88. $\sqrt{29}$ feet;   5.4 feet

 CONCEPTUAL   WRITING   ▲ CHALLENGING   SCIENTIFIC CALCULATOR   GRAPHICS CALCULATOR

89. .003   90. 581
91. $22x^4 - 10x^3$
92. cannot be simplified
further   93. $8q^2 - 3q$
94. $15m^5 - 3m^3$

 In Exercises 89 and 90, use a calculator to find the answer correct to the decimal place specified.

89. A formula from electronics is

$$I = \frac{E}{\sqrt{R^2 + \omega^2 L^2}}$$

where the variables are in appropriate units. Find I if $E = 282$, $R = 100$, $L = 264$ and $\omega = 120\pi$. Give your answer to the nearest thousandth.

90. In the study of sound, one version of the law of tensions is

$$f_1 = f_2 \sqrt{\frac{F_1}{F_2}}$$

If $F_1 = 300$, $F_2 = 60$, and $f_2 = 260$, find f_1 to the nearest unit.

REVIEW EXERCISES

Combine like terms. See Section 3.3.

91. $13x^4 - 12x^3 + 9x^4 + 2x^3$

92. $-15z^3 - z^2 + 4z^4 + 12z^8$

93. $9q^2 + 2q - 5q - q^2$

94. $7m^5 - 2m^3 + 8m^5 - m^3$

5.4 ADDING AND SUBTRACTING RADICAL EXPRESSIONS

FOR EXTRA HELP	OBJECTIVES
SSG pp. 150–151 **SSM** pp. 222–224	**1 ▶** Define a radical expression. **2 ▶** Simplify radical expressions involving addition and subtraction.
Video 6	
Tutorial IBM MAC	

1 ▶ Define a radical expression.

▶ A **radical expression** is an algebraic expression that contains radicals. For example,

$$\sqrt[4]{3} + \sqrt{6}, \qquad \sqrt{x + 2y} - 1, \qquad \text{and} \quad \sqrt{8} - \sqrt{2r}$$

are radical expressions. The examples in the previous section discussed simplifying radical expressions that involve multiplication and division. Now we show how to simplify radical expressions that involve addition and subtraction.

2 ▶ Simplify radical expressions involving addition and subtraction.

▶ An expression such as $4\sqrt{2} + 3\sqrt{2}$ can be simplified by using the distributive property.

$$4\sqrt{2} + 3\sqrt{2} = (4 + 3)\sqrt{2} = 7\sqrt{2}$$

As another example, $2\sqrt{3} - 5\sqrt{3} = (2 - 5)\sqrt{3} = -3\sqrt{3}$. This is very similar to simplifying $2x + 3x$ to $5x$ or $5y - 8y$ to $-3y$.

INSTRUCTOR'S RESOURCES

ITM pp. 399–401 **ISM** pp. 337–342 **IAM** pp. 21	**TEST GENERATOR** DOS Windows MAC	**TRANSPARENCIES**

CAUTION Only radical expressions with the same index and the same radicand may be combined. Expressions such as $5\sqrt{3} + 2\sqrt{2}$ and $3\sqrt{3} + 2\sqrt[3]{3}$ cannot be simplified.

<table>
<tr><td>

EXAMPLE 1

Adding and Subtracting Radical Expressions

</td></tr>
</table>

Add or subtract the following radical expressions.

(a) $3\sqrt{24} + \sqrt{54}$

Begin by simplifying each radical. Then use the distributive property.

$$3\sqrt{24} + \sqrt{54} = 3\sqrt{4} \cdot \sqrt{6} + \sqrt{9} \cdot \sqrt{6} \qquad \text{Product rule}$$
$$= 3 \cdot 2\sqrt{6} + 3\sqrt{6}$$
$$= 6\sqrt{6} + 3\sqrt{6}$$
$$= 9\sqrt{6} \qquad\qquad\qquad \text{Combine terms.}$$

(b) $2\sqrt{20x} - \sqrt{45x} = 2\sqrt{4}\sqrt{5x} - \sqrt{9}\sqrt{5x} \qquad \text{Product rule}$
$$= 2 \cdot 2\sqrt{5x} - 3\sqrt{5x}$$
$$= 4\sqrt{5x} - 3\sqrt{5x}$$
$$= \sqrt{5x} \quad (\text{if } x \geq 0)$$

(c) $2\sqrt{3} - 4\sqrt{5}$

Here the radicals differ, and are already simplified, so that $2\sqrt{3} - 4\sqrt{5}$ cannot be simplified further. ■

Chalkboard Exercise

Simplify
$3\sqrt{8} - 6\sqrt{50} + 2\sqrt{200}$.

Answer: $-4\sqrt{2}$

▶ **TEACHING TIP**

Explain how adding and subtracting "like radicals" is done exactly the same way as adding and subtracting like terms. For example, $5\sqrt{6} + 7\sqrt{6} = 12\sqrt{6}$, in the same way that $5x + 7x = 12x$. ◀

CAUTION Do not confuse the product rule with combining like terms. The root of a sum **does not equal** the sum of the roots. That is,

$$\sqrt{25} = \sqrt{9 + 16} \neq \sqrt{9} + \sqrt{16}.$$
$$\sqrt{25} = 5 \qquad \text{but} \qquad \sqrt{9} + \sqrt{16} = 3 + 4 = 7$$

◆ 1.618033989

◇ **CONNECTIONS** ◇

A triangle that has whole number measures for the lengths of two sides may have an irrational number as the measure of the third side. For example, a right triangle with the two shorter sides measuring 1 and 2 units will have a longest side measuring $\sqrt{5}$ units. The ratio of the dimensions of the golden rectangle, considered to have the most pleasing dimensions of any rectangle, is irrational. To sketch a golden rectangle, begin with the square *ONRS*. Divide it into two equal parts by segment *MK*, as shown in the figure. Let *M* be the center of a circle with radius *MN*. Sketch the rectangle *PQRS*. This is a golden rectangle, with the property that if the original square is taken away, *PQNO* is still a golden rectangle. If the square with side *OP* is taken away, another golden rectangle results, and so on.

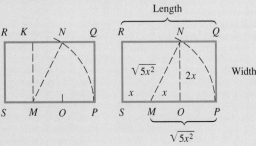

If the sides of the generating square have measure $2x$, then by the Pythagorean formula,

$$MN = \sqrt{x^2 + (2x)^2} = \sqrt{x^2 + 4x^2} = \sqrt{5x^2}.$$

Since NP is an arc of the circle with radius MN,

$$MP = MN = \sqrt{5x^2}.$$

The ratio of length to width is

$$\frac{\text{length}}{\text{width}} = \frac{x + \sqrt{5x^2}}{2x} = \frac{x + x\sqrt{5}}{2x} = \frac{x(1 + \sqrt{5})}{2x} = \frac{1 + \sqrt{5}}{2},$$

which is an irrational number.

FOR DISCUSSION OR WRITING

The golden rectangle has been widely used in art and architecture. See whether you can find some examples of its use. Use a calculator to approximate the ratio found above, called the *golden ratio*. Compare the approximation to the decimal number that the ratios of the successive terms of the Fibonacci sequence approach, as identified in Section 4.3.

EXAMPLE 2

Adding and Subtracting Radicals with Higher Indexes

Add or subtract the following radical expressions. Assume that all variables represent positive real numbers.

(a)
$$\begin{aligned}
2\sqrt[3]{16} - 5\sqrt[3]{54} &= 2\sqrt[3]{8 \cdot 2} - 5\sqrt[3]{27 \cdot 2} \\
&= 2\sqrt[3]{8} \cdot \sqrt[3]{2} - 5\sqrt[3]{27} \cdot \sqrt[3]{2} \\
&= 2 \cdot 2 \cdot \sqrt[3]{2} - 5 \cdot 3 \cdot \sqrt[3]{2} \\
&= 4\sqrt[3]{2} - 15\sqrt[3]{2} \\
&= -11\sqrt[3]{2}
\end{aligned}$$

(b)
$$\begin{aligned}
2\sqrt[3]{x^2y} + \sqrt[3]{8x^5y^4} &= 2\sqrt[3]{x^2y} + \sqrt[3]{(8x^3y^3)x^2y} \\
&= 2\sqrt[3]{x^2y} + 2xy\sqrt[3]{x^2y} \\
&= (2 + 2xy)\sqrt[3]{x^2y} \qquad \text{Distributive property} \blacksquare
\end{aligned}$$

CAUTION It is a common error to forget to write the index when working wih cube roots, fourth roots, and so on. Be careful not to forget to write these indexes.

The next example shows how to simplify radical expressions involving fractions.

EXAMPLE 3

Adding and Subtracting Radical Expressions with Fractions

Perform the indicated operations.

(a)
$$\begin{aligned}
2\sqrt{\frac{75}{16}} + 4\frac{\sqrt{8}}{\sqrt{32}} &= 2\sqrt{\frac{25 \cdot 3}{16}} + 4\frac{\sqrt{4 \cdot 2}}{\sqrt{16 \cdot 2}} \\
&= 2\left(\frac{5\sqrt{3}}{4}\right) + 4\left(\frac{2\sqrt{2}}{4\sqrt{2}}\right) \\
&= \frac{5\sqrt{3}}{2} + 2 \qquad \text{Multiply.}
\end{aligned}$$

Chalkboard Exercise

Simplify. Assume all variables represent positive real numbers.

$$\sqrt{\frac{80}{y^4}} + \sqrt{\frac{81}{y^{10}}}$$

Answer: $\dfrac{4y^3\sqrt{5} + 9}{y^5}$

$$= \frac{5\sqrt{3}}{2} + \frac{4}{2} \qquad \text{Find the common denominator.}$$

$$= \frac{5\sqrt{3} + 4}{2}$$

(b) $10\sqrt[3]{\dfrac{5}{x^6}} - 3\sqrt[3]{\dfrac{4}{x^9}} = 10\dfrac{\sqrt[3]{5}}{\sqrt[3]{x^6}} - 3\dfrac{\sqrt[3]{4}}{\sqrt[3]{x^9}}$

$$= \frac{10\sqrt[3]{5}}{x^2} - \frac{3\sqrt[3]{4}}{x^3}$$

$$= \frac{10x\sqrt[3]{5}}{x^3} - \frac{3\sqrt[3]{4}}{x^3} \qquad \text{Find the common denominator.}$$

$$= \frac{10x\sqrt[3]{5} - 3\sqrt[3]{4}}{x^3} \qquad (\text{if } x \neq 0) \quad \blacksquare$$

◇ **CONNECTIONS** ◇

To get a sense of the value of a radical expression we can use a calculator. When using a calculator to evaluate some expressions it is important to be careful to use parentheses. For example, to approximate

$$\frac{5\sqrt{3} + 4}{2}$$

it is necessary to use parentheses around the numerator before dividing by 2: enter it as (5 * √3 + 4) ÷ 2.

5.4 EXERCISES

◉ **1.** Which one of the following sums could be simplified without first simplifying the individual radical expressions?
   (a) $\sqrt{50} + \sqrt{32}$ **(b)** $3\sqrt{6} + 9\sqrt{6}$ **(c)** $\sqrt[3]{32} - \sqrt[3]{108}$ **(d)** $\sqrt[5]{6} - \sqrt[5]{192}$

◉ **2.** Let $a = 1$ and let $b = 64$.
   (a) Evaluate $\sqrt{a} + \sqrt{b}$. Then find $\sqrt{a + b}$. Are they equal?
   (b) Evaluate $\sqrt[3]{a} + \sqrt[3]{b}$. Then find $\sqrt[3]{a + b}$. Are they equal?
   (c) Complete the following: In general, $\sqrt[n]{a} + \sqrt[n]{b} \neq$ _____, based on the observations in parts (a) and (b) of this exercise.

◉ **3.** Even though the root indexes of the terms are not equal, the sum $\sqrt{64} + \sqrt[3]{125} + \sqrt[4]{16}$ can be simplified quite easily. What is this sum? Why can we add them so easily?

◉ **4.** There is an old saying "You can't add apples and oranges." How does this apply to the expression $\sqrt{a} + \sqrt[3]{a}$?

Simplify. Assume that all variables represent positive real numbers. See Examples 1–3.

5. $\sqrt{36} - \sqrt{100}$

6. $\sqrt{25} - \sqrt{81}$

7. $-2\sqrt{48} + 3\sqrt{75}$

8. $4\sqrt{32} - 2\sqrt{8}$

9. $6\sqrt{18} - \sqrt{32} + 2\sqrt{50}$

10. $5\sqrt{8} + 3\sqrt{72} - 3\sqrt{50}$

11. $-2\sqrt{63} + 2\sqrt{28} + 2\sqrt{7}$

12. $-\sqrt{27} + 2\sqrt{48} - \sqrt{75}$

1. (b)   2. (a) 9; $\sqrt{65}$; no   (b) 5; $\sqrt[3]{65}$;   no
(c) $\sqrt[n]{a + b}$
3. 15;   Each radical represents a whole number.   4. Since the indexes are different, we cannot combine $\sqrt{a} + \sqrt[3]{a}$ into a single radical using the methods of this section.
5. -4 6. -4 7. $7\sqrt{3}$
8. $12\sqrt{2}$ 9. $24\sqrt{2}$
10. $13\sqrt{2}$ 11. 0   12. 0

◉ CONCEPTUAL   ✐ WRITING   ▲ CHALLENGING   ▦ SCIENTIFIC CALCULATOR   ▦ GRAPHICS CALCULATOR

13. $20\sqrt{5}$ 14. $3\sqrt{6}$
15. $12\sqrt{2x}$ 16. $11\sqrt{2k}$
17. $-11m\sqrt{2}$
18. $-49p\sqrt{3}$
19. $\sqrt[3]{2}$ 20. $37\sqrt[3]{3}$
21. $2\sqrt[3]{x}$ 22. $30\sqrt[3]{2m}$
23. $19\sqrt[4]{2}$ 24. $16\sqrt[4]{2}$
25. $x\sqrt[4]{xy}$
26. $-m^2p\sqrt[4]{mp^2}$
27. $(4 + 3xy)\sqrt[3]{xy^2}$
28. $(5 - 3st)\sqrt[4]{s^3t}$
29. $\dfrac{7\sqrt{2}}{6}$ 30. $\sqrt{3}$
31. $\dfrac{5\sqrt{2}}{3}$ 32. $\dfrac{\sqrt{3}}{2}$
33. $\dfrac{m\sqrt[3]{m^2}}{2}$ 34. $-\dfrac{2a\sqrt[4]{a}}{3}$
35. Both are approximately 11.3137085.
36. Both are approximately -22.5166605.
37. Both are approximately 31.6227766.
38. Both are approximately 23.8117618.   39. (a)
40. (a)
41. $12\sqrt{5} + 5\sqrt{3}$ inches
42. $24\sqrt{3}$ meters

13. $2\sqrt{5} + 3\sqrt{20} + 4\sqrt{45}$

14. $5\sqrt{54} - 2\sqrt{24} - 2\sqrt{96}$

15. $8\sqrt{2x} - \sqrt{8x} + \sqrt{72x}$

16. $4\sqrt{18k} - \sqrt{72k} + \sqrt{50k}$

17. $3\sqrt{72m^2} - 5\sqrt{32m^2} - 3\sqrt{18m^2}$

18. $9\sqrt{27p^2} - 14\sqrt{108p^2} + 2\sqrt{48p^2}$

19. $-\sqrt[3]{54} + 2\sqrt[3]{16}$

20. $15\sqrt[3]{81} - 4\sqrt[3]{24}$

21. $2\sqrt[3]{27x} - 2\sqrt[3]{8x}$

22. $6\sqrt[3]{128m} + 3\sqrt[3]{16m}$

23. $5\sqrt[4]{32} + 3\sqrt[4]{162}$

24. $2\sqrt[4]{512} + 4\sqrt[4]{32}$

25. $3\sqrt[4]{x^5y} - 2x\sqrt[4]{xy}$

26. $2\sqrt[4]{m^9p^6} - 3m^2p\sqrt[4]{mp^2}$

27. $\sqrt[3]{64xy^2} + \sqrt[3]{27x^4y^5}$

28. $\sqrt[4]{625s^3t} - \sqrt[4]{81s^7t^5}$

29. $\sqrt{\dfrac{8}{9}} + \sqrt{\dfrac{18}{36}}$

30. $\sqrt{\dfrac{12}{16}} + \sqrt{\dfrac{48}{64}}$

31. $\dfrac{\sqrt{32}}{3} + \dfrac{2\sqrt{2}}{3} - \dfrac{\sqrt{2}}{\sqrt{9}}$

32. $\dfrac{\sqrt{27}}{2} - \dfrac{3\sqrt{3}}{2} + \dfrac{\sqrt{3}}{\sqrt{4}}$

33. $3\sqrt[3]{\dfrac{m^5}{27}} - 2m\sqrt[3]{\dfrac{m^2}{64}}$

34. $2a\sqrt[4]{\dfrac{a}{16}} - 5a\sqrt[4]{\dfrac{a}{81}}$

In Example 1(a) we show that $3\sqrt{24} + \sqrt{54} = 9\sqrt{6}$. All these terms are expressed as square roots using radical symbols. To support this result, one method is to find a calculator approximation of $3\sqrt{24}$, find a calculator approximation of $\sqrt{54}$, and then add these two approximations. Next, we find a calculator approximation of $9\sqrt{6}$. It should correspond to the sum we found. (For this example, both approximations are 22.04540769. Due to rounding procedures, there may be a discrepancy in the final digit if you try to duplicate this work.)

Follow this procedure to support the statements in Exercises 35–38.

35. $3\sqrt{32} - 2\sqrt{8} = 8\sqrt{2}$

36. $4\sqrt{12} - 7\sqrt{27} = -13\sqrt{3}$

37. $2\sqrt{40} + 6\sqrt{90} - 3\sqrt{160} = 10\sqrt{10}$

38. $5\sqrt{28} - 3\sqrt{63} + 2\sqrt{112} = 9\sqrt{7}$

⊚ **39.** A rectangular yard has a length of $\sqrt{192}$ meters and a width of $\sqrt{48}$ meters. Choose the best estimate of its dimensions. (All measures are in meters.)
   (a) 14 by 7   **(b)** 5 by 7   **(c)** 14 by 8   **(d)** 15 by 8

⊚ **40.** If the base of a triangle is $\sqrt{65}$ inches and its height is $\sqrt{26}$ inches, which one of the following is the best estimate for its area?
   (a) 20 square inches   **(b)** 26 square inches   **(c)** 40 square inches
   (d) 52 square inches

▲ *Work the problem. Give the answer as a simplified radical expression.*

41. Find the perimeter of the triangle shown here.

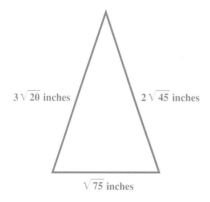

$3\sqrt{20}$ inches $2\sqrt{45}$ inches

$\sqrt{75}$ inches

42. Find the perimeter of the rectangle shown here.

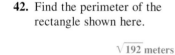
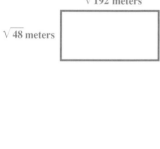

$\sqrt{192}$ meters

$\sqrt{48}$ meters

43. What is the perimeter of a computer graphic with sides measuring $4\sqrt{18}$, $4\sqrt{32}$, $6\sqrt{50}$, and $5\sqrt{12}$ centimeters?

44. Find the area of a trapezoid with bases $\sqrt{288}$ and $\sqrt{72}$ inches and height $\sqrt{24}$ inches.

REVIEW EXERCISES

Find the product. See Section 3.4.

45. $5xy(2x^2y^3 - 4x)$ **46.** $(3x + 7)(2x - 6)$ **47.** $(a^2 + b)(a^2 - b)$

48. $(2p - 7)^2$ **49.** $(4x^3 + 3)^3$ **50.** $(2 + 3y)(2 - 3y)$

Write in lowest terms. See Section 4.1.

51. $\dfrac{8x^2 - 10x}{6x^2}$ **52.** $\dfrac{15y^3 - 9y^2}{6y}$

43. $58\sqrt{2} + 10\sqrt{3}$ centimeters
44. $36\sqrt{3}$ square inches
45. $10x^3y^4 - 20x^2y$
46. $6x^2 - 4x - 42$
47. $a^4 - b^2$
48. $4p^2 - 28p + 49$
49. $64x^9 + 144x^6 + 108x^3 + 27$ **50.** $4 - 9y^2$
51. $\dfrac{4x - 5}{3x}$ **52.** $\dfrac{y(5y - 3)}{2}$

5.5 MULTIPLYING AND DIVIDING RADICAL EXPRESSIONS

FOR EXTRA HELP

 SSG pp. 157–160
SSM pp. 229–234

 Video 7

 Tutorial IBM MAC

OBJECTIVES

1 ▶ Multiply radical expressions.
2 ▶ Rationalize denominators with one radical term.
3 ▶ Rationalize denominators with binomials involving radicals.
4 ▶ Write radical quotients in lowest terms.

1 ▶ Multiply radical expressions.

▶ We multiply binomial expressions involving radicals by using the FOIL (first, outside, inside, last) method, which depends on the distributive property. For example, the product of the binomials $(\sqrt{5} + 3)(\sqrt{6} + 1)$ is found as follows.

$$
\begin{array}{cccc}
\text{First} & \text{Outside} & \text{Inside} & \text{Last} \\
\downarrow & \downarrow & \downarrow & \downarrow
\end{array}
$$

$$(\sqrt{5} + 3)(\sqrt{6} + 1) = \sqrt{5} \cdot \sqrt{6} + \sqrt{5} \cdot 1 + 3 \cdot \sqrt{6} + 3 \cdot 1$$
$$= \sqrt{30} + \sqrt{5} + 3\sqrt{6} + 3$$

This result cannot be simplified further.

E X A M P L E 1
Multiplying Binomials Involving Radical Expressions

Multiply.

(a) $(7 - \sqrt{3})(\sqrt{5} + \sqrt{2}) = 7\sqrt{5} + 7\sqrt{2} - \sqrt{3} \cdot \sqrt{5} - \sqrt{3} \cdot \sqrt{2}$
$$= 7\sqrt{5} + 7\sqrt{2} - \sqrt{15} - \sqrt{6}$$

(b) $(\sqrt{10} + \sqrt{3})(\sqrt{10} - \sqrt{3}) = \sqrt{10}\sqrt{10} - \sqrt{10}\sqrt{3} + \sqrt{10}\sqrt{3} - \sqrt{3}\sqrt{3}$
$$= 10 - 3$$
$$= 7$$

Notice that this is an example of the kind of product that results in the difference of two squares:

$$(a + b)(a - b) = a^2 - b^2.$$

Here, $a = \sqrt{10}$ and $b = \sqrt{3}$.

Chalkboard Exercise

Multiply $(\sqrt{6} + \sqrt{2})(\sqrt{5} - \sqrt{7})$.

Answer:
$\sqrt{30} - \sqrt{42} + \sqrt{10} - \sqrt{14}$

INSTRUCTOR'S RESOURCES

 ITM pp. 402–405 **ISM** pp. 342–350
IAM pp. 21–22

 TEST GENERATOR
DOS Windows MAC

 TRANSPARENCIES

 ● CONCEPTUAL ✍ WRITING ▲ CHALLENGING ▦ SCIENTIFIC CALCULATOR ▨ GRAPHICS CALCULATOR

▶ **TEACHING TIP**
Review the FOIL method with
polynomials at this time. Some
examples can be $(5x + 3)(6x + 1)$,
$(x + 3y)(x - 3y)$, and $(7x - 3)^2$.

   Review: $\sqrt{5} \cdot \sqrt{5} = 5$;
            $\sqrt{5} \cdot \sqrt{6} = \sqrt{30}$;
         $2\sqrt{3} \cdot 4\sqrt{2} = 8\sqrt{6}$. ◀

(c) $(\sqrt{7} - 3)^2 = (\sqrt{7} - 3)(\sqrt{7} - 3)$

$$= \sqrt{7} \cdot \sqrt{7} - \sqrt{7} \cdot 3 - 3\sqrt{7} + 3 \cdot 3$$

$$= 7 - 6\sqrt{7} + 9$$

$$= 16 - 6\sqrt{7}$$

▶ **TEACHING TIP**
Explain why $16 - 6\sqrt{7} \neq$
$10\sqrt{7}$. ◀

(d) $(5 - \sqrt[3]{3})(5 + \sqrt[3]{3}) = 5 \cdot 5 + 5\sqrt[3]{3} - 5\sqrt[3]{3} - \sqrt[3]{3} \cdot \sqrt[3]{3}$

$$= 25 - \sqrt[3]{3^2}$$

$$= 25 - \sqrt[3]{9}$$

(e) $(\sqrt{k} + \sqrt{y})(\sqrt{k} - \sqrt{y}) = (\sqrt{k})^2 - (\sqrt{y})^2$

$$= k - y \quad (\text{if } k \geq 0 \quad \text{and} \quad y \geq 0) \quad ■$$

2 ▶ Rationalize
denominators with one
radical term.

▶   As defined earlier, a simplified radical expression should have no radical in the denominator. For example,

$$\frac{3}{\sqrt{7}}$$

is not simplified; it will not be simplified until the radical is eliminated from the denominator. Before hand-held calculators were easily available, in order to find a decimal approximation for a quotient such as $3/\sqrt{7}$, it was advantageous to rewrite the expression so that it was not necessary to divide by a decimal approximation. Although calculators are now available, there are other reasons to learn how to *rationalize* a radical denominator in a quotient. The process of removing radicals from the denominator so that the denominator contains only rational numbers is called **rationalizing the denominator.** This process is explained in the following examples.

EXAMPLE 2

Rationalizing
Denominators with
Square Roots

Rationalize each denominator.

(a) $\dfrac{3}{\sqrt{7}}$

   Multiply numerator and denominator by $\sqrt{7}$. This is, in effect, multiplying by 1.

$$\frac{3}{\sqrt{7}} = \frac{3 \cdot \sqrt{7}}{\sqrt{7} \cdot \sqrt{7}}$$

Since $\sqrt{7} \cdot \sqrt{7} = \sqrt{7 \cdot 7} = \sqrt{49} = 7$,

$$\frac{3}{\sqrt{7}} = \frac{3\sqrt{7}}{7}.$$

The denominator is now a rational number.

Chalkboard Exercise ·

Rationalize the denominator.

$$\frac{3}{\sqrt{48}}$$

Answer: $\dfrac{\sqrt{3}}{4}$

(b) $\dfrac{5\sqrt{2}}{\sqrt{5}}$

Multiply the numerator and denominator by $\sqrt{5}$.

$$\frac{5\sqrt{2}}{\sqrt{5}} = \frac{5\sqrt{2} \cdot \sqrt{5}}{\sqrt{5} \cdot \sqrt{5}} = \frac{5\sqrt{10}}{5} = \sqrt{10}$$

(c) $\dfrac{6}{\sqrt{12}}$

Less work is involved if the radical in the denominator is simplified first.

$$\frac{6}{\sqrt{12}} = \frac{6}{\sqrt{4 \cdot 3}} = \frac{6}{2\sqrt{3}} = \frac{3}{\sqrt{3}}$$

Now rationalize the denominator by multiplying the numerator and the denominator by $\sqrt{3}$.

$$\frac{3 \cdot \sqrt{3}}{\sqrt{3} \cdot \sqrt{3}} = \frac{3\sqrt{3}}{3} = \sqrt{3} \quad \blacksquare$$

▶ **TEACHING TIP**
Show that rationalizing directly or simplifying the denominator first will yield the same result. Explain that radicals can often be simplified in several ways, but regardless of what method is used, the result will be the same. ◀

The next example shows how to rationalize denominators in expressions involving quotients under a radical sign.

EXAMPLE 3
Rationalizing Denominators in Roots of Fractions

Simplify each of the following.

(a) $\sqrt{\dfrac{18}{125}}$

$$\sqrt{\frac{18}{125}} = \frac{\sqrt{18}}{\sqrt{125}} \qquad \text{Quotient rule}$$

$$= \frac{\sqrt{9 \cdot 2}}{\sqrt{25 \cdot 5}} \qquad \text{Factor.}$$

$$= \frac{3\sqrt{2}}{5\sqrt{5}} \qquad \text{Product rule}$$

$$= \frac{3\sqrt{2} \cdot \sqrt{5}}{5\sqrt{5} \cdot \sqrt{5}} \qquad \text{Multiply by } \sqrt{5} \text{ in numerator and denominator.}$$

$$= \frac{3\sqrt{10}}{5 \cdot 5} \qquad \text{Product rule}$$

$$= \frac{3\sqrt{10}}{25}$$

Chalkboard Exercise

Simplify. Assume all variables represent positive real numbers.

$$\sqrt{\frac{72}{y^3}}$$

Answer: $\dfrac{6\sqrt{2y}}{y^2}$

▶ **TEACHING TIP**
Show in Example 3(b) that, alternatively, multiplying directly by

$$\frac{\sqrt{p^5}}{\sqrt{p^5}}$$

will yield the same result after much simplification.   ◀

(b) $\sqrt{\dfrac{50m^4}{p^5}}, \quad p > 0$

$$\sqrt{\frac{50m^4}{p^5}} = \frac{\sqrt{50m^4}}{\sqrt{p^5}} \qquad \text{Quotient rule}$$

$$= \frac{\sqrt{25m^4 \cdot 2}}{\sqrt{p^4 \cdot p}}$$

$$= \frac{5m^2\sqrt{2}}{p^2\sqrt{p}} \qquad \text{Product rule}$$

$$= \frac{5m^2\sqrt{2} \cdot \sqrt{p}}{p^2\sqrt{p} \cdot \sqrt{p}} \qquad \text{Multiply by } \sqrt{p} \text{ in numerator and denominator.}$$

$$= \frac{5m^2\sqrt{2p}}{p^2 \cdot p} \qquad \text{Product rule}$$

$$= \frac{5m^2\sqrt{2p}}{p^3} \qquad ∎$$

EXAMPLE 4

Rationalizing a Denominator with a Cube Root

Simplify $\sqrt[3]{\dfrac{27}{16}}$.

Use the quotient rule and simplify the numerator and denominator.

$$\sqrt[3]{\frac{27}{16}} = \frac{\sqrt[3]{27}}{\sqrt[3]{16}} = \frac{3}{\sqrt[3]{8} \cdot \sqrt[3]{2}} = \frac{3}{2\sqrt[3]{2}}$$

Chalkboard Exercise

Simplify $\sqrt[3]{\dfrac{15}{32}}$.

Answer: $\dfrac{\sqrt[3]{30}}{4}$

To get a rational denominator multiply the numerator and denominator by a number that will result in a perfect cube in the radicand in the denominator. Since $2 \cdot 4 = 8$, a perfect cube, multiply the numerator and denominator by $\sqrt[3]{4}$.

$$\frac{3}{2\sqrt[3]{2}} = \frac{3 \cdot \sqrt[3]{4}}{2 \cdot \sqrt[3]{2} \cdot \sqrt[3]{4}} = \frac{3\sqrt[3]{4}}{2\sqrt[3]{8}} = \frac{3\sqrt[3]{4}}{2 \cdot 2} = \frac{3\sqrt[3]{4}}{4} \qquad ∎$$

CAUTION   It is easy to make mistakes in problems like the one in Example 4. A typical error is to multiply numerator and denominator by $\sqrt[3]{2}$, forgetting that

$$\sqrt[3]{2} \cdot \sqrt[3]{2} \neq 2.$$

You need *three* factors of 2 to get 2^3 under the radical. As shown in Example 4,

$$\sqrt[3]{2} \cdot \sqrt[3]{2} \cdot \sqrt[3]{2} = 2.$$

---

3 ▶ Rationalize denominators with binomials involving radicals.

▶ **TEACHING TIP**
Starting with the example

$$\frac{3}{1 + \sqrt{2}},$$

show how trying to rationalize with $\sqrt{2}$ or $1 + \sqrt{2}$ will not eliminate radicals in the denominator.   ◀

▶   Recall the special product

$$(a + b)(a - b) = a^2 - b^2.$$

In order to rationalize a denominator that contains a binomial expression (one that contains exactly two terms) involving radicals, such as

$$\frac{3}{1 + \sqrt{2}},$$

we must use conjugates. The **conjugate** of $1 + \sqrt{2}$ is $1 - \sqrt{2}$. In general, $a + b$ and $a - b$ are conjugates.

Rationalizing Binomial Denominators

Whenever a radical expression has a binomial with square root radicals in the denominator, rationalize by multiplying both the numerator and the denominator by the conjugate of the denominator.

For the expression $\dfrac{3}{1 + \sqrt{2}}$, rationalize the denominator by multiplying both the numerator and denominator by the conjugate of the denominator, $1 - \sqrt{2}$.

$$\frac{3}{1 + \sqrt{2}} = \frac{3(1 - \sqrt{2})}{(1 + \sqrt{2})(1 - \sqrt{2})}$$

Then $(1 + \sqrt{2})(1 - \sqrt{2}) = 1^2 - (\sqrt{2})^2 = 1 - 2 = -1$. Placing -1 in the denominator gives

$$\frac{3}{1 + \sqrt{2}} = \frac{3(1 - \sqrt{2})}{-1} = -3(1 - \sqrt{2}) \qquad \text{or} \qquad -3 + 3\sqrt{2}.$$

EXAMPLE 5
Rationalizing a Binomial Denominator

Rationalize the denominator in the following expressions. Assume that all variables represent positive real numbers.

(a) $\dfrac{5}{4 - \sqrt{3}}$

Multiply numerator and denominator by the conjugate of the denominator, $4 + \sqrt{3}$.

$$\frac{5}{4 - \sqrt{3}} = \frac{5(4 + \sqrt{3})}{(4 - \sqrt{3})(4 + \sqrt{3})}$$
$$= \frac{5(4 + \sqrt{3})}{16 - 3}$$
$$= \frac{5(4 + \sqrt{3})}{13}$$

> **Chalkboard Exercise**
>
> Rationalize the denominator.
> $$\frac{7}{\sqrt{2} + \sqrt{13}}$$
> Answer: $\dfrac{-7(\sqrt{2} - \sqrt{13})}{11}$

▶ **TEACHING TIP**
Emphasize that by leaving the numerator in factored form, it is easier to determine whether the expression can be reduced to lowest terms. ◀

Notice that the numerator is left in factored form. Doing this makes it easier to determine whether the expression can be reduced to lowest terms.

(b) $\dfrac{\sqrt{2} - \sqrt{3}}{\sqrt{5} + \sqrt{3}}$

Multiplication of both numerator and denominator by $\sqrt{5} - \sqrt{3}$ will rationalize the denominator.

$$\frac{\sqrt{2} - \sqrt{3}}{\sqrt{5} + \sqrt{3}} = \frac{(\sqrt{2} - \sqrt{3})(\sqrt{5} - \sqrt{3})}{(\sqrt{5} + \sqrt{3})(\sqrt{5} - \sqrt{3})}$$
$$= \frac{\sqrt{10} - \sqrt{6} - \sqrt{15} + 3}{5 - 3}$$
$$= \frac{\sqrt{10} - \sqrt{6} - \sqrt{15} + 3}{2}$$

▶ **TEACHING TIP**
Many students feel uncomfortable about answers that are not "easy numbers" like 2, −6, 1/2, and so on. Warn them that many of the answers in the exercise section are messier than usual. ◀

(c) $\dfrac{3}{\sqrt{5m} - \sqrt{p}} = \dfrac{3(\sqrt{5m} + \sqrt{p})}{(\sqrt{5m} - \sqrt{p})(\sqrt{5m} + \sqrt{p})}$
$$= \frac{3(\sqrt{5m} + \sqrt{p})}{5m - p} \qquad (5m \neq p) \quad \blacksquare$$

◆ 1. $\dfrac{319}{6(8\sqrt{5}+1)}$

2. $\dfrac{9a-b}{b(3\sqrt{a}-\sqrt{b})}$

3. $\dfrac{9a-b}{(\sqrt{b}-\sqrt{a})(3\sqrt{a}-\sqrt{b})}$

4. $\dfrac{(3\sqrt{a}+\sqrt{b})(\sqrt{b}+\sqrt{a})}{b-a}$

◆ **CONNECTIONS** ◆

Sometimes it is desirable to rationalize the *numerator* in an expression. The procedure is similar to rationalizing the denominator. For example, to rationalize the numerator of

$$\frac{6-\sqrt{2}}{4},$$

we multiply by the conjugate of the numerator, $6+\sqrt{2}$.

$$\frac{6-\sqrt{2}}{4}=\frac{(6-\sqrt{2})(6+\sqrt{2})}{4(6+\sqrt{2})}=\frac{36-2}{4(6+\sqrt{2})}=\frac{34}{4(6+\sqrt{2})}$$

In the final expression, the numerator is rationalized.

FOR DISCUSSION OR WRITING

Rationalize the numerator of the following expressions, assuming a and b are nonnegative real numbers.

1. $\dfrac{8\sqrt{5}-1}{6}$ 2. $\dfrac{3\sqrt{a}+\sqrt{b}}{b}$ 3. $\dfrac{3\sqrt{a}+\sqrt{b}}{\sqrt{b}-\sqrt{a}}$

4. Rationalize the denominator of the expression in Item 3, and then describe the difference in the procedure you used from what you did in Item 3.

4 ▶ Write radical quotients in lowest terms.

▶ In the next example we show how to write radical quotients in lowest terms.

EXAMPLE 6
Writing a Radical Quotient in Lowest Terms

Write each expression in lowest terms.

(a) $\dfrac{6+2\sqrt{5}}{4}$

Factor the numerator, and then simplify.

$$\frac{6+2\sqrt{5}}{4}=\frac{2(3+\sqrt{5})}{2\cdot2}=\frac{3+\sqrt{5}}{2}$$

Here is an alternative method for writing this expression in lowest terms.

$$\frac{6+2\sqrt{5}}{4}=\frac{6}{4}+\frac{2\sqrt{5}}{4}=\frac{3}{2}+\frac{\sqrt{5}}{2}=\frac{3+\sqrt{5}}{2}$$

(b) $\dfrac{5y-\sqrt{8y^2}}{6y}=\dfrac{5y-2y\sqrt{2}}{6y}=\dfrac{y(5-2\sqrt{2})}{6y}=\dfrac{5-2\sqrt{2}}{6}$ (if $y>0$)

Notice that the final fraction cannot be reduced further because there is no common factor of 2 in the numerator. ■

Chalkboard Exercise

Write in lowest terms.

$$\frac{24-36\sqrt{7}}{16}$$

Answer: $\dfrac{6-9\sqrt{7}}{4}$

▶ **TEACHING TIP**
Expressions like the one in Example 6(a) occur when the quadratic formula is used, as in the next chapter. Errors often occur when students try to reduce to lowest terms before factoring. Warn them here (and whenever the opportunity arises) that we reduce to lowest terms *only* after the numerator and denominator have been factored. ◀

CAUTION Refer to Example 6(a). A common error occurs when students try to write in lowest terms *before* factoring. Be careful to factor before writing a quotient in lowest terms.

5.5 EXERCISES

◆———————◆ **MATHEMATICAL CONNECTIONS** (Exercises 1–6) ◇————◇

The following operations apply to the exercises of this section. You should be able to perform them without writing intermediate steps. Fill in the blank with the correct response. Assume all variables represent positive real numbers.

1. $\sqrt{a} \cdot \sqrt{b} =$ _____

2. $(x + y)(x - y) =$ _____

3. $(x + \sqrt{y})(x - \sqrt{y}) =$ _____

4. $(\sqrt{x} + \sqrt{y})(\sqrt{x} - \sqrt{y}) =$ _____

5. $(x + y)^2 =$ _____

6. $(\sqrt{x} + \sqrt{y})^2 =$ _____

————————————◆————————————

◉ **7.** Explain why $\sqrt[3]{x} \cdot \sqrt[3]{x}$ is not equal to x. What is it equal to?

◉ **8.** Explain why $\sqrt[4]{x} \cdot \sqrt[4]{x}$ is not equal to x, but *is* equal to $\sqrt{x}$, for $x \geq 0$.

Multiply and then simplify the products. Assume that all variables represent positive real numbers. See Example 1.

9. $\sqrt{3}(\sqrt{12} - 4)$

10. $\sqrt{5}(\sqrt{125} - 6)$

11. $\sqrt{2}(\sqrt{18} - \sqrt{3})$

12. $\sqrt{5}(\sqrt{15} + \sqrt{5})$

13. $(\sqrt{6} + 2)(\sqrt{6} - 2)$

14. $(\sqrt{7} + 8)(\sqrt{7} - 8)$

15. $(\sqrt{12} - \sqrt{3})(\sqrt{12} + \sqrt{3})$

16. $(\sqrt{18} + \sqrt{8})(\sqrt{18} - \sqrt{8})$

17. $(\sqrt{3} + 2)(\sqrt{6} - 5)$

18. $(\sqrt{7} + 1)(\sqrt{2} - 4)$

19. $(\sqrt{3x} + 2)(\sqrt{3x} - 2)$

20. $(\sqrt{6y} - 4)(\sqrt{6y} + 4)$

21. $(2\sqrt{x} + \sqrt{y})(2\sqrt{x} - \sqrt{y})$

22. $(\sqrt{p} + 5\sqrt{s})(\sqrt{p} - 5\sqrt{s})$

23. $(4\sqrt{x} + 3)^2$

24. $(5\sqrt{p} - 6)^2$

25. $(9 - \sqrt[3]{2})(9 + \sqrt[3]{2})$

26. $(7 + \sqrt[3]{6})(7 - \sqrt[3]{6})$

◉ **27.** The correct answer to Exercise 9 is $6 - 4\sqrt{3}$. Explain why this is not equal to $2\sqrt{3}$.

◉ **28.** When we rationalize the denominator in the radical expression $\dfrac{1}{\sqrt{2}}$, we multiply both the numerator and the denominator by $\sqrt{2}$. What property of real numbers covered in Section 1.3 justifies this procedure?

Rationalize the denominator in the expression. Assume that all variables represent positive real numbers. See Examples 2 and 3.

29. $\dfrac{7}{\sqrt{7}}$

30. $\dfrac{11}{\sqrt{11}}$

31. $\dfrac{15}{\sqrt{3}}$

32. $\dfrac{12}{\sqrt{6}}$

33. $\dfrac{\sqrt{3}}{\sqrt{2}}$

34. $\dfrac{\sqrt{7}}{\sqrt{6}}$

35. $\dfrac{9\sqrt{3}}{\sqrt{5}}$

36. $\dfrac{3\sqrt{2}}{\sqrt{11}}$

37. $\dfrac{-6}{\sqrt{18}}$

38. $\dfrac{-5}{\sqrt{24}}$

39. $\sqrt{\dfrac{7}{2}}$

40. $\sqrt{\dfrac{10}{3}}$

41. $-\sqrt{\dfrac{7}{50}}$

42. $-\sqrt{\dfrac{13}{75}}$

43. $\sqrt{\dfrac{24}{x}}$

44. $\sqrt{\dfrac{52}{y}}$

45. $\dfrac{-8\sqrt{3}}{\sqrt{k}}$

46. $\dfrac{-4\sqrt{13}}{\sqrt{m}}$

47. $-\sqrt{\dfrac{150m^5}{n^3}}$

48. $-\sqrt{\dfrac{98r^3}{s^5}}$

49. $\sqrt{\dfrac{288x^7}{y^9}}$

50. $\sqrt{\dfrac{242t^9}{u^{11}}}$

1. $\sqrt{ab}$ 2. $x^2 - y^2$
3. $x^2 - y$ 4. $x - y$
5. $x^2 + 2xy + y^2$
6. $x + 2\sqrt{xy} + y$
9. $6 - 4\sqrt{3}$
10. $25 - 6\sqrt{5}$
11. $6 - \sqrt{6}$
12. $5\sqrt{3} + 5$
13. 2   14. -57
15. 9   16. 10
17. $3\sqrt{2} - 5\sqrt{3} + 2\sqrt{6} - 10$
18. $\sqrt{14} - 4\sqrt{7} + \sqrt{2} - 4$
19. $3x - 4$ 20. $6y - 16$
21. $4x - y$ 22. $p - 25s$
23. $16x + 24\sqrt{x} + 9$
24. $25p - 60\sqrt{p} + 36$
25. $81 - \sqrt[3]{4}$
26. $49 - \sqrt[3]{36}$
28. identity property for multiplication
29. $\sqrt{7}$ 30. $\sqrt{11}$
31. $5\sqrt{3}$ 32. $2\sqrt{6}$
33. $\dfrac{\sqrt{6}}{2}$ 34. $\dfrac{\sqrt{42}}{6}$
35. $\dfrac{9\sqrt{15}}{5}$ 36. $\dfrac{3\sqrt{22}}{11}$
37. $-\sqrt{2}$ 38. $\dfrac{-5\sqrt{6}}{12}$
39. $\dfrac{\sqrt{14}}{2}$ 40. $\dfrac{\sqrt{30}}{3}$
41. $-\dfrac{\sqrt{14}}{10}$ 42. $-\dfrac{\sqrt{39}}{15}$
43. $\dfrac{2\sqrt{6x}}{x}$ 44. $\dfrac{2\sqrt{13y}}{y}$
45. $\dfrac{-8\sqrt{3k}}{k}$
46. $\dfrac{-4\sqrt{13m}}{m}$
47. $\dfrac{-5m^2\sqrt{6mn}}{n^2}$
48. $\dfrac{-7r\sqrt{2rs}}{s^3}$
49. $\dfrac{12x^3\sqrt{2xy}}{y^5}$
50. $\dfrac{11t^4\sqrt{2tu}}{u^6}$

 CONCEPTUAL   WRITING   ▲ CHALLENGING   SCIENTIFIC CALCULATOR    GRAPHICS CALCULATOR

53. $\dfrac{\sqrt[3]{18}}{3}$ 54. $\dfrac{\sqrt[3]{100}}{5}$

55. $\dfrac{\sqrt[3]{12}}{3}$ 56. $\dfrac{\sqrt[3]{20}}{4}$

57. $-\dfrac{\sqrt[3]{2pr}}{r}$

58. $-\dfrac{\sqrt[3]{6xy}}{y}$

59. $\dfrac{2\sqrt[4]{x^3}}{x}$ 60. $\dfrac{3\sqrt[4]{y^3}}{y}$

63. $\dfrac{2(4-\sqrt{3})}{13}$

64. $\dfrac{6(5-\sqrt{2})}{23}$

65. $3(\sqrt{5}-\sqrt{3})$
66. $4(\sqrt{6}-\sqrt{3})$
67. $\sqrt{3}+\sqrt{7}$
68. $\sqrt{2}-\sqrt{5}$
69. $\sqrt{7}-\sqrt{6}-\sqrt{14}+2\sqrt{3}$
70. $-\sqrt{6}+\sqrt{5}-3\sqrt{2}+\sqrt{15}$

71. $\dfrac{4\sqrt{x}(\sqrt{x}+2\sqrt{y})}{x-4y}$

72. $\dfrac{5\sqrt{r}(3\sqrt{r}-\sqrt{s})}{9r-s}$

73. $\dfrac{x-2\sqrt{xy}+y}{x-y}$

74. $\dfrac{a+2\sqrt{ab}+b}{a-b}$

75. Square both sides to show that each is equal to $\dfrac{2-\sqrt{3}}{4}$. 76. Each original expression is approximately equal to .2588190451.

77. $\dfrac{5+2\sqrt{6}}{4}$

78. $\dfrac{2-\sqrt{2}}{4}$

79. $\dfrac{4+2\sqrt{2}}{3}$

80. $\dfrac{2+9\sqrt{2}}{3}$

81. $\dfrac{6+2\sqrt{6x}}{3}$

82. $\dfrac{1+y\sqrt{2y}}{2}$

⦿ **51.** Look again at the expression in Exercise 44. After writing as a quotient of radicals, multiply both the numerator and the denominator by $\sqrt{y}$, and then obtain the final answer. Next, start over, multiplying both the numerator and the denominator by $\sqrt{y^3}$, to obtain the same answer. Which method do you prefer? Why?

⦿ **52.** Explain why $\dfrac{1}{\sqrt[3]{2}}$ cannot be written with denominator rationalized if you begin by multiplying both the numerator and the denominator by $\sqrt[3]{2}$. By what can you multiply them both to achieve the desired result?

Simplify. See Example 4.

53. $\sqrt[3]{\dfrac{2}{3}}$ **54.** $\sqrt[3]{\dfrac{4}{5}}$ **55.** $\sqrt[3]{\dfrac{4}{9}}$ **56.** $\sqrt[3]{\dfrac{5}{16}}$

57. $-\sqrt[3]{\dfrac{2p}{r^2}}$ **58.** $-\sqrt[3]{\dfrac{6x}{y^2}}$ **59.** $\sqrt[4]{\dfrac{16}{x}}$ **60.** $\sqrt[4]{\dfrac{81}{y}}$

⦿ **61.** Explain the procedure you will use to rationalize the denominator of the expression in Exercise 63: $\dfrac{2}{4+\sqrt{3}}$.

⦿ **62.** Would multiplying both the numerator and the denominator of $\dfrac{2}{4+\sqrt{3}}$ by $4+\sqrt{3}$ lead to a rationalized denominator? Why or why not?

Rationalize the denominator in each expression. Assume that all variables represent positive real numbers and that no denominators are zero. See Example 5.

63. $\dfrac{2}{4+\sqrt{3}}$ **64.** $\dfrac{6}{5+\sqrt{2}}$ **65.** $\dfrac{6}{\sqrt{5}+\sqrt{3}}$

66. $\dfrac{12}{\sqrt{6}+\sqrt{3}}$ **67.** $\dfrac{-4}{\sqrt{3}-\sqrt{7}}$ **68.** $\dfrac{-3}{\sqrt{2}+\sqrt{5}}$

69. $\dfrac{1-\sqrt{2}}{\sqrt{7}+\sqrt{6}}$ **70.** $\dfrac{-1-\sqrt{3}}{\sqrt{6}+\sqrt{5}}$ **71.** $\dfrac{4\sqrt{x}}{\sqrt{x}-2\sqrt{y}}$

72. $\dfrac{5\sqrt{r}}{3\sqrt{r}+\sqrt{s}}$ **73.** $\dfrac{\sqrt{x}-\sqrt{y}}{\sqrt{x}+\sqrt{y}}$ **74.** $\dfrac{\sqrt{a}+\sqrt{b}}{\sqrt{a}-\sqrt{b}}$

▲ **75.** If a and b are both positive numbers and $a^2=b^2$, then $a=b$. Use this fact to show that $\dfrac{\sqrt{6}-\sqrt{2}}{4}=\dfrac{\sqrt{2}-\sqrt{3}}{2}$.

▦ **76.** Use a calculator approximation to support your result for Exercise 75.

Write the expression in lowest terms. Assume that all variables represent positive real numbers. See Example 6.

77. $\dfrac{25+10\sqrt{6}}{20}$ **78.** $\dfrac{12-6\sqrt{2}}{24}$ **79.** $\dfrac{16+4\sqrt{8}}{12}$

80. $\dfrac{12+9\sqrt{72}}{18}$ **81.** $\dfrac{6x+\sqrt{24x^3}}{3x}$ **82.** $\dfrac{11y+\sqrt{242y^5}}{22y}$

83. The following expression occurs in a certain standard problem in trigonometry:

$$\frac{1}{\sqrt{2}} \cdot \frac{\sqrt{3}}{2} - \frac{1}{\sqrt{2}} \cdot \frac{1}{2}.$$

Show that it simplifies to $\dfrac{\sqrt{6} - \sqrt{2}}{4}$. Then verify using a calculator approximation.

84. The following expression occurs in a certain standard problem in trigonometry:

$$\frac{\sqrt{3} + 1}{1 - \sqrt{3}}.$$

Show that it simplifies to $-2 - \sqrt{3}$. Then verify using a calculator approximation.

83. Each expression is approximately equal to .2588190451.
84. Each expression is approximately equal to -3.732050808.
85. $\left\{\dfrac{3}{8}\right\}$ 86. $\left\{\dfrac{19}{3}\right\}$
87. $\left\{-\dfrac{1}{3}, \dfrac{3}{2}\right\}$ 88. $\left\{\dfrac{1}{3}, \dfrac{2}{5}\right\}$
89. $4x^2 + 20x + 25$
90. $9x^2 - 6x + 1$
91. $x^4 + 2x^2 + 5$
92. $4x^2 - 2x + 3$
93. true   94. false
95. true   96. false

REVIEW EXERCISES

Solve the equation. See Sections 2.1 and 3.9.

85. $-8t + 7 = 4$

86. $3x - 7 = 12$

87. $6x^2 - 7x = 3$

88. $m(15m - 11) = -2$

Apply the exponent in each expression to give a polynomial as a result. See Section 3.4.

89. $(2x + 5)^2$

90. $(3x - 1)^2$

91. $(\sqrt{x^4 + 2x^2 + 5})^2$

92. $(\sqrt[3]{4x^2 - 2x + 3})^3$

Determine whether the statement is true or false for the value of x given. See Section 1.2.

93. $\sqrt{4 - x} = x + 2$ for $x = 0$

94. $\sqrt{4 - x} = x + 2$ for $x = -5$

95. $\sqrt{x^2 - 4x + 9} = x - 1$ for $x = 4$

96. $\sqrt{5x + 6} + \sqrt{3x + 4} = 2$ for $x = 15$

5.6 EQUATIONS WITH RADICALS

FOR EXTRA HELP

SSG pp. 157–160
SSM pp. 229–234

Video 7

Tutorial
IBM   MAC

OBJECTIVES

1 ▶ Learn the power rule.
2 ▶ Solve radical equations, such as $\sqrt{3x + 4} = 8$.
3 ▶ Solve radical equations that require the square of a binomial, such as $\sqrt{m^2 - 4m + 9} = m - 1$.
4 ▶ Solve radical equations that require squaring twice, such as $\sqrt{5m + 6} + \sqrt{3m + 4} = 2$.
5 ▶ Solve radical equations with indexes greater than 2.

Now that we have learned how to perform operations on radical expressions, we can use the same concepts to solve equations that include radical expressions.

INSTRUCTOR'S RESOURCES

 ITM pp. 405–409 **ISM** pp. 350–362
IAM p. 22

 TEST GENERATOR
DOS Windows MAC

 **TRANSPARENCIES**

 CONCEPTUAL   WRITING    CHALLENGING   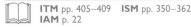 SCIENTIFIC CALCULATOR    GRAPHICS CALCULATOR

◆ 62.5 meters;
155 meters

◆ **CONNECTIONS** ◆

A number of useful formulas involve radicals or radical expressions. Many occur in the mathematics needed for working with objects in space. In *Spacemathematics,* the formula

$$N = \frac{1}{2\pi} \sqrt{\frac{a}{r}}$$

is used to find the rotational rate N of a space station.* Here a is the acceleration and r represents the radius of the space station. To find the value of r that will make N simulate the effect of gravity on Earth, the equation must be solved for r, using the required value of N.

FOR DISCUSSION OR WRITING
After reading through this section, find the value of r that makes $N = .063$ rotations per second, if a is to equal 9.8 (the gravity on Earth in meters per second per second). What radius is needed if N is to equal .04 rotations per second and $a = 9.8$ meters per second per second? What units will r have? Show how you can determine the appropriate units for r.

1 ▶ Learn the power rule.

▶ The equation $x = 1$ has only one solution. Its solution set is $\{1\}$. If we square both sides of this equation, another equation is obtained: $x^2 = 1$. This new equation has two solutions: -1 and 1. Notice that the solution of the original equation is also a solution of the squared equation. However, the squared equation has another solution, -1, that is *not* a solution of the original equation. When solving equations with radicals, we will use this idea of raising both sides to a power. It is an application of the *power rule*.

Power Rule If both sides of an equation are raised to the same power, all solutions of the original equation are also solutions of the new equation.

Read the power rule carefully; it does *not* say that all solutions to the new equation are solutions to the original equation. They may or may not be.

NOTE When the power rule is used to solve an equation, **every solution to the new equation *must* be checked in the original equation.**

Solutions that do not satisfy the original equation are called **extraneous;** they must be discarded.

2 ▶ Solve radical equations, such as $\sqrt{3x + 4} = 8$.

▶ The first example shows how to use the power rule in solving an equation.

*Source: Formula, from *Spacemathematics* by Bernice Kastner, Ph.D. Copyright © 1972 by the National Aeronautics and Space Administration. Reprinted by permission of the National Aeronautics and Space Administration.

EXAMPLE 1

Using the Power Rule

Solve $\sqrt{3x + 4} = 8$.

Use the power rule and square both sides of the equation to get

$$(\sqrt{3x + 4})^2 = 8^2$$
$$3x + 4 = 64$$
$$3x = 60$$
$$x = 20.$$

Check this proposed solution in the *original* equation.

$$\sqrt{3x + 4} = 8$$
$$\sqrt{3 \cdot 20 + 4} = 8 \quad ? \quad \text{Let } x = 20.$$
$$\sqrt{64} = 8 \quad ?$$
$$8 = 8 \quad \text{True}$$

► **TEACHING TIP**
Review: For $x \geq 0$,
$(\sqrt{x})^2 = \sqrt{x} \cdot \sqrt{x} = x$. In words:
When squaring the square root of a number, the result is the number. ◄

Since 20 satisfies the *original* equation, the solution set is {20}. ∎

The solution of the equation in Example 1 can be generalized to give a method for solving equations with radicals.

Solving an Equation with Radicals

Step 1 **Isolate the radical.** Make sure that one radical term is alone on one side of the equation.

Step 2 **Apply the power rule.** Raise each side of the equation to a power that is the same as the index of the radical.

Step 3 **Solve.** Solve the resulting equation; if it still contains a radical, repeat Steps 1 and 2.

Step 4 **Check.** It is essential that all potential solutions be checked in the original equation.

CAUTION Be careful not to skip Step 4 or you may get an incorrect solution set.

EXAMPLE 2

Using the Power Rule

Solve $\sqrt{5q - 1} + 3 = 0$.

To get the radical alone on one side, subtract 3 from both sides.

$$\sqrt{5q - 1} = -3$$

Now square both sides.

$$(\sqrt{5q - 1})^2 = (-3)^2$$
$$5q - 1 = 9$$
$$5q = 10$$
$$q = 2$$

► **TEACHING TIP**
Emphasize that after raising both sides of an equation to a power, it is essential to check all potential solutions, since some may not be actual solutions. Emphasize that we must go back to the *original* equation to check. ◄

We must check the potential solution, 2, by substituting it in the original equation.

$$\sqrt{5q - 1} + 3 = 0$$
$$\sqrt{5 \cdot 2 - 1} + 3 = 0 \quad ? \quad \text{Let } q = 2.$$
$$3 + 3 = 0 \quad \text{False}$$

This false result shows that 2 is not a solution of the original equation; it is extraneous. The solution set is ∅. ∎

NOTE We could have determined after the first step that the equation in Example 2 had no solution. The equation $\sqrt{5q - 1} = -3$ has no solution because the expression on the left cannot be negative.

3 ▶ Solve radical equations that require the square of a binomial, such as $\sqrt{m^2 - 4m + 9} = m - 1$.

▶ The next examples involve finding the square of a binomial. Recall that $(x + y)^2 = x^2 + 2xy + y^2$.

EXAMPLE 3
Using the Power Rule; Squaring a Binomial

Solve $\sqrt{4 - x} = x + 2$.

Square both sides; the square of $x + 2$ is $(x + 2)^2 = x^2 + 4x + 4$. Thus, the new equation is quadratic, and we need to get 0 on one side.

$$(\sqrt{4 - x})^2 = (x + 2)^2$$
$$4 - x = x^2 + 4x + 4$$

↑———— Twice the product of 2 and x

$$0 = x^2 + 5x \qquad \text{Subtract 4 and add } x.$$
$$0 = x(x + 5) \qquad \text{Factor.}$$
$$x = 0 \qquad \text{or} \qquad x + 5 = 0 \qquad \text{Zero-factor property}$$
$$x = -5$$

Check each potential solution in the original equation.

If $x = 0$,
$$\sqrt{4 - x} = x + 2$$
$$\sqrt{4 - 0} = 0 + 2 \qquad ?$$
$$\sqrt{4} = 2 \qquad ?$$
$$2 = 2. \qquad \text{True}$$

If $x = -5$,
$$\sqrt{4 - x} = x + 2$$
$$\sqrt{4 - (-5)} = -5 + 2 \qquad ?$$
$$\sqrt{9} = -3 \qquad ?$$
$$3 = -3. \qquad \text{False}$$

The solution set is {0}. The other potential solution, -5, is extraneous. ■

CAUTION When a radical equation requires squaring a binomial as in Example 3, the middle term is often omitted in error. Remember:

$$(x + 2)^2 \neq x^2 + 4 \qquad (x + 2)^2 = x^2 + 4x + 4.$$

EXAMPLE 4
Using the Power Rule; Squaring a Binomial

Solve $\sqrt{m^2 - 4m + 9} = m - 1$.

Square both sides. The square of the binomial $m - 1$ is $(m - 1)^2 = m^2 - 2(m)(1) + 1^2$.

$$(\sqrt{m^2 - 4m + 9})^2 = (m - 1)^2$$
$$m^2 - 4m + 9 = m^2 - 2m + 1$$

↑———— Twice the product of m and 1

Subtract m^2 and 1 from both sides. Then add $4m$ to both sides, to get

$$8 = 2m$$
$$4 = m.$$

Check this potential solution in the original equation.

$$\sqrt{m^2 - 4m + 9} = m - 1$$
$$\sqrt{4^2 - 4 \cdot 4 + 9} = 4 - 1 \qquad ? \qquad \text{Let } m = 4.$$
$$3 = 3 \qquad\qquad\qquad \text{True}$$

The solution set of the original equation is $\{4\}$. ∎

4 ▶ Solve radical equations that require squaring twice, such as $\sqrt{5m + 6} + \sqrt{3m + 4} = 2$.

▶ In the next example, we show an equation where both sides must be squared twice.

E X A M P L E 5

Using the Power Rule; Squaring Twice

Solve $\sqrt{5m + 6} + \sqrt{3m + 4} = 2$.

Start by getting one radical alone on one side of the equation. Do this by subtracting $\sqrt{3m + 4}$ from both sides.

$$\sqrt{5m + 6} = 2 - \sqrt{3m + 4}$$

Now square both sides.

$$(\sqrt{5m + 6})^2 = (2 - \sqrt{3m + 4})^2$$
$$5m + 6 = 4 - 4\sqrt{3m + 4} + (3m + 4)$$

⎯ Twice the product of 2 and $\sqrt{3m + 4}$

This equation still contains a radical, so it will be necessary to square both sides again. Before doing this, isolate the radical term on the right.

$$5m + 6 = 8 + 3m - 4\sqrt{3m + 4}$$
$$2m - 2 = -4\sqrt{3m + 4} \qquad \text{Subtract 8 and } 3m.$$
$$m - 1 = -2\sqrt{3m + 4} \qquad \text{Divide by 2.}$$

Now square both sides again.

$$(m - 1)^2 = (-2\sqrt{3m + 4})^2$$
$$(m - 1)^2 = (-2)^2(\sqrt{3m + 4})^2 \qquad (ab)^2 = a^2b^2$$
$$m^2 - 2m + 1 = 4(3m + 4)$$
$$m^2 - 2m + 1 = 12m + 16 \qquad \text{Distributive property}$$

This equation is quadratic and may be solved with the zero-factor property. Start by getting 0 on one side of the equation; then factor.

$$m^2 - 14m - 15 = 0$$
$$(m - 15)(m + 1) = 0$$

By the zero-factor property,

$$m - 15 = 0 \qquad \text{or} \qquad m + 1 = 0$$
$$m = 15 \qquad \text{or} \qquad m = -1.$$

Check each of these potential solutions in the original equation. Only -1 works, so the solution set, $\{-1\}$, has only one element. ∎

5 ▶ Solve radical equations with indexes greater than 2.

▶ The power rule also works for powers greater than 2, as the next example shows.

Chalkboard Exercise

Solve $\sqrt{2y + 3} + \sqrt{y + 1} = 1$.

Answer: $\{-1\}$

▶ TEACHING TIP

A common error occurs when students try to square both sides by squaring term by term. Explain how the right side of the equation is a binomial and must be squared by using the identity

$$(a + b)^2 = a^2 + 2ab + b^2. \quad ◀$$

EXAMPLE 6

Using the Power Rule for Powers Greater than 2

Solve $\sqrt[4]{z + 5} = \sqrt[4]{2z - 6}$.

Raise both sides to the fourth power.

$$(\sqrt[4]{z + 5})^4 = (\sqrt[4]{2z - 6})^4$$
$$z + 5 = 2z - 6$$
$$11 = z$$

Check this result in the original equation.

$$\sqrt[4]{z + 5} = \sqrt[4]{2z - 6}$$
$$\sqrt[4]{11 + 5} = \sqrt[4]{2 \cdot 11 - 6} \quad ? \quad \text{Let } z = 11.$$
$$\sqrt[4]{16} = \sqrt[4]{16} \qquad\qquad\quad \text{True}$$

The solution set is $\{11\}$. ∎

5.6 EXERCISES

1. (a) yes   (b) no
2. (a) yes   (b) no
3. (a) yes   (b) no
4. (a) no   (b) yes
5. no;   There is no solution.   6. Since the radical on the left-hand side cannot be negative, and it must equal x, x cannot be negative.
7. $\{19\}$ 8. $\{23\}$ 9. $\left\{\dfrac{38}{3}\right\}$
10. $\left\{\dfrac{37}{2}\right\}$ 11. $\emptyset$
12. $\emptyset$ 13. $\{5\}$ 14. $\{2\}$
15. $\{1\}$ 16. $\{2\}$ 17. $\{9\}$
18. $\{4\}$ 19. $\{3\}$ 20. $\{-3\}$
21. $\{-3, 3\}$ 22. $\{3\}$
23. $\{-3, 3\}$ 24. (a) more than   (b) the same as

Check the equation to see if the given value for x is a solution.

1. $\sqrt{3x + 18} = x$
   (a) 6   (b) -3

2. $\sqrt{2x + 3} - x = 0$
   (a) 3   (b) -1

3. $\sqrt{x + 2} = \sqrt{9x - 2} - 2\sqrt{x - 1}$
   (a) 2   (b) 7

4. $\sqrt[3]{3x + 4} = \sqrt[3]{2x - 1}$
   (a) $-\dfrac{3}{5}$ (b) -5

5. Is 9 a solution of the equation $\sqrt{x} = -3$? If not, what is the solution of this equation?

 6. Before even attempting to solve $\sqrt{3x + 18} = x$, how can you be sure that the equation cannot have a negative solution?

Solve the equation. See Examples 1 and 2.

7. $\sqrt{x - 3} = 4$

8. $\sqrt{y + 2} = 5$

9. $\sqrt{3k - 2} = 6$

10. $\sqrt{4t + 7} = 9$

11. $\sqrt{x + 9} = 0$

12. $\sqrt{w + 4} = 0$

13. $\sqrt{3x - 6} - 3 = 0$

14. $\sqrt{7y + 11} - 5 = 0$

15. $\sqrt{6x + 2} - \sqrt{5x + 3} = 0$

16. $\sqrt{3 + 5x} - \sqrt{x + 11} = 0$

17. $3\sqrt{x} = \sqrt{8x + 9}$

18. $6\sqrt{p} = \sqrt{30p + 24}$

◆ **MATHEMATICAL CONNECTIONS** (Exercises 19–24) ◆

Solve the equation.

19. $x = 3$

20. $x = -3$

21. $x^2 = 9$

22. $x^3 = 27$

23. $x^4 = 81$

24. Fill in the blanks in the following generalization.

   Suppose both sides of $x = k$ are raised to the nth power.
   (a) If n is even, the number of solutions of the new equation is
   _____ the number of solutions of the original equation.
   (more than/the same as/fewer than)
   (b) If n is odd, the number of solutions of the new equation is
   _____ the number of solutions of the original equation.
   (more than/the same as/fewer than)

◆

 CONCEPTUAL   WRITING   CHALLENGING   SCIENTIFIC CALCULATOR    GRAPHICS CALCULATOR

◉ **25.** What is *wrong* with this first step in the solution process for $\sqrt{3x + 4} = 8 - x$?

$$3x + 4 = 64 + x^2$$

◉ **26.** What is *wrong* with this first step in the solution process for
$\sqrt{5y + 6} - \sqrt{y + 3} = 3$?

$$(5y + 6) + (y + 3) = 9$$

Solve the equation. See Examples 3 and 4.

27. $\sqrt{3x + 4} = 8 - x$ **28.** $\sqrt{5x + 1} = 2x - 2$

29. $\sqrt{13 + 4t} = t + 4$ **30.** $\sqrt{50 + 7k} = k + 8$

31. $\sqrt{r^2 - 15r + 15} + 5 = r$ **32.** $\sqrt{p^2 + 12p - 4} + 4 = p$

Solve the equation. See Example 5.

33. $\sqrt{r + 4} - \sqrt{r - 4} = 2$ **34.** $\sqrt{m + 1} - \sqrt{m - 2} = 1$

▲ **35.** $\sqrt{11 + 2q} + 1 = \sqrt{5q + 1}$ ▲ **36.** $\sqrt{6 + 5y} - 3 = \sqrt{y + 3}$

▲ **37.** $\sqrt{3 - 3p} - \sqrt{3p + 2} = 3$ ▲ **38.** $\sqrt{3x + 4} - \sqrt{2x - 4} = 2$

◉ **39.** What is the smallest power to which you can raise both sides of the radical equation $\sqrt[3]{x + 3} = \sqrt[3]{5 + 4x}$ so that the radicals are eliminated?

◉ **40.** What is the smallest power to which you can raise both sides of the radical equation $\sqrt{x + 3} = \sqrt[3]{10x + 14}$ so that the radicals are eliminated?

Solve the equation. See Example 6.

41. $\sqrt[3]{2x^2 + 3x - 7} = \sqrt[3]{2x^2 + 4x + 6}$ **42.** $\sqrt[3]{3y^2 - 4y + 6} = \sqrt[3]{3y^2 - 2y + 8}$

43. $\sqrt[3]{1 - 2k} - \sqrt[3]{-k - 13} = 0$ **44.** $\sqrt[3]{11 - 2t} - \sqrt[3]{-1 - 5t} = 0$

45. $\sqrt[4]{x - 1} + 2 = 0$ **46.** $\sqrt[3]{2k + 3} + 1 = 0$

47. $\sqrt[4]{x + 7} = \sqrt[4]{2x}$ **48.** $\sqrt[4]{y + 8} = \sqrt[4]{3y}$

▲ *For the equation, rewrite the expressions with rational exponents as radical expressions, and then solve using the procedures explained in this section.*

49. $(5r - 6)^{1/2} = 2 + (3r - 6)^{1/2}$ **50.** $(3w + 7)^{1/2} = 1 + (w + 2)^{1/2}$

51. $(2w - 1)^{2/3} - w^{1/3} = 0$ **52.** $(x^2 - 2x)^{1/3} - x^{1/3} = 0$

If x is the number of years since 1900, the equation $y = x^{.7}$ approximates the timber grown in the United States in billions of cubic feet. Let x = 20 represent 1920, x = 52 represent 1952, and so on.

▦ **53.** Replace *x* in the equation for each year shown in the figure and give the value of *y*. (Round answers to the nearest billion.)

54. Use the figure to estimate the amount of timber grown for each year shown.

55. Compare the values found from the equation with your estimates from the figure. Does the equation give a good approximation of the data from the figure? In which year is the approximation best?

56. From the figure, estimate the amount of timber harvested in each year shown.

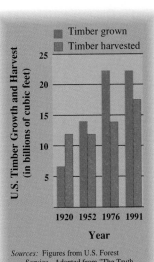

■ Timber grown
■ Timber harvested

U.S. Timber Growth and Harvest (in billions of cubic feet)

1920 1952 1976 1991

Year

Sources: Figures from U.S. Forest Service. Adapted from "The Truth about America's Forests," Evergreen, 4025 Crater Lake Hwy., Medford, Ore. 97504.

Answers:

25. You cannot just square each term. The right-hand side should be $(8 - x)^2 = 64 - 16x + x^2$.
26. You cannot just square each term. The left-hand side should be $5y + 6 - 2\sqrt{(5y + 6)(y + 3)} + y + 3$.   27. {4}
28. {3}   29. {-3, -1}
30. {-7, -2}   31. $\emptyset$
32. $\emptyset$ 33. {5}   34. {3}
35. {7}   36. {6}   37. $\emptyset$
38. {4, 20}   39. 3   40. 6
41. {-13}   42. {-1}
43. {14}   44. {-4}
45. $\emptyset$ 46. $\emptyset$ 47. {7}
48. {4}   49. {2, 14}

50. {-2, -1}   51. $\left\{\dfrac{1}{4}, 1\right\}$

52. {0, 3}   53. 8 billion cubic feet;
16 billion cubic feet;
21 billion cubic feet;
24 billion cubic feet
54. 7 billion cubic feet;
14 billion cubic feet;
22.5 billion cubic feet;
22.5 billion cubic feet
55. fairly good;   1976
56. 12 billion cubic feet;
12 billion cubic feet;
14 billion cubic feet;
17.5 billion cubic feet

◉ CONCEPTUAL   🖉 WRITING   ▲ CHALLENGING   ▦ SCIENTIFIC CALCULATOR   ▦ GRAPHICS CALCULATOR

57. 6 billion cubic feet;
12 billion cubic feet;
15 billion cubic feet;
16 billion cubic feet
58. fairly good; 1920
59. $(x - 6)^2$
60. $(2x + 5)^2$
61. $-14, 14$ **62.** $-75, 75$
63. $\{-6\}$ **64.** $\left\{-\dfrac{5}{2}\right\}$
65. $\dfrac{6 - 3\sqrt{2}}{4}$
66. $2 + \sqrt{6}$

57. Use the equation $y = x^{.62}$ to approximate the amount of timber harvested in each of the given years. (Round answers to the nearest billion.)

58. Compare your answers from Exercises 56 and 57. Does the equation give a good approximation? For which year is it poorest?

REVIEW EXERCISES

Factor the polynomial. See Section 3.7.

59. $x^2 - 12x + 36$ **60.** $4x^2 + 20x + 25$

Give all square roots of the number. See Section 5.1.

61. 196   **62.** 5625

Solve the equation. See Section 3.9.

63. $x^2 + 12x = -36$ **64.** $4x^2 + 20x + 25 = 0$

Express in lowest terms. See Section 5.5.

65. $\dfrac{12 - 6\sqrt{2}}{8}$ **66.** $\dfrac{4 + \sqrt{24}}{2}$

5.7 COMPLEX NUMBERS

FOR EXTRA HELP	OBJECTIVES
SSG pp. 160–165 **SSM** pp. 234–239	**1** ▶ Simplify numbers of the form $\sqrt{-b}$, where $b > 0$.
Video 7	**2** ▶ Recognize imaginary complex numbers.
Tutorial IBM MAC	**3** ▶ Add and subtract complex numbers.
	4 ▶ Multiply complex numbers.
	5 ▶ Divide complex numbers.
	6 ▶ Find powers of i.

As discussed in Chapter 1, the set of real numbers includes many other number sets (the rational numbers, integers, and natural numbers, for example). In this section a new set of numbers is introduced that includes the set of real numbers, as well as numbers that are even roots of negative numbers, like $\sqrt{-2}$.

1 ▶ Simplify numbers of the form $\sqrt{-b}$, where $b > 0$.

▶ The equation $x^2 + 1 = 0$ has no real number solutions, since any solution must be a number whose square is -1. In the set of real numbers all squares are nonnegative numbers, because the product of either two positive numbers or two negative numbers is positive and $0^2 = 0$. To provide a solution for the equation $x^2 + 1 = 0$, a new number i is defined so that

$$i^2 = -1.$$

That is, i is a number whose square is -1, so $i = \sqrt{-1}$. This definition of i makes it possible to define any square root of a negative number as follows.

INSTRUCTOR'S RESOURCES

ITM pp. 409–412 **ISM** pp. 362–370 **IAM** pp. 22–23	**TEST GENERATOR** DOS Windows MAC **TRANSPARENCIES**

For any positive real number b, $\quad \sqrt{-b} = i\sqrt{b}$.

Write each number as a product of a real number and i.

(a) $\sqrt{-100} = i\sqrt{100} = 10i$

(b) $\sqrt{-2} = \sqrt{2}i = i\sqrt{2}$ ■

CAUTION It is easy to mistake $\sqrt{2}i$ for $\sqrt{2i}$, with the i under the radical. For this reason, we often write $\sqrt{2}i$ as $i\sqrt{2}$.

When finding a product such as $\sqrt{-4} \cdot \sqrt{-9}$, we cannot use the product rule for radicals since that rule applies only when both radicals represent real numbers. For this reason, we always change $\sqrt{-b}$ ($b > 0$) to the form $i\sqrt{b}$ before performing any multiplications or divisions. For example,

$$\sqrt{-4} \cdot \sqrt{-9} = i\sqrt{4} \cdot i\sqrt{9} = i \cdot 2 \cdot i \cdot 3 = 6i^2.$$

Since $i^2 = -1$,

$$6i^2 = 6(-1) = -6.$$

An *incorrect* use of the product rule for radicals would give a wrong answer.

$$\sqrt{-4} \cdot \sqrt{-9} = \sqrt{(-4)(-9)} = \sqrt{36} = 6 \quad \text{INCORRECT}$$

Multiply.

(a) $\sqrt{-3} \cdot \sqrt{-7} = i\sqrt{3} \cdot i\sqrt{7} = i^2\sqrt{3 \cdot 7} = (-1)\sqrt{21} = -\sqrt{21}$

(b) $\sqrt{-2} \cdot \sqrt{-8} = i\sqrt{2} \cdot i\sqrt{8} = i^2\sqrt{2 \cdot 8} = (-1)\sqrt{16} = (-1)4 = -4$

(c) $\sqrt{-5} \cdot \sqrt{6} = i\sqrt{5} \cdot \sqrt{6} = i\sqrt{30}$ ■

The methods used to find products also apply to quotients, as the next example shows.

Divide.

(a) $\dfrac{\sqrt{-75}}{\sqrt{-3}} = \dfrac{i\sqrt{75}}{i\sqrt{3}} = \sqrt{\dfrac{75}{3}} = \sqrt{25} = 5$

(b) $\dfrac{\sqrt{-32}}{\sqrt{8}} = \dfrac{i\sqrt{32}}{\sqrt{8}} = i\sqrt{\dfrac{32}{8}} = i\sqrt{4} = 2i$ ■

2 ▶ Recognize imaginary complex numbers.

▶ With the new number i and the real numbers, a new set of numbers can be formed that includes the real numbers as a subset. The *complex numbers* are defined as follows.

| Complex Number | If a and b are real numbers, then any number of the form $a + bi$ is called a **complex number**. |

▶ **TEACHING TIP**
Show examples of real numbers and imaginary numbers written in standard form. For example, $3 = 3 + 0i$ and $4i = 0 + 4i$. ◀

In the complex number $a + bi$, the number a is called the **real part** and b is called the **imaginary part**. When $b = 0$, $a + bi$ is a real number, so the real numbers are a subset of the complex numbers. Complex numbers with $b \neq 0$ are called **imaginary numbers.*** In spite of their name, imaginary numbers are very useful in applications, particularly in work with electricity.

The relationships among the various sets of numbers discussed in this book are shown in Figure 2.

▶ **TEACHING TIP**
Relate the exercises in this section to simplifying algebraic expressions in earlier chapters. However, emphasize that $i^2 = -1$. Stress that when i^2 appears, simply replace it with -1. ◀

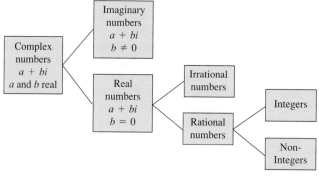

FIGURE 2

◈ Answers will vary.

◈ **CONNECTIONS** ◈

Girolamo Cardano first introduced the concept of the square root of a negative number in 1545. Complex numbers have since been shown to have many applications in such areas as atomic physics, electronics, and the study of forces. For example, a car going around a banked curve is subject to the forces shown in the figure. Each of these forces has magnitude, shown by the length of the arrow, and direction, indicated by its direction in the figure. Imagine

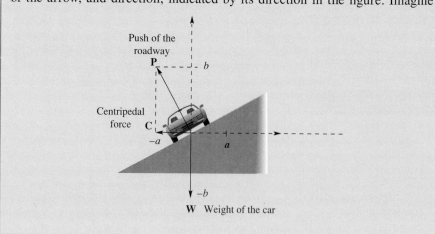

---

* Imaginary numbers are sometimes defined as complex numbers with $a = 0$ and $b \neq 0$.

axes (dashed lines) with the origin at the point where the three arrows meet. Then the end of each arrow can be located on the plane with reference to the horizontal and vertical axes. Complex numbers are used to indicate the position of each arrowhead. In the figure, force **P** is described as $-a + bi$, force **C** as $-a + 0i$ (or just $-a$), and force **W** as $0 - bi$ (or just $-bi$).

FOR DISCUSSION OR WRITING
Elsewhere in this text we discuss graphing points on the plane. From your previous experience with graphing, what relationship do you see between the points on the plane and the forces as described above?

3 ▶ Add and subtract complex numbers.

▶ The commutative, associative, and distributive properties for real numbers are also valid for complex numbers. To add complex numbers, add their real parts and add their imaginary parts.

EXAMPLE 4
Adding Complex Numbers

Add.

(a) $(2 + 3i) + (6 + 4i) = (2 + 6) + (3 + 4)i$ Commutative, associative, and distributive properties

$$= 8 + 7i$$

(b) $5 + (9 - 3i) = (5 + 9) - 3i$
$$= 14 - 3i$$ ∎

We subtract complex numbers by subtracting their real parts and subtracting their imaginary parts.

EXAMPLE 5
Subtracting Complex Numbers

Subtract.

(a) $(6 + 5i) - (3 + 2i) = (6 - 3) + (5 - 2)i = 3 + 3i$

(b) $(7 - 3i) - (8 - 6i) = (7 - 8) + [-3 - (-6)]i = -1 + 3i$

(c) $(-9 + 4i) - (-9 + 8i) = [-9 - (-9)] + (4 - 8)i$
$$= 0 - 4i \quad \text{or} \quad -4i$$ ∎

In Example 5(c), the answer was written as $0 - 4i$ and then as just $-4i$. A complex number written in the form $a + bi$, like $-4i$, is in **standard form.** In this section, most answers will be given in standard form, but if b or a is 0, we consider answers such as a or bi to be in standard form.

4 ▶ Multiply complex numbers.

▶ Complex numbers of the form $a + bi$ have the same form as a binomial, so we multiply two complex numbers by using the FOIL method for multiplying binomials. (Recall that FOIL stands for *First-Outside-Inside-Last.*)

EXAMPLE 6
Multiplying Complex Numbers

(a) Multiply $3 + 5i$ and $4 - 2i$.
Use the FOIL method.

$$(3 + 5i)(4 - 2i) = 3(4) + 3(-2i) + 5i(4) + 5i(-2i)$$

First   Outside   Inside   Last

Now simplify. (Remember that $i^2 = -1$.)

$$(3 + 5i)(4 - 2i) = 12 - 6i + 20i - 10i^2$$
$$= 12 + 14i - 10(-1) \qquad \text{Let } i^2 = -1.$$
$$= 12 + 14i + 10$$
$$= 22 + 14i$$

► **TEACHING TIP**

Remind students to combine the
real terms in their results.   ◄

(b) $(2 + 3i)(1 - 5i) = 2(1) + 2(-5i) + 3i(1) + 3i(-5i)$
$$= 2 - 10i + 3i - 15i^2$$
$$= 2 - 7i - 15(-1)$$
$$= 2 - 7i + 15$$
$$= 17 - 7i$$

(c) $4i(2 + 3i)$

Use the distributive property.

$$4i(2 + 3i) = 4i(2) + 4i(3i) = 8i + 12i^2 = 8i + 12(-1) = -12 + 8i \quad ■$$

The two complex numbers $a + bi$ and $a - bi$ are called *conjugates* of each other. The product of a complex number and its conjugate is always a real number, as shown here.

$$(a + bi)(a - bi) = a \cdot a - abi + abi - b^2i^2$$
$$= a^2 - b^2(-1)$$
$$= a^2 + b^2$$
$$(a + bi)(a - bi) = a^2 + b^2$$

For example, $(3 + 7i)(3 - 7i) = 3^2 + 7^2 = 9 + 49 = 58$.

5 ► Divide complex
numbers.

► The quotient of two complex numbers should be a complex number. To write the quotient as a complex number, we need to eliminate i in the denominator. We use conjugates to do this, as the next example shows.

EXAMPLE 7
Dividing Complex
Numbers

Find the quotients.

(a) $\dfrac{4 - 3i}{5 + 2i}$

Multiply the numerator and the denominator by the conjugate of the denominator. The conjugate of $5 + 2i$ is $5 - 2i$.

Chalkboard Exercise

Divide $\dfrac{6 + 2i}{4 - 3i}$.

Answer: $\dfrac{18}{25} + \dfrac{26}{25}i$

$$\frac{4 - 3i}{5 + 2i} = \frac{(4 - 3i)(5 - 2i)}{(5 + 2i)(5 - 2i)}$$
$$= \frac{20 - 8i - 15i + 6i^2}{5^2 + 2^2}$$
$$= \frac{14 - 23i}{29} \qquad \text{or} \qquad \frac{14}{29} - \frac{23}{29}i$$

Notice that this is just like rationalizing the denominator. The final result is given in standard form.

(b) $\dfrac{1 + i}{i}$

The conjugate of i is $-i$. Multiply the numerator and the denominator by $-i$.

$$\frac{1 + i}{i} = \frac{(1 + i)(-i)}{i(-i)}$$

$$= \frac{-i - i^2}{-i^2}$$

$$= \frac{-i - (-1)}{-(-1)}$$

$$= \frac{-i + 1}{1}$$

$$= 1 - i \quad \blacksquare$$

6 ▶ Find powers of i.   ▶   The fact that $i^2 = -1$ can be used to find higher powers of i, as shown in the following examples.

$$i^3 = i \cdot i^2 = i(-1) = -i \qquad i^6 = i^2 \cdot i^4 = (-1) \cdot 1 = -1$$
$$i^4 = i^2 \cdot i^2 = (-1)(-1) = 1 \qquad i^7 = i^3 \cdot i^4 = (-i) \cdot 1 = -i$$
$$i^5 = i \cdot i^4 = i \cdot 1 = i \qquad i^8 = i^4 \cdot i^4 = 1 \cdot 1 = 1$$

A few powers of i are listed here.

Powers of i

$i^1 = i$	$i^5 = i$	$i^9 = i$	$i^{13} = i$
$i^2 = -1$	$i^6 = -1$	$i^{10} = -1$	$i^{14} = -1$
$i^3 = -i$	$i^7 = -i$	$i^{11} = -i$	$i^{15} = -i$
$i^4 = 1$	$i^8 = 1$	$i^{12} = 1$	$i^{16} = 1$

As these examples suggest, the powers of i rotate through the four numbers i, -1, $-i$, and 1. Larger powers of i can be simplified by using the fact that $i^4 = 1$. For example, $i^{75} = (i^4)^{18} \cdot i^3 = 1^{18} \cdot i^3 = 1 \cdot i^3 = i^3 = -i$. This example suggests a quick method for simplifying large powers of i.

EXAMPLE 8

Simplifying Powers of i

Find each power of i.

(a) $i^{12} = (i^4)^3 = 1^3 = 1$

(b) $i^{39} = i^{36} \cdot i^3 = (i^4)^9 \cdot i^3 = 1^9 \cdot (-i) = -i$

(c) $i^{-2} = \dfrac{1}{i^2} = \dfrac{1}{-1} = -1$

(d) $i^{-1} = \dfrac{1}{i}$

To simplify this quotient, multiply the numerator and the denominator by the conjugate of i, which is $-i$.

$$i^{-1} = \frac{1}{i} = \frac{1(-i)}{i(-i)} = \frac{-i}{-i^2} = \frac{-i}{-(-1)} = \frac{-i}{1} = -i \quad \blacksquare$$

Chalkboard Exercise

Find the power of i.

i^{50}

Answer: -1

308 CHAPTER 5 ROOTS AND RADICALS

5.7 EXERCISES

Answers (left column):

1. false 2. true 3. true
4. false 5. true 6. true
7. $13i$ 8. $15i$ 9. $-12i$
10. $-14i$ 11. $i\sqrt{5}$
12. $i\sqrt{21}$ 13. $4i\sqrt{3}$
14. $4i\sqrt{6}$ 15. -15
16. -19 17. -10
18. -27 19. $\sqrt{3}$
20. 2 21. $5i$ 22. $4i$
25. $10 + 8i$ 26. $3 - 2i$
27. $-1 + 7i$
28. $-4 + 29i$ 29. 0
30. 0 31. $7 + 3i$
32. $6 - i$ 33. -2 34. 3
35. $1 + 13i$ 36. $4 + 8i$
37. $6 + 6i$ 38. $1 - 4i$
39. $4 + 2i$ 40. -5
41. -81 42. -625
43. -16 44. -64
45. $-10 - 30i$
46. $-27 + 12i$
47. $10 - 5i$ 48. $23 + i$
49. $-9 + 40i$

Decide whether the statement is true or false.

1. $i = -1$ **2.** $i^2 = -1$

3. $-i = -\sqrt{-1}$ **4.** If $a > 0$, $\sqrt{-a} = -1\sqrt{a}$.

5. If $a > 0$, $\sqrt{-a} = i\sqrt{a}$.

6. If $a > 0$ and $b > 0$, then $\sqrt{-ab} = i\sqrt{ab}$.

Write the number as a product of a real number and i. Simplify all radical expressions. See Example 1.

7. $\sqrt{-169}$ **8.** $\sqrt{-225}$ **9.** $-\sqrt{-144}$ **10.** $-\sqrt{-196}$

11. $\sqrt{-5}$ **12.** $\sqrt{-21}$ **13.** $\sqrt{-48}$ **14.** $\sqrt{-96}$

Multiply or divide as indicated. See Examples 2 and 3.

15. $\sqrt{-15} \cdot \sqrt{-15}$ **16.** $\sqrt{-19} \cdot \sqrt{-19}$ **17.** $\sqrt{-4} \cdot \sqrt{-25}$ **18.** $\sqrt{-9} \cdot \sqrt{-81}$

19. $\dfrac{\sqrt{-300}}{\sqrt{-100}}$ **20.** $\dfrac{\sqrt{-40}}{\sqrt{-10}}$ **21.** $\dfrac{\sqrt{-75}}{\sqrt{3}}$ **22.** $\dfrac{\sqrt{-160}}{\sqrt{10}}$

23. Every real number is a complex number. Explain why this is so.

24. Not every complex number is a real number. Give an example of this, and explain why this statement is true.

Add or subtract as indicated. Write your answers in the form a + bi. See Examples 4 and 5.

25. $(6 + 2i) + (4 + 6i)$ **26.** $(1 + i) + (2 - 3i)$

27. $(3 + 2i) + (-4 + 5i)$ **28.** $(7 + 15i) + (-11 + 14i)$

29. $(5 - i) + (-5 + i)$ **30.** $(-2 + 6i) + (2 - 6i)$

31. $(4 + i) - (-3 - 2i)$ **32.** $(9 + i) - (3 + 2i)$

33. $(-3 - 4i) - (-1 - 4i)$ **34.** $(-2 - 3i) - (-5 - 3i)$

35. $(-4 + 11i) + (-2 - 4i) + (7 + 6i)$ **36.** $(-1 + i) + (2 + 5i) + (3 + 2i)$

37. $[(7 + 3i) - (4 - 2i)] + (3 + i)$ **38.** $[(7 + 2i) + (-4 - i)] - (2 + 5i)$

39. Fill in the blank with the correct response:

Because $(4 + 2i) - (3 + i) = 1 + i$, using the definition of subtraction, we can check this to find that $(1 + i) + (3 + i) =$ _____ .

40. Fill in the blank with the correct response:

Because $\dfrac{-5}{2 - i} = -2 - i$, using the definition of division, we can check this to find that $(-2 - i)(2 - i) =$ _____ .

Multiply. See Example 6.

41. $(3i)(27i)$ **42.** $(5i)(125i)$ **43.** $(-8i)(-2i)$

44. $(-32i)(-2i)$ **45.** $5i(-6 + 2i)$ **46.** $3i(4 + 9i)$

47. $(4 + 3i)(1 - 2i)$ **48.** $(7 - 2i)(3 + i)$ **49.** $(4 + 5i)^2$

⊙ CONCEPTUAL   ✎ WRITING   ▲ CHALLENGING   ▦ SCIENTIFIC CALCULATOR   ▦ GRAPHICS CALCULATOR

50. $(3 + 2i)^2$

51. $(-4 - i)^2$

52. $(-3 - i)^2$

53. $(12 + 3i)(12 - 3i)$

54. $(6 + 7i)(6 - 7i)$

55. $(4 + 9i)(4 - 9i)$

56. $(7 + 2i)(7 - 2i)$

◉ **57.** **(a)** What is the conjugate of $a + bi$?

(b) If we multiply $a + bi$ by its conjugate, we get _____ + _____ , which is always a real number.

◉ **58.** Explain the procedure you would use to find the quotient

$$\frac{-1 + 5i}{3 + 2i}.$$

Write the expression in the form $a + bi$. See Example 7.

59. $\dfrac{2}{1 - i}$

60. $\dfrac{29}{5 + 2i}$

61. $\dfrac{-7 + 4i}{3 + 2i}$

62. $\dfrac{-38 - 8i}{7 + 3i}$

63. $\dfrac{8i}{2 + 2i}$

64. $\dfrac{-8i}{1 + i}$

65. $\dfrac{2 - 3i}{2 + 3i}$

66. $\dfrac{-1 + 5i}{3 + 2i}$

◆ **MATHEMATICAL CONNECTIONS** (Exercises 67–72) ◆

Consider the following expressions:

Binomials	Complex Numbers
$x + 2$, $\quad 3x - 1$	$1 + 2i$, $\quad 3 - i$

When we add, subtract, or multiply complex numbers in standard form, the rules are the same as those for the corresponding operations on binomials. That is, we add or subtract like terms, and we use FOIL to multiply. Division, however, is comparable to division by the sum or difference of radicals, where we multiply by the conjugate to get a rational denominator. To express the quotient of two complex numbers in standard form, we also multiply by the conjugate of the denominator.

The following exercises illustrate these ideas. Work them in order.

67. **(a)** Add the two binomials. **(b)** Add the two complex numbers.

68. **(a)** Subtract the second binomial from the first. **(b)** Subtract the second complex number from the first.

69. **(a)** Multiply the two binomials. **(b)** Multiply the two complex numbers.

70. **(a)** Rationalize the denominator: $\dfrac{\sqrt{3} - 1}{1 + \sqrt{2}}$. **(b)** Write in standard form: $\dfrac{3 - i}{1 + 2i}$.

✎ **71.** Explain why the answers for (a) and (b) in Exercise 69 do not correspond as the answers in Exercises 67–68 do.

✎ **72.** Explain why the answers for (a) and (b) in Exercise 70 do not correspond as the answers in Exercises 67–68 do.

◆

◉ **73.** Recall that if $a \neq 0$, $\dfrac{1}{a}$ is called the reciprocal of a. Use this definition to express the reciprocal of $5 - 4i$ in the form $a + bi$.

◉ **74.** Recall that if $a \neq 0$, a^{-1} is defined to be $\dfrac{1}{a}$. Use this definition to express $(4 - 3i)^{-1}$ in the form $a + bi$.

50. $5 + 12i$ 51. $15 + 8i$
52. $8 + 6i$ 53. 153
54. 85 55. 97 56. 53
57. (a) $a - bi$ (b) a^2; b^2
59. $1 + i$ 60. $5 - 2i$
62. $-1 + 2i$ 62. $-5 + i$
63. $2 + 2i$ 64. $-4 - 4i$
65. $-\dfrac{5}{13} - \dfrac{12}{13}i$
66. $\dfrac{7}{13} + \dfrac{17}{13}i$
67. (a) $4x + 1$ (b) $4 + i$
68. (a) $-2x + 3$
(b) $-2 + 3i$
69. (a) $3x^2 + 5x - 2$
(b) $5 + 5i$
70. (a) $-\sqrt{3} + \sqrt{6} + 1 - \sqrt{2}$ (b) $\dfrac{1}{5} - \dfrac{7}{5}i$
73. $\dfrac{5}{41} + \dfrac{4}{41}i$
74. $\dfrac{4}{25} + \dfrac{3}{25}i$

 ◉ CONCEPTUAL ✎ WRITING ▲ CHALLENGING ▦ SCIENTIFIC CALCULATOR ▦ GRAPHICS CALCULATOR

75. -1 76. -1 77. i
78. 1   79. $-i$ 80. $-i$
82. $\dfrac{37}{10} - \dfrac{19}{10}i$
83. $\dfrac{1}{2} + \dfrac{1}{2}i$ 84. 4
85. $(1 + 5i)^2 - 2(1 + 5i) + 26$ will simplify to 0 when the operations are applied.
86. $(3 + 2i)^2 - 6(3 + 2i) + 13$ will simplify to 0 when the operations are applied.
87. $\left\{ -\dfrac{13}{6} \right\}$ 88. $\left\{ \dfrac{7}{4} \right\}$
89. $\{-8, 5\}$ 90. $\left\{ -1, \dfrac{7}{2} \right\}$

Find each power of i. See Example 8.

75. i^{18} **76.** i^{26} **77.** i^{89} **78.** i^{48} **79.** i^{-5} **80.** i^{-17}

▲**81.** A student simplified i^{-18} as follows:

$$i^{-18} = i^{-18} \cdot i^{20} = i^{-18+20} = i^2 = -1.$$

Explain the mathematical justification for this correct work.

▲**82.** Add: $3(2 - i)^{-1} + 5(1 + i)^{-1}$.

▲ *Ohm's law for the current I in a circuit with voltage E, resistance R, capacitance reactance X_c, and inductive reactance X_L is*

$$I = \frac{E}{R + (X_L - X_c)i}.$$

83. Find I if $E = 2 + 3i$, $R = 5$, $X_L = 4$, and $X_c = 3$.

84. Using the law given for Exercise 83, find E if $I = 1 - i$, $R = 2$, $X_L = 3$, and $X_c = 1$.

85. Show that $1 + 5i$ is a solution of $x^2 - 2x + 26 = 0$.

86. Show that $3 + 2i$ is a solution of $x^2 - 6x + 13 = 0$.

REVIEW EXERCISES

Solve the equation. See Sections 2.1 and 3.9.

87. $6x + 13 = 0$ **88.** $4x - 7 = 0$

89. $x(x + 3) = 40$ **90.** $2x^2 - 5x - 7 = 0$

CHAPTER 5 SUMMARY

KEY TERMS

5.1 principal nth root
radicand
index (order)
radical

5.3 Pythagorean
formula
hypotenuse
legs

5.4 radical expression

5.5 rationalizing the
denominator
conjugate

5.6 extraneous solution

5.7 complex number
real part
imaginary part
imaginary number
standard form

NEW SYMBOLS

$a^{1/n}$	a to the power $\dfrac{1}{n}$
$\sqrt[n]{a}$	principal nth root of a
$\pm$	positive or negative
$a^{m/n}$	a to the power $\dfrac{m}{n}$
$\approx$	is approximately equal to
i	a number whose square is -1

QUICK REVIEW

CONCEPTS	EXAMPLES
5.1 RATIONAL EXPONENTS AND RADICALS	
$a^{1/n} = b$ means $b^n = a$. $a^{1/n}$ is the principal nth root of a. $a^{1/n} = \sqrt[n]{a}$ whenever $\sqrt[n]{a}$ is a real number.	The two square roots of 64 are $\sqrt{64} = 8$ and $-\sqrt{64} = -8$. Of these, 8 is the principal square root of 64. $$25^{1/2} = \sqrt{25} = 5$$ $$(-64)^{1/3} = \sqrt[3]{-64} = -4$$

CONCEPTS	EXAMPLES
$\sqrt[n]{a^n} = \lvert a \rvert$ if n is even. $\sqrt[n]{a^n} = a$ if n is odd.	$\sqrt[3]{-27} = \sqrt[3]{(-3)^3} = -3 \qquad \sqrt[4]{(-2)^4} = \lvert -2 \rvert = 2$
If m and n are positive integers with m/n in lowest terms, then $$a^{m/n} = (a^{1/n})^m$$ provided that $a^{1/n}$ is a real number.	$8^{5/3} = (8^{1/3})^5 = 2^5 = 32$

5.2 MORE ABOUT RATIONAL EXPONENTS AND RADICALS

All the usual rules for exponents are valid for rational exponents.	$5^{-1/2} \cdot 5^{1/4} = 5^{-1/2+1/4} = 5^{-1/4} = \dfrac{1}{5^{1/4}}$ $$(y^{2/5})^{10} = y^4$$ $$\frac{x^{-1/3}}{x^{-1/2}} = x^{-1/3+1/2} = x^{1/6} \quad (x > 0)$$

5.3 SIMPLIFYING RADICALS

Product and Quotient Rules for Radicals If $\sqrt[n]{a}$ and $\sqrt[n]{b}$ are real numbers, and if n is a natural number, $$\sqrt[n]{a} \cdot \sqrt[n]{b} = \sqrt[n]{ab}$$ and $$\sqrt[n]{\frac{a}{b}} = \frac{\sqrt[n]{a}}{\sqrt[n]{b}} \quad (b \neq 0).$$	Apply the product and quotient rules. $$\sqrt{3} \cdot \sqrt{7} = \sqrt{21}$$ $$\sqrt[5]{x^3y} \cdot \sqrt[5]{xy^2} = \sqrt[5]{x^4y^3}$$ $$\frac{\sqrt{x^5}}{\sqrt{x^4}} = \sqrt{\frac{x^5}{x^4}} = \sqrt{x} \quad (x > 0)$$
Simplified Radical 1. The radicand has no factor raised to a power greater than or equal to the index. 2. The radicand has no fractions. 3. No denominator contains a radical. 4. Exponents in the radicand and the index of the radical have no common factor (except 1). That is, $$\sqrt[kn]{a^{km}} = \sqrt[n]{a^m}$$ if all roots exist.	Simplify each radical. $$\sqrt{18} = \sqrt{9 \cdot 2} = 3\sqrt{2}$$ $$\sqrt[3]{54x^5y^3} = \sqrt[3]{27x^3y^3 \cdot 2x^2} = 3xy\sqrt[3]{2x^2}$$ $$\sqrt{\frac{7}{4}} = \frac{\sqrt{7}}{\sqrt{4}} = \frac{\sqrt{7}}{2}$$ $$\sqrt[9]{x^3} = x^{3/9} = x^{1/3} \quad \text{or} \quad \sqrt[3]{x}$$ $$\sqrt[8]{64} = \sqrt[8]{2^6} = \sqrt[4]{2^3} = \sqrt[4]{8}$$

5.4 ADDING AND SUBTRACTING RADICAL EXPRESSIONS

Only radical expressions with the same index and the same radicand may be combined.	Perform the indicated operations. $$3\sqrt{17} + 2\sqrt{17} - 8\sqrt{17} = (3 + 2 - 8)\sqrt{17}$$ $$= -3\sqrt{17}$$ $$\sqrt[3]{2} - \sqrt[3]{250} = \sqrt[3]{2} - 5\sqrt[3]{2}$$ $$= -4\sqrt[3]{2}$$ $\left.\begin{array}{l} \sqrt{15} + \sqrt{30} \\ \sqrt{3} + \sqrt[3]{3} \end{array}\right\}$ cannot be further simplified.

CONCEPTS	EXAMPLES

5.5 MULTIPLYING AND DIVIDING RADICAL EXPRESSIONS

Radical expressions are multiplied by using the distributive property or the FOIL method. Special products from Section 3.7 may apply.	Multiply. $$(\sqrt{2} + \sqrt{7})(\sqrt{3} - \sqrt{6}) = \sqrt{6} - \sqrt{12} + \sqrt{21} - \sqrt{42}$$ $$= \sqrt{6} - 2\sqrt{3} + \sqrt{21} - \sqrt{42}$$ $$(\sqrt{5} - \sqrt{10})(\sqrt{5} + \sqrt{10}) = 5 - 10 = -5$$ $$(\sqrt{3} - \sqrt{2})^2 = 3 - 2\sqrt{3} \cdot \sqrt{2} + 2$$ $$= 5 - 2\sqrt{6}$$
Rationalize the denominator by multiplying both the numerator and denominator by the same expression.	Rationalize the denominator. $$\frac{\sqrt{7}}{\sqrt{5}} = \frac{\sqrt{7}}{\sqrt{5}} \cdot \frac{\sqrt{5}}{\sqrt{5}} = \frac{\sqrt{35}}{5}$$ $$\frac{\sqrt[3]{2}}{\sqrt[3]{4}} = \frac{\sqrt[3]{2}}{\sqrt[3]{4}} \cdot \frac{\sqrt[3]{2}}{\sqrt[3]{2}} = \frac{\sqrt[3]{4}}{\sqrt[3]{8}} = \frac{\sqrt[3]{4}}{2}$$ $$\frac{4}{\sqrt{5} - \sqrt{2}} = \frac{4}{\sqrt{5} - \sqrt{2}} \cdot \frac{\sqrt{5} + \sqrt{2}}{\sqrt{5} + \sqrt{2}}$$ $$= \frac{4(\sqrt{5} + \sqrt{2})}{5 - 2}$$ $$= \frac{4(\sqrt{5} + \sqrt{2})}{3}$$
To write a quotient involving radicals, such as $$\frac{5 + 15\sqrt{6}}{10}$$ in lowest terms, factor the numerator and denominator, and then divide both by the greatest common factor.	$$\frac{5 + 15\sqrt{6}}{10} = \frac{5(1 + 3\sqrt{6})}{5 \cdot 2}$$ $$= \frac{1 + 3\sqrt{6}}{2}$$

5.6 EQUATIONS WITH RADICALS

Solving Equations with Radicals *Step 1* Isolate one radical on one side of the equals sign. *Step 2* Raise each side of the equation to a power that is the same as the index of the radical. *Step 3* Solve the resulting equation; if it still contains a radical, repeat Steps 1 and 2. *Step 4* Check all potential solutions in the *original* equation. Potential solutions that do not check are *extraneous;* they are not part of the solution set.	Solve $\sqrt{2x + 3} - x = 0$. $$\sqrt{2x + 3} = x$$ $$2x + 3 = x^2$$ $$x^2 - 2x - 3 = 0$$ $$(x - 3)(x + 1) = 0$$ $$x - 3 = 0 \quad \text{or} \quad x + 1 = 0$$ $$x = 3 \qquad\qquad x = -1$$ A check shows that 3 is a solution, but -1 is extraneous. The solution set is $\{3\}$.

CONCEPTS	EXAMPLES

5.7   COMPLEX NUMBERS

$i^2 = -1$

For any positive number b,

$$\sqrt{-b} = i\sqrt{b}.$$

To perform a multiplication such as $\sqrt{-3} \cdot \sqrt{-27}$, first change each factor to the form $i\sqrt{b}$, then multiply. The same procedure applies to quotients such as

$$\frac{\sqrt{-18}}{\sqrt{-2}}.$$

Simplify.

$$\sqrt{-3} \cdot \sqrt{-27} = i\sqrt{3} \cdot i\sqrt{27}$$
$$= i^2\sqrt{81}$$
$$= -1 \cdot 9 = -9$$

$$\frac{\sqrt{-18}}{\sqrt{-2}} = \frac{i\sqrt{18}}{i\sqrt{2}} = \frac{\sqrt{18}}{\sqrt{2}} = \sqrt{9} = 3$$

Adding and Subtracting Complex Numbers

$(a + bi) + (c + di) = (a + c) + (b + d)i$

$(a + bi) - (c + di) = (a - c) + (b - d)i$

Perform the indicated operations.

$(5 + 3i) + (8 - 7i) = 13 - 4i$

$(5 + 3i) - (8 - 7i) = (5 + 3i) + (-8 + 7i)$
$$= -3 + 10i$$

Multiplying and Dividing Complex Numbers
Multiply complex numbers by using the FOIL method. Divide complex numbers by multiplying the numerator and the denominator by the conjugate of the denominator.

$(2 + i)(5 - 3i) = 10 - 6i + 5i - 3i^2$
$$= 10 - i - 3(-1)$$
$$= 10 - i + 3$$
$$= 13 - i$$

$$\frac{2}{3 + i} = \frac{2}{3 + i} \cdot \frac{3 - i}{3 - i}$$
$$= \frac{2(3 - i)}{9 - i^2}$$
$$= \frac{2(3 - i)}{10}$$
$$= \frac{3 - i}{5} = \frac{3}{5} - \frac{1}{5}i$$

CHAPTER 5 REVIEW EXERCISES

———◆ **MATHEMATICAL CONNECTIONS** (Exercises 1–6) ◆———

In the first three chapters of this book, we worked with polynomials. In Chapters 4 and 5, we extended the rules developed in the earlier chapters to rational expressions and radical expressions.

1. Evaluate the following expressions for $x = 2$.

  (a) $2x^2 - 5x + 6$ **(b)** $\dfrac{x^3 - 4}{x + 1}$ **(c)** $\sqrt{5x - 2}$

2. Which of the expressions in Exercise 1 is a
    (a) rational expression?   **(b)** polynomial?   **(c)** radical expression?

1. (a) 4   (b) $\dfrac{4}{3}$ (c) $\sqrt{8}$ or $2\sqrt{2}$

2. (a) b   (b) a   (c) c

6. integer—polynomial; rational numbers—rational expression; irrational numbers—radical expression
7. 32   8. −4
9. $-\dfrac{216}{125}$ 10. −32
11. $\dfrac{1000}{27}$ 12. 15   14. 42
15. −17   16. 6   17. −5
18. $-3z^4$ 19. −2
20. if n is even and a is negative
21. (a) $|x|$ (b) not a real number   (c) $-|x|$ (d) x
22. $\sqrt[3]{5}$
23. $(\sqrt[5]{2k})^2$ or $\sqrt[5]{4k^2}$
24. $\sqrt{m + 3n}$
25. $\dfrac{1}{(\sqrt[3]{3a + b})^5}$ or $\dfrac{1}{\sqrt[3]{(3a + b)^5}}$ 26. 3^9
27. $7^{9/2}$ 28. $5^{3/4}$
29. $p^{4/5}$ 30. 5^2 or 25
31. 96   32. $a^{2/3}$ 33. $\dfrac{1}{y^{1/2}}$
34. $\dfrac{z^{1/2}x^{8/5}}{4}$ 35. $r^{1/2} + r$
36. $s^{1/2}$ 37. $r^{3/2}$
38. $p^{1/2}$ 39. $k^{9/4}$
40. $m^{13/3}$ 41. $z^{1/12}$
42. −6.856   43. 6.164
44. −5.053   45. 4.960
46. 10.903   47. .009
48. −3968.503   49. −.189

3. In your own words, describe the integers.

4. In your own words, describe the rational numbers.

5. In your own words, describe the irrational numbers.

6. Match each number set with the appropriate expression.

integers	rational expression
rational numbers	radical expression
irrational numbers	polynomial

[5.1] *Simplify.*

7. $16^{5/4}$

8. $-8^{2/3}$

9. $-\left(\dfrac{36}{25}\right)^{3/2}$

10. $\left(-\dfrac{1}{8}\right)^{-5/3}$

11. $\left(\dfrac{81}{10,000}\right)^{-3/4}$

12. $\dfrac{15^{1/4}}{15^{-3/4}}$

13. Explain the relationship between the expressions $a^{m/n}$ and $\sqrt[n]{a^m}$.

Find the root. Use a calculator as necessary.

14. $\sqrt{1764}$

15. $-\sqrt{289}$

16. $\sqrt[3]{216}$

17. $\sqrt[3]{-125}$

18. $-\sqrt[3]{27z^{12}}$

19. $\sqrt[5]{-32}$

20. Under what conditions is $\sqrt[n]{a}$ not a real number?

21. Find any root that represents a real number.

   (a) $\sqrt{x^2}$ **(b)** $\sqrt{-x^2}$ **(c)** $-\sqrt{x^2}$ **(d)** $\sqrt[3]{x^3}$

Write the expression as a radical.

22. $5^{1/3}$

23. $(2k)^{2/5}$

24. $(m + 3n)^{1/2}$

25. $(3a + b)^{-5/3}$

Write the expression with a rational exponent.

26. $\sqrt{3^{18}}$

27. $\sqrt{7^9}$

28. $\sqrt[4]{5^3}$

29. $\sqrt[5]{p^4}$

[5.2] *Use the rules for exponents to simplify the expression. Write the answer with only positive exponents. Assume that all variables represent positive real numbers.*

30. $5^{1/4} \cdot 5^{7/4}$

31. $\dfrac{96^{2/3}}{96^{-1/3}}$

32. $\dfrac{(a^{1/3})^4}{a^{2/3}}$

33. $\dfrac{y^{-1/3} \cdot y^{5/6}}{y}$

34. $\left(\dfrac{z^{-1}x^{-3/5}}{2^{-2}z^{-1/2}x}\right)^{-1}$

35. $r^{-1/2}(r + r^{3/2})$

Simplify by first writing the radical in exponential form. Leave the answer in exponential form. Assume all variables represent positive real numbers.

36. $\sqrt[8]{s^4}$

37. $\sqrt[6]{r^9}$

38. $\dfrac{\sqrt{p^5}}{p^2}$

39. $\sqrt[4]{k^3} \cdot \sqrt{k^3}$

40. $\sqrt[3]{m^5} \cdot \sqrt[3]{m^8}$

41. $\sqrt[4]{\sqrt[3]{z}}$

Use a calculator to find a decimal approximation for the number. Give the answer to the nearest thousandth.

42. $-\sqrt{47}$

43. $\sqrt{38}$

44. $\sqrt[3]{-129}$

45. $\sqrt[4]{605}$

46. $36^{2/3}$

47. $500^{-3/4}$

48. $-500^{4/3}$

49. $-28^{-1/2}$

50. A triangle has a base of $\sqrt{38}$ inches and a height of $\sqrt{99}$ inches. What is the best estimate of its area?

(a) 3600   (b) 30   (c) 60   (d) 360

[5.3] *Simplify the radical. Assume that all variables represent positive real numbers.*

51. $\sqrt{6} \cdot \sqrt{11}$

52. $\sqrt{5} \cdot \sqrt{r}$

53. $\sqrt[3]{6} \cdot \sqrt[3]{5}$

54. $\sqrt[4]{7} \cdot \sqrt[4]{3}$

55. $\sqrt{20}$

56. $\sqrt{75}$

57. $-\sqrt{125}$

58. $\sqrt[3]{-108}$

59. $\sqrt{100y^7}$

60. $\sqrt[3]{64p^4q^6}$

61. $\sqrt[3]{108a^8b^5}$

62. $\sqrt[3]{632r^8t^4}$

63. $\sqrt{\dfrac{y^3}{144}}$

64. $\sqrt[3]{\dfrac{m^{15}}{27}}$

65. $\sqrt[3]{\dfrac{r^2}{8}}$

66. $\sqrt[4]{\dfrac{a^9}{81}}$

Simplify the radical.

67. $\sqrt[6]{15^3}$

68. $\sqrt[4]{p^6}$

69. $\sqrt[3]{2} \cdot \sqrt[4]{5}$

70. $\sqrt{x} \cdot \sqrt[5]{x}$

71. Find the missing length in the right triangle. Simplify the answer if necessary.

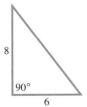

[5.4] *Perform the indicated operations. Assume that all variables represent positive real numbers.*

72. $2\sqrt{8} - 3\sqrt{50}$

73. $8\sqrt{80} - 3\sqrt{45}$

74. $-\sqrt{27y} + 2\sqrt{75y}$

75. $2\sqrt{54m^3} + 5\sqrt{96m^3}$

76. $3\sqrt[3]{54} + 5\sqrt[3]{16}$

77. $-6\sqrt[4]{32} + \sqrt[4]{512}$

78. $\dfrac{3}{\sqrt{16}} - \dfrac{\sqrt{5}}{2}$

79. $\dfrac{4}{\sqrt{25}} + \dfrac{\sqrt{5}}{4}$

In Exercises 80 and 81, leave answers as simplified radicals.

80. Find the perimeter of a lot with sides measuring $3\sqrt{18}$, $2\sqrt{32}$, $4\sqrt{50}$, and $5\sqrt{12}$ yards.

81. What is the perimeter of a triangle with sides measuring $3\sqrt{54}$, $4\sqrt{24}$, and $\sqrt{80}$ meters?

[5.5] *Multiply.*

82. $(\sqrt{3} + 1)(\sqrt{3} - 2)$

83. $(\sqrt{7} + \sqrt{5})(\sqrt{7} - \sqrt{5})$

84. $(3\sqrt{2} + 1)(2\sqrt{2} - 3)$

85. $(\sqrt{13} - \sqrt{2})^2$

86. $(\sqrt[3]{2} + 3)(\sqrt[3]{4} - 3\sqrt[3]{2} + 9)$

87. $(\sqrt[3]{4y} - 1)(\sqrt[3]{4y} + 3)$

88. Use a calculator to show that the answer to Exercise 85, $15 - 2\sqrt{26}$, is not equal to $13\sqrt{26}$.

Rationalize the denominator. Assume that all variables represent positive real numbers.

89. $\dfrac{\sqrt{6}}{\sqrt{5}}$

90. $\dfrac{-6\sqrt{3}}{\sqrt{2}}$

91. $\dfrac{3\sqrt{7p}}{\sqrt{y}}$

92. $\sqrt{\dfrac{11}{8}}$

93. $-\sqrt[3]{\dfrac{9}{25}}$

94. $\sqrt[3]{\dfrac{108m^3}{n^5}}$

95. $\dfrac{1}{\sqrt{2} + \sqrt{7}}$

96. $\dfrac{-5}{\sqrt{6} - 3}$

Answers column:

50. (b)   51. $\sqrt{66}$
52. $\sqrt{5r}$ 53. $\sqrt[3]{30}$
54. $\sqrt[4]{21}$ 55. $2\sqrt{5}$
56. $5\sqrt{3}$ 57. $-5\sqrt{5}$
58. $-3\sqrt[3]{4}$ 59. $10y^3\sqrt{y}$
60. $4pq^2\sqrt[3]{p}$
61. $3a^2b\sqrt[3]{4a^2b^2}$
62. $2r^2t\sqrt[3]{79r^2t}$ 63. $\dfrac{y\sqrt{y}}{12}$
64. $\dfrac{m^5}{3}$ 65. $\dfrac{\sqrt[3]{r^2}}{2}$
66. $\dfrac{a^2\sqrt[4]{a}}{3}$ 67. $\sqrt{15}$
68. $p\sqrt{p}$ 69. $\sqrt[12]{2000}$
70. $\sqrt[10]{x^7}$ 71. 10
72. $-11\sqrt{2}$ 73. $23\sqrt{5}$
74. $7\sqrt{3y}$ 75. $26m\sqrt{6m}$
76. $19\sqrt[3]{2}$ 77. $-8\sqrt[4]{2}$
78. $\dfrac{3 - 2\sqrt{5}}{4}$
79. $\dfrac{16 + 5\sqrt{5}}{20}$
80. $37\sqrt{2} + 10\sqrt{3}$ yards
81. $17\sqrt{6} + 4\sqrt{5}$ meters
82. $1 - \sqrt{3}$ 83. 2
84. $9 - 7\sqrt{2}$
85. $15 - 2\sqrt{26}$ 86. 29
87. $2\sqrt[3]{2y^2} + 2\sqrt[3]{4y} - 3$
88. $4.801960973 \neq$
66.28725368
89. $\dfrac{\sqrt{30}}{5}$ 90. $-3\sqrt{6}$
91. $\dfrac{3\sqrt{7py}}{y}$ 92. $\dfrac{\sqrt{22}}{4}$
93. $-\dfrac{\sqrt[3]{45}}{5}$ 94. $\dfrac{3m\sqrt[3]{4n}}{n^2}$
95. $\dfrac{\sqrt{2} - \sqrt{7}}{-5}$
96. $\dfrac{5(\sqrt{6} + 3)}{3}$

99. $\dfrac{1 - \sqrt{5}}{4}$

100. $\dfrac{1 - 4\sqrt{2}}{3}$

101. $2 + \sqrt{3k}$ 102. $\{2\}$

103. $\{6\}$ 104. $\emptyset$

105. $\{0, 5\}$ 106. $\{9\}$

107. $\{3\}$ 108. $\{7\}$

109. $\left\{-\dfrac{1}{2}\right\}$ 110. $\{6\}$

111. 0 does not satisfy the equation.   112. $5i$

113. $10i\sqrt{2}$ 114. no

115. $-10 - 2i$

116. $14 + 7i$ 117. $-\sqrt{35}$

118. -45 119. 3

120. $5 + i$ 121. $32 - 24i$

122. $1 - i$ 123. $4 + i$

124. $-i$ 125. -1

126. -4 127. $\dfrac{1}{100}$

128. $\dfrac{1}{z^{3/5}}$ 129. k^6

130. $3z^3t^2\sqrt[3]{2t^2}$

131. $57\sqrt{2}$ 132. $6x\sqrt[3]{y^2}$

133. $\sqrt{35} + \sqrt{15} - \sqrt{21} - 3$

⊙ 🖉 **97.** A friend wants to rationalize the denominator of the fraction $\dfrac{5 + \sqrt{7}}{2 + \sqrt{3}}$, and he decides to multiply the numerator and denominator by $\sqrt{3}$. Why will his plan not work?

⊙ 🖉 **98.** Another friend wants to rationalize the denominator of the fraction $\dfrac{5}{\sqrt[3]{6}}$, and she decides to multiply the numerator and denominator by $\sqrt[3]{6}$. Why will her plan not work?

Write in lowest terms.

99. $\dfrac{2 - 2\sqrt{5}}{8}$

100. $\dfrac{4 - 8\sqrt{8}}{12}$

101. $\dfrac{6k + \sqrt{27k^3}}{3k}, \quad k > 0$

[5.6] *Solve the equation.*

102. $\sqrt{8y + 9} = 5$

103. $\sqrt{2z - 3} - 3 = 0$

104. $\sqrt{3m + 1} = -1$

105. $\sqrt{7z + 1} = z + 1$

106. $3\sqrt{m} = \sqrt{10m - 9}$

107. $\sqrt{p^2 + 3p + 7} = p + 2$

108. $\sqrt{a + 2} - \sqrt{a - 3} = 1$

109. $\sqrt[3]{5m - 1} = \sqrt[3]{3m - 2}$

110. $\sqrt[4]{b + 6} = \sqrt[4]{2b}$

⊙ **111.** In Exercise 104, when we square both sides of the equation and solve, the result is $m = 0$. Explain why the correct answer is $\emptyset$.

[5.7] *Write the expression as a product of a real number and i.*

112. $\sqrt{-25}$

113. $\sqrt{-200}$

⊙ **114.** If a is a positive real number, is $-\sqrt{-a}$ a real number?

Perform the indicated operations. Write each imaginary number answer in the form $a + bi$.

115. $(-2 + 5i) + (-8 - 7i)$

116. $(5 + 4i) - (-9 - 3i)$

117. $\sqrt{-5} \cdot \sqrt{-7}$

118. $\sqrt{-25} \cdot \sqrt{-81}$

119. $\dfrac{\sqrt{-72}}{\sqrt{-8}}$

120. $(2 + 3i)(1 - i)$

121. $(6 - 2i)^2$

122. $\dfrac{3 - i}{2 + i}$

123. $\dfrac{5 + 14i}{2 + 3i}$

Find the power of i.

124. i^{11}

125. i^{-10}

MIXED REVIEW EXERCISES

Simplify. Assume all variables represent positive real numbers.

126. $-\sqrt[4]{256}$

127. $1000^{-2/3}$

128. $\dfrac{z^{-1/5} \cdot z^{3/10}}{z^{7/10}}$

129. $\sqrt[4]{k^{24}}$

130. $\sqrt[3]{54z^9t^8}$

131. $-5\sqrt{18} + 12\sqrt{72}$

132. $8\sqrt[3]{x^3y^2} - 2x\sqrt[3]{y^2}$

133. $(\sqrt{5} - \sqrt{3})(\sqrt{7} + \sqrt{3})$

134. $\dfrac{-1}{\sqrt{12}}$

135. $\sqrt[3]{\dfrac{12}{25}}$

136. $\dfrac{2\sqrt{z}}{\sqrt{z}-2}$

137. $\sqrt{-49}$

138. $(4-9i)+(-1+2i)$

139. $\dfrac{\sqrt{50}}{\sqrt{-2}}$

140. i^{-1000}

Solve the equation.

141. $\sqrt{x+4}=x-2$

142. $\sqrt[3]{2x-9}=\sqrt[3]{5x+3}$

143. $\sqrt{6+2y}-1=\sqrt{7-2y}$

Work the following problem. Express your answer to the nearest tenth of a foot.

144. Carpenters stabilize wall frames with a diagonal brace as shown in the figure. The length of the brace is given by $L=\sqrt{H^2+W^2}$. If the bottom of the brace is attached 9 feet from the corner and the brace is 12 feet long, how far up the corner post should it be nailed?

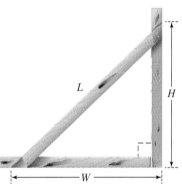

134. $-\dfrac{\sqrt{3}}{6}$

135. $\dfrac{\sqrt[3]{60}}{5}$

136. $\dfrac{2\sqrt{z}(\sqrt{z}+2)}{z-4}$

137. $7i$
138. $3-7i$ **139.** $-5i$
140. 1   **141.** $\{5\}$

142. $\{-4\}$ **143.** $\left\{\dfrac{3}{2}\right\}$

144. 7.9 feet

CHAPTER 5 TEST

Simplify the expression. Assume that all variables represent positive numbers.

1. $\left(\dfrac{16}{25}\right)^{-3/2}$

2. $(-64)^{-4/3}$

3. $\dfrac{3^{2/5}x^{-1/4}y^{2/5}}{3^{-8/5}x^{7/4}y^{1/10}}$

4. $7^{3/4}\cdot 7^{-1/4}$

5. $\sqrt[3]{a^4}\cdot\sqrt[3]{a^7}$

Find the root. Use a calculator as necessary.

6. $-\sqrt{841}$

7. $\sqrt[3]{3375}$

8. Find the closest estimate of $\sqrt{146.25}$ from the choices given.
   (a) 10   **(b)** 11   **(c)** 12   **(d)** 13

9. Give a calculator approximation of $\sqrt{146.25}$ to the nearest hundredth.

Simplify the expression. Assume that all variables represent positive real numbers.

10. $\sqrt{54x^5y^6}$

11. $\sqrt[4]{32a^7b^{13}}$

12. $\sqrt{2}\cdot\sqrt[3]{5}$

13. $3\sqrt{20}-5\sqrt{80}+4\sqrt{500}$

14. $(7\sqrt{5}+4)(2\sqrt{5}-1)$

15. $(\sqrt{3}-2\sqrt{5})^2$

16. $\dfrac{-5}{\sqrt{40}}$

[5.1] **1.** $\dfrac{125}{64}$ **2.** $\dfrac{1}{256}$

[5.2] **3.** $\dfrac{9y^{3/10}}{x^2}$ **4.** $7^{1/2}$

5. $a^3\sqrt[3]{a^2}$ or $a^{11/3}$
[5.1] **6.** -29 **7.** 15
[5.2] **8.** (c)   **9.** 12.09
[5.3] **10.** $3x^2y^3\sqrt{6x}$
11. $2ab^3\sqrt[4]{2a^3b}$
12. $\sqrt[6]{200}$
[5.4] **13.** $26\sqrt{5}$
[5.5] **14.** $66+\sqrt{5}$
15. $23-4\sqrt{15}$
16. $-\dfrac{\sqrt{10}}{4}$

RESOURCES FOR TEST

ITM pp. 101–116 **ISM** pp. 381–383 | **SSG** pp. 166–167   **TEST GENERATOR**
IAM p. 23   **SSM** pp. 250–252   **DOS** **Windows** **MAC**

⦿ CONCEPTUAL   WRITING   ▲ CHALLENGING   SCIENTIFIC CALCULATOR    GRAPHICS CALCULATOR

⊙ **17.** In rationalizing the denominator in the expression $\dfrac{2}{\sqrt[3]{5}}$, by what expression should we multiply both the numerator and denominator?

18. Rationalize the denominator of $\dfrac{-4}{\sqrt{7} + \sqrt{5}}$.

19. The following formula is used in physics, relating the velocity of sound V to the temperature T.

$$V = \frac{V_0}{\sqrt{1 - kT}}$$

Find an approximation of V to the nearest tenth if $V_0 = 50$, $k = .01$, and $T = 30$. Use a calculator.

Solve for x.

20. $3\sqrt{x} = \sqrt{2x + 7}$

21. $x - 5 = \sqrt{7 - x}$

Perform the indicated operations. Express the answer in the form $a + bi$.

22. $(-2 + 5i) - (3 + 6i) - 7i$

23. $\dfrac{7 + i}{1 - i}$

24. $(1 + 5i)(3 + i)$

25. Simplify i^{37}.

CUMULATIVE REVIEW (Chapters 1–5)

Evaluate the expression if $a = -3$, $b = 5$, and $c = -4$.

1. $|2a^2 - 3b + c|$

2. $\dfrac{(a + b)(a + c)}{3b - 6}$

Perform the indicated operations.

3. $(3x - y)(x + 2y)$

4. $\left(\dfrac{2}{3}t^2 - 6\right)^2$

5. $(x + 2y)(x - 2y)$

6. $5x^2y^3(-2x^3y + 5x^2y - 3)$

7. $(4x^3 - 3x - 1) \div (x - 1)$

8. $(a + b)^3$

Factor the polynomial completely.

9. $8x^6 - 27$

10. $144y^2 - 49x^2$

11. $10x^2 + x - 3$

12. $2xy + 8x + 3y + 12$

13. $1000a^3 + 1$

14. $36a^2 - 84a + 49$

Write the expression in simplest form, using only positive exponents. Assume all variables represent positive real numbers.

15. $27^{-2/3}$

16. $\dfrac{(a^{-3}b^4)^{-1}}{a^{-2}b^{-1}}$

17. $\sqrt{48x^5y^2}$

18. $\sqrt{50} + \sqrt{8}$

19. $\dfrac{1}{\sqrt{10} - \sqrt{8}}$

20. $(2\sqrt{x} + \sqrt{y})(-3\sqrt{x} - 4\sqrt{y})$

Perform the indicated operations and express the answer in lowest terms.

21. $\dfrac{z^2}{z - y} - \dfrac{y^2}{z - y}$

22. $\dfrac{12}{x^2 - 9} + \dfrac{2}{x + 3}$

23. $\dfrac{x^2 + 3x + 2}{x^2 - 3x - 4} \div \dfrac{x^2 + 5x + 6}{x^2 - 10x + 24}$

24. $\dfrac{4}{2 - a} + \dfrac{7}{a - 2}$

25. $\dfrac{x + \dfrac{1}{x}}{\dfrac{3}{x} - x}$

26. $\dfrac{x^{-1}}{y - x^{-1}}$

Solve the equation.

27. $3(x + 2) - 4(2x + 3) = -3x + 2$

28. $\dfrac{1}{3}x + \dfrac{1}{4}(x + 8) = x + 7$

29. $.04y + .06(100 - y) = 5.88$

30. $|6x + 7| = 13$

31. $|-2x + 4| = |-2x - 3|$

32. $\dfrac{p + 1}{p - 3} = \dfrac{4}{p - 3} + 6$

33. $4t^2 = 3t + 10$

34. $\sqrt{3r - 8} = r - 2$

Solve the inequality. Give the solution set in interval form.

35. $3x - 2(2x + 1) \le 0$

36. $|4 - 2x| \ge 3$

Solve the problem.

37. The *fall speed* of a vehicle running off the road into a ditch is given by

$$S = \dfrac{2.74D}{\sqrt{h}},$$

where D is the horizontal distance traveled from the level surface to the bottom of the ditch and h is the height (or depth) of the ditch. What is the fall speed of a vehicle that traveled 32 feet horizontally into a 5-foot deep ditch?

38. How many liters of pure alcohol must be mixed with 40 liters of 18% alcohol to obtain a 22% alcohol solution?

39. Natalie can ride her bike 4 miles per hour faster than her husband, Chuck. If Natalie can ride 48 miles in the same time that Chuck can ride 24 miles, what are their speeds?

40. Find the measures of the marked angles.

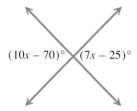

$(10x - 70)°$ $(7x - 25)°$

21. $z + y$ **22.** $\dfrac{2}{x - 3}$

23. $\dfrac{x - 6}{x + 3}$

24. $\dfrac{3}{a - 2}$ or $\dfrac{-3}{2 - a}$

25. $\dfrac{x^2 + 1}{3 - x^2}$

26. $\dfrac{1}{xy - 1}$ **27.** $\{-4\}$

28. $\{-12\}$ **29.** $\{6\}$

30. $\left\{-\dfrac{10}{3}, 1\right\}$ **31.** $\left\{\dfrac{1}{4}\right\}$

32. $\emptyset$ **33.** $\left\{-\dfrac{5}{4}, 2\right\}$

34. $\{3, 4\}$ **35.** $[-2, \infty)$

36. $\left(-\infty, \dfrac{1}{2}\right] \cup \left[\dfrac{7}{2}, \infty\right)$

37. 39.2 miles per hour

38. $\dfrac{80}{39}$ or $2\dfrac{2}{39}$ liters

39. Natalie: 8 miles per hour; Chuck: 4 miles per hour **40.** Both angles measure 80°.

QUADRATIC EQUATIONS AND INEQUALITIES

CONNECTIONS

As mentioned earlier, quadratic equations are good models for problems from everyday life. For example, union membership in the United States rose from 1930 to 1960 and then declined. The quadratic equation

$$M = -.011x^2 + 1.22x - 8.5$$

approximates the number of union members (in millions) in the years 1950 to 1990, where x is the number of years since 1930 (when unions were just starting out).* In this chapter we learn methods to solve this equation for any given value of M.

FOR DISCUSSION OR WRITING

Use the quadratic equation (and a calculator) to approximate the union membership for the years 1950 ($x = 20$), 1970, 1980, and 1990. How did the membership change from 1950 to 1980? from 1980 to 1990? Compare this with the description of the union membership above.

◈ 1950: 11.5 million; 1970: 22.7 million; 1980: 25 million; 1990: 25.1 million. From 1950 to 1980, membership increased by 14.5 million. From 1980 to 1990, it increased by only .1 million.

*Source: U.S. Union Membership, 1930–1990 Table. Bureau of Labor Statistics, U.S. Department of Labor.

In this chapter we continue the study of equations we began in Chapter 2 and touched on again in each succeeding chapter. Earlier chapters dealt primarily with linear equations. Now in this chapter, several methods are introduced for solving second-degree equations.

6.1 COMPLETING THE SQUARE

FOR EXTRA HELP	OBJECTIVES
SSG pp. 168–172 **SSM** pp. 257–262	**1** ▶ Review the zero-factor property.
Video **7**	**2** ▶ Learn the square root property. **3** ▶ Solve quadratic equations of the form $(ax - b)^2 = c$ by using the square root property.
Tutorial IBM MAC	**4** ▶ Solve quadratic equations by completing the square. **5** ▶ Solve quadratic equations with complex solutions.

We solved quadratic equations in Section 3.9 by using the zero-factor property. In this section we introduce additional methods for solving quadratic equations. Recall that a quadratic equation is a second-degree equation. For example, $m^2 + 4m - 5 = 0$ and $3q^2 = 4q - 8$ are quadratic equations, with the first equation in standard form.

1 ▶ Review the zero-factor property.

▶   The zero-factor property is now stated again.

Zero-Factor Property

If two numbers have a product of 0, then at least one of the numbers must be 0. That is, if $ab = 0$, then $a = 0$ or $b = 0$.

The first example is a review of the method of solving quadratic equations first illustrated in Section 3.9.

EXAMPLE 1

Using the Zero-Factor Property to Solve an Equation

Use the zero-factor property to solve $3x^2 - 5x = 28$.

$$3x^2 - 5x = 28 \qquad \text{Given.}$$
$$3x^2 - 5x - 28 = 0 \qquad \text{Standard form}$$
$$(3x + 7)(x - 4) = 0 \qquad \text{Factor.}$$
$$3x + 7 = 0 \quad \text{or} \quad x - 4 = 0 \qquad \text{Use the zero-factor property.}$$
$$3x = -7 \qquad\qquad x = 4 \qquad \text{Solve each linear equation.}$$
$$x = -\frac{7}{3}$$

The solution set is $\left\{ -\frac{7}{3}, 4 \right\}$. ■

Chalkboard Exercise

Use the zero-factor property to solve $2x^2 - 3x = -1$.

Answer: $\left\{ \frac{1}{2}, 1 \right\}$

2 ▶ Learn the square root property.

▶   Although factoring is the simplest way to solve quadratic equations, not every quadratic equation can be solved easily by factoring. In this section and the next, we develop other methods of solving quadratic equations based on the following property.

INSTRUCTOR'S RESOURCES

 **ITM** pp. 413–415   **ISM** pp. 389–397
IAM p. 24

 **TEST GENERATOR**
DOS   Windows   MAC

 TRANSPARENCIES

| **Square Root Property** | If $x^2 = b$, $b \geq 0$, then $x = \sqrt{b}$ or $x = -\sqrt{b}$. |

CAUTION Do not forget that if $b \neq 0$, using the square root property always produces *two* square roots, one positive and one negative.

EXAMPLE 2

Using the Square Root Property

Solve $r^2 = 5$.

From the square root property, since $(\sqrt{5})^2 = 5$,

$$r = \sqrt{5} \quad \text{or} \quad r = -\sqrt{5}$$

and the solution set is $\{\sqrt{5}, -\sqrt{5}\}$. ∎

Recall from Chapter 5 that roots such as those in Example 2 are sometimes abbreviated with the symbol $\pm$ (read "positive or negative"); with this symbol the solutions in Example 2 would be written $\pm\sqrt{5}$.

3 ▶ Solve quadratic equations of the form $(ax - b)^2 = c$ by using the square root property.

▶ The square root property can be used to solve more complicated equations, such as

$$(x - 5)^2 = 36,$$

by substituting $(x - 5)^2$ for x^2 and 36 for b in the square root property to get

$$x - 5 = 6 \quad \text{or} \quad x - 5 = -6$$
$$x = 11 \qquad\qquad x = -1.$$

Check that both 11 and -1 satisfy the given equation so that the solution set is $\{11, -1\}$.

EXAMPLE 3

Using the Square Root Property

Solve $(2a - 3)^2 = 18$.

By the square root property,

$$2a - 3 = \sqrt{18} \quad \text{or} \quad 2a - 3 = -\sqrt{18},$$

from which

$$2a = 3 + \sqrt{18} \quad \text{or} \quad 2a = 3 - \sqrt{18}$$

$$a = \frac{3 + \sqrt{18}}{2} \quad \text{or} \quad a = \frac{3 - \sqrt{18}}{2}.$$

Since $\sqrt{18} = \sqrt{9 \cdot 2} = 3\sqrt{2}$, the solution set can be written

$$\left\{ \frac{3 + 3\sqrt{2}}{2}, \frac{3 - 3\sqrt{2}}{2} \right\}. \quad ∎$$

4 ▶ Solve quadratic equations by completing the square.

▶ We can use the square root property to solve any quadratic equation by writing it in the form $(x + k)^2 = n$. That is, the left side of the equation must be rewritten as a perfect square trinomial that can be factored as $(x + k)^2$ and the right side must be a constant. Rewriting the equation in this form is called **completing the square.**

For example,

$$m^2 + 8m + 10 = 0$$

is a quadratic equation that cannot be solved easily by factoring. To get a perfect square trinomial on the left side of the equation $m^2 + 8m + 10 = 0$, we first subtract 10 from both sides:

$$m^2 + 8m = -10.$$

The left side should be a perfect square, say $(m + k)^2$. Since $(m + k)^2 = m^2 + 2mk + k^2$, comparing $m^2 + 8m$ with $m^2 + 2mk$ shows that

$$2mk = 8m$$
$$k = 4.$$

If $k = 4$, then $(m + k)^2$ becomes $(m + 4)^2$, or $m^2 + 8m + 16$. We get the necessary $+16$ by adding 16 on both sides so that

$$m^2 + 8m = -10$$

becomes

$$m^2 + 8m + 16 = -10 + 16.$$

Now we factor the left side, which should be a perfect square. Since $m^2 + 8m + 16$ factors as $(m + 4)^2$, the equation becomes

$$(m + 4)^2 = 6.$$

We can solve this equation with the square root property:

$$m + 4 = \sqrt{6} \qquad \text{or} \qquad m + 4 = -\sqrt{6},$$

leading to the solution set $\{-4 + \sqrt{6}, -4 - \sqrt{6}\}$.

Based on the work of this example, we convert an equation of the form $x^2 + px = q$ into an equation of the form $(x + k)^2 = n$ by adding the square of half the coefficient of the first-degree term to both sides of the equation.

In summary, to find the solutions of a quadratic equation by completing the square, proceed as follows.

Completing the Square

To solve $ax^2 + bx + c = 0$ $(a \neq 0)$, use the following steps.

Step 1 If $a \neq 1$, divide both sides by a.
Step 2 Rewrite the equation so that both terms containing variables are on one side of the equals sign and the constant is on the other side.
Step 3 Take half the coefficient of x (the first-degree term) and square it.
Step 4 Add the square to both sides of the equation.
Step 5 One side should now be a perfect square trinomial. Factor that side and write it as the square of a binomial. Simplify the other side.
Step 6 Use the square root property to complete the solution.

▶ **TEACHING TIP** Keeping track of the steps for completing the square is difficult for students. Emphasize that they need to know the mechanics presented in this section in order to understand material presented later in the book (finding the vertex of a parabola, the center and radius of a circle). In Step 6, remind students to take the square root of both sides; do not forget the ± sign on the right side; solve for the unknown and reduce the result. ◀

NOTE Steps 1 and 2 can be done in either order. With some problems, it is more convenient to do Step 2 first.

◆ Original square: x^2; each strip: x;   total area of the strips: $6x$;   each small square: 1;   total area of the small squares: 9;   area of new larger square is $(x + 3)^2$ or $x^2 + 6x + 9$.

◆——————◆ **CONNECTIONS** ◆——————◆

The Greeks had a method of completing the square geometrically in which they literally changed a figure into a square. For example, to complete the square for $x^2 + 6x$, we begin with a square of side x, as in the figure. We add three rectangles of width 1 to the right side and the bottom to get a region with area $x^2 + 6x$. To fill in the corner (complete the square) we must add 9 1-by-1 squares as shown.

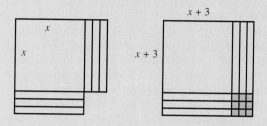

FOR DISCUSSION OR WRITING
Find the following areas: the original square; each strip; the total area of the strips; each small square; the total area of the small squares. What is the area of the new larger square?

Example 4 illustrates how to determine the number that must be added to complete the square.

EXAMPLE 4
Determining the Number to Be Used to Complete the Square

If the equation $2x^2 + 8x - 3 = 0$ is to be solved by completing the square, what number would be added to both sides after Step 2 in the procedure box above?

To solve this equation by completing the square, we would apply Steps 1 and 2 as follows.

$$2x^2 + 8x - 3 = 0 \qquad \text{Given equation}$$

$$x^2 + 4x - \frac{3}{2} = 0 \qquad \text{Step 1}$$

$$x^2 + 4x = \frac{3}{2} \qquad \text{Step 2}$$

We would now add $\left(\frac{1}{2} \cdot 4\right)^2 = 2^2 = 4$ to both sides to complete the square. The left side would then be $x^2 + 4x + 4 = (x + 2)^2$. ■

EXAMPLE 5
Solving a Quadratic Equation with $a = 1$ by Completing the Square

Solve $k^2 + 5k - 1 = 0$ by completing the square.

Follow the steps listed earlier. Since $a = 1$, Step 1 is not needed here. Begin by adding 1 to both sides.

$$k^2 + 5k = 1 \qquad \text{Step 2}$$

Take half of 5 (the coefficient of the first-degree term) and square the result.

$$\frac{1}{2} \cdot 5 = \frac{5}{2} \qquad \text{and} \qquad \left(\frac{5}{2}\right)^2 = \frac{25}{4} \qquad \text{Step 3}$$

Add the square to each side of the equation to get

$$k^2 + 5k + \frac{25}{4} = 1 + \frac{25}{4}. \qquad \text{Step 4}$$

▶ **TEACHING TIP**
After working through several examples, have students work through an exercise from the book; allow them to refer to the steps listed in the book. ◀

Write the left side as a perfect square and add on the right. Then use the square root property.

$$\left(k + \frac{5}{2} \right)^2 = \frac{29}{4} \qquad \text{Step 5}$$

$$k + \frac{5}{2} = \sqrt{\frac{29}{4}} \qquad \text{or} \qquad k + \frac{5}{2} = -\sqrt{\frac{29}{4}} \qquad \text{Step 6}$$

$$k + \frac{5}{2} = \frac{\sqrt{29}}{2} \qquad \text{or} \qquad k + \frac{5}{2} = \frac{-\sqrt{29}}{2} \qquad \text{Simplify.}$$

$$k = -\frac{5}{2} + \frac{\sqrt{29}}{2} \qquad \text{or} \qquad k = -\frac{5}{2} - \frac{\sqrt{29}}{2} \qquad \text{Solve.}$$

$$k = \frac{-5 + \sqrt{29}}{2} \qquad \text{or} \qquad k = \frac{-5 - \sqrt{29}}{2} \qquad \text{Combine terms.}$$

Check that the solution set is $\left\{ \dfrac{-5 + \sqrt{29}}{2}, \dfrac{-5 - \sqrt{29}}{2} \right\}$. ■

EXAMPLE 6

Solving a Quadratic Equation with $a \neq 1$ by Completing the Square

Chalkboard Exercise

Solve by completing the square.
$$3a^2 + 6a - 2 = 0$$
Answer:
$$\left\{ \frac{-3 + \sqrt{15}}{3}, \frac{-3 - \sqrt{15}}{3} \right\}$$

Solve $2a^2 - 4a - 5 = 0$.

Go through the steps for completing the square. First divide both sides of the equation by 2 to make the coefficient of the second-degree term equal to 1, getting

$$a^2 - 2a - \frac{5}{2} = 0. \qquad \text{Step 1}$$

$$a^2 - 2a = \frac{5}{2} \qquad \text{Step 2}$$

$$\frac{1}{2}(-2) = -1 \text{ and } (-1)^2 = 1. \qquad \text{Step 3}$$

$$a^2 - 2a + 1 = \frac{5}{2} + 1 \qquad \text{Step 4}$$

$$(a - 1)^2 = \frac{7}{2} \qquad \text{Step 5}$$

$$a - 1 = \sqrt{\frac{7}{2}} \qquad \text{or} \qquad a - 1 = -\sqrt{\frac{7}{2}} \qquad \text{Step 6}$$

$$a = 1 + \sqrt{\frac{7}{2}} \qquad\qquad a = 1 - \sqrt{\frac{7}{2}} \qquad \text{Solve.}$$

$$a = 1 + \frac{\sqrt{14}}{2} \qquad \text{or} \qquad a = 1 - \frac{\sqrt{14}}{2} \qquad \text{Rationalize denominators.}$$

Add the two terms in each solution as follows:

$$1 + \frac{\sqrt{14}}{2} = \frac{2}{2} + \frac{\sqrt{14}}{2} = \frac{2 + \sqrt{14}}{2}$$

$$1 - \frac{\sqrt{14}}{2} = \frac{2}{2} - \frac{\sqrt{14}}{2} = \frac{2 - \sqrt{14}}{2}.$$

The solution set is

$$\left\{ \frac{2 + \sqrt{14}}{2}, \frac{2 - \sqrt{14}}{2} \right\}. \quad ■$$

5 ▶ Solve quadratic equations with complex solutions.

▶   In the equation $x^2 = b$, if $b < 0$, there will be two imaginary solutions. The square root property can be extended to complex numbers as follows.

If $x^2 = -b, b > 0,$ then $x = i\sqrt{b}$ or $x = -i\sqrt{b}.$

EXAMPLE 7 ■
Extending the Methods of Solving Quadratic Equations

Find all solutions of each equation.

(a) $x^2 = -15$

By the extension of the square root property, we have

$$x^2 = -15$$
$$x = \sqrt{-15} \quad \text{or} \quad x = -\sqrt{-15}$$
$$x = i\sqrt{15} \quad \text{or} \quad x = -i\sqrt{15}$$

The solution set is $\{-i\sqrt{15}, i\sqrt{15}\}$.

(b) $(b + 2)^2 = -16$

$$b + 2 = \sqrt{-16} \quad \text{or} \quad b + 2 = -\sqrt{-16} \quad \text{Square root property}$$
$$b + 2 = 4i \quad \text{or} \quad b + 2 = -4i \quad \sqrt{-16} = 4i$$
$$b = -2 + 4i \quad \text{or} \quad b = -2 - 4i$$

The solution set is $\{-2 + 4i, -2 - 4i\}$.

(c) $x^2 + 2x + 7 = 0$

We will solve by completing the square.

$$x^2 + 2x = -7 \qquad \text{Add } -7.$$
$$x^2 + 2x + 1 = -7 + 1 \qquad [\tfrac{1}{2} \cdot 2]^2 = 1^2 = 1; \text{ add 1 to both sides.}$$
$$(x + 1)^2 = -6 \qquad \text{Factor on the left, combine terms on the right.}$$
$$x + 1 = \pm i\sqrt{6} \qquad \text{Square root property}$$
$$x = -1 \pm i\sqrt{6} \qquad \text{Subtract 1 from both sides.}$$

The solution set is $\{-1 + i\sqrt{6}, -1 - i\sqrt{6}\}$. ■

6.1 **EXERCISES**

📝 **1.** Give a one-sentence description or explanation of each of the following.
   (a) quadratic equation in standard form
   (b) zero-factor property
   (c) square root property

◎ **2.** Of the two equations

$$(2x + 1)^2 = 5 \qquad \text{and} \qquad x^2 + 4x = 12,$$

one is more suitable for solving by the square root property, and the other is more suitable for solving by completing the square. Which method do you think most students would use for each equation?

2. square root property for $(2x + 1)^2 = 5$; completing the square for $x^2 + 4x = 12$

 CONCEPTUAL   WRITING   CHALLENGING    SCIENTIFIC CALCULATOR   GRAPHICS CALCULATOR

3. $\{-9, 9\}$
4. $\{-15, 15\}$
5. $\{-\sqrt{17}, \sqrt{17}\}$
6. $\{-\sqrt{19}, \sqrt{19}\}$
7. $\{-4\sqrt{2}, 4\sqrt{2}\}$
8. $\{-8\sqrt{2}, 8\sqrt{2}\}$
9. $\{-7, 3\}$ 10. $\{5, 11\}$
11. $\left\{\dfrac{1 + \sqrt{7}}{3}, \dfrac{1 - \sqrt{7}}{3}\right\}$
12. $\left\{\dfrac{-4 + \sqrt{10}}{2}, \dfrac{-4 - \sqrt{10}}{2}\right\}$
13. $\left\{\dfrac{-1 + 2\sqrt{6}}{4}, \dfrac{-1 - 2\sqrt{6}}{4}\right\}$
14. $\left\{\dfrac{2 + 2\sqrt{3}}{5}, \dfrac{2 - 2\sqrt{3}}{5}\right\}$
15. (b)   16. (d)
17. $\{-2i\sqrt{3}, 2i\sqrt{3}\}$
18. $\{-3i\sqrt{2}, 3i\sqrt{2}\}$
19. $\{5 + i\sqrt{3}, 5 - i\sqrt{3}\}$
20. $\{-6 + i\sqrt{5}, -6 - i\sqrt{5}\}$
21. $\left\{\dfrac{1 + 2i\sqrt{2}}{6}, \dfrac{1 - 2i\sqrt{2}}{6}\right\}$
22. $\left\{\dfrac{7 + 3i\sqrt{3}}{4}, \dfrac{7 - 3i\sqrt{3}}{4}\right\}$
23. Divide both sides by 2.
25. $\{-4, 6\}$ 26. $\{-4, 8\}$
27. $\left\{-3, \dfrac{8}{3}\right\}$
28. $\left\{-3, \dfrac{13}{4}\right\}$
29. $\left\{\dfrac{-5 + \sqrt{41}}{4}, \dfrac{-5 - \sqrt{41}}{4}\right\}$
30. $\left\{\dfrac{-1 + \sqrt{7}}{3}, \dfrac{-1 - \sqrt{7}}{3}\right\}$

Use the square root property to solve the equation. (All solutions for these equations are real numbers.) See Examples 2 and 3.

3. $x^2 = 81$ **4.** $y^2 = 225$ **5.** $t^2 = 17$

6. $k^2 = 19$ **7.** $m^2 = 32$ **8.** $w^2 = 128$

9. $(x + 2)^2 = 25$ **10.** $(8 - y)^2 = 9$ **11.** $(1 - 3k)^2 = 7$

12. $(2x + 4)^2 = 10$ **13.** $(4p + 1)^2 = 24$ **14.** $(5k - 2)^2 = 12$

◉ **15.** Which one of the following is another form of $\sqrt{-5}$?
    (a) $5i$ **(b)** $i\sqrt{5}$ **(c)** $-\sqrt{5}$ **(d)** $-5i$

◉ **16.** Which one of the following equations has imaginary solutions?
    (a) $x^2 = 37$ **(b)** $-x^2 = -37$ **(c)** $x^2 - 37 = 0$ **(d)** $x^2 = -37$

Find the imaginary number solutions of the equation. See Example 7.

17. $x^2 = -12$ **18.** $y^2 = -18$ **19.** $(r - 5)^2 = -3$

20. $(t + 6)^2 = -5$ **21.** $(6k - 1)^2 = -8$ **22.** $(4m - 7)^2 = -27$

◉ **23.** What would be your first step in solving $2x^2 + 8x = 9$ by completing the square?

◉ **24.** Why would most students find the equation $x^2 + 4x = 20$ easier to solve by completing the square than the equation $5x^2 + 2x = 3$?

Solve the equation by completing the square. (All solutions for these equations are real numbers.) See Examples 4–6.

25. $x^2 - 2x - 24 = 0$ **26.** $m^2 - 4m - 32 = 0$ **27.** $3y^2 + y = 24$

28. $4z^2 - z = 39$ **29.** $2k^2 + 5k - 2 = 0$ **30.** $3r^2 + 2r - 2 = 0$

Solve the equation by completing the square. (Some solutions for these equations are imaginary numbers.) See Examples 4, 5, 6, and 7(c).

31. $m^2 + 4m + 13 = 0$
31. $\{-2 + 3i, -2 - 3i\}$

32. $t^2 + 6t + 10 = 0$
32. $\{-3 + i, -3 - i\}$

33. $9x^2 - 24x = -13$
33. $\left\{\dfrac{4 + \sqrt{3}}{3}, \dfrac{4 - \sqrt{3}}{3}\right\}$

34. $25n^2 - 20n = 1$
34. $\left\{\dfrac{2 + \sqrt{5}}{5}, \dfrac{2 - \sqrt{5}}{5}\right\}$

35. $z^2 - \dfrac{4}{3}z = -\dfrac{1}{9}$
35. $\left\{\dfrac{2 + \sqrt{3}}{3}, \dfrac{2 - \sqrt{3}}{3}\right\}$

36. $p^2 - \dfrac{8}{3}p = -1$
36. $\left\{\dfrac{4 + \sqrt{7}}{3}, \dfrac{4 - \sqrt{7}}{3}\right\}$

37. $3r^2 + 4r + 4 = 0$
37. $\left\{\dfrac{-2 + 2i\sqrt{2}}{3}, \dfrac{-2 - 2i\sqrt{2}}{3}\right\}$

38. $4x^2 + 5x + 5 = 0$
38. $\left\{\dfrac{-5 + i\sqrt{55}}{8}, \dfrac{-5 - i\sqrt{55}}{8}\right\}$

39. $.1x^2 - .2x - .1 = 0$
39. $\{1 + \sqrt{2}, 1 - \sqrt{2}\}$

40. $.1p^2 - .4p + .1 = 0$
40. $\{2 + \sqrt{3}, 2 - \sqrt{3}\}$

41. $-m^2 - 6m - 12 = 0$
41. $\{-3 + i\sqrt{3}, -3 - i\sqrt{3}\}$

42. $-k^2 - 5k - 10 = 0$
42. $\left\{\dfrac{-5 + i\sqrt{15}}{2}, \dfrac{-5 - i\sqrt{15}}{2}\right\}$

43. What is wrong with the following "solution" using the zero-factor property?

$$(x - 2)(x + 1) = 5$$
$$x - 2 = 5 \quad \text{or} \quad x + 1 = 1$$
$$x = 7 \qquad\qquad x = 0$$

44. A student was asked to solve the quadratic equation $x^2 = 16$ and did not get full credit for the solution set {4}. What was the reason for this?

45. Why is the zero-factor property insufficient for solving *all* quadratic equations?

46. Can *any* quadratic equation be solved by completing the square?

▲ *Solve for x. Assume that a and b represent positive real numbers.*

47. $x^2 - b = 0$ **48.** $x^2 = 4b$ **49.** $4x^2 = b^2 + 16$

50. $9x^2 - 25a = 0$ **51.** $(5x - 2b)^2 = 3a$ **52.** $x^2 - a^2 - 36 = 0$

 **MATHEMATICAL CONNECTIONS** (Exercises 53–58) ◆

Read the Connections material at the beginning of this section. Based on information from the U.S. Justice Statistics, the number of prisoners in the United States under sentence of death is defined by the quadratic equation

$$P = -6.5x^2 + 132.5x + 2117$$

for the years 1988, 1989, and 1990, where $x = 0$ corresponds to 1988, $x = 1$ corresponds to 1989, and $x = 2$ corresponds to 1990. The accompanying bar graph illustrates this in another way.

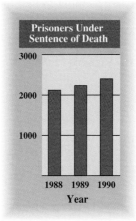

Now work Exercises 53–58 in order, using a calculator as necessary.

53. How many such prisoners were there in 1988?

54. How many such prisoners were there in 1989?

55. How many such prisoners were there in 1990?

56. The equation given was based only on information for the three years named. If we wanted to make a prediction for the number of prisoners in 1991, how would we do it based on this quadratic equation?

57. If we wanted to use this equation to predict the year in which the number of such prisoners would be 2543, what would be the standard form of the equation? (Do not try to solve it.) Now verify that one positive solution for this equation is 4. What year would this correspond to?

58. Suppose that the number of people in a small town grew according to the equation $P = 5x^2 + 45x + 350$, where $x = 0$ corresponds to 1994, $x = 1$ to 1995, and so on. In what year would the population be 700? (Solve the equation by completing the square.)

◆

REVIEW EXERCISES

Evaluate $\sqrt{b^2 - 4ac}$ for the given values of a, b, and c. See Section 1.2.

59. $a = 3, b = 1, c = -1$ **60.** $a = 4, b = 11, c = -3$

61. $a = 6, b = 7, c = 2$ **62.** $a = 1, b = -6, c = 9$

---

44. The equation is also true for −4.   **45.** Some quadratic polynomials cannot easily be factored.
46. yes
47. $\{-\sqrt{b}, \sqrt{b}\}$
48. $\{-2\sqrt{b}, 2\sqrt{b}\}$
49.
$$\left\{ \frac{-\sqrt{b^2 + 16}}{2}, \frac{\sqrt{b^2 + 16}}{2} \right\}$$
50.
$$\left\{ \frac{-5\sqrt{a}}{3}, \frac{5\sqrt{a}}{3} \right\}$$
51.
$$\left\{ \frac{2b + \sqrt{3a}}{5}, \frac{2b - \sqrt{3a}}{5} \right\}$$
52.
$$\{-\sqrt{a^2 + 36}, \sqrt{a^2 + 36}\}$$
53. 2117   **54.** 2243
55. 2356   **56.** We would evaluate P for $x = 3$.
57. $-6.5x^2 + 132.5x - 426 = 0$. Let $x = 4$ and show that 4 is a solution. This corresponds to the year 1992.   **58.** 1999
59. $\sqrt{13}$ **60.** 13   **61.** 1
62. 0

 CONCEPTUAL   WRITING   ▲ CHALLENGING   SCIENTIFIC CALCULATOR   GRAPHICS CALCULATOR

6.2 THE QUADRATIC FORMULA

FOR EXTRA HELP

SSG pp. 173–181
SSM pp. 262–272

Video
7

Tutorial
IBM   MAC

OBJECTIVES

1 ▶ Solve quadratic equations by using the quadratic formula.
2 ▶ Solve applications by using the quadratic formula.
3 ▶ Use the discriminant to determine the number and type of solutions.
4 ▶ Use the discriminant to decide whether a quadratic trinomial can be factored.

▶ **TEACHING TIP**

Going through the derivation of the quadratic formula is a good review of the previous section. Students should be able to follow the individual steps (although they may not be able to present the derivation in its entirety). ◀

▶ **TEACHING TIP**

After getting one side equal to zero, students can circle the coefficients representing *a, b, c.* This formality ensures that the minus sign is not omitted when these coefficients are substituted into the quadratic formula. ◀

The examples in the previous section showed that any quadratic equation can be solved by completing the square. However, completing the square can be tedious and time consuming. In this section we use the method of completing the square to solve the general quadratic equation $ax^2 + bx + c = 0$, where a, b, and c are complex numbers and $a \neq 0$. The solution of this general equation can then be used as a formula to find the solution of any specific quadratic equation.

To solve $ax^2 + bx + c = 0$ by completing the square (assuming $a > 0$ for now), we follow the steps given in Section 6.1.

$$ax^2 + bx + c = 0$$

$$x^2 + \frac{b}{a}x + \frac{c}{a} = 0 \qquad \text{Divide each side by } a. \text{ (Step 1)}$$

$$x^2 + \frac{b}{a}x = -\frac{c}{a} \qquad \text{Subtract } \frac{c}{a} \text{ from each side. (Step 2)}$$

$$\frac{1}{2}\left(\frac{b}{a}\right) = \frac{b}{2a}; \quad \left(\frac{b}{2a}\right)^2 = \frac{b^2}{4a^2} \qquad \text{(Step 3)}$$

$$x^2 + \frac{b}{a}x + \frac{b^2}{4a^2} = -\frac{c}{a} + \frac{b^2}{4a^2} \qquad \text{Add } \frac{b^2}{4a^2} \text{ to each side. (Step 4)}$$

$$x^2 + \frac{b}{a}x + \frac{b^2}{4a^2} = \left(x + \frac{b}{2a}\right)^2 \qquad \text{Write the left side as a square.}$$

$$\left(x + \frac{b}{2a}\right)^2 = \frac{b^2}{4a^2} + \frac{-c}{a}. \qquad \text{Substitute on the left and rearrange the right side. (Step 5)}$$

$$= \frac{b^2}{4a^2} + \frac{-4ac}{4a^2} \qquad \text{Find a common denominator.}$$

$$\left(x + \frac{b}{2a}\right)^2 = \frac{b^2 - 4ac}{4a^2}. \qquad \text{Add fractions.}$$

$$x + \frac{b}{2a} = \sqrt{\frac{b^2 - 4ac}{4a^2}} \qquad \text{or} \qquad x + \frac{b}{2a} = -\sqrt{\frac{b^2 - 4ac}{4a^2}} \qquad \text{Square root property (Step 6)}$$

Since

$$\sqrt{\frac{b^2 - 4ac}{4a^2}} = \frac{\sqrt{b^2 - 4ac}}{\sqrt{4a^2}} = \frac{\sqrt{b^2 - 4ac}}{2a},$$

INSTRUCTOR'S RESOURCES

ITM pp. 415–419   **ISM** pp. 397–415
IAM pp. 24–25

TEST GENERATOR
DOS   Windows   MAC

TRANSPARENCIES

the result can be expressed as

$$x + \frac{b}{2a} = \frac{\sqrt{b^2 - 4ac}}{2a} \qquad \text{or} \qquad x + \frac{b}{2a} = \frac{-\sqrt{b^2 - 4ac}}{2a}$$

$$x = \frac{-b}{2a} + \frac{\sqrt{b^2 - 4ac}}{2a} \qquad \text{or} \qquad x = \frac{-b}{2a} - \frac{\sqrt{b^2 - 4ac}}{2a}$$

$$x = \frac{-b + \sqrt{b^2 - 4ac}}{2a} \qquad \text{or} \qquad x = \frac{-b - \sqrt{b^2 - 4ac}}{2a}.$$

If $a < 0$, it can be shown that the same two solutions are obtained. The result is the **quadratic formula,** often abbreviated as follows.

| Quadratic Formula | The solutions of $ax^2 + bx + c = 0$ $(a \neq 0)$ are $$x = \frac{-b \pm \sqrt{b^2 - 4ac}}{2a}.$$ |

CAUTION Notice in the quadratic formula that the square root is added to or subtracted from the value of $-b$ *before* dividing by $2a$.

1 ▶ Solve quadratic equations by using the quadratic formula.

▶ To use the quadratic formula, first write the given equation in the form $ax^2 + bx + c = 0$; then identify the values of a, b, and c and substitute them into the quadratic formula, as shown in the next examples.

EXAMPLE 1

Using the Quadratic Formula to Find Rational Solutions

Solve $6x^2 - 5x - 4 = 0$.

First identify the letters a, b, and c of the general quadratic equation. Here a, the coefficient of the second-degree term, is 6, while b, the coefficient of the first-degree term, is -5, and the constant, c, is -4. Substitute these values into the quadratic formula.

$$x = \frac{-b \pm \sqrt{b^2 - 4ac}}{2a}$$

$$x = \frac{-(-5) \pm \sqrt{(-5)^2 - 4(6)(-4)}}{2(6)} \qquad a = 6, b = -5, c = -4$$

$$= \frac{5 \pm \sqrt{25 + 96}}{12}$$

$$= \frac{5 \pm \sqrt{121}}{12} = \frac{5 \pm 11}{12}$$

This last statement leads to two solutions, one from $+$ and one from $-$. Using $+$ and $-$ in turn gives

$$x = \frac{5 + 11}{12} = \frac{16}{12} = \frac{4}{3} \qquad \text{or} \qquad x = \frac{5 - 11}{12} = \frac{-6}{12} = -\frac{1}{2}.$$

Check each of these solutions by substitution in the original equation. The solution set is $\{4/3, -1/2\}$. ■

Chalkboard Exercise

Solve by using the quadratic formula.

$$4y^2 - 11y - 3 = 0$$

Answer: $\left\{3, -\frac{1}{4}\right\}$

▶ **TEACHING TIP**

Have students evaluate b^2 and $-4ac$, then write these numbers in their proper place under the radical sign. By doing this first, one may avoid possible sign errors under the radical sign. ◀

▶ **TEACHING TIP**

Students often forget to carry out the arithmetic in the case of two rational answers. ◀

E X A M P L E   2

Using the Quadratic Formula to Find Irrational Solutions

Solve $4r^2 = 8r - 1$.

Rewrite the equation in standard form as $4r^2 - 8r + 1 = 0$ and identify $a = 4$, $b = -8$, and $c = 1$. Now use the quadratic formula.

$$r = \frac{-b \pm \sqrt{b^2 - 4ac}}{2a}$$

$$r = \frac{-(-8) \pm \sqrt{(-8)^2 - 4(4)(1)}}{2(4)} \qquad a = 4, b = -8, c = 1$$

$$= \frac{8 \pm \sqrt{64 - 16}}{8} = \frac{8 \pm \sqrt{48}}{8}$$

$$= \frac{8 \pm 4\sqrt{3}}{8} = \frac{4(2 \pm \sqrt{3})}{8} = \frac{2 \pm \sqrt{3}}{2}$$

The solution set for $4r^2 = 8r - 1$ is $\left\{ \dfrac{2 + \sqrt{3}}{2}, \dfrac{2 - \sqrt{3}}{2} \right\}$. ∎

Chalkboard Exercise

Solve by using the quadratic formula.

$$2k^2 + 19 = 14k$$

Answer: $\left\{ \dfrac{7 + \sqrt{11}}{2}, \dfrac{7 - \sqrt{11}}{2} \right\}$

The solutions of the equations in the next two examples are imaginary numbers.

E X A M P L E   3

Using the Quadratic Formula to Find Imaginary Solutions

Solve $(9q + 3)(q - 1) = -8$.

Every quadratic equation must be in standard form to begin its solution, whether we are factoring or using the quadratic formula. To put this equation in standard form, we first multiply on the left, then collect all nonzero terms on the left.

$$(9q + 3)(q - 1) = -8$$
$$9q^2 - 6q - 3 = -8$$
$$9q^2 - 6q + 5 = 0$$

After writing the equation in the standard form, as $9q^2 - 6q + 5 = 0$, identify $a = 9$, $b = -6$, and $c = 5$. Then use the quadratic formula.

$$q = \frac{-(-6) \pm \sqrt{(-6)^2 - 4(9)(5)}}{2(9)}$$

$$= \frac{6 \pm \sqrt{-144}}{18} = \frac{6 \pm 12i}{18} \qquad \sqrt{-144} = 12i$$

$$= \frac{6(1 \pm 2i)}{18} = \frac{1 \pm 2i}{3}$$

The solution set is $\left\{ \dfrac{1}{3} + \dfrac{2}{3}i, \dfrac{1}{3} - \dfrac{2}{3}i \right\}$. ∎

Chalkboard Exercise

Solve by using the quadratic formula.

$$z^2 = 4z - 5$$

Answer: $\{2 + i, 2 - i\}$

▶ **TEACHING TIP**

Remind them to *factor first, cancel second* when reducing their answers. ◀

▶ **TEACHING TIP**

Imaginary answers are to be written in standard form, where possible. ◀

The solutions in Example 3 were written as $a + bi$, the standard form for complex numbers, as follows:

$$\frac{1 \pm 2i}{3} = \frac{1}{3} \pm \frac{2}{3}i.$$

EXAMPLE 4

Solving a Quadratic Equation with Imaginary Coefficients

Solve $ix^2 - 5x + 2i = 0$.

Here $a = i, b = -5, c = 2i$. Substituting these values into the quadratic formula gives

$$x = \frac{5 \pm \sqrt{25 - (4)(i)(2i)}}{2i} \qquad a = i, b = -5, c = 2i$$

$$= \frac{5 \pm \sqrt{25 - 8i^2}}{2i}$$

$$= \frac{5 \pm \sqrt{25 + 8}}{2i} \qquad i^2 = -1$$

$$= \frac{5 \pm \sqrt{33}}{2i}.$$

Write the solutions without i in the denominator by multiplying the numerator and the denominator by $-i$ as follows.

$$\frac{5 \pm \sqrt{33}}{2i} \cdot \frac{-i}{-i} = \frac{-5i \mp i\sqrt{33}}{-2i^2} = \frac{-5i \pm i\sqrt{33}}{2}$$

The solution set is $\left\{ \dfrac{-5i + i\sqrt{33}}{2}, \dfrac{-5i - i\sqrt{33}}{2} \right\}$. ■

2 ▶ Solve applications by using the quadratic formula.

▶ We can now use the quadratic formula to solve applied problems that produce quadratic equations that cannot be solved by factoring.

PROBLEM SOLVING

In Chapter 4 we solved problems about work rates. Recall, a person's work rate is $\frac{1}{t}$ part of the job per hour, where t is the time in hours required to do the complete job. Thus, the fractional part of the job the person will do in x hours is $\frac{1}{t}x$.

EXAMPLE 5

Solving a Work Problem

Two mechanics take 4 hours to repair a car. If each worked alone, one of them could do the job in 1 hour less time than the other. How long would it take the slower one to complete the job alone?

Let x represent the number of hours for the slower mechanic to complete the job alone. Then the faster mechanic could do the entire job in $x - 1$ hours. Together, they do the job in 4 hours. This information is shown in the following chart.

Worker	Rate	Time Working Together (in hours)	Fractional Part of the Job Done
Faster mechanic	$\dfrac{1}{x - 1}$	4	$\dfrac{1}{x - 1} \cdot 4$
Slower mechanic	$\dfrac{1}{x}$	4	$\dfrac{1}{x} \cdot 4$

The sum of the fractional parts done by each should equal 1 (the whole job).

part done by slower mechanic	+	part done by faster mechanic	=	1 whole job
$\dfrac{4}{x}$	+	$\dfrac{4}{x - 1}$	=	1

Multiply both sides by the common denominator, $x(x - 1)$.

$$4(x - 1) + 4x = x(x - 1)$$

$$4x - 4 + 4x = x^2 - x \qquad \text{Distributive property}$$

$$0 = x^2 - 9x + 4 \qquad \text{Standard form}$$

$$x = \frac{9 \pm \sqrt{81 - 16}}{2} \qquad \text{Quadratic formula}$$

$$= \frac{9 \pm \sqrt{65}}{2}$$

$$x \approx 8.5 \qquad \text{or} \qquad x \approx .5$$

Only the solution 8.5 makes sense in the original equation. (Why?) Thus, the slower mechanic can do the job in about 8.5 hours and the faster in about 8.5 − 1 = 7.5 hours. ■

◆ **C O N N E C T I O N S** ◆

In applied problems, we often prefer approximations to exact values. In Example 5 the calculator values are more useful. Using a calculator, we find

$$\frac{9 + \sqrt{65}}{2} \approx 8.5 \qquad \text{or} \qquad \frac{9 - \sqrt{65}}{2} \approx .5$$

rounded to the nearest tenth.

---

3 ▶ Use the discriminant to determine the number and type of solutions.

▶ The solutions of the quadratic equation $ax^2 + bx + c = 0$ are

$$x = \frac{-b \pm \sqrt{b^2 - 4ac}}{2a}.$$

▶ **TEACHING TIP**
Due to time constraints, this topic is often omitted. Consider including the discriminant table in the text. Give a simple example illustrating each case in the table. ◀

If a, b, and c are integers, the type of solutions of a quadratic equation (that is, rational, irrational, or imaginary) is determined by the quantity under the square root sign, $b^2 - 4ac$. Because it distinguishes among the three types of solutions, the quantity $b^2 - 4ac$ is called the **discriminant.** By calculating the discriminant before solving a quadratic equation, we can predict whether the solutions will be rational numbers, irrational numbers, or imaginary numbers. This can be useful in an applied problem, for example, where irrational or imaginary number solutions are not acceptable. If the discriminant is a perfect square (including 0), the equation can be solved by factoring. Otherwise, the quadratic formula should be used.

Discriminant

The discriminant of $ax^2 + bx + c = 0$ is given by $b^2 - 4ac$. If a, b and c are integers, then the type of solution is determined as follows.

Discriminant	Type of Solutions
Positive, and the square of an integer	Two different rational solutions
Positive, but not the square of an integer	Two different irrational solutions
Zero	One rational solution
Negative	Two different imaginary solutions

In the next example, we show how to use the discriminant to decide what type of solution a quadratic equation has.

EXAMPLE 6

Using the Discriminant

Predict the number and type of solutions for the following equations.

(a) $6x^2 - x - 15 = 0$

We find the discriminant by evaluating $b^2 - 4ac$.

$$b^2 - 4ac = (-1)^2 - 4(6)(-15) \qquad a = 6, b = -1, c = -15$$
$$= 1 + 360 = 361$$

A calculator shows that $361 = 19^2$, a perfect square. Since a, b, and c are integers, the solutions will be two different rational numbers, and the equation can be solved by factoring.

(b) $3m^2 - 4m = 5$

Rewrite the equation as $3m^2 - 4m - 5 = 0$ to find $a = 3$, $b = -4$, $c = -5$. The discriminant is

$$b^2 - 4ac = (-4)^2 - 4(3)(-5) = 16 + 60 = 76.$$

Because 76 is not the square of an integer, $\sqrt{76}$ is irrational. From this and from the fact that a, b, and c integers, the equation will have two different irrational solutions, one using $\sqrt{76}$ and one using $-\sqrt{76}$.

(c) $4x^2 + x + 1 = 0$

Since $a = 4$, $b = 1$, and $c = 1$, the discriminant is

$$(1)^2 - 4(4)(1) = -15.$$

Since the discriminant is negative and a, b, and c are integers, this quadratic equation will have two imaginary number solutions. ■

EXAMPLE 7

Using the Discriminant

Find k so that $9x^2 + kx + 4 = 0$ will have only one rational number solution.

The equation will have only one rational number solution if the discriminant is 0. Here, since $a = 9$, $b = k$, and $c = 4$, the discriminant is

$$b^2 - 4ac = k^2 - 4(9)(4) = k^2 - 144.$$

We set this result equal to 0 and solve for k.

$$k^2 - 144 = 0$$
$$k^2 = 144$$
$$k = 12 \qquad \text{or} \qquad k = -12$$

The equation will have only one rational number solution if $k = 12$ or $k = -12$. ■

4 ▶ Use the discriminant to decide whether a quadratic trinomial can be factored.

▶ A quadratic trinomial can be factored with rational coefficients only if the corresponding quadratic equation has rational solutions. Thus, the discriminant can be used to decide whether or not a given trinomial is factorable.

EXAMPLE 8

Deciding Whether a Trinomial Is Factorable

Decide whether or not the following trinomials can be factored.

(a) $24x^2 + 7x - 5$

To decide whether the solutions of $24x^2 + 7x - 5 = 0$ are rational numbers, we evaluate the discriminant.

$$b^2 - 4ac = 7^2 - 4(24)(-5) = 49 + 480 = 529 = 23^2$$

Since 529 is a perfect square, the solutions are rational numbers and the trinomial can be factored. Verify that

$$24x^2 + 7x - 5 = (3x - 1)(8x + 5).$$

(b) $11m^2 - 9m + 12$

The discriminant is $b^2 - 4ac = (-9)^2 - 4(11)(12) = -447$. This number is negative, so the corresponding quadratic equation has imaginary number solutions and therefore the trinomial cannot be factored. ∎

◆ 1. Since $\dfrac{1 + \sqrt{41}}{5} +$

$\dfrac{1 - \sqrt{41}}{5} = \dfrac{2}{5} = -\dfrac{b}{a}$ and

$\dfrac{1 + \sqrt{41}}{5} \cdot \dfrac{1 - \sqrt{41}}{5} =$

$-\dfrac{8}{5} = \dfrac{c}{a}$, the solutions are

correct.

2. Since $\dfrac{3i}{2} + (-4i) =$

$-\dfrac{5i}{2} = -\dfrac{b}{a}$ and $\dfrac{3i}{2}(-4i) =$

$6 = \dfrac{c}{a}$, the solutions are

correct.

◇ **CONNECTIONS** ◇

We can develop two interesting and useful properties of the solutions of a quadratic equation $ax^2 + bx + c = 0$, $a \neq 0$. If

$$x_1 = \frac{-b + \sqrt{b^2 - 4ac}}{2a} \quad \text{and} \quad x_2 = \frac{-b - \sqrt{b^2 - 4ac}}{2a},$$

then the sum of the solutions is

$$x_1 + x_2 = \frac{-b + \sqrt{b^2 - 4ac}}{2a} + \frac{-b - \sqrt{b^2 - 4ac}}{2a} = -\frac{b}{a}.$$

The product of the solutions is

$$x_1 x_2 = \left(\frac{-b + \sqrt{b^2 - 4ac}}{2a}\right)\left(\frac{-b - \sqrt{b^2 - 4ac}}{2a}\right).$$

Using the rule $(x + y)(x - y) = x^2 - y^2$ with $x = -b$ and $y = \sqrt{b^2 - 4ac}$ gives the product of the numerators.

$$x_1 x_2 = \frac{(-b)^2 - (\sqrt{b^2 - 4ac})^2}{(2a)^2} = \frac{c}{a}$$

FOR DISCUSSION OR WRITING

A good way to use the sum and product of the solutions of a quadratic equation is to check that the solutions you find are correct.

1. Use the sum and product of the solutions to verify that the solutions of
$$5x^2 - 2x - 8 = 0 \text{ are } \frac{1 + \sqrt{41}}{5} \text{ and } \frac{1 - \sqrt{41}}{5}.$$

2. Verify that the solutions of $-2x^2 - 5ix = 12$ are $3i/2$ and $-4i$.

3. Fill in the missing steps above that show that the sum of the solutions is $-b/a$ and the product of the solutions is c/a.

6.2 EXERCISES

1. In both cases, the solutions are -2 and 3.

2. $x = \dfrac{-q \pm \sqrt{q^2 - 4pr}}{2p}$

◉ **1.** If we begin with the quadratic equation $x^2 - x - 6 = 0$, another quadratic equation with the same solution set can be obtained by multiplying both sides by a nonzero constant. Suppose we multiply both sides by -1 to obtain the equation $-x^2 + x + 6 = 0$. Show that if we use $a = 1$, $b = -1$, and $c = -6$ (values from the first equation), we get the same solutions as with $a = -1$, $b = 1$, and $c = 6$ (values from the second equation).

2. What would the quadratic formula be if we chose $px^2 + qx + r = 0$ ($p \neq 0$) as the standard form of a quadratic equation?

◉ CONCEPTUAL   ✐ WRITING   ▲ CHALLENGING   ▦ SCIENTIFIC CALCULATOR   ▦ GRAPHICS CALCULATOR

Use the quadratic formula to solve the equation. (All solutions for these equations are real numbers.) See Examples 1 and 2.

3. $m^2 - 8m + 15 = 0$ **4.** $x^2 + 3x - 28 = 0$ **5.** $2k^2 + 4k + 1 = 0$

6. $2y^2 + 3y - 1 = 0$ **7.** $2x^2 - 2x = 1$ **8.** $9t^2 + 6t = 1$

9. $x^2 + 18 = 10x$ **10.** $x^2 - 4 = 2x$ **11.** $-2t(t + 2) = -3$

12. $-3x(x + 2) = -4$ **13.** $(r - 3)(r + 5) = 2$ **14.** $(k + 1)(k - 7) = 1$

15. $p^2 + \dfrac{p}{3} = \dfrac{2}{3}$ **16.** $\dfrac{x^2}{4} - \dfrac{x}{2} = 1$ **17.** $4k(k + 1) = 1$

18. $(r - 1)(4r) = 19$ **19.** $(g + 2)(g - 3) = 1$ **20.** $(y - 5)(y + 2) = 6$

21. $3x^2 + 2x = 2$ **22.** $26r - 2 = 3r^2$

23. $y = \dfrac{5(5 - y)}{3(y + 1)}$ **24.** $k = \dfrac{k + 15}{3(k - 1)}$

◉ **25.** What is wrong with the following solution of $5x^2 - 5x + 1 = 0$?

$$x = \frac{5 \pm \sqrt{25 - 4(5)(1)}}{2(5)} \qquad a = 5, b = -5, c = 1$$

$$= \frac{5 \pm \sqrt{5}}{10}$$

$$= \frac{1}{2} \pm \sqrt{5}$$

◉ **26.** Is $\dfrac{-b \pm \sqrt{b^2 - 4ac}}{2a}$ equal to $-b \pm \dfrac{\sqrt{b^2 - 4ac}}{2a}$? Explain.

◉ **27.** Can the quadratic formula be used to solve $2y^2 - 5 = 0$? Explain.

◉ **28.** Can $4m^2 + 3m = 0$ be solved by the quadratic formula? Explain.

Use the quadratic formula to solve the equation. (All solutions for these equations are imaginary numbers.) See Examples 3 and 4.

29. $k^2 + 47 = 0$ **30.** $x^2 + 19 = 0$ **31.** $r^2 - 6r + 14 = 0$

32. $t^2 + 4t + 11 = 0$ **33.** $4x^2 - 4x = -7$ **34.** $9x^2 - 6x = -7$

35. $x(3x + 4) = -2$ **36.** $y(2y + 3) = -2$ **37.** $\dfrac{x + 5}{2x - 1} = \dfrac{x - 4}{x - 6}$

38. $\dfrac{3x - 4}{2x - 5} = \dfrac{x + 5}{x + 2}$ **39.** $\dfrac{1}{x^2} + 1 = -\dfrac{1}{x}$ **40.** $\dfrac{4}{r^2} + 3 = \dfrac{1}{r}$

41. $-3t^2 + 4it = 0$ **42.** $5x^2 - 8ix = 0$

43. $2iz^2 - 3z = -2i$ **44.** $r^2 - ir = -12$

▦ *Solve the problem and use a calculator as necessary. Round your answer to the nearest tenth. See Example 5.*

45. Working together, two people can cut a large lawn in 2 hours. One person can do the job alone in 1 hour less than the other. How long would it take the faster person to do the job?

46. A janitorial service provides two people to clean an office building. Working together, the two can clean the building in 5 hours. One person is new to the job and would take 2 hours longer than the other person to clean the building working alone. How long would it take the new worker to clean the building working alone?

3. $\{3, 5\}$ **4.** $\{-7, 4\}$

5. $\left\{\dfrac{-2 + \sqrt{2}}{2}, \dfrac{-2 - \sqrt{2}}{2}\right\}$

6. $\left\{\dfrac{-3 + \sqrt{17}}{4}, \dfrac{-3 - \sqrt{17}}{4}\right\}$

7. $\left\{\dfrac{1 + \sqrt{3}}{2}, \dfrac{1 - \sqrt{3}}{2}\right\}$

8. $\left\{\dfrac{-1 + \sqrt{2}}{3}, \dfrac{-1 - \sqrt{2}}{3}\right\}$

9. $\{5 + \sqrt{7}, 5 - \sqrt{7}\}$

10. $\{1 + \sqrt{5}, 1 - \sqrt{5}\}$

11. $\left\{\dfrac{-2 + \sqrt{10}}{2}, \dfrac{-2 - \sqrt{10}}{2}\right\}$

12. $\left\{\dfrac{-3 + \sqrt{21}}{3}, \dfrac{-3 - \sqrt{21}}{3}\right\}$

13. $\{-1 + 3\sqrt{2}, -1 - 3\sqrt{2}\}$

14. $\{3 + \sqrt{17}, 3 - \sqrt{17}\}$

15. $\left\{-1, \dfrac{2}{3}\right\}$

16. $\{1 + \sqrt{5}, 1 - \sqrt{5}\}$

17. $\left\{\dfrac{-1 + \sqrt{2}}{2}, \dfrac{-1 - \sqrt{2}}{2}\right\}$

18. $\left\{\dfrac{1 + 2\sqrt{5}}{2}, \dfrac{1 - 2\sqrt{5}}{2}\right\}$

19. $\left\{\dfrac{1 + \sqrt{29}}{2}, \dfrac{1 - \sqrt{29}}{2}\right\}$

20. $\left\{\dfrac{3 + \sqrt{73}}{2}, \dfrac{3 - \sqrt{73}}{2}\right\}$

21. $\left\{\dfrac{-1 + \sqrt{7}}{3}, \dfrac{-1 - \sqrt{7}}{3}\right\}$

22. $\left\{\dfrac{13 + \sqrt{163}}{3}, \dfrac{13 - \sqrt{163}}{3}\right\}$

23. $\left\{\dfrac{-4 + \sqrt{91}}{3}, \dfrac{-4 - \sqrt{91}}{3}\right\}$

24. $\left\{-\dfrac{5}{3}, 3\right\}$

29. $\{-i\sqrt{47}, i\sqrt{47}\}$
30. $\{-i\sqrt{19}, i\sqrt{19}\}$
31. $\{3 + i\sqrt{5}, 3 - i\sqrt{5}\}$
32. $\{-2 + i\sqrt{7}, -2 - i\sqrt{7}\}$

◉ CONCEPTUAL   🖉 WRITING   ▲ CHALLENGING   ▦ SCIENTIFIC CALCULATOR   ▦ GRAPHICS CALCULATOR

33. $\left\{\dfrac{1 + i\sqrt{6}}{2}, \dfrac{1 - i\sqrt{6}}{2}\right\}$

34. $\left\{\dfrac{1 + i\sqrt{6}}{3}, \dfrac{1 - i\sqrt{6}}{3}\right\}$

35. $\left\{\dfrac{-2 + i\sqrt{2}}{3}, \dfrac{-2 - i\sqrt{2}}{3}\right\}$

36. $\left\{\dfrac{-3 + i\sqrt{7}}{4}, \dfrac{-3 - i\sqrt{7}}{4}\right\}$

37. $\{4 + 3i\sqrt{2}, 4 - 3i\sqrt{2}\}$

38. $\left\{\dfrac{3 + i\sqrt{59}}{2}, \dfrac{3 - i\sqrt{59}}{2}\right\}$

39. $\left\{\dfrac{-1 + i\sqrt{3}}{2}, \dfrac{-1 - i\sqrt{3}}{2}\right\}$

40. $\left\{\dfrac{1 + i\sqrt{47}}{6}, \dfrac{1 - i\sqrt{47}}{6}\right\}$

41. $\left\{0, \dfrac{4}{3}i\right\}$ 42. $\left\{0, \dfrac{8}{5}i\right\}$

43. $\left\{-2i, \dfrac{1}{2}i\right\}$

44. $\{-3i, 4i\}$

45. 3.6 hours

46. 11.1 hours

47. Rusty: 25.0 hours; Nancy: 23.0 hours

48. Young: 2.6 hours; Parker: 3.6 hours

49. 2.4 seconds and 5.6 seconds

50. 3.7 seconds and 9.3 seconds   51. It reaches its maximum height at 5 seconds, since this is the only time it reaches 400 feet.

52. Because a negative number appears as the discriminant, it never reaches a height of 425 feet.

53. (d)   54. (d)   55. (a)

56. (a)   57. (b)   58. (b)

59. (c)   60. (c)

61. −10 or 10

62. −14 or 14   63. 16

64. 9   65. 25   66. 9

47. Rusty and Nancy Brauner are planting flats of spring flowers. Working alone, Rusty would take 2 hours longer than Nancy to plant the flowers. Working together, they do the job in 12 hours. How long would it have taken each person working alone?

48. John Young can work through a stack of invoices in 1 hour less time than Arnold Parker can. Working together they take 3/2 hours. How long would it take each person working alone?

49. A ball is thrown vertically upward from the ground. Its distance in feet from the ground in t seconds is $s = -16t^2 + 128t$. At what times will the ball be 213 feet from the ground?

50. A toy rocket is launched from ground level. Its distance in feet from the ground in t seconds is $s = -16t^2 + 208t$. At what times will the rocket be 550 feet from the ground?

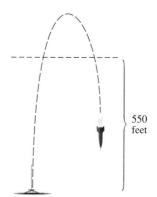

⊙ *A rock is thrown upward from ground level, and its distance from the ground in t seconds is $s = -16t^2 + 160t$. Use algebra and a short explanation to answer Exercises 51 and 52.*

51. After how many seconds does it reach a height of 400 feet? How would you describe in words its position at this height?

52. After how many seconds does it reach a height of 425 feet? How would you interpret the mathematical result here?

Use the discriminant to determine whether the equation has solutions that are (a) two different rational numbers, (b) exactly one rational number, (c) two different irrational numbers, or (d) two different imaginary numbers. See Example 6.

53. $2x^2 - x + 1 = 0$ **54.** $4x^2 - 4x + 3 = 0$ **55.** $6m^2 + 7m - 3 = 0$

56. $7x^2 - 32x - 15 = 0$ **57.** $x^2 + 4x = -4$ **58.** $4y^2 + 36y = -81$

59. $9t^2 = 30t - 15$ **60.** $25k^2 = -20k + 2$

Find the value of a, b, or c so that the equation will have exactly one rational solution. See Example 7.

61. $p^2 + bp + 25 = 0$ **62.** $r^2 - br + 49 = 0$ **63.** $am^2 + 8m + 1 = 0$

64. $ay^2 + 24y + 16 = 0$ **65.** $9y^2 - 30y + c = 0$ **66.** $4m^2 + 12m + c = 0$

⊙ **67.** Is it possible for the solution of a quadratic equation with integer coefficients to include just one irrational number? Why?

⊙ **68.** Can the solution of a quadratic equation with integer coefficients include one real and one imaginary number? Why?

Use the discriminant to tell if the polynomial can be factored (using integer coefficients). If a polynomial can be factored, factor it. See Example 8.

69. $24x^2 - 34x - 45$

70. $36y^2 + 69y + 28$

71. $36x^2 + 21x - 24$

72. $18k^2 + 13k - 12$

73. $12x^2 - 83x - 7$

74. $16y^2 - 61y - 12$

────────◆ **MATHEMATICAL CONNECTIONS** (Exercises 75–78) ◆────────

When studying quadratic equations, we are usually interested in solving an equation for its solutions. In these Connections, we will look at the opposite procedure: given two solutions, how do we go about finding an equation that will yield them?

There is a theorem that states that if a quadratic equation has real coefficients and $a + bi$ is a solution, then its conjugate, $a - bi$, must also be a solution.

Work Exercises 75–78 in order to see how we can find a quadratic equation with $1 + 5i$ and $1 - 5i$ as solutions.

75. The equation $(x - r)(x - s) = 0$ has solutions r and s. Use $r = 1 + 5i$ and $s = 1 - 5i$, and write the equation making these substitutions. Do not multiply out the factors.

76. Distribute the negative sign in each factor in the equation from Exercise 75 so that the left-hand side is the product of the sum of two terms and the difference of two terms. One of these terms should be written as a binomial within parentheses.

77. Use the pattern $(m + n)(m - n) = m^2 - n^2$ and the fact that $i^2 = -1$ to write the equation from Exercise 76 in the form $ax^2 + bx + c = 0$.

78. Solve the equation from Exercise 77 using the quadratic formula to verify that the solutions are indeed $1 + 5i$ and $1 - 5i$.

▲**79.** One solution of $4x^2 + bx - 3 = 0$ is $-\dfrac{5}{2}$. Find b and the other solution.

▲**80.** One solution of $3x^2 - 7x + c = 0$ is $\dfrac{1}{3}$. Find c and the other solution.

REVIEW EXERCISES

Rewrite the expression making the indicated substitution. Then factor the trinomial obtained after the substitution has been made. See Section 3.8.

81. $(7z + 3)^2 + 4(7z + 3) - 5$;   let $7z + 3 = u$

82. $4\left(\dfrac{1}{2}w + 8\right)^2 - 7\left(\dfrac{1}{2}w + 8\right) - 2$;   let $\dfrac{1}{2}w + 8 = u$

Solve the equation. See Section 2.1.

83. $\dfrac{3}{4}x + \dfrac{1}{2}x = -10$

84. $\dfrac{x}{5} + \dfrac{3x}{4} = -19$

Solve the equation. See Section 5.6.

85. $\sqrt{2x + 6} = x - 1$

86. $\sqrt{2x + 1} + \sqrt{x + 3} = 0$

69. The discriminant is 74^2, so it can be factored; $(6x + 5)(4x - 9)$
70. The discriminant is 27^2, so it can be factored; $(12y + 7)(3y + 4)$
71. The discriminant, 3897, is not a perfect square, so it cannot be factored.
72. The discriminant, 1033, is not a perfect square, so it cannot be factored.
73. The discriminant is 85^2, so it can be factored; $(12x + 1)(x - 7)$
74. The discriminant is 67^2, so it can be factored; $(16y + 3)(y - 4)$
75. $[x - (1 + 5i)] \cdot [x - (1 - 5i)] = 0$
76. $[(x - 1) - 5i] \cdot [(x - 1) + 5i] = 0$
77. $x^2 - 2x + 26 = 0$
78. Use $a = 1$, $b = -2$, and $c = 26$. The solutions are $1 + 5i$ and $1 - 5i$.
79. $b = \dfrac{44}{5}$; $x_2 = \dfrac{3}{10}$
80. $c = 2$; $x_2 = 2$
81. $u^2 + 4u - 5$; $(u + 5)(u - 1)$
82. $4u^2 - 7u - 2$; $(4u + 1)(u - 2)$
83. $\{-8\}$ **84.** $\{-20\}$
85. $\{5\}$ **86.** $\emptyset$

6.3 EQUATIONS QUADRATIC IN FORM

FOR EXTRA HELP

 SSG pp. 181–188
SSM pp. 272–281

 Video
8

Tutorial
IBM   MAC

OBJECTIVES

1 ▶ Solve an equation with fractions by writing it in quadratic form.
2 ▶ Use quadratic equations to solve applied problems.
3 ▶ Solve an equation with radicals by writing it in quadratic form.
4 ▶ Solve an equation that is quadratic in form by substitution.

We have now introduced four methods for solving quadratic equations written in the form $ax^2 + bx + c = 0$. The following chart gives some advantages and disadvantages of each method.

Methods for Solving Quadratic Equations

Method	Advantages	Disadvantages
Factoring	Usually the fastest method	Not all polynomials are factorable; some factorable polynomials are hard to factor.
Square root property	Simplest method for solving equations of the form $(x + a)^2 = b$	Few equations are given in this form.
Completing the square	None for solving equations (the procedure is useful in other areas of mathematics)	It requires more steps than other methods.
Quadratic formula	Can always be used	It is more difficult than factoring because of the square root.

▶ **TEACHING TIP**
Spend some lecture time briefly reviewing the methods outlined in this table. Emphasize that these two methods are most often used: *factoring* (look for this first), and the *quadratic formula* (used when we cannot factor). ◀

◈ $x = 2$

◈ **C O N N E C T I O N S** ◈

Methods of solving linear and quadratic equations have been known since the time of the Babylonians. However, for centuries mathematicians wrestled with finding a formula that could solve cubic (third-degree) equations. In the sixteenth century, Nicolo Tartaglia developed the following formula for finding one real solution of a cubic equation of the form $x^3 + mx = n$.

$$x = \sqrt[3]{\frac{n}{2} + \sqrt{\left(\frac{n}{2}\right)^2 + \left(\frac{m}{3}\right)^3}} - \sqrt[3]{-\left(\frac{n}{2}\right) + \sqrt{\left(\frac{n}{2}\right)^2 + \left(\frac{m}{3}\right)^3}}$$

FOR DISCUSSION OR WRITING
Try solving for one solution of the equation $x^3 + 9x = 26$ using this formula. Check your answer to see if it is correct. Why do you think it is so much more difficult to solve a cubic equation than a quadratic equation? (*Hint:* Consider how we developed the formula for the quadratic formula.)

INSTRUCTOR'S RESOURCES

 ITM pp. 419–422   **ISM** pp. 415–432
IAM pp. 25–26

TEST GENERATOR
DOS   Windows   MAC

TRANSPARENCIES

1 ▶ Solve an equation with fractions by writing it in quadratic form.

▶ A variety of nonquadratic equations can be written in the form of a quadratic equation and solved by using one of the methods in the chart. For example, some equations with fractions lead to quadratic equations. As you solve the equations in this section try to decide which is the best method for each equation.

E X A M P L E   1

Writing an Equation with Fractions in Quadratic Form

■ Solve $\dfrac{1}{x} + \dfrac{1}{x-1} = \dfrac{7}{12}$.

Clear the equation of fractions by multiplying each term by the common denominator, $12x(x-1)$. (We note that $x \neq 0$ and $x \neq 1$.)

$$12x(x-1)\frac{1}{x} + 12x(x-1)\frac{1}{x-1} = 12x(x-1)\frac{7}{12}$$

$$12(x-1) + 12x = 7x(x-1)$$

$$12x - 12 + 12x = 7x^2 - 7x \qquad \text{Distributive property}$$

$$24x - 12 = 7x^2 - 7x \qquad \text{Combine terms.}$$

A quadratic equation must be in the standard form $ax^2 + bx + c = 0$ before it can be solved by factoring. Combine terms and arrange them so that one side of the equation is zero. Then factor.

$$0 = 7x^2 - 31x + 12 \qquad \text{Standard form}$$

$$0 = (7x - 3)(x - 4) \qquad \text{Factor.}$$

Setting each factor equal to 0 and solving the two linear equations gives the solutions 3/7 and 4. Check by substituting these solutions in the original equation. The solution set is {3/7, 4}. ■

Chalkboard Exercise

Solve $\dfrac{4}{m-1} + 9 = -\dfrac{7}{m}$.

Answer: $\left\{\dfrac{7}{9}, -1\right\}$

▶ TEACHING TIP

Remind students that we usually expect two different answers from a second-degree equation. ◀

▶ TEACHING TIP

Try factoring *first*. Point out that the quadratic formula will work in case one has difficulty with a factorable trinomial. ◀

---

2 ▶ Use quadratic equations to solve applied problems.

▶ Earlier we solved distance-rate-time (or motion) problems that led to linear equations or rational equations. Now we can extend that work to motion problems that lead to quadratic equations. Writing an equation to solve the problem will be done just as it was earlier. Distance-rate-time applications often lead to equations with fractions, as in the next example.

E X A M P L E   2

Solving a Motion Problem

■ A riverboat for tourists averages 12 miles per hour in still water. It takes the boat 1 hour, 4 minutes to go 6 miles upstream and return. Find the speed of the current. See Figure 1.

Chalkboard Exercise

In $1\dfrac{3}{4}$ hours Khe rows his boat 5 miles upriver and comes back. The speed of the current is 3 miles per hour. How fast does Khe row?

Answer: 7 miles per hour

FIGURE 1

For a problem about rate (or speed), we use the distance formula, $d = rt$.

Let
$$x = \text{the speed of the current;}$$
$$12 - x = \text{the rate upstream;}$$
$$12 + x = \text{the rate downstream.}$$

The rate upstream is the difference of the speed of the boat in still water and the speed of the current, or $12 - x$. The speed downstream is, in the same way, $12 + x$. To find the time, we rewrite the formula $d = rt$ as

$$t = \frac{d}{r}.$$

This information was used to complete the following chart.

	d	r	t
Upstream	6	$12 - x$	$\dfrac{6}{12 - x}$
Downstream	6	$12 + x$	$\dfrac{6}{12 + x}$

Times in hours

We can write the total time, 1 hour and 4 minutes, as

$$1 + \frac{4}{60} = 1 + \frac{1}{15} = \frac{16}{15} \text{ hours.}$$

Since the time upstream plus the time downstream equals 16/15 hours,

$$\frac{6}{12 - x} + \frac{6}{12 + x} = \frac{16}{15}.$$

Now multiply both sides of the equation by the common denominator $15(12 - x)(12 + x)$ and solve the resulting quadratic equation.

$$15(12 + x)6 + 15(12 - x)6 = 16(12 - x)(12 + x)$$
$$90(12 + x) + 90(12 - x) = 16(144 - x^2)$$
$$1080 + 90x + 1080 - 90x = 2304 - 16x^2 \qquad \text{Distributive property}$$
$$2160 = 2304 - 16x^2 \qquad \text{Combine terms.}$$
$$16x^2 = 144$$
$$x^2 = 9$$

▶ **TEACHING TIP**
Alternatively, a student may wish to write this equation as $x^2 - 9 = 0$ and then solve by factoring. ◀

Solve $x^2 = 9$ by using the square root property to get the two solutions

$$x = 3 \qquad \text{or} \qquad x = -3.$$

The speed of the current cannot be -3, so the solution is $x = 3$ miles per hour. ■

3 ▶ Solve an equation with radicals by writing it in quadratic form.

▶ In Section 5.6 we saw that some equations with radicals lead to quadratic equations.

EXAMPLE 3

Solving a Radical Equation that Leads to a Quadratic Equation

Solve $x + \sqrt{x} = 6$.

This equation is not quadratic. However, squaring both sides of the equation gives a quadratic equation that can be solved by factoring.

$$\sqrt{x} = 6 - x \qquad \text{Isolate the radical on one side.}$$
$$x = 36 - 12x + x^2 \qquad \text{Square both sides.}$$
$$0 = x^2 - 13x + 36 \qquad \text{Get } 0 \text{ on one side.}$$
$$0 = (x - 4)(x - 9) \qquad \text{Factor.}$$
$$x - 4 = 0 \qquad \text{or} \qquad x - 9 = 0 \qquad \text{Set each factor equal to } 0.$$
$$x = 4 \qquad\qquad x = 9$$

Chalkboard Exercise

Solve $2x = \sqrt{x} + 1$.
Answer: $\{1\}$

▶ **TEACHING TIP**

As with rational equations, answers to equations with radicals must be checked. Remind students that square roots of a negative number are imaginary numbers. ◀

Check both potential solutions in the *original* equation.

If $x = 4$,		If $x = 9$,	
$4 + \sqrt{4} = 6$	?	$9 + \sqrt{9} = 6$	?
$6 = 6$.	True	$12 = 6$.	False

The solution set is $\{4\}$. ∎

4 ▶ Solve an equation that is quadratic in form by substitution.

▶ An equation that can be written in the form $au^2 + bu + c = 0$, for $a \neq 0$ and u an algebraic expression, is called **quadratic in form**.

EXAMPLE 4

Solving Equations that Are Quadratic in Form

Solve each equation.

(a) $2(4m - 3)^2 + 7(4m - 3) + 5 = 0$

Because of the repeated quantity $4m - 3$, this equation is quadratic in form with $u = 4m - 3$. (Any letter, except m, could be used instead of u.) Write

$$2(4m - 3)^2 + 7(4m - 3) + 5 = 0$$

Chalkboard Exercise

Solve
$5(r + 3)^2 + 9(r + 3) = 2$.

Answer: $\left\{-\dfrac{14}{5}, -5\right\}$

as
$$2u^2 + 7u + 5 = 0. \qquad \text{Let } 4m - 3 = u.$$
$$(2u + 5)(u + 1) = 0 \qquad \text{Factor.}$$

By the zero-factor property, the solutions of $2u^2 + 7u + 5 = 0$ are

$$u = -\frac{5}{2} \qquad \text{or} \qquad u = -1.$$

▶ **TEACHING TIP**

Although substitution is the preferable method, one could simplify the expression and then solve by factoring directly. ◀

To find m, substitute $4m - 3$ for u.

$$4m - 3 = -\frac{5}{2} \qquad \text{or} \qquad 4m - 3 = -1$$

$$4m = \frac{1}{2} \qquad \text{or} \qquad 4m = 2$$

$$m = \frac{1}{8} \qquad \text{or} \qquad m = \frac{1}{2}$$

The solution set of the original equation is $\left\{\frac{1}{8}, \frac{1}{2}\right\}$.

(b) $y^4 = 6y^2 - 3$.

First write the equation as

$$y^4 - 6y^2 + 3 = 0.$$

Then substitute u for y^2 and u^2 for y^4 to get $u^2 - 6u + 3 = 0$, with $a = 1$, $b = -6$, and $c = 3$. By the quadratic formula,

$$u = y^2 = \frac{6 \pm \sqrt{36 - 12}}{2} \qquad a = 1, b = -6, c = 3$$

$$= \frac{6 \pm \sqrt{24}}{2}$$

$$= \frac{6 \pm 2\sqrt{6}}{2}$$

$$= 3 \pm \sqrt{6}.$$

Find y by using the square root property as follows.

$$y^2 = 3 + \sqrt{6} \qquad \text{or} \qquad y^2 = 3 - \sqrt{6}$$

$$y = \pm\sqrt{3 + \sqrt{6}} \qquad \text{or} \qquad y = \pm\sqrt{3 - \sqrt{6}}$$

► **TEACHING TIP**
Point out that we can obtain up to four distinct solutions to a fourth-degree equation. ◄

The solution set contains four numbers:

$$\{\sqrt{3 + \sqrt{6}}, \ -\sqrt{3 + \sqrt{6}}, \ \sqrt{3 - \sqrt{6}}, \ -\sqrt{3 - \sqrt{6}}\}. \ ■$$

EXAMPLE 5

Solving Equations that Are Quadratic in Form

Solve each equation.

(a) Solve $4x^6 + 1 = 5x^3$.

Let $x^3 = u$. Then $x^6 = u^2$. Substitute into the given equation.

$$4x^6 + 1 = 5x^3$$

$$4u^2 + 1 = 5u \qquad \text{Let } x^6 = u^2, x^3 = u.$$

$$4u^2 - 5u + 1 = 0 \qquad \text{Get 0 on one side.}$$

$$(4u - 1)(u - 1) = 0 \qquad \text{Factor.}$$

$$u = \frac{1}{4} \qquad \text{or} \qquad u = 1$$

Chalkboard Exercise

Solve $4m^{2/3} = 3m^{1/3} + 1$.

Answer: $\left\{-\dfrac{1}{64}, 1\right\}$

Replace u with x^3 to get

$$x^3 = \frac{1}{4} \qquad \text{or} \qquad x^3 = 1.$$

► **TEACHING TIP**
Show several specific examples of *quadratic forms* such as:

$$3x^4 + 4x^2 + 2$$

or

$$5x^6 + 2x^3 - 7. \ ◄$$

From these equations,

$$x = \sqrt[3]{\frac{1}{4}} = \frac{\sqrt[3]{1}}{\sqrt[3]{4}} = \frac{1}{\sqrt[3]{4}} \cdot \frac{\sqrt[3]{2}}{\sqrt[3]{2}} = \frac{\sqrt[3]{2}}{2} \qquad \text{or} \qquad x = \sqrt[3]{1} = 1.$$

This method of substitution will give only the real number solutions for equations with polynomials of degree $2n$ where n is odd. However, it gives all complex number solutions for equations with polynomials of degree $2n$ where n is even. The real number solution set of $4x^6 + 1 = 5x^3$ is $\{\sqrt[3]{2}/2, 1\}$.

(b) $2a^{2/3} - 11a^{1/3} + 12 = 0$

Let $a^{1/3} = u$; then $a^{2/3} = u^2$. Substitute into the given equation.

$$2u^2 - 11u + 12 = 0 \qquad \text{Let } a^{1/3} = u, \, a^{2/3} = u^2.$$

$$(2u - 3)(u - 4) = 0 \qquad \text{Factor.}$$

$$2u - 3 = 0 \qquad\qquad \text{or} \qquad u - 4 = 0$$

$$u = \frac{3}{2} \qquad\qquad \text{or} \qquad u = 4$$

$$a^{1/3} = \frac{3}{2} \qquad\qquad \text{or} \qquad a^{1/3} = 4 \qquad u = a^{1/3}$$

$$a = \left(\frac{3}{2}\right)^3 = \frac{27}{8} \qquad \text{or} \qquad a = 4^3 = 64 \qquad \text{Cube both sides.}$$

(Recall that equations with fractional exponents were solved in Section 5.6 by raising both sides to the same power.) Check that the solution set is $\{27/8, 64\}$. ∎

6.3 EXERCISES

⊚ **1.** Of the four methods for solving quadratic equations, it is often said that the quadratic formula is the most efficient method. Do you agree with this statement? Explain why or why not.

⊚ **2.** What is the relationship between the method of completing the square in solving quadratic equations and the quadratic formula?

2. The quadratic formula is derived using the method of completing the square for the standard form $ax^2 + bx + c = 0, a \neq 0$.

Solve by first clearing the equation of fractions. Check your answers. See Example 1.

3. $\dfrac{14}{x} = x - 5$

3. $\{-2, 7\}$

4. $\dfrac{-12}{x} = x + 8$

4. $\{-6, -2\}$

5. $1 - \dfrac{3}{x} - \dfrac{28}{x^2} = 0$

5. $\{-4, 7\}$

6. $4 - \dfrac{7}{r} - \dfrac{2}{r^2} = 0$

6. $\left\{-\dfrac{1}{4}, 2\right\}$

7. $3 - \dfrac{1}{t} = \dfrac{2}{t^2}$

7. $\left\{-\dfrac{2}{3}, 1\right\}$

8. $1 + \dfrac{2}{k} = \dfrac{3}{k^2}$

8. $\{-3, 1\}$

9. $\dfrac{1}{x} + \dfrac{2}{x + 2} = \dfrac{17}{35}$

9. $\left\{-\dfrac{14}{17}, 5\right\}$

10. $\dfrac{2}{m} + \dfrac{3}{m + 9} = \dfrac{11}{4}$

10. $\left\{-8, \dfrac{9}{11}\right\}$

11. $\dfrac{2}{x + 1} + \dfrac{3}{x + 2} = \dfrac{7}{2}$

11. $\left\{-\dfrac{11}{7}, 0\right\}$

12. $\dfrac{4}{3 - y} + \dfrac{2}{5 - y} = \dfrac{26}{15}$

12. $\left\{0, \dfrac{59}{13}\right\}$

13. $\dfrac{3}{2x} - \dfrac{1}{2(x + 2)} = 1$

13. $\left\{\dfrac{-1 + \sqrt{13}}{2}, \dfrac{-1 - \sqrt{13}}{2}\right\}$

14. $\dfrac{4}{3x} - \dfrac{1}{2(x + 1)} = 1$

14. $\left\{\dfrac{-1 + \sqrt{193}}{12}, \dfrac{-1 - \sqrt{193}}{12}\right\}$

⊚ CONCEPTUAL   🖉 WRITING   ▲ CHALLENGING   ▦ SCIENTIFIC CALCULATOR   ▦ GRAPHICS CALCULATOR

15. $\dfrac{1}{m}$ job per hour
16. (a) $20 - t$ miles per hour
(b) $20 + t$ miles per hour
17. 25 miles per hour
18. 255 miles per hour
19. 80 kilometers per hour
20. 50 miles per hour
21. 9 minutes
22. 3 hours; 6 hours
23. $\{3\}$ 24. $\left\{\dfrac{1}{2}\right\}$
25. $\left\{\dfrac{8}{9}\right\}$ 26. $\left\{\dfrac{3}{4}\right\}$
27. $\{16\}$ 28. $\{9\}$
29. $\left\{\dfrac{2}{5}\right\}$ 30. $\left\{\dfrac{5}{6}\right\}$
31. $\{-3, 3\}$ 32. $\{-2, 2\}$
33. $\left\{-\dfrac{3}{2}, -1, 1, \dfrac{3}{2}\right\}$
34. $\left\{-\dfrac{4}{3}, -1, 1, \dfrac{4}{3}\right\}$
35. $\{-6, -5\}$ 36. $\{-1, 8\}$
37. $\{-4, 1\}$
38. $\left\{-\dfrac{16}{3}, -2\right\}$
39. $\left\{-\dfrac{1}{3}, \dfrac{1}{6}\right\}$
40. $\left\{-\dfrac{4}{3}, \dfrac{1}{2}\right\}$
41. $\left\{-\dfrac{1}{2}, 3\right\}$ 42. $\left\{\dfrac{2}{3}, 2\right\}$

⊙ **15.** If it takes m hours to grade a set of papers, what is the grader's rate (in job per hour)?

⊙ **16.** If a boat goes 20 miles per hour in still water, and the rate of the current is t miles per hour,
   (a) what is the rate of the boat when it travels upstream?
   (b) what is the rate of the boat when it travels downstream?

Solve the problem by writing an equation with fractions and solving it. See Example 2.

17. On a windy day Yoshiaki found that he could go 16 miles downstream and then 4 miles back upstream at top speed in a total of 48 minutes. What was the top speed of Yoshiaki's boat if the current was 15 miles per hour?

18. Lekesha flew her plane for 6 hours at a constant speed. She traveled 810 miles with the wind, then turned around and traveled 720 miles against the wind. The wind speed was a constant 15 miles per hour. Find the speed of the plane.

19. In Canada, Medicine Hat and Cranbrook are 300 kilometers apart. Harry rides his Honda 20 kilometers per hour faster than Karen rides her Yamaha. Find Harry's average speed if he travels from Cranbrook to Medicine Hat in 1 1/4 hours less time than Karen.

20. The distance from Jackson to Lodi is about 40 miles, as is the distance from Lodi to Manteca. Rico drove from Jackson to Lodi during the rush hour, stopped in Lodi for a root beer, and then drove on to Manteca at 10 miles per hour faster. Driving time for the entire trip was 88 minutes. Find his speed from Jackson to Lodi.

21. A washing machine can be filled in 6 minutes if both the hot and cold water taps are fully opened. To fill the washer with hot water alone takes 9 minutes longer than filling with cold water alone. How long does it take to fill the tank with cold water?

22. Two pipes together can fill a large tank in 2 hours. One of the pipes, used alone, takes 3 hours longer than the other to fill the tank. How long would each pipe take to fill the tank alone?

Find all solutions by first squaring. Check your answers. See Example 3.

23. $2x = \sqrt{11x + 3}$

24. $4x = \sqrt{6x + 1}$

25. $3y = (16 - 10y)^{1/2}$

26. $4t = (8t + 3)^{1/2}$

27. $p - 2\sqrt{p} = 8$

28. $k + \sqrt{k} = 12$

29. $m = \sqrt{\dfrac{6 - 13m}{5}}$

30. $r = \sqrt{\dfrac{20 - 19r}{6}}$

Find all solutions to the equation. Check your answers. See Examples 4 and 5.

31. $t^4 - 18t^2 + 81 = 0$

32. $y^4 - 8y^2 + 16 = 0$

33. $4k^4 - 13k^2 + 9 = 0$

34. $9x^4 - 25x^2 + 16 = 0$

35. $(x + 3)^2 + 5(x + 3) + 6 = 0$

36. $(k - 4)^2 + (k - 4) - 20 = 0$

37. $(t + 5)^2 + 6 = 7(t + 5)$

38. $3(m + 4)^2 - 8 = 2(m + 4)$

39. $2 + \dfrac{5}{3k - 1} = \dfrac{-2}{(3k - 1)^2}$

40. $3 - \dfrac{7}{2p + 2} = \dfrac{6}{(2p + 2)^2}$

41. $2 - 6(m - 1)^{-2} = (m - 1)^{-1}$

42. $3 - 2(x - 1)^{-1} = (x - 1)^{-2}$

Use substitution to solve the equation. Check your answers. See Example 5.

43. $x^{2/3} + x^{1/3} - 2 = 0$

44. $3x^{2/3} - x^{1/3} - 24 = 0$

45. $2(1 + \sqrt{y})^2 = 13(1 + \sqrt{y}) - 6$

46. $(k^2 + k)^2 + 12 = 8(k^2 + k)$

47. $2x^4 + x^2 - 3 = 0$

48. $4k^4 + 5k^2 + 1 = 0$

49. What is wrong with the following "solution"?

$$2(m - 1)^2 - 3(m - 1) + 1 = 0$$
$$2u^2 - 3u + 1 = 0 \quad \text{Let } u = m - 1.$$
$$(2u - 1)(u - 1) = 0$$
$$2u - 1 = 0 \quad \text{or} \quad u - 1 = 0$$
$$u = \frac{1}{2} \qquad\qquad u = 1$$

Solution set: $\{\frac{1}{2}, 1\}$

50. Explain why the equation

$$y + \sqrt{y} - 6 = 0$$

can be solved either by using the method of Example 3 or the method of Example 4.

▲ *The equations in Exercises 51–58 are not grouped by type. Decide which method of solution applies, and then solve the equation. Give only real solutions. See Examples 1, 3, 4, and 5.*

51. $x^4 - 16x^2 + 48 = 0$

52. $\left(x - \frac{1}{2}\right)^2 + 5\left(x - \frac{1}{2}\right) - 4 = 0$

53. $\sqrt{2x + 3} = 2 + \sqrt{x - 2}$

54. $\sqrt{m + 1} = -1 + \sqrt{2m}$

55. $2m^6 + 11m^3 + 5 = 0$

56. $8x^6 + 513x^3 + 64 = 0$

57. $2 - (y - 1)^{-1} = 6(y - 1)^{-2}$

58. $3 = 2(p - 1)^{-1} + (p - 1)^{-2}$

──────◆ **MATHEMATICAL CONNECTIONS** (Exercises 59–64) ◆──────

Consider the following equation, which contains variable expressions in the denominators.

$$\frac{x^2}{(x - 3)^2} + \frac{3x}{x - 3} - 4 = 0.$$

Work Exercises 59–64 in order. They all pertain to this equation.

59. Why can 3 not possibly be a solution for this equation?

60. Multiply both sides of the equation by the LCD, $(x - 3)^2$, and solve. There is only one solution—what is it?

61. Write the equation in a different manner so that it is quadratic in form, with the rational expression $\frac{x}{x - 3}$ as the expression that you will substitute another variable for.

62. In your own words, explain why the expression $\frac{x}{x - 3}$ cannot equal 1.

---

43. $\{-8, 1\}$

44. $\left\{-\dfrac{512}{27}, 27\right\}$ **45.** $\{25\}$

46. $\{-3, -2, 1, 2\}$

47. $\left\{-1, 1, -\dfrac{\sqrt{6}}{2}i, \dfrac{\sqrt{6}}{2}i\right\}$

48. $\left\{-i, i, -\dfrac{1}{2}i, \dfrac{1}{2}i\right\}$

51. $\{-2\sqrt{3}, -2, 2, 2\sqrt{3}\}$

52. $\left\{\dfrac{-4 + \sqrt{41}}{2}, \dfrac{-4 - \sqrt{41}}{2}\right\}$

53. $\{3, 11\}$ **54.** $\{8\}$

55. $\left\{-\sqrt[3]{5}, -\dfrac{\sqrt[3]{4}}{2}\right\}$

56. $\left\{-4, -\dfrac{1}{2}\right\}$

57. $\left\{-\dfrac{1}{2}, 3\right\}$ **58.** $\left\{\dfrac{2}{3}, 2\right\}$

59. It would cause both denominators to be 0, and division by 0 is undefined.

60. The solution is $\dfrac{12}{5}$.

61. $\left(\dfrac{x}{x - 3}\right)^2 +$ $3\left(\dfrac{x}{x - 3}\right) - 4 = 0$

 CONCEPTUAL   WRITING   ▲ CHALLENGING   SCIENTIFIC CALCULATOR    GRAPHICS CALCULATOR

63. $\left\{\dfrac{12}{5}\right\}$; The values for t are -4 and 1. The value 1 is impossible because it leads to a contradiction (since $\dfrac{x}{x-3}$ is never equal to 1).

64. $\left\{\dfrac{12}{5}\right\}$; The values for s are $\dfrac{1}{x}$ and $\dfrac{-4}{x}$. The value $\dfrac{1}{x}$ is impossible, since $\dfrac{1}{x} \neq \dfrac{1}{x-3}$ for all x.

⊙ **63.** Solve the equation from Exercise 61 by making the substitution $t = \dfrac{x}{x-3}$. You should get two values for t. Why is one of them impossible for this equation?

⊙ **64.** Solve the equation $x^2(x-3)^{-2} + 3x(x-3)^{-1} - 4 = 0$ by letting $s = (x-3)^{-1}$. You should get two values for s. Why is this impossible for this equation?

REVIEW EXERCISES

Solve the equation for the specified variable. See Section 2.2.

65. $P = 2L + 2W$; for W

65. $W = \dfrac{P - 2L}{2}$ or $W = \dfrac{P}{2} - L$

66. $A = \dfrac{1}{2}bh$; for h

66. $h = \dfrac{2A}{b}$

67. $F = \dfrac{9}{5}C + 32$; for C

67. $C = \dfrac{5}{9}(F - 32)$

68. $S = 2\pi rh + 2\pi r^2$; for h

68. $h = \dfrac{S - 2\pi r^2}{2\pi r}$ or $h = \dfrac{S}{2\pi r} - r$

6.4 FORMULAS AND APPLICATIONS

FOR EXTRA HELP

 SSG pp. 188–193
SSM pp. 281–288

 Video 8

 Tutorial IBM MAC

OBJECTIVES

1 ▶ Solve formulas for variables involving squares and square roots.
2 ▶ Solve applied problems about motion along a straight line.
3 ▶ Solve applied problems using the Pythagorean formula.
4 ▶ Solve applied problems using formulas for area.

1 ▶ Solve formulas for variables involving squares and square roots.

▶ Many useful formulas have a variable that is squared or under a radical. The methods presented earlier in this chapter and the previous one can be used to solve for such variables.

EXAMPLE 1
Solving for a Variable Involving a Square or a Square Root

■ **(a)** Solve $w = \dfrac{kFr}{v^2}$ for v.

Begin by multiplying each side by v^2 to clear the equation of fractions. Then solve first for v^2.

Chalkboard Exercise

Solve $A = \pi r^2$ for r.

Answer: $r = \dfrac{\pm\sqrt{A\pi}}{\pi}$

$$w = \dfrac{kFr}{v^2}$$

$$v^2 w = kFr \qquad \text{Multiply each side by } v^2.$$

$$v^2 = \dfrac{kFr}{w} \qquad \text{Divide each side by } w.$$

$$v = \pm\sqrt{\dfrac{kFr}{w}} \qquad \text{Square root property}$$

$$v = \dfrac{\pm\sqrt{kFr}}{\sqrt{w}} \cdot \dfrac{\sqrt{w}}{\sqrt{w}} = \dfrac{\pm\sqrt{kFrw}}{w} \qquad \text{Rationalize the denominator.}$$

INSTRUCTOR'S RESOURCES

ITM pp. 422–426 **ISM** pp. 433–446
IAM p. 26

TEST GENERATOR
DOS Windows MAC

TRANSPARENCIES

(b) Solve $d = \sqrt{\dfrac{4A}{\pi}}$ for A.

First square both sides to eliminate the radical.

$$d = \sqrt{\dfrac{4A}{\pi}}$$

$$d^2 = \dfrac{4A}{\pi} \qquad \text{Square both sides.}$$

$$\pi d^2 = 4A \qquad \text{Multiply both sides by } \pi.$$

$$\dfrac{\pi d^2}{4} = A \qquad \text{Divide both sides by } 4.$$

Check the solution in the original equation. ▪

E X A M P L E   2

Solving for a Squared Variable

Solve $s = 2t^2 + kt$ for t.

Since the equation has terms with t^2 and t, write it in the standard quadratic form $ax^2 + bx + c = 0$, with t as the variable instead of x.

$$s = 2t^2 + kt$$

$$0 = 2t^2 + kt - s$$

Now use the quadratic formula with $a = 2$, $b = k$, and $c = -s$.

$$t = \dfrac{-k \pm \sqrt{k^2 - 4(2)(-s)}}{2(2)} \qquad a = 2, b = k, c = -s$$

$$= \dfrac{-k \pm \sqrt{k^2 + 8s}}{4}$$

The solutions are $t = \dfrac{-k + \sqrt{k^2 + 8s}}{4}$ and $t = \dfrac{-k - \sqrt{k^2 + 8s}}{4}$. ▪

Chalkboard Exercise

Solve for t.

$$2t^2 - 5t + k = 0$$

Answer:

$t = \dfrac{5 + \sqrt{25 - 8k}}{4}$,

$t = \dfrac{5 - \sqrt{25 - 8k}}{4}$

▶ TEACHING TIP

For this example, after getting one side equal to zero, circle the coefficients of t, then label these coefficients with a, b, c. Finally, substitute these circled values into the quadratic formula. ◀

The following examples show that it is important to check all proposed solutions of applied problems against the information of the original problem. Numbers that are valid solutions of the equation may not satisfy the physical conditions of the problem.

2 ▶ Solve applied problems about motion along a straight line.

▶   In the next example, we show how quadratic equations can be used to solve problems about motion in a straight line.

EXAMPLE 3

Solving a Straight Line Motion Problem

If a rock is dropped from a 144-foot building, its position (in feet above the ground) is given by $s = -16t^2 + 112t + 144$, where t is time in seconds after it was dropped. When does it hit the ground?

When the rock hits the ground, its distance above the ground is zero. Find t when s is zero by solving the equation

$$0 = -16t^2 + 112t + 144. \qquad \text{Let } s = 0.$$

$$0 = t^2 - 7t - 9 \qquad \text{Divide both sides by } -16.$$

$$t = \frac{7 \pm \sqrt{49 + 36}}{2} \qquad \text{Quadratic formula}$$

$$t = \frac{7 \pm \sqrt{85}}{2}$$

$$t \approx 8.1 \qquad \text{or} \qquad t \approx -1.1$$

Discard the negative solution. The rock will hit the ground about 8.1 seconds after it is dropped. ■

◊ approximately 46 feet

◊ **CONNECTIONS** ◊

Sometimes it is necessary to find the height of an object that cannot be measured directly. One way to solve such a problem is to use an old friend, the Pythagorean formula, which was introduced in Chapter 5. Using a special instrument surveyors can measure the distance to the top of the object from a point a known distance (horizontally) from the object. As shown in the figure, these distances give the lengths of the hypotenuse and one leg of a right triangle. The Pythagorean formula can then be used to find the length of the other leg, which is the unknown height.

If the length of the hypotenuse cannot be found, the surveyor can measure the angle between the hypotenuse and the leg whose length is known. Then methods of trigonometry are needed to find the unknown height.

FOR DISCUSSION OR WRITING

Suppose the measurements referred to above are determined to be 32 feet and 56 feet for a tower. Use a calculator to find the height of the tower. You may want to investigate further the kind of work done by surveyors.

3 ▶ Solve applied problems using the Pythagorean formula.

▶ The Pythagorean formula is used again in the solution of the next example.

E X A M P L E 4

Using the Pythagorean Formula

Two cars left an intersection at the same time, one heading due north, and the other due west. Some time later, they were exactly 100 miles apart. The car headed north had gone 20 miles farther than the car headed west. How far had each car traveled?

Let x be the distance traveled by the car headed west. Then $x + 20$ is the distance traveled by the car headed north. These distances are shown in Figure 2. The cars are 100 miles apart, so the hypotenuse of the right triangle equals 100 and the two legs are equal to x and $x + 20$. By the Pythagorean formula,

$$c^2 = a^2 + b^2.$$

$$100^2 = x^2 + (x + 20)^2$$

$$10{,}000 = x^2 + x^2 + 40x + 400 \qquad \text{Square the binomial.}$$

$$0 = 2x^2 + 40x - 9600 \qquad \text{Get } 0 \text{ on one side.}$$

$$0 = 2(x^2 + 20x - 4800) \qquad \text{Factor out the common factor.}$$

$$0 = x^2 + 20x - 4800 \qquad \text{Divide both sides by } 2.$$

Use the quadratic formula to find x.

$$x = \frac{-20 \pm \sqrt{400 - 4(1)(-4800)}}{2} \qquad a = 1, b = 20, c = -4800$$

$$= \frac{-20 \pm \sqrt{19{,}600}}{2}$$

$$x = 60 \qquad \text{or} \qquad x = -80 \qquad \text{Use a calculator.}$$

Discard the negative solution. The required distances are 60 miles and $60 + 20 = 80$ miles. ■

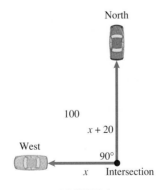

FIGURE 2

4 ▶ Solve applied problems using formulas for area.

▶ Formulas for area often lead to quadratic equations, as shown next.

EXAMPLE 5 ■

Solving an Area Problem

A rectangular reflecting pool in a park is 20 feet wide and 30 feet long. The park gardener wants to plant a strip of grass of uniform width around the edge of the pool. She has enough seed to cover 336 square feet. How wide will the strip be?

The pool is shown in Figure 3. If x represents the unknown width of the grass strip, the width of the large rectangle is given by $20 + 2x$ (the width of the pool plus two grass strips), and the length is given by $30 + 2x$. The area of the large rectangle is given by the product of its length and width, $(20 + 2x)(30 + 2x)$. The area of the pool is $20 \cdot 30 = 600$ square feet. The area of the large rectangle, minus the area of the pool, should equal the area of the grass strip. Since the area of the grass strip is to be 336 square feet, the equation is

$$\underset{\substack{\text{Area of} \\ \text{rectangle}}}{(20 + 2x)(30 + 2x)} - \underset{\substack{\text{Area of} \\ \text{pool}}}{600} = \underset{\substack{\text{Area of grass}}}{336}$$

$$600 + 100x + 4x^2 - 600 = 336 \qquad \text{Multiply.}$$

$$4x^2 + 100x - 336 = 0 \qquad \text{Combine terms.}$$

$$x^2 + 25x - 84 = 0 \qquad \text{Divide by 4.}$$

$$(x + 28)(x - 3) = 0 \qquad \text{Factor.}$$

$$x = -28 \qquad \text{or} \qquad x = 3.$$

The width cannot be -28 feet, so the grass strip should be 3 feet wide. ■

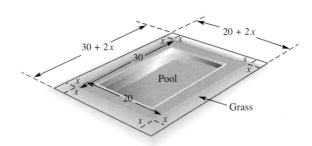

FIGURE 3

6.4 EXERCISES

In Exercises 1 and 2, solve for m in terms of the other variables. Remember that m must be positive.

1.

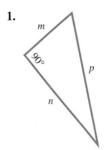

2.

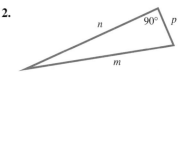

 CONCEPTUAL WRITING ▲ CHALLENGING  SCIENTIFIC CALCULATOR GRAPHICS CALCULATOR

Solve the equation for the indicated variable. (While in practice we would often reject a negative value due to the physical nature of the quantity represented by the variable, leave ± in your answers here.) See Examples 1 and 2.

3. $d = kt^2$; for t

4. $s = kwd^2$; for d

5. $I = \dfrac{ks}{d^2}$; for d

6. $R = \dfrac{k}{d^2}$; for d

7. $F = \dfrac{kA}{v^2}$; for v

8. $L = \dfrac{kd^4}{h^2}$; for h

9. $V = \dfrac{1}{3}\pi r^2 h$; for r

10. $V = \pi(r^2 + R^2)h$; for r

11. $At^2 + Bt = -C$; for t

12. $S = 2\pi rh + \pi r^2$; for r

13. $D = \sqrt{kh}$; for h

14. $F = \dfrac{k}{\sqrt{d}}$; for d

15. $p = \sqrt{\dfrac{k\ell}{g}}$; for ℓ

16. $p = \sqrt{\dfrac{k\ell}{g}}$; for g

17. In the formula of Exercise 15, if g is a positive number explain why k and ℓ must have the same sign in order for p to be a real number.

18. Refer to Example 2 of this section. Suppose that k and s both represent positive numbers.
   (a) Which one of the two solutions given is positive?
   (b) Which one is negative?
   (c) How can you tell?

Solve the problem by using a quadratic equation. Use a calculator as necessary, and round the answer to the nearest tenth. See Example 3.

19. The Mart Hotel in Dallas, Texas, is 400 feet high. Suppose that a ball is projected upward from the top of the Mart, and its position s in feet above the ground is given by the equation $s = -16t^2 + 45t + 400$, where t is the number of seconds elapsed. How long will it take for the ball to reach a height of 200 feet above the ground?

20. The Toronto Dominion Center in Winnipeg, Manitoba, is 407 feet high. Suppose that a ball is projected upward from the top of the Center, and its position s in feet above the ground is given by the equation $s = -16t^2 + 75t + 407$, where t is the number of seconds elapsed. How long will it take for the ball to reach a height of 450 feet above the ground?

21. Refer to the equations in Exercises 19 and 20. Suppose that the first sentence in each problem did not give the height of the building. How could you use the equation to determine the height of the building?

22. In Exercises 19 and 20, one problem has only one solution while the other has two solutions. Explain why this is so.

23. A search light moves horizontally back and forth along a wall with the distance of the light from a starting point at t minutes given by $s = 100t^2 - 300t$. How long will it take before the light returns to the starting point?

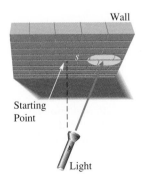

Answers:

3. $t = \dfrac{\pm\sqrt{dk}}{k}$

4. $d = \dfrac{\pm\sqrt{skw}}{kw}$

5. $d = \dfrac{\pm\sqrt{skI}}{I}$

6. $d = \dfrac{\pm\sqrt{kR}}{R}$

7. $v = \dfrac{\pm\sqrt{kAF}}{F}$

8. $h = \dfrac{\pm d^2\sqrt{kL}}{L}$

9. $r = \dfrac{\pm\sqrt{3\pi Vh}}{\pi h}$

10. $r = \dfrac{\pm\sqrt{V\pi h - \pi^2 R^2 h^2}}{\pi h}$

11. $t = \dfrac{-B \pm \sqrt{B^2 - 4AC}}{2A}$

12. $r = \dfrac{-\pi h \pm \sqrt{\pi^2 h^2 + \pi S}}{\pi}$

13. $h = \dfrac{D^2}{k}$ **14.** $d = \dfrac{k^2}{F^2}$

15. $\ell = \dfrac{p^2 g}{k}$ **16.** $g = \dfrac{k\ell}{p^2}$

18. (a) $\dfrac{-k + \sqrt{k^2 + 8s}}{4}$ is positive

(b) $\dfrac{-k - \sqrt{k^2 + 8s}}{4}$ is negative

19. 5.2 seconds

20. .7 second and 4.0 seconds

21. Find s when $t = 0$.

23. 3 minutes

24. (a) 1 second and
8 seconds   (b) 9 seconds
after it is projected
25. 3.4 seconds
26. 2.4 seconds
27. 2.3, 5.3, 5.8   28. 7.9,
8.9, 11.9   29. 412.3 feet
30. 470.0 feet

24. An object is projected directly upward from the ground. After t seconds its distance in feet above the ground is $s = 144t - 16t^2$.
(a) After how many seconds will the object be 128 feet above the ground? (*Hint:* Look for a common factor before solving the equation.)
(b) When does the object strike the ground?

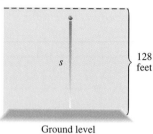

s 128 feet

Ground level

25. The formula $D = 100t - 13t^2$ gives the distance in feet a car going approximately 68 miles per hour will skid in t seconds. Find the time it would take for the car to skid 190 feet. (*Hint:* Your answer must be less than the time it takes the car to stop, which is 3.8 seconds.)

D

26. The formula in Exercise 25 becomes $D = 73t - 13t^2$ for a car going 50 miles per hour. Find the time for such a car to skid 100 feet. (*Hint:* The car will stop at about 2.8 seconds.)

Use the Pythagorean formula to solve the problem. Use a calculator as necessary, and when appropriate, round answers to the nearest tenth. See Example 4.

27. Find the lengths of the sides of the triangle.

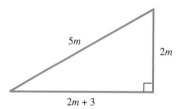

$5m$ $2m$
$2m + 3$

28. Find the lengths of the sides of the triangle.

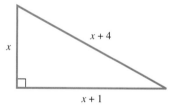

x $x + 4$
$x + 1$

29. Refer to Exercise 19. Suppose that a wire is attached to the top of the Mart, and pulled tight. It is attached to the ground 100 feet from the base of the building, as shown in the figure. How long is the wire?

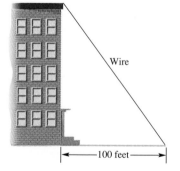

Wire

100 feet

30. Refer to Exercise 20. Suppose that a wire is attached to the top of the Center, and pulled tight. The length of the wire is twice the distance between the base of the Center and the point on the ground where the wire is attached. How long is the wire?

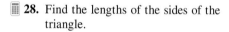

31. Two ships leave port at the same time, one heading due south and the other heading due east. Several hours later, they are 170 miles apart. If the ship traveling south had traveled 70 miles farther than the other, how many miles had they each traveled?

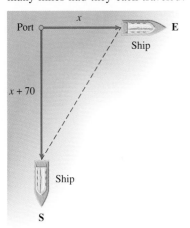

32. Peggy Cardella is flying a kite that is 30 feet farther above her hand than its horizontal distance from her. The string from her hand to the kite is 150 feet long. How high is the kite?

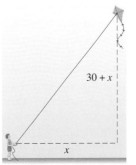

31. eastbound ship: 80 miles; southbound ship: 150 miles
32. 120 feet
33. 5 centimeters, 12 centimeters, 13 centimeters
34. 20 meters, 21 meters, 29 meters
35. length: 2 centimeters; width: 1.5 centimeters
36. 5 feet   37. 1 foot

33. A toy manufacturer needs a piece of plastic in the shape of a right triangle with the longer leg 2 centimeters more than twice as long as the shorter leg, and the hypotenuse 1 centimeter more than the longer leg. How long should the three sides of the triangular piece be?

34. Michael, a developer, owns a piece of land enclosed on three sides by streets, giving it the shape of a right triangle. The hypotenuse is 8 meters longer than the longer leg, and the shorter leg is 9 meters shorter than the hypotenuse. Find the lengths of the three sides of the property.

35. Two pieces of a large wooden puzzle fit together to form a rectangle with a length 1 centimeter less than twice the width. The diagonal, where the two pieces meet, is 2.5 centimeters in length. Find the length and width of the rectangle.

36. A 13-foot ladder is leaning against a house. The distance from the bottom of the ladder to the house is 7 feet less than the distance from the top of the ladder to the ground. How far is the bottom of the ladder from the house?

Use a quadratic equation to solve the problem. See Example 5.

37. Catarina and José want to buy a rug for a room that is 15 by 20 feet. They want to leave an even strip of flooring uncovered around the edges of the room. How wide a strip will they have if they buy a rug with an area of 234 square feet?

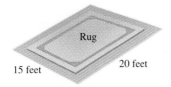

38. 2 feet   39. length:
26 meters;   width:
16 meters
40. length: 20 inches;
width: 12 inches
41. 1955   42. 1987
43. .035 or 3.5%
44. 18 inches
45. $.80
46. $11.93

38. A club swimming pool is 30 feet wide and 40 feet long. The club members want an exposed aggregate border in a strip of uniform width around the pool. They have enough material for 296 square feet. How wide can the strip be?

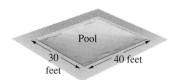

39. Arif's backyard is 20 by 30 meters. He wants to put a flower garden in the middle of the backyard, leaving a strip of grass of uniform width around the flower garden. To be happy, Arif must have 184 square meters of grass. Under these conditions what will the length and width of the garden be?

40. A rectangular piece of sheet metal has a length that is 4 inches less than twice the width. A square piece 2 inches on a side is cut from each corner. The sides are then turned up to form an uncovered box of volume 256 cubic inches. Find the length and width of the original piece of metal.

▲ *Solve the problem using a quadratic equation. Use a calculator as necessary.*

41. Refer to the Connections that introduce this chapter. According to the quadratic equation that models union membership described there, in what year did union membership reach 15.125 million?

42. The adjusted poverty threshold for a single person from the year 1984 to the year 1990 is approximated by the quadratic model $T = 18.7x^2 + 105.3x + 4814.1$, where $x = 0$ corresponds to 1984, and T is in dollars. In what year during this period was the threshold T approximately $5300? (*Source:* Congressional Budget Office.)

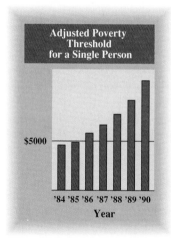

43. The formula $A = P(1 + r)^2$ gives the amount A in dollars that P dollars will grow to in 2 years at interest rate r (where r is given as a decimal), using compound interest. What interest rate will cause $2000 to grow to $2142.25 in 2 years?

44. If a square piece of cardboard has 3-inch squares cut from its corners and then has the flaps folded up to form an open-top box, the volume of the box is given by the formula $V = 3(x - 6)^2$, where x is the length of each side of the original piece of cardboard in inches. What original length would yield a box with a volume of 432 cubic inches?

45. A certain bakery has found that the daily demand for bran muffins is $\frac{3200}{p}$, where p is the price of a muffin in cents. The daily supply is $3p - 200$. Find the price at which supply and demand are equal.

46. In one area the demand for compact discs is $\frac{700}{P}$ per day, where P is the price in dollars per disc. The supply is $5P - 1$ per day. At what price does supply equal demand?

▲ *Recall that the corresponding sides of similar triangles are proportional. Use this fact to find the lengths of the indicated sides of the pair of similar triangles. Check all possible solutions in both triangles. Sides of a triangle cannot be negative.*

47. side AC

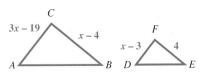

48. side RQ

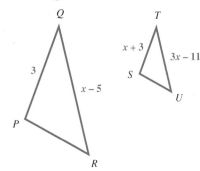

▲ *Solve the equation for the indicated variable.*

49. $p = \dfrac{E^2 R}{(r + R)^2}$;   for R $(E > 0)$

50. $S(6S - t) = t^2$;   for S

51. $10p^2 c^2 + 7pcr = 12r^2$;   for r

52. $S = vt + \dfrac{1}{2} gt^2$;   for t

53. $LI^2 + RI + \dfrac{1}{c} = 0$;   for I

54. $P = EI - RI^2$;   for I

47. 5 or 14
48. 4
49. $R = \dfrac{E^2 - 2pr \pm E\sqrt{E^2 - 4pr}}{2p}$
50. $S = \dfrac{-t}{3}$ or $S = \dfrac{t}{2}$
51. $r = \dfrac{5pc}{4}$ or $r = -\dfrac{2pc}{3}$
52. $t = \dfrac{-v \pm \sqrt{v^2 + 2gS}}{g}$
53. $I = \dfrac{-cR \pm \sqrt{c^2 R^2 - 4cL}}{2cL}$
54. $I = \dfrac{E \pm \sqrt{E^2 - 4RP}}{2R}$
55.
56.
57.
58. $[-2, \infty)$ **59.** $\left(-\dfrac{3}{2}, \infty\right)$
60. $(-\infty, -16)$

REVIEW EXERCISES

Graph the interval on a number line. See Section 2.5.

55. $[1, 5]$ **56.** $(-6, 1]$ **57.** $(-\infty, 1] \cup [5, \infty)$

Solve the inequality. See Section 2.5.

58. $3 - x \le 5$ **59.** $-2x + 1 < 4$ **60.** $-\dfrac{1}{2}x - 3 > 5$

6.5 NONLINEAR INEQUALITIES

FOR EXTRA HELP

📖 **SSG** pp. 194–199
  SSM pp. 288–299

📼 **Video**
  8

💾 **Tutorial**
  IBM   MAC

OBJECTIVES

1 ▶ Solve quadratic inequalities.
2 ▶ Solve polynomial inequalities of degree **3** or more.
3 ▶ Solve rational inequalities.

One of the most important parts of algebra is the study of methods of equation solving. Once we study a method for solving an equation, it is natural to extend the method to solving a related inequality.

◆ The solutions of $-.011x^2 + 1.22x - 23.5 = 0$ are, to the nearest whole numbers, 25 and 86. The solutions to $-.011x^2 + 1.22x - 8.5 \geq 15$ are the years between year 25 and year 86, or between 1955 and 2016. The larger endpoint is not appropriate since we have not yet reached 2016. It should be the current year (in terms of this application).

◆ **C O N N E C T I O N S** ◆

The Connections box at the beginning of this chapter gave an equation relating union membership (in millions) to the years from 1950 to 1990: $M = -.011x^2 + 1.22x - 8.5$. Recall x represents the number of years since 1930. Suppose we want to know in what years union membership was greater than or equal to some number, say 15 million. That is, we want to solve the inequality

$$-.011x^2 + 1.22x - 8.5 \geq 15.$$

The quadratic formula and Objective 1 in this section can be used to solve this inequality.

FOR DISCUSSION OR WRITING

After studying this section, solve the inequality given above. (You will need to use a calculator to solve the corresponding equation.) Round solutions of the equation to the nearest year. What is wrong in terms of the applied problem with the larger endpoint you found as a solution of the inequality? In terms of the application what should the upper endpoint be?

We have discussed methods of solving linear inequalities (earlier) and methods of solving quadratic equations (in this chapter). Now this work can be extended to include solving *quadratic inequalities*.

Quadratic Inequality

A **quadratic inequality** can be written in the form

$$ax^2 + bx + c < 0 \qquad \text{or} \qquad ax^2 + bx + c > 0,$$

where a, b, and c are real numbers, with $a \neq 0$.

As before, $<$ and $>$ may be replaced with $\leq$ and $\geq$ as necessary.

1 ▶ Solve quadratic inequalities.

▶ A method for solving quadratic inequalities is shown in the next example.

EXAMPLE 1

Solving a Quadratic Inequality

Solve $x^2 - x - 12 > 0$.

We first solve the quadratic *equation*

$$x^2 - x - 12 = 0.$$
$$(x - 4)(x + 3) = 0 \qquad \text{Factor.}$$
$$x - 4 = 0 \qquad \text{or} \qquad x + 3 = 0 \qquad \text{Zero-factor property}$$
$$x = 4 \qquad \text{or} \qquad x = -3$$

The numbers 4 and -3 divide the number line into three regions, as shown in Figure 4. (Be careful to put the smaller number on the left.)

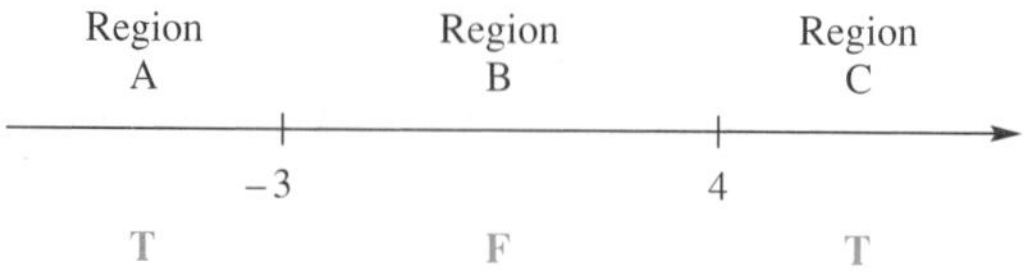

Chalkboard Exercise

Solve and graph the solution.
$$2p^2 + 3p \geq 2$$

Answer: $(-\infty, -2] \cup \left[\dfrac{1}{2}, \infty\right)$

Region A   Region B   Region C

-3 4

T   F   T

FIGURE 4

▶ **TEACHING TIP**

The solutions to the quadratic *equation* indicate the values that make the left side equal to zero. The solutions to the quadratic *inequality* give values of the variable which cause the left side to become positive (in this example).   ◀

The numbers 4 and -3 are the only numbers that make the expression $x^2 - x - 12$ equal to zero. All other numbers either make the expression positive or negative and the value of the expression can change from positive to negative or from negative to positive only on either side of a number that makes it zero. Therefore, if one number in a region satisfies the inequality, then all the numbers in that region will satisfy the inequality. Choose any number from Region A in Figure 4 (any number less than -3). Substitute this number for x in the inequality $x^2 - x - 12 > 0$. If the result is true, then all numbers in Region A satisfy the original inequality.

Let us choose -5 from Region A. Substitute -5 into $x^2 - x - 12 > 0$, getting

$$(-5)^2 - (-5) - 12 > 0 \qquad ?$$
$$25 + 5 - 12 > 0 \qquad ?$$
$$18 > 0. \qquad \text{True}$$

Since -5 from Region A satisfies the inequality, all numbers from Region A are solutions.

We try 0 from Region B. If $x = 0$, then

$$0^2 - 0 - 12 > 0 \qquad ?$$
$$-12 > 0. \qquad \text{False}$$

▶ **TEACHING TIP**

Pay attention to the symbols at the endpoints (brackets or parentheses), to the careful graphing of the solution set (including the arrows), and to properly writing the solution set in interval notation.   ◀

The numbers in Region B are *not* solutions. Verify that the number 5 satisfies the inequality, so that the numbers in Region C are also solutions to the inequality.

Based on these results (shown by the colored letters in Figure 4), the solution set includes the numbers in Regions A and C, as shown on the graph in Figure 5. The solution set is written

$$(-\infty, -3) \cup (4, \infty). \quad \blacksquare$$

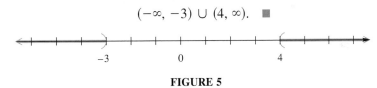

FIGURE 5

In summary, we solve a quadratic inequality by following these steps.

Solving a Quadratic Inequality	*Step 1* Replace the inequality symbol with $=$, and solve the equation.
	Step 2 Place the numbers found in Step 1 on a number line. These numbers divide the number line into regions.
	Step 3 Substitute a number from each region into the inequality to determine the intervals that make the inequality true. All numbers in those intervals that make the inequality true are in the solution set. A graph of the solution set will usually look like one of these.

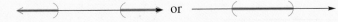

	Step 4 The numbers found in Step 1 are included in the solution set if the symbol is $\leq$ or $\geq$; they are not included if it is $<$ or $>$.

Special cases of quadratic inequalities may occur, such as those discussed in the next example.

EXAMPLE 2

Solving Special Cases

Solve $(2y - 3)^2 > -1$.

Since $(2y - 3)^2$ is never negative, it is always greater than -1. Thus, the solution is the set of all real numbers $(-\infty, \infty)$. In the same way, there is no solution for $(2y - 3)^2 < -1$ and the solution set is $\emptyset$. ■

2 ▶ Solve polynomial inequalities of degree 3 or more.

▶ Higher-degree polynomial inequalities that are factorable can be solved in the same way as quadratic inequalities.

EXAMPLE 3

Solving a Third-degree Polynomial Inequality

Solve $(x - 1)(x + 2)(x - 4) \leq 0$.

This is a *cubic* (third-degree) inequality rather than a quadratic inequality, but it can be solved by the method shown above and by extending the zero-factor property to more than two factors. Begin by setting the factored polynomial *equal* to 0 and solving the equation.

$$(x - 1)(x + 2)(x - 4) = 0$$

$$x - 1 = 0 \quad \text{or} \quad x + 2 = 0 \quad \text{or} \quad x - 4 = 0$$

$$x = 1 \quad \text{or} \quad x = -2 \quad \text{or} \quad x = 4$$

Locate the numbers -2, 1, and 4 on a number line as in Figure 6 to determine the regions A, B, C, and D.

For Example 2

Chalkboard Exercise

Solve $(3k - 2)^2 > -2$.
Answer: $(-\infty, \infty)$

Chalkboard Exercise

Solve and graph the solution.

$(2k + 1)(3k - 1)(k + 4) > 0$

Answer: $\left(-4, -\dfrac{1}{2}\right) \cup \left(\dfrac{1}{3}, \infty\right)$

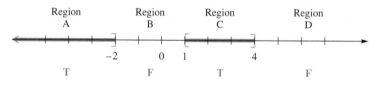

FIGURE 6

Substitute a number from each region into the original inequality to determine which regions satisfy the inequality. These results are shown below the number line in Figure 6. For example, in Region A, using $x = -3$ gives

$$(-3 - 1)(-3 + 2)(-3 - 4) \leq 0 \qquad ?$$

$$(-4)(-1)(-7) \leq 0 \qquad ?$$

$$-28 \leq 0. \qquad \text{True}$$

The numbers in Region A are in the solution set. Verify that the numbers in Region C are also in the solution set, which is written

$$(-\infty, -2] \cup [1, 4].$$

Notice that the three endpoints are included in the solution set, graphed in Figure 6. ■

▶ TEACHING TIP

Using a *table* is a way to organize the information needed to graph the solution set. Include the following column headings for this table: *Region* (such as A, B, C), *Test Number* and *Value* (of the expression with *T* or *F* at the test number). ◀

3 ▶ Solve rational inequalities.

▶ Inequalities involving fractions, called **rational inequalities,** are solved in a similar manner by going through the following steps.

Solving Rational Inequalities	
Step 1	Write the inequality as an equation and solve the equation.
Step 2	Set the denominator equal to zero and solve that equation.
Step 3	Use the solutions from Steps 1 and 2 to divide a number line into regions.
Step 4	Test a number from each region by substitution in the inequality to determine the intervals that satisfy the inequality.
Step 5	Be sure to exclude any values that make the denominator equal to zero.

CAUTION Don't forget Step 2. Any number that makes the denominator zero *must* separate two regions on the number line, since there will be no point on the number line for that value of the variable.

EXAMPLE 4
Solving a Rational Inequality

Solve the inequality $\dfrac{-1}{p-3} \geq 1$.

Write the corresponding equation and solve it. (Step 1)

$$\frac{-1}{p-3} = 1$$

$$-1 = p - 3 \qquad \text{Multiply by the common denominator.}$$

$$2 = p$$

Find the number that makes the denominator 0. (Step 2)

$$p - 3 = 0$$
$$p = 3$$

These two numbers, 2 and 3, divide a number line into three regions. (Step 3) (See Figure 7.)

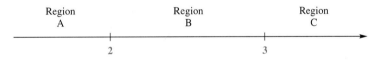

FIGURE 7

Testing one number from each region in the given inequality shows that the solution set is the interval $[2, 3)$. (Step 4) Note that this interval does not include the number 3, because 3 makes the denominator of the original inequality equal to zero. (Step 5) A graph of the solution set is given in Figure 8. ■

FIGURE 8

6.5 EXERCISES

1. Explain how you determine whether to include or exclude endpoints when solving a quadratic or higher degree inequality.

2. Explain why the number 7 cannot possibly be a solution of a rational inequality that has $x - 7$ as the denominator of a fraction.

3. The solution set of the inequality $x^2 + x - 12 < 0$ is the interval $(-4, 3)$. Without actually performing any work, give the solution set of the inequality $x^2 + x - 12 \geq 0$.   3. $(-\infty, -4] \cup [3, \infty)$

4. Without actually performing any work, give the solution set of the rational inequality $\dfrac{3}{x^2 + 1} > 0$.

   (*Hint:* Determine the sign of the numerator. Determine what the sign of the denominator *must* be. Then consider the inequality symbol.)   4. $(-\infty, \infty)$

 CONCEPTUAL   WRITING   CHALLENGING    SCIENTIFIC CALCULATOR   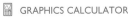 GRAPHICS CALCULATOR

Solve the inequality and graph the solution set. See Example 1. (Hint: In Exercises 17 and 18, use the quadratic formula.)

5. $(x + 1)(x - 5) > 0$
5. $(-\infty, -1) \cup (5, \infty)$

6. $(m + 6)(m - 2) > 0$
6. $(-\infty, -6) \cup (2, \infty)$

7. $(r + 4)(r - 6) < 0$
7. $(-4, 6)$

8. $(y + 4)(y - 8) < 0$
8. $(-4, 8)$

9. $x^2 - 4x + 3 \geq 0$
9. $(-\infty, 1] \cup [3, \infty)$

10. $m^2 - 3m - 10 \geq 0$
10. $(-\infty, -2] \cup [5, \infty)$

11. $10a^2 + 9a \geq 9$
11. $\left(-\infty, -\dfrac{3}{2}\right] \cup \left[\dfrac{3}{5}, \infty\right)$

12. $3r^2 + 10r \geq 8$
12. $(-\infty, -4] \cup \left[\dfrac{2}{3}, \infty\right)$

13. $9p^2 + 3p < 2$
13. $\left(-\dfrac{2}{3}, \dfrac{1}{3}\right)$

14. $2y^2 + y < 15$
14. $\left(-3, \dfrac{5}{2}\right)$

15. $6x^2 + x \geq 1$
15. $\left(-\infty, -\dfrac{1}{2}\right] \cup \left[\dfrac{1}{3}, \infty\right)$

16. $4y^2 + 7y \geq -3$
16. $(-\infty, -1] \cup \left[-\dfrac{3}{4}, \infty\right)$

▲ **17.** $y^2 - 6y + 6 \geq 0$
17. $(-\infty, 3 - \sqrt{3}] \cup [3 + \sqrt{3}, \infty)$

▲ **18.** $3k^2 - 6k + 2 \leq 0$
18. $\left[\dfrac{3 - \sqrt{3}}{3}, \dfrac{3 + \sqrt{3}}{3}\right]$

Solve the inequality. See Example 2.

19. $(4 - 3x)^2 \geq -2$
19. $(-\infty, \infty)$

20. $(6y + 7)^2 \geq -1$
20. $(-\infty, \infty)$

21. $(3x + 5)^2 \leq -4$
21. $\emptyset$

22. $(8t + 5)^2 \leq -5$
22. $\emptyset$

Solve the inequality and graph the solution set. See Example 3.

23. $(p - 1)(p - 2)(p - 4) < 0$
23. $(-\infty, 1) \cup (2, 4)$

24. $(2r + 1)(3r - 2)(4r + 7) < 0$
24. $\left(-\infty, -\dfrac{7}{4}\right) \cup \left(-\dfrac{1}{2}, \dfrac{2}{3}\right)$

25. $(a - 4)(2a + 3)(3a - 1) \geq 0$
25. $\left[-\dfrac{3}{2}, \dfrac{1}{3}\right] \cup [4, \infty)$

26. $(z + 2)(4z - 3)(2z + 7) \geq 0$
26. $\left[-\dfrac{7}{2}, -2\right] \cup \left[\dfrac{3}{4}, \infty\right)$

Solve the inequality and graph the solution set. See Example 4.

27. $\dfrac{x - 1}{x - 4} > 0$
27. $(-\infty, 1) \cup (4, \infty)$

28. $\dfrac{x + 1}{x - 5} > 0$
28. $(-\infty, -1) \cup (5, \infty)$

29. $\dfrac{2y + 3}{y - 5} \leq 0$
29. $\left[-\dfrac{3}{2}, 5\right)$

30. $\dfrac{3t + 7}{t - 3} \leq 0$
30. $\left[-\dfrac{7}{3}, 3\right)$

31. $\dfrac{8}{x - 2} \geq 2$
31. $(2, 6]$

32. $\dfrac{20}{y - 1} \geq 1$
32. $(1, 21]$

33. $\dfrac{3}{2t - 1} < 2$
33. $\left(-\infty, \dfrac{1}{2}\right) \cup \left(\dfrac{5}{4}, \infty\right)$

34. $\dfrac{6}{m - 1} < 1$
34. $(-\infty, 1) \cup (7, \infty)$

35. $\dfrac{a}{a + 2} \geq 2$
35. $[-4, -2)$

36. $\dfrac{m}{m + 5} \geq 2$
36. $[-10, -5)$

37. $\dfrac{4k}{2k - 1} < k$
37. $\left(0, \dfrac{1}{2}\right) \cup \left(\dfrac{5}{2}, \infty\right)$

38. $\dfrac{r}{r + 2} < 2r$
38. $\left(-2, -\dfrac{3}{2}\right) \cup (0, \infty)$

▲ **39.** $\dfrac{2x - 3}{x^2 + 1} \geq 0$

39. $\left[\dfrac{3}{2}, \infty\right)$

▲ **40.** $\dfrac{9x - 8}{4x^2 + 25} < 0$

40. $\left(-\infty, \dfrac{8}{9}\right)$

▲ **41.** $\dfrac{(3x - 5)^2}{x + 2} > 0$

41. $\left(-2, \dfrac{5}{3}\right) \cup \left(\dfrac{5}{3}, \infty\right)$

▲ **42.** $\dfrac{(5x - 3)^2}{2x + 1} \leq 0$

42. $\left(-\infty, -\dfrac{1}{2}\right) \cup \left\{\dfrac{3}{5}\right\}$

◆ **MATHEMATICAL CONNECTIONS** (Exercises 43–46) ◆

A rock is projected vertically upward from the ground. Its distance s in feet above the ground after t seconds is given by the equation $s = -16t^2 + 256t$. *Work Exercises 43–46 in order to see how quadratic equations and inequalities are connected.*

43. At what times will the rock be 624 feet above the ground? (*Hint*: Set $s = 624$ and solve the quadratic *equation.*)

43. 3 seconds and 13 seconds

44. At what times will the rock be more than 624 feet above the ground? (*Hint*: Set $s > 624$ and solve the quadratic *inequality.*)

44. between 3 seconds and 13 seconds

45. At what times will the rock be at ground level? (*Hint*: Set $s = 0$ and solve the quadratic *equation.*)

45. at 0 seconds (the time when it is initially projected) and at 16 seconds (the time when it hits the ground)

46. At what times will the rock be less than 624 feet above the ground? (*Hint*: Set $s < 624$, solve the quadratic *inequality,* and observe the solutions in Exercise 45 to determine the smallest and largest possible values of *t.*)

46. between 0 and 3 seconds and also between 13 and 16 seconds

◆

REVIEW EXERCISES

For the equation $3x + 2y = 24$, *find y for the given value of x. See Section 1.2.*

47. $x = 0$ **48.** $x = -2$ **49.** $x = 8$ **50.** $x = 1.5$

47. 12   48. 15   49. 0   50. 9.75

CHAPTER 6 SUMMARY

KEY TERMS

6.2 **quadratic formula**
discriminant

6.3 **quadratic in form**

6.5 **quadratic inequality**
rational inequality

QUICK REVIEW

CONCEPTS	EXAMPLES
6.1 COMPLETING THE SQUARE	
Square Root Property **(a)** If $x^2 = b$, $b \geq 0$, then $x = \sqrt{b}$ or $x = -\sqrt{b}$.	Solve. **(a)** $(x - 1)^2 = 8$ $\qquad x - 1 = \pm\sqrt{8} = \pm 2\sqrt{2}$ $\qquad\qquad x = 1 \pm 2\sqrt{2}$
(b) If $x^2 = -b$, $b > 0$, then $x = i\sqrt{b}$ or $x = -i\sqrt{b}$.	**(b)** $x^2 = -5$ $\qquad x = i\sqrt{5}$ or $x = -i\sqrt{5}$

 CONCEPTUAL    WRITING   CHALLENGING    SCIENTIFIC CALCULATOR    GRAPHICS CALCULATOR

CONCEPTS	EXAMPLES
Completing the Square To solve $ax^2 + bx + c = 0$: If $a \neq 1$, divide both sides by a. Write the equation with variable terms on one side and the constant on the other. Find half the coefficient of x and square it. Add the square to both sides. One side should now be a perfect square. Write it as the square of a binomial. Use the square root property to complete the solution.	Solve. $\qquad$ $$2x^2 - 4x - 18 = 0$$ $$x^2 - 2x - 9 = 0$$ $$x^2 - 2x = 9$$ $$\left[\frac{1}{2}(-2)\right]^2 = (-1)^2 = 1$$ $$x^2 - 2x + 1 = 9 + 1$$ $$(x - 1)^2 = 10$$ $$x - 1 = \pm\sqrt{10}$$ $$x = 1 \pm \sqrt{10}$$

6.2 THE QUADRATIC FORMULA

Quadratic Formula The solutions of $ax^2 + bx + c = 0$ $(a \neq 0)$ are $$x = \frac{-b \pm \sqrt{b^2 - 4ac}}{2a}.$$	Solve $3x^2 + 5x + 2 = 0$. $$x = \frac{-5 \pm \sqrt{5^2 - 4(3)(2)}}{2(3)}$$ $$x = -1 \quad \text{or} \quad x = -\frac{2}{3}$$ Solution set: $\left\{-1, -\frac{2}{3}\right\}$
The Discriminant If a, b, and c are integers, then the discriminant, $b^2 - 4ac$, of $ax^2 + bx + c = 0$ determines the type of solutions as follows. | Discriminant	Solutions
Positive square of an integer	2 rational solutions
Positive, not square of an integer	2 irrational solutions
Zero	1 rational solution
Negative	2 imaginary solutions		For $x^2 + 3x - 10 = 0$, the discriminant is $$3^2 - 4(1)(-10) = 49.$$ There are **2 rational** solutions. For $2x^2 + 5x + 1 = 0$, the discriminant is $$5^2 - 4(2)(1) = 17.$$ There are **2 irrational** solutions. For $9x^2 - 6x + 1 = 0$, the discriminant is $$(-6)^2 - 4(9)(1) = 0.$$ There is **1 rational** solution. For $4x^2 + x + 1 = 0$, the discriminant is $$1^2 - 4(4)(1) = -15.$$ There are **2 imaginary** solutions.

6.4 FORMULAS AND APPLICATIONS

To solve a formula for a squared variable, proceed as follows. **(a)** The variable appears only to the second degree. Isolate the squared variable on one side of the equation, then use the square root property.	Solve $A = \dfrac{2mp}{r^2}$ for r. $$r^2 A = 2mp \qquad \text{Multiply by } r^2.$$ $$r^2 = \frac{2mp}{A} \qquad \text{Divide by } A.$$

CONCEPTS	EXAMPLES
	$$r = \pm \sqrt{\frac{2mp}{A}} \qquad \text{Take square roots.}$$ $$r = \pm \frac{\sqrt{2mpA}}{A} \qquad \text{Rationalize.}$$
(b) The variable appears to the first and second degree. Write the equation in standard quadratic form, then use the quadratic formula to solve.	Solve $m^2 + rm = t$ for m. $$m^2 + rm - t = 0 \qquad \text{Standard form}$$ $$m = \frac{-r \pm \sqrt{r^2 - 4(1)(-t)}}{2(1)} \qquad \begin{array}{l} a = 1, b = r, \\ c = -t \end{array}$$ $$m = \frac{-r \pm \sqrt{r^2 + 4t}}{2}$$

6.5 NONLINEAR INEQUALITIES

Solving a Quadratic (or Higher Degree Polynomial) Inequality	Solve $2x^2 + 5x + 2 < 0$.
Step 1 Write the inequality as an equation and solve.	$$2x^2 + 5x + 2 = 0$$ $$x = -\frac{1}{2}, \; x = -2$$
Step 2 Place the numbers found in Step 1 on a number line. These numbers divide the line into regions.	$\qquad -2 \qquad -\frac{1}{2}$
Step 3 Substitute a number from each region into the inequality to determine the intervals that belong in the solution set—those intervals containing numbers that make the inequality true.	$x = -3$ makes it false; $x = -1$ makes it true; $x = 0$ makes it false. Solution: $\left(-2, -\frac{1}{2} \right)$
Solving a Rational Inequality	Solve $\dfrac{x}{x + 2} \geq 4$.
Step 1 Write the inequality as an equation and solve the equation.	$$\frac{x}{x + 2} = 4 \text{ leads to } -\frac{8}{3}.$$
Step 2 Set the denominator equal to zero and solve that equation.	$$x + 2 = 0$$ $$x = -2$$
Step 3 Use the solutions from Steps 1 and 2 to divide a number line into regions.	$\qquad -\frac{8}{3} \qquad -2$
Step 4 Test a number from each region in the inequality to determine the regions that satisfy the inequality.	-4 makes it false; $-\frac{7}{3}$ makes it true; 0 makes it false.
Step 5 Exclude any values that make the denominator zero.	The solution is $\left[-\frac{8}{3}, -2 \right)$, since -2 makes the denominator 0.

CHAPTER 6 REVIEW EXERCISES

Answers column (left):

1. $\{-11, 11\}$

2. $\{-\sqrt{3}, \sqrt{3}\}$

3. $\left\{-\dfrac{15}{2}, \dfrac{5}{2}\right\}$

4. $\left\{\dfrac{2 + 5i}{3}, \dfrac{2 - 5i}{3}\right\}$

5. $\{-2 + \sqrt{19}, -2 - \sqrt{19}\}$

6. $\left\{\dfrac{1}{2}, 1\right\}$ 7. $\left\{-\dfrac{7}{2}, 3\right\}$

8. $\left\{\dfrac{-5 + \sqrt{53}}{2}, \dfrac{-5 - \sqrt{53}}{2}\right\}$

9. $\left\{\dfrac{1 + \sqrt{41}}{2}, \dfrac{1 - \sqrt{41}}{2}\right\}$

10. $\left\{\dfrac{7}{3}\right\}$

11. $\left\{\dfrac{2 + i\sqrt{2}}{3}, \dfrac{2 - i\sqrt{2}}{3}\right\}$

12. $\left\{\dfrac{-7 + \sqrt{37}}{2}, \dfrac{-7 - \sqrt{37}}{2}\right\}$

13. $\{(-2 + \sqrt{5})i, (-2 - \sqrt{5})i\}$

15. 4 seconds and 12 seconds

17. 5.5 hours and 6.5 hours

18. 4.6 hours   19. (c)

20. (a)   21. (d)   22. (b)

23. The discriminant is 34^2, so it can be factored; $(6x - 5)(4x - 9)$.

24. The discriminant, 9657, is not a perfect square, so it cannot be factored.

25. $\left\{-\dfrac{5}{2}, 3\right\}$

26. $\left\{-\dfrac{1}{2}, 1\right\}$

27. $\left\{-\dfrac{11}{6}, -\dfrac{19}{12}\right\}$

28. $\{-4\}$

29. $\left\{-\dfrac{343}{8}, 64\right\}$

30. $\left\{-1, 1, -\dfrac{\sqrt{15}}{5}i, \dfrac{\sqrt{15}}{5}i\right\}$

[6.1] *Solve the equation by either the square root property or the method of completing the square.*

1. $t^2 = 121$ **2.** $p^2 = 3$ **3.** $(2x + 5)^2 = 100$

*4. $(3k - 2)^2 = -25$ **5.** $x^2 + 4x = 15$ **6.** $2m^2 - 3m = -1$

[6.2] *Solve the equation by the quadratic formula.*

7. $2y^2 + y - 21 = 0$ **8.** $k^2 + 5k = 7$ **9.** $(t + 3)(t - 4) = -2$

10. $9p^2 = 42p - 49$ ***11.** $3p^2 = 2(2p - 1)$

12. $m(2m - 7) = 3m^2 + 3$ ***13.** $ix^2 - 4x = -i$

14. A student wrote the following as the quadratic formula for solving $ax^2 + bx + c = 0$, $a \neq 0$: $x = -b \pm \dfrac{\sqrt{b^2 - 4ac}}{2a}$. Was this correct? If not, what is wrong with it?

15. A rock is projected vertically upward from the ground. Its distance s in feet above the ground after t seconds is $s = -16t^2 + 256t$. At what times will the rock be 768 feet above the ground?

16. Explain why the problem in Exercise 15 has two answers.

Solve the problem. Use a calculator and round answers to the nearest tenth.

17. A paint-mixing machine has two inlet pipes. One takes 1 hour less than the other to fill the tank. Together they fill the tank in 3 hours. How long would it take each of them alone to fill the tank?

18. An old machine processes a batch of checks in one hour more time than a new one. How long would it take the old machine to process a batch of checks that the two machines together process in 2 hours?

*Use the discriminant to predict whether the solutions to the equation are (**a**) two distinct rational numbers; (**b**) exactly one rational number; (**c**) two distinct irrational numbers; (**d**) two distinct imaginary numbers.*

19. $a^2 + 5a + 2 = 0$ **20.** $4x^2 = 3 - 4x$

21. $4x^2 = 6x - 8$ **22.** $9z^2 + 30z + 25 = 0$

Use the discriminant to tell if the polynomial can be factored using integer coefficients. If a polynomial can be factored, factor it.

23. $24x^2 - 74x + 45$ **24.** $36x^2 + 69x - 34$

[6.3] *Solve the equation.*

25. $\dfrac{15}{x} = 2x - 1$ **26.** $\dfrac{1}{y} + \dfrac{2}{y + 1} = 2$

27. $8(3x + 5)^2 + 2(3x + 5) - 1 = 0$ **28.** $-2r = \sqrt{\dfrac{48 - 20r}{2}}$

29. $2x^{2/3} - x^{1/3} - 28 = 0$ ***30.** $5x^4 - 2x^2 - 3 = 0$

Exercises identified with asterisks have imaginary number solutions.

RESOURCES FOR REVIEW

 ITM pp. 413–428   **ISM** pp. 466–485
IAM pp. 27–28

 SSG pp. 168–199
SSM pp. 299–319

 **TEST GENERATOR**
DOS   **Windows**   **MAC**

◉ CONCEPTUAL   ✎ WRITING   ▲ CHALLENGING   ▦ SCIENTIFIC CALCULATOR   ▥ GRAPHICS CALCULATOR

Solve the problem.

31. Lisa Wunderle drove 8 miles to pick up her friend Laurie, and then drove 11 miles to a mall at a speed 15 miles per hour faster. If Lisa's total travel time was 24 minutes, what was her speed on the trip to pick up Laurie?

 32. It takes Lisa De Mol 2 hours longer to complete a project for her boss than it takes Ed Moura. Working together, it would take them 3 hours. How long would it take each one to do the job alone? Round your answer to the nearest tenth of an hour.

33. Why can't the equation $x = \sqrt{2x + 4}$ have a negative solution?

34. If you were to use the quadratic formula to solve $x^4 - 5x^2 + 6 = 0$, with $a = 1$, $b = -5$, and $c = 6$, what would you have to remember after you applied the formula?

[6.4] *Solve the formula for the indicated variable. (Give answers with ±.)*

35. $S = \dfrac{Id^2}{k}$; for d

36. $k = \dfrac{rF}{wv^2}$; for v

37. $2\pi R^2 + 2\pi RH - S = 0$; for R

38. $mt^2 = 3mt + 6$; for t

Solve the problem.

 39. The equation $s = 16t^2 + 15t$ gives the distance s in feet an object dropped off a building has fallen in t seconds. Find the time t when the object has fallen 25 feet. Round the answer to the nearest hundredth.

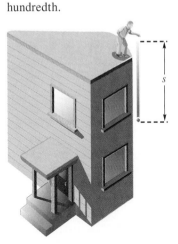

40. A large machine requires a part in the shape of a right triangle with a hypotenuse 9 feet less than twice the length of the longer leg. The shorter leg must be 3/4 the length of the longer leg. Find the lengths of the three sides of the part.

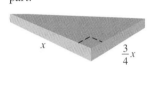

41. The hypotenuse of a right triangle is 13 feet shorter than twice the length of the longer leg. The shorter leg is 1 foot shorter than the longer leg. Find the lengths of the three sides of the triangle.

42. A rectangle has a length 2 meters more than its width. If one meter is cut from the length and one meter is added to the width, the resulting figure is a square with an area of 121 square meters. Find the dimensions of the original rectangle.

43. The product of the page numbers of two pages facing each other in this book is 4692. What are the page numbers?

31. 40 miles per hour
32. Lisa: 7.2 hours; Ed: 5.2 hours
33. Because x appears on the left side alone, and because it is equal to the nonnegative square root of $2x + 4$, it cannot be negative. 34. Take the positive and negative square roots of each value obtained from the formula.
35. $d = \dfrac{\pm\sqrt{SkI}}{I}$
36. $v = \dfrac{\pm\sqrt{rFkw}}{kw}$
37. $R = \dfrac{-\pi H \pm \sqrt{\pi^2 H^2 + 2\pi S}}{2\pi}$
38. $t = \dfrac{3m \pm \sqrt{9m^2 + 24m}}{2m}$
39. .87 second   40. 9 feet, 12 feet, and 15 feet
41. 20 feet, 21 feet, and 29 feet
42. 10 meters by 12 meters
43. 68 and 69

44. 1 inch   45. 1.1 seconds
and 2.9 seconds
46. 24.1 miles per hour
47. 4.5%   48. 120 meters
49. $\left(-\infty, -\dfrac{3}{2}\right) \cup (4, \infty)$

50. $[-4, 3]$

51. $\left(-\infty, -\dfrac{1}{2}\right) \cup (3, \infty)$

52. $[-3, 2)$

53. $(-\infty, -5] \cup [-2, 3]$

54. 0

55. $R = \dfrac{\pm\sqrt{Vh - r^2h}}{h}$

56. $\left\{\dfrac{3 + i\sqrt{3}}{3}, \dfrac{3 - i\sqrt{3}}{3}\right\}$

57. $\{-2, -1, 3, 4\}$

58. $(-\infty, -6) \cup \left(-\dfrac{3}{2}, 1\right)$

59.
$\left\{\dfrac{-11 + \sqrt{7}}{3}, \dfrac{-11 - \sqrt{7}}{3}\right\}$

60. $y = \dfrac{6p^2}{z}$ 61. $\left\{\dfrac{3}{5}, 1\right\}$

62. $\left\{-\dfrac{5}{3}, -\dfrac{3}{2}\right\}$

63. $\left(-5, -\dfrac{23}{5}\right]$

64. 3 hours and 6 hours
65. $\{-i, i, -1, 1\}$

66. $\left\{\dfrac{4}{11}, 5\right\}$

67. $\{-11i\sqrt{2}, 11i\sqrt{2}\}$
68. $(-\infty, \infty)$

44. Nancy wants to buy a mat for a photograph that measures 14 inches by 20 inches. She wants to have an even border around the picture when it is mounted on the mat. If the area of the mat she chooses is 352 square inches, how wide will the border be?

▦ **45.** The formula $s = -16t^2 + 64t + 100$ gives the height (in feet) an object propelled upward from the top of a building has reached in t seconds. At what time or times will the object have reached a height (from the ground) of 150 feet? Round to the nearest tenth of a second.

▦ **46.** An excursion boat traveled 20 miles upriver and then traveled back. If the current was 10 miles per hour and the entire trip took 2 hours, find the speed of the boat in still water. Use a calculator, and round your answer to the nearest tenth.

▦ **47.** Use the formula $A = P(1 + r)^2$ to find the interest rate r at which a principal P of $10,000 will increase to $10,920.25 in 2 years.

48. A lot is in the shape of a right triangle. The shortest side measures 50 meters. The longest side is 110 meters less than twice the middle side. How long is the middle side?

[6.5] *Solve the inequality and graph the solution set.*

49. $(x - 4)(2x + 3) > 0$

50. $x^2 + x \le 12$

51. $2k^2 > 5k + 3$

52. $\dfrac{3y + 4}{y - 2} \le 1$

53. $(x + 2)(x - 3)(x + 5) \le 0$

54. $(4m + 3)^2 \le -4$

MIXED REVIEW EXERCISES

Solve.

55. $V = r^2 + R^2h$;   for R

*56. $3t^2 - 6t = -4$

57. $(b^2 - 2b)^2 = 11(b^2 - 2b) - 24$

58. $(r - 1)(2r + 3)(r + 6) < 0$

59. $(3k + 11)^2 = 7$

60. $p = \sqrt{\dfrac{yz}{6}}$;   for y

61. $-5x^2 = -8x + 3$

62. $6 + \dfrac{15}{s^2} = -\dfrac{19}{s}$

63. $\dfrac{-2}{x + 5} \le -5$

64. Two pipes together can fill a large tank in 2 hours. One of the pipes, used alone, takes 3 hours longer than the other to fill the tank. How long would each pipe take to fill the tank alone?

*65. $x^4 - 1 = 0$

66. $\dfrac{2}{x - 4} + \dfrac{1}{x} = \dfrac{11}{5}$

*67. $y^2 = -242$

68. $(8k - 7)^2 \ge -1$

◉ **69.** A student gave the following solution to the equation $b^2 = 12$.

$$b^2 = 12$$
$$b = \sqrt{12}$$
$$b = 2\sqrt{3}$$

What is wrong with this solution?

◉ **70.** Explain how you would go about writing a quadratic equation whose solutions are -5 and 6. What is the standard form of one such equation?

━━━━━━━◆ **MATHEMATICAL CONNECTIONS** (Exercises 71–78) ◆━━━━━━━

Work Exercises 71–78 in order, so that the connections between equations and inequalities can be seen.

71. Use the methods of Chapter 2 to solve the equation or inequality, and graph the solution set.
(a) $3x - (4x + 2) = 0$ (b) $3x - (4x + 2) > 0$ (c) $3x - (4x + 2) < 0$

72. Use the methods of this chapter to solve the equation or inequality, and graph the solution set.
(a) $x^2 - 6x + 5 = 0$ (b) $x^2 - 6x + 5 > 0$ (c) $x^2 - 6x + 5 < 0$

73. Use the methods of Section 4.6 and Section 6.5 to solve the equation or inequality, and graph the solution set.

(a) $\dfrac{-5x + 20}{x - 2} = 0$ (b) $\dfrac{-5x + 20}{x - 2} > 0$ (c) $\dfrac{-5x + 20}{x - 2} < 0$

Review the definition of the union of two sets in Section 2.6, and then answer the questions in Exercises 74–76.

74. Form the union of the solution sets in Exercise 71. What is their union?

75. Repeat Exercise 74 for the solution sets in Exercise 72.

76. For the equation and inequalities in Exercise 73, what value of x cannot possibly be part of any of the solution sets? What is the union of the solution sets of the equation and inequalities in Exercise 73?

77. Fill in the blanks in the following statement: If we solve a linear, quadratic, or rational equation and the two inequalities associated with it, the union of the three solution sets will be _____; the only exception will be in the case of the rational equation and inequalities, where the number or numbers that cause the _____ to be zero will be deleted.

78. Suppose that the solution set of a quadratic equation is $\{-5, 3\}$ and the solution set of one of the associated inequalities is $(-\infty, -5) \cup (3, \infty)$. What is the solution set of the other associated inequality?

━━━━━━━━━━━━◆━━━━━━━━━━━━

71. (a) $\{-2\}$

(b) $(-\infty, -2)$

(c) $(-2, \infty)$

72. (a) $\{1, 5\}$

(b) $(-\infty, 1) \cup (5, \infty)$

(c) $(1, 5)$

73. (a) $\{4\}$

(b) $(2, 4)$

(c) $(-\infty, 2) \cup (4, \infty)$

74. $(-\infty, \infty)$ **75.** $(-\infty, \infty)$
76. 2; $(-\infty, 2) \cup (2, \infty)$
77. $(-\infty, \infty)$;
denominator
78. $(-5, 3)$

CHAPTER 6 TEST

*Items marked * require knowledge of imaginary numbers.*

Solve by the square root property.

1. $t^2 = 54$

2. $(7x + 3)^2 = 25$

Solve by completing the square.

3. $x^2 + 2x = 1$

Solve by the quadratic formula.

4. $2x^2 - 3x - 1 = 0$ ***5.** $3t^2 - 4t = -5$ **6.** $3x = \sqrt{\dfrac{9x + 2}{2}}$

[6.1] **1.** $\{-3\sqrt{6}, 3\sqrt{6}\}$

2. $\left\{-\dfrac{8}{7}, \dfrac{2}{7}\right\}$

3. $\{-1 + \sqrt{2}, -1 - \sqrt{2}\}$
[6.2]

4. $\left\{\dfrac{3 + \sqrt{17}}{4}, \dfrac{3 - \sqrt{17}}{4}\right\}$

5. $\left\{\dfrac{2 + i\sqrt{11}}{3}, \dfrac{2 - i\sqrt{11}}{3}\right\}$

[6.3] **6.** $\left\{\dfrac{2}{3}\right\}$

🖥 **7.** Maretha and Lillaana are typesetters. For a certain report, Lillaana can set the type in 2 hours less than Maretha. If they work together, they can do the entire report in 5 hours. How long will it take each of them working alone to prepare the report? Give your answers to the nearest tenth of an hour.

◉ ***8.** If k is a negative number, then which one of the following equations will have two imaginary solutions?

    (a) $x^2 = 4k$ **(b)** $x^2 = -4k$ **(c)** $(x + 2)^2 = -k$ **(d)** $x^2 + k = 0$

Use the discriminant to predict the number and type of solutions. Do not solve.

9. $2x^2 - 8x - 3 = 0$ **10.** $24x^2 + 42x - 27 = 0$

Solve by any method.

11. $3 - \dfrac{16}{x} - \dfrac{12}{x^2} = 0$ **12.** $4x^2 + 7x - 3 = 0$

13. $9x^4 + 4 = 37x^2$ **14.** $12 = (2d + 1)^2 + (2d + 1)$

15. Sandi Goldstein paddled her canoe 10 miles upstream, and then paddled back to her starting point. If the rate of the current was 3 miles per hour and the entire trip took $3\frac{1}{2}$ hours, what was Sandi's rate?

16. Solve for r: $S = 4\pi r^2$ (Leave $\pm$ in your answer.)

Solve the problem by writing a quadratic equation. Use any method to solve the equation.

17. Adam Bryer has a pool 24 feet long and 10 feet wide. He wants to construct a concrete walk around the pool. If he plans for the walk to be of uniform width and cover 152 square feet, what will be the width of the walk?

18. At a point 30 meters from the base of a tower, the distance to the top of the tower is 2 meters more than twice the height of the tower. Find the height of the tower.

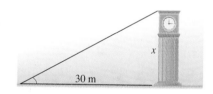

Solve and graph the solution set.

19. $2x^2 + 7x > 15$ **20.** $\dfrac{5}{t - 4} \le 1$

CUMULATIVE REVIEW (Chapters 1–6)

Let $S = \left\{-\dfrac{7}{3}, -2, -\sqrt{3}, 0, .7, \sqrt{12}, \sqrt{-8}, 7, \dfrac{32}{3}\right\}$. *List the elements of S that are elements of the set.*

  1. integers   **2.** rational numbers   **3.** real numbers   **4.** complex numbers

Simplify the expression.

  5. $|-3| + 8 - |-9| - (-7 + 3)$ **6.** $2(-3)^2 + (-8)(-5) + (-17)$

Solve the equation.

7. $7 - (4 + 3t) + 2t = -6(t - 2) - 5$

8. $|-3x + 5| = 9$

9. $|6x - 9| = |-4x + 2|$

10. $2x = \sqrt{\dfrac{5x + 2}{3}}$

11. $\dfrac{3}{x - 3} - \dfrac{2}{x - 2} = \dfrac{3}{x^2 - 5x + 6}$

12. $10x^2 + x = 2$

13. $(r - 5)(2r + 3) = 1$

14. $b^4 - 5b^2 + 4 = 0$

Solve the inequality.

15. $-2x + 4 \le -x + 3$

16. $|3y - 7| \le 1$

17. $|4z + 2| > 7$

18. $x^2 - 4x + 3 < 0$

19. $\dfrac{3}{y + 2} > 1$

20. $(9z + 4)^2 \ge -4$

Perform the indicated operations.

21. $(7x + 4)(2x - 3)$

22. $\left(\dfrac{2}{3}t + 9\right)^2$

23. $(3t^3 + 5t^2 - 8t + 7) - (6t^3 + 4t - 8)$

24. Divide $4x^3 + 2x^2 - x + 26$ by $x + 2$.

Factor completely.

25. $16x - x^3$

26. $24m^2 + 2m - 15$

27. $1000p^6 - 1$

28. $(3x + 2)^2 - 4(3x + 2) - 5$

29. $8x^3 + 27y^3$

30. $9x^2 - 30xy + 25y^2$

Perform the operations and express the answer in lowest terms. Assume denominators are nonzero.

31. $\dfrac{x^2 - 3x - 10}{x^2 + 3x + 2} \cdot \dfrac{x^2 - 2x - 3}{x^2 + 2x - 15}$

32. $\dfrac{5t + 2}{-6} \div \dfrac{15t + 6}{5}$

33. $\dfrac{3}{2 - k} - \dfrac{5}{k} + \dfrac{6}{k^2 - 2k}$

34. $\dfrac{\dfrac{r}{s} - \dfrac{s}{r}}{\dfrac{r}{s} + 1}$

Simplify the radical expression.

35. $\sqrt[3]{\dfrac{27}{16}}$

36. $\dfrac{2}{\sqrt{7} - \sqrt{5}}$

Solve the problem.

37. The perimeter of a rectangle is 20 inches and its area is 21 square inches. What are the dimensions of the rectangle?

38. Tri rode his bicycle for 12 miles and then walked an additional 8 miles. The total time for the trip was 5 hours. If his rate while walking was 10 miles per hour less than his rate while riding, what was each rate?

7. $\left\{\dfrac{4}{5}\right\}$ **8.** $\left\{-\dfrac{4}{3}, \dfrac{14}{3}\right\}$

9. $\left\{\dfrac{11}{10}, \dfrac{7}{2}\right\}$ **10.** $\left\{\dfrac{2}{3}\right\}$

11. 0 **12.** $\left\{-\dfrac{1}{2}, \dfrac{2}{5}\right\}$

13. $\left\{\dfrac{7 + \sqrt{177}}{4}, \dfrac{7 - \sqrt{177}}{4}\right\}$

14. $\{-2, -1, 1, 2\}$

15. $[1, \infty)$ **16.** $\left[2, \dfrac{8}{3}\right]$

17. $\left(-\infty, -\dfrac{9}{4}\right) \cup \left(\dfrac{5}{4}, \infty\right)$

18. $(1, 3)$ **19.** $(-2, 1)$

20. $(-\infty, \infty)$

21. $14x^2 - 13x - 12$

22. $\dfrac{4}{9}t^2 + 12t + 81$

23. $-3t^3 + 5t^2 - 12t + 15$

24. $4x^2 - 6x + 11 + \dfrac{4}{x + 2}$

25. $x(4 + x)(4 - x)$

26. $(4m - 3)(6m + 5)$

27. $(10p^2 - 1) \cdot (100p^4 + 10p^2 + 1)$

28. $9(x - 1)(x + 1)$

29. $(2x + 3y) \cdot (4x^2 - 6xy + 9y^2)$

30. $(3x - 5y)^2$

31. $\dfrac{x - 5}{x + 5}$ **32.** $-\dfrac{5}{18}$

33. $-\dfrac{8}{k}$ **34.** $\dfrac{r - s}{r}$

35. $\dfrac{3\sqrt[3]{4}}{4}$ **36.** $\sqrt{7} + \sqrt{5}$

37. 7 inches by 3 inches
38. biking: 12 miles per hour; walking: 2 miles per hour

39. southbound car:
57 miles; eastbound car:
76 miles   40. 40 miles per hour

39. Two cars left an intersection at the same time, one heading due south and the other due east. Later they were exactly 95 miles apart. The car heading east had gone 38 miles less than twice as far as the car heading south. How far had each car traveled?

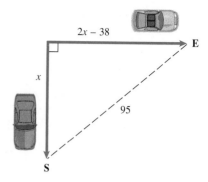

40. Two cars traveled from Elmhurst to Oakville, a distance of 100 miles. One car traveled 10 miles per hour faster than the other and arrived $\frac{5}{6}$ of an hour sooner. Find the speed of the faster car.

◆ Answers will vary.

CONNECTIONS

René Descartes, a French mathematician of the seventeenth century, is credited with giving us an indispensible method of locating a point in a plane. His attempts to unify algebra and geometry influenced the creation of what became coordinate geometry (analytic geometry).

Howard Eves, in *An Introduction to the History of Mathematics* (sixth edition) relates how the idea of rectangular coordinates, which we discuss in this chapter, supposedly came to Descartes:

Another story, perhaps on a par with the story of Isaac Newton and the falling apple, says that the initial flash of analytic geometry came to Descartes when watching a fly crawling about on the ceiling near a corner of his room. It struck him that the path of the fly on the ceiling could be described . . . if only one knew the relationship connecting the fly's distances from two adjacent walls.

FOR DISCUSSION OR WRITING

In your mind picture the path of a fly wandering on a ceiling. Think of the intersection of two adjacent walls with the ceiling as two number lines intersecting at their origins. How might we relate points on the path of the fly to these number lines?

373

7.1 THE RECTANGULAR COORDINATE SYSTEM

FOR EXTRA HELP

📖 **SSG** pp. 202–209
SSM pp. 332–339

📼 **Video**
8

💾 **Tutorial**
IBM   MAC

OBJECTIVES

1 ▶ Plot ordered pairs.
2 ▶ Find ordered pairs that satisfy a given equation.
3 ▶ Graph lines.
4 ▶ Find x- and y-intercepts.
5 ▶ Recognize equations of vertical or horizontal lines.
6 ▶ Use the distance formula.

The *rectangular, or Cartesian (for Descartes), coordinate system* of locating a point using two number lines intersecting at right angles is introduced in this chapter. We also study methods of graphing equations in two variables.

Graphs are widely used in the media. Newspapers, magazines, television, reports to stockholders, and newsletters often present information in graph form. The bar graph in Figure 1(a) shows the number of enrollees, in millions, in California health maintenance organizations (HMOs). The pie graph in Figure 1(b) shows sources of credit card fraud. Figure 1(c) on the facing page shows a line graph representing the increase in concentrations of atmospheric CO_2. Graphs are widely used because they show a lot of information in a form that makes it easy to understand. As the saying goes, "A picture is worth a thousand words." In this section we show how to graph equations of lines.

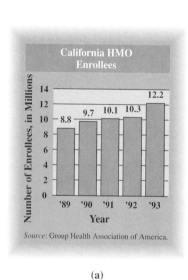

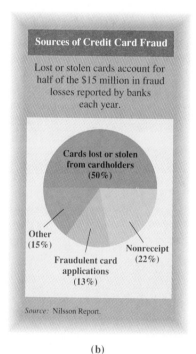

(a)   (b)

FIGURE 1

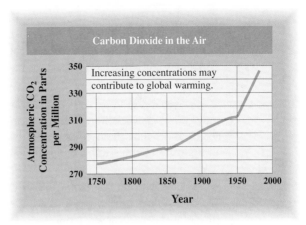

(c)

FIGURE 1 (continued)

1 ▶ Plot ordered pairs.

▶   Each of the pairs of numbers (1, 2), (−1, 5), and (3, 7) is an example of an **ordered pair**; that is, a pair of numbers written within parentheses in which the order of the numbers is important. We graph an ordered pair using two perpendicular number lines that intersect at their zero points, as shown in Figure 2. The common zero point is called the **origin.** The horizontal line, the **x-axis,** represents the first number in an ordered pair, and the vertical line, the **y-axis,** represents the second. The x-axis and the y-axis make up a **rectangular** (or Cartesian) **coordinate system.**

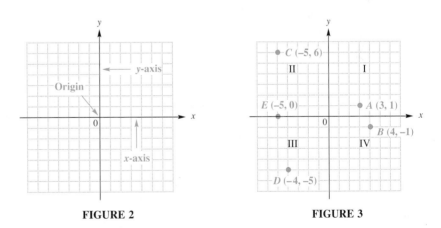

FIGURE 2   **FIGURE 3**

To locate, or **plot,** the point on the graph that corresponds to the ordered pair (3, 1) we move three units from zero to the right along the x-axis, and then one unit up parallel to the y-axis. The point corresponding to the ordered pair (3, 1) is labeled A in Figure 3. The point (4, −1) is labeled B, (−5, 6) is labeled C, and (−4, −5) is labeled D. Point E corresponds to (−5, 0). The phrase "the point corresponding to the ordered pair (3, 1)" is often abbreviated "the point (3, 1)." The numbers in an ordered pair are called the **coordinates** of the corresponding point.

CAUTION The parentheses used to represent an ordered pair are also used to represent an open interval (introduced in Chapter 1). In general, there is no confusion between these symbols because the context of the discussion tells us whether we are discussing ordered pairs or open intervals.

The four regions of the graph shown in Figure 3 are called **quadrants I, II, III, and IV,** reading counterclockwise from the upper right quadrant. The points of the x-axis and y-axis themselves do not belong to any quadrant. For example, point E in Figure 3 belongs to no quadrant.

2 ► Find ordered pairs that satisfy a given equation.

► Each solution to an equation with two variables includes two numbers, one for each variable. To keep track of which number goes with which variable we write the solutions as ordered pairs, with the x-value given first. For example, we can show that $(6, -2)$ is a solution of $2x + 3y = 6$ by substitution.

$$2x + 3y = 6$$
$$2(6) + 3(-2) = 6 \qquad ?$$
$$12 - 6 = 6 \qquad ?$$
$$6 = 6 \qquad \text{True}$$

Since the pair of numbers $(6, -2)$ makes the equation true, it is a solution. On the other hand, since

$$2(5) + 3(1) = 10 + 3 = 13 \neq 6,$$

$(5, 1)$ is not solution of the equation.

To find ordered pairs that satisfy an equation we select any number for one of the variables, substitute it into the equation for that variable, and then solve for the other variable.

Some other ordered pairs satisfying $2x + 3y = 6$ are $(0, 2)$ and $(3, 0)$. Since every real number could be selected for one variable and would lead to a real number for the other variable, equations with two variables usually have an infinite number of solutions.

EXAMPLE 1
Completing Ordered Pairs

Complete the following ordered pairs for $2x + 3y = 6$.

(a) $(-3, \quad)$

Let $x = -3$. Substitute into the equation.

$$2(-3) + 3y = 6 \qquad \text{Let } x = -3.$$
$$-6 + 3y = 6$$
$$3y = 12$$
$$y = 4$$

The ordered pair is $(-3, 4)$.

Chalkboard Exercise

Complete the following ordered pairs for $3x - 4y = 12$.

$(0, \quad), (\quad, 0), (\quad, -2),$
$(-4, \quad), (-6, \quad)$

Answer: $(0, -3)$, $(4, 0)$, $\left(\dfrac{4}{3}, -2\right)$,

$(-4, -6)$, $\left(-6, -\dfrac{15}{2}\right)$

(b) (, −4)

Replace y with −4.

$$2x + 3y = 6$$
$$2x + 3(-4) = 6 \qquad \text{Let } y = -4.$$
$$2x - 12 = 6$$
$$2x = 18$$
$$x = 9$$

The ordered pair is $(9, -4)$. ∎

3 ▶ Graph lines.

▶ The **graph of an equation** is the set of points that correspond to all the ordered pairs that satisfy the equation. It gives a "picture" of the equation. Since most equations with two variables have an infinite set of ordered pairs, their graphs include an infinite number of points. To graph an equation, we plot a number of ordered pairs that satisfy the equation until we have enough points to suggest the shape of the graph. For example, we graph $2x + 3y = 6$ by first plotting all the ordered pairs mentioned above. These are shown in Figure 4(a). The resulting points appear to lie on a straight line. If all the ordered pairs that satisfy the equation $2x + 3y = 6$ were graphed, they would form a straight line. In fact, the graph of any first-degree equation in two variables is a straight line. The graph of $2x + 3y = 6$ is the line shown in Figure 4(b).

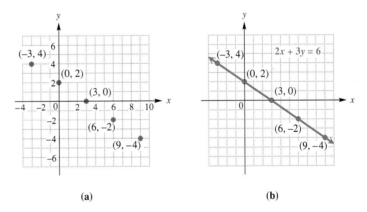

(a)   (b)

FIGURE 4

Since first-degree equations with two variables have straight-line graphs, they are called **linear equations in two variables.** (We discussed linear equations in one variable in Chapter 2.)

Standard Form of a Linear Equation in Two Variables	An equation that can be written in the form $$Ax + By = C \quad (A \text{ and } B \text{ not both } 0)$$ is a linear equation. This form is called the **standard form.**

4 ▶ Find x- and y-intercepts.

▶ A straight line is determined if any two different points on the line are known, so finding two different points is enough to graph the line. Two points that are useful for graphing are the x- and y-intercepts. The **x-intercept** is the point (if any) where the line crosses the x-axis; likewise, the **y-intercept** is the point (if any) where the line crosses the y-axis. In Figure 4(b), the y-value of the point where the line crosses the x-axis is 0. Similarly, the x-value of the point where the line crosses the y-axis is 0. This suggests a method for finding the x- and y-intercepts.

Intercepts	Let $y = 0$ to find the x-intercept; let $x = 0$ to find the y-intercept.

EXAMPLE 2

Finding Intercepts

■ Find the x- and y-intercepts of $4x - y = -3$, and graph the equation.
Find the x-intercept by letting $y = 0$.

$$4x - 0 = -3 \quad \text{Let } y = 0.$$
$$4x = -3$$
$$x = -\frac{3}{4} \quad x\text{-intercept is } \left(-\frac{3}{4}, 0\right).$$

For the y-intercept, let $x = 0$.

$$4(0) - y = -3 \quad \text{Let } x = 0.$$
$$-y = -3$$
$$y = 3 \quad y\text{-intercept is } (0, 3).$$

The intercepts are the two points $(-3/4, 0)$ and $(0, 3)$. Use these two points to draw the graph, as shown in Figure 5. ■

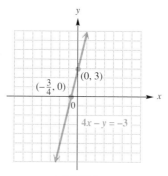

FIGURE 5

NOTE While two points, such as the two intercepts in Figure 5, are sufficient to graph a straight line, it is a good idea to use a third point to guard against errors. Verify that $(-1, -1)$ also lies on the graph of $4x - y = -3$.

5 ▶ Recognize equations of vertical or horizontal lines.

▶ The next example shows that a graph can fail to have an x-intercept, which is why the phrase "if any" was added when discussing intercepts.

E X A M P L E   3

Graphing a Horizontal Line

Graph $y = 2$.

Writing $y = 2$ as $0x + 1y = 2$ shows that any value of x, including $x = 0$, gives $y = 2$, making the y-intercept $(0, 2)$. Since y is always 2, there is no value of x corresponding to $y = 0$, and so the graph has no x-intercept. The graph, shown in Figure 6, is a horizontal line. ■

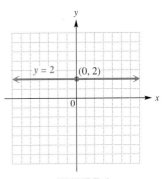

▶ **TEACHING TIP**

Helpful hints:

1. An equation with only the variable x will always intersect the x-axis and thus will be *vertical*.
2. An equation with only the variable y will always intersect the y-axis and thus will be *horizontal*. ◀

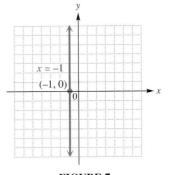

FIGURE 6

It is also possible for a graph to have no y-intercept, as in the next example.

E X A M P L E   4

Graphing a Vertical Line

Graph $x + 1 = 0$.

The form $1x + 0y = -1$ shows that every value of y leads to $x = -1$, and so no value of y makes $x = 0$. The graph, therefore, has no y-intercept. The only way a straight line can have no y-intercept is to be vertical, as shown in Figure 7. ■

FIGURE 7

The line graphed in the next example has both x-intercept and y-intercept at the origin.

EXAMPLE 5

Graphing a Line that Passes Through the Origin

Graph $x + 2y = 0$.

Find the x-intercept by letting $y = 0$.

$$x + 2y = 0$$
$$x + 2(0) = 0 \qquad \text{Let } y = 0.$$
$$x + 0 = 0$$
$$x = 0 \qquad x\text{-intercept is } (0, 0).$$

To find the y-intercept, let $x = 0$.

$$x + 2y = 0$$
$$0 + 2y = 0 \qquad \text{Let } x = 0.$$
$$y = 0 \qquad y\text{-intercept is } (0, 0).$$

Both intercepts are the same ordered pair, $(0, 0)$. We need another point to graph the line. We can choose any number for x (or for y), say $x = 4$, and solve for y.

$$x + 2y = 0$$
$$4 + 2y = 0 \qquad \text{Let } x = 4.$$
$$2y = -4$$
$$y = -2$$

This gives the ordered pair $(4, -2)$. These two points lead to the graph shown in Figure 8. ■

Chalkboard Exercise

Graph $3x - y = 0$.

Answer:

▶ **TEACHING TIP**

Students often have trouble finding points for a line passing through the origin. Inform students that a line of the form $Ax + By = 0$ will always pass through $(0, 0)$. A second point with integer coordinates can be found as follows:

1. Find a multiple of the coefficients of x and y.
2. Substitute this multiple for x.
3. Solve for y.
4. Use the results for steps (2) and (3) above as the second ordered pair. ◄

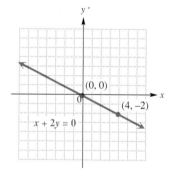

FIGURE 8

---

6 ▶ Use the distance formula.

▶ Figure 9 shows the points $(3, -4)$ and $(-5, 3)$. To find the distance between these points, we use the Pythagorean formula from geometry, first mentioned in Chapter 5.

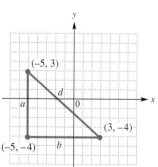

FIGURE 9

The vertical line through $(-5, 3)$ and the horizontal line through $(3, -4)$ intersect at the point $(-5, -4)$. Thus, the point $(-5, -4)$ becomes the vertex of the right angle in a right triangle. By the Pythagorean formula, the square of the length of the hypotenuse, d, of the right triangle in Figure 9 is equal to the sum of the squares of the lengths of the two legs a and b:

$$d^2 = a^2 + b^2.$$

The length a is the difference between the coordinates of the endpoints. Since the x-coordinate of both points is -5, the side is vertical, and we can find a by finding the difference between the y-coordinates. Subtract -4 from 3 to get a positive value for a.

$$a = 3 - (-4) = 7$$

Similarly, find b by subtracting -5 from 3.

$$b = 3 - (-5) = 8.$$

Substituting these values into the formula we have

$$d^2 = a^2 + b^2$$
$$d^2 = 7^2 + 8^2 \qquad \text{Let } a = 7 \text{ and } b = 8.$$
$$d^2 = 49 + 64$$
$$d^2 = 113$$
$$d = \sqrt{113}. \qquad \text{Use the square root property.}$$

Therefore, the distance between $(-5, 3)$ and $(3, -4)$ is $\sqrt{113}$.

NOTE It is customary to leave the distance in radical form. Do not use a calculator to get an approximation unless you are specifically directed to do so.

This result can be generalized. Figure 10 shows the two different points (x_1, y_1) and (x_2, y_2). To find a formula for the distance d between these two points, notice that the distance between (x_2, y_2) and (x_2, y_1) is given by $a = y_2 - y_1$, and the distance between (x_1, y_1) and (x_2, y_1) is given by $b = x_2 - x_1$. From the Pythagorean formula,

$$d^2 = (x_2 - x_1)^2 + (y_2 - y_1)^2,$$

and by using the square root property, we get the distance formula.

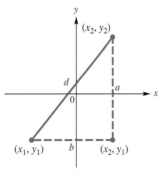

FIGURE 10

Distance Formula

The distance between the points (x_1, y_1) and (x_2, y_2) is

$$d = \sqrt{(x_2 - x_1)^2 + (y_2 - y_1)^2}.$$

This result is called the **distance formula.**

The small numbers 1 and 2 in the ordered pairs (x_1, y_1) and (x_2, y_2) are called *subscripts*. We read x_1 as "x sub 1." Subscripts are used to distinguish between different values of a variable, here, the x-values in two ordered pairs.

E X A M P L E   6

Using the Distance Formula

Find the distance between $(-3, 5)$ and $(6, 4)$.

When using the distance formula to find the distance between two points, designating the points as (x_1, y_1) and (x_2, y_2) is arbitrary. Let us choose $(x_1, y_1) = (-3, 5)$ and $(x_2, y_2) = (6, 4)$.

$$\begin{aligned} d &= \sqrt{(x_2 - x_1)^2 + (y_2 - y_1)^2} \\ &= \sqrt{(6 - (-3))^2 + (4 - 5)^2} \qquad x_2 = 6, y_2 = 4, x_1 = -3, y_1 = 5 \\ &= \sqrt{9^2 + (-1)^2} \\ &= \sqrt{82} \quad \blacksquare \end{aligned}$$

Chalkboard Exercise

Find the distance between $(-3, 2)$ and $(0, -4)$.

Answer: $\sqrt{45}$ or $3\sqrt{5}$

◇ 1. the figure on the right   2. $y = -\dfrac{x}{2}$

◇ **C O N N E C T I O N S** ◇

*Graphics calculators** are the latest development in the evolution of scientific calculators. Of their many advanced features, the most obvious is the capability of graphing equations on their screens. Beginning in this chapter we address the basic features of such calculators. Because of the ever-changing nature of makes and models, we do not attempt to give specific keystrokes, but will address in a generic fashion the features common to graphics calculators. Therefore, readers should refer to the owner's manual for specific keystrokes and instructions.

Choosing an appropriate viewing screen for a particular equation is of utmost importance. We will refer to a screen having minimum x- and y-values of -10 and maximum x- and y-values of 10 as the *standard viewing window*. The *scale* on each axis determines the distance between tick marks; in a standard window, the scale will be 1 for both axes. See the figure for an example of the standard viewing window.

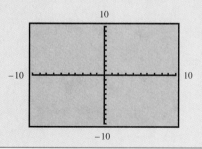

*See the discussion *An Introduction to Graphics Calculators* included at the front of this book.

To graph a linear equation such as $4x - y = 3$, it is usually necessary to solve the equation for y before inputting it into the calculator. The following figures show the equation $y = 4x - 3$ graphed in the standard viewing window. Notice that graphics calculators allow us to find the intercepts of the line.

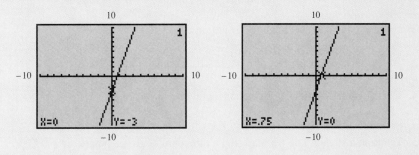

FOR WRITING OR DISCUSSION

1. The figures each show the graph of $x + y = 15$ (which was entered as $y = -x + 15$). However, different viewing windows are used. Which one of the two windows do you think would be more useful for this graph? Why?

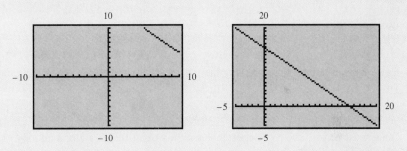

2. The figure shows the graph of $x + 2y = 0$. It was sketched in a traditional manner in Example 5. Suppose that your calculator requires that you solve the equation first for y. What would be the desired form of the equation?

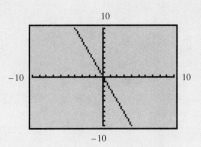

7.1 EXERCISES

◉ *Fill in the blank with the correct response.*

1. The point with coordinates (0, 0) is called the _____ of a rectangular coordinate system.

1. origin

2. For any value of x, the point $(x, 0)$ lies on the _____ -axis.

2. x

3. To find the x-intercept of a line, we let _____ equal 0 and solve for _____ .

3. y; x

4. The equation _____ = 4 has a horizontal line as its graph.
            (*x* or *y*)

4. y

5. To graph a straight line we must find a minimum of _____ points.

5. two

6. The point (_____ , 4) is on the graph of $2x - 3y = 0$.

6. 6

Name the quadrant, if any, in which the point is located.

7. (a) $(1, 6)$
(b) $(-4, -2)$
(c) $(-3, 6)$
(d) $(7, -5)$
(e) $(-3, 0)$

8. (a) $(-2, -10)$
(b) $(4, 8)$
(c) $(-9, 12)$
(d) $(3, -9)$
(e) $(0, -8)$

7. (a) **I**   (b) **III**   (c) **II**   (d) **IV**   (e) **none**

8. (a) **III**   (b) **I**   (c) **II**   (d) **IV**   (e) **none**

◉ **9.** Use the given information to determine the possible quadrants in which the point (x, y) must lie.

(a) $xy > 0$ (b) $xy < 0$ (c) $\dfrac{x}{y} < 0$ (d) $\dfrac{x}{y} > 0$

9. (a) **I or III**   (b) **II or IV**   (c) **II or IV**   (d) **I or III**

◉ **10.** What must be true about the coordinates of any point that lies along an axis?

10. One of the coordinates must be zero.

Locate the following points on the rectangular coordinate system.

11. $(2, 3)$ **12.** $(-1, 2)$ **11–20.**

13. $(-3, -2)$ **14.** $(1, -4)$

15. $(0, 5)$ **16.** $(-2, -4)$

17. $(-2, 4)$ **18.** $(3, 0)$

19. $(-2, 0)$ **20.** $(3, -3)$

Complete the given ordered pairs for the equation, and then graph the equation. See Example 1.

21. $x - y = 3$
    $(0, \underline{\quad}), (\underline{\quad}, 0)$
    $(5, \underline{\quad}), (2, \underline{\quad})$

22. $x - y = 5$
    $(0, \underline{\quad}), (\underline{\quad}, 0)$
    $(1, \underline{\quad}), (3, \underline{\quad})$

23. $x + 2y = 5$
    $(0, \underline{\quad}), (\underline{\quad}, 0)$
    $(2, \underline{\quad}), (\underline{\quad}, 2)$

21. -3; 3; 2; -1

22. -5; 5; -4; -2

23. $\dfrac{5}{2}$; 5; $\dfrac{3}{2}$; 1

24. $x + 3y = -5$
$(0, \underline{\quad}), (\underline{\quad}, 0)$
$(1, \underline{\quad}), (\underline{\quad}, -1)$
24. $-\dfrac{5}{3}$; -5; -2; -2

25. $4x - 5y = 20$
$(0, \underline{\quad}), (\underline{\quad}, 0)$
$(2, \underline{\quad}), (\underline{\quad}, -3)$
25. -4; 5; $-\dfrac{12}{5}$; $\dfrac{5}{4}$

26. $6x - 5y = 30$
$(0, \underline{\quad}), (\underline{\quad}, 0)$
$(3, \underline{\quad}), (\underline{\quad}, -2)$
26. -6; 5; $-\dfrac{12}{5}$; $\dfrac{10}{3}$

⊙ **27.** Explain why the graph of $x + y = k$ cannot pass through quadrant III if $k > 0$.

⊙ **28.** Explain why the graph of $ax + by = 0$ must pass through the origin for any values of a and b.

Find the x-intercept and the y-intercept. Then graph the equation. See Examples 2–5.

29. $2x + 3y = 12$
29. $(6, 0)$; $(0, 4)$

30. $5x + 2y = 10$
30. $(2, 0)$; $(0, 5)$

31. $x - 3y = 6$
31. $(6, 0)$; $(0, -2)$

32. $x - 2y = -4$
32. $(-4, 0)$; $(0, 2)$

33. $3x - 7y = 9$
33. $(3, 0)$; $\left(0, -\dfrac{9}{7}\right)$

34. $5x + 6y = -10$
34. $(-2, 0)$; $\left(0, -\dfrac{5}{3}\right)$

35. $y = 5$
35. none; $(0, 5)$

36. $y = -3$
36. none; $(0, -3)$

37. $x = 2$
37. $(2, 0)$; none

38. $x = -3$
38. $(-3, 0)$; none

39. $x + 5y = 0$
39. $(0, 0)$; $(0, 0)$

40. $x - 3y = 0$
40. $(0, 0)$; $(0, 0)$

Find the distance between the pair of points. See Example 6.

41. $(-8, 2)$ and $(-4, 1)$
41. $\sqrt{17}$

42. $(5, -3)$ and $(-1, -1)$
42. $2\sqrt{10}$

43. $(-1, 4)$ and $(5, 3)$
43. $\sqrt{37}$

44. $(-6, 5)$ and $(3, -4)$
44. $9\sqrt{2}$

45. $(\sqrt{2}, \sqrt{6})$ and $(-2\sqrt{2}, 4\sqrt{6})$
45. $6\sqrt{2}$

46. $(\sqrt{7}, 9\sqrt{3})$ and $(-\sqrt{7}, 4\sqrt{3})$
46. $\sqrt{103}$

47. $(x + y, y)$ and $(x - y, x)$
47. $\sqrt{5y^2 - 2xy + x^2}$

48. $(c, c - d)$ and $(d, c + d)$
48. $\sqrt{c^2 - 2cd + 5d^2}$

⊙ **49.** As given in the text, the distance formula is expressed with a radical. Write the distance formula using rational exponents. 49. $d = [(x_2 - x_1)^2 + (y_2 - y_1)^2]^{1/2}$

⊙ **50.** An alternate form of the distance formula is

$$d = \sqrt{(x_1 - x_2)^2 + (y_1 - y_2)^2}.$$

Compare this to the form given in this section, and explain why the two forms are equivalent.

⊙ CONCEPTUAL   ✎ WRITING   ▲ CHALLENGING   ▦ SCIENTIFIC CALCULATOR   ▨ GRAPHICS CALCULATOR

Find the perimeter of the triangle.

51.

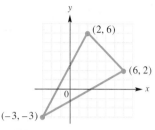

51. $2\sqrt{106} + 4\sqrt{2}$

52.

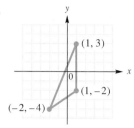

52. $5 + \sqrt{58} + \sqrt{13}$

The midpoint of the line segment joining the two points (x_1, y_1) and (x_2, y_2) is the point $(\bar{x}, \bar{y})$ where

$$(\bar{x}, \bar{y}) = \left(\frac{x_1 + x_2}{2}, \frac{y_1 + y_2}{2}\right).$$

Find the midpoint of the line segment PQ.

53. $P(3, 5)$, $Q(-4, 7)$

53. $\left(-\frac{1}{2}, 6\right)$

54. $P(8, -6)$, $Q(2, 3)$

54. $\left(5, -\frac{3}{2}\right)$

55. $P(0, 4)$, $Q(-3, -1)$

55. $\left(-\frac{3}{2}, \frac{3}{2}\right)$

56. $P(1, -7)$, $Q(-8, 4)$

56. $\left(-\frac{7}{2}, -\frac{3}{2}\right)$

57. $P(7, 2)$, $Q(-3, -8)$

57. $(2, -3)$

58. $P(10, -4)$, $Q(6, 5)$

58. $\left(8, \frac{1}{2}\right)$

Use the graph to answer the questions.

59. The graph indicates the percentage of all U.S. workers without any private or public health insurance. (*Source:* Census Bureau, Employee Benefit Research Institute)
   (a) Between which two years was the percentage approximately the same?
   (b) Between which two years was the increase the greatest?
   (c) In what year was the percent about 16.5%?

60. The graph indicates the production of handguns, rifles, and shotguns in the U.S. (*Source:* Bureau of Alcohol, Tobacco, and Firearms)
   (a) In what year between 1983 and 1992 was the production the greatest?
   (b) In what year was the production the least?
   (c) In what two successive years was the production less than the previous year's?

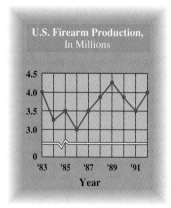

60. (a) 1989   (b) 1986   (c) 1990 and 1991

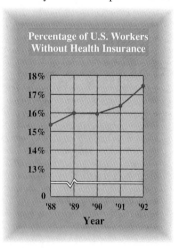

59. (a) between 1989 and 1990   (b) between 1991 and 1992   (c) 1991

61. Between the years 1980 and 1992, the linear model $y = 2.503x + 198.729$ approximated the winning speed for the Indianapolis 500 race. In the model, $x = 0$ corresponds to 1980, $x = 1$ to 1981, and so on, and y is in miles per hour. Use this model to approximate the speed of the 1988 winner, Rick Mears.

61. **218.753 miles per hour**

62. According to information provided by Families USA Foundation, the national average family health care cost in dollars between 1980 and 2000 (projected) can be approximated by the linear model $y = 382.75x + 1742$, where $x = 0$ corresponds to 1980 and $x = 20$ corresponds to 2000. Based on this model, what would be the expected national average health care cost in 1998?

62. **$8631.50**

 63. The screen shows the graph of one of the equations below, along with the coordinates of a point on the graph. Which one of the equations is it?

   (a) $x + 2y = 4$ **(b)** $-3x + 5y = 15$
   (c) $y = 4x - 2$ **(d)** $y = -2$

63. (c)

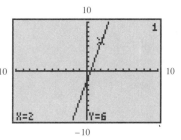

64. The screen shows the graph of one of the equations below. Two views of the graph are given, along with the intercepts. Which one of the equations is it?

   (a) $3x + 2y = 6$ **(b)** $-3x + 2y = 6$ **(c)** $-3x - 2y = 6$ **(d)** $3x - 2y = 6$

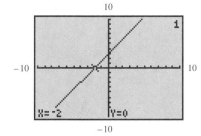   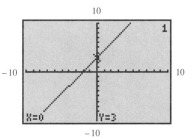

64. (b)

65. The table of points shown was generated by a graphics calculator with a *table* feature. Which one of the equations below corresponds to this table of points?

   (a) $y_1 = 2x - 3$ **(b)** $y_1 = -2x - 3$
   (c) $y_1 = 2x + 3$ **(d)** $y_1 = -2x + 3$

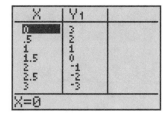

65. (d)

66. Refer to the model equation in Exercise 62. A portion of its graph is shown on the accompanying screen, along with the coordinates of a point on the line displayed at the bottom. How is this point interpreted in the context of the model?

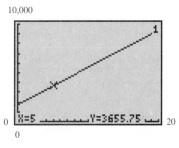

66. It means that when $x = 5$ (year 1985), the national average family health care cost was approximately $3655.75.

◆ **MATHEMATICAL CONNECTIONS** (Exercises 67–76) ◆

A statement of the form "if p, then q" is called a conditional statement. The letter p represents the antecedent and q represents the consequent. If we interchange p and q to obtain "if q, then p," the resulting conditional statement is called the converse of the original one.

Use this idea in working Exercises 67–76 in order.

67. Given a true conditional statement, the converse may or may not be necessarily true. What is the converse of "If $3x = 0$, then $x = 0$"? Is it a true statement?

67. If $x = 0$, then $3x = 0$. It is a true statement.

68. What is the converse of "If $x > 4$, then $x > 3$"? Why is it not necessarily true?

68. If $x > 3$, then $x > 4$. It is not necessarily true—for example, if $x = 3\frac{1}{2}$, $x > 3$ but $x \not> 4$.

69. The converse of the Pythagorean theorem *is* a true statement. Fill in the blank with the correct response: If a and b are the shorter sides of a triangle and d is the longest side, with $d^2 = a^2 + b^2$, then _____ .

69. the triangle is a right triangle

70. Sketch the triangle with vertices at $(3, 2)$, $(12, -10)$, and $(0, -4)$.

70.

71. Find the length of the side of the triangle joining the vertices $(3, 2)$ and $(12, -10)$.

71. 15

72. Find the length of the side of the triangle joining the vertices $(12, -10)$ and $(0, -4)$.

72. $6\sqrt{5}$

73. Find the length of the side of the triangle joining the vertices $(3, 2)$ and $(0, -4)$.

73. $3\sqrt{5}$

74. Which one of the sides of the triangle is the longest?

74. The longest side joins the vertices $(3, 2)$ and $(12, -10)$.

75. Use the converse of the Pythagorean theorem (Exercise 69) to determine whether the triangle you sketched in Exercise 70 is a right triangle.

75. $15^2 = (\sqrt{180})^2 + (\sqrt{45})^2$ is a true statement, so the triangle is a right triangle.

76. Without sketching, show that the triangle with vertices at $(5, 6)$, $(0, -3)$, and $(9, 2)$ is not a right triangle.

76. $(\sqrt{32})^2 + (\sqrt{106})^2 \neq (\sqrt{106})^2$

◆

REVIEW EXERCISES

Find the quotient. See Section 1.2.

77. $\dfrac{6 - 2}{5 - 3}$ 77. 2

78. $\dfrac{5 - 7}{-4 - 2}$ 78. $\dfrac{1}{3}$

79. $\dfrac{4 - (-1)}{-3 - (-5)}$ 79. $\dfrac{5}{2}$

80. $\dfrac{-6 - 0}{0 - (-3)}$ 80. -2

81. $\dfrac{-5 - (-5)}{3 - 2}$ 81. 0

82. $\dfrac{7 - (-2)}{-3 - (-3)}$ 82. undefined

7.2 THE SLOPE OF A LINE

FOR EXTRA HELP

 SSG pp. 209–213
SSM pp. 339–345

 Video
8

 Tutorial
IBM   MAC

OBJECTIVES

1 ► Find the slope of a line given two points on the line.
2 ► Find the slope of a line given the equation of the line.
3 ► Graph a line given its slope and a point on the line.
4 ► Use slope to decide whether two lines are parallel or perpendicular.
5 ► Solve problems involving average rate of change.

► **TEACHING TIP**

Begin this section by showing the four cases for slopes of a line:

Positive   Negative
slope ↗   slope ↘

Undefined   Zero
slope ↕   slope ↔

Emphasize that zero *slope* is not the same as *undefined* slope. ◄

Slope is used in many ways in our everyday world. The slope of a hill (sometimes called the *grade*) is often given in the form of percent. For example, a 10% (or $10/100 = 1/10$) slope means the hill rises 1 unit for every 10 horizontal units. Stairs and roofs have slopes too, as shown in Figure 11.

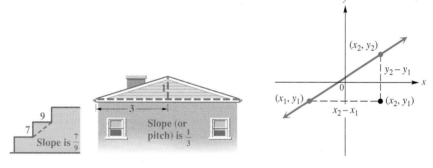

FIGURE 11   **FIGURE 12**

1 ► Find the slope of a line given two points on the line.

► To get a formal definition of the slope of a line, suppose (x_1, y_1) and (x_2, y_2) are two different points on a line. Then, as we move along the line from (x_1, y_1) to (x_2, y_2), the y-value changes from y_1 to y_2, an amount equal to $y_2 - y_1$. As y changes from y_1 to y_2, the value of x changes from x_1 to x_2 by the amount $x_2 - x_1$. See Figure 12. The ratio of the change in y to the change in x is called the **slope** of the line, with the letter m used for the slope.

Slope

If $x_1 \neq x_2$, the slope of the line through the distinct points (x_1, y_1) and (x_2, y_2) is

$$m = \frac{\text{change in } y}{\text{change in } x} = \frac{y_2 - y_1}{x_2 - x_1}.$$

► **TEACHING TIP**

Students should remember the familiar definition:
slope = rise/run. ◄

EXAMPLE 1
Using the Definition of Slope

Find the slope of the line through the points $(2, -1)$ and $(-5, 3)$.
   If $(2, -1) = (x_1, y_1)$ and $(-5, 3) = (x_2, y_2)$, then

$$m = \frac{y_2 - y_1}{x_2 - x_1}$$

$$= \frac{3 - (-1)}{-5 - (2)} = \frac{4}{-7} = -\frac{4}{7}.$$

INSTRUCTOR'S RESOURCES

 ITM pp. 432–436   **ISM** pp. 510–523
IAM p. 30

 **TEST GENERATOR**
DOS Windows MAC

 TRANSPARENCIES

Find the slope of the line through the points $(-6, 9)$ and $(3, -5)$.

Answer: $-\dfrac{14}{9}$

▶ **TEACHING TIP**
Accompany each use of the slope formula with a graph that shows the unit changes in y and x (a "slope triangle"). A graph is used to check for the correct sign of the slope resulting from the formula. ◀

See Figure 13. On the other hand, if $(2, -1) = (x_2, y_2)$ and $(-5, 3) = (x_1, y_1)$, the slope would be

$$m = \frac{-1 - 3}{2 - (-5)} = \frac{-4}{7} = -\frac{4}{7},$$

the same answer. ■

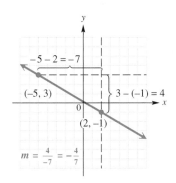

FIGURE 13

Example 1 suggests that the slope is the same no matter which point is considered first. Also, using similar triangles from geometry, it can be shown that the slope is the same no matter which two different points on the line are chosen.

CAUTION   In calculating the slope, be careful to subtract the y-values and the x-values in the *same* order.

	Correct			Incorrect	
$\dfrac{y_2 - y_1}{x_2 - x_1}$	or	$\dfrac{y_1 - y_2}{x_1 - x_2}$	$\dfrac{y_2 - y_1}{x_1 - x_2}$	or	$\dfrac{y_1 - y_2}{x_2 - x_1}$

Also, remember that the change in y is in the *numerator* and the change in x is in the *denominator*.

---

2 ▶ Find the slope of a line given the equation of the line.

▶   When the equation of a line is given, one way to find the slope is to use the definition of slope by first finding two different points on the line.

E X A M P L E   2

Finding the Slope of a Line

Find the slope of the line $4x - y = 8$.
　　The intercepts can be used as the two different points needed to find the slope. Let $y = 0$ to find that the x-intercept is $(2, 0)$. Then let $x = 0$ to find that the y-intercept is $(0, -8)$. Use these two points in the slope formula. The slope is

Find the slope of the line $3x - 4y = 12$.

Answer: $\dfrac{3}{4}$

$$m = \frac{-8 - 0}{0 - 2} = \frac{-8}{-2} = 4. \quad ■$$

EXAMPLE 3

Finding the Slope of a Line

Chalkboard Exercise

Find the slope of the line
$y + 3 = 0$.

Answer: 0

▶ **TEACHING TIP**

Review the cases for the slopes of vertical and horizontal lines. An equation of the form $x = k$ always intersects the *x-axis*; thus it is vertical and has an undefined slope. An equation of the form $y = k$ always intersects the *y-axis*; thus it is horizontal and has a zero slope.

Avoid using *no slope* in place of *undefined slope* as students confuse *no slope* with *zero slope*. ◀

◆ **48 miles per hour**

Find the slope of each of the following lines.

(a) $x = -3$

By inspection, $(-3, 5)$ and $(-3, -4)$ are two points that satisfy the equation $x = -3$. Use these two points to find the slope.

$$m = \frac{-4 - 5}{-3 - (-3)} = \frac{-9}{0}$$

Since division by zero is undefined, the slope is undefined. This is why the definition of slope includes the restriction that $x_1 \neq x_2$.

(b) $y = 5$

Find the slope by selecting two different points on the line, such as $(3, 5)$ and $(-1, 5)$, and by using the definition of slope.

$$m = \frac{5 - 5}{3 - (-1)} = \frac{0}{4} = 0 \ ■$$

As shown in Section 7.1, $x = -3$ has a graph that is a vertical line, and $y = 5$ has a graph that is a horizontal line. Generalizing from those results and the results of Example 3 above, we can make the following statements about vertical and horizontal lines.

Slope of Vertical and Horizontal Lines

The slope of a vertical line is undefined; the slope of a horizontal line is 0.

◆ **CONNECTIONS** ◆

On a trip from San Francisco, a passenger kept track every half hour of the distance traveled, with the following results for the first three hours.

Time in Hours	0	.5	1	1.5	2	2.5	3
Distance in Miles	0	20	48	80	104	126	150

From the distance formula, $d = rt$, solving for rate (or speed) gives

$$\text{Average speed} = \frac{\text{Distance}}{\text{Time}}.$$

For example, the average speed over the time interval from $t = 0$ to $t = 3$ is

$$\text{Average speed} = \frac{150 - 0}{3 - 0} = 50,$$

or 50 miles per hour. Similarly, the average speed from $t = 1$ to $t = 2$ is

$$\text{Average speed} = \frac{104 - 48}{2 - 1} = 56,$$

or 56 miles per hour. If the time in hours is represented by x and the distance is represented by y, then we can write the following formula for average speed.

$$\text{Average speed} = \frac{y_2 - y_1}{x_2 - x_1}.$$

Of course, x and y may represent other quantities, so this formula can be used to find the rate of change of any quantity with respect to another quantity. As we see in this section, one application of this formula is the slope of a line, where it represents the average change in y as x changes by 1 unit.

FOR DISCUSSION OR WRITING
Use the formula developed above to find the average speed from $t = 2$ to $t = 3$. Why do you think the average speed keeps changing? How might we find the speed at some particular instant?

3 ▶ Graph a line given its slope and a point on the line.

▶ Examples 4 and 5 show how to graph a straight line by using the slope and one point that the line contains.

E X A M P L E   4

Using the Slope and a Point to Graph a Line

Graph the line that has slope 2/3 and goes through the point $(-1, 4)$.
First locate the point $(-1, 4)$ on a graph as shown in Figure 14. Then, from the definition of slope,

$$m = \frac{\text{change in } y}{\text{change in } x} = \frac{2}{3}.$$

Move *up* 2 units in the y-direction and then 3 units to the *right* in the x-direction to locate another point on the graph (labeled P). The line through $(-1, 4)$ and P is the required graph.   ■

Chalkboard Exercise

Graph the line through $(-3, -2)$ that has slope $\frac{1}{2}$.

Answer:

▶ TEACHING TIP

To graph a line requires a minimum of two points. Here, the first point is given. From the first point, use the given slope (a positive or negative change in y followed by a positive change in x) to get to the second point.

Be sure that the graph corresponds to the sign of the slope.   ◀

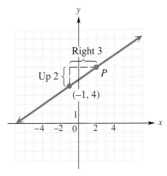

FIGURE 14

E X A M P L E   5

Using the Slope and a Point to Graph a Line

Graph the line through $(2, 1)$ that has slope $-4/3$.
Start by locating the point $(2, 1)$ on the graph. Find a second point on the line by using the definition of slope.

$$\text{slope} = \frac{\text{change in } y}{\text{change in } x} = \frac{-4}{3}$$

Move *down* 4 units from $(2, 1)$ and then 3 units to the *right*. Draw a line through this second point and $(2, 1)$, as shown in Figure 15. The slope also could be written as

$$\frac{\text{change in } y}{\text{change in } x} = \frac{4}{-3}.$$

Chalkboard Exercise

Graph the line through $(1, -3)$ that has slope $-\dfrac{3}{4}$.

Answer:

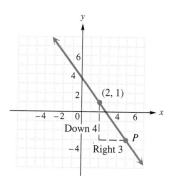

FIGURE 15

In that case the second point is located *up* 4 units and 3 units to the *left*. Verify that this approach produces the same line. ■

In Example 4, the slope of the line is the *positive* number 2/3. The graph of the line in Figure 14 goes up from left to right. The line in Example 5 has a *negative* slope, $-4/3$. As Figure 15 shows, its graph goes down from left to right. These facts suggest the following generalization.

> A positive slope indicates that the line goes up from left to right; a negative slope indicates that the line goes down from left to right.

Figure 16 shows lines of positive, zero, negative, and undefined slopes.

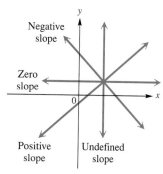

FIGURE 16

4 ▶ Use slope to decide whether two lines are parallel or perpendicular.

▶   The slopes of a pair of parallel or perpendicular lines are related in a special way. The slope of a line measures the steepness of a line. Since parallel lines have equal steepness, their slopes also must be equal. Also, lines with the same slope are parallel.

Slopes of Parallel Lines

Two nonvertical lines with the same slope are parallel; two nonvertical parallel lines have the same slope.

EXAMPLE 6

Determining Whether Two Lines Are Parallel

▶ **TEACHING TIP**

For the *parallel line* case: graph two different lines with y-intercepts at $(0, 0)$ and $(0, 2)$, but with the same slope $2/3$.

For the *perpendicular line* case: graph a line with y-intercept $(0, 0)$, slope $2/3$. Graph a second line with slope $-3/2$. ◀

▶ **TEACHING TIP**

It is helpful to say that perpendicular lines have "different slopes," then proceed to get more specific.

Inform students of a *third case:* skewed lines are two lines that are neither parallel nor perpendicular. ◀

Are the lines L_1, through $(-2, 1)$ and $(4, 5)$, and L_2, through $(3, 0)$ and $(0, -2)$, parallel?

The slope of L_1 is

$$m_1 = \frac{5 - 1}{4 - (-2)} = \frac{4}{6} = \frac{2}{3}.$$

The slope of L_2 is

$$m_2 = \frac{-2 - 0}{0 - 3} = \frac{-2}{-3} = \frac{2}{3}.$$

Since the slopes are equal, the lines are parallel. ■

To see how the slopes of perpendicular lines are related, consider a nonvertical line with slope a/b. If this line is rotated $90°$, the vertical change and the horizontal change are exchanged and the slope is $-(b/a)$, since the horizontal change is now negative. See Figure 17. Thus, the slopes of perpendicular lines have a product of -1 and are negative reciprocals of each other. For example, if the slopes of two lines are $\frac{3}{4}$ and $-\frac{4}{3}$, then the lines are perpendicular because $\left(\frac{3}{4}\right)\left(-\frac{4}{3}\right) = -1$.

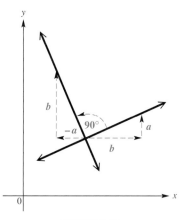

FIGURE 17

Slopes of Perpendicular Lines

If neither is vertical, perpendicular lines have slopes that are negative reciprocals; that is, their product is -1. Also, lines with slopes that are negative reciprocals are perpendicular.

EXAMPLE 7

Determining Whether Two Lines Are Perpendicular

Are the lines with equations $2y = 3x - 6$ and $2x + 3y = -6$ perpendicular?

Find the slope of each line by first finding two points on the line. The points $(0, -3)$ and $(2, 0)$ are on the first line. The slope is

$$m_1 = \frac{0 - (-3)}{2 - 0} = \frac{3}{2}.$$

The second line goes through $(-3, 0)$ and $(0, -2)$ and has slope

$$m_2 = \frac{-2 - 0}{0 - (-3)} = -\frac{2}{3}.$$

Since the product of the slopes of the two lines is $(3/2)(-2/3) = -1$, the lines are perpendicular. ■

5 ▶ Solve problems involving average rate of change.

▶ We have seen how the slope of a line is the ratio of the change in y (vertical change) to the change in x (horizontal change). This idea can be extended to real-life situations as follows: the slope gives the average rate of change in y per unit of change in x, where the value of y depends on the value of x. The next example illustrates this idea of average rate of change. We assume a linear relationship between x and y.

E X A M P L E   8 ■

Interpreting Slope as Average Rate of Change

The bar graph in Figure 18* shows the number of multimedia personal computers (PCs), in millions, in U.S. homes. (The figures for 1995 to 1997 are estimates.) Find the average rate of change in the number of multimedia PCs per year.

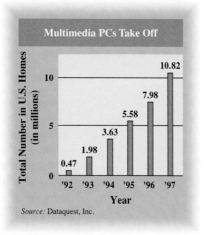

FIGURE 18

Since connecting the tops of the bars would closely approach a straight line, we can use the slope formula. We need two pairs of data. If we let 1992 represent 0, then 1993 represents 1, 1994 represents 2, and so on. Then the ordered pair for 1993 is $(1, 1.98)$ and the pair for 1996 is $(4, 7.98)$. The average rate of change in the number of multimedia PCs is found by using the slope formula.

$$\text{Average rate of change} = \frac{y_2 - y_1}{x_2 - x_1} = \frac{7.98 - 1.98}{4 - 1} = 2$$

The result, 2, indicates that the number of multimedia PCs increases by 2 million each year. ■

---

* Graph for Figure 18, "Multimedia PC's Take Off," from *The Wall Street Journal*, March 21, 1994. Reprinted by permission of The Wall Street Journal, Copyright © 1994 Dow Jones & Company, Inc. All Rights Reserved Worldwide.

◆ Answers will vary.

◆ C O N N E C T I O N S ◆

While graphics calculators provide us with a powerful method of analyzing graphs, we must never forget that an understanding of the concepts of mathematics is necessary in interpreting what we see on their screens. Otherwise, we may reach incorrect conclusions.

For example, consider the graphs of the lines $2x - 3y = -3$ and $7x - 10y = 30$, shown in the standard viewing window in the figure. Based on the graphs alone, one might conclude that the lines are parallel. However, it can be shown algebraically that their slopes are not equal and thus the lines are *not* parallel.

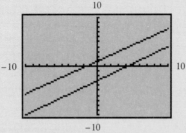

The graphs are *not* parallel,
though they may appear to be.

In Example 7 of this section we show algebraically that the lines with equations $2y = 3x - 6$ and $2x + 3y = -6$ are perpendicular. If we graph them in the standard viewing window of the typical graphics calculator, they do not appear to be perpendicular. This is because the horizontal-to-vertical aspect ratio of such a calculator is usually about 3 to 2. See the first figure below. In order to obtain a more realistic picture, we can use the *square window,* as seen in the second figure below. Again, we cannot rely completely on what we see on the screen—we must understand the mathematical theory to draw the correct conclusion.

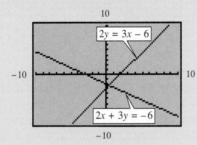

In the standard window,
the lines *do not* appear
to be perpendicular.

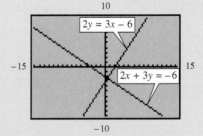

In the square window,
the lines *do* appear
to be perpendicular.

FOR WRITING OR DISCUSSION

Comment on the following statement by Peg Crider of Tomball College: *The most powerful tool in the entire process of understanding mathematics is your brain.*

7.2 EXERCISES

1. If a walkway rises 2 feet for every 10 feet on the horizontal, which of the following express its slope (or grade)? (There are several correct choices.)

(a) .2   (b) $\dfrac{2}{10}$ (c) $\dfrac{1}{5}$ (d) 20%

(e) 5   (f) $\dfrac{20}{100}$ (g) 500%   (h) $\dfrac{10}{2}$

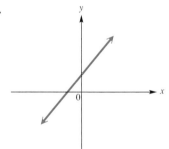

1. (a), (b), (c), (d), (f)

2. If the pitch of a roof is $\dfrac{1}{4}$, how many feet in the horizontal direction corresponds to a rise of 3 feet?

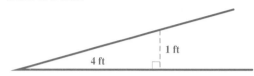

2. 12 feet

Find the slope of the line through the pair of points by using the slope formula. See Example 1.

3. $(-2, -3)$ and $(-1, 5)$

3. 8

4. $(-4, 3)$ and $(-3, 4)$

4. 1

5. $(-4, 1)$ and $(2, 6)$

5. $\dfrac{5}{6}$

6. $(-3, -3)$ and $(5, 6)$

6. $\dfrac{9}{8}$

7. $(2, 4)$ and $(-4, 4)$

7. 0

8. $(-6, 3)$ and $(2, 3)$

8. 0

◉ *Tell whether the slope of the line is positive, negative, zero, or undefined.*

9.

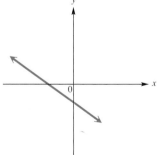

9. positive

10.

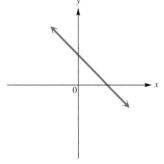

10. positive

11.

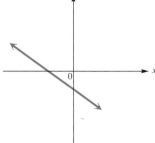

11. negative

12.

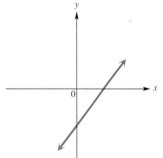

12. negative

 CONCEPTUAL   WRITING   ▲ CHALLENGING   SCIENTIFIC CALCULATOR   GRAPHICS CALCULATOR

13.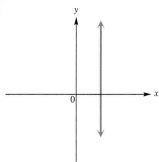

14.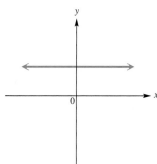

13. undefined 14. zero

Find the slope of the line and sketch the graph. See Examples 1–3.

15. $x + 2y = 4$

15. $-\dfrac{1}{2}$

16. $x + 3y = -6$

16. $-\dfrac{1}{3}$

17. $-x + y = 4$

17. 1

18. $-x + y = 6$

18. 1

19. $6x + 5y = 30$

19. $-\dfrac{6}{5}$

20. $3x + 4y = 12$

20. $-\dfrac{3}{4}$

21. $5x - 2y = 10$

21. $\dfrac{5}{2}$

22. $4x - y = 4$

22. 4

23. $y = 4x$

23. 4

24. $y = -3x$

24. -3

25. $y - 3 = 0$

25. 0

26. $y + 5 = 0$

26. 0

⊙ **27.** A vertical line has equation _____ = k for some constant k; a horizontal line has equation _____ = k for some constant k.

27. x; y

⊙ **28.** Explain the procedure for graphing a straight line using its slope and a point on the line.

Use the methods shown in Examples 4 and 5 to graph the line described.

29. through $(-4, 2)$; $m = \dfrac{1}{2}$ 29.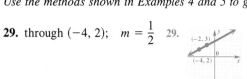

30. through $(-2, -3)$; $m = \dfrac{5}{4}$ 30.

31. through $(0, -2)$; $m = -\dfrac{2}{3}$ 31.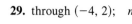

32. through $(0, -4)$; $m = -\dfrac{3}{2}$ 32.

33. through $(-1, -2)$; $m = 3$ **33.**

34. through $(-2, -4)$; $m = 4$ **34.**

35. $m = 0$; through $(2, -5)$ **35.**

36. undefined slope; through $(-3, 1)$ **36.**

37. undefined slope; through $(-4, 1)$ **37.**

38. $m = 0$; through $(5, 3)$ **38.**

◉ **39.** If a line has slope $-4/9$, then any line parallel to it has slope _____ , and any line perpendicular to it has slope _____ .

39. $-\dfrac{4}{9}$; $\dfrac{9}{4}$

◉ **40.** What is the slope of any line perpendicular to a line with undefined slope?

40. zero

Decide whether the two lines are parallel. See Example 6.

41. $3x = y$ and $2y - 6x = 5$

41. parallel

42. $2x + 5y = -8$ and $6 + 2x = 5y$

42. not parallel

43. $4x + y = 0$ and $5x - 8 = 2y$

43. not parallel

44. $x = 6$ and $6 - x = 8$

44. parallel

45. The line through $(4, 6)$ and $(-8, 7)$ and the line through $(7, 4)$ and $(-5, 5)$

45. parallel

46. The line through $(9, 15)$ and $(-7, 12)$ and the line through $(-4, 8)$ and $(-20, 5)$

46. parallel

Decide whether the two lines are perpendicular. See Example 7.

47. $4x - 3y = 8$ and $4y + 3x = 12$

47. perpendicular

48. $2x = y + 3$ and $2y + x = 3$

48. perpendicular

49. $4x - 3y = 5$ and $3x - 4y = 2$

49. not perpendicular

50. $5x - y = 7$ and $5x = 3 + y$

50. not perpendicular

51. $2x + y = 1$ and $x - y = 2$

51. not perpendicular

52. $2y - x = 3$ and $y + 2x = 1$

52. perpendicular

▦ *Use the concept of slope to solve the problem.*

53. The upper deck at the new Comiskey Park in Chicago has produced, among other complaints, displeasure with its steepness. It's been compared to a ski jump. It is 160 feet from home plate to the front of the upper deck and 250 feet from home plate to the back. The top of the upper deck is 63 feet above the bottom. What is its slope?

53. $\dfrac{7}{10}$

54. When designing the new arena in Boston to replace the old Boston Garden, architects were careful to design the ramps leading up to the entrances so that circus elephants would be able to march up the ramps. The maximum grade (or slope) that an elephant will walk on is 13%. Suppose that such a ramp was constructed with a horizontal run of 150 feet. What would be the maximum vertical rise the architects could use?

54. 19.5 feet

◉ CONCEPTUAL   🖉 WRITING   ▲ CHALLENGING   ▦ SCIENTIFIC CALCULATOR   ▧ GRAPHICS CALCULATOR

Use the idea of average rate of change to solve the problem. See Example 8.

55. The graph shows how average monthly rates for cable television increased from 1980 to 1992. The graph can be approximated by a straight line.

    (a) Use the information provided for 1980 and 1992 to determine the average rate of change in price per year.

⊙ **(b)** In your own words, explain how a *positive* rate of change affects the consumer in a situation such as the one illustrated by this graph.

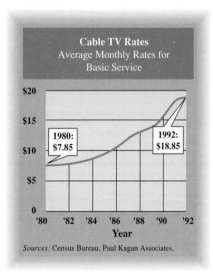

Sources: Census Bureau, Paul Kagan Associates.

56. Assuming a linear relationship, what is the average rate of change for cable industry revenues over the period from 1990 to 1992?

Sources: Census Bureau, Paul Kagan Associates.

56. $1.6 billion

55. (a) $.92   (b) It means an *increase* in price.

57. The 1993 Annual Report of AT&T cites the following figures concerning the global growth of international telephone calls: From 47.5 billion minutes in 1993, traffic is expected to rise to 60 billion minutes in 1995. Assuming a linear relationship, what is the average rate of change for this time period? (*Source:* TeleGeography, 1993, Washington, D.C.)

57. 6.25 billion minutes per year

58. The market for international phone calls during the ten-year period from 1986 to 1995 is depicted in the accompanying bar graph. The tops of the bars approximate a straight line. Assuming that the traffic volume at the beginning of this period was 18 billion minutes and at the end was 60 billion minutes, what was the average rate of change for the ten-year period?

58. 4.2 billion minutes per year

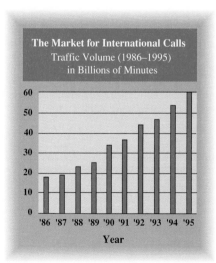

59. In one state the fine for driving 10 miles per hour over the speed limit is $35 and the fine for driving 15 miles per hour over the speed limit is $42. What is the average rate of change in the fine for each mile per hour over the limit?

59. **$1.40 per mile**

60. On the third day of a rotation diet, Lynn Elliott weighed 92.5 kilograms. By the eleventh day, she weighed 90.9 kilograms. What was her average rate of weight loss per day?

60. **.2 kilogram per day**

61. The graphics calculator screen shows a line with the coordinates of a point displayed at the bottom. The slope of the line is 3. What is the y-corrdinate of the point on the line whose x-coordinate is 4?

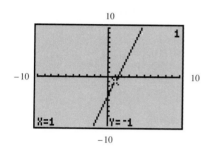

61. **8**

63. The table shown was generated by a graphics calculator. It gives several points that lie on the graph of a line. What is the slope of the line?

63. **1.75**

62. The graphics calculator screen shows two lines. One is the graph of $y_1 = -2x + 3$ and the other is the graph of $y_2 = 3x - 4$. Which is which?

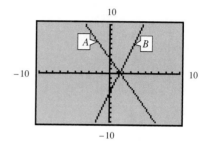

62. *A* is y_1 and *B* is y_2.

64. The graphics calculator screen shows two lines. One is the graph of $y_1 = 2x - 5$ and the other is the graph of $y_2 = 4x - 5$. Which is which?

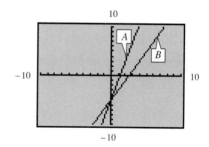

64. *A* is y_2 and *B* is y_1.

▲ *Solve the problem using your knowledge of the slopes of parallel and perpendicular lines.*

65. Show that $(-13, -9)$, $(-11, -1)$, $(2, -2)$, and $(4, 6)$ are the vertices of a parallelogram. (*Hint:* A parallelogram is a four-sided figure with opposite sides parallel.)

65. **Since the slopes of both pairs of opposite sides are equal, the figure is a parallelogram.**

66. Is the figure with vertices at $(-11, -5)$, $(-2, -19)$, $(12, -10)$, and $(3, 4)$ a parallelogram? Is it a rectangle? (A rectangle is a parallelogram with a right angle.)

66. **yes; yes**

◆ **MATHEMATICAL CONNECTIONS** (Exercises 67–76) ◆

*In these exercises we investigate two methods of determining whether three points lie on the same straight line. (Such points are said to be **collinear**.) The points we consider are A(3, 1), B(6, 2), and C(9, 3). Work Exercises 67–76 in order.*

67. Find the distance between *A* and *B*.

67. $\sqrt{10}$

68. Find the distance between *B* and *C*.

68. $\sqrt{10}$

69. Find the distance between *A* and *C*.

69. $2\sqrt{10}$

70. If $AB + BC = AC$, then the points are collinear. Use the results of Exercises 67–69 to show that this statement is satisfied.

70. $\sqrt{10} + \sqrt{10} = 2\sqrt{10}$ is true.

71. Find the slope of segment *AB*.

71. $\dfrac{1}{3}$

72. Find the slope of segment *BC*.

72. $\dfrac{1}{3}$

73. Find the slope of segment *AC*.

73. $\dfrac{1}{3}$

74. If slope of AB = slope of BC = slope of AC, then A, B, and C are collinear. Use the results of Exercises 71–73 to show that this statement is satisfied.

74. $\dfrac{1}{3} = \dfrac{1}{3} = \dfrac{1}{3}$ is true.

75. Use either the distance formula or the slope formula to determine whether the points $(1, -2)$, $(3, -1)$, and $(5, 0)$ are collinear.

75. They are collinear.

76. Repeat Exercise 75 for the points $(0, 6)$, $(4, -5)$, and $(-2, 12)$.

76. They are not collinear.

◆

REVIEW EXERCISES

Solve the equation for y. See Section 2.2.

77. $3x + 2y = 8$

77. $y = \dfrac{-3x + 8}{2}$

78. $4x + 3y = 0$

78. $y = -\dfrac{4}{3}x$

79. $y - 2 = 4(x + 3)$

79. $y = 4x + 14$

Other correct forms are possible for the answers in Exercises 77–82.

Write the equation without fractions in the form Ax + By = C. Combine terms if possible. See Section 2.1.

80. $y - (-2) = \dfrac{3}{2}(x - 5)$

80. $3x - 2y = 19$

81. $y - (-1) = \dfrac{5}{3}[x - (-4)]$

81. $5x - 3y = -17$

82. $y - 7 = -\dfrac{1}{4}[x - (-3)]$

82. $x + 4y = 25$

7.3 LINEAR EQUATIONS

FOR EXTRA HELP

 **SSG** pp. 214–219
SSM pp. 345–352

 Video
9

 Tutorial
IBM   MAC

OBJECTIVES

1 ▶ Write the equation of a line given its slope and a point on the line.
2 ▶ Write the equation of a line given two points on the line.
3 ▶ Write the equation of a line given its slope and y-intercept.
4 ▶ Find the slope and y-intercept of a line given its equation.
5 ▶ Write the equation of a line parallel or perpendicular to a given line.
6 ▶ Use a graphics calculator to solve linear equations.

Many real-world situations can be described by straight-line graphs. In this section we see how to write a linear equation in such situations.

1 ▶ Write the equation of a line given its slope and a point on the line.

▶ **TEACHING TIP**

Start with the graph of a line through (x_1, y_1) and (x, y), complete the "slope triangle," then show the slope ratio: $m = (y - y_1)/(x - x_1)$. Proceed to the formula $y - y_1 = m(x - x_1)$, stating that the boldface quantities are the only ones to be replaced (x and y are left as variables). ◀

▶   A straight line is a set of points in the plane such that the slope between any two points is the same. In Figure 19, point P is on the line through P_1 and P_2 if the slope of the line through points P_1 and P equals the slope of the line through points P and P_2. If these slopes are equal to m, then

$$\frac{y - y_1}{x - x_1} = \frac{y - y_2}{x - x_2} = m,$$

$$\frac{y - y_1}{x - x_1} = m$$

$$y - y_1 = m(x - x_1) \qquad \text{Multiply both sides by } x - x_1.$$

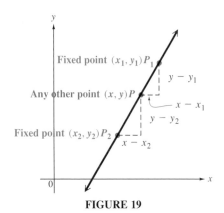

FIGURE 19

This last equation gives the *point-slope form* of the equation of the line, which shows the coordinates of a point (x_1, y_1) on the line and the slope of the line.

Point-Slope Form

The point-slope form of the equation of a line is

$$\underset{\text{Given point}}{\underset{\uparrow}{y} - y_1 = \overset{\overset{\text{Slope}}{\downarrow}}{m} (x - x_1).}$$

INSTRUCTOR'S RESOURCES

 ITM pp. 436–440   **ISM** pp. 523–534
IAM p. 31

 TEST GENERATOR
DOS   Windows   MAC

 TRANSPARENCIES

Find the equation of the line with slope 1/3, going through the point $(-2, 5)$.

Use the point-slope form of the equation of a line, with $(x_1, y_1) = (-2, 5)$ and $m = 1/3$.

$$y - y_1 = m(x - x_1)$$

$$y - 5 = \frac{1}{3}[(x - (-2)]$$ Let $y_1 = 5$, $m = \frac{1}{3}$, $x_1 = -2$.

$$y - 5 = \frac{1}{3}(x + 2)$$

$$3y - 15 = x + 2$$ Multiply by 3.

or $$x - 3y = -17$$ ■

Chalkboard Exercise

Find the equation of the line

with slope $\frac{2}{5}$, going through the

point $(3, -4)$. Write the equation in standard form.

Answer: $2x - 5y = 26$

▶ **TEACHING TIP**

It is helpful to restate the restriction on standard form as follows: A, B, and C cannot be fractions and A should be nonnegative. Proceed with several examples. Include slopes of 3 and $-2/3$. ◀

It is convenient to have an agreement on the form in which a linear equation should be written. In Section 7.1, we defined *standard form* for a linear equation as

$$Ax + By = C.$$

In addition, from now on, let us agree that A, B and C will be integers with no common factor (except 1) and $A \geq 0$. For example, the final equation found in Example 1, $x - 3y = -17$, is written in standard form.

CAUTION The definition of "standard form" is not standard from one text to another. Any linear equation can be written in many different (all equally correct) forms. For example, the equation $2x + 3y = 8$ can be written as $2x = 8 - 3y$, $3y = 8 - 2x$, $x + (3/2)y = 4$, $4x + 6y = 16$, and so on. In addition to writing it in the form $Ax + By = C$ (with $A \geq 0$), let us agree that the form $2x + 3y = 8$ is preferred over any multiples of both sides, such as $4x + 6y = 16$.

◈ approximately 12 million; 1992; Since $-.02$ is close to 0, the number of PCs doubles each year in the indicated years.

◈ **CONNECTIONS** ◈

Many of the earlier Connections boxes in this book have given equations that describe real-world data. Now we are able to show how such equations can be found. The process of writing an equation to describe a set of data is called *data fitting*.

The data shown in the bar graph in Example 8 in the previous section increased "linearly". That is, a line through the data pairs would be very close to a straight line. This indicates that we can write a linear equation to approximate the data, allowing us to estimate the number of PCs for years other than those shown in the graph. In Example 8, we found that the slope of the line is 2, and that $(1, 1.98)$ and $(4, 7.98)$ are two points on the line. Using $(1, 1.98)$, we get the following equation of the line.

$$y - y_1 = m(x - x_1)$$
$$y - 1.98 = 2(x - 1)$$ Let $m = 2$, $x_1 = 1$, $y_1 = 1.98$.
$$y - 1.98 = 2x - 2$$ Distributive property
$$y = 2x - .02$$ Add 1.98.

The equation tells us that the number of multimedia PCs (in millions) in year x is given by $2x - .02$.

> **FOR DISCUSSION OR WRITING**
> Use the equation found above to predict the number of multimedia PCs in 1998. (Recall that 1992 is represented by 0.) According to the equation, in what year were there no multimedia PCs? What does the $-.02$ in the equation indicate about the number of multimedia PCs? What might cause this equation to become very inaccurate in future years?

2 ▶ Write the equation of a line given two points on the line.

▶ When two points on a line are known, we can find an equation of the line by first using the slope formula to find the slope of the line. Then we use the slope with either one of the given points in the point-slope form.

E X A M P L E   2

Finding an Equation of a Line When Two Points Are Known

Find an equation of the line through the points $(-4, 3)$ and $(5, -7)$.

First find the slope, using the definition.

$$m = \frac{-7 - 3}{5 - (-4)} = -\frac{10}{9}$$

Use either $(-4, 3)$ or $(5, -7)$ as (x_1, y_1) in the point-slope form of the equation of a line. If $(-4, 3)$ is used, then $-4 = x_1$ and $3 = y_1$.

$y - y_1 = m(x - x_1)$	Point-slope form
$y - 3 = -\dfrac{10}{9}[x - (-4)]$	Let $y_1 = 3$, $m = -\frac{10}{9}$, $x_1 = -4$.
$y - 3 = -\dfrac{10}{9}(x + 4)$	
$9(y - 3) = -10(x + 4)$	Multiply by 9.
$9y - 27 = -10x - 40$	Distributive property
$10x + 9y = -13$	Standard form

Verify that if $(5, -7)$ were used, the same equation would be found. ∎

Chalkboard Exercise

Find an equation of the line through the points $(-2, 6)$ and $(1, 4)$.

Answer: $2x + 3y = 14$

▶ TEACHING TIP
Write the point-slope form of the equation, leaving the locations for y_1, m, and x_1 blank.

Label the given points as (x_1, y_1), (x_2, y_2), respectively. Use the slope formula. Fill in the blanks left in the formula. Put into standard form. ◀

Notice that the point-slope form does not apply to a vertical line, since the slope of a vertical line is undefined. A vertical line through the point (k, y), where k is a constant and y represents any real number, has equation $x = k$.

A horizontal line has slope 0. From the point-slope form, the equation of a horizontal line through the point (x, k), where x is any real number and k is a constant, is

$$y - y_1 = m(x - x_1)$$
$$y - k = 0(x - x) \qquad y_1 = k, x_1 = x$$
$$y - k = 0$$
$$y = k$$

▶ TEACHING TIP
The point-slope form of the line is used for cases involving two variables.

For one-variable cases passing through, for example, $(5, 3)$: the vertical line equation is $x = 5$; the horizontal line equation is $y = 3$. ◀

In summary, horizontal and vertical lines have the following special equations.

Equations of Vertical and Horizontal Lines

If k is a constant, the vertical line through (k, y) has equation $x = k$, and the horizontal line through (x, k) has equation $y = k$.

3 ▶ Write the equation of a line given its slope and y-intercept.

▶ Suppose a line has slope m and we know that the y-intercept is $(0, b)$. Using the point-slope form gives

$$y - y_1 = m(x - x_1)$$
$$y - b = m(x - 0) \qquad x_1 = 0, y_1 = b$$
$$y = mx + b. \qquad \text{Add } b.$$

When the equation is solved for y, the coefficient of x is the slope, m, and the constant b is the y-value of the y-intercept. Because this form of the equation shows the slope and the y-intercept, it is called the *slope-intercept form.*

Slope-Intercept Form	The equation of a line with slope m and y-intercept $(0, b)$ is written in **slope-intercept form** as

$$y = mx + b.$$

Slope y-intercept is $(0, b)$

EXAMPLE 3

Using the Slope-Intercept Form

Find an equation of the line with slope $-4/5$ and y-intercept $(0, -2)$ in standard form.

Here $m = -4/5$ and $b = -2$. Substitute these values into the slope-intercept form.

$$y = mx + b$$
$$y = -\frac{4}{5}x - 2$$
$$5y = -4x - 10 \qquad \text{Multiply by 5.}$$
$$4x + 5y = -10 \qquad \text{Standard form} \blacksquare$$

Chalkboard Exercise

Write an equation in standard form of the line with slope 2 and y-intercept $(0, -3)$.

Answer: $2x - y = 3$

4 ▶ Find the slope and y-intercept of a line given its equation.

▶ In the previous section, the slope of the line $4x - y = 8$ was found to be 4 by first getting two points on the line and then using the definition of slope. For an alternative method of finding the slope of this line, transform $4x - y = 8$ into the form $y = mx + b$.

$$4x - y = 8$$
$$-y = -4x + 8 \qquad \text{Subtract } 4x.$$
$$y = 4x - 8 \qquad \text{Multiply by } -1.$$

When the equation is solved for y, the coefficient of x is the slope. Writing the equation in this form also tells us the y-intercept. In this case, it is $(0, -8)$.

EXAMPLE 4

Finding the Slope and y-intercept from the Equation

Write $3y + 2x = 9$ in slope-intercept form, and find the slope and y-intercept. Put the equation in slope-intercept form by solving for y.

$$3y + 2x = 9$$
$$3y = -2x + 9$$
$$y = -\frac{2}{3}x + 3$$

Slope ____↑   ↑____ y-intercept is $(0, 3)$.

Chalkboard Exercise

Find the slope and y-intercept of the line $2x - 5y = 1$.

Answer: $m = \frac{2}{5}, \left(0, -\frac{1}{5}\right)$

From the slope-intercept form, the slope is $-2/3$ and the y-intercept is $(0, 3)$. ■

NOTE The importance of the slope-intercept form of a linear equation cannot be overemphasized. First, every linear equation (of a non-vertical line) has a *unique* (one and only one) slope-intercept form. Second, in Section 7.5 we will study linear *functions*, where the slope-intercept form is necessary in specifying such functions.

5 ▶ Write the equation of a line parallel or perpendicular to a given line.

▶ As mentioned in the last section, parallel lines have the same slope and perpendicular lines have slopes with a product of -1. These results are used in the next example.

EXAMPLE 5

Finding Equations of Parallel or Perpendicular Lines

Find an equation of the line passing through the point $(-4, 5)$, and **(a)** parallel to the line $2x + 3y = 6$; **(b)** perpendicular to the line $2x + 3y = 6$.

(a) The slope of the graph of $2x + 3y = 6$ can be found by solving for y.

$$2x + 3y = 6$$
$$3y = -2x + 6 \qquad \text{Subtract } 2x \text{ on both sides.}$$
$$y = -\frac{2}{3}x + 2 \qquad \text{Divide both sides by } 3.$$

The slope is given by the coefficient of x, so $m = -2/3$.

$$y = -\overset{\text{Slope}}{\underset{\downarrow}{\tfrac{2}{3}}}x + 2$$

This means that the required equation of the line through $(-4, 5)$ and parallel to $2x + 3y = 6$ has slope $-2/3$. Now use the point-slope form, with $(x_1, y_1) = (-4, 5)$ and $m = -2/3$.

$$y - 5 = -\frac{2}{3}[x - (-4)] \qquad y_1 = 5, \, m = -\tfrac{2}{3}, \, x_1 = -4$$
$$y - 5 = -\frac{2}{3}(x + 4)$$
$$3(y - 5) = -2(x + 4) \qquad \text{Multiply by } 3.$$
$$3y - 15 = -2x - 8 \qquad \text{Distributive property}$$
$$2x + 3y = 7 \qquad \text{Standard form}$$

(b) To be perpendicular to the line $2x + 3y = 6$, a line must have a slope that is the negative reciprocal of $-2/3$, which is $3/2$. Use the point $(-4, 5)$ and slope $3/2$ in the point-slope form of the equation.

$$y - 5 = \frac{3}{2}[x - (-4)] \qquad y_1 = 5, \, m = \tfrac{3}{2}, \, x_1 = -4$$
$$2y - 10 = 3[x + 4] \qquad \text{Multiply by } 2.$$
$$2y - 10 = 3x + 12 \qquad \text{Distributive property}$$
$$-3x + 2y = 22$$
$$3x - 2y = -22 \qquad \text{Standard form} \quad ■$$

A summary of the various forms of linear equations follows.

Summary of Forms of Linear Equations	$Ax + By = C$	**Standard form** (Neither A nor B is 0.) Slope is $-\dfrac{A}{B}$. x-intercept is $\left(\dfrac{C}{A}, 0\right)$. y-intercept is $\left(0, \dfrac{C}{B}\right)$.
	$x = k$	**Vertical line** Undefined slope x-intercept is $(k, 0)$.
	$y = k$	**Horizontal line** Slope is 0. y-intercept is $(0, k)$.
	$y = mx + b$	**Slope-intercept form** Slope is m. y-intercept is $(0, b)$.
	$y - y_1 = m(x - x_1)$	**Point-slope form** Slope is m. Line passes through (x_1, y_1).

6 ▶ Use a graphics calculator to solve linear equations.

▶ We saw how to solve linear equations in one variable in Chapter 2. Now, we can use a graphics calculator to solve a linear equation.* First, we observe that the graph of $y = mx + b$ is a straight line, and if $m \neq 0$, the x-intercept of the graph is the solution of the equation $mx + b = 0$. For a simple example, suppose that we wish to solve $-4x + 7 = 0$ graphically. We begin by graphing $y_1 = -4x + 7$ in a viewing window that shows the x-intercept. Then we use the capability of the calculator to find the x-intercept. As seen in Figure 20, the x-intercept of the graph, which is also the solution or root of the equation, is 1.75.† Therefore, the solution set of $-4x + 7 = 0$ is {1.75}. This can easily be verified using strictly algebraic methods.

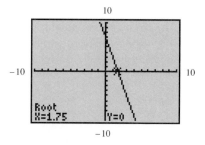

FIGURE 20

---

* The graphics calculator exposition and exercises may also be used with a computer that has appropriate graphing software.

† For convenience, here we refer to the *number* as the x-intercept rather than the point.

E X A M P L E   6

Solving a More
Complicated Equation

Solve $-2x - 4(2 - x) = 3x + 4$.

Begin by writing the equation as an equivalent equation with 0 on one side. If we subtract $3x$ and subtract 4 from both sides, we get

$$-2x - 4(2 - x) - 3x - 4 = 0.$$

Then we graph $y_1 = -2x - 4(2 - x) - 3x - 4$ and find the x-intercept. Notice that the viewing window must be altered from the one shown in Figure 20 because the x-intercept does not lie in the interval $[-10, 10]$. As seen in Figure 21, the x-intercept of the line, and thus the solution or root of the equation, is -12. The solution set is $\{-12\}$. ■

Chalkboard Exercise

Use a graphics calculator to solve

$4(x - 3) - x = x - 6$.

Answer: {3}

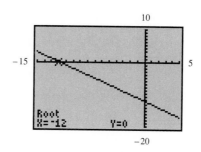

FIGURE 21

7.3 EXERCISES

◉ *Match the equation with the figure that most closely resembles its graph. The equation is given in the form $y = mx + b$, so consider the signs of m and b in making your choice.*

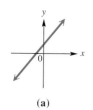

(a)	**(b)**	**(c)**	**(d)**

1. (a)   2. (a)   3. (c)
4. (c)   5. (b)   6. (b)
7. (d)   8. (d)   9. (c)
10. (c) 11. (a) 12. (a)
13. (d) 14. (d) 15. (b)
16. (b)

1. $y = 3x + 6$ **2.** $y = 4x + 5$ **3.** $y = -3x + 6$ **4.** $y = -4x + 5$

5. $y = 3x - 6$ **6.** $y = 4x - 5$ **7.** $y = -3x - 6$ **8.** $y = -4x - 5$

◉ *Match the equation with the figure that most closely resembles its graph.*

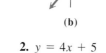
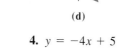

(a)	**(b)**	**(c)**	**(d)**

9. $y + 2 = 0$ **10.** $y + 4 = 0$ **11.** $x + 3 = 0$ **12.** $x + 7 = 0$

13. $y - 2 = 0$ **14.** $y - 4 = 0$ **15.** $x - 3 = 0$ **16.** $x - 7 = 0$

◉ CONCEPTUAL   ✎ WRITING   ▲ CHALLENGING   ▦ SCIENTIFIC CALCULATOR   ▦ GRAPHICS CALCULATOR

17. $3x + 4y = 10$
18. $5x + 6y = 31$
19. $2x + y = 18$
20. $x - y = 2$
21. $x - 2y = -13$
22. $x - 4y = 15$
23. $y = 12$ 24. $y = 5$
25. $4x - y = 12$
26. $5x + y = -10$
28. (d) 29. $x = 9$
30. $x = -2$ 31. $x = .5$
32. $x = \dfrac{5}{8}$ 33. $y = 8$
34. $y = \sqrt{7}$ 35. $2x - y = 2$
36. $2x + y = 8$
37. $x + 2y = 8$
38. $2x - 3y = -19$
39. $2x - 13y = -6$
40. $120x - 21y = 34$
41. $y = 5$ 42. $y = 2$
43. $x = 7$ 44. $x = 13$
45. $\sqrt{5}x - 2y = -\sqrt{5}$
46. $2\sqrt{2}x + 9y = \sqrt{2}$
47. $y = 5x + 15$
48. $y = -2x + 12$
49. $y = -\dfrac{2}{3}x + \dfrac{4}{5}$
50. $y = -\dfrac{5}{8}x - \dfrac{1}{3}$
51. $y = \dfrac{2}{5}x + 5$
52. $y = -\dfrac{3}{4}x + 7$
53. (a) $y = -x + 12$
(b) -1 (c) $(0, 12)$
54. (a) $y = x - 14$
(b) 1 (c) $(0, -14)$
55. (a) $y = -\dfrac{5}{2}x + 10$
(b) $-\dfrac{5}{2}$ (c) $(0, 10)$
56. (a) $y = -\dfrac{6}{5}x + 8$
(b) $-\dfrac{6}{5}$ (c) $(0, 8)$
57. (a) $y = \dfrac{2}{3}x - \dfrac{10}{3}$
(b) $\dfrac{2}{3}$ (c) $\left(0, -\dfrac{10}{3}\right)$
58. (a) $y = \dfrac{4}{3}x - \dfrac{10}{3}$
(b) $\dfrac{4}{3}$ (c) $\left(0, -\dfrac{10}{3}\right)$

Write the equation of the line satisfying the given conditions. See Example 1.

17. through $(-2, 4)$; $m = -\dfrac{3}{4}$

18. through $(-1, 6)$; $m = -\dfrac{5}{6}$

19. through $(5, 8)$; $m = -2$

20. through $(12, 10)$; $m = 1$

21. through $(-5, 4)$; $m = \dfrac{1}{2}$

22. through $(7, -2)$; $m = \dfrac{1}{4}$

23. through $(-4, 12)$; horizontal

24. through $(1, 5)$; horizontal

25. x-intercept $(3, 0)$; $m = 4$

26. x-intercept $(-2, 0)$; $m = -5$

27. Explain why the point-slope form of an equation cannot be used to find the equation of a vertical line.

28. Which one of the following equations is in standard form, according to the definition of standard form given in this text?
   (a) $3x + 2y - 6 = 0$ (b) $y = 5x - 12$
   (c) $2y = 3x + 4$ (d) $6x - 5y = 12$

Write an equation in the form $x = k$ or $y = k$ for some constant k for the line described.

29. through $(9, 10)$; undefined slope

30. through $(-2, 8)$; undefined slope

31. through $(.5, .2)$; vertical

32. through $\left(\dfrac{5}{8}, \dfrac{2}{9}\right)$; vertical

33. through $(-7, 8)$; horizontal

34. through $(\sqrt{2}, \sqrt{7})$; horizontal

Write the standard form of the equation of the line passing through the two points. See Example 2.

35. $(3, 4)$ and $(5, 8)$

36. $(5, -2)$ and $(-3, 14)$

37. $(6, 1)$ and $(-2, 5)$

38. $(-2, 5)$ and $(-8, 1)$

▲ **39.** $\left(-\dfrac{2}{5}, \dfrac{2}{5}\right)$ and $\left(\dfrac{4}{3}, \dfrac{2}{3}\right)$

▲ **40.** $\left(\dfrac{3}{4}, \dfrac{8}{3}\right)$ and $\left(\dfrac{2}{5}, \dfrac{2}{3}\right)$

41. $(2, 5)$ and $(1, 5)$

42. $(-2, 2)$ and $(4, 2)$

43. $(7, 6)$ and $(7, -8)$

44. $(13, 5)$ and $(13, -1)$

▲ **45.** $(1, \sqrt{5})$ and $(3, 2\sqrt{5})$

▲ **46.** $(-4, \sqrt{2})$ and $(5, -\sqrt{2})$

Find the equation of the line satisfying the given conditions. Write it in slope-intercept form. See Example 3.

47. $m = 5$; $b = 15$

48. $m = -2$; $b = 12$

49. $m = -\dfrac{2}{3}$; $b = \dfrac{4}{5}$

50. $m = -\dfrac{5}{8}$; $b = -\dfrac{1}{3}$

51. slope $\dfrac{2}{5}$; y-intercept $(0, 5)$

52. slope $-\dfrac{3}{4}$; y-intercept $(0, 7)$

For the given equation (a) write in slope-intercept form, (b) give the slope of the line, and (c) give the y-intercept. See Example 4.

53. $x + y = 12$

54. $x - y = 14$

55. $5x + 2y = 20$

56. $6x + 5y = 40$

57. $2x - 3y = 10$

58. $4x - 3y = 10$

Write an equation in standard form of the line satisfying the given conditions. See Example 5.

59. through $(7, 2)$; parallel to $3x - y = 8$

60. through $(4, 1)$; parallel to $2x + 5y = 10$

61. through $(-2, -2)$; parallel to $-x + 2y = 10$

62. through $(-1, 3)$; parallel to $-x + 3y = 12$

63. through $(8, 5)$; perpendicular to $2x - y = 7$

64. through $(2, -7)$; perpendicular to $5x + 2y = 18$

65. through $(-2, 7)$; perpendicular to $x = 9$

66. through $(8, 4)$; perpendicular to $x = -3$

59. $3x - y = 19$
60. $2x + 5y = 13$
61. $x - 2y = 2$
62. $x - 3y = -10$
63. $x + 2y = 18$
64. $2x - 5y = 39$
65. $y = 7$ 66. $y = 4$
67. $-22.125x + 220$
68. $y = 158.5x + 575$
69. $y = -\dfrac{791}{6}x + 29,750$

Many real-world situations can be modeled approximately by straight-line graphs. One way to find the equation of such a line is to use two typical data points from the information provided, and then apply the point-slope form of the equation of a line. Because of the usefulness of the slope-intercept form, such equations are often given in the form $y = mx + b$. Assume that the situation described can be modeled by a straight-line graph, and use the information to find the $y = mx + b$ form of the equation of the line.

67. Let $x = 0$ represent the year 1985. In 1985, U.S. sales of Volkswagen vehicles were about 220,000. In 1993, sales had declined to 43,000. Let y represent sales in thousands. (Thus, the data points for this model are $(0, 220)$ and $(8, 43)$.) (*Source:* Autodata)

68. Let $x = 0$ represent the year 1989. In 1989, the number of auto accidents resulting in "catastrophic claims"—those amounting to one hundred thousand dollars or more—led to an average expected payment of \$575,000. In 1991, the average expected payment had risen to \$892,000. Let y represent the average expected payment in thousands of dollars. (*Source:* Insurance Research Council)

69. The number of post offices in the United States has been declining in recent years. Use the information given on the bar graph for the years 1984 and 1990, letting $x = 0$ represent the year 1984 and letting y represent the number of post offices. (*Source:* U.S. Postal Service, Annual Report of the Postmaster General)

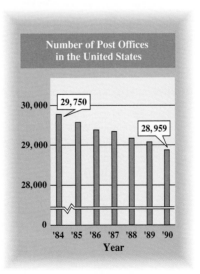

70. $y = -1545.5x + 49,391$
71. $y = 4x - 160$
72. $45°F$
73. (a) $-3x + 9 = 0$
(b) $\{3\}$
(c)

74. (a) $2x - 2 = 0$
(b) $\{1\}$
(c)

75. (a) $4x + 2 = 0$
(b) $\{-.5\}$
(c)

76. (a) $2x - 8 = 0$
(b) $\{4\}$
(c)

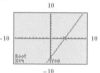

77. (d)
78. (a) $\{12\}$
(c)

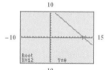

70. The number of motor vehicle deaths in the United States declined from 1988 to 1990 as seen in the accompanying bar graph. Use the information given, with $x = 0$ representing 1988 and y representing the number of deaths. (*Source*: National Safety Council)

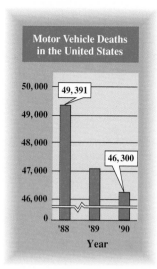

71. Erin is a biology student. She has heard that the number of times a cricket chirps in one minute can be used to find the temperature. In an experiment, she finds that a cricket chirps 40 times per minute when the temperature is 50°F, and 80 times per minute when the temperature is 60°F. Let x represent the temperature in degrees Fahrenheit, and let y represent the number of chirps per minute.

72. Refer to Exercise 71. What is the temperature when the cricket chirps 20 times per minute?

A linear equation in the variable x is given. Do each of the following.
(a) *Collect like terms and write the equation in the form* $y_1 = 0$, *where* y_1 *is an expression in x.*
(b) *Solve the equation found in part (a) using the methods of Chapter 2.*
(c) *Graph* y_1 *in the standard window of your graphics calculator, and show that the x-intercept corresponds to the solution you found in part (b).*
See Example 6.

73. $2x + 7 - x = 4x - 2$ **74.** $7x - 2x + 4 - 5 = 3x + 1$
75. $3(2x + 1) - 2(x - 2) = 5$ **76.** $4x - 3(4 - 2x) = 2(x - 3) + 6x + 2$

77. The graph of y_1 is shown in the *standard viewing window*. Which is the only choice that could possibly be the solution of the equation $y_1 = 0$?
(a) -15 **(b)** 0 **(c)** 5 **(d)** 15

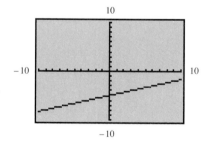

78. **(a)** Solve the equation $-2(x - 5) = -x - 2$ using the methods of Chapter 2.
   (b) Explain why the standard viewing window of a graphics calculator cannot give a graphical support of the solution found in part (a).
   (c) Use a window with x-minimum of -10 and x-maximum of 15 to show a graphical support for the solution found in part (a).

──────── ◈ **MATHEMATICAL CONNECTIONS** (Exercises 79–84) ◈ ────────

In Section 2.2 we learned how formulas can be applied to problem solving. In Exercises 79–84, we will see how the formula that relates the Celsius and Fahrenheit temperatures is derived. Work Exercises 79–84 in order.

79. There is a linear relationship between Celsius and Fahrenheit temperatures. When $C = 0°$, $F = $ _____ $°$, and when $C = 100°$, $F = $ _____ $°$.

80. Think of ordered pairs of temperatures (C, F), where C and F represent corresponding Celsius and Fahrenheit temperatures. The equation that relates the two scales has a straight-line graph that contains the two points determined in Exercise 79. What are these two points?

81. Find the slope of the line described in Exercise 80.

82. Now think of the point-slope form of the equation in terms of C and F, where C replaces x and F replaces y. Use the slope you found in Exercise 81 and one of the two points determined earlier, and find the equation that gives F in terms of C.

83. To obtain another form of the formula, use the equation you found in Exercise 82 and solve for C in terms of F.

84. The equation found in Exercise 82 is graphed on the graphics calculator screen shown here. Observe the display at the bottom, and interpret it in the context of this group of exercises.

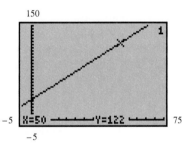

79. 32;   212   80. (0, 32)

and (100, 212)   81. $\dfrac{9}{5}$

82. $F = \dfrac{9}{5}C + 32$

83. $C = \dfrac{5}{9}(F - 32)$

84. When the Celsius temperature is 50°, the Fahrenheit temperature is 122°.   85. $(-\infty, 2)$

86. $(-\infty, 1)$

87. $\left(-\infty, -\dfrac{4}{3}\right]$ 88. $[0, \infty)$

──────── ◈ ────────

REVIEW EXERCISES

Solve the inequality. See Section 2.5.

85. $2x + 5 < 9$ **86.** $-x + 4 > 3$ **87.** $5 - 3x \geq 9$ **88.** $-x \leq 0$

7.4 LINEAR INEQUALITIES

FOR EXTRA HELP	OBJECTIVES
📖 **SSG** pp. 219–223 **SSM** pp. 352–356	**1** ▶ Graph linear inequalities.
📼 **Video** 9	**2** ▶ Graph the intersection of two linear inequalities.
💾 **Tutorial** IBM MAC	**3** ▶ Use a graphics calculator to solve linear inequalities. 🖩

1 ▶ Graph linear inequalities.

▶   Linear inequalities with one variable were graphed on the number line in Chapter 2. In this section linear inequalities in two variables are graphed on a rectangular coordinate system.

INSTRUCTOR'S RESOURCES

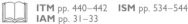 **ITM** pp. 440–442   **ISM** pp. 534–544   **TEST GENERATOR**   **TRANSPARENCIES**
IAM pp. 31–33   DOS   Windows   MAC

◉ CONCEPTUAL   ✎ WRITING   ▲ CHALLENGING   ▦ SCIENTIFIC CALCULATOR    GRAPHICS CALCULATOR

Linear Inequality	An inequality that can be written as

$$Ax + By < C \quad \text{or} \quad Ax + By > C,$$

where A, B, and C are real numbers and A and B are not both 0, is called a **linear inequality in two variables.**

Also, $\leq$ and $\geq$ may replace $<$ and $>$ in the definition.

A line divides the plane into three regions: the line itself and the two half-planes on either side of the line. Recall that the graphs of linear inequalities in one variable are intervals on the number line that sometimes include an endpoint. The graphs of linear inequalities in two variables are *regions* in the real number plane and may include a *boundary line*. The **boundary line** for the inequality $Ax + By < C$ or $Ax + By > C$ is the graph of the *equation $Ax + By = C$.* To graph a linear inequality, we go through the following steps.

Graphing a Linear Inequality	*Step 1* **Draw the boundary.** Draw the graph of the straight line that is the boundary. Make the line solid if the inequality involves $\leq$ or $\geq$; make the line dashed if the inequality involves $<$ or $>$.
	Step 2 **Choose a test point.** Choose any point not on the line as a test point.
	Step 3 **Shade the appropriate region.** Shade the region that includes the test point if it satisfies the original inequality; otherwise, shade the region on the other side of the boundary line.

E X A M P L E 1

Graphing a Linear Inequality

Graph $3x + 2y \geq 6$.

First graph the straight line $3x + 2y = 6$. The graph of this line, the boundary of the graph of the inequality, is shown in Figure 22. The graph of the inequality $3x + 2y \geq 6$ includes the points of the line $3x + 2y = 6$, and either the points *above* the line $3x + 2y = 6$ or the points *below* that line. To decide which, select any point not on the line $3x + 2y = 6$ as a test point. The origin, $(0, 0)$, is often a good choice. Substitute the values from the test point $(0, 0)$ for x and y in the inequality $3x + 2y > 6$.

$$3(0) + 2(0) > 6 \qquad ?$$
$$0 > 6 \qquad \text{False}$$

Chalkboard Exercise

Graph the solutions of
$x + y \leq 4.$

Answer:

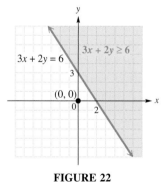

FIGURE 22

Since the result is false (0, 0) does not satisfy the inequality, and so the solution includes all points on the other side of the line. This region is shaded in Figure 22. ■

If the inequality is written in the form $y > mx + b$ or $y < mx + b$, the inequality symbol indicates which half-plane to shade.

If $y > mx + b$, shade *above* the boundary line;

if $y < mx + b$, shade *below* the boundary line.

EXAMPLE 2

Graphing a Linear Inequality

Graph the solutions of $x - 3y > 4$.

First graph the boundary line, shown in Figure 23. The points of the boundary line do not belong to the inequality $x - 3y > 4$ (since the inequality symbol is $>$, not $\geq$). For this reason, the line is dashed. Now solve the equation for y.

$$x - 3y > 4$$
$$-3y > -x + 4$$
$$y < \frac{x}{3} - \frac{4}{3} \qquad \text{Multiply by } -\frac{1}{3}; \text{ change the inequality.}$$

Because of the less than symbol, we should shade *below* the line. As a check, we can choose any point not on the line, say (1, 2) and substitute for x and y in the original inequality.

$$1 - 3(2) > 4 \qquad ?$$
$$-5 > 4 \qquad \text{False}$$

This result agrees with our decision to shade below the line. The solutions, graphed in Figure 23, include only those points in the shaded half-plane (not those on the line). ■

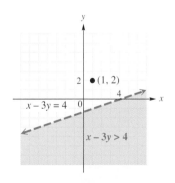

FIGURE 23

2 ▶ Graph the intersection of two linear inequalities.

▶ In Section 2.6 we discussed how the words "and" and "or" are used with compound inequalities. In that section, the inequalities were in a single variable. Those ideas can be extended to include inequalities in two variables. If a pair of inequalities is joined with the word "and," it is interpreted as the intersection of the solutions of the inequalities. The graph of the intersection of two or more inequalities is the region of the plane where all points satisfy all of the inequalities at the same time.

E X A M P L E   3

Graphing the Intersection of Two Inequalities

Graph the solutions of $2x + 4y \geq 5$ and $x \geq 1$.

To begin, we graph each of the two inequalities $2x + 4y \geq 5$ and $x \geq 1$ separately. The graph of $2x + 4y \geq 5$ is shown in Figure 24(a), and the graph of $x \geq 1$ is shown in Figure 24(b).

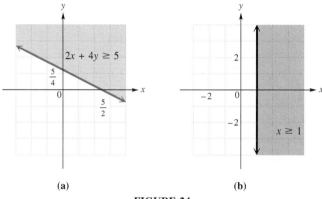

(a)   (b)

FIGURE 24

In practice, the two graphs in Figure 24 are graphed on the same axes. Then we use heavy shading to identify the intersection of the graphs, as shown in Figure 25.

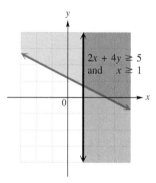

FIGURE 25

1. $x \leq 200$, $x \geq 100$, $y \geq 3000$

2.

3. $C = 50x + 100y$

4. Some examples are (100, 5000), (150, 3000), and (150, 5000). The corner points are (100, 3000) and (200, 3000).

◆ **C O N N E C T I O N S** ◆

Many realistic problems involve inequalities. For example, suppose a factory can have *no more than* 200 workers on a shift, but must have *at least* 100 and must manufacture *at least* 3000 units at minimum cost. The managers need to know how many workers should be on a shift in order to produce the required units at minimal cost. *Linear programming* is a method for finding the optimal (best possible) solution that meets all the conditions for such problems. The first step in solving linear programming problems with two variables is to express the conditions (constraints) as inequalities, graph the system of inequalities, and identify the region that satisfies all the inequalities at once.

5. The least cost occurs when $x = 100$ and $y = 3000$. The company should use 100 workers and manufacture 3000 units to achieve the lowest possible cost.

FOR DISCUSSION OR WRITING

Let x represent the number of workers and y represent the number of units manufactured.

1. Write three inequalities expressing the conditions in the problem given above.
2. Graph the inequalities from Item 1 and shade the intersection.
3. The cost per worker is \$50 per day and the cost to manufacture 1 unit is \$100. Write an expression representing the total daily cost, C.
4. Find values of x and y for several points in or on the boundary of the shaded region. Include any "corner points."
5. Of the values of x and y that you chose in Item 4, which gives the least cost when substituted in the cost equation from Item 3? What does your answer mean in terms of the given problem? Is your answer reasonable? Explain why.

---

3 ▶ Use a graphics calculator to solve linear inequalities.

▶ In Section 7.3 we observed that the x-intercept of the graph of the line $y = mx + b$ is the solution of the equation $mx + b = 0$. We can extend our observations to find solutions of the associated inequalities $mx + b > 0$ and $mx + b < 0$. The solution set of $mx + b > 0$ is the set of all x-values for which the graph of $y = mx + b$ is *above* the x-axis. (We consider points above because the symbol is $>$.) On the other hand, the solution set of $mx + b < 0$ is the set of all x-values for which the graph of $y = mx + b$ is *below* the x-axis. (We consider points below because the symbol is $<$.) Therefore, once we know the solution set of the equation and have the graph of the line, we can determine the solution sets of the corresponding inequalities.

In Figure 26, we see that the x-intercept of $y_1 = 3x - 9$ is 3. Therefore, as shown in Section 7.3, the solution set of $3x - 9 = 0$ is $\{3\}$. Because the graph of y_1 lies above the x-axis for x-values greater than 3, the solution set of $3x - 9 > 0$ is $(3, \infty)$. Because the graph lies below the x-axis for x-values less than 3, the solution set of $3x - 9 < 0$ is $(-\infty, 3)$.

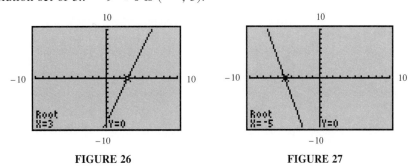

FIGURE 26   FIGURE 27

Suppose that we wish to solve the equation $-2(3x + 1) = -2x + 18$, and the associated inequalities $-2(3x + 1) > -2x + 18$ and $-2(3x + 1) < -2x + 18$. We begin by considering the equation rewritten so that the right side is equal to 0: $-2(3x + 1) + 2x - 18 = 0$. Graphing $y_1 = -2(3x + 1) + 2x - 18$ yields the x-intercept -5 as shown in Figure 27. The first inequality listed above is equivalent to $y_1 > 0$. Because the line lies *above* the x-axis for x-values less than -5, the solution set of $-2(3x + 1) > -2x + 18$ is $(-\infty, -5)$. Because the line lies *below* the x-axis for x-values greater than -5, the solution set of $-2(3x + 1) < -2x + 18$ is $(-5, \infty)$.

7.4 EXERCISES

Write the word solid *if the boundary line is included in the graph of the linear inequality in two variables. Write* dashed *if the boundary line is not included.*

1. $x + y \leq 2$ **2.** $-x + y \leq 3$ **3.** $4x - y < 5$ **4.** $3x + y > 6$
1. solid   2. solid   3. dashed   4. dashed

Graph the linear inequality in two variables. See Examples 1 and 2.

5. $x + y \leq 2$ **6.** $x + y \leq -3$ **7.** $4x - y < 4$

5.   6.   7.

8. $3x - y < 3$ **9.** $x + 3y \geq -2$ **10.** $x + 4y \geq -3$

8.   9.   10.

11. $x + y > 0$ **12.** $x + 2y > 0$ **13.** $x - 3y \leq 0$

11.   12.   13.

14. $x - 5y \leq 0$ **15.** $y < x$ **16.** $y \leq 4x$

14.   15.   16.

Graph the compound inequality. See Example 3.

17. $x + y \leq 1$ and $x \geq 1$ **18.** $x - y \geq 2$ and $x \geq 3$

17.   18.

19. $2x - y \geq 2$ and $y < 4$ **20.** $3x - y \geq 3$ and $y < 3$

19.   20.

21. $x + y > -5$ and $y < -2$ **22.** $6x - 4y < 10$ and $y > 2$

21.   22.

 CONCEPTUAL   WRITING   CHALLENGING   SCIENTIFIC CALCULATOR   GRAPHICS CALCULATOR

▲ *Use the method described in Section 2.7 to write the absolute value inequality as an "and" statement. Then solve the compound inequality and graph its solution set in the rectangular coordinate plane.*

23. $|x| < 3$
23. $-3 < x < 3$

24. $|y| < 5$
24. $-5 < y < 5$

25. $|x + 1| < 2$
25. $-2 < x + 1 < 2$

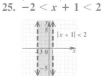

26. $|y - 3| < 2$
26. $-2 < y - 3 < 2$

▲ *When a compound inequality involves the word* or, *the graph is found by graphing each individual inequality and then taking the* union *of the two. For example, the graph of*

$$2x + 4y \geq 5 \quad \text{or} \quad x \geq 1$$

is shown here. Use this idea to graph the compound inequality.

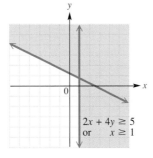

27. $x - y \geq 1$ or $y \geq 2$

28. $x + y \leq 2$ or $y \geq 3$

29. $x - 2 > y$ or $x < 1$

30. $x + 3 < y$ or $x > 3$

31. $3x + 2y < 6$ or $x - 2y > 2$

32. $x - y \geq 1$ or $x + y \leq 4$

33. (a) $\{-4\}$
(b) $(-\infty, -4)$ (c) $(-4, \infty)$
34. (a) $\{6\}$ (b) $(-\infty, 6)$
(c) $(6, \infty)$
35. (a) $\{3.5\}$ (b) $(3.5, \infty)$
(c) $(-\infty, 3.5)$
36. $\{-1.25\}$ (b) $(-1.25, \infty)$
(c) $(-\infty, -1.25)$
37. (a) $\{-.6\}$ (b) $(-.6, \infty)$
(c) $(-\infty, -.6)$
The graph of $y_1 = 5x + 3$ has x-intercept $-.6$, supporting the result of part (a). The graph of y_1 lies *above* the x-axis for values of x greater than $-.6$, supporting the result of part (b). The graph of y_1 lies *below* the x-axis for values of x less than $-.6$, supporting the result of part (c).

38. (a) $\{-.5\}$
(b) $(-.5, \infty)$
(c) $(-\infty, -.5)$
The graph of $y_1 = 6x + 3$ has x-intercept $-.5$, supporting the result of part (a). The graph of y_1 lies *above* the x-axis for values of x *greater than* $-.5$, supporting the result of part (b). The graph of y_1 lies *below* the x-axis for values of x *less than* $-.5$, supporting the result of part (c).

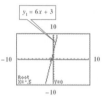

39. (a) $\{-1.2\}$
(b) $(-\infty, -1.2]$
(c) $[-1.2, \infty)$
The graph of $y_1 = -8x - (2x + 12)$ has x-intercept -1.2, supporting the result of part (a). The graph of y_1 lies *above or on* the x-axis for values of x *less than or equal to* -1.2, supporting the result of part (b). The graph of y_1 lies *below or on* the x-axis for values of x *greater than or equal to* -1.2, supporting the result of part (c).

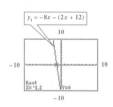

The graph of y_1 is shown on a graphics calculator screen, along with the x-intercept of the line. Use the screen to solve (a) the equation $y_1 = 0$, (b) the inequality $y_1 < 0$, and (c) the inequality $y_1 > 0$.

33.

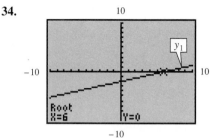

34.

35.

36.

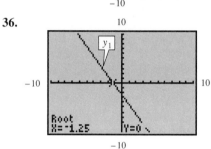

Solve the equation in part (a) and the associated inequalities in parts (b) and (c) using the methods of Chapter 2. Then graph the left side as y_1 in the standard viewing window of a graphics calculator, and explain how the graph supports your answers in parts (a), (b), and (c).

37. (a) $5x + 3 = 0$
(b) $5x + 3 > 0$
(c) $5x + 3 < 0$

38. (a) $6x + 3 = 0$
(b) $6x + 3 > 0$
(c) $6x + 3 < 0$

39. (a) $-8x - (2x + 12) = 0$
(b) $-8x - (2x + 12) \geq 0$
(c) $-8x - (2x + 12) \leq 0$

40. (a) $-4x - (2x + 18) = 0$
(b) $-4x - (2x + 18) \geq 0$
(c) $-4x - (2x + 18) \leq 0$

Some graphics calculators have the capability of graphing inequalities by shading above or below the boundary. For example, the graph of $y > -2x + 3$ appears in the standard viewing window as shown here.

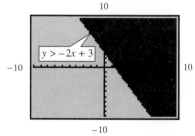

Use a graphics calculator to graph the inequality in the standard window.

41. $y > -3x + 2$

42. $y > 3x - 2$

41.

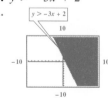

42.

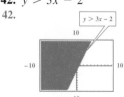

43. $y < .5x + 3$

43.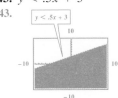

44. $y < .75x - 5$

44.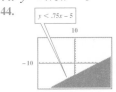

40. (a) $\{-3\}$
(b) $(-\infty, -3]$
(c) $[-3, \infty)$
The graph of $y_1 = -4x - (2x + 18)$ has x-intercept -3, supporting the result of part (a). The graph of y_1 lies *above or on* the x-axis for values of x less than or equal to -3, supporting the result of part (b). The graph of y_1 lies *below or on* the x-axis for values of x *greater than or equal to -3*, supporting the result of part (c).

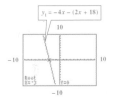

REVIEW EXERCISES

Evaluate y when x = 3. See Section 7.1.

45. $y = -7x + 12$
45. -9

46. $y = -5x - 4$
46. -19

47. $y = 3x - 8$
47. 1

48. $y = 2x - 9$
48. -3

Solve for y. See Section 2.2.

49. $3x - 7y = 8$
49. $y = \dfrac{3}{7}x - \dfrac{8}{7}$

50. $2x - 4y = 7$
50. $y = \dfrac{1}{2}x - \dfrac{7}{4}$

51. $\dfrac{1}{2}x - 4y = 5$
51. $y = \dfrac{1}{8}x - \dfrac{5}{4}$

52. $\dfrac{3}{4}x + 2y = 9$
52. $y = -\dfrac{3}{8}x + \dfrac{9}{2}$

7.5 INTRODUCTION TO FUNCTIONS

FOR EXTRA HELP	OBJECTIVES
SSG pp. 224–229 **SSM** pp. 356–362 **Video** 9 **Tutorial** IBM MAC	**1 ▶** Define and identify relations and functions. **2 ▶** Find the domain and range of a relation. **3 ▶** Use the vertical line test. **4 ▶** Use $f(x)$ notation. **5 ▶** Identify linear functions.

It is often useful to describe one quantity in terms of another. For example, the growth of a plant is related to the amount of light it receives, the demand for a product is related to the price of the product, the cost of a trip is related to the distance traveled, and so on. To represent these corresponding quantities, it is helpful to use ordered pairs.

For example, we can indicate the relationship between the demand for a product and its price by writing ordered pairs in which the first number represents the price and the second number represents the demand. The ordered pair (5, 1000) then could indicate a demand for 1000 items when the price of the item is $5. Since the demand depends on the price charged, we place the price first and the demand second. The ordered pair is an abbreviation for the sentence "If the price is 5 (dollars), then the demand is for 1000 (items)." Similarly, the ordered pairs (3, 5000) and (10, 250) show that a price of $3 produces a demand for 5000 items, and a price of $10 produces a demand for 250 items.

INSTRUCTOR'S RESOURCES

 ITM pp. 442–447 **ISM** pp. 544–555
 IAM pp. 33–34

 TEST GENERATOR
 DOS Windows MAC

 TRANSPARENCIES

 CONCEPTUAL WRITING ▲ CHALLENGING SCIENTIFIC CALCULATOR GRAPHICS CALCULATOR

In this example, the demand depends on the price of the item. For this reason, demand is called the *dependent variable,* and price the *independent variable.* Generalizing, if the value of the variable y depends on the variable x, then y is the **dependent variable** and x the **independent variable.**

1 ▶ Define and identify relations and functions.

▶ Since related quantities can be written using ordered pairs, the concept of *relation* can be defined as follows.

Relation	A **relation** is a set of ordered pairs.

A special kind of relation, called a *function,* is very important in mathematics and its applications.

Function	A **function** is a relation in which, for each value of the first component of the ordered pairs, there is exactly one value of the second component.

Functions can be expressed in many ways—for example, as sets of ordered pairs, mappings, graphs, or equations.

EXAMPLE 1

Determining Whether a Relation Is a Function

Determine whether each of the following relations is a function.

$$F = \{(1, 2), (-2, 5), (3, -1)\}$$
$$G = \{(-2, 1), (-1, 0), (0, 1), (1, 2), (2, 2)\}$$
$$H = \{(-4, 1), (-2, 1), (-2, 0)\}$$

Relations F and G are functions, because for each x-value, there is only one y-value. Notice that in G, the last two ordered pairs have the same y-value. This does not violate the definition of function, since each first component (x-value) has only one second component (x-value).

Relation H is not a function, because the last two ordered pairs have the same x-value, but different y-values. ■

EXAMPLE 2

Recognizing a Function Expressed as a Mapping

A function can also be expressed as a correspondence or *mapping* from one set to another. The mapping in Figure 28 is a function that assigns to a state its population (in millions) expected by the year 2000. ■

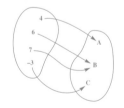
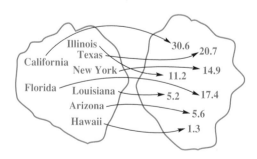

FIGURE 28

◆ Answers will vary.

◆ **C O N N E C T I O N S** ◆

Many of the Connections boxes throughout this book have discussed models of real situations or formulas. In each case we have been able to predict (or find the exact value of) one variable given the value of one or more other variables. In Chapter 3, we gave the formula $h = -16t^2 + 128t$ for the distance in feet an object will fall in t seconds under certain conditions. In Chapter 4, the expression for the number of gallons of heating oil,

$$\frac{125,000 - 25x}{125 + 2x},$$

produced for every x gallons of gasoline was discussed. Then in Chapter 6, the expression $-.011x^2 + 1.22x - 8.5$ approximated union membership (in millions) for year x. In the last two examples, we can set the expression equal to y, so that each of these examples has two variables. In each case, substituting a value for x (or t) gives exactly one corresponding value for y (or h). Thus, each of these is an example of a function that applies to a real situation. These examples illustrate the importance of the function concept, one of the most important concepts in mathematics.

FOR DISCUSSION OR WRITING
Look back through the first six chapters of this book to find other examples of the function concept. Remember, if an algebraic expression represents a real quantity, it can be set equal to y, so that it is expressed as a function.

2 ► Find the domain and range of a relation.

► **TEACHING TIP**

Emphasize that the domain and the range are *sets*. ◄

► The set of all first components (x-values) of the ordered pairs of a relation is called the **domain** of the relation, and the set of all second components (y-values) is called the **range**. For example, the domain of function F in Example 1 is $\{1, -2, 3\}$; the range is $\{2, 5, -1\}$. Also, the domain of function G is $\{-2, -1, 0, 1, 2\}$ and the range is $\{0, 1, 2\}$. Domains and ranges can also be defined in terms of independent and dependent variables.

Domain and Range

In a relation, the set of all values of the independent variable (x) is the **domain;** the set of all values of the dependent variable (y) is the **range.**

E X A M P L E   3

Finding Domains and Ranges

Give the domain and range of each function.

(a) $\{(3, -1), (4, 2), (0, 5)\}$
  The domain, the set of x-values, is $\{3, 4, 0\}$; the range is the set of y-values, $\{-1, 2, 5\}$.

(b) The function in Figure 28
  The domain is {Illinois, Texas, California, New York, Florida, Louisiana, Arizona, Hawaii} and the range is $\{1.3, 5.2, 5.6, 11.2, 14.9, 17.4, 20.7, 30.6\}$. ■

The **graph of a relation** is the graph of its ordered pairs. The graph gives a picture of the relation.

E X A M P L E   4
Finding Domains and
Ranges from Graphs

Three relations are graphed in Figure 29. Give the domain and range of each.

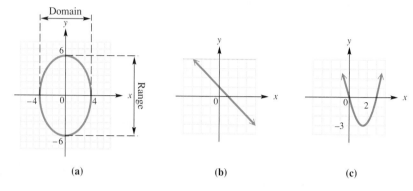

(a)   (b)   (c)

FIGURE 29

Chalkboard Exercise

Give the domain and range of
the function.

Answer: domain: $(-\infty, \infty)$;
range: $(-\infty, 4]$

(a) In Figure 29(a), the x-values of the points on the graph include all numbers between -4 and 4, inclusive. The y-values include all numbers between -6 and 6, inclusive. Using interval notation,

the domain is $[-4, 4]$;
the range is $[-6, 6]$.

(b) In Figure 29(b), the arrowheads indicate that the line extends indefinitely left and right, as well as up and down. Therefore, both the domain and the range are the set of all real numbers, written $(-\infty, \infty)$.

(c) In Figure 29(c), the arrowheads indicate that the graph extends indefinitely left and right, as well as upward. The domain is $(-\infty, \infty)$. Because there is a least y-value, -3, the range includes all numbers greater than or equal to -3, written $[-3, \infty)$.   ∎

Relations are often defined by equations such as $y = 2x + 3$ and $y^2 = x$. It is sometimes necessary to determine the domain of a relation from its equation. In this book, the following agreement on the domain of a relation is assumed.

Agreement on Domain   The domain of a relation is assumed to be all real numbers that produce real numbers when substituted for the independent variable.

To illustrate this agreement, since any real number can be used as a replacement for x in $y = 2x + 3$, the domain of this function is the set of real numbers. As another example, the function defined by $y = 1/x$ has all real numbers except 0 as a domain, since y is undefined if $x = 0$. In general, the domain of a function defined by an algebraic expression is all real numbers except those numbers that lead to division by zero or an even root of a negative number.

EXAMPLE 5

Identifying Functions and Determining Domains

For each of the following, decide whether it defines a function and give the domain.

(a) $y = \sqrt{2x - 1}$

Here, for any choice of x in the domain, there is exactly one corresponding value for y (the radical is a nonnegative number), so this equation defines a function. Since the radicand cannot be negative,

$$2x - 1 \geq 0$$
$$2x \geq 1$$
$$x \geq \frac{1}{2}.$$

The domain is $\left[\frac{1}{2}, \infty\right)$.

(b) $y^2 = x$

The ordered pairs $(16, 4)$ and $(16, -4)$ both satisfy this equation. Since one value of x, 16, corresponds to two values of y, 4 and -4, this equation does not define a function. Solving $y^2 = x$ for y gives $y = \sqrt{x}$ or $y = -\sqrt{x}$, which shows that two values of y correspond to each positive value of x. Because x is equal to the square of y, the values of x must always be nonnegative. The domain is $[0, \infty)$.

(c) $y \leq x - 1$

By definition, y is a function of x if a value of x leads to exactly one value of y. In this example, a particular value of x, say 1, corresponds to many values of y. The ordered pairs $(1, 0), (1, -1), (1, -2), (1, -3)$, and so on, all satisfy the inequality. For this reason, this inequality does not define a function. Any number can be used for x, so the domain is the set of real numbers $(-\infty, \infty)$.

(d) $y = \dfrac{5}{x^2 - 1}$

Given any value of x in the domain, we find y by squaring x, subtracting 1, then dividing the result into 5. This process produces exactly one value of y for each value in the domain, so this equation defines a function. The domain includes all real numbers except those that make the denominator zero. We find these numbers by setting the denominator equal to zero and solving for x.

$$x^2 - 1 = 0$$
$$x^2 = 1$$
$$x = 1 \quad \text{or} \quad x = -1 \qquad \text{Square root property}$$

Thus, the domain includes all real numbers except 1 and -1. In interval notation this is written as

$$(-\infty, -1) \cup (-1, 1) \cup (1, \infty). \quad \blacksquare$$

In summary, three variations of the definition of function are given here.

Variations of the Definition of Function

1. A **function** is a relation in which, for each value of the first component of the ordered pairs, there is exactly one value of the second component.
2. A **function** is a set of ordered pairs in which no first component is repeated.
3. A **function** is a rule or correspondence that assigns exactly one range value to each domain value.

3 ▶ Use the vertical line test.

▶ In a function each value of *x* leads to only one value of *y*, so that any vertical line drawn through the graph of a function would have to cut the graph in at most one point. This is the **vertical line test for a function.**

Vertical Line Test	If a vertical line cuts the graph of a relation in more than one point, then the relation does not represent a function.

For example, the graph shown in Figure 30(a) is not the graph of a function, since a vertical line can cut the graph in more than one point, while the graph in Figure 30(b) does represent a function.

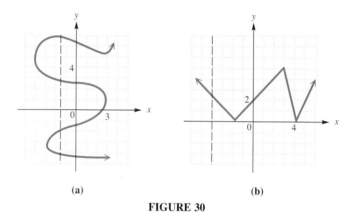

(a) (b)

FIGURE 30

4 ▶ Use $f(x)$ notation.

▶ **TEACHING TIP**

Emphasize that y and $f(x)$ are equivalent.

It is helpful to review the order of operations before giving examples using functional notation. ◀

▶ To say that *y* is a function of *x* means that for each value of *x* from the domain of the function, there is exactly one value of *y*. To emphasize that *y is a function of x,* or that *y* depends on *x,* it is common to write

$$y = f(x),$$

with $f(x)$ read "*f* of *x*." (In this special notation, the parentheses do not indicate multiplication.) The letter *f* stands for *function.* For example, if $y = 9x - 5$, we emphasize that *y* is a function of *x* by writing $y = 9x - 5$ as

$$f(x) = 9x - 5.$$

This **functional notation** can be used to simplify certain statments. For example, if $y = 9x - 5$, then replacing *x* with 2 gives

$$y = 9 \cdot 2 - 5$$
$$= 18 - 5$$
$$= 13.$$

The statement "if $x = 2$, then $y = 13$" is abbreviated with functional notation as

$$f(2) = 13.$$

Also, $f(0) = 9 \cdot 0 - 5 = -5$, and $f(-3) = -32$.

These ideas and the symbols used to represent them can be explained as follows.

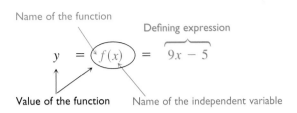

Name of the function

Defining expression

$$y \;=\; \boxed{f(x)} \;=\; \overbrace{9x - 5}$$

Value of the function   Name of the independent variable

CAUTION The symbol $f(x)$ *does not* indicate "f times x," but represents the y-value for the indicated x-value. As shown above, $f(2)$ is the y-value that corresponds to the x-value 2.

EXAMPLE 6
Using Functional Notation

Let $f(x) = -x^2 + 5x - 3$. Find the following.

(a) $f(2)$
   Replace x with 2.
$$f(2) = -2^2 + 5 \cdot 2 - 3 = -4 + 10 - 3 = 3$$

(b) $f(-1)$
$$f(-1) = -(-1)^2 + 5(-1) - 3 = -1 - 5 - 3 = -9$$

(c) $f(q)$
   Replace x with q.
$$f(q) = -q^2 + 5q - 3$$

The replacement of one variable with another is important in later courses. ■

If a function is defined by an equation with x and y, not with functional notation, use the following steps to find $f(x)$.

Finding an Expression for $f(x)$

If an equation that defines a function is given with x and y, to find $f(x)$:

1. solve the equation for y;
2. replace y with $f(x)$.

EXAMPLE 7
Writing Equations Using Functional Notation

Rewrite each equation using functional notation; then find $f(-2)$ and $f(a)$.

(a) $y = x^2 + 1$
   This equation is already solved for y. Since $y = f(x)$,
$$f(x) = x^2 + 1.$$

To find $f(-2)$, we let $x = -2$:
$$f(-2) = (-2)^2 + 1$$
$$= 4 + 1$$
$$= 5.$$

We find $f(a)$ by letting $x = a$: $f(a) = a^2 + 1$.

(b) $x - 4y = 5$

First solve $x - 4y = 5$ for y.

$$x - 4y = 5$$
$$x - 5 = 4y$$
$$y = \frac{x - 5}{4} \quad \text{so} \quad f(x) = \frac{1}{4}x - \frac{5}{4}$$

Now find $f(-2)$ and $f(a)$.

$$f(-2) = \frac{1}{4}(-2) - \frac{5}{4} = -\frac{7}{4}$$

and

$$f(a) = \frac{1}{4}a - \frac{5}{4} \quad ■$$

5 ▶ Identify linear functions.

▶ Our first two-dimensional graphing was of straight lines. Straight-line graphs (except for vertical lines) are graphs of *linear functions*.

Linear Function

A function that can be written in the form

$$f(x) = mx + b$$

for real numbers m and b is a **linear function.**

Recall from Section 7.3 that m is the slope of the line and $(0, b)$ is the y-intercept. A linear function of the form $f(x) = k$ is sometimes called a **constant function.** The domain of any linear function is $(-\infty, \infty)$. The range of a nonconstant linear function is $(-\infty, \infty)$, while the range of the constant function $f(x) = k$ is $\{k\}$.

In later chapters of this book, we learn about several other types of functions.

7.5 EXERCISES

2. $\{(-3, 4), (2, 4), (2, 6), (6, 4)\}$ (There are other correct answers.)
3. function; domain: $\{5, 3, 4, 7\}$; range: $\{1, 2, 9, 3\}$
4. function; domain: $\{8, 5, 9, 3\}$; range: $\{0, 4, 3, 9\}$
5. not a function; domain: $\{2, 0\}$; range: $\{4, 2, 6\}$
6. not a function; domain: $\{9, -3\}$; range: $\{-2, 5, 1\}$
7. not a function; domain: $\{1, 2, 3, 5\}$; range: $\{10, 15, 19, -27\}$
8. function; domain: $\{9, 11, 4, 17, 25\}$; range: $\{32, 47, -69, 14\}$

◉ **1.** Explain the meaning of the term.
 (a) relation **(b)** domain of a relation **(c)** range of a relation **(d)** function

◉ **2.** Give an example of a relation that is not a function having domain $\{-3, 2, 6\}$ and range $\{4, 6\}$. (There are many possible correct answers.)

Decide whether the relation is a function and give the domain and the range of the relation. Use the vertical line test in Exercises 9–12. See Examples 1–4.

3. $\{(5, 1), (3, 2), (4, 9), (7, 3)\}$

4. $\{(8, 0), (5, 4), (9, 3), (3, 9)\}$

5. $\{(2, 4), (0, 2), (2, 6)\}$

6. $\{(9, -2), (-3, 5), (9, 1)\}$

7.
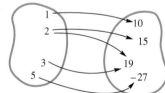

8.

9.
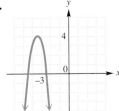

10.
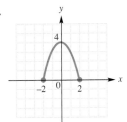

11.
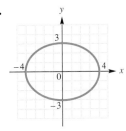

12.
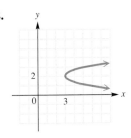

Decide whether the given equation defines y as a function of x. Give the domain. Identify any linear functions. See Example 5.

13. $y = x^2$ **14.** $y = x^3$ **15.** $x = y^2$ **16.** $x = y^4$

17. $x + y < 4$ **18.** $x - y < 3$ **19.** $y = \sqrt{x}$ **20.** $y = -\sqrt{x}$

21. $xy = 1$ **22.** $xy = -3$ **23.** $y = 2x - 6$ **24.** $y = -6x + 8$

25. $y = \sqrt{4x + 2}$ **26.** $y = \sqrt{9 - 2x}$ **27.** $y = \dfrac{2}{x^2 - 9}$ **28.** $y = \dfrac{-7}{x^2 - 16}$

Let $f(x) = -3x + 4$ and $g(x) = -x^2 + 4x + 1$. Find the following. See Example 6.

29. $f(0)$ **30.** $f(-3)$ ▲ **31.** $f(-x)$ ▲ **32.** $f(x + 2)$

33. $g(10)$ **34.** $g(-2)$ **35.** $g(-4)$ **36.** $g(1)$

▲ **37.** $g(k)$ ▲ **38.** $g(2p)$ ▲ **39.** $f(g(1))$ ▲ **40.** $g(f(1))$

▲ ◉ **41.** Compare the answers to Exercises 39 and 40. Do you think that $f(g(x))$ is, in general, equal to $g(f(x))$?

▲ ◉ **42.** Make up two linear functions f and g such that $f(g(2)) = 4$. (There are many ways to do this.)

An equation that defines y as a function of x is given. (a) Solve for y in terms of x, and replace y with the functional notation $f(x)$. (b) Find $f(3)$. See Examples 6 and 7.

43. $x + 3y = 12$ **44.** $x - 4y = 8$ **45.** $y + 2x^2 = 3$

46. $y - 3x^2 = 2$ **47.** $4x - 3y = 8$ **48.** $-2x + 5y = 9$

◉ **49.** Fill in the blanks with the correct responses.

The equation $2x + y = 4$ has a straight _____ as its graph. One point that lies on the line is (3, _____). If we solve the equation for y and use functional notation, we have a linear function $f(x) = $ _____. For this function, $f(3) = $ _____, meaning that the point (_____, _____) lies on the graph of the function.

◉ **50.** Which one of the following defines a linear function?

(a) $y = \dfrac{x - 5}{4}$ **(b)** $y = \sqrt[3]{x}$ **(c)** $y = x^2$ **(d)** $y = x^{1/2}$

◉ **51.** Which one of the functions in Exercise 50 has domain $[0, \infty)$?

◉ **52.** Which one of the functions in Exercise 50 does not have negative numbers in its domain?

◉ CONCEPTUAL   ▢ WRITING   ▲ CHALLENGING   ▦ SCIENTIFIC CALCULATOR   ▦ GRAPHICS CALCULATOR

9. function; domain: $(-\infty, \infty)$; range: $(-\infty, 4]$
10. function; domain: $[-2, 2]$; range: $[0, 4]$
11. not a function; domain: $[-4, 4]$; range: $[-3, 3]$
12. not a function; domain: $[3, \infty)$; range: $(-\infty, \infty)$
13. function; domain: $(-\infty, \infty)$ 14. function; domain: $(-\infty, \infty)$
15. not a function; domain: $[0, \infty)$ 16. not a function; domain: $[0, \infty)$
17. not a function; domain: $(-\infty, \infty)$
18. not a function; domain: $(-\infty, \infty)$
19. function; domain: $[0, \infty)$ 20. function; domain: $[0, \infty)$
21. function; domain: $(-\infty, 0) \cup (0, \infty)$
22. function; domain: $(-\infty, 0) \cup (0, \infty)$
23. function (also a linear function); domain: $(-\infty, \infty)$
24. function (also a linear function); domain: $(-\infty, \infty)$
25. function; domain: $\left[-\dfrac{1}{2}, \infty\right)$
26. function; domain: $\left(-\infty, \dfrac{9}{2}\right]$
27. function; domain: $(-\infty, -3) \cup (-3, 3) \cup (3, \infty)$
28. function; domain: $(-\infty, -4) \cup (-4, 4) \cup (4, \infty)$
29. 4   30. 13
31. $3x + 4$ 32. $-3x - 2$
33. -59 34. -11
35. -31 36. 4
37. $-k^2 + 4k + 1$
38. $-4p^2 + 8p + 1$
39. -8 40. 4
41. No—in general, $f(g(x)) \neq g(f(x))$.
42. One example is $f(x) = 12 - x$ and $g(x) = 2x + 4$.

43. (a) $f(x) = \dfrac{12 - x}{3}$

(b) 3

44. (a) $f(x) = \dfrac{8 - x}{-4}$

(b) $-\dfrac{5}{4}$

45. (a) $f(x) = 3 - 2x^2$
(b) -15

46. (a) $f(x) = 2 + 3x^2$
(b) 29   47. (a) $f(x) =$
$\dfrac{8 - 4x}{-3}$ (b) $\dfrac{4}{3}$

48. (a) $f(x) = \dfrac{9 + 2x}{5}$

(b) 3   49. line; -2;
$-2x + 4$; -2;   3; -2
50. (a)   51. (d)   52. (d)
53. domain: $(-\infty, \infty)$;
range: $(-\infty, \infty)$

54. domain: $(-\infty, \infty)$;
range: $(-\infty, \infty)$

55. domain: $(-\infty, \infty)$;
range: $(-\infty, \infty)$

56. domain: $(-\infty, \infty)$;
range: $(-\infty, \infty)$

57. domain: $(-\infty, \infty)$;
range: $(-\infty, \infty)$

58. domain: $(-\infty, \infty)$;
range: $(-\infty, \infty)$

Sketch the graph of the linear function. Give the domain and range.

53. $f(x) = -2x + 5$

54. $g(x) = 4x - 1$

55. $h(x) = \dfrac{1}{2}x + 2$

56. $F(x) = -\dfrac{1}{4}x + 1$

57. $G(x) = 2x$

58. $H(x) = -3x$

59. $f(x) = 5$

60. $g(x) = -4$

61. (a) Suppose that a taxicab driver charges \$.50 per mile. Fill in the chart with the correct response for the price $R(x)$ she charges for a trip of x miles.

x	$R(x)$
0	
1	
2	
3	

(b) The linear function that gives a rule for the amount charged is $R(x) =$ _____ .
(c) Graph this function for x, where x is an element of $\{0, 1, 2, 3\}$.

62. Suppose that a package weighing x pounds costs $C(x)$ dollars to mail to a given location, where

$$C(x) = 2.75x.$$

(a) What is the value of $C(3)$?
(b) In your own words, describe what 3 and the value $C(3)$ mean in part (a), using the terminology domain and range.

63. The graph shown here depicts spot prices in dollars per barrel for West Texas intermediate crude oil over a 10-day period during October, 1990.
(a) Is this the graph of a function?
(b) What is the domain?
(c) What is the range? (Round to nearest half dollar.)
(d) Estimate the price on October 24.
(e) On what day was oil at its highest price? Its lowest price?

64. The graph shown here depicts gasoline prices (per gallon) in the New Orleans metropolitan area for one month in late 1990.
(a) Is this the graph of a function?
(b) What is the domain?
(c) What is the range?
(d) Estimate the price on October 12.
(e) By how much had gasoline prices risen from the invasion of Kuwait on August 2 to October 25? (*Hint:* Look at the inserts.)

Oil Prices Crash
Spot prices for West Texas intermediate crude oil for last 10 days in October (per barrel)

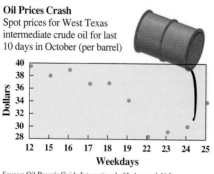

Source: Oil Buyer's Guide International of Lakewood, N.J.

Thursday's Gas Price: Down
Average prices (per gallon)*

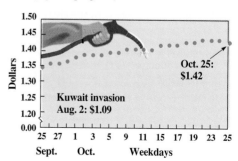

*Regular unleaded, average from 14 stations in the metro area.

Forensic scientists use the lengths of certain bones to calculate the height of a person. Two bones often used are the tibia (t), the bone from the ankle to the knee, and the femur (f), the bone from the knee to the hip socket. A person's height (h) is determined from the lengths of these bones using functions defined by the following formulas. All measurements are in centimeters.

$$\text{For men:} \quad h = 69.09 + 2.24f \quad \text{or} \quad h = 81.69 + 2.39t$$
$$\text{For women:} \quad h = 61.41 + 2.32f \quad \text{or} \quad h = 72.57 + 2.53t$$

65. Find the height of a man with a femur measuring 56 centimeters.

66. Find the height of a man with a tibia measuring 40 centimeters.

67. Find the height of a woman with a femur measuring 50 centimeters.

68. Find the height of a woman with a tibia measuring 36 centimeters.

 *Federal regulations set standards for the size of the quarters of marine mammals. A pool to house sea otters must have a volume of "the square of the sea otter's average adult length (in meters) multiplied by 3.14 and by .91 meters." If x represents the sea otter's average adult length and f(x) represents the volume of the corresponding pool size, this formula can be written as $f(x) = (.91)(3.14)x^2$. Find the volume of the pool for each of the following adult lengths (in meters). Round answers to the nearest hundredth.*

69. .8   **70.** 1.0   **71.** 1.2   **72.** 1.5

73. The linear function $f(x) = -123x + 29{,}685$ provides a model for the number of post offices in the United States from 1984 to 1990, where $x = 0$ corresponds to 1984, $x = 1$ corresponds to 1985, and so on. Use this model to give the approximate number of post offices during the given year. (*Source:* U.S. Postal Service, Annual Report of the Postmaster General)
(a) 1985
(b) 1987
(c) 1990
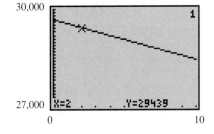 **(d)** The graphics calculator screen shows a portion of the graph of $y = f(x)$ with the coordinates of a point on the graph displayed at the bottom of the screen. Interpret the meaning of the display in the context of this exercise.

74. The linear function $f(x) = 1650x + 3817$ provides a model for the United States defense budget for the decade of the 1980's, where $x = 0$ corresponds to 1980, $x = 1$ corresponds to 1981, and so on, with $f(x)$ representing the budget in millions of dollars. Use this model to approximate the defense budget during the given year. (*Source:* U.S. Office of Management and Budget)
(a) 1983
(b) 1985
(c) 1988
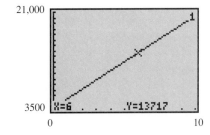 **(d)** The graphics calculator screen shows a portion of the graph of $y = f(x)$ with the coordinates of a point on the graph displayed at the bottom of the screen. Interpret the meaning of the display in the context of this exercise.

59. domain: $(-\infty, \infty)$; range: $\{5\}$

60. domain: $(-\infty, \infty)$; range: $\{-4\}$

61. (a) $0; $.50; $1.00; $1.50 **(b)** $.50x
(c)

62. (a) 8.25 (dollars)
63. (a) yes **(b)** {Oct. 12, 15, 16, 17, 18, 19, 22, 23, 24, 25} **(c)** {$39.50, $38.00, $39.00, $37.00, $34.00, $28.00, $29.00, $30.00}
(d) $30.00 **(e)** October 12; October 22 **64. (a)** yes
(b) {Sept. 25, 26, 27, 28; Oct. 1, 2, 3, 4, 5, 8, 9, 10, 11, 12, 15, 16, 17, 18, 19, 22, 23, 24, 25} **(c)** {$1.34, $1.35, $1.36, $1.37, $1.38, $1.39, $1.40, $1.41, $1.42, $1.43} **(d)** $1.40 **(e)** $.33
65. 194.53 centimeters
66. 177.29 centimeters
67. 177.41 centimeters
68. 163.65 centimeters
69. 1.83 cubic meters
70. 2.86 cubic meters
71. 4.11 cubic meters
72. 6.43 cubic meters
73. (a) 29,562 **(b)** 29,316
(c) 28,947 **(d)** In 1986 (when $x = 2$), the number of post offices was approximately 29,439.
74. (a) $8767 million
(b) $12,067 million
(c) $17,017 million **(d)** In 1986 (when $x = 6$), the defense budget was approximately $13,717 million.

75. $f(3) = 7$
76. (a) $f(2) = 1.1$
(b) $x = 6$ (c) -1.2 (d) 3.5
(e) $f(x) = -1.2x + 3.5$
77. $f(x) = -3x + 5$
78. When the Celsius
temperature is $-40°$, the
Fahrenheit temperature is
also $-40°$.
79. $\dfrac{1}{3}$ 80. $\dfrac{1}{9}$ 81. 3
82. 9

75. The graphics calculator screen shows the graph of a linear function $y = f(x)$, along with the display of coordinates of a point on the graph. Use functional notation to write what the display indicates.

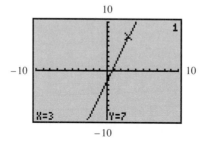

76. The table was generated by a graphics calculator for a linear function $y_1 = f(x)$. Use the table to answer.
   (a) What is $f(2)$?
   (b) If $f(x) = -3.7$, what is the value of x?
   (c) What is the slope of the line?
   (d) What is the y-intercept of the line?
   (e) Find the expression for $f(x)$.

77. The two screens show the graph of the same linear function $y = f(x)$. Find the expression for $f(x)$.

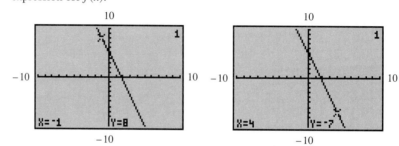

78. The formula for converting Celsius to Fahrenheit is $F = 1.8C + 32$. If we graph $y = f(x) = 1.8x + 32$ on a graphics calculator screen, we obtain the accompanying picture. (We used the interval $[-50, 50]$ on the x-axis and $[-75, 50]$ on the y-axis.) The point $(-40, -40)$ lies on the graph, as indicated by the display. Interpret the meaning of this in the context of this exercise.

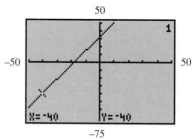

REVIEW EXERCISES

Find k, given that y = 1 and x = 3. See Section 2.1.

79. $y = kx$

80. $y = kx^2$

81. $y = \dfrac{k}{x}$

82. $y = \dfrac{k}{x^2}$

7.6 VARIATION

FOR EXTRA HELP

SSG pp. 230–235
SSM pp. 362–367

Video
9

Tutorial
IBM   MAC

OBJECTIVES

1 ▶ Write an equation expressing direct variation.
2 ▶ Find the constant of variation and solve direct variation problems.
3 ▶ Solve inverse variation problems.
4 ▶ Solve joint variation problems.
5 ▶ Solve combined variation problems.

◈ In the distance formula $d = rt$, distance varies directly as rate and time. In the simple interest formula $I = prt$, interest earned varies directly as the principal, the interest rate, and the time.

◈ **CONNECTIONS** ◈

Certain types of functions are very common, especially in the physical sciences. These are functions where y depends on a multiple of x, or y depends on a number divided by x. In such situations, y is said to *vary directly* as x (in the first case) or *vary inversely* as x (in the second case). For example, the period of a pendulum varies directly as the square root of the length of the pendulum and inversely as the square root of the acceleration due to gravity.

FOR DISCUSSION OR WRITING

Find some other common examples of direct and inverse variation. As a start, consider the distance formula, the simple interest formula, and formulas for area and volume.

1 ▶ Write an equation expressing direct variation.

▶ The circumference of a circle is given by the formula $C = 2\pi r$, where r is the radius of the circle. As the formula shows, the circumference is always a constant multiple of the radius (C is always found by multiplying r by the constant 2π). Because of this, the circumference is said to *vary directly* as the radius.

Direct Variation

y varies directly as x if there exists some constant k such that

$$y = kx.$$

▶ **TEACHING TIP**

Some simple examples of variation include:
Direct variation: the harder one pushes on a car's gas pedal, the faster the car goes.
Inverse variation: the harder one pushes on a car's brake pedal, the slower the car goes. ◀

Also, y is said to be **proportional** to x. The number k is called the **constant of variation.** In direct variation, for $k > 0$, as the value of x increases, the value of y also increases. Similarly, as x decreases, y also decreases.

The direct variation equation defines a linear function. In applications, functions are often defined by variation equations. For example, if Tom earns \$8 per hour, his wages vary directly as, or are proportional to, the number of hours he works. If y represents his total wages and x the number of hours he has worked, then

$$y = 8x.$$

Here k, the constant of variation, is 8.

2 ▶ Find the constant of variation and solve direct variation problems.

▶ The following example shows how to find the value of the constant k.

INSTRUCTOR'S RESOURCES

ITM pp. 447–451   **ISM** pp. 555–564
IAM p. 34

TEST GENERATOR
DOS   Windows   MAC

TRANSPARENCIES

E X A M P L E   1

Finding the Constant of Variation

Steven is paid an hourly wage. One week he worked 43 hours and was paid $795.50. How much does he earn per hour?

Let h represent the number of hours he works and T represent his corresponding pay. Then, T varies directly as h, and

$$T = kh.$$

Here k represents Steven's hourly wage. Since $T = 795.50$ when $h = 43$,

$$795.50 = 43k$$
$$k = 18.50 \qquad \text{Use a calculator.}$$

His hourly wage is $18.50. Thus, T and h are related by

$$T = 18.50h. \quad \blacksquare$$

E X A M P L E   2

Solving a Direct Variation Problem

Hooke's law for an elastic spring states that the distance a spring stretches is proportional to the force applied. If a force of 150 newtons* stretches a certain spring 8 centimeters, how much will a force of 400 newtons stretch the spring?

If d is the distance the spring stretches and f is the force applied, then $d = kf$ for some constant k. Since a force of 150 newtons stretches the spring 8 centimeters,

$$d = kf \qquad \text{Formula}$$
$$8 = k \cdot 150 \qquad d = 8, f = 150$$
$$k = \frac{8}{150} = \frac{4}{75}, \qquad \text{Find } k.$$

and $d = \frac{4}{75}f.$

For a force of 400 newtons,

$$d = \frac{4}{75}(400) \qquad \text{Let } f = 400.$$
$$= \frac{64}{3}.$$

The spring will stretch 64/3 centimeters if a force of 400 newtons is applied. $\quad \blacksquare$

▶ **TEACHING TIP**

The problems in this section follow a consistent pattern:

1. Write the equation of variation.
2. Substitute the initial values and solve for k.
3. Rewrite the equation of variation with the new k value.
4. Substitute the remaining values, solve for the unknown, and find the final answer. ◀

The direct variation equation $y = kx$ is a linear equation. However, other kinds of variation involve polynomial equations of higher degree. That is, one variable can be directly proportional to a power of another variable.

Direct Variation as a Power

y **varies directly as the nth power of x** if there exists a real number k such that

$$y = kx^n.$$

---

*A newton is a unit of measure of force used in physics.

An example of direct variation as a power involves the area of a circle. The formula for the area of a circle is

$$A = \pi r^2.$$

Here, π is the constant of variation, and the area varies directly as the square of the radius.

EXAMPLE 3

Solving a Problem Involving Direct Variation as a Power

The distance a body falls from rest varies directly as the square of the time it falls (here we disregard air resistance). If an object falls 64 feet in 2 seconds, how far will it fall in 8 seconds?

If d represents the distance the object falls and t the time it takes to fall,

$$d = kt^2$$

for some constant k. To find the value of k, use the fact that the object falls 64 feet in 2 seconds.

$$d = kt^2 \qquad \text{Formula}$$
$$64 = k(2)^2 \qquad \text{Let } d = 64 \text{ and } t = 2.$$
$$k = 16 \qquad \text{Find } k.$$

With this result, the variation equation becomes

$$d = 16t^2.$$

Now let $t = 8$ to find the number of feet the object will fall in 8 seconds.

$$d = 16(8)^2 \qquad \text{Let } t = 8.$$
$$= 1024$$

The object will fall 1024 feet in 8 seconds. ∎

3 ▶ Solve inverse variation problems.

▶ In direct variation where $k > 0$, as x increases, y increases, and similarly as x decreases, y decreases. Another type of variation is *inverse variation*.

Inverse Variation

y varies inversely as x if there exists a real number k such that

$$y = \frac{k}{x}.$$

Also, **y varies inversely as the nth power of x** if there exists a real number k such that

$$y = \frac{k}{x^n}.$$

With inverse variation, for $k > 0$, as x increases, y decreases, and as x decreases, y increases. Notice that the inverse variation equation also defines a function. Since x is in the denominator, these functions are called *rational functions*.

An example of inverse variation can be found by looking at the formula for the area of a parallelogram. In its usual form, the formula is

$$A = bh.$$

Dividing both sides by b gives

$$h = \frac{A}{b}.$$

Here, h (height) varies inversely as b (base), with A (the area) serving as the constant of variation. For example, if a parallelogram has an area of 72 square inches, the values of b and h might be any of the following:

$$b = 2, h = 36 \qquad b = 12, h = 6$$
$$b = 3, h = 24 \qquad b = 9, \; h = 8$$
$$b = 4, h = 18 \qquad b = 8, \; h = 9.$$

Notice that in the first group, as b increases, h decreases. In the second group, as b decreases, h increases.

EXAMPLE 4

Solving an Inverse Variation Problem

Chalkboard Exercise

Suppose p varies inversely as the cube of q and $p = 100$ when $q = 3$. Find p, given $q = 5$.

Answer: $\dfrac{108}{5}$

The weight of a object above the earth varies inversely as the square of its distance from the center of the earth. A space vehicle in an elliptical orbit has a maximum distance from the center of the earth (apogee) of 6700 miles. Its minimum distance from the center of the earth (perigee) is 4090 miles. See Figure 31. If an astronaut in the vehicle weighs 57 pounds at its apogee, what does the astronaut weigh at its perigee?

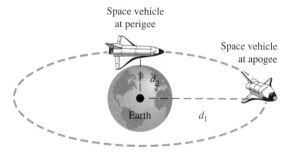

FIGURE 31

If w is the weight and d is the distance from the center of the earth, then

$$w = \frac{k}{d^2}$$

for some constant k. At the apogee the astronaut weighs 57 pounds and the distance from the center of the earth is 6700 miles. Use these values to find k.

$$57 = \frac{k}{(6700)^2} \qquad \text{Let } w = 57 \text{ and } d = 6700.$$
$$k = 57(6700)^2$$

Then the weight at the perigee with $d = 4090$ miles is

$$w = \frac{57(6700)^2}{(4090)^2} \approx 153 \text{ pounds.} \quad \blacksquare$$

NOTE The approximate answer in Example 4, 153 pounds, was obtained by using a calculator. A calculator will often be helpful in performing operations required in variation problems.

4 ▶ Solve joint variation problems.

▶ It is common for one variable to depend on several others. For example, if one variable varies as the product of several other variables (perhaps raised to powers), the first variable is said to **vary jointly** as the others.

EXAMPLE 5

Solving a Joint Variation Problem

The strength of a rectangular beam varies jointly as its width and the square of its depth. If the strength of a beam 2 inches wide by 10 inches deep is 1000 pounds per square inch, what is the strength of a beam 4 inches wide and 8 inches deep?

If S represents the strength, w the width, and d the depth, then

$$S = kwd^2$$

for some constant, k. Since $S = 1000$ if $w = 2$ and $d = 10$,

$$1000 = k(2)(10)^2. \qquad \text{Let } S = 1000, w = 2, \text{ and } d = 10.$$

Solving this equation for k gives

$$1000 = k \cdot 2 \cdot 100$$
$$1000 = 200k$$
$$k = 5,$$

so

$$S = 5wd^2.$$

Find S when $w = 4$ and $d = 8$ by substitution in $S = 5wd^2$.

$$S = 5(4)(8)^2 \qquad \text{Let } w = 4 \text{ and } d = 8.$$
$$= 1280$$

The strength of the beam is 1280 pounds per square inch. ∎

5 ▶ Solve combined variation problems.

▶ There are many combinations of direct and inverse variation. The final example shows a typical **combined variation** problem.

EXAMPLE 6

Solving a Combined Variation Problem

The body–mass index, or BMI, is used by physicians to assess a person's level of fatness.* The BMI varies directly as an individual's weight in pounds and inversely as the square of the individual's height in inches. A person who weighs 118 pounds and is 64 inches tall has a BMI of 20. (The BMI is rounded to the nearest whole number.) Find the BMI of a person who weighs 165 pounds with a height of 70 inches.

Let B represent the BMI, w the weight, and h the height. Then

$$B = \frac{kw}{h^2}. \qquad \begin{array}{l} \leftarrow \text{BMI varies directly as the weight.} \\ \leftarrow \text{BMI varies inversely as the square of the height.} \end{array}$$

_____
* From _Reader's Digest,_ October 1993.

To find k, let $B = 20$, $w = 118$, and $h = 64$.

$$20 = \frac{k(118)}{64^2}$$

$$k = \frac{20(64^2)}{118} \qquad \text{Multiply by } 64^2; \text{ divide by } 118.$$

$$= 694 \qquad \text{Use a calculator.}$$

Now find B when $k = 694$, $w = 165$ and $h = 70$.

$$B = \frac{694(165)}{70^2} = 23 \quad \text{(rounded)}$$

The required BMI is 23. A BMI from 20 through 26 is considered desirable. ■

7.6 EXERCISES

⊙ *Determine whether the equation represents direct, inverse, joint, or combined variation.*

1. $y = \dfrac{3}{x}$ 　　 **2.** $y = \dfrac{8}{x}$ 　　 **3.** $y = 10x^2$ 　　 **4.** $y = 2x^3$

5. $y = 3xz^4$ 　　 **6.** $y = 6x^3z^2$ 　　 **7.** $y = \dfrac{4x}{wz}$ 　　 **8.** $y = \dfrac{6x}{st}$

Solve the problem.

9. If x varies directly as y, and $x = 9$ when $y = 3$, find x when $y = 12$.

10. If x varies directly as y, and $x = 10$ when $y = 7$, find y when $x = 50$.

11. If z varies inversely as w, and $z = 10$ when $w = .5$, find z when $w = 8$.

12. If t varies inversely as s, and $t = 3$ when $s = 5$, find s when $t = 5$.

13. p varies jointly as q and r^2, and $p = 200$ when $q = 2$ and $r = 3$. Find p when $q = 5$ and $r = 2$.

14. f varies jointly as g^2 and h, and $f = 50$ when $g = 4$ and $h = 2$. Find f when $g = 3$ and $h = 6$.

⊙ **15.** For $k > 0$, if y varies directly as x, when x increases, y _____ , and when x decreases, y _____ .

⊙ **16.** For $k > 0$, if y varies inversely as x, when x increases, y _____ , and when x decreases, y _____ .

⊙ **17.** Explain the difference between inverse variation and direct variation.

⊙ **18.** What is meant by the constant of variation in a direct variation problem? If you were to graph the linear equation $y = kx$ for some nonnegative constant k, what role does the value of k play in the graph?

⊙ CONCEPTUAL 　 🖉 WRITING 　 ▲ CHALLENGING 　 ▦ SCIENTIFIC CALCULATOR 　 ▦ GRAPHICS CALCULATOR

Solve the variation problem. Use a calculator as necessary. See Examples 1–6.

19. Todd bought 8 gallons of gasoline and paid $8.79. To the nearest tenth of a cent, what is the price of gasoline per gallon?

20. Melissa gives horseback rides at Shadow Mountain Ranch. A 2.5-hour ride costs $50.00. What is the price per hour?

21. The weight of an object on earth is directly proportional to the weight of that same object on the moon. A 200-pound astronaut would weigh 32 pounds on the moon. How much would a 50-pound dog weigh on the moon?

22. In the study of electricity, the resistance of a conductor of uniform cross-sectional area is directly proportional to its length. Suppose that the resistance of a certain type of copper wire is .640 ohm per 1000 feet. What is the resistance of 2500 feet of the wire?

23. The amount of water emptied by a pipe varies directly as the square of the diameter of the pipe. For a certain constant water flow, a pipe emptying into a canal will allow 200 gallons of water to escape in an hour. The diameter of the pipe is 6 inches. How much water would a 12-inch pipe empty into the canal in an hour, assuming the same water flow?

24. For a body falling freely from rest (disregarding air resistance), the distance the body falls varies directly as the square of the time. If an object is dropped from the top of a tower 576 feet high and hits the ground in 6 seconds, how far did it fall in the first 4 seconds?

25. The pressure exerted by a certain liquid at a given point varies directly as the depth of the point beneath the surface of the liquid. The pressure at 30 meters is 80 newtons per square centimeter. What pressure is exerted at 50 meters?

26. The force required to compress a spring is proportional to the change in length of the spring. If a force of 20 newtons is required to compress a certain spring 2 centimeters, how much force is required to compress the spring from 20 centimeters to 8 centimeters?

27. The illumination produced by a light source varies inversely as the square of the distance from the source. If the illumination produced 4 meters from a certain light source is 48 footcandles, find the illumination produced 16 meters from the same source.

28. A meteorite approaching the earth has a velocity inversely proportional to the square root of its distance from the center of the earth. If the velocity is 5 kilometers per second when the distance is 8100 kilometers from the center of the earth, find the velocity at a distance of 6400 kilometers.

29. The frequency of a vibrating string varies inversely as its length. That is, a longer string vibrates fewer times in a second than a shorter string. Suppose a piano string 2 feet long vibrates 250 cycles per second. What frequency would a string 5 feet long have?

30. The force with which the earth attracts an object above the earth's surface varies inversely with the square of the distance of the object from the center of the earth. If an object 4000 miles from the center of the earth is attracted with a force of 160 pounds, find the force of attraction if the object were 6000 miles from the center of the earth.

31. The distance that a person can see to the horizon from a point above the surface of the earth varies directly as the square root of the height of that point (disregarding mountains, smog, and haze). If a person 144 meters above the surface of the earth can see for 18 kilometers to the horizon, how far can a person see to the horizon from a point 1600 meters high?

19. 1.09\frac{9}{10}$

20. $20.00   **21.** 8 pounds
22. 1.600 ohms
23. 800 gallons

24. 256 feet   **25.** $133\frac{1}{3}$ newtons per square centimeter
26. 120 newtons
27. 3 footcandles

28. $5\frac{5}{8}$ kilometers per second
29. 100 cycles per second

30. $71\frac{1}{9}$ pounds

31. 60 kilometers

32. 1.105 liters

33. 800 pounds   34. $\frac{8}{9}$ ton

35. .71π seconds
36. 448.1 pounds
37. 480 kilograms
38. about 68,600 calls
39. 25   40. Answers will
vary.   41. 9

32. The volume of a gas varies inversely as the pressure and directly as the temperature. (Temperature must be measured in *degrees Kelvin* (K), a unit of measurement used in physics.) If a certain gas occupies a volume of 1.3 liters at 300 K and a pressure of 18 newtons per square centimeter, find the volume at 340 K and a pressure of 24 newtons per square centimeter.

33. The force of the wind blowing on a vertical surface varies jointly as the area of the surface and the square of the velocity. If a wind of 40 miles per hour exerts a force of 50 pounds on a surface of $\frac{1}{2}$ square foot, how much force will a wind of 80 miles per hour place on a surface of 2 square feet?

34. It is shown in engineering that the maximum load a cylindrical column of circular cross section can hold varies directly as the fourth power of the diameter and inversely as the square of the height. If a column 9 feet high and 3 feet in diameter will support a load of 8 tons, how great a load will be supported by a column 12 feet high and 2 feet in diameter?

35. The period of a pendulum varies directly as the square root of the length of the pendulum and inversely as the square root of the acceleration due to gravity. Find the period when the length is 4 feet and the acceleration due to gravity is 32 feet per second per second, if the period is 1.06π seconds when the length is 9 feet and the acceleration due to gravity is 32 feet per second per second.

36. The force needed to keep a car from skidding on a curve varies inversely as the radius of the curve and jointly as the weight of the car and the square of the speed. If 242 pounds of force keep a 2000-pound car from skidding on a curve of radius 500 feet at 30 miles per hour, what force would keep the same car from skidding on a curve of radius 750 feet at 50 miles per hour?

37. The maximum load of a horizontal beam that is supported at both ends varies directly as the width and the square of the height and inversely as the length between the supports. A beam 6 meters long, .1 meter wide, and .06 meter high supports a load of 360 kilograms. What is the maximum load supported by a beam 16 meters long, .2 meter wide, and .08 meter high?

38. The number of long distance phone calls between two cities in a certain time period varies directly as the populations p_1 and p_2 of the cities, and inversely as the distance between them. If 80,000 calls are made between two cities 400 miles apart, with populations of 70,000 and 100,000, how many calls are made between cities with populations of 50,000 and 75,000 that are 250 miles apart?

39. Key Griffey, Jr. weighs 205 pounds and is 6 feet, 3 inches tall. Use the information given in Example 6 to find his body–mass index.

40. A body–mass index from 27 through 29 carries a slight risk of weight-related health problems, while one of 30 or more indicates a great increase in risk. Use your own height and weight and the information in Example 6 to determine whether you are at risk.

41. The graphics calculator screen shows a portion of the graph of a function $y = f(x)$ that satisfies the conditions for direct variation. What is $f(36)$?

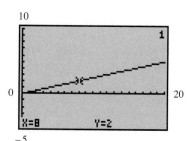

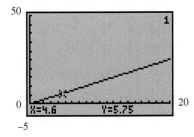

42. The accompanying table of points was generated by a graphics calculator. The points lie on the graph of a function $y_1 = f(x)$ that satisfies the conditions for direct variation. What is $f(36)$?

X	Y1
0	0
1	1.5
2	3
3	4.5
4	6
5	7.5
6	9

X=0

◆ **MATHEMATICAL CONNECTIONS** (Exercises 43–50) ◆

A routine activity such as pumping gasoline can be related to many of the concepts studied in this chapter. Suppose that premium unleaded costs $1.25 per gallon. Work Exercises 43–50 in order.

43. 0 gallons of gasoline cost $0.00, while 1 gallon costs $1.25. Represent these two pieces of information as ordered pairs of the form (gallons, price).

44. Use the information from Exercise 43 to find the slope of the line on which the two points lie.

45. Write the slope-intercept form of the equation of the line on which the two points lie.

46. Using function notation, if $f(x) = ax + b$ represents the line from Exercise 45, what are the values of a and b?

47. How does the value of a from Exercise 46 relate to gasoline in this situation? With relationship to the line, what do we call this number?

48. Why does the equation from Exercise 46 satisfy the conditions for direct variation? In the context of variation, what do we call the value of a?

49. The graph of the equation from Exercise 46 is shown in the accompanying graphics calculator screen, along with a display at the bottom of the screen. How is this display interpreted in the context of these exercises?

50
1
0
X=4.6   Y=5.75
20
−5

50. The accompanying table was generated by a graphics calculator, with y_1 entered as the equation from Exercise 46. Notice that when $x = 12$, $y_1 = 15$. How is this interpreted in the context of these exercises?

X	Y1
7	8.75
8	10
9	11.25
10	12.5
11	13.75
12	15
13	16.25

X=12

Answers (margin)

REVIEW EXERCISES

Solve the equation. See Section 2.1.

51. $2\left(\dfrac{3x - 1}{2}\right) - 2x = -4$

52. $3\left(\dfrac{5p + 4}{3}\right) + p = 6$

53. $-4\left(\dfrac{t - 3}{3}\right) + 2t = 8$

54. $-2\left(\dfrac{3n + 1}{5}\right) + n = -4$

○ CONCEPTUAL   ✎ WRITING   ▲ CHALLENGING   ▦ SCIENTIFIC CALCULATOR   ▦ GRAPHICS CALCULATOR

CHAPTER 7 SUMMARY

KEY TERMS

7.1 ordered pair
origin
x-axis
y-axis
rectangular
(Cartesian)
coordinate system
plot
coordinate
quadrant
graph of an
equation
linear equation in
two variables
standard form
x-intercept
y-intercept

7.2 slope

7.4 linear inequality in
two variables
boundary line

7.5 dependent variable
independent
variable
relation
function
domain
range
graph of a relation
vertical line test
functional notation
linear function

7.6 proportional
constant of variation
vary jointly
combined variation

NEW SYMBOLS

(a, b) ordered pair
x_1 a specific value of the variable x (read "x sub one")
m slope
$f(x)$ function of x (read "f of x")

QUICK REVIEW

CONCEPTS	EXAMPLES
7.1 THE RECTANGULAR COORDINATE SYSTEM	
Finding Intercepts	The graph of $2x + 3y = 12$ has
To find the x-intercept, let $y = 0$.	x-intercept (6, 0)
To find the y-intercept, let $x = 0$.	and y-intercept (0, 4).
Distance Formula	
The distance between (x_1, y_1) and (x_2, y_2) is	The distance between $(3, -2)$ and $(-1, 1)$ is
$$d = \sqrt{(x_2 - x_1)^2 + (y_2 - y_1)^2}.$$	$$\sqrt{(-1 - 3)^2 + [1 - (-2)]^2}$$ $$= \sqrt{(-4)^2 + 3^2} = \sqrt{16 + 9}$$ $$= \sqrt{25} = 5.$$
7.2 THE SLOPE OF A LINE	
If $x_2 \neq x_1$, then	For $2x + 3y = 12$,
$$m = \frac{\text{change in } y}{\text{change in } x} = \frac{y_2 - y_1}{x_2 - x_1}.$$	$$m = \frac{4 - 0}{0 - 6} = -\frac{2}{3}.$$
A vertical line has undefined slope.	$x = 3$ has undefined slope.
A horizontal line has 0 slope.	$y = -5$ has $m = 0$.
Parallel lines have equal slopes.	$y = 2x + 3$ $4x - 2y = 6$ $m = 2$ $m = 2$ These lines are **parallel**.
The slopes of perpendicular lines are negative reciprocals with a product of -1.	$y = 3x - 1$ $x + 3y = 4$ $m = 3$ $m = -\dfrac{1}{3}$ These lines are **perpendicular**.

CONCEPTS	EXAMPLES

7.3 LINEAR EQUATIONS

Standard Form $\quad Ax + By = C$	$2x - 5y = 8$
Vertical Line $\quad x = k$	$x = -1$
Horizontal Line $\quad y = k$	$y = 4$
Slope-Intercept Form $\quad y = mx + b$	$y = 2x + 3$ $m = 2$, y-intercept is $(0, 3)$.
Point-Slope Form $\quad y - y_1 = m(x - x_1)$	$y - 3 = 4(x - 5)$ $(5, 3)$ is on the line, $m = 4$.

7.4 LINEAR INEQUALITIES

Graphing a Linear Inequality *Step 1* Draw the graph of the line that is the boundary. Make the line solid if the inequality involves $\leq$ or $\geq$; make the line dashed if the inequality involves $<$ or $>$.	Graph $2x - 3y \leq 6$. Draw the graph of $2x - 3y = 6$. Use a solid line because of $\leq$.
Step 2 Choose any point not on the line as a test point.	Choose $(1, 2)$. $\qquad 2(1) - 3(2) = 2 - 6 \leq 6 \qquad$ True
Step 3 Shade the region that includes the test point if the test point satisfies the original inequality; otherwise, shade the region on the other side of the boundary line.	Shade the side of the line that includes $(1, 2)$. 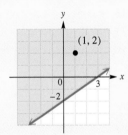

7.5 INTRODUCTION TO FUNCTIONS

To evaluate a function using functional notation (that is, $f(x)$ notation) for a given value of x, substitute the value wherever x appears.	If $f(x) = x^2 - 7x + 12$, then $\qquad f(1) = 1^2 - 7(1) + 12 = 6.$
To write the equation that defines a function in functional notation, solve the equation for y. Then replace y with $f(x)$.	Given: $2x + 3y = 12$. $\qquad 3y = -2x + 12$ $\qquad y = -\dfrac{2}{3}x + 4$ $\qquad f(x) = -\dfrac{2}{3}x + 4$

7.6 VARIATION

If there is some constant k such that: $y = kx^n$, then y varies directly as x^n, $y = \dfrac{k}{x^n}$, then y varies inversely as x^n.	The area of a circle **varies directly** as the square of the radius. $\qquad A = kr^2$ Pressure **varies inversely** as volume. $\qquad p = \dfrac{k}{V}$

CHAPTER 7 REVIEW EXERCISES

1. (0, 5); $\left(\frac{10}{3}, 0\right)$;

(2, 2); $\left(\frac{14}{3}, -2\right)$

2. (2, −6); (5, −3);
(3, −5); (6, −2)

3. (3, 0); (0, −4)

4. $\left(\frac{28}{5}, 0\right)$; (0, 4)

5. (10, 0); (0, 4)

6. (8, 0); (0, −2)

7. $\sqrt{13}$ 8. $\sqrt{53}$
9. $\sqrt{65} + \sqrt{181} + 2\sqrt{17}$

11. $-\frac{7}{5}$ 12. $-\frac{1}{2}$

13. 2   14. $\frac{3}{4}$

15. undefined   16. $\frac{2}{3}$

17. $-\frac{1}{3}$ 18. undefined

[7.1] *Complete the given ordered pairs for the equation. Then graph the equation.*

1. $3x + 2y = 10$;   (0,), (, 0), (2,), (, −2)
2. $x - y = 8$;   (2,), (, −3), (3,), (, −2)

Find the x- and y-intercepts and then graph the equation.

3. $4x - 3y = 12$ **4.** $5x + 7y = 28$
5. $2x + 5y = 20$ **6.** $x - 4y = 8$

Find the distance between the pair of points.

7. (1, 5) and (−2, 3)

8. (−4, −2) and (3, 0)

9. Find the perimeter of the triangle whose vertices are the points (−1, 3), (7, 2), and (9, −6).

⊙ **10.** Explain how the signs of the x- and y-coordinates of a point determine the quadrant in
✍   which the point lies.

[7.2] *Find the slope of the line.*

11. through (−1, 2) and (4, −5)   **12.** through (0, 3) and (−2, 4)
13. $y = 2x + 3$ **14.** $3x - 4y = 5$
15. $x = 5$ **16.** parallel to $3y = 2x + 5$
17. perpendicular to $3x - y = 4$ **18.** through (−1, 5) and (−1, −4)

▦ **19.** the line containing the points shown in this table

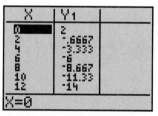

▦ **20.** the line pictured in the two screens

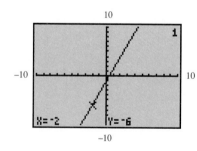
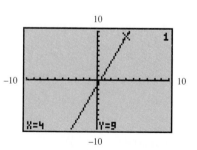

RESOURCES FOR REVIEW

| 📖 | **ITM** pp. 429–451   **ISM** pp. 564–576 | 📖 | **SSG** pp. 202–235 | 💾 | **TEST GENERATOR** |
| | **IAM** p. 35 | | **SSM** pp. 367–380 | | **DOS   Windows   MAC** |

⊙ CONCEPTUAL   ✍ WRITING   ▲ CHALLENGING   ▦ SCIENTIFIC CALCULATOR   ▦ GRAPHICS CALCULATOR

Tell whether the line has positive, negative, zero, or undefined slope.

21.

22.

23.

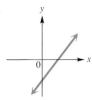

24.

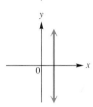

25. Family income in the United States has steadily increased for many years (primarily due to inflation). In 1970 the median family income was about $10,000 a year. In 1985 it was about $25,000 a year. Find the average rate of change of family income over that period.

[7.3] *Write an equation in standard form for the line.*

26. slope $-\dfrac{1}{3}$, y-intercept $(0, -1)$

27. slope 0, y-intercept $(0, -2)$

28. slope $-\dfrac{4}{3}$, through $(2, 7)$

29. slope 3, through $(-1, 4)$

30. vertical, through $(2, 5)$

31. through $(2, -5)$ and $(1, 4)$

32. through $(-3, -1)$ and $(2, 6)$

33. parallel to $4x - y = 3$ and through $(7, -1)$

34. perpendicular to $2x - 5y = 7$ and through $(4, 3)$

35. the line containing the points in the table accompanying Exercise 19

36. the line pictured in Exercise 20

The equation $y = 1.7x + 230$ is a model that was recently used to predict the U.S. population, where $x = 0$ represents the year 1980 and y is the population in millions.

37. According to this equation, what was the population in 1982?

38. According to this equation, in what year would the population have reached 247 million?

[7.4] *Graph the solutions of the inequality or compound inequality.*

39. $3x - 2y \le 12$

40. $5x - y > 6$

41. $x \ge 2$

42. $2x + y \le 1$ and $x \ge 2y$

[7.5] *Give the domain and range of the relation. Identify any functions.*

43. $\{(-4, 2), (-4, -2), (1, 5), (1, -5)\}$

44.

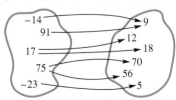

45.

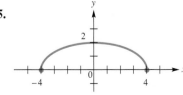

---

19. $-\dfrac{4}{3}$ **20.** $\dfrac{5}{2}$

21. positive   **22.** negative
23. zero   **24.** undefined
25. $1000 per year
26. $x + 3y = -3$
27. $y = -2$
28. $4x + 3y = 29$
29. $3x - y = -7$
30. $x = 2$
31. $9x + y = 13$
32. $7x - 5y = -16$
33. $4x - y = 29$
34. $5x + 2y = 26$
35. $4x + 3y = 6$
36. $5x - 2y = 2$
37. 233.4 million
38. 1990
39.

40.

41.

42.

43. domain: $\{-4, 1\}$;
range: $\{2, -2, 5, -5\}$; not a function
44. domain:
$\{-14, 91, 17, 75, -23\}$;
range: $\{9, 12, 18, 70, 56, 5\}$;
not a function
45. domain: $[-4, 4]$;
range: $[0, 2]$; function

---

46. function;
domain: $(-\infty, \infty)$; linear
function 47. not a
function; domain:
$(-\infty, \infty)$ 48. function;
domain: $(-\infty, \infty)$
49. function; domain:
$\left[-\dfrac{7}{4}, \infty\right)$ 50. not a
function; domain: $[0, \infty)$
51. function; domain:
$(-\infty, -6) \cup (-6, 6) \cup$
$(6, \infty)$ 53. -6
54. -15 55. -8
56. $-2k^2 + 3k - 6$
57. -96
58. $-8p^2 + 6p - 6$
59. (a); $y = 3 - 2x^2$;
$f(x) = 3 - 2x^2$; $f(3) =$
-15 60. (c)
62. (c) 63. 15 64. 5
65. .850 ohm
66. 430 millimeters
67. approximately 5600
pounds (to the nearest
hundred)
68. approximately 28 (to
the nearest unit)

Determine whether the equation defines y as a function of x. Give the domain in each case. Identify any linear functions.

46. $y = 3x - 3$

47. $y < x + 2$

48. $y = |x - 4|$

49. $y = \sqrt{4x + 7}$

50. $x = y^2$

51. $y = \dfrac{7}{x^2 - 36}$

52. Explain the test that allows us to determine whether a graph is that of a function.

Given $f(x) = -2x^2 + 3x - 6$, find the function value.

53. $f(0)$

54. $f(3)$

55. $f\left(-\dfrac{1}{2}\right)$

56. $f(k)$

57. $f[f(0)]$

58. $f(2p)$

59. Only one of the equations below defines y as a function of x. Determine which one it is, solve for y in terms of x, rewrite it using $f(x)$ notation, and find $f(3)$.
  (a) $2x^2 + y = 3$ **(b)** $2x^2 + y^2 = 8$ **(c)** $x = |y|$ **(d)** $x + y^2 = 4$

60. Suppose that $2x - 5y = 7$ defines a function. If $y = f(x)$, which one of the following defines the same function?

  (a) $f(x) = \dfrac{7 - 2x}{5}$ **(b)** $f(x) = \dfrac{-7 - 2x}{5}$

  (c) $f(x) = \dfrac{-7 + 2x}{5}$ **(d)** $f(x) = \dfrac{7 + 2x}{5}$

61. Can the graph of a linear function have undefined slope? Explain.

[7.6]

62. In which one of the following does x vary inversely as y?

  (a) $y = 2x$ **(b)** $y = \dfrac{x}{3}$ **(c)** $y = \dfrac{3}{x}$ **(d)** $y = x^2$

63. m varies directly as p^2 and inversely as q, and $m = 32$ when $p = 8$ and $q = 10$. Find q when $p = 12$ and $m = 48$.

64. x varies jointly as y and z and inversely as $\sqrt{w}$. If $x = 12$, when $y = 3$, $z = 8$, and $w = 36$, find y when $x = 12$, $z = 4$, and $w = 25$.

Solve the problem.

65. The resistance in ohms of a platinum wire temperature sensor varies directly as the temperature in *degrees Kelvin* (K). If the resistance is .646 ohm at a temperature of 190 K, find the resistance at a temperature of 250 K.

66. For the subject in a photograph to appear in the same perspective in the photograph as in real life, the viewing distance must be properly related to the amount of enlargement. For a particular camera, the viewing distance varies directly as the amount of enlargement. A picture taken with this camera that is enlarged 5 times should be viewed from a distance of 250 millimeters. Suppose a print 8.6 times the size of the negative is made. From what distance should it be viewed?

67. The force needed to keep a car from skidding on a curve varies inversely as the radius of the curve and jointly as the weight of the car and the square of the speed. If 3000 pounds of force keep a 2000-pound car from skidding as it travels at 30 miles per hour on a curve of radius 500 feet, what force would keep the same car from skidding on a curve of radius 750 feet at 50 miles per hour?

68. The frequency of vibration, f, of a guitar string varies directly as the square root of the tension, t, and inversely as the length, L, of the string. If the frequency is 20 when the tension is 9 (in appropriate units) and the length is 30 inches, find f when the tension is doubled and the length remains the same.

———◆ **MATHEMATICAL CONNECTIONS** (Exercises 69–82) ◆———

Refer to the straight–line graph shown and work Exercises 69–82 in order.

◉ **69.** By just looking at the graph, how can you tell whether the slope is positive, negative, zero, or undefined?

70. Use the slope formula to find the slope of the line.

71. Use the point-slope form to find the equation of the line, and write it in slope-intercept form.

72. Use the result of Exercise 71 to find the *x*-intercept of the graph.

73. Use the result of Exercise 71 to find the *y*-intercept of the graph.

74. Use functional notation to write the equation of the line. Use *f* to designate the function.

75. Find $f(8)$.

76. If $f(x) = -8$, what is the value of *x*?

77. Graph the solution set of $f(x) \geq 0$.

78. What is the solution set of $f(x) = 0$?

79. What is the solution set of $f(x) < 0$? (Use the graph and the result of Exercise 78.)

80. What is the solution set of $f(x) > 0$? (Use the graph and the result of Exercise 78.)

81. What is the slope of any line perpendicular to the line shown?

82. What is the equation of the line perpendicular to the line shown, passing through the origin?

Graph: points $(-1, 5)$ and $(3, -1)$ on a line falling from left to right.

69. Because it falls from left to right, the slope is negative. **70.** $-\dfrac{3}{2}$

71. $y = -\dfrac{3}{2}x + \dfrac{7}{2}$

72. $\left(\dfrac{7}{3}, 0\right)$ **73.** $\left(0, \dfrac{7}{2}\right)$

74. $f(x) = -\dfrac{3}{2}x + \dfrac{7}{2}$

75. $f(8) = -\dfrac{17}{2}$

76. $x = \dfrac{23}{3}$

77. ←———┤———→
    $\frac{7}{3}$

78. $\left\{\dfrac{7}{3}\right\}$ **79.** $\left(\dfrac{7}{3}, \infty\right)$

80. $\left(-\infty, \dfrac{7}{3}\right)$ **81.** $\dfrac{2}{3}$

82. $y = \dfrac{2}{3}x$

———◆———

CHAPTER 7 TEST

Consider points $A(6, 4)$ and $B(-4, -1)$.

1. Find the distance between the points *A* and *B*.

2. Find the slope of the line through *A* and *B*.

Find the x- and y-intercepts, and graph the equation.

3. $3x - 2y = 20$ **4.** $y = 5$ **5.** $x = 2$

◉ **6.** Describe how the graph of a line with undefined slope is situated in a rectangular coordinate system.

Determine whether the pair of lines are parallel, perpendicular, or neither.

7. $5x - y = 8$ and $5y = -x + 3$ **8.** $2y = 3x + 12$ and $3y = 2x - 5$

Find the equation of the line, and write it in standard form.

9. through $(4, -1)$; $m = -5$ **10.** through $(-3, 14)$; horizontal

[7.1] **1.** $5\sqrt{5}$

[7.2] **2.** $\dfrac{1}{2}$

[7.1, 7.3] **3.** $\left(\dfrac{20}{3}, 0\right)$; $(0, -10)$

4. none; $(0, 5)$

5. (2, 0);   none

[7.2] 6. It is a vertical
line.   7. perpendicular
8. neither
[7.3] 9. $5x + y = 19$
10. $y = 14$
11. $3x + 5y = -11$
12. $x + 2y = -3$
13. $x + 2y = 4$ 14. (b)
15. 11,989
16. (a) It is the slope of the
line.   (b) It is the annual
increase in the number of cases
served.
[7.4] 17.

18.

[7.5] 19. (d)   20. (d)
21. 0
22. domain: $(-\infty, \infty)$;
range: $(-\infty, \infty)$

11. through $(-7, 2)$ and parallel to $3x + 5y = 6$

12. through $(-7, 2)$ and perpendicular to $y = 2x$

13. the line shown in the figures (Look at the displays at the bottom.)

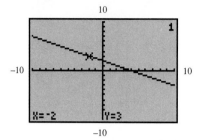
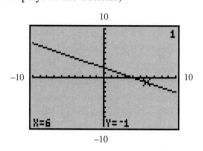

14. Which one of the following has positive slope and negative y-coordinate for its y-intercept?

(a)   (b)   (c)   (d)

15. The linear equation $y = 825x + 8689$ provides a model for the number of cases served by the Child Support Enforcement program from 1985 to 1990, where $x = 0$ corresponds to 1985, $x = 1$ corresponds to 1986, and so on. Use this model to approximate the number of cases served during 1989. (*Source:* Office of Child Support Enforcement)

16. What does the number 825 in the equation in Exercise 15 refer to
(a) with respect to the graph?   (b) in the context of the problem?

Graph the inequality or compound inequality.

17. $3x - 2y > 6$ **18.** $y < 2x - 1$ and $x - y < 3$

19. Which one of the following is the graph of a function?

(a)   (b)   (c)   (d)

20. Which of the following does not define a function?
(a) $\{(0, 1), (-2, 3), (4, 8)\}$ (b) $y = 2x - 6$ (c) $y = \sqrt{x + 2}$
(d)

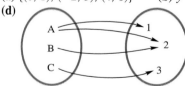

21. If $f(x) = -x^2 + 2x - 1$, find $f(1)$.

22. Graph the linear function $f(x) = \frac{2}{3}x - 1$. What is its domain? What is its range?

Solve the problem.

23. Suppose that y varies directly as x and inversely as z. If $y = \dfrac{3}{2}$ when $x = 3$ and $z = 10$, find y when $x = 4$ and $z = 12$.

24. The current in a simple electrical circuit is inversely proportional to the resistance. If the current is 80 amps when the resistance is 30 ohms, find the current when the resistance is 12 ohms.

▦ **25.** The collision impact of an automobile varies jointly as its mass and the square of its speed. Suppose a 2000-pound car traveling at 55 miles per hour has a collision impact of 6.1. What is the collision impact of the same car at 65 miles per hour? Round to the nearest tenth.

[7.6] 23. $\dfrac{5}{3}$

24. 200 amps   25. 8.5

CUMULATIVE REVIEW (Chapters 1–7)

1. Solve for x. $\dfrac{2x + 3}{2} - \dfrac{x + 12}{8} = -7$

2. The sale price of a compact disc player is $280. This represents 30% off the regular price. What was the regular price?

3. Solve and graph the solution set.
$$4 - 7(q + 4) < -3q$$

4. Solve the compound inequality and graph the solution set.
$$3x - 2 < 10 \quad \text{and} \quad -2x < 10$$

5. Solve the absolute value equation.
$$|5x - 3| = |2x + 6|$$

6. Solve the absolute value inequality and graph the solution set.
$$|4x + 3| \geq 11$$

Perform the indicated operations.

7. $(3k^3 - 5k^2 + 8k - 2) - (4k^3 + 11k + 7) + (2k^2 - 5k)$

8. $(8x - 7)(x + 3)$

9. $\dfrac{8z^3 - 16z^2 + 24z}{8z^2}$

10. $\dfrac{6y^4 - 3y^3 + 5y^2 + 6y - 9}{2y + 1}$

Factor the polynomial completely.

11. $2p^2 - 5pq + 3q^2$ **12.** $18k^4 + 9k^2 - 20$ **13.** $x^3 + 512$

Solve by factoring.

14. $2x^2 + 11x + 15 = 0$ **15.** $5t(t - 1) = 2(1 - t)$

Perform the operation and express the answer in lowest terms.

16. $\dfrac{y^2 + y - 12}{y^3 + 9y^2 + 20y} \div \dfrac{y^2 - 9}{y^3 + 3y^2}$ **17.** $\dfrac{1}{x + y} + \dfrac{3}{x - y}$

1. $\{-8\}$ 2. $400
3. $(-6, \infty)$ ⟵—————⟶
                    $^{-6}$
4. $(-5, 4)$ ⟵————⟶
              $^{-5}$ ^{4}
5. $\left\{-\dfrac{3}{7}, 3\right\}$
6. $\left(-\infty, -\dfrac{7}{2}\right] \cup [2, \infty)$
7. $-k^3 - 3k^2 - 8k - 9$
8. $8x^2 + 17x - 21$
9. $z - 2 + \dfrac{3}{z}$
10. $3y^3 - 3y^2 + 4y + 1 + \dfrac{-10}{2y + 1}$
11. $(2p - 3q)(p - q)$
12. $(3k^2 + 4)(6k^2 - 5)$
13. $(x + 8)(x^2 - 8x + 64)$
14. $\left\{-3, -\dfrac{5}{2}\right\}$
15. $\left\{-\dfrac{2}{5}, 1\right\}$ 16. $\dfrac{y}{y + 5}$
17. $\dfrac{4x + 2y}{(x + y)(x - y)}$

18. $-\dfrac{9}{4}$ 19. $\dfrac{-1}{a+b}$

20. $\emptyset$ 21. $m = \dfrac{5zy}{2y-z}$

22. $3\dfrac{3}{14}$ hours

23. 15 miles per hour

24. $\dfrac{1}{243}$ 25. $x^{1/12}$

26. $8\sqrt{5}$ 27. $\dfrac{-9\sqrt{5}}{20}$

28. $4(\sqrt{6}+\sqrt{5})$ 29. $6\sqrt[3]{4}$

30. $\{6\}$ 31. $\left\{-\dfrac{4}{3}, 0\right\}$

32. $\{-3, 5\}$

33. $\left\{\dfrac{2+\sqrt{10}}{2}, \dfrac{2-\sqrt{10}}{2}\right\}$

34. $\left\{-\dfrac{3}{4}, 1\right\}$

35. $\left(-5, \dfrac{3}{2}\right)$

36. 6 inches, 8 inches, 10 inches   37. x-intercept: $(-2, 0)$; y-intercept: $(0, 4)$

38. $-\dfrac{3}{2}$ 39. $-\dfrac{3}{4}$

40.

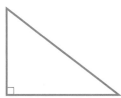

41. (a) $(-\infty, \infty)$ (b) 24
42. $9.92

Simplify the complex fraction.

18. $\dfrac{\dfrac{-6}{x-2}}{\dfrac{8}{3x-6}}$

19. $\dfrac{\dfrac{1}{a}-\dfrac{1}{b}}{\dfrac{a}{b}-\dfrac{b}{a}}$

Solve the equation.

20. $\dfrac{2}{x-3} - \dfrac{3}{x+3} = \dfrac{12}{x^2-9}$

21. $\dfrac{2}{z} - \dfrac{5}{m} = \dfrac{1}{y};$ for m

Solve the problem.

22. Wayne can do a job in 9 hours, while Susan can do the same job in 5 hours. How long would it take them to do the job if they worked together?

23. The current of a river runs at 3 miles per hour. Brent's boat can go 36 miles downstream in the same time that it takes to go 24 miles upstream. Find the speed of the boat in still water.

Simplify.

24. $27^{-5/3}$

25. $\dfrac{x^{-2/3}}{x^{-3/4}}$ $(x \neq 0)$

26. $8\sqrt{20} + 3\sqrt{80} - 2\sqrt{500}$

27. $\dfrac{-9}{\sqrt{80}}$

28. $\dfrac{4}{\sqrt{6}-\sqrt{5}}$

29. $\dfrac{12}{\sqrt[3]{2}}$

30. Solve for x. $\sqrt{8x-4} - \sqrt{7x+2} = 0$

Solve the equation by the method stated.

31. $(3m+2)^2 = 4$ (square root method)

32. $z^2 - 2z = 15$ (completing the square)

33. $2x^2 - 4x - 3 = 0$ (quadratic formula)

34. $4 = \dfrac{1}{t} + \dfrac{3}{t^2}$ (method of your choice)

35. Solve the inequality and graph its solution. $2x^2 + 7x < 15$

36. In a right triangle, the shorter leg is 4 inches less than the hypotenuse, while the hypotenuse is 2 inches longer than the longer leg. What are the lengths of the sides of the triangle?

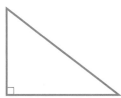

37. Graph $-4x + 2y = 8$ and give the intercepts.

Find the slope of the line described.

38. through $(-5, 8)$ and $(-1, 2)$

39. perpendicular to $4x - 3y = 12$

40. Graph the inequality $-2x + y < -6$.

41. For the function $f(x) = -3x + 6$,
  (a) what is the domain?   **(b)** what is $f(-6)$?

42. The cost of a pizza varies directly as the square of its radius. If a pizza with a 7-inch radius costs $6.00, how much should a pizza with a 9-inch radius cost?

SYSTEMS OF LINEAR EQUATIONS

CONNECTIONS

A few years ago, the United States' industrial inferiority to Japan and Germany was an accepted fact. In the last few years, however, the U.S. has made a strong comeback. The figure shows the economic growth rate from the beginning of 1990 to the end of 1993 for the U.S., Japan, and Germany.* As the graphs indicate, the U.S. has pulled ahead of these two countries economically. The three graphs represent three functions. The equations that correspond to the graphs (which are unknown here) form a system of equations. The points where two of the graphs intersect are frequently points of interest in a system of equations.

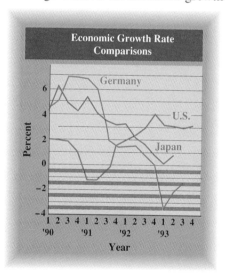

◆ fourth quarter of 1991: about 1.5%; at the end of the first quarter of 1992: 2%; All three graphs would be best fit with a polynomial function. None are linear.

FOR DISCUSSION OR WRITING

Estimate from the figure the approximate year and quarter of the year when the U.S. economic growth rate matched Germany's economic growth rate. What was the growth rate at that time? When did the U.S. growth rate coincide with Japan's? What was the growth rate then? So far in this text, we have studied only linear functions. Are any of these graphs the graph of a linear function?

* Graph, "The U.S. Economy Is Growing," from *The New York Times*, February 27, 1994. Copyright © 1994 by The New York Times Company. Reprinted by permission.

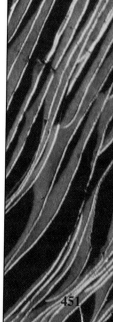

Many of the applied problems studied earlier are more easily solved by writing two or more equations and finding a solution common to this set of equations. When two quantities are to be found, it is natural to use two variables and use the information in the problem to write two equations involving those variables. In this chapter we show how to solve such problems in this way.

8.1 LINEAR SYSTEMS OF EQUATIONS IN TWO VARIABLES

FOR EXTRA HELP	OBJECTIVES
SSG pp. 239–246 **SSM** pp. 391–399	**1** ▶ Decide whether an ordered pair is a solution of a linear system.
Video 9	**2** ▶ Solve linear systems by graphing.
Tutorial IBM MAC	**3** ▶ Solve linear systems with two equations and two unknowns by the elimination method. **4** ▶ Solve linear systems with two equations and two unknowns by the substitution method. **5** ▶ Solve linear systems using a graphics calculator.

Methods shown in this chapter help to identify numbers that make two or more equations true at the same time. Such a set of equations is called a **system of equations.**

1 ▶ **Decide whether an ordered pair is a solution of a linear system.**

▶ Recall from the previous chapter that the graph of a first-degree equation of the form $Ax + By = C$ is a straight line. For this reason, such an equation is called a linear equation. Two or more linear equations form a **linear system.** The solution set of a linear system of equations contains all ordered pairs that satisfy all the equations of the system at the same time.

EXAMPLE 1

Deciding Whether an Ordered Pair Is a Solution

Decide whether the given ordered pair is a solution of the system.

(a) $x + y = 6$; $(4, 2)$
$4x - y = 14$

Replace x with 4 and y with 2 in each equation of the system.

$x + y = 6$		$4x - y = 14$	
$4 + 2 = 6$	?	$4(4) - 2 = 14$	?
$6 = 6$ True		$14 = 14$ True	

Since $(4, 2)$ makes both equations true, $(4, 2)$ is a solution of the system.

(b) $3x + 2y = 11$; $(-1, 7)$
$x + 5y = 36$

$3x + 2y = 11$		$x + 5y = 36$	
$3(-1) + 2(7) = 11$	?	$-1 + 5(7) = 36$	?
$-3 + 14 = 11$ True		$-1 + 35 = 36$ False	

The ordered pair $(-1, 7)$ is not a solution of the system, since it does not make *both* equations true. ■

INSTRUCTOR'S RESOURCES

ITM pp. 452–456   **ISM** pp. 587–602   **TEST GENERATOR**   **TRANSPARENCIES**
IAM pp. 36–37   DOS   Windows   MAC

2 ▶ Solve linear systems by graphing.

▶ Sometimes we can estimate the solution set of a linear system of equations by graphing the equations of the system on the same axes and then estimating the coordinates of any point of intersection.

E X A M P L E 2

Solving a System by Graphing

■ Solve the following system by graphing.

$$x + y = 5$$
$$2x - y = 4$$

The graphs of these linear equations are shown in Figure 1. The graph suggests that the point of intersection is the ordered pair (3, 2). Check this by substituting these values for x and y in both of the equations. As the check shows, the solution set of the system is $\{(3, 2)\}$. ■

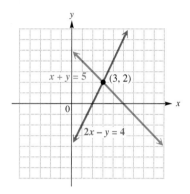

FIGURE 1

Since the graph of a linear equation is a straight line, there are three possibilities for the solution of a system of two linear equations, as shown in Figure 2.

Graphs of a Linear System

1. The two graphs intersect in a single point. The coordinates of this point give the only solution of the system. This is the most common case. See Figure 2(a).
2. The graphs are parallel lines. In this case the system is **inconsistent;** that is, there is no solution common to both equations of the system, and the solution set is Ø. See Figure 2(b).
3. The graphs are the same line. In this case the equations are **dependent,** since any solution of one equation of the system is also a solution of the other. The solution set is an infinite set of ordered pairs representing the points on the line. See Figure 2(c).

▶ TEACHING TIP
Review the procedure for carefully graphing lines by determining the *x*- and *y*-intercepts (where possible). Students *must* use a straight edge to solve systems of equations graphically. ◀

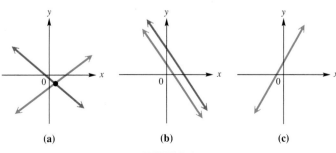

(a) (b) (c)

FIGURE 2

3 ▶ Solve linear systems with two equations and two unknowns by the elimination method.

▶ It is possible to find the solution of a system of equations by graphing. However, since it can be hard to read exact coordinates from a graph, an algebraic method is usually used to solve a system. (See Objective 5, where we discuss using a graphics calculator to solve a system by graphing.) One such algebraic method, called the **elimination method** (or the **addition method**), is explained in the following examples.

E X A M P L E   3

Solving a System by Elimination

Solve the system

$$5x - 2y = 4 \tag{1}$$
$$2x + 3y = 13. \tag{2}$$

The elimination method involves combining the two equations so that one variable is eliminated. This is done using the following fact.

$$\text{If } a = b \text{ and } c = d, \text{ then } a + c = b + d.$$

Suppose we wish to eliminate the variable x. Since the coefficients of x are *not* opposites, we must first transform one or both equations so that the coefficients *are* opposites. Then, we combine the equations, the term with x will have a coefficient of 0, and we will be able to solve for y. We begin by multiplying equation (1) by 2 and equation (2) by -5.

$$10x - \ 4y = 8 \qquad \text{2 times each side of equation (1)}$$
$$-10x - 15y = -65 \qquad \text{-5 times each side of equation (2)}$$

Now we add.

$$\begin{aligned} 10x - \ 4y &= \ \ \ 8 \\ -10x - 15y &= -65 \\ \hline -19y &= -57 \\ y &= 3 \end{aligned}$$

To find x, we substitute 3 for y in either equation (1) or equation (2). Substituting in equation (2) gives

$$\begin{aligned} 2x + 3y &= 13 \\ 2x + 3(\mathbf{3}) &= 13 \qquad \text{Let } y = 3. \\ 2x + 9 &= 13 \\ 2x &= 4 \qquad \text{Subtract 9.} \\ x &= 2. \qquad \text{Divide by 2.} \end{aligned}$$

The solution set of the system is $\{(2, 3)\}$. Check this solution in both equations of the given system. ■

The general method of solving a system by the elimination method is summarized as follows.

Solving Linear Systems by Elimination	
Step 1	Write both equations in the form $Ax + By = C$.
Step 2	Multiply one or both equations by appropriate numbers so that the sum of the coefficients of either x or y is zero.
Step 3	Add the new equations. The sum should be an equation with just one variable.
Step 4	Solve the equation from Step 3.
Step 5	Substitute the result of Step 4 into either of the given equations and solve for the other variable.
Step 6	Check the solution in both of the given equations.

EXAMPLE 4

Solving a System of Dependent Equations

▶ **TEACHING TIP**

The solution set for a system of dependent equations is written in *set builder notation,* first introduced in Chapter 1. ◀

Solve the system

$$2x - y = 3 \qquad \textbf{(3)}$$
$$6x - 3y = 9. \qquad \textbf{(4)}$$

Multiply both sides of equation (3) by -3, and then add the result to equation (4).

$$\begin{array}{ll} -6x + 3y = -9 & \text{-3 times each side of equation (3)} \\ \underline{6x - 3y = 9} & \text{(4)} \\ 0 = 0 & \text{True} \end{array}$$

Adding these equations gave the true statement $0 = 0$. In the original system, equation (4) could be obtained from equation (3) by multiplying both sides of equation (3) by 3. Because of this, equations (3) and (4) are equivalent and have the same line as their graph. The solution set, the infinite set of ordered pairs on the line with equation $2x - y = 3$, is written as

$$\{(x, y) \mid 2x - y = 3\}. \quad ■$$

EXAMPLE 5

Solving an Inconsistent System

Solve the system

$$x + 3y = 4 \qquad \textbf{(5)}$$
$$-2x - 6y = 3. \qquad \textbf{(6)}$$

Multiply both sides of equation (5) by 2, and then add the result to equation (6).

$$\begin{array}{ll} 2x + 6y = 8 & \text{2 times each side of equation (5)} \\ \underline{-2x - 6y = 3} & \text{(6)} \\ 0 = 11 & \text{False} \end{array}$$

The result of the addition step here is a false statement, which shows that the system is inconsistent. The graphs of the equations of the system are parallel lines similar to the graphs in Figure 2(b). There are no ordered pairs that satisfy both equations, so the solution set of the system is $\emptyset$. ■

The results of Examples 4 and 5 are generalized as follows.

> If both variables are eliminated when a system of linear equations is solved,
>
> **1.** there is no solution if the resulting statement is *false;*
> **2.** there are infinitely many solutions if the resulting statement is *true.*

Slopes and y-intercepts can be used to decide if the graphs of a system of equations are parallel lines or if they coincide. For the system of Example 4, writing each equation in slope-intercept form shows that both lines have a slope of 2 and a y-intercept of $(0, -3)$, so the graphs are the same line and the system has an infinite solution set.

In Example 5, both equations have a slope of $-1/3$, but the y-intercepts are $(0, 4/3)$ and $(0, -1/2)$, showing that the graphs are two distinct parallel lines. Thus, the system has no solution.

4 ▶ Solve linear systems with two equations and two unknowns by the substitution method.

▶ Linear systems can also be solved by the **substitution method.** The substitution method is most useful in solving linear systems in which one variable has a coefficient of 1. However, as shown in the next chapter, the substitution method is the best choice for solving many *nonlinear* systems.

The method of solving a system by substitution is summarized as follows.

Solving Linear Systems by Substitution

Step 1 Solve one of the equations for either variable. (If one of the variables has coefficient 1 or −1, choose it, since the substitution method is usually easier this way.)

Step 2 Substitute for that variable in the other equation. The result should be an equation with just one variable.

Step 3 Solve the equation from Step 2.

Step 4 Substitute the result from Step 3 into the equation from Step 1 to find the value of the other variable.

Step 5 Check the solution in both of the given equations.

The next two examples illustrate this method.

EXAMPLE 6

Solving a System by Substitution

Solve the system

$$\frac{x}{2} + \frac{y}{3} = \frac{13}{6} \tag{7}$$

$$4x - y = -1. \tag{8}$$

Write the first equation without fractions by multiplying both sides by the common denominator, 6.

$$6 \cdot \frac{x}{2} + 6 \cdot \frac{y}{3} = 6 \cdot \frac{13}{6} \qquad \text{Multiply by 6.}$$

$$3x + 2y = 13 \tag{9}$$

The new system is

$$3x + 2y = 13 \qquad (9)$$

$$4x - y = -1. \qquad (8)$$

To use the substitution method, first solve one of the equations for either x or y. Since the coefficient of y in equation (8) is −1, it is easiest to solve for y in equation (8).

$$-y = -1 - 4x$$

$$y = 1 + 4x$$

Substitute $1 + 4x$ for y in equation (9), and solve for x.

$$3x + 2y = 13 \qquad (9)$$

$$3x + 2(1 + 4x) = 13 \qquad \text{Let } y = 1 + 4x.$$

$$3x + 2 + 8x = 13 \qquad \text{Distributive property}$$

$$11x = 11 \qquad \text{Combine terms; subtract 2.}$$

$$x = 1 \qquad \text{Divide by 11.}$$

Since $y = 1 + 4x$, $y = 1 + 4(1) = 5$. Check that the solution set is $\{(1, 5)\}$. ■

Chalkboard Exercise

Solve by substitution.

$$\frac{x}{5} + \frac{2y}{3} = -\frac{8}{5}$$

$$\frac{3x}{4} - \frac{y}{3} = \frac{5}{2}$$

Answer: $\{(2, -3)\}$

▶ **TEACHING TIP**

Emphasize that one should multiply *each* term by the common denominator. ◀

EXAMPLE 7

Solving a System by Substitution

Solve the system

$$4x - 3y = 7 \qquad \textbf{(10)}$$
$$3x - 2y = 6. \qquad \textbf{(11)}$$

If the substitution method is to be used, one equation must be solved for one of the two variables. Let us solve equation (11) for x.

$$3x = 2y + 6$$
$$x = \frac{2y + 6}{3}$$

Now substitute $\dfrac{2y + 6}{3}$ for x in equation (10).

$$4x - 3y = 7 \qquad (10)$$
$$4\left(\frac{2y + 6}{3}\right) - 3y = 7 \qquad \text{Let } x = \tfrac{2y + 6}{3}.$$

Multiply both sides of the equation by the common denominator 3 to eliminate the fraction.

$$4(2y + 6) - 9y = 21 \qquad \text{Multiply by 3.}$$
$$8y + 24 - 9y = 21 \qquad \text{Distributive property}$$
$$24 - y = 21 \qquad \text{Combine terms.}$$
$$-y = -3 \qquad \text{Add } -24.$$
$$y = 3 \qquad \text{Divide by } -1.$$

Since $x = \dfrac{2y + 6}{3}$ and $y = 3$,

$$x = \frac{2(3) + 6}{3} = \frac{6 + 6}{3} = 4,$$

and the solution set is $\{(4, 3)\}$. ∎

The substitution method is usually not the best choice for a system like the one in Example 7. However, it is sometimes necessary when solving a system of *nonlinear* equations to proceed as shown in this example.

5 ▶ Solve linear systems using a graphics calculator.

▶ In Example 3 we solved the system

$$5x - 2y = 4$$
$$2x + 3y = 13$$

by elimination. As shown there, the solution set of the system is $\{(2, 3)\}$, meaning that the graphs of the two equations are lines that intersect at $(2, 3)$. This can be supported with a graphics calculator by graphing the two equations and using the capabilities of the calculator to find the point of intersection. First, however, we must write each equation in linear function form—that is, solved for y—so that the equations can be entered. Solving $5x - 2y = 4$ for y, we get $y = (5/2)x - 2$.

Solving $2x + 3y = 13$ for y gives $y = -(2/3)x + (13/3)$. If we enter these equations as y_1 and y_2, respectively, we get the graphs shown in Figure 3. The display at the bottom of the figure indicates that the point of intersection is $(2, 3)$, as we determined algebraically in Example 3.

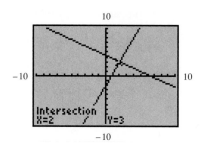

FIGURE 3

8.1 EXERCISES

2. ordered pair
3. yes   4. yes   5. no
6. no
7. $\{(2, 2)\}$

8. $\{(-2, -3)\}$

9. $\{(3, -1)\}$ 10. $\{(1, 1)\}$
11. $\{(2, -3)\}$
12. $\{(4, -5)\}$
13. $\left\{\left(\dfrac{3}{2}, -\dfrac{3}{2}\right)\right\}$
14. $\left\{\left(\dfrac{1}{4}, -\dfrac{1}{2}\right)\right\}$
15. $\{(x, y)\,|\,7x + 2y = 6\}$; dependent equations
16. $\{(x, y)\,|\,x - 4y = 2\}$; dependent equations
17. $\{(2, -4)\}$
18. $\left\{\left(13, -\dfrac{7}{5}\right)\right\}$
19. $\emptyset$;   inconsistent system
20. $\emptyset$;   inconsistent system

1. What is meant by a system of equations?

2. A solution of a system of linear equations in two variables is a(n) _____ _____ of real numbers.

Decide whether the ordered pair is a solution of the given system. See Example 1.

3. $x + y = 6$ $(5, 1)$
$x - y = 4$

4. $x - y = 17$ $(8, -9)$
$x + y = -1$

5. $2x - y = 8$ $(5, 2)$
$3x + 2y = 20$

6. $3x - 5y = -12$ $(-1, 2)$
$x - y = 1$

Solve the system by graphing. See Example 2.

7. $x + y = 4$
$2x - y = 2$

8. $x + y = -5$
$-2x + y = 1$

Solve the system by elimination. See Examples 3–5.

9. $2x - 5y = 11$
$3x + y = 8$

10. $-2x + 3y = 1$
$-4x + y = -3$

11. $3x + 4y = -6$
$5x + 3y = 1$

12. $4x + 3y = 1$
$3x + 2y = 2$

13. $3x + 3y = 0$
$4x + 2y = 3$

14. $8x + 4y = 0$
$4x - 2y = 2$

15. $7x + 2y = 6$
$-14x - 4y = -12$

16. $x - 4y = 2$
$4x - 16y = 8$

17. $\dfrac{x}{2} + \dfrac{y}{3} = -\dfrac{1}{3}$

$\dfrac{x}{2} + 2y = -7$

18. $\dfrac{x}{5} + y = \dfrac{6}{5}$

$\dfrac{x}{10} + \dfrac{y}{3} = \dfrac{5}{6}$

19. $5x - 5y = 3$
$x - y = 12$

20. $2x - 3y = 7$
$-4x + 6y = 14$

21. Suppose that two linear equations are graphed on the same set of coordinate axes. Sketch what the graph might look like if the system has the given description.
   (a) The system has a single solution.   **(b)** The system has no solution.
   (c) The system has infinitely many solutions.

 CONCEPTUAL   ✎ WRITING   ▲ CHALLENGING   ▦ SCIENTIFIC CALCULATOR   ▦ GRAPHICS CALCULATOR

22. How many solutions will a system of linear equations have if
  (a) the lines have the same slope but different y-intercepts?
  (b) the lines have different slopes?
  (c) the lines have the same slope and the same y-intercept?

Write the two equations of the system in slope-intercept form and then tell how many solutions the system has. Do not actually solve the system.

23. $3x + 7y = 4$
    $6x + 14y = 3$

24. $-x + 2y = 8$
    $4x - 8y = 1$

25. $2x = -3y + 1$
    $6x = -9y + 3$

26. $5x = -2y + 1$
    $10x = -4y + 2$

Solve the system by substitution. See Examples 6 and 7.

27. $4x + y = 6$
    $y = 2x$

28. $2x - y = 6$
    $y = 5x$

29. $\quad 3x - 4y = -22$
    $-3x + \ y = 0$

30. $-3x + \ y = -5$
    $\quad x + 2y = 0$

31. $-x - 4y = -14$
    $\ 2x = y + 1$

32. $-3x - 5y = -17$
    $\ 4x = y - 8$

33. $5x - 4y = 9$
    $3 - 2y = -x$

34. $6x - y = -9$
    $4 + 7x = -y$

35. $\ x = 3y + 5$
    $\ x = \dfrac{3}{2}y$

36. $x = 6y - 2$
    $x = \dfrac{3}{4}y$

37. $\dfrac{1}{2}x + \dfrac{1}{3}y = 3$
    $y = 3x$

38. $\dfrac{1}{4}x - \dfrac{1}{5}y = 9$
    $y = 5x$

39. Refer to Example 3. What other numbers might equations (1) and (2) be multiplied by to eliminate y by adding the two equations?

40. What makes a system of equations inconsistent?

41. What must be true for a system of two linear equations (in two variables) to have no solution?

42. What must be true for a system of two linear equations (in two variables) to have an infinite number of solutions?

▲ *In the system let $p = 1/x$ and $q = 1/y$. Substitute, solve for p and q, and then find x and y. (Hint: $3/x = 3 \cdot 1/x = 3p$.)*

43. $\dfrac{3}{x} + \dfrac{4}{y} = \dfrac{5}{2}$

  $\dfrac{5}{x} - \dfrac{3}{y} = \dfrac{7}{4}$

44. $\dfrac{4}{x} - \dfrac{9}{y} = -1$

  $-\dfrac{7}{x} + \dfrac{6}{y} = -\dfrac{3}{2}$

45. $\dfrac{2}{x} - \dfrac{5}{y} = \dfrac{3}{2}$

  $\dfrac{4}{x} + \dfrac{1}{y} = \dfrac{4}{5}$

46. $\dfrac{2}{x} + \dfrac{3}{y} = \dfrac{11}{2}$

  $-\dfrac{1}{x} + \dfrac{2}{y} = -1$

▲ *Solve by any method. Assume that a and b represent nonzero constants.*

47. $\quad ax + \ by = 2$
    $-ax + 2by = 1$

48. $2ax - y = 3$
    $\quad y = 5ax$

49. $\ 3ax + 2y = 1$
    $-ax + \ y = 2$

50. $ax + by = c$
    $bx + ay = c$

21. Answers will vary.
(a)

(b)

(c)

22. (a) 0 **(b)** 1
(c) infinitely many

23. $y = -\dfrac{3}{7}x + \dfrac{4}{7}$;

$y = -\dfrac{3}{7}x + \dfrac{3}{14}$; 0

24. $y = \dfrac{1}{2}x + 4$;

$y = \dfrac{1}{2}x - \dfrac{1}{8}$; 0

25. both are

$y = -\dfrac{2}{3}x + \dfrac{1}{3}$;

infinitely many
26. both are

$y = -\dfrac{5}{2}x + \dfrac{1}{2}$; infinitely

many
27. $\{(1, 2)\}$
28. $\{(-2, -10)\}$

29. $\left\{\left(\dfrac{22}{9}, \dfrac{22}{3}\right)\right\}$

30. $\left\{\left(\dfrac{10}{7}, -\dfrac{5}{7}\right)\right\}$

31. $\{(2, 3)\}$ **32.** $\{(-1, 4)\}$
33. $\{(5, 4)\}$ **34.** $\{(-1, 3)\}$

35. $\left\{\left(-5, -\dfrac{10}{3}\right)\right\}$

36. $\left\{\left(\dfrac{2}{7}, \dfrac{8}{21}\right)\right\}$

37. $\{(2, 6)\}$
38. $\{(-12, -60)\}$
39. Multiply (1) by 3 and (2) by 2.
43. $\{(2, 4)\}$ **44.** $\{(2, 3)\}$
45. $\{(4, -5)\}$

46. $\left\{\left(\dfrac{1}{2}, 2\right)\right\}$

47. $\left\{\left(\dfrac{1}{a},\dfrac{1}{b}\right)\right\}$

48. $\left\{\left(-\dfrac{1}{a},-5\right)\right\}$

49. $\left\{\left(-\dfrac{3}{5a},\dfrac{7}{5}\right)\right\}$

50. $\left\{\left(\dfrac{c}{a+b},\dfrac{c}{a+b}\right)\right\}$

51. $\{(1,3)\}$
52. $f(x)=-3x+6$; linear

53. $g(x)=\dfrac{2}{3}x+\dfrac{7}{3}$; linear

54. one; 1; 3; 1; 3; 1; 3 55. $(3,-4)$
56. (a) $y_1=4x+4$
(b) $y_2=-2x+7$

(c) $\left\{\left(\dfrac{1}{2},6\right)\right\}$

57. (a)

◆─────◆ **MATHEMATICAL CONNECTIONS** (Exercises 51–54) ◆─────◆

Work Exercises 51–54 in order to see the connections between systems of linear equations and the graphs of linear functions.

51. Solve the system

$$3x + y = 6$$
$$-2x + 3y = 7.$$

Use elimination or substitution.

52. For the first equation in the system of Exercise 51, solve for y and rename it $f(x)$. What special kind of function is f?

53. For the second equation in the system of Exercise 51, solve for y and rename it $g(x)$. What special kind of function is g?

⊙ **54.** Use the result of Exercise 51 to fill in the blanks with the appropriate responses:

Because the graphs of f and g are straight lines that are neither parallel nor coincide, they intersect in exactly ____ point. The coordinates of the point are (____ , ____). Using functional notation, this is given by $f($____$) =$ ____ and $g($____$) =$ ____ .

◆────────◆────────

55. The table shown was generated by a graphics calculator. The functions defined by y_1 and y_2 are linear. Based on the table, what are the coordinates of the point of intersection of the graphs?

X	Y1	Y2
0	-7	-1
1	-6	-2
2	-5	-3
3	-4	-4
4	-3	-5
5	-2	-6
6	-1	-7

X=0

56. The table shown was generated by a graphics calculator. The functions defined by y_1 and y_2 are linear.
(a) Use the methods of Chapter 7 to find the equation for y_1.
(b) Use the methods of Chapter 7 to find the equation for y_2.
(c) Solve the system of equations formed by y_1 and y_2.

X	Y1	Y2
0	4	7
1	8	5
2	12	3
3	16	1
4	20	-1
5	24	-3
6	28	-5

X=0

57. The solution set of the system

$$y_1 = 3x - 5$$
$$y_2 = -4x + 2$$

is $\{(1, -2)\}$. Using slopes and y-intercepts, determine which one of the two calculator-generated graphs is the appropriate one for this system.

(a)

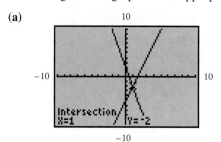

(b)

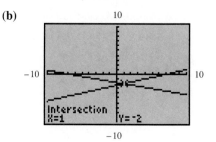

58. Which one of the ordered pairs listed could be the only possibility for the solution of the system whose graphs are shown in the standard viewing window of a graphics calculator?
(a) $(15, -15)$ (b) $(15, 15)$
(c) $(-15, 15)$ (d) $(-15, -15)$

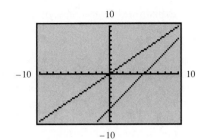

For the system **(a)** *solve by elimination or substitution and* **(b)** *use a graphics calculator to support your result. In part* **(b)***, be sure to solve each equation for y first. See Objective 5.*

59. $x + y = 10$
$2x - y = 5$

60. $6x + y = 5$
$-x + y = -9$

61. $3x - 2y = 4$
$3x + y = -2$

62. $2x - 3y = 3$
$2x + 2y = 8$

Answer the questions in Exercises 63 and 64 by observing the graphs provided.

63. Eboni Perkins compared the monthly payments she would incur for two types of mortgages: fixed-rate and variable-rate. Her observations led to this graph.
(a) For which years would the monthly payment be more for the fixed-rate mortgage than the variable-rate mortgage?
(b) In what year would the payments be the same, and what would those payments be?

64. The figure shows graphs that represent the supply and demand for a certain brand of low-fat frozen yogurt at various prices per half gallon.
(a) At what price does supply equal demand?
(b) For how many half-gallons does supply equal demand?
(c) What are the supply and the demand at a price of $2 per half-gallon?

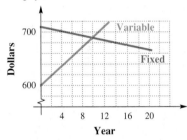

Year

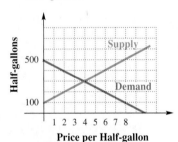

Price per Half-gallon

65. The accompanying graph shows the trends during the years 1966–1990 relating to bachelor's degrees awarded in the United States. (*Source:* National Science Foundation)
(a) Between what years shown on the horizontal axis did the number of degrees in all fields for men and women reach equal numbers?
(b) When the number of degrees for men and women reached equal numbers, what was that number (approximately)?

58. (b)
59. (a) $\{(5, 5)\}$
(b)
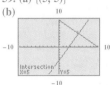

60. (a) $\{(2, -7)\}$
(b)
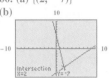

61. (a) $\{(0, -2)\}$
(b)
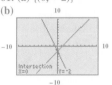

62. (a) $\{(3, 1)\}$
(b)
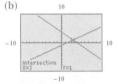

63. (a) years 0 to 10
(b) year 10; about $690
64. (a) $4 **(b)** 300 half-gallons **(c)** supply: 200 half-gallons; demand: 400 half-gallons
65. (a) 1978 and 1982
(b) just less than 500,000

○ CONCEPTUAL   ✎ WRITING   ▲ CHALLENGING   ▦ SCIENTIFIC CALCULATOR   ▦ GRAPHICS CALCULATOR

66. (a) In 1991 they both reached the level of about 360 million.
(b) (1987, 100 million)
67. $8x - 12y + 4z = 20$
68. $9x - 24y + 3z = 0$
69. 4   70. 2

 (c) Explain why, if trends continue, the number of degrees for women in science and engineering will eventually equal the number of similar degrees for men.

66. The accompanying graph shows how the production of vinyl LPs, audiocassettes, and compact discs (CDs) changed over the years from 1983 to 1993. (*Source*: Recording Industry of America)

(a) In what year did cassette production and CD production reach equal levels? What was that level?

(b) Express as an ordered pair of the form (year, production level) the point of intersection of the graphs of LP production and CD production.

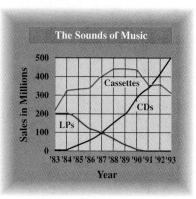

REVIEW EXERCISES

Multiply both sides of the equation by the given number. See Section 2.1.

67. $2x - 3y + z = 5$ by 4

68. $-3x + 8y - z = 0$ by -3

Solve for z if $x = 1$ and $y = -2$.

69. $x + 2y + 3z = 9$

70. $-3x - y + z = 1$

8.2 LINEAR SYSTEMS OF EQUATIONS IN THREE VARIABLES

FOR EXTRA HELP

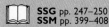 **SSG** pp. 247–250
SSM pp. 399–408

 Video 10

Tutorial IBM   MAC

OBJECTIVES

1 ▶ Solve linear systems with three equations and three unknowns by the elimination method.

2 ▶ Solve linear systems with three equations and three unknowns where some of the equations have missing terms.

3 ▶ Solve linear systems with three equations and three unknowns that are inconsistent or that include dependent equations.

A solution of an equation in three variables, such as $2x + 3y - z = 4$, is called an **ordered triple** and is written (x, y, z). For example, the ordered triples $(1, 1, 1)$ and $(10, -3, 7)$ are each solutions of $2x + 3y - z = 4$, since the numbers in these ordered triples satisfy the equation when used as replacements for x, y, and z, respectively.

In the rest of this chapter, the term *linear equation* is extended to first-degree equations of the form $Ax + By + Cz + \cdots + Dw = K$. For example, $2x + 3y - 5z = 7$ and $x - 2y - z + 3w - 2v = 8$ are linear equations, the first having three variables, and the second having five variables.

INSTRUCTOR'S RESOURCES

ITM pp. 456–459   **ISM** pp. 602–617
IAM p. 37

TEST GENERATOR
DOS   Windows   MAC

TRANSPARENCIES

In this section we discuss the solution of a system of linear equations in three variables such as

$$4x + 8y + z = 2$$
$$x + 7y - 3z = -14$$
$$2x - 3y + 2z = 3.$$

Theoretically, a system of this type can be solved by graphing. However, the graph of a linear equation with three variables is a *plane* and not a line. Since the graph of each equation of the system is a plane, which requires three-dimensional graphing, this method is not practical. However, it does serve to illustrate the number of solutions possible for such systems, as Figure 4 shows.

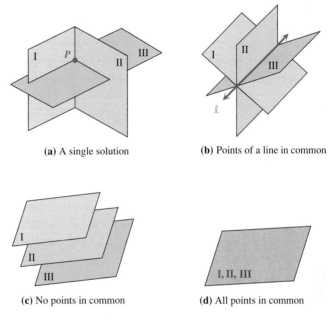

(a) A single solution   (b) Points of a line in common

(c) No points in common   (d) All points in common

FIGURE 4

Graphs of Linear Systems in Three Variables	1. The three planes may meet at a single, common point that is the solution of the system. See Figure 4(a).

1. The three planes may meet at a single, common point that is the solution of the system. See Figure 4(a).
2. The three planes may have the points of a line in common so that the set of points that satisfy the equation of the line is the solution of the system. See Figure 4(b).
3. The planes may have no points common to all three so that there is no solution for the system. See Figure 4(c).
4. The three planes may coincide so that the solution of the system is the set of all points on a plane. See Figure 4(d).

▶ **TEACHING TIP**
Use the walls, floor, and ceiling of your classroom as examples of intersecting and parallel planes. ◀

1 ▶ Solve linear systems with three equations and three unknowns by the elimination method.

▶   Since a graphic solution of a system of three equations in three variables is impractical, these systems are solved with an extension of the elimination method, summarized as follows.

Solving Linear Systems in Three Variables by Elimination

Step 1 Use the elimination method to eliminate any variable from any two of the given equations. The result is an equation in two variables.

Step 2 Eliminate the *same* variable from any *other* two equations. The result is an equation in the same two variables as in Step 1.

Step 3 Use the elimination method to eliminate a second variable from the two equations in two variables that result from Steps 1 and 2. The result is an equation in one variable that gives the value of that variable.

Step 4 Substitute the value of the variable found in Step 3 into either of the equations in two variables to find the value of the second variable.

Step 5 Use the values of the two variables from Steps 3 and 4 to find the value of the third variable by substituting into any of the original equations.

▶ **TEACHING TIP**

It helps to outline this procedure as going from *three equations* (in three unknowns) to *two equations* (in two unknowns) to *one equation* (in one unknown). Once this last equation is solved, proceed to substitute this answer into a two-variable equation to find the value of a second unknown. Finally, substitute the two known values into a three-variable equation. ◄

EXAMPLE 1

Solving a System in Three Variables

Solve the system

$$4x + 8y + z = 2 \tag{1}$$
$$x + 7y - 3z = -14 \tag{2}$$
$$2x - 3y + 2z = 3. \tag{3}$$

As before, the elimination method involves eliminating a variable from the sum of two equations. The choice of which variable to eliminate is arbitrary. Suppose we decide to begin by eliminating z. To do this, multiply both sides of equations (1) by 3 and then add the result to equation (2).

$$\begin{array}{ll} 12x + 24y + 3z = 6 & \text{Multiply both sides of (1) by 3.} \\ \underline{x + 7y - 3z = -14} & \text{(2)} \\ 13x + 31y = -8 & \text{(4)} \end{array}$$

Equation (4) has only two variables. To get another equation without z, multiply both sides of equation (1) by -2 and add the result to equation (3). It is essential at this point to eliminate the *same variable, z*.

$$\begin{array}{ll} -8x - 16y - 2z = -4 & \text{Mutliply both sides of (1) by } -2. \\ \underline{2x - 3y + 2z = 3} & \text{(3)} \\ -6x - 19y = -1 & \text{(5)} \end{array}$$

Now solve the system of equations (4) and (5) for x and y. This step is possible only if the *same* variable is eliminated in the first two steps.

$$\begin{array}{ll} 78x + 186y = -48 & \text{Multiply both sides of (4) by 6.} \\ \underline{-78x - 247y = -13} & \text{Multiply both sides of (5) by 13.} \\ -61y = -61 \\ y = 1 \end{array}$$

Substitute 1 for y in either equation (4) or (5). Choosing (5) gives

$$-6x - 19y = -1 \quad \text{(5)}$$
$$-6x - 19(1) = -1 \quad \text{Let } y = 1.$$
$$-6x - 19 = -1$$
$$-6x = 18$$
$$x = -3.$$

Substitute -3 for x and 1 for y in equation (1) to find z. (Any one of the three given equations could have been used.)

$$4x + 8y + z = 2 \quad (1)$$
$$4(-3) + 8(1) + z = 2 \quad \text{Let } x = -3 \text{ and } y = 1.$$
$$z = 6$$

The ordered triple $(-3,\ 1,\ 6)$ is the only solution of the system. Check that the solution satisfies all three equations of the system, so the solution set is $\{(-3,\ 1,\ 6)\}$. ∎

2 ▶ Solve linear systems with three equations and three unknowns where some of the equations have missing terms.

▶   When one or more of the equations of a system has a missing term, one elimination step can be omitted.

E X A M P L E   2
Solving a System of Equations with Missing Terms

■ Solve the system

$$6x - 12y = -5 \tag{6}$$
$$8y + z = 0 \tag{7}$$
$$9x - z = 12. \tag{8}$$

Since equation (8) is missing the variable y, one way to begin the solution is to eliminate y again with equations (6) and (7).

$$
\begin{array}{ll}
12x - 24y \quad\quad = -10 & \text{Multiply both sides of (6) by 2.} \\
\underline{\quad\quad 24y + 3z = \quad 0} & \text{Multiply both sides of (7) by 3.} \\
12x \quad\quad + 3z = -10 & \tag{9}
\end{array}
$$

Use this result, together with equation (8), to eliminate z. Multiply both sides of equation (8) by 3. This gives

$$
\begin{array}{ll}
27x - 3z = \quad 36 & \text{Multiply both sides of (8) by 3.} \\
\underline{12x + 3z = -10} & \text{(9)} \\
39x \quad\quad = \quad 26 &
\end{array}
$$
$$x = \frac{26}{39} = \frac{2}{3}.$$

Substitution into equation (8) gives

$$9x - z = 12 \quad (8)$$
$$9\left(\frac{2}{3}\right) - z = 12 \quad \text{Let } x = \tfrac{2}{3}.$$
$$6 - z = 12$$
$$z = -6.$$

Substitution of -6 for z in equation (7) gives

$$8y + z = 0 \quad (7)$$
$$8y - 6 = 0 \quad \text{Let } z = -6.$$
$$8y = 6$$
$$y = \frac{3}{4}.$$

Check in each of the original equations of the system to verify that the solution set of the system is $\{(2/3,\ 3/4,\ -6)\}$. ∎

3 ▶ Solve linear systems with three equations and three unknowns that are inconsistent or that include dependent equations.

▶ Linear systems with three variables may be inconsistent or may include dependent equations. The next examples illustrate these cases.

E X A M P L E 3

Solving an Inconsistent System with Three Variables

Solve the system

$$2x - 4y + 6z = 5 \qquad \textbf{(10)}$$
$$-x + 3y - 2z = -1 \qquad \textbf{(11)}$$
$$x - 2y + 3z = 1. \qquad \textbf{(12)}$$

Eliminate x by adding equations (11) and (12) to get the equation

$$y + z = 0.$$

Now, *eliminate x again,* using equations (10) and (12).

$$
\begin{array}{ll}
-2x + 4y - 6z = -2 & \text{Multiply both sides of } (12) \text{ by } -2. \\
\underline{2x - 4y + 6z = 5} & (10) \\
\ 0 = 3 & \text{False}
\end{array}
$$

The resulting false statement indicates that equations (10) and (12) have no common solution. Thus, the system is inconsistent and the solution set is ∅. The graph of this system would show at least two of the planes parallel to one another. (See Figure 4.) ■

NOTE If you get a false statement from the addition step as in Example 3, you do not need to go any further with the solution. Since two of the three planes are parallel, it is not possible for the three planes to have any common points.

E X A M P L E 4

Solving a System of Dependent Equations with Three Variables

Solve the system

$$2x - 3y + 4z = 8 \qquad \textbf{(13)}$$
$$-x + \frac{3}{2}y - 2z = -4 \qquad \textbf{(14)}$$
$$6x - 9y + 12z = 24. \qquad \textbf{(15)}$$

Multiplying both sides of equation (13) by 3 gives equation (15). Multiplying both sides of equation (14) by -6 also results in equation (15). Because of this, the equations are dependent. All three equations have the same graph, as illustrated in Figure 4(d). The solution set is written $\{(x, y, z) \mid 2x - 3y + 4z = 8\}$. Although any one of the three equations could be used to write the solution set, we prefer to use the equation with coefficients that are integers with no common factor (except 1). This is similar to our choice of a standard form for a linear equation earlier. ■

The method discussed in this section can be extended to solve larger systems. For example, to solve a system of four equations in four variables, eliminate a variable from three pairs of equations to get a system of three equations in three unknowns. Then proceed as shown above.

8.2 EXERCISES

1. Explain what the following statement means: The solution set of the system

$$2x + y + z = 3$$
$$3x - y + z = -2 \quad \text{is } \{(-1, 2, 3)\}.$$
$$4x - y + 2z = 0$$

2. Write a system of three linear equations in three variables that has solution set $\{(3, 1, 2)\}$. Then solve the system. (*Hint:* Start with the solution and make up three equations that are satisfied by the solution. There are many ways to do this.)

Solve the system of equations. See Example 1.

3. $3x + 2y + z = 8$
   $2x - 3y + 2z = -16$
   $x + 4y - z = 20$

4. $-3x + y - z = -10$
   $-4x + 2y + 3z = -1$
   $2x + 3y - 2z = -5$

5. $2x + 5y + 2z = 0$
   $4x - 7y - 3z = 1$
   $3x - 8y - 2z = -6$

6. $5x - 2y + 3z = -9$
   $4x + 3y + 5z = 4$
   $2x + 4y - 2z = 14$

7. $x + y - z = -2$
   $2x - y + z = -5$
   $-x + 2y - 3z = -4$

8. $x + 2y + 3z = 1$
   $-x - y + 3z = 2$
   $-6x + y + z = -2$

Solve the system of equations. See Example 2.

9. $2x - 3y + 2z = -1$
   $x + 2y + z = 17$
   $2y - z = 7$

10. $2x - y + 3z = 6$
    $x + 2y - z = 8$
    $2y + z = 1$

11. $4x + 2y - 3z = 6$
    $x - 4y + z = -4$
    $-x + 2z = 2$

12. $2x + 3y - 4z = 4$
    $x - 6y + z = -16$
    $-x + 3z = 8$

13. $2x + y = 6$
    $3y - 2z = -4$
    $3x - 5z = -7$

14. $4x - 8y = -7$
    $4y + z = 7$
    $-8x + z = -4$

15. Using your immediate surroundings give an example of three planes that
    (a) intersect in a single point   **(b)** do not intersect
    (c) intersect in infinitely many points.

16. Explain how you can determine algebraically that a system of three linear equations in three variables has no solution. Then do the same for infinitely many solutions.

Solve the system of equations. See Examples 1, 3, and 4.

17. $2x + 2y - 6z = 5$
    $-3x + y - z = -2$
    $-x - y + 3z = 4$

18. $-2x + 5y + z = -3$
    $5x + 14y - z = -11$
    $7x + 9y - 2z = -5$

19. $-5x + 5y - 20z = -40$
    $x - y + 4z = 8$
    $3x - 3y + 12z = 24$

20. $x + 4y - z = 3$
    $-2x - 8y + 2z = -6$
    $3x + 12y - 3z = 9$

21. $2x + y - z = 6$
    $4x + 2y - 2z = 12$
    $-x - \dfrac{1}{2}y + \dfrac{1}{2}z = -3$

22. $2x - 8y + 2z = -10$
    $-x + 4y - z = 5$
    $\dfrac{1}{8}x - \dfrac{1}{2}y + \dfrac{1}{8}z = -\dfrac{5}{8}$

23. $x + y - 2z = 0$
    $3x - y + z = 0$
    $4x + 2y - z = 0$

24. $2x + 3y - z = 0$
    $x - 4y + 2z = 0$
    $3x - 5y - z = 0$

---

2. Answers will vary. One such system is
   $x + y + z = 6$
   $2x + 3y - z = 7$
   $3x - y - z = 6.$
3. $\{(1, 4, -3)\}$
4. $\{(2, -1, 3)\}$
5. $\{(0, 2, -5)\}$
6. $\{(0, 3, -1)\}$
7. $\left\{\left(-\dfrac{7}{3}, \dfrac{22}{3}, 7\right)\right\}$
8. $\left\{\left(\dfrac{20}{59}, -\dfrac{33}{59}, \dfrac{35}{59}\right)\right\}$
9. $\{(4, 5, 3)\}$
10. $\{(5, 1, -1)\}$
11. $\{(2, 2, 2)\}$
12. $\{(4, 4, 4)\}$
13. $\left\{\left(\dfrac{8}{3}, \dfrac{2}{3}, 3\right)\right\}$
14. $\left\{\left(\dfrac{3}{4}, \dfrac{5}{4}, 2\right)\right\}$
15. Answers will vary. Some possible answers are (a) two perpendicular walls and the ceiling in a normal room (b) the floors of three different levels of an office building (c) three pages of this book (since they intersect in the spine).   17. ∅
18. ∅ The solution sets in Exercises 19–22 may be given in other equivalent forms.
19. $\{(x, y, z) \mid x - y + 4z = 8\}$
20. $\{(x, y, z) \mid x + 4y - z = 3\}$
21. $\{(x, y, z) \mid 2x + y - z = 6\}$
22. $\{(x, y, z) \mid -x + 4y - z = 5\}$
23. $\{(0, 0, 0)\}$
24. $\{(0, 0, 0)\}$

---

25. $\{(2, 1, 5, 3)\}$

26. $\{(4, 1, 2, -1)\}$

27. $128 = a + b + c$

28. $140 = 2.25a + 1.5b + c$

29. $80 = 9a + 3b + c$

30. $\quad a + \quad b + c = 128$
$\quad 2.25a + 1.5b + c = 140$
$\quad\quad 9a + \quad 3b + c = 80;$
$\{(-32, 104, 56)\}$

31. $f(x) = -32x^2 +$
$104x + 56$ 32. 56 feet

33. after aproximately 3.72
seconds 34. after 3.25
seconds

35. $a = 3, b = 1, c = -2;$
$f(x) = 3x^2 + x - 2$

36. $a = 1, b = 4, c = 3;$
$y_1 = x^2 + 4x + 3$

39. 100 inches, 103 inches,
120 inches

▲ *Extend the method of this section to solve the system.*

25. $\quad x + \quad y + \quad z - \quad w = 5$
$\quad 2x + \quad y - \quad z + \quad w = 3$
$\quad\quad x - 2y + 3z + \quad w = 18$
$\quad -x - \quad y + \quad z + 2w = 8$

26. $\quad 3x + y - \quad z + 2w = 9$
$\quad\quad x + y + 2z - \quad w = 10$
$\quad\quad x - y - \quad z + 3w = -2$
$\quad -x + y - \quad z + \quad w = -6$

◆ **MATHEMATICAL CONNECTIONS** (Exercises 27–36) ◆

Suppose that on a distant planet a function of the form

$$f(x) = ax^2 + bx + c \quad (a \neq 0)$$

describes the height in feet of a projectile x seconds after it has been projected upward. Work through Exercises 27–36 in order to see how this can be related to a system of three equations in three variables a, b, and c.

27. After 1 second, the height of a certain projectile is 128 feet. Thus, $f(1) = 128$. Use this information to find one equation in the variables a, b, and c. (*Hint:* Substitute 1 for x and 128 for $f(x)$.)

28. After 1.5 seconds, the height is 140 feet. Find a second equation in a, b, and c.

29. After 3 seconds, the height is 80 feet. Find a third equation in a, b, and c.

30. Write a system of three equations in a, b, and c, based on your answers in Exercises 27–29. Solve the system.

31. What is the function f for this particular projectile?

32. What was the initial height of the projectile? (*Hint:* Find $f(0)$.)

33. When will the projectile hit the ground? (*Hint:* Find the positive solution of $f(x) = 0$.)

34. At what other time was the projectile the same height as its initial position? (*Hint:* Solve $f(x) = c$ for the value of c you found earlier.)

35. In Chapter 9 we discuss graphs of functions of the form $f(x) = ax^2 + bx + c$ $(a \neq 0)$. Use a system of equations to find the values of a, b, and c for the function of this form that satisfies $f(1) = 2$, $f(-1) = 0$, and $f(-2) = 8$. Then write the expression for $f(x)$.

36. The accompanying table was generated by a graphics calculator for a function $y_1 = ax^2 + bx + c$. Use any three points shown to find the values of a, b, and c. Then write the expression for y_1.

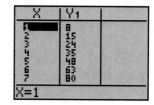

◎ **37.** Discuss why it is necessary to eliminate the same variable in the first two steps of the elimination method with three equations and three variables.

◎ **38.** In Step 3 of the elimination method for solving systems in three variables, does it matter which variable is eliminated? Explain.

REVIEW EXERCISES

Solve the problem. See Sections 2.3 and 2.4.

39. The perimeter of a triangle is 323 inches. The shortest side measures five-sixths the length of the longest side, and the medium side measures 17 inches less than the longest side. Find the lengths of the sides of the triangle.

40. The sum of the three angles of a triangle is 180°. The largest angle is twice the measure of the smallest, and the third angle measures 10° less than the largest. Find the measures of the three angles.

41. The sum of three numbers is 16. The largest number is −3 times the smallest, while the middle number is 4 less than the largest. Find the three numbers.

42. Margaret Maggio has a collection of cents, dimes, and quarters. The number of dimes is one less than twice the number of cents. If there are 27 coins in all worth a total of $4.20, how many of each denomination of coin is in the collection?

40. 38°, 66°, 76°
41. −4, 8, 12
42. 5 cents, 9 dimes, 13 quarters

8.3 APPLICATIONS OF LINEAR SYSTEMS OF EQUATIONS

FOR EXTRA HELP

📖 **SSG** pp. 250–254
SSM pp. 408–418

📼 **Video**
10

💾 **Tutorial**
IBM   MAC

OBJECTIVES

1 ▶ Solve geometry problems using two variables.
2 ▶ Solve money problems using two variables.
3 ▶ Solve mixture problems using two variables.
4 ▶ Solve distance-rate-time problems using two variables.
5 ▶ Solve problems with three unknowns using a system of three equations.

◈ "Mixed price" refers to the price of a mixture of the two products. The system is $9x + 7y = 107$, $7x + 9y = 101$, where x represents the price of a citron and y represents the price of a wood apple.

◈ **C O N N E C T I O N S** ◈

Problems that can be solved by writing a system of equations have been of interest historically. The following problem appeared in a Hindu work that dates back to about 850 A.D.

> The mixed price of 9 citrons and 7 fragrant wood apples is 107; again, the mixed price of 7 citrons and 9 fragrant wood apples is 101. O you arithmetician, tell me quickly the price of a citron and the price of a wood apple here, having distinctly separated those prices well.

(*Answer:* 8 for a citron and 5 for a wood apple.)

FOR DISCUSSION OR WRITING
What do you think is meant by "the mixed price" in the problem quoted above? Use the method discussed in this section to write a system of equations for this problem. Solve the system and compare your answer with the one given above.

PROBLEM SOLVING   Many applied problems involve more than one unknown quantity. Although most problems with two unknowns can be solved using just one variable, many times it is easier to use two variables. To solve a problem with two unknowns, we must write two equations that relate the unknown quantities. The system formed by the pair of equations then can be solved using the methods of Section 8.1.

The following steps give a strategy for solving problems using more than one variable.

INSTRUCTOR'S RESOURCES

📖 **ITM** pp. 460–466   **ISM** pp. 617–633
IAM p. 38

💾 **TEST GENERATOR**
DOS   Windows   MAC

✳ **TRANSPARENCIES**

 CONCEPTUAL   WRITING   CHALLENGING    SCIENTIFIC CALCULATOR    GRAPHICS CALCULATOR

Solving an Applied Problem by Writing a System of Equations

Step 1 **Determine what you are to find.** Assign a variable for each unknown and *write down* what it represents.

Step 2 **Write down other information.** If appropriate, draw a figure or a diagram and label it using the variables from Step 1. Use a chart or a box diagram to summarize the information.

Step 3 **Write a system of equations.** Write as many equations as there are unknowns.

Step 4 **Solve the system.**

Step 5 **Answer the question(s).** Be sure you have answered all questions posed.

Step 6 **Check.** Check your solution(s) in the original problem. Be sure your answer makes sense.

1 ▶ Solve geometry problems using two variables.

▶ Problems about the perimeter of a geometric figure often involve two unknowns. The next example shows how to write a system of equations to solve such a problem.

EXAMPLE 1
Solving a Geometry Problem

The length of the foundation of a rectangular house is to be 6 meters more than its width. Find the length and width of the house if the perimeter must be 48 meters.

Begin by sketching a rectangle to represent the foundation of the house.

Let
$$x = \text{the length},$$
$$y = \text{the width}.$$

See Figure 5. The length, x, is 6 meters more than the width, y. Therefore,

$$x = 6 + y.$$

The formula for the perimeter of a rectangle is $P = 2L + 2W$. Here $P = 48$, $L = x$, and $W = y$, so

$$48 = 2x + 2y.$$

The length and width can now be found by solving the system

$$x = 6 + y$$
$$48 = 2x + 2y.$$

Solve this system to find that the width is 9 meters and the length 15 meters. Be sure to check the solution in the words of the original problem. ■

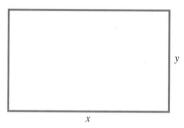

FIGURE 5

2 ▶ Solve money problems using two variables.

▶   Another type of problem that often leads to a system of equations is one about different amounts of money.

EXAMPLE   2

Solving a Problem About Money

For an art project Kay bought 8 sheets of colored paper and 3 marker pens for $6.50. She later needed 2 more sheets of colored paper and 2 different colored pens. These items cost $3.00. Find the cost of 1 marker pen and 1 sheet of colored paper.

Let x represent the cost of a sheet of paper and y represent the cost of a pen. For the first purchase $8x$ represents the cost of the paper and $3y$ the cost of the pens. The total cost was $6.50, so

$$8x + 3y = 6.50.$$

For the second purchase,

$$2x + 2y = 3.00.$$

(A box diagram or a table could be used to organize this information as we did in Chapter 2.) We can solve the system by multiplying both sides of the second equation by -4 and adding the result to the first equation.

$$
\begin{array}{rcr}
8x + 3y &=& 6.50 \\
-8x - 8y &=& -12.00 \\
\hline
-5y &=& -5.50 \\
y &=& 1.10
\end{array}
$$

By substituting 1.10 for y in either of the equations, verify that $x = .40$. Kay paid $.40 for a sheet of colored paper and $1.10 for a pen.   ■

NOTE In Example 2, x and y represented costs in *dollars*, because the right side of each equation was in dollars, so the left side had to agree. Therefore, $x = .40$ represents $.40, not .40¢.

---

3 ▶ Solve mixture problems using two variables.

▶   We solved mixture problems earlier using one variable. For many mixture problems it seems more natural to use more than one variable and a system of equations.

EXAMPLE 3

Solving a Mixture
Problem

How many ounces of 5% hydrochloric acid and of 20% hydrochloric acid must be combined to get 10 ounces of solution that is 12.5% hydrochloric acid?

Let x represent the number of ounces of 5% solution and y represent the number of ounces of 20% solution. Again, a table summarizes the given information.

Kind of Solution	Ounces of Solution	Ounces of Acid
5%	x	$.05x$
20%	y	$.20y$
12.5%	10	$(.125)10$

When the x ounces of 5% solution and the y ounces of 20% solution are combined, the total number of ounces is 10, so that

$$x + y = 10. \tag{1}$$

The ounces of acid in the 5% solution, $.05x$, plus the ounces of acid in the 20% solution, $.20y$, should equal the total number of ounces of acid in the mixture, which is $(.125)10$. That is,

$$.05x + .20y = (.125)10. \tag{2}$$

Eliminate x by first multiplying both sides of equation (2) by 100 to clear it of decimals, and then multiplying both sides of equation (1) by -5. Then add the results.

$$\begin{array}{rl}
5x + 20y = & 125 \qquad \text{Multiply both sides of (2) by 100.} \\
-5x - 5y = & -50 \qquad \text{Multiply both sides of (1) by } -5. \\
\hline
15y = & 75 \\
y = & 5
\end{array}$$

Since $y = 5$ and $x + y = 10$, x is also 5. Therefore, 5 ounces each of the 5% and the 20% solutions is required. ∎

Chalkboard Exercise

A grocer has some $4-per-pound coffee and some $8-per-pound coffee which she will mix to make 50 pounds of $5.60-per-pound coffee. How many pounds of each should be used?

Answer: 30 pounds at $4, 20 pounds at $8

4 ▶ Solve distance-rate-time problems using two variables.

▶ Constant rate applications require the distance formula, $d = rt$, where d is distance, r is rate (or speed), and t is time. These applications often lead to a system of equations, as in the next example.

EXAMPLE 4

Solving a Motion Problem

A car travels 250 kilometers in the same time that a truck travels 225 kilometers. If the speed of the car is 8 kilometers per hour faster than the speed of the truck, find both speeds.

A table is useful for organizing the information in problems about distance, rate, and time. Fill in the given information for each vehicle (in this case, distance) and use variables for the unknown speeds (rates) as follows.

	d	r	t
Car	250	x	
Truck	225	y	

The problem states that the car travels 8 kilometers per hour faster than the truck. Since the two speeds are x and y,

$$x = y + 8.$$

Chalkboard Exercise

A train travels 600 miles in the same time that a truck travels 520 miles. Find the speed of each vehicle if the train's average speed is 8 miles per hour faster than the truck's.

Answer: The train travels at 60 miles per hour and the truck at 52 miles per hour.

The table shows nothing about time. To get an expression for time, solve the distance formula, $d = rt$, for t to get

$$\frac{d}{r} = t.$$

The two times can be written as $250/x$ and $225/y$. Since both vehicles travel for the same time,

$$\frac{250}{x} = \frac{225}{y}.$$

This is not a linear equation. However, multiplying both sides by xy gives

$$250y = 225x,$$

which is linear. Now solve the system.

$$x = y + 8 \tag{3}$$
$$250y = 225x \tag{4}$$

The substitution method can be used. Replace x with $y + 8$ in equation (4).

$$250y = 225(y + 8) \qquad \text{Let } x = y + 8.$$
$$250y = 225y + 1800 \qquad \text{Distributive property}$$
$$25y = 1800$$
$$y = 72$$

Since $x = y + 8$, the value of x is $72 + 8 = 80$. It is important to check the solution in the original problem since one of the equations had variable denominators. Checking verifies that the speeds are 80 kilometers per hour for the car and 72 kilometers per hour for the truck. ∎

5 ▶ Solve problems with three unknowns using a system of three equations.

▶ Some applications involve three unknowns.

PROBLEM SOLVING

To solve applied problems with three or more unknowns, we extend the method given earlier for problems with two unknowns. When three variables are used, three equations are necessary to find a solution. The next two examples illustrate the procedure.

EXAMPLE 5

Solving a Mixture Problem

A plant food is to be made from three chemicals. The mix must include 60% of the first two chemicals. The other two chemicals must be in a ratio of 4 to 3 by weight. How much of each chemical is needed to make 750 kilograms of the plant food?
First, choose variables to represent the three unknowns.

Let $x =$ the number of kilograms of the first chemical;

$y =$ the number of kilograms of the second chemical;

$z =$ the number of kilograms of the third chemical.

Chalkboard Exercise

How many ounces of 5% hydrochloric acid, 20% hydrocholoric acid, and water must be combined to get 10 ounces of solution that is 8.5% hydrochloric acid, if the amount of water used must equal the total amount of the other two solutions?

Answer: 1 ounce of the 5% solution, 4 ounces of the 20% solution, and 5 ounces of water

Next, use the information in the problem to write three equations. To make 750 kilograms of the mix will require 60% of 750 kilograms of the first two chemicals, so

$$x + y = .60(750) = 450.$$

Since the ratio of the second and third chemicals is to be 4 to 3,

$$\frac{y}{z} = \frac{4}{3}.$$

Finally, the total amount of mix is to be 750 kilograms, so

$$x + y + z = 750.$$

Now, we must solve the system

$$x + y = 450$$
$$\frac{y}{z} = \frac{4}{3}$$
$$x + y + z = 750.$$

Use the method shown earlier to find the solution (50, 400, 300). The plant food should contain 50 kilograms of the first chemical, 400 kilograms of the second chemical, and 300 kilograms of the third chemical. ■

Business problems involving production sometimes require the solution of a system of equations. The final example shows how to set up such a system.

EXAMPLE 6
Solving a Business
Production Problem

Chalkboard Exercise

A paper mill makes newsprint, bond, and copy machine paper. Each ton of newsprint requires 3 tons of recycled paper and 1 ton of wood pulp. Each ton of bond requires 2 tons of recycled paper, 4 tons of wood pulp, and 3 tons of rags. A ton of copy machine paper requires 2 tons of recycled paper, 3 tons of wood pulp, and 2 tons of rags. The mill has 4200 tons of recycled paper, 5800 tons of wood pulp, and 3900 tons of rags. How much of each kind of paper can be made from these supplies?

Answer: 400 tons of newsprint, 900 tons of bond, and 600 tons of copy machine paper

A company produces three color television sets, models X, Y, and Z. Each model X set requires 2 hours of electronics work, 2 hours of assembly time, and 1 hour of finishing time. Each model Y requires 1, 3, and 1 hours of electronics, assembly, and finishing time, respectively. Each model Z requires 3, 2, and 2 hours of the same work, respectively. There are 100 hours available for electronics, 100 hours available for assembly, and 65 hours available for finishing per week. How many of each model should be produced each week if all available time must be used?

Let $x =$ the number of model X produced per week;
          $y =$ the number of model Y produced per week;
          $z =$ the number of model Z produced per week.

A table is useful for organizing the information in a problem of this type.

	Each Model X	Each Model Y	Each Model Z	Totals
Hours of electronics work	2	1	3	100
Hours of assembly time	2	3	2	100
Hours of finishing time	1	1	2	65

The x model X sets require $2x$ hours of electronics, the y model Y sets require $1y$ (or y) hours of electronics, and the z model Z sets require $3z$ hours of electronics. Since 100 hours are available for electronics,

$$2x + y + 3z = 100.$$

▶ **TEACHING TIP**

Point out that a problem involving three unknowns must produce three equations for a unique solution to exist.   ◀

Similarly, from the fact that 100 hours are available for assembly,

$$2x + 3y + 2z = 100,$$

and the fact that 65 hours are available for finishing leads to the equation

$$x + y + 2z = 65.$$

Solve the system

$$2x + y + 3z = 100$$
$$2x + 3y + 2z = 100$$
$$x + y + 2z = 65$$

to find $x = 15$, $y = 10$, and $z = 20$. The company should produce 15 model X, 10 model Y, and 20 model Z sets per week.   ■

Notice the advantage of setting up the table as in Example 6. By reading across, we can easily determine the coefficients and the constants in the system.

8.3 EXERCISES

For each application in this exercise set select variables to represent the unknown quantitites, write equations using the variables, and solve the resulting systems. The applications in Exercises 1–34 require solving systems with two variables, while the ones that follow require solving systems with three variables.

1. length: 78 feet;   width: 36 feet   2. length: 94 feet;   width: 50 feet
3. length: 12 feet;   width: 5 feet   4. square: 12 centimeters;   triangle: 8 centimeters

Solve the problem. See Example 1.

1. Andre and Monica measured the perimeter of a tennis court and found that it was 42 feet longer than it was wide, and had a perimeter of 228 feet. What were the length and the width of the tennis court?

2. Kareem and Manute found that the width of their basketball court was 44 feet less than the length. If the perimeter was 288 feet, what were the length and the width of their court?

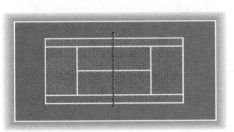

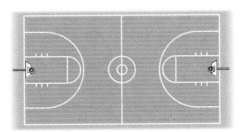

3. The length of a rectangle is 7 feet more than the width. If the length were decreased by 3 feet and the width were increased by 2 feet, the perimeter would be 32 feet. Find the length and width of the original rectangle.

4. The side of a square is 4 centimeters longer than the side of an equilateral triangle. The perimeter of the square is 24 centimeters more than the perimeter of the triangle. Find the lengths of a side of the square and a side of the triangle.

 CONCEPTUAL   WRITING   ▲ CHALLENGING   SCIENTIFIC CALCULATOR   GRAPHICS CALCULATOR

5. $x = 40, y = 50$, so the angles measure 40° and 50°.   6. $x = 100$ and $y = 80$, so the angles measure 100° and 80°.
7. CGA monitor: $400; VGA monitor: $500
8. cap: $20;   jersey: $110
9. 6 units of yarn;   2 units of thread
10. 20 standard;   24 top-of-the-line
11. dark clay: $5 per kilogram;   light clay: $4 per kilogram   12. 4 vats of green algae;   7 vats of brown algae
13. (a) 3.2 ounces
(b) 8 ounces
(c) 12.8 ounces
(d) 16 ounces

5. Find the measures of the angles marked x and y.

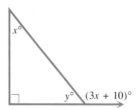

6. Find the measures of the angles marked x and y.

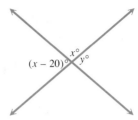

Solve the problem. See Examples 2 and 5.

7. San Jacinto College has decided to supply its mathematics labs with color monitors. A trip to the local electronics outlet leads to the following information: 4 CGA monitors and 6 VGA monitors can be purchased for $4600, while 6 CGA monitors and 4 VGA monitors will cost $4400. What are the prices of a single CGA monitor and a single VGA monitor?

8. The Cleveland Indians gift shop will sell you 5 New Era baseball caps and 3 Diamond Collection jerseys for $430, or 3 New Era caps and 4 Diamond Collection jerseys for $500. What are the prices of a single cap and a single jersey?

9. A factory makes use of two basic machines, *A* and *B,* which turn out two different products, yarn and thread. Each unit of yarn requires 1 hour on machine *A* and 2 hours on machine *B,* while each unit of thread requires 1 hour on *A* and 1 hour on *B.* Machine *A* runs 8 hours per day, while machine *B* runs 14 hours per day. How many units each of yarn and thread should the factory make to keep its machines running at capacity?

10. A company that makes personal computers has found that each standard model requires 4 hours to manufacture electronics and 2 hours for the case. The top-of-the-line model requires 5 hours for the eltronics and 1.5 hours for the case. On a particular production run, the company has available 200 hours in the electronics department and 76 hours in the cabinet department. How many of each model can be made?

11. Theodis bought 2 kilograms of dark clay and 3 kilograms of light clay, paying $22 for the clay. He later needed 1 kilogram of dark clay and 2 kilograms of light clay, costing $13 altogether. How much did he pay for each type of clay?

12. A biologist wants to grow two types of algae, green and brown. She has 15 kilograms of nutrient X and 26 kilograms of nutrient Y. A vat of green algae needs 2 kilograms of nutrient X and 3 kilograms of nutrient Y, while a vat of brown algae needs 1 kilogram of nutrient X and 2 kilograms of nutrient Y. How many vats of each type of algae should the biologist grow in order to use all the nutrients?

The formulas $p = br$ (percentage = base × rate) and $I = prt$ (simple interest = principal × rate × time) are used in the applications found in Exercises 17–28. To prepare for the use of these formulas, answer the questions in Exercises 13 and 14.

13. If a container of liquid contain 32 ounces of solution, what is the number of ounces of pure acid if the given solution contains the following acid concentrations?
(a) 10%   **(b)** 25%   **(c)** 40%   **(d)** 50%

14. If $2000 is invested in an account paying simple annual interest, how much interest will be earned during the first year at the following rates?
(**a**) 2%   (**b**) 3%   (**c**) 4%   (**d**) 3.5%

15. If a pound of oranges costs $.69, how much will *x* pounds cost?

16. If a ticket to a movie costs $3.50, and *y* tickets are sold, how much is collected from the sale?

Solve the problem. See Example 3.

17. How many gallons each of 25% alcohol and 35% alcohol should be mixed to get 20 gallons of 32% alcohol?

18. How many liters each of 15% acid and 33% acid should be mixed to get 40 liters of 21% acid?

19. Pure acid is to be added to a 10% acid solution to obtain 27 liters of a 20% acid solution. What amounts of each should be used?

20. A truck radiator holds 18 liters of fluid. How much pure antifreeze must be added to a mixture that is 4% antifreeze in order to fill the radiator with a mixture that is 20% antifreeze?

21. A party mix is made by adding nuts that sell for $2.50 a kilogram to a cereal mixture that sells for $1 a kilogram. How much of each should be added to get 30 kilograms of a mix that will sell for $1.70 a kilogram?

22. A popular fruit drink is made by mixing fruit juices. Such a mixture with 50% juice is to be mixed with another mixture that is 30% juice to get 200 liters of a mixture that is 45% juice. How much of each should be used?

23. Tickets to a production of *Othello* at Nicholls State University cost $2.50 for general admission or $2.00 with a student identification. If 184 people paid to see a performance and $406 was collected, how many of each type of admission were sold?

24. A grocer plans to mix candy that sells for $1.20 a pound with candy that sells for $2.40 a pound to get a mixture that he plans to sell for $1.65 a pound. How much of the $1.20 and $2.40 candy should he use if he wants 80 pounds of the mix?

25. Stacy Gregg has been saving dimes and quarters. She has 94 coins in all. If the total value is $19.30, how many dimes and how many quarters does she have?

26. A teller at the Bank of New Roads received a checking account deposit in twenty-dollar bills and fifty-dollar bills. She received a total of 70 bills, and the amount of the deposit was $3200. How many of each denomination were deposited?

27. A total of $3000 is invested, part at 2% simple interest and part at 4%. If the total annual return from the two investments is $100, how much is invested at each rate?

28. An investor must invest a total of $15,000 in two accounts, one paying 4% annual simple interest, and the other 3%. If he wants to earn $550 annual interest, how much should he invest at each rate?

The formula d = rt (distance = rate × time) is used in the applications found in Exercises 31–34. To prepare for the use of this formula, answer the questions in Exercises 29 and 30.

29. If the speed of a train is 65 miles per hour and the train travels *y* hours, how many miles does the train travel?

30. If the speed of a boat in still water is 10 miles per hour, and the speed of the current of a river is *x* miles per hour, what is the speed of the boat (**a**) going upstream (that is, against the current) and   (**b**) going downstream (that is, with the current)?

14. (a) $40 (b) $60
(c) $80 (d) $70
15. $.69x   16. $3.50y
17. 6 gallons of
25%; 14 gallons of 35%
18. $26\frac{2}{3}$ liters of 15%;
$13\frac{1}{3}$ liters of 33%
19. 3 liters of pure
acid; 24 liters of 10%
acid
20. 3 liters of pure
antifreeze
21. 14 kilograms of nuts;
16 kilograms of cereal
22. 150 liters of
50% juice; 50 liters of
30% juice 23. 76 general
admission; 108 with
student identification
24. 50 pounds of $1.20
candy; 30 pounds of
$2.40 candy
25. 28 dimes; 66 quarters
26. 10 twenties; 60 fifties
27. $1000 at 2%; $2000 at
4% 28. $10,000 at
4%; $5000 at 3%
29. 65y miles
30. (a) $10 - x$ miles per
hour (b) $10 + x$ miles per
hour

31. freight train: 50 kilometers per hour; express train: 80 kilometers per hour
32. train: 60 kilometers per hour; plane: 160 kilometers per hour
33. boat: 21 miles per hour; current: 3 miles per hour   34. plane: 600 miles per hour; wind: 50 miles per hour
35. $x + y + z = 180$; angle measures: 70°, 30°, 80°   36. $x = y - 10$, $x = z - 20$, $x + y + z = 180$; 50°, 60°, 70°
37. first: 20°; second: 70°; third: 90°   38. largest: 84°; middle: 70°; smallest: 26°   39. shortest: 12 centimeters; middle: 25 centimeters; longest: 33 centimeters
40. shortest: 10 inches; middle: 20 inches; longest: 26 inches
41. A: 180 cases; B: 60 cases;   C: 80 cases

Solve the problem. See Example 4.

31. A freight train and an express train leave towns 390 kilometers apart, traveling toward one another. The freight train travels 30 kilometers per hour slower than the express train. They pass one another 3 hours later. What are their speeds?

32. A train travels 150 kilometers in the same time that a plane covers 400 kilometers. If the speed of the plane is 20 kilometers per hour less than 3 times the speed of the train, find both speeds.

33. In his motorboat, Nguyen travels upstream at top speed to his favorite fishing spot, a distance of 36 miles, in two hours. Returning, he finds that the trip downstream, still at top speed, takes only 1.5 hours. Find the speed of Nguyen's boat and the speed of the current.

34. Traveling for three hours into a steady headwind, a plane makes a trip of 1650 miles. The pilot determines that flying with the same wind for two hours, he could make a trip of 1300 miles. What is the speed of the plane and the speed of the wind?

Solve the problem involving three unknowns. See Examples 5 and 6. (In Exercises 35–38, remember that the sum of the measures of the angles of a triangle is 180°.)

35. In the figure shown, $z = x + 10$ and $x + y = 100$. Determine a third equation involving x, y, and z, and then find the measures of the three angles.

36. In the figure shown, x is 10 less than y and 20 less than z. Write a system of equations and find the measures of the three angles.

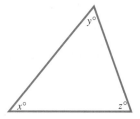

37. In a certain triangle, the measure of the second angle is 10° more than three times the first. The third angle measure is equal to the sum of the measures of the other two. Find the measures of the three angles.

38. The measure of the largest angle of a triangle is 12° less than the sum of the measures of the other two. The smallest angle measures 58° less than the largest. Find the measures of the angles.

39. The perimeter of a triangle is 70 centimeters. The longest side is 4 centimeters less than the sum of the other two sides. Twice the shortest side is 9 centimeters less than the longest side. Find the length of each side of the triangle.

40. The perimeter of a triangle is 56 inches. The longest side measures 4 inches less than the sum of the other two sides. Three times the shortest side is 4 inches more than the longest side. Find the lengths of the three sides.

41. A Mardi Gras trinket manufacturer supplies three wholesalers, A, B, and C. The output from a day's production is 320 cases of trinkets. She must send wholesaler A three times as many cases as she sends B, and she must send wholesaler C 160 cases less than she provides A and B together. How many cases should she send to each wholesaler to distribute the entire day's production to them?

42. A hardware supplier manufactures three kinds of clamps, types A, B, and C. Production restrictions require them to make 10 units more type C clamps than the total of the other types and twice as many type B clamps as type A. The shop must produce a total of 490 units of clamps per day. How many units of each type can be made per day?

43. The manager of a candy store wants to feature a special Easter candy mixture of jelly beans, small chocolate eggs, and marshmallow chicks. She plans to make 15 pounds of mix to sell at $1 a pound. Jelly beans sell for $.80 a pound, chocolate eggs for $2 a pound, and marshmallow chicks for $1 a pound. She will use twice as many pounds of jelly beans as eggs and chicks combined and fives times as many pounds of jelly beans as chocolate eggs. How many pounds of each candy should she use?

44. Three kinds of tickets are available for a Rhonda Rock concert: "up close," "in the middle," and "far out." "Up close" tickets cost $2 more than "in the middle" tickets, while "in the middle" tickets cost $1 more than "far out" tickets. Twice the cost of an "up close" ticket is $1 less than 3 times the cost of a "far out" seat. Find the price of each kind of ticket.

MATHEMATICAL CONNECTIONS (Exercises 45–48)

In the next chapter we will see that an equation of the form

$$x^2 + y^2 + ax + by + c = 0$$

may have a circle as its graph. It is a fact from geometry that given three noncollinear points (that is, points that do not all lie on the same straight line), there will be a circle that contains them. For example, the points $(4, 2)$, $(-5, -2)$, and $(0, 3)$ lie on the circle whose equation is

$$x^2 + y^2 - (7/5)x + (27/5)y - 126/5 = 0.$$

The circle is shown in the figure.

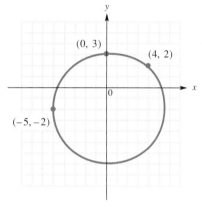

Work Exercises 45–48 in order, so that the equation of the circle passing through the points $(2, 1)$, $(-1, 0)$, and $(3, 3)$ can be found.

45. Let $x = 2$ and $y = 1$ in the equation $x^2 + y^2 + ax + by + c = 0$ to find an equation in a, b, and c.

46. Let $x = -1$ and $y = 0$ to find a second equation in a, b, and c.

47. Let $x = 3$ and $y = 3$ to find a third equation in a, b, and c.

48. Solve the system of equations formed by your answers in Exercises 45–47 to find the values of a, b, and c. What is the equation of the circle?

REVIEW EXERCISES

*Give **(a)** the additive inverse and **(b)** the multiplicative inverse (reciprocal) of the number. See Sections 1.1 and 1.3.*

49. -6 **50.** $.2$ **51.** $\dfrac{7}{8}$ **52.** 2.25

Answers (margin):

42. type A: 80; type B: 160; type C: 250
43. 10 pounds of jelly beans; 2 pounds of chocolate eggs; 3 pounds of marshmallow chicks
44. up close: $10; in the middle: $8; far out: $7
45. $2a + b + c = -5$
46. $-a + c = -1$
47. $3a + 3b + c = -18$
48. $a = 1, b = -7, c = 0$; $x^2 + y^2 + x - 7y = 0$
49. (a) 6   (b) $-\dfrac{1}{6}$
50. (a) $-.2$ (b) 5
51. (a) $-\dfrac{7}{8}$ (b) $\dfrac{8}{7}$
52. (a) -2.25
(b) $\dfrac{4}{9}$ or $.\overline{4}$

 CONCEPTUAL   WRITING   CHALLENGING   SCIENTIFIC CALCULATOR   GRAPHICS CALCULATOR

8.4 SOLUTION OF LINEAR SYSTEMS OF EQUATIONS BY MATRIX METHODS

FOR EXTRA HELP	OBJECTIVES
📖 **SSG** pp. 254–260 **SSM** pp. 418–424	**1 ▶** Define a matrix.
📼 **Video** 10	**2 ▶** Write the augmented matrix for a system. **3 ▶** Use row operations to solve a system with two equations.
💾 **Tutorial** IBM MAC	**4 ▶** Use row operations to solve a system with three equations. **5 ▶** Use row operations to solve inconsistent systems or systems with dependent equations.

1 ▶ Define a matrix.

▶ An ordered array of numbers such as

$$\begin{bmatrix} 2 & 3 & 5 \\ 7 & 1 & 2 \end{bmatrix}$$

▶ **TEACHING TIP**

This chapter is a good introduction to the Gauss-Jordan method used in finite mathematics and linear algebra courses.

Students may approach the material in this section with resistance. Mention that this method is just a variation of the elimination method: the variables are removed at the beginning and then replaced in the final answer. ◀

is called a **matrix.** Matrices (the plural of matrix) are named according to the number of rows and columns they contain. The rows are read horizontally and the columns are read vertically. For example, the first row in the matrix above is 2 3 5 and the first column is $\dfrac{2}{7}$. The matrix above is a 2 × 3 (read "two by three") matrix because it has 2 rows and 3 columns. The number of rows is given first, and then the number of columns.

The matrix

$$\begin{bmatrix} -1 & 0 \\ 1 & -2 \end{bmatrix}$$

is a 2 × 2 matrix, and the matrix

$$\begin{bmatrix} 8 & -1 & -3 \\ 2 & 1 & 6 \\ 0 & 5 & -3 \\ 5 & 9 & 7 \end{bmatrix}$$

is a 4 × 3 matrix. A **square matrix** is one that has the same number of rows as columns.

◈ The zero 2 × 2 matrix would be $\begin{bmatrix} 0 & 0 \\ 0 & 0 \end{bmatrix}$. The additive inverse of $\begin{bmatrix} a & b \\ c & d \end{bmatrix}$ would be $\begin{bmatrix} -a & -b \\ -c & -d \end{bmatrix}$.

◈ **C O N N E C T I O N S** ◈

The use of matrices has gained increasing importance in the fields of management, natural science, and social science because matrices provide a convenient way to organize data. Matrices can be treated as mathematical entities and added, subtracted, and multiplied. Many square matrices have an inverse matrix, similar to the multiplicative inverse for a real number. One of the most important uses of matrices is to solve linear systems of equations with many variables. In fact, matrix methods of solution of such systems are the most suitable for use with computers and graphics calculators.

INSTRUCTOR'S RESOURCES

📖 **ITM** pp. 467–471   **ISM** pp. 634–645
IAM p. 38

💾 **TEST GENERATOR**
DOS Windows MAC

 TRANSPARENCIES

FOR DISCUSSION OR WRITING

Suppose we consider the set of all 2 × 2 matrices. Two matrices are added or subtracted by adding or subtracting corresponding elements. With that definition, what would the zero matrix be? What would the *additive* inverse of the matrix

$$\begin{bmatrix} a & b \\ c & d \end{bmatrix}$$

be? (*Hint:* What is the additive inverse of the real number a?)

In this section we discuss a method of solving linear systems that uses matrices. This method is really just a very structured way of using the elimination method to solve a linear system. The advantage of this new method is that it can be done by computer, allowing large systems of equations to be solved more easily.

2 ▶ Write the augmented matrix for a system.

▶ To begin, we write an *augmented matrix* for the system. An **augmented matrix** has a vertical bar that separates the columns of the matrix into two groups. For example, to solve the system

$$x - 3y = 1$$
$$2x + y = -5,$$

start with the augmented matrix

$$\left[\begin{array}{cc|c} 1 & -3 & 1 \\ 2 & 1 & -5 \end{array}\right].$$

Place the coefficients of the variables to the left of the bar, and the constants to the right. The bar itself is used only to separate the coefficients from the constants. The matrix is just a shorthand way of writing the system of equations, so the rows of the augmented matrix can be treated the same as the equations of a system of equations.

We know that exchanging the position of two equations in a system does not change the system. Also, multiplying any equation in a system by a nonzero number does not change the system. Comparable changes to the augmented matrix of a system of equations produce new matrices that correspond to systems with the same solutions as the original system.

The following **row operations** produce new matrices that lead to systems having the same solutions as the original system.

Matrix Row Operations

1. Any two rows of the matrix may be interchanged.
2. The numbers in any row may be multiplied by any nonzero real number.
3. Any row may be changed by adding to the numbers of the row the product of a real number and the corresponding numbers of another row.

Examples of these row operation follow.

Using row operation 1,

$$\begin{bmatrix} 2 & 3 & 9 \\ 4 & 8 & -3 \\ 1 & 0 & 7 \end{bmatrix} \quad \text{becomes} \quad \begin{bmatrix} 1 & 0 & 7 \\ 4 & 8 & -3 \\ 2 & 3 & 9 \end{bmatrix}$$

by interchanging row 1 and row 3.

Using row operation 2,

$$\begin{bmatrix} 2 & 3 & 9 \\ 4 & 8 & -3 \\ 1 & 0 & 7 \end{bmatrix} \quad \text{becomes} \quad \begin{bmatrix} 6 & 9 & 27 \\ 4 & 8 & -3 \\ 1 & 0 & 7 \end{bmatrix}$$

by multiplying the numbers in row 1 by 3.

Using row operation 3,

$$\begin{bmatrix} 2 & 3 & 9 \\ 4 & 8 & -3 \\ 1 & 0 & 7 \end{bmatrix} \quad \text{becomes} \quad \begin{bmatrix} 0 & 3 & -5 \\ 4 & 8 & -3 \\ 1 & 0 & 7 \end{bmatrix}$$

by multiplying the numbers in row 3 by -2 and adding them to the corresponding numbers in row 1.

The third row operation corresponds to the way we eliminated a variable from a pair of equations in the previous sections.

3 ▶ Use row operations to solve a system with two equations.

▶ Row operations can be used to rewrite the matrix until it is the matrix of a system where the solution is easy to find. The goal is a matrix in the form

$$\left[\begin{array}{cc|c} 1 & a & b \\ 0 & 1 & c \end{array}\right] \quad \text{or} \quad \left[\begin{array}{ccc|c} 1 & a & b & c \\ 0 & 1 & d & e \\ 0 & 0 & 1 & f \end{array}\right]$$

for systems with two or three equations respectively. Notice that there are 1s down the diagonal from upper left to lower right and 0s below the 1s. When these matrices are rewritten as systems of equations, the value of one variable is known, and the rest can be found by substitution. The following examples illustrate the method.

▶ TEACHING TIP

It might interest students that computer solutions to systems of equations follow the steps outlined in this section. Of course, computers can process these steps much quicker and with greater accuracy. ◀

E X A M P L E   1

Using Row Operations to Solve a System with Two Variables

Use row operations to solve the system

$$\begin{aligned} x - 3y &= 1 \\ 2x + y &= -5. \end{aligned}$$

We start by writing the augmented matrix of the system.

$$\left[\begin{array}{cc|c} 1 & -3 & 1 \\ 2 & 1 & -5 \end{array}\right]$$

Now we use the various row operations to change this matrix into one that leads to a system that is easier to solve.

It is best to work by columns. We start with the first column and make sure that there is a 1 in the first row, first column position. There already is a 1 in this position. Next, we get 0s in every position below the first. To get a 0 in row two, column one, we use the third row operation and add to the numbers in row two the result of multiplying each number in row one by -2. (We abbreviate this as $-2R_1 + R_2$.) Row one remains unchanged.

$$\begin{bmatrix} 1 & -3 & \bigm| & 1 \\ 2 + 1(-2) & 1 + -3(-2) & \bigm| & -5 + 1(-2) \end{bmatrix}$$

Original number -2 times number
from row two   from row one

$$\begin{bmatrix} 1 & -3 & \bigm| & 1 \\ 0 & 7 & \bigm| & -7 \end{bmatrix} \quad -2R_1 + R_2$$

The matrix now has a 1 in the first position of column one, with 0s in every position below the first.

Now we go to column two. A 1 is needed in row two, column two. We get this 1 by using the second row operation, multiplying each number of row two by $1/7$.

$$\begin{bmatrix} 1 & -3 & \bigm| & 1 \\ 0 & 1 & \bigm| & -1 \end{bmatrix} \quad \tfrac{1}{7}R_2$$

This augmented matrix leads to the system of equations

$$\begin{aligned} 1x - 3y &= 1 \\ 0x + 1y &= -1 \end{aligned} \qquad \text{or} \qquad \begin{aligned} x - 3y &= 1 \\ y &= -1. \end{aligned}$$

From the second equation, $y = -1$. We can substitute -1 for y in the first equation to get

$$\begin{aligned} x - 3y &= 1 \\ x - 3(-1) &= 1 \\ x + 3 &= 1 \\ x &= -2. \end{aligned}$$

The solution set of the system is $\{(-2, -1)\}$. Check this solution by substitution in both equations. ∎

4 ▶ Use row operations to solve a system with three equations.

▶ A linear system with three equations is solved in a similar way. We use row operations to get 1s down the diagonal from left to right and all 0s below each 1.

E X A M P L E   2

Using Row Operations to Solve a System with Three Variables

Use matrix methods to solve the system

$$\begin{aligned} x - y + 5z &= -6 \\ 3x + 3y - z &= 10 \\ x + 3y + 2z &= 5. \end{aligned}$$

Start by writing the augmented matrix of the system.

$$\begin{bmatrix} 1 & -1 & 5 & \bigm| & -6 \\ 3 & 3 & -1 & \bigm| & 10 \\ 1 & 3 & 2 & \bigm| & 5 \end{bmatrix}$$

This matrix already has 1 in row one, column one. Next get 0s in the rest of column one. First, add to row two the results of multiplying each number of row one by -3. This gives the matrix

$$\begin{bmatrix} 1 & -1 & 5 & | & -6 \\ 0 & 6 & -16 & | & 28 \\ 1 & 3 & 2 & | & 5 \end{bmatrix}. \quad -3R_1 + R_2$$

Now add to the numbers in row three the results of multiplying each number of row one by -1.

$$\begin{bmatrix} 1 & -1 & 5 & | & -6 \\ 0 & 6 & -16 & | & 28 \\ 0 & 4 & -3 & | & 11 \end{bmatrix} \quad -1R_1 + R_3$$

We get 1 in row two, column two by multiplying each number in row two by $1/6$.

$$\begin{bmatrix} 1 & -1 & 5 & | & -6 \\ 0 & 1 & -\dfrac{8}{3} & | & \dfrac{14}{3} \\ 0 & 4 & -3 & | & 11 \end{bmatrix} \quad \tfrac{1}{6}R_2$$

Get 0 in row three, column two by adding to row three the results of multiplying each number in row two by -4.

$$\begin{bmatrix} 1 & -1 & 5 & | & -6 \\ 0 & 1 & -\dfrac{8}{3} & | & \dfrac{14}{3} \\ 0 & 0 & \dfrac{23}{3} & | & -\dfrac{23}{3} \end{bmatrix} \quad -4R_2 + R_3$$

Finally, get 1 in row three, column three by multiplying each number in row three by $3/23$.

$$\begin{bmatrix} 1 & -1 & 5 & | & -6 \\ 0 & 1 & -\dfrac{8}{3} & | & \dfrac{14}{3} \\ 0 & 0 & 1 & | & -1 \end{bmatrix} \quad \tfrac{3}{23}R_3$$

This final matrix gives the system of equations

$$x - y + 5z = -6$$
$$y - \frac{8}{3}z = \frac{14}{3}$$
$$z = -1.$$

Substitute -1 for z in the second equation, to get

$$y - \frac{8}{3}z = \frac{14}{3}$$
$$y - \frac{8}{3}(-1) = \frac{14}{3}$$
$$y + \frac{8}{3} = \frac{14}{3}$$
$$y = 2.$$

Finally, substitute 2 for y and -1 for z in the first equation.

$$x - y + 5z = -6$$
$$x - 2 + 5(-1) = -6$$
$$x - 2 - 5 = -6$$
$$x = 1$$

The solution set of the original system is $\{(1, 2, -1)\}$. This solution should be checked by substitution in the system. ∎

5 ▶ Use row operations to solve inconsistent systems or systems with dependent equations.

▶ In the final example we show how to recognize inconsistent systems or systems with dependent equations when solving these systems with row operations.

E X A M P L E   3

Recognizing Inconsistent Systems or Dependent Equations

Use row operations to solve each system.

(a) $2x - 3y = 8$
  $-6x + 9y = 4$
  Write the augmented matrix.

$$\begin{bmatrix} 2 & -3 & \bigm| & 8 \\ -6 & 9 & \bigm| & 4 \end{bmatrix}$$

Multiply the first row by $1/2$ to get 1 in row one column one.

$$\begin{bmatrix} 1 & -\dfrac{3}{2} & \bigm| & 4 \\ -6 & 9 & \bigm| & 4 \end{bmatrix} \qquad \tfrac{1}{2}R_1$$

Multiply row one by 6 and add the results to row two.

$$\begin{bmatrix} 1 & -\dfrac{3}{2} & \bigm| & 4 \\ 0 & 0 & \bigm| & 28 \end{bmatrix} \qquad 6R_1 + R_2$$

The corresponding system of equations is

$$x - \frac{3}{2}y = 4$$
$$0 = 28,$$

which has no solution and so is inconsistent. The solution set is ∅.

(b) $-10x + 12y = 30$
   $5x - 6y = -15$
  The augmented matrix is

$$\begin{bmatrix} -10 & 12 & \bigm| & 30 \\ 5 & -6 & \bigm| & -15 \end{bmatrix}.$$

Multiply the first row by $-1/10$.

$$\begin{bmatrix} 1 & -\dfrac{6}{5} & \bigm| & -3 \\ 5 & -6 & \bigm| & -15 \end{bmatrix} \qquad -\tfrac{1}{10}R_1$$

Multiply the first row by -5 and add the products to row two to get

$$\begin{bmatrix} 1 & -\dfrac{6}{5} & \bigm| & -3 \\ 0 & 0 & \bigm| & 0 \end{bmatrix}. \qquad -5R_1 + R_2$$

The corresponding system is

$$x - \frac{6}{5}y = -3$$
$$0 = 0,$$

which has dependent equations. Clearing fractions in the first equation, we write the solution set as $\{(x, y)\,|\,5x - 6y = -15\}$. ■

◇ **CONNECTIONS** ◇

While modern graphics calculators are probably best known for their graphing capabilities, they possess other powerful features as well. One of these is the capability to handle work with matrices. Many graphics calculators can perform row operations on matrices, making it possible to solve systems of linear equations using the method presented in this section. Refer to the owner's manual for your calculator to learn how to use its matrix capabilities.

8.4 EXERCISES

✎ **1.** Write a short explanation of the meaning of the term.
  (a) matrix **(b)** square matrix **(c)** augmented matrix
  (d) row of a matrix **(e)** column of a matrix **(f)** row operations on a matrix

◉ **2.** Which one of the following is an example of a 3×2 matrix?

  (a) $\begin{bmatrix} 6 & -1 & 5 \\ 0 & 3 & 7 \end{bmatrix}$ **(b)** $\begin{bmatrix} 6 & 0 \\ -1 & 3 \\ 5 & 7 \end{bmatrix}$ **(c)** $\begin{bmatrix} 3 & 3 & 3 & 3 \\ 2 & 2 & 2 & 2 \end{bmatrix}$ **(d)** $\begin{bmatrix} 3 & 2 \\ 3 & 2 \\ 3 & 2 \\ 3 & 2 \end{bmatrix}$ 2. (b)

Complete the steps in the matrix solution of the system by filling in the boxes. Give the final system and the solution. See Example 1.

3. $4x + 8y = 44$
  $2x - y = -3$

$$\begin{bmatrix} 4 & 8 & \bigm| & 44 \\ 2 & -1 & \bigm| & -3 \end{bmatrix}$$

$$\begin{bmatrix} 1 & ▨ & \bigm| & ▨ \\ 2 & -1 & \bigm| & -3 \end{bmatrix} \tfrac{1}{4}R_1$$

$$\begin{bmatrix} 1 & 2 & \bigm| & 11 \\ 0 & ▨ & \bigm| & ▨ \end{bmatrix} -2R_1 + R_2$$

$$\begin{bmatrix} 1 & 2 & \bigm| & 11 \\ 0 & 1 & \bigm| & ▨ \end{bmatrix} -\tfrac{1}{5}R_2$$

3. $\begin{bmatrix} 1 & 2 & \bigm| & 11 \\ 2 & -1 & \bigm| & -3 \end{bmatrix};$

$\begin{bmatrix} 1 & 2 & \bigm| & 11 \\ 0 & -5 & \bigm| & -25 \end{bmatrix};$

$\begin{bmatrix} 1 & 2 & \bigm| & 11 \\ 0 & 1 & \bigm| & 5 \end{bmatrix}$

$x + 2y = 11,$
$y = 5;$
$\{(1, 5)\}$

4. $2x - 5y = -1$
  $3x + y = 7$

$$\begin{bmatrix} 2 & -5 & \bigm| & -1 \\ 3 & 1 & \bigm| & 7 \end{bmatrix}$$

$$\begin{bmatrix} 1 & -\dfrac{5}{2} & \bigm| & ▨ \\ 3 & 1 & \bigm| & 7 \end{bmatrix} \tfrac{1}{2}R_1$$

$$\begin{bmatrix} 1 & -\dfrac{5}{2} & \bigm| & -\dfrac{1}{2} \\ 0 & ▨ & \bigm| & ▨ \end{bmatrix} -3R_1 + R_2$$

$$\begin{bmatrix} 1 & -\dfrac{5}{2} & \bigm| & -\dfrac{1}{2} \\ 0 & 1 & \bigm| & ▨ \end{bmatrix} \tfrac{2}{17}R_2$$

4. $\begin{bmatrix} 1 & -\dfrac{5}{2} & \bigm| & -\dfrac{1}{2} \\ 3 & 1 & \bigm| & 7 \end{bmatrix};$

$\begin{bmatrix} 1 & -\dfrac{5}{2} & \bigm| & -\dfrac{1}{2} \\ 0 & \dfrac{17}{2} & \bigm| & \dfrac{17}{2} \end{bmatrix};$

$\begin{bmatrix} 1 & -\dfrac{5}{2} & \bigm| & -\dfrac{1}{2} \\ 0 & 1 & \bigm| & 1 \end{bmatrix};$

$x - \dfrac{5}{2}y = -\dfrac{1}{2}, y = 1;$
$\{(2, 1)\}$

 CONCEPTUAL   WRITING   CHALLENGING    SCIENTIFIC CALCULATOR   GRAPHICS CALCULATOR

Use row operations to solve the system. See Examples 1 and 3.

5. $x + y = 5$
$x - y = 3$
5. $\{(4, 1)\}$

6. $x + 2y = 7$
$x - y = -2$
6. $\{(1, 3)\}$

7. $2x + 4y = 6$
$3x - y = 2$
7. $\{(1, 1)\}$

8. $4x + 5y = -7$
$x - y = 5$
8. $\{(2, -3)\}$

9. $3x + 4y = 13$
$2x - 3y = -14$
9. $\{(-1, 4)\}$

10. $5x + 2y = 8$
$3x - y = 7$
10. $\{(2, -1)\}$

11. $-4x + 12y = 36$
$x - 3y = 9$
11. $\emptyset$

12. $2x - 4y = 8$
$-3x + 6y = 5$
12. $\emptyset$

⊙ 13. Compare the use of an augmented matrix as a shorthand way of writing a system of linear equations and the use of synthetic division as a shorthand way to divide polynomials.

⊙ 14. Compare the use of the third row operation on a matrix and the elimination method of solving a system of linear equations.

Complete the steps in the matrix solution of the system by filling in the boxes. Give the final system and the solution. See Example 2.

15. $x + y - z = -3$
$2x + y + z = 4$
$5x - y + 2z = 23$

$$\begin{bmatrix} 1 & 1 & -1 & -3 \\ 2 & 1 & 1 & 4 \\ 5 & -1 & 2 & 23 \end{bmatrix}$$

$$\begin{bmatrix} 1 & 1 & -1 & -3 \\ 0 & \blacksquare & \blacksquare & \blacksquare \\ 0 & \blacksquare & \blacksquare & \blacksquare \end{bmatrix} \begin{matrix} \\ -2R_1 + R_2 \\ -5R_1 + R_3 \end{matrix}$$

$$\begin{bmatrix} 1 & 1 & -1 & -3 \\ 0 & 1 & \blacksquare & \blacksquare \\ 0 & -6 & 7 & 38 \end{bmatrix} \begin{matrix} \\ -1R_2 \\ \\ \end{matrix}$$

$$\begin{bmatrix} 1 & 1 & -1 & -3 \\ 0 & 1 & -3 & -10 \\ 0 & 0 & \blacksquare & \blacksquare \end{bmatrix} \begin{matrix} \\ \\ 6R_2 + R_3 \end{matrix}$$

$$\begin{bmatrix} 1 & 1 & -1 & -3 \\ 0 & 1 & -3 & -10 \\ 0 & 0 & 1 & \blacksquare \end{bmatrix} \begin{matrix} \\ \\ -\frac{1}{11}R_3 \end{matrix}$$

15.
$$\begin{bmatrix} 1 & 1 & -1 & -3 \\ 0 & -1 & 3 & 10 \\ 0 & -6 & 7 & 38 \end{bmatrix};$$

$$\begin{bmatrix} 1 & 1 & -1 & -3 \\ 0 & 1 & -3 & -10 \\ 0 & -6 & 7 & 38 \end{bmatrix};$$

$$\begin{bmatrix} 1 & 1 & -1 & -3 \\ 0 & 1 & -3 & -10 \\ 0 & 0 & -11 & -22 \end{bmatrix};$$

$$\begin{bmatrix} 1 & 1 & -1 & -3 \\ 0 & 1 & -3 & -10 \\ 0 & 0 & 1 & 2 \end{bmatrix};$$

$x + y - z = -3,$
$y - 3z = -10, z = 2;$
$\{(3, -4, 2)\}$

16. $2x + y + 2z = 11$
$2x - y - z = -3$
$3x + 2y + z = 9$

$$\begin{bmatrix} 2 & 1 & 2 & 11 \\ 2 & -1 & -1 & -3 \\ 3 & 2 & 1 & 9 \end{bmatrix}$$

$$\begin{bmatrix} 1 & \blacksquare & \blacksquare & \blacksquare \\ 2 & -1 & -1 & -3 \\ 3 & 2 & 1 & 9 \end{bmatrix} \begin{matrix} \frac{1}{2}R_1 \\ \\ \end{matrix}$$

$$\begin{bmatrix} 1 & \frac{1}{2} & 1 & \frac{11}{2} \\ 0 & \blacksquare & \blacksquare & \blacksquare \\ 0 & \blacksquare & \blacksquare & \blacksquare \end{bmatrix} \begin{matrix} \\ -2R_1 + R_2 \\ -3R_1 + R_3 \end{matrix}$$

$$\begin{bmatrix} 1 & \frac{1}{2} & 1 & \frac{11}{2} \\ 0 & 1 & \blacksquare & \blacksquare \\ 0 & \frac{1}{2} & -2 & -\frac{15}{2} \end{bmatrix} \begin{matrix} \\ -\frac{1}{2}R_2 \\ \\ \end{matrix}$$

$$\begin{bmatrix} 1 & \frac{1}{2} & 1 & \frac{11}{2} \\ 0 & 1 & \frac{3}{2} & 7 \\ 0 & 0 & \blacksquare & \blacksquare \end{bmatrix} \begin{matrix} \\ \\ -\frac{1}{2}R_2 + R_3 \end{matrix}$$

$$\begin{bmatrix} 1 & \frac{1}{2} & 1 & \frac{11}{2} \\ 0 & 1 & \frac{3}{2} & 7 \\ 0 & 0 & 1 & \blacksquare \end{bmatrix} \begin{matrix} \\ \\ -\frac{4}{11}R_3 \end{matrix}$$

16.
$$\begin{bmatrix} 1 & \frac{1}{2} & 1 & \frac{11}{2} \\ 2 & -1 & -1 & -3 \\ 3 & 2 & 1 & 9 \end{bmatrix};$$

$$\begin{bmatrix} 1 & \frac{1}{2} & 1 & \frac{11}{2} \\ 0 & -2 & -3 & -14 \\ 0 & \frac{1}{2} & -2 & -\frac{15}{2} \end{bmatrix};$$

$$\begin{bmatrix} 1 & \frac{1}{2} & 1 & \frac{11}{2} \\ 0 & 1 & \frac{3}{2} & 7 \\ 0 & \frac{1}{2} & -2 & -\frac{15}{2} \end{bmatrix};$$

$$\begin{bmatrix} 1 & \frac{1}{2} & 1 & \frac{11}{2} \\ 0 & 1 & \frac{3}{2} & 7 \\ 0 & 0 & -\frac{11}{4} & -11 \end{bmatrix};$$

$$\begin{bmatrix} 1 & \frac{1}{2} & 1 & \frac{11}{2} \\ 0 & 1 & \frac{3}{2} & 7 \\ 0 & 0 & 1 & 4 \end{bmatrix};$$

$x + \frac{1}{2}y + z = \frac{11}{2},$
$y + \frac{3}{2}z = 7, z = 4;$
$\{(1, 1, 4)\}$

Use row operations to solve the system. See Examples 2 and 3.

17. $x + y - 3z = 1$
$2x - y + z = 9$
$3x + y - 4z = 8$ 17. $\{(4, 0, 1)\}$

18. $2x + 4y - 3z = -18$
$3x + y - z = -5$
$x - 2y + 4z = 14$ 18. $\{(0, -3, 2)\}$

19. $x + y - z = 6$
$2x - y + z = -9$
$x - 2y + 3z = 1$ 19. $\{(-1, 23, 16)\}$

20. $x + 3y - 6z = 7$
$2x - y + 2z = 0$
$x + y + 2z = -1$ 20. $\{(1, 0, -1)\}$

21. $x - y = 1$
$y - z = 6$
$x + z = -1$ 21. $\{(3, 2, -4)\}$

22. $x + y = 1$
$2x - z = 0$
$y + 2z = -2$ 22. $\{(-1, 2, -2)\}$

23. $x - 2y + z = 4$
$3x - 6y + 3z = 12$
$-2x + 4y - 2z = -8$

24. $4x + 8y + 4z = 9$
$x + 3y + 4z = 10$
$5x + 10y + 5z = 12$ 24. $\emptyset$

23. $\{(x, y, z) \mid x - 2y + z = 4\}$

The screen shown here is a typical graphics calculator augmented matrix display for the system of equations found in Exercise 3.

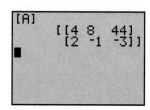

After using the calculator to perform the necessary row operations, the transformed matrix appears as follows.

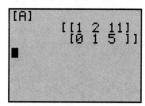

Compare these to the first matrix shown in Exercise 3 and the final matrix shown for the system in the answer section.

Use a graphics calculator with row operation capability to solve the system.

25. $4x + y = 5$
$2x + y = 3$ 25. $\{(1, 1)\}$

26. $5x + 3y = 7$
$7x - 3y = -19$ 26. $\{(-1, 4)\}$

27. $5x + y - 3z = -6$
$2x + 3y + z = 5$
$-3x - 2y + 4z = 3$ 27. $\{(-1, 2, 1)\}$

28. $x + y + z = 3$
$3x - 3y - 4z = -1$
$x + y + 3z = 11$ 28. $\{(2, -3, 4)\}$

29. $x + z = -3$
$y + z = 3$
$x + y = 8$ 29. $\{(1, 7, -4)\}$

30. $x - y = -1$
$-y + z = -2$
$x + z = -2$ 30. $\{(-.5, .5, -1.5)\}$

REVIEW EXERCISES

Evaluate the expression. See Section 1.2.

31. $5(-6) - 8(-2)$ 31. -14

32. $ad - bc$, if $a = 0, b = 3, c = -2, d = 1$ 32. 6

33. $[(-1)(4) + (-4)(-2)] - [(-2)(3) - (-3)(1)]$
33. 7

34. $[(5)(-4) + (-1)(-6)] + [(-2)(-4) - (-1)(3)]$
34. -3

8.5 DETERMINANTS

FOR EXTRA HELP	OBJECTIVES
SSG pp. 260–264 **SSM** pp. 424–427 **Video** 10 **Tutorial** IBM MAC	**1** ▶ Evaluate 2×2 determinants. **2** ▶ Use expansion by minors about the first column to evaluate 3×3 determinants. **3** ▶ Use expansion by minors about any row or column to evaluate determinants. **4** ▶ Evaluate larger determinants.

▶ **TEACHING TIP**

Sections 8.5 and 8.6 are possible optional sections. An understanding of Section 8.5 is required before attempting Section 8.6.

Most students who continue in mathematics will see the topic of determinants again in calculus when studying vectors. The topic is also studied in detail in a linear algebra course. ◀

Three methods for solving linear systems have now been presented: the elimination method, substitution, and a matrix method. A method of solving linear systems by using *determinants* is introduced in the next section. In this section we introduce determinants and learn how to evaluate them.

Associated with every square matrix is a real number called the **determinant** of the matrix. A determinant is symbolized by the entries of the matrix placed between two vertical lines, such as

$$\begin{vmatrix} 2 & 3 \\ 7 & 1 \end{vmatrix} \quad \text{or} \quad \begin{vmatrix} 7 & 4 & 3 \\ 0 & 1 & 5 \\ 6 & 0 & 1 \end{vmatrix}.$$

Like matrices, determinants are named according to the number of rows and columns they contain. For example, the first determinant shown is a 2×2 (read "two by two") determinant. The second is a 3×3 determinant.

1 ▶ Evaluate 2×2 determinants.

▶ The value of the 2×2 determinant

$$\begin{vmatrix} a & b \\ c & d \end{vmatrix}$$

is defined as follows.

Value of a 2×2 Determinant	$\begin{vmatrix} a & b \\ c & d \end{vmatrix} = ad - bc$

EXAMPLE 1
Evaluating a 2×2 Determinant

Evaluate the determinant

$$\begin{vmatrix} -1 & -3 \\ 4 & -2 \end{vmatrix}.$$

Here $a = -1$, $b = -3$, $c = 4$, and $d = -2$, and the determinant equals

$$\begin{vmatrix} -1 & -3 \\ 4 & -2 \end{vmatrix} = (-1)(-2) - (-3)(4) = 2 + 12 = 14. \quad ■$$

A 3×3 determinant can be evaluated in a similar way.

Chalkboard Exercise

Evaluate the determinant.

$$\begin{vmatrix} -4 & 6 \\ 2 & 3 \end{vmatrix}$$

Answer: -24

INSTRUCTOR'S RESOURCES

ITM pp. 471–474 **ISM** pp. 646–651
IAM p. 39

TEST GENERATOR
DOS Windows MAC

TRANSPARENCIES

Value of a 3 × 3 Determinant

$$\begin{vmatrix} a_1 & b_1 & c_1 \\ a_2 & b_2 & c_2 \\ a_3 & b_3 & c_3 \end{vmatrix} = (a_1b_2c_3 + b_1c_2a_3 + c_1a_2b_3) - (a_3b_2c_1 + b_3c_2a_1 + c_3a_2b_1)$$

This rule for evaluating a 3 × 3 determinant is hard to remember. A method for calculating a 3 × 3 determinant that is easier to use is based on the rule above. Rearranging terms and factoring gives

$$\begin{vmatrix} a_1 & b_1 & c_1 \\ a_2 & b_2 & c_2 \\ a_3 & b_3 & c_3 \end{vmatrix} = a_1(b_2c_3 - b_3c_2) - a_2(b_1c_3 - b_3c_1) + a_3(b_1c_2 - b_2c_1). \qquad \textbf{(1)}$$

Each of the quantities in parentheses represents a 2 × 2 determinant, which is that part of the 3 × 3 determinant remaining when the row and column of the multiplier are eliminated, as shown below.

$$a_1(b_2c_3 - b_3c_2) \qquad \begin{vmatrix} a_1 & b_1 & c_1 \\ a_2 & b_2 & c_2 \\ a_3 & b_3 & c_3 \end{vmatrix}$$

$$a_2(b_1c_3 - b_3c_1) \qquad \begin{vmatrix} a_1 & b_1 & c_1 \\ a_2 & b_2 & c_2 \\ a_3 & b_3 & c_3 \end{vmatrix}$$

$$a_3(b_1c_2 - b_2c_1) \qquad \begin{vmatrix} a_1 & b_1 & c_1 \\ a_2 & b_2 & c_2 \\ a_3 & b_3 & c_3 \end{vmatrix}$$

These 2 × 2 determinants are called **minors** of the elements in the 3 × 3 determinant. In the determinant above, the minors of a_1, a_2, and a_3 are, respectively,

$$\begin{vmatrix} b_2 & c_2 \\ b_3 & c_3 \end{vmatrix}, \quad \begin{vmatrix} b_1 & c_1 \\ b_3 & c_3 \end{vmatrix}, \quad \begin{vmatrix} b_1 & c_1 \\ b_2 & c_2 \end{vmatrix}.$$

2 ▶ Use expansion by minors about the first column to evaluate 3 × 3 determinants.

▶ A 3 × 3 determinant can be evaluated by multiplying each element in the first column by its minor and combining the products as shown in equation (1). This is called the **expansion of the determinant by minors** about the first column.

EXAMPLE 2
Evaluating a 3 × 3 Determinant

Evaluate the determinant shown below by expanding by minors about the first column.

In this determinant, $a_1 = 1$, $a_2 = -1$, and $a_3 = 1$. Multiply each of these numbers by its minor and combine the three terms using the definition. Notice that the second term in the definition is *subtracted*.

$$\begin{vmatrix} 1 & 3 & -2 \\ -1 & -2 & -3 \\ 1 & 1 & 2 \end{vmatrix} = 1\begin{vmatrix} -2 & -3 \\ 1 & 2 \end{vmatrix} - (-1)\begin{vmatrix} 3 & -2 \\ 1 & 2 \end{vmatrix} + 1\begin{vmatrix} 3 & -2 \\ -2 & -3 \end{vmatrix}$$

$$= 1[(-2)(2) - (-3)(1)] + 1[(3)(2) - (-2)(1)]$$
$$\quad + 1[(3)(-3) - (-2)(-2)]$$
$$= 1(-1) + 1(8) + 1(-13)$$
$$= -1 + 8 - 13$$
$$= -6 \quad \blacksquare$$

Chalkboard Exercise

Evaluate the determinant by expansion by minors about the first column.

$$\begin{vmatrix} 0 & -1 & 0 \\ 2 & 4 & 2 \\ 3 & 1 & 5 \end{vmatrix}$$

Answer: 4

3 ▶ Use expansion by minors about any row or column to evaluate determinants.

▶ To get equation (1) we could have rearranged terms in the definition of the determinant and factored out the three elements of the second or third columns or of any of the three rows. Therefore, expanding by minors about any row or any column results in the same value for a 3 × 3 determinant. To determine the correct signs for the terms of other expansions, the following **array of signs** is helpful.

Array of Signs for a **3 × 3 Determinant**		

$$
\begin{array}{ccc}
+ & - & + \\
- & + & - \\
+ & - & +
\end{array}
$$

The signs alternate for each row and column beginning with $+$ in the first row, first column position. For example, if the expansion is to be about the second column, the first term would have a minus sign associated with it, the second term a plus sign, and the third term a minus sign.

E X A M P L E   3

Evaluating a 3 × 3 Determinant by a Different Expansion

Evaluate the determinant of Example 2 by expansion by minors about the second column.

$$
\begin{vmatrix} 1 & 3 & -2 \\ -1 & -2 & -3 \\ 1 & 1 & 2 \end{vmatrix} = -3\begin{vmatrix} -1 & -3 \\ 1 & 2 \end{vmatrix} + (-2)\begin{vmatrix} 1 & -2 \\ 1 & 2 \end{vmatrix} - 1\begin{vmatrix} 1 & -2 \\ -1 & -3 \end{vmatrix}
$$

$$
= -3(1) - 2(4) - 1(-5)
$$

$$
= -3 - 8 + 5
$$

$$
= -6
$$

As expected, the result is the same as in Example 2. ■

◆ Your answer should agree with the answer for Exercise 9, but not with the answer in Example 4.

◇ **C O N N E C T I O N S** ◇

Another method is available for evaluating a 3 × 3 determinant. Referring to Example 2: carry over the first two columns to the right of the original determinant to get

$$
\begin{vmatrix} 1 & 3 & -2 \\ -1 & -2 & -3 \\ 1 & 1 & 2 \end{vmatrix} \begin{matrix} 1 & 3 \\ -1 & -2 \\ 1 & 1 \end{matrix}.
$$

Multiply along the diagonals as shown below, placing the product at the end of the arrow:

Add the top numbers:

$$-4 - 9 + 2 = -11.$$

Add the bottom numbers:

$$4 - 3 - 6 = -5.$$

Find the *difference* between these sums to obtain the final answer:

$$-11 - (-5) = -6.$$

FOR DISCUSSION OR WRITING

Use this method to find the determinant in Exercise 9. Which method do you prefer? Why? Try an extension of this method to find the determinant given in Example 4. After checking to be sure your work is correct, do you get the same result? How does this affect your answer if you preferred this method for finding a 3×3 determinant?

4 ▶ Evaluate larger determinants.

▶   The method of expansion by minors can be extended to evaluate larger determinants, such as 4×4 or 5×5. For a larger determinant, the sign array also is extended. For example, the signs for a 4×4 determinant are arranged as follows.

Array of Signs for a 4×4 Determinant

$$
\begin{array}{cccc}
+ & - & + & - \\
- & + & - & + \\
+ & - & + & - \\
- & + & - & +
\end{array}
$$

EXAMPLE 4

Evaluating a 4×4 Determinant

Evaluate the determinant.

$$
\begin{bmatrix}
-1 & -2 & 3 & 2 \\
0 & 1 & 4 & -2 \\
3 & -1 & 4 & 0 \\
2 & 1 & 0 & 3
\end{bmatrix}
$$

The work can be reduced by choosing a row or column with zeros, say the fourth row. Expand by minors about the fourth row using the elements of the fourth row and the signs from the fourth row of the sign array given above. The minors are 3×3 determinants.

$$
-2 \begin{vmatrix} -2 & 3 & 2 \\ 1 & 4 & -2 \\ -1 & 4 & 0 \end{vmatrix} + 1 \begin{vmatrix} -1 & 3 & 2 \\ 0 & 4 & -2 \\ 3 & 4 & 0 \end{vmatrix} - 0 \begin{vmatrix} -1 & -2 & 2 \\ 0 & 1 & -2 \\ 3 & -1 & 0 \end{vmatrix}
$$

$$
+ 3 \begin{vmatrix} -1 & -2 & 3 \\ 0 & 1 & 4 \\ 3 & -1 & 4 \end{vmatrix}
$$

$$= -2(6) + 1(-50) - 0 + 3(-41)$$

$$= -185 \ \blacksquare$$

Chalkboard Exercise

Evaluate.

$$
\begin{vmatrix}
1 & 0 & 2 & 0 \\
3 & 0 & 0 & 4 \\
0 & -1 & 1 & 0 \\
2 & 0 & -1 & 0
\end{vmatrix}
$$

Answer: 20

Each of the four 3×3 determinants in Example 4 is evaluated by expansion of three 2×2 minors. Thus, a great deal of work is needed to evaluate a 4×4 or larger determinant. Such large determinants can be evaluated quickly, however, with the aid of a computer or a graphics calculator.

◆ C O N N E C T I O N S ◆

Graphics calculators with matrix capabilities can be used to find determinants of square matrices. Refer to your owner's manual to learn how to evaluate determinants with your calculator.

8.5 EXERCISES

◎ **1.** Give the expression for evaluating the determinant $\begin{vmatrix} p & q \\ r & s \end{vmatrix}$.

◎ **2.** What is the value of the determinant $\begin{vmatrix} a & 0 \\ b & 0 \end{vmatrix}$ for any values of a and b?

Evaluate the determinant. See Example 1.

3. $\begin{vmatrix} -2 & 5 \\ -1 & 4 \end{vmatrix}$

4. $\begin{vmatrix} 3 & -6 \\ 2 & -2 \end{vmatrix}$

5. $\begin{vmatrix} 1 & -2 \\ 7 & 0 \end{vmatrix}$

6. $\begin{vmatrix} -5 & -1 \\ 1 & 0 \end{vmatrix}$

7. $\begin{vmatrix} 0 & 4 \\ 0 & 4 \end{vmatrix}$

8. $\begin{vmatrix} 8 & -3 \\ 0 & 0 \end{vmatrix}$

Evaluate the determinant by expansion by minors about the first column. See Example 2.

9. $\begin{vmatrix} -1 & 2 & 4 \\ -3 & -2 & -3 \\ 2 & -1 & 5 \end{vmatrix}$

10. $\begin{vmatrix} 2 & -3 & -5 \\ 1 & 2 & 2 \\ 5 & 3 & -1 \end{vmatrix}$

11. $\begin{vmatrix} 1 & 0 & -2 \\ 0 & 2 & 3 \\ 1 & 0 & 5 \end{vmatrix}$

12. $\begin{vmatrix} 2 & -1 & 0 \\ 0 & -1 & 1 \\ 1 & 2 & 0 \end{vmatrix}$

13. $\begin{vmatrix} 1 & 0 & 0 \\ 0 & 1 & 0 \\ 0 & 0 & 1 \end{vmatrix}$

14. $\begin{vmatrix} 0 & 0 & 1 \\ 0 & 1 & 0 \\ 1 & 0 & 0 \end{vmatrix}$

◎ **15.** Explain in your own words how to evaluate a 2×2 determinant.

◎ **16.** Explain in your own words the method of evaluating a 3×3 determinant.

Evaluate the determinant by expansion by minors about any row or column. (Hint: The work is easier if you choose a row or a column with zeros.) See Example 3.

17. $\begin{vmatrix} 4 & 4 & 2 \\ 1 & -1 & -2 \\ 1 & 0 & 2 \end{vmatrix}$

18. $\begin{vmatrix} 3 & -1 & 2 \\ 1 & 5 & -2 \\ 0 & 2 & 0 \end{vmatrix}$

19. $\begin{vmatrix} 3 & 5 & -2 \\ 1 & -4 & 1 \\ 3 & 1 & -2 \end{vmatrix}$

20. $\begin{vmatrix} 1 & 3 & 2 \\ 3 & -1 & -2 \\ 1 & 10 & 20 \end{vmatrix}$

21. $\begin{vmatrix} 0 & 0 & 3 \\ 4 & 0 & -2 \\ 2 & -1 & 3 \end{vmatrix}$

22. $\begin{vmatrix} 3 & 0 & -2 \\ 1 & -4 & 1 \\ 3 & 1 & -2 \end{vmatrix}$

23. $\begin{vmatrix} 1 & 1 & 2 \\ 5 & 5 & 7 \\ 3 & 3 & 1 \end{vmatrix}$

24. $\begin{vmatrix} 2 & 4 & -1 \\ 1 & 0 & 1 \\ 2 & 4 & -1 \end{vmatrix}$

25. $\begin{vmatrix} 2 & 4 & 0 \\ 3 & -5 & 0 \\ 6 & -7 & 0 \end{vmatrix}$

◎ **26.** Explain why a determinant with a row or column of zeros has a value of zero.

1. $ps - rq$ 2. 0 3. −3
4. 6 5. 14 6. 1 7. 0
8. 0 9. 59 10. −14
11. 14 12. −5 13. 1
14. −1 17. −22 18. 16
19. 20 20. −124
21. −12 22. −5 23. 0
24. 0 25. 0

◎ CONCEPTUAL ✑ WRITING ▲ CHALLENGING ▦ SCIENTIFIC CALCULATOR ▦ GRAPHICS CALCULATOR

27. $\dfrac{y_2 - y_1}{x_2 - x_1}$

28.
$y - y_1 = \dfrac{y_2 - y_1}{x_2 - x_1}(x - x_1)$

29. $x_2y - x_1y - x_2y_1 - xy_2 + x_1y_2 + xy_1 = 0$

30. The result is the same as in Exercise 29.

31. {2}   32. {10}

33. 6.078

34. −22.04285452

35. −32   36. −6

———————◆ **MATHEMATICAL CONNECTIONS** (Exercises 27–30) ◇————————

Recall the formula for slope and the point-slope form of the equation of a line, as found in Chapter 7. Use these formulas in working Exercises 27–30 in order, so that you can see how a determinant can be used in writing the equation of a line.

27. Write the expression for the slope of a line passing through the points (x_1, y_1) and (x_2, y_2).

28. Using the expression from Exercise 27 as m, and the point (x_1, y_1), write the point-slope form of the equation of the line.

29. Using the equation obtained in Exercise 28, multiply both sides by $x_2 - x_1$, and write the equation so that 0 is on the right side.

30. Consider the *determinant equation*

$$\begin{vmatrix} x & y & 1 \\ x_1 & y_1 & 1 \\ x_2 & y_2 & 1 \end{vmatrix} = 0.$$

Expand by minors on the left and show that this determinant equation yields the same result that you obtained in Exercise 29.

———————◆———————

▲ *Solve the equation by finding an expression for the determinant on the left, and then solving using the methods of Chapter 2.*

31. $\begin{vmatrix} 4 & x \\ 2 & 3 \end{vmatrix} = 8$

32. $\begin{vmatrix} 5 & 3 \\ x & x \end{vmatrix} = 20$

In Example 2 we showed how to evaluate the determinant

$$\begin{vmatrix} 1 & 3 & -2 \\ -1 & -2 & -3 \\ 1 & 1 & 2 \end{vmatrix}$$

by expanding by minors. Modern graphics calculators have the capability of finding determinants at the stroke of a key. The display on the left is a graphics calculator-generated depiction of the matrix with the same entries as shown in Example 2, and the display on the right shows that its determinant is indeed −6.

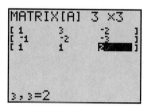 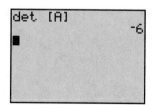

🖼 *Use a graphics calculator with matrix capabilities to find the determinant.*

33. $\begin{vmatrix} 1.5 & 2.6 & 9.3 \\ 5.2 & -1.4 & 8.6 \\ 0 & .7 & 1.2 \end{vmatrix}$

34. $\begin{vmatrix} \sqrt{5} & \sqrt{2} & -\sqrt{3} \\ \sqrt{7} & -\sqrt{6} & \sqrt{10} \\ -\sqrt{5} & -\sqrt{2} & \sqrt{17} \end{vmatrix}$ (To as many places as the calculator shows)

35. $\begin{vmatrix} 1 & 0 & 2 & 2 \\ 2 & 4 & 1 & -1 \\ 1 & -3 & 1 & 0 \\ 1 & 1 & 0 & 1 \end{vmatrix}$

36. $\begin{vmatrix} 2 & -1 & 1 & 0 \\ 1 & 1 & 0 & 1 \\ 0 & -1 & 1 & 1 \\ 1 & 2 & 1 & 2 \end{vmatrix}$

Evaluate the determinant by expansion by minors about the second row. See Example 4.

37.
$$\begin{vmatrix} 1 & 4 & 2 & 0 \\ 0 & 2 & 0 & -1 \\ 3 & -1 & 2 & 0 \\ 1 & 4 & -1 & 2 \end{vmatrix}$$

38.
$$\begin{vmatrix} 4 & 1 & 0 & 2 \\ 1 & 0 & 0 & -2 \\ 3 & 4 & 1 & -3 \\ -2 & 1 & 1 & -1 \end{vmatrix}$$

39.
$$\begin{vmatrix} 3 & 5 & 1 & 9 \\ 0 & 5 & 2 & 0 \\ 2 & -1 & -1 & -1 \\ -4 & 2 & 2 & 2 \end{vmatrix}$$

40.
$$\begin{vmatrix} 2 & 3 & 2 & 2 \\ 1 & -1 & 0 & 0 \\ 2 & 1 & 1 & 1 \\ 0 & 0 & -3 & -2 \end{vmatrix}$$

37. -55 38. -22 39. 0
40. 1   41. $\dfrac{7}{5}$ 42. -37
43. undefined   44. 0

REVIEW EXERCISES

Evaluate the expression. See Section 1.2.

41. $\dfrac{-3(-4) - 5(-6)}{-2(-3) - 4(-6)}$

42. $\dfrac{-9(6) - 4(5)}{6(-1) - 2(-4)}$

43. $\dfrac{-5(-5) - 2(3)}{(-10)(-2) - (-4)(-5)}$

44. $\dfrac{4(9) - 12(3)}{3(-1) - 4(-2)}$

8.6 SOLUTION OF LINEAR SYSTEMS OF EQUATIONS BY DETERMINANTS—CRAMER'S RULE

FOR EXTRA HELP

 SSG pp. 264–269
SSM pp. 427–433

 Video 10

💾 **Tutorial** IBM   MAC

OBJECTIVES

1 ▶ Understand the derivation of Cramer's rule.
2 ▶ Apply Cramer's rule to a linear system with two equations and two unknowns.
3 ▶ Apply Cramer's rule to a linear system with three equations and three unknowns.
4 ▶ Determine when Cramer's rule does not apply.

1 ▶ Understand the derivation of Cramer's rule.

▶ In this section we use the elimination method to solve the general system of two equations with two variables,

$$a_1x + b_1y = c_1 \tag{1}$$
$$a_2x + b_2y = c_2. \tag{2}$$

The result will be a formula that can be used for any system of two equations with two unknowns. To get this general solution, we eliminate y and solve for x by first multiplying both sides of equation (1) by b_2 and both sides of equation (2) by $-b_1$. Then we add these results and solve for x.

$$
\begin{array}{ll}
a_1b_2x + b_1b_2y = c_1b_2 & \quad b_2 \text{ times both sides of equation } (1) \\
\underline{-a_2b_1x - b_1b_2y = -c_2b_1} & \quad -b_1 \text{ times both sides of equation } (2) \\
(a_1b_2 - a_2b_1)x = c_1b_2 - c_2b_1 &
\end{array}
$$

$$x = \frac{c_1b_2 - c_2b_1}{a_1b_2 - a_2b_1} \quad (\text{if } a_1b_2 - a_2b_1 \neq 0)$$

INSTRUCTOR'S RESOURCES

 ITM pp. 475–478   **ISM** pp. 651–663
IAM p. 39

 TEST GENERATOR
DOS Windows MAC

 TRANSPARENCIES

 ⊙ CONCEPTUAL    ✍ WRITING   ▲ CHALLENGING   🔢 SCIENTIFIC CALCULATOR    🖥 GRAPHICS CALCULATOR

To solve for y, we multiply both sides of equation (1) by $-a_2$ and both sides of equation (2) by a_1 and add.

$$\begin{array}{ll} -a_1a_2x - a_2b_1\,y = -a_2c_1 & \quad -a_2 \text{ times both sides of } (1) \\ \underline{a_1a_2x + a_1b_2\,y = a_1c_2} & \quad a_1 \text{ times both sides of } (2) \\ (a_1b_2 - a_2b_1)y = a_1c_2 - a_2c_1 & \end{array}$$

$$y = \frac{a_1c_2 - a_2c_1}{a_1b_2 - a_2b_1}$$

Both numerators and the common denominator of these values for x and y can be written as determinants, since

$$a_1c_2 - a_2c_1 = \begin{vmatrix} a_1 & c_1 \\ a_2 & c_2 \end{vmatrix},$$

$$c_1b_2 - c_2b_1 = \begin{vmatrix} c_1 & b_1 \\ c_2 & b_2 \end{vmatrix},$$

and

$$a_1b_2 - a_2b_1 = \begin{vmatrix} a_1 & b_1 \\ a_2 & b_2 \end{vmatrix}.$$

Using these results, the solutions for x and y become

$$x = \frac{\begin{vmatrix} c_1 & b_1 \\ c_2 & b_2 \end{vmatrix}}{\begin{vmatrix} a_1 & b_1 \\ a_2 & b_2 \end{vmatrix}} \quad \text{and} \quad y = \frac{\begin{vmatrix} a_1 & c_1 \\ a_2 & c_2 \end{vmatrix}}{\begin{vmatrix} a_1 & b_1 \\ a_2 & b_2 \end{vmatrix}}, \quad \text{if} \begin{vmatrix} a_1 & b_1 \\ a_2 & b_2 \end{vmatrix} \neq 0.$$

For convenience, we denote the three determinants in the solution as

$$\begin{vmatrix} a_1 & b_1 \\ a_2 & b_2 \end{vmatrix} = D, \quad \begin{vmatrix} c_1 & b_1 \\ c_2 & b_2 \end{vmatrix} = D_x, \quad \begin{vmatrix} a_1 & c_1 \\ a_2 & c_2 \end{vmatrix} = D_y.$$

Note that the elements of D are the four coefficients of the variables in the given system; the elements of D_x are obtained by replacing the coefficients of x by the respective constants; the elements of D_y are obtained by replacing the coefficients of y by the respective constants.

These results are summarized as **Cramer's rule.**

Cramer's Rule for 2 × 2 Systems

▶ **TEACHING TIP**
Students may reach the conclusion that Cramer's rule is a replacement for the elimination method. When $D = 0$, one should use the elimination method to determine whether the system is inconsistent or has dependent equations. ◀

Given the system

$$a_1x + b_1 y = c_1$$
$$a_2x + b_2 y = c_2,$$

with

$$a_1b_2 - a_2b_1 \neq 0,$$

then

$$x = \frac{\begin{vmatrix} c_1 & b_1 \\ c_2 & b_2 \end{vmatrix}}{\begin{vmatrix} a_1 & b_1 \\ a_2 & b_2 \end{vmatrix}} = \frac{D_x}{D} \quad \text{and} \quad y = \frac{\begin{vmatrix} a_1 & c_1 \\ a_2 & c_2 \end{vmatrix}}{\begin{vmatrix} a_1 & b_1 \\ a_2 & b_2 \end{vmatrix}} = \frac{D_y}{D}.$$

---

2 ▶ Apply Cramer's rule to a linear system with two equations and two unknowns.

▶ To use Cramer's rule to solve a system of linear equations, find the three determinants, D, D_x, and D_y, and then write the necessary quotients for x and y.

CAUTION As indicated above, Cramer's rule does not apply if $D = a_1 b_2 - a_2 b_1$ is 0. When $D = 0$, the system is inconsistent or has dependent equations. For this reason, it is a good idea to evaluate D first.

EXAMPLE 1

Using Cramer's Rule for a 2 × 2 System

Use Cramer's rule to solve the system

$$5x + 7y = -1$$
$$6x + 8y = 1.$$

By Cramer's rule, $x = D_x/D$ and $y = D_y/D$. We will find D first, since if $D = 0$, Cramer's rule does not apply. If $D \neq 0$, then we will find D_x and D_y.

$$D = \begin{vmatrix} 5 & 7 \\ 6 & 8 \end{vmatrix} = 5(8) - 6(7) = -2$$

$$D_x = \begin{vmatrix} -1 & 7 \\ 1 & 8 \end{vmatrix} = (-1)8 - 7(1) = -15$$

$$D_y = \begin{vmatrix} 5 & -1 \\ 6 & 1 \end{vmatrix} = 5(1) - (-1)6 = 11$$

From Cramer's rule,

$$x = \frac{D_x}{D} = \frac{-15}{-2} = \frac{15}{2}$$

and

$$y = \frac{D_y}{D} = \frac{11}{-2} = -\frac{11}{2}.$$

The solution set is $\{(15/2, -11/2)\}$, as we can verify by checking in the given system. ∎

Chalkboard Exercise

Solve by Cramer's rule.
$$2x - 3y = -26$$
$$3x + 4y = 12$$
Answer: $\{(-4, 6)\}$

3 ▶ Apply Cramer's rule to a linear system with three equations and three unknowns.

▶ In a similar manner, Cramer's rule can be applied to systems of three linear equations with three variables.

Cramer's Rule for 3 × 3 Systems

Given the system

$$a_1 x + b_1 y + c_1 z = d_1$$
$$a_2 x + b_2 y + c_2 z = d_2$$
$$a_3 x + b_3 y + c_3 z = d_3,$$

with

$$D_x = \begin{vmatrix} d_1 & b_1 & c_1 \\ d_2 & b_2 & c_2 \\ d_3 & b_3 & c_3 \end{vmatrix}, \qquad D_y = \begin{vmatrix} a_1 & d_1 & c_1 \\ a_2 & d_2 & c_2 \\ a_3 & d_3 & c_3 \end{vmatrix},$$

$$D_z = \begin{vmatrix} a_1 & b_1 & d_1 \\ a_2 & b_2 & d_2 \\ a_3 & b_3 & d_3 \end{vmatrix}, \qquad D = \begin{vmatrix} a_1 & b_1 & c_1 \\ a_2 & b_2 & c_2 \\ a_3 & b_3 & c_3 \end{vmatrix} \neq 0,$$

then

$$x = \frac{D_x}{D}, \qquad y = \frac{D_y}{D}, \qquad z = \frac{D_z}{D}.$$

EXAMPLE 2

Using Cramer's Rule for a 3 × 3 System

Use Cramer's rule to solve the system

$$x + y - z + 2 = 0$$
$$2x - y + z + 5 = 0$$
$$x - 2y + 3z - 4 = 0.$$

To use Cramer's rule, we must rewrite the system in the form

$$x + y - z = -2$$
$$2x - y + z = -5$$
$$x - 2y + 3z = 4.$$

We expand by minors about row 1 to find D.

$$D = \begin{vmatrix} 1 & 1 & -1 \\ 2 & -1 & 1 \\ 1 & -2 & 3 \end{vmatrix} = 1\begin{vmatrix} -1 & 1 \\ -2 & 3 \end{vmatrix} - 1\begin{vmatrix} 2 & 1 \\ 1 & 3 \end{vmatrix} + (-1)\begin{vmatrix} 2 & -1 \\ 1 & -2 \end{vmatrix}$$

$$= 1(-1) - 1(5) - 1(-3) = -3$$

Expanding D_x by minors about row 1 gives

$$D_x = \begin{vmatrix} -2 & 1 & -1 \\ -5 & -1 & 1 \\ 4 & -2 & 3 \end{vmatrix} = -2\begin{vmatrix} -1 & 1 \\ -2 & 3 \end{vmatrix} - 1\begin{vmatrix} -5 & 1 \\ 4 & 3 \end{vmatrix} + (-1)\begin{vmatrix} -5 & -1 \\ 4 & -2 \end{vmatrix}$$

$$= -2(-1) - 1(-19) - 1(14) = 7.$$

In the same way, $D_y = -22$ and $D_z = -21$, so that

$$x = \frac{D_x}{D} = \frac{7}{-3} = -\frac{7}{3}, \qquad y = \frac{D_y}{D} = \frac{-22}{-3} = \frac{22}{3}, \qquad z = \frac{D_z}{D} = \frac{-21}{-3} = 7.$$

The solution set is $\left\{\left(-\dfrac{7}{3}, \dfrac{22}{3}, 7\right)\right\}$. ■

4 ▶ Determine when Cramer's rule does not apply.

▶ As mentioned earlier, Cramer's rule does not apply when $D = 0$. The next example illustrates this case.

EXAMPLE 3

Determining when Cramer's Rule Does not Apply

Use Cramer's rule to solve the system

$$2x - 3y + 4z = 8$$
$$6x - 9y + 12z = 24$$
$$x + 2y - 3z = 5.$$

We need to find D, D_x, D_y, and D_z. Here, expanding about column 1 gives

$$D = \begin{vmatrix} 2 & -3 & 4 \\ 6 & -9 & 12 \\ 1 & 2 & -3 \end{vmatrix} = 2\begin{vmatrix} -9 & 12 \\ 2 & -3 \end{vmatrix} - 6\begin{vmatrix} -3 & 4 \\ 2 & -3 \end{vmatrix} + 1\begin{vmatrix} -3 & 4 \\ -9 & 12 \end{vmatrix}$$

$$= 2(3) - 6(1) + 1(0)$$
$$= 0.$$

Since $D = 0$ here, Cramer's rule does not apply and we must use another method to solve the system. Multiplying the first equation on both sides by 3 shows that the first two equations have the same solutions, so this system has dependent equations and an infinite solution set. ■

Cramer's rule can be extended to 4×4 or larger systems. See a standard college algebra text for details.

8.6 EXERCISES

◉ **1.** For the system

$$8x - 4y = 8$$
$$x + 3y = 22,$$

$D_x = 112$, $D_y = 168$, and $D = 28$. What is the solution set of the system?

◉ **2.** For the system

$$x + 3y - 6z = 7$$
$$2x - y + z = 1$$
$$x + 2y + 2z = -1,$$

the solution set is $\{(1, 0, -1)\}$ and $D = -43$. Find the values of D_x, D_y, and D_z.

Use Cramer's rule to solve the linear system in two variables. See Example 1.

3. $3x + 5y = -5$
    $-2x + 3y = 16$

4. $5x + 2y = -3$
    $4x - 3y = -30$

5. $3x + 2y = 3$
    $2x - 4y = 2$

6. $7x - 2y = 6$
    $4x - 5y = 15$

7. $8x + 3y = 1$
    $6x - 5y = 2$

8. $3x - y = 9$
    $2x + 5y = 8$

9. $2x + 3y = 4$
    $5x + 6y = 7$

10. $4x + 5y = 6$
    $7x + 8y = 9$

◉ **11.** Look at the coefficients in the systems in Exercises 9 and 10. Notice that in both cases, the six coefficients are consecutive integers. Make up a system having this same pattern for its coefficients, and solve it using Cramer's rule.

◉ **12. (a)** Compare the solutions in Exercises 9, 10, and 11. What do you notice?
▲ **(b)** Use Cramer's rule to prove that the system

$$ax + (a + 1)y = a + 2$$
$$(a + 3)x + (a + 4)y = a + 5, \quad \text{where } D \neq 0,$$

has solution set $\{(-1, 2)\}$.

Use Cramer's rule where applicable to solve the linear system in three variables. See Examples 2 and 3.

13. $2x + 3y + 2z = 15$
    $x - y + 2z = 5$
    $x + 2y - 6z = -26$

14. $x - y + 6z = 19$
    $3x + 3y - z = 1$
    $x + 9y + 2z = -19$

15. $2x + 2y + z = 10$
    $4x - y + z = 20$
    $-x + y - 2z = -5$

16. $x + 3y - 4z = -12$
    $3x + y - z = -5$
    $5x - y + z = -3$

1. $\{(4, 6)\}$ 2. $D_x = -43$, $D_y = 0$, $D_z = 43$
3. $\{(-5, 2)\}$ 4. $\{(-3, 6)\}$
5. $\{(1, 0)\}$ 6. $\{(0, -3)\}$
7. $\left\{\left(\dfrac{11}{58}, -\dfrac{5}{29}\right)\right\}$
8. $\left\{\left(\dfrac{53}{17}, \dfrac{6}{17}\right)\right\}$
9. $\{(-1, 2)\}$ 10. $\{(-1, 2)\}$
11. Answers will vary. One example is
$6x + 7y = 8$
$9x + 10y = 11$.
12. **(a)** Each has the solution set $\{(-1, 2)\}$.
13. $\{(-2, 3, 5)\}$
14. $\{(4, -3, 2)\}$
15. $\{(5, 0, 0)\}$
16. $\{(-1, 3, 5)\}$

 CONCEPTUAL   WRITING    CHALLENGING    SCIENTIFIC CALCULATOR   GRAPHICS CALCULATOR

17. Cramer's rule does not apply.   18. Cramer's rule does not apply.
19. $\{(20, -13, -12)\}$
20. $\{(-2, 1, 3)\}$
21. $\left\{\left(\dfrac{62}{5}, -\dfrac{1}{5}, \dfrac{27}{5}\right)\right\}$
22. $\left\{\left(\dfrac{49}{9}, -\dfrac{155}{9}, \dfrac{136}{9}\right)\right\}$
23. $\{(-1, 2, 5, 1)\}$
24. $\{(7, 2, 3, 0)\}$
25.

26. $\dfrac{1}{2}\begin{vmatrix} 0 & 0 & 1 \\ -3 & -4 & 1 \\ 2 & -2 & 1 \end{vmatrix}$

27. 7   28. $a = \sqrt{29}$, $b = 2\sqrt{2}$, $c = 5$,
$s = \dfrac{1}{2}(\sqrt{29} + 2\sqrt{2} + 5)$;
The area is 7.

17.
$$2x - 3y + 4z = 8$$
$$6x - 9y + 12z = 24$$
$$-4x + 6y - 8z = -16$$

18.
$$7x + y - z = 4$$
$$2x - 3y + z = 2$$
$$-6x + 9y - 3z = -6$$

19.
$$3x + 5z = 0$$
$$2x + 3y = 1$$
$$-y + 2z = -11$$

20.
$$-x + 2y = 4$$
$$3x + y = -5$$
$$2x + z = -1$$

21.
$$x - 3y = 13$$
$$2y + z = 5$$
$$-x + z = -7$$

22.
$$-5x - y = -10$$
$$3x + 2y + z = -3$$
$$-y - 2z = -13$$

▲ **23.**
$$3x + 2y - w = 0$$
$$2x + z + 2w = 5$$
$$x + 2y - z = -2$$
$$2x - y + z + w = 2$$

▲ **24.**
$$x + 2y - z + w = 8$$
$$2x - y - w = 12$$
$$y + 3z = 11$$
$$x - z - w = 4$$

◆ **MATHEMATICAL CONNECTIONS** (Exercises 25–28) ◆

In this section we have seen how determinants can be used to solve systems of equations. In the Mathematical Connections in the exercises for Section 8.5, we saw how a determinant can be used to write the equation of a line given two points on a line. Here, we show how a determinant can be used to find the area of a triangle if we know the coordinates of its vertices.

Suppose that $A(x_1, y_1)$, $B(x_2, y_2)$ and $C(x_3, y_3)$ are the coordinates of the vertices of triangle ABC in the coordinate plane. Then it can be shown that the area of the triangle is given by the absolute value of

$$\frac{1}{2}\begin{vmatrix} x_1 & y_1 & 1 \\ x_2 & y_2 & 1 \\ x_3 & y_3 & 1 \end{vmatrix}.$$

Work Exercises 25–28 in order.

25. Sketch triangle ABC in the coordinate plane, given that the coordinates of A are $(0, 0)$, of B are $(-3, -4)$, and of C are $(2, -2)$.

26. Write the determinant expression as described above that gives the area of triangle ABC described in Exercise 25.

27. Evaluate the absolute value of the determinant expression in Exercise 26 to find the area.

▦ **28.** *Heron's formula* states that if a, b, and c are the lengths of the sides of a triangle and $s = \frac{1}{2}(a + b + c)$, then the area of the triangle is given by the expression $\sqrt{s(s - a)(s - b)(s - c)}$. Use the distance formula to find the lengths of the three sides of the triangle of Exercise 25, find s, and then use Heron's formula to find the area of the triangle. Use a calculator as necessary. It must agree with the answer in Exercise 27.

◆

Because graphics calculators with matrix capabilities can evaluate determinants, they can be used to apply Cramer's rule. For example, the system

$$x + y - z = -2$$
$$2x - y + z = -5$$
$$x - 2y + 3z = 4,$$

solved in a traditional manner in Example 2, had $D_x = 7$ and $D = -3$. Using $[A]$ to represent the matrix for D_x, $[B]$ to represent the matrix for D, and *det* to represent determinant (as shown on the calculator screen), typical graphics calculator displays are shown.

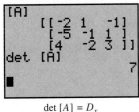

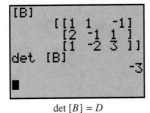

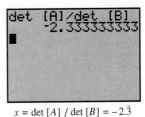

det $[A] = D_x$ det $[B] = D$ $x = \text{det } [A] \, / \, \text{det } [B] = -2.\overline{3}$

The values of y and z can be found similarly. Graphics calculators can be programmed to solve systems using Cramer's rule as well.

Use a graphics calculator with matrix capabilities to solve the system.

29. $\begin{aligned} x + 2y + z &= 10 \\ 2x - y - 3z &= -20 \\ -x + 4y + z &= 18 \end{aligned}$

30. $\begin{aligned} 2x + y + 3z &= 1 \\ x - 2y + z &= -3 \\ -3x + y - 2z &= -4 \end{aligned}$

31. $\begin{aligned} -8w + 4x - 2y + z &= -28 \\ -w + x - y + z &= -10 \\ w + x + y + z &= -4 \\ 27w + 9x + 3y + z &= 2 \end{aligned}$

32. $\begin{aligned} 5w + 2x - 3y + z &= 4.7 \\ -2w + x + 2y - z &= -3.2 \\ w + 3x - y + 2z &= 2.1 \\ 2w + x - 5y + 3z &= 3.4 \end{aligned}$

REVIEW EXERCISES

Find the function value. See Section 7.5.

33. $f(x) = x^2 + 4x - 3$. Find $f(2)$.

34. $f(x) = 2(x - 3)^2 + 5$. Find $f(3)$.

35. $f(x) = ax^2 + bx + c$. Find $f\left(\dfrac{b}{-2a}\right)$.

See Sections 3.9 and 6.2.

36. If $f(x) = 8x^2 - 2x - 3$, find all values of x for which $f(x) = 0$.

29. $\{(-1, 3, 5)\}$
30. $\{(4, 2, -3)\}$
31. $\{(1, -3, 2, -4)\}$
32. $\{(1.2, -.5, .6, 1.5)\}$
33. 9   34. 5
35. $\dfrac{4ac - b^2}{4a}$ 36. $\dfrac{3}{4}, -\dfrac{1}{2}$

CHAPTER 8 SUMMARY

KEY TERMS

8.1 system of equations
linear system
inconsistent system
dependent equations
elimination (or
addition) method
substitution method

8.2 ordered triple

8.4 matrix
square matrix
augmented matrix

8.5 determinant

8.6 Cramer's rule

NEW SYMBOLS

(x, y, z) ordered triple

$\begin{bmatrix} a & b & c \\ d & e & f \end{bmatrix}$ matrix with two rows, three columns

$\begin{vmatrix} a & b \\ c & d \end{vmatrix}$ determinant of a 2×2 matrix

$\begin{vmatrix} a & b & c \\ d & e & f \\ g & h & i \end{vmatrix}$ determinant of a 3×3 matrix

 CONCEPTUAL   WRITING   ▲ CHALLENGING   SCIENTIFIC CALCULATOR    GRAPHICS CALCULATOR

QUICK REVIEW

CONCEPTS	EXAMPLES
8.1 LINEAR SYSTEMS OF EQUATIONS IN TWO VARIABLES	

Solving Linear Systems by Elimination	Solve by elimination.
Step 1 Write both equations in the form $Ax + By = C$.	$$5x + y = 2$$ $$2x - 3y = 11$$
Step 2 Multiply one or both equations by appropriate numbers so that the sum of the coefficients of either x or y is zero.	To eliminate y, multiply the top equation by 3, and add.
Step 3 Add the new equations. The sum should be an equation with just one variable.	$$\begin{array}{r} 15x + 3y = 6 \\ 2x - 3y = 11 \\ \hline 17x \phantom{{}- 3y} = 17 \end{array}$$
Step 4 Solve the equation from Step 3.	$$x = 1$$
Step 5 Substitute the result of Step 4 into either of the given equations and solve for the other variable.	Let $x = 1$ in the top equation, and solve for y. $$5(1) + y = 2$$ $$y = -3$$
Step 6 Check the solution in both of the given equations.	Check to verify that $\{(1, -3)\}$ is the solution set.
Solving Linear Systems by Substitution	Solve by substitution. $$4x - y = 7$$ $$3x + 2y = 30$$
Step 1 Solve one of the equations for either variable.	Solve for y in the top equation. $$y = 4x - 7$$
Step 2 Substitute for that variable in the other equation. The result should be an equation with just one variable.	Substitute $4x - 7$ for y in the bottom equation, and solve for x. $$3x + 2(4x - 7) = 30$$ $$3x + 8x - 14 = 30$$
Step 3 Solve the equation from Step 2.	$$11x = 44$$ $$x = 4$$
Step 4 Substitute the result from Step 3 into the equation from Step 1 to find the value of the other variable.	Substitute 4 for x in the equation $y = 4x - 7$ to find that $y = 9$.
Step 5 Check the solution in both of the given equations.	Check to see that $\{(4, 9)\}$ is the solution set.
8.2 LINEAR SYSTEMS OF EQUATIONS IN THREE VARIABLES	
Solving Linear Systems in Three Variables	Solve the system
Step 1 Use the elimination method to eliminate any variable from any two of the given equations. The result is an equation in two variables.	$$x + 2y - z = 6$$ $$x + y + z = 6$$ $$2x + y - z = 7.$$ Add the first and second equations; z is eliminated and the result is $2x + 3y = 12$.

CONCEPTS	EXAMPLES
Step 2 Eliminate the *same* variable from any *other* two equations. The result is an equation in the same two variables as in Step 1.	Eliminate z again by adding the second and third equations to get $3x + 2y = 13$. Now solve the system $$2x + 3y = 12 \qquad (*)$$ $$3x + 2y = 13.$$
Step 3 Use the elimination method to eliminate a second variable from the two equations in two variables that result from Steps 1 and 2. The result is an equation in one variable that gives the value of that variable.	To eliminate x, multiply the top equation by -3 and the bottom equation by 2. $$\begin{array}{r} -6x - 9y = -36 \\ 6x + 4y = 26 \\ \hline -5y = -10 \\ y = 2 \end{array}$$
Step 4 Substitute the value of the variable found in Step 3 into either of the equations in two variables to find the value of the second variable.	Let $y = 2$ in equation $(*)$. $$2x + 3(2) = 12$$ $$2x = 6$$ $$x = 3$$
Step 5 Use the values of the two variables from Steps 3 and 4 to find the value of the third variable by substituting into any of the original equations.	Let $y = 2$ and $x = 3$ in any of the original equations to find $z = 1$. The solution set is $\{(3, 2, 1)\}$.

8.3   APPLICATIONS OF LINEAR SYSTEMS OF EQUATIONS

To solve an applied problem with two (three) unknowns, write two (three) equations that relate the unknowns. Then solve the system.	The perimeter of a rectangle is 18 feet. The length is 3 feet more than twice the width. Find the dimensions of the rectangle. Let x represent the length and y represent the width. From the perimeter formula, one equation is $2x + 2y = 18$. From the problem, another equation is $x = 3 + 2y$. Now solve the system $$2x + 2y = 18$$ $$x = 3 + 2y.$$ The solution of the system is $(7, 2)$. Therefore, the length is 7 feet and the width is 2 feet.

8.4   SOLUTION OF LINEAR SYSTEMS OF EQUATIONS BY MATRIX METHODS

Matrix Row Operations 1. Any two rows of the matrix may be interchanged.	$\begin{bmatrix} 1 & 5 & 7 \\ 3 & 9 & -2 \\ 0 & 6 & 4 \end{bmatrix}$ becomes $\begin{bmatrix} 3 & 9 & -2 \\ 1 & 5 & 7 \\ 0 & 6 & 4 \end{bmatrix}$ Interchange R_1 and R_2.

CONCEPTS	EXAMPLES
2. The numbers in any row may be multiplied by any nonzero real number.	$\begin{bmatrix} 1 & 5 & 7 \\ 3 & 9 & -2 \\ 0 & 6 & 4 \end{bmatrix}$ becomes $\begin{bmatrix} 1 & 5 & 7 \\ 1 & 3 & -\dfrac{2}{3} \\ 0 & 6 & 4 \end{bmatrix}$ $\frac{1}{3}R_2$
3. Any row may be changed by adding to the numbers of the row the product of a real number and the numbers of another row.	$\begin{bmatrix} 1 & 5 & 7 \\ 3 & 9 & -2 \\ 0 & 6 & 4 \end{bmatrix}$ becomes $\begin{bmatrix} 1 & 5 & 7 \\ 0 & -6 & -23 \\ 0 & 6 & 4 \end{bmatrix}$ $-3R_1 + R_2$

8.5 DETERMINANTS

Value of a 2 × 2 Determinant $$\begin{vmatrix} a & b \\ c & d \end{vmatrix} = ad - bc.$$ Determinants larger than 2 × 2 are evaluated by **expansion by minors** about a column or row. **Array of Signs for a 3 × 3 Determinant** $$\begin{matrix} + & - & + \\ - & + & - \\ + & - & + \end{matrix}$$	$$\begin{vmatrix} 3 & 4 \\ -2 & 6 \end{vmatrix} = (3)(6) - (4)(-2) = 26$$ Evaluate $$\begin{vmatrix} 2 & -3 & -2 \\ -1 & -4 & -3 \\ -1 & 0 & 2 \end{vmatrix}$$ by expanding about the second column. $$\begin{vmatrix} 2 & -3 & -2 \\ -1 & -4 & -3 \\ -1 & 0 & 2 \end{vmatrix} = -(-3)(-5) + (-4)(2) - (0)(-8)$$ $$= -15 - 8 + 0$$ $$= -23$$

8.6 SOLUTION OF LINEAR SYSTEMS BY DETERMINANTS—CRAMER'S RULE

Cramer's Rule for 2 × 2 Systems Given the system $$a_1 x + b_1 y = c_1$$ $$a_2 x + b_2 y = c_2$$ with $a_1 b_2 - a_2 b_1 = D \neq 0$, then $\quad x = \dfrac{\begin{vmatrix} c_1 & b_1 \\ c_2 & b_2 \end{vmatrix}}{\begin{vmatrix} a_1 & b_1 \\ a_2 & b_2 \end{vmatrix}} = \dfrac{D_x}{D}$ and $\quad y = \dfrac{\begin{vmatrix} a_1 & c_1 \\ a_2 & c_2 \end{vmatrix}}{\begin{vmatrix} a_1 & b_1 \\ a_2 & b_2 \end{vmatrix}} = \dfrac{D_y}{D}.$	Solve using Cramer's rule. $$x - 2y = -1$$ $$2x + 5y = 16$$ $$x = \dfrac{\begin{vmatrix} -1 & -2 \\ 16 & 5 \end{vmatrix}}{\begin{vmatrix} 1 & -2 \\ 2 & 5 \end{vmatrix}} = \dfrac{-5 + 32}{5 + 4} = \dfrac{27}{9} = 3$$ $$y = \dfrac{\begin{vmatrix} 1 & -1 \\ 2 & 16 \end{vmatrix}}{\begin{vmatrix} 1 & -2 \\ 2 & 5 \end{vmatrix}} = \dfrac{16 + 2}{5 + 4} = \dfrac{18}{9} = 2$$ The solution set is $\{(3, 2)\}$.

CONCEPTS	EXAMPLES
Cramer's Rule for 3 × 3 Systems Given the system $$a_1x + b_1y + c_1z = d_1$$ $$a_2x + b_2y + c_2z = d_2$$ $$a_3x + b_3y + c_3z = d_3$$ with $$D_x = \begin{vmatrix} d_1 & b_1 & c_1 \\ d_2 & b_2 & c_2 \\ d_3 & b_3 & c_3 \end{vmatrix},$$ $$D_y = \begin{vmatrix} a_1 & d_1 & c_1 \\ a_2 & d_2 & c_2 \\ a_3 & d_3 & c_3 \end{vmatrix},$$ $$D_z = \begin{vmatrix} a_1 & b_1 & d_1 \\ a_2 & b_2 & d_2 \\ a_3 & b_3 & d_3 \end{vmatrix},$$ $$D = \begin{vmatrix} a_1 & b_1 & c_1 \\ a_2 & b_2 & c_2 \\ a_3 & b_3 & c_3 \end{vmatrix} \neq 0,$$ then $$x = \frac{D_x}{D}, \quad y = \frac{D_y}{D}, \quad z = \frac{D_z}{D}.$$	Solve using Cramer's rule. $$3x + 2y + z = -5$$ $$x - y + 3z = -5$$ $$2x + 3y + z = 0$$ Using the methods of expansion by minors, it can be shown that $D_x = 45$, $D_y = -30$, $D_z = 0$, and $D = -15$. Therefore, $$x = \frac{D_x}{D} = \frac{45}{-15} = -3,$$ $$y = \frac{D_y}{D} = \frac{-30}{-15} = 2,$$ $$z = \frac{D_z}{D} = \frac{0}{-15} = 0.$$ The solution set is $\{(-3, 2, 0)\}$.

CHAPTER 8 REVIEW EXERCISES

[8.1] *Solve the system of equations by the elimination method. In Exercises 1 and 2, also graph the system.*

1. $x + 3y = 8$
$\quad 2x - y = 2$

2. $x - 4y = -4$
$\quad 3x + y = 1$

3. $6x + 5y = 4$
$\quad -4x + 2y = 8$

4. $\dfrac{x}{6} + \dfrac{y}{6} = -\dfrac{1}{2}$
$\quad x - y = -9$

5. $4x + 5y = 9$
$\quad 3x + 7y = -1$

6. $9x - y = -4$
$\quad y = x + 4$

7. $-3x + y = 6$
$\quad y = 6 + 3x$

8. $\quad 5x - 4y = 2$
$\quad -10x + 8y = 7$

Solve the following systems by the substitution method.

9. $3x + y = -4$
$\quad x = \dfrac{2}{3}y$

10. $-5x + 2y = -2$
$\quad x + 6y = 26$

11. $\dfrac{1}{x} + \dfrac{1}{y} = \dfrac{7}{12}$
$\quad \dfrac{2}{x} - \dfrac{3}{y} = -\dfrac{1}{12}$

1. $\{(2, 2)\}$

2. $\{(0, 1)\}$

3. $\{(-1, 2)\}$ 4. $\{(-6, 3)\}$

5. $\left\{ \left(\dfrac{68}{13}, -\dfrac{31}{13} \right) \right\}$

6. $\{(0, 4)\}$

7. $\{(x, y) \mid -3x + y = 6\}$

8. ∅   9. $\left\{ \left(-\dfrac{8}{9}, -\dfrac{4}{3} \right) \right\}$

10. $\{(2, 4)\}$ 11. $\{(3, 4)\}$

12. (a) inconsistent
(b) dependent equations
14. (a) $y_1 = -3x + 5$
(b) $y_2 = .5x + 1.5$
(c) $\{(1, 2)\}$
(d)

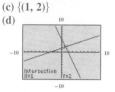

16. $\{(1, -5, 3)\}$
17. $\{(1, 2, 3)\}$ **18. 0**
19. length: 6 feet;   width:
4 feet   **20.** 3 weekend
days;   5 weekdays
21. plane: 300 miles per
hour;   wind: 20 miles per
hour

12. State whether the system graphed is inconsistent or has dependent equations.

(a)

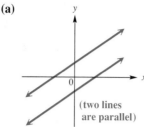

(two lines
are parallel)

(b)

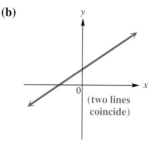

(two lines
coincide)

13. Explain why the system

$$y = 3x + 2$$
$$y = 3x - 4$$

has 0 as its solution set without doing any algebraic work but answering based on your knowledge of the graphs of the two lines.

14. Consider the system

$$3x + y = 5$$
$$-x + 2y = 3.$$

(a) Solve the first equation for y and designate it y_1.
(b) Solve the second equation for y and designate it y_2.
(c) Solve the system algebraically by solving $y_1 = y_2$ for x, and then finding the corresponding value of y.
(d) Support your solution with a graphics calculator by graphing both y_1 and y_2 in the standard viewing window and finding the point of intersection.

15. Explain why the point $(-5, 15)$ cannot possibly be the solution of the system of linear equations whose graphs are shown in the standard viewing window of a graphics calculator. What would be a good approximation for the solution?

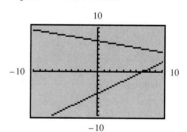

[8.2] *Solve the system.*

16.
$$2x + 3y - z = -16$$
$$x + 2y + 2z = -3$$
$$-3x + y + z = -5$$

17. $4x - y = 2$
$$3y + z = 9$$
$$x + 2z = 7$$

18.
$$3x - y - z = -8$$
$$4x + 2y + 3z = 15$$
$$-6x + 2y + 2z = 10$$

[8.3] *Solve the problem by writing a system of equations and then solving the system.*

19. A rectangular table top is 2 feet longer than it is wide, and its perimeter is 20 feet. Find the length and the width of the table top.

20. On an 8-day business trip Clay rented a car for $32 per day at weekday rates and $19 per day at weekend rates. If his total rental bill was $217, how many days did he rent at each rate?

21. A plane flies 560 miles in 1.75 hours traveling with the wind. The return trip later against the same wind takes the plane 2 hours. Find the speed of the plane and the speed of the wind.

22. Sweet's Candy Store is offering a special mix for Valentine's Day. Ms. Sweet will mix some $2-a-pound candy with some $1-a-pound candy to get 100 pounds of mix which she will sell at $1.30 a pound. How many pounds of each should she use?

23. The sum of the measures of the angles of a triangle is 180°. One angle measures 10° less than the sum of the other two. The measure of the middle-sized angle is the average of the other two. Find the measures of the three angles.

24. David Zerangue sells real estate. On three recent sales, he made 10% commission, 6% commission, and 5% commission. His total commissions on these sales were $17,000, and he sold property worth $280,000. If the 5% sale amounted to the sum of the other two, what were the three sales prices?

25. How many liters each of 8%, 10%, and 20% hydrogen peroxide should be mixed together to get 8 liters of 12.5% solution, if the amount of 8% solution used must be 2 liters more than the amount of 20% solution used?

 26. A farmer wishes to satisfy her fertilizer needs with three brands, A, B, and C. She needs to apply a total of 26.4 pounds of nitrogen, 28 pounds of potash, and 26.8 pounds of sulfate of ammonia. Brand A contains 8% nitrogen, 5% potash, and 10% sulfate of ammonia. Brand B contains 6% nitrogen, 10% potash, and 6% sulfate of ammonia. Brand C contains 10% nitrogen, 8% potash, and 5% sulfate of ammonia. How much of each fertilizer should be used? Round answers to the nearest tenth.

27. Find the values of a, b, and c so that the points $(1, 1)$, $(0, -4)$, and $(-2, 4)$ lie on the graph of the equation $y = ax^2 + bx + c$.

[8.4] *Solve the system of equations by matrix methods.*

28. $2x + 5y = -4$
$4x - y = 14$

29. $6x + 3y = 9$
$-7x + 2y = 17$

30. $x + 2y - z = 1$
$3x + 4y + 2z = -2$
$-2x - y + z = -1$

31. $x + 3y = 7$
$3x + z = 2$
$y - 2z = 4$

[8.5]

32. Which one of the following determinants is equal to 0?

(a) $\begin{vmatrix} 3 & 2 \\ 2 & 3 \end{vmatrix}$ **(b)** $\begin{vmatrix} 4 & 2 \\ -3 & 2 \end{vmatrix}$ **(c)** $\begin{vmatrix} -1 & 1 \\ 8 & 8 \end{vmatrix}$ **(d)** $\begin{vmatrix} 1 & 2 \\ 6 & 12 \end{vmatrix}$

▲**33.** If $\begin{vmatrix} x & 2 \\ 5 & -2 \end{vmatrix} = 12$, what is the value of x?

Evaluate the determinant.

34. $\begin{vmatrix} 2 & -9 \\ 8 & 4 \end{vmatrix}$

35. $\begin{vmatrix} 7 & 0 \\ 5 & -3 \end{vmatrix}$

36. $\begin{vmatrix} 2 & 10 & 4 \\ 0 & 1 & 3 \\ 0 & 6 & -1 \end{vmatrix}$

37. $\begin{vmatrix} 0 & 0 & 0 \\ 0 & 2 & 5 \\ -1 & 3 & 6 \end{vmatrix}$

38. $\begin{vmatrix} 0 & 0 & 2 \\ 2 & 1 & 0 \\ -1 & 0 & 0 \end{vmatrix}$

39. $\begin{vmatrix} -1 & 7 & 2 \\ 3 & 0 & 5 \\ -1 & 2 & 6 \end{vmatrix}$

[8.6]

◉ **40.** Under what conditions can a system *not* be solved using Cramer's rule?

◉ **41.** Why can't the system $\begin{array}{l} 3x + 2y + z = 0 \\ -x + y - 3z = 1 \end{array}$ be solved using Cramer's rule?

22. 30 pounds of $2-a-pound candy; 70 pounds of $1-a-pound candy **23.** 85°, 60°, 35°
24. $40,000 at 10%; $100,000 at 6%; $140,000 at 5%
25. 5 liters of 8%; 3 liters of 20%; none of 10%
26. 141.5 pounds of Brand A; 170.4 pounds of Brand B; 48.6 pounds of Brand C
27. $a = 3, b = 2, c = -4$
28. $\{(3, -2)\}$
29. $\{(-1, 5)\}$
30. $\{(0, 0, -1)\}$
31. $\{(1, 2, -1)\}$
32. (d) **33.** -11
34. 80 **35.** -21 **36.** -38
37. 0 **38.** 2 **39.** -139
40. Cramer's rule does not apply if $D = 0$.
41. For three unknowns we need three equations.

42. $\left\{\left(\dfrac{37}{11}, \dfrac{14}{11}\right)\right\}$

43. $\left\{\left(\dfrac{39}{10}, \dfrac{6}{5}\right)\right\}$

44. $\{(3, -2, 1)\}$

45. $\left\{\left(\dfrac{172}{67}, -\dfrac{14}{67}, -\dfrac{87}{67}\right)\right\}$

46. $\{(12, 9)\}$ 47. $\emptyset$

48. $\{(3, -1)\}$ 49. $\{(5, 3)\}$

50. $\{(0, 4)\}$

51. $\left\{\left(\dfrac{82}{23}, -\dfrac{4}{23}\right)\right\}$

52. (a) after about 11 years
(b) fixed: about $600;
graduated: about $560
(c) in the eleventh year

53. $5y + z = 7$;
$-5y + z = -13$

54. $\{(2, -3)\}$

55. $\{(4, 2, -3)\}$

56. $\{(4, 2, -3)\}$

Use Cramer's rule to solve the system of equations.

42. $3x - 4y = 5$
$2x + y = 8$

43. $-4x + 3y = -12$
$2x + 6y = 15$

44. $4x + y + z = 11$
$x - y - z = 4$
$y + 2z = 0$

45. $-x + 3y - 4z = 2$
$2x + 4y + z = 3$
$3x - z = 9$

MIXED REVIEW EXERCISES

Solve by any method.

46. $\dfrac{2}{3}x + \dfrac{1}{6}y = \dfrac{19}{2}$

$\dfrac{1}{3}x - \dfrac{2}{9}y = 2$

47. $2x + 5y - z = 12$
$-x + y - 4z = -10$
$-8x - 20y + 4z = 31$

48. $x = 7y + 10$
$2x + 3y = 3$

49. $x + 4y = 17$
$-3x + 2y = -9$

50. $-7x + 3y = 12$
$5x + 2y = 8$

51. $2x - 5y = 8$
$3x + 4y = 10$

52. Julio Nicolai compared the monthly interest costs for two types of home mortgages; a fixed-rate mortgage and a graduated-payment mortgage. His results are shown in the figure.

(a) In how many years will the monthly interest costs be equal for the two plans?

(b) What is the monthly interest in the sixth year for each plan?

(c) In what year will the monthly interest cost be $590 for each plan?

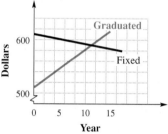

◆ **MATHEMATICAL CONNECTIONS** (Exercises 53–58) ◆

Consider the system of equations

$$2x + y + 3z = 1$$
$$x - 2y + z = -3$$
$$-3x + y - 2z = -4.$$

In Exercises 53–58 we examine several different ways of solving this system. Work the exercises in order.

53. Eliminate x from the first and second equations, and eliminate x from the second and third equations. Write the resulting system of two equations in y and z.

54. Solve the system of two equations found in Exercise 53 using elimination.

55. Complete the solution of the original system using the results of Exercise 54.

56. Solve the original system using matrix row operations (Section 8.4). Verify that your solution is the same as the one found in Exercise 55.

57. Evaluate these determinants.

(a) $D = \begin{vmatrix} 2 & 1 & 3 \\ 1 & -2 & 1 \\ -3 & 1 & -2 \end{vmatrix}$ (b) $D_x = \begin{vmatrix} 1 & 1 & 3 \\ -3 & -2 & 1 \\ -4 & 1 & -2 \end{vmatrix}$

(c) $D_y = \begin{vmatrix} 2 & 1 & 3 \\ 1 & -3 & 1 \\ -3 & -4 & -2 \end{vmatrix}$ (d) $D_z = \begin{vmatrix} 2 & 1 & 1 \\ 1 & -2 & -3 \\ -3 & 1 & -4 \end{vmatrix}$

58. Use the results of Exercise 57 and Cramer's rule to solve the original system. Verify that your solution is the same as the one found in Exercises 55 and 56.

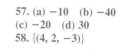

57. (a) -10 (b) -40
(c) -20 (d) 30
58. $\{(4, 2, -3)\}$

CHAPTER 8 TEST

1. Use a graph to solve the system

$$x + y = 7$$
$$x - y = 5.$$

Solve the system by elimination.

2. $3x + y = 12$
$2x - y = 3$

3. $-5x + 2y = -4$
$6x + 3y = -6$

4. $3x + 4y = 8$
$8y = 7 - 6x$

5. $3x + 5y + 3z = 2$
$6x + 5y + z = 0$
$3x + 10y - 2z = 6$

Solve the system by substitution.

6. $2x - 3y = 24$
$y = -\dfrac{2}{3}x$

7. $12x - 5y = 8$
$3x = \dfrac{5}{4}y + 2$

Solve the problem by writing a system of equations.

8. Two cars start from points 420 miles apart and travel toward each other. They meet after 3.5 hours. Find the average speed of each car if one travels 30 miles per hour slower than the other.

9. A chemist needs 12 liters of a 40% alcohol solution. She must mix a 20% solution and a 50% solution. How many liters of each will be required to obtain what she needs?

10. A local electronics store will sell 7 AC adaptors and 2 rechargeable flashlights for $86, or 3 AC adaptors and 4 rechargeable flashlights for $84. What is the price of a single AC adaptor and a single rechargeable flashlight?

 11. Why would it probably be easier to solve the system

$$6x - 8y = 4$$
$$y = \dfrac{1}{2}x$$

by substitution than by elimination? Would you get the same solution set in both cases?

[8.1] 1. $\{(6, 1)\}$

2. $\{(3, 3)\}$
3. $\{(0, -2)\}$ 4. ∅
[8.2] 5. $\left\{\left(-\dfrac{2}{3}, \dfrac{4}{5}, 0\right)\right\}$

[8.1] 6. $\{(6, -4)\}$
7. $\{(x, y) \mid 12x - 5y = 8\}$
[8.3] 8. slower car: 45 miles per hour; faster car: 75 miles per hour
9. 4 liters of 20%; 8 liters of 50% 10. AC adaptor: $8; rechargeable flashlight: $15
[8.1] 11. The second equation is already solved for y, so substitution would be easier. Yes, the solution set would be the same.

[8.4] 12. $\left\{\left(\dfrac{2}{5}, \dfrac{7}{5}\right)\right\}$

13. $\{(-1, 2, 3)\}$

[8.2, 8.4, 8.6]

14. $\{(-3, -2, -4)\}$

[8.5] 15. 3   16. 0

[8.6] 17. $\left\{\left(-\dfrac{9}{4}, \dfrac{5}{4}\right)\right\}$

18. $\{(-1, 2, 3)\}$

[8.1] 19. $x = 8$ or 800
items; $3000

20. about $500

Solve the system by matrix methods.

12. $3x + 2y = 4$
$5x + 5y = 9$

13. $x + 3y + 2z = 11$
$3x + 7y + 4z = 23$
$5x + 3y - 5z = -14$

14. Use any method described in this chapter to solve the system $\begin{aligned} 4x - 2y &= -8 \\ 3y - 5z &= 14 \\ 2x + z &= -10. \end{aligned}$

Evaluate the determinant.

15. $\begin{vmatrix} 6 & -3 \\ 5 & -2 \end{vmatrix}$

16. $\begin{vmatrix} 4 & 1 & 0 \\ -2 & 7 & 3 \\ 0 & 5 & 2 \end{vmatrix}$

Solve the system by Cramer's rule.

17. $3x - y = -8$
$2x + 6y = 3$

18. $x + y + z = 4$
$-2x + z = 5$
$3y + z = 9$

The graph shows a company's costs to produce computer parts and the revenue from the sale of computer parts.

⊙ 19. At what production level does the cost equal the revenue? What is the revenue at that point?

20. Profit is revenue less cost. Estimate the profit on the sale of 1100 parts.

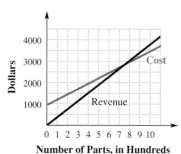

Number of Parts, in Hundreds

CUMULATIVE REVIEW (Chapters 1–8)

1. $(-\infty, \infty)$ 2. $[-3, 3]$
3. $4000 at 4%; $8000 at
3%   4. 78 or greater
5. 0   6. 0   7. yes
8. $(2x - 1)(4x^2 + 2x + 1) \cdot$
$(2x + 1)(4x^2 - 2x + 1)$

1. Solve for x: $3x + 2(x - 4) = 5x - 8$.

2. Solve the inequality: $-3 \le \dfrac{2}{3}x - 1 \le 1$.

3. Earl Karn invested some money at 4% interest and twice as much at 3% interest. His interest for the first year was $400. How much did he invest at each rate?

4. A student must have an average grade of at least 80 on the four tests in a course to earn a grade of B. David Hingle had grades of 79, 75, and 88 on the first three tests. What possible scores can he make on the fourth test so that he can get a B in the course?

5. Solve for t: $|6t - 4| + 8 = 3$.

6. Evaluate: $-4^2 + (-4)^2$.

7. Use synthetic division to decide whether 4 is a solution of
$$x^4 - 8x^3 + 21x^2 - 14x - 24 = 0.$$

8. Factor completely: $64x^6 - 1$ (*Hint:* Factor as the difference of two squares *first.*)

9. Simplify: $\left(\dfrac{a^{-2}b^{-1}}{3a^2}\right)^{-2}\left(\dfrac{b^{-2}\cdot 3a^4}{2b^{-3}}\right)^{-2}$. $(a \neq 0,\, b \neq 0)$

10. Solve: $6z^2 = 5z + 50$.

Perform the indicated operations and express answers in lowest terms.

11. $\dfrac{2}{x+2} - \dfrac{3}{4x+8}$

12. $\dfrac{2}{x^2+x-2} + \dfrac{3}{x^2+5x+6}$

13. $\dfrac{3p(p-2)}{p+5} \div \dfrac{p^2-4}{4p+20}$

14. $\dfrac{\dfrac{2}{r}+\dfrac{5}{s}}{\dfrac{25r^2-4s^2}{3rs}}$

15. Solve for m: $\dfrac{m-4}{3} - \dfrac{m+1}{2} = -\dfrac{11}{6}$.

Write the radical expression in simplified form.

16. $3\sqrt{40} + 6\sqrt{90} - 8\sqrt{160}$

17. $\dfrac{3}{12+\sqrt{5}}$

18. $\sqrt[3]{\dfrac{x^{13}}{y}}$ $(y \neq 0)$

19. $\dfrac{12-9\sqrt{7}}{6}$

20. Multiply: $(4+\sqrt{7})(5-\sqrt{7})$.

21. Solve for x: $\sqrt{4x^2+2x-3} - 2x - 1 = 6$.

22. Solve by factoring, making a substitution for $3m-1$:
$$2(3m-1)^2 - 7(3m-1) = 15.$$

23. The formula $A = P(1+r)^2$ can be used to find the amount A that P dollars will grow to in 2 years at rate r. If $P = \$10{,}000$ is invested for 2 years and grows to $A = \$11{,}664$, what is the interest rate r?

24. Find the solution set: $2x^2 \leq -5x + 3$.

25. On a 30-mile bicycle trip, Joanna took $1/2$ hour less time than Ben. If Ben's speed was 5 miles per hour slower than Joanna's, what was Ben's speed?

In Exercises 26–31, point A has coordinates $(-2, 6)$ and point B has coordinates $(4, -2)$.

26. What is the equation of the horizontal line through A?

27. What is the equation of the vertical line through B?

28. What is the slope of line AB?

29. What is the slope of a line perpendicular to line AB?

30. What is the standard form of the equation of line AB?

31. Graph line AB.

32. Find the standard form of the equation of the line with x-intercept $(-3, 0)$ and y-intercept $(0, 5)$.

33. Graph the line having slope $\dfrac{2}{3}$ and passing through the point $(-1, -3)$.

34. Graph the inequality: $-3x - 2y \leq 6$.

9. 4   **10.** $\left\{-\dfrac{5}{2}, \dfrac{10}{3}\right\}$

11. $\dfrac{5}{4(x+2)}$

12. $\dfrac{5x+3}{(x+2)(x-1)(x+3)}$

13. $\dfrac{12p}{p+2}$

14. $\dfrac{-3}{2s-5r}$ or $\dfrac{3}{5r-2s}$

15. $\{0\}$

16. $-8\sqrt{10}$

17. $\dfrac{3(12-\sqrt{5})}{139}$

18. $\dfrac{x^4\sqrt[3]{xy^2}}{y}$ **19.** $\dfrac{4-3\sqrt{7}}{2}$

20. $13+\sqrt{7}$ **21.** $\{-2\}$

22. $\left\{-\dfrac{1}{6}, 2\right\}$ **23.** 8%

24. $\left[-3, \dfrac{1}{2}\right]$ **25.** 15 miles per hour   **26.** $y = 6$

27. $x = 4$ **28.** $-\dfrac{4}{3}$

29. $\dfrac{3}{4}$ **30.** $4x + 3y = 10$

31.

32. $5x - 3y = -15$

33.

34.

 CONCEPTUAL   WRITING   ▲ CHALLENGING   SCIENTIFIC CALCULATOR   GRAPHICS CALCULATOR

35. 5 pounds of oranges; 1 pound of apples 36. length: 12 meters; width: 9 meters 37. small: $1.50; large: $2.50 38. peanuts: $2 per pound; cashews: $4 per pound 39. video rental firm: $60,000; bond and money market fund: $10,000 each 40. $20,000 at 8%; $50,000 at 10%; $5000 at 7%

In Exercises 35–40, solve the problem using a system of equations. Use the method of your choice: elimination, substitution, row operations with matrices, or Cramer's rule.

35. Mabel Johnston bought apples and oranges at DeVille's Grocery. She bought 6 pounds of fruit. Oranges cost $.90 per pound, while apples cost $.70 per pound. If she spent a total of $5.20, how many pounds of each kind of fruit did she buy?

36. The length of a rectangle is 3 meters more than the width. The perimeter is 42 meters. Find the length and width of the rectangle.

37. Kenneth and Peggy are planning to move, and they need some cardboard boxes. They can buy 10 small and 20 large boxes for $65, or 6 small and 10 large boxes for $34. Find the cost of each size of box.

38. At the Chalmette Nut Shop, 6 pounds of peanuts and 12 pounds of cashews cost $60, while 3 pounds of peanuts and 4 pounds of cashews cost $22. Find the cost of each type of nut.

39. Alexis has inherited $80,000 from her aunt. She invests part of the money in a video rental firm which produces a return of 7% per year, and divides the rest equally between a tax-free bond at 6% a year and a money market fund at 4% a year. Her annual return on these investments is $5200. How much is invested in each?

40. Jeff's Rental Properties, Inc., borrowed in three loans for major renovations. The company borrowed a total of $75,000. Some of the money was borrowed at 8% interest, and $30,000 more than that amount was borrowed at 10%. The rest was borrowed at 7%. How much was borrowed at each rate if the total annual simple interest was $6950?

QUADRATIC FUNCTIONS AND THE CONIC SECTIONS

◆ Answers will vary.

CONNECTIONS

When a plane intersects an infinite cone (Figure 1 on the next page), a conic section is produced. Conic sections appear in countless places in our world (and beyond). For example, cross-sections of the backyard satellite dishes that bring programming into living rooms are examples of *parabolas.*

Johann Kepler (1571–1630) established the importance of the *ellipse* in 1609 when he discovered that the orbits of the planets around the sun were elliptical, not circular. The orbits of the planets are nearly circular, while Halley's comet has an elliptical orbit that is long and narrow.

Ships and planes often use a location finding system called LORAN, based on sending out a series of pulses. A branch of a *hyperbola* is crucial in the use of this system. And *circles* are found just about everywhere.

FOR DISCUSSION OR WRITING

Additional examples of applications in daily life of conic sections are mentioned throughout this chapter. See how many you can locate. You may be able to find additional ones as well. At the end of the chapter, compare your list with those of your classmates.

In this chapter we study the properties and the graphs of conic sections.

9.1 QUADRATIC FUNCTIONS: PARABOLAS

FOR EXTRA HELP	OBJECTIVES
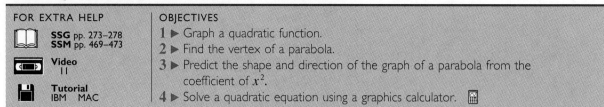 **SSG** pp. 273–278 **SSM** pp. 469–473	**1** ▶ Graph a quadratic function.
Video II	**2** ▶ Find the vertex of a parabola.
Tutorial IBM MAC	**3** ▶ Predict the shape and direction of the graph of a parabola from the coefficient of x^2.
	4 ▶ Solve a quadratic equation using a graphics calculator.

▶ **TEACHING TIP**

Use three-dimensional models to help students visualize each conic section. ◀

As we saw in Chapter 7, the graphs of first-degree equations are straight lines. In this chapter the graphs of second-degree equations, which are equations with one or more second-degree terms, are discussed. These graphs result from cutting an infinite cone with a plane, as shown in Figure 1. Because of this, the graphs are called **conic sections.**

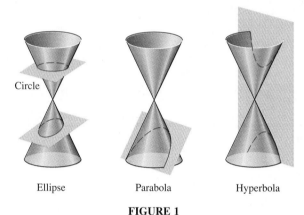

Circle

Ellipse   Parabola   Hyperbola

FIGURE 1

1 ▶ Graph a quadratic function.

▶ Let us begin by graphing the equation $y = x^2$. First, make a table of ordered pairs satisfying the equation.

x	-2	$-\dfrac{3}{2}$	-1	$-\dfrac{1}{2}$	0	$\dfrac{1}{2}$	1	$\dfrac{3}{2}$	2
y	4	$\dfrac{9}{4}$	1	$\dfrac{1}{4}$	0	$\dfrac{1}{4}$	1	$\dfrac{9}{4}$	4

Plot these points and draw a smooth curve through them to get the graph shown in Figure 2. This graph is called a **parabola.** The point (0, 0), with the smallest y-value of any point on the curve, is the **vertex** of this parabola. The vertical line through the vertex is the **axis** of this parabola. The parabola is symmetric about its axis; that is, if the graph were folded along the axis, the two portions of the curve would coincide.

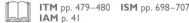 

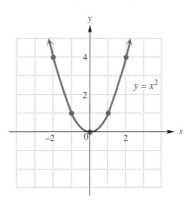

$y = x^2$

FIGURE 2

Because the graph of $y = x^2$ satisfies the conditions of the graph of a function (Section 7.5), we may write its equation as $f(x) = x^2$. This is the simplest example of a quadratic function.

Qauadratic Function	A function that can be written in the form

$$f(x) = ax^2 + bx + c$$

for real numbers a, b, and c, with $a \neq 0$, is a **quadratic function.**

The graph of any quadratic function is a parabola with a vertical axis.

Each conic section is defined in terms of distance. A parabola is a set of all points in a plane equally distant from a given point and a given line not containing the point. (See Exercise 53.) Parabolas have many applications. For example, if an object is thrown upward, then (disregarding air resistance) the path it follows is a parabola. The large disks seen on the sidelines of televised football games, which are used by television crews to pick up the shouted signals of the players on the field, have cross sections that are parabolas. Cross sections of radar dishes and automobile headlights also form parabolas. Applications of the other conic sections are mentioned in later sections.

For the rest of this section and the next, we use the symbols y and $f(x)$ interchangeably in discussing parabolas.

2 ▶ Find the vertex of a parabola.

▶ Parabolas need not have their vertices at the origin, as does $f(x) = x^2$. For example, to graph a parabola of the form $f(x) = x^2 + k$, we start by selecting the sample values of x that were used to graph $f(x) = x^2$. The corresponding values of $f(x)$ in $f(x) = x^2 + k$ differ by k from those of $f(x) = x^2$. For this reason, the graph of $f(x) = x^2 + k$ is shifted k units vertically compared with that of $f(x) = x^2$.

EXAMPLE 1

Graphing a Parabola with
a Vertical Shift

Graph $f(x) = x^2 - 2$.

As we mentioned before, this graph has the same shape as $f(x) = x^2$, but since k here is -2, the graph is shifted 2 units downward, with vertex at $(0, -2)$. Every function value is 2 less than the corresponding function value of $f(x) = x^2$. Plotting points gives the graph in Figure 3. ∎

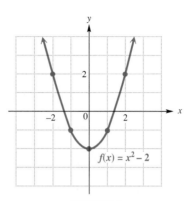

FIGURE 3

Vertical Shifts

The graph of $f(x) = x^2 + k$ is a parabola with the same shape as the graph of $f(x) = x^2$. The parabola is shifted k units upward if $k > 0$, and $|k|$ units downward if $k < 0$. The vertex is $(0, k)$.

The graph of $f(x) = (x - h)^2$ is also a parabola with the same shape as $f(x) = x^2$. The vertex of the parabola $f(x) = (x - h)^2$ is the lowest point on the parabola. The lowest point occurs here when $f(x)$ is 0. To get $f(x)$ equal to 0, we let $x = h$, so the vertex of $f(x) = (x - h)^2$ is at $(h, 0)$. Based on this, the graph of $f(x) = (x - h)^2$ is shifted h units horizontally compared with that of $f(x) = x^2$.

EXAMPLE 2

Graphing a Parabola with
a Horizontal Shift

Graph $f(x) = (x - 2)^2$.

When $x = 2$, then $f(x) = 0$, giving the vertex $(2, 0)$. The parabola $f(x) = (x - 2)^2$ has the same shape as $f(x) = x^2$ but is shifted 2 units to the right, as shown in Figure 4. ∎

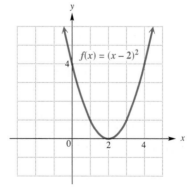

FIGURE 4

Horizontal Shifts

The graph of $f(x) = (x - h)^2$ is a parabola with the same shape as the graph of $f(x) = x^2$. The parabola is shifted h units horizontally: h units to the right if $h > 0$, and $|h|$ units to the left if $h < 0$. The vertex is $(h, 0)$.

CAUTION Errors frequently occur when horizontal shifts are involved. In order to determine the direction and magnitude of horizontal shifts, find the value that would cause the expression $x - h$ to equal 0. For example, the graph of $f(x) = (x - 5)^2$ would be shifted 5 units to the *right,* because $+5$ would cause $x - 5$ to equal 0. On the other hand, the graph of $f(x) = (x + 4)^2$ would be shifted 4 units to the *left,* because -4 would cause $x + 4$ to equal 0.

A parabola can have both a horizontal and a vertical shift, as in Example 3.

EXAMPLE 3

Graphing a Parabola with Horizontal and Vertical Shifts

■ Graph $f(x) = (x + 3)^2 - 2$.

This graph has the same shape as $f(x) = x^2$, but is shifted 3 units to the left (since $x + 3 = 0$ if $x = -3$), and 2 units downward (because of the -2). As shown in Figure 5, the vertex is at $(-3, -2)$. ■

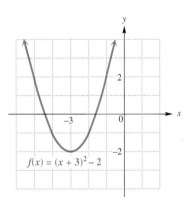

FIGURE 5

The characteristics of the graph of a parabola of the form $f(x) = (x - h)^2 + k$ are summarized as follows.

Vertex and Axis

The graph of $f(x) = (x - h)^2 + k$ is a parabola with the same shape as $f(x) = x^2$ and with vertex at (h, k). The axis is the vertical line $x = h$.

3 ▶ Predict the shape and direction of the graph of a parabola from the coefficient of x^2.

▶ Not all parabolas open upward, and not all parabolas have the same shape as $f(x) = x^2$. In the next example we show how to identify parabolas opening downward and having a different shape from that of $f(x) = x^2$.

E X A M P L E   4

Graphing a Parabola that
Opens Downward

Graph $f(x) = -\dfrac{1}{2}x^2$.

This parabola is shown in Figure 6. Some ordered pairs that satisfy the equation are $(0, 0)$, $(1, -\frac{1}{2})$, $(2, -2)$, $(-1, -\frac{1}{2})$, and $(-2, -2)$. The coefficient $-1/2$ affects the shape of the graph; the $1/2$ makes the parabola wider (since the values of $f(x)$ grow more slowly than they would for $f(x) = x^2$), and the negative sign makes the parabola open downward. The graph is not shifted in any direction; the vertex is still at $(0, 0)$. Here, the vertex has the *largest* function value of any point on the graph. ∎

Chalkboard Exercise

Graph $f(x) = -2x^2 - 3$.

Answer:

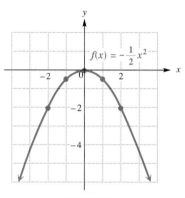

FIGURE 6

Some general principles concerning the graph of $f(x) = a(x - h)^2 + k$ are summarized as follows.

General Principles

1. The graph of the quadratic function

$$f(x) = a(x - h)^2 + k, \qquad a \neq 0$$

   is a parabola with vertex at (h, k) and the vertical line $x = h$ as axis.
2. The graph opens upward if a is positive and downward if a is negative.
3. The graph is wider than $f(x) = x^2$ if $0 < |a| < 1$. The graph is narrower than $f(x) = x^2$ if $|a| > 1$.

E X A M P L E   5

Using the General
Principles to Graph a
Parabola

Graph $f(x) = -2(x + 3)^2 + 4$.

The parabola opens downward (because $a < 0$), and is narrower than the graph of $f(x) = x^2$, since $|-2| = 2 > 1$, causing the values of $f(x)$ to grow more quickly than they would for $f(x) = x^2$. This parabola has vertex at $(-3, 4)$ as shown in Figure 7. To complete the graph, we plotted the ordered pairs $(-4, 2)$ and $(-2, 2)$. Notice that these two points are symmetric to the axis of the parabola. This symmetry is very useful for finding additional ordered pairs that satisfy the equation. ∎

Chalkboard Exercise

Graph $f(x) = \dfrac{1}{2}(x - 2)^2 + 1$.

Answer:

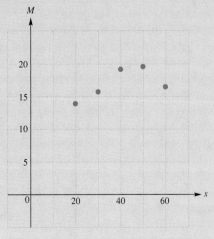

$f(x) = -2(x + 3)^2 + 4$

FIGURE 7

◈ 1. It must be negative.
2. approximately 45
3. approximately 20

◇ **CONNECTIONS** ◇

In a Connections box at the beginning of Chapter 6, we gave an equation that described union membership in the United States: $M = -.011x^2 + 1.22x - 8.5$.* Recall that in the equation x represented the number of years since 1930, while M gave the corresponding number of union members (in millions). You may wonder how that equation was determined. The necessary data, shown below, are from the Bureau of Labor Statistics.

Year	1950	1960	1970	1980	1990
x	20	30	40	50	60
M	14	16	19.4	19.8	16.7

A graph of these ordered pairs is shown below. A straight line cannot be drawn through these points, so we know that an equation describing the union membership cannot be linear. As the graphs in this section suggest, a parabola should closely approximate these points. In a later Connections box in this chapter, we shall see how the equation of a parabola that closely fits a set of points can be found.

*Source: World Almanac and Book of Facts, 1992, "U.S. Union Membership, 1930–1990", p. 180.

> **FOR DISCUSSION OR WRITING**
> After working through this section, you should be able to answer the following questions about the parabola that will "fit" the ordered pairs graphed on the previous page.
>
> **1.** What can be said about the coefficient of x^2?
> **2.** What should be the x-value of the vertex?
> **3.** What should be the y-value of the vertex?

4 ▶ Solve a quadratic equation using a graphics calculator.

▶ In Chapter 6 we saw how to solve quadratic equations such as $2x^2 + 3x - 2 = 0$. The graph of the function $y = 2x^2 + 3x - 2$ is a parabola. Recall that the x-intercept of the line $y = mx + b$ is the solution of the equation $mx + b = 0$. Extending this idea to quadratic equations, we may say that the x-intercept(s), if any, of the graph of $y = ax^2 + bx + c$, indicate the real solutions of the equation $ax^2 + bx + c = 0$. Therefore, from the graph and displays shown in Figure 8, the solutions of $2x^2 + 3x - 2 = 0$ are -2 and $.5$. We could verify that these solutions are correct (and exact) by solving the equation using one of the methods given in Chapter 6.

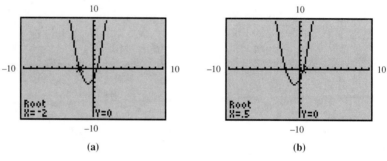

(a)　　　　　　　　　　　(b)

FIGURE 8

9.1 EXERCISES

1. Explain in your own words the meaning of the term.
   (a) vertex of a parabola　　**(b)** axis of a parabola

2. Explain why the axis of the graph of a quadratic function cannot be a horizontal line.

Identify the vertex of the graph of the quadratic function. See Examples 1–3.

3. $f(x) = -3x^2$ **4.** $f(x) = -.5x^2$ **5.** $f(x) = x^2 + 4$ **6.** $f(x) = x^2 - 4$
3. **(0, 0)**   4. **(0, 0)**   5. **(0, 4)**   6. **(0, −4)**

7. $f(x) = (x - 1)^2$ **8.** $f(x) = (x + 3)^2$ **9.** $f(x) = (x + 3)^2 - 4$ **10.** $f(x) = (x - 5)^2 - 8$
7. **(1, 0)**   8. **(−3, 0)**   9. **(−3, −4)**   10. **(5, −8)**

For the quadratic function, tell whether the graph opens upward or downward, and tell whether the graph is wider, narrower, or the same as the graph of $f(x) = x^2$. See Example 4.

11. $f(x) = -2x^2$ **12.** $f(x) = -3x^2 + 1$ **13.** $f(x) = .5x^2$ **14.** $f(x) = \frac{2}{3}x^2 - 4$

11. downward; narrower   12. downward; narrower   13. upward; wider   14. upward; wider

15. Describe how the graph of each parabola in Exercises 9 and 10 is shifted compared to the graph of $y = x^2$.

◎ CONCEPTUAL   ✎ WRITING   ▲ CHALLENGING   ▦ SCIENTIFIC CALCULATOR   ▦ GRAPHICS CALCULATOR

16. Describe how the sign and the absolute value of a in $f(x) = a(x - h)^2 + k$ affects the graph of the function when compared to the graph of $y = x^2$.

17. For $f(x) = a(x - h)^2 + k$, in what quadrant is the vertex if:
   (a) $h > 0, k > 0$; **(b)** $h > 0, k < 0$; **(c)** $h < 0, k > 0$; **(d)** $h < 0, k < 0$?
   17. **(a) I** **(b) IV** **(c) II** **(d) III**

18. (a) What is the value of h if the graph of $f(x) = a(x - h)^2 + k$ has vertex on the y-axis?
   (b) What is the value of k if the graph of $f(x) = a(x - h)^2 + k$ has vertex on the x-axis?
   18. **(a) 0** **(b) 0**

In Exercises 19–24, match the equation with the figure that most closely resembles its graph.

19. $g(x) = x^2 - 5$
19. **F**

20. $h(x) = -x^2 + 4$
20. **B**

21. $F(x) = (x - 1)^2$
21. **C**

22. $G(x) = (x + 1)^2$
22. **A**

23. $H(x) = (x - 1)^2 + 1$
23. **E**

24. $K(x) = (x + 1)^2 + 1$
24. **D**

A.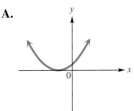

B.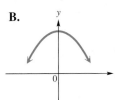

C.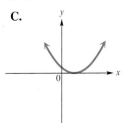

D.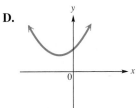

E.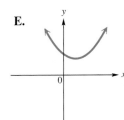

F.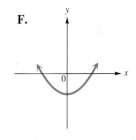

Sketch the graph of the parabola. Plot at least two points in addition to the vertex. See Examples 1–5.

25. $f(x) = -2x^2$
25.

26. $f(x) = \frac{1}{3}x^2$
26.

27. $f(x) = x^2 - 1$
27.

28. $f(x) = x^2 + 3$
28.

29. $f(x) = -x^2 + 2$
29.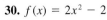

30. $f(x) = 2x^2 - 2$
30.

31. $f(x) = .5(x - 4)^2$
31.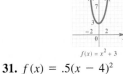

32. $f(x) = -2(x + 1)^2$
32.

33. $f(x) = (x + 2)^2 - 1$
33.

34. $f(x) = (x - 1)^2 + 2$
34.

◉ CONCEPTUAL   ✎ WRITING   ▲ CHALLENGING   ▥ SCIENTIFIC CALCULATOR   ▦ GRAPHICS CALCULATOR

35. $f(x) = 2(x - 2)^2 - 4$

35.

36. $f(x) = -2(x + 3)^2 + 4$

36. $f(x) = -2(x + 3)^2 + 4$

37. $f(x) = -.5(x + 1)^2 + 2$

37. $f(x) = -.5(x + 1)^2 + 2$

38. $f(x) = -\dfrac{2}{3}(x + 2)^2 + 1$

38. $f(x) = -\frac{2}{3}(x + 2)^2 + 1$

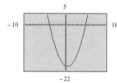

39. $f(x) = 2(x - 2)^2 - 3$

39.

40. $f(x) = \dfrac{4}{3}(x - 3)^2 - 2$

40.

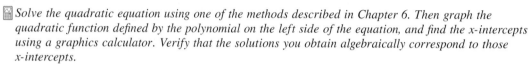

Solve the quadratic equation using one of the methods described in Chapter 6. Then graph the quadratic function defined by the polynomial on the left side of the equation, and find the x-intercepts using a graphics calculator. Verify that the solutions you obtain algebraically correspond to those x-intercepts.

41. $x^2 - x - 20 = 0$

41. $\{-4, 5\}$ The x values of the intercepts are -4 and 5.

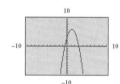

42. $x^2 + 9x + 14 = 0$

42. $\{-7, -2\}$ The x values of the intercepts are -7 and -2.

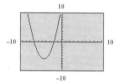

43. $-2x^2 + 5x + 3 = 0$

43. $\{-.5, 3\}$ The x values of the intercepts are $-.5$ and 3.

44. $-4x^2 - 11x + 3 = 0$

44. $\{-3, .25\}$ The x values of the intercepts are -3 and $.25$.

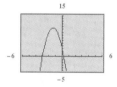

The graph of a quadratic function $y = f(x)$ is shown in the calculator-generated graph to the right. It is shown in the standard viewing window, with no tick marks on the x-axis. Refer to it to answer Exercises 45 and 46.

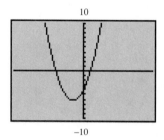

45. Which one of the following choices would be the only possible solution set for the equation $f(x) = 0$?

    (a) $\{-4, 1\}$ **(b)** $\{1, 4\}$ **(c)** $\{-1, -4\}$

45. (a)

46. Explain why only one choice in Exercise 45 is possible.

────────────◇ **MATHEMATICAL CONNECTIONS** (Exercises 47–52) ◇────────────

The procedures described in this section that allow the graph of $y = x^2$ to be shifted vertically and horizontally are applicable to other types of functions as well. In Section 7.5 we introduced linear functions (functions of the form $f(x) = ax + b$). Consider the graph of the simplest linear function, $f(x) = x$, shown here, and then work through Exercises 47–52 in order.

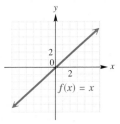

47. Based on the concepts of this section, how does the graph of $y = x^2 + 6$ compare to the graph of $y = x^2$ if a *vertical* shift is considered?

47. **It is shifted 6 units upward.**

48. Graph the linear function $y = x + 6$.

48.

49. Based on the concepts of Chapter 7, how does the graph of $y = x + 6$ compare to the graph of $y = x^2$ if a vertical shift is considered? (*Hint:* Look at the y-intercept.)

49. **It is shifted 6 units upward.**

50. Based on the concepts of this section, how does the graph of $y = (x - 6)^2$ compare to the graph of $y = x^2$ if a *horizontal* shift is considered?

50. **It is shifted 6 units to the right.**

51. Graph the linear function $y = x - 6$.

51.

52. Based on the concepts of Chapter 7, how does the graph of $y = x - 6$ compare to the graph of $y = x$ if a horizontal shift is considered? (*Hint:* Look at the x-intercept.)

52. **It is shifted 6 units to the right.**

───────────────◆───────────────

In the following exercise, the distance formula is used to develop the equation of a parabola.

53. *A parabola can be defined as the set of all points in a plane equally distant from a given point and a given line not containing the point. (The point is called the *focus* and the line is called the *directrix*.) See the figure.*

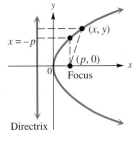

  (a) Suppose (x, y) is to be on the parabola. Suppose the directrix has equation $x = -p$. Find the distance between (x, y) and the directrix. (The distance from a point to a line is the length of the perpendicular from the point to the line.)
  (b) If $x = -p$ is the equation of the directrix, why should the focus have coordinates $(p, 0)$? (*Hint:* See the figure.)
  (c) Find an expression for the distance from (x, y) to $(p, 0)$.
  (d) Find an equation for the parabola of the figure. (*Hint:* Use the results of parts (a) and (c) and the fact that (x, y) is equally distant from the focus and the directrix.)

53. (a) $|x + p|$ (b) **The distance from the focus to the origin should equal the distance from the directrix to the origin.**   (c) $\sqrt{(x - p)^2 + y^2}$ (d) $y^2 = 4px$

54. Use the equation derived in Exercise 53 to find an equation for a parabola with focus $(3, 0)$ and directrix with the equation $x = -3$.

54. $y^2 = 12x$

⊙ CONCEPTUAL   ✎ WRITING   ▲ CHALLENGING   ▦ SCIENTIFIC CALCULATOR   ▨ GRAPHICS CALCULATOR

REVIEW EXERCISES

Solve the quadratic equation by completing the square. See Section 6.1.

55. $x^2 + 6x - 3 = 0$ **56.** $x^2 + 8x - 4 = 0$ **57.** $2x^2 - 12x = 5$ **58.** $3x^2 - 12x = 10$

55. $\{-3 + 2\sqrt{3}, -3 - 2\sqrt{3}\}$ 56. $\{-4 + 2\sqrt{5}, -4 - 2\sqrt{5}\}$ 57. $\left\{\dfrac{6 + \sqrt{46}}{2}, \dfrac{6 - \sqrt{46}}{2}\right\}$ 58. $\left\{\dfrac{6 + \sqrt{66}}{3}, \dfrac{6 - \sqrt{66}}{3}\right\}$

9.2 MORE ABOUT PARABOLAS AND THEIR APPLICATIONS

FOR EXTRA HELP	OBJECTIVES

SSG pp. 278–286
 SSM pp. 473–479
  Video
 11
  Tutorial
 IBM MAC
	1 ▶ Find the vertex of a vertical parabola.
	 2 ▶ Graph a quadratic function.
	 3 ▶ Use the discriminant to find the number of x-intercepts of a vertical parabola.
	 4 ▶ Use quadratic functions to solve problems involving maximum or minimum value.
	 5 ▶ Graph horizontal parabolas.

In the previous section we saw how to graph a parabola with an equation in the form $f(x) = a(x - h)^2 + k$ (where $a \neq 0$). Suppose we want to graph the function $f(x) = 3x^2 - 4x + 2$. We could start by finding ordered pairs. Some ordered pairs are shown in the table below.

x	-2	-1	0	1	2	3
y	22	9	2	1	6	17

We can plot these ordered pairs and sketch the graph through them, but to get an accurate graph we need to know the vertex.

1 ▶ Find the vertex of a vertical parabola.

▶ We can find the vertex of a parabola in two ways. The first is by completing the square, as shown in Examples 1 and 2. The second is by using a formula that may be derived by completing the square.

EXAMPLE 1

Completing the Square to Find the Vertex

Find the vertex of the graph of $f(x) = x^2 - 4x + 5$.

To find the vertex, we need to express $x^2 - 4x + 5$ in the form $(x - h)^2 + k$. This is done by completing the square. (Recall that this process was introduced in Section 6.1.) To simplify the notation, we replace $f(x)$ by y.

$y = x^2 - 4x + 5$	
$y - 5 = x^2 - 4x$	Get the constant term on the left.
$y - 5 + 4 = x^2 - 4x + 4$	Half of -4 is -2; $(-2)^2 = 4$. Add 4 to both sides.
$y - 1 = (x - 2)^2$	Combine terms on the left and factor on the right.
$y = (x - 2)^2 + 1$	Add 1 to both sides.

Now write the original equation as $f(x) = (x - 2)^2 + 1$. The method of Section 9.1 shows that the vertex of this parabola is $(2, 1)$. ∎

Chalkboard Exercise

Find the vertex of the parabola by completing the square.

$$f(x) = x^2 + 4x - 9$$

Answer: $(-2, -13)$

EXAMPLE 2

Completing the Square to Find the Vertex when $a \neq 1$

Find the vertex of the graph of $y = -3x^2 + 6x - 1$.

We must complete the square on $-3x^2 + 6x$. Because the x^2 term has a coefficient other than 1, divide both sides by this coefficient, and then proceed as in Example 1.

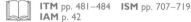 

▶ **TEACHING TIP**

Work through one example showing the "completing the square" process, then show the vertex formula

$$\left(\frac{-b}{2a}, \frac{4ac - b^2}{4a} \right).$$

Substitute the values for a, b, c into this formula to verify the result. ◀

$$y = -3x^2 + 6x - 1$$

$$\frac{y}{-3} = x^2 - 2x + \frac{1}{3} \qquad \text{Divide both sides by } -3.$$

$$\frac{y}{-3} - \frac{1}{3} = x^2 - 2x \qquad \text{Get the constant term on the left.}$$

$$\frac{y}{-3} - \frac{1}{3} + 1 = x^2 - 2x + 1 \qquad \text{Half of } -2 \text{ is } -1; (-1)^2 = 1.$$
$$\text{Add 1 to both sides.}$$

$$\frac{y}{-3} + \frac{2}{3} = (x - 1)^2 \qquad \text{Combine terms on the left and factor} \\ \text{on the right.}$$

$$\frac{y}{-3} = (x - 1)^2 - \frac{2}{3} \qquad \text{Subtract } \tfrac{2}{3} \text{ from both sides.}$$

$$y = -3(x - 1)^2 + 2 \qquad \text{Multiply by } -3 \text{ to get into desired form.}$$

The vertex is $(1, 2)$.   ∎

We can derive a formula for the vertex of the graph of the quadratic function $y = ax^2 + bx + c$ by completing the square for the general form of the equation. Going through the same steps as in Example 2 gives the equation in the form

$$y = a\left[x - \underbrace{\left(\frac{-b}{2a} \right)}_{h} \right]^2 + \underbrace{\frac{4ac - b^2}{4a}}_{k}.$$

This equation shows that the vertex (h, k) can be expressed in terms of a, b, and c. However, it is not necessary to memorize the expression for k, since it can be obtained by replacing x by $\frac{-b}{2a}$. Using function notation, if $y = f(x)$, the y-value of the vertex is $f\left(\frac{-b}{2a}\right)$.

Vertex Formula

The graph of the quadratic function $f(x) = ax^2 + bx + c$ has its vertex at

$$\left(\frac{-b}{2a}, f\left(\frac{-b}{2a} \right) \right),$$

and the axis of the parabola is the line $x = \frac{-b}{2a}$.

EXAMPLE 3

Using the Formula to Find the Vertex

Use the vertex formula to find the vertex of the graph of the function

$$f(x) = x^2 - x - 6.$$

For this function, $a = 1$, $b = -1$, and $c = -6$. The x-coordinate of the vertex of the parabola is given by

$$\frac{-b}{2a} = \frac{-(-1)}{2(1)} = \frac{1}{2}.$$

The y-coordinate is $f\left(\dfrac{-b}{2a} \right) = f\left(\dfrac{1}{2} \right).$

$$f\left(\frac{1}{2} \right) = \left(\frac{1}{2} \right)^2 - \frac{1}{2} - 6 = \frac{1}{4} - \frac{1}{2} - 6 = -\frac{25}{4}$$

Finally, the vertex is $(1/2, -25/4)$.   ∎

2 ▶ Graph a quadratic function.

▶ Graphing quadratic functions was introduced in Section 9.1. A more general approach involving intercepts and the vertex is given here.

Graphing a Quadratic Function $f(x) = ax^2 + bx + c$

Step 1 **Find the y-intercept.** Find the y-intercept by evaluating $f(0)$.

Step 2 **Find the x-intercepts.** Find the x-intercepts, if any, by solving $f(x) = 0$.

Step 3 **Find the vertex.** Find the vertex either by using the formula or by completing the square.

▶ **TEACHING TIP**
A minimum of two points on either side of the vertex is needed for an accurate graph. ◀

Step 4 **Complete the graph.** Find and plot additional points as needed, using the symmetry about the axis.

Verify that the graph opens upward (if $a > 0$) or opens downward (if $a < 0$).

EXAMPLE 4

Using the Steps for Graphing a Quadratic Function

Chalkboard Exercise

Graph the quadratic function.

$$f(x) = x^2 - 6x + 5$$

Answer:

■ Graph the quadratic function $f(x) = x^2 - x - 6$.

Begin by finding the y-intercept.

$$f(x) = x^2 - x - 6$$
$$f(0) = 0^2 - 0 - 6 \qquad \text{Find } f(0).$$
$$f(0) = -6$$

The y-intercept is $(0, -6)$. Now find any x-intercepts.

$$f(x) = x^2 - x - 6$$
$$0 = x^2 - x - 6 \qquad \text{Let } f(x) = 0.$$
$$0 = (x - 3)(x + 2) \qquad \text{Factor.}$$
$$x - 3 = 0 \quad \text{or} \quad x + 2 = 0 \qquad \text{Set each factor equal to 0 and solve.}$$
$$x = 3 \quad \text{or} \quad x = -2$$

The x-intercepts are $(3, 0)$ and $(-2, 0)$. The vertex, found in Example 3, is $(1/2, -25/4)$. Plot the points found so far, and plot any additional points as needed. The symmetry of the graph is helpful here. The graph is shown in Figure 9. ■

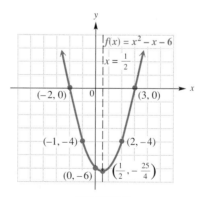

FIGURE 9

3 ▶ Use the discriminant to find the number of x-intercepts of a vertical parabola.

▶ The graph of a quadratic function may have one x-intercept, two x-intercepts, or no x-intercepts, as shown in Figure 10. Recall from Section 6.2 that the value of $b^2 - 4ac$ is called the *discriminant* of the quadratic equation $ax^2 + bx + c = 0$. It can be used to determine the number of real solutions of a quadratic equation. In a similar way, we can use the discriminant of a quadratic *function* to determine the number of x-intercepts of its graph. If the discriminant is positive, the parabola will have two x-intercepts. If the discriminant is 0, there will be only one x-intercept, and it will be the vertex of the parabola. If the discriminant is negative, the graph will have no x-intercepts.

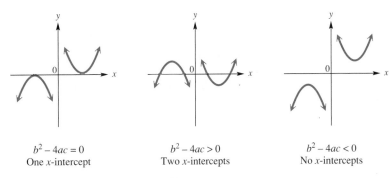

$b^2 - 4ac = 0$ $b^2 - 4ac > 0$ $b^2 - 4ac < 0$
One x-intercept   Two x-intercepts   No x-intercepts

FIGURE 10

EXAMPLE 5

Using the Discriminant to Determine the Number of x-intercepts

Determine the number of x-intercepts of the graph of each quadratic function. Use the discriminant.

(a) $f(x) = 2x^2 + 3x - 5$
The discriminant is $b^2 - 4ac$. Here $a = 2$, $b = 3$, and $c = -5$, so

$$b^2 - 4ac = 9 - 4(2)(-5) = 49.$$

Since the discriminant is positive, the parabola has two x-intercepts.

(b) $f(x) = -3x^2 - 1$
In this equation, $a = -3$, $b = 0$, and $c = -1$. The discriminant is

$$b^2 - 4ac = 0 - 4(-3)(-1) = -12.$$

The discriminant is negative and so the graph has no x-intercepts.

(c) $f(x) = 9x^2 + 6x + 1$
Here, $a = 9$, $b = 6$, and $c = 1$. The discriminant is

$$b^2 - 4ac = 36 - 4(9)(1) = 0.$$

The parabola has only one x-intercept (its vertex) since the value of the discriminant is 0. ■

4 ▶ Use quadratic functions to solve problems involving maximum or minimum value.

▶ As we have seen, the vertex of a parabola is either the highest or the lowest point on the parabola. The y-value of the vertex gives the maximum or minimum value of y, while the x-value tells where that maximum or minimum occurs.

◆ 4 seconds; 256 feet;
Let $h = 0$ and find t.
8 second

◆ C O N N E C T I O N S ◆

Quadratic functions are particularly good models for applications where a quantity increases up to a point and then decreases, or decreases and then increases. In such applications, we are often most interested in the maximum or minimum value of the function. For example, in an earlier Connections box, we discussed the quadratic function $h = -16t^2 + 128t$, which described the height h in feet of a toy rocket at time t in seconds. Since the rocket goes up, stops, and then falls to the ground, a quadratic equation is a good model for its height. Now we are able to find the exact time when the rocket reaches its maximum height by finding the vertex of the parabola that is the graph of the function.

FOR DISCUSSION OR WRITING

Find the time when the rocket reaches its maximum height. What is the maximum height? How can you determine how long it takes for the rocket to return to the ground? How long does it take?

PROBLEM SOLVING

In many practical problems we want to know the largest or smallest value of some quantity. When that quantity can be expressed using a quadratic function $y = ax^2 + bx + c$, as in the next example, the vertex can be used to find the desired value.

E X A M P L E   6

Finding the Maximum Area of a Rectangular Region

A farmer has 120 feet of fencing. He wants to put a fence around a rectangular plot of land next to a river. See Figure 11. Find the maximum area he can enclose.

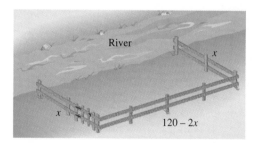

FIGURE 11

Figure 11 shows the plot. Let x represent the width of the plot. Then, since there are 120 feet of fencing,

$$x + x + \text{length} = 120 \qquad \text{Sum of the sides is 120 feet.}$$
$$2x + \text{length} = 120 \qquad \text{Combine terms.}$$
$$\text{length} = 120 - 2x. \qquad \text{Subtract } 2x.$$

The area is given by the product of the width and length, so

$$A = x(120 - 2x) = 120x - 2x^2.$$

To make the area (and thus $120x - 2x^2$) as large as possible, first find the vertex of the parabola $A = 120x - 2x^2$. We can do this by using the vertex formula. Writing the equation in standard form as $A = -2x^2 + 120x$ shows that $a = -2$, $b = 120$, and $c = 0$, so

$$h = -\frac{b}{2a} = -\frac{120}{2(-2)} = -\frac{120}{-4} = 30$$

$$f(30) = -2(30)^2 + 120(30) = -2(900) + 3600 = 1800.$$

The graph is a parabola that opens downward, and its vertex is (30, 1800). The vertex shows that the maximum area will be 1800 square feet. This area will occur if x, the width of the plot, is 30 feet.   ∎

CAUTION   Be careful when interpreting the meanings of the coordinates of the vertex in problems involving maximum or minimum values. The first coordinate, x, gives the value for which the *function value* is a maximum or a minimum. It is always necessary to read the problem carefully to determine whether you are asked to find the value of the independent variable, the function value, or both.

---

5 ▶ Graph horizontal parabolas.

▶   If x and y are exchanged in the equation $y = ax^2 + bx + c$, the equation becomes $x = ay^2 + by + c$. Because of the interchange of the roles of x and y, these parabolas are horizontal (with horizontal lines as axes), compared with the vertical ones graphed previously.

Graph of a Horizontal Parabola

The graph of $x = ay^2 + by + c$ or $x = a(y - k)^2 + h$ is a parabola with vertex at (h, k) and the horizontal line $y = k$ as axis. The graph opens to the right if a is positive and to the left if a is negative.

EXAMPLE 7

Graphing a Horizontal Parabola

Graph $x = (y - 2)^2 - 3$.

This graph has its vertex at $(-3, 2)$, since the roles of x and y are reversed. It opens to the right, the positive x-direction, and has the same shape as $y = x^2$. Plotting a few additional points gives the graph shown in Figure 12.   ∎

Chalkboard Exercise

Graph the parabola.

$x = (y + 1)^2 - 4$

Answer:

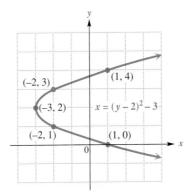

FIGURE 12

When a quadratic equation is given in the form $x = ay^2 + by + c$, completing the square on y will put the equation into a form in which the vertex can be identified.

EXAMPLE 8

Completing the Square to Graph a Horizontal Parabola

■ Graph $x = -2y^2 + 4y - 3$.

Complete the square on the right to express the equation in the form $x = a(y - k)^2 + h$.

$$\frac{x}{-2} = y^2 - 2y + \frac{3}{2} \qquad \text{Divide by } -2.$$

$$\frac{x}{-2} - \frac{3}{2} = y^2 - 2y \qquad \text{Subtract } \tfrac{3}{2}.$$

$$\frac{x}{-2} - \frac{3}{2} + 1 = y^2 - 2y + 1 \qquad \text{Add } 1.$$

$$\frac{x}{-2} - \frac{1}{2} = (y - 1)^2 \qquad \text{Factor on the right; add on the left.}$$

$$\frac{x}{-2} = (y - 1)^2 + \frac{1}{2} \qquad \text{Add } \tfrac{1}{2}.$$

$$x = -2(y - 1)^2 - 1 \qquad \text{Multiply by } -2.$$

Because the coefficient is -2, the graph opens to the left (the negative x direction) and is narrower than $y = x^2$. As shown in Figure 13, the vertex is $(-1, 1)$. ■

▶ **TEACHING TIP**
You might want to mention that the vertex formula for quadratic relations of the form

$$x = ay^2 + by + c$$

is given by

$$\left(\frac{4ac - b^2}{4a}, \frac{-b}{2a} \right). \; \blacktriangleleft$$

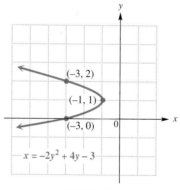

FIGURE 13

CAUTION Only quadratic equations that are solved for y are examples of functions. The graphs of the equations in Examples 7 and 8 are not graphs of functions. They do not satisfy the conditions of the vertical line test. Furthermore, the vertex formula given earlier in the section does not apply to parabolas with horizontal axes.

In summary, the graphs of parabolas studied in this section and the previous one fall into the following categories.

Graphs of Parabolas

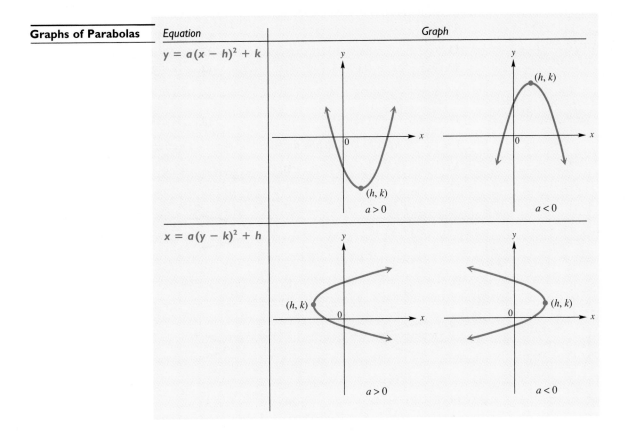

Equation

$y = a(x - h)^2 + k$

Graph

(h, k)
$a > 0$

(h, k)
$a < 0$

$x = a(y - k)^2 + h$

(h, k)
$a > 0$

(h, k)
$a < 0$

9.2 EXERCISES

1. How can you determine just by looking at the equation of a parabola whether it has a vertical or a horizontal axis?

2. Why can't the graph of a quadratic function be a parabola that opens to the left or to the right?

3. How can you determine the number of x-intercepts of the graph of a quadratic function without graphing the function?

4. If the vertex of the graph of a quadratic function is $(1, -3)$, and the graph opens downward, how many x-intercepts does the graph have?

Find the vertex of the parabola. Decide whether the graph opens upward, downward, to the left, or to the right, and state whether it is wider, narrower, or the same shape as the graph of $y = x^2$. If it is a vertical parabola, use the discriminant to determine the number of x-intercepts. See Examples 1–3, 5, 7, and 8.

5. $y = 2x^2 + 4x + 5$

6. $y = 3x^2 - 6x + 4$

7. $y = -x^2 + 5x + 3$

8. $x = -y^2 + 7y + 2$

9. $x = \frac{1}{3}y^2 + 6y + 24$

10. $x = .5y^2 + 10y - 5$

4. none
5. $(-1, 3)$; upward; narrower; no x-intercepts
6. $(1, 1)$; upward; narrower; no x-intercepts
7. $\left(\frac{5}{2}, \frac{37}{4}\right)$; downward; same; two x-intercepts
8. $\left(\frac{57}{4}, \frac{7}{2}\right)$; to the left; same   9. $(-3, -9)$; to the right; wider
10. $(-55, -10)$; to the right; wider

 CONCEPTUAL   WRITING   ▲ CHALLENGING    SCIENTIFIC CALCULATOR    GRAPHICS CALCULATOR

11. $f(x) = x^2 + 8x + 10$

12.

$f(x) = x^2 + 10x + 23$

13.

$y = -2x^2 + 4x - 5$

14. $y = -3x^2 + 12x - 8$

15. $x = -\frac{1}{5}y^2 + 2y - 4$

16.

$x = -.5y^2 - 4y - 6$

17.

$x = 3y^2 + 12y + 5$

18.

$x = 4y^2 + 16y + 11$

19. F **20.** A **21.** C
22. B **23.** D **24.** E
25. in 1995; 68 million
barrels daily
26. in 1993; 10%
27. 16 feet; 2 seconds

Graph the parabola using the techniques described in this section. See Examples 3, 4, 7, and 8.

11. $f(x) = x^2 + 8x + 10$

12. $f(x) = x^2 + 10x + 23$

13. $y = -2x^2 + 4x - 5$

14. $y = -3x^2 + 12x - 8$

15. $x = -\frac{1}{5}y^2 + 2y - 4$

16. $x = -.5y^2 - 4y - 6$

17. $x = 3y^2 + 12y + 5$

18. $x = 4y^2 + 16y + 11$

Use the concepts of this section to match the equation with its graph in Exercises 19–24.

A.

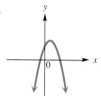

B.

C.

D.

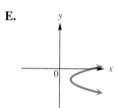

E.

F.

19. $y = 2x^2 + 4x - 3$ **20.** $y = -x^2 + 3x + 5$ **21.** $y = -\frac{1}{2}x^2 - x + 1$

22. $x = y^2 + 6y + 3$ **23.** $x = -y^2 - 2y + 4$ **24.** $x = 3y^2 + 6y + 5$

Solve the problem. See Example 6.

25. The world consumption of oil in millions of barrels daily is given by

$$f(x) = -.1x^2 + 2x + 58,$$

where x is the number of years since 1985 (1985 corresponds to 0). According to this equation, when will oil consumption reach a maximum? What will the maximum consumption be? Discuss how realistically this equation describes oil consumption.

26. Recent annual inflation rates (in percent) in Mexico are given by the function

$$f(x) = 4x^2 - 48x + 154,$$

where x represents the number of years since 1987. In what year does this equation indicate that the inflation rate will be a minimum? What is the minimum rate? Discuss how realistic this equation may or may not be.

27. If an object is thrown upward with an initial velocity of 32 feet per second, then its height after t seconds is given by

$$h = 32t - 16t^2.$$

Find the maximum height attained by the object and the number of seconds it takes to hit the ground.

28. A projectile is fired straight upward so that its distance (in feet) above the ground t seconds after firing is

$$s = -16t^2 + 400t.$$

Find the maximum height it reaches and the number of seconds it takes to reach that height.

29. Keisha Hughes has 100 meters of fencing material to enclose a rectangular exercise run for her dog. What width will give the enclosure the maximum area?

30. Morgan's Department Store wants to construct a rectangular parking lot on land bordered on one side by a highway. It has 280 feet of fencing that is to be used to fence off the other three sides. What should be the dimensions of the lot if the enclosed area is to be a maximum? What is the maximum area?

31. If air resistance is neglected, a projectile shot straight upward with an initial velocity of 40 meters per second will be at a height s in meters given by the function

$$s(t) = -4.9t^2 + 40t,$$

where t is the number of seconds elapsed after projection. After how many seconds will it reach its maximum height, and what is this maximum height? Round your answers to the nearest tenth.

▲ **32.** For a trip to a resort, a charter bus company charges a fare of $48 per person, plus $2 per person for each unsold seat on the bus. If the bus has 42 seats and x represents the number of unsold seats, find the following:
 (a) a function that defines the total revenue, R, from the trip (*Hint:* Multiply the total number riding, $42 - x$, by the price per ticket, $48 + 2x$);
 (b) the graph of the function from part (a);
 (c) the number of unsold seats that produces the maximum revenue;
 (d) the maximum revenue.

The accompanying bar graph shows the annual average number of nonfarm payroll jobs in California for the years 1988 through 1992. If the tops of the bars were joined by a smooth curve, the curve would resemble the graph of a quadratic function (that is, a parabola). Using a technique from statistics it can be determined that this function can be described approximately as

$$f(x) = -.10x^2 + .42x + 11.90,$$

where $x = 0$ corresponds to 1988, $x = 1$ corresponds to 1989, and so on, and $f(x)$ represents the number of payroll jobs in millions. (Source: California Employment Development Department)

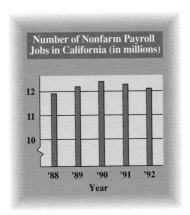

33. Explain why the coefficient of x^2 in the function is negative, based on the graph formed by joining the tops of the bars.

34. Determine the coordinates of the vertex of the graph using algebraic methods.

35. How does the x-coordinate of the vertex of the parabola indicate that during the time period under consideration, the maximum number of payroll jobs was in 1990?

36. What does the y-coordinate of the vertex of the parabola indicate?

28. 2500 feet; 12.5 seconds
29. 25 meters
30. 140 feet by 70 feet; 9800 square feet
31. 4.1 seconds; 81.6 meters
32. (a) $R(x) = (42 - x)(48 + 2x) = -2x^2 + 36x + 2016$
(b)

(c) 9   **(d)** $2178
33. The parabola opens downward, indicating $a < 0$.   **34.** (2.1, 12.34)
35. Since $x = 2.1 \approx 2$, the year corresponding to 2, which is 1990, is the year of maximum number of jobs.   **36.** It indicates the maximum number of jobs, which was approximately 12.34 million.

37. (a) $\{1, 3\}$
(b) $(-\infty, 1) \cup (3, \infty)$
(c) the open interval $(1, 3)$

38. (a) $\left\{-4, \dfrac{2}{3}\right\}$

(b) $(-\infty, -4] \cup \left[\dfrac{2}{3}, \infty\right)$

(c) the open interval
$\left(-4, \dfrac{2}{3}\right)$ 39. (a) $\{-2, 5\}$

(b) $[-2, 5]$

(c) $(-\infty, -2] \cup [5, \infty)$

40. (a) $\left\{-3, \dfrac{5}{2}\right\}$

(b) $\left[-3, \dfrac{5}{2}\right]$

(c) $(-\infty, -3] \cup \left[\dfrac{5}{2}, \infty\right)$

◆ **MATHEMATICAL CONNECTIONS** (Exercises 37–40) ◇

In Example 1 of Section 6.5, we determined the solution set of the quadratic *inequality* $x^2 - x - 12 > 0$ by using regions on a number line and testing values in the inequality. We know that if we graph $f(x) = x^2 - x - 12$, the *x*-intercepts will determine the solutions of the quadratic *equation* $x^2 - x - 12 = 0$. The *x*-values of the points on the graph that are *above* the *x*-axis form the solution set of $x^2 - x - 12 > 0$. As seen in the figure, this solution set is $(-\infty, -3) \cup (4, \infty)$, which supports the result found in Section 6.5. Similarly, the solution set of the quadratic inequality $x^2 - x - 12 < 0$ is found by locating the points on the graph that lie *below* the *x*-axis. Those *x*-values belong to the open interval $(-3, 4)$.

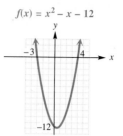

$f(x) = x^2 - x - 12$

The graph is *above* the *x*-axis for $(-\infty, -3) \cup (4, \infty)$.

In Exercises 37–40, the graph of a quadratic function f is given. Use only the graph to find the solution set of the equation or inequality. Work through parts (a)–(c) in order each time.

37. $f(x) = x^2 - 4x + 3$

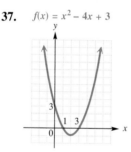

(a) $x^2 - 4x + 3 = 0$
(b) $x^2 - 4x + 3 > 0$
(c) $x^2 - 4x + 3 < 0$

38. $f(x) = 3x^2 + 10x - 8$

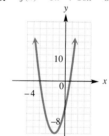

(a) $3x^2 + 10x - 8 = 0$
(b) $3x^2 + 10x - 8 \geq 0$
(c) $3x^2 + 10x - 8 < 0$

39. $f(x) = -x^2 + 3x + 10$

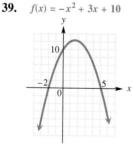

(a) $-x^2 + 3x + 10 = 0$
(b) $-x^2 + 3x + 10 \geq 0$
(c) $-x^2 + 3x + 10 \leq 0$

40. $f(x) = -2x^2 - x + 15$

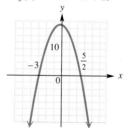

(a) $-2x^2 - x + 15 = 0$
(b) $-2x^2 - x + 15 \geq 0$
(c) $-2x^2 - x + 15 \leq 0$

The accompanying graphics calculator-generated graph is of the function $y = f(x) = x^2 - x - 6$. Notice that modern graphics calculators not only are capable of finding x-intercepts, but are also capable of finding the vertex of the graph of a quadratic function. Since this graph opens upward, its vertex is a *minimum,* as designated on the screen. (For a parabola that opens downward, it is called a *maximum.*) Now compare this figure to Figure 9 in this section, which is a traditional graph of this function. Because $\frac{1}{2} = .5$ and $-\frac{25}{4} = -6.25$, we see that the calculator graph supports our algebraic work in Example 3. (The long string of decimal places for the value of x is typical of how graphics calculators compute such values. If you obtain such a display, do not be overly concerned. This is an instance where we see the necessity for being able to compute using strictly algebraic methods; we must not rely totally on what we see on the screen!)

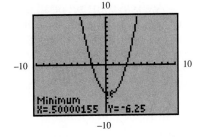

 For the quadratic function **(a)** *determine the coordinates of the vertex using the methods of this section and* **(b)** *use a graphics calculator to support your answer.*

41. $f(x) = x^2 + 4x - 3$

42. $f(x) = x^2 + 2x - 4$

43. $f(x) = -3x^2 + 12x - 4$

44. $f(x) = -.5x^2 + x + 6$

REVIEW EXERCISES

Find the distance between the pair of points. See Section 7.1.

45. $(2, -1)$ and $(4, 3)$

46. $(5, 6)$ and $(-2, -3)$

47. (x, y) and $(-2, 5)$

48. (x, y) and (h, k)

41. (a) $(-2, -7)$
(b)
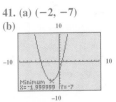

42. (a) $(-1, -5)$
(b)
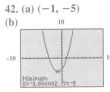

43. (a) $(2, 8)$
(b)
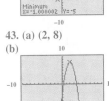

44. (a) $\left(1, \frac{13}{2}\right)$
(b)
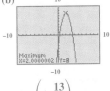

45. $2\sqrt{5}$ 46. $\sqrt{130}$
47. $\sqrt{(x + 2)^2 + (y - 5)^2}$
48. $\sqrt{(x - h)^2 + (y - k)^2}$

9.3 THE CIRCLE AND THE ELLIPSE

FOR EXTRA HELP

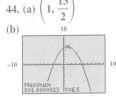

SSG pp. 286–291
SSM pp. 479–494

Video
11

Tutorial
IBM MAC

OBJECTIVES

1 ► Find the equation of a circle given the center and radius.
2 ► Determine the center and radius of a circle given its equation.
3 ► Recognize the equation of an ellipse.
4 ► Graph ellipses.
5 ► Graph circles and ellipses using a graphics calculator.

In Sections 9.1 and 9.2 the second-degree equations $y = ax^2 + bx + c$ and $x = ay^2 + by + c$ $(a \neq 0)$ were discussed. These equations have just one second-degree term. This section begins a discussion of second-degree equations that have both x^2 and y^2 terms, starting with those second-degree equations that have a circle for a graph.

INSTRUCTOR'S RESOURCES

 ITM pp. 484–487 ISM pp. 719–729
IAM p. 43

 TEST GENERATOR
DOS Windows MAC

TRANSPARENCIES

CONCEPTUAL   WRITING   ▲ CHALLENGING   SCIENTIFIC CALCULATOR   GRAPHICS CALCULATOR

1 ▶ Find the equation of a circle given the center and radius.

▶   A **circle** is the set of all points in a plane that lie a fixed distance from a fixed point. The fixed point is called the **center** and the fixed distance is called the **radius.** We use the distance formula derived in Section 7.1 to find the equation of a circle.

E X A M P L E   1

Finding the Equation of a Circle and Graphing It

Find an equation of the circle with radius 3 and center at $(0, 0)$, and graph the circle.

If the point (x, y) is on the circle, the distance from (x, y) to the center $(0, 0)$ is 3. By the distance formula,

$$\sqrt{(x_2 - x_1)^2 + (y_2 - y_1)^2} = d$$
$$\sqrt{(x - 0)^2 + (y - 0)^2} = 3$$
$$x^2 + y^2 = 9.$$

An equation of this circle is $x^2 + y^2 = 9$. The graph is shown in Figure 14. ■

▶ **TEACHING TIP**

One can accurately graph any circle without using a compass. First, lightly draw horizontal and vertical work lines through the center. Count out the number of units corresponding to the radius and plot four points: above, below, to the left, and to the right of the center. Connect these points as quarter circles to obtain the graph.

(This same procedure can be used to plot ellipses.)   ◀

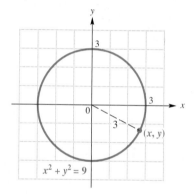

FIGURE 14

E X A M P L E   2

Finding the Equation of a Circle and Graphing It

Find an equation for the circle that has its center at $(4, -3)$ and radius 5, and graph the circle.

Again we use the distance formula.

$$\sqrt{(x - 4)^2 + (y + 3)^2} = 5$$
$$(x - 4)^2 + (y + 3)^2 = 25$$

The graph of this circle is shown in Figure 15. ■

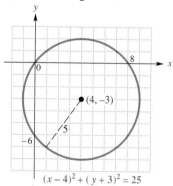

FIGURE 15

Examples 1 and 2 can be generalized to get an equation of a circle with radius r and center at (h, k). If (x, y) is a point on the circle, the distance from the center (h, k) to the point (x, y) is r. Then by the distance formula,

$$\sqrt{(x - h)^2 + (y - k)^2} = r.$$

Squaring both sides gives us the following equation of a circle.

Equation of a Circle

$$(x - h)^2 + (y - k)^2 = r^2$$

is an equation of a circle of radius r with center at (h, k).

E X A M P L E   3

Using the Form of the Equation of a Circle

Find an equation of the circle with center at $(-1, 2)$ and radius 4.

We use the last result, with $h = -1$, $k = 2$, and $r = 4$ to get

$$(x - h)^2 + (y - k)^2 = r^2$$
$$[x - (-1)]^2 + (y - 2)^2 = (4)^2$$
$$(x + 1)^2 + (y - 2)^2 = 16. \quad ■$$

2 ▶ Determine the center and radius of a circle given its equation.

▶ In the equation found in Example 2, multiplying out $(x - 4)^2$ and $(y + 3)^2$ and then combining like terms gives

$$(x - 4)^2 + (y + 3)^2 = 25$$
$$x^2 - 8x + 16 + y^2 + 6y + 9 = 25$$
$$x^2 + y^2 - 8x + 6y = 0.$$

This result suggests that an equation that has both x^2 and y^2 terms may represent a circle. The next example shows how to tell, using the method of completing the square.

For Example 3

Chalkboard Exercise

Find an equation of the circle with center at $(2, -1)$ and radius 3.

Answer:
$(x - 2)^2 + (y + 1)^2 = 9$

E X A M P L E   4

Completing the Square to Find the Center and Radius

Graph $x^2 + y^2 + 2x + 6y - 15 = 0$.

Since the equation has x^2 and y^2 terms with equal coefficients, its graph might be that of a circle. To find the center and radius, we complete the square on x and y as follows.

Chalkboard Exercise

Find the center and radius of the circle.

$x^2 + y^2 + 6x - 4y - 51 = 0$

Answer: center at $(-3, 2)$; radius 8

$x^2 + y^2 + 2x + 6y = 15$	Get the constant on the right.
$(x^2 + 2x \quad) + (y^2 + 6y \quad) = 15$	Rewrite in anticipation of completing the square.
$(x^2 + 2x + 1) + (y^2 + 6y + 9) = 15 + 1 + 9$	Complete the square in both x and y.
$(x + 1)^2 + (y + 3)^2 = 25$	Factor on the left and add on the right.

The last equation shows that the graph is a circle with center at $(-1, -3)$ and radius 5. The graph is shown in Figure 16. ■

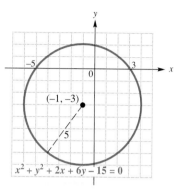

FIGURE 16

NOTE If the procedure of Example 4 leads to an equation of the form $(x - h)^2 + (y - k)^2 = 0$, the graph is the single point (h, k). If the constant on the right side is negative, the equation has no graph.

3 ▶ Recognize the equation of an ellipse.

▶ An **ellipse** is the set of all points in a plane the sum of whose distances from two fixed points is constant. These fixed points are called **foci** (singular: *focus*). Figure 17 shows an ellipse whose foci are $(c, 0)$ and $(-c, 0)$, with x-intercepts $(a, 0)$ and $(-a, 0)$ and y-intercepts $(0, b)$ and $(0, -b)$. The origin is the **center of the ellipse**. From the definition above, it can be shown by the distance formula that an ellipse has the following equation.

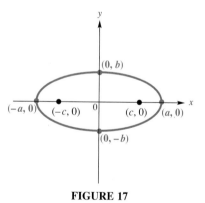

FIGURE 17

Equation of an Ellipse

The ellipse whose x-intercepts are $(a, 0)$ and $(-a, 0)$ and whose y-intercepts are $(0, b)$ and $(0, -b)$ has an equation of the form

$$\frac{x^2}{a^2} + \frac{y^2}{b^2} = 1.$$

The proof of this is outlined in the exercises. (See Exercise 49.) Note that a circle is a special case of an ellipse, where $a^2 = b^2$.

The paths of the earth and other planets around the sun are approximately ellipses; the sun is at one focus and a point in space is at the other. The orbits of communication satellites and other space vehicles are elliptical.

◆ Answers will vary.

◆ CONNECTIONS ◆

An interesting and perhaps surprising application of ellipses in our everyday life appears in gears. Elliptical bicycle gears are designed to respond to the legs' natural strengths and weaknesses. At the top and bottom of the powerstroke where the legs have the least leverage, the gear offers little resistance, but as the gear rotates, the resistance increases. This allows the legs to apply more power where it is most naturally available.

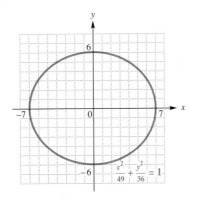

FOR DISCUSSION OR WRITING

A circle can be thought of as an ellipse in which $a = b$ in the equation. Explain how this fact distinguishes the operation of a circular gear from an elliptical gear.

4 ▶ Graph ellipses.

▶ To graph an ellipse, we plot the four intercepts and then sketch an ellipse through those points.

EXAMPLE 5

Graphing an Ellipse

■ Graph $\dfrac{x^2}{49} + \dfrac{y^2}{36} = 1$.

The x-intercepts for this ellipse are $(7, 0)$ and $(-7, 0)$. The y-intercepts are $(0, 6)$ and $(0, -6)$. Plotting the intercepts and sketching the ellipse through them gives the graph in Figure 18.   ■

Chalkboard Exercise

Graph $\dfrac{x^2}{4} + \dfrac{y^2}{25} = 1$.

Answer:

▶ TEACHING TIP

Students may have difficulty properly graphing an ellipse. Show students several examples of what an ellipse is *not* supposed to look like (in particular, a rhombus). Remind them that curve sketching takes a great deal of practice.   ◀

FIGURE 18

As with the graphs of parabolas and circles, the graph of an ellipse may be shifted horizontally and vertically, as in the next example.

E X A M P L E   6 ■

Graphing an Ellipse
Shifted Horizontally and
Vertically

Graph $\dfrac{(x - 2)^2}{25} + \dfrac{(y + 3)^2}{49} = 1$.

Just as $(x - 2)^2$ and $(y + 3)^2$ would indicate that the center of a circle would be $(2, -3)$, so is it with this ellipse. Figure 19 shows that the graph goes through the four points $(2, 4), (7, -3), (2, -10)$, and $(-3, -3)$. The x-values of these points are found by adding $\pm a = \pm 5$ to 2, and the y-values come from adding $\pm b = \pm 7$ to -3. ■

Chalkboard Exercise

Graph $\dfrac{(x + 4)^2}{16} + \dfrac{(y - 1)^2}{36} = 1$.

Answer:

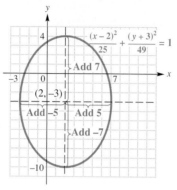

FIGURE 19

5 ▶ Graph circles and
ellipses using a graphics
calculator.

▶   A graphics calculator in the function mode cannot directly graph a circle or an ellipse. We must first solve the equation for y, getting two functions y_1 and y_2. The union of these two graphs is the graph of the entire figure. To get an undistorted screen a *square window* must be used. See your instruction manual for details. For example, to graph $(x + 3)^2 + (y + 2)^2 = 25$, we begin by solving for y.

$$(x + 3)^2 + (y + 2)^2 = 25$$
$$(y + 2)^2 = 25 - (x + 3)^2 \qquad \text{Subtract } (x + 3)^2.$$
$$y + 2 = \pm\sqrt{25 - (x + 3)^2} \qquad \text{Take square roots.}$$
$$y = -2 \pm \sqrt{25 - (x + 3)^2} \qquad \text{Subtract 2.}$$

The two functions to be graphed are

$$y_1 = -2 + \sqrt{25 - (x + 3)^2} \qquad \text{and} \qquad y_2 = -2 - \sqrt{25 - (x + 3)^2}.$$

See Figure 20. (*Note:* The two semicircles seem to be disconnected. This is because the graphs are nearly vertical at those points, and the calculator cannot show a true picture of the behavior there.)

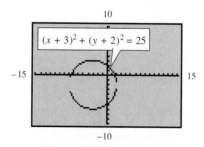

Square viewing window

FIGURE 20

9.3 EXERCISES

Match the equation with the correct graph. See Examples 1–3.

1. $(x - 3)^2 + (y - 2)^2 = 25$

1. B

2. $(x - 3)^2 + (y + 2)^2 = 25$

2. C

3. $(x + 3)^2 + (y - 2)^2 = 25$

3. D

4. $(x + 3)^2 + (y + 2)^2 = 25$

4. A

A.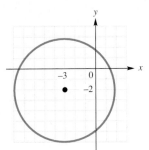

B.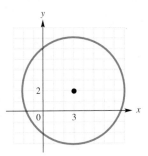

C.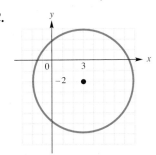

D.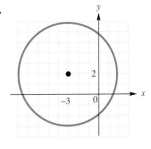

Write an equation of the circle with the given center and radius. See Examples 1–3.

5. $(0, 0)$; $r = 6$

5. $x^2 + y^2 = 36$

6. $(0, 0)$; $r = 5$

6. $x^2 + y^2 = 25$

7. $(-1, 3)$; $r = 4$

7. $(x + 1)^2 + (y - 3)^2 = 16$

8. $(2, -2)$; $r = 3$

8. $(x - 2)^2 + (y + 2)^2 = 9$

9. $(0, 4)$; $r = \sqrt{3}$

9. $x^2 + (y - 4)^2 = 3$

10. $(-2, 0)$; $r = \sqrt{5}$

10. $(x + 2)^2 + y^2 = 5$

11. Suppose that a circle has an equation of the form $x^2 + y^2 = r^2$, $r > 0$. What is the center of the circle? What is the radius of the circle?

11. center: $(0, 0)$; radius: r

12. **(a)** How many points are there on the graph of $(x - 4)^2 + (y - 1)^2 = 0$? Explain your answer.
  (b) How many points are there on the graph of $(x - 4)^2 + (y - 1)^2 = -1$? Explain your answer.

Find the center and the radius of the circle. (Hint: In Exercises 17 and 18 divide both sides by the greatest common factor.)

13. $x^2 + y^2 + 4x + 6y + 9 = 0$

13. center: $(-2, -3)$; radius: 2

14. $x^2 + y^2 - 8x - 12y + 3 = 0$

14. center: $(4, 6)$; radius: 7

15. $x^2 + y^2 + 10x - 14y - 7 = 0$

15. center: $(-5, 7)$; radius: 9

16. $x^2 + y^2 - 2x + 4y - 4 = 0$

16. center: $(1, -2)$; radius: 3

17. $3x^2 + 3y^2 - 12x - 24y + 12 = 0$

17. center: $(2, 4)$; radius: 4

18. $2x^2 + 2y^2 + 20x + 16y + 10 = 0$

18. center: $(-5, -4)$; radius: 6

CONCEPTUAL   WRITING   CHALLENGING   SCIENTIFIC CALCULATOR   GRAPHICS CALCULATOR

Graph the circle. Identify the center if it is not at the origin. See Examples 1, 2 and 4.

19. $x^2 + y^2 = 9$

19.

20. $x^2 + y^2 = 4$

20.

21. $2y^2 = 10 - 2x^2$

21.

22. $3x^2 = 48 - 3y^2$

22.

23. $(x + 3)^2 + (y - 2)^2 = 9$

23. center: $(-3, 2)$

24. $(x - 1)^2 + (y + 3)^2 = 16$

24. center: $(1, -3)$

25. $x^2 + y^2 - 4x - 6y + 9 = 0$

25. center: $(2, 3)$

26. $x^2 + y^2 + 8x + 2y - 8 = 0$

26. center: $(-4, -1)$

27. A circle can be drawn on a piece of posterboard by fastening one end of a string, pulling the string taut with a pencil, and tracing a curve as shown in the figure. Explain why this methods works.

28. It is possible to sketch an ellipse on a piece of posterboard by fastening two ends of a length of string, pulling the string taut with a pencil, and tracing a curve, as shown in the drawing. Explain why this method works.

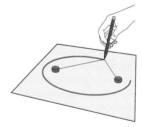

29. This figure shows the crawfish race held at the Crawfish Festival in Breaux Bridge, Louisiana. Explain why a circular "racetrack" is appropriate for such a race.

30. A *lithotripter* is a machine used to crush kidney stones using shock waves. It is effective because it uses a special property of ellipses. Look up this term in an encyclopedia and write a short paragraph on how it works.

Graph the ellipse. See Examples 5 and 6.

31. $\dfrac{x^2}{9} + \dfrac{y^2}{25} = 1$ 　　　 **31.**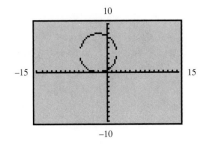

32. $\dfrac{x^2}{9} + \dfrac{y^2}{16} = 1$ 　　　 **32.**

33. $\dfrac{x^2}{36} = 1 - \dfrac{y^2}{16}$ 　　　 **33.**

34. $\dfrac{x^2}{9} = 1 - \dfrac{y^2}{4}$ 　　　 **34.**

35. $\dfrac{y^2}{25} = 1 - \dfrac{x^2}{49}$ 　　　 **35.**

36. $\dfrac{y^2}{9} = 1 - \dfrac{x^2}{16}$ 　　　 **36.**

37. $\dfrac{(x + 1)^2}{64} + \dfrac{(y - 2)^2}{49} = 1$ 　　 **37.**

38. $\dfrac{(x - 4)^2}{9} + \dfrac{(y + 2)^2}{4} = 1$ 　 **38.**

39. $\dfrac{(x - 2)^2}{16} + \dfrac{(y - 1)^2}{9} = 1$ 　 **39.**

40. $\dfrac{(x + 3)^2}{25} + \dfrac{(y + 2)^2}{36} = 1$ 　 **40.**

41. The circle shown in the calculator-generated graph was created using the function mode with a square viewing window. It is the graph of $(x + 2)^2 + (y - 4)^2 = 16$. What are the two functions y_1 and y_2 that were used to obtain this graph?

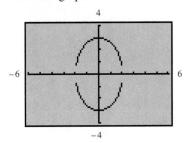

42. The ellipse shown in the calculator-generated graph was graphed using a graphics calculator in the function mode, with a square viewing window. It is the graph of $\dfrac{x^2}{4} + \dfrac{y^2}{9} = 1$. What are the two functions y_1 and y_2 that were used to obtain this graph?

41. $y_1 = 4 + \sqrt{16 - (x + 2)^2}$,
$y_2 = 4 - \sqrt{16 - (x + 2)^2}$

42. $y_1 = 3\sqrt{1 - \dfrac{x^2}{4}}$, $y_2 = -3\sqrt{1 - \dfrac{x^2}{4}}$

 CONCEPTUAL 　 WRITING 　 CHALLENGING 　 SCIENTIFIC CALCULATOR 　 GRAPHICS CALCULATOR

Use a graphics calculator in the function mode to graph the circle or ellipse. Use a square viewing window.

43. $x^2 + y^2 = 36$ 43.

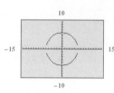

44. $(x - 2)^2 + y^2 = 49$ 44.

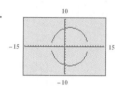

45. $\dfrac{x^2}{16} + \dfrac{y^2}{4} = 1$ 45.

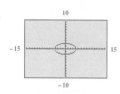

46. $\dfrac{(x - 3)^2}{25} + \dfrac{y^2}{9} = 1$ 46.

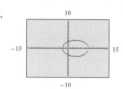

▲ **47.** The orbit of Venus around the sun (one of the foci) is an ellipse with equation

$$\frac{x^2}{5013} + \frac{y^2}{4970} = 1,$$

where x and y are measured in millions of miles.
(a) Find the farthest distance between Venus and the sun.
(b) Find the smallest distance between Venus and the sun. (*Hint:* See Figure 17 and use the fact that $c^2 = a^2 - b^2$.)

47. (a) 77.4 million miles   (b) 64.2 million miles (Both answers are rounded to the nearest tenth.)

▲ **48.** A one-way road passes under an overpass in the form of half of an ellipse, 15 feet high at the center and 20 feet wide. Assuming a truck is 12 feet wide, what is the tallest truck that can pass under the overpass?

48. 12 feet

▲ **49. (a)** Suppose that $(c, 0)$ and $(-c, 0)$ are the foci of an ellipse and that the sum of the distances from any point (x, y) on the ellipse to the two foci is $2a$. See the figure. Show that the equation of the resulting ellipse is

$$\frac{x^2}{a^2} + \frac{y^2}{a^2 - c^2} = 1.$$

(b) Show that in the equation in part (a), the x-intercepts are $(a, 0)$ and $(-a, 0)$.
(c) Let $b^2 = a^2 - c^2$, and show that $(0, b)$ and $(0, -b)$ are the y-intercepts in the equation in part (a).

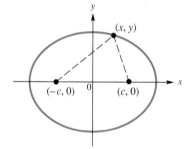

▲ **50.** Use the result of Exercise 49(a) to find an equation of an ellipse with foci $(3, 0)$ and $(-3, 0)$, where the sum of the distances from any point of the ellipse to the two foci is 10.

50. $\dfrac{x^2}{25} + \dfrac{y^2}{16} = 1$

REVIEW EXERCISES

51. Plot the points $(3, 4)$, $(-3, 4)$, $(3, -4)$ and $(-3, -4)$. See Section 7.1.

51.

52. Sketch the graphs of $y = \frac{4}{3}x$ and $y = -\frac{4}{3}x$ on the same axes. See Section 7.1.

52.

53. Find the x- and y-intercepts of the graph of $4x + 3y = 12$. See Section 7.1.
53. x-intercept: $(3, 0)$; y-intercept: $(0, 4)$

54. Solve the equation $x^2 = 121$. See Section 6.1.
54. $\{-11, 11\}$

9.4 THE HYPERBOLA AND SQUARE ROOT FUNCTIONS

FOR EXTRA HELP

📖 **SSG** pp. 291–298
SSM pp. 484–489

📼 **Video**
11

💾 **Tutorial**
IBM MAC

OBJECTIVES

1 ▶ Recognize the equation of a hyperbola.
2 ▶ Graph hyperbolas by using the asymptotes.
3 ▶ Identify conic sections by their equations.
4 ▶ Graph square root functions.

This section begins by introducing the final conic section, the *hyperbola*.

1 ▶ Recognize the equation of a hyperbola.

▶ A **hyperbola** is the set of all points in a plane such that the absolute value of the *difference* of the distances from two fixed points (called *foci*) is constant. Figure 21 shows a hyperbola; using the distance formula and the definition above, we can show that this hyperbola is given by the equation

$$\frac{x^2}{16} - \frac{y^2}{12} = 1.$$

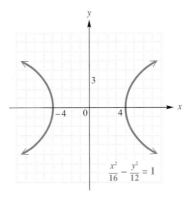

FIGURE 21

INSTRUCTOR'S RESOURCES

📖 **ITM** pp. 487–489 **ISM** pp. 730–741
IAM pp. 44–45

💾 **TEST GENERATOR**
DOS Windows MAC

✳️ **TRANSPARENCIES**

◎ CONCEPTUAL   ✎ WRITING   ▲ CHALLENGING   🖩 SCIENTIFIC CALCULATOR   🖩 GRAPHICS CALCULATOR

To find the x-intercepts, let $y = 0$.

$$\frac{x^2}{16} - \frac{0^2}{12} = 1$$

$$\frac{x^2}{16} = 1.$$

$$x^2 = 16$$

$$x = \pm 4$$

The x-intercepts are $(4, 0)$ and $(-4, 0)$. To find any y-intercepts, let $x = 0$.

$$\frac{0^2}{16} - \frac{y^2}{12} = 1$$

$$\frac{-y^2}{12} = 1$$

$$y^2 = -12$$

Because there are no *real* solutions to $y^2 = -12$, the graph has no y-intercepts. In the same way, we can show that Figure 22 gives the graph of

$$\frac{y^2}{25} - \frac{x^2}{9} = 1.$$

Here the y-intercepts are $(0, 5)$ and $(0, -5)$, and there are no x-intercepts.

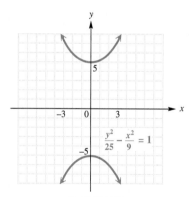

FIGURE 22

These examples suggest the following summary.

Equations of Hyperbolas

A hyperbola with x-intercepts $(a, 0)$ and $(-a, 0)$ has an equation of the form

$$\frac{x^2}{a^2} - \frac{y^2}{b^2} = 1,$$

and a hyperbola with y-intercepts $(0, b)$ and $(0, -b)$ has an equation of the form

$$\frac{y^2}{b^2} - \frac{x^2}{a^2} = 1.$$

2 ▶ Graph hyperbolas by using the asymptotes.

▶ The two branches of the graph of a hyperbola approach a pair of intersecting straight lines called **asymptotes.** See Figure 23. These lines are useful for sketching the graph of the hyperbola. We find the asymptotes as follows.

Asymptotes of Hyperbolas	The extended diagonals of the rectangle with corners at the points (a, b), $(-a, b)$, $(-a, -b)$, and $(a, -b)$ are the **asymptotes** of either of the hyperbolas $$\frac{x^2}{a^2} - \frac{y^2}{b^2} = 1 \quad \text{or} \quad \frac{y^2}{b^2} - \frac{x^2}{a^2} = 1.$$

▶ **TEACHING TIP**
Have students graph the fundamental rectangle and asymptotes as *barely visible* work lines. The graph of the resulting hyperbola will then stand out. ◀

This rectangle is called the **fundamental rectangle.** Using the methods of Chapter 7 we could show that the equations of these asymptotes are

$$y = \frac{b}{a}x \quad \text{and} \quad y = -\frac{b}{a}x.$$

EXAMPLE 1

Graphing a Horizontal Hyperbola

Chalkboard Exercise

Graph $\frac{x^2}{81} - \frac{y^2}{64} = 1.$

Answer:

For Example 2

Chalkboard Exercise

Graph $\frac{y^2}{4} - \frac{x^2}{25} = 1.$

Answer:

■ Graph $\frac{x^2}{16} - \frac{y^2}{25} = 1.$

Here $a = 4$ and $b = 5$. The x-intercepts are $(4, 0)$ and $(-4, 0)$. The four points $(4, 5)$, $(-4, 5)$, $(-4, -5)$, and $(4, -5)$ are the corners of the rectangle that determine the asymptotes, as shown in Figure 23. The equations of the asymptotes are $y = \pm\frac{5}{4}x$, and the hyperbola approaches these lines as x and y get larger and larger in absolute value. ■

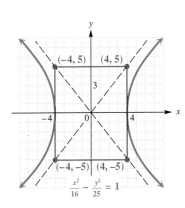

FIGURE 23

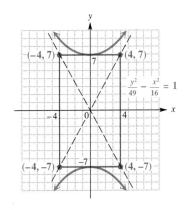

FIGURE 24

EXAMPLE 2

Graphing a Vertical Hyperbola

■ Graph $\frac{y^2}{49} - \frac{x^2}{16} = 1.$

This hyperbola has y-intercepts $(0, 7)$ and $(0, -7)$. The asymptotes are the extended diagonals of the rectangle with corners at $(4, 7)$, $(-4, 7)$, $(-4, -7)$, and $(4, -7)$, having equations $y = \pm\frac{7}{4}x$. See Figure 24. ■

In summary, to graph either of the two forms of hyperbolas, $\frac{x^2}{a^2} - \frac{y^2}{b^2} = 1$ or $\frac{y^2}{b^2} - \frac{x^2}{a^2} = 1$, follow these steps.

Graphing a Hyperbola

Step 1 **Find the intercepts.** Locate the intercepts: at $(a, 0)$ and $(-a, 0)$ if the x^2 term has a positive coefficient, or at $(0, b)$ and $(0, -b)$ if the y^2 term has a positive coefficient.

Step 2 **Find the fundamental rectangle.** Locate the corners of the fundamental rectangle at (a, b), $(a, -b)$, $(-a, -b)$, and $(-a, b)$.

Step 3 **Sketch the asymptotes.** The extended diagonals of the rectangle are the asymptotes of the hyperbola, and have equations $y = \pm \frac{b}{a} x$.

Step 4 **Draw the graph.** Sketch each branch of the hyperbola through an intercept and approaching (but not touching) the asymptotes.

CAUTION When sketching the graph of a hyperbola, be sure that the branches do not touch the asymptotes.

◆ Answers will vary.

◆ CONNECTIONS ◆

A hyperbola and a parabola are used together in one kind of microwave antenna system. The cross-sections of the system consist of a parabola and a hyperbola with the focus of the parabola coinciding with one focus of the hyperbola.

The incoming microwaves that are parallel to the axis of the parabola are reflected from the parabola up toward the hyperbola and back to the other focus of the hyperbola, where the cone of the antenna is located to capture the signal.

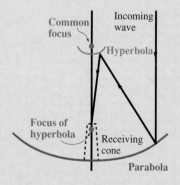

FOR DISCUSSION OR WRITING

The property of the parabola and the hyperbola that is used here is a "reflection property" of the foci. Explain why this name is appropriate.

3 ▶ Identify conic sections by their equations.

▶ Rewriting a second-degree equation in one of the forms given for ellipses, hyperbolas, circles, or parabolas makes it possible to determine when the graph is one of these figures. A summary of the equations and graphs of the conic sections is given on the next page.

Equation	Graph	Description	Identification
$y = a(x - h)^2 + k$	Parabola	Opens upward if $a > 0$, downward if $a < 0$. Vertex is at (h, k).	x^2 term y is not squared.
$x = a(y - k)^2 + h$	Parabola	Opens to right if $a > 0$, to left if $a < 0$. Vertex is at (h, k).	y^2 term x is not squared.
$(x - h)^2$ $+ (y - k)^2 = r^2$	Circle	Center at (h, k), radius r	x^2 and y^2 terms have the same positive coefficient.
$\dfrac{x^2}{a^2} + \dfrac{y^2}{b^2} = 1$	Ellipse	x-intercepts are $(a, 0)$ and $(-a, 0)$. y-intercepts are $(0, b)$ and $(0, -b)$.	x^2 and y^2 terms have different positive coefficients.
$\dfrac{x^2}{a^2} - \dfrac{y^2}{b^2} = 1$	Hyperbola	x-intercepts are $(a, 0)$ and $(-a, 0)$. Asymptotes are found from (a, b), $(a, -b)$, $(-a, -b)$. and $(-a, b)$.	x^2 has a positive coefficient. y^2 has a negative coefficient.
$\dfrac{y^2}{b^2} - \dfrac{x^2}{a^2} = 1$	Hyperbola	y-intercepts are $(0, b)$ and $(0, -b)$. Asymptotes are found from (a, b), $(a, -b)$, $(-a, -b)$, and $(-a, b)$.	y^2 has a positive coefficient. x^2 has a negative coefficient.

4 ▶ Graph square root functions.

▶ Recall that no vertical line will intersect the graph of a function in more than one point. Thus, horizontal parabolas and all circles, ellipses, and the hyperbolas discussed in this section are examples of graphs that do not satisfy the conditions of a function. However, by considering only a part of the graph of each of these we can have the graph of a function, as seen in Figure 25.

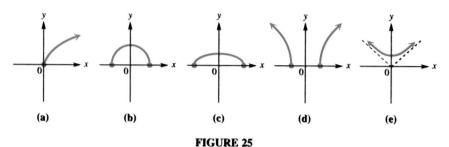

(a) (b) (c) (d) (e)

FIGURE 25

In parts (a), (b), (c) and (e) of Figure 25, the top portion of a conic section is shown (parabola, circle, ellipse, and hyperbola, respectively). In part (d), the top two portions of a hyperbola are shown. In each case, the graph is the graph of a function, since the graph satisfies the conditions of the vertical line test.

In order to obtain equations for the graphs shown in Figure 25, the idea of a square root function is introduced.

Square Root Function	A function of the form $$f(x) = \sqrt{u}$$ for an algebraic expression u, with $u \geq 0$, is called a **square root function**.

◈ For example, let $x = 2$. Then $y \approx 69$ using the equation and this appears to be close to the y-value of the point on the curve with $x = 2$.

◈▷ **C O N N E C T I O N S** ◁◈

Some applied problems can be modeled by a square root function. The graph shows the number of gun-related deaths per 100,000 for young African-American males starting in 1985. The shape of the graph looks like half of the upper branch of a hyperbola, indicating that a square root function would be a good model. Since the intercept is on the vertical axis, in the equation for a vertical hyperbola, $b = 50$. To find a, we choose another point on the curve and determine its coordinates. Let 0 correspond to 1985. Then 1990 corresponds to 5, and we can use the ordered pair (5, 130). Substituting these values for x, y, and b we get

$$\frac{y^2}{50^2} - \frac{x^2}{a^2} = 1$$

$$\frac{130^2}{50^2} - \frac{5^2}{a^2} = 1.$$

Solving this equation, we find $a \approx 2.1$. Now we replace a with 2.1 and solve the equation for y, choosing the positive square root. The resulting root function that approximates the graph is

$$y = 23.8\sqrt{4.41 + x^2}.$$

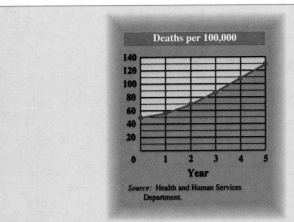

Source: Health and Human Services Department.

FOR DISCUSSION OR WRITING

To see how well this function fits the curve, choose some x-values, substitute them into the function, and note how closely the resulting y-values match the coordinates of the corresponding points on the curve. What do you find? Is the function a good fit?

EXAMPLE 3

Graphing a Square Root Function

Chalkboard Exercise

Graph $\dfrac{y}{3} = \sqrt{1 - \dfrac{x^2}{4}}$.

Answer:

Graph $f(x) = \sqrt{25 - x^2}$.

Replace $f(x)$ with y and square both sides to get the equation

$$y^2 = 25 - x^2, \quad \text{or} \quad x^2 + y^2 = 25.$$

This is the equation of a circle with center at $(0, 0)$ and radius 5. Since $f(x)$, or y, represents a principal square root in the original equation, $f(x)$ must be nonnegative. This restricts the graph to the upper half of the circle, as shown in Figure 26. Use the graph and the vertical line test to verify that it is indeed a function. ∎

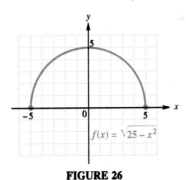

FIGURE 26

NOTE Refer to Figure 26. Had we wanted to graph the function

$$g(x) = -\sqrt{25 - x^2},$$

the graph would simply be obtained by reflecting the graph of $f(x)$ across the x-axis. This would ensure nonpositive y-values, as indicated by the $-$ sign in the rule for $g(x)$.

◆ C O N N E C T I O N S ◆

 Hyperbolas may be graphed with a graphics calculator in much the same way as circles and ellipses, by first writing the equations of two root functions that combined are equivalent to the equation of the hyperbola. A square window gives a truer shape for hyperbolas, too. Root functions, since they are functions, can be entered and graphed directly.

9.4 EXERCISES

Match the equation with the correct graph. See Examples 1 and 2, and Example 5 in Section 9.3.

1. $\dfrac{x^2}{25} + \dfrac{y^2}{9} = 1$ **A.**

1. C

2. $\dfrac{x^2}{9} + \dfrac{y^2}{25} = 1$

2. B

3. $\dfrac{x^2}{9} - \dfrac{y^2}{25} = 1$

3. D

4. $\dfrac{x^2}{25} - \dfrac{y^2}{9} = 1$ **C.**

4. A

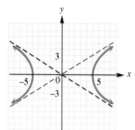

B.

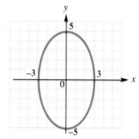

D.

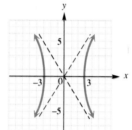

◎ **5.** Write an explanation of how you can tell from the equation whether the branches of a hyperbola
🖉   open up and down or open left and right.

◎ **6.** Explain why the graph of a hyperbola of the type discussed in this section does not satisfy the
🖉   conditions for the graph of a function.

Graph the hyperbola. See Examples 1 and 2.

7. $\dfrac{x^2}{16} - \dfrac{y^2}{9} = 1$ **8.** $\dfrac{y^2}{4} - \dfrac{x^2}{25} = 1$ **9.** $\dfrac{y^2}{9} - \dfrac{x^2}{9} = 1$

7. 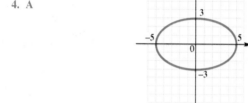   8.   9.

10. $\dfrac{x^2}{49} - \dfrac{y^2}{16} = 1$ **11.** $\dfrac{x^2}{25} - \dfrac{y^2}{36} = 1$ **12.** $\dfrac{y^2}{9} - \dfrac{x^2}{4} = 1$

10.   11.   12.

◎ CONCEPTUAL   🖉 WRITING   ▲ CHALLENGING   ▦ SCIENTIFIC CALCULATOR   GRAPHICS CALCULATOR

In this chapter we have studied the four conic sections: parabolas, circles, ellipses, and hyperbolas. Identify the graph of the equation as one of the four conic sections. (It may be necessary to transform the equation into a more recognizable form.) Then sketch the graph of the equation.

13. $x^2 - y^2 = 16$

13. hyperbola

14. $x^2 + y^2 = 16$

14. circle

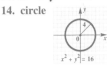

15. $4x^2 + y^2 = 16$

15. ellipse

16. $x^2 - 2y = 0$

16. parabola

17. $y^2 = 36 - x^2$

17. circle

18. $9x^2 + 25y^2 = 225$

18. ellipse

19. $9x^2 = 144 + 16y^2$

19. hyperbola

20. $x^2 + 9y^2 = 9$

20. ellipse

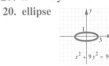

21. $y^2 = 4 + x^2$

21. hyperbola

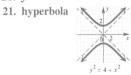

22. State in your own words the major difference between the definitions of *ellipse* and *hyperbola*.

Graph the square root function. See Example 3.

23. $f(x) = -\sqrt{36 - x^2}$

23.

24. $f(x) = 3\sqrt{1 + \dfrac{x^2}{9}}$

24.

25. $f(x) = \sqrt{\dfrac{x + 4}{2}}$

25.

26. $f(x) = -2\sqrt{\dfrac{9 - x^2}{9}}$

26.

▲ *Recall from Section 9.3 that the center of an ellipse may be shifted away from the origin. (See Example 6 in Section 9.3, for instance.) The same shifting process can be applied to hyperbolas. For example, the hyperbola*

$$\frac{(x + 5)^2}{4} - \frac{(y - 2)^2}{9} = 1$$

would have the same graph as $\dfrac{x^2}{4} - \dfrac{y^2}{9} = 1$, *but centered at* $(-5, 2)$. *Graph the hyperbola with center shifted away from the origin.*

27. $\dfrac{(x - 2)^2}{4} - \dfrac{(y + 1)^2}{9} = 1$

27.

28. $\dfrac{(x + 3)^2}{16} - \dfrac{(y - 2)^2}{25} = 1$

28.

 CONCEPTUAL   WRITING   ▲ CHALLENGING    SCIENTIFIC CALCULATOR   GRAPHICS CALCULATOR

29. $\dfrac{y^2}{36} - \dfrac{(x-2)^2}{49} = 1$

29.

30. $\dfrac{(y-5)^2}{9} - \dfrac{x^2}{25} = 1$

30.

▲ **31.** An arch has the shape of half an ellipse. The equation of the ellipse is $100x^2 + 324y^2 = 32{,}400$, where x and y are in meters.
 (a) How high is the center of the arch?
 (b) How wide is the arch across the bottom? (See the figure.)

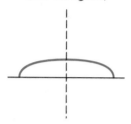

31. (a) 10 meters   (b) 36 meters

▲ **32.** Two buildings in a sports complex are shaped and positioned like a portion of the branches of the hyperbola $400x^2 - 625y^2 = 250{,}000$, where x and y are in meters.
 (a) How far apart are the buildings at their closest point?
 (b) Find the distance d in the figure.

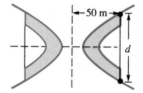

32. (a) 50 meters   (b) 69.3 meters

▦ **33.** The hyperbola shown in the calculator-generated graph in the figure was graphed in function mode with a square viewing window. It is the graph of $\dfrac{x^2}{9} - y^2 = 1$. What are the two functions y_1 and y_2 that were used to obtain this graph?

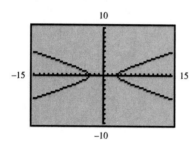

33. $y_1 = \sqrt{\dfrac{x^2}{9} - 1}$, $y_2 = -\sqrt{\dfrac{x^2}{9} - 1}$

▦ **34.** Repeat Exercise 34 for the graph of $\dfrac{y^2}{9} - x^2 = 1$, shown in the accompanying figure.

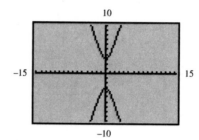

34. $y_1 = 3\sqrt{1 + x^2}$, $y_2 = -3\sqrt{1 + x^2}$

▦ *Use a graphics calcuator in function mode to graph the hyperbola. Use a square viewing window.*

35. $\dfrac{x^2}{25} - \dfrac{y^2}{49} = 1$

35.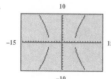

36. $\dfrac{x^2}{4} - \dfrac{y^2}{16} = 1$

36.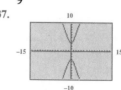

37. $\dfrac{y^2}{9} - x^2 = 1$

37.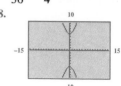

38. $\dfrac{y^2}{36} - \dfrac{x^2}{4} = 1$

38.

─────────◇ **MATHEMATICAL CONNECTIONS** (Exercises 39–44) ◇─────────

From the discussion in this section, we know that the graph of $\dfrac{x^2}{4} - y^2 = 1$ is a hyperbola.

We know that the graph of this hyperbola approaches its asymptotes as x gets larger and larger. Work through Exercises 39–44 in order to see the relationship between the hyperbola and one of its asymptotes.

39. Solve $\dfrac{x^2}{4} - y^2 = 1$ for y, and choose the positive square root.

39. $y = \sqrt{\dfrac{x^2}{4} - 1}$

40. Find the equation of the asymptote with positive slope.

40. $y = \dfrac{1}{2}x$

41. Use a calculator to evaluate the y-coordinate of the point where $x = 50$ on the graph of the portion of the hyperbola represented by the equation obtained in Exercise 39. Round your answer to the nearest hundredth.

41. $y \approx 24.98$

42. Find the y-coordinate of the point where $x = 50$ on the graph of the asymptote found in Exercise 40.

42. $y = 25$

43. Compare your results in Exercises 41 and 42. How do they support the following statement? When $x = 50$, the graph of the function defined by the equation found in Exercise 39 lies *below* the graph of the asymptote found in Exercise 40.

43. Because $24.98 < 25$, the graph of $y = \sqrt{\dfrac{x^2}{4} - 1}$ lies below the graph of $y = \dfrac{1}{2}x$.

44. What do you think will happen if we choose x-values larger than 50?

44. The y-values on the hyperbola will approach the y-values on the line, but will always be less.

───────────◆───────────

45. Suppose that a hyperbola has center at the origin, foci at $(-c, 0)$ and $(c, 0)$, and the absolute value of the difference between the distances from any point (x, y) of the hyperbola to the two foci is $2a$. See the figure. Let $b^2 = c^2 - a^2$, and show that an equation of the hyperbola is

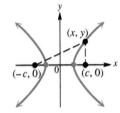

$$\frac{x^2}{a^2} - \frac{y^2}{b^2} = 1.$$

46. Use the result of Exercise 45 to find an equation of a hyperbola with center at the origin, foci at $(-2, 0)$ and $(2, 0)$, and the absolute value of the difference between the distances from any point of the hyperbola to the two foci equal to 2.

46. $x^2 - \dfrac{y^2}{3} = 1$

 CONCEPTUAL   WRITING   ▲ CHALLENGING   SCIENTIFIC CALCULATOR    GRAPHICS CALCULATOR

REVIEW EXERCISES

Solve the system by substitution. See Section 8.1.

47. $2x + y = 13$
$y = 3x + 3$
47. $\{(2, 9)\}$

48. $3x + y = 4$
$x = y - 8$
48. $\{(-1, 7)\}$

49. $9x + 2y = 10$
$x - y = -5$
49. $\{(0, 5)\}$

50. $5x + 2y = 15$
$x - y = 3$
50. $\{(3, 0)\}$

Solve the equation. See Section 6.3.

51. $2x^4 - 5x^2 - 3 = 0$
51. $\left\{ -\sqrt{3}, \sqrt{3}, -\dfrac{\sqrt{2}}{2}i, \dfrac{\sqrt{2}}{2}i \right\}$

52. $3x^4 + 26x^2 + 35 = 0$
52. $\left\{ -\dfrac{\sqrt{15}}{3}i, \dfrac{\sqrt{15}}{3}i, \sqrt{7}i, -\sqrt{7}i \right\}$

53. $r^4 - 7r^2 + 12 = 0$
53. $\{-2, 2, -\sqrt{3}, \sqrt{3}\}$

54. $p^4 - 14p^2 + 45 = 0$
54. $\{-3, 3, -\sqrt{5}, \sqrt{5}\}$

9.5 NONLINEAR SYSTEMS OF EQUATIONS

FOR EXTRA HELP	OBJECTIVES

SSG pp. 298–305 **SSM** pp. 489–499
  Video 11
  Tutorial IBM MAC
	$1 \blacktriangleright$ Solve a nonlinear system by substitution.
	 $2 \blacktriangleright$ Use the elimination method to solve a system with two second-degree equations.
	 $3 \blacktriangleright$ Solve a system that requires a combination of methods.

An equation in which some terms have more than one variable or a variable of degree two or higher is called a **nonlinear equation**. A **nonlinear system of equations** includes at least one nonlinear equation.

When solving nonlinear systems, it is helpful to visualize the types of graphs of the equations of the system in order to determine the possible number of points of intersection. For example, if a system contains two equations where the graph of one is a parabola and the graph of the other is a line, then there may be 0, 1, or 2 points of intersection. This is illustrated in Figure 27.

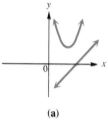
(a)
No points of intersection

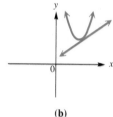
(b)
One point of intersection

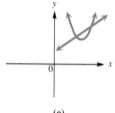
(c)
Two points of intersection

FIGURE 27

INSTRUCTOR'S RESOURCES

 ITM pp. 489–492   **ISM** pp. 741–758
IAM p. 45–46

 TEST GENERATOR
DOS   Windows   MAC

TRANSPARENCIES

1 ▶ Solve a nonlinear system by substitution.

▶ We solve nonlinear systems by the elimination method, the substitution method, or a combination of the two. The following examples show the use of these methods for solving nonlinear systems. The substitution method is usually the most useful when one of the equations is linear. The first two examples illustrate this kind of system.

E X A M P L E   1

Using Substitution When One Equation Is Linear

Solve the system

$$x^2 + y^2 = 9 \qquad \textbf{(1)}$$
$$2x - y = 3. \qquad \textbf{(2)}$$

The graph of (1) is a circle and the graph of (2) is a line. Visualizing the possibilities indicates that there may be 0, 1, or 2 points of intersection. When solving a system of this type, it is best to solve the linear equation for one of the two variables; then substitute the resulting expression into the nonlinear equation to obtain an equation in one variable. Solving equation (2) for y gives

$$2x - y = 3 \qquad \text{(2)}$$
$$y = 2x - 3. \qquad \textbf{(3)}$$

Substitute $2x - 3$ for y in equation (1) to get

$$x^2 + (2x - 3)^2 = 9$$
$$x^2 + 4x^2 - 12x + 9 = 9$$
$$5x^2 - 12x = 0$$
$$x(5x - 12) = 0 \qquad \text{Common factor is } x.$$
$$x = 0 \qquad \text{or} \qquad x = \frac{12}{5}.$$

▶ TEACHING TIP

It helps to provide a set of empty ordered pairs, like (,), before solving the nonlinear system. Provide as many ordered pairs as might be expected in the solution of the problem. Then, as the solutions for x are found, there is a place for them. The blank spaces in the ordered pairs will indicate that y-values need to be found. ◀

Let $x = 0$ in equation (3) to get $y = -3$. If $x = 12/5$, then $y = 9/5$. The solution set of the system is $\{(0, -3), (12/5, 9/5)\}$. The graph of the system, shown in Figure 28 confirms the two points of intersection. ■

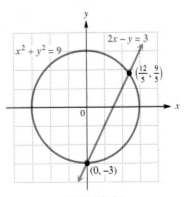

FIGURE 28

◈ quadratic; a parabola opening upward; linear; a straight line; 0, 1, or 2, because a line may intersect a parabola at two points, be tangent to the curve at one point, or not intersect at all. during 1980 ($x \approx 6.9$) and 2000 ($x \approx 26.7$); In 1990, men received about 517 thousand degrees and women received about 589 thousand degrees.

◈ **C O N N E C T I O N S** ◈

Historically, the number of bachelor's degrees earned by men has been greater than the number earned by women.* However, in the 1970s the situation in the United States began to change as the number earned by men decreased and then in the 1980s slowly began to increase again. Meanwhile, the number of degrees earned by women continued to rise steadily throughout that period. These changes can be expressed as functions and defined as shown below, where y is the number of degrees (in thousands) granted in year x, with $x = 0$ corresponding to 1974.

Men: $\quad y = .75(x - 9)^2 + 480$

Women: $\quad y = 11.7x + 402$

These two equations form a system of nonlinear equations that can be solved by the methods of this section to find the year in which the number of bachelor's degrees awarded to men and women were the same.

FOR DISCUSSION OR WRITING

What type of equation is the one for men? What kind of graph would it have? What type of equation is the one for women? What kind of graph would it have? How many intersections could these two graphs have? How do you know? In what year were the number of degrees granted to men and women the same? (A graphics calculator would help here.) How many degrees were awarded to each gender in 1990?

E X A M P L E 2

Using Substitution When One Equation Is Linear

Solve the system

$$6x - y = 5 \qquad (4)$$
$$xy = 4. \qquad (5)$$

The graph of (4) is a line, and although we have not specifically mentioned equations like (5), it can be shown by plotting points that its graph is a hyperbola. Once again, visualizing a line and a hyperbola indicates that there may be 0, 1, or 2 points of intersection. Since neither equation has a squared term here, solve either equation for one of the variables and then substitute the result into the other equation. Solving $xy = 4$ for x gives $x = 4/y$. Substituting $4/y$ for x in equation (4) gives

$$6\left(\frac{4}{y}\right) - y = 5.$$

Clear fractions by multiplying both sides by y, noting the restriction that y cannot be 0. Then solve for y.

$$\frac{24}{y} - y = 5$$
$$24 - y^2 = 5y \qquad \text{Multiply by } y.$$
$$0 = y^2 + 5y - 24$$
$$0 = (y - 3)(y + 8) \qquad \text{Factor.}$$
$$y = 3 \quad \text{or} \quad y = -8$$

---

*Source: National Science Foundation.

Substitute these results into $x = 4/y$ to obtain the corresponding values of x.

$$\text{If } y = 3, \text{ then } x = \frac{4}{3}.$$

$$\text{If } y = -8, \text{ then } x = -\frac{1}{2}.$$

The solution set has two ordered pairs: $\{(4/3, 3), (-1/2, -8)\}$. The graph in Figure 29 shows that there are two points of intersection. ∎

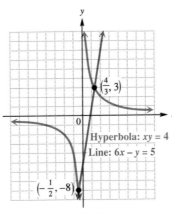

FIGURE 29

2 ▶ **Use the elimination method to solve a system with two second-degree equations.**

▶ The elimination method is often useful when both equations are second-degree equations. This method is shown in the following example.

E X A M P L E 3
Solving a Nonlinear System by Elimination

Solve the system

$$x^2 + y^2 = 9 \qquad (6)$$
$$2x^2 - y^2 = -6. \qquad (7)$$

The graph of (6) is a circle, while the graph of (7) is a hyperbola. Analyzing the possibilities leads to the conclusion that there may be 0, 1, 2, 3, or 4 points of intersection. Adding the two equations will eliminate y, leaving an equation that can be solved for x.

$$\begin{array}{rcl} x^2 + y^2 &=& 9 \\ 2x^2 - y^2 &=& -6 \\ \hline 3x^2 &=& 3 \end{array}$$

$$x^2 = 1$$

$$x = 1 \qquad \text{or} \qquad x = -1$$

Each value of x gives corresponding values for y when substituted into one of the original equations. Using equation (6) gives the following.

If $x = 1$,

$$(1)^2 + y^2 = 9$$
$$y^2 = 8$$
$$y = \sqrt{8} \qquad \text{or} \qquad -\sqrt{8}$$
$$y = 2\sqrt{2} \qquad \text{or} \qquad -2\sqrt{2}.$$

If $x = -1$,

$$(-1)^2 + y^2 = 9$$
$$y^2 = 8$$
$$y = 2\sqrt{2} \qquad \text{or} \qquad -2\sqrt{2}.$$

Chalkboard Exercise

Solve the system.

$$x^2 - 5y^2 = 4$$
$$x^2 - 3y^2 = 6$$

Answer: $\{(3, 1), (-3, 1), (-3, -1), (3, -1)\}$

▶ **TEACHING TIP**

Before working out an example of this type, review the following:

1. Square root property: For real numbers a and b,

   if $a^2 = b$ then $a = \pm\sqrt{b}$.

2. Definition of i: $i = \sqrt{-1}$. ◀

▶ **TEACHING TIP**
Two intersecting conics with centers at (0, 0) will always have points of intersection that are symmetric about the origin. ◀

The solution set has four ordered pairs: $\{(1, 2\sqrt{2}), (1, -2\sqrt{2}), (-1, 2\sqrt{2}),$ $(-1, -2\sqrt{2})\}$. Figure 30 shows the four points of intersection. ■

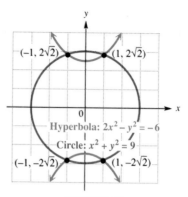

FIGURE 30

3 ▶ Solve a system that requires a combination of methods.

▶The next example shows a system of second-degree equations that requires a combination of methods to solve.

E X A M P L E   4

Solving a Nonlinear System by a Combination of Methods

Solve the system

$$x^2 + 2xy - y^2 = 7 \qquad (8)$$
$$x^2 - y^2 = 3. \qquad (9)$$

While we have not graphed equations like (8), its graph is a hyperbola. The graph of (9) is also a hyperbola. Two hyperbolas may have 0, 1, 2, 3, or 4 points of intersection. The elimination method can be used here in combination with the substitution method. We begin by eliminating the squared terms by multiplying both sides of equation (9) by -1 and then adding the result to equation (8).

$$\begin{array}{rcr} x^2 + 2xy - y^2 = & 7 \\ -x^2 \qquad\quad + y^2 = & -3 \\ \hline 2xy \qquad\quad = & 4 \end{array}$$

Next, we solve $2xy = 4$ for y. (Either variable would do.)

$$2xy = 4$$
$$y = \frac{2}{x} \qquad (10)$$

Now substitute $y = 2/x$ into one of the original equations. It is easier to do this with equation (9).

$$x^2 - y^2 = 3 \qquad \text{\color{gray}(9)}$$
$$x^2 - \left(\frac{2}{x}\right)^2 = 3$$
$$x^2 - \frac{4}{x^2} = 3$$
$$x^4 - 4 = 3x^2 \qquad \text{Multiply by } x^2.$$
$$x^4 - 3x^2 - 4 = 0 \qquad \text{Subtract } 3x^2.$$
$$(x^2 - 4)(x^2 + 1) = 0 \qquad \text{Factor.}$$

$$x^2 - 4 = 0 \quad \text{or} \quad x^2 + 1 = 0$$
$$x^2 = 4 \quad \text{or} \quad x^2 = -1$$
$$x = 2 \quad \text{or} \quad x = -2 \quad x = i \quad \text{or} \quad x = -i$$

Substituting the four values of x from above into equation (10) gives the corresponding values for y.

If $x = 2$, then $y = 1$.   If $x = i$, then $y = -2i$.

If $x = -2$, then $y = -1$.   If $x = -i$, then $y = 2i$.

Note that if you substitute the x-values found above into equations (8) or (9) instead of into equation (10), you get extraneous solutions. It is always wise to check all solutions in both of the given equations. There are four ordered pairs in the solution set, two with real values and two with imaginery values:

$$\{(2, 1), (-2, -1), (i, -2i), (-i, 2i)\}.$$

The graph of the system, shown in Figure 31, shows only the two real intersection points because the graph is in the real number plane. The two ordered pairs with imaginary components are solutions of the system, but do not show up on the graph. ■

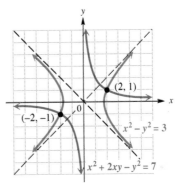

FIGURE 31

NOTE In the examples of this section, we analyzed the possible number of points of intersection of the graphs in each system. However, in Examples 2 and 4, we worked with equations whose graphs had not been studied. Keep in mind that it is not absolutely essential to visualize the number of points of intersection in order to solve the system. Without having studied equations of the types of (5) and (8), it would be unrealistic to expect you to know that these graphs are hyperbolas. Visualizing the geometry of the graphs is only an aid to solving these systems, and not an essential part of the solution process.

◆ **CONNECTIONS** ◆

If the equations in a nonlinear system can be solved for y (perhaps as the union of two equations), then we can graph the equations of the system with a graphics calculator and use the capabilities of the calculator to find all the intersection points. For instance, the two equations in Example 3 would require graphing four separate functions, two for each of the original equations.

9.5 EXERCISES

Suppose that a nonlinear system is composed of equations whose graphs are those described, and the number of points of intersection of the two graphs is as given. Make a sketch satisfying these conditions. (There may be more than one way to do this.)

1. a line and a circle;   no points

2. a line and a circle;   one point

3. a line and a circle;   two points

4. a line and an ellipse;   no points

5. a line and an ellipse;   one point

6. a line and an ellipse;   two points

7. a line and a hyperbola;   no points

8. a line and a hyperbola;   one point

9. a line and a hyperbola;   two points

10. a circle and an ellipse;   one point

11. a circle and an ellipse;   four points

12. a parabola and an ellipse;   one point

13. a parabola and an ellipse;   four points

14. a parabola and a hyperbola;   two points

◎ **15.** Write an explanation of the steps you would use to solve the system

$$x^2 + y^2 = 25$$
$$y = x - 1$$

by the substitution method.

⊙ **16.** Why would the elimination method not be appropriate for the system in Exercise 15?

16. In the first equation both variables are squared, and in the second both variables are to the first power. There are no like terms to eliminate.

Solve the system by the substitution method. See Examples 1 and 2.

17. $y = 4x^2 - x$
$y = x$
17. $\left\{ (0, 0), \left(\dfrac{1}{2}, \dfrac{1}{2} \right) \right\}$

18. $y = x^2 + 6x$
$3y = 12x$
18. $\{(0, 0), (-2, -8)\}$

19. $y = x^2 + 6x + 9$
$x + y = 3$
19. $\{(-6, 9), (-1, 4)\}$

20. $y = x^2 + 8x + 16$
$x - y = -4$
20. $\{(-3, 1), (-4, 0)\}$

21. $x^2 + y^2 = 2$
$2x + y = 1$
21. $\left\{ \left(-\dfrac{1}{5}, \dfrac{7}{5} \right), (1, -1) \right\}$

22. $2x^2 + 4y^2 = 4$
$x = 4y$
22. $\left\{ \left(\dfrac{4}{3}, \dfrac{1}{3} \right), \left(-\dfrac{4}{3}, -\dfrac{1}{3} \right) \right\}$

23. $xy = 4$
$3x + 2y = -10$
23. $\left\{ (-2, -2), \left(-\dfrac{4}{3}, -3 \right) \right\}$

24. $xy = -5$
$2x + y = 3$
24. $\left\{ \left(\dfrac{5}{2}, -2 \right), (-1, 5) \right\}$

25. $xy = -3$
$x + y = -2$
25. $\{(-3, 1), (1, -3)\}$

26. $xy = 12$
$x + y = 8$
26. $\{(6, 2), (2, 6)\}$

27. $y = 3x^2 + 6x$
$y = x^2 - x - 6$
27. $\left\{ \left(-\dfrac{3}{2}, -\dfrac{9}{4} \right), (-2, 0) \right\}$

28. $y = 2x^2 + 1$
$y = 5x^2 + 2x - 7$
28. $\left\{ \left(\dfrac{4}{3}, \dfrac{41}{9} \right), (-2, 9) \right\}$

29. $2x^2 - y^2 = 6$
$y = x^2 - 3$
29. $\{(-\sqrt{3}, 0), (\sqrt{3}, 0), (-\sqrt{5}, 2), (\sqrt{5}, 2)\}$

30. $x^2 + y^2 = 4$
$y = x^2 - 2$
30. $\{(0, -2), (\sqrt{3}, 1), (-\sqrt{3}, 1)\}$

⊙ **31.** Write an explanation of the steps you would use to solve the system

$$x^2 + 2y^2 = 9$$
$$4x^2 + 3y^2 = 16$$

by the elimination method.

⊙ **32.** If you were to solve the system in Exercise 31 by substitution, which variable in which equation would be the best choice to solve for to avoid introducing fractions as coefficients?

32. Solving for x^2 in the first equation would be the best choice.

Solve the system by the elimination method or a combination of the elimination and substitution methods. See Examples 3 and 4.

33. $3x^2 + 2y^2 = 12$
$x^2 + 2y^2 = 4$
33. $\{(-2, 0), (2, 0)\}$

34. $2x^2 + y^2 = 28$
$4x^2 - 5y^2 = 28$
34. $\{(-2\sqrt{3}, -2), (-2\sqrt{3}, 2), (2\sqrt{3}, -2), (2\sqrt{3}, 2)\}$

35. $2x^2 + 3y^2 = 6$
$x^2 + 3y^2 = 3$
35. $\{(\sqrt{3}, 0), (-\sqrt{3}, 0)\}$

36. $6x^2 + y^2 = 9$
$3x^2 + 4y^2 = 36$
36. $\{(0, 3), (0, -3)\}$

37. $2x^2 = 8 - 2y^2$
$3x^2 = 24 - 4y^2$
37. $\{(-2i\sqrt{2}, -2\sqrt{3}), (-2i\sqrt{2}, 2\sqrt{3}), (2i\sqrt{2}, -2\sqrt{3}), (2i\sqrt{2}, 2\sqrt{3})\}$

38. $5x^2 = 20 - 5y^2$
$2y^2 = 2 - x^2$
38. $\{(-\sqrt{6}, -i\sqrt{2}), (-\sqrt{6}, i\sqrt{2}), (\sqrt{6}, -i\sqrt{2}), (\sqrt{6}, i\sqrt{2})\}$

39. $x^2 + xy + y^2 = 15$
$x^2 + y^2 = 10$
39. $\{(-\sqrt{5}, -\sqrt{5}), (\sqrt{5}, \sqrt{5})\}$

40. $2x^2 + 3xy + 2y^2 = 21$
$x^2 + y^2 = 6$
40. $\{(\sqrt{3}, \sqrt{3}), (-\sqrt{3}, -\sqrt{3})\}$

⊙ CONCEPTUAL   ✐ WRITING   ▲ CHALLENGING   ▦ SCIENTIFIC CALCULATOR   ▦ GRAPHICS CALCULATOR

41. $3x^2 + 2xy - 3y^2 = 5$
$\quad\;\; -x^2 - 3xy + y^2 = 3$

42. $-2x^2 + 7xy - 3y^2 = 4$
$\qquad\quad 2x^2 - 3xy + 3y^2 = 4$

41. $\{(i, 2i), (-i, -2i), (2, -1), (-2, 1)\}$

42. $\left\{(\sqrt{2}, \sqrt{2}), (-\sqrt{2}, -\sqrt{2}), \left(\sqrt{3}, \dfrac{2\sqrt{3}}{3}\right), \left(-\sqrt{3}, -\dfrac{2\sqrt{3}}{3}\right)\right\}$

In Example 1 of this section, we solved the system

$$x^2 + y^2 = 9$$
$$2x - y = 3$$

algebraically by using the substitution method. As shown in Figure 28, the points of intersection of the graphs of the circle and the line are $\left(\dfrac{12}{5}, \dfrac{9}{5}\right)$ and $(0, -3)$. This can be supported with a graphics calculator by graphing the circle as the union of two functions whose graphs are semicircles, graphing the line (by first solving for y), and then finding the points of intersection using the capabilities of the calculator. The accompanying graphs show $y_1 = \sqrt{9 - x^2}$, $y_2 = -\sqrt{9 - x^2}$, and $y = 2x - 3$ all on the same square screen. Notice that the points of intersection as indicated at the bottoms of the screens correspond to the results found in Example 1.

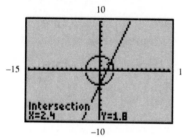 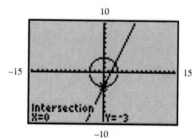

📷 *Use a graphics calculator to solve the system. Then confirm your answer algebraically by solving the system using substitution, elimination, or a combination of the methods.*

43. $y = x^2 - 4x + 4$ **43.** $\{(2, 0), (1, 1)\}$
$\quad\; x + y = 2$

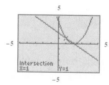 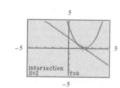

44. $x^2 + y^2 = 25$ **44.** $\{(4, 3), (-3, -4)\}$
$\quad\; x - y = 1$

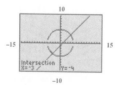 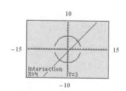

45. $xy = -6$ **45.** $\{(2, -3), (-3, 2)\}$
$\quad\; x + y = -1$

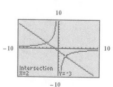 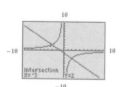

46. $y = 2x^2 + 4x$ **46.** $\left\{\left(-\dfrac{1}{3}, -\dfrac{10}{9}\right), (-1, -2)\right\}$
$\quad\; y = -x^2 - 1$

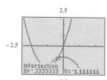 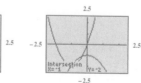

---◆ **MATHEMATICAL CONNECTIONS** (Exercises 47–52) ◆---

In order to see how solving quadratic equations in one variable is related to solving a nonlinear system involving one quadratic equation and one linear equation, work through Exercises 47–52 in order.

47. Use one of the methods of Chapter 3 or Chapter 6 to solve the quadratic equation $x^2 = 3x + 10$.

47. $\{-2, 5\}$

48. Graph the equations $y = x^2$ and $y = 3x + 10$ on the same set of coordinate axes.

48.

49. Solve the nonlinear system

$$y = x^2$$
$$y = 3x + 10$$

using substitution.

49. $\{(-2, 4), (5, 25)\}$

50. How do the x-coordinates of the solutions in Exercise 49 compare to the solutions of the equation in Exercise 47?

50. They are the same.

51. Graph the quadratic function $y = x^2 - 3x - 10$.

51.

52. How do the x-intercepts of the quadratic function in Exercise 51 compare to the x-coordinates of the points of intersection of the graphs in Exercise 48?

52. They are the same.

---◆---

Solve the problem by using a nonlinear system.

53. The sum of the squares of two numbers is 8. The product of the two numbers is 4. Find the numbers.

53. 2 and 2 or -2 and -2

54. The sum of the squares of two numbers is 26. The difference of the squares of the same two numbers is 24. Find the numbers.

54. 5 and 1, or -5 and 1, or 5 and -1, or -5 and -1

55. The area of a rectangular rug is 84 square feet and its perimeter is 38 feet. Find the length and width of the rug.

55. length: 12 feet;   width: 7 feet

56. Find the length and width of a rectangular room whose perimeter is 50 meters and whose area is 100 square meters.

56. length: 20 meters;   width: 5 meters

57. A company has found that the price p (in dollars) of its scientific calculator is related to the supply x (in thousands) by the equation $px = 16$. Also, the price is related to the demand x (in thousands) for the calculator by the equation $p = 10x + 12$. The *equilibrium price* is the value of p where demand equals supply. Find the equilibrium price and the supply/demand at that price by solving a system of equations.

57. $20;   800 calculators

58. The calculator company in Exercise 57 has also determined that the cost y to make x (thousand) calculators is $y = 4x^2 + 36x + 20$, while the revenue y from the sale of x (thousand) calculators is $36x^2 - 3y = 0$. Find the *break-even point*, where cost just equals revenue, by solving a system of equations. (Values of y are in dollars.)

58. 5000 calculators; $300

 CONCEPTUAL   WRITING   ▲ CHALLENGING   SCIENTIFIC CALCULATOR    GRAPHICS CALCULATOR

REVIEW EXERCISES

Graph the inequality. See Section 7.4.

59. $2x - y \leq 4$

60. $-x + 3y > 9$

61. $-5x + 3y \leq 15$

62. $2x \leq y$

59.

60.

61.

62.

9.6 SECOND-DEGREE INEQUALITIES AND SYSTEMS OF INEQUALITIES

FOR EXTRA HELP

 SSG pp. 305–309
SSM pp. 499–504

 Video
12

 Tutorial
IBM   MAC

OBJECTIVES

1 ▶ Graph second-degree inequalities.
2 ▶ Graph the solution set of a system of inequalities.

We can now extend the discussion of inequalities in Chapter 7 to nonlinear inequalities.

1 ▶ Graph second-degree inequalities.

▶ The linear inequality $3x + 2y \leq 5$ is graphed by first graphing the boundary line $3x + 2y = 5$. **Second-degree inequalities** such as $x^2 + y^2 \leq 36$ are graphed in much the same way. The boundary of the inequality $x^2 + y^2 \leq 36$ is the graph of the equation $x^2 + y^2 = 36$, a circle with radius 6 and center at the origin, as shown in Figure 32. As with linear inequalities, the inequality $x^2 + y^2 \leq 36$ will include either the points outside the circle or the points inside the circle. We decide which region to shade by substituting any point not on the circle, such as $(0, 0)$, into the inequality. Since $0^2 + 0^2 < 36$ is a true statement, the inequality includes the points inside the circle, the shaded region in Figure 32.

▶ **TEACHING TIP**

This section provides an excellent review of graph recognition and graphing techniques for lines and conic sections. ◀

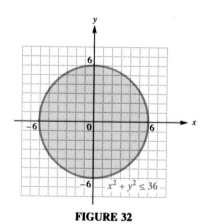

FIGURE 32

EXAMPLE 1

Graphing a Second-Degree Inequality

Graph $y < -2(x - 4)^2 - 3$.

The boundary, $y = -2(x - 4)^2 - 3$, is a parabola opening downward with vertex at $(4, -3)$. Using the point $(0, 0)$ as a test point gives

$$0 < -2(0 - 4)^2 - 3 \qquad ?$$
$$0 < -32 - 3 \qquad ?$$
$$0 < -35. \qquad \text{False}$$

Because the final inequality is a false statement, the points in the region containing $(0, 0)$ do not satisfy the inequality. Figure 33 shows the final graph; the parabola is drawn with a dashed line since the points of the parabola itself do not satisfy the inequality, and the region inside (or below) the parabola is shaded. ■

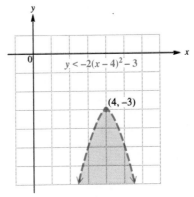

FIGURE 33

EXAMPLE 2

Graphing a Second-Degree Inequality

Graph $16y^2 \leq 144 + 9x^2$.

First rewrite the inequality as follows.

$$16y^2 - 9x^2 \leq 144$$
$$\frac{y^2}{9} - \frac{x^2}{16} \leq 1 \qquad \text{Divide by } 144.$$

This form of the inequality shows that the boundary is the hyperbola

$$\frac{y^2}{9} - \frac{x^2}{16} = 1.$$

The desired region will either be the region between the branches of the hyperbola or the regions above the top branch and below the bottom branch. Test the region between the branches by choosing $(0, 0)$ as a test point. Substitute into the original inequality.

$$16y^2 \leq 144 + 9x^2$$
$$16(0)^2 \leq 144 + 9(0)^2 \qquad ?$$
$$0 \leq 144 + 0 \qquad ?$$
$$0 \leq 144 \qquad \text{True}$$

Since the test point $(0, 0)$ satisfies the inequality $16y^2 \leq 144 + 9x^2$, the region between the branches containing $(0, 0)$ is shaded, as shown in Figure 34. ■

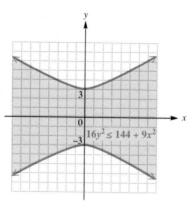

FIGURE 34

2 ▶ Graph the solution set of a system of inequalities.

▶ If two or more inequalities are considered at the same time, at least one of which is nonlinear, we have a **nonlinear system of inequalities.** In order to find the solution set of a nonlinear system, we find the intersection of the graphs of each inequality in the system. This is just an extension of the method used in Section 7.4 to graph pairs of linear inequalities joined by the word *and*.

E X A M P L E   3 ■

Graphing a Nonlinear System of Two Inequalities

Graph the solution set of

$$2x + 3y > 6$$
$$x^2 + y^2 < 16.$$

We begin by graphing the solution of $2x + 3y > 6$. Using the methods of Section 7.4, we find that the boundary line is the graph of $2x + 3y = 6$, and is a dashed line because of the symbol $>$. The test point $(0, 0)$ leads to a false statement in the inequality $2x + 3y > 6$, so we shade the region above the line, as shown in Figure 35.

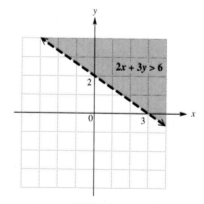

FIGURE 35

Using the methods of Examples 1 and 2, we find that the graph of $x^2 + y^2 < 16$ is the interior of a dashed circle centered at the origin with radius 4. This is shown in Figure 36.

Finally, to get the graph of the solution set of the system, we find the intersection of the graphs of the two inequalities. The overlapping region in Figure 37 is the solution set. ■

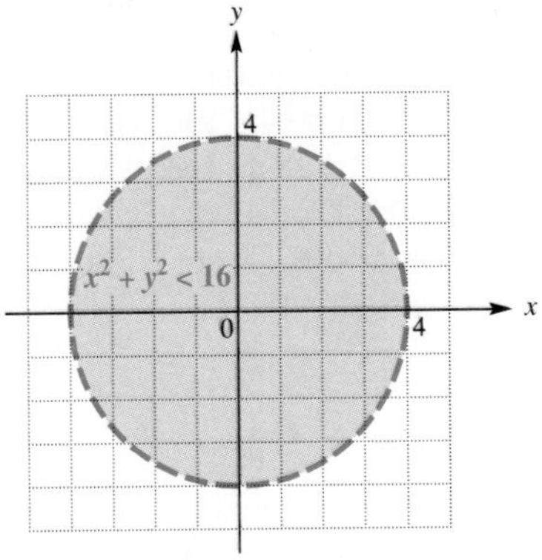

FIGURE 36

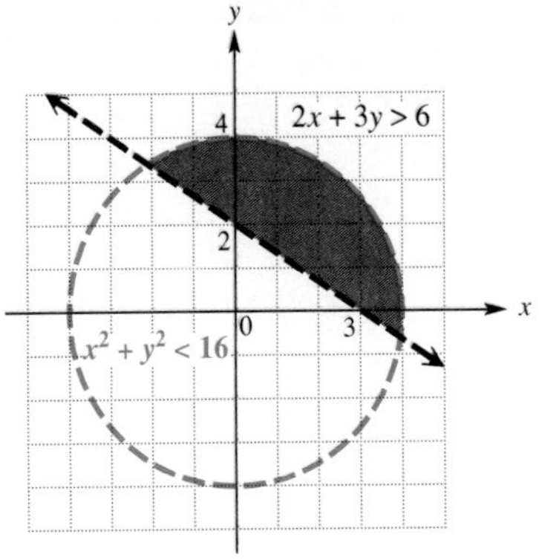

FIGURE 37

While the system in the following example does not contain a nonlinear inequality, it is different from those that we have solved in both this section and Section 7.4. It contains three linear inequalities.

EXAMPLE 4

Graphing a Linear System with Three Inequalities

Graph the solutions of the system

$$x + y < 1$$
$$y \leq 2x + 3$$
$$y \geq -2.$$

We graph each inequality separately, on the same axes. The graph of $x + y < 1$ consists of all points below the dashed line $x + y = 1$. The graph of $y \leq 2x + 3$ is the region below the solid line $y = 2x + 3$. Finally, the graph of $y \geq -2$ is the region above the solid horizontal line $y = -2$. The graph of the system is the triangular region enclosed by the three boundary lines in Figure 38 including two of its boundaries. ■

Chalkboard Exercise

Graph the solutions of the system

$$2x - 5y \geq 10$$
$$x + 3y \geq 6$$
$$y \leq 2.$$

Answer:

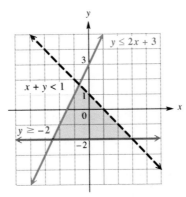

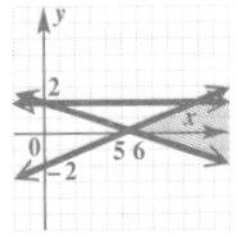

FIGURE 38

EXAMPLE 5

Graphing a Nonlinear System with Three Inequalities

Graph the solutions of the system

$$y \geq x^2 - 2x + 1$$
$$2x^2 + y^2 > 4$$
$$y < 4.$$

The graph of $y = x^2 - 2x + 1$ is a parabola with vertex at $(1, 0)$. Those points above (or in the interior of) the parabola satisfy the condition $y > x^2 - 2x + 1$. Thus points on the parabola or in the interior are in the solution of $y \geq x^2 - 2x + 1$. The graph of the equation $2x^2 + y^2 = 4$ is an ellipse. We draw it with a dashed line. To satisfy the inequality $2x^2 + y^2 > 4$, a point must lie outside the ellipse. The graph of $y < 4$ includes all points below the dashed line $y = 4$. Finally, the graph of the system is the shaded region in Figure 39 that lies outside the ellipse, inside or on the boundary of the parabola, and below the line $y = 4$. ■

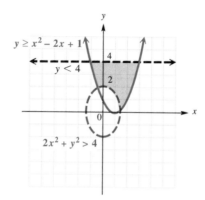

FIGURE 39

◆ CONNECTIONS ◆

Graphics calculators can show the graph of the solutions of a system of inequalities. We can use the techniques for graphing lines, parabolas, and root functions to graph the boundaries of inequalities with a graphics calculator. Many of these calculators have a draw function that can be used to shade the appropriate region.

9.6 EXERCISES

1. Write an explanation of how to graph the solution set of a nonlinear inequality.
2. Write an explanation of how to graph the solution set of a system of nonlinear inequalities.

Match the nonlinear inequality with its graph.

A.
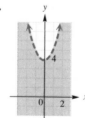

B.
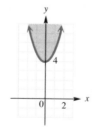

C.
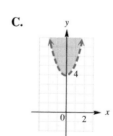

D.
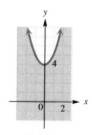

3. $y \geq x^2 + 4$
3. B

4. $y \leq x^2 + 4$
4. D

5. $y < x^2 + 4$
5. A

6. $y > x^2 + 4$
6. C

Graph the nonlinear inequality. See Examples 1 and 2.

7. $y^2 > 4 + x^2$

7.

8. $y^2 \le 4 - 2x^2$

8.

9. $y + 2 \ge x^2$

9.

10. $x^2 \le 16 - y^2$

10.

11. $2y^2 \ge 8 - x^2$

11.

12. $x^2 \le 16 + 4y^2$

12.

13. $y \le x^2 + 4x + 2$

13.

14. $9x^2 < 16y^2 - 144$

14.

15. $9x^2 > 16y^2 + 144$

15.

16. $4y^2 \le 36 - 9x^2$

16.

17. $x^2 - 4 \ge -4y^2$

17.

18. $x \ge y^2 - 8y + 14$

18.

19. $x \le -y^2 + 6y - 7$

19.

20. $y^2 - 16x^2 \le 16$

20.

⊚ 21. Which one of the following is a description of the graph of the solution set of the system below?

$$x^2 + y^2 < 25$$
$$y > -2$$

(a) all points outside the circle $x^2 + y^2 = 25$ and above the line $y = -2$
(b) all points outside the circle $x^2 + y^2 = 25$ and below the line $y = -2$
(c) all points inside the circle $x^2 + y^2 = 25$ and above the line $y = -2$
(d) all points inside the circle $x^2 + y^2 = 25$ and below the line $y = -2$

21. (c)

⊚ 22. Fill in the blank with the appropriate response. The graph of the system

$$y > x^2 + 1$$
$$\frac{x^2}{9} + \frac{y^2}{4} > 1$$
$$y < 5$$

consists of all points _____?_____ the parabola $y = x^2 + 1$, _____?_____ the ellipse
(above/below)   (inside/outside)

$\frac{x^2}{9} + \frac{y^2}{4} = 1$, and _____?_____ the line $y = 5$.
(above/below)

22. above; outside; below

Graph the system of inequalities. See Examples 3–5.

23. $2x + 5y < 10$
$x - 2y < 4$

23.

24. $3x - y > -6$
$4x + 3y > 12$

24.

25. $5x - 3y \leq 15$
$4x + y \geq 4$

25.

26. $4x - 3y \leq 0$
$x + y \leq 5$

26.

27. $x \leq 5$
$y \leq 4$

27.

28. $x \geq -2$
$y \leq 4$

28.

29. $y > x^2 - 4$
$y < -x^2 + 3$

29.

30. $x^2 - y^2 \geq 9$
$\dfrac{x^2}{16} + \dfrac{y^2}{9} \leq 1$

30.

31. $y^2 - x^2 \geq 4$
$-5 \leq y \leq 5$

31.

32. $x^2 + y^2 \geq 4$
$x + y \leq 5$
$x \geq 0$
$y \geq 0$

32.

33. $y \leq -x^2$
$y \geq x - 3$
$y \leq -1$
$x < 1$

33.

34. $y < x^2$
$y > -2$
$x + y < 3$
$3x - 2y > -6$

34.

▲ *For each nonlinear inequality, a restriction is placed on one or both variables. For example, the graph of*

$$x^2 + y^2 \leq 4, \quad x \geq 0$$

would be as shown in the figure. Only the right half of the interior of the circle is shaded, because of the restriction that x must be nonnegative. Graph the nonlinear inequality with restrictions.

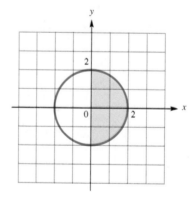

35. $x^2 + y^2 > 36, \quad x \geq 0$

35.

36. $4x^2 + 25y^2 < 100, \quad y < 0$

36.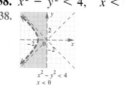

37. $x < y^2 - 3, \quad x < 0$

37.

38. $x^2 - y^2 < 4, \quad x < 0$

38.

39. $4x^2 - y^2 > 16$, $x < 0$

39.

40. $x^2 + y^2 > 4$, $y < 0$

40.

41. $x^2 + 4y^2 \geq 1$, $x \geq 0, y \geq 0$

41.

42. $2x^2 - 32y^2 \leq 8$, $x \leq 0, y \geq 0$

42.

The calculator-generated graph shown here is a graph of the nonlinear system

$$y \geq x^2 - 6$$
$$y \leq x + 1.$$

The system contains all points that are on or above (inside) the graph of the parabola $y = x^2 - 6$ and all points that are on or below the graph of the line $y = x + 1$. As mentioned in the Connections box in this section, some graphics calculators have the

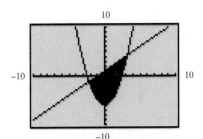

capability of shading above or below a curve. Of course, we cannot tell whether the curve itself is part of the solution due to the limited resolution of the screen. We must rely on the inequality symbols in the system to make such a determination. Once again, we see the need for understanding the mathematical concepts behind the technology.

 Use the shading feature of a graphics calculator to graph the system.

43. $y \geq x - 3$
    $y \leq -x + 4$

43.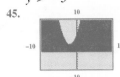

44. $y \geq -x^2 + 5$
    $y \leq x^2 - 3$

44.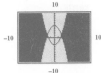

45. $y < x^2 + 4x + 4$
    $y > -3$

45.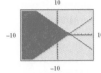

46. $y > (x - 4)^2 - 3$
    $y < 5$

46.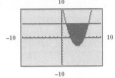

REVIEW EXERCISES

Graph the two linear functions on the same set of axes. See Section 7.5.

47. (a) $f(x) = 2x - 3$
    (b) $g(x) = \dfrac{x + 3}{2}$

47.

48. (a) $f(x) = x - 5$
    (b) $g(x) = x + 5$

48.

⊙ CONCEPTUAL   ✐ WRITING   ▲ CHALLENGING   ▦ SCIENTIFIC CALCULATOR   ▧ GRAPHICS CALCULATOR

Give the domain and the range of the function. See Section 7.5.

49. $\left\{\left(-3, \frac{1}{8}\right), \left(-2, \frac{1}{4}\right), \left(-1, \frac{1}{2}\right), (0, 1), (1, 2), (2, 4), (3, 8)\right\}$

49. domain: $\{-3, -2, -1, 0, 1, 2, 3\}$;   range: $\left\{\frac{1}{8}, \frac{1}{4}, \frac{1}{2}, 1, 2, 4, 8\right\}$

50. $f(x) = x^2$

50. domain: $(-\infty, \infty)$;   range: $[0, \infty)$

Decide whether the graph is that of a function. See Section 7.5.

51.

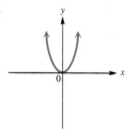

51. function

52.

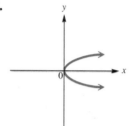

52. not a function

Solve for y in terms of x. See Section 2.2.

53. $x - 5y = 4$

53. $y = \frac{4 - x}{-5}$ or $y = \frac{x - 4}{5}$

54. $2x^2 + 3y = 5$

54. $y = \frac{5 - 2x^2}{3}$

CHAPTER 9 SUMMARY

KEY TERMS

9.1 conic sections
parabola
vertex
axis
quadratic function

9.3 circle
center
radius
ellipse
foci

9.4 hyperbola
asymptotes of a
hyperbola
fundamental
rectangle
square root function

9.5 nonlinear equation
nonlinear system of
equations

9.6 second-degree
inequality
system of
inequalities

QUICK REVIEW

CONCEPTS	EXAMPLES
9.1 QUADRATIC FUNCTIONS: PARABOLAS	
1. The graph of the quadratic function $f(x) = a(x - h)^2 + k,\quad a \neq 0$ is a parabola with vertex at (h, k) and the vertical line $x = h$ as axis. **2.** The graph opens upward if a is positive and downward if a is negative. **3.** The graph is wider than $f(x) = x^2$ if $0 < \|a\| < 1$ and narrower if $\|a\| > 1$.	Graph $f(x) = -(x + 3)^2 + 1$. 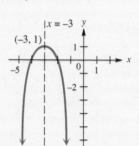

CONCEPTS	EXAMPLES

9.2   MORE ABOUT PARABOLAS AND THEIR APPLICATIONS

The vertex of the graph of $f(x) = ax^2 + bx + c$, $a \neq 0$, may be found by completing the square. The vertex has coordinates $$\left(-\frac{b}{2a}, f\left(-\frac{b}{2a}\right)\right).$$	Graph $f(x) = x^2 + 4x + 3$. The vertex is $(-2, -1)$. Since $f(0) = 3$, the y-intercept is $(0, 3)$. The solutions of $x^2 + 4x + 3 = 0$ are -1 and -3, so the x-intercepts are $(-1, 0)$ and $(-3, 0)$.
To graph a quadratic function: Find the y-intercept by evaluating $f(0)$. Find any x-intercepts by solving $f(x) = 0$. Find the vertex either by using the formula or by completing the square. Find and plot any additional points as needed, using the symmetry about the axis. Verify that the graph opens upward (if $a > 0$) or opens downward (if $a < 0$).	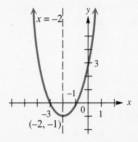
If the discriminant, $b^2 - 4ac$, is positive, the graph of $f(x) = ax^2 + bx + c$ has two x-intercepts; if zero, one x-intercept; if negative, no x-intercepts.	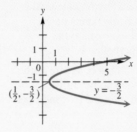
The graph of $x = ay^2 + by + c$ is a horizontal parabola, opening to the right if $a > 0$, or to the left if $a < 0$.	Graph $x = 2y^2 + 6y + 5$. The graph is shown above.

9.3   THE CIRCLE AND THE ELLIPSE

The circle with radius r and center at (h, k) has an equation of the form $$(x - h)^2 + (y - k)^2 = r^2.$$	The circle $(x + 2)^2 + (y - 3)^2 = 25$ has center $(-2, 3)$ and radius 5. 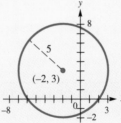
The ellipse whose x-intercepts are $(a, 0)$ and $(-a, 0)$ and whose y-intercepts are $(0, b)$ and $(0, -b)$ has an equation of the form $$\frac{x^2}{a^2} + \frac{y^2}{b^2} = 1.$$	Graph $\dfrac{x^2}{9} + \dfrac{y^2}{4} = 1$.

CONCEPTS	EXAMPLES
9.4 THE HYPERBOLA AND SQUARE ROOT FUNCTIONS	

A hyperbola with x-intercepts $(a, 0)$ and $(-a, 0)$ has an equation of the form $$\frac{x^2}{a^2} - \frac{y^2}{b^2} = 1$$ and a hyperbola with y-intercepts $(0, b)$ and $(0, -b)$ has an equation of the form $$\frac{y^2}{b^2} - \frac{x^2}{a^2} = 1.$$	Graph $\dfrac{x^2}{4} - \dfrac{y^2}{4} = 1.$ The graph has x-intercepts $(2, 0)$ and $(-2, 0)$. 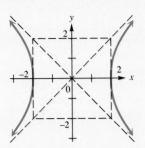
The extended diagonals of the fundamental rectangle with corners at the points (a, b), $(-a, b)$, $(-a, -b)$, and $(a, -b)$ are the asymptotes of these hyperbolas.	The fundamental rectangle has corners at $(2, 2)$, $(-2, 2)$, $(-2, -2)$, and $(2, -2)$.
To graph a square root function, square both sides so that the equation can be easily recognized. Then graph only the part indicated by the original equation.	Graph $y = -\sqrt{4 - x^2}.$ Square both sides and rearrange terms to get $$x^2 + y^2 = 4.$$ This equation has a circle as its graph. However, graph only the lower half of the circle, since the original equation indicates that y cannot be positive. 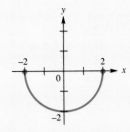

9.5 NONLINEAR SYSTEMS OF EQUATIONS	

Nonlinear systems can be solved by the substitution method, the elimination method, or a combination of the two.	Solve the system $$x^2 + 2xy - y^2 = 14$$ $$x^2 - y^2 = -16. \qquad (*)$$ Multiply equation $(*)$ by -1 and use elimination. $$\begin{aligned} x^2 + 2xy - y^2 &= 14 \\ -x^2 \qquad\quad + y^2 &= 16 \\ \hline 2xy \qquad\quad &= 30 \\ xy \qquad\quad &= 15 \end{aligned}$$

CONCEPTS	EXAMPLES
	Solve for y to obtain $y = 15/x$, and substitute into equation (*). $$x^2 - \left(\frac{15}{x}\right)^2 = -16$$ $$x^2 - \frac{225}{x^2} = -16.$$ $x^4 + 16x^2 - 225 = 0$ Multiply by x^2; add $16x^2$. $(x^2 - 9)(x^2 + 25) = 0$ Factor. $x = \pm 3$ $x = \pm 5i$ Zero-factor property. Find corresponding y values to get the solution set $$\{(3, 5), (-3, -5), (5i, -3i), (-5i, 3i)\}.$$

9.6   SECOND-DEGREE INEQUALITIES AND SYSTEMS OF INEQUALITIES

To graph a second-degree inequality, graph the corresponding equation as a boundary and use test points to determine which region(s) form the solution. Shade the appropriate region(s).	Graph $y \geq x^2 - 2x + 3$. 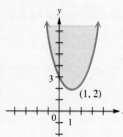
The solution set of a system of inequalities is the intersection of the solution sets of the individual inequalities.	Graph the solution set of the system $$3x - 5y > -15$$ $$x^2 + y^2 \leq 25.$$ 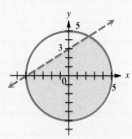

CHAPTER 9 REVIEW EXERCISES

[9.1–9.2] *Identify the vertex of the parabola.*

1. $y = 3x^2 - 2$

1. $(0, -2)$

2. $y = 6 - 2x^2$

2. $(0, 6)$

3. $f(x) = -(x - 1)^2$

3. $(1, 0)$

4. $f(x) = (x + 2)^2$

4. $(-2, 0)$

5. $y = (x - 3)^2 + 7$

5. $(3, 7)$

6. $y = \frac{4}{3}(x - 2)^2 + 1$

6. $(2, 1)$

7. $y = -3x^2 + 4x - 2$

7. $\left(\dfrac{2}{3}, -\dfrac{2}{3}\right)$

8. $x = 2(y + 3)^2 - 4$

8. $(-4, -3)$

9. $x = 2y^2 + 5y + 4$

9. $\left(\dfrac{7}{8}, -\dfrac{5}{4}\right)$

Graph the parabola.

10. $y = 4x^2 + 4x - 2$

10.

11. $f(x) = -2x^2 + 8x - 5$

11.

12. $x = -\dfrac{1}{2}y^2 + 6y - 14$

12.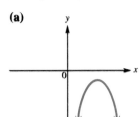

13. $x = 2y^2 + 8y + 3$

13.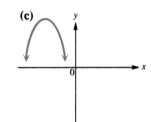

⊚ **14.** Explain how the discriminant can be used to determine the number of x-intercepts of the graph of a quadratic function.

⊚ **15.** Which one of the following would most closely resemble the graph of $f(x) = a(x - h)^2 + k$ if $a < 0$, $h > 0$, and $k < 0$?

(a)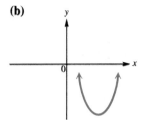

(b)

(c)

(d)

15. (a)

16. From 1982 to 1990, sales (in billions of dollars) of video games were approximated by the quadratic function

$$f(x) = .2x^2 - 1.6x + 3.3,$$

where $x = 0$ corresponds to 1982, $x = 1$ corresponds to 1983, and so on.
(a) In what year during that period were sales at a minimum?
(b) What were the minimum sales during that period?

16. (a) 1986   (b) $.1 billion

17. The height (in feet) of a projectile t seconds after being fired into the air is given by the function
$s(t) = -16t^2 + 160t$.
(a) Find the number of seconds required for the projectile to reach its maximum height.
(b) What is the maximum height?

17. (a) 5 seconds   (b) 400 feet

18. Find the length and the width of a rectangle having a perimeter of 600 meters if the area is to be a maximum.

18. length: 150 meters;   width: 150 meters

19. Find the two numbers whose sum is 40 and whose product is a maximum.

19. 20 and 20

The function $f(x) = 2x^2 - 7x - 4$ was graphed in the standard viewing window of a graphics calculator and the two x-intercepts were located. See the figures.

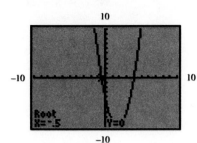

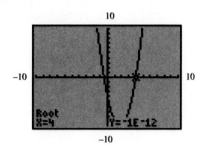

20. What is the solution set of $2x^2 - 7x - 4 = 0$?
20. $\{-.5, 4\}$

21. What is the solution set of $2x^2 - 7x - 4 > 0$?
21. $(-\infty, -.5) \cup (4, \infty)$

22. What is the solution set of $2x^2 - 7x - 4 \leq 0$?
22. $[-.5, 4]$

[9.3] *Write an equation of the circle described or graphed.*

23. center $(-2, 4)$, radius 3
23. $(x + 2)^2 + (y - 4)^2 = 9$

24.

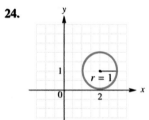

24. $(x - 2)^2 + (y - 1)^2 = 1$

Find the center and radius of the circle.

25. $x^2 + y^2 + 6x - 4y - 3 = 0$
25. center: $(-3, 2)$; radius: 4

26. $x^2 + y^2 - 8x - 2y + 13 = 0$
26. center: $(4, 1)$; radius: 2

27. $2x^2 + 2y^2 + 4x + 20y = -34$
27. center: $(-1, -5)$; radius: 3

28. $4x^2 + 4y^2 - 24x + 16y = 48$
28. center: $(3, -2)$; radius: 5

[9.3–9.4] *Graph the following.*

29. $\dfrac{x^2}{16} + \dfrac{y^2}{9} = 1$
29.

30. $\dfrac{x^2}{49} + \dfrac{y^2}{25} = 1$
30.

31. $\dfrac{x^2}{16} - \dfrac{y^2}{25} = 1$
31.

32. $\dfrac{y^2}{25} - \dfrac{x^2}{4} = 1$
32.

33. $x^2 + 9y^2 = 9$
33.

34. $f(x) = \sqrt{4 + x^2}$
34.

35. $(x - 2)^2 + (y + 3)^2 = 16$
35.

⊙ CONCEPTUAL   ✎ WRITING   ▲ CHALLENGING   ▦ SCIENTIFIC CALCULATOR   ▦ GRAPHICS CALCULATOR

Identify the graph of the equation as a parabola, circle, ellipse, or hyperbola.

36. $y = 2x^2 - 3$

36. parabola

37. $y^2 = 2x^2 - 8$

37. hyperbola

38. $y^2 = 8 - 2x^2$

38. ellipse

39. $x = y^2 + 4$

39. parabola

40. $x^2 + y^2 = 64$

40. circle

[9.5] *Solve the system.*

41. $2y = 3x - x^2$
$x + 2y = -12$

41. $\{(6, -9), (-2, -5)\}$

42. $y + 1 = x^2 + 2x$
$y + 2x = 4$

42. $\{(1, 2), (-5, 14)\}$

43. $x^2 + 3y^2 = 28$
$y - x = -2$

43. $\{(4, 2), (-1, -3)\}$

44. $xy = 8$
$x - 2y = 6$

44. $\{(-2, -4), (8, 1)\}$

45. $x^2 + y^2 = 6$
$x^2 - 2y^2 = -6$

45. $\{(-\sqrt{2}, 2), (-\sqrt{2}, -2),$
$(\sqrt{2}, -2), (\sqrt{2}, 2)\}$

46. $3x^2 - 2y^2 = 12$
$x^2 + 4y^2 = 18$

46. $\{(-\sqrt{6}, -\sqrt{3}) (-\sqrt{6}, \sqrt{3}),$
$(\sqrt{6}, -\sqrt{3}), (\sqrt{6}, \sqrt{3})\}$

⊙ **47.** How many solutions are possible for a system of two equations whose graphs are a circle and a line?

47. 0, 1, or 2

⊙ **48.** How many solutions are possible for a system of two equations whose graphs are a parabola and a hyperbola?

48. 0, 1, 2, 3, or 4

[9.6] *Graph the nonlinear inequality.*

49. $9x^2 \geq 16y^2 + 144$

49.

50. $4x^2 + y^2 \geq 16$

50.

51. $y < -(x + 2)^2 + 1$

51.

Graph the system of inequalities.

52. $2x + 5y \leq 10$
$3x - y \leq 6$

52.

53. $|x| \leq 2$
$|y| > 1$
$4x^2 + 9y^2 \leq 36$

53.

54. $9x^2 \leq 4y^2 + 36$
$x^2 + y^2 \leq 16$

54.

MIXED REVIEW EXERCISES

Graph.

55. $\dfrac{x^2}{64} + \dfrac{y^2}{25} = 1$

55.

56. $\dfrac{y^2}{4} - 1 = \dfrac{x^2}{9}$

56.

57. $x^2 + y^2 = 25$

57.

58. $y = 2(x - 2)^2 - 3$

58.

$y = 2(x - 2)^2 - 3$

59. $f(x) = -\sqrt{16 - x^2}$

59.

$f(x) = -\sqrt{16 - x^2}$

60. $f(x) = \sqrt{4 - x}$

60.

$f(x) = \sqrt{4 - x}$

61. $3x + 2y \geq 0$
$y \leq 4$
$x \leq 4$

61.

$3x + 2y \geq 0$
$y \leq 4$
$x \leq 4$

62. $4y > 3x - 12$
$x^2 < 16 - y^2$

62.

$4y > 3x - 12$
$x^2 < 16 - y^2$

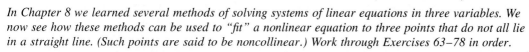

◆ **MATHEMATICAL CONNECTIONS** (Exercises 63–78) ◆

In Chapter 8 we learned several methods of solving systems of linear equations in three variables. We now see how these methods can be used to "fit" a nonlinear equation to three points that do not all lie in a straight line. (Such points are said to be noncollinear.) Work through Exercises 63–78 in order.

63. Let us find the equation of the quadratic function $f(x)$ such that $f(-3) = 16$, $f(-1) = -4$, and $f(2) = 11$. We know that $f(x) = ax^2 + bx + c$ is true for some values of a, b, and c. Since $f(-3) = 16$, write one equation in a, b, and c by using the fact that when $x = -3$, $f(x) = 16$.

63. $9a - 3b + c = 16$

64. Write a second equation in a, b, and c, using the fact that $f(-1) = -4$.

64. $a - b + c = -4$

65. Write a third equation in a, b, and c, using the fact that $f(2) = 11$.

65. $4a + 2b + c = 11$

66. Solve the system of the three equations found in Exercises 63–65, and give the function satisfying the conditions stated in Exercise 63.

66. $\{(3, 2, -5)\}$; $f(x) = 3x^2 + 2x - 5$

67. Use the methods of this chapter to find the vertex and intercepts of the parabola which is the graph of the function in Exercise 66. Sketch the graph.

67. vertex: $\left(-\dfrac{1}{3}, -\dfrac{16}{3}\right)$; x-intercepts: $\left(-\dfrac{5}{3}, 0\right)$, $\left(1, 0\right)$; y-intercept: $(0, -5)$

$f(x) = 3x^2 + 2x - 5$

68. Let us find the equation of the horizontal parabola $x = ay^2 + by + c$ that contains the points $(3, 1)$, $(-15, -2)$, and $(-5, 3)$. Let $x = 3$ and $y = 1$ to find one equation in a, b, and c.

68. $a + b + c = 3$

 CONCEPTUAL   ✎ WRITING   ▲ CHALLENGING   ▦ SCIENTIFIC CALCULATOR   ▦ GRAPHICS CALCULATOR

69. Let $x = -15$ and $y = -2$ to find a second equation in a, b, and c.

69. $4a - 2b + c = -15$

70. Let $x = -5$ and $y = 3$ to find a third equation in a, b, and c.

70. $9a + 3b + c = -5$

71. Solve the system of the three equations found in Exercises 68–70, and give the equation of the horizontal parabola that satisfies the conditions of Exercise 68.

71. $\{(-2, 4, 1)\}$; $x = -2y^2 + 4y + 1$

72. Graph the horizontal parabola found in Exercise 71.   **72.**

$x = -2y^2 + 4y + 1$

73. Given three noncollinear points in a plane, there is one and only one circle containing them. The equation of a circle can be written in the form $x^2 + y^2 + ax + by + c = 0$ for some values of a, b, and c. We will find the equation of the circle that contains the points $(2, 4)$, $(5, 1)$, and $(-1, 1)$. Determine one equation in a, b, and c by letting $x = 2$ and $y = 4$ in the general form given. Write it in the form with a, b, and c on the left and the constant on the right.

73. $2a + 4b + c = -20$

74. Repeat Exercise 73 for the point $(5, 1)$.

74. $5a + b + c = -26$

75. Repeat Exercise 73 for the point $(-1, 1)$.

75. $-a + b + c = -2$

76. Solve the system of equations formed by the equations found in Exercises 73–75, and give the equation of the circle that satisfies the conditions described in Exercise 73.

76. $\{(-4, -2, -4)\}$; $x^2 + y^2 - 4x - 2y - 4 = 0$

77. Use the methods of this chapter to find the center and the radius of the circle in Exercise 76.

77. center: $(2, 1)$; radius: 3

78. Graph the circle found in Exercise 76.   **78.**

$x^2 + y^2 - 4x - 2y - 4 = 0$

CHAPTER 9 TEST

1. Graph the quadratic function $f(x) = .5x^2 - 2$. Identify the vertex.   [9.1] **1.** vertex: $(0, -2)$

2. Identify the vertex of the graph of $f(x) = -x^2 + 4x - 1$. Sketch the graph.   [9.2] **2.** vertex: $(2, 3)$

3. The distance in feet that an object moves from a fixed point in t seconds is given by the function $f(t) = -2t^2 + 12t + 64$. Find the maximum distance it reaches and the time it takes to reach that distance.

3. 82 feet; 3 seconds

4. Give an equation of the circle shown in the figure.
[9.3] **4.** $(x + 4)^2 + (y - 4)^2 = 16$

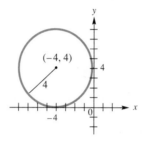

5. Find the coordinates of the center and the radius of the circle with equation $x^2 + y^2 + 8x - 2y = 8$.
5. center: $(-4, 1)$;   radius: 5

Identify the graph of the equation as one of the following: parabola, hyperbola, ellipse, or circle.

6. $3x^2 + 3y^2 = 27$ **7.** $9x^2 + 4y^2 = 36$ **8.** $9x^2 = 36 + 4y^2$ **9.** $x = 36 - 4y^2$

6. circle   **7.** ellipse   [9.4] **8. hyperbola**   [9.2] **9. parabola**

Sketch the graph of the equation.

10. $y = x^2 - 4x + 5$ **11.** $x = -(y - 2)^2 + 2$ **12.** $x^2 + y^2 = 64$

10.   **11.**   [9.3] **12.**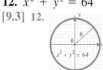

13. $4x^2 + 9y^2 = 36$ **14.** $16y^2 - 4x^2 = 64$ **15.** $f(x) = \sqrt{16 - x^2}$

13.   [9.4] **14.**   **15.**

16. If a parabola and an ellipse are graphed in the same plane, what are the possible number of points of intersection of the graphs?
[9.5] **16.** 0, 1, 2, 3, or 4

RESOURCES FOR TEST

ITM pp. 173–198   ISM pp. 788–793   SSG pp. 310–312   **TEST GENERATOR**
IAM pp. 49–50   SSM pp. 521–526   DOS   Windows   MAC

◉ CONCEPTUAL   ✎ WRITING   ▲ CHALLENGING   ▦ SCIENTIFIC CALCULATOR   ▦ GRAPHICS CALCULATOR

Solve the nonlinear system.

17. $2x - y = 9$
$xy = 5$

18. $x - 4 = 3y$
$x^2 + y^2 = 8$

19. $x^2 + y^2 = 25$
$x^2 - 2y^2 = 16$

17. $\left\{\left(-\dfrac{1}{2}, -10\right), (5, 1)\right\}$

18. $\left\{(-2, -2), \left(\dfrac{14}{5}, -\dfrac{2}{5}\right)\right\}$

19. $\{(-\sqrt{22}, -\sqrt{3}), (-\sqrt{22}, \sqrt{3}),$
$(\sqrt{22}, -\sqrt{3}), (\sqrt{22}, \sqrt{3})\}$

◉ **20.** Sketch a parabola and an ellipse on the same set of axes so that there are two points of intersection of the graphs.

20.

Graph the inequality or system of inequalities.

21. $y \le x^2 - 2$

22. $2x - 5y \ge 12$
$3x + 4y \le 12$

23. $x^2 + 25y^2 \le 25$
$x^2 + y^2 \le 9$

[9.6] 21.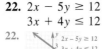

22.

23.

The number of chinook salmon that returned each year through the Lower Granite Dam in the state of Washington during the period from 1982 to 1987 can be approximated by the quadratic function

$$f(x) = -2.6x^2 + 11.7x + 22.5,$$

where $x = 0$ corresponds to 1982, $x = 1$ corresponds to 1983, and so on, and $f(x)$ is in thousands. (*Source: Idaho Department of Fish and Game*)

24. Based on this model, how many returned in 1983?
[9.2] 24. 31,600

25. In what year did the return reach a maximum? To the nearest thousand, how many salmon returned that year?
25. 1984; 36,000

CUMULATIVE REVIEW (Chapters 1–9)

1. Simplify $-10 + |-5| - |3| + 4$.
1. -4

Solve.

2. $4 - (2x + 3) + x = 5x - 3$
2. $\left\{\dfrac{2}{3}\right\}$

3. $-4k + 7 \ge 6k + 1$
3. $\left(-\infty, \dfrac{3}{5}\right]$

4. $|5m| - 6 = 14$
4. $\{-4, 4\}$

5. $|2p - 5| > 15$
5. $(-\infty, -5) \cup (10, \infty)$

Perform the operations.

6. $(5y - 3)^2$
6. $25y^2 - 30y + 9$

7. $(2r + 7)(6r - 1)$
7. $12r^2 + 40r - 7$

8. $(8x^4 - 4x^3 + 2x^2 + 13x + 8) \div (2x + 1)$
8. $4x^3 - 4x^2 + 3x + 5 + \dfrac{3}{2x + 1}$

Factor.

9. $12x^2 - 7x - 10$
9. $(3x + 2)(4x - 5)$

10. $2y^4 + 5y^2 - 3$
10. $(2y^2 - 1)(y^2 + 3)$

11. $z^4 - 1$
11. $(z^2 + 1)(z + 1)(z - 1)$

12. $a^3 - 27b^3$
12. $(a - 3b)(a^2 + 3ab + 9b^2)$

Simplify.

13. $\dfrac{5x - 15}{24} \cdot \dfrac{64}{3x - 9}$
13. $\dfrac{40}{9}$

14. $\dfrac{y^2 - 4}{y^2 - y - 6} \div \dfrac{y^2 - 2y}{y - 1}$
14. $\dfrac{y - 1}{y(y - 3)}$

15. $\dfrac{5}{c + 5} - \dfrac{2}{c + 3}$
15. $\dfrac{3c + 5}{(c + 5)(c + 3)}$

16. $\dfrac{p}{p^2 + p} + \dfrac{1}{p^2 + p}$
16. $\dfrac{1}{p}$

17. Kareem and Jamal want to clean an office they share. Kareem can do the job alone in 3 hours, while Jamal can do it alone in 2 hours. How long will it take them if they work together?
17. $\dfrac{6}{5}$ or $1\dfrac{1}{5}$ hours

Simplify.

18. $\left(\dfrac{4}{3}\right)^{-1}$
18. $\dfrac{3}{4}$

19. $\dfrac{(2a)^{-2}a^4}{a^{-3}}$
$\dfrac{a^5}{4}$

20. $4\sqrt[3]{16} - 2\sqrt[3]{54}$
20. $2\sqrt[3]{2}$

21. $\dfrac{3\sqrt{5x}}{\sqrt{2x}}, \quad x > 0$
21. $\dfrac{3\sqrt{10}}{2}$

22. $\dfrac{5 + 3i}{2 - i}$
22. $\dfrac{7}{5} + \dfrac{11}{5}i$

Solve for real values of the variable.

23. $2\sqrt{k} = \sqrt{5k + 3}$
23. $\emptyset$

24. $10q^2 + 13q = 3$
24. $\left\{\dfrac{1}{5}, -\dfrac{3}{2}\right\}$

25. $(4x - 1)^2 = 8$
25. $\left\{\dfrac{1 - 2\sqrt{2}}{4}, \dfrac{1 + 2\sqrt{2}}{4}\right\}$

26. $3k^2 - 3k - 2 = 0$
26. $\left\{\dfrac{3 - \sqrt{33}}{6}, \dfrac{3 + \sqrt{33}}{6}\right\}$

27. $2(x^2 - 3)^2 - 5(x^2 - 3) = 12$
27. $\left\{-\dfrac{\sqrt{6}}{2}, \dfrac{\sqrt{6}}{2}, -\sqrt{7}, \sqrt{7}\right\}$

28. $F = \dfrac{kwv^2}{r}$; for v
28. $v = \dfrac{\pm\sqrt{rFkw}}{kw}$

29. Find the slope of the line through $(2, 5)$ and $(-4, 1)$.
29. $\dfrac{2}{3}$

30. Find the equation of the line through $(-3, -2)$ and perpendicular to $2x - 3y = 7$.
30. $3x + 2y = -13$

31. $3x + y = 5$

32. $f(x) = -2(x - 1)^2 + 3$

33. $y = \sqrt{x - 2}$

31.

32.

33.

34. $\dfrac{x^2}{4} - \dfrac{y^2}{16} = 1$

35. $\dfrac{x^2}{25} + \dfrac{y^2}{16} \le 1$

34.

35.

Solve the system.

36. $3x - y = 12$
$2x + 3y = -3$
36. $\{(3, -3)\}$

37. $x + y - 2z = 9$
$2x + y + z = 7$
$3x - y - z = 13$
37. $\{(4, 1, -2)\}$

38. $xy = -5$
$2x + y = 3$
38. $\left\{ (-1, 5), \left(\dfrac{5}{2}, -2 \right) \right\}$

Solve the problem.

39. Al and Bev traveled from their apartment to a picnic 20 miles away. Al traveled on his bike; Bev, who left later, took her car. Al's average speed was half of Bev's average speed. The trip took Al one half hour longer than Bev. What was Bev's average speed?
39. **40 miles per hour**

40. A cash drawer contains only fives and twenties. There are eight more fives than twenties. The total value of the money is $215. How many of each type of bill is in the drawer?
40. **15 fives and 7 twenties**

INVERSE, EXPONENTIAL, AND LOGARITHMIC FUNCTIONS

CONNECTIONS

Exponential and logarithmic functions, the subject of this chapter, are used as models to describe growth and decay. Growth or decay that change at a faster and faster pace are described by exponential functions. Some examples of exponential growth or decay are compound interest paid on deposits or for loans, uninhibited population growth, and the decay of radioactive materials. Logarithmic functions describe growth or decay that is changing at a decreasing rate. Some populations grow in this way due to inhibiting factors, as do sales for some types of businesses. The percent of moisture that falls as snow in the Sierra Nevada Mountains of California increases logarithmically with the height of the location.

◆ For a growth function, as x increases, y also increases. If y is a decay function, as x increases, y decreases.

FOR DISCUSSION OR WRITING

As you work through Section 10.2, consider these questions: If y is a growth function of x, what happens to y as x increases? What if y is a decay function of x?

587

In this chapter we study two important types of functions, *exponential* and *logarithmic* functions. These functions are related in a special way. They are *inverses* of one another. We begin by discussing inverse functions in general.

10.1 INVERSE FUNCTIONS

FOR EXTRA HELP	OBJECTIVES
SSG pp. 313–317 **SSM** pp. 534–538	**1** ▶ Decide whether a function is one-to-one and, if it is, find its inverse.
Video 12	**2** ▶ Use the horizontal line test to determine whether a function is one-to-one. **3** ▶ Find the equation of the inverse of a function.
Tutorial IBM MAC	**4** ▶ Graph the inverse f^{-1} from the graph of f. **5** ▶ Use a graphics calculator to graph inverse functions.

1 ▶ Decide whether a function is one-to-one and, if it is, find its inverse.

▶   Suppose that G is the function $\{(-2, 2), (-1, 1), (0, 0), (1, 3), (2, 5)\}$. Another set of ordered pairs can be formed from G by exchanging the x- and y-values of each pair in G. Call this set F, with

$$F = \{(2, -2), (1, -1), (0, 0), (3, 1), (5, 2)\}.$$

To show that these two sets are related, F is called the *inverse* of G. For a function f to have an inverse, f must be *one-to-one*. In a **one-to-one function** each x-value corresponds to only one y-value and each y-value corresponds to just one x-value.

The function shown in Figure 1(a) is not one-to-one because the y-value 7 corresponds to *two* x-values, 2 and 3. That is, the ordered pairs $(2, 7)$ and $(3, 7)$ both appear in the function. The function in Figure 1(b) is one-to-one.

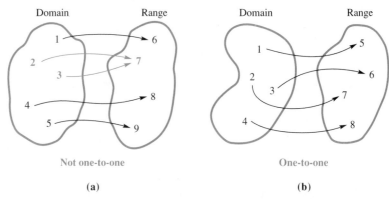

FIGURE 1

The *inverse* of any one-to-one function f is found by exchanging the components of the ordered pairs of f. The inverse of f is written f^{-1}. Read f^{-1} as "the inverse of f" or "f-inverse."

CAUTION   The symbol $f^{-1}(x)$ does not represent $\dfrac{1}{f(x)}$.

The definition of the inverse of a function follows.

Inverse of a Function

The **inverse** of a one-to-one function, f, written f^{-1}, is the set of all ordered pairs of the form (y, x) where (x, y) belongs to f. Since the inverse is formed by interchanging x and y, the domain of f becomes the range of f^{-1} and the range of f becomes the domain of f^{-1}.

E X A M P L E 1

Deciding Whether a Function Is One-To-One

Decide whether each function is one-to-one. If it is, find the inverse function.

(a) $F = \{(-2, 1), (-1, 0), (0, 1), (1, 2), (2, 2)\}$

Each x-value in F corresponds to just one y-value. However, the y-value 2 corresponds to two x-values, 1 and 2. Also, the y-value 1 corresponds to both -2 and 0. Because some y-values corresponds to more than one x-value, F is not one-to-one.

(b) $G = \{(3, 1), (0, 2), (2, 3), (4, 0)\}$

Every x-value in G corresponds to only one y-value, and every y-value corresponds to only one x-value, so G is a one-to-one function. The inverse function is found by exchanging the numbers in each ordered pair.

$$G^{-1} = \{(1, 3), (2, 0), (3, 2), (0, 4)\}$$

Notice how the domain and range of G become the range and domain, respectively, of G^{-1}.

(c) *Minutes Needed to Burn 100 Calories by Exercising*

vacuuming	16
walking	27
jogging	13
running	6
skiing	13
bicycling	16

Let f be the function defined by the correspondence in the table. Then f is not one-to-one because 16 minutes corresponds to vacuuming and bicycling. Also, 13 minutes corresponds to jogging and skiing. ■

2 ▶ Use the horizontal line test to determine whether a function is one-to-one.

▶ It may be difficult to decide whether a function is one-to-one just by looking at the equation that defines the function. However, by graphing the function and observing the graph, we can use the following *horizontal line test* to tell whether it is one-to-one.

Horizontal Line Test

A function is one-to-one if every horizontal line intersects the graph of the function at most once.

The horizontal line test follows from the definition of a one-to-one function. Any two points that lie on the same horizontal line have the same y-coordinate. No two ordered pairs that belong to a one-to-one function may have the same y-coordinate, and therefore no horizontal line will intersect the graph of a one-to-one function more than once.

EXAMPLE 2

Using the Horizontal Line Test

Use the horizontal line test to determine whether the graphs in Figures 2 and 3 are graphs of one-to-one functions.

(a)

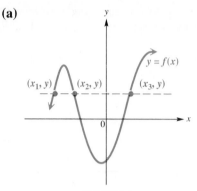

FIGURE 2

Because the horizontal line shown in Figure 2 intersects the graph in more than one point (actually three points in this case), the function is not one-to-one.

(b)

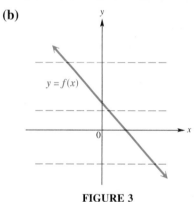

FIGURE 3

Every horizontal line will intersect the graph in Figure 3 in exactly one point. This function is one-to-one. ∎

3 ▶ Find the equation of the inverse of a function.

▶ By definition, the inverse of a function is found by exchanging the x- and y-values of the ordered pairs of the function. The equation of the inverse of a function defined by $y = f(x)$ is found in the same way.

Finding the Equation of the Inverse of $y = f(x)$

For a one-to-one function f defined by an equation $y = f(x)$, find the defining equation of the inverse as follows.

Step 1 Exchange x and y.
Step 2 Solve for y.
Step 3 Replace y with $f^{-1}(x)$.

This procedure is illustrated in the following example.

EXAMPLE 3
Finding the Equation of the Inverse

Chalkboard Exercise

Decide whether the equation defines a one-to-one function. If so, find the equation of the inverse.

$$f(x) = 3x - 4$$

Answer: Yes, it is one-to-one.

$$f^{-1}(x) = \frac{x + 4}{3}$$

▶ **TEACHING TIP**

For $f(x) = 2x + 5$, find

$$f(3) = 2 \cdot 3 + 5 = 11.$$

Now, for $f^{-1}(x) = \frac{x - 5}{2}$ find

$$f^{-1}(11) = \frac{11 - 5}{2} = 3.$$

This shows numerically that when the original y-value is evaluated in $f^{-1}(x)$, we obtain the original x-value. ◀

Decide whether each of the following defines a one-to-one function. If so, find the equation of the inverse.

(a) $f(x) = 2x + 5$

By the definition, this is a one-to-one function. Use the steps given above to find the inverse. Let $y = f(x)$ so that

$$y = 2x + 5$$
$$x = 2y + 5 \qquad \text{Exchange } x \text{ and } y.$$
$$2y = x - 5 \qquad \text{Solve for } y.$$
$$y = \frac{x - 5}{2}.$$

From the last equation,

$$f^{-1}(x) = \frac{x - 5}{2},$$

which is a linear function. In the function $y = 2x + 5$, the value of y is found by starting with a value of x, multiplying by 2, and adding 5. The equation for the inverse has us *subtract* 5, and then *divide* by 2. This shows how an inverse is used to "undo" what a function does to the variable x.

(b) $f(x) = (x - 2)^3$

Because of the cube, this is a one-to-one function. Find the inverse by replacing $f(x)$ with y and then exchanging x and y.

$$y = (x - 2)^3$$
$$x = (y - 2)^3$$

Take the cube root on each side to solve for y.

$$\sqrt[3]{x} = \sqrt[3]{(y - 2)^3}$$
$$\sqrt[3]{x} = y - 2$$
$$\sqrt[3]{x} + 2 = y$$
$$f^{-1}(x) = \sqrt[3]{x} + 2 \qquad \text{Replace } y \text{ with } f^{-1}(x).$$

(c) $y = x^2 + 2$

Both $x = 3$ and $x = -3$ correspond to $y = 11$. Because of the x^2 term, there are many pairs of x-values that correspond to the same y-value. This means that the function defined by $y = x^2 + 2$ is not one-to-one.

If this were not noticed, following the steps given above for finding the equation of an inverse leads to

$$y = x^2 + 2$$
$$x = y^2 + 2 \qquad \text{Exchange } x \text{ and } y.$$
$$x - 2 = y^2$$
$$\pm\sqrt{x - 2} = y. \qquad \text{Square-root property}$$

The last step shows that there are two y-values for each choice of $x > 2$, so the given function is not one-to-one and cannot have an inverse. ■

4 ▶ Graph the inverse f^{-1} from the graph of f.

▶ Suppose the point (a, b) shown in Figure 4 belongs to a one-to-one function f. Then the point (b, a) would belong to f^{-1}. The line segment connecting (a, b) and (b, a) is perpendicular to and cut in half by the line $y = x$. The points (a, b) and (b, a) are "mirror images" of each other with respect to $y = x$. For this reason the graph of f^{-1} can be found from the graph of f by locating the mirror image of each point of f with respect to $y = x$.

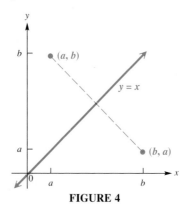

FIGURE 4

E X A M P L E 4

Graphing Inverses

Graph the inverses of the functions shown in Figure 5.

In Figure 5 the graphs of two functions are shown in blue and their inverses are shown in red. In each case, the graph of f^{-1} is symmetric to the graph of f with respect to the line $y = x$. ■

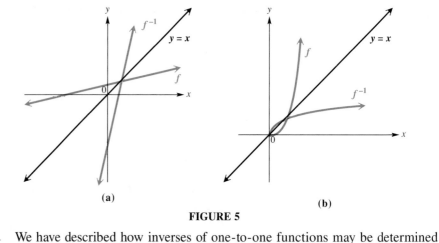

(a)   **(b)**

FIGURE 5

5 ▶ Use a graphics calculator to graph inverse functions.

▶ We have described how inverses of one-to-one functions may be determined algebraically. We also explained how the graph of a one-to-one function f compares to the graph of its inverse f^{-1}: it is a reflection of the graph of f^{-1} across the line $y = x$. In Example 3 we showed that the inverse of the one-to-one function $f(x) = 2x + 5$ is given by $f^{-1}(x) = \dfrac{x - 5}{2}$. If we use a "square" viewing window and graph $y_1 = f(x) = 2x + 5$, $y_2 = f^{-1}(x) = \dfrac{x - 5}{2}$, and $y_3 = x$, we can see how this reflection appears on the screen. See Figure 6.

▶ TEACHING TIP

Point out that the graphs for $f(x)$ and $f^{-1}(x)$ will intersect along the line $y = x$ at the x-value for which $f(x) = f^{-1}(x)$. ◀

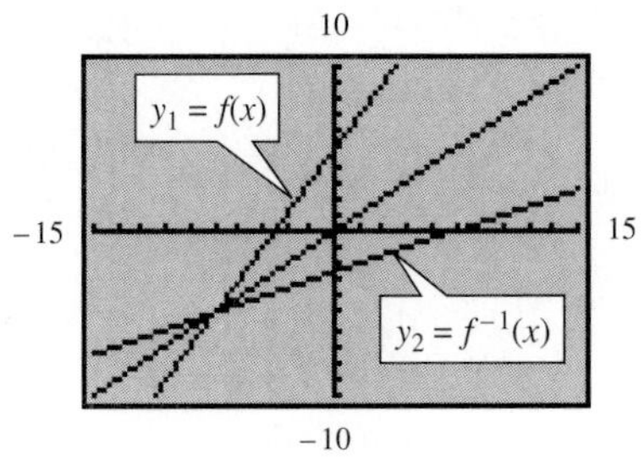

FIGURE 6

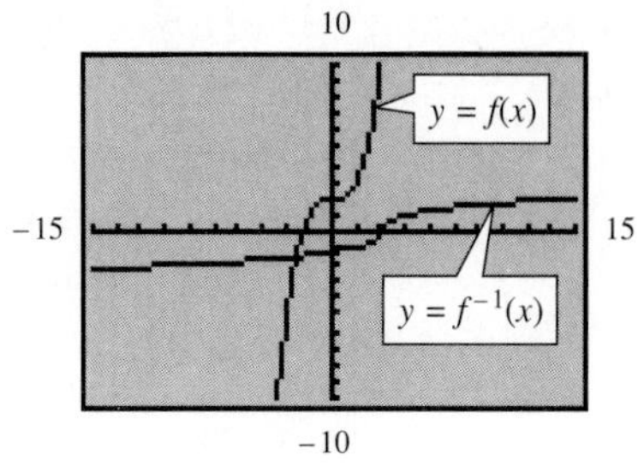

FIGURE 7

Some graphics calculators have the capability to "draw" the inverse of a function. Figure 7 shows the graphs of $f(x) = x^3 + 2$ and its inverse in a square viewing window.

10.1 EXERCISES

◎*Decide whether the statement is true or false.*

1. If a function is made up of ordered pairs in such a way that the same y-value appears in correspondence with two different x-values, then the function is not one-to-one.

2. The function $f(x) = x^2$ is a one-to-one function.

3. The function $f(x) = x^3$ has an inverse and it is $f^{-1}(x) = \sqrt[3]{x}$.

4. A *linear* function, defined by $f(x) = mx + b$, has an inverse provided $m \neq 0$.

◎ **5.** Suppose that you consider the set of ordered pairs (x, y) such that x represents a
✍ person in your mathematics class and y represents that person's mother. Explain how this function might not be a one-to-one function.

◎ **6.** The road mileage between Denver, Colorado and several selected U.S. cities is shown
✍ in the table below.

City	Distance to Denver in Miles
Atlanta	1398
Dallas	781
Indianapolis	1058
Kansas City, MO	600
Los Angeles	1059
San Francisco	1235

If we consider this as a function that pairs a city with a distance, is it a one-to-one function? How could we change the answer to this question by adding 1 mile to one of the distances shown?

If the function is one-to-one, find its inverse. See Examples 1–3.

7. $\{(3, 6), (2, 10), (5, 12)\}$

8. $\left\{(-1, 3), (0\ 5), (5, 0), \left(7, -\dfrac{1}{2}\right)\right\}$

9. $\{(-1, 3), (2, 7), (4, 3), (5, 8)\}$

10. $\{(-8, 6), (-4, 3), (0, 6), (5, 10)\}$

1. true   2. false   3. true
4. true
7. $\{(6, 3), (10, 2), (12, 5)\}$
8. $\{(3, -1), (5, 0), (0, 5),$
$\left(-\dfrac{1}{2}, 7\right)\}$ 9. not a one-
to-one function
10. not a one-to-one function

11. $f^{-1}(x) = \dfrac{x-4}{2}$

12. $f^{-1}(x) = \dfrac{x-1}{3}$

13. $g^{-1}(x) = x^2 + 3, x \ge 0$
14. $g^{-1}(x) = x^2 - 2, x \ge 0$
15. not one-to-one
16. not one-to-one
17. $f^{-1}(x) = \sqrt[3]{x+4}$
18. $f^{-1}(x) = \sqrt[3]{x+3}$
19. (a)   20. (a) 3   (b) −2
(c) −19   (d) a 21. (a) 8
(b) 3   22. (a) 16   (b) 4
23. (a) 1   (b) 0

24. (a) $\dfrac{1}{4}$ (b) −2

25. (a) one-to-one
(b)

26. (a) one-to-one
(b)

27. (a) not one-to-one
28. (a) not one-to-one
29. (a) one-to-one
(b)

30. (a) one-to-one
(b)

31.   32.

33.   34.

11. $f(x) = 2x + 4$
▲ **13.** $g(x) = \sqrt{x-3}, \quad x \ge 3$
15. $f(x) = 3x^2 + 2$
17. $f(x) = x^3 - 4$

12. $f(x) = 3x + 1$
▲ **14.** $g(x) = \sqrt{x+2}, \quad x \ge -2$
16. $f(x) = -4x^2 - 1$
18. $f(x) = x^3 - 3$

◉ **19.** If a function f is one-to-one and the point (p, q) lies on the graph of f, then which one of the following *must* lie on the graph of f^{-1}?
    (a) (q, p) **(b)** $(-p, q)$ **(c)** $(-q, -p)$ **(d)** $(p, -q)$

◉ **20.** Suppose that f is a one-to-one function.
    (a) If $f(3) = 5$, then $f^{-1}(5) =$ _____ .   **(b)** If $f(-2) = 4$, then $f^{-1}(4) =$ _____ .
    (c) If $f(-19) = 3$, then $f^{-1}(3) =$ _____ .   **(d)** If $f(a) = b$, then $f^{-1}(b) =$ _____ .

▲ *Let $f(x) = 2^x$. We will see in the next section that this function is one-to-one. Find each of the following, always working part (a) before part (b).*

21. (a) $f(3)$ **(b)** $f^{-1}(8)$

23. (a) $f(0)$ **(b)** $f^{-1}(1)$

22. (a) $f(4)$ **(b)** $f^{-1}(16)$

24. (a) $f(-2)$ **(b)** $f^{-1}\!\left(\dfrac{1}{4}\right)$

*The graphs of some functions are given in Exercises 25–30. (**a**) Use the horizontal line test to determine whether the function is one-to-one.   (**b**) If the function is one-to-one, graph the inverse of the function. (Remember that if f is one-to-one and $f(a) = b$, then $f^{-1}(b) = a$.) See Example 4.*

25.
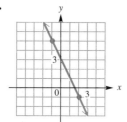

26.
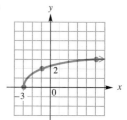

27.
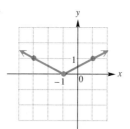

28.
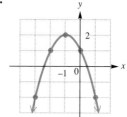

29.
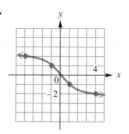

30.
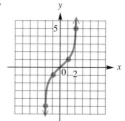

Each function in Exercises 31–38 is a one-to-one function. Graph the function as a solid line (or curve) and then graph its inverse on the same set of axes as a dashed line (or curve). In Exercises 35–38 you are given a table to complete so that graphing the function will be a bit easier. See Example 4.

31. $f(x) = 2x - 1$

32. $f(x) = 2x + 3$

33. $g(x) = -4x$

34. $g(x) = -2x$

35. $f(x) = \sqrt{x}, \quad x \geq 0$

x	$f(x)$
0	
1	
4	

36. $f(x) = -\sqrt{x}, \quad x \geq 0$

x	$f(x)$
0	
1	
4	

37. $y = x^3 - 2$

x	y
-1	
0	
1	
2	

38. $y = x^3 + 3$

x	y
-2	
-1	
0	
1	

 Each function is one-to-one. Find the inverse algebraically, and then graph both the function and its inverse on the same square graphics calculator screen.

39. $f(x) = 2x - 7$

40. $f(x) = -3x + 2$

41. $f(x) = x^3 + 5$

42. $f(x) = \sqrt[3]{x + 2}$

Some graphics calculators have the capability to draw the "inverse" of a function even if the function is not one-to-one; therefore, the inverse is not technically a function, but is a relation. For example, the graphs of $y = x^2$ and $x = y^2$ are shown in the accompanying square window.

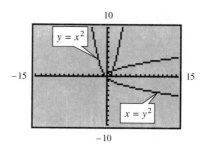

 Read your instruction manual to see if your model has this capability. Draw both y_1 and its inverse in the same square window.

43. $y_1 = x^2 + 3x + 4$

44. $y_1 = x^3 - 9x$

◉ **45.** Explain why the "inverse" of the function in Exercise 43 does not actually satisfy the definition of inverse as given in this section.

◉ **46.** Which one of the following graphs does not have an inverse as defined in this section?

(a)

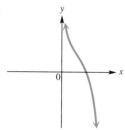

(b)

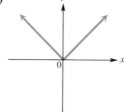

35.

x	$f(x)$
0	0
1	1
4	2

36.

x	$f(x)$
0	0
1	-1
4	-2

37.

x	y
-1	-3
0	-2
1	-1
2	6

38.

x	y
-2	-5
-1	2
0	3
1	4

39. $f^{-1}(x) = \dfrac{x + 7}{2}$

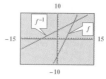

40. $f^{-1}(x) = \dfrac{x - 2}{-3}$

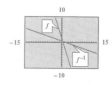

41. $f^{-1}(x) = \sqrt[3]{x} - 5$

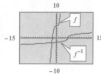

42. $f^{-1}(x) = x^3 - 2$

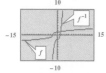

43.

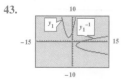

44.

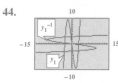

46. (b)

47. $f^{-1}(x) = \dfrac{x + 5}{4}$

48. My graphics calculator is the greatest thing since sliced bread.

50. Answers will vary. For example, Jane Doe is 1004 5 2748 129 68 3379 129.

51. 64   52. 1

53. 2   54. $\dfrac{1}{2}$ 55. 44.02

56. 10.27

━━━━━◆ **MATHEMATICAL CONNECTIONS** (Exercises 47–50) ◇━━━━━

Inverse functions are used by government agencies and other businesses to send and receive coded information. The functions they use are usually very complicated. A simple example might use the function $f(x) = 2x + 5$. (Note that it is one-to-one.) Suppose that each letter of the alphabet is assigned a numerical value according to its position, as follows:

A	1	G	7	L	12	Q	17	V	22
B	2	H	8	M	13	R	18	W	23
C	3	I	9	N	14	S	19	X	24
D	4	J	10	O	15	T	20	Y	25
E	5	K	11	P	16	U	21	Z	26
F	6								

Using the function, the word ALGEBRA would be encoded as

$$7 \quad 29 \quad 19 \quad 15 \quad 9 \quad 41 \quad 7,$$

because $f(A) = f(1) = 2(1) + 5 = 7, f(L) = f(12) = 2(12) + 5 = 29$, and so on. The message would then be decoded by using the inverse of f, $f^{-1}(x) = \dfrac{x - 5}{2}$. For example,

$$f^{-1}(7) = \frac{7 - 5}{2} = 1 = A, f^{-1}(29) = \frac{29 - 5}{2} = 12 = L, \text{ and so on.}$$

Work Exercises 47–50 in order.

47. Suppose that you are an agent for a detective agency and you know that today's function for your code is $f(x) = 4x - 5$. Find the rule for f^{-1} algebraically.

48. You receive the following coded message today.

47 95 23 67 −1 59 27 31   7 71   7 −1 43   7 79 43 −1   75 55 67

31 71 75 27   15 23 67 15   −1 75 15 71   75 75 27 31   51

23 71 31 51   7 15 71 43   31   7 15 11   3 67 15 −1   11

Use the letter/number assignment described earlier to decode the message.

◉ **49.** Why is a one-to-one function essential in this coding/decoding process?

50. Use the function $f(x) = x^3 + 4$ to encode your name, using the letter/number assignment described earlier.

━━━━━━◆━━━━━━

REVIEW EXERCISES

If $f(x) = 4^x$, find the value indicated. In Exercises 55 and 56, use a calculator, and give answers to the nearest hundredth. See Section 7.5.

51. $f(3)$ **52.** $f(0)$ **53.** $f\left(\dfrac{1}{2}\right)$

54. $f\left(-\dfrac{1}{2}\right)$ ▦ **55.** $f(2.73)$ ▦ **56.** $f(1.68)$

10.2 EXPONENTIAL FUNCTIONS

FOR EXTRA HELP	OBJECTIVES
SSG pp. 318–322 **SSM** pp. 538–542 ▭ **Video** 12 ▭ **Tutorial** IBM MAC	1 ▶ Identify exponential functions. 2 ▶ Graph exponential functions. 3 ▶ Solve exponential equations of the form $a^x = a^k$ for x. 4 ▶ Use exponential functions in applications.

A calculator with the following keys will be very helpful in this chapter.

We will explain how these keys are used at appropriate places in the chapter.

1 ▶ Identify exponential functions.

▶ In Section 5.1, we showed how to evaluate 2^x for rational values of x. For example,

$$2^3 = 8,$$

$$2^{-1} = \frac{1}{2},$$

$$2^{1/2} = \sqrt{2}$$

$$2^{3/4} = \sqrt[4]{2^3} = \sqrt[4]{8}.$$

In more advanced courses it is shown that 2^x exists for all real number values of x, both rational and irrational. (Later in the chapter, methods are given for approximating the value of 2^x for irrational x.) The following definition of an exponential function assumes that a^x exists for all real numbers x.

Exponential Function For $a > 0$ and $a \neq 1$, and all real numbers x,

$$F(x) = a^x$$

is an **exponential function.**

NOTE The two restrictions on a in the definition of exponential function are important. The restriction that a must be positive is necessary so that the function can be defined for all real numbers x. For example, letting a be negative ($a = -2$, for instance) and letting $x = 1/2$ would give the expression $(-2)^{1/2}$, which is not real. The other restriction, $a \neq 1$, is necessary because 1 raised to any power is equal to 1, and the function would then be the linear function $F(x) = 1$.

2 ▶ Graph exponential functions.

▶ We can graph exponential functions by finding several ordered pairs that belong to the function. Plotting these points and connecting them with a smooth curve gives the graph.

CAUTION Be sure to plot enough points to see how rapidly the graph rises.

E X A M P L E   1

Graphing an Exponential Function

■ Graph the exponential function $f(x) = 2^x$.

Choose some values of x and find the corresponding values of $f(x)$.

x	-3	-2	-1	0	1	2	3	4
$f(x) = 2^x$	$\dfrac{1}{8}$	$\dfrac{1}{4}$	$\dfrac{1}{2}$	1	2	4	8	16

Plotting these points and drawing a smooth curve through them gives the graph shown in Figure 8. ■

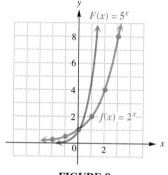

FIGURE 8

The graph in Figure 8 is typical of the graphs of exponential functions of the form $F(x) = a^x$, where $a > 1$. The larger the value of a, the faster the graph rises. To see this, compare the graph of $F(x) = 5^x$ with the graph of $f(x) = 2^x$ in Figure 8.

By the vertical line test, the graphs in Figure 8 represent functions. As these graphs suggest, the domain of an exponential function includes all real numbers. Since y is always positive, the range is $(0, \infty)$.

E X A M P L E   2

Graphing an Exponential Function with $a < 1$

■ Graph $g(x) = \left(\dfrac{1}{2}\right)^x$.

Again, find some points on the graph.

x	-3	-2	-1	0	1	2	3	4
$g(x) = \left(\dfrac{1}{2}\right)^x$	8	4	2	1	$\dfrac{1}{2}$	$\dfrac{1}{4}$	$\dfrac{1}{8}$	$\dfrac{1}{16}$

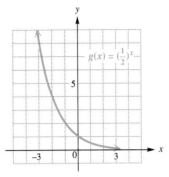

FIGURE 9

The graph, shown in Figure 9, is very similar to that of $f(x) = 2^x$, shown in Figure 8, except that here as x gets larger, y *decreases.* This graph is typical of the graph of a function of the form $F(x) = a^x$, where $0 < a < 1$. ■

Based on Examples 1 and 2, we can make the following generalizations about the graphs of exponential functions of the form $F(x) = a^x$.

Graph of $F(x) = a^x$

1. The graph will always contain the point $(0, 1)$.
2. When $a > 1$, the graph will *rise* from left to right. When $0 < a < 1$, the graph will *fall* from left to right. In both cases, the graph goes from the second quadrant to the first.
3. The graph will approach the x-axis, but never touch it. (Such a line is called an an **asymptote.**)
4. The domain is $(-\infty, \infty)$ and the range is $(0, \infty)$.

To graph a more complicated exponential function, we plot carefully selected points, as shown in the next example.

E X A M P L E   3
Graphing a More Complicated Exponential Function

Graph $y = 3^{2x-4}$.

Find some ordered pairs. For example, if $x = 0$,

$$y = 3^{2(0)-4}$$
$$= 3^{-4}$$
$$= \frac{1}{81}.$$

Also, for $x = 2$,

$$y = 3^{2(2)-4}$$
$$= 3^0$$
$$= 1.$$

Chalkboard Exercise

Graph $y = 2^{4x-3}$.

Answer:

These ordered pairs, $(0, 1/81)$ and $(2, 1)$, along with the ordered pairs $(1, 1/9)$ and $(3, 9)$, lead to the graph shown in Figure 10. The graph is similar to the graph of $f(x) = 2^x$ except that it is shifted to the right and rises more rapidly. ■

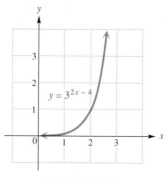

$y = 3^{2x-4}$

FIGURE 10

3 ▶ Solve exponential equations of the form $a^x = a^k$ for x.

▶ Until now in this book, all equations that we have solved have had the variable as a base; all exponents have been constants. An **exponential equation** is an equation that has a variable in an exponent, such as

$$9^x = 27.$$

▶ **TEACHING TIP**

It helps students to remember this property if it is stated in words: "If bases are equal, then exponents are equal." ◀

Because the exponential function $F(x) = a^x$ is a one-to-one function, the following property can be used to solve many exponential equations.

For $a > 0$ and $a \neq 1$, if $a^x = a^y$ then $x = y$.

This property would not necessarily be true if $a = 1$.

To solve an exponential equation using this property, go through the following steps.

Solving Exponential Equations

Step 1 **Each side must have the same base.** If the two sides of the equation do not have the same base, express each as a power of the same base.

Step 2 **Simplify exponents.** If necessary, use the rules of exponents to simplify the exponents.

Step 3 **Set exponents equal.** Use the property to set the exponents equal to each other.

Step 4 **Solve.** Solve the equation obtained in Step 3.

NOTE The steps above cannot be applied to an exponential equation like

$$3^x = 12,$$

since Step 1 cannot easily be done. A method for solving such equations is given in Section 10.6.

E X A M P L E   4
Solving an Exponential
Equation

Solve the equation $9^x = 27$.

We can use the property given above if both sides are changed to the same base. Since $9 = 3^2$ and $27 = 3^3$, the equation $9^x = 27$ is solved as follows.

$$(3^2)^x = 3^3 \qquad \text{Substitute.} \qquad \textbf{(Step 1)}$$

$$3^{2x} = 3^3 \qquad \text{Power rule for exponents} \qquad \textbf{(Step 2)}$$

$$2x = 3 \qquad \text{If } a^x = a^y, \text{ then } x = y. \qquad \textbf{(Step 3)}$$

$$x = \frac{3}{2} \qquad \qquad \textbf{(Step 4)}$$

Check that the solution set is $\{3/2\}$ by substituting 3/2 for x in the given equation. ▪

Chalkboard Exercise

Solve $25^x = 125$.

Answer: $\left\{\dfrac{3}{2}\right\}$

---

4 ▶ Use exponential functions in applications.

▶ Exponential functions frequently occur in applications describing growth or decay of some quantity. In particular, they are used to describe the growth and decay of populations.

E X A M P L E   5
Solving an Application of
an Exponential Function

The air pollution, y, in appropriate units, in a large industrial city has been growing according to the equation

$$y = 1000(2)^{.3x},$$

where x is time in years from 1990. That is, $x = 0$ represents 1990, $x = 2$ represents 1992, and so on.

(a) Find the amount of pollution in 1990.

Let $x = 0$, and solve for y.

$$y = 1000(2)^{.3x}$$
$$= 1000(2)^{(.3)(0)} \qquad \text{Let } x = 0.$$
$$= 1000(2)^0$$
$$= 1000(1)$$
$$= 1000$$

The pollution in 1990 was 1000 units.

(b) Assuming that air pollution continues to increase at the same rate, estimate the pollution in 2010.

To represent 2010, let $x = 20$, since $2010 - 1990 = 20$.

$$y = 1000(2)^{.3x}$$
$$= 1000(2)^{(.3)(20)} \qquad \text{Let } x = 20.$$
$$= 1000(2)^6$$
$$= 1000(64)$$
$$= 64,000$$

In 2010 the pollution will be about 64,000 units.

Chalkboard Exercise

The amount of a radioactive substance, in grams, present at time t is $A = 100(3)^{-.5t}$, where t is in months. Find the amount present at

(a) $t = 0$;
(b) $t = 2$;
(c) $t = 10$.
(d) Graph the equation.

Answers: (a) 100 grams;

(b) $33\dfrac{1}{3}$ grams; (c) .41 grams;

(d)

(c) Graph the equation $y = 1000(2)^{.3x}$.

The scale on the y-axis must be quite large to allow for the very large y-values. A calculator can be used to find a few more ordered pairs. (The use of the exponential key was discussed in Section 5.2.) The graph is shown in Figure 11. ∎

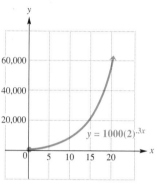

FIGURE 11

◈ about 341 parts per million in 1995; about 343 parts per million in 2000

◈ **CONNECTIONS** ◈

One result of rapidly increasing world population is an increase of carbon dioxide in the air, which scientists believe may be contributing to global warming. Both population and carbon dioxide in the air are increasing exponentially. This means that the growth rate is continually increasing. The graph in the figure shows the concentration of carbon dioxide (in parts per million) in the air. The data fit the function $y = 278(1.00084)^x$, where x is the number of years since 1750. The effects on the environment result not only from the increase in population, but also from increasing consumption as poorer nations strive to catch up to wealthier nations.

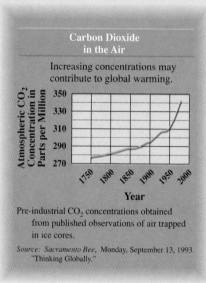

FOR DISCUSSION OR WRITING
Use the exponential function given above to predict the carbon dioxide concentration in the years 1995 and 2000. Most population growth is occuring in those countries that are developing. Discuss the implications of this for environmental changes.

10.2 EXERCISES

⦿ *Fill in the blank with the correct response.*

1. For an exponential function $f(x) = a^x$, if $a > 1$, the graph _____ from left to right. If $0 < a < 1$, the graph _____ from left to right.
   (rises/falls)   (rises/falls)

2. The *y*-intercept of the graph of $y = a^x$ is _____ .

3. The graph of the exponential function $f(x) = a^x$ _____ have an *x*-intercept.
   (does/does not)

4. The point $(2,$ _____ $)$ is on the graph of $f(x) = 3^{4x-3}$.

Graph the exponential function. See Examples 1–3.

5. $f(x) = 3^x$ **6.** $f(x) = 5^x$ **7.** $g(x) = \left(\dfrac{1}{3}\right)^x$ **8.** $g(x) = \left(\dfrac{1}{5}\right)^x$

9. $y = 4^{-x}$ **10.** $y = 6^{-x}$ **11.** $y = 2^{2x-2}$ **12.** $y = 2^{2x+1}$

⦿ **13.** Based on your answer to Exercise 1, make a conjecture (an educated guess) concerning whether an exponential function $f(x) = a^x$ is one-to-one. Then decide whether it has an inverse based on the concepts of Section 10.1.

14. What is the domain of an exponential function $f(x) = a^x$? What is its range?

Solve the equation. See Example 4.

15. $100^x = 1000$ **16.** $8^x = 4$ **17.** $16^{2x+1} = 64^{x+3}$ **18.** $9^{2x-8} = 27^{x-4}$

19. $5^x = \dfrac{1}{125}$ **20.** $3^x = \dfrac{1}{81}$ **21.** $5^x = .2$ **22.** $10^x = .1$

23. $\left(\dfrac{3}{2}\right)^x = \dfrac{8}{27}$ **24.** $\left(\dfrac{4}{3}\right)^x = \dfrac{27}{64}$

🖩 *Use the exponential key of a calculator to find an approximation to the nearest thousandth.*

25. $12^{2.6}$ **26.** $13^{1.8}$ **27.** $.5^{3.921}$ **28.** $.6^{4.917}$ **29.** $2.718^{2.5}$ **30.** $2.718^{-3.1}$

🖩 **31.** Try to evaluate $(-2)^4$ on a scientific calculator. You may get an error message, since the exponential function key on many calculators does not allow negative bases. Discuss the concept introduced in this section that is closely related to this "peculiarity" of many scientific calculators.

⦿📝

⦿ **32.** Explain why the exponential equation $4^x = 6$ cannot be solved using the method explained in this section.

📝

---

1. rises;   falls   2. (0, 1)
3. does not   4. 243
5.   6.

7.   8.

9.   10.

11.   12.

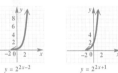

13. It is one-to-one and thus has an inverse.
14. domain: $(-\infty, \infty)$; range: $(0, \infty)$

15. $\left\{\dfrac{3}{2}\right\}$ 16. $\left\{\dfrac{2}{3}\right\}$

17. $\{7\}$ 18. $\{4\}$ 19. $\{-3\}$
20. $\{-4\}$ 21. $\{-1\}$
22. $\{-1\}$ 23. $\{-3\}$
24. $\{-3\}$
25. 639.545   26. 101.181
27. .066   28. .081
29. 12.179   30. .045

⦿ CONCEPTUAL   📝 WRITING   ▲ CHALLENGING   🖩 SCIENTIFIC CALCULATOR   🖩 GRAPHICS CALCULATOR

33. (a) .5°C (b) .35°C
34. (a) 1.0°C (b) .4°C
35. (a) 1.6°C (b) .5°C
36. (a) 3.0°C (b) .7°C
37. (a) 172.3 million
(b) 310.6 million
(c) It will be twice as large.
38. (a) 126.4 million
(b) 252.8 million   (c) It
will be twice as large.
39. (a) 100 grams
(b) 31.25 grams
(c) .30 gram (to the nearest
hundredth)
(d)

The figure shown here accompanied the article "Is Our World Warming?" which appeared in the October 1990 issue of National Geographic. *It shows projected temperature increases using two graphs: one an exponential-type curve, and the other linear. From the figure, approximate the increase* **(a)** *for the exponential curve, and* **(b)** *for the linear graph for each of the following years.*

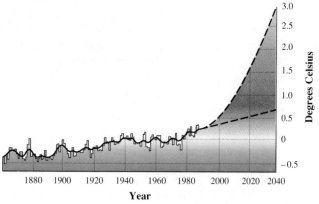

Graph, "Zero Equals Average Global Temperature for the Period 1950–1979." Dale D. Glasgow, © National Geographic Society. Reprinted by permission.

33. 2000

34. 2010

35. 2020

36. 2040

▦ *Solve the problem. See Example 5. Use a calculator as needed. The equations in Exercises 37 and 38 are models based on the assumption that current growth rates remain the same.*

37. The population of Brazil, in millions, is approximated by the function

$$f(x) = 155.3(2)^{.025x},$$

where $x = 0$ corresponds to 1994, $x = 1$ corresponds to 1995, and so on.
(a) What will be the population of Brazil in the year 2000 according to this model?
(b) What will be the population in 2034?
(c) How will the population in 2034 compare to the population in 1994?

38. The population of Pakistan, in millions, is approximated by the function

$$f(x) = 126.4(2)^{.04x},$$

where $x = 0$ corresponds to 1994, $x = 1$ corresponds to 1995, and so on.
(a) What was the population of Pakistan in the year 1994?
(b) What will the population be in 2019?
(c) How will the population in 2019 compare to the population in 1994?

39. The amount of radioactive material in an ore sample is given by the function

$$A(t) = 100(3.2)^{-.5t},$$

where $A(t)$ is the amount present, in grams, of the sample t months after the initial measurement.
(a) How much was present at the initial measurement? (*Hint:* $t = 0$.)
(b) How much was present 2 months later?
(c) How much was present 10 months later?
(d) Graph the function.

40. A small business estimates that the value $V(t)$ of a copy machine is decreasing according to the function

$$V(t) = 5000(2)^{-.15t},$$

where t is the number of years that have elapsed since the machine was purchased, and $V(t)$ is in dollars.

(a) What was the original value of the machine?

(b) What is the value of the machine 5 years after purchase? Give your answer to the nearest dollar.

(c) What is the value of the machine 10 years after purchase? Give your answer to the nearest dollar.

(d) Graph the function.

41. Refer to the function in Exercise 40. When will the value of the machine be $2500? (*Hint:* Let $V(t) = 2500$, divide both sides by 5000, and use the method of Example 4.)

42. Refer to the function in Exercise 40. When will the value of the machine be $1250?

The bar graph shows the average annual major league baseball player's salary for each year since free agency began. Using a technique from statistics, it was determined that the function

$$S(x) = 74{,}741(1.17)^x$$

approximates the salary, where $x = 0$ corresponds to 1976, and so on, up to $x = 18$ representing 1994. (Salary is in dollars.)

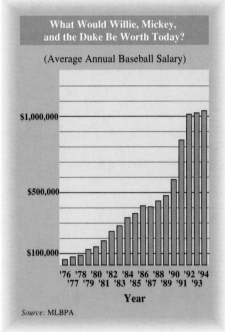

What Would Willie, Mickey, and the Duke Be Worth Today?

(Average Annual Baseball Salary)

Year

Source: MLBPA

43. Based on this model, what was the average salary in 1986?

44. Based on the graph, in what year did the average salary first exceed $1,000,000?

45. The accompanying graphics calculator screen shows this function graphed from $x = 0$ to $x = 20$. Interpret the display at the bottom of the screen.

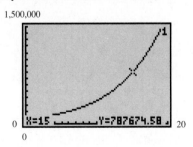

1,500,000

X=15   Y=787674.58

0   20

0

46. In the bar graph, we see that the tops of the bars rise from left to right, and in the calculator-generated graph, the curve rises from left to right. What in the model equation $S(x) = 74{,}741(1.17)^x$ indicates that during the time period, baseball salaries were *rising*?

40. (a) $5000 (b) $2973 (c) $1768 (d)

41. 6.67 years after it was purchased   42. 13.33 years after it was purchased   43. about $360,000   44. 1992 45. In 1991, the average salary was about $800,000.   46. Because the base of the exponential function is greater than 1 (1.17 > 1), the function value increases as x gets larger.

◉ CONCEPTUAL   ✎ WRITING   ▲ CHALLENGING   ▦ SCIENTIFIC CALCULATOR   ▦ GRAPHICS CALCULATOR

47. $(\sqrt[4]{16})^3$;　8
48. $\sqrt[4]{16^3}$;　8　49. 8
50. Because $\sqrt{\sqrt{x}} = (x^{1/2})^{1/2} = x^{1/4} = \sqrt[4]{x}$, the fourth root of 16^3 can be found by taking the square root twice.
51. $f(.75) = 8$

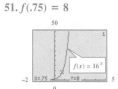

52. $(\sqrt[40]{16})^{30} = 8$; $16^{30/40} = 16^{3/4}$ since $\frac{30}{40} = \frac{3}{4}$.　53. 4　54. -4

55. 0　56. $\frac{1}{2}$

◆ **MATHEMATICAL CONNECTIONS** (Exercises 47–52) ◆

In these exercises we examine several methods of simplifying the expression $16^{3/4}$. Work Exercises 47–52 in order.

47. Write $16^{3/4}$ as a radical expression with the exponent outside the radical. Then simplify the expression.

48. Write $16^{3/4}$ as a radical expression with the exponent under the radical. Then simplify the expression.

49. Use a calculator to find the square root of 16^3. Now find the square root of that result.

50. Explain why the result in Exercise 49 is equal to $16^{3/4}$.

51. The accompanying calculator-generated graph is that of $f(x) = 16^x$. If $x = .75$, what is the value of $f(x)$? Support this answer by using a graphics calculator.

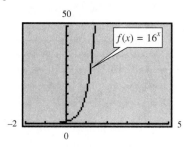

52. Use a calculator to approximate $\sqrt[40]{16}$. Then raise the result to the 30th power. Why is the result equal to $16^{3/4}$?

REVIEW EXERCISES

Determine what number would have to be placed in the box for the statement to be true. See Sections 3.1 and 5.1.

53. $2^\square = 16$　**54.** $2^\square = \frac{1}{16}$　**55.** $2^\square = 1$　**56.** $2^\square = \sqrt{2}$

10.3 LOGARITHMIC FUNCTIONS

FOR EXTRA HELP

 SSG pp. 322–326
SSM pp. 542–546

 Video 12

 Tutorial IBM　MAC

OBJECTIVES
1 ▶ Define a logarithm.
2 ▶ Write exponential statements in logarithmic form and logarithmic statements in exponential form.
3 ▶ Solve logarithmic equations of the form $\log_a b = k$ for a, b, or k.
4 ▶ Graph logarithmic functions.
5 ▶ Use logarithmic functions in applications.

The graph of $y = 2^x$ is the curve shown in **blue** in Figure 12. Since $y = 2^x$ is a one-to-one function, it has an inverse. Interchanging x and y gives $x = 2^y$, the inverse of $y = 2^x$. As we saw in Section 10.1, the graph of the inverse is found by reflecting the graph of $y = 2^x$ about the line $y = x$. The graph of $x = 2^y$ is shown as a **red** curve in Figure 12.

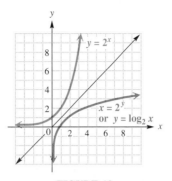

FIGURE 12

$1 \blacktriangleright$ Define a logarithm. $\blacktriangleright$ We cannot solve the equation $x = 2^y$ for the dependent variable y with the methods presented up to now. The following definition is used to solve $x = 2^y$ for y.

Definition of Logarithm	For all positive numbers a, where $a \neq 1$, and all positive numbers x, $$y = \log_a x \quad \text{means the same as} \quad x = a^y.$$

This key statement should be memorized. The abbreviation **log** is used for **logarithm.** Read $\log_a x$ as "the logarithm of x to the base a." To remember the location of the base and the exponent in each form, refer to the diagram that follows.

$$\text{Exponent}$$
$$\downarrow$$
$$\text{Logarithmic form: } y = \log_a x$$
$$\uparrow$$
$$\text{Base}$$

$$\text{Exponent}$$
$$\downarrow$$
$$\text{Exponential form: } x = a^y$$
$$\uparrow$$
$$\text{Base}$$

In working with logarithmic form and exponential form, remember the following.

Meaning of $\log_a x$	A **logarithm** is an exponent; $\log_a x$ is the exponent on the base a that yields the number x.

$2 \blacktriangleright$ Write exponential statements in logarithmic form and logarithmic statements in exponential form. $\blacktriangleright$ We can use the definition of logarithm to write exponential statements in logarithmic form and logarithmic statements in exponential form.

The list below shows several pairs of equivalent statements. The same statement is written in both exponential and logarithmic form.

Exponential Form	Logarithmic Form
$3^2 = 9$	$\log_3 9 = 2$
$\left(\dfrac{1}{5}\right)^{-2} = 25$	$\log_{1/5} 25 = -2$
$10^5 = 100{,}000$	$\log_{10} 100{,}000 = 5$
$4^{-3} = \dfrac{1}{64}$	$\log_4 \dfrac{1}{64} = -3$	■

3 ▶ Solve logarithmic equations of the form $\log_a b = k$ for a, b, or k.

▶ A **logarithmic equation** is an equation with a logarithm in at least one term. We solve logarithmic equations of the form $\log_a b = k$ for any of the three variables by first writing the equation in exponential form.

Solve the following equations.

(a) $\log_4 x = -2$

By the definition of logarithm, $\log_4 x = -2$ is equivalent to $4^{-2} = x$. Then

$$x = 4^{-2} = \frac{1}{4^2} = \frac{1}{16}.$$

The solution set is {1/16}.

(b) $\log_{1/2} 16 = y$

First write the statement in exponential form.

$$\log_{1/2} 16 = y$$

$$\left(\frac{1}{2}\right)^y = 16 \qquad \text{Convert to exponential form.}$$

$$(2^{-1})^y = 2^4 \qquad \text{Write with the same base.}$$

$$2^{-y} = 2^4 \qquad \text{Property of exponents}$$

$$-y = 4 \qquad \text{Set exponents equal.}$$

$$y = -4 \qquad \text{Multiply by } -1.$$

The solution set is {−4}. ■

For any positive real number b, we know that $b^1 = b$ and $b^0 = 1$. Writing these two statements in logarithmic form gives the following two properties of logarithms.

For any positive real number b, $b \neq 1$,

$$\log_b b = 1 \qquad \text{and} \qquad \log_b 1 = 0$$

(a) $\log_7 7 = 1$ **(b)** $\log_{\sqrt{2}} \sqrt{2} = 1$

(c) $\log_9 1 = 0$ **(d)** $\log_{.2} 1 = 0$ ■

Now we can define the logarithmic function with base a as follows.

| Logarithmic Function | If a and x are positive numbers, with $a \neq 1$, then |

$$f(x) = \log_a x$$

defines the **logarithmic function with base a.**

4 ▶ Graph logarithmic functions.

▶ To graph a logarithmic function, it is helpful to write it in exponential form first. Then plot selected ordered pairs to determine the graph.

E X A M P L E   4

Graphing a Logarithmic Function

■ Graph $y = \log_{1/2} x$.

By writing $y = \log_{1/2} x$ in its exponential form as $x = \left(\dfrac{1}{2}\right)^y$, we can identify ordered pairs that satisfy the equation. Here it is easier to choose values for y and find the corresponding values of x. Doing this gives the following pairs.

Chalkboard Exercise

Graph $y = \log_3 x$.

Answer:

x	$\dfrac{1}{4}$	$\dfrac{1}{2}$	1	2	4	8
y	2	1	0	-1	-2	-3

Plotting these points (be careful to get them in the right order) and connecting them with a smooth curve gives the graph in Figure 13. This graph is typical of logarithmic functions with $0 < a < 1$. The graph of $x = 2^y$ in Figure 12, which is equivalent to $y = \log_2 x$, is typical of graphs of logarithmic functions with base $a > 1$. ■

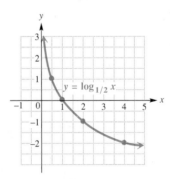

FIGURE 13

Based on the graphs of the functions $y = \log_2 x$ in Figure 12 and $y = \log_{1/2} x$ in Figure 13 we can make the following generalizations about the graphs of logarithmic functions of the form $G(x) = \log_a x$.

| Graph of $G(x) = \log_a x$ |

1. The graph will always contain the point $(1, 0)$.
2. When $a > 1$, the graph will *rise* from left to right, from the fourth quadrant to the first. When $0 < a < 1$, the graph will *fall* from left to right, from the first quadrant to the fourth.
3. The graph will approach the y-axis, but never touch it. (It is an asymptote.)
4. The domain is $(0, \infty)$ and the range is $(-\infty, \infty)$.

▶ **TEACHING TIP**
To emphasize that the domain of the logarithm function is strictly positive, have students try to find *log 0* and *log (−2)* on their calculators and comment on the results. ◀

Compare these generalizations to the similar ones for exponential functions in Section 10.2.

◊ almost 4 times as powerful; about 300 times as powerful

◊ **CONNECTIONS** ◊

In the United States, the intensity of an earthquake is rated using the *Richter scale*. The Richter scale rating of an earthquake of intensity x is given by

$$R = \log_{10} \frac{x}{x_0},$$

where x_0 is the intensity of an earthquake of a certain (small) size. The figure shows Richter scale ratings for major Southern California earthquakes since 1920. As the figure indicates, earthquakes "come in bunches" and the 1990s have been an especially busy time.

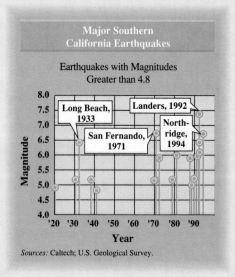

Major Southern California Earthquakes

Earthquakes with Magnitudes Greater than 4.8

Long Beach, 1933 · Landers, 1992 · Northridge, 1994 · San Fernando, 1971

Sources: Caltech; U.S. Geological Survey.

FOR DISCUSSION OR WRITING

Writing the logarithmic equation given above in exponential form, we get

$$10^R = \frac{x}{x_0} \qquad \text{or} \qquad x = 10^R x_0.$$

The 1994 Northridge earthquake had a Richter scale rating of 6.7; the Landers earthquake had a rating of 7.3. How much more powerful was the Landers earthquake than the Northridge earthquake? Compare the smallest rated earthquake in the figure (at 4.8) with the Landers quake. How much more powerful was the Landers quake?

5 ▶ Use logarithmic functions in applications.

▶ Logarithmic functions, like exponential functions, are used in applications to describe growth and decay.

EXAMPLE 5

Solving an Application of a Logarithmic Function

Sales (in thousands of units) of a new product are approximated by

$$S = 100 + 30 \log_3(2t + 1),$$

where t is the number of years after the product is introduced.

(a) What were the sales after 1 year?
  Let $t = 1$ and find S.

$$S = 100 + 30 \log_3(2t + 1)$$
$$= 100 + 30 \log_3(2 \cdot 1 + 1) \qquad \text{Let } t = 1.$$
$$= 100 + 30 \log_3 3$$
$$= 100 + 30(1) \qquad \log_3 3 = 1$$
$$S = 130$$

Sales were 130 thousand units after 1 year.

(b) Find the sales after 13 years.
  Substitute $t = 13$ into the expression for S.

$$S = 100 + 30 \log_3(2t + 1)$$
$$= 100 + 30 \log_3(2 \cdot 13 + 1) \qquad \text{Let } t = 13.$$
$$= 100 + 30 \log_3 27$$
$$= 100 + 30(3) \qquad \log_3 27 = 3$$
$$S = 190$$

After 13 years, sales had increased to 190 thousand units.

(c) Graph S.
  Use the two ordered pairs (1, 130) and (13, 190) found above. Check that (0, 100) and (40, 220) also satisfy the equation. Use these ordered pairs and a knowledge of the general shape of the graph of a logarithmic function to get the graph in Figure 14. ■

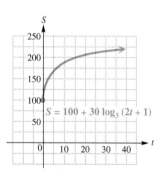

FIGURE 14

10.3 EXERCISES

◉ **1.** By definition $\log_a x$ is the exponent to which the base a must be raised in order to obtain x. Use this to simplify each of the following, without doing any written work. (Example: $\log_3 9$ is 2, because 2 is the exponent to which 3 must be raised in order to obtain 9.)

1. (a) 2   (b) 4   (c) −1
(d) −2   (e) $\frac{1}{2}$ (f) 0

  (a) $\log_4 16$ (b) $\log_3 81$ (c) $\log_3\left(\dfrac{1}{3}\right)$ (d) $\log_{10} .01$
  (e) $\log_5 \sqrt{5}$ (f) $\log_{12} 1$

◉ **2.** Explain why $\log_a 1$ is 0 for any value of a that is allowed as the base of a logarithm. Use a rule of exponents introduced in Chapter 3 in your explanation.

◉ CONCEPTUAL   🖉 WRITING   ▲ CHALLENGING   ▦ SCIENTIFIC CALCULATOR   ▨ GRAPHICS CALCULATOR

3. $\log_4 1024 = 5$
4. $\log_3 729 = 6$
5. $\log_{1/2} 8 = -3$
6. $\log_{1/6} 216 = -3$
7. $\log_{10} .001 = -3$
8. $\log_{36} 6 = \dfrac{1}{2}$ 9. $4^3 = 64$

10. $2^9 = 512$
11. $10^{-4} = \dfrac{1}{10,000}$
12. $100^1 = 100$
13. $6^0 = 1$ 14. $\pi^0 = 1$
17. $\left\{\dfrac{1}{3}\right\}$ 18. $\left\{\dfrac{1}{3}\right\}$
19. $\{81\}$ 20. $\{25\}$
21. $\left\{\dfrac{1}{5}\right\}$ 22. $\left\{\dfrac{1}{2}\right\}$
23. $\{1\}$ 24. $\{1\}$
25. $\{x \mid x > 0, x \ne 1\}$
26. $\{x \mid x > 0, x \ne 1\}$
27. $\{5\}$ 28. $\{10\}$
29. $\left\{\dfrac{5}{3}\right\}$ 30. $\left\{\dfrac{3}{4}\right\}$
31. $\{4\}$ 32. $\{9\}$ 33. $\left\{\dfrac{3}{2}\right\}$
34. $\left\{\dfrac{3}{2}\right\}$

35.
$y = \log_3 x$

36.
$y = \log_5 x$

37.
$y = \log_{1/3} x$

38.
$y = \log_{1/5} x$

39.
$y = \log_{2.718} x$

Write in logarithmic form. See Example 1.

3. $4^5 = 1024$ **4.** $3^6 = 729$ **5.** $\left(\dfrac{1}{2}\right)^{-3} = 8$

6. $\left(\dfrac{1}{6}\right)^{-3} = 216$ **7.** $10^{-3} = .001$ **8.** $36^{1/2} = 6$

Write in exponential form. See Example 1.

9. $\log_4 64 = 3$ **10.** $\log_2 512 = 9$ **11.** $\log_{10} \dfrac{1}{10,000} = -4$

12. $\log_{100} 100 = 1$ **13.** $\log_6 1 = 0$ **14.** $\log_\pi 1 = 0$

◉ **15.** When a student asked his teacher to explain to him how to evaluate $\log_9 3$ without showing any work, his teacher told him, "Think radically." Explain what the teacher meant by this hint.

◉ **16.** A student told her teacher "I know that $\log_2 1$ is the exponent to which 2 must be raised in order to obtain 1, but I can't think of any such number." How would you explain to the student that the value of $\log_2 1$ is 0?

Solve the equation for x. See Examples 2 and 3.

17. $x = \log_{27} 3$ **18.** $x = \log_{125} 5$ **19.** $\log_x 9 = \dfrac{1}{2}$ **20.** $\log_x 5 = \dfrac{1}{2}$

21. $\log_x 125 = -3$ **22.** $\log_x 64 = -6$ **23.** $\log_{12} x = 0$ **24.** $\log_4 x = 0$

25. $\log_x x = 1$ **26.** $\log_x 1 = 0$ **27.** $\log_x \dfrac{1}{25} = -2$ **28.** $\log_x \dfrac{1}{10} = -1$

29. $\log_8 32 = x$ **30.** $\log_{81} 27 = x$ **31.** $\log_\pi \pi^4 = x$ **32.** $\log_{\sqrt{2}} \sqrt{2}^9 = x$

33. $\log_6 \sqrt{216} = x$ **34.** $\log_4 \sqrt{64} = x$

If the point (p, q) is on the graph of $f(x) = a^x$, (for $a > 0$ and $a \ne 1$) then the point (q, p) is on the graph of $f^{-1}(x) = \log_a x$. Use this fact, and refer to the graphs required in Exercises 5–8 in Section 10.2 to graph the logarithmic function. See Example 4.

35. $y = \log_3 x$ **36.** $y = \log_5 x$ **37.** $y = \log_{1/3} x$ **38.** $y = \log_{1/5} x$

▦ **39.** Graph the function $y = \log_{2.718} x$ using the exponential key of a calculator, and ▲ rewriting it as $2.718^y = x$. Choose -1, 0, and 1 as y values, and approximate x to the nearest tenth.

▦ **40.** Use the exponential key of your calculator to find approximations for the expression ◉ $\left(1 + \dfrac{1}{x}\right)^x$, using x values of 1, 10, 100, 1000, and 10,000. Explain what seems to be happening as x gets larger and larger. (*Hint:* Look at the base in Exercise 39.)

Solve the application of a logarithmic function. See Example 5.

41. A study showed that the number of mice in an old abandoned house was approximated by the function

$$M(t) = 6 \log_4(2t + 4),$$

where t is measured in months and $t = 0$ corresponds to January 1993. Find the number of mice in the house in
(a) January 1993
(b) July 1993
(c) July 1995.
(d) Graph the function.

42. A supply of hybrid striped bass were introduced into a lake in January 1980. Biologists researching the bass population over the next decade found that the number of bass in the lake was approximated by the function

$$B(t) = 500 \log_3(2t + 3),$$

where $t = 0$ corresponds to January 1980, $t = 1$ to January 1981, $t = 2$ to January 1982, and so on. Use this function to find the bass population in

(a) January 1980
(b) January 1983
(c) January 1992.
(d) Graph the function for $0 \le t \le 12$.

Use the graph at the right to predict the value of $f(t)$ for the given value of t.

43. $t = 0$

44. $t = 10$

45. $t = 60$

46. Show that the points determined in Exercises 43–45 lie on the graph of $f(t) = 8 \log_5(2t + 5)$.

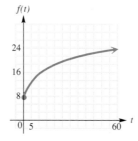

$f(t)$

◉ **47.** Explain why 1 is not allowed as a base for a logarithmic function.

◉ **48.** Compare the summary of facts about the graph of $F(x) = a^x$ in Section 10.2 with the similar summary of facts about the graph of $G(x) = \log_a x$ in this section. Make a list of the facts that reinforce the concept that F and G are inverse functions.

▲ **49.** The domain of $F(x) = a^x$ is $(-\infty, \infty)$, while the range is $(0, \infty)$. Therefore, since
◉ $G(x) = \log_a x$ is the inverse of F, the domain of G is _____ , while the range of G is _____ .

◉ **50.** The graphs of both $F(x) = 3^x$ and $G(x) = \log_3 x$ rise from left to right. Which one rises at a faster rate?

As mentioned in Section 10.1, some graphics calculators have the capability of drawing the inverse of a function. For example, the two screens that follow show the graphs of $f(x) = 2^x$ and $g(x) = \log_2 x$. The graph of g was obtained by drawing the graph of f^{-1}, since $g(x) = f^{-1}(x)$. (Compare to Figure 12 in this section.)

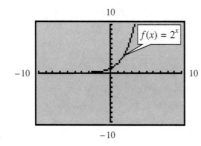
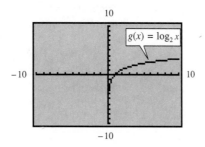

▦ *Use a graphics calculator with the capability of drawing the inverse of a function to draw the graph of the logarithmic function. Use the standard viewing window.*

51. $g(x) = \log_3 x$ (Compare to Exercise 35.)
52. $g(x) = \log_5 x$ (Compare to Exercise 36.)

41. (a) 6   (b) 12
(c) 18
(d) $M(t) = 6 \log_4(2t + 4)$

Months since January 1993

42. (a) **500**   (b) **1000**
(c) **1500**
(d) $B(t) = 500 \log_3(2t + 3)$

Years since January 1980

43. **8**   44. **16**   45. **24**
46. $f(0) = 8$; $f(10) = 16$;
$f(60) = 24$ 49. $(0, \infty)$;
$(-\infty, \infty)$ 50. $F(x) = 3^x$
51.

52.

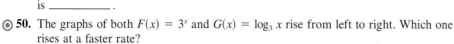
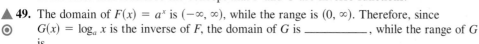

◉ CONCEPTUAL   ▢ WRITING   ▲ CHALLENGING   ▦ SCIENTIFIC CALCULATOR   ▦ GRAPHICS CALCULATOR

53.

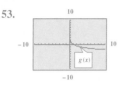

54.

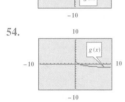

53. $g(x) = \log_{1/3} x$ (Compare to Exercise 37.)

54. $g(x) = \log_{1/5} x$ (Compare to Exercise 38.)

REVIEW EXERCISES

Use the properties of exponents to simplify the expression. Write answers using only positive exponents. See Sections 3.1 and 3.2.

55. $4^7 \cdot 4^2$ **56.** $3^{-3} \cdot 3^{16}$ **57.** $\dfrac{5^{-3}}{5^8}$ **58.** $\dfrac{7^8}{7^{-4}}$ **59.** $(9^3)^{-2}$ **60.** $(6x^5)^5$

55. 4^9 **56.** 3^{13} **57.** $\dfrac{1}{5^{11}}$ **58.** 7^{12} **59.** $\dfrac{1}{9^6}$ **60.** $6^5 x^{25}$

10.4 PROPERTIES OF LOGARITHMS

FOR EXTRA HELP	OBJECTIVES

SSG pp. 326–329
 SSM pp. 546–549
 	
  Video
 12
 	
  Tutorial
 IBM   MAC
	1 ▶ Use the product rule for logarithms.
	 2 ▶ Use the quotient rule for logarithms.
	 3 ▶ Use the power rule for logarithms.
	 4 ▶ Use the properties of logarithms to write logarithmic expressions in alternative forms.

Logarithms have been used as an aid to numerical calculation for several hundred years. Today the widespread use of calculators has made the use of logarithms for calculation obsolete. However, logarithms are very important in applications and in further work in mathematics. The properties that make logarithms so useful are given in this section.

1 ▶ Use the product rule for logarithms.

▶ One way in which logarithms simplify problems is by changing a problem of multiplication into one of addition. This is done with the product rule for logarithms.

Product Rule For Logarithms

If x, y, and b are positive numbers, where $b \neq 1$, then

$$\log_b xy = \log_b x + \log_b y.$$

(The logarithm of a product is the sum of the logarithms of the factors.)

NOTE The word statement of the product rule can also be stated by replacing "logarithm" with "exponent," and the rule then becomes the familiar rule for multiplying exponential expressions: The *exponent* of a product is equal to the sum of the *exponents* of the factors.

To prove this rule, let $m = \log_b x$ and $n = \log_b y$, and recall that

$$\log_b x = m \qquad \text{means} \qquad b^m = x,$$
$$\log_b y = n \qquad \text{means} \qquad b^n = y.$$

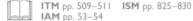

Now consider the product xy.

$$xy = b^m \cdot b^n \qquad \text{Substitution}$$
$$xy = b^{m+n} \qquad \text{Product rule for exponents}$$
$$\log_b xy = m + n \qquad \text{Convert to logarithmic form.}$$
$$\log_b xy = \log_b x + \log_b y \qquad \text{Substitution}$$

The last statement is the result we wish to prove.

EXAMPLE 1
Using the Product Rule

Use the product rule for logarithms to rewrite the following. Assume $x > 0$.

(a) $\log_5 6 \cdot 9$

By the product rule, $\log_5 6 \cdot 9 = \log_5 6 + \log_5 9$.

(b) $\log_7 8 + \log_7 12$

$$\log_7 8 + \log_7 12 = \log_7 8 \cdot 12 = \log_7 96$$

(c) $\log_3 3x$

$$\log_3 3x = \log_3 3 + \log_3 x$$
$$\log_3 3x = 1 + \log_3 x \qquad \log_3 3 = 1$$

(d) $\log_4 x^3$

Since $x^3 = x \cdot x \cdot x$,

$$\log_4 x^3 = \log_4(x \cdot x \cdot x)$$
$$= \log_4 x + \log_4 x + \log_4 x$$
$$= 3 \log_4 x. \quad \blacksquare$$

Chalkboard Exercise

Use the product rule to rewrite the following logarithm.

$$\log_8 8k \quad (k > 0)$$

Answer: $1 + \log_8 k$

2 ▸ Use the quotient rule for logarithms.

▸ The rule for division is similar to the rule for multiplication.

Quotient Rule For Logarithms

If x, y, and b are positive numbers, where $b \neq 1$, then

$$\log_b \frac{x}{y} = \log_b x - \log_b y.$$

(The logarithm of a quotient is the difference between the logarithm of the numerator and the logarithm of the denominator.)

The proof of this rule is similar to the proof of the product rule.

EXAMPLE 2
Using the Quotient Rule

Use the quotient rule for logarithms to rewrite the following.

(a) $\log_4 \dfrac{7}{9} = \log_4 7 - \log_4 9$

(b) $\log_5 \dfrac{6}{x} = \log_5 6 - \log_5 x$, for $x > 0$.

(c) $\log_3 \dfrac{27}{5} = \log_3 27 - \log_3 5$

$$= 3 - \log_3 5 \qquad \log_3 27 = 3 \quad \blacksquare$$

Chalkboard Exercise

Use the quotient rule to rewrite the following logarithm.

$$\log_4 \frac{3}{16}$$

Answer: $\log_4 3 - 2$

3 ▶ Use the power rule for logarithms.

▶ The next rule gives a method for evaluating powers and roots of numbers such as

$$2^{\sqrt{2}}, \quad (\sqrt{2})^{3/4}, \quad (.032)^{5/8}, \quad \text{and} \quad \sqrt[5]{12}.$$

This rule makes it possible to find approximations for numbers that could not be evaluated before. Example 1(d) showed that $\log_4 x^3 = 3 \log_4 x$, which suggests the following generalization.

Power Rule for Logarithms	If x and b are positive real numbers, where $b \neq 1$, and if r is any real number, then $$\log_b x^r = r(\log_b x)$$ (The logarithm of a number to a power equals the exponent times the logarithm of the number.)

As examples of this result,

$$\log_b m^5 = 5 \log_b m \quad \text{and} \quad \log_3 5^{3/4} = \frac{3}{4} \log_3 5.$$

To prove the power rule, let

$\log_b x = m$	
$b^m = x$	Convert to exponential form.
$(b^m)^r = x^r$	Raise to the power r.
$b^{mr} = x^r$	Power rule for exponents
$\log_b x^r = mr$	Convert to logarithmic form.
$\log_b x^r = rm$	
$\log_b x^r = r \log_b x.$	$m = \log_b x$

This is the statement to be proved.

As a special case of the rule above, let $r = 1/p$, so that

$$\log_b \sqrt[p]{x} = \log_b x^{1/p} = \frac{1}{p} \log_b x.$$

For example, using this result, with $x > 0$,

$$\log_b \sqrt[5]{x} = \frac{1}{5} \log_b x \quad \text{and} \quad \log_b \sqrt[3]{x^4} = \frac{4}{3} \log_b x.$$

EXAMPLE 3
Using the Power Rule

Use the power rule to rewrite each of the following. Assume $a > 0$, $b > 0$, $x > 0$, $a \neq 1$, and $b \neq 1$.

(a) $\log_3 5^2 = 2 \log_3 5$ **(b)** $\log_a x^4 = 4 \log_a x$

(c) $\log_b \sqrt{7}$

When using the power rule with logarithms of expressions involving radicals, begin by rewriting the radical expression with a rational exponent, as shown in Section 5.1.

$$\log_b \sqrt{7} = \log_b 7^{1/2} \qquad \sqrt{x} = x^{1/2}$$

$$= \frac{1}{2} \log_b 7 \qquad \text{Power rule}$$

(d) $\log_2 \sqrt[5]{x^2} = \log_2 x^{2/5} = \dfrac{2}{5} \log_2 x$ ∎

Two special properties involving both exponential and logarithmic expressions come directly from the fact that logarithmic and exponential functions are inverses of each other.

Special Properties   If $b > 0$ and $b \neq 1$, then

$$b^{\log_b x} = x \quad (x > 0) \qquad \text{and} \qquad \log_b b^x = x.$$

To prove the first statement, let

$$y = \log_b x.$$
$$b^y = x \qquad \qquad \text{Convert to exponential form.}$$
$$b^{\log_b x} = x \qquad \qquad \text{Replace } y \text{ with } \log_b x.$$

The proof of the second statement is similar.

E X A M P L E   4
Using the Special Properties

Find the value of the following logarithmic expressions.

(a) $\log_5 5^4$
Since $\log_b b^x = x$,
$$\log_5 5^4 = 4.$$

(b) $\log_3 9$
Since $9 = 3^2$,
$$\log_3 9 = \log_3 3^2 = 2.$$

The property $\log_b b^x = x$ was used in the last step.

(c) $4^{\log_4 10} = 10$ ∎

4 ▶ Use the properties of logarithms to write logarithmic expressions in alternative forms.

▶   The properties of logarithms are useful for writing expressions in an alternative form. This use of logarithms is important in solving equations with logarithms and in calculus.

E X A M P L E   5
Writing Logarithms in Alternative Forms

Use the properties of logarithms to rewrite each expression. Assume all variables represent positive real numbers.

(a) $\log_4 4x^3 = \log_4 4 + \log_4 x^3$ Product rule
$$= 1 + 3 \log_4 x \qquad \log_4 4 = 1; \text{ power rule}$$

(b) $\log_7 \sqrt{\dfrac{p}{q}} = \log_7 \left(\dfrac{p}{q}\right)^{1/2}$

$$= \frac{1}{2} \log_7 \frac{p}{q} \qquad \qquad \text{Power rule}$$

$$= \frac{1}{2} (\log_7 p - \log_7 q) \qquad \text{Quotient rule}$$

(c) $\log_5 \dfrac{a}{bc} = \log_5 a - \log_5 bc$ Quotient rule

$= \log_5 a - (\log_5 b + \log_5 c)$ Product rule

$= \log_5 a - \log_5 b - \log_5 c$

Notice the careful use of parentheses in the second step. Since we are subtracting the logarithm of a product, and it is being rewritten as a sum of two terms, parentheses *must* be placed around the sum.

(d) $3 \log_b x + \dfrac{1}{2} \log_b y = \log_b x^3 + \log_b y^{1/2}$ Power rule

$= \log_b x^3 \sqrt{y}$ Product rule; $y^{1/2} = \sqrt{y}$

(e) $\log_8(2p + 3r)$ cannot be rewritten by the properties of logarithms. ■

CAUTION Remember that there is no property of logarithms to rewrite the logarithm of a *sum* or *difference*. For example, we *cannot* write $\log_b(x + y)$ in terms of $\log_b x$ and $\log_b y$. Also, $\log_b \dfrac{x}{y} \neq \dfrac{\log_b x}{\log_b y}$.

In the next example, we use numerical values for $\log_2 5$ and $\log_2 3$. While we use the equals sign to give these values, they are actually just approximations, since most logarithms of this type are irrational numbers. While it would be more correct to use the symbol $\approx$, we will simply use $=$ with the understanding that the values are correct to four decimal places.

EXAMPLE 6

Using the Properties of Logarithms with Numerical Values

Given that $\log_2 5 = 2.3219$ and $\log_2 3 = 1.5850$, evaluate the following.

(a) $\log_2 15$

$\log_2 15 = \log_2 3 \cdot 5$

$= \log_2 3 + \log_2 5$ Product rule

$= 1.5850 + 2.3219$

$= 3.9069$

(b) $\log_2 .6$

$\log_2 .6 = \log_2 \dfrac{3}{5}$ $.6 = \frac{6}{10} = \frac{3}{5}$

$= \log_2 3 - \log_2 5$ Quotient rule

$= 1.5850 - 2.3219$

$= -.7369$

(c) $\log_2 27$

$\log_2 27 = \log_2 3^3$

$= 3 \log_2 3$ Power rule

$= 3(1.5850)$

$= 4.7550$ ■

EXAMPLE 7

Deciding Whether
Statements about
Logarithms Are True

Decide whether each of the following statements is true or false.

(a) $\log_2 8 - \log_2 4 = \log_2 4$
Evaluate both sides.

$$\log_2 8 - \log_2 4 = \log_2 2^3 - \log_2 2^2 = 3 - 2 = 1$$
$$\log_2 4 = \log_2 2^2 = 2$$

The statement is false because $2 \neq 1$.

Chalkboard Exercise

True or false.

$$\log_6(\log_2 16) = \frac{\log_6 6}{\log_6 36}$$

Answer: false

(b) $\log_3(\log_2 8) = \dfrac{\log_7 49}{\log_8 64}$
Evaluate both sides.

$$\log_3(\log_2 8) = \log_3(3) = 1$$
$$\frac{\log_7 49}{\log_8 64} = \frac{\log_7 7^2}{\log_8 8^2} = \frac{2}{2} = 1$$

The statement is true. ∎

$\log_{10} 458.3 \approx 2.661149857$
$+ \log_{10} 294.6 \approx 2.469232743$
$\underline{\hspace{3cm}} \approx 5.130382600$
$10^{5.130382600} \approx 135,015.18$
A calculator gives
$(458.3)(294.6) = 135,015.18$.

◆ **CONNECTIONS** ◆

Long before the days of calculators and computers, the search for making calculations easier was an ongoing process. Machines built by Charles Babbage and Blaise Pascal, a system of "rods" used by John Napier, and slide rules were the forerunners of today's electronic marvels. The invention of logarithms by John Napier in the sixteenth century was a great breakthrough in the search for easier methods of calculation.

Since logarithms are exponents, their properties allowed users of tables of common logarithms to multiply by adding, divide by subtracting, raise to powers by multiplying, and take roots by dividing. Although logarithms are no longer used for computations, they play an important part in higher mathematics.

FOR DISCUSSION OR WRITING
To multiply 458.3 by 294.6 using logarithms, we must add $\log_{10} 458.3$ and $\log_{10} 294.6$, then find 10 to the sum. Perform this multiplication using the log* key and the 10^x key on your calculator. Check your answer by multiplying directly with your calculator. Try division, raising to a power, and taking a root by this method.

10.4 EXERCISES

◉ *Fill in the blank with the correct response.*

1. sum   2. difference

1. The logarithm of the product of two numbers is equal to the _____ of the logarithms of the numbers.

2. The logarithm of the quotient of two numbers is equal to the _____ of the logarithms of the numbers.

---

*In this text, the notation log x is used to mean $\log_{10} x$. This is also the meaning of the log key on calculators.

 CONCEPTUAL   WRITING   CHALLENGING    SCIENTIFIC CALCULATOR    GRAPHICS CALCULATOR

3. k 4. $\dfrac{1}{2}$

5. $\log_7 4 - \log_7 5$

6. $\log_8 9 - \log_8 11$

7. $\dfrac{1}{4}\log_2 8$ or $\dfrac{3}{4}$

8. $\dfrac{3}{4}\log_3 9$ or $\dfrac{3}{2}$

9. $\log_4 3 + \dfrac{1}{2}\log_4 x -$
$\log_4 y$

10. $\log_5 6 + \dfrac{1}{2}\log_5 z -$
$\log_5 w$

11. $\dfrac{1}{3}\log_3 4 - 2\log_3 x -$
$\log_3 y$

12. $\dfrac{1}{3}\log_7 13 - \log_7 p -$
$2\log_7 q$

13. $\dfrac{1}{2}\log_3 x + \dfrac{1}{2}\log_3 y -$
$\dfrac{1}{2}\log_3 5$

14. $\dfrac{1}{2}\log_6 p + \dfrac{1}{2}\log_6 q -$
$\dfrac{1}{2}\log_6 7$

15. $\dfrac{1}{3}\log_2 x + \dfrac{1}{5}\log_2 y -$
$2\log_2 r$

16. $\dfrac{1}{4}\log_4 z + \dfrac{1}{5}\log_4 w -$
$2\log_4 s$

19. $\log_b xy$ 20. $\log_b 2z$

21. $\log_a \dfrac{m^3}{n}$ 22. $\log_b \dfrac{x^5}{y}$

23. $\log_a \dfrac{rt^3}{s}$ 24. $\log_a \dfrac{pr^2}{q}$

25. $\log_a \dfrac{125}{81}$ 26. $\log_a 375$

27. $\log_{10}(x^2 - 9)$

28. $\log_{10}(y^2 - 16)$

29. $\log_p \dfrac{x^3 y^{1/2}}{z^{3/2}a^3}$

30. $\log_b \dfrac{x^{1/3}y^{2/3}}{s^{3/4}t^{2/3}}$

31. 1.2552   32. .6532

33. 1.5562   34. 2.2094

35. .4771   36. 1.5050

37. true   38. true

39. false   40. false

---

3. $\log_a b^k = $ _____ $\log_a b$ $(a > 0, a \neq 1, b > 0)$

4. The logarithm of the square root of a number is equal to _____ times the logarithm of the number.

Use the properties of logarithms introduced in this section to express the logarithm as a sum or difference of logarithms, or as a single number if possible. Assume that all variables represent positive real numbers. See Examples 1–5.

5. $\log_7 \dfrac{4}{5}$ 6. $\log_8 \dfrac{9}{11}$ 7. $\log_2 8^{1/4}$

8. $\log_3 9^{3/4}$ 9. $\log_4 \dfrac{3\sqrt{x}}{y}$ 10. $\log_5 \dfrac{6\sqrt{z}}{w}$

11. $\log_3 \dfrac{\sqrt[3]{4}}{x^2 y}$ 12. $\log_7 \dfrac{\sqrt[3]{13}}{pq^2}$ 13. $\log_3 \sqrt{\dfrac{xy}{5}}$

14. $\log_6 \sqrt{\dfrac{pq}{7}}$ 15. $\log_2 \dfrac{\sqrt[3]{x} \cdot \sqrt[5]{y}}{r^2}$ 16. $\log_4 \dfrac{\sqrt[4]{z} \cdot \sqrt[5]{w}}{s^2}$

17. A student erroneously wrote $\log_a(x + y) = \log_a x + \log_a y$. When his teacher explained that this was indeed wrong, the student claimed that he had used the distributive property. Write a few sentences explaining why the distributive property does not apply in this case.

18. Write a few sentences explaining how the rules for multiplying and dividing powers of the same base are similar to the rules for finding logarithms of products and quotients.

Use the properties of logarithms introduced in this section to express each of the following as a single logarithm. Assume that all variables are defined in such a way that the variable expressions are positive, and bases are positive numbers not equal to 1. See Examples 1–5.

19. $\log_b x + \log_b y$ 20. $\log_b 2 + \log_b z$

21. $3\log_a m - \log_a n$ 22. $5\log_b x - \log_b y$

23. $(\log_a r - \log_a s) + 3\log_a t$ 24. $(\log_a p - \log_a q) + 2\log_a r$

25. $3\log_a 5 - 4\log_a 3$ 26. $3\log_a 5 + \dfrac{1}{2}\log_a 9$

27. $\log_{10}(x + 3) + \log_{10}(x - 3)$ 28. $\log_{10}(y + 4) + \log_{10}(y - 4)$

29. $3\log_p x + \dfrac{1}{2}\log_p y - \dfrac{3}{2}\log_p z - 3\log_p a$

30. $\dfrac{1}{3}\log_b x + \dfrac{2}{3}\log_b y - \dfrac{3}{4}\log_b s - \dfrac{2}{3}\log_b t$

Given that $\log_{10} 2 \approx .3010$ and $\log_{10} 9 \approx .9542$, evaluate the following by applying the appropriate rule or rules from this section. **DO NOT USE A CALCULATOR FOR THESE EXERCISES.** *See Example 6.*

31. $\log_{10} 18$ 32. $\log_{10} 4.5$ $\left(Hint: 4.5 = \dfrac{9}{2}\right)$

33. $\log_{10} 36$ 34. $\log_{10} 162$ 35. $\log_{10} 3$ 36. $\log_{10} 32$

Decide whether the statement is true or false. See Example 7.

37. $\log_6 60 - \log_6 10 = 1$ 38. $\log_3 7 + \log_3 \dfrac{1}{7} = 0$

39. $\dfrac{\log_{10} 7}{\log_{10} 14} = \dfrac{1}{2}$ 40. $\dfrac{\log_{10} 10}{\log_{10} 100} = \dfrac{1}{10}$

📝 **41.** Refer to the "NOTE" following the word statement of the product rule for logarithms in this section. Now, state the quotient rule in words, replacing "logarithm" with "exponent."

◉ 📝 **42.** Explain why the statement for the power rule for logarithms requires that x be a positive real number.

◉ **43.** Refer to Example 7(a). Change the right side of the equation using the quotient rule so that the statement becomes true, and simplify.

◉ **44.** What is wrong with the following "proof" that $\log_2 16$ does not exist?

$$\log_2 16 = \log_2(-4)(-4)$$
$$= \log_2(-4) + \log_2(-4)$$

Since the logarithm of a negative number is not defined, the final step cannot be evaluated, and so $\log_2 16$ does not exist.

◆ **MATHEMATICAL CONNECTIONS** (Exercises 45–50) ◆

Work Exercises 45–50 in order.

45. Evaluate $\log_3 81$.

46. Write the *meaning* of the expression $\log_3 81$.

47. Evaluate $3^{\log_3 81}$.

48. Write the *meaning* of the expression $\log_2 19$.

49. Evaluate $2^{\log_2 19}$.

50. Keeping in mind that a logarithm is an exponent, and using the results from Exercises 45–49, what is the simplest form of the expression $k^{\log_k m}$?

REVIEW EXERCISES

Write the exponential statement in logarithmic form. See Section 10.3.

51. $10^4 = 10,000$ **52.** $10^{1/2} = \sqrt{10}$ **53.** $10^{-2} = .01$

Write the logarithmic statement in exponential form. See Section 10.3.

54. $\log_{10} .001 = -3$ **55.** $\log_{10} 1 = 0$ **56.** $\log_{10} \sqrt[3]{10} = \dfrac{1}{3}$

43. $\log_2 8 - \log_2 4 =$
$\log_2 \dfrac{8}{4} = \log_2 2 = 1$
44. We cannot apply the product rule, since
$\log_b xy = \log_b x + \log_b y$
only if x and y are positive numbers.
45. 4   **46.** It is the exponent to which 3 must be raised in order to obtain 81.   **47.** 81
48. It is the exponent to which 2 must be raised in order to obtain 19.
49. 19   **50.** m
51. $\log_{10} 10,000 = 4$
52. $\log_{10} \sqrt{10} = \dfrac{1}{2}$
53. $\log_{10} .01 = -2$
54. $10^{-3} = .001$
55. $10^0 = 1$
56. $10^{1/3} = \sqrt[3]{10}$

10.5 EVALUATING LOGARITHMS

FOR EXTRA HELP	OBJECTIVES
📖 **SSG** pp. 330–334 **SSM** pp. 549–554	**1** ► Evaluate common logarithms by using a calculator.
📼 **Video** 13	**2** ► Use common logarithms in an application.
💾 **Tutorial** IBM MAC	**3** ► Evaluate natural logarithms using a calculator.
	4 ► Use natural logarithms in applications.
	5 ► Use the change-of-base rule.

As mentioned earlier, logarithms are important in many applications of mathematics to everyday problems, particularly in biology, engineering, economics, and social science. In this section we show how to find numerical approximations for

INSTRUCTOR'S RESOURCES

 ITM pp. 512–515 **ISM** pp. 830–838
IAM pp. 54

 TEST GENERATOR
DOS   Windows   MAC

 TRANSPARENCIES

 CONCEPTUAL   📝 WRITING   ▲ CHALLENGING   SCIENTIFIC CALCULATOR   GRAPHICS CALCULATOR

logarithms. Traditionally, base 10 logarithms have been used most extensively, since our number system is base 10. Logarithms to base 10 are called **common logarithms** and $\log_{10} x$ is abbreviated as simply $\log x$, where the base is understood to be 10.

1 ▶ Evaluate common logarithms by using a calculator.

▶ We use calculators to evaluate common logarithms. In the next example we give the results of evaluating some common logarithms using a calculator with a log key. (This may be a second function key on some calculators.) For simple scientific calculators, just enter the number, then touch the log key. For graphics calculators, these steps are reversed. We will give all logarithms to four decimal places.

E X A M P L E   1

Evaluating Common Logarithms

Evaluate each logarithm using a calculator.

(a) $\log 327.1 = 2.5147$

(b) $\log 437{,}000 = 5.6405$

(c) $\log .0615 = -1.2111$

In part (c), $\log .0615$ is found to be -1.2111, a negative result. The common logarithm of a number between 0 and 1 is always negative because the logarithm is the exponent on 10 that produces the number. For example,

$$10^{-1.2111} = .0615.$$

If the exponent (the logarithm) were positive, the result would be greater than 1, since $10^0 = 1$. ■

2 ▶ Use common logarithms in an application.

▶ In chemistry, the **pH** of a solution is defined as follows.

Definition of pH

$$\text{pH} = -\log[\text{H}_3\text{O}^+],$$

where $[\text{H}_3\text{O}^+]$ is the hydronium ion concentration in moles per liter.

The pH is a measure of the acidity or alkalinity of a solution, with water, for example, having a pH of 7. In general, acids have pH numbers less than 7, and alkaline solutions have pH values greater than 7.

E X A M P L E   2

Finding pH

Find the pH of grapefruit with a hydronium ion concentration of 6.3×10^{-4}. Use the definition of pH.

$$\begin{aligned}
\text{pH} &= -\log(6.3 \times 10^{-4}) \\
&= -(\log 6.3 + \log 10^{-4}) \quad \text{Product rule} \\
&= -[.7993 - 4(1)] \\
&= -.7993 + 4 \approx 3.2
\end{aligned}$$

It is customary to round pH values to the nearest tenth. ■

E X A M P L E   3

Finding Hydronium Ion Concentration

Find the hydronium ion concentration of drinking water with a pH of 6.5.

$$\begin{aligned}
\text{pH} = 6.5 &= -\log[\text{H}_3\text{O}^+] \\
\log[\text{H}_3\text{O}^+] &= -6.5 \quad \text{Multiply by } -1.
\end{aligned}$$

Solve for $[H_3O^+]$ by writing the equation in exponential form, remembering that the base is 10.

$$[H_3O^+] = 10^{-6.5} = 3.2 \times 10^{-7}. \quad \blacksquare$$

3 ▶ Evaluate natural logarithms using a calculator.

Chalkboard Exercise

Find the hydronium ion concentration of a solution with a pH of 4.6.

Answer: 2.5×10^{-5}

▶ The most important logarithms used in applications are **natural logarithms,** which have as base the number e. The number e is irrational, like π: $e \approx 2.7182818$. Logarithms to base e are called natural logarithms because they occur in biology and the social sciences in natural situations that involve growth or decay. The base e logarithm of x is written $\ln x$ (read "el en x"). A graph of $y = \ln x$, the natural logarithm function, is given in Figure 15.

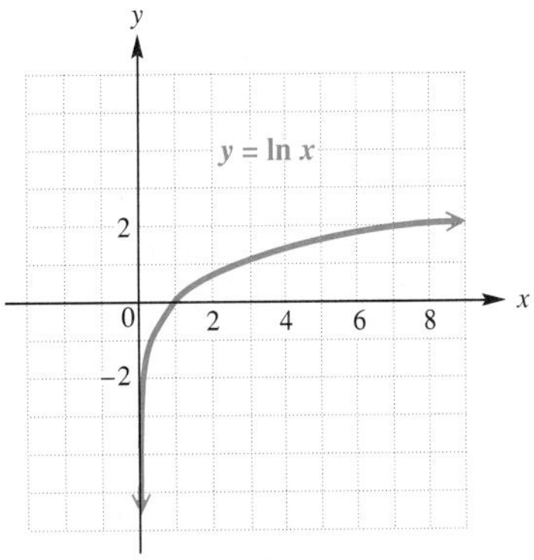

FIGURE 15

◆ 2; 2.5; $2.\overline{6}$; $2.708\overline{3}$; $2.716\overline{6}$; The difference is .0016151618. It approaches e fairly quickly.

▶ **TEACHING TIP**
With many scientific calculators, the hydronium ion concentration in Examples 2 and 3 can be entered directly using the | EE | key or | Exp | key. One need only press the | log | key followed by the | +/− | key to obtain the answer. ◀

◆ **CONNECTIONS** ◆

The number e is a fundamental number in our universe. For this reason, e, like π, is called a *universal constant*. If there are intelligent beings elsewhere, they too will have to use e to do higher mathematics.

The letter e is used to honor Leonhard Euler, who published extensive results on the number in 1748. The first few digits of the decimal value of e are 2.7182818. Since it is an irrational number, its decimal expansion never terminates and never repeats.

The properties of e are used in calculus and in higher mathematics extensively. In Section 10.6 we see how it applies to growth and decay in the physical world.

FOR DISCUSSION OR WRITING
The value of e can be expressed as

$$e = 1 + \frac{1}{1} + \frac{1}{1 \cdot 2} + \frac{1}{1 \cdot 2 \cdot 3} + \frac{1}{1 \cdot 2 \cdot 3 \cdot 4} + \cdots.$$

Approximate e using 2 terms of this expression, then 3 terms, 4 terms, 5 terms, and 6 terms. How close is the approximation to the value of e given above with 6 terms? Does this infinite sum approach the value of e very quickly?

◆ **C O N N E C T I O N S** ◆

🖩 A calculator key labeled ln x is used to evaluate natural logarithms. If your calculator has an e^x key, but not a key labeled ln x, find natural logarithms by entering the number, touching the INV key, and then touching the e^x key. This works because $y = e^x$ is the inverse function of $y = \ln x$ (or $y = \log_e x$).

E X A M P L E   4

Finding Natural Logarithms

■ Find each of the following logarithms to four significant digits.

(a) ln .5841

Enter .5841 and touch the ln key to get ln .5841 $= -.5377$. (Again, with some calculators, the steps are reversed.) As with common logarithms, a number between 0 and 1 has a negative natural logarithm.

(b) ln 192.7 $= 5.261$

(c) ln 10.84 $= 2.383$ ■

4 ▶ Use natural logarithms in applications.

▶ One of the most common applications of exponential functions depends on the fact that in many situations involving growth or decay of a population, the amount or number of some quantity present at time t can be closely approximated by

$$y = y_0 e^{kt},$$

where y_0 is the amount or number present at time $t = 0$, k is a constant, and e is the base of natural logarithms mentioned earlier.

E X A M P L E   5

Applying Natural Logarithms

■ Suppose that the population of a small town in the Southwest is

$$P = 10{,}000e^{.04t},$$

where t represents time, measured in years. The population at time $t = 0$ is

$$P = 10{,}000e^{(.04)(0)}$$
$$= 10{,}000e^0$$
$$= 10{,}000(1) \qquad e^0 = 1$$
$$= 10{,}000.$$

The population of the town is 10,000 at time $t = 0$, that is, at present. The population of the town 5 years from now, at $t = 5$, is

$$P = 10{,}000e^{(.04)(5)}$$
$$= 10{,}000e^{.2}.$$

The number $e^{.2}$ can be found with the e^x key of a calculator. Enter .2, then touch e^x to get 1.22. Alternatively, use the INV and ln x keys for the same result. Use this result to evaluate P.

$$P \approx (10{,}000)(1.22)$$
$$= 12{,}200$$

In 5 years the population of the town will be about 12,200.   ■

E X A M P L E   6

Applying Natural
Logarithms

E X A M P L E   6

Applying Natural
Logarithms

The number of years, $N(r)$, since two independently evolving languages split off from a common ancestral language is approximated by

$$N(r) = -5000 \ln r$$

where r is the percent of words from the ancestral language common to both languages now. Find N if $r = 70\%$.

Write 70% as .7 and find $N(.7)$.

$$N(.7) = -5000 \ln .7$$
$$\approx -5000(-.3567)$$
$$\approx 1783$$

Approximately 1800 years have passed since the two languages separated. ■

**5 ▶ Use the
change-of-base rule.**

▶ A calculator can be used to approximate the values of common logarithms (base 10) or natural logarithms (base e). However, sometimes we need to use logarithms to other bases. The following rule is used to convert logarithms from one base to another.

Change-of-Base Rule

If $a > 0$, $a \neq 1$, $b > 0$, $b \neq 1$, and $x > 0$, then

$$\log_a x = \frac{\log_b x}{\log_b a}.$$

NOTE As an aid in remembering the change-of-base rule, notice that x is "above" a on both sides of the equation.

Any positive number other than 1 can be used for base b in the change-of-base rule, but usually the only practical bases are e and 10, since calculators give logarithms only for these two bases.

To prove the formula for change of base, let $\log_a x = m$.

$$\log_a x = m$$
$$a^m = x \qquad \text{Change to exponential form.}$$

Since logarithmic functions are one-to-one, if all variables are positive and if $x = y$, then $\log_b x = \log_b y$.

$$\log_b(a^m) = \log_b x \qquad \text{Take logarithms on both sides.}$$
$$m \log_b a = \log_b x \qquad \text{Use the power rule.}$$
$$(\log_a x)(\log_b a) = \log_b x \qquad \text{Substitute for } m.$$
$$\log_a x = \frac{\log_b x}{\log_b a} \qquad \text{Divide both sides by } \log_b a.$$

The next example shows how the change-of-base rule is used to find logarithms to bases other than 10 or e with a calculator.

EXAMPLE 7
Using the
Change-of-Base Rule

■ Find each logarithm using a calculator.

(a) $\log_5 12$

Use common logarithms and the rule for change-of-base.

$$\log_5 12 = \frac{\log 12}{\log 5} \approx 1.5440$$

(b) $\log_2 134$

Use natural logarithms and the change-of-base rule.

$$\log_2 134 = \frac{\ln 134}{\ln 2} \approx 7.0661 \ \blacksquare$$

Chalkboard Exercise

Find the logarithm.

$\log_3 17$

Answer: 2.5789

NOTE In Example 7, the final answers were obtained *without* rounding off the intermediate values. In general, it is best to wait until the final step to round off the answer; otherwise, a build-up of round-off error may cause the final answer to have an incorrect final decimal place digit.

10.5 EXERCISES

1. 10   2. *e*   3. 0 and 1
4. 1 and 2   5. 19.2
6. $\sqrt{2}$ 7. 1.6335
8. 1.9912   9. 2.5164
10. 2.6601   11. −1.4868
12. −.7592   13. 9.6776
14. 4.3284   15. 2.0592
16. 2.1187   17. −2.8896
18. −3.8304   19. 5.9613
20. 6.8486   21. 4.1506
22. 5.0096   23. 2.3026
24. .4343
25. (a) 2.552424846
(b) 1.552424846
(c) 0.552424846
(d) The whole number parts will vary but the decimal parts are the same.
26. Answers will vary. Suppose the last name is Miller, with $k = 6$.
(a) .7781512504   (b) 6
(c) Because $y = \log x$ and $y = 10^x$ are inverses, we will always obtain k as our answer.

◉ **1.** What base is understood in the expression $\log x$?

◉ **2.** What base is understood in the expression $\ln x$?

◉ **3.** Since $10^0 = 1$ and $10^1 = 10$, between what two consecutive integers is the value of $\log 5$?

◉ **4.** Since $e^1 \approx 2.718$ and $e^2 \approx 7.389$, between what two consecutive integers is the value of $\ln 4$?

◉ **5.** Without using a calculator, give the value of $\log 10^{19.2}$.

◉ **6.** Without using a calculator, give the value of $\ln e^{\sqrt{2}}$.

You may need a calculator for the remaining exercises in this set.

▦ *Find the logarithm. Give an approximation to the nearest ten-thousandth. See Examples 1 and 4.*

7. $\log 43$ **8.** $\log 98$ **9.** $\log 328.4$

10. $\log 457.2$ **11.** $\log .0326$ **12.** $\log .1741$

13. $\log(4.76 \times 10^9)$ **14.** $\log(2.13 \times 10^4)$ **15.** $\ln 7.84$

16. $\ln 8.32$ **17.** $\ln .0556$ **18.** $\ln .0217$

19. $\ln 388.1$ **20.** $\ln 942.6$ **21.** $\ln(8.59 \times e^2)$

22. $\ln(7.46 \times e^3)$ **23.** $\ln 10$ **24.** $\log e$

▦ **25.** Use your calculator to find approximations of the following logarithms:
(a) $\log 356.8$ (b) $\log 35.68$ (c) $\log 3.568$.
(d) Observe your answers and make a conjecture concerning the decimal values of the common logarithms of numbers greater than 1 that have the same digits.

▦ **26.** Let k represent the number of letters in your last name.
(a) Use your calculator to find $\log k$.
(b) Raise 10 to the power indicated by the number you found in part (a). What is your result?
(c) Use the concepts of Section 10.1 to explain why you obtained the answer you found in part (b). Would it matter what number you used for k to observe the same result?

◎ CONCEPTUAL   ✐ WRITING   ▲ CHALLENGING   ▦ SCIENTIFIC CALCULATOR   ▦ GRAPHICS CALCULATOR

27. Enter -1 into a scientific calculator and touch the log key. What happens? Explain why this happens.

28. Repeat Exercise 27 but use the ln key instead.

29. Which one of the following is not equal to 1? Do not use a calculator.
(a) log 1   (b) log 10   (c) ln e (d) $(\ln 10)^0$

30. Which one of the following is a negative number? Do not use a calculator.
(a) log 125   (b) ln 5   (c) log 1.348   (d) ln .322

Use a calculator to find **(a)** 10^x *and* **(b)** e^x *for the value of x. Give the approximation to the nearest ten-thousandth.*

31. $x = 1.7486$ **32.** $x = 2.9137$ **33.** $x = -2.8476$ **34.** $x = -1.9938$

Use the change-of-base rule (with either common or natural logarithms) to find the following logarithms. Give approximations to the nearest ten-thousandth. See Example 7.

35. $\log_6 13$ **36.** $\log_7 19$ **37.** $\log_{\sqrt{2}} \pi$ **38.** $\log_\pi \sqrt{2}$

39. Let m be the number of letters in your first name, and let n be the number of letters in your last name.
(a) In your own words, explain what $\log_m n$ means.
(b) Use your calculator to find $\log_m n$.
(c) Raise m to the power indicated by the number you found in part (b). What is your result?

40. The equation $5^x = 7$ cannot be solved using the methods described in Section 10.2. However, in solving this equation, we must find the exponent to which 5 must be raised in order to obtain 7: this is $\log_5 7$.
(a) Use the change-of-base rule and your calculator to find $\log_5 7$.
(b) Raise 5 to the number you found in part (a). What is your result?
(c) Using as many decimal places as your calculator gives, write the solution set of $5^x = 7$. (Equations of this type will be studied in more detail in Section 10.6.)

Use the formula $\text{pH} = -\log[H_3O^+]$ *to find the pH of the substance with the given hydronium ion concentration. See Example 2.*

41. ammonia, 2.5×10^{-12} **42.** sodium bicarbonate, 4.0×10^{-9}

43. grapes, 5.0×10^{-5} **44.** tuna, 1.3×10^{-6}

Use the formula for pH *to find the hydronium ion concentration of the substance with the given pH. See Example 3.*

45. human blood plasma,   7.4   **46.** human gastric contents,   2.0

47. spinach,   5.4   **48.** bananas,   4.6

Solve the problem. See Examples 5 and 6.

49. Suppose that the amount, in grams, of plutonium-241 present in a given sample is determined by the function

$$A(t) = 2.00e^{-.053t},$$

where t is measured in years. Find the amount present in the sample after the given number of years.
(a) 4   (b) 10   (c) 20
(d) What was the initial amount present?

29. (a)   30. (d)
31. (a) 56.0531   (b) 5.7466
32. (a) 819.7851
(b) 18.4248
33. (a) .0014   (b) .0580
34. (a) .0101   (b) .1362
35. 1.4315   36. 1.5131
37. 3.3030   38. .3028
39. Answers will vary.
Suppose the name is Paul
Bunyan, with $m = 4$ and
$n = 6$.   (a) $\log_4 6$ is the
exponent to which 4 must
be raised in order to
obtain 6.   (b) 1.29248125
(c) 6 (the value of n)
40. (a) 1.209061955   (b) 7
(c) {1.209061955}
41. 11.6   42. 8.4
43. 4.3   44. 5.9
45. 4.0×10^{-8}
46. 1.0×10^{-2}
47. 4.0×10^{-6}
48. 2.5×10^{-5}
49. (a) 1.62 grams
(b) 1.18 grams
(c) .69 gram
(d) 2.00 grams

50. (a) 3.22 grams
(b) 3.11 grams
(c) 2.62 grams
(d) 3.25 grams
51. 2228 million (or
appoximately 2.2 billion)
52. 333.5 billion dollars
53. 527 years
54. 6020 years
55. 3466 years
56. 1116 years
57. 789.3 thousand (or
789,300)
58. 355.6 thousand (or
3,556,000)
59. In 1989 (when $x = 10$),
the number of births was
about 730,000.

50. Suppose that the amount, in grams, of radium-226 present in a given sample is determined by the function

$$A(t) = 3.25e^{-.00043t},$$

where t is measured in years. Find the amount present in the sample after the given number of years.
(a) 20   **(b)** 100   **(c)** 500
(d) What was the initial amount present?

51. The number of books, in millions, sold per year in the United States between 1985 and 1990 can be approximated by the function

$$N(t) = 1757e^{.0264t},$$

where $t = 0$ corresponds to the year 1985. Based on this model, how many books were sold in 1994? (*Source:* Book Industry Study Group)

52. Personal consumption expenditures for recreation in billions of dollars in the United States during the years 1984 through 1990 can be approximated by the function

$$C(t) = 185.4e^{.0587t},$$

where $t = 0$ corresponds to the year 1984. Based on this model, how much were personal consumption expenditures in 1994? (*Source:* U.S. Bureau of Economic Analysis)

For Exercises 53–56, refer to the function in Example 6.

53. Find $N(.9)$.   **54.** Find $N(.3)$.   **55.** Find $N(.5)$.

56. How many years have elapsed since the split if 80% of the words of the ancestral language are common to both languages today?

57. The number of cesarean section deliveries in the United States has increased over the years. According to statistics provided by the U.S. National Center for Health Statistics, between the years 1980 and 1989, the number of such births, in thousands, can be approximated by the function

$$B(t) = 624.6e^{.0156t},$$

where $t = 1$ corresponds to the year 1980. Based on this model, how many such births were there in 1994?

58. According to an article in *The AMATYC Review* (Spring, 1993), the number of students enrolled, in thousands, in intermediate algebra in two-year colleges since 1966 can be approximated by the function

$$E(t) = 39.8e^{.073t},$$

where $t = 1$ corresponds to 1966. If these trends continue, how many students could we expect to be enrolled in intermediate algebra at the two-year college level in 1995?

59. The function $B(x) = 624.6e^{.0156x}$, described in Exercise 57 with $x = t$, is graphed in a graphics calculator-generated window in the accompanying figure. Interpret the meanings of x and y in the display at the bottom in the context of Exercise 57.

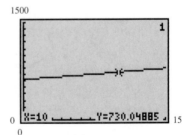

60. The function $E(x) = 39.8e^{.073x}$, described in Exercise 58 with $x = t$, is graphed in a graphics calculator-generated window in the accompanying figure. Interpret the meanings of x and y in the display at the bottom in the context of Exercise 58.

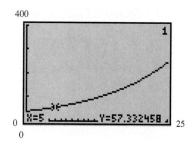

60. In 1970 (when $x = 5$), the number of students was about 57,300.

61.

62.

63.

64.

Because graphics calculators are equipped with log x and ln x keys, it is possible to graph the functions $f(x) = \log x$ and $g(x) = \ln x$ directly, as shown in the figures that follow.

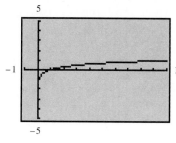

$$f(x) = \log x$$

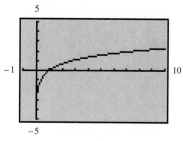

$$g(x) = \ln x$$

65. $\left\{-\dfrac{3}{5}\right\}$ **66.** $\left\{-\dfrac{4}{3}\right\}$

67. {5}   **68.** {8}   **69.** {−3}

70. $\{a \mid a > 0, a \neq 1\}$

71. $\log(x + 2)(x - 3)$ or $\log(x^2 - x - 6)$

72. $\log_4 \dfrac{x + 4}{(3x + 1)^2}$ or $\log_4 \dfrac{x + 4}{9x^2 + 6x + 1}$

In order to graph functions defined by logarithms to bases other than 10 or e, however, we must use the change-of-base rule. For example, to graph $y = \log_2 x$, we may enter y_1 as $(\log x)/(\log 2)$ or $(\ln x)/(\ln 2)$. This is shown in the figure that follows. (Compare it to the figure in the exercises of Section 10.3, where it was drawn using the fact that $y = \log_2 x$ is the inverse of $y = 2^x$.)

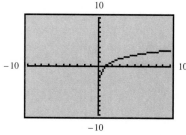

$$y_1 = \log_2 x$$

Use the change-of-base rule to graph the logarithmic function with a graphics calculator. Use a viewing window with $X\mathrm{min} = -1$, $X\mathrm{max} = 10$, $Y\mathrm{min} = -5$, and $Y\mathrm{max} = 5$.

61. $g(x) = \log_3 x$ **62.** $g(x) = \log_5 x$ **63.** $g(x) = \log_{1/3} x$ **64.** $g(x) = \log_{1/5} x$

REVIEW EXERCISES

Solve the equation. See Sections 10.2 and 10.3.

65. $4^{2x} = 8^{3x+1}$ **66.** $2^{5x} = \left(\dfrac{1}{16}\right)^{x+3}$ **67.** $\log_3(x + 4) = 2$

68. $\log_x 64 = 2$ **69.** $\log_{1/2} 8 = x$ **70.** $\log_a 1 = 0$

Write as a single logarithm. See Section 10.4.

71. $\log(x + 2) + \log(x - 3)$ **72.** $\log_4(x + 4) - 2\log_4(3x + 1)$

⊙ CONCEPTUAL   ✐ WRITING   ▲ CHALLENGING   ▦ SCIENTIFIC CALCULATOR   ▦ GRAPHICS CALCULATOR

10.6 EXPONENTIAL AND LOGARITHMIC EQUATIONS AND THEIR APPLICATIONS

FOR EXTRA HELP

SSG pp. 335–342
SSM pp. 554–559

Video 13

Tutorial IBM MAC

OBJECTIVES

1 ▶ Solve equations involving variables in the exponents.
2 ▶ Solve equations involving logarithms.
3 ▶ Solve applications involving compound interest.
4 ▶ Solve applications involving exponential growth and decay.
5 ▶ Use a graphics calculator to solve exponential and logarithmic equations.

◆ about 6.96 grams;
about 4.84 grams;
5730 years

◆ **CONNECTIONS** ◆

Carbon-14 is a radioactive form of carbon that is found in all living plants and animals. After a plant or animal dies, the radioactive carbon-14 disintegrates according to the function

$$y = y_0 e^{-.000121t},$$

where t is time in years, and y is the amount of the sample at time t.

You have probably heard of the carbon-14 dating process used to determine the age of fossils. The method used is based on the exponential decay function.

FOR DISCUSSION OR WRITING

If an initial sample contains $y_0 = 10$ grams of carbon-14, how many grams will be present after 3000 years? How many grams of the initial 10-gram sample will be present after 6000 years? The *half-life* of a radioactive substance is the time until half the sample has disintegrated. What is the half-life of carbon-14?

▶ **TEACHING TIP**

Also, state the properties for equation solving in words as follows:

1. If exponents are equal, then the bases are equal.
2. If bases are equal, then the exponents are equal.
3. If two numbers are equal, then the logarithms of these numbers are equal.
4. If the logarithms of two numbers are equal, then the numbers are equal. ◀

As mentioned earlier, exponential and logarithmic functions are important in many applications of mathematics. Using these functions in applications requires solving exponential and logarithmic equations. Some simple equations were solved in Sections 10.2 and 10.3. More general methods for solving these equations depend on the following properties.

Properties for Equation Solving

For all real numbers $b > 0$, $b \neq 1$, and any real numbers x and y:

1. If $x = y$, then $b^x = b^y$.
2. If $b^x = b^y$, then $x = y$.
3. If $x = y$, and $x > 0$, $y > 0$, then $\log_b x = \log_b y$.
4. If $x > 0$, $y > 0$, and $\log_b x = \log_b y$, then $x = y$.

We used property 2 to solve exponential equations in Section 10.2 and property 3 was used in the proof of the change-of-base rule in the previous section. We will refer to these properties by number throughout this section.

1 ▶ Solve equations involving variables in the exponents.

▶ The first examples illustrate a general method for solving exponential equations using property 3.

INSTRUCTOR'S RESOURCES

 ITM pp. 515–518 ISM pp. 838–848
IAM pp. 54–55

 TEST GENERATOR
DOS Windows MAC

TRANSPARENCIES

EXAMPLE 1

Solving an Exponential Equation

▶ **TEACHING TIP**
An alternate approach to this example is as follows:

$3^m = 12$ Exponential form

becomes

$\log_3 12 = m$ Log form

so $m = \dfrac{\log 12}{\log 3}$

$= 2.262$

using the change-of-base formula. ◀

Solve the equation $3^m = 12$. Give the answer in decimal form.

$$3^m = 12$$

$$\log 3^m = \log 12 \qquad \text{Property 3}$$

$$m \log 3 = \log 12 \qquad \text{Power rule for logarithms}$$

$$m = \frac{\log 12}{\log 3}$$

This quotient is the exact solution. To get a decimal approximation for the solution, we use a calculator. Correct to three decimal places, a calculator gives

$$m = 2.262,$$

and the solution set is {2.262}. ∎

CAUTION Be careful: $\dfrac{\log 12}{\log 3}$ is *not* equal to log 4, since log 4 = .6021, but

$$\frac{\log 12}{\log 3} = 2.262.$$

When an exponential equation has e as the base, it is easiest to use base e logarithms.

EXAMPLE 2

Solving an Exponential Equation with Base e

▶ **TEACHING TIP**
Start with a simple example of this type:
Solve $x^3 = 8$.

$x^3 = 8$

$(x^3)^{1/3} = 8^{1/3}$ Raise both sides to the reciprocal power.

$x = 2$ ◀

Solve $e^{.003x} = 40$.

Take base e logarithms on both sides.

$$\ln e^{.003x} = \ln 40$$

$$.003x \ln e = \ln 40 \qquad \text{Power rule for logarithms}$$

$$.003x = \ln 40 \qquad \ln e = \ln e^1 = 1$$

$$x = \frac{\ln 40}{.003} \qquad \text{Divide by .003.}$$

$$x \approx 1230 \qquad \text{Use a calculator.}$$

The solution set is {1230}. Check that $e^{.003(1230)} \approx 40$. ∎

In summary, exponential equations can be solved by one of the following methods. (The method used depends on the form of the equation.) Examples 1 and 2 illustrate Method 1. We gave examples of Method 2 in Section 10.2.

Solving an Exponential Equation

1. Using property 3, take logarithms to the same base on each side; then use the power rule of logarithms.
2. Using property 2, write both sides as exponentials with the same base; then set the exponents equal.

2 ▶ Solve equations involving logarithms.

▶   The next three examples illustrate ways to solve equations with logarithms. The properties of logarithms from Section 10.4 are useful here, as is using the definition of a logarithm to change to exponential form.

EXAMPLE 3

Solving a Logarithmic Equation

Solve $\log_2(x + 5)^3 = 4$.

$$(x + 5)^3 = 2^4 \qquad \text{Convert to exponential form.}$$
$$(x + 5)^3 = 16$$
$$x + 5 = \sqrt[3]{16} \qquad \text{Take cube roots on both sides.}$$
$$x = -5 + \sqrt[3]{16}$$

Verify that the solution satisfies the equation, so the solution set is $\{-5 + \sqrt[3]{16}\}$.

CAUTION Recall that the domain of $y = \log_b x$ is $(0, \infty)$. For this reason, it is always necessary to check that the solution of an equation with logarithms yields only logarithms of positive numbers in the original equation.

EXAMPLE 4

Solving a Logarithmic Equation

Solve $\log_2(x + 1) - \log_2 x = \log_2 8$.

$$\log_2(x + 1) - \log_2 x = \log_2 8$$
$$\log_2 \frac{x + 1}{x} = \log_2 8 \qquad \text{Quotient rule}$$
$$\frac{x + 1}{x} = 8 \qquad \text{Property 4}$$
$$8x = x + 1 \qquad \text{Multiply by } x.$$
$$x = \frac{1}{7} \qquad \text{Subtract } x; \text{ divide by 7.}$$

Check this solution by substitution in the given equation. Here, both $x + 1$ and x must be positive. If $x = 1/7$, this condition is satisfied, and the solution set is $\{1/7\}$.

EXAMPLE 5

Solving a Logarithmic Equation

▶ **TEACHING TIP**
Before working an example like Examples 4 and 5:

1. Review the product, quotient, and power rules for logarithms.
2. Inform students that a logarithmic equation cannot be solved with *two* logarithms on one side of the equation (the equation must be translated into an expression with only one logarithm). ◀

Solve $\log x + \log(x - 21) = 2$.

For this equation, write the left side as a single logarithm. Then write in exponential form and solve the equation.

$$\log x + \log(x - 21) = 2$$
$$\log x(x - 21) = 2 \qquad \text{Product rule}$$
$$x(x - 21) = 10^2 \qquad \log x = \log_{10} x; \text{ write in exponential form}$$
$$x^2 - 21x = 100$$
$$x^2 - 21x - 100 = 0 \qquad \text{Standard form}$$
$$(x - 25)(x + 4) = 0 \qquad \text{Factor.}$$
$$x - 25 = 0 \quad \text{or} \quad x + 4 = 0 \qquad \text{Set each factor equal to 0.}$$
$$x = 25 \quad \text{or} \quad x = -4$$

The value -4 must be rejected as a solution, since it leads to the logarithm of a negative number in the original equation:

$$\log(-4) + \log(-4 - 21) = 2. \qquad \text{The left side is not defined.}$$

The only solution, therefore, is 25, and the solution set is $\{25\}$.

CAUTION Do not reject a potential solution just because it is nonpositive. Reject any value that *leads to* the logarithm of a nonpositive number.

In summary, use the following steps to solve a logarithmic equation.

Solving a Logarithmic Equation

Step 1 Use the product or quotient rules of logarithms to get a single logarithm on one side.

Step 2 **(a)** Use property 4: If $\log_b x = \log_b y$, then $x = y$. (See Example 4.)

(b) Write the equation in exponential form: if $\log_b x = k$, then $x = b^k$. (See Example 3 and Example 5.)

3 ▶ Solve applications involving compound interest.

▶ Logarithms can be used to solve problems involving compound interest. So far in this book, problems involving applications of interest have been limited to the use of the simple interest formula, $I = prt$. In most cases, banks pay compound interest (interest paid on both principal and interest). The formula for compound interest is an important application of exponential functions.

Compound Interest Formula

If P dollars is deposited in an account paying an annual rate of interest r compounded (paid) n times per year, the account will contain

$$A = P\left(1 + \frac{r}{n}\right)^{nt}$$

dollars after t years.

In the formula above, r is usually expressed as a decimal.

EXAMPLE 6

Solving a Compound Interest Problem

How much money will there be in an account at the end of 5 years if $1000 is deposited at 6% compounded quarterly? (Assume no withdrawals are made.)

Since interest is compounded quarterly, $n = 4$. The other values given in the problem are $A = 1000$, $r = .06$ (since $6\% = .06$), and $t = 5$. Substitute into the compound interest formula to get the value of A.

$$A = 1000\left(1 + \frac{.06}{4}\right)^{4 \cdot 5}$$

$$A = 1000(1.015)^{20}$$

Now use the y^x key on a calculator, and round the answer to the nearest cent.

$$A = 1346.86$$

The account will contain $1346.86. (The actual amount of interest earned is $1346.86 - \$1000 = \$346.86$. Do you see why?) ■

◆ $1221.40; about 17.3 years; about 13.9 years; as *r* increases, doubling time decreases.

▶ **TEACHING TIP**

Refer to *P* as the amount at "present" and *A* as the amount "after." It should make sense to students that e^{rt} is a multiplier. ◀

◆ **CONNECTIONS** ◆

Interest can be compounded annually, semiannually, quarterly, daily, and so on. The number of compounding periods can get larger and larger. If the value of *n* is allowed to approach infinity, we have an example of *continuous compounding*. However, the compound interest formula above cannot be used for continuous compounding, since there is no finite value for *n*. The formula for continuous compounding is an example of exponential growth involving the number *e*. If a principal of *P* dollars is deposited at a rate of interest *r* compounded continuously for *t* years, the final amount on deposit is $A = Pe^{rt}$.

FOR DISCUSSION OR WRITING

If $1000 is deposited at 4% annual interest compounded continuously for 5 years, find the final amount. Find the time it would take for the amount to double at 4% annual interest. This is called the **doubling time.** What is the doubling time at 5%? What happens to the doubling time as *r* increases?

---

4 ▶ Solve applications involving exponential growth and decay.

▶ We saw some applications involving exponential growth and decay in Section 10.2. In many cases, quantities grow or decay according to a function defined by an exponential expression with base *e*. The next example illustrates this.

E X A M P L E 7

Solving an Exponential Decay Problem

Nuclear energy derived from radioactive isotopes can be used to supply power to space vehicles. The output of the radioactive power supply for a certain satellite is given by the function

$$y = 40e^{-.004t},$$

where *y* is in watts and *t* is the time in days.

(a) How much power will be available at the end of 180 days?

Let $t = 180$ in the formula.

$$y = 40e^{-.004(180)}$$
$$y \approx 19.5 \qquad \text{Use a calculator.}$$

About 19.5 watts will be left.

(b) How long will it take for the amount of power to be half of its original strength?

The original amount of power is 40 watts. (Why?) Since half of 40 is 20, replace *y* with 20 in the formula, and solve for *t*.

$$20 = 40e^{-.004t}$$
$$.5 = e^{-.004t} \qquad \text{Divide by 40.}$$
$$\ln .5 = -.004t \qquad \ln e^k = k$$
$$t = \frac{\ln .5}{-.004}$$
$$t \approx 173 \qquad \text{Use a calculator.}$$

After about 173 days, the amount of available power will be half of its original amount. ■

▶ **TEACHING TIP**

An alternative approach is to write $.5 = e^{-.004t}$ in logarithmic form.

$$\log_e .5 = -.004t$$

or $\ln .5 = -.004t$

Thus, $t = \dfrac{\ln .5}{-.004}$

$$\approx 173 \text{ days.} \quad ◀$$

In Example 7(b), we found the amount of time that it would take for an amount to decrease by half. This is an example of half-life. The **half-life** of a quantity that decays exponentially is the amount of time that it takes for any initial amount to decay to half its value.

5 ▶ Use a graphics calculator to solve exponential and logarithmic equations.

▶ Earlier, we saw how the x-intercepts of the graph of a function f correspond to the real solutions of the equation $f(x) = 0$. The ideas presented there dealt with linear and quadratic functions. We can extend those ideas to exponential and logarithmic functions. For example, consider the equation $8^x = 4$. Using traditional methods, we can determine that the solution set of this equation is $\{2/3\}$. Now if we write the equation with 0 on one side, we get $8^x - 4 = 0$. Graphing $y_1 = 8^x - 4$ and finding the x-intercept supports this solution, as seen in Figure 16. The x-intercept is a decimal approximation for the solution $2/3$.

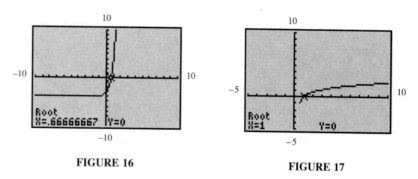

FIGURE 16   **FIGURE 17**

To find the solution of the logarithmic equation $\log(2x - 1) + \log 10x = \log 10$, we graph $y_1 = \log(2x - 1) + \log 10x - \log 10$. The x-intercept is 1, supporting the result obtained when the equation is solved algebraically. See Figure 17.

10.6 EXERCISES

A calculator with log *and* ln *keys will be essential in working many of the exercises in this section. Different models may give answers that differ by a few cents in answers involving money.*

1. Using the method described in Objective 3 of Section 10.2, the equation $5^x = 125$ is solved by writing 125 as 5^3, and then solving $5^x = 5^3$ by setting exponents equal to get the solution set $\{3\}$. Solve this same equation using the method of Example 1 in this section. Use your calculator, and do not round off in intermediate steps.

2. Repeat Exercise 1 above for the equation $9^x = 27$, which was solved in Example 4 of Section 10.2.

3. By inspection, determine the solution set of $2^x = 64$. Then without using a calculator, give the exact value of $\dfrac{\log 64}{\log 2}$.

4. Which one of the following is *not* a solution of $7^x = 23$?
   (a) $\dfrac{\log 23}{\log 7}$ (b) $\dfrac{\ln 23}{\ln 7}$ (c) $\log_7 23$ (d) $\log_{23} 7$

1. $\{3\}$ 2. $\left\{\dfrac{3}{2}\right\}$

3. $\{6\}$; $\dfrac{\log 64}{\log 2} = 6$

4. (d)

 CONCEPTUAL   ✎ WRITING   ▲ CHALLENGING   ▦ SCIENTIFIC CALCULATOR   ▦ GRAPHICS CALCULATOR

5. {.827}   6. {.792}
7. {.833}   8. {−.725}
9. {2.269}   10. {−13.257}
11. {566.866}
12. {261.291}
13. {−18.892}
14. {−10.718}
15. {43.301}   16. {5.879}
19. $\left\{\dfrac{2}{3}\right\}$ 20. $\left\{\dfrac{143}{12}\right\}$
21. $\{-1 + \sqrt[3]{49}\}$
22. $\{3 + 4\sqrt[3]{4}\}$
25. $\left\{\dfrac{1}{3}\right\}$ 26. {−5}
27. {2}   28. {4}   29. 0
30. 0   31. {8}   32. {1}
33. $\left\{\dfrac{4}{3}\right\}$ 34. $\left\{\dfrac{1}{2}\right\}$
35. {8}   36. {4}
37. $2539.47
38. $ 3828.78
39. about 180 grams
40. about 21.66 years
41. (a) $11,260.96
(b) $11,416.64
(c) $11,497.99
(d) $11,580.90
(e) $11,581.83

▦ *Solve the equation. Give solutions to the nearest thousandth. See Example 1.*

5. $7^x = 5$

6. $4^x = 3$

7. $9^{-x+2} = 13$

8. $6^{-t+1} = 22$

▲ **9.** $2^{y+3} = 5^y$

▲ **10.** $6^{m+3} = 4^m$

▦ *Use natural logarithms to solve the equation. Give solutions to the nearest thousandth. See Example 2.*

11. $e^{.006x} = 30$

12. $e^{.012x} = 23$

13. $e^{-.103x} = 7$

14. $e^{-.205x} = 9$

▲ **15.** $\ln e^{.04x} = \sqrt{3}$

▲ **16.** $\ln e^{.45x} = \sqrt{7}$

◉ **17.** Try solving one of the equations in Exercises 11–14 using common logarithms rather than natural logarithms. (You should get the same solution.) Explain why using natural logarithms is a better choice.

◉ **18.** If you were asked to solve $10^{.0025x} = 75$, would natural or common logarithms be a better choice? Explain your answer.

Solve the equation. Give the exact solution. See Example 3.

19. $\log_3(6x + 5) = 2$

20. $\log_5(12x − 8) = 3$

21. $\log_7(x + 1)^3 = 2$

22. $\log_4(y − 3)^3 = 4$

◉ **23.** Suppose that in solving a logarithmic equation having the term $\log_4(x − 3)$ you obtain an apparent solution of 2. All algebraic work is correct. Explain why 2 must be rejected as a solution of the equation.

◉ **24.** Suppose that in solving a logarithmic equation having the term $\log_7(3 − x)$, you obtain an apparent solution of −4. All algebraic work is correct. Should you reject −4 as a solution of the equation? Explain why or why not.

Solve the equation. Give the exact solution. See Examples 4 and 5.

25. $\log(6x + 1) = \log 3$

26. $\log(7 − x) = \log 12$

27. $\log_5(3t + 2) − \log_5 t = \log_5 4$

28. $\log_2(x + 5) − \log_2(x − 1) = \log_2 3$

29. $\log 4x − \log(x − 3) = \log 2$

30. $\log(−x) + \log 3 = \log(2x − 15)$

31 $\log_2 x + \log_2(x − 7) = 3$

32. $\log(2x − 1) + \log 10x = \log 10$

33. $\log 5x − \log(2x − 1) = \log 4$

34. $\log_3 x + \log_3(2x + 5) = 1$

35. $\log_2 x + \log_2(x − 6) = 4$

36. $\log_2 x + \log_2(x + 4) = 5$

▦ *Solve the problem. See Examples 6 and 7.*

37. How much money will there be in an account at the end of 6 years if $2000.00 is deposited at 4% compounded quarterly? (Assume no withdrawals are made.)

38. How much money will there be in an account at the end of 7 years if $3000.00 is deposited at 3.5% compounded quarterly? (Assume no withdrawals are made.)

39. A sample of 400 grams of lead-210 decays to polonium-210 according to the function

$$A(t) = 400e^{-.032t},$$

where t is time in years. How much lead will be left in the sample after 25 years?

40. How long will it take the initial sample of lead in Exercise 39 to decay to half of its original amount?

41. Find the amount of money in an account after 12 years if $5000 is deposited at 7% annual interest compounded as follows.
(a) annually   **(b)** semiannually   **(c)** quarterly
(d) daily (Use $n = 365$.)   **(e)** continuously

42. How much money will be in an account at the end of 8 years if $4500 is deposited at 6% annual interest compounded as follows?
(a) annually   **(b)** semiannually   **(c)** quarterly
(d) daily (Use $n = 365$.)   **(e)** continuously

43. How much money must be deposited today to become $1850 in 40 years at 6.5% compounded continuously?

44. How much money must be deposited today to amount to $1000 in 10 years at 5% compounded continuously?

45. The concentration y of a drug in a person's system decreases according to the relationship

$$y = 2e^{-.125t},$$

where y is in appropriate units, and t is in hours. Find the amount of time that it will take for the concentration to be half of its original value.

46. The number y of ants in an anthill grows according to the function

$$y = 300e^{.4t}$$

where t is time measured in days. Find the time it will take for the number of ants to double.

47. Carbon-14 is a radioactive form of carbon that is found in all living plants and animals. After a plant or animal dies, the radioactive carbon-14 disintegrates according to the function

$$y = y_0 e^{-.000121t},$$

where t is time in years, and y is the amount of the sample at time t.
(a) If an initial sample contains $y_0 = 20$ grams of carbon-14, how many grams will be present after 4000 years?
(b) How many grams of the initial 20-gram sample will be present after 8000 years?
(c) What is the half-life of carbon-14?

48. Radioactive strontium decays according to the function

$$y = y_0 e^{-.0239t},$$

where t is time in years.
(a) If an initial sample contains $y_0 = 5$ grams of radioactive strontium, how many grams will be present after 20 years?
(b) How many grams of the initial 5-gram sample will be present after 60 years?
(c) What is the half-life of radioactive strontium?

 As explained in Objective 5, we can find the solution set of either an exponential equation or a logarithmic equation of the type found in this section using a graphics calculator. Provide a graphical solution to the equation given in the form $y_1 = y_2$ by entering the function $y_1 - y_2$ and finding the x-intercept of the graph. Note that the equation was solved using algebraic methods earlier in this section.

49. $7^x = 5$ (Exercise 5)   **50.** $4^x = 3$ (Exercise 6)

51. $\log(6x + 1) = \log 3$ (Exercise 25)   **52.** $\log(7 - x) = \log 12$ (Exercise 26)

 **53.** Refer to Figure 17. Explain how the graph supports the solution found algebraically in Exercise 32 of this section.

 54. Use a graphics calculator to show why the equations

$$\log x^2 = 1 \quad \text{and} \quad 2 \log x = 1$$

do not have the same solution set. Why does the power rule from Section 10.4 not apply here?

Answers (right column):

42. (a) $7172.32
(b) $7221.18
(c) $7246.46
(d) $7272.05
(e) $7272.33
43. $137.41   44. $606.53
45. 5.55 hours
46. 1.73 days
47. (a) 12.33 grams
(b) 7.60 grams
(c) about 5730 years
48. (a) 3.10 grams
(b) 1.19 grams
(c) 29 years
49. {.827}

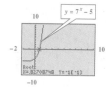

50. {.792}

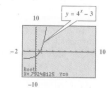

51. $\left\{\dfrac{1}{3}\right\}$ or {.333}

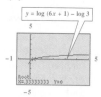

52. {−5}

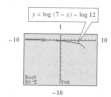

54. The soution set for the first equation has two elements, while the solution set for the second equation has one element. Log x^2 is not equivalent to $2 \log x$ if $x < 0$.

55. (a) 6 (b) $\dfrac{7}{2}$ (c) $\dfrac{8}{3}$

(d) $\dfrac{9}{4}$ 56. (a) 0 (b) $\dfrac{1}{3}$

(c) $\dfrac{1}{2}$ (d) $\dfrac{3}{5}$ 57. (a) 0

(b) 2 (c) 6 (d) 12

58. (a) -2 (b) -2

(c) 0 (d) 4

REVIEW EXERCISES

*Evaluate the expression for (**a**) $n = 1$, (**b**) $n = 2$, (**c**) $n = 3$, and (**d**) $n = 4$. See Section 1.2.*

55. $\dfrac{n + 5}{n}$ **56.** $\dfrac{n - 1}{n + 1}$ **57.** $n^2 - n$ **58.** $n(n - 3)$

CHAPTER 10 SUMMARY

KEY TERMS

10.1 one-to-one function
inverse of a
function

10.2 asymptote
exponential
equation

10.3 logarithm
logarithmic
equation

10.5 common logarithm
natural logarithm

10.6 doubling time
half-life

NEW SYMBOLS

$f^{-1}(x)$ the inverse of $f(x)$
$\log_a x$ the logarithm of x to the base a
$\log x$ common (base 10) logarithm of x
$\ln x$ natural (base e) logarithm of x
e a constant, approximately 2.7182818

QUICK REVIEW

CONCEPTS	EXAMPLES
10.1 INVERSE FUNCTIONS	
Horizontal Line Test If a horizontal line intersects the graph of a function in no more than one point, then the function is one-to-one.	Find f^{-1} if $f(x) = 2x - 3$. The graph of f is a straight line, so f is one-to-one by the horizontal line test.
Inverse Functions For a one-to-one function f defined by an equation $y = f(x)$, the defining equation of the inverse function f^{-1} is found by exchanging x and y, solving for y, and replacing y with $f^{-1}(x)$.	Exchange x and y in the equation $y = 2x - 3$. $$x = 2y - 3$$ Solve for y to get $\quad y = \dfrac{1}{2}x + \dfrac{3}{2}.$ Therefore, $\quad f^{-1}(x) = \dfrac{1}{2}x + \dfrac{3}{2}.$
The graph of f^{-1} is a mirror image of the graph of f with respect to the line $y = x$.	The graphs of a function f and its inverse f^{-1} are given here. 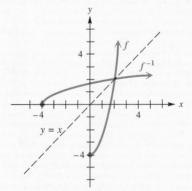

CONCEPTS	EXAMPLES
10.2 EXPONENTIAL FUNCTIONS	

For $a > 0$, $a \neq 1$, $f(x) = a^x$ is an exponential function with base a.

Graph of $f(x) = a^x$
The graph contains the point $(0, 1)$. When $a > 1$, the graph rises from left to right. When $0 < a < 1$, the graph falls from left to right. The x-axis is an asymptote. The domain is $(-\infty, \infty)$; the range is $(0, \infty)$.

$f(x) = 3^x$ is an exponential function with base 3. Its graph is shown here.

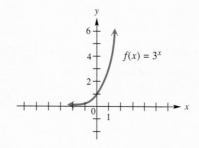

| **10.3 LOGARITHMIC FUNCTIONS** | |

$y = \log_a x$ has the same meaning as $a^y = x$.

For $b > 0$, $b \neq 1$, $\log_b b = 1$ and $\log_b 1 = 0$.

For $a > 0$, $a \neq 1$, $x > 0$, $g(x) = \log_a x$ is the logarithmic function with base a.

Graph of $g(x) = \log_a x$
The graph contains the point $(1, 0)$. When $a > 1$, the graph rises from left to right. When $0 < a < 1$, the graph falls from left to right. The y-axis is an asymptote. The domain is $(0, \infty)$; the range is $(-\infty, \infty)$.

$y = \log_2 x$ means $x = 2^y$.

$$\log_3 3 = 1, \qquad \log_5 1 = 0$$

$g(x) = \log_3 x$ is the logarithmic function with base 3. Its graph is shown here.

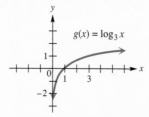

| **10.4 PROPERTIES OF LOGARITHMS** | |

Product Rule
$$\log_a xy = \log_a x + \log_a y$$

$$\log_2 3m = \log_2 3 + \log_2 m \quad (m > 0)$$

Quotient Rule
$$\log_a \frac{x}{y} = \log_a x - \log_a y$$

$$\log_5 \frac{9}{4} = \log_5 9 - \log_5 4$$

Power Rule
$$\log_a x^r = r \log_a x$$

$$\log_{10} 2^3 = 3 \log_{10} 2$$

Special Properties
$$b^{\log_b x} = x \quad \text{and} \quad \log_b b^x = x$$

$$6^{\log_6 10} = 10 \quad \text{and} \quad \log_3 3^4 = 4$$

CONCEPTS	EXAMPLES
10.5 EVALUATING LOGARITHMS	
Change-of-Base Rule If $a > 0$, $a \neq 1$, $b > 0$, $b \neq 1$, $x > 0$, then $$\log_a x = \frac{\log_b x}{\log_b a}.$$	$$\log_3 17 = \frac{\ln 17}{\ln 3} = \frac{\log 17}{\log 3} \approx 2.579$$
10.6 EXPONENTIAL AND LOGARITHMIC EQUATIONS AND THEIR APPLICATIONS	
To solve exponential equations, use these properties ($b > 0$, $b \neq 1$). **1.** If $b^x = b^y$, then $x = y$.	Solve. $\quad\quad 2^{3x} = 2^5$ $$3x = 5$$ $$x = \frac{5}{3}$$ The solution set is $\left\{ \dfrac{5}{3} \right\}$.
2. If $x = y$, $(x > 0, y > 0)$, then $\log_b x = \log_b y$.	Solve. $\quad\quad 5^m = 8$ $$\log 5^m = \log 8$$ $$m \log 5 = \log 8$$ $$m = \frac{\log 8}{\log 5} \approx 1.29$$ The solution set is $\{1.29\}$.
To solve logarithmic equations, use these properties, where $b > 0$, $b \neq 1$, $x > 0$, $y > 0$. First use the properties of Section 10.4, if necessary, to get the equation in the proper form. **1.** If $\log_b x = \log_b y$, then $x = y$.	Solve. $\quad\quad \log_3 2x = \log_3(x + 1)$ $$2x = x + 1$$ $$x = 1$$ The solution set is $\{1\}$.
2. If $\log_b x = y$, then $b^y = x$.	Solve. $\quad\quad \log_2(3a - 1) = 4$ $$3a - 1 = 2^4 = 16$$ $$3a = 17$$ $$a = \frac{17}{3}$$ The solution set is $\left\{ \dfrac{17}{3} \right\}$.

CHAPTER 10 REVIEW EXERCISES

[10.1] *Determine whether the graph is the graph of a one-to-one function.*

1.

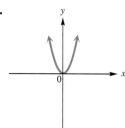

2.

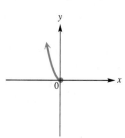

Determine whether the function is one-to-one. If it is, find its inverse.

3. $f(x) = -3x + 7$ **4.** $f(x) = \sqrt[3]{6x - 4}$ **5.** $f(x) = -x^2 + 3$ **6.** $f(x) = x$

The function graphed is one-to-one. Graph its inverse.

7.

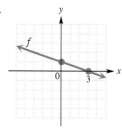

8.

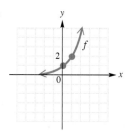

[10.2] *Graph the function.*

9. $f(x) = 3^x$ **10.** $f(x) = \left(\dfrac{1}{3}\right)^x$ **11.** $y = 3^{x+1}$ **12.** $y = 2^{2x+3}$

Solve the equation.

13. $4^{3x} = 8^{x+4}$ **14.** $\left(\dfrac{1}{27}\right)^{x-1} = 9^{2x}$

◉ **15.** What is the *y*-intercept of the graph of $y = a^x$ $(a > 0,\ a \neq 1)$?

◉ **16.** How does the answer to Exercise 15 reinforce the definition of 0 as an exponent?

🖉

17. The production of an oil well, in millions of barrels, is decreasing according to the function

$$f(t) = 2.40^{-.4t},$$

where *t* is time in years. Find the production at the following times.
(a) $t = 0$ **(b)** $t = 2$ **(c)** $t = 5$ **(d)** $t = 10$
(e) Graph the function.

[10.3] *Graph the function.*

18. $g(x) = \log_3 x$ (*Hint:* See Exercise 9.) **19.** $g(x) = \log_{1/3} x$ (*Hint:* See Exercise 10.)

---

1. not one-to-one
2. one-to-one
3. $f^{-1}(x) = \dfrac{x - 7}{-3}$ or $\dfrac{7 - x}{3}$
4. $f^{-1}(x) = \dfrac{x^3 + 4}{6}$
5. not one-to-one
6. $f^{-1}(x) = x$
7.

8.

9.

10.

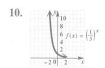

11.

12.

13. $\{4\}$ 14. $\left\{\dfrac{3}{7}\right\}$

15. $(0, 1)$
17. (a) 1 million barrels
(b) .50 million barrels
(c) .17 million barrels
(d) .03 million barrels
(e)

RESOURCES FOR REVIEW

📖 **ITM** pp. 495–518 **ISM** pp. 848–861
 IAM pp. 55–56

📖 **SSG** pp. 313–342
 SSM pp. 559–571

💾 **TEST GENERATOR**
 DOS Windows MAC

◉ CONCEPTUAL 🖉 WRITING ▲ CHALLENGING ▦ SCIENTIFIC CALCULATOR ▦ GRAPHICS CALCULATOR

18.

$g(x) = \log_3 x$

19.

$g(x) = \log_{1/3} x$

20. $\{2\}$ **21.** $\left\{\dfrac{3}{2}\right\}$

22. $\{-2\}$

23. $\{8\}$ **24.** $\{4\}$

25. $\{b \mid b > 0, b \ne 1\}$

27. a

28. $300{,}000

400 $S(x)$
300
200
100 $S(x) = 100 \log_2(x + 2)$
0 | 2 4 6 8 | x
Weeks
Sales in Thousands of Dollars

29. $\log_4 3 + 2 \log_4 x$

30. $2 \log_2 p + \log_2 r -$

$\dfrac{1}{2} \log_2 z$ **31.** $\log_b \dfrac{3x}{y^2}$

32. $\log_3 \dfrac{x + 7}{4x + 6}$

33. 1.4609 **34.** $-.5901$

35. 4.8613 **36.** 3.3638

37. -1.3587 **38.** 4.8613

39. $.9251$ **40.** 1.7925

41. 1.4315 **42.** 6.4

43. 8.4 **44.** (a) $1{,}000{,}000$

(b) about $1{,}040{,}800$

45. (a) 500 grams

(b) about 409 grams

46. $\{2.042\}$ **47.** $\{18.310\}$

Solve the equation.

20. $\log_8 64 = x$

21. $\log_2 \sqrt{8} = x$

22. $\log_7\left(\dfrac{1}{49}\right) = x$

23. $\log_4 x = \dfrac{3}{2}$

24. $\log_k 4 = 1$

25. $\log_b b^2 = 2$

26. In your own words, explain the meaning of $\log_b a$.

27. Based on the meaning of $\log_b a$, what is the simplest form of $b^{\log_b a}$?

28. A company has found that total sales, in thousands of dollars, are given by the function

$$S(x) = 100 \log_2(x + 2),$$

where x is the number of weeks after a major advertising campaign was introduced. What were the total sales 6 weeks after the campaign was introduced? Graph the function.

[10.4] *Apply the properties of logarithms introduced in Section 10.4 to express the logarithm as a sum or difference of logarithms, or as a single number if possible. Assume that all variables represent positive real numbers.*

29. $\log_4 3x^2$

30. $\log_2 \dfrac{p^2 r}{\sqrt{z}}$

Use the properties of logarithms introduced in Section 10.4 to write the expression as a single logarithm. Assume that all variables represent positive real numbers, $b \ne 1$.

31. $\log_b 3 + \log_b x - 2 \log_b y$

32. $\log_3(x + 7) - \log_3(4x + 6)$

[10.5] *Find the logarithm. Give approximations to the nearest ten-thousandth.*

33. $\log 28.9$

34. $\log .257$

35. $\log 10^{4.8613}$

36. $\ln 28.9$

37. $\ln .257$

38. $\ln e^{4.8613}$

Use the change-of-base rule (either with common or natural logarithms) to find the logarithm. Give an approximation to the nearest ten-thousandth.

39. $\log_{16} 13$

40. $\log_4 12$

41. $\log_{\sqrt{6}} \sqrt{13}$

Use the formula $\text{pH} = -\log[H_3O^+]$ *to find the pH of the substance with the given hydronium ion concentration.*

42. milk, 4.0×10^{-7}

43. crackers, 3.8×10^{-9}

44. Suppose that the population of a city is given by

$$P(t) = 1{,}000{,}000 e^{.02t}$$

where t represents time measured in years. Find the following values.

(a) $P(0)$ **(b)** $P(2)$

45. Suppose the quantity, measured in grams, of a radioactive substance present at time t is given by

$$Q(t) = 500 e^{-.05t}$$

where t is measured in days. Find the quantity present at the following times.

(a) $t = 0$ **(b)** $t = 4$

[10.6] *Solve the equation. Give the solution to the nearest thousandth.*

46. $3^x = 9.42$

47. $e^{.06x} = 3$

Solve the equation. Give the exact solution.

48. $\log_3(9x + 8) = 2$

49. $\log_5(y + 6)^3 = 2$

50. $\log_3(p + 2) - \log_3 p = \log_3 2$

51. $\log(2x + 3) = \log 3x + 2$

52. $\log_4 x + \log_4(8 - x) = 2$

53. $\log_2 x + \log_2(x + 15) = 4$

54. Consider the logarithmic equation

$$\log(2x + 3) = \log x + 1.$$

(a) Solve the equation using properties of logarithms.

(b) If $y_1 = \log(2x + 3)$ and $y_2 = \log x + 1$, then the graph of $y_1 - y_2$ in a selected window of a graphics calculator looks like this. Explain how the display at the bottom of the screen confirms the solution set found in part (a).

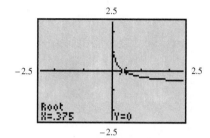

 Solve the problem. Use a calculator as necessary.

55. If $20,000 is deposited at 7% annual interest compounded quarterly, how much will be in the account after 5 years, assuming no withdrawals are made?

56. How much will $10,000 compounded continuously at 6% annual interest amount to in three years?

57. Which is a better plan?

> Plan A: Invest $1000.00 at 4% compounded quarterly for 3 years
> Plan B: Invest $1000.00 at 3.9% compounded monthly for 3 years

58. Suppose that a certain collection of termites is growing according to the function

$$y = 3000e^{.04t},$$

where t is time measured in months. How long will it take for the collection to double?

59. Recall from Example 6 in Section 10.5 that the number of years, $N(r)$, since two independently evolving languages split off from a common ancestral language is approximated by

$$N(r) = -5000 \ln r,$$

where r is the percent of words from the ancestral language common to both languages now. Find r if the split occurred 2000 years ago.

60. A machine purchased for business use *depreciates,* or loses value, over a period of years. The value of the machine at the end of its useful life is called its scrap value. By one method of depreciation (where it is assumed a constant percentage of the value depreciates annually), the scrap value, S, is given by

$$S = C(1 - r)^n,$$

where C is the original cost, n is the useful life in years, and r is the constant percent of depreciation.

(a) Find the scrap value of a machine costing $30,000, having a useful life of 12 years and a constant annual rate of depreciation of 15%.

(b) A machine has a "half-life" of 6 years. Find the constant annual rate of depreciation.

Answers (right column):

48. $\left\{\dfrac{1}{9}\right\}$ **49.** $\{-6 + \sqrt[3]{25}\}$

50. $\{2\}$ **51.** $\left\{\dfrac{3}{298}\right\}$

52. $\{4\}$ **53.** $\{1\}$

54. (a) $\left\{\dfrac{3}{8}\right\}$

(b) The *x*-value of the *x*-intercept is .375, the decimal equivalent of $\dfrac{3}{8}$.

55. $28,295.56

56. $11,972.17

57. Plan A is better, since it would pay $2.92 more.

58. about 17.3 months

59. 67%   **60.** (a) $4267

(b) about 11%

61. $\{72\}$ **62.** $\{5\}$

63. $\left\{\dfrac{1}{9}\right\}$ **64.** $\left\{\dfrac{4}{3}\right\}$

65. $\{3\}$ **66.** $\left\{\dfrac{1}{8}\right\}$

67. $\left\{\dfrac{11}{3}\right\}$ **68.** (a) 500

(b) 1000   (c) 1500

69. $\dfrac{1}{4}, \dfrac{1}{2}, 1, 2, 4, 8$

$f(x) = 2^x$

70. $-2, -1, 0, 1, 2, 3$

$g(x) = \log_2 x$

71. The roles of x and y are reversed. They are inverses.

72. horizontal;   vertical

73. 5;   2;   3;   5

74. 4;   8 (or 8; 4)

75. 3.700439718 (The number of displayed digits may vary.)   **77.** 13

78. 13;   The number in Exercise 75 is the exponent to which 2 must be raised in order to obtain 13.

79. 3.700439718

80. $\left\{-\dfrac{8}{5}\right\}$ **81.** $\{4\}$

MIXED REVIEW EXERCISES

Solve.

61. $\log_3(x + 9) = 4$

62. $\log_2 32 = x$

63. $\log_x \dfrac{1}{81} = 2$

64. $27^x = 81$

65. $2^{2x-3} = 8$

66. $\log_3(x + 1) - \log_3 x = 2$

67. $\log(3x - 1) = \log 10$

68. A population of hares in a specific area is growing according to the function

$$H(t) = 500 \log_3(2t + 3),$$

where t is time in years after the population was introduced into the area. Find the number of hares for the following times.
(a) $t = 0$ **(b)** $t = 3$ **(c)** $t = 12$

◆ **MATHEMATICAL CONNECTIONS** (Exercises 69–81) ◆

Work Exercises 69–81 in order, so that you can see some of the relationships between exponential and logarithmic properties and functions.

69. Complete the table, and graph the function $f(x) = 2^x$.

x	$f(x)$
-2	
-1	
0	
1	
2	
3	

70. Complete the table, and graph the function $g(x) = \log_2 x$.

x	$g(x)$
$\dfrac{1}{4}$	
$\dfrac{1}{2}$	
1	
2	
4	
8	

71. What do you notice about the ordered pairs found in Exercises 69 and 70? What do we call the functions f and g in relationship to each other?

72. Fill in the blank with the word *vertical* or *horizontal:* The graph of f in Exercise 69 has a _____ asymptote, while the graph of g in Exercise 70 has a _____ asymptote.

73. Using properties of exponents, $2^2 \cdot 2^3 = 2^?$, because ___?___ + ___?___ = ___?___ .

74. It is a fact that $32 = 4 \cdot 8$. Therefore, using properties of logarithms, $\log_2 32 = \log_2$ _____ $+ \log_2$ _____ .

75. Use the change-of-base rule to find an approximation for $\log_2 13$. Give as many digits as your calculator displays, and store this approximation in memory.

76. In your own words, explain what $\log_2 13$ means.

77. Simplify without the use of a calculator: $2^{\log_2 13}$.

78. Use the exponential key of your calculator to raise 2 to the power obtained in Exercise 75. What is the result? Why is this so?

79. Based on your result in Exercise 75, the point $(13, \underline{\quad})$ lies on the graph of $g(x) = \log_2 13$.

80. Use the method of Section 10.2 to solve the equation $2^{x+1} = 8^{2x+3}$.

81. Use the method of Section 10.6 to solve the equation $\log_2(x + 4) + \log_2 x = 5$.

CHAPTER 10 TEST

1. Decide whether the function is one-to-one.

    (a) $f(x) = x^2 + 9$ **(b)**

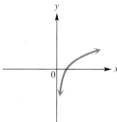

2. Find $f^{-1}(x)$ for the one-to-one function $f(x) = \sqrt[3]{x + 7}$.

3. Graph the inverse of f, given the graph of f at the right.

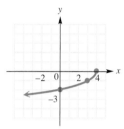

Graph the function.

    4. $y = 6^x$ **5.** $y = \log_6 x$

     6. Explain how the graph of the function in Exercise 5 can be obtained from the graph of the function in Exercise 4.

Solve the equation. Give the exact solution.

    7. $5^x = \dfrac{1}{625}$ **8.** $2^{3x-7} = 8^{2x+2}$

Solve the problem.

    9. A small business estimates that the value y of a copy machine is decreasing according to the function

$$y = 5000(2)^{-.15t},$$

where t is the number of years that have elapsed since the machine was purchased, and y is in dollars.

    (a) What was the original value of the machine? (*Hint:* Let $t = 0$.)

    (b) What is the value of the machine 5 years after purchase? Give your answer to the nearest dollar.

    (c) What is the value of the machine 10 years after purchase? Give your answer to the nearest dollar.

    (d) Graph the function.

10. Write in logarithmic form: $4^{-2} = .0625$.

11. Write in exponential form: $\log_7 49 = 2$.

[10.1] 1. (a) not one-to-one **(b)** one-to-one

2. $f^{-1}(x) = x^3 - 7$

3.

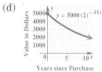

[10.2] 4.

[10.3] 5.

6. Simply exchange the x and y values in the ordered pairs.

[10.2] 7. $\{-4\}$ **8.** $\left\{-\dfrac{13}{3}\right\}$

9. (a) $5000 **(b)** $2973 **(c)** $1768 **(d)**

[10.3] 10. $\log_4 .0625 = -2$

11. $7^2 = 49$

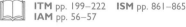

12. $\{32\}$ 13. $\left\{\dfrac{1}{2}\right\}$ 14. $\{2\}$

15. 5;   2; 5th;   32

[10.4] 16. $2 \log_3 x + \log_3 y$

17. $\dfrac{1}{2} \log_5 x - \log_5 y -$

$\log_5 z$ 18. $\log_b \dfrac{s^3}{t}$

19. $\log_b \dfrac{r^{1/4} s^2}{t^{2/3}}$

[10.5] 20. 1.3284

21. $-.8440$

22. (a) 2 and 3   (b) 2.1245

[10.6] 23. $\{3.9656\}$

24. $\{3\}$ 25. (a) $2973

(b) after $15.48 \approx 15$ years

Solve the equation.

12. $\log_{1/2} x = -5$ **13.** $x = \log_9 3$ **14.** $\log_x 16 = 4$

⦿ **15.** Fill in the blanks with the correct responses: The value of $\log_2 32$ is _____ . This means that if we raise _____ to the _____ power, the result is _____ .

Use properties of logarithms to write the expressions in Exercises 16 and 17 as sums or differences of logarithms. Assume variables represent positive numbers.

16. $\log_3 x^2 y$ **17.** $\log_5\left(\dfrac{\sqrt{x}}{yz}\right)$

Use properties of logarithms to write the expressions in Exercises 18 and 19 as single logarithms. Assume variables represent positive real numbers, $b \neq 1$.

18. $3 \log_b s - \log_b t$ **19.** $\dfrac{1}{4} \log_b r + 2 \log_b s - \dfrac{2}{3} \log_b t$

▦ *Use a calculator to find an approximation to the nearest ten-thousandth for the logarithm.*

20. $\log 21.3$ **21.** $\ln .43$

⦿ **22.** (a) Between what two consecutive integers must the value of $\log_6 45$ be?
   ▦ (b) Use a calculator to find an approximation of $\log_6 45$ to four decimal places.

▦ **23.** Solve for x, and give the solution correct to the nearest ten-thousandth.

$$3^x = 78$$

24. Solve: $\log_8(x + 5) + \log_8(x - 2) = \log_8 8$.

▦ **25.** The accompanying graph is that of a function equivalent to the one in Exercise 9, describing the value of the office copy machine, entered as $y = 5000e^{-.104x}$.
   (a) Based on the display at the bottom of the screen, what will be the value of the machine after 5 years? Give your answer to the nearest dollar.
   (b) To the nearest whole number, after how many years will the value of the machine be $1000? Determine the answer either graphically or using natural logarithms.

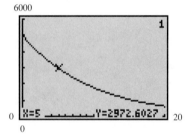

CUMULATIVE REVIEW (Chapters 1–10)

1. $-2, 0, 6, \dfrac{30}{3}$ (or 10)

2. $-\dfrac{9}{4}, -2, 0, .6, 6, \dfrac{30}{3}$
(or 10)   3. $-\sqrt{2}, \sqrt{11}$

4. all except $\sqrt{-8}$

Let $S = \left\{-\dfrac{9}{4}, -2, -\sqrt{2}, 0, .6, \sqrt{11}, \sqrt{-8}, 6, \dfrac{30}{3}\right\}$. *List the elements of S that are members of the set.*

1. integers   **2.** rational numbers

3. irrational numbers   **4.** real numbers

Simplify the expression.

5. $|-8| + 6 - |-2| - (-6 + 2)$

6. $-12 - |-3| - 7 - |-5|$

7. $2(-5) + (-8)(4) - (-3)$

Solve the equation or inequality.

8. $7 - (3 + 4a) + 2a = -5(a - 1) - 3$

9. $2m + 2 \leq 5m - 1$

10. $|2x - 5| = 9$

11. $|3p| - 4 = 12$

12. $|3k - 8| \leq 1$

13. $|4m + 2| > 10$

Perform the indicated operations.

14. $(2p + 3)(3p - 1)$

15. $(4k - 3)^2$

16. $(3m^3 + 2m^2 - 5m) - (8m^3 + 2m - 4)$

17. Divide $6t^4 + 17t^3 - 4t^2 + 9t + 4$ by $3t + 1$.

Factor.

18. $8x + x^3$

19. $24y^2 - 7y - 6$

20. $5z^3 - 19z^2 - 4z$

21. $16a^2 - 25b^4$

22. $8c^3 + d^3$

23. $16r^2 + 56rq + 49q^2$

Simplify.

24. $\dfrac{(5p^3)^4(-3p^7)}{2p^2(4p^4)}$

25. $\dfrac{x^2 - 9}{x^2 + 7x + 12} \div \dfrac{x - 3}{x + 5}$

26. $\dfrac{2}{k + 3} - \dfrac{5}{k - 2}$

27. $\dfrac{3}{p^2 - 4p} - \dfrac{4}{p^2 + 2p}$

28. Candy worth $1.00 per pound is to be mixed with 10 pounds of candy worth $1.96 per pound to get a mixture that will be sold for $1.60 per pound. How many pounds of the $1.00 candy should be used?

Simplify.

29. $\left(\dfrac{5}{4}\right)^{-2}$

30. $\dfrac{6^{-3}}{6^2}$

31. $2\sqrt{32} - 5\sqrt{98}$

32. Multiply: $(5 + 4i)(5 - 4i)$.

Solve the equation or inequality.

33. $3x^2 - x - 1 = 0$

34. $k^2 + 2k - 8 > 0$

35. Recently the U.S. population has been growing according to the equation

$$y = 1.7x + 230$$

where y gives the population (in millions) in year x, measured from year 1980. For example, in 1980 $x = 0$ and $y = 1.7(0) + 230 = 230$. This means that the population was about 230 million in 1980. To find the population in 1985, let $x = 5$, and so on. Find the population in each of the following years.

(a) 1982   **(b)** 1985   **(c)** 1990

(d) In what year will the population reach 315 million, based on this equation?

Answers (right margin):

5. 16   6. −27

7. −39   8. $\left\{-\dfrac{2}{3}\right\}$

9. $[1, \infty)$

10. $\{-2, 7\}$

11. $\left\{-\dfrac{16}{3}, \dfrac{16}{3}\right\}$

12. $\left[\dfrac{7}{3}, 3\right]$

13. $(-\infty, -3) \cup (2, \infty)$

14. $6p^2 + 7p - 3$

15. $16k^2 - 24k + 9$

16. $-5m^3 + 2m^2 - 7m + 4$

17. $2t^3 + 5t^2 - 3t + 4$

18. $x(8 + x^2)$

19. $(3y - 2)(8y + 3)$

20. $z(5z + 1)(z - 4)$

21. $(4a + 5b^2)(4a - 5b^2)$

22. $(2c + d) \cdot (4c^2 - 2cd + d^2)$

23. $(4r + 7q)^2$

24. $-\dfrac{1875p^{13}}{8}$ 25. $\dfrac{x + 5}{x + 4}$

26. $\dfrac{-3k - 19}{(k + 3)(k - 2)}$

27. $\dfrac{22 - p}{p(p - 4)(p + 2)}$

28. 6 pounds   29. $\dfrac{16}{25}$

30. $\dfrac{1}{6^5}$ 31. $-27\sqrt{2}$

32. 41

33. $\left\{\dfrac{1 + \sqrt{13}}{6}, \dfrac{1 - \sqrt{13}}{6}\right\}$

34. $(-\infty, -4) \cup (2, \infty)$

35. (a) 233.4 million
(b) 238.5 million
(c) 247 million   (d) 2030

 CONCEPTUAL   ✐ WRITING   ▲ CHALLENGING   ▦ SCIENTIFIC CALCULATOR   ▦ GRAPHICS CALCULATOR

36. −5,587,500
37. $3x - 4y = 19$
38.

39.

40.

41.

42.

43. $\{(4, 2)\}$
44. $\{(1, -1, 4)\}$ 45. −1
46. −9
47.

48. $\{-1\}$
49.

50. (a) 25,000   (b) 30,500
(c) 37,300   (d) 68,000

36. The graph indicates that the long-term debt of the Port of New Orleans has dropped from $70,000,000 in 1986 to $25,300,000 in 1994. What is the slope of the line graphed? (*Source:* Division of Finance and Accounting, Port of New Orleans)

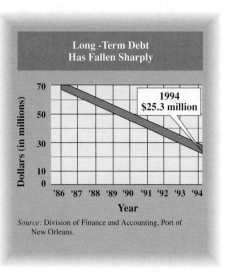

Source: Division of Finance and Accounting, Port of New Orleans.

37. Find the standard form of the equation of the line through $(5, -1)$ and parallel to the line with equation $3x - 4y = 12$.

Graph.

38. $5x + 2y = 10$ **39.** $-4x + y \le 5$ **40.** $f(x) = \dfrac{1}{3}(x - 1)^2 + 2$

41. $\dfrac{x^2}{9} + \dfrac{y^2}{16} = 1$ **42.** $25x^2 - 16y^2 = 400$

Solve the system.

43. $5x - 3y = 14$
$\quad\ 2x + 5y = 18$

44. $\quad x + 2y + 3z = 11$
$\quad 3x - \ \ y + \ \ z = 8$
$\quad 2x + 2y - 3z = -12$

Evaluate the determinant.

45. $\begin{vmatrix} -2 & -1 \\ 5 & 3 \end{vmatrix}$

46. $\begin{vmatrix} 2 & 4 & 5 \\ 1 & 3 & 0 \\ 0 & -1 & -2 \end{vmatrix}$

47. Graph $f(x) = 2^x$.

48. Solve $5^{x+3} = \left(\dfrac{1}{25}\right)^{3x+2}$.

49. Graph $f(x) = \log_3 x$.

50. Let the number of bacteria present in a certain culture be given by

$$B = 25{,}000e^{.2t},$$

where t is time measured in hours, and $t = 0$ corresponds to noon. Find, to the nearest hundred, the number of bacteria present at:

(a) noon;   **(b)** 1 P.M.;   **(c)** 2 P.M.;   **(d)** 5 P.M.

SEQUENCES AND SERIES

CONNECTIONS

A *sequence* is a list of numbers, usually written in some sort of order. One of the most famous sequences in mathematics is the **Fibonacci sequence**:

$$1, 1, 2, 3, 5, 8, 13, 21, 34, 55, \ldots.$$

This sequence is named for the Italian mathematician Leonardo of Pisa (1170–1250), who was also known as Fibonacci. The Fibonacci sequence is found in numerous places in nature. For example, male honeybees hatch from eggs that have not been fertilized, so a male bee has only one parent, a female. On the other hand, female honeybees hatch from fertilized eggs, so a female has two parents, one male and one female. The number of ancestors in consecutive generations of bees follows the Fibonacci sequence. Successive terms in the sequence also appear in plants: in the daisy head, the pineapple, and the pine cone, for instance.

FOR DISCUSSION OR WRITING

1. See if you can discover the pattern in the Fibonacci sequence. (*Hint:* Can you explain how to find the next term, given the preceding two terms?)
2. Draw a tree showing the number of ancestors of a male bee in each generation following the description given above.

◆ 1. After the first term, each term is the sum of the two preceding terms. 2. We show only four generations in this sketch.

Intuitively, a **sequence** is a list of numbers in which order is important. Sequences, a special type of function, are useful in our daily lives, as well as in higher mathematics. For instance, the interest portion of monthly payments made to pay off an automobile or home loan and the list of maximum daily temperatures in one area for a month are sequences.

Some polynomials are sums of sequences, called series. These sums of sequences are used in our calculators and computers to calculate logarithms and powers and roots of numbers. The binomial theorem, presented in Section 11.4, is a series that has a wide range of applications.

11.1 SEQUENCES AND SERIES

FOR EXTRA HELP	OBJECTIVES
SSG pp. 347–353 **SSM** pp. 584–587 **Video** 13 **Tutorial** IBM MAC	**1** ▶ Find the terms of a sequence given the general term. **2** ▶ Find the general term of a sequence. **3** ▶ Use sequences to solve applied problems. **4** ▶ Use summation notation to evaluate a series. **5** ▶ Write a series using summation notation. **6** ▶ Find an arithmetic mean.

SEQUENCES   In the Palace of the Alhambra, residence of the Moorish rulers of Granada, Spain, the Sultana's quarters feature an interesting architectural pattern. There are 2 matched marble slabs inlaid in the floor, 4 walls, an octagon (8-sided) ceiling, 16 windows, 32 arches, and so on. This ordered list of numbers,

$$2, 4, 8, 16, 32, \ldots,$$

is an example of a sequence.

Infinite Sequence

An **infinite sequence** is a function with the set of positive integers as the domain.

---

1 ▶ Find the terms of a sequence given the general term.

▶   For any positive integer n, the function value (y-value) of a sequence is written as a_n (read "a sub-n") instead of $a(n)$ or $f(n)$. The function values $a_1, a_2, a_3, \ldots$, written in order, are the **terms** of the sequence, with a_1 the first term, a_2 the second term, and so on. The expression a_n, which defines the sequence, is called the **general term** of the sequence.

In the example given above, the first five terms of the sequence are

$$a_1 = 2 \qquad a_2 = 4 \qquad a_3 = 8 \qquad a_4 = 16 \qquad \text{and} \qquad a_5 = 32.$$

The general term for this sequence is $a_n = 2^n$.

EXAMPLE 1

Writing the Terms of a Sequence from the General Term

■ Given an infinite sequence with $a_n = n + \dfrac{1}{n}$, find the following.

(a) The second term of the sequence

To get a_2, the second term, replace n by 2.

$$a_2 = 2 + \frac{1}{2} = \frac{5}{2}$$

▶ **TEACHING TIP**
Students have difficulty finding the
general term of a sequence from
the first few terms. Illustrate with
simple sequences like 2, 4, 6, 8, . . .
or 1/2, 2/3, 3/4, 4/5, ... Have
students come up with the general
term as a class participation
exercise. Often the expression they
come up with will work for the first
few terms but not a_3, a_4, and so
forth. ◀

2 ▶ Find the general
term of a sequence.

E X A M P L E   2
Finding the General Term
of a Sequence

(b) The fifth term
Replace n by 5.

$$a_5 = 5 + \frac{1}{5} = \frac{26}{5}$$

(c) $a_{10} = 10 + \dfrac{1}{10} = \dfrac{101}{10}$

(d) $a_{12} = 12 + \dfrac{1}{12} = \dfrac{145}{12}$ ■

◆ **C O N N E C T I O N S** ◆

Sequences can be graphed by using the sequence mode and dot mode of a graphics calculator. In the figure, we show a calculator screen with the graph of $a_n = n + \dfrac{1}{n}$.

▶ Sometimes we need to find a general term to fit the first few terms of a given sequence. There are no rules for finding the general term of a sequence from the first few terms. In fact, it is possible to give more than one general term that produce the same first three or four terms. However, in many examples, the terms may suggest a general term.

Find an expression for the general term a_n of the sequence

$$5, \quad 10, \quad 15, \quad 20, \quad 25, \ldots.$$

By inspection, $a_n = 5n$ will produce the given first five terms. ■

CAUTION One problem with using just a few terms to suggest a general term, as in Example 2, is that there may be more than one general term that gives the same first few terms.

3 ▶ Use sequences to solve applied problems.

▶ Practical problems often involve finite sequences.

Finite Sequence

A **finite sequence** has a domain that includes only the first n positive integers.

For example, if n is 5, the domain is $\{1, 2, 3, 4, 5\}$, and the sequence has five terms.

■

PROBLEM SOLVING

As mentioned in the introduction to this chapter, there are many applications of sequences. To solve problems involving sequences, a good strategy is to list the first few terms and look for a pattern that suggests a general term. When the general term is known, we can find any term in the sequence without writing all the preceding terms.

■

EXAMPLE 3

Using a Sequence in an Application

A colony of bacteria doubles in weight every hour. If the colony weighs 1 gram at the beginning of an experiment, find the weight after ten hours.

At the end of the first hour, the colony will weigh 2 grams. At the end of the second hour, the weight will be $2 \cdot 2$ or $2^2 = 4$ grams. After three hours, the weight will be $2 \cdot 2 \cdot 2$ or $2^3 = 8$ grams, and so on. Continuing in this way gives the sequence shown in the chart below.

Time	a_1 End hour 1	a_2 End hour 2	a_3 End hour 3	a_4 End hour 4
Weight	2	4	8	16

In general, the colony will weigh 2^n grams at the end of n hours, so after 10 hours, the colony should weigh $2^{10} = 1024$ grams. ■

SERIES The indicated sum of the terms of a sequence is called a **series.** Since a sequence can be finite or infinite, there are finite or infinite series. One type of infinite series is discussed in Section 11.3, and the binomial theorem discussed in Section 11.4 defines an important finite series. In this section we discuss only finite series.

4 ▶ Use summation notation to evaluate a series.

▶ TEACHING TIP

Students will use *sigma notation* again in statistics, finite math, and calculus courses. ◀

▶ We use a compact notation, called **summation notation,** to write a series from the general term of the corresponding sequence. For example, the sum of the first six terms of the sequence with general term $a_n = 3n + 2$ is written with the Greek letter Σ (sigma) as

$$\sum_{i=1}^{6} (3i + 2).$$

We read this as "the sum from $i = 1$ to 6 of $3i + 2$." To find this sum, we replace the letter i in $3i + 2$ with 1, 2, 3, 4, 5, and 6, as follows.

► **TEACHING TIP**
It is helpful to write the value of
the index i above each term as
follows:

$$i = 1 \qquad i = 2$$
$$(3 \cdot 1 + 2) \quad + \quad (3 \cdot 2 + 2)$$
$$i = 3$$
$$+ \quad (3 \cdot 3 + 2) \quad + \quad \cdots \quad ◄$$

$$\sum_{i=1}^{6} (3i + 2) = (3 \cdot 1 + 2) + (3 \cdot 2 + 2) + (3 \cdot 3 + 2)$$
$$+ (3 \cdot 4 + 2) + (3 \cdot 5 + 2) + (3 \cdot 6 + 2)$$
$$= 5 + 8 + 11 + 14 + 17 + 20$$
$$= 75$$

The letter i is called the **index of summation.**

CAUTION This use of i has no connection with the use of i to represent a complex number.

EXAMPLE 4

Evaluating a Series
Written in Summation
Notation

Write out the terms and evaluate each of the following.

(a) $\displaystyle\sum_{i=1}^{5} (i - 4) = (1 - 4) + (2 - 4) + (3 - 4) + (4 - 4) + (5 - 4)$
$$= -3 - 2 - 1 + 0 + 1$$
$$= -5$$

(b) $\displaystyle\sum_{i=3}^{7} 3i^2 = 3(3)^2 + 3(4)^2 + 3(5)^2 + 3(6)^2 + 3(7)^2$
$$= 27 + 48 + 75 + 108 + 147$$
$$= 405 \quad ∎$$

Chalkboard Exercise

Write out the terms and evaluate the following.
$$\sum_{i=2}^{6} (i + 1)(i - 2)$$
Answer:
$0 + 4 + 10 + 18 + 28 = 60$

◇ **CONNECTIONS** ◇

The list feature of a graphics calculator can be used to find the sum of the terms of a finite series. First, we save the definition of the sequence in a list. Then we can use the capability of the calculator to get the sum of the list.

Typical screens for the series $\displaystyle\sum_{i=0}^{3} [1000(1.06)^i]$ are shown in the figure.

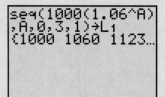

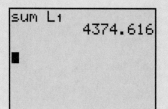

5 ► Write a series using
summation notation.

► Sometimes we want to express a sum in compact form using summation notation.

EXAMPLE 5

Writing a Series with Summation Notation

Write the following sums with summation notation.

(a) $2 + 5 + 8 + 11$

First, find a general term a_n that will give these four terms for a_1, a_2, a_3, and a_4. Inspection (and trial and error) shows that $3i - 1$ will work for these four terms, since

$$3(1) - 1 = 2$$
$$3(2) - 1 = 5$$
$$3(3) - 1 = 8$$
$$3(4) - 1 = 11.$$

(Remember, there may be other expressions that also work. These four terms may be the first terms of more than one sequence.) Since i ranges from 1 to 4, write the sum as

$$2 + 5 + 8 + 11 = \sum_{i=1}^{4} (3i - 1).$$

(b) $8 + 27 + 64 + 125 + 216$

Since these numbers are the cubes of 2, 3, 4, 5, and 6,

$$8 + 27 + 64 + 125 + 216 = \sum_{i=2}^{6} i^3. \quad \blacksquare$$

6 ▶ Find an arithmetic mean.

▶   In statistics, the **arithmetic mean** of a list of numbers is defined as the sum of all the numbers, divided by the number of numbers.

Arithmetic Mean

If $\bar{x}$ represents the mean, then

$$\bar{x} = \frac{\sum_{i=1}^{n} x_i}{n}.$$

Here the values of x_i represent the individual numbers in the collection, and n represents the number of numbers.

EXAMPLE 6

Finding the Arithmetic Mean

Find the arithmetic mean for the following collection of numbers.

$$4, \quad 7, \quad 8, \quad 10, \quad 12, \quad 13$$

Let $x_1 = 4$, $x_2 = 7$, $x_3 = 8$, and so on. Since there are 6 numbers in the collection, n is 6. By definition,

$$\bar{x} = \frac{\sum_{i=1}^{6} x_i}{6}$$

$$= \frac{4 + 7 + 8 + 10 + 12 + 13}{6}$$

$$= \frac{54}{6}$$

$$= 9. \quad \blacksquare$$

11.1 EXERCISES

◉ **1.** Suppose the general term of a sequence is $a_n = mn + b$, for some constants m and b. What kind of function does this suggest?

◉ **2.** If the general term of a sequence is $a_n = k^n$, for some constant k, what kind of function is suggested?

Write out the first five terms of the sequence. See Example 1.

3. $a_n = \dfrac{n + 3}{n}$ **4.** $a_n = \dfrac{n + 2}{n + 1}$ **5.** $a_n = 3^n$ **6.** $a_n = 1^{n-1}$

7. $a_n = \dfrac{1}{n^2}$ **8.** $a_n = \dfrac{n^2}{n + 1}$ **9.** $a_n = (-1)^n$ **10.** $a_n = (-1)^{2n-1}$

Find the indicated term for the sequence. See Example 2.

11. $a_n = -9n + 2;\quad a_8$ **12.** $a_n = 3n - 7;\quad a_{12}$

13. $a_n = \dfrac{3n + 7}{2n - 5};\quad a_{14}$ **14.** $a_n = \dfrac{5n - 9}{3n + 8};\quad a_{16}$

15. $a_n = (n + 1)(2n + 3);\quad a_8$ **16.** $a_n = (5n - 2)(3n + 1);\quad a_{10}$

Find a general term, a_n, for the given terms of the sequence. See Example 1.

17. 4, 8, 12, 16, . . .   **18.** $-10, -20, -30, -40, \ldots$

19. $\dfrac{1}{3}, \dfrac{1}{9}, \dfrac{1}{27}, \dfrac{1}{81}, \ldots$ **20.** $\dfrac{1}{2}, \dfrac{2}{3}, \dfrac{3}{4}, \dfrac{4}{5}, \ldots$

Solve the applied problem by writing the first few terms of a sequence. See Example 3.

21. Anne borrows $1000 and agrees to pay $100 plus interest of 1% on the unpaid balance each month. Find the payments for the first six months and the remaining debt at the end of this period.

22. Suppose that an automobile loses 1/5 of its value each year; that is, at the end of any given year, the value is 4/5 of the value at the beginning of that year. If a car cost $20,000 new, what is its value at the end of 5 years?

23. A package of supplies is dropped to an isolated work site. The package falls $15n$ meters in n seconds. Write a sequence showing how far it falls in each of the first 4 seconds. How far will it fall in 10 seconds?

24. Ms. Perez is offered a new job with a salary of $20,000 + 2500n$ dollars per year at the end of the nth year. Write a sequence showing her salary at the end of each of the first five years. If she continues in this way, what will her salary be at the end of the tenth year?

Write out the series and evaluate it. See Example 4.

25. $\displaystyle\sum_{i=1}^{3} (i^2 + 2)$ **26.** $\displaystyle\sum_{i=1}^{4} i(i + 3)$ **27.** $\displaystyle\sum_{i=2}^{5} \dfrac{1}{i}$ **28.** $\displaystyle\sum_{i=0}^{4} \dfrac{i}{i + 1}$

29. $\displaystyle\sum_{i=1}^{6} (-1)^i$ **30.** $\displaystyle\sum_{i=1}^{5} (-1)^i \cdot i$ **31.** $\displaystyle\sum_{i=3}^{7} (i - 3)(i + 2)$ **32.** $\displaystyle\sum_{i=2}^{6} \dfrac{i^2 + 1}{2}$

Write out the terms of the series.

33. $\displaystyle\sum_{i=1}^{5} 2x \cdot i$ **34.** $\displaystyle\sum_{i=1}^{6} x^i$ **35.** $\displaystyle\sum_{i=1}^{5} i \cdot x^i$ **36.** $\displaystyle\sum_{i=2}^{6} \dfrac{x + i}{x - i}$

---

1. linear
2. exponential
3. $4, \dfrac{5}{2}, 2, \dfrac{7}{4}, \dfrac{8}{5}$
4. $\dfrac{3}{2}, \dfrac{4}{3}, \dfrac{5}{4}, \dfrac{6}{5}, \dfrac{7}{6}$
5. 3, 9, 27, 81, 243
6. 1, 1, 1, 1, 1
7. $1, \dfrac{1}{4}, \dfrac{1}{9}, \dfrac{1}{16}, \dfrac{1}{25}$
8. $\dfrac{1}{2}, \dfrac{4}{3}, \dfrac{9}{4}, \dfrac{16}{5}, \dfrac{25}{6}$
9. $-1, 1, -1, 1, -1$
10. $-1, -1, -1, -1, -1$
11. -70 12. 29
13. $\dfrac{49}{23}$ 14. $\dfrac{71}{56}$
15. 171   16. 1488
17. $4n$ 18. $-10n$
19. $\dfrac{1}{3^n}$ 20. $\dfrac{n}{n + 1}$
21. $110, $109, $108, $107, $106, $105; $400
22. $6554   23. 15 meters, 30 meters, 45 meters, 60 meters;   150 meters
24. $22,500, $25,000, $27,500, $30,000, $32,500; $45,000
25. $3 + 6 + 11 = 20$
26. $4 + 10 + 18 + 28 = 60$
27. $\dfrac{1}{2} + \dfrac{1}{3} + \dfrac{1}{4} + \dfrac{1}{5} = \dfrac{77}{60}$
28. $0 + \dfrac{1}{2} + \dfrac{2}{3} + \dfrac{3}{4} + \dfrac{4}{5} = \dfrac{163}{60}$
29. $-1 + 1 - 1 + 1 - 1 + 1 = 0$
30. $-1 + 2 - 3 + 4 - 5 = -3$
31. $0 + 6 + 14 + 24 + 36 = 80$
32. $\dfrac{5}{2} + \dfrac{10}{2} + \dfrac{17}{2} + \dfrac{26}{2} + \dfrac{37}{2} = \dfrac{95}{2}$
33. $2x + 4x + 6x + 8x + 10x$
34. $x + x^2 + x^3 + x^4 + x^5 + x^6$
35. $x + 2x^2 + 3x^3 + 4x^4 + 5x^5$

◉ CONCEPTUAL   ✐ WRITING   ▲ CHALLENGING   ▦ SCIENTIFIC CALCULATOR   ▦ GRAPHICS CALCULATOR

36. $\dfrac{x+2}{x-2} + \dfrac{x+3}{x-3} +$
$\dfrac{x+4}{x-4} + \dfrac{x+5}{x-5} + \dfrac{x+6}{x-6}$

37. $\displaystyle\sum_{i=1}^{5}(i+2)$

38. $\displaystyle\sum_{i=1}^{4} i^2$ 39. $\displaystyle\sum_{i=1}^{5} \dfrac{1}{i+1}$

40. $\displaystyle\sum_{i=1}^{6}(-1)^i \cdot i$

43. $\dfrac{59}{7}$ 44. $\dfrac{72}{5}$

45. 5 46. $\dfrac{25}{6}$

47. $\displaystyle\sum_{i=1}^{6} i^2 + \sum_{i=1}^{6} 3i + \sum_{i=1}^{6} 5$

48. $3\displaystyle\sum_{i=1}^{6} i$ 49. $6 \cdot 5 = 30$

50. $\displaystyle\sum_{i=1}^{n} i = \dfrac{n(n+1)}{2}$

51.
$\displaystyle\sum_{i=1}^{n} i^2 = \dfrac{n(n+1)(2n+1)}{6}$

52. $91 + 3(21) +$
$6(5) = 184$
53. 572   54. -2620
55. $a = 6, d = 2$
56. $a = 5, d = 1$
57. 10   58. 14

Write in summation notation. See Example 5.

37. $3 + 4 + 5 + 6 + 7$

38. $1 + 4 + 9 + 16$

39. $\dfrac{1}{2} + \dfrac{1}{3} + \dfrac{1}{4} + \dfrac{1}{5} + \dfrac{1}{6}$

40. $-1 + 2 - 3 + 4 - 5 + 6$

⊙ 41. Does $\displaystyle\sum_{i=1}^{4} i^2 = \left(\sum_{i=1}^{4} i\right)^2$? Explain.

⊙ 42. Does $\displaystyle\sum_{i=1}^{3} 5i = 5\sum_{i=1}^{3} i$? Explain.

Find the arithmetic mean for the collection of numbers. See Example 6.

43. 8, 11, 14, 9, 3, 6, 8

44. 10, 12, 8, 19, 23

45. 5, 9, 8, 2, 4, 7, 3, 2

46. 2, 1, 4, 8, 3, 7

───◆ **MATHEMATICAL CONNECTIONS** (Exercises 47–54) ◆───

The following properties of series provide useful shortcuts for evaluating series.

If $a_1, a_2, a_3, \ldots, a_n$ and $b_1, b_2, b_3, \ldots, b_n$ are two sequences, and c is a constant, then for every positive integer n,

(a) $\displaystyle\sum_{i=1}^{n} c = nc$

(b) $\displaystyle\sum_{i=1}^{n} ca_i = c\sum_{i=1}^{n} a_i$

(c) $\displaystyle\sum_{i=1}^{n}(a_i + b_i) = \sum_{i=1}^{n} a_i + \sum_{i=1}^{n} b_i$

(d) $\displaystyle\sum_{i=1}^{n}(a_i - b_i) = \sum_{i=1}^{n} a_i - \sum_{i=1}^{n} b_i.$

47. Use property (c) to write $\displaystyle\sum_{i=1}^{6}(i^2 + 3i + 5)$ as the sum of three summations.

48. Use property (b) to rewrite the second summation from Exercise 47.

49. Use property (a) to rewrite the third summation from Exercise 47.

50. Rewrite $1 + 2 + 3 + 4 + \cdots + n = \dfrac{n(n+1)}{2}$ in summation notation.

51. Rewrite $1^2 + 2^2 + 3^2 + 4^2 + \cdots + n^2 = \dfrac{n(n+1)(2n+1)}{6}$ in summation notation.

52. Use the summations you wrote in Exercises 50 and 51, and the properties given above to evaluate the three summations from Exercises 47–49. This gives the value of
$\displaystyle\sum_{i=1}^{6}(i^2 + 3i + 5)$ without writing out all six terms.

53. Use the properties and summations given above to evaluate $\displaystyle\sum_{i=1}^{12}(i^2 - i)$.

54. Use the properties and summations given above to evaluate $\displaystyle\sum_{i=1}^{20}(2 + i - i^2)$.

───────◆───────

REVIEW EXERCISES

Find the values of a and d by solving the system. See Section 8.1.

55. $a + 3d = 12$
$a + 8d = 22$

56. $a + 7d = 12$
$a + 2d = 7$

Evaluate $A = a + (n - 1)d$ for the given values. See Section 1.2.

57. $a = -2, n = 5, d = 3$

58. $a = \dfrac{1}{2}, n = 10, d = \dfrac{3}{2}$

11.2 ARITHMETIC SEQUENCES

FOR EXTRA HELP

 SSG pp. 353–359
SSM pp. 588–592

 Video
13

Tutorial
IBM   MAC

OBJECTIVES

1 ▶ Find the common difference for an arithmetic sequence.
2 ▶ Find the general term of an arithmetic sequence.
3 ▶ Find any specified term or the number of terms of an arithmetic sequence.
4 ▶ Find the sum of a specified number of terms of an arithmetic sequence.

1 ▶ Find the common difference for an arithmetic sequence.

▶ In this section we introduce a special type of sequence that has many applications.

Arithmetic Sequence

A sequence in which each term after the first differs from the preceding term by a constant amount is called an **arithmetic sequence** or **arithmetic progression.**

▶ **TEACHING TIP**
We can refer to the *common difference d* as the *spacing* between the numbers. ◀

▶ **TEACHING TIP**
Students have trouble with the formula $d = a_{n+1} - a_n$. One can find the value of d by computing $a_2 - a_1$ then checking to be sure that the remaining terms in the sequence differ by this amount d.

Also, an increasing sequence will have a positive d value; a decreasing sequence will have a negative d value. ◀

For example, the sequence

$$6, 11, 16, 21, 26, \ldots$$

is an arithmetic sequence, since the difference between any two adjacent terms is always 5. The number 5 is called the **common difference** of the arithmetic sequence. The common difference, d, is found by subtracting any pair of terms a_n and a_{n+1}. That is

$$d = a_{n+1} - a_n.$$

EXAMPLE 1

Finding the Common Difference

Find d for the arithmetic sequence

$$-11, -4, 3, 10, 17, 24, \ldots.$$

Since the sequence is arithmetic, d is the difference between any two adjacent terms. Choosing the terms 10 and 17 gives

$$d = 17 - 10$$
$$= 7.$$

The terms -11 and -4 would give $d = -4 - (-11) = 7$, the same result. ■

Chalkboard Exercise

Find d for the arithmetic sequence.

$$1, \frac{4}{3}, \frac{5}{3}, 2, \frac{7}{3}, \frac{8}{3}, 3, \ldots$$

Answer: $d = \frac{1}{3}$

EXAMPLE 2

Writing the Terms of a Sequence From the First Term and Common Difference

Write the first five terms of the arithmetic sequence with first term 3 and common difference -2.

The second term is found by adding -2 to the first term 3, getting 1. For the next term, add -2 to 1, and so on. The first five terms are

$$3, 1, -1, -3, -5. \quad ■$$

INSTRUCTOR'S RESOURCES

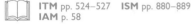 **ITM** pp. 524–527 **ISM** pp. 880–889
IAM p. 58

 **TEST GENERATOR**
DOS   Windows   MAC

 TRANSPARENCIES

2 ▶ Find the general term of an arithmetic sequence.

▶ Generalizing from Example 2, if we know the first term, a_1, and the common difference, d, of an arithmetic sequence, then the sequence is completely defined as

$$a_1, \quad a_2 = a_1 + d, \quad a_3 = a_1 + 2d, \quad a_4 = a_1 + 3d, \ldots.$$

Writing the terms of the sequence in this way suggests the following rule.

General Term of an Arithmetic Sequence

The general term of an arithmetic sequence with first term a_1 and common difference d is

$$a_n = a_1 + (n - 1)d.$$

Since $a_n = a_1 + (n - 1)d = dn + (a_1 - d)$ is a linear function in n, any linear expression of the form $kn + c$, where k and c are real numbers, defines an arithmetic sequence.

E X A M P L E   3

Finding the General Term of an Arithmetic Sequence

(a) Find the general term of the arithmetic sequence with $a_1 = -4$ and $d = 3$. Then find a_{10} and a_{20}.

Use the formula shown above with $a_1 = -4$ and $d = 3$.

$$
\begin{aligned}
a_n &= a_1 + (n - 1)d \\
a_n &= -4 + (n - 1)(3) & \text{Let } a_1 = -4, d = 3. \\
a_n &= -4 + 3n - 3 & \text{Distributive property} \\
a_n &= -7 + 3n & \text{Combine terms.}
\end{aligned}
$$

Now use $a_n = -7 + 3n$ to find a_{10} and a_{20}.

$$
\begin{aligned}
a_{10} &= -7 + 3(10) = 23 \\
a_{20} &= -7 + 3(20) = 53
\end{aligned}
$$

(b) Find the general term for the arithmetic sequence

$$-9, -6, -3, 0, 3, 6, \ldots.$$

Here the first term is $a_1 = -9$. To find d, subtract any two adjacent terms. For example,

$$d = -3 - (-6) = 3.$$

Now find a_n.

$$
\begin{aligned}
a_n &= a_1 + (n - 1)d \\
&= -9 + (n - 1)(3) & \text{Let } a_1 = -9, d = 3. \\
&= -9 + 3n - 3 & \text{Distributive property} \\
a_n &= 3n - 12 & \text{Combine terms.} \quad ■
\end{aligned}
$$

For Example 2

Chalkboard Exercise

Write the first five terms of the arithmetic sequence with first term 5 and common difference $\frac{1}{2}$.

Answer: $5, 5\frac{1}{2}, 6, 6\frac{1}{2}, 7$

Chalkboard Exercise

Find the general term of the arithmetic sequence with $a_1 = 4$ and $d = -2$. Then find a_8.

Answer: $a_n = 6 - 2n$; $a_8 = -10$

3 ▶ Find any specified term or the number of terms of an arithmetic sequence.

▶ The formula for the general term has four variables: $a_n, a_1, n,$ and d. If we know any three of these, the formula can be used to find the value of the fourth variable. The next example shows how we find a particular term.

EXAMPLE 4

Finding a Specified Term

▶ **TEACHING TIP**

One can get too involved with formulas when completing each part of this example. Specifically, part (b) may be more easily understood if the terms from a_5 to a_{11} are listed.

Value: 2 -10

Term: a_5 a_6 a_7 a_8 a_9 a_{10} a_{11}

Find the difference between a_{11} and a_5: $-10 - 2 = -12$. Divide -12 by 6 (since there are six spaces between a_5 and a_{11}) to get -2. So, $d = -2$. Now, between a_{11} and a_{17} there are 6 spaces, so

$$a_{17} = a_{11} + 6d$$
$$= -10 + 6(-2)$$
$$= -22. \quad ◀$$

Find the indicated term for each of the following arithmetic sequences.

(a) $a_1 = -6$, $d = 12$; a_{15}

Use the formula $a_n = a_1 + (n - 1)d$. Since we want $a_n = a_{15}$, $n = 15$.

$$a_{15} = a_1 + (15 - 1)d \qquad \text{Let } n = 15.$$
$$= -6 + 14(12) \qquad \text{Let } a_1 = -6, d = 12.$$
$$= 162$$

(b) $a_5 = 2$ and $a_{11} = -10$; a_{17}

Any term can be found if a_1 and d are known. Use the formula for a_n with the two given terms.

$$a_5 = a_1 + (5 - 1)d \qquad\qquad a_{11} = a_1 + (11 - 1)d$$
$$a_5 = a_1 + 4d \qquad\qquad\qquad a_{11} = a_1 + 10d$$
$$2 = a_1 + 4d \quad {\scriptstyle a_5 = 2} \qquad\qquad -10 = a_1 + 10d \quad {\scriptstyle a_{11} = -10}$$

This gives a system of two equations with two variables, a_1 and d. Find d by adding -1 times one equation to the other to eliminate a_1.

$$-10 = \quad a_1 + 10d$$
$$\underline{-2 = -a_1 - \quad 4d} \qquad \text{Multiply } 2 = a_1 + 4d \text{ by } -1.$$
$$-12 = \qquad\quad 6d \qquad \text{Add.}$$
$$-2 = d \qquad\qquad\qquad \text{Divide by 6.}$$

Now find a_1 by substituting -2 for d into either equation.

$$-10 = a_1 + 10(-2) \qquad \text{Let } d = -2.$$
$$-10 = a_1 - 20$$
$$10 = a_1$$

Use the formula for a_n to find a_{17}.

$$a_{17} = a_1 + (17 - 1)d \qquad \text{Use } n = 17.$$
$$= a_1 + 16d$$
$$= 10 + 16(-2) \qquad \text{Let } a_1 = 10, d = -2.$$
$$= -22 \quad ■$$

Sometimes we need to find out how many terms are in a sequence as shown in the following example.

EXAMPLE 5

Finding the Number of Terms in a Sequence

Find the number of terms in the arithmetic sequence

$$-8, -2, 4, 10, \ldots, 52.$$

Let n represent the number of terms in the sequence. Since $a_n = 52$, $a_1 = -8$, and $d = -2 - (-8) = 6$, use the formula $a_n = a_1 + (n - 1)d$ to find n. Substituting the known values into the formula gives

$$a_n = a_1 + (n - 1)d$$
$$52 = -8 + (n - 1)6 \qquad \text{Let } a_n = 52, a_1 = -8, d = 6.$$
$$52 = -8 + 6n - 6 \qquad \text{Distributive property}$$
$$66 = 6n \qquad\qquad\qquad \text{Combine terms.}$$
$$n = 11. \qquad\qquad\qquad \text{Divide by 6.}$$

The sequence has 11 terms. ■

4 ▶ Find the sum of a specified number of terms of an arithmetic sequence.

▶ To find a formula for the sum, S_n, of the first n terms of an arithmetic sequence, we can write out the terms as

$$S_n = a_1 + (a_1 + d) + (a_1 + 2d) + \cdots + [a_1 + (n - 1)d].$$

This same sum can be written in reverse as

$$S_n = a_n + (a_n - d) + (a_n - 2d) + \cdots + [a_n - (n - 1)d].$$

Now add the corresponding terms of these two expressions for S_n to get

$$2S_n = (a_1 + a_n) + (a_1 + a_n) + (a_1 + a_n) + \cdots + (a_1 + a_n).$$

The right-hand side of this expression contains n terms, each equal to $a_1 + a_n$, so that

$$2S_n = n(a_1 + a_n)$$

$$S_n = \frac{n}{2}(a_1 + a_n).$$

E X A M P L E 6
Finding the Sum of the First n Terms

Find the sum of the first five terms of the arithmetic sequence in which $a_n = 2n - 5$.

We can use the formula $S_n = \frac{n}{2}(a_1 + a_n)$ to find the sum of the first five terms. Here $n = 5$, $a_1 = 2(1) - 5 = -3$, and $a_5 = 2(5) - 5 = 5$. From the formula,

$$S_5 = \frac{5}{2}(-3 + 5) = \frac{5}{2}(2) = 5. \quad \blacksquare$$

It is sometimes useful to express the sum of an arithmetic sequence, S_n, in terms of a_1 and d, the quantities that define the sequence. We can do this as follows. Since

$$S_n = \frac{n}{2}(a_1 + a_n) \qquad \text{and} \qquad a_n = a_1 + (n - 1)d,$$

by substituting the expression for a_n into the expression for S_n, we get

$$S_n = \frac{n}{2}(a_1 + [a_1 + (n - 1)d])$$

$$S_n = \frac{n}{2}[2a_1 + (n - 1)d].$$

The following summary gives both of the alternative forms that may be used to find the sum of the first n terms of an arithmetic sequence.

Chalkboard Exercise

Find the sum of the first nine terms of the arithmetic sequence in which $a_n = 5 + 2n$.

Answer: $S_9 = 135$

▶ **TEACHING TIP**
For an example like this, first write out the five terms, then add these directly. Finally, apply the formula to get the same result.

Again, emphasize that this formula only works for arithmetic sequences in which a common difference between the terms exists. ◀

Sum of the First n Terms of an Arithmetic Sequence

The sum of the first n terms of the arithmetic sequence with first term a_1, nth term a_n, and common difference d is

$$S_n = \frac{n}{2}(a_1 + a_n)$$

or

$$S_n = \frac{n}{2}[2a_1 + (n - 1)d].$$

Find the sum of the first 8 terms of the arithmetic sequence having first term 3 and common difference -2.

Since the known values, $a_1 = 3$, $d = -2$, and $n = 8$, appear in the second formula for S_n, we use it.

$$S_n = \frac{n}{2}[2a_1 + (n - 1)d]$$

$$S_8 = \frac{8}{2}[2(3) + (8 - 1)(-2)] \qquad \text{Let } a_1 = 3, d = -2, n = 8.$$

$$= 4[6 - 14]$$

$$= -32 \quad \blacksquare$$

As mentioned above, linear expressions of the form $kn + c$, where k and c are real numbers, define an arithmetic sequence. For example, the sequences defined by $a_n = 2n + 5$ and $a_n = n - 3$ are arithmetic sequences. For this reason,

$$\sum_{i=1}^{n} (ki + c)$$

represents the sum of the first n terms of an arithmetic sequence having first term $a_1 = k(1) + c = k + c$ and general term $a_n = k(n) + c = kn + c$. We can find this sum with the first formula for S_n given above, as shown in the next example.

Find $\displaystyle\sum_{i=1}^{12} (2i - 1)$.

This is the sum of the first 12 terms of the arithmetic sequence having $a_n = 2n - 1$. This sum, S_{12}, is found with the formula for S_n,

$$S_n = \frac{n}{2}(a_1 + a_n).$$

Here $n = 12$, $a_1 = 2(1) - 1 = 1$, $a_{12} = 2(12) - 1 = 23$. Substitute these values into the formula to get

$$S_{12} = \frac{12}{2}(1 + 23) = 6(24) = 144. \quad \blacksquare$$

11.2 EXERCISES

◎ **1.** Can any two terms of an arithmetic sequence be used to find the common difference?
🖉 Explain.

If a sequence is arithmetic, find the common difference, d. If a sequence is not arithmetic, say so. See Example 1.

2. 1, 2, 3, 4, 5, . . .

4. 2, -4, 6, -8, 10, -12, . . .

6. -10, -5, 0, 5, 10, . . .

8. 3.42, 5.57, 7.72, 9.87, . . .

10. $-\dfrac{5}{3}$, -1, $-\dfrac{1}{3}$, $\dfrac{1}{3}$, ...

3. 2, 5, 8, 11, . . .

5. -6, -10, -14, -18, . . .

7. 1, 2, 4, 7, 11, 16, . . .

9. 1, $\dfrac{3}{2}$, 2, $\dfrac{5}{2}$, 3, $\dfrac{7}{2}$, ...

11. $\dfrac{1}{2}$, $\dfrac{1}{3}$, $\dfrac{1}{4}$, $\dfrac{1}{5}$, $\dfrac{1}{6}$, ...

2. $d = 1$ 3. $d = 3$
4. not arithmetic
5. $d = -4$ 6. $d = 5$
7. not arithmetic

8. $d = 2.15$ 9. $d = \dfrac{1}{2}$

10. $d = \dfrac{2}{3}$

11. not arithmetic

 CONCEPTUAL   WRITING   ▲ CHALLENGING    SCIENTIFIC CALCULATOR   GRAPHICS CALCULATOR

13. $a_n = 5n - 3$
14. $a_n = 8 - 3n$
15. $a_n = \dfrac{9}{4} + \dfrac{3}{4}n$
16. $a_n = 10n - 6$
17. $a_n = 3n - 6$
18. $a_n = -9.4721 + 4n$
19. $m + b, 2m + b,$
$3m + b$
20. yes 21. m
22. $a_n = mn + b$
23. 76 24. $-\dfrac{9}{2}$ 25. 48
26. 197 27. -1 28. $\dfrac{32}{5}$
29. 16 30. 15 31. 6
32. 13 34. 81 35. 90
36. -3 37. -90
38. 87 39. 159 40. 390
41. 136 42. 320 43. $\dfrac{35}{2}$
44. 31,375 45. 2,001,000
46. 281,625 47. 2500
48. no; 3; 9 49. 6

⊙ **12.** What does the letter n represent in the formula for the general term?

Use the formula for a_n to find the general term for the arithmetic sequence. See Example 3.

13. $a_1 = 2, d = 5$ **14.** $a_1 = 5, d = -3$

15. $a_1 = 3, d = \dfrac{3}{4}$ **16.** 4, 14, 24, . . .

17. $-3, 0, 3, \ldots$ ⊞**18.** $-5.4721, -1.4721, \ldots$

 **MATHEMATICAL CONNECTIONS** (Exercises 19–22)

Let $f(x) = mx + b$.

19. Find $f(1), f(2),$ and $f(3)$.

20. Consider the sequence $f(1), f(2), f(3), \ldots$. Is it an arithmetic sequence?

21. If the sequence is arithmetic, what is the common difference?

22. What is a_n for the sequence described in Exercise 20?

Find the indicated term for the arithmetic sequence. See Examples 2 and 4.

23. $a_1 = 4, d = 3;\quad a_{25}$ **24.** $a_1 = 1, d = -\dfrac{1}{2};\quad a_{12}$

25. 2, 4, 6, . . . ; a_{24} **26.** 1, 5, 9, . . . ; a_{50}

27. $a_{12} = -45, a_{10} = -37;\quad a_1$ **28.** $a_{10} = -2, a_{15} = -8;\quad a_3$

Find the number of terms in the arithmetic sequence. See Example 5.

29. 3, 5, 7, . . . , 33 **30.** $2, \dfrac{3}{2}, 1, \dfrac{1}{2}, \ldots, -5$

31. $\dfrac{3}{4}, 3, \dfrac{21}{4}, \ldots, 12$ **32.** 4, 1, $-2, \ldots, -32$

✎ **33.** Explain when you would use each of the two formulas for S_n.

Find S_6 for the arithmetic sequence. See Examples 6 and 7.

34. $a_1 = 6, d = 3$ **35.** $a_1 = 5, d = 4$ **36.** $a_1 = 7, d = -3$
37. $a_1 = -5, d = -4$ **38.** $a_n = 4 + 3n$ **39.** $a_n = 9 + 5n$

Use a formula for S_n to evaluate the series. See Example 8.

40. $\displaystyle\sum_{i=1}^{10} (8i - 5)$ **41.** $\displaystyle\sum_{i=1}^{17} (i - 1)$ **42.** $\displaystyle\sum_{i=1}^{20} (2i - 5)$

43. $\displaystyle\sum_{i=1}^{10} \left(\dfrac{1}{2}i - 1\right)$ **44.** $\displaystyle\sum_{i=1}^{250} i$ **45.** $\displaystyle\sum_{i=1}^{2000} i$

Solve the applied problem.

46. Find the sum of the first 750 positive integers.

47. Find the sum of the odd integers from 1 through 99.

▲ **48.** A child builds with blocks, placing 35 blocks in the first row, 31 in the second row, 27 in the third row, and so on. Continuing this pattern, can she end with a row containing exactly 1 block? If not, how many blocks will the last row contain? How many rows can she build this way?

▲ **49.** A stack of firewood has 28 pieces on the bottom, 24 on top of those, then 20, and so on. If there are 108 pieces of wood, how many rows are there? (*Hint: $n \leq 7$.*)

▲ **50.** Suppose that you are offered a job at $1600 a month with a guranteed increase of $50 every 6 months for 5 years. What will your salary be at the end of this period of time?

▲ **51.** Jay's grandfather promised to give him 1¢ on the first day of his birthday month, 2¢ on the second day, 3¢ on the third day, and so on for thirty days. How much will that amount to?

▲ **52.** A seating section in a theater-in-the-round has 20 seats in the first row, 22 in the second row, 24 in the third row, and so on for 25 rows. How many seats are there in the last row? How many seats are there in the section?

50. $2100 a month
51. $4.65   52. 68;   1100
53. 18   54. 48   55. $\frac{1}{2}$
56. $\frac{5}{16}$

REVIEW EXERCISES

Evaluate ar^n for the given values of a, r, and n. See Section 3.1.

53. $a = 2, r = 3, n = 2$

54. $a = 3, r = 2, n = 4$

55. $a = 4, r = \frac{1}{2}, n = 3$

56. $a = 5, r = \frac{1}{4}, n = 2$

11.3 GEOMETRIC SEQUENCES

FOR EXTRA HELP	OBJECTIVES
SSG pp. 360–366 **SSM** pp. 592–597	**1** ▶ Find the common ratio of a geometric sequence.
Video 13	**2** ▶ Find the general term of a geometric sequence.
	3 ▶ Find any specified term of a geometric sequence.
Tutorial IBM MAC	**4** ▶ Find the sum of a specified number of terms of a geometric sequence.
	5 ▶ Find the sum of an infinite number of terms of certain geometric sequences.

In an arithmetic sequence, each term after the first is found by *adding* a fixed number to the previous term. A *geometric sequence* is defined as follows.

Geometric Sequence   A **geometric sequence** or **geometric progression** is a sequence in which each term after the first is a constant multiple of the preceding term.

1 ▶ Find the common ratio of a geometric sequence.

▶   We find the constant multiplier, called the **common ratio,** by dividing any term after the first by the preceding term. That is, the common ratio is

$$r = \frac{a_{n+1}}{a_n}.$$

For example,

$$2, 6, 18, 54, 162, \ldots$$

is a geometric sequence in which the first term, a_1, is 2 and the common ratio is

$$r = \frac{6}{2} = \frac{18}{6} = \frac{54}{18} = \frac{162}{54} = 3.$$

▶ **TEACHING TIP**

Show the contrast between *arithmetic* and *geometric* sequences with the following examples.

2, 4, 6, 8, 10, . . .
Arithmetic: $d = a_2 - a_1 = 2$

2, 4, 8, 16, 32, . . .
Geometric: $r = \frac{a_2}{a_1} = 2$

Note that the terms in a geometric sequence get quite large early on in the sequence (for $r > 1$). ◀

INSTRUCTOR'S RESOURCES

 ITM pp. 527–531   **ISM** pp. 889–898
IAM p. 58

TEST GENERATOR
DOS   **Windows**   **MAC**

TRANSPARENCIES

 CONCEPTUAL   WRITING   CHALLENGING   SCIENTIFIC CALCULATOR    GRAPHICS CALCULATOR

EXAMPLE 1

Finding the
Common Ratio

Find r for the geometric sequence

$$15, \frac{15}{2}, \frac{15}{4}, \frac{15}{8}, \dots$$

To find r choose any two adacent terms and divide the second one by the first. Choosing the second and third terms of the sequence,

$$r = \frac{a_3}{a_2} = \frac{15}{4} \div \frac{15}{2} = \frac{1}{2}.$$

Any other two adjacent terms could have been used to find r. Additional terms of the sequence can be found by multiplying each successive term by $1/2$. ■

2 ▶ Find the general term of a geometric sequence.

▶ The general term a_n of a geometric sequence $a_1, a_2, a_3, \dots$ is expressed in terms of a_1 and r by writing the first few terms as

$$a_1, \quad a_2 = a_1 r, \quad a_3 = a_1 r^2, \quad a_4 = a_1 r^3, \dots,$$

which suggests the following rule.

General Term of a Geometric Sequence

The general term of the geometric sequence with first term a_1 and common ratio r is

$$a_n = a_1 r^{n-1}.$$

CAUTION Be careful to use the correct order of operations when finding $a_1 r^{n-1}$. The value of r^{n-1} must be found first. Then multiply the result by a_1.

EXAMPLE 2

Finding the General Term

Find the general term of the sequence in Example 1.

The first term is $a_1 = 15$ and the common ratio is $r = 1/2$. Substituting into the formula for the general term gives

$$a_n = a_1 r^{n-1} = 15\left(\frac{1}{2}\right)^{n-1},$$

the required general term. Notice that it is not possible to simplify further, because the exponent must be applied before the multiplication can be done. ■

3 ▶ Find any specified term of a geometric sequence.

▶ We can use the formula for the general term to find any particular term.

EXAMPLE 3

Finding a Specified Term

Find the indicated term for each geometric sequence.

(a) $a_1 = 4, r = -3; \ a_6$

Let $n = 6$. From the general term $a_n = a_1 r^{n-1}$,

$$\begin{aligned} a_6 &= a_1 \cdot r^{6-1} &&\text{Let } n = 6. \\ &= 4 \cdot (-3)^5 &&\text{Let } a_1 = 4, r = -3. \\ &= -972 &&\text{Evaluate } (-3)^5 \text{ first.} \end{aligned}$$

(b) $\dfrac{3}{4}, \dfrac{3}{8}, \dfrac{3}{16}, \ldots; \quad a_7$

Here, $r = 1/2$, $a_1 = 3/4$, and n is 7.

$$a_7 = \dfrac{3}{4} \cdot \left(\dfrac{1}{2}\right)^6 = \dfrac{3}{4} \cdot \dfrac{1}{64} = \dfrac{3}{256} \quad \blacksquare$$

EXAMPLE 4
Writing the Terms of a Sequence
■

Write the first five terms of the geometric sequence whose first term is 5 and whose common ratio is 1/2.

Using the formula $a_n = a_1 r^{n-1}$,

$$a_1 = 5,$$

$$a_2 = 5\left(\dfrac{1}{2}\right) = \dfrac{5}{2},$$

$$a_3 = 5\left(\dfrac{1}{2}\right)^2 = \dfrac{5}{4},$$

$$a_4 = 5\left(\dfrac{1}{2}\right)^3 = \dfrac{5}{8},$$

$$a_5 = 5\left(\dfrac{1}{2}\right)^4 = \dfrac{5}{16}. \quad \blacksquare$$

4 ▶ Find the sum of a specified number of terms of a geometric sequence.

▶ It is useful to have a formula for the sum of the first n terms of a geometric sequence, S_n. We can develop a formula by first writing out S_n.

$$S_n = a_1 + a_1 r + a_1 r^2 + a_1 r^3 + \cdots + a_1 r^{n-1}$$

Next, we multiply both sides by r.

$$rS_n = a_1 r + a_1 r^2 + a_1 r^3 + a_1 r^4 + \cdots + a_1 r^n$$

We subtract the first result from the second.

$$rS_n - S_n = (a_1 r - a_1) + (a_1 r^2 - a_1 r) + (a_1 r^3 - a_1 r^2)$$
$$+ (a_1 r^4 - a_1 r^3) + \cdots + (a_1 r^n - a_1 r^{n-1})$$

Using the commutative and associative properties to rearrange the terms on the right, we get

$$rS_n - S_n = (a_1 r - a_1 r) + (a_1 r^2 - a_1 r^2)$$
$$+ (a_1 r^3 - a_1 r^3) + \cdots + (a_1 r^n - a_1)$$
$$(r - 1)S_n = a_1 r^n - a_1, \qquad \text{Distributive property}$$

so, if $r \neq 1$,

$$S_n = \dfrac{a_1 r^n - a_1}{r - 1}. \qquad \text{Divide by } r - 1.$$

A summary of this discussion follows.

Sum of the First n Terms of a Geometric Sequence

The sum of the first n terms of the geometric sequence with first term a_1 and common ratio r is

$$S_n = \dfrac{a_1(r^n - 1)}{r - 1} \qquad (r \neq 1).$$

If $r = 1$, $S_n = a_1 + a_1 + a_1 + \cdots + a_1 = na_1$.

Multiplying the formula for S_n by $-1/(-1)$ gives us an alternative form that is sometimes preferable.

$$S_n = \frac{a_1(r^n - 1)}{r - 1} \cdot \frac{-1}{-1} = \frac{a_1(1 - r^n)}{1 - r}$$

EXAMPLE 5

Finding the Sum of the First n Terms

■ Find the sum of the first six terms of the geometric sequence with first term -2 and common ratio 3.

Substitute $n = 6$, $a_1 = -2$, and $r = 3$ into the formula for S_n.

$$S_n = \frac{a_1(r^n - 1)}{r - 1}$$

$$S_6 = \frac{-2(3^6 - 1)}{3 - 1} \qquad \text{Let } n = 6, a_1 = -2, r = 3.$$

$$= \frac{-2(729 - 1)}{2} \qquad \text{Evaluate the exponential.}$$

$$= -728 \quad ■$$

> **Chalkboard Exercise**
>
> Find the sum of the first six terms of the geometric sequence with first term 6 and common ratio 3.
>
> Answer: 2184

▶ **TEACHING TIP**
To point out the utility of this formula, list the first six terms of the geometric sequence

$-2, -6, -18, -54, -162, -486$

and add them together. Next, apply the formula directly to obtain the same answer. Emphasize that the formula applies only to geometric sequences. Consider showing a summation example with a fraction for the common ratio, such as the geometric series:

$$\frac{1}{2} + \frac{1}{4} + \frac{1}{8} + \frac{1}{16} + \frac{1}{32}. \quad ◀$$

A series of the form

$$\sum_{i=1}^{n} a \cdot b^i$$

represents the sum of the first n terms of a geometric sequence having first term $a_1 = a \cdot b^1 = ab$ and common ratio b. For example,

$$\sum_{i=1}^{4} 3 \cdot 2^i = 3 \cdot 2^1 + 3 \cdot 2^2 + 3 \cdot 2^3 + 3 \cdot 2^4$$

is a geometric sequence with $a_1 = 3 \cdot 2^1 = 6$ and common ratio $r = 2$. We can find this sum using the formula for S_n, as shown in the next example.

EXAMPLE 6

Using the Formula for S_n to Find a Summation

■ Find $\sum_{i=1}^{4} 3 \cdot 2^i$.

Since the series is in the form

$$\sum_{i=1}^{n} a \cdot b^i,$$

it represents the sum of the first n terms of the geometric sequence with $a_1 = a \cdot b^1$ and $r = b$. The sum is found by using the formula

$$S_n = \frac{a_1(r^n - 1)}{r - 1}.$$

Here $n = 4$. Also, $a_1 = 6$ and $r = 2$. Now substitute into the formula for S_n.

$$S_4 = \frac{6(2^4 - 1)}{2 - 1} \qquad \text{Let } n = 4, a_1 = 6, r = 2.$$

$$= \frac{6(16 - 1)}{1} \qquad \text{Evaluate } 2^4.$$

$$= 90 \quad ■$$

> **Chalkboard Exercise**
>
> Find $\sum_{i=1}^{6} 3\left(\frac{1}{4}\right)^i$.
>
> Answer: .9998 or $\dfrac{4095}{4096}$

▶ **TEACHING TIP**
Write down the first few terms of the series:

Index: $i = 1$ $i = 2$ $i = 3$
Values: 6   +   12   +   24   + $\cdots$
Terms: a_1 a_2 a_3

Now, $a_1 = 6$, $r = a_2/a_1 = 12/6 = 2$. Substitute a_1 and r into the summation formula and simplify. ◀

5 ▶ Find the sum of an infinite number of terms of certain geometric sequences.

▶ Now, consider an infinite geometric sequence such as

$$\frac{1}{3}, \frac{1}{6}, \frac{1}{12}, \frac{1}{24}, \frac{1}{48}, \ldots$$

Can the sum of the terms of such a sequence be found somehow? The sum of the first two terms is

$$S_2 = \frac{1}{3} + \frac{1}{6} = \frac{1}{2} = .5.$$

In a similar manner,

$$S_3 = S_2 + \frac{1}{12} = \frac{1}{2} + \frac{1}{12} = \frac{7}{12} \approx .583,$$

$$S_4 = S_3 + \frac{1}{24} = \frac{7}{12} + \frac{1}{24} = \frac{15}{24} = .625,$$

$$S_5 = \frac{31}{48} \approx .64583,$$

$$S_6 = \frac{21}{32} = .65625,$$

$$S_7 = \frac{127}{192} \approx .6614583.$$

Each term of the geometric sequence is smaller than the preceding one, so that each additional term is contributing less and less to the sum. In decimal form (to the nearest thousandth) the first seven terms and the tenth term are given below.

Term	a_1	a_2	a_3	a_4	a_5	a_6	a_7	a_{10}
Value	.333	.167	.083	.042	.021	.010	.005	.001

As the table suggests, the value of a term gets closer and closer to zero as the number of the term increases. To express this idea, we say that as n increases without bound (written $n \to \infty$), the limit of the term a_n is zero, written

$$\lim_{n \to \infty} a_n = 0.$$

A number that can be defined as the sum of an infinite number of terms of a geometric sequence can be found by starting with the expression for the sum of a finite number of terms:

$$S_n = \frac{a_1(r^n - 1)}{r - 1}.$$

If $|r| < 1$, then as n increases without bound the value of r^n gets closer and closer to zero. For example, in the infinite sequence discussed above, $r = 1/2 = .5$. The chart below shows how $r^n = (.5)^n$, given to nearest thousandth, gets smaller as n increases.

n	1	2	3	4	5	6	7	10
r^n	.5	.25	.125	.063	.031	.016	.008	.001

As r^n approaches 0, $r^n - 1$ approaches $0 - 1 = -1$, and S_n approaches the quotient $\dfrac{-a_1}{r-1}$. Thus,

$$\lim_{r^n \to 0} S_n = \lim_{r^n \to 0} \frac{a_1(r^n - 1)}{r - 1}$$

$$= \frac{a_1(0 - 1)}{r - 1}$$

$$= \frac{-a_1}{r - 1} = \frac{a_1}{1 - r}.$$

This limit is defined to be the sum of the infinite geometric sequence:

$$a_1 + a_1 r + a_1 r^2 + a_1 r^3 + \cdots = \frac{a_1}{1 - r}, \quad \text{if } |r| < 1.$$

What if $|r| > 1$? For example, suppose the sequence is

$$6, 12, 24, \ldots, 3(2)^n, \ldots.$$

In this kind of sequence, as n increases, the value of r^n also increases and so does the sum S_n. Since each new term adds a larger and larger amount to the sum, there is no limit to the value of S_n, and the sum S_n does not exist. A similar situation exists if $r = 1$.

In summary, the sum of the terms of an infinite geometric sequence is as follows.

Sum of the Terms of an Infinite Geometric Sequence	The sum of the terms of an infinite geometric sequence with first term a_1 and common ratio r, where $|r| < 1$, is $$\frac{a_1}{1 - r}.$$ If $|r| \geq 1$, the sum does not exist.

EXAMPLE 7

Finding the Sum of the Terms of an Infinite Geometric Sequence

Find the sum of the terms of the infinite geometric sequence with $a_1 = 3$ and $r = -1/3$.

From the rule just above, the sum is

$$\frac{a_1}{1 - r} = \frac{3}{1 - (-1/3)}$$

$$= \frac{3}{4/3}$$

$$= \frac{9}{4}. \quad \blacksquare$$

In summation notation, the sum of an infinite geometric sequence is written as

$$\sum_{i=1}^{\infty} a_i.$$

E X A M P L E   8

Finding the Sum of the Terms of an Infinite Geometric Series

■ Find $\sum_{i=1}^{\infty} \left(\frac{1}{2}\right)^i$.

This is the infinite geometric series

$$\frac{1}{2} + \frac{1}{4} + \frac{1}{8} + \cdots,$$

with $a_1 = 1/2$ and $r = 1/2$. Since $|r| < 1$, we find the sum as follows.

$$\frac{a_1}{1-r} = \frac{\frac{1}{2}}{1 - \frac{1}{2}} = \frac{\frac{1}{2}}{\frac{1}{2}} = 1 \quad ■$$

Chalkboard Exercise

Find $\sum_{i=1}^{\infty} \left(\frac{1}{5}\right)\left(\frac{5}{7}\right)^i$.

Answer: $\frac{1}{2}$

11.3 EXERCISES

◉ **1.** Explain in your own words how to evaluate $2(3)^5$.

◉ **2.** If $2, x, 9/2$ is a geometric sequence, how can the value of x be found?

If a sequence is geometric, find the common ratio, r. If a sequence is not geometric, say so. See Example 1.

3. $4, 8, 16, 32, \ldots$

4. $5, 15, 45, 135, \ldots$

5. $\frac{1}{3}, \frac{2}{3}, \frac{3}{3}, \frac{4}{3}, \frac{5}{3}, \ldots$

6. $\frac{1}{3}, \frac{2}{3}, \frac{4}{3}, \frac{8}{3}, \ldots$

7. $1, -3, 9, -27, 81, \ldots$

8. $1, -3, 7, -11, \ldots$

9. $1, -\frac{1}{2}, \frac{1}{4}, -\frac{1}{8}, \frac{1}{16}, \ldots$

10. $\frac{2}{3}, \frac{2}{15}, \frac{2}{75}, \frac{2}{375}, \ldots$

Find a general term for the geometric sequence. See Example 2.

11. $5, 10, \ldots$

12. $-2, -6, \ldots$

13. $\frac{1}{9}, \frac{1}{3}, \ldots$

14. $-3, \frac{3}{2}, \ldots$

15. $10, -2, \ldots$

16. $-4, 8, \ldots$

—◆ **MATHEMATICAL CONNECTIONS** (Exercises 17–20) ◆—

Let $g(x) = ab^x$.

17. Find $g(1)$, $g(2)$, and $g(3)$.

18. Consider the sequence $g(1), g(2), g(3), \ldots$. Is it a geometric sequence? If so, what is the common ratio?

19. What is the general term of the sequence in Exercise 18?

◉ **20.** Explain how geometric sequences are related to exponential functions.

————◆————

Find the indicated term for the geometric sequence. See Example 3.

21. $2, 10, 50, \ldots;$ a_{10}

22. $-1, -3, -9, \ldots;$ a_{15}

23. $\frac{1}{2}, \frac{1}{6}, \frac{1}{18}, \ldots;$ a_{12}

24. $\frac{2}{3}, -\frac{1}{3}, \frac{1}{6}, \ldots;$ a_{18}

▲ **25.** $a_3 = \frac{1}{2}, a_7 = \frac{1}{32};$ a_{25}

▲ **26.** $a_5 = 48, a_8 = -384;$ a_{10}

3. $r = 2$ **4.** $r = 3$
5. not geometric
6. $r = 2$ **7.** $r = -3$
8. not geometric
9. $r = -\frac{1}{2}$ **10.** $r = \frac{1}{5}$
11. $a_n = 5(2)^{n-1}$
12. $a_n = -2(3)^{n-1}$
13. $a_n = \frac{3^{n-1}}{9}$
14. $a_n = -3\left(-\frac{1}{2}\right)^{n-1}$
15. $a_n = 10\left(-\frac{1}{5}\right)^{n-1}$
16. $a_n = -4(-2)^{n-1}$
17. ab, ab^2, ab^3
18. yes; b **19.** $a_n = ab^n$
21. $2(5)^9$ **22.** -3^{14}
23. $\frac{1}{2}\left(\frac{1}{3}\right)^{11}$
24. $\frac{2}{3}\left(-\frac{1}{2}\right)^{17} = -\frac{1}{3 \cdot 2^{16}}$
25. $2\left(\frac{1}{2}\right)^{24} = \frac{1}{2^{23}}$
26. $3(-2)^9 = -1536$

◉ CONCEPTUAL   ✐ WRITING   ▲ CHALLENGING   ▦ SCIENTIFIC CALCULATOR   ▣ GRAPHICS CALCULATOR

27. $\dfrac{121}{243}$ 28. 84

29. -1.997 30. $.195$

31. 2.662 32. 9.610

33. -2.982 34. $.656$

35. 9   36. $\dfrac{25}{2}$ 37. $\dfrac{10,000}{11}$

38. $21,250$ 39. $-\dfrac{9}{20}$

40. 3   41. does not exist

42. does not exist

43. $10\left(\dfrac{3}{5}\right)^4 \approx 1.3$ feet

44. 3 days, $\dfrac{1}{4}$ gram

45. $\left(\dfrac{2}{3}\right)^6 = \dfrac{64}{729}$ units

46. (a) $1.1(1.06)^5 \approx$ 1.5 billion units   (b) about 12 years

47. (a) $1.1(1.02)^5 \approx$ 1.21 billion units (b) about 35 years

48. $\$50,000\left(\dfrac{3}{4}\right)^8 \approx \$5000$

Use the formula for S_n to find the sum for the geometric sequence. See Examples 5 and 6. In Exercises 29–34, give the answer to the nearest thousandth.

27. $\dfrac{1}{3}, \dfrac{1}{9}, \dfrac{1}{27}, \dfrac{1}{81}, \dfrac{1}{243}$

28. $\dfrac{4}{3}, \dfrac{8}{3}, \dfrac{16}{3}, \dfrac{32}{3}, \dfrac{64}{3}, \dfrac{128}{3}$

29. $-\dfrac{4}{3}, -\dfrac{4}{9}, -\dfrac{4}{27}, -\dfrac{4}{81}, -\dfrac{4}{243}, -\dfrac{4}{729}$

30. $\dfrac{5}{16}, -\dfrac{5}{32}, \dfrac{5}{64}, -\dfrac{5}{128}$

31. $\displaystyle\sum_{i=1}^{7} 4\left(\dfrac{2}{5}\right)^i$

32. $\displaystyle\sum_{i=1}^{8} 5\left(\dfrac{2}{3}\right)^i$

33. $\displaystyle\sum_{i=1}^{10} (-2)\left(\dfrac{3}{5}\right)^i$

34. $\displaystyle\sum_{i=1}^{6} (-2)\left(-\dfrac{1}{2}\right)^i$

Find the sum, if it exists, of the infinite geometric sequence. See Examples 7 and 8.

35. $a_1 = 6, r = \dfrac{1}{3}$

36. $a_1 = 10, r = \dfrac{1}{5}$

37. $a_1 = 1000, r = -\dfrac{1}{10}$

38. $a_1 = 8500, r = \dfrac{3}{5}$

39. $\displaystyle\sum_{i=1}^{\infty} \dfrac{9}{8}\left(-\dfrac{2}{3}\right)^i$

40. $\displaystyle\sum_{i=1}^{\infty} \dfrac{3}{5}\left(\dfrac{5}{6}\right)^i$

41. $\displaystyle\sum_{i=1}^{\infty} \dfrac{12}{5}\left(\dfrac{5}{4}\right)^i$

42. $\displaystyle\sum_{i=1}^{\infty} \left(-\dfrac{16}{3}\right)\left(-\dfrac{9}{8}\right)^i$

▲ *Solve the applied problem.*

43. A certain ball when dropped from a height rebounds 3/5 of the original height. How high will the ball rebound after the fourth bounce if it was dropped from a height of 10 feet?

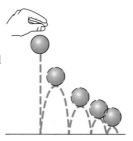

44. A particular substance decays in such a way that it loses half its weight each day. In how many days will 256 grams of the substance be reduced to 32 grams? How much of the substance is left after 10 days?

45. A tracer dye is injected into a system with an input and an excretion. After one hour 2/3 of the dye is left. At the end of the second hour 2/3 of the remaining dye is left, and so on. If one unit of the dye is injected, how much is left after 6 hours?

46. Historically, the consumption of electricity has increased about 6% per year.
    (a) If a community uses 1.1 billion units of electricity now, how much will it use five years from now?
    (b) Find how many years it will take for the consumption to double.

47. Suppose the community in Exercise 46 reduces its increase in consumption to 2% per year.
    (a) How much will it use five years from now?
    (b) Find the number of years for the consumption to double.

48. A machine depreciates by 1/4 of its value each year. If it cost $50,000 new, what is its value after 8 years?

REVIEW EXERCISES

Multiply. See Section 3.4.

49. $(3x + 2y)^2$

50. $(4x - 3y)^2$

51. $(a - b)^3$

52. $(x + y)^4$

49. $9x^2 + 12xy + 4y^2$
50. $16x^2 - 24xy + 9y^2$
51. $a^3 - 3a^2b + 3ab^2 - b^3$
52. $x^4 + 4x^3y + 6x^2y^2 + 4xy^3 + y^4$

11.4 THE BINOMIAL THEOREM

FOR EXTRA HELP	OBJECTIVES
SSG pp. 367–369 **SSM** pp. 597–599	**1** ▶ Expand a binomial raised to a power.
Video 14	**2** ▶ Find any specified term of the expansion of a binomial.
Tutorial IBM MAC	

1 ▶ Expand a binomial raised to a power.

▶ TEACHING TIP

This is an important section with applications in statistics, finite math, and calculus. ◀

▶ Writing out the binomial expression $(x + y)^n$ for nonnegative integer values of n gives a family of expressions that is important in the study of mathematics and its applications. For example,

$$(x + y)^0 = 1$$
$$(x + y)^1 = x + y,$$
$$(x + y)^2 = x^2 + 2xy + y^2,$$
$$(x + y)^3 = x^3 + 3x^2y + 3xy^2 + y^3,$$
$$(x + y)^4 = x^4 + 4x^3y + 6x^2y^2 + 4xy^3 + y^4,$$
$$(x + y)^5 = x^5 + 5x^4y + 10x^3y^2 + 10x^2y^3 + 5xy^4 + y^5.$$

Inspection shows that these expansions follow a pattern. By identifying the pattern we can write a general expression for $(x + y)^n$.

First, if n is a positive integer, each expansion after $(x + y)^0$ begins with x raised to the same power to which the binomial is raised. That is, the expansion of $(x + y)^1$ has a first term of x^1, the expansion of $(x + y)^2$ has a first term of x^2, the expansion of $(x + y)^3$ has a first term of x^3, and so on. Also, the last term in each expansion is y to this same power, so that the expansion of $(x + y)^n$ should begin with the term x^n and end with the term y^n.

The exponents on x decrease by one in each term after the first, while the exponents on y, beginning with y in the second term, increase by one in each succeeding term. Thus the *variables* in the expansion of $(x + y)^n$ have the following pattern.

$$x^n, \quad x^{n-1}y, \quad x^{n-2}y^2, \quad x^{n-3}y^3, \ldots, xy^{n-1}, \quad y^n$$

This pattern suggests that the sum of the exponents on x and y in each term is n. For example, in the third term above, the variable part is $x^{n-2}y^2$ and the sum of the exponents, $n - 2$ and 2, is n.

Now examine the pattern for the *coefficients* in the terms of the expansions shown above. Writing the coefficients alone in a triangular pattern gives the following.

INSTRUCTOR'S RESOURCES

ITM pp. 531–533 **ISM** pp. 898–901 **IAM** p. 58	**TEST GENERATOR** DOS Windows MAC	**TRANSPARENCIES**

◉ CONCEPTUAL   ✍ WRITING   ▲ CHALLENGING   ▦ SCIENTIFIC CALCULATOR   ▦ GRAPHICS CALCULATOR

Pascal's Triangle

$$
\begin{array}{ccccccccccccc}
 & & & & & & 1 & & & & & & \\
 & & & & & 1 & & 1 & & & & & \\
 & & & & 1 & & 2 & & 1 & & & & \\
 & & & 1 & & 3 & & 3 & & 1 & & & \\
 & & 1 & & 4 & & 6 & & 4 & & 1 & & \\
 & 1 & & 5 & & 10 & & 10 & & 5 & & 1 & \quad\text{and so on}
\end{array}
$$

▶ **TEACHING TIP**
Note that the number of terms in $(x + y)^n$ is $n + 1$, so for $(x + y)^6$, one needs to find the row in Pascal's triangle with $6 + 1 = 7$ numbers in it. ◀

Arranging the coefficients in this way shows that each number in the triangle is the sum of the two numbers just above it (one to the right and one to the left). For example, starting with 1 as the first row, in the fifth row from the top, 1 is the sum of 1 (the only number above it), 4 is the sum of 1 and 3, while 6 is the sum of 3 and 3, and so on.

We get the coefficients for $(x + y)^6$ by attaching the seventh row to the table by adding pairs of numbers from the sixth row.

$$1 \quad 6 \quad 15 \quad 20 \quad 15 \quad 6 \quad 1$$

Use these coefficients to expand $(x + y)^6$ as

$$(x + y)^6 = x^6 + 6x^5y + 15x^4y^2 + 20x^3y^3 + 15x^2y^4 + 6xy^5 + y^6.$$

This triangular array of numbers is known as **Pascal's triangle**, in honor of the seventeenth-century mathematician Blaise Pascal, one of the first to use it extensively.

◈ 1. 21 and 34
2. 15, 21, 28, 36, 45
3. Answers will vary.

◈ **CONNECTIONS** ◈

Over the years, many interesting patterns have been discovered in Pascal's triangle. In the figure below, the triangular array is written in a different form. The indicated sums along the diagonals shown are the terms of the *Fibonacci sequence,* mentioned at the beginning of the chapter. The presence of this sequence in the triangle apparently was not recognized by Pascal.

$$
\begin{array}{ccccccccc}
1 & & & & & & & & \\
1 & 1 & & & & & & & \\
1 & 1 & 2 & & & & & & \\
1 & 2 & 1 & 3 & & & & & \\
1 & 3 & 3 & 1 & 5 & & & & \\
1 & 4 & 6 & 4 & 1 & 8 & & & \\
1 & 5 & 10 & 10 & 5 & 1 & 13 & & \\
1 & 6 & 15 & 20 & 15 & 6 & 1 & &
\end{array}
$$

Triangular numbers are found by counting the number of points in triangular arrangements of points. The first few triangular numbers are shown below.

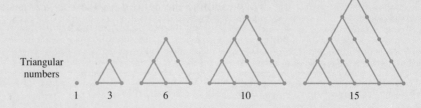

Triangular numbers   1   3   6   10   15

The number of points in these figures form the sequence, 1, 3, 6, 10, . . . , a sequence that is found in Pascal's triangle as shown in the next figure.

FOR DISCUSSION OR WRITING

1. Predict the next two numbers in the sequence of sums of the diagonals of Pascal's triangle.
2. Predict the next five numbers in the list of triangular numbers.
3. Describe other sequences that can be found in Pascal's triangle.

Although it is possible to use Pascal's triangle to find the coefficients of $(x + y)^n$ for any positive integer value of n, it is impractical for large values of n. A more efficient way to determine these coefficients is needed. It is helpful to use the following notational shorthand. The symbol $\boldsymbol{n!}$ (read "n factorial") is defined as follows.

n **Factorial ($n!$)**   For any positive integer n,

$$n(n - 1)(n - 2)(n - 3) \cdots (2)(1) = n!.$$

▶ **TEACHING TIP**

Students can use the *factorial* key on their calculators. For example, 5!/(3!2!) can be evaluated on the calculator as

$5! \div 3! \div 2! = 10.$ ◀

For example,

$$3! = 3 \cdot 2 \cdot 1 = 6 \quad \text{and} \quad 5! = 5 \cdot 4 \cdot 3 \cdot 2 \cdot 1 = 120.$$

From the definition of n factorial, $n[(n - 1)!] = n!$. If $n = 1$, then $1(0!) = 1! = 1$. Because of this, 0! is defined as

$$0! = 1.$$

◆ **CONNECTIONS** ◆

Many calculators have the capability of finding $n!$. A calculator with a 10-digit display will give the exact value of $n!$ for $n \leq 13$ and approximate values of $n!$ for $14 \leq n \leq 69$. The figure shows the display for 13!, 25!, and 69!.

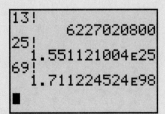

EXAMPLE 1

Evaluating Expressions with $n!$

▶ **TEACHING TIP**
The binomial coefficient

$$\frac{n!}{r!(n-r)!}$$

is calculated by using the $\boxed{_nC_r}$ key on a scientific calculator.

For example, to find the fourth term in the expansion of $(x+y)^9$, the variable part is x^6y^3, so the binomial coefficient is $9!/(6!3!)$. Use the $\boxed{_nC_r}$ key with $n=9$ and $r=6$ to find the value of the coefficient. Students should refer to the owner's manual to determine how to use this key. ◀

Find the value of each of the following.

(a) $\dfrac{5!}{4!\ 1!} = \dfrac{5\cdot4\cdot3\cdot2\cdot1}{(4\cdot3\cdot2\cdot1)(1)} = 5$

(b) $\dfrac{5!}{3!\ 2!} = \dfrac{5\cdot4\cdot3\cdot2\cdot1}{(3\cdot2\cdot1)(2\cdot1)} = \dfrac{5\cdot4}{2\cdot1} = 10$

(c) $\dfrac{6!}{3!\ 3!} = \dfrac{6\cdot5\cdot4\cdot3\cdot2\cdot1}{(3\cdot2\cdot1)(3\cdot2\cdot1)} = \dfrac{6\cdot5\cdot4}{3\cdot2\cdot1} = 20$

(d) $\dfrac{4!}{4!\ 0!} = \dfrac{4\cdot3\cdot2\cdot1}{(4\cdot3\cdot2\cdot1)(1)} = 1$ ∎

Now look again at the coefficients of the expansion

$$(x+y)^5 = x^5 + 5x^4y + 10x^3y^2 + 10x^2y^3 + 5xy^4 + y^5.$$

The coefficient of the second term is 5 and the exponents on the variables in that term are 4 and 1. From Example 1(a), $5!/(4!\ 1!) = 5$. The coefficient of the third term is 10, and the exponents are 3 and 2. From Example 1(b), $5!/(3!\ 2!) = 10$. Similar results hold true for the remaining terms. The first term can be written as $1x^5y^0$ and the last term can be written as $1x^0y^5$. Then the coefficient of the first term should be $5!/(5!\ 0!) = 1$, and the coefficient of the last term would be $5!/(0!\ 5!) = 1$. Generalizing, the coefficient for a term of $(x+y)^n$ in which the variable part is x^ry^{n-r} will be

$$\frac{n!}{r!(n-r)!}. \quad *$$

NOTE The denominator factorials in the coefficient of a term are the same as the exponents on the variables in that term.

Summarizing this work gives the **binomial theorem,** or the **general binomial expansion.**

Binomial Theorem

For any positive integer n,

$$(x+y)^n = x^n + \frac{n!}{(n-1)!\ 1!}x^{n-1}y + \frac{n!}{(n-2)!\ 2!}x^{n-2}y^2$$

$$+ \frac{n!}{(n-3)!\ 3!}x^{n-3}y^3 + \cdots + \frac{n!}{1!(n-1)!}xy^{n-1} + y^n.$$

The binomial theorem can be written in summation notation as

$$\sum_{i=0}^{n} \frac{n!}{(n-i)!\ i!}x^{n-i}y^i.$$

NOTE The letter i is used here instead of r because we are using summation notation. It is not the imaginary number i.

---

* This quantity may be symbolized as $\dbinom{n}{r}$ or $_nC_r$.

EXAMPLE 2

Using the Binomial Theorem

Expand $(2m + 3)^4$.

$$(2m + 3)^4 = (2m)^4 + \frac{4!}{3!\,1!}(2m)^3(3) + \frac{4!}{2!\,2!}(2m)^2(3)^2 + \frac{4!}{1!\,3!}(2m)(3)^3 + (3)^4$$

$$= 16m^4 + 4(8m^3)(3) + 6(4m^2)(9) + 4(2m)(27) + 81$$

$$= 16m^4 + 96m^3 + 216m^2 + 216m + 81 \quad \blacksquare$$

EXAMPLE 3

Using the Binomial Theorem

Expand $\left(a - \dfrac{b}{2}\right)^5$.

$$\left(a - \frac{b}{2}\right)^5 = a^5 + \frac{5!}{4!\,1!}a^4\left(-\frac{b}{2}\right) + \frac{5!}{3!\,2!}a^3\left(-\frac{b}{2}\right)^2 + \frac{5!}{2!\,3!}a^2\left(-\frac{b}{2}\right)^3$$

$$+ \frac{5!}{1!\,4!}a\left(-\frac{b}{2}\right)^4 + \left(-\frac{b}{2}\right)^5$$

$$= a^5 + 5a^4\left(-\frac{b}{2}\right) + 10a^3\left(\frac{b^2}{4}\right) + 10a^2\left(-\frac{b^3}{8}\right)$$

$$+ 5a\left(\frac{b^4}{16}\right) + \left(-\frac{b^5}{32}\right)$$

$$= a^5 - \frac{5}{2}a^4b + \frac{5}{2}a^3b^2 - \frac{5}{4}a^2b^3 + \frac{5}{16}ab^4 - \frac{1}{32}b^5 \quad \blacksquare$$

For Example 2

Chalkboard Exercise

Expand $(x^2 + 3)^5$.

Answer: $x^{10} + 15x^8 + 90x^6 + 270x^4 + 405x^2 + 243$

Chalkboard Exercise

Expand $\left(\dfrac{x}{4} - 3y\right)^4$.

Answer: $\dfrac{1}{256}x^4 - \dfrac{3}{16}x^3y + \dfrac{27}{8}x^2y^2 - 27xy^3 + 81y^4$

CAUTION When the binomial is the *difference* of two terms as in Example 3, the signs of the terms in the expansion will alternate. Those terms with odd exponents on the second variable expression ($-b/2$ in Example 3) will be negative, while those with even exponents on the second variable expression will be positive.

2 ▶ Find any specified term of the expansion of a binomial.

▶ Any single term of a binomial expansion can be determined without writing out the whole expansion. For example, if $n \geq 10$, the tenth term of $(x + y)^n$ has y raised to the ninth power (since y has the power of 1 in the second term, the power of 2 in the third term, and so on). Since the exponents on x and y in any term must have a sum of n, the exponent on x in the tenth term is $n - 9$. These quantities, 9 and $n - 9$, determine the factorials in the denominator of the coefficient. Thus,

$$\frac{n!}{(n - 9)!\,9!}x^{n-9}y^9$$

is the tenth term of $(x + y)^n$. A generalization of this idea follows.

rth Term of the Binomial Expansion

If $n \geq r$, the rth term of the expansion of $(x + y)^n$ is

$$\frac{n!}{[n - (r - 1)]!(r - 1)!}x^{n-(r-1)}y^{r-1}.$$

This general expression is confusing. Remember to start with the exponent on y, which is 1 less than the term number r. Then subtract that exponent from n to get the exponent on x: $n - (r - 1)$. The two exponents are then used as the factorials in the denominator of the coefficient.

EXAMPLE 4

Finding a Single Term of a Binomial Expansion

■ Find the fourth term of $(a + 2b)^{10}$.

In the fourth term, $2b$ has an exponent of $4 - 1 = 3$ and a has an exponent of $10 - 3 = 7$. The fourth term is

$$\frac{10!}{7! \, 3!}(a^7)(2b)^3 = \frac{10 \cdot 9 \cdot 8}{3 \cdot 2 \cdot 1}(a^7)(8b^3)$$

$$= 120a^7(8b^3)$$

$$= 960a^7b^3. \quad ■$$

Chalkboard Exercise

Find the fifth term of $\left(\dfrac{x}{2} - y\right)^9$.

Answer: $\dfrac{63}{16}x^5y^4$

11.4 EXERCISES

◉ **1.** What is the "binomial" referred to in the binomial theorem?

1. The quantity raised to the nth power is a binomial.

Use the binomial theorem to expand. See Examples 2 and 3.

2. $(m + n)^4$

2. $m^4 + 4m^3n + 6m^2n^2 + 4mn^3 + n^4$

3. $(x + r)^5$

3. $x^5 + 5x^4r + 10x^3r^2 + 10x^2r^3 + 5xr^4 + r^5$

4. $(a - b)^5$

4. $a^5 - 5a^4b + 10a^3b^2 - 10a^2b^3 + 5ab^4 - b^5$

5. $(p - q)^4$

5. $p^4 - 4p^3q + 6p^2q^2 - 4pq^3 + q^4$

6. $(2x + 3)^3$

6. $8x^3 + 36x^2 + 54x + 27$

7. $\left(\dfrac{x}{2} - y\right)^4$

7. $\dfrac{x^4}{16} - \dfrac{x^3y}{2} + \dfrac{3x^2y^2}{2} - 2xy^3 + y^4$

8. $\left(\dfrac{x}{3} + 2y\right)^5$

8. $\dfrac{x^5}{243} + \dfrac{10x^4y}{81} + \dfrac{40x^3y^2}{27} + \dfrac{80x^2y^3}{9} + \dfrac{80xy^4}{3} + 32y^5$

▲ **9.** $(x^2 + 1)^4$

9. $x^8 + 4x^6 + 6x^4 + 4x^2 + 1$

▲ **10.** $(mx - n^2)^3$

10. $m^3x^3 - 3m^2n^2x^2 + 3mn^4x - n^6$

▲ **11.** $(2p^2 - q^2)^3$

11. $8p^6 - 12p^4q^2 + 6p^2q^4 - q^6$

Write the first four terms of the binomial expansion.

12. $(r + 2s)^{12}$

12. $r^{12} + 24r^{11}s + 264r^{10}s^2 + 1760r^9s^3$

13. $(m - n)^{20}$

13. $m^{20} - 20m^{19}n + 190m^{18}n^2 - 1140m^{17}n^3$

14. $(3x - y)^{14}$

14. $3^{14}x^{14} - 14(3^{13})x^{13}y + 91(3^{12})x^{12}y^2 - 364(3^{11})x^{11}y^3$

15. $(2p + 3q)^{11}$

15. $2^{11}p^{11} + 33(2^{10})p^{10}q + 495(2^9)p^9q^2 + 4455(2^8)p^8q^3$

16. $(t^2 + u^2)^{10}$

16. $t^{20} + 10t^{18}u^2 + 45t^{16}u^4 + 120t^{14}u^6$

17. $(x^2 - y^2)^{15}$

17. $x^{30} - 15x^{28}y^2 + 105x^{26}y^4 - 455x^{24}y^6$

✏ **18.** Explain why the exponent on y in the rth term of $(x + y)^n$ is $r - 1$ and not r.

Find the indicated term of the binomial expansion. See Example 4.

19. $(2m + n)^{10}$; fourth term

19. $120(2^7)m^7n^3$

20. $(a - 3b)^{12}$; fifth term

20. $495(3^4)a^8b^4$

21. $\left(x + \dfrac{y}{2}\right)^8$; seventh term

21. $\dfrac{7x^2y^6}{16}$

22. $(3p - 2q)^{15}$; eighth term

22. $-6435(3^8)(2^7)p^8q^7$

◉ CONCEPTUAL   ✏ WRITING   ▲ CHALLENGING   ▦ SCIENTIFIC CALCULATOR   ▦ GRAPHICS CALCULATOR

23. $(k - 1)^9$; third term

23. $36k^7$

24. $(-4 - S)^{11}$; fourth term

24. $-165(4^8)S^3$

▲ **25.** the middle term of $(x^2 - 2y)^6$

25. $-160x^6y^3$

▲ **26.** the middle term of $(m^3 + 3)^8$

26. $5670m^{12}$

▲ **27.** the term with x^9y^4 in $(3x^3 - 4y^2)^5$

27. $4320x^9y^4$

▲ **28.** the term with x^{10} in $\left(x^3 - \dfrac{2}{x}\right)^6$

28. $60x^{10}$

CHAPTER 11 SUMMARY

KEY TERMS

11.1 sequence
infinite sequence
terms of a
sequence
general term
finite sequence
series
summation
notation
index of
summation
arithmetic mean

11.2 arithmetic
sequence
(arithmetic
progression)
common difference

11.3 geometric sequence
(geometric
progression)
common ratio

11.4 Pascal's triangle
binomial theorem
(general
binomial
expansion)

NEW SYMBOLS

a_n nth term of a sequence

$\displaystyle\sum_{i=1}^{n} a_i$ summation notation

S_n sum of first n terms of a sequence

$\displaystyle\lim_{n\to\infty} a_n$ limit of a_n as n gets larger and larger

$\displaystyle\sum_{i=1}^{\infty} a_i$ sum of an infinite number of terms

$n!$ n factorial

QUICK REVIEW

CONCEPTS	EXAMPLES
11.1 SEQUENCES AND SERIES	
Sequence	$1, \dfrac{1}{2}, \dfrac{1}{3}, \dfrac{1}{4}, \ldots, \dfrac{1}{n}$
General Term a_n	has general term $\dfrac{1}{n}$.
Series	The corresponding series is the *sum* $$1 + \dfrac{1}{2} + \dfrac{1}{3} + \dfrac{1}{4} + \cdots + \dfrac{1}{n}.$$
11.2 ARITHMETIC SEQUENCES	
Assume a_1 is the first term, a_n is the nth term, and d is the common difference.	The arithmetic sequence 2, 5, 8, 11, . . . has $a_1 = 2$.
Common Difference $d = a_{n+1} - a_n$	$= 5 - 2 = 3$ (Any two successive terms could have been used.)
General Term $a_n = a_1 + (n - 1)d$	The 10th term is $$a_{10} = 2 + (10 - 1)3$$ $$= 2 + 9 \cdot 3 = 29.$$

 CONCEPTUAL   WRITING   CHALLENGING  SCIENTIFIC CALCULATOR GRAPHICS CALCULATOR

CONCEPTS	EXAMPLES
Sum of the First n Terms $$S_n = \frac{n}{2}(a_1 + a_n) \quad \text{or}$$ $$S_n = \frac{n}{2}[2a_1 + (n-1)d]$$	The sum of the first 10 terms is $$S_{10} = \frac{10}{2}(2 + a_{10})$$ $$= 5(2 + 29) = 5(31) = 155$$ or $\quad S_{10} = \frac{10}{2}[2(2) + (10 - 1)3]$ $$= 5(4 + 9 \cdot 3)$$ $$= 5(4 + 27) = 5(31) = 155.$$

11.3 GEOMETRIC SEQUENCES

a_1 is the first term, a_n is the nth term, and r is the common ratio. **Common Ratio** $$r = \frac{a_{n+1}}{a_n}$$	The geometric sequence 1, 2, 4, 8, . . . has $a_1 = 1$. $$r = \frac{8}{4} = 2$$ (Any two successive terms could have been used.)
General Term $$a_n = a_1 r^{n-1}$$	The 6th term is $$a_6 = (1)(2)^{6-1} = 1(2)^5 = 32.$$
Sum of the First n Terms $$S_n = \frac{a_1(r^n - 1)}{r - 1} \quad (r \neq 1)$$	The sum of the first 6 terms is $$S_6 = \frac{1(2^6 - 1)}{2 - 1} = \frac{64 - 1}{1} = 63.$$
Sum of the Terms of an Infinite Geometric Sequence with $|r| < 1$ $$\frac{a_1}{1 - r}$$	The sum of the terms of an infinite geometric sequence with $a_1 = 1$ and $r = 1/2$ is $$\frac{1}{1 - 1/2} = \frac{1}{1/2} = 2.$$

11.4 THE BINOMIAL THEOREM

General Binomial Expansion For any positive integer n, $$(x + y)^n = x^n + \frac{n!}{(n-1)! \, 1!} x^{n-1}y$$ $$+ \frac{n!}{(n-2)! \, 2!} x^{n-2}y^2$$ $$+ \frac{n!}{(n-3)! \, 3!} x^{n-3}y^3 + \cdots$$ $$+ \frac{n!}{1!(n-1)!} xy^{n-1} + y^n.$$	$(2m + 3)^4 = (2m)^4 + \dfrac{4!}{3! \, 1!}(2m)^3(3) + \dfrac{4!}{2! \, 2!}(2m)^2(3)^2$ $\qquad + \dfrac{4!}{1! \, 3!}(2m)(3)^3 + 3^4$ $\quad = 2^4 m^4 + 4(2)^3 m^3(3)$ $\qquad + 6(2)^2 m^2(9) + 4(2m)(27) + 81$ $\quad = 16m^4 + 12(8)m^3 + 54(4)m^2 + 216m + 81$ $\quad = 16m^4 + 96m^3 + 216m^2 + 216m + 81$

CONCEPTS	EXAMPLES
rth Term of the Binomial Expansion of $(x + y)^n$ $$\frac{n!}{[n - (r - 1)]!\,(r - 1)!}x^{n-(r-1)}y^{r-1}$$	The 8th term of $(a - 2b)^{10}$ is $$\frac{10!}{3!\,7!}a^3(-2b)^7 = \frac{10 \cdot 9 \cdot 8}{3 \cdot 2 \cdot 1}a^3(-2)^7b^7$$ $$= 120(-128)a^3b^7$$ $$= -15{,}360a^3b^7.$$

CHAPTER 11 REVIEW EXERCISES

[11.1] *Write out the first four terms of the sequence.*

1. $a_n = 2n - 3$ **2.** $a_n = \dfrac{n - 1}{n}$ **3.** $a_n = n^2$

4. $a_n = \left(\dfrac{1}{2}\right)^n$ **5.** $a_n = (n + 1)(n - 1)$

Write the series as a sum of terms.

6. $\displaystyle\sum_{i=1}^{5} i^2 x$ **7.** $\displaystyle\sum_{i=1}^{6} (i + 1)x^i$

Evaluate the series.

8. $\displaystyle\sum_{i=1}^{4} (i + 2)$ **9.** $\displaystyle\sum_{i=1}^{6} 2^i$ **10.** $\displaystyle\sum_{i=4}^{7} \dfrac{i}{i + 1}$ **11.** $\displaystyle\sum_{i=1}^{5} (-1)^i(i + 1)$

[11.2–11.3] *Decide whether the sequence is arithmetic, geometric, or neither. If the sequence is arithmetic, find the common difference, d. If it is geometric, find the common ratio, r.*

12. $2, 5, 8, 11, \ldots$ **13.** $-6, -2, 2, 6, 10, \ldots$

14. $\dfrac{2}{3}, -\dfrac{1}{3}, \dfrac{1}{6}, -\dfrac{1}{12}, \ldots$ **15.** $-1, 1, -1, 1, -1, \ldots$

16. $64, 32, 8, \dfrac{1}{2}, \ldots$ **17.** $64, 32, 16, 8, \ldots$

18. $10, 8, 6, 4, \ldots$ **19.** $2, \dfrac{5}{2}, 3, \dfrac{7}{2}, \ldots$

[11.2] *Find the indicated term for the arithmetic sequence.*

20. $a_1 = -2, d = 5;\quad a_{16}$ **21.** $a_6 = 12, a_8 = 18;\quad a_{25}$

Find the general term for the arithmetic sequence.

22. $a_1 = -4, d = -5$ **23.** $6, 3, 0, -3, \ldots$

Find the number of terms in the arithmetic sequence.

24. $7, 10, 13, \ldots, 49$ **25.** $5, 1, -3, \ldots, -79$

Answers (right column):

1. $-1, 1, 3, 5$
2. $0, \dfrac{1}{2}, \dfrac{2}{3}, \dfrac{3}{4}$ 3. $1, 4, 9, 16$
4. $\dfrac{1}{2}, \dfrac{1}{4}, \dfrac{1}{8}, \dfrac{1}{16}$
5. $0, 3, 8, 15$
6. $x + 4x + 9x + 16x + 25x$
7. $2x + 3x^2 + 4x^3 + 5x^4 + 6x^5 + 7x^6$
8. 18 9. 126
10. $\dfrac{2827}{840}$ 11. -4
12. arithmetic; $d = 3$
13. arithmetic; $d = 4$
14. geometric; $r = -\dfrac{1}{2}$
15. geometric; $r = -1$
16. neither
17. geometric; $r = \dfrac{1}{2}$
18. arithmetic; $d = -2$
19. arithmetic; $d = \dfrac{1}{2}$
20. 73 21. 69
22. $a_n = 1 - 5n$
23. $a_n = 9 - 3n$
24. 15 25. 22

RESOURCES FOR REVIEW

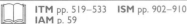 **ITM** pp. 519–533 **ISM** pp. 902–910
IAM p. 59

 **SSG** pp. 347–369
SSM pp. 599–607

 **TEST GENERATOR**
DOS Windows MAC

 CONCEPTUAL   WRITING   ▲ CHALLENGING    SCIENTIFIC CALCULATOR    GRAPHICS CALCULATOR

26. 152   27. 164
28. 116
29. $a_n = -1(4)^{n-1} = -(4)^{n-1}$

30. $a_n = \dfrac{2}{3}\left(\dfrac{1}{5}\right)^{n-1}$

31. $2(-3)^{10} = 118{,}098$
32. $5(2)^9 = 2560$ or $5(-2)^9 = -2560$

33. $\dfrac{341}{1024}$ 34. 0

35. $\dfrac{242}{81}$ 36. 1

37. does not exist
44. 235;   280
45. $-3(-2)^9 = 1536$; $2^{10} - 1 = 1023$
46. $(-3)^8 = 6561$; $-\dfrac{1}{4}(3^{10} - 1) = -14{,}762$
47. 38;   95
48. $a_n = 5n - 3$
49. $a_n = 2(4)^{n-1}$

50. $a_n = 27\left(\dfrac{1}{3}\right)^{n-1}$

51. $a_n = 15 - 3n$
52. 10 seconds
53. $78

Find S_8 for the arithmetic sequence or series.

26. $a_1 = -2, d = 6$ **27.** $a_n = -2 + 5n$ **28.** $\displaystyle\sum_{i=1}^{8} (3i + 1)$

[11.3] *Find the general term for the geometric sequence.*

29. $-1, -4, \ldots$ **30.** $\dfrac{2}{3}, \dfrac{2}{15}, \ldots$

Find the indicated term for the geometric sequence.

31. $2, -6, 18, \ldots$; a_{11} **32.** $a_3 = 20, a_5 = 80$; a_{10}

Find the sum, if it exists, of the geometric sequence.

33. $\displaystyle\sum_{i=1}^{5} \left(\dfrac{1}{4}\right)^i$ **34.** $\displaystyle\sum_{i=1}^{8} \dfrac{3}{4}(-1)^i$ **35.** $2, \dfrac{2}{3}, \dfrac{2}{9}, \dfrac{2}{27}, \dfrac{2}{81}$

36. $\displaystyle\sum_{i=1}^{\infty} 4\left(\dfrac{1}{5}\right)^i$ **37.** $\displaystyle\sum_{i=1}^{\infty} 2(3)^i$

[11.4] *Use the binomial theorem to expand the binomial.*

38. $(2p - q)^5$
38. $32p^5 - 80p^4q + 80p^3q^2 - 40p^2q^3 + 10pq^4 - q^5$
39. $(x^2 + 3y)^4$
39. $x^8 + 12x^6y + 54x^4y^2 + 108x^2y^3 + 81y^4$
40. $(\sqrt{m} + \sqrt{n})^4$
40. $m^2 + 4m\sqrt{mn} + 6mn + 4n\sqrt{mn} + n^2$

41. Write the first four terms of the expansion of $\left(m + \dfrac{n}{2}\right)^{20}$.

41. $m^{20} + 10m^{19}n + \dfrac{95m^{18}n^2}{2} + \dfrac{285m^{17}n^3}{2}$

42. Write the fourth term of the expansion of $(3a + 2b)^{19}$.

42. $7752(3)^{16}a^{16}b^3$

43. Write the twenty-third term of the expansion of $(-2k + 3)^{25}$.

43. $-18{,}400(3)^{22}k^3$

MIXED REVIEW EXERCISES

Find the indicated term and S_{10} for the sequence.

44. a_{40}: arithmetic; $1, 7, 13, \ldots$ **45.** a_{10}: geometric; $-3, 6, -12, \ldots$

46. a_9: geometric; $a_1 = 1, r = -3$ **47.** a_{15}: arithmetic; $a_1 = -4, d = 3$

Find the general term for the arithmetic or geometric sequence.

48. $2, 7, 12, \ldots$ **49.** $2, 8, 32, \ldots$

50. $27, 9, 3, \ldots$ **51.** $12, 9, 6, \ldots$

Solve the problem.

52. When Mary's sled goes down the hill near her home, she covers 3 feet in the first second, then for each second after that she goes 4 feet more than in the preceding second. If the distance she covers going down is 210 feet, how long does it take her to reach the bottom?

53. Karl saved $1 in January, $2 in February, $3 in March, and so on for a year. How much did he save over the year?

54. A piece of paper .05 centimeter thick is folded in half, then in half again, and so on, until it has been folded 10 times. How thick is the result?

55. The school population in Scoville has been dropping 3% per year. The current population is 50,000. If this trend continues, what will the population be in 6 years?

56. A pump removes $1/2$ of the liquid in a container with each stroke. What fraction of the liquid is left in the container after 7 strokes?

57. Can the sum of the terms of the infinite geometric sequence with $a_n = 5(2)^n$ be found? Explain.

58. Can any two terms of a geometric sequence be used to find the common ratio? Explain.

54. 51.2 centimeters
55. $50,000(.97)^6 \approx 42,000$
56. $\dfrac{1}{128} = .0078125$

CHAPTER 11 TEST

Write the first five terms for the sequence.

1. $a_n = (-1)^n + 1$

2. $a_n = (n + 1)(n + 2)$

3. arithmetic; $a_1 = 4, d = 2$

4. arithmetic; $a_3 = 5, d = -4$

5. geometric; $a_1 = 2, r = -3$

6. geometric; $a_4 = 6, r = \dfrac{1}{2}$

Find a_4 for the sequence.

7. arithmetic; $a_1 = 6, d = -2$

8. arithmetic; $a_2 = 11, a_5 = 38$

9. geometric; $a_2 = 9, a_3 = 4$

10. geometric; $a_5 = 16, a_7 = 9$

Find S_5 for the sequence.

11. arithmetic; $a_1 = 8, a_5 = 12$

12. arithmetic; $a_2 = 12, a_3 = 15$

13. geometric; $a_2 = 9, a_3 = 18$

14. geometric; $a_5 = 4, a_7 = 1$

Find the sum if it exists.

15. $\displaystyle\sum_{i=1}^{5} (2i + 8)$

16. $\displaystyle\sum_{i=1}^{6} (3i - 5)$

17. $\displaystyle\sum_{i=1}^{500} i$

18. $\displaystyle\sum_{i=1}^{3} \frac{1}{2}(4^i)$

19. $\displaystyle\sum_{i=1}^{\infty} \left(\frac{1}{4}\right)^i$

20. $\displaystyle\sum_{i=1}^{\infty} 6\left(\frac{3}{2}\right)^i$

Solve the problem.

21. Write the fifth term of $\left(2x - \dfrac{y}{3}\right)^{12}$.

22. Expand $(3k - 5)^4$.

23. During the summer months, the population of a certain insect colony triples each week. If there are 20 insects in the colony at the end of the first week in July, how many are present by the end of September? (Assume exactly four weeks in a month.)

24. Cheryl bought a new sewing machine for $300. She agreed to pay $20 a month for 15 months plus interest of 1% each month on the unpaid balance. Find the total cost of the machine.

25. Under what conditions does an infinite geometric series have a sum?

[11.1] **1.** 0, 2, 0, 2, 0
2. 6, 12, 20, 30, 42
[11.2–11.3] **3.** 4, 6, 8, 10, 12
4. 13, 9, 5, 1, −3
5. 2, −6, 18, −54, 162
6. 48, 24, 12, 6, 3
7. 0   **8.** 29   **9.** $\dfrac{16}{9}$
10. $\dfrac{64}{3}$ or $-\dfrac{64}{3}$
11. 50   **12.** 75   **13.** $\dfrac{279}{2}$
14. 124 or 44   **15.** 70
16. 33   **17.** 125,250
18. 42   **19.** $\dfrac{1}{3}$
20. does not exist
[11.4] **21.** $\dfrac{14,080x^8y^4}{9}$
22. $81k^4 - 540k^3 + 1350k^2 - 1500k + 625$
[11.3] **23.** $20(3^{11})$
[11.2] **24.** $324
25. It has a sum if $|r| < 1$.

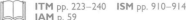

CUMULATIVE REVIEW (Chapters 1–11)

1. $10, 0, \dfrac{45}{15}, -3$

2. $-\dfrac{8}{3}, 10, 0, \dfrac{45}{15}, .82, -3$

3. $\sqrt{13}, -\sqrt{3}$ **4.** all except $\sqrt{-7}$ **5.** 8

6. -35 **7.** -55

8. $\left\{\dfrac{1}{6}\right\}$ **9.** $[10, \infty)$

10. $\left\{-\dfrac{9}{2}, 6\right\}$ **11.** $\{9\}$

12. $\left(-3, \dfrac{3}{2}\right)$

13. $(-\infty, -3] \cup [8, \infty)$

14. $20p^2 - 2p - 6$

15. $9k^2 - 42k + 49$

16. $-5m^3 - 3m^2 + 3m + 8$ **17.** $2t^3 + 3t^2 - 4t + 2 + \dfrac{3}{3t - 2}$

18. $x(7 + x^2)$

19. $(7y - 4)(2y + 3)$

20. $z(3z + 4)(2z - 1)$

21. $(7a^2 + 3b)(7a^2 - 3b)$

22. $(c + 3d) \cdot (c^2 - 3cd + 9d^2)$

23. $(8r + 3q)^2$

24. $-\dfrac{27p^2}{10}$ **25.** $\dfrac{x + 7}{x - 2}$

26. $\dfrac{-2k - 23}{(k + 4)(k - 1)}$

27. $\dfrac{3p - 26}{p(p + 3)(p - 4)}$

28. $\emptyset$ **29.** 2 pounds

30. $\dfrac{9}{4}$ **31.** $\dfrac{1}{5^5}$ or $\dfrac{1}{3125}$

32. $10\sqrt{2}$ **33.** 73

34. $\left\{\dfrac{-5 + \sqrt{217}}{12}, \dfrac{-5 - \sqrt{217}}{12}\right\}$

This set of exercises may be considered a final examination for the course.

Let $P = \left\{-\dfrac{8}{3}, 10, 0, \sqrt{13}, -\sqrt{3}, \dfrac{45}{15}, \sqrt{-7}, .82, -3\right\}$. *List the elements of P that are members of the set.*

1. integers

2. rational numbers

3. irrational numbers

4. real numbers

Simplify the expression.

5. $|-7| + 6 - |-10| - (-8 + 3)$

6. $-15 - |-4| - 10 - |-6|$

7. $4(-6) + (-8)(5) - (-9)$

Solve the equation or inequality.

8. $9 - (5 + 3a) + 5a = -4(a - 3) - 7$

9. $7m + 18 \le 9m - 2$

10. $|4x - 3| = 21$

11. $\dfrac{x + 3}{12} - \dfrac{x - 3}{6} = 0$

12. $|4k + 3| < 9$

13. $|2m - 5| \ge 11$

Perform the indicated operations.

14. $(4p + 2)(5p - 3)$

15. $(3k - 7)^2$

16. $(2m^3 - 3m^2 + 8m) - (7m^3 + 5m - 8)$

17. Divide $6t^4 + 5t^3 - 18t^2 + 14t - 1$ by $3t - 2$.

Factor.

18. $7x + x^3$

19. $14y^2 + 13y - 12$

20. $6z^3 + 5z^2 - 4z$

21. $49a^4 - 9b^2$

22. $c^3 + 27d^3$

23. $64r^2 + 48rq + 9q^2$

Simplify.

24. $\dfrac{(3p^2)^3(-2p^6)}{4p^3(5p^7)}$

25. $\dfrac{x^2 - 16}{x^2 + 2x - 8} \div \dfrac{x - 4}{x + 7}$

26. $\dfrac{3}{k + 4} + \dfrac{-5}{k - 1}$

27. $\dfrac{5}{p^2 + 3p} - \dfrac{2}{p^2 - 4p}$

Solve.

28. $\dfrac{4}{x - 3} - \dfrac{6}{x + 3} = \dfrac{24}{x^2 - 9}$

29. Nuts worth \$3 per pound are to be mixed with 8 pounds of nuts worth \$4.25 per pound to get a mixture that will be sold for \$4 per pound. How many pounds of the \$3 nuts should be used?

Simplify.

30. $\left(\dfrac{2}{3}\right)^{-2}$

31. $\dfrac{5^{-4}}{5}$

32. $5\sqrt{72} - 4\sqrt{50}$

33. Multiply: $(8 + 3i)(8 - 3i)$.

Solve the equation or inequality.

34. $6x^2 + 5x = 8$ **35.** $k^2 - k - 6 \le 0$

36. It is estimated that y, the number of items of a particular commodity (in millions) sold in the United States in year x, where x represents the number of years since 1996, is given by

$$y = 1.71x + 2.98.$$

That is, $x = 0$ represents 1996, $x = 1$ represents 1997, and so on. Find the number of items sold in the following years.
 (a) 1996   **(b)** 1997   **(c)** 1998
 (d) In what year will about 10 million items be sold?

37. Find the slope of the line through $(4, -5)$ and $(-12, -17)$.

38. Find the standard form of the equation of the line through $(-2, 10)$ and parallel to the line with equation $3x + y = 7$.

Graph.

39. $x - 3y = 6$ **40.** $4x - y < 4$

41. $f(x) = 2(x - 2)^2 - 3$ **42.** $\dfrac{x^2}{9} + \dfrac{y^2}{25} = 1$

43. Find the equation of a circle with center at $(-5, 12)$ and radius 9.

Solve the system of equations.

44. $2x + 5y = -19$
     $-3x + 2y = -19$

45. $x + 2y + z = 8$
     $2x - y + 3z = 15$
     $-x + 3y - 3z = -11$

Evaluate the determinant.

46. $\begin{vmatrix} -3 & -2 \\ 6 & 9 \end{vmatrix}$ **47.** $\begin{vmatrix} 2 & 4 & 1 \\ 1 & 3 & 6 \\ 2 & 3 & -1 \end{vmatrix}$

48. Find $f^{-1}(x)$, if $f(x) = 9x + 5$.   **49.** Graph $g(x) = \left(\dfrac{1}{3}\right)^x$.

50. Solve $3^{2x-1} = 81$.   **51.** Graph $y = \log_{1/3} x$.

52. Solve $\log_8 x + \log_8(x + 2) = 1$.

53. Write the first five terms of the sequence defined by $a_n = 5n - 12$.

54. Find the sum of the first six terms of the arithmetic sequence with $a_1 = 8$ and $d = 2$.

55. Find the sum of the geometric series $15 - 6 + \dfrac{12}{5} - \dfrac{24}{25} + \cdots$.

56. Find the sum: $\displaystyle\sum_{i=1}^{4} 3i$.

57. Use the binomial theorem to expand $(2a - 1)^5$.

---

35. $[-2, 3]$
36. (a) 2.98 million
(b) 4.69 million
(c) 6.4 million   **(d)** 2000
37. $\dfrac{3}{4}$ **38.** $3x + y = 4$
39.

40.

41.

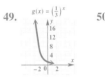

42.

$$\frac{x^2}{9} + \frac{y^2}{25} = 1$$

43. $(x + 5)^2 + (y - 12)^2 = 81$
44. $\{(3, -5)\}$
45. $\{(2, 1, 4)\}$
46. -15 **47.** 7
48. $f^{-1}(x) = \dfrac{x - 5}{9}$
49. $g(x) = \left(\frac{1}{3}\right)^x$ **50.** $\left\{\dfrac{5}{2}\right\}$

51. $y = \log_{1/3} x$

52. $\{2\}$
53. $-7, -2, 3, 8, 13$
54. 78   **55.** $\dfrac{75}{7}$ **56.** 30
57. $32a^5 - 80a^4 + 80a^3 - 40a^2 + 10a - 1$

ANSWERS TO SELECTED EXERCISES

In this section we provide the answers that we think most students will obtain when they work the exercises using the methods explained in the text. If your answer does not look exactly like the one given here, it is not necessarily wrong. In many cases there are equivalent forms of the answer. For example, if the answer section shows 3/4 and your answer is .75, you have obtained the correct answer but written it in a different (yet equivalent) form. Unless the directions specify otherwise, .75 is just as valid an answer as 3/4. In general, if your answer does not agree with the one given in the text, see whether it can be transformed into the other form. If it can, then it is the correct answer. If you still have doubts, talk with your instructor.

CHAPTER 1 THE REAL NUMBERS

SECTION 1.1 (PAGE 10)

EXERCISES **1.** false **3.** true **5.** true **7.** $\{1, 2, 3, 4, 5\}$ **9.** $\{5, 6, 7, 8, \ldots\}$ **11.** $\{10, 12, 14, 16, \ldots\}$
13. $\emptyset$ **15.** $\{-4, 4\}$ **17.** $\{0, 3, 6, 9, \ldots\}$ **19.** yes **21.** $\{x \mid x \text{ is a multiple of 4 greater than 0}\}$ **23.** $\{x \mid x \text{ is an even}$

natural number less than or equal to 8$\}$ **25. (a)** $4, 5, 17, \frac{40}{2}$ (or 20) **(b)** $0, 4, 5, 17, \frac{40}{2}$ **(c)** $-8, 0, 4, 5, 17, \frac{40}{2}$

(d) $-8, -.6, 0, \frac{3}{4}, 4, 5, \frac{13}{2}, 17, \frac{40}{2}$ **(e)** $-\sqrt{5}, \sqrt{3}$ **(f)** All are real numbers except $\frac{1}{0}$. **(g)** $\frac{1}{0}$ **27.**

29. $-\frac{2}{3}$ $\frac{4}{5}$ $\frac{12}{5}$ $\frac{9}{2}$ -4.8 **33.** 8 **35.** -5 **37.** -2 **39.** -4.5 **41.** 5 **43.** 6 **45.** 22 **47.** 0 **49.** true

51. false **53.** true **55.** true **59.** $6 < 11$ **61.** $4 > x$ **63.** $3t - 4 \le 10$ **65.** $5 \ge 5$ **67.** $-3 < t < 5$
69. $-3 \le 3x < 4$ **71.** $5x + 3 \ne 0$ **73.** $3 \ge 2$ **75.** $-3 \le -3$ **77.** $5 \not< 3$ **79.** Pacific Ocean, Indian
Ocean, Caribbean Sea, South China Sea, Gulf of California **81.** true **83.** $(-2, \infty)$

85. $(-\infty, 6]$ **87.** $(0, 3.5)$ **89.** $[2, 7]$

91. $(-4, 3]$ **93.** $(0, 3]$
95. October 1991, November 1991, February 1992, and March 1992 **97.** December 1991, February 1992, and March 1992

SECTION 1.2 (PAGE 22)

◆ **CONNECTIONS** **Page 19:** Answers will vary. **Page 20:** A calculator will show an error message for $\sqrt{-25}$. Some calculators will also show an error message for $\sqrt[3]{-125}$, even though $\sqrt[3]{-125} = -5$.

EXERCISES   **1.** always true   **3.** sometimes true; for example, $-3 + 4 = 1$ is positive, but $-4 + 3 = -1$ is negative. **5.** always true   **7.** sometimes true; for example, $4 - 5 = -1$ is negative, but $5 - 4 = 1$ is positive. **9.** sometimes true; for example, $1 + (-1) = 0$, but $-3 + 2 = -1 \neq 0$.   **11.** 9   **13.** -19 **15.** $-\dfrac{19}{12}$ **17.** .187   **19.** 8   **21.** $-\dfrac{7}{4}$ **23.** 3.018   **25.** 6   **27.** $\dfrac{13}{2}$ or $6\dfrac{1}{2}$ **29.** Answers will vary.   **30.** It is less than 0.   **31.** Answers will vary.   **32.** It is less than 0; $<$ **33.** Answers will vary. One example is $-4 - (-9) = -4 + 9 = 5.$ **37.** 45   **39.** 180   **41.** 7.383   **43.** 4   **45.** $-\dfrac{9}{13}$ **47.** undefined   **51.** true   **53.** true   **55.** false   **57.** true   **59.** false   **61.** 11   **63.** .021952   **65.** $\dfrac{49}{100}$ **67.** -30 **69.** -7 **71.** 10   **The number of digits displayed will vary in Exercises 75–79.**   **75.** 136.011029   **77.** 4.534873083   **79.** 1.659870127   **81.** not a real number   **83.** 29   **85.** -79 **87.** 39   **89.** -2 **91.** 2   **93.** undefined   **95.** 13   **97.** $-\dfrac{32}{5}$ **99.** -1 **101.** 13   **103.** -96 **105.** $-\dfrac{15}{238}$ **107.** 112° Fahrenheit   **109.** 11,331 feet   **111.** \$266 million

SECTION 1.3 (PAGE 31)

◆ **CONNECTIONS**   **Page 26:** Answers will vary.

EXERCISES   **1.** The identity element for addition is 0.   **3.** Like terms are terms with exactly the same variables raised to exactly the same powers.   **5.** The commutative properties state that the *order* in which the terms are operated on does not affect the answer, while the associative properties state that the *grouping* of the terms does not affect the answer.   **7.** $8k$ **9.** $-2r$ **11.** $-8z + 4w$ (cannot be simplified)   **13.** $6a$ **15.** $2m + 2p$ **17.** $-12x + 12y$ **19.** $-10d - 5f$ **21.** $7x + 26$ **23.** $-6y + 3$ **25.** $p + 11$ **27.** $-2k + 15$ **29.** $m - 14$ **31.** -1 **33.** $2p + 7$ **35.** $-6z - 39$ **37.** $(5 + 8)x = 13x$ **39.** $(5 \cdot 9)r = 45r$ **41.** $9y + 5x$ **43.** 7   **45.** 0   **47.** $8(-4) + 8x = -32 + 8x$ **49.** 0   **51.** associative property of addition   **52.** associative property of addition   **53.** commutative property of addition   **54.** associative property of addition   **55.** distributive property   **56.** arithmetic facts   **57.** Answers will vary. One example is washing your face and brushing your teeth.   **59.** $2 + 6 \cdot 5 = 2 + 30 = 32$, which does not equal $8 \cdot 5 = 40.$ **61.** 1900   **63.** 75   **65.** 431   **67.** Yes. Any nonzero numbers a and b that have the same absolute value satisfy $a/b = b/a$.   **69.** No. One example is $7 + (5 \cdot 3) = (7 + 5)(7 + 3)$, which is false.

CHAPTER 1 REVIEW EXERCISES (PAGE 35)

1. $\dfrac{9}{4}$

3. 16   **5.** -4 **7.** 0, 4   **9.** $-9, -\dfrac{4}{3}, 0, \dfrac{5}{3}, 4$ **11.** $\{4, 5, 6, 7, 8\}$ **13.** true   **15.** true   **17.** $(-2, 3]$ **19.** $-\dfrac{1}{2}$ **21.** -17.09 **23.** -1 **25.** $-\dfrac{5}{18}$ **29.** -90 **31.** -11.408 **33.** -15 **35.** true   **37.** $\dfrac{5}{7 - 7}$ is undefined.   **39.** -125 **41.** 2.89   **43.** 3   **45.** 3   **47.** -4 **49.** $\dfrac{7}{3}$ **51.** -30 **53.** -116 **55.** Work within the parentheses first.   **57.** $21q$ **59.** $5m$ **61.** $-2k - 6$ **63.** $18m + 27n$ **65.** $-p - 3q$ **67.** $y + 1$ **69.** 2   **71.** $(2 + 3)x = 5x$ **73.** $(2 \cdot 4)x = 8x$ **75.** 0   **77.** 7   **79.** $(3 + 5 + 6)a = 14a$ **81.** $\dfrac{256}{625}$ **83.** 31   **85.** 0   **87.** $\dfrac{4}{3}$ **89.** -9 **91.** $-\dfrac{47}{3}$ **92.** It is greater than $-16.$ **93. (a)** no **(b)** yes   **94.** $\dfrac{47}{3}$ **95.** no   **96.** yes   **97.** $\dfrac{47}{3}$ **98.** $-\dfrac{3}{47}$ **99.** yes   **100.** No, the new answer is $-\dfrac{29}{3}.$

CHAPTER 1 TEST (PAGE 39)

[1.1] 1.   **2.** 0, 3, 12   **3.** $-1, 0, 3, 12$ **4.** $-1, -.5, 0, 3, 7.5, 12$ **5.** All are real numbers.   **6.** $(-\infty, -3)$ **7.** $(-4, 2]$ **[1.2] 8.** 0   **9.** -26 **10.** 19   **11.** 1

12. $\dfrac{16}{7}$ **13.** $\dfrac{11}{23}$ **14.** 14 **15.** -15 **16.** -3 **17.** not a real number **18. (a)** a must be positive.

(b) n must be odd and a must be negative. **(c)** a must be zero. **19.** 2 **20.** $-\dfrac{6}{23}$ **[1.3] 21.** $10k - 10$

22. It changes the sign of each term. The simplified form is $7r + 2$. **23.** B **24.** E **25.** D **26.** A **27.** F
28. C **29.** C **30.** E

CHAPTER 2 LINEAR EQUATIONS AND INEQUALITIES

SECTION 2.1 (PAGE 47)

◆ **CONNECTIONS** **Page 44: 1. (a)** 1940 **(b)** 1964 **2.** men: 37.22 seconds; women: 39.06 seconds

EXERCISES **1.** yes **3.** no **5.** yes **7.** (b) **9.** $\{-1\}$ **11.** $\{3\}$ **13.** $\{-7\}$ **15.** $\{0\}$

17. $\left\{-\dfrac{5}{3}\right\}$ **19.** $\left\{-\dfrac{1}{2}\right\}$ **21.** $\{2\}$ **23.** $\{-2\}$ **25.** $\{7\}$ **27.** $\{-5\}$ **31.** 2 (that is, 10^2) **33.** $\{4\}$

35. $\{0\}$ **37.** $\{0\}$ **39.** $\{2000\}$ **41.** $\{25\}$ **43.** $\{40\}$ **45.** identity; contradiction **47.** contradiction; $\emptyset$
49. conditional; $\{0\}$ **51.** identity; {all real numbers} **53.** $\{-8\}$ **54.** $\{-8\}$ **55.** 33 **56.** 33 **57.** $\{-8\}$
58. -7 **59.** equivalent **61.** not equivalent **63.** not equivalent **65.** 2820 million dollars; 1996 (when $x = 7$)
67. (a) 230 billion dollars **(b)** 275 billion dollars **(c)** 300 billion dollars **(d)** 335 billion dollars **69.** 36 **71.** 72

73. 50 **75.** $\dfrac{39}{2}$ or $19\dfrac{1}{2}$

SECTION 2.2 (PAGE 56)

◆ **CONNECTIONS** **Page 50:** .10 or 10%; .0676 or 6.76%; .1125 or 11.25%

EXERCISES **3.** L should not appear on both sides of the final equation. **5.** $r = \dfrac{d}{t}$ **7.** $b = \dfrac{A}{h}$ **9.** $a = P - b - c$

11. $h = \dfrac{2A}{b}$ **13.** $h = \dfrac{S - 2\pi r^2}{2\pi r}$ or $h = \dfrac{S}{2\pi r} - r$ **15.** $F = \dfrac{9}{5}C + 32$ **17.** $H = \dfrac{A - 2LW}{2W + 2L}$ **19. (a)** $3x = 5x + 8$

(b) $ct = bt + k$ **20. (a)** $3x - 5x = 8$ **(b)** $ct - bt = k$ **21. (a)** $-2x = 8$; distributive property **(b)** $(c - b)t = k$;

distributive property **22. (a)** $x = -4$ **(b)** $t = \dfrac{k}{c - b}$ **23.** $b \neq c$; If $b = c$, the denominator becomes zero, and zero is not

allowed as the denominator of a fraction. **25.** about 3.7 hours **27.** -40 degrees Fahrenheit **29.** 230 meters **31.** 52
miles per hour **33.** 15 meters **35.** 6 inches **37.** 8 feet **39.** perimeter **41.** 75% water, 25% alcohol
43. 63,120 **45.** 455 **47.** 3% **49.** $10.51 **51.** $45.66 **53.** 10% **55.** $x = -8$; $49°$; $49°$ **57.** $x = 27$;

$49°$; $49°$ **59.** $x = 47$; $48°$; $132°$ **61.** $\{12\}$ **63.** $\left\{\dfrac{1}{2}\right\}$ **65.** -3

SECTION 2.3 (PAGE 68)

◆ **CONNECTIONS** **Page 62:** Our steps 1, 3, 4, and 6 correspond to Polya's steps.

EXERCISES **3.** $x - 13$ **5.** $7 + x$ **7.** $8(x + 12)$ **9.** $\dfrac{x}{6}$ **11.** $\dfrac{x}{12}$ **13.** $\dfrac{6}{7}x$ **15.** (d)

17. Labrador retrievers: 120,879; Rottweilers: 95,445 **19.** Eisner: 40.1 million dollars; Horrigan: 21.7 million dollars
21. Ruth: 2873 hits; Hornsby: 2930 hits **23.** The leading state, North Carolina, employs 105,633 people, and the second
leading state, Kentucky, employs 61,648 people. **27.** 7.0% **29.** $2008 **31.** $122.28 **33. (a)** $800 - x$
(b) $800 - y$ **34. (a)** $.05x$; $.10(800 - x)$ **(b)** $.05y$; $.10(800 - y)$ **35. (a)** $.05x + .10(800 - x) = 800(.0875)$
(b) $.05y + .10(800 - y) = 800(.0875)$ **36. (a)** $200 at 5%; $600 at 10% **(b)** 200 liters of 5% acid;

600 liters of 10% acid **39.** 5 liters **41.** 4 liters **43.** $18\dfrac{2}{11}$ liters **45.** 5 liters **47.** $4000 at 3%; $8000 at 4%

49. $10,000 at 4.5%; $19,000 at 3% **51.** $58,000 **53.** 180 **55.** 180 **57.** $20°$; $30°$; $130°$ **59.** $65°$; $115°$
61. $180°$ **63.** 200 **65.** 6 **67.** 12

SECTION 2.4 (PAGE 77)

EXERCISES   **1.** $2.45   **3.** 52 miles per hour   **5.** (c)   **7.** 13 quarters, 23 half dollars
9. 17 pennies, 17 dimes, 10 quarters   **11.** 11 two-cent pieces, 33 three-cent pieces   **13.** 305 students, 105 nonstudents
17. $1\frac{3}{4}$ hours   **19.** 10:00 A.M.   **21.** 18 miles   **23.** 8 hours   **25.** $\frac{5}{6}$ hour   **27.** width: 165 feet;   length: 265 feet
29. 850 miles, 925 miles, 1300 miles   **31.** length: 60 meters;   width: 30 meters   **33.** 76 and 77   **35.** 19, 20, 21
37. 40°, 80°   **38.** 120°   **39.** The sum is equal to the measure of the angle found in Exercise 38.   **40.** The sum of the
measures of angles ① and ② is equal to the measure of angle ③.   **41.** $425   **43.** length: 8 inches;   width: 5 inches
45. 65 heads   **47.** ⟶ 4   **49.** ⟶ −2 6   **51.** ⟶ −4 9

SECTION 2.5 (PAGE 87)

◊ **CONNECTIONS**   **Page 81:** Answers will vary.

EXERCISES   **3.** $(2, \infty)$ ⟶ 2   **5.** $(-\infty, -3]$ ⟶ −3   **7.** $[5, \infty)$ ⟶ 5
9. $(7, \infty)$ ⟶ 7   **11.** $(-4, \infty)$ ⟶ −4   **13.** $(-\infty, -40]$ ⟶ −40
15. $[3, \infty)$ ⟶ 3   **17.** $(-\infty, 4]$ ⟶ 4   **19.** $\left(-\infty, -\frac{15}{2}\right)$ ⟶ −$\frac{15}{2}$
21. $\left[\frac{1}{2}, \infty\right)$ ⟶ $\frac{1}{2}$ **23.** $(3, \infty)$ ⟶ 3   **25.** $(-\infty, 4)$ ⟶ 4
27. $\left(-\infty, \frac{23}{6}\right]$ ⟶ $\frac{23}{6}$ **29.** $\left(-\infty, \frac{76}{11}\right)$ ⟶ $\frac{76}{11}$ **31.** $(-\infty, \infty)$ ⟶ 0
33. ∅   **37.** {−9} ⟶ −9   **38.** $(-9, \infty)$ ⟶ −9   **39.** $(-\infty, -9)$ ⟶ −9
40. We obtain the set of all real numbers. ⟶ −9   **41.** $(-\infty, -3)$ **43.** $(1, 11)$ ⟶ 1 11
45. $[-14, 10]$ ⟶ −14 10   **47.** $[-5, 6]$ ⟶ −5 6   **49.** $\left[-\frac{14}{3}, 2\right]$ ⟶ −$\frac{14}{3}$ 2
51. $\left[-\frac{1}{2}, \frac{35}{2}\right]$ ⟶ −$\frac{1}{2}$ $\frac{35}{2}$ **53.** $\left(-\frac{1}{3}, \frac{1}{9}\right]$ ⟶ −$\frac{1}{3}$ $\frac{1}{9}$ **55.** April, May, June, July
57. January, February, March, August, September, October, November, December   **59.** 2 miles   **61.** at least 80   **63.** 50 miles
65. 26 tapes   **67.** $-5 < -x < 1$ **71. (a)** ⟶ 4 5 **(b)** ⟶ 4 5
73. (a) ⟶ 4 5 **(b)** ⟶ 4 5

SECTION 2.6 (PAGE 96)

EXERCISES   **1.** true   **3.** true   **5.** false;   6 is not included in the union.   **7.** {1, 3, 5} or B   **9.** {4} or D
11. ∅   **13.** {1, 2, 3, 4, 5, 6} or A   **15.** {1, 3, 5, 6}   **17.** {1, 4, 6}   **19.** Each is equal to {1}. This illustrates the
associative property of set intersection.   **23.** ⟶ −3 2   **25.** ⟶ 2
27. $[5, 9]$ ⟶ 5 9   **29.** $(-3, -1)$ ⟶ −3 −1   **31.** $(-\infty, 4]$ ⟶ 4   **33.** ⟶ 2 4

35. **37.** $(-\infty, -5) \cup (5, \infty)$ **39.** $(-\infty, -1) \cup (2, \infty)$

41. $[-4, -1]$ **43.** $[-9, -6]$ **45.** $[6, 11]$ **47.** $(4, 6]$ **49.** intersection; $(-5, -1)$

51. union; $(-\infty, 4)$ **53.** intersection; $[4, 12]$

55. union; $(-\infty, 0] \cup [2, \infty)$ **57.** 1988, 1989, 1990, 1991, 1992 **59.** Mario, Joe **60.** none of them

61. none of them **62.** Luigi, Than **63.** $[-6, \infty)$ **65.** $(-3, 2)$ **67.** -21

SECTION 2.7 (PAGE 104)

◆ **CONNECTIONS** **Page 99:** Answers will vary.

EXERCISES **1.** E; C; D; B; A **5.** $\{-12, 12\}$ **7.** $\{-5, 5\}$ **9.** $\{-6, 12\}$ **11.** $\{-4, 3\}$

13. $\left\{-3, \dfrac{11}{2}\right\}$ **15.** $\left\{-\dfrac{19}{2}, \dfrac{9}{2}\right\}$ **17.** $\{-10, -2\}$ **19.** $\left\{-8, \dfrac{32}{3}\right\}$ **21.** (a)

(b) **23.** $(-\infty, -3) \cup (3, \infty)$ **25.** $(-\infty, -4] \cup [4, \infty)$

27. $(-\infty, -12) \cup (8, \infty)$ **29.** $\left(-\infty, -\dfrac{7}{3}\right] \cup [3, \infty)$

31. $(-\infty, -2) \cup (8, \infty)$ **33.** $[-3, 3]$ **35.** $(-4, 4)$

37. $[-12, 8]$ **39.** $\left(-\dfrac{7}{3}, 3\right)$ **41.** $[-2, 8]$

43. $(-\infty, -5) \cup (13, \infty)$ **45.** $\{-6, -1\}$ **47.** $\left[-\dfrac{10}{3}, 4\right]$

49. $\left[-\dfrac{7}{6}, -\dfrac{5}{6}\right]$ **51.** $|x - 4| = 9$ (or $|4 - x| = 9$) **53.** $\{-5, -3\}$ **55.** $(-\infty, -3) \cup (2, \infty)$

57. $[-10, 0]$ **59.** $\{-1, 3\}$ **61.** $\left\{-3, \dfrac{5}{3}\right\}$ **63.** $\left\{-\dfrac{1}{3}, -\dfrac{1}{15}\right\}$ **65.** $\left\{-\dfrac{5}{4}\right\}$ **67.** $\emptyset$ **69.** $\left\{-\dfrac{1}{4}\right\}$ **71.** $\emptyset$

73. $(-\infty, \infty)$ **75.** $\left\{-\dfrac{3}{7}\right\}$ **77.** $(-\infty, \infty)$ **79.** $\left(-\infty, -\dfrac{7}{10}\right) \cup \left(-\dfrac{7}{10}, \infty\right)$ **81.** 460.2 feet **82.** Federal Office

Building, City Hall, Kansas City Power and Light, Hyatt Regency **83.** Southwest Bell Telephone, City Center Square, Commerce Tower, Federal Office Building, City Hall, Kansas City Power and Light, Hyatt Regency **84.** (a) $|x - 460.2| \geq 75$ **(b)** $x \geq 535.2$ or $x \leq 385.2$ **(c)** Pershing Road Associates, AT&T Town Pavillion, One Kansas City Place **(d)** It makes sense because it includes all buildings *not* listed earlier. **85.** $[30.4, 33.6]$ **87.** 64 **89.** 625 **91.** $\dfrac{81}{256}$

SUMMARY: Solving Linear and Absolute Value Equations and Inequalities (PAGE 108)

1. $\{12\}$ **3.** $\{7\}$ **5.** $\emptyset$ **7.** $\left[-\dfrac{2}{3}, \infty\right)$ **9.** $\{-3\}$ **11.** $(-\infty, 5]$ **13.** $\{2\}$ **15.** $\emptyset$ **17.** $(-5.5, 5.5)$

19. $\left\{-\dfrac{96}{5}\right\}$ **21.** $(-\infty, -24)$ **23.** $\left\{\dfrac{7}{2}\right\}$ **25.** $(-\infty, \infty)$ **27.** $(-\infty, -4) \cup (7, \infty)$

29. $\left\{-\dfrac{1}{5}\right\}$ **31.** $\left[-\dfrac{1}{3}, 3\right]$ **33.** $\left\{-\dfrac{1}{6}, 2\right\}$ **35.** $(-\infty, -1] \cup \left[\dfrac{5}{3}, \infty\right)$ **37.** $\left\{-\dfrac{5}{2}\right\}$

39. $\left[-\dfrac{9}{2}, \dfrac{15}{2}\right]$ **41.** $(-\infty, \infty)$ **43.** $(-\infty, \infty)$ **45.** $\{-2\}$ **47.** $(-\infty, -1) \cup (2, \infty)$

CHAPTER 2 REVIEW EXERCISES (PAGE 113)

1. $\left\{-\dfrac{9}{5}\right\}$ **3.** {10} **5.** $\emptyset$ **7.** {16} **9.** identity; $(-\infty, \infty)$ **11.** conditional; {0} **13.** $H = \dfrac{V}{LW}$ **15.** $d = \dfrac{C}{\pi}$

17. $x = -4M - 3y$ **19.** 11.6 centimeters **21.** 8 feet **23.** 6.5% **25.** 100°C **27.** −40°F **29.** 66.6%

31. 105°; 105° **33.** 37.4% **35.** $3x$ **37.** $5 + \dfrac{1}{2}x$ **39.** 2 **41.** 30 **43.** 12 kilograms **45.** 30 liters

47. 25° **49.** 40°; 45°; 95° **51.** 850 reserved; 246 general admission **53.** 15 pounds **55.** 6 inches; 12 inches;

16 inches **57.** 6 inches **59.** 52, 53 **61.** $(-9, \infty)$ **63.** $\left(\dfrac{3}{2}, \infty\right)$ **65.** $[-3, \infty)$ **67.** $[3, 5)$ **69.** $\left(\dfrac{59}{31}, \infty\right)$

71. $(-\infty, 1]$ **73.** 87 points **75.** {a, c} **77.** {a, c, e, f, g} **79.** $(-\infty, 3)$ **81.** $\emptyset$

83. $(-\infty, \infty)$ **85.** $(-3, 4)$ **87.** $(4, \infty)$ **89.** managerial and professional specialty **91.** {−7, 7}

93. $\left\{-\dfrac{1}{3}, 5\right\}$ **95.** {0, 7} **97.** $\left\{-\dfrac{3}{4}, \dfrac{1}{2}\right\}$ **99.** $(-14, 14)$ **101.** $[-3, -2]$ **103.** $\left(-\infty, -\dfrac{8}{5}\right) \cup (2, \infty)$ **105.** $\emptyset$

107. $|x - 14| > 12$ (or $|14 - x| > 12$) **109.** $(-2, \infty)$ **111.** $[-2, 3)$ **113.** 15 centimeters **115.** $\left(-\infty, -\dfrac{13}{5}\right) \cup (3, \infty)$

117. 5 liters **119.** {30} **121.** $\left\{1, \dfrac{11}{3}\right\}$ **123.** $\emptyset$ **125.** $\dfrac{21}{2} = 10.5$; $.05x + .25(50 - x) = 10.5$ **126.** 10^2 or 100

127. {10} **128.** $(-\infty, 10)$ **129.** $(10, \infty)$

130. $(-\infty, \infty)$ **131.** $\emptyset$ **132.** The union will be $(-\infty, \infty)$ and the intersection will be $\emptyset$.

CHAPTER 2 TEST (PAGE 119)

[2.1] 1. {−19} **2.** {5} **3.** $(-\infty, \infty)$ **4.** contradiction; $\emptyset$ **[2.2] 5.** $v = \dfrac{S + 16t^2}{t}$ **[2.3, 2.4] 6.** 3.2 hours

7. 3.75% **8.** 9696 residents **9.** $8000 at 3%; $20,000 at 5% **10.** 40°; 40°; 100° **11.** $15,345,000

[2.5] 12. We must reverse the direction of the inequality symbol. **13.** $[1, \infty)$

14. $(-\infty, 28)$ **15.** $[-3, 3]$ **16.** (c) **17.** 82%

[2.6] 18. (a) {1, 5} (b) {1, 2, 5, 7, 9, 12} **19.** (a) $[2, 9)$ (b) $(-\infty, 3) \cup [6, \infty)$ **[2.7] 20.** $\left\{-1, \dfrac{5}{2}\right\}$

21. $(-\infty, -1) \cup \left(\dfrac{5}{2}, \infty\right)$ **22.** $\left(-1, \dfrac{5}{2}\right)$ **23.** $\left\{-\dfrac{5}{7}, \dfrac{11}{3}\right\}$ **24.** $\left(\dfrac{1}{3}, \dfrac{7}{3}\right)$ **25.** (a) $\emptyset$ (b) $(-\infty, \infty)$ (c) $\emptyset$

CUMULATIVE REVIEW (Chapters 1–2) (PAGE 120)

1. 9, 6 **2.** 0, 9, 6 **3.** −8, 0, 9, 6 **4.** $-8, -\dfrac{2}{3}, 0, \dfrac{4}{5}, 9, 6$ **5.** $-\sqrt{6}$ **6.** All are real numbers. **7.** $-\dfrac{22}{21}$

8. 8 **9.** 8 **10.** 0 **11.** −243 **12.** $\dfrac{216}{343}$ **13.** $-\dfrac{8}{27}$ **14.** −4096 **15.** $\sqrt{-36}$ is not a real number.

16. $\dfrac{4 + 4}{4 - 4}$ is undefined. **17.** −16 **18.** −34 **19.** 184 **20.** $\dfrac{27}{16}$ **21.** $-20r + 17$ **22.** $13k + 42$

23. commutative property **24.** distributive property **25.** inverse property **26.** $-\dfrac{3}{2}$ **27.** {5} **28.** {30}

29. {15} **30.** $b = P - a - c$ **31.** $[-14, \infty)$ **32.** $\left[\dfrac{5}{3}, 3\right)$

33. $(-\infty, 0) \cup (2, \infty)$ <img ...> 0 2   **34.** $\left(-\infty, -\dfrac{1}{7}\right] \cup [1, \infty)$ <img ...> $-\dfrac{1}{7}$ 1

35. $5000   **36.** $6\dfrac{1}{3}$ grams   **37.** 74 or greater   **38.** $\dfrac{1}{8}$ hour   **39.** 2 liters   **40.** 9 cents, 12 nickels, 8 quarters

41. 44 milligrams   **42.** 20 drops   **43. (a)** $317,000 **(b)** 55%   **44.** 25.7

CHAPTER 3 EXPONENTS AND POLYNOMIALS

SECTION 3.1 (PAGE 129)

EXERCISES **1.** a^8 **3.** y^6 **5.** $-18x^5y^8$ **7.** p^{14} **9.** z^6 **11.** r^{16} **13.** $-\dfrac{56}{k^2}$ **15.** $-24x^4$ **17.** $\dfrac{1}{2pq}$

19. 1   **23.** $\dfrac{4}{9}$ **25.** $\dfrac{1}{64}$ **27.** $-\dfrac{1}{64}$ **29.** $-\dfrac{1}{64}$ **31.** 9   **33.** $-\dfrac{16}{3}$ **35.** $\dfrac{27}{8}$ **37.** $\dfrac{5}{6}$ **39.** $-\dfrac{1}{12}$ **41.** $\dfrac{1}{2}$

43. $\dfrac{1}{27}$ **45.** $-\dfrac{1}{81}$ **47.** 1   **48.** 2   **49.** $\dfrac{5^2}{5^2}$ **50.** 5^0 **51.** $1 = 5^0$ because they are both equal to $\dfrac{25}{25}$.

52. If $a \neq 0$, $a^0 = 1$.   **53.** 1   **55.** -1 **57.** 1   **59.** 1   **61.** -2 **63.** (c)   **65.** 15   **67.** $-\dfrac{1}{k^3}$

69. $\dfrac{1}{(3x^4)^2}$ **71.** $-\dfrac{12}{r^8}$ **73.** $\dfrac{1}{5}$ **75.** -1 **79.** p^{8q} **81.** a^{3r+3} **83.** m^a **85.** $\dfrac{1}{5}$; $\dfrac{5}{6}$; $\dfrac{1}{5} \neq \dfrac{5}{6}$

87. 25; 13; $25 \neq 13$ **89.** $\left(\dfrac{1}{2}\right)^7$ **91.** -2^6

SECTION 3.2 (PAGE 137)

◊ **CONNECTIONS** **Page 134:** Answers will vary.   **Page 136:** Answers will vary.

EXERCISES **1.** (d)   **3.** incorrect; a^2b^2 **5.** correct   **7.** incorrect; $\dfrac{4^5}{a^5}$ **9.** incorrect; z^{20} **11.** $\left(\dfrac{4}{3}\right)^2$

13. $\dfrac{5}{6}$ **15.** $\dfrac{1}{2^9 \cdot 5^3}$ **17.** $5^{12} \cdot 6^6$ **19.** $\dfrac{1}{k^2}$ **21.** $-4r^6$ **23.** $\dfrac{5^4}{a^{10}}$ **25.** $\dfrac{z^4}{x^3}$ **27.** $\dfrac{1}{5p^{10}}$ **29.** $\dfrac{4}{a^2}$ **31.** $\dfrac{1}{6y^{13}}$

33. $\dfrac{2^2k^5}{m^2}$ **35.** $\dfrac{2^2k^{17}}{5^3}$ **37.** $\dfrac{2k^5}{3}$ **39.** $\left(\dfrac{b^6}{a^3}\right)^{-2}$; $\dfrac{b^{-12}}{a^{-6}} = \dfrac{a^6}{b^{12}}$ **40.** $\dfrac{a^{16}b^{-4}}{a^{10}b^8}$; $\dfrac{a^{16-10}}{b^{8+4}} = \dfrac{a^6}{b^{12}}$ **41.** They are the same.

42. Both methods are correct.   **43.** 5.3×10^2 **45.** 8.3×10^{-1} **47.** 6.92×10^{-6} **49.** -3.85×10^4
51. 72,000   **53.** .00254   **55.** $-60,000$ **57.** .000012   **59.** .06   **61.** .0000025   **63.** 200,000   **65.** 3000
67. $3838.38   **69.** 63,360 inches in a mile   **71.** approximately $3.2 \times 10^4 = 32,000$ hours (about 3.7 years)
73. 500 seconds, which is approximately 8.3 minutes   **75.** 20,000 hours   **77. (a)** .7 astronomical unit **(b)** 1 astronomical
unit **(c)** 1.6 astronomical units   **79.** 2×10^{30} bacteria   **81.** 1987   **83.** p^{2y+2} **85.** a^{-4-2k} **87.** r^{p^2+3p-6}
89. $-\dfrac{3}{32m^8p^4}$ **91.** $\dfrac{2}{3y^4}$ **93.** $\dfrac{3p^8}{16q^{14}}$ **95.** $9x$ **97.** $4 - 11y$ **99.** $2x - 2$

SECTION 3.3 (PAGE 146)

◊ **CONNECTIONS** **Page 145: 1.** $P(1) = 1$, $P(2) = 2$ **2.** 4, 8, 16   **3.** $P(3) = 4$, $P(4) = 8$, $P(5) = 16$ **4.** Because
the pattern 1, 2, 4, 8, 16 emerges, most people will predict 32, since the terms are doubling each time. However, $P(6) = 31$ (not 32).
See the figure.

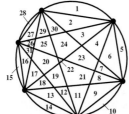

EXERCISES **1.** 7; 1   **3.** -15; 2   **5.** 1; 4   **7.** -1; 6   **11.** neither   **13.** ascending   **15.** descending
17. monomial; 0   **19.** binomial; 1   **21.** trinomial; 3   **23.** none of these; 5   **25.** (a)   **27.** $12p - 4$
29. $-9p^2 + 11p - 9$ **31.** $14m^2 - 13m + 6$ **33.** $8q^3 - 7q + 11$ **35.** $(2 \times 10^2) + (4 \times 10^1) + (1 \times 10^0)$

36. Corresponding place values are not aligned in columns. **38.** Corresponding powers of the variable are not aligned in columns. **41.** $8z^4$ **43.** $7m^3$ **45.** $5x$ **47.** $7y - 3y^2$ **49.** $8k^2 + 2k - 7$ **51.** $-2n^4 - n^3 + n^2$ **53.** $3m + 11$ **55.** $-p - 4$ **57.** $-4p^3 + 10p^2 + 5p - 5$ **59.** $2x^5 - x^4 - 2x^3 + 1$ **61.** $6a + 3$ **63. (a)** -10 **(b)** 8 **65. (a)** 8 **(b)** 2 **67. (a)** 8 **(b)** 74 **69.** 22.5 **71.** 187.5 **73.** -137.5 **75.** \$1.05 billion **77.** \$1.48 billion **79.** $3y^2 - 4y + 2$ **81.** $-4m^2 + 4n^2 - 7n$ **83.** $y^4 - 4y^2 - 4$ **85.** $-3xy - 11z^2$ **87.** $10z^2 - 16z$ **89.** $5a^{3x} + 2a^{2x} + 3a^x + 2$ **91.** $-8k^{4p} - 9k^{2p} + 2$ **93.** $3p^{2k} + 8p^k - 7$ **95.** 0 **97.** $8z^2 + 4z + 2$ **99.** $12m^5$ **101.** $-6a^3b^9$ **103.** $60x^3y^4$

SECTION 3.4 (PAGE 154)

◆ **CONNECTIONS** **Page 152:** Answers will vary. **Page 153:**

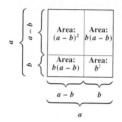

EXERCISES **3.** $-24m^5$ **5.** $-6x^2 + 15x$ **7.** $-2q^3 - 3q^4$ **9.** $18k^4 + 12k^3 + 6k^2$ **11.** $6m^3 + m^2 - 14m - 3$ **13.** $-d^4 + 6d^3 + 2d^2 - 13d + 6$ **15.** $6y^2 + y - 12$ **17.** $-2b^3 + 2b^2 + 18b + 12$ **19.** $25m^2 - 9n^2$ **21.** $m^2 - 3m - 40$ **23.** $6z^2 - 5z - 4$ **25.** $15k^2 - 7k - 2$ **27.** $12m^3 + 22m^2 + 4m - 3$ **29.** $8z^4 - 14z^3 + 17z^2 + 20z - 3$ **31.** $-2x^5 + 16x^4 - 11x^3 + 40x^2 - 15x$ **33.** $6p^4 + p^3 + 4p^2 - 27p - 6$

37. $4p^2 - 9$ **39.** $25m^2 - 1$ **41.** $9a^2 - 4c^2$ **43.** $16x^2 - \dfrac{4}{9}$ **45.** $16m^2 - 49n^4$ **47.** $25y^6 - 4$

49. $y^2 - 10y + 25$ **51.** $4p^2 + 28p + 49$ **53.** $16n^2 - 24nm + 9m^2$ **55.** $k^2 - \dfrac{10}{7}kp + \dfrac{25}{49}p^2$ **59.** $25x^2 + 10x + 1 + 60xy + 12y + 36y^2$ **61.** $4a^2 + 4ab + b^2 - 12a - 6b + 9$ **63.** $4a^2 + 4ab + b^2 - 9$ **65.** $6a^3 + 7a^2b + 4ab^2 + b^3$ **67.** $4z^4 - 17z^3x + 12z^2x^2 - 6zx^3 + x^4$ **69.** $m^4 - 4m^2p^2 + 4mp^3 - p^4$ **71.** $a^3 - 7ab^2 - 6b^3$ **73.** $y^3 + 6y^2 + 12y + 8$ **75.** $q^4 - 8q^3 + 24q^2 - 32q + 16$ **77.** $49; \ 25; \ 49 \neq 25$ **79.** $2401; \ 337; \ 2401 \neq 337$ **81.** Although they are equal for this *particular* case, they are not equal *in general*. **83.** $\dfrac{9}{2}x^2 - 2y^2$ **85.** $15x^2 - 2x - 24$ **87.** $a - b$ **88.** $A = s^2; \ (a - b)^2$ **89.** $(a - b)b$ or $ab - b^2; \ 2ab - 2b^2$ **90.** b^2 **91.** $a^2; \ a$ **92.** $a^2 - (2ab - 2b^2) - b^2 = a^2 - 2ab + b^2$ **93.** They must be equal to each other. **94.** $(a - b)^2 = a^2 - 2ab + b^2;$ This reinforces the special product for the square of a binomial difference. **95.** $12b^{2p-2} + 6b^{3p}$ **97.** $3z^{2m-2} - 4z^{-4}$ **99.** $12k^{2n} - 16k^n - 3$ **101.** $16k^{3m} + 56k^my^m - 6y^{3m}k^{2m} - 21y^{4m}$ **103.** $y^{4n} - 1$ **105.** $9(6 + r^2)$

107. $7(2x - 3z)$ **109.** $(3x + 4)(x + 1)$ **111.** $18z^7w^9$ **113.** $\dfrac{12p^3}{5q^3}$

SECTION 3.5 (PAGE 161)

EXERCISES **3.** $z^2(m + n)^4$ **5.** $12(m + 5)$ **7.** $8k(k^2 + 3)$ **9.** $xy(1 - 5y)$ **11.** $-2p^2q^4(2p + q)$ **13.** $7x^3(3x^2 + 5x - 2)$ **15.** $5ac(3ac^2 - 5c + 1)$ **17.** $-9m^3p^3(3p^2 - 4m + 8m^2p)$ **19.** $(m - 4)(2m + 5)$ **21.** $11(2z - 1)$ **23.** $-y^5(r + w + yz + yk)$ **25.** $(2 - x)(10 - x - x^2)$ **27.** $(3 - x)(6 + 2x - x^2)$ **29.** $20z(2z + 1)(3z + 4)$ **31.** $5(m + p)^2(m + p - 2 - 3m^2 - 6mp - 3p^2)$ **33. (a)** **35.** $2x^2(-x^3 + 3x + 2)$ or $-2x^2(x^3 - 3x - 2)$ **37.** $16a^2m^3(-2a^2m^2 - 1 - 4a^3m^3)$ or $-16a^2m^3(2a^2m^2 + 1 + 4a^3m^3)$ **39.** $2^2 \cdot 3 \cdot 5$ **40.** $2^2 \cdot 3 \cdot 5 \cdot 7$ **41.** $2^3 \cdot 3^2 \cdot 5 + 2^2 \cdot 3 \cdot 5$ **42.** $2^2 \cdot 3 \cdot 5$ **43.** $2^2 \cdot 3 \cdot 5(2 \cdot 3 + 1)$ **44.** It represents 7. **45.** $2^2 \cdot 3 \cdot 5 \cdot 7$ **46.** Yes, the answers are the same. **47.** $(m + 3q)(x + y)$ **49.** $(5m + n)(2 + k)$ **51.** $(m - 3)(m + 5)$ **53.** $(p + q)(p - 4z)$ **55.** $(a + 5)(3a - 2)$ **57.** $(-3p + q)(5p + 2q)$ **59.** $(a^2 + b^2)(-3a + 2b)$ **61.** $(y - 2)(x - 2)$ **63.** $(3y - 2)(3y^3 - 4)$ **65.** $(1 - a)(1 - b)$ **67. (d)** **69.** $p^{4m}(p^{2m} - 2)$ **71.** $q^k(q^{2k} + 2q^k + 3)$ **73.** $y^{r+2}(y^3 + y^2 + 1)$ **75.** $(r^p + q^p)(m^p - z^p)$ **77.** $m^{-5}(3 + m^2)$ or $\dfrac{m^2 + 3}{m^5}$ **79.** $p^{-3}(3 + 2p - 4p^2)$ or $\dfrac{-4p^2 + 2p + 3}{p^3}$ **81.** $m^2 - m - 6$ **83.** $10m^2 - 3m - 27$ **85.** $64z^2 - 48az + 9a^2$ **87.** $2x^4 - 17x^2 + 30$ **89.** $2x^3 - 11x^2 - 21x$

SECTION 3.6 (PAGE 170)

EXERCISES **1. (a)** $x^2 + x - 12; \ x^2 - x - 12$ **(b)** $2x^2 - 3x - 35; \ 2x^2 + 3x - 35$ **(c)** $12x^2 + 23x - 24; \ 12x^2 - 23x - 24$ **(d)** The signs are opposites. **(e)** reverse the signs of the second terms of the binomials **3.** $(a - 5)(a + 3)$ **5.** $(p - 8)(p + 7)$ **7.** $(-m + 6)(m - 10)$ **9.** $(a + 5b)(a - 7b)$ **11.** prime

13. $(xy + 9)(xy + 2)$ **15.** $(-6m + 5)(m + 3)$ **17.** $(5x - 6)(2x + 3)$ **19.** $(4k + 3)(5k + 8)$
21. $(3a - 2b)(5a - 4b)$ **23.** $(6m - 5)^2$ **25.** prime **27.** $(2xz - 1)(3xz + 4)$ **29.** $3(4x + 5)(2x + 1)$
31. $5(a + 6)(3a - 4)$ **33.** $4m(m + 5)(m - 2)$ **35.** $11x(x - 6)(x - 4)$ **37.** $2xy^3(x - 12y)(x - 12y)$
39. $3(3a + 2)(2a - 3)$ **41.** $6a(a - 3)(a + 5)$ **43.** $13y(y + 4)(y - 1)$ **45.** $8mn^2(m + n)(m - 4n)$
47. $-(x - 9)(x + 2)$ **49.** $-(9a + 5)(2a - 3)$ **51.** $-r(7r + 1)(2r - 3)$ **55.** $-18a^3 + 27a^2b + 315ab^2$ **57.** no
58. 1, 3, 5, 9, 15, 45; no **59.** no **60.** 1, 2, 4, 5, 10, 20; no **61.** no **62.** $(5x + 2)(2x + 5)$; no **63.** Since k is
odd, 2 is not a factor of $2x^2 + kx + 8$, and because 2 is a factor of $2x + 4$, the binomial $2x + 4$ cannot be a factor.
65. $(p^2 - 8)(p^2 - 2)$ **67.** $(2x^2 + 3)(x^2 - 6)$ **69.** $(4x^2 + 3)(4x^2 + 1)$ **71.** $(6p^3 - r)(2p^3 - 5r)$
73. $(5k + 4)(2k + 1)$ **75.** $(3m + 3p + 5)(m + p - 4)$ **77.** $(a + b)^2(a - 3b)(a + 2b)$ **79.** $(p + q)^2(p + 3q)$
81. $(z - x)^2(z + 2x)$ **83.** $(p^n - 3)(p^n + 2)$ **85.** $(2z^{2r} + 1)(3z^{2r} - 4)$ **87.** $2k^r(6k^r + 1)(3k^r + 2)$ **89.** $9x^2 - 25$
91. $p^2 + 6pq + 9q^2$ **93.** $y^3 + 27$

SECTION 3.7 (PAGE 175)

◈ **CONNECTIONS** **Page 172:** Answers will vary.

EXERCISES **1.** (a), (d) **3.** (b), (c) **5.** $(p + 4)(p - 4)$ **7.** $(5x + 2)(5x - 2)$ **9.** $(3a + 7b)(3a - 7b)$
11. $4(4m^2 + y^2)(2m + y)(2m - y)$ **13.** $(y + z + 9)(y + z - 9)$ **15.** $(4 + x + 3y)(4 - x - 3y)$ **17.** $4pq$
19. $(k - 3)^2$ **21.** $(2z + w)^2$ **23.** $(4m - 1 + n)(4m - 1 - n)$ **25.** $(2r - 3 + s)(2r - 3 - s)$
27. $(x + y - 1)(x - y + 1)$ **29.** $2(7m + 3n)^2$ **31.** $(p + q + 1)^2$ **33.** $(a - b + 4)^2$
35. $(2x - y)(4x^2 + 2xy + y^2)$ **37.** $(4g + 3h)(16g^2 - 12gh + 9h^2)$ **39.** $3(2n + 3p)(4n^2 - 6np + 9p^2)$
41. $(y + z - 4)(y^2 + 2yz + z^2 + 4y + 4z + 16)$ **43.** $(x^3 - y^3)(x^3 + y^3)$; $(x - y)(x^2 + xy + y^2)(x + y)(x^2 - xy + y^2)$
44. $(x^2 + xy + y^2)(x^2 - xy + y^2)$ **45.** $(x^2 - y^2)(x^4 + x^2y^2 + y^4)$; $(x - y)(x + y)(x^4 + x^2y^2 + y^4)$
46. $x^4 + x^2y^2 + y^4$ **47.** The product must equal $x^4 + x^2y^2 + y^4$. Multiply $(x^2 + xy + y^2)(x^2 - xy + y^2)$ to verify this.
48. Start by factoring as the difference of squares. **49.** 16 **51.** 25 **53.** -56 or 56
55. $(3x + y)(9x^2 - 3xy + y^2 + 3x - y)$ **57.** $(10k - m)(100k^2 + 10km + m^2 + 2)$
59. $(y + 1)^2(y^2 - y + 1)$ **61.** $5(x + y)(2x - 2y + x^2 - xy + y^2)$ **63.** $(4m^{2x} + 3)(4m^{2x} - 3)$
65. $(8r^{4z} + 1)(8r^{4z} - 1)$ **67.** $(10m^z + 3p^{4z})(10m^z - 3p^{4z})$ **69.** $(3a^{2z} - 5)^2$ **71.** $(x^n - 2)(x^{2n} + 2x^n + 4)$
73. $8y^3(3 - 2y^2 + 8y^4)$ **75.** $(2x + y)(a - b)$ **77.** $(y - 2)(y + 1)$ **79.** $(3t - s)(2t + 7s)$

SECTION 3.8 (PAGE 180)

EXERCISES **1.** $(10a + 3b)(10a - 3b)$ **3.** $3p^2(p - 6)(p + 5)$ **5.** $3pq(a + 6b)(a - 5b)$ **7.** prime
9. $(6b + 1)(b - 3)$ **11.** $3mn(3m + 2n)(2m - n)$ **13.** $(2p + 5q)(p + 3q)$ **15.** $9m(m - 5 + 2m^2)$
17. $2(3m - 10)(9m^2 + 30m + 100)$ **19.** $(2a + 1)(a - 4)$ **21.** $(k - 9)(q + r)$ **23.** prime
25. $9(x^2 + 4y^2)$; A sum of two squares can be factored when the greatest common factor is not 1.
27. $(x - 5)(x + 5)(x^2 + 25)$ **29.** $(p + 4)(p^2 - 4p + 16)$ **31.** $(8m + 25)(8m - 25)$
33. $6z(2z^2 - z + 3)$ **35.** $16(4b + 5c)(4b - 5c)$ **37.** $8(5z + 4)(25z^2 - 20z + 16)$ **39.** $(5r - s)(2r + 5s)$
41. $4pq(2p + q)(3p + 5q)$ **43.** $3(4k^2 + 9)(2k + 3)(2k - 3)$ **45.** $(m - n)(m^2 + mn + n^2 + m + n)$
47. $(x - 2m - n)(x + 2m + n)$ **49.** $2W + 9$ **51.** $2(x + 4)(x - 5)$ **53.** $8mn$ **55.** $2(5p + 9)(5p - 9)$
57. $4rx(3m^2 + mn + 10n^2)$ **59.** $(7a - 4b)(3a + b)$ **61.** prime **63.** $(p + 8q - 5)^2$ **65.** $(7m^2 + 1)(3m^2 - 5)$
67. $(2r - t)(r^2 - rt + 19t^2)$ **69.** $(x + 3)(x^2 + 1)(x + 1)(x - 1)$ **71.** $(m + n - 5)(m - n + 1)$ **73.** $a(1 + y)$
75. $(x^2 + 4)(1 + y)$ **77.** $(3m^k + 2)(m^k - 3)$ **79.** $3c^y(5c^{3y} + c^y - 2)$ **81.** $(z^{2x} + w^x)(z^{2x} - w^x)$

83. $\left\{ -\dfrac{2}{3} \right\}$ **85.** $\{0\}$ **87.** $\{-10\}$

SECTION 3.9 (PAGE 186)

◈ **CONNECTIONS** **Page 185:** 8 seconds; 4 seconds

EXERCISES **3.** $\{5, -10\}$ **5.** $\left\{ \dfrac{5}{2}, -\dfrac{8}{3} \right\}$ **7.** $\left\{ -6, \dfrac{3}{4}, 1 \right\}$ **11.** $\{-2, 5\}$ **13.** $\{-6, -3\}$ **15.** $\left\{ -\dfrac{1}{2}, 4 \right\}$

17. $\left\{ -\dfrac{1}{3}, \dfrac{4}{5} \right\}$ **19.** $\{-3, 4\}$ **21.** $\{-3, 3\}$ **23.** $\left\{ -\dfrac{3}{5}, 0 \right\}$ **25.** $\left\{ -\dfrac{3}{4} \right\}$ **27.** $\left\{ -\dfrac{3}{2}, \dfrac{3}{2} \right\}$ **29.** $x + 4$ **30.** $x - 8$
31. $(x + 4)(x - 8)$ **32.** $x^2 - 4x - 32$ **33.** $x^2 - 4x - 32 = 0$; The solutions are -4 and 8. **34.** $x^2 - 4x - 45 = 0$
35. $18x^2 - 3x - 10 = 0$ **36.** $x^2 + 18x + 81 = 0$ **37.** $\{-4, 2\}$ **39.** $\left\{ -\dfrac{1}{2}, 6 \right\}$ **41.** $\{1, 6\}$ **45.** $\left\{ -\dfrac{1}{2}, 0, 5 \right\}$

47. $\left\{ -\dfrac{4}{3}, 0, \dfrac{4}{3} \right\}$ **49.** $\left\{ -\dfrac{5}{2}, -1, 1 \right\}$ **51.** By dividing both sides by a variable expression, she "lost" the solution 0.

53. $\left\{-\dfrac{1}{2}, 6\right\}$ **55.** $\left\{-\dfrac{2}{3}, \dfrac{4}{15}\right\}$ **57.** $\{2, 4\}$ **59.** $\left\{-\dfrac{3}{2}, \dfrac{1}{2}\right\}$ **61.** width: 16 feet;   length: 20 feet   **63.** base: 8 meters;
height: 11 meters   **65.** after 3 seconds and 5 seconds   **67.** 5 seconds   **69.** Each side measures 7 inches.
71. -9 and -8 or 8 and 9   **73.** 100 feet by 300 feet   **75.** $4p$ **77.** $-\dfrac{3}{4m^4n^3}$ **79.** $\dfrac{36}{75}$

CHAPTER 3 REVIEW EXERCISES (PAGE 192)

1. $-12x^2y^8$ **3.** $\dfrac{10p^8}{q^7}$ **5.** 64   **7.** -125 **9.** $\dfrac{81}{16}$ **11.** $\dfrac{11}{30}$ **13.** 0   **15.** $\dfrac{1}{3^8}$ **17.** $\dfrac{y^6}{x^2}$ **19.** $\dfrac{25}{m^{18}}$ **21.** $\dfrac{25}{z^4}$
23. $\dfrac{2025}{8r^4}$ **25.** m^{8q-3} **27.** For example, let $x = 2$ and $y = 3$. Then $(x^2 + y^2)^2 = (2^2 + 3^2)^2 = 169$.
$x^4 + y^4 = 2^4 + 3^4 = 97 \neq 169$.   **29.** 7.65×10^{-8} **31.** 1,210,000   **33.** .63   **35.** 1.5×10^3 or 1500
37. 2.7×10^{-2} or .027   **39.** 5.449×10^3 **41.** 14   **43.** 504   **45.** (a) $9m^7 + 14m^6$ (b) binomial (c) 7
47. (a) $-7q^5r^3$ (b) monomial (c) 8   **49.** (a) 27 (b) -3 **51.** $-x^2 - 3x + 1$ **53.** $-5y^3 - 4y^2 + 6y - 12$
55. $8y^2 - 9y + 5$ **57.** $-15b^3 - 50b$ **59.** $15m^2 - 7m - 2$ **61.** $6w^2 - 13wt + 6t^2$ **63.** $3q^3 - 13q^2 - 14q + 20$
65. $36r^4 - 1$ **67.** $16m^2 + 24m + 9$ **69.** $18y^{2q-1} - 45$ **71.** $6z^{2r} - 11x^{2r}z^r - 10x^{4r}$ **73.** $6p(2p - 1)$
75. $4qb(3q + 2b - 5q^2b)$ **77.** $(x + 3)(x - 3)$ **79.** $(m + q)(4 + n)$ **81.** $(m + 3)(2 - a)$ **83.** $(3p - 4)(p + 1)$
85. $(3r + 1)(4r - 3)$ **87.** $(2k - h)(5k - 3h)$ **89.** $2x(4 + x)(3 - x)$ **91.** $(y^2 + 4)(y^2 - 2)$
93. $(p + 2)^2(p + 3)(p - 2)$ **95.** $(5m^{2q} - 3)(2m^{2q} + 1)$ **99.** $(4x + 5)(4x - 5)$ **101.** $(x + 7)^2$
103. $(r + 3)(r^2 - 3r + 9)$ **105.** $(m + 1)(m^2 - m + 1)(m - 1)(m^2 + m + 1)$ **107.** $(x + 3 + 5y)(x + 3 - 5y)$
109. $(x + 1)(x - 1)(x - 2)(x^2 + 2x + 4)$ **111.** $(x^{2y} + 3)^2$ **113.** $(5r^k + 1)(25r^{2k} - 5r^k + 1)$
115. (a) $(x - 3)(6x + 16) = 0$ (b) $\left\{3, -\dfrac{8}{3}\right\}$ **117.** $\{2, 3\}$ **119.** $\left\{-\dfrac{5}{2}, \dfrac{10}{3}\right\}$ **121.** $\left\{-\dfrac{3}{2}, -\dfrac{1}{4}\right\}$ **123.** $\left\{-\dfrac{3}{2}, 0\right\}$
125. $\{4\}$ **127.** $\{-3, -2, 2\}$ **129.** 3 feet   **131.** after 16 seconds   **135.** $\dfrac{y^4}{36}$ **137.** $\dfrac{1}{16y^{18}}$ **139.** $21p^9 + 7p^8 + 14p^7$
141. -9 **143.** $-3k^2 + 4k - 7$ **145.** $k(11 + 12k)$ **147.** prime   **149.** $(5z - 3m)^2$ **151.** $\{-1, 0, 1\}$
153. $(x^{14} - x^2) - (4x^{13} - 4x) + (4x^{12} - 4)$ **154.** $x^2(x^{12} - 1) - 4x(x^{12} - 1) + 4(x^{12} - 1)$ **155.** $(x^{12} - 1)(x^2 - 4x + 4)$
156. $(x^6 - 1)(x^6 + 1)(x - 2)^2$ **157.** $(x^3 - 1)(x^3 + 1)(x^2 + 1)(x^4 - x^2 + 1)(x - 2)^2$
158. $(x - 1)(x^2 + x + 1)(x + 1)(x^2 - x + 1)(x^2 + 1)(x^4 - x^2 + 1)(x - 2)^2$

CHAPTER 3 TEST (PAGE 197)

[3.1–3.2] 1. For example, if $a = 4$, $(2 \cdot 4)^{-3} = 8^{-3} = \dfrac{1}{512}$, and $\dfrac{2}{a^3} = \dfrac{2}{4^3} = \dfrac{1}{32}$. $\dfrac{1}{512} \neq \dfrac{1}{32}$.   **2.** $\dfrac{2^3}{3^3}$ **3.** $-\dfrac{m^2}{4n^4}$ **4.** $\dfrac{a^7}{5^3 \cdot 2}$
5. 1   **6.** $\dfrac{a^8}{3^3 \cdot 2^4}$ **7.** .0000037   **8.** 3×10^{-4} or .0003   **9.** 980 square miles   **[3.3] 10.** $x^3 - 2x^2 - 10x - 13$
[3.4] 11. $10x^2 - x - 3$ **12.** $6m^3 - 7m^2 - 30m + 25$ **13.** $36x^2 - y^2$ **14.** $9k^2 + 6kq + q^2$
15. $4y^2 - 9z^2 + 6zx - x^2$ **[3.3] 16.** -66 **[3.5] 17.** $11z(z - 4)$ **18.** $2(h - 1)(h + 1)$ **19.** $(x + y)(3 + b)$
[3.6–3.7] 20. $(4p - q)(p + q)$ **21.** $(4a + 5b)^2$ **22.** $(y - 6)(y^2 + 6y + 36)$ **23.** $(3k + 11j)(3k - 11j)$
24. $(2k^2 - 5)(3k^2 + 7)$ **25.** (d)   **[3.9] 26.** $\left\{-2, -\dfrac{2}{3}\right\}$ **27.** $\left\{\dfrac{1}{5}, \dfrac{3}{2}\right\}$ **28.** $\left\{-\dfrac{2}{5}, 1\right\}$ **29.** length: 8 inches;
width: 5 inches   **30.** 2 seconds and 4 seconds

CUMULATIVE REVIEW (Chapters 1–3) (PAGE 198)

1. $-2m + 6$ **2.** $4m - 3$ **3.** $2x^2 + 5x + 4$ **4.** $(2, \infty)$ **5.** $(-\infty, 1]$ **6.** $(-3, 5]$ **7.** no   **8.** -24
9. 204   **10.** 56   **11.** undefined   **12.** 10   **13.** $\left\{\dfrac{7}{6}\right\}$ **14.** $\{-1\}$ **15.** $\left(-\infty, \dfrac{15}{4}\right]$ **16.** $\left(-\dfrac{1}{2}, \infty\right)$ **17.** $(2, 3)$
18. $(-\infty, 2) \cup (3, \infty)$ **19.** $\left\{-\dfrac{16}{5}, 2\right\}$ **20.** $(-11, 7)$ **21.** $(-\infty, -2] \cup [7, \infty)$ **22.** 2 hours   **23.** 4 glazed pots
24. 343   **25.** -1 **26.** 16   **27.** $\dfrac{y}{18x}$ **28.** $\dfrac{5my^4}{3}$ **29.** $10p^3 + 7p^2 - 28p - 24$ **30.** $x^3 + 12x^2 - 3x - 7$
31. $(2w + 7z)(8w - 3z)$ **32.** $(2x - 1 + y)(2x - 1 - y)$ **33.** $(2y - 9)^2$ **34.** $(10x^2 + 9)(10x^2 - 9)$
35. $(2p + 3)(4p^2 - 6p + 9)$ **36.** $\left\{-4, -\dfrac{3}{2}, 1\right\}$ **37.** $\left\{\dfrac{1}{3}\right\}$ **38.** $\{-3, 3\}$
39. longer sides: 18 inches;   distance between: 16 inches   **40.** 4 meters

CHAPTER 4   RATIONAL EXPRESSIONS

SECTION 4.1 (PAGE 208)

EXERCISES **3.** 7   **5.** $-\dfrac{1}{7}$ **7.** 0   **9.** $-2, \dfrac{3}{2}$ **11.** none   **13.** none

15. (a) numerator: $x^2, 4x$;   denominator: $x, 4$ **17.** $\dfrac{16.5}{73.2}$, .23   **19.** $\dfrac{26}{705}$, .04   **21.** $\dfrac{4x}{3y}$ **23.** $\dfrac{x-3}{x+5}$ **25.** $\dfrac{x+3}{2x(x-3)}$

27. already in lowest terms   **29.** $\dfrac{6}{7}$ **31.** $\dfrac{z}{6}$ **33.** $\dfrac{2}{t-3}$ **35.** $\dfrac{x-3}{x+1}$ **37.** $\dfrac{4x+1}{4x+3}$ **39.** $a^2 - ab + b^2$ **41.** $\dfrac{c+6d}{c-d}$

43. $\dfrac{a+b}{a-b}$ **45. (b)**   **47.** -1 **In Exercises 49–53, there are several other acceptable ways to express the answer.**

49. $-(x+y)$ **51.** $-\dfrac{x+y}{x-y}$ **53.** $-\dfrac{1}{2}$ **55.** already in lowest terms   **59.** $\dfrac{3y}{x^2}$ **61.** $\dfrac{3a^3b^2}{4}$ **63.** $\dfrac{7x}{6}$

65. $-\dfrac{p+5}{2p}$ (There are other ways.)   **67.** $\dfrac{-m(m+7)}{m+1}$ (There are other ways.)   **69.** -2 **71.** $\dfrac{x+4}{x-4}$

73. $\dfrac{2x+3y}{2x-3y}$ **75.** $\dfrac{k+5p}{2k+5p}$ **77.** $(k-1)(k-2)$ **79.** $\dfrac{(a+5)(a-1)}{3a^2-2a+1}$ **81.** $(3x+5)(x-1)$

82. $(3x+5)(3x-5)$ **83.** $\dfrac{x-1}{3x-5}$ **84.** Many answers are possible. One example is

$\dfrac{(3x^2+2x-5)(4x)}{(9x^2-25)(4x)} = \dfrac{12x^3+8x^2-20x}{36x^3-100x}$.   **85. (a)** 1   **(b)** 1   **86.** They are the same. $\dfrac{R}{S}, \dfrac{S}{R}$ **87.** $\dfrac{17}{42}$ **89.** $-\dfrac{2}{3}$

SECTION 4.2 (PAGE 216)

EXERCISES **3.** $\dfrac{9}{t}$ **5.** $\dfrac{2}{x}$ **7.** 1   **9.** $x-5$ **11.** $\dfrac{5}{p+3}$ **13.** $a-b$ **15.** $72x^4y^5$

17. $z(z-2)$ **19.** $2(y+4)$ **21.** $30(x+3)$ **23.** $(m+n)(m-n)$ **25.** $x(x-4)(x+1)$

27. $(t+5)(t-2)(2t-3)$ **29.** $2y(y+3)(y-3)$ **31.** $6x^2(x+1)$ **35.** $\dfrac{31}{3t}$ **37.** $\dfrac{5-22x}{12x^2y}$

39. $\dfrac{1}{x(x-1)}$ **41.** $\dfrac{5a^2-7a}{(a+1)(a-3)}$ **43.** $\dfrac{-6x+3}{x-4}$ or $\dfrac{6x-3}{4-x}$ **45.** $\dfrac{w+z}{w-z}$ or $\dfrac{-w-z}{z-w}$ **47.** $\dfrac{2(2x-1)}{x-1}$

49. $\dfrac{6}{x-2}$ **51.** $\dfrac{3x-2}{x-1}$ **53.** $\dfrac{4x-7}{x^2-x+1}$ **55.** $\dfrac{2x+1}{x}$ **57.** $\dfrac{2x^2+21xy-10y^2}{(x+2y)(x-y)(x+6y)}$ **59.** $\dfrac{3r-2s}{(2r-s)(3r-s)}$

61. If $x=4$ and $y=2$, then $\dfrac{1}{x} + \dfrac{1}{y} = \dfrac{1}{4} + \dfrac{1}{2} = \dfrac{3}{4}$ which does not equal $\dfrac{1}{x+y} = \dfrac{1}{4+2} = \dfrac{1}{6}$.

63. (a)–(d) All equal $\dfrac{7m+5}{m(m+1)}$.   **65. (a)** $\dfrac{-x}{y(4x+3y)} + \dfrac{8x+6y}{(4x+3y)(4x-3y)}$ **(b)** $\dfrac{-x}{y(4x+3y)} \cdot \dfrac{2(4x+3y)}{(4x+3y)(4x-3y)}$

66. (a) $\dfrac{-x(4x-3y)}{y(4x+3y)(4x-3y)} + \dfrac{y(8x+6y)}{y(4x+3y)(4x-3y)}$ **(b)** $\dfrac{-2x(4x+3y)}{y(4x+3y)^2(4x-3y)}$ **67. (a)** $\dfrac{-4x^2+11xy+6y^2}{y(4x+3y)(4x-3y)}$

(b) $\dfrac{-2x}{y(4x+3y)(4x-3y)}$ **69.** $\dfrac{4}{15}$ **71.** $\dfrac{7}{17}$

SECTION 4.3 (PAGE 222)

◆ **CONNECTIONS**   **Page 220: 1.** Answers will vary.   **2.** $\dfrac{34}{21}$; 1, 2, 1.5, 1.$\overline{6}$, 1.6, 1.625, 1.615384615, 1.619047619;

They seem to approach a number close to 1.62.

EXERCISES **3.** $\dfrac{2x}{x-1}$ **5.** $\dfrac{2(k+1)}{3k-1}$ **7.** $\dfrac{5x^2}{9z^3}$ **9.** $\dfrac{1+x}{-1+x}$ **11.** $\dfrac{y+x}{y-x}$ **13.** $4x$ **15.** $x+4y$ **17.** $\dfrac{3y}{2}$

19. $\dfrac{x^2+5x+4}{x^2+5x+10}$ **21.** $\dfrac{m^2+6m-4}{m(m-1)}$ **22.** $\dfrac{m^2-m-2}{m(m-1)}$ **23.** $\dfrac{m^2+6m-4}{m^2-m-2}$ **24.** $m(m-1)$

25. $\dfrac{m^2+6m-4}{m^2-m-2}$ **27.** $\dfrac{x^2y^2}{y^2+x^2}$ **29.** $\dfrac{y^2+x^2}{xy^2+x^2y}$ or $\dfrac{y^2+x^2}{xy(y+x)}$ **31.** $\dfrac{rs}{s+r}$ **33. (a)** $\dfrac{\dfrac{3}{mp} - \dfrac{4}{p} + \dfrac{8}{m}}{\dfrac{2}{m} - \dfrac{3}{p}}$

(c) $\dfrac{3 - 4m + 8p}{2p - 3m}$ **35.** $\dfrac{-1}{6y - 1}$ or $\dfrac{1}{1 - 6y}$ **37.** $\dfrac{p + p^2 + 1}{p^3 + p^2 + 2p + 1}$ **39.** $2p^4$ **41.** $\dfrac{-4b^6}{3a^2}$

43. $-8a^2 + a + 4$ **45.** $-2x^3 - 13x^2 + 11x$

SECTION 4.4 (PAGE 227)

EXERCISES   **1.** exponents   **3.** $r^2 - 7r + 6$ **5.** $3y + 4 - \dfrac{5}{y}$ **7.** $3m + 5 + \dfrac{6}{m}$ **9.** $n - \dfrac{3n^2}{2m} + 2$

11. $\dfrac{2y}{x} + \dfrac{3}{4} + \dfrac{3w}{x}$ **13.** $y - 3$ **15.** $t + 5$ **17.** $z^2 + 3$ **19.** $x^2 + 2x - 3 + \dfrac{6}{4x + 1}$ **21.** $2x - 5 + \dfrac{-4x + 5}{3x^2 - 2x + 4}$

23. $2k^2 + 3k - 1$ **25.** $9z^2 - 4z + 1 + \dfrac{-z + 6}{z^2 - z + 2}$ **27.** $t^3 + 6t^2 + 5t + 4$ **28.** $t + 4$ **29.** $t^2 + 2t - 3 + \dfrac{16}{t + 4}$

30. $118\dfrac{1}{7}$ **31.** $118\dfrac{1}{7}$ **32.** They are the same.   **33.** The correct quotient is $a^2 + 1$.   **35.** $\dfrac{2}{3}x - 1$

37. $\dfrac{3}{4}a - 2 + \dfrac{1}{4a + 3}$ **39.** $2p + 7$ **43.** -6 **45.** -21 **47.** -6

SECTION 4.5 (PAGE 232)

EXERCISES   **3.** $x - 5$ **5.** $4m - 1$ **7.** $2a + 4 + \dfrac{5}{a + 2}$ **9.** $p - 4 + \dfrac{9}{p + 1}$ **11.** $4a^2 + a + 3$

13. $x^4 + 2x^3 + 2x^2 + 7x + 10 + \dfrac{18}{x - 2}$ **15.** $-4r^5 - 7r^4 - 10r^3 - 5r^2 - 11r - 8 + \dfrac{-5}{r - 1}$

17. $-3y^4 + 8y^3 - 21y^2 + 36y - 72 + \dfrac{143}{y + 2}$ **19.** $y^2 + y + 1 + \dfrac{2}{y - 1}$ **21.** 7 **23.** -2 **25.** 0

28. $(2x - 3)(x + 4)$ **29.** $\left\{\dfrac{3}{2}, -4\right\}$ **30.** $P(-4) = 0, P\left(\dfrac{3}{2}\right) = 0$ **31.** a **32.** Yes, $x - 3$ is a factor.

$Q(x) = (x - 3)(3x - 1)(x + 2)$ **33.** yes   **35.** no   **37.** no   **39.** yes   **41.** $5(8 + x)$ **43.** $6(5m - 3n)$

45. $(x + 1)(x + y)$

SECTION 4.6 (PAGE 237)

◆ **CONNECTIONS**   **Page 236: 1.** 14,000 gallons   **2.** It decreases.

EXERCISES   **1.** $-1, 2$ **3.** $-\dfrac{5}{3}, 0, -\dfrac{3}{2}$ **5.** $4, \dfrac{7}{2}$ **7.** There are no numbers that would have to be rejected.

9. $\{8\}$ **11.** $\{-3\}$ **13.** $\left\{-\dfrac{7}{12}\right\}$ **15.** $\varnothing$ **17.** $\{1\}$ **19.** $\{-6, 4\}$ **21.** $\left\{-\dfrac{23}{5}\right\}$ **23.** $\{0\}$ **25.** $\{5\}$ **27.** $\varnothing$

29. $\{-2\}$ **31.** $\varnothing$ **33.** $\left\{x \mid x \neq -\dfrac{3}{2}, x \neq \dfrac{3}{2}\right\}$ **35.** 6 **36. (a)** $3x + 2x = -30$ **(b)** $\dfrac{3x}{6} + \dfrac{2x}{6}$

37. (a) $\{-6\}$ **(b)** $\dfrac{5x}{6}$ **41.** $t = \dfrac{d}{r}$ **43.** $c = P - a - b$ **45.** $\dfrac{1}{2}$ or 4 **47.** $\dfrac{60}{7}$

Summary of Rational Expressions: Equations Versus Expressions (PAGE 239)

1. $\{20\}$ **3.** $\dfrac{2(x + 5)}{5}$ **5.** $\dfrac{y + x}{y - x}$ **7.** $\{7\}$ **9.** $\{1\}$ **11.** $\dfrac{25}{4(r + 2)}$ **13.** $\dfrac{24p}{p + 2}$ **15.** $\{0\}$ **17.** $\dfrac{5}{3z}$ **19.** $\{2\}$

21. $\dfrac{-x}{3x + 5y}$ **23.** $\dfrac{3}{2s - 5r}$ **25.** $\left\{\dfrac{5}{4}\right\}$ **27.** $\dfrac{2z - 3}{2z + 3}$ **29.** $\dfrac{t - 2}{8}$ **31.** $\dfrac{13x + 28}{2x(x + 4)(x - 4)}$ **33.** $\dfrac{k(2k^2 - 2k + 5)}{(k - 1)(3k^2 - 2)}$

SECTION 4.7 (PAGE 246)

◆ **CONNECTIONS**   **Page 245:** 48 miles per hour

EXERCISES   **1. (a)**   **3. (a)**   **5.** 1.349 **7.** 24 **9.** $G = \dfrac{Fd^2}{Mm}$ **11.** $a = \dfrac{bc}{c + b}$ **13.** $v = \dfrac{PVt}{pT}$

15. $e = \dfrac{Er}{R + r}$ **17.** $b = \dfrac{2A}{h} - B$ or $b = \dfrac{2A - Bh}{h}$ **19.** $R = \dfrac{Ar}{r - A}$ **21.** Multiply both sides by $a - b$.

23. 15 girls, 5 boys   **25.** $\frac{1}{2}$ job per hour   **27.** $\frac{4}{5}$ **29.** $\frac{5}{13}$; $\frac{5}{12}$ **31.** \$300   **33.** 25,000

There are two ways to set up this problem.   **35.** Find the distance from Dr. Dawson's office to his home.   *Method 1:* Let x represent his time riding his bike to the office.   *Method 2:* Let x represent the distance.

36. *Method 1:*

	d	r	t
bike	$12x$	12	x
car	$36\left(x - \frac{1}{4}\right)$	36	$x - \frac{1}{4}$

Method 2:

	d	r	t
bike	x	12	$\frac{x}{12}$
car	x	36	$\frac{x}{36}$

37. *Method 1:* $12x = 36\left(x - \frac{1}{4}\right)$;

Method 2: $\frac{x}{36} = \frac{x}{12} - \frac{1}{4}$ **38.** *Method 1:* $x = \frac{3}{8}$ hour;   *Method 2:* $x = \frac{9}{2}$ miles   **39.** The distance is $\frac{9}{2}$ or 4.5 miles.

40. $\frac{9}{2}$ miles at 12 miles per hour takes $\frac{\frac{9}{2}}{12} = \frac{3}{8}$ hour; $\frac{9}{2}$ miles at 36 miles per hour takes $\frac{\frac{9}{2}}{36} = \frac{1}{8}$ hour. $\frac{3}{8} - \frac{1}{4} = \frac{1}{8}$ as required.

41. *Step 1:* Find the distance from Tulsa to Detroit. Let x represent that distance.   *Step 2:*

d	r	t
x	50	$\frac{x}{50}$
x	60	$\frac{x}{60}$

Step 3: $\frac{x}{60} = \frac{x}{50} - 3$ *Step 4:* $x = 900$ *Step 5:* The distance is 900 miles.   *Step 6:* Check: 900 miles at 50 miles per hour takes 18 hours;   900 miles at 60 miles per hour takes 15 hours; $15 = 18 - 3$ as required.   **43.** 150 miles

45. 200 miles   **49.** $\frac{40}{13}$ or $3\frac{1}{13}$ hours   **51.** $17\frac{1}{2}$ hours   **53.** 36 hours   **55.** 24 hours

57. $x = \frac{7}{2}$; $AC = 8$; $DF = 12$ **59.** 9   **61.** 4   **63.** 3

CHAPTER 4 REVIEW EXERCISES (PAGE 253)

3. -6 **7.** $\frac{x}{2}$ **9.** $\frac{5m + n}{5m - n}$ **13.** $\frac{3y^2(2y + 3)}{2y - 3}$ **15.** $\frac{y(y + 5)}{y - 5}$ **17.** 1   **21.** $96b^5$ **23.** $45(2k + 1)$

25. $\frac{16z - 3}{2z^2}$ **27.** 12   **29.** $\frac{3x + 2}{x - 5}$ **31.** $\frac{17}{12}$ **32.** $\frac{5}{7}$ **33.** no;   denominator;   numerators;   least common

denominator   **34.** .67, .71, .75; $\frac{2}{3} < \frac{5}{7} < \frac{3}{4}$;   It lies between the original two fractions.   **35.** numerators;   denominators

36. $\frac{2}{5}$ **37.** $\frac{3 + 2t}{4 - 7t}$ **39.** $\frac{1}{32}$ **41.** $\frac{1}{3q + 2p}$ **43.** $y^2 - 3y + \frac{5}{4}$ **45.** $p^2 + 3p - 6$ **47.** $4y^2 + 1 + \frac{-2y}{3y^2 + 1}$

49. $10k - 23 + \frac{31}{k + 2}$ **51.** $-a^3 + 4a^2 + 3a + 6 + \frac{-9}{a + 4}$ **53.** no   **55.** -5 **57.** $\{-3\}$ **59.** $\{0\}$ **61.** $\emptyset$

65. $m = \frac{Fd^2}{GM}$ **67.** $m = \frac{Mv - \mu M}{\mu}$ or $m = \frac{Mv}{\mu} - M$ **69.** \$21.06   **71.** bus: 50 miles per hour;

train: 60 miles per hour   **73.** $\frac{24}{5}$ or $4\frac{4}{5}$ minutes   **75.** $\frac{6m + 5}{3m^2}$ **77.** $\frac{x^2 - 6}{2(2x + 1)}$ **79.** $k^2 - 7k + 6$ **81.** $\frac{1}{3}$

83. $\frac{5a^2 + 4ab + 12b^2}{(a + 3b)(a - 2b)(a + b)}$ **85.** $r = \frac{AR}{R - A}$ or $r = \frac{-AR}{A - R}$ **87.** $\left\{-\frac{14}{3}\right\}$

CHAPTER 4 TEST (PAGE 257)

[4.1] 1. $-2, \frac{4}{3}$ **2.** $\frac{2x - 5}{x(3x - 1)}$ **3.** $\frac{3x}{2y^8}$ **4.** $\frac{y + 4}{y - 5}$ **5.** $\frac{x + 5}{x}$ **[4.2] 6.** $t^2(t + 3)(t - 2)$

7. $\frac{7 - 2t}{6t^2}$ **8.** $\frac{13x + 35}{(x - 7)(x + 7)}$ **9.** $\frac{4}{x + 2}$ **[4.3] 10.** $\frac{72}{11}$ **11.** $\frac{-1}{a + b}$ **[4.4] 12.** $4p - 8 + \frac{6}{p}$

13. $3q^3 - 4q^2 + q + 4 + \frac{-2}{3q - 2}$ **14.** $3y^2 - 2y - 2 + \frac{12y - 3}{2y^2 + 3}$ **[4.5] 15.** yes

16. $9x^4 - 5x^3 + 2x^2 - 2x + 4 + \frac{2}{x + 5}$ **[4.2, 4.6] 17. (a)** $\frac{11(x - 6)}{12}$ **(b)** $\{6\}$ **[4.6] 18.** $\left\{\frac{1}{2}\right\}$ **19.** $\{5\}$

20. A solution cannot make a denominator zero.   **[4.7] 21.** $r = \dfrac{En - IRn}{I}$ or $r = \dfrac{En}{I} - Rn$ **22.** $\dfrac{48}{5}$ **23.** $\dfrac{45}{14}$ or $3\dfrac{3}{14}$ hours
24. 15 miles per hour   **25.** 48,000

CUMULATIVE REVIEW   (Chapters 1–4)   (PAGE 258)

1. -199 **2.** 455   **3.** 14   **4.** $\left\{-\dfrac{15}{4}\right\}$ **5.** $\left\{\dfrac{2}{3}, 2\right\}$ **6.** $x = \dfrac{d - by}{a - c}$ or $x = \dfrac{by - d}{c - a}$ **7.** $\{11\}$

8. $\left(-\infty, \dfrac{240}{13}\right]$ **9.** $\left[-2, \dfrac{2}{3}\right]$ **10.** $(-\infty, \infty)$ **11.** 180 votes, 225 votes

12. 6 meters   **13.** 35 cents, 29 nickels, 30 dimes   **14.** $46°, 46°, 88°$ **15.** 3.5×10^4 **16.** 7.6×10^{-9}
17. 5,600,000,000   **18.** .00000489   **19.** $\dfrac{21y^7}{x^9}$ **20.** $\dfrac{a^{10}}{b^{10}}$ **21.** $\dfrac{m}{n}$ **22.** $4y^2 - 7y - 6$
23. $-25x^3 - 2x^2 - 36x + 114$ **24.** $-6x^6 + 18x^5 - 12x^4$ **25.** $12f^2 + 5f - 3$ **26.** $x^3 + y^3$ **27.** $49t^6 - 64$
28. $\dfrac{1}{16}x^2 + \dfrac{5}{2}x + 25$ **29.** $4x^4 - 3x^2 + 2x + 1$ **30.** $x^2 + 4x - 7$ **31.** 20   **32.** $2x^3 + 5x^2 - 3x - 2$

33. $(2x + 5)(x - 9)$ **34.** $25(2t^2 + 1)(2t^2 - 1)$ **35.** $(2p + 5)(4p^2 - 10p + 25)$ **36.** $\left\{-\dfrac{7}{3}, 1\right\}$ **37.** $\dfrac{y + 4}{y - 4}$
38. $\dfrac{2x - 3}{2(x - 1)}$ **39.** 3   **40.** $\dfrac{-a - 5b}{(a + b)(a - b)}$ **41.** $\dfrac{2(x + 2)}{2x - 1}$ **42.** $\dfrac{5 + 3x - 3y}{(x - y)(x^2 + xy + y^2)}$ **43.** $\{-4\}$

44. $q = \dfrac{fp}{p - f}$ or $q = \dfrac{-fp}{f - p}$ **45.** 150 miles per hour   **46.** $\dfrac{6}{5}$ or $1\dfrac{1}{5}$ hours

CHAPTER 5   ROOTS AND RADICALS

SECTION   5.1   (PAGE 266)

◆ **CONNECTIONS**   **Page 263:** Yes;   No, because T depends on h^3, not h.

EXERCISES   1. cube (or third); 8; 2; second; 4; 4   **3. (a)** m must be even   **(b)** m must be odd

7. 13   **9.** 9   **11.** 2   **13.** $\dfrac{8}{9}$ **15.** -3 **17.** 1000   **19.** -32 **21.** not a real number

23. $\dfrac{1}{512}$ **25.** $\dfrac{9}{4}$ **27.** 6   **29.** $\dfrac{8}{9}$ **31.** not a real number   **33.** 6   **35.** -4 **37.** -8 **39.** -3
41. not a real number   **43.** 2   **45.** -9 **47.** $|x|$ **49.** x **51.** x^5 **53.** 4 and -4 **54.** 4   **55.** 4, -4; 4
56. $\{4, -4\}$ **57.** 3 and -3 **58.** 3   **59.** 3, -3; 3   **60.** $\{3, -3\}$ **63. (a)** not a real number   **(b)** negative

(c) zero   **65.** 5; 12; 3; 4; 625   **67.** $\sqrt{12}$ **69.** $(\sqrt[4]{8})^3$ **71.** $(\sqrt[8]{9q})^5$ **73.** $\dfrac{1}{(\sqrt{2m})^3}$ **75.** $(\sqrt[3]{2y + x})^2$

77. $\dfrac{1}{(\sqrt[3]{3m^4 + 2k^2})^2}$ **79.** $\sqrt{a^2 + b^2} = \sqrt{3^2 + 4^2} = 5$; $a + b = 3 + 4 = 7$; $5 \neq 7$ **81.** 64   **83.** 64   **85.** x^{10}
87. $\sqrt[6]{x^5}$ **89.** $y\sqrt{7y}$ **91.** $\sqrt[15]{t^8}$ **93.** 10 miles   **95.** 1.6 seconds   **97.** $\dfrac{1}{x^{11}}$ **99.** $r^8 s^{12}$ **101.** $\dfrac{-6x^3}{y^{13}}$

SECTION   5.2   (PAGE 271)

EXERCISES   3. The product rule for exponents applies only if the bases are the same.   **5.** 9   **7.** 4   **9.** y
11. $k^{2/3}$ **13.** $9x^8 y^{10}$ **15.** $\dfrac{1}{x^{10/3}}$ **17.** $\dfrac{1}{m^{1/4} n^{3/4}}$ **19.** $\dfrac{c^{11/3}}{b^{11/4}}$ **21.** $p + 2p^2$ **23.** $k^{7/4} - k^{3/4}$ **25.** $6 + 18a$
27. $x^{8/5} - 4x^{2/5}$ **29.** $y^{5a/6}$ **31.** $x^{3a/2}$ **33.** $x^{(a-b)/(ab)} \cdot y$ **35.** $x^{-1/2}$ **36.** $m^{5/2}$ **37.** $k^{-3/4}$ **38.** $x^{-1/2}(3 - 4x)$
39. $m^{5/2}(m^{1/2} - 3)$ **40.** $k^{-3/4}(9 + 2k^{1/2})$ **41.** $t^{-1/2}(4 + 7t^2)$ **42.** $x^{-1/3}(8x - 5)$
43. In the definition of $a^{1/n}$, if n is even, we must have $a \geq 0$. Thus $(-1)^{2/2} = [(-1)^2]^{1/2}$ is not a valid step.
45. For example, let $x = 3$ and $y = 4$. $(x^2 + y^2)^{1/2} = (3^2 + 4^2)^{1/2} = 25^{1/2} = 5$; $x + y = 3 + 4 = 7$; $5 \neq 7$ **47.** $x^{17/20}$
49. $\dfrac{1}{x^{3/2}}$ **51.** $y^{5/6} z^{1/3}$ **53.** $m^{1/12}$ **55. (a)**   **57. (c)**   **59. (c)**   **61.** 3, 7   **63.** 97.381   **65.** 16.863
67. 2.646   **69.** -4.359 **71.** -9.055 **73.** 7.507   **75.** 3.162   **77.** 1.885   **79.** 15.155   **81.** .272
87. 392,000 square miles   **89.** r^3 **91.** $\dfrac{q^8}{112}$

SECTION 5.3 (PAGE 280)

◆ **CONNECTIONS** **Page 279:** no; no

EXERCISES **1.** true; Both are equal to $4\sqrt{3}$ and approximately 6.92820323. **3.** true; Both are equal to $6\sqrt{2}$ and approximately 8.485281374. **5.** (d) **7.** $\sqrt{30}$ **9.** $\sqrt[3]{14xy}$ **11.** $\sqrt[4]{33}$ **13.** $\dfrac{8}{11}$ **15.** $\dfrac{\sqrt{3}}{5}$ **17.** $\dfrac{\sqrt{x}}{5}$

19. $\dfrac{p^3}{9}$ **21.** $\dfrac{3}{4}$ **23.** $-\dfrac{\sqrt[3]{r^2}}{2}$ **25.** $2\sqrt{7}$ **27.** $-4\sqrt{2}$ **29.** $10\sqrt{3}$ **31.** $4\sqrt[3]{2}$ **33.** $-2\sqrt[3]{2}$ **35.** $2\sqrt[3]{5}$

37. $-4\sqrt[4]{2}$ **39.** $2\sqrt[5]{2}$ **41.** His reasoning was incorrect. Here 8 is a term and not a factor. **43.** $6k\sqrt{2}$ **45.** $\dfrac{3\sqrt[3]{3}}{4}$

47. $11x^3$ **49.** $-3t^4$ **51.** $-10m^4z^2$ **53.** $5a^2b^3c^4$ **55.** $\dfrac{1}{2}r^2t^5$ **57.** $-x^3y^4\sqrt{13x}$ **59.** $2z^2w^3$

61. $-2zt^2\sqrt[3]{2z^2t}$ **63.** $3x^3y^4$ **65.** $-3r^3s^2\sqrt[4]{2r^3s}$ **67.** $\dfrac{y^5\sqrt{y}}{6}$ **69.** $\dfrac{x^5\sqrt[3]{x}}{3}$ **71.** $4\sqrt{3}$ **73.** $\sqrt{5}$ **75.** $x^2\sqrt{x}$

77. $\sqrt[6]{432}$ **79.** $\sqrt[12]{6912}$ **81.** $\sqrt[6]{x^5}$ **83.** 5 **85.** $8\sqrt{2}$ **87.** $8\sqrt{5}$ feet; 17.9 feet **89.** .003
91. $22x^4 - 10x^3$ **93.** $8q^2 - 3q$

SECTION 5.4 (PAGE 285)

◆ **CONNECTIONS** **Page 283:** 1.618033989

EXERCISES **1.** (b) **3.** 15; Each radical represents a whole number. **5.** -4 **7.** $7\sqrt{3}$ **9.** $24\sqrt{2}$ **11.** 0
13. $20\sqrt{5}$ **15.** $12\sqrt{2x}$ **17.** $-11m\sqrt{2}$ **19.** $\sqrt[3]{2}$ **21.** $2\sqrt[3]{x}$ **23.** $19\sqrt[4]{2}$ **25.** $x\sqrt[4]{xy}$

27. $(4 + 3xy)\sqrt[3]{xy^2}$ **29.** $\dfrac{7\sqrt{2}}{6}$ **31.** $\dfrac{5\sqrt{2}}{3}$ **33.** $\dfrac{m\sqrt[3]{m^2}}{2}$ **35.** Both are approximately 11.3137085. **37.** Both are approximately 31.6227766. **39.** (a) **41.** $12\sqrt{5} + 5\sqrt{3}$ inches **43.** $58\sqrt{2} + 10\sqrt{3}$ centimeters

45. $10x^3y^4 - 20x^2y$ **47.** $a^4 - b^2$ **49.** $64x^9 + 144x^6 + 108x^3 + 27$ **51.** $\dfrac{4x - 5}{3x}$

SECTION 5.5 (PAGE 293)

◆ **CONNECTIONS** **Page 292:** **1.** $\dfrac{319}{6(8\sqrt{5} + 1)}$ **2.** $\dfrac{9a - b}{b(3\sqrt{a} - \sqrt{b})}$ **3.** $\dfrac{9a - b}{(\sqrt{b} - \sqrt{a})(3\sqrt{a} - \sqrt{b})}$
4. $\dfrac{(3\sqrt{a} + \sqrt{b})(\sqrt{b} + \sqrt{a})}{b - a}$

EXERCISES **1.** $\sqrt{ab}$ **2.** $x^2 - y^2$ **3.** $x^2 - y$ **4.** $x - y$ **5.** $x^2 + 2xy + y^2$ **6.** $x + 2\sqrt{xy} + y$
9. $6 - 4\sqrt{3}$ **11.** $6 - \sqrt{6}$ **13.** 2 **15.** 9 **17.** $3\sqrt{2} - 5\sqrt{3} + 2\sqrt{6} - 10$ **19.** $3x - 4$ **21.** $4x - y$
23. $16x + 24\sqrt{x} + 9$ **25.** $81 - \sqrt[3]{4}$ **29.** $\sqrt{7}$ **31.** $5\sqrt{3}$ **33.** $\dfrac{\sqrt{6}}{2}$ **35.** $\dfrac{9\sqrt{15}}{5}$ **37.** $-\sqrt{2}$ **39.** $\dfrac{\sqrt{14}}{2}$

41. $-\dfrac{\sqrt{14}}{10}$ **43.** $\dfrac{2\sqrt{6x}}{x}$ **45.** $\dfrac{-8\sqrt{3k}}{k}$ **47.** $\dfrac{-5m^2\sqrt{6mn}}{n^2}$ **49.** $\dfrac{12x^3\sqrt{2xy}}{y^5}$ **53.** $\dfrac{\sqrt[3]{18}}{3}$ **55.** $\dfrac{\sqrt[3]{12}}{3}$

57. $-\dfrac{\sqrt[3]{2pr}}{r}$ **59.** $\dfrac{2\sqrt[4]{x^3}}{x}$ **63.** $\dfrac{2(4 - \sqrt{3})}{13}$ **65.** $3(\sqrt{5} - \sqrt{3})$ **67.** $\sqrt{3} + \sqrt{7}$ **69.** $\sqrt{7} - \sqrt{6} - \sqrt{14} + 2\sqrt{3}$

71. $\dfrac{4\sqrt{x}(\sqrt{x} + 2\sqrt{y})}{x - 4y}$ **73.** $\dfrac{x - 2\sqrt{xy} + y}{x - y}$ **75.** Square both sides to show that each is equal to $\dfrac{2 - \sqrt{3}}{4}$.

77. $\dfrac{5 + 2\sqrt{6}}{4}$ **79.** $\dfrac{4 + 2\sqrt{2}}{3}$ **81.** $\dfrac{6 + 2\sqrt{6x}}{3}$ **83.** Each expression is approximately equal to .2588190451.

85. $\left\{\dfrac{3}{8}\right\}$ **87.** $\left\{-\dfrac{1}{3}, \dfrac{3}{2}\right\}$ **89.** $4x^2 + 20x + 25$ **91.** $x^4 + 2x^2 + 5$ **93.** true **95.** true

SECTION 5.6 (PAGE 300)

◆ **CONNECTIONS** **Page 296:** 62.5 meters; 155 meters

EXERCISES **1.** (a) yes (b) no **3.** (a) yes (b) no **5.** no; There is no solution. **7.** $\{19\}$ **9.** $\left\{\dfrac{38}{3}\right\}$
11. $\emptyset$ **13.** $\{5\}$ **15.** $\{1\}$ **17.** $\{9\}$ **19.** $\{3\}$ **20.** $\{-3\}$ **21.** $\{-3, 3\}$ **22.** $\{3\}$ **23.** $\{-3, 3\}$

24. (a) more than **(b)** the same as **25.** You cannot just square each term. The right-hand side should be $(8 - x)^2 = 64 - 16x + x^2$. **27.** $\{4\}$ **29.** $\{-3, -1\}$ **31.** $\emptyset$ **33.** $\{5\}$ **35.** $\{7\}$ **37.** $\emptyset$ **39.** 3 **41.** $\{-13\}$ **43.** $\{14\}$ **45.** $\emptyset$ **47.** $\{7\}$ **49.** $\{2, 14\}$ **51.** $\left\{\dfrac{1}{4}, 1\right\}$ **53.** 8 billion cubic feet; 16 billion cubic feet; 21 billion cubic feet; 24 billion cubic feet **55.** fairly good; 1976 **57.** 6 billion cubic feet; 12 billion cubic feet; 15 billion cubic feet; 16 billion cubic feet **59.** $(x - 6)^2$ **61.** $-14, 14$ **63.** $\{-6\}$ **65.** $\dfrac{6 - 3\sqrt{2}}{4}$

SECTION 5.7 (PAGE 308)

◆ **CONNECTIONS** **Page 304:** Answers will vary.

EXERCISES **1.** false **3.** true **5.** true **7.** $13i$ **9.** $-12i$ **11.** $i\sqrt{5}$ **13.** $4i\sqrt{3}$ **15.** -15 **17.** -10 **19.** $\sqrt{3}$ **21.** $5i$ **25.** $10 + 8i$ **27.** $-1 + 7i$ **29.** 0 **31.** $7 + 3i$ **33.** -2 **35.** $1 + 13i$ **37.** $6 + 6i$ **39.** $4 + 2i$ **41.** -81 **43.** -16 **45.** $-10 - 30i$ **47.** $10 - 5i$ **49.** $-9 + 40i$ **51.** $15 + 8i$ **53.** 153 **55.** 97 **57. (a)** $a - bi$ **(b)** a^2; b^2 **59.** $1 + i$ **61.** $-1 + 2i$ **63.** $2 + 2i$ **65.** $-\dfrac{5}{13} - \dfrac{12}{13}i$ **67. (a)** $4x + 1$ **(b)** $4 + i$ **68. (a)** $-2x + 3$ **(b)** $-2 + 3i$ **69. (a)** $3x^2 + 5x - 2$ **(b)** $5 + 5i$ **70. (a)** $-\sqrt{3} + \sqrt{6} + 1 - \sqrt{2}$ **(b)** $\dfrac{1}{5} - \dfrac{7}{5}i$ **73.** $\dfrac{5}{41} + \dfrac{4}{41}i$ **75.** -1 **77.** i **79.** $-i$ **83.** $\dfrac{1}{2} + \dfrac{1}{2}i$ **85.** $(1 + 5i)^2 - 2(1 + 5i) + 26$ will simplify to 0 when the operations are applied. **87.** $\left\{-\dfrac{13}{6}\right\}$ **89.** $\{-8, 5\}$

CHAPTER 5 REVIEW EXERCISES (PAGE 313)

1. (a) 4 **(b)** $\dfrac{4}{3}$ **(c)** $\sqrt{8}$ or $2\sqrt{2}$ **2. (a)** (b) **(b)** (a) **(c)** (c) **6.** integer—polynomial; rational numbers—rational expression; irrational numbers—radical expression **7.** 32 **9.** $-\dfrac{216}{125}$ **11.** $\dfrac{1000}{27}$ **15.** -17 **17.** -5 **19.** -2 **21. (a)** $|x|$ **(b)** not a real number **(c)** $-|x|$ **(d)** x **23.** $(\sqrt[5]{2k})^2$ or $\sqrt[5]{4k^2}$ **25.** $\dfrac{1}{(\sqrt[3]{3a + b})^5}$ or $\dfrac{1}{\sqrt[3]{(3a + b)^5}}$ **27.** $7^{9/2}$ **29.** $p^{4/5}$ **31.** 96 **33.** $\dfrac{1}{y^{1/2}}$ **35.** $r^{1/2} + r$ **37.** $r^{3/2}$ **39.** $k^{9/4}$ **41.** $z^{1/12}$ **43.** 6.164 **45.** 4.960 **47.** .009 **49.** $-.189$ **51.** $\sqrt{66}$ **53.** $\sqrt[3]{30}$ **55.** $2\sqrt{5}$ **57.** $-5\sqrt{5}$ **59.** $10y^3\sqrt{y}$ **61.** $3a^2b\sqrt[3]{4a^2b^2}$ **63.** $\dfrac{y\sqrt{y}}{12}$ **65.** $\dfrac{\sqrt[3]{r^2}}{2}$ **67.** $\sqrt{15}$ **69.** $\sqrt[12]{2000}$ **71.** 10 **73.** $23\sqrt{5}$ **75.** $26m\sqrt{6m}$ **77.** $-8\sqrt[4]{2}$ **79.** $\dfrac{16 + 5\sqrt{5}}{20}$ **81.** $17\sqrt{6} + 4\sqrt{5}$ meters **83.** 2 **85.** $15 - 2\sqrt{26}$ **87.** $2\sqrt[3]{2y^2} + 2\sqrt[3]{4y} - 3$ **89.** $\dfrac{\sqrt{30}}{5}$ **91.** $\dfrac{3\sqrt{7py}}{y}$ **93.** $-\dfrac{\sqrt[3]{45}}{5}$ **95.** $\dfrac{\sqrt{2} - \sqrt{7}}{-5}$ **99.** $\dfrac{1 - \sqrt{5}}{4}$ **101.** $2 + \sqrt{3k}$ **103.** $\{6\}$ **105.** $\{0, 5\}$ **107.** $\{3\}$ **109.** $\left\{-\dfrac{1}{2}\right\}$ **111.** 0 does not satisfy the equation. **113.** $10i\sqrt{2}$ **115.** $-10 - 2i$ **117.** $-\sqrt{35}$ **119.** 3 **121.** $32 - 24i$ **123.** $4 + i$ **125.** -1 **127.** $\dfrac{1}{100}$ **129.** k^6 **131.** $57\sqrt{2}$ **133.** $\sqrt{35} + \sqrt{15} - \sqrt{21} - 3$ **135.** $\dfrac{\sqrt[3]{60}}{5}$ **137.** $7i$ **139.** $-5i$ **141.** $\{5\}$ **143.** $\left\{\dfrac{3}{2}\right\}$

CHAPTER 5 TEST (PAGE 317)

[5.1] 1. $\dfrac{125}{64}$ **2.** $\dfrac{1}{256}$ **[5.2] 3.** $\dfrac{9y^{3/10}}{x^2}$ **4.** $7^{1/2}$ **5.** $a^3\sqrt[3]{a^2}$ or $a^{11/3}$ **[5.1] 6.** -29 **7.** 15 **[5.2] 8.** (c) **9.** 12.09 **[5.3] 10.** $3x^2y^3\sqrt{6x}$ **11.** $2ab^3\sqrt[4]{2a^3b}$ **12.** $\sqrt[6]{200}$ **[5.4] 13.** $26\sqrt{5}$ **[5.5] 14.** $66 + \sqrt{5}$ **15.** $23 - 4\sqrt{15}$ **16.** $-\dfrac{\sqrt{10}}{4}$ **17.** $\sqrt[3]{25}$ **18.** $-2(\sqrt{7} - \sqrt{5})$ **19.** 59.8 **[5.6] 20.** $\{1\}$ **21.** $\{6\}$ **[5.7] 22.** $-5 - 8i$ **23.** $3 + 4i$ **24.** $-2 + 16i$ **25.** i

CUMULATIVE REVIEW (Chapters 1–5) (PAGE 318)

1. 1   **2.** $-\dfrac{14}{9}$ **3.** $3x^2 + 5xy - 2y^2$ **4.** $\dfrac{4}{9}t^4 - 8t^2 + 36$ **5.** $x^2 - 4y^2$ **6.** $-10x^5y^4 + 25x^4y^4 - 15x^2y^3$

7. $4x^2 + 4x + 1$ **8.** $a^3 + 3a^2b + 3ab^2 + b^3$ **9.** $(2x^2 - 3)(4x^4 + 6x^2 + 9)$ **10.** $(12y + 7x)(12y - 7x)$

11. $(5x + 3)(2x - 1)$ **12.** $(2x + 3)(y + 4)$ **13.** $(10a + 1)(100a^2 - 10a + 1)$ **14.** $(6a - 7)^2$ **15.** $\dfrac{1}{9}$

16. $\dfrac{a^5}{b^3}$ **17.** $4x^2y\sqrt{3x}$ **18.** $7\sqrt{2}$ **19.** $\dfrac{\sqrt{10} + 2\sqrt{2}}{2}$ **20.** $-6x - 11\sqrt{xy} - 4y$ **21.** $z + y$ **22.** $\dfrac{2}{x - 3}$

23. $\dfrac{x - 6}{x + 3}$ **24.** $\dfrac{3}{a - 2}$ or $\dfrac{-3}{2 - a}$ **25.** $\dfrac{x^2 + 1}{3 - x^2}$ **26.** $\dfrac{1}{xy - 1}$ **27.** $\{-4\}$ **28.** $\{-12\}$ **29.** $\{6\}$

30. $\left\{-\dfrac{10}{3}, 1\right\}$ **31.** $\left\{\dfrac{1}{4}\right\}$ **32.** $\emptyset$ **33.** $\left\{-\dfrac{5}{4}, 2\right\}$ **34.** $\{3, 4\}$ **35.** $[-2, \infty)$ **36.** $\left(-\infty, \dfrac{1}{2}\right] \cup \left[\dfrac{7}{2}, \infty\right)$

37. 39.2 miles per hour   **38.** $\dfrac{80}{39}$ or $2\dfrac{2}{39}$ liters   **39.** Natalie: 8 miles per hour;   Chuck: 4 miles per hour

40. Both angles measure 80°.

CHAPTER 6   QUADRATIC EQUATIONS AND INEQUALITIES

SECTION 6.1 (PAGE 327)

◆ **CONNECTIONS** **Page 321:** 1950: 11.5 million;   1970: 22.7 million;   1980: 25 million;   1990: 25.1 million. From 1950 to 1980, membership increased by 14.5 million. From 1980 to 1990, it increased by only .1 million. **Page 325:** Original square: x^2;   each strip: x;   total area of the strips: $6x$;   each small square: 1;   total area of the small squares: 9;   area of new larger square is $(x + 3)^2$ or $x^2 + 6x + 9$.

EXERCISES **3.** $\{-9, 9\}$ **5.** $\{-\sqrt{17}, \sqrt{17}\}$ **7.** $\{-4\sqrt{2}, 4\sqrt{2}\}$ **9.** $\{-7, 3\}$ **11.** $\left\{\dfrac{1 + \sqrt{7}}{3}, \dfrac{1 - \sqrt{7}}{3}\right\}$

13. $\left\{\dfrac{-1 + 2\sqrt{6}}{4}, \dfrac{-1 - 2\sqrt{6}}{4}\right\}$ **15.** (b)   **17.** $\{-2i\sqrt{3}, 2i\sqrt{3}\}$ **19.** $\{5 + i\sqrt{3}, 5 - i\sqrt{3}\}$

21. $\left\{\dfrac{1 + 2i\sqrt{2}}{6}, \dfrac{1 - 2i\sqrt{2}}{6}\right\}$ **23.** Divide both sides by 2.   **25.** $\{-4, 6\}$ **27.** $\left\{-3, \dfrac{8}{3}\right\}$

29. $\left\{\dfrac{-5 + \sqrt{41}}{4}, \dfrac{-5 - \sqrt{41}}{4}\right\}$ **31.** $\{-2 + 3i, -2 - 3i\}$ **33.** $\left\{\dfrac{4 + \sqrt{3}}{3}, \dfrac{4 - \sqrt{3}}{3}\right\}$ **35.** $\left\{\dfrac{2 + \sqrt{3}}{3}, \dfrac{2 - \sqrt{3}}{3}\right\}$

37. $\left\{\dfrac{-2 + 2i\sqrt{2}}{3}, \dfrac{-2 - 2i\sqrt{2}}{3}\right\}$ **39.** $\{1 + \sqrt{2}, 1 - \sqrt{2}\}$ **41.** $\{-3 + i\sqrt{3}, -3 - i\sqrt{3}\}$ **45.** Some quadratic

polynomials cannot easily be factored.   **47.** $\{-\sqrt{b}, \sqrt{b}\}$ **49.** $\left\{\dfrac{-\sqrt{b^2 + 16}}{2}, \dfrac{\sqrt{b^2 + 16}}{2}\right\}$ **51.** $\left\{\dfrac{2b + \sqrt{3a}}{5}, \dfrac{2b - \sqrt{3a}}{5}\right\}$

53. 2117   **54.** 2243   **55.** 2356   **56.** We would evaluate P for $x = 3$.   **57.** $-6.5x^2 + 132.5x - 426 = 0$. Let $x = 4$ and show that 4 is a solution. This corresponds to the year 1992.   **58.** 1999   **59.** $\sqrt{13}$ **61.** 1

SECTION 6.2 (PAGE 336)

◆ **CONNECTIONS** **Page 336: 1.** Since $\dfrac{1 + \sqrt{41}}{5} + \dfrac{1 - \sqrt{41}}{5} = \dfrac{2}{5} = -\dfrac{b}{a}$ and $\dfrac{1 + \sqrt{41}}{5} \cdot \dfrac{1 - \sqrt{41}}{5} = -\dfrac{8}{5} = \dfrac{c}{a}$, the

solutions are correct.   **2.** Since $\dfrac{3i}{2} + (-4i) = -\dfrac{5i}{2} = -\dfrac{b}{a}$ and $\dfrac{3i}{2}(-4i) = 6 = \dfrac{c}{a}$, the solutions are correct.

EXERCISES **1.** In both cases, the solutions are -2 and 3.   **3.** $\{3, 5\}$ **5.** $\left\{\dfrac{-2 + \sqrt{2}}{2}, \dfrac{-2 - \sqrt{2}}{2}\right\}$

7. $\left\{\dfrac{1 + \sqrt{3}}{2}, \dfrac{1 - \sqrt{3}}{2}\right\}$ **9.** $\{5 + \sqrt{7}, 5 - \sqrt{7}\}$ **11.** $\left\{\dfrac{-2 + \sqrt{10}}{2}, \dfrac{-2 - \sqrt{10}}{2}\right\}$ **13.** $\{-1 + 3\sqrt{2}, -1 - 3\sqrt{2}\}$

15. $\left\{-1, \dfrac{2}{3}\right\}$ **17.** $\left\{\dfrac{-1 + \sqrt{2}}{2}, \dfrac{-1 - \sqrt{2}}{2}\right\}$ **19.** $\left\{\dfrac{1 + \sqrt{29}}{2}, \dfrac{1 - \sqrt{29}}{2}\right\}$ **21.** $\left\{\dfrac{-1 + \sqrt{7}}{3}, \dfrac{-1 - \sqrt{7}}{3}\right\}$

23. $\left\{\dfrac{-4 + \sqrt{91}}{3}, \dfrac{-4 - \sqrt{91}}{3}\right\}$ **29.** $\{-i\sqrt{47}, i\sqrt{47}\}$ **31.** $\{3 + i\sqrt{5}, 3 - i\sqrt{5}\}$ **33.** $\left\{\dfrac{1 + i\sqrt{6}}{2}, \dfrac{1 - i\sqrt{6}}{2}\right\}$

35. $\left\{\dfrac{-2 + i\sqrt{2}}{3}, \dfrac{-2 - i\sqrt{2}}{3}\right\}$ **37.** $\{4 + 3i\sqrt{2}, 4 - 3i\sqrt{2}\}$ **39.** $\left\{\dfrac{-1 + i\sqrt{3}}{2}, \dfrac{-1 - i\sqrt{3}}{2}\right\}$ **41.** $\left\{0, \dfrac{4}{3}i\right\}$

43. $\left\{-2i, \dfrac{1}{2}i\right\}$ **45.** 3.6 hours   **47.** Rusty: 25.0 hours;   Nancy: 23.0 hours   **49.** 2.4 seconds and 5.6 seconds
51. It reaches its maximum height at 5 seconds, since this is the only time it reaches 400 feet.   **53.** (d)   **55.** (a)
57. (b)   **59.** (c)   **61.** -10 or 10 **63.** 16   **65.** 25   **69.** The discriminant is 74^2, so it can be factored;
$(6x + 5)(4x - 9)$ **71.** The discriminant, 3897, is not a perfect square, so it cannot be factored.   **73.** The discriminant
is 85^2, so it can be factored; $(12x + 1)(x - 7)$ **75.** $[x - (1 + 5i)][x - (1 - 5i)] = 0$
76. $[(x - 1) - 5i][(x - 1) + 5i] = 0$ **77.** $x^2 - 2x + 26 = 0$ **78.** Use $a = 1$, $b = -2$, and $c = 26$. The solutions are
$1 + 5i$ and $1 - 5i$.   **79.** $b = \dfrac{44}{5}$; $x_2 = \dfrac{3}{10}$ **81.** $u^2 + 4u - 5$; $(u + 5)(u - 1)$ **83.** $\{-8\}$ **85.** $\{5\}$

SECTION 6.3 (PAGE 345)

◆ **CONNECTIONS**   **Page 340:** $x = 2$

EXERCISES   **3.** $\{-2, 7\}$ **5.** $\{-4, 7\}$ **7.** $\left\{-\dfrac{2}{3}, 1\right\}$ **9.** $\left\{-\dfrac{14}{17}, 5\right\}$ **11.** $\left\{-\dfrac{11}{7}, 0\right\}$

13. $\left\{\dfrac{-1 + \sqrt{13}}{2}, \dfrac{-1 - \sqrt{13}}{2}\right\}$ **15.** $\dfrac{1}{m}$ job per hour   **17.** 25 miles per hour   **19.** 80 kilometers per hour

21. 9 minutes   **23.** $\{3\}$ **25.** $\left\{\dfrac{8}{9}\right\}$ **27.** $\{16\}$ **29.** $\left\{\dfrac{2}{5}\right\}$ **31.** $\{-3, 3\}$ **33.** $\left\{-\dfrac{3}{2}, -1, 1, \dfrac{3}{2}\right\}$ **35.** $\{-6, -5\}$

37. $\{-4, 1\}$ **39.** $\left\{-\dfrac{1}{3}, \dfrac{1}{6}\right\}$ **41.** $\left\{-\dfrac{1}{2}, 3\right\}$ **43.** $\{-8, 1\}$ **45.** $\{25\}$ **47.** $\left\{-1, 1, -\dfrac{\sqrt{6}}{2}i, \dfrac{\sqrt{6}}{2}i\right\}$

51. $\{-2\sqrt{3}, -2, 2, 2\sqrt{3}\}$ **53.** $\{3, 11\}$ **55.** $\left\{-\sqrt[3]{5}, -\dfrac{\sqrt[3]{4}}{2}\right\}$ **57.** $\left\{-\dfrac{1}{2}, 3\right\}$ **59.** It would cause both denominators to

be 0, and division by 0 is undefined.   **60.** The solution is $\dfrac{12}{5}$.   **61.** $\left(\dfrac{x}{x - 3}\right)^2 + 3\left(\dfrac{x}{x - 3}\right) - 4 = 0$ **63.** $\left\{\dfrac{12}{5}\right\}$;   The

values for t are -4 and 1. The value 1 is impossible because it leads to a contradiction (since $\dfrac{x}{x - 3}$ is never equal to 1).

64. $\left\{\dfrac{12}{5}\right\}$;   The values for s are $\dfrac{1}{x}$ and $\dfrac{-4}{x}$. The value $\dfrac{1}{x}$ is impossible, since $\dfrac{1}{x} \neq \dfrac{1}{x - 3}$ for all x.

65. $W = \dfrac{P - 2L}{2}$ or $W = \dfrac{P}{2} - L$ **67.** $C = \dfrac{5}{9}(F - 32)$

SECTION 6.4 (PAGE 352)

◆ **CONNECTIONS**   **Page 350:** approximately 46 feet

EXERCISES   **1.** $m = \sqrt{p^2 - n^2}$ **3.** $t = \dfrac{\pm\sqrt{dk}}{k}$ **5.** $d = \dfrac{\pm\sqrt{skI}}{I}$ **7.** $v = \dfrac{\pm\sqrt{kAF}}{F}$ **9.** $r = \dfrac{\pm\sqrt{3\pi Vh}}{\pi h}$

11. $t = \dfrac{-B \pm \sqrt{B^2 - 4AC}}{2A}$ **13.** $h = \dfrac{D^2}{k}$ **15.** $\ell = \dfrac{p^2 g}{k}$ **19.** 5.2 seconds   **21.** Find s when $t = 0$.   **23.** 3 minutes

25. 3.4 seconds   **27.** 2.3, 5.3, 5.8   **29.** 412.3 feet   **31.** eastbound ship: 80 miles;   southbound ship: 150 miles
33. 5 centimeters, 12 centimeters, 13 centimeters   **35.** length: 2 centimeters;   width: 1.5 centimeters   **37.** 1 foot
39. length: 26 meters;   width: 16 meters   **41.** 1955   **43.** .035 or 3.5%   **45.** \$.80   **47.** 5 or 14

49. $R = \dfrac{E^2 - 2pr \pm E\sqrt{E^2 - 4pr}}{2p}$ **51.** $r = \dfrac{5pc}{4}$ or $r = -\dfrac{2pc}{3}$ **53.** $I = \dfrac{-cR \pm \sqrt{c^2R^2 - 4cL}}{2cL}$

55. [diagram: number line with bracket segment from 1 to 5]   **57.** [diagram: number line]   **59.** $\left(-\dfrac{3}{2}, \infty\right)$

SECTION 6.5 (PAGE 361)

◊ **CONNECTIONS   Page 358:** The solutions of $-.011x^2 + 1.22x - 23.5 = 0$ are, to the nearest whole numbers, 25 and 86. The solutions to $-.011x^2 + 1.22x - 8.5 \geq 15$ are the years between year 25 and year 86, or between 1955 and 2016. The larger endpoint is not appropriate since we have not yet reached 2016. It should be the current year (in terms of this application).

EXERCISES   3. $(-\infty, -4] \cup [3, \infty)$ **5.** $(-\infty, -1) \cup (5, \infty)$ **7.** $(-4, 6)$

9. $(-\infty, 1] \cup [3, \infty)$ **11.** $\left(-\infty, -\dfrac{3}{2}\right] \cup \left[\dfrac{3}{5}, \infty\right)$ **13.** $\left(-\dfrac{2}{3}, \dfrac{1}{3}\right)$

15. $\left(-\infty, -\dfrac{1}{2}\right] \cup \left[\dfrac{1}{3}, \infty\right)$ **17.** $(-\infty, 3 - \sqrt{3}] \cup [3 + \sqrt{3}, \infty)$

19. $(-\infty, \infty)$ **21.** $\emptyset$ **23.** $(-\infty, 1) \cup (2, 4)$ **25.** $\left[-\dfrac{3}{2}, \dfrac{1}{3}\right] \cup [4, \infty)$

27. $(-\infty, 1) \cup (4, \infty)$ **29.** $\left[-\dfrac{3}{2}, 5\right)$ **31.** $(2, 6]$

33. $\left(-\infty, \dfrac{1}{2}\right) \cup \left(\dfrac{5}{4}, \infty\right)$ **35.** $[-4, -2)$ **37.** $\left(0, \dfrac{1}{2}\right) \cup \left(\dfrac{5}{2}, \infty\right)$

39. $\left[\dfrac{3}{2}, \infty\right)$ **41.** $\left(-2, \dfrac{5}{3}\right) \cup \left(\dfrac{5}{3}, \infty\right)$

43. 3 seconds and 13 seconds   **44.** between 3 seconds and 13 seconds   **45.** at 0 seconds (the time when it is initially projected) and at 16 seconds (the time when it hits the ground)   **46.** between 0 and 3 seconds and also between 13 and 16 seconds   **47.** 12   **49.** 0

CHAPTER 6 REVIEW EXERCISES (PAGE 366)

1. $\{-11, 11\}$ **3.** $\left\{-\dfrac{15}{2}, \dfrac{5}{2}\right\}$ **5.** $\{-2 + \sqrt{19}, -2 - \sqrt{19}\}$ **7.** $\left\{-\dfrac{7}{2}, 3\right\}$ **9.** $\left\{\dfrac{1 + \sqrt{41}}{2}, \dfrac{1 - \sqrt{41}}{2}\right\}$

11. $\left\{\dfrac{2 + i\sqrt{2}}{3}, \dfrac{2 - i\sqrt{2}}{3}\right\}$ **13.** $\{(-2 + \sqrt{5})i, (-2 - \sqrt{5})i\}$ **15.** 4 seconds and 12 seconds   **17.** 5.5 hours and

6.5 hours   **19.** (c)   **21.** (d)   **23.** The discriminant is 34^2, so it can be factored; $(6x - 5)(4x - 9)$.   **25.** $\left\{-\dfrac{5}{2}, 3\right\}$

27. $\left\{-\dfrac{11}{6}, -\dfrac{19}{12}\right\}$ **29.** $\left\{-\dfrac{343}{8}, 64\right\}$ **31.** 40 miles per hour   **33.** Because x appears on the left side alone, and because

it is equal to the nonnegative square root of $2x + 4$, it cannot be negative.   **35.** $d = \dfrac{\pm\sqrt{SkI}}{I}$

37. $R = \dfrac{-\pi H \pm \sqrt{\pi^2 H^2 + 2\pi S}}{2\pi}$ **39.** .87 second   **41.** 20 feet, 21 feet, and 29 feet   **43.** 68 and 69   **45.** 1.1 seconds

and 2.9 seconds   **47.** 4.5%   **49.** $\left(-\infty, -\dfrac{3}{2}\right) \cup (4, \infty)$ **51.** $\left(-\infty, -\dfrac{1}{2}\right) \cup (3, \infty)$

53. $(-\infty, -5] \cup [-2, 3]$ **55.** $R = \dfrac{\pm\sqrt{Vh - r^2 h}}{h}$ **57.** $\{-2, -1, 3, 4\}$ **59.** $\left\{\dfrac{-11 + \sqrt{7}}{3}, \dfrac{-11 - \sqrt{7}}{3}\right\}$

61. $\left\{\dfrac{3}{5}, 1\right\}$ **63.** $\left(-5, -\dfrac{23}{5}\right]$ **65.** $\{-i, i, -1, 1\}$ **67.** $\{-11i\sqrt{2}, 11i\sqrt{2}\}$ **71. (a)** $\{-2\}$

(b) $(-\infty, -2)$ **(c)** $(-2, \infty)$ **72. (a)** $\{1, 5\}$

(b) $(-\infty, 1) \cup (5, \infty)$ **(c)** $(1, 5)$ **73. (a)** $\{4\}$

(b) $(2, 4)$ **(c)** $(-\infty, 2) \cup (4, \infty)$

74. $(-\infty, \infty)$ **75.** $(-\infty, \infty)$ **76.** 2; $(-\infty, 2) \cup (2, \infty)$ **77.** $(-\infty, \infty)$; denominator **78.** $(-5, 3)$

CHAPTER 6 TEST (PAGE 369)

[6.1] 1. $\{-3\sqrt{6}, 3\sqrt{6}\}$ **2.** $\left\{-\dfrac{8}{7}, \dfrac{2}{7}\right\}$ **3.** $\{-1 + \sqrt{2}, -1 - \sqrt{2}\}$ **[6.2] 4.** $\left\{\dfrac{3 + \sqrt{17}}{4}, \dfrac{3 - \sqrt{17}}{4}\right\}$

5. $\left\{\dfrac{2 + i\sqrt{11}}{3}, \dfrac{2 - i\sqrt{11}}{3}\right\}$ **[6.3] 6.** $\left\{\dfrac{2}{3}\right\}$ **[6.4] 7.** Maretha: 11.1 hours; Lillaana: 9.1 hours

[6.1] 8. (a) **[6.2] 9.** two irrational solutions **10.** two rational solutions **[6.1–6.3] 11.** $\left\{-\dfrac{2}{3}, 6\right\}$

12. $\left\{\dfrac{-7 + \sqrt{97}}{8}, \dfrac{-7 - \sqrt{97}}{8}\right\}$ **13.** $\left\{-2, -\dfrac{1}{3}, \dfrac{1}{3}, 2\right\}$ **14.** $\left\{-\dfrac{5}{2}, 1\right\}$ **[6.4] 15.** 7 miles per hour

16. $r = \dfrac{\pm\sqrt{\pi S}}{2\pi}$ **17.** 2 feet **18.** 16 meters **[6.5] 19.** $(-\infty, -5) \cup \left(\dfrac{3}{2}, \infty\right)$

20. $(-\infty, 4) \cup [9, \infty)$

CUMULATIVE REVIEW (Chapters 1–6) (PAGE 370)

1. $-2, 0, 7$ **2.** $-\dfrac{7}{3}, -2, 0, .7, 7, \dfrac{32}{3}$ **3.** all except $\sqrt{-8}$ **4.** All are complex numbers. **5.** 6 **6.** 41 **7.** $\left\{\dfrac{4}{5}\right\}$

8. $\left\{-\dfrac{4}{3}, \dfrac{14}{3}\right\}$ **9.** $\left\{\dfrac{11}{10}, \dfrac{7}{2}\right\}$ **10.** $\left\{\dfrac{2}{3}\right\}$ **11.** $\emptyset$ **12.** $\left\{-\dfrac{1}{2}, \dfrac{2}{5}\right\}$ **13.** $\left\{\dfrac{7 + \sqrt{177}}{4}, \dfrac{7 - \sqrt{177}}{4}\right\}$

14. $\{-2, -1, 1, 2\}$ **15.** $[1, \infty)$ **16.** $\left[2, \dfrac{8}{3}\right]$ **17.** $\left(-\infty, -\dfrac{9}{4}\right) \cup \left(\dfrac{5}{4}, \infty\right)$ **18.** $(1, 3)$ **19.** $(-2, 1)$ **20.** $(-\infty, \infty)$

21. $14x^2 - 13x - 12$ **22.** $\dfrac{4}{9}t^2 + 12t + 81$ **23.** $-3t^3 + 5t^2 - 12t + 15$ **24.** $4x^2 - 6x + 11 + \dfrac{4}{x + 2}$

25. $x(4 + x)(4 - x)$ **26.** $(4m - 3)(6m + 5)$ **27.** $(10p^2 - 1)(100p^4 + 10p^2 + 1)$ **28.** $9(x - 1)(x + 1)$

29. $(2x + 3y)(4x^2 - 6xy + 9y^2)$ **30.** $(3x - 5y)^2$ **31.** $\dfrac{x - 5}{x + 5}$ **32.** $-\dfrac{5}{18}$ **33.** $-\dfrac{8}{k}$ **34.** $\dfrac{r - s}{r}$ **35.** $\dfrac{3\sqrt[3]{4}}{4}$

36. $\sqrt{7} + \sqrt{5}$ **37.** 7 inches by 3 inches **38.** biking: 12 miles per hour; walking: 2 miles per hour **39.** southbound car: 57 miles; eastbound car: 76 miles **40.** 40 miles per hour

CHAPTER 7 THE STRAIGHT LINE

SECTION 7.1 (PAGE 384)

◈ **CONNECTIONS** **Page 373:** Answers will vary. **Page 382: 1.** the figure on the right **2.** $y = -\dfrac{x}{2}$

EXERCISES **1.** origin **3.** y; x **5.** two **7. (a)** I **(b)** III **(c)** II **(d)** IV **(e)** none **9. (a)** I or III **(b)** II or IV **(c)** II or IV **(d)** I or III **11–20.**

21. -3; 3; 2; -1 **23.** $\dfrac{5}{2}$; 5; $\dfrac{3}{2}$; 1 **25.** -4; 5; $-\dfrac{12}{5}$; $\dfrac{5}{4}$ **29.** $(6, 0)$; $(0, 4)$ **31.** $(6, 0)$; $(0, -2)$

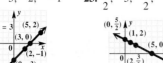   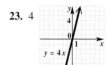

33. $(3, 0)$; $\left(0, -\dfrac{9}{7}\right)$ **35.** none; $(0, 5)$ **37.** $(2, 0)$; none   **39.** $(0, 0)$; $(0, 0)$ **41.** $\sqrt{17}$ **43.** $\sqrt{37}$ **45.** $6\sqrt{2}$

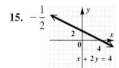   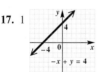   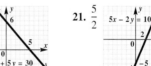

47. $\sqrt{5y^2 - 2xy + x^2}$ **49.** $d = [(x_2 - x_1)^2 + (y_2 - y_1)^2]^{1/2}$ **51.** $2\sqrt{106} + 4\sqrt{2}$ **53.** $\left(-\dfrac{1}{2}, 6\right)$ **55.** $\left(-\dfrac{3}{2}, \dfrac{3}{2}\right)$

57. $(2, -3)$ **59. (a)** between 1989 and 1990 **(b)** between 1991 and 1992 **(c)** 1991   **61.** 218.753 miles per hour
63. (c)   **65.** (d)   **67.** If $x = 0$, then $3x = 0$. It is a true statement.   **68.** If $x > 3$, then $x > 4$. It is not necessarily true—for
example, if $x = 3\frac{1}{2}$, $x > 3$ but $x \not> 4$.   **69.** the triangle is a right triangle   **70.**

71. 15   **72.** $6\sqrt{5}$ **73.** $3\sqrt{5}$ **74.** The longest side joins the vertices $(3, 2)$ and $(12, -10)$.   **75.** $15^2 = (\sqrt{180})^2 + (\sqrt{45})^2$
is a true statement, so the triangle is a right triangle.   **76.** $(\sqrt{32})^2 + (\sqrt{106})^2 \neq (\sqrt{106})^2$ **77.** 2   **79.** $\dfrac{5}{2}$ **81.** 0

SECTION 7.2 (PAGE 397)

◈ **CONNECTIONS**   **Page 391:** 48 miles per hour   **Page 396:** Answers will vary.

EXERCISES   **1.** (a), (b), (c), (d), (f)   **3.** 8   **5.** $\dfrac{5}{6}$ **7.** 0   **9.** positive   **11.** negative   **13.** undefined

15. $-\dfrac{1}{2}$ 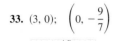   **17.** 1 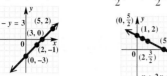   **19.** $-\dfrac{6}{5}$ **21.** $\dfrac{5}{2}$ **23.** 4

25. 0   **27.** x; y **29.** 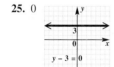   **31.** 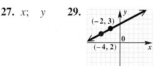   **33.**

35.   **37.**   **39.** $-\dfrac{4}{9}$; $\dfrac{9}{4}$ **41.** parallel   **43.** not parallel   **45.** parallel

47. perpendicular   **49.** not perpendicular   **51.** not perpendicular   **53.** $\dfrac{7}{10}$ **55. (a)** \$.92 **(b)** It means an *increase* in
price.   **57.** 6.25 billion minutes per year   **59.** \$1.40 per mile   **61.** 8   **63.** 1.75   **65.** Since the slopes of both pairs of
opposite sides are equal, the figure is a parallelogram.   **67.** $\sqrt{10}$ **68.** $\sqrt{10}$ **69.** $2\sqrt{10}$ **70.** $\sqrt{10} + \sqrt{10} = 2\sqrt{10}$ is
true.   **71.** $\dfrac{1}{3}$ **72.** $\dfrac{1}{3}$ **73.** $\dfrac{1}{3}$ **74.** $\dfrac{1}{3} = \dfrac{1}{3} = \dfrac{1}{3}$ is true.   **75.** They are collinear.   **76.** They are not collinear.

Other correct forms are possible for the answers in Exercises 77–81.

77. $y = \dfrac{-3x + 8}{2}$ **79.** $y = 4x + 14$ **81.** $5x - 3y = -17$

SECTION 7.3 (PAGE 409)

◆ **CONNECTIONS** **Page 404:** approximately 12 million; 1992; Since $-.02$ is close to 0, the number of PCs doubles each year in the indicated years.

EXERCISES **1.** (a) **3.** (c) **5.** (b) **7.** (d) **9.** (c) **11.** (a) **13.** (d) **15.** (b) **17.** $3x + 4y = 10$
19. $2x + y = 18$ **21.** $x - 2y = -13$ **23.** $y = 12$ **25.** $4x - y = 12$ **29.** $x = 9$ **31.** $x = .5$ **33.** $y = 8$
35. $2x - y = 2$ **37.** $x + 2y = 8$ **39.** $2x - 13y = -6$ **41.** $y = 5$ **43.** $x = 7$ **45.** $\sqrt{5}x - 2y = -\sqrt{5}$
47. $y = 5x + 15$ **49.** $y = -\dfrac{2}{3}x + \dfrac{4}{5}$ **51.** $y = \dfrac{2}{5}x + 5$ **53.** (a) $y = -x + 12$ (b) -1 (c) $(0, 12)$
55. (a) $y = -\dfrac{5}{2}x + 10$ (b) $-\dfrac{5}{2}$ (c) $(0, 10)$ **57.** (a) $y = \dfrac{2}{3}x - \dfrac{10}{3}$ (b) $\dfrac{2}{3}$ (c) $\left(0, -\dfrac{10}{3}\right)$ **59.** $3x - y = 19$
61. $x - 2y = 2$ **63.** $x + 2y = 18$ **65.** $y = 7$ **67.** $y = -22.125x + 220$ **69.** $y = -\dfrac{791}{6}x + 29{,}750$
71. $y = 4x - 160$ **73.** (a) $-3x + 9 = 0$ (b) $\{3\}$ (c) **75.** (a) $4x + 2 = 0$ (b) $\{-.5\}$

(c) 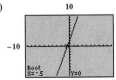 **77.** (d) **79.** 32; 212 **80.** $(0, 32)$ and $(100, 212)$ **81.** $\dfrac{9}{5}$ **82.** $F = \dfrac{9}{5}C + 32$

83. $C = \dfrac{5}{9}(F - 32)$ **84.** When the Celsius temperature is 50°, the Fahrenheit temperature is 122°. **85.** $(-\infty, 2)$ **87.** $\left(-\infty, -\dfrac{4}{3}\right]$

SECTION 7.4 (PAGE 418)

◆ **CONNECTIONS** **Page 416: 1.** $x \leq 200,\ x \geq 100,\ y \geq 3000$ **2.** 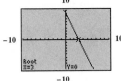 **3.** $C = 50x + 100y$

4. Some examples are $(100, 5000)$, $(150, 3000)$, and $(150, 5000)$. The corner points are $(100, 3000)$ and $(200, 5000)$.
5. The least cost occurs when $x = 100$ and $y = 3000$. The company should use 100 workers and manufacture 3000 units to achieve the lowest possible cost.

EXERCISES **1.** solid **3.** dashed **5.** **7.** **9.**

11. **13.** **15.** **17.** **19.**

21.

23. $-3 < x < 3$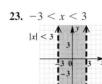

25. $-2 < x + 1 < 2$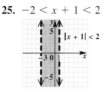

27.

29.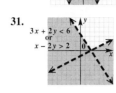

31.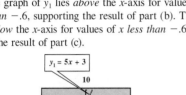

33. (a) $\{-4\}$ **(b)** $(-\infty, -4)$ **(c)** $(-4, \infty)$
35. (a) $\{3.5\}$ **(b)** $(3.5, \infty)$ **(c)** $(-\infty, 3.5)$

37. (a) $\{-.6\}$ **(b)** $(-.6, \infty)$ **(c)** $(-\infty, -.6)$ The graph of $y_1 = 5x + 3$ has x-intercept $-.6$, supporting the result of part (a). The graph of y_1 lies *above* the x-axis for values of x *greater than* $-.6$, supporting the result of part (b). The graph of y_1 lies *below* the x-axis for values of x *less than* $-.6$, supporting the result of part (c).

39. (a) $\{-1.2\}$ **(b)** $(-\infty, -1.2]$ **(c)** $[-1.2, \infty)$ The graph of $y_1 = -8x - (2x + 12)$ has x-intercept -1.2, supporting the result of part (a). The graph of y_1 lies *above or on* the x-axis for values of x *less than or equal to* -1.2, supporting the result of part (b). The graph of y_1 lies *below or on* the x-axis for values of x *greater than or equal to* -1.2, supporting the result of part (c).

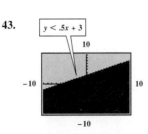

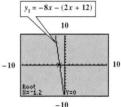

41.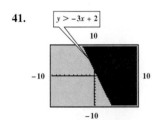

43.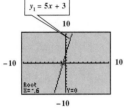

45. -9 **47.** 1 **49.** $y = \dfrac{3}{7}x - \dfrac{8}{7}$ **51.** $y = \dfrac{1}{8}x - \dfrac{5}{4}$

SECTION 7.5 (PAGE 428)

◆ **CONNECTIONS** **Page 423:** Answers will vary.

EXERCISES **3.** function; domain: $\{5, 3, 4, 7\}$; range: $\{1, 2, 9, 3\}$ **5.** not a function; domain: $\{2, 0\}$; range: $\{4, 2, 6\}$ **7.** not a function; domain: $\{1, 2, 3, 5\}$; range: $\{10, 15, 19, -27\}$ **9.** function; domain: $(-\infty, \infty)$; range: $(-\infty, 4]$ **11.** not a function; domain: $[-4, 4]$; range: $[-3, 3]$ **13.** function; domain: $(-\infty, \infty)$ **15.** not a function; domain: $[0, \infty)$ **17.** not a function; domain: $(-\infty, \infty)$ **19.** function; domain: $[0, \infty)$ **21.** function; domain: $(-\infty, 0) \cup (0, \infty)$ **23.** function (also a linear function); domain: $(-\infty, \infty)$ **25.** function; domain: $\left[-\dfrac{1}{2}, \infty\right)$

27. function; domain: $(-\infty, -3) \cup (-3, 3) \cup (3, \infty)$ **29.** 4 **31.** $3x + 4$ **33.** -59 **35.** -31

37. $-k^2 + 4k + 1$ **39.** -8 **41.** No—in general, $f(g(x)) \neq g(f(x))$. **43. (a)** $f(x) = \dfrac{12 - x}{3}$ **(b)** 3

45. (a) $f(x) = 3 - 2x^2$ **(b)** -15 **47. (a)** $f(x) = \dfrac{8 - 4x}{-3}$ **(b)** $\dfrac{4}{3}$ **49.** line; -2; $-2x + 4$; -2; 3; -2

51. (d) **53.** domain: $(-\infty, \infty)$; range: $(-\infty, \infty)$ **55.** domain: $(-\infty, \infty)$; range: $(-\infty, \infty)$

57. domain: $(-\infty, \infty)$;   range: $(-\infty, \infty)$ **59.** domain: $(-\infty, \infty)$;   range: $\{5\}$ **61. (a)** \$0;   \$.50;   \$1.00;   \$1.50   **(b)** \$.50x

(c)

63. (a) yes   **(b)** {Oct. 12, 15, 16, 17, 18, 19, 22, 23, 24, 25}
(c) {\$39.50, \$38.00, \$39.00, \$37.00, \$34.00, \$28.00, \$29.00, \$30.00}   **(d)** \$30.00   **(e)** October 12;   October 22
65. 194.53 centimeters   **67.** 177.41 centimeters   **69.** 1.83 cubic meters   **71.** 4.11 cubic meters   **73. (a)** 29,562
(b) 29,316   **(c)** 28,947   **(d)** In 1986 (when $x = 2$), the number of post offices was approximately 29,439.   **75.** $f(3) = 7$
77. $f(x) = -3x + 5$ **79.** $\dfrac{1}{3}$ **81.** 3

SECTION 7.6   (PAGE 438)

◊ **CONNECTIONS**   **Page 433:** In the distance formula $d = rt$, distance varies directly as rate and time. In the simple interest
formula $I = prt$, interest earned varies directly as the principal, the interest rate, and the time.

EXERCISES   **1.** inverse   **3.** direct   **5.** joint   **7.** combined   **9.** 36   **11.** .625   **13.** $222\dfrac{2}{9}$

15. increases;   decreases   **19.** $\$1.09\dfrac{9}{10}$ **21.** 8 pounds   **23.** 800 gallons   **25.** $133\dfrac{1}{3}$ newtons per square centimeter
27. 3 footcandles   **29.** 100 cycles per second   **31.** 60 kilometers   **33.** 800 pounds   **35.** $.71\pi$ seconds
37. 480 kilograms   **39.** 25   **41.** 9   **43.** (0, 0), (1, 1.25)   **44.** 1.25   **45.** $y = 1.25x + 0$ or $y = 1.25x$
46. $a = 1.25, b = 0$ **47.** It is the price per gallon, and it is the slope of the line.   **48.** It can be written in the form $y = kx$
(where $k = a$). The value of a is called the constant of variation.   **49.** It means that 4.6 gallons cost \$5.75.   **50.** It means that
12 gallons cost \$15.00.   **51.** $\{-3\}$ **53.** $\{6\}$

CHAPTER 7 REVIEW EXERCISES   (PAGE 444)

1. $(0, 5)$; $\left(\dfrac{10}{3}, 0\right)$; $(2, 2)$; $\left(\dfrac{14}{3}, -2\right)$ **3.** $(3, 0)$; $(0, -4)$ **5.** $(10, 0)$; $(0, 4)$ **7.** $\sqrt{13}$

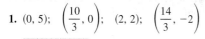

9. $\sqrt{65} + \sqrt{181} + 2\sqrt{17}$ **11.** $-\dfrac{7}{5}$ **13.** 2   **15.** undefined   **17.** $-\dfrac{1}{3}$ **19.** $-\dfrac{4}{3}$ **21.** positive   **23.** zero
25. \$1000 per year   **27.** $y = -2$ **29.** $3x - y = -7$ **31.** $9x + y = 13$ **33.** $4x - y = 29$ **35.** $4x + 3y = 6$
37. 233.4 million   **39.**   **41.**   **43.** domain: $\{-4, 1\}$;   range: $\{2, -2, 5, -5\}$;   not a function

45. domain: $[-4, 4]$;   range: $[0, 2]$;   function   **47.** not a function;   domain: $(-\infty, \infty)$ **49.** function;   domain: $\left[-\dfrac{7}{4}, \infty\right)$
51. function;   domain: $(-\infty, -6) \cup (-6, 6) \cup (6, \infty)$ **53.** -6 **55.** -8 **57.** -96
59. (a); $y = 3 - 2x^2$; $f(x) = 3 - 2x^2$; $f(3) = -15$ **63.** 15   **65.** .850 ohm   **67.** approximately 5600 pounds (to the
nearest hundred)   **69.** Because it falls from left to right, the slope is negative.
70. $-\dfrac{3}{2}$ **71.** $y = -\dfrac{3}{2}x + \dfrac{7}{2}$ **72.** $\left(\dfrac{7}{3}, 0\right)$ **73.** $\left(0, \dfrac{7}{2}\right)$ **74.** $f(x) = -\dfrac{3}{2}x + \dfrac{7}{2}$ **75.** $f(8) = -\dfrac{17}{2}$

76. $x = \dfrac{23}{3}$ **77.** 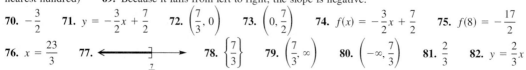   **78.** $\left\{\dfrac{7}{3}\right\}$ **79.** $\left(\dfrac{7}{3}, \infty\right)$ **80.** $\left(-\infty, \dfrac{7}{3}\right)$ **81.** $\dfrac{2}{3}$ **82.** $y = \dfrac{2}{3}x$

CHAPTER 7 TEST (PAGE 447)

[7.1] 1. $5\sqrt{5}$ **[7.2] 2.** $\dfrac{1}{2}$ **[7.1, 7.3] 3.** $\left(\dfrac{20}{3}, 0\right)$; $(0, -10)$ **4.** none; $(0, 5)$ **5.** $(2, 0)$;   none

[7.2] 6. It is a vertical line.   **7.** perpendicular   **8.** neither   **[7.3] 9.** $5x + y = 19$ **10.** $y = 14$
11. $3x + 5y = -11$ **12.** $x + 2y = -3$ **13.** $x + 2y = 4$ **14. (b)**   **15.** 11,989   **16. (a)** It is the slope of the line.
(b) It is the annual increase in the number of cases served.   **[7.4] 17.**   **18.**

[7.5] 19. (d)   **20. (d)**   **21.** 0   **22.** domain: $(-\infty, \infty)$;   range: $(-\infty, \infty)$ **[7.6] 23.** $\dfrac{5}{3}$ **24.** 200 amps   **25.** 8.5

CUMULATIVE REVIEW (Chapters 1–7) (PAGE 449)

1. $\{-8\}$ **2.** \$400   **3.** $(-6, \infty)$ **4.** $(-5, 4)$ **5.** $\left\{-\dfrac{3}{7}, 3\right\}$ **6.** $\left(-\infty, -\dfrac{7}{2}\right] \cup [2, \infty)$

7. $-k^3 - 3k^2 - 8k - 9$ **8.** $8x^2 + 17x - 21$ **9.** $z - 2 + \dfrac{3}{z}$ **10.** $3y^3 - 3y^2 + 4y + 1 + \dfrac{-10}{2y + 1}$

11. $(2p - 3q)(p - q)$ **12.** $(3k^2 + 4)(6k^2 - 5)$ **13.** $(x + 8)(x^2 - 8x + 64)$ **14.** $\left\{-3, -\dfrac{5}{2}\right\}$ **15.** $\left\{-\dfrac{2}{5}, 1\right\}$

16. $\dfrac{y}{y + 5}$ **17.** $\dfrac{4x + 2y}{(x + y)(x - y)}$ **18.** $-\dfrac{9}{4}$ **19.** $\dfrac{-1}{a + b}$ **20.** $\emptyset$ **21.** $m = \dfrac{5zy}{2y - z}$ **22.** $3\dfrac{3}{14}$ hours

23. 15 miles per hour   **24.** $\dfrac{1}{243}$ **25.** $x^{1/12}$ **26.** $8\sqrt{5}$ **27.** $\dfrac{-9\sqrt{5}}{20}$ **28.** $4(\sqrt{6} + \sqrt{5})$ **29.** $6\sqrt[3]{4}$ **30.** $\{6\}$

31. $\left\{-\dfrac{4}{3}, 0\right\}$ **32.** $\{-3, 5\}$ **33.** $\left\{\dfrac{2 + \sqrt{10}}{2}, \dfrac{2 - \sqrt{10}}{2}\right\}$ **34.** $\left\{-\dfrac{3}{4}, 1\right\}$ **35.** $\left(-5, \dfrac{3}{2}\right)$

36. 6 inches, 8 inches, 10 inches   **37.** x-intercept: $(-2, 0)$; y-intercept: $(0, 4)$ **38.** $-\dfrac{3}{2}$ **39.** $-\dfrac{3}{4}$ **40.**

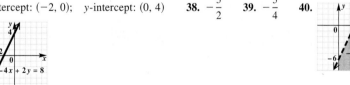

41. (a) $(-\infty, \infty)$ **(b)** 24   **42.** \$9.92

CHAPTER 8   SYSTEMS OF LINEAR EQUATIONS

◆ **CONNECTIONS**   **Page 451:** fourth quarter of 1991: about 1.5%;   at the end of the first quarter of 1992: 2%;   All three graphs would be best fit with a polynomial function. None are linear.

EXERCISES   **3.** yes   **5.** no   **7.** $\{(2, 2)\}$ **9.** $\{(3, -1)\}$ **11.** $\{(2, -3)\}$ **13.** $\left\{\left(\dfrac{3}{2}, -\dfrac{3}{2}\right)\right\}$

15. $\{(x, y) \mid 7x + 2y = 6\}$;   dependent equations   **17.** $\{(2, -4)\}$ **19.** ∅;   inconsistent system   **21.** Answers will

vary.   **(a)**   **(b)**   **(c)**   **23.** $y = -\dfrac{3}{7}x + \dfrac{4}{7}$; $y = -\dfrac{3}{7}x + \dfrac{3}{14}$;   0

25. both are $y = -\dfrac{2}{3}x + \dfrac{1}{3}$;   infinitely many   **27.** $\{(1, 2)\}$ **29.** $\left\{\left(\dfrac{22}{9}, \dfrac{22}{3}\right)\right\}$ **31.** $\{(2, 3)\}$ **33.** $\{(5, 4)\}$

35. $\left\{\left(-5, -\dfrac{10}{3}\right)\right\}$ **37.** $\{(2, 6)\}$ **39.** Multiply (1) by 3 and (2) by 2.   **43.** $\{(2, 4)\}$ **45.** $\{(4, -5)\}$ **47.** $\left\{\left(\dfrac{1}{a}, \dfrac{1}{b}\right)\right\}$

49. $\left\{\left(-\dfrac{3}{5a}, \dfrac{7}{5}\right)\right\}$ **51.** $\{(1, 3)\}$ **52.** $f(x) = -3x + 6$;   linear   **53.** $g(x) = \dfrac{2}{3}x + \dfrac{7}{3}$;   linear   **54.** one;   1;
3;   1;   3;   1;   3   **55.** $(3, -4)$ **57.** (a)   **59. (a)** $\{(5, 5)\}$ **(b)**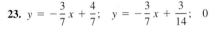

61. (a) $\{(0, -2)\}$ **(b)** 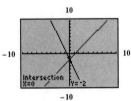   **63. (a)** years 0 to 10   **(b)** year 10;   about \$690   **65. (a)** 1978 and 1982

(b) just less than 500,000   **67.** $8x - 12y + 4z = 20$ **69.** 4

EXERCISES   **3.** $\{(1, 4, -3)\}$ **5.** $\{(0, 2, -5)\}$ **7.** $\left\{\left(-\dfrac{7}{3}, \dfrac{22}{3}, 7\right)\right\}$ **9.** $\{(4, 5, 3)\}$ **11.** $\{(2, 2, 2)\}$

13. $\left\{\left(\dfrac{8}{3}, \dfrac{2}{3}, 3\right)\right\}$ **15.** Answers will vary. Some possible answers are   **(a)** two perpendicular walls and the ceiling in a normal room   **(b)** the floors of three different levels of an office building   **(c)** three pages of this book (since they intersect in the spine).
17. ∅   **The solution sets in Exercises 19 and 21 may be given in other equivalent forms.**   **19.** $\{(x, y, z) \mid x - y + 4z = 8\}$
21. $\{(x, y, z) \mid 2x + y - z = 6\}$ **23.** $\{(0, 0, 0)\}$ **25.** $\{(2, 1, 5, 3)\}$ **27.** $128 = a + b + c$
28. $140 = 2.25a + 1.5b + c$ **29.** $80 = 9a + 3b + c$
30. $a + b + c = 128$
 $2.25a + 1.5b + c = 140$
 $9a + 3b + c = 80$; $\{(-32, 104, 56)\}$
31. $f(x) = -32x^2 + 104x + 56$ **32.** 56 feet   **33.** after approximately 3.72 seconds   **34.** after 3.25 seconds
35. $a = 3, b = 1, c = -2$; $f(x) = 3x^2 + x - 2$ **36.** $a = 1, b = 4, c = 3$; $y_1 = x^2 + 4x + 3$ **39.** 100 inches, 103
inches, 120 inches   **41.** $-4, 8, 12$

SECTION 8.3 (PAGE 475)

◈ **CONNECTIONS**   **Page 469:** "Mixed price" refers to the price of a mixture of the two products. The system is $9x + 7y = 107, 7x + 9y = 101$, where x represents the price of a citron and y represents the price of a wood apple.

EXERCISES   **1.** length: 78 feet; width: 36 feet   **3.** length: 12 feet; width: 5 feet   **5.** $x = 40, y = 50$, so the angles measure 40° and 50°.   **7.** CGA monitor: $400; VGA monitor: $500   **9.** 6 units of yarn; 2 units of thread   **11.** dark clay: $5 per kilogram; light clay: $4 per kilogram   **13. (a)** 3.2 ounces **(b)** 8 ounces **(c)** 12.8 ounces **(d)** 16 ounces   **15.** $.69x   **17.** 6 gallons of 25%; 14 gallons of 35%   **19.** 3 liters of pure acid; 24 liters of 10% acid   **21.** 14 kilograms of nuts; 16 kilograms of cereal   **23.** 76 general admission; 108 with student identification   **25.** 28 dimes; 66 quarters   **27.** $1000 at 2%; $2000 at 4%   **29.** $65y$ miles   **31.** freight train: 50 kilometers per hour; express train: 80 kilometers per hour   **33.** boat: 21 miles per hour; current: 3 miles per hour   **35.** $x + y + z = 180$; angle measures: 70°, 30°, 80°   **37.** first: 20°; second: 70°; third: 90°   **39.** shortest: 12 centimeters; middle: 25 centimeters; longest: 33 centimeters   **41.** A: 180 cases; B: 60 cases; C: 80 cases   **43.** 10 pounds of jelly beans; 2 pounds of chocolate eggs; 3 pounds of marshmallow chicks   **45.** $2a + b + c = -5$ **46.** $-a + c = -1$ **47.** $3a + 3b + c = -18$ **48.** $a = 1, b = -7, c = 0$; $x^2 + y^2 + x - 7y = 0$ **49. (a)** 6 **(b)** $-\dfrac{1}{6}$ **51. (a)** $-\dfrac{7}{8}$ **(b)** $\dfrac{8}{7}$

SECTION 8.4 (PAGE 486)

◈ **CONNECTIONS**   **Page 480:** The zero 2 × 2 matrix would be $\begin{bmatrix} 0 & 0 \\ 0 & 0 \end{bmatrix}$. The additive inverse of $\begin{bmatrix} a & b \\ c & d \end{bmatrix}$ would be $\begin{bmatrix} -a & -b \\ -c & -d \end{bmatrix}$.

EXERCISES   **3.** $\begin{bmatrix} 1 & 2 & | & 11 \\ 2 & -1 & | & -3 \end{bmatrix}$; $\begin{bmatrix} 1 & 2 & | & 11 \\ 0 & -5 & | & -25 \end{bmatrix}$; $\begin{bmatrix} 1 & 2 & | & 11 \\ 0 & 1 & | & 5 \end{bmatrix}$; $x + 2y = 11, y = 5$; $\{(1, 5)\}$ **5.** $\{(4,1)\}$

7. $\{(1, 1)\}$ **9.** $\{(-1, 4)\}$ **11.** $\emptyset$ **15.** $\begin{bmatrix} 1 & 1 & -1 & | & -3 \\ 0 & -1 & 3 & | & 10 \\ 0 & -6 & 7 & | & 38 \end{bmatrix}$; $\begin{bmatrix} 1 & 1 & -1 & | & -3 \\ 0 & 1 & -3 & | & -10 \\ 0 & -6 & 7 & | & 38 \end{bmatrix}$; $\begin{bmatrix} 1 & 1 & -1 & | & -3 \\ 0 & 1 & -3 & | & -10 \\ 0 & 0 & -11 & | & -22 \end{bmatrix}$;

$\begin{bmatrix} 1 & 1 & -1 & | & -3 \\ 0 & 1 & -3 & | & -10 \\ 0 & 0 & 1 & | & 2 \end{bmatrix}$; $x + y - z = -3, y - 3z = -10, z = 2$; $\{(3, -4, 2)\}$ **17.** $\{(4, 0, 1)\}$ **19.** $\{(-1, 23, 16)\}$
21. $\{(3, 2, -4)\}$ **23.** $\{(x, y, z) \mid x - 2y + z = 4\}$ **25.** $\{(1, 1)\}$ **27.** $\{(-1, 2, 1)\}$ **29.** $\{(1, 7, -4)\}$ **31.** -14
33. 7

SECTION 8.5 (PAGE 493)

◈ **CONNECTIONS**   **Page 491:** Your answer should agree with the answer for Exercise 9, but not with the answer in Example 4.

EXERCISES   **1.** $ps - rq$ **3.** -3 **5.** 14   **7.** 0   **9.** 59   **11.** 14   **13.** 1   **17.** -22 **19.** 20

21. -12 **23.** 0   **25.** 0   **27.** $\dfrac{y_2 - y_1}{x_2 - x_1}$ **28.** $y - y_1 = \dfrac{y_2 - y_1}{x_2 - x_1}(x - x_1)$ **29.** $x_2y - x_1y - x_2y_1 - xy_2 + x_1y_2 + xy_1 = 0$

30. The result is the same as in Exercise 29.   **31.** $\{2\}$ **33.** 6.078   **35.** -32 **37.** -55 **39.** 0   **41.** $\dfrac{7}{5}$
43. undefined

SECTION 8.6 (PAGE 499)

EXERCISES   **1.** $\{(4, 6)\}$ **3.** $\{(-5, 2)\}$ **5.** $\{(1, 0)\}$ **7.** $\left\{\left(\dfrac{11}{58}, -\dfrac{5}{29}\right)\right\}$ **9.** $\{(-1, 2)\}$ **11.** Answers will vary.
One example is $6x + 7y = 8$ **13.** $\{(-2, 3, 5)\}$ **15.** $\{(5, 0, 0)\}$ **17.** Cramer's rule does not apply.   **19.** $\{(20, -13, -12)\}$
$9x + 10y = 11$.

21. $\left\{\left(\dfrac{62}{5}, -\dfrac{1}{5}, \dfrac{27}{5}\right)\right\}$ **23.** $\{(-1, 2, 5, 1)\}$ **25.**

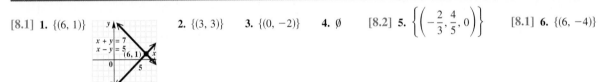

26. $\dfrac{1}{2}\begin{vmatrix} 0 & 0 & 1 \\ -3 & -4 & 1 \\ 2 & -2 & 1 \end{vmatrix}$ **27.** 7

28. $a = \sqrt{29}, b = 2\sqrt{2}, c = 5, s = \dfrac{1}{2}(\sqrt{29} + 2\sqrt{2} + 5)$;   The area is 7.   **29.** $\{(-1, 3, 5)\}$ **31.** $\{(1, -3, 2, -4)\}$

33. 9   **35.** $\dfrac{4ac - b^2}{4a}$

CHAPTER 8 REVIEW EXERCISES   (PAGE 505)

1. $\{(2, 2)\}$ **3.** $\{(-1, 2)\}$ **5.** $\left\{\left(\dfrac{68}{13}, -\dfrac{31}{13}\right)\right\}$ **7.** $\{(x, y) \mid -3x + y = 6\}$ **9.** $\left\{\left(-\dfrac{8}{9}, -\dfrac{4}{3}\right)\right\}$ **11.** $\{(3, 4)\}$

17. $\{(1, 2, 3)\}$ **19.** length: 6 feet;   width: 4 feet   **21.** plane: 300 miles per hour;   wind: 20 miles per hour

23. 85°, 60°, 35°   **25.** 5 liters of 8%;   3 liters of 20%;   none of 10%   **27.** $a = 3, b = 2, c = -4$ **29.** $\{(-1, 5)\}$

31. $\{(1, 2, -1)\}$ **33.** -11 **35.** -21 **37.** 0   **39.** -139 **41.** For three unknowns we need three equations.

43. $\left\{\left(\dfrac{39}{10}, \dfrac{6}{5}\right)\right\}$ **45.** $\left\{\left(\dfrac{172}{67}, -\dfrac{14}{67}, -\dfrac{87}{67}\right)\right\}$ **47.** ∅   **49.** $\{(5, 3)\}$ **51.** $\left\{\left(\dfrac{82}{23}, -\dfrac{4}{23}\right)\right\}$ **53.** $5y + z = 7$;

$-5y + z = -13$ **54.** $\{(2, -3)\}$ **55.** $\{(4, 2, -3)\}$ **56.** $\{(4, 2, -3)\}$ **57. (a)** -10 **(b)** -40 **(c)** -20 **(d)** 30

58. $\{(4, 2, -3)\}$

CHAPTER 8 TEST   (PAGE 509)

[8.1] 1. $\{(6, 1)\}$ **2.** $\{(3, 3)\}$ **3.** $\{(0, -2)\}$ **4.** ∅   **[8.2] 5.** $\left\{\left(-\dfrac{2}{3}, \dfrac{4}{5}, 0\right)\right\}$ **[8.1] 6.** $\{(6, -4)\}$

7. $\{(x, y) \mid 12x - 5y = 8\}$ **[8.3] 8.** slower car: 45 miles per hour;   faster car: 75 miles per hour   **9.** 4 liters of 20%;

8 liters of 50%   **10.** AC adaptor: $8;   rechargeable flashlight: $15   **[8.1] 11.** The second equation is already solved for y, so

substitution would be easier. Yes, the solution set would be the same.   **[8.4] 12.** $\left\{\left(\dfrac{2}{5}, \dfrac{7}{5}\right)\right\}$ **13.** $\{(-1, 2, 3)\}$

[8.2, 8.4, 8.6] 14. $\{(-3, -2, -4)\}$ **[8.5] 15.** 3   **16.** 0   **[8.6] 17.** $\left\{\left(-\dfrac{9}{4}, \dfrac{5}{4}\right)\right\}$ **18.** $\{(-1, 2, 3)\}$

[8.1] 19. $x = 8$ or 800 items; $3000   **20.** about $500

CUMULATIVE REVIEW   (Chapters 1–8)   (PAGE 510)

1. $(-\infty, \infty)$ **2.** $[-3, 3]$ **3.** $4000 at 4%; $8000 at 3%   **4.** 78 or greater   **5.** ∅   **6.** 0   **7.** yes

8. $(2x - 1)(4x^2 + 2x + 1)(2x + 1)(4x^2 - 2x + 1)$ **9.** 4   **10.** $\left\{-\dfrac{5}{2}, \dfrac{10}{3}\right\}$ **11.** $\dfrac{5}{4(x + 2)}$

12. $\dfrac{5x + 3}{(x + 2)(x - 1)(x + 3)}$ **13.** $\dfrac{12p}{p + 2}$ **14.** $\dfrac{-3}{2s - 5r}$ or $\dfrac{3}{5r - 2s}$ **15.** $\{0\}$ **16.** $-8\sqrt{10}$ **17.** $\dfrac{3(12 - \sqrt{5})}{139}$

18. $\dfrac{x^4 \sqrt[3]{xy^2}}{y}$ **19.** $\dfrac{4 - 3\sqrt{7}}{2}$ **20.** $13 + \sqrt{7}$ **21.** $\{-2\}$ **22.** $\left\{-\dfrac{1}{6}, 2\right\}$ **23.** 8%   **24.** $\left[-3, \dfrac{1}{2}\right]$

25. 15 miles per hour   **26.** $y = 6$ **27.** $x = 4$ **28.** $-\dfrac{4}{3}$ **29.** $\dfrac{3}{4}$ **30.** $4x + 3y = 10$

31.   **32.** $5x - 3y = -15$ **33.**   **34.**

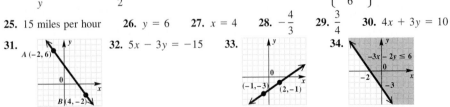

35. 5 pounds of oranges;   1 pound of apples   **36.** length: 12 meters;   width: 9 meters   **37.** small: $1.50;   large: $2.50
38. peanuts: $2 per pound;   cashews: $4 per pound   **39.** video rental firm: $60,000;   bond and money market fund: $10,000 each
40. $20,000 at 8%; $50,000 at 10%; $5000 at 7%

CHAPTER 9   QUADRATIC FUNCTIONS AND THE CONIC SECTIONS

SECTION 9.1   (PAGE 520)

◆ **CONNECTIONS**   **Page 513:** Answers will vary.   **Page 519: 1.** It must be negative.   **2.** approximately 45
3. approximately 20

EXERCISES   3. (0, 0)   **5.** (0, 4)   **7.** (1, 0)   **9.** $(-3, -4)$ **11.** downward;   narrower   **13.** upward;   wider
17. (a) I **(b)** IV   **(c)** II   **(d)** III   **19.** F   **21.** C   **23.** E   **25.** 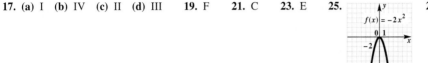   **27.**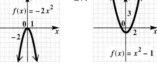

29.   **31.**   **33.**   **35.**   **37.** $f(x) = -.5(x + 1)^2 + 2$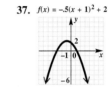

39.   **41.** $\{-4, 5\}$ The x-values of the x-intercepts are -4 and 5.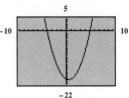

43. $\{-.5, 3\}$ The x-values of the x-intercepts are $-.5$ and 3. 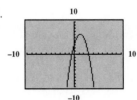   **45. (a)**

47. It is shifted 6 units upward.   **48.** 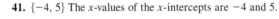   **49.** It is shifted 6 units upward.   **50.** It is shifted 6 units to the right.

51.   **52.** It is shifted 6 units to the right.   **53. (a)** $|x + p|$ **(b)** The distance from the focus to the origin

should equal the distance from the directrix to the origin.   **(c)** $\sqrt{(x - p)^2 + y^2}$ **(d)** $y^2 = 4px$ **55.** $\{-3 + 2\sqrt{3}, -3 - 2\sqrt{3}\}$
57. $\left\{ \dfrac{6 + \sqrt{46}}{2}, \dfrac{6 - \sqrt{46}}{2} \right\}$

SECTION 9.2 (PAGE 531)

◆ **CONNECTIONS** **Page 528:** 4 seconds; 256 feet; Let $h = 0$ and find t. 8 seconds

EXERCISES **5.** $(-1, 3)$; upward; narrower; no x-intercepts

7. $\left(\dfrac{5}{2}, \dfrac{37}{4}\right)$; downward; same; two x-intercepts **9.** $(-3, -9)$; to the right; wider

11. $f(x) = x^2 + 8x + 10$ **13.** **15.** $x = -\dfrac{1}{5}y^2 + 2y - 4$ **17.**

19. F **21.** C **23.** D **25.** in 1995; 68 million barrels daily **27.** 16 feet; 2 seconds **29.** 25 meters
31. 4.1 seconds; 81.6 meters **33.** The parabola opens downward, indicating $a < 0$.
35. Since $x = 2.1 \approx 2$, the year corresponding to 2, which is 1990, is the year of maximum number of jobs.
37. (a) $\{1, 3\}$ (b) $(-\infty, 1) \cup (3, \infty)$ (c) the open interval $(1, 3)$

38. (a) $\left\{-4, \dfrac{2}{3}\right\}$ (b) $(-\infty, -4] \cup \left[\dfrac{2}{3}, \infty\right)$ (c) the open interval $\left(-4, \dfrac{2}{3}\right)$

39. (a) $\{-2, 5\}$ (b) $[-2, 5]$ (c) $(-\infty, -2] \cup [5, \infty)$

40. (a) $\left\{-3, \dfrac{5}{2}\right\}$ (b) $\left[-3, \dfrac{5}{2}\right]$ (c) $(-\infty, -3] \cup \left[\dfrac{5}{2}, \infty\right)$

41. (a) $(-2, -7)$ (b) 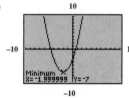 **43.** (a) $(2, 8)$ (b) 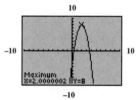 **45.** $2\sqrt{5}$

47. $\sqrt{(x + 2)^2 + (y - 5)^2}$

SECTION 9.3 (PAGE 541)

◆ **CONNECTIONS** **Page 539:** Answers will vary.

EXERCISES **1.** B **3.** D **5.** $x^2 + y^2 = 36$ **7.** $(x + 1)^2 + (y - 3)^2 = 16$ **9.** $x^2 + (y - 4)^2 = 3$
11. center: $(0, 0)$; radius: r **13.** center: $(-2, -3)$; radius: 2 **15.** center: $(-5, 7)$; radius: 9
17. center: $(2, 4)$; radius: 4 **19.** **21.** **23.** center: $(-3, 2)$ **25.** center: $(2, 3)$

31. **33.** **35.** **37.** **39.**

41. $y_1 = 4 + \sqrt{16 - (x + 2)^2}$, $y_2 = 4 - \sqrt{16 - (x + 2)^2}$ **43.** **45.**

47. (a) 77.4 million miles **(b)** 64.2 million miles (Both answers are rounded to the nearest tenth.)
51.   **53.** x-intercept: $(3, 0)$; y-intercept: $(0, 4)$

SECTION 9.4 (PAGE 552)

◆ **CONNECTIONS**   **Page 548:** Answers will vary.   **Page 550:** For example, let $x = 2$. Then $y \approx 69$ using the equation and this appears to be close to the y-value of the point on the curve with $x = 2$.

EXERCISES   **1.** C   **3.** D   **7.**   **9.**   **11.**   **13.** hyperbola

15. ellipse   **17.** circle   **19.** hyperbola   **21.** hyperbola   **23.**

25.   **27.** $\dfrac{(x-2)^2}{4} - \dfrac{(y+1)^2}{9} = 1$ **29.** $\dfrac{y^2}{36} - \dfrac{(x-2)^2}{49} = 1$ **31. (a)** 10 meters **(b)** 36 meters

33. $y_1 = \sqrt{\dfrac{x^2}{9} - 1},\ y_2 = -\sqrt{\dfrac{x^2}{9} - 1}$ **35.**   **37.**

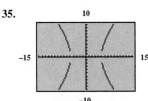 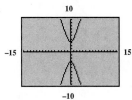

39. $y = \sqrt{\dfrac{x^2}{4} - 1}$ **40.** $y = \dfrac{1}{2}x$ **41.** $y \approx 24.98$ **42.** $y = 25$

43. Because $24.98 < 25$, the graph of $y = \sqrt{\dfrac{x^2}{4} - 1}$ lies below the graph of $y = \dfrac{1}{2}x$.

44. The y-values on the hyperbola will approach the y-values on the line, but will always be less.

47. $\{(2, 9)\}$ **49.** $\{(0, 5)\}$ **51.** $\left\{-\sqrt{3}, \sqrt{3}, -\dfrac{\sqrt{2}}{2}i, \dfrac{\sqrt{2}}{2}i\right\}$ **53.** $\{-2, 2, -\sqrt{3}, \sqrt{3}\}$

SECTION 9.5 (PAGE 562)

◆ **CONNECTIONS**   **Page 558:** quadratic;   a parabola opening upward;   linear;   a straight line;   0, 1, or 2, because a line may intersect a parabola at two points, be tangent to the curve at one point, or not intersect at all.   during 1980 ($x \approx 6.9$) and 2000 ($x \approx 26.7$);   In 1990, men received about 517 thousand degrees and women received about 589 thousand degrees.

EXERCISES 1. 3. 5. 7. 9.

11. 13. 17. $\left\{ (0, 0), \left(\frac{1}{2}, \frac{1}{2} \right) \right\}$ 19. $\{(-6, 9), (-1, 4)\}$ 21. $\left\{ \left(-\frac{1}{5}, \frac{7}{5} \right), (1, -1) \right\}$

23. $\left\{ (-2, -2), \left(-\frac{4}{3}, -3 \right) \right\}$ 25. $\{(-3, 1), (1, -3)\}$ 27. $\left\{ \left(-\frac{3}{2}, -\frac{9}{4} \right), (-2, 0) \right\}$

29. $\{(-\sqrt{3}, 0), (\sqrt{3}, 0), (-\sqrt{5}, 2), (\sqrt{5}, 2)\}$ 33. $\{(-2, 0), (2, 0)\}$ 35. $\{(\sqrt{3}, 0), (-\sqrt{3}, 0)\}$ 37. $\{(-2i\sqrt{2}, -2\sqrt{3}),$
$(-2i\sqrt{2}, 2\sqrt{3}), (2i\sqrt{2}, -2\sqrt{3}), (2i\sqrt{2}, 2\sqrt{3})\}$ 39. $\{(-\sqrt{5}, -\sqrt{5}), (\sqrt{5}, \sqrt{5})\}$ 41. $\{(i, 2i), (-i, -2i), (2, -1), (-2, 1)\}$
43. $\{(2, 0), (1, 1)\}$ 45. $\{(2, -3), (-3, 2)\}$

 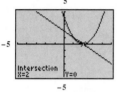 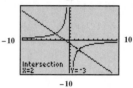 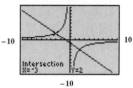

47. $\{-2, 5\}$ 48. 49. $\{(-2, 4), (5, 25)\}$ 50. They are the same. 51.

52. They are the same. 53. 2 and 2 or -2 and -2 55. length: 12 feet; width: 7 feet 57. $20; 800 calculators

59. 61.

SECTION 9.6 (PAGE 570)

EXERCISES 3. B 5. A 7. 9. 11. 13.

15. 17. 19. 21. (c) 23. 25.

27. 29. 31. 33. 35.

37.

39.

41.

43.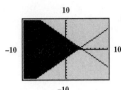

45.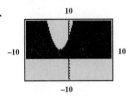

47.

49. domain: $\{-3, -2, -1, 0, 1, 2, 3\}$;　range: $\left\{\dfrac{1}{8}, \dfrac{1}{4}, \dfrac{1}{2}, 1, 2, 4, 8\right\}$

51. function　**53.** $y = \dfrac{4 - x}{-5}$ or $y = \dfrac{x - 4}{5}$

CHAPTER 9 REVIEW EXERCISES (PAGE 577)

1. $(0, -2)$　**3.** $(1, 0)$　**5.** $(3, 7)$　**7.** $\left(\dfrac{2}{3}, -\dfrac{2}{3}\right)$　**9.** $\left(\dfrac{7}{8}, -\dfrac{5}{4}\right)$　**11.** $f(x) = -2x^2 + 8x - 5$ 　**13.**

15. (a)　**17.** (a) 5 seconds　(b) 400 feet　**19.** 20 and 20　**21.** $(-\infty, -.5) \cup (4, \infty)$　**23.** $(x + 2)^2 + (y - 4)^2 = 9$

25. center: $(-3, 2)$;　radius: 4　**27.** center: $(-1, -5)$;　radius: 3　**29.**

31. 　**33.** 　**35.** 　**37.** hyperbola　**39.** parabola　**41.** $\{(6, -9), (-2, -5)\}$

43. $\{(4, 2), (-1, -3)\}$　**45.** $\{(-\sqrt{2}, 2), (-\sqrt{2}, -2), (\sqrt{2}, -2), (\sqrt{2}, 2)\}$　**47.** 0, 1, or 2　**49.**

51. 　**53.** 　**55.** 　**57.** 　**59.**

61.
$3x + 2y \geq 0$
$y \leq 4$
$x \leq 4$

63. $9a - 3b + c = 16$ **64.** $a - b + c = -4$ **65.** $4a + 2b + c = 11$

66. $\{(3, 2, -5)\}$; $f(x) = 3x^2 + 2x - 5$ **67.** vertex: $\left(-\dfrac{1}{3}, -\dfrac{16}{3}\right)$;   x-intercepts: $\left(-\dfrac{5}{3}, 0\right)$, $(1, 0)$;   y-intercept: $(0, -5)$

68. $a + b + c = 3$ **69.** $4a - 2b + c = -15$ **70.** $9a + 3b + c = -5$ **71.** $\{(-2, 4, 1)\}$; $x = -2y^2 + 4y + 1$

72.
$x = -2y^2 + 4y + 1$

73. $2a + 4b + c = -20$ **74.** $5a + b + c = -26$ **75.** $-a + b + c = -2$

76. $\{(-4, -2, -4)\}$; $x^2 + y^2 - 4x - 2y - 4 = 0$ **77.** center: $(2, 1)$;   radius: 3   **78.**
$x^2 + y^2 - 4x - 2y - 4 = 0$

CHAPTER 9 TEST (PAGE 583)

[9.1] 1. vertex: $(0, -2)$ **[9.2] 2.** vertex: $(2, 3)$ **3.** 82 feet;   3 seconds   **[9.3] 4.** $(x + 4)^2 + (y - 4)^2 = 16$

$f(x) = .5x^2 - 2$

$f(x) = -x^2 + 4x - 1$

5. center: $(-4, 1)$;   radius: 5   **6.** circle   **7.** ellipse   **[9.4] 8.** hyperbola   **[9.2] 9.** parabola

10.
$y = x^2 - 4x + 5$

11.
$x = -(y - 2)^2 + 2$

[9.3] 12.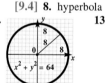
$x^2 + y^2 = 64$

13.
$4x^2 + 9y^2 = 36$

[9.4] 14.
$16y^2 - 4x^2 = 64$

15.
$f(x) = \sqrt{16 - x^2}$

[9.5] 16. 0, 1, 2, 3, or 4   **17.** $\left\{\left(-\dfrac{1}{2}, -10\right), (5, 1)\right\}$ **18.** $\left\{(-2, -2), \left(\dfrac{14}{5}, -\dfrac{2}{5}\right)\right\}$

19. $\{(-\sqrt{22}, -\sqrt{3}), (-\sqrt{22}, \sqrt{3}), (\sqrt{22}, -\sqrt{3}), (\sqrt{22}, \sqrt{3})\}$ **20.**   **[9.6] 21.**
$y \leq x^2 - 2$

22.
$2x - 5y \geq 12$
$3x + 4y \leq 12$

23.
$x^2 + 25y^2 \leq 25$
$x^2 + y^2 \leq 9$

[9.2] 24. 31,600   **25.** 1984;   36,000

CUMULATIVE REVIEW   (Chapters 1–9)   (PAGE 584)

1. -4 **2.** $\left\{\dfrac{2}{3}\right\}$ **3.** $\left(-\infty, \dfrac{3}{5}\right]$ **4.** $\{-4, 4\}$ **5.** $(-\infty, -5) \cup (10, \infty)$ **6.** $25y^2 - 30y + 9$ **7.** $12r^2 + 40r - 7$

8. $4x^3 - 4x^2 + 3x + 5 + \dfrac{3}{2x + 1}$ **9.** $(3x + 2)(4x - 5)$ **10.** $(2y^2 - 1)(y^2 + 3)$ **11.** $(z^2 + 1)(z + 1)(z - 1)$

12. $(a - 3b)(a^2 + 3ab + 9b^2)$ **13.** $\dfrac{40}{9}$ **14.** $\dfrac{y - 1}{y(y - 3)}$ **15.** $\dfrac{3c + 5}{(c + 5)(c + 3)}$ **16.** $\dfrac{1}{p}$ **17.** $\dfrac{6}{5}$ or $1\dfrac{1}{5}$ hours

18. $\dfrac{3}{4}$ **19.** $\dfrac{a^5}{4}$ **20.** $2\sqrt[3]{2}$ **21.** $\dfrac{3\sqrt{10}}{2}$ **22.** $\dfrac{7}{5} + \dfrac{11}{5}i$ **23.** $\varnothing$ **24.** $\left\{\dfrac{1}{5}, -\dfrac{3}{2}\right\}$ **25.** $\left\{\dfrac{1 - 2\sqrt{2}}{4}, \dfrac{1 + 2\sqrt{2}}{4}\right\}$

26. $\left\{\dfrac{3 - \sqrt{33}}{6}, \dfrac{3 + \sqrt{33}}{6}\right\}$ **27.** $\left\{-\dfrac{\sqrt{6}}{2}, \dfrac{\sqrt{6}}{2}, -\sqrt{7}, \sqrt{7}\right\}$ **28.** $v = \dfrac{\pm\sqrt{rFkw}}{kw}$ **29.** $\dfrac{2}{3}$ **30.** $3x + 2y = -13$

31.   **32.** $f(x) = -2(x - 1)^2 + 3$ **33.**   **34.**   **35.**

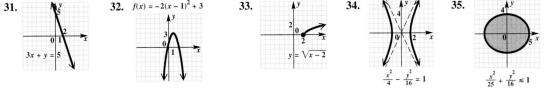

36. $\{(3, -3)\}$ **37.** $\{(4, 1, -2)\}$ **38.** $\left\{(-1, 5), \left(\dfrac{5}{2}, -2\right)\right\}$ **39.** 40 miles per hour   **40.** 15 fives and 7 twenties

CHAPTER 10   INVERSE, EXPONENTIAL, AND LOGARITHMIC FUNCTIONS

SECTION 10.1   (PAGE 593)

◊ **CONNECTIONS**   **Page 587:** For a growth function, as x increases, y also increases. If y is a decay function, as x increases, y decreases.

EXERCISES   **1.** true   **3.** true   **7.** $\{(6, 3), (10, 2), (12, 5)\}$ **9.** not a one-to-one function   **11.** $f^{-1}(x) = \dfrac{x - 4}{2}$

13. $g^{-1}(x) = x^2 + 3, \; x \geq 0$ **15.** not one-to-one   **17.** $f^{-1}(x) = \sqrt[3]{x + 4}$ **19.** (a)   **21.** (a) 8   (b) 3

23. (a) 1   (b) 0   **25.** (a) one-to-one   (b)   **27.** (a) not one-to-one

29. (a) one-to-one   (b) 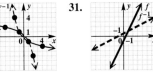   **31.** 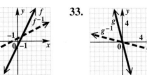   **33.** 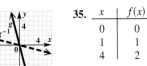   **35.**

x	$f(x)$
0	0
1	1
4	2

37.

x	y
-1	-3
0	-2
1	-1
2	6

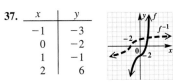   **39.** $f^{-1}(x) = \dfrac{x + 7}{2}$ 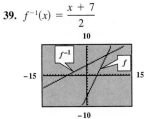   **41.** $f^{-1}(x) = \sqrt[3]{x - 5}$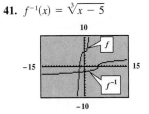

43.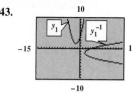

47. $f^{-1}(x) = \dfrac{x + 5}{4}$ **48.** My graphics calculator is the greatest thing since sliced bread.

50. Answers will vary. For example, Jane Doe is   1004 5 2748 129 68 3379 129.
51. 64   **53.** 2   **55.** 44.02

S E C T I O N   10.2   (PAGE 603)

◆ **CONNECTIONS**   **Page 602:** about 341 parts per million in 1995;   about 343 parts per million in 2000

EXERCISES   **1.** rises;   falls   **3.** does not   **5.**   **7.**   **9.**

11.   **13.** It is one-to-one and thus has an inverse.   **15.** $\left\{\dfrac{3}{2}\right\}$ **17.** $\{7\}$ **19.** $\{-3\}$ **21.** $\{-1\}$

23. $\{-3\}$ **25.** 639.545   **27.** .066   **29.** 12.179   **33. (a)** .5°C **(b)** .35°C   **35. (a)** 1.6°C **(b)** .5°C
37. (a) 172.3 million **(b)** 310.6 million **(c)** It will be twice as large.   **39. (a)** 100 grams **(b)** 31.25 grams
(c) .30 gram (to the nearest hundredth) **(d)** 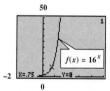   **41.** 6.67 years after it was purchased   **43.** about $360,000

45. In 1991, the average salary was about $800,000.   **47.** $(\sqrt[4]{16})^3$;   8   **48.** $\sqrt[4]{16^3}$;   8   **49.** 8
50. Because $\sqrt{\sqrt{x}} = (x^{1/2})^{1/2} = x^{1/4} = \sqrt[4]{x}$, the fourth root of 16^3 can be found by taking the square root twice.
51. $f(.75) = 8$ **52.** $(\sqrt[40]{16})^{30} = 8$; $16^{30/40} = 16^{3/4}$ since $\dfrac{30}{40} = \dfrac{3}{4}$.   **53.** 4   **55.** 0

S E C T I O N   10.3   (PAGE 611)

◆ **CONNECTIONS**   **Page 610:** almost 4 times as powerful;   about 300 times as powerful

EXERCISES   **1. (a)** 2 **(b)** 4 **(c)** −1 **(d)** −2 **(e)** $\dfrac{1}{2}$ **(f)** 0   **3.** $\log_4 1024 = 5$ **5.** $\log_{1/2} 8 = -3$

7. $\log_{10} .001 = -3$ **9.** $4^3 = 64$ **11.** $10^{-4} = \dfrac{1}{10,000}$ **13.** $6^0 = 1$ **17.** $\left\{\dfrac{1}{3}\right\}$ **19.** $\{81\}$ **21.** $\left\{\dfrac{1}{5}\right\}$ **23.** $\{1\}$

25. $\{x \mid x > 0, x \neq 1\}$ **27.** $\{5\}$ **29.** $\left\{\dfrac{5}{3}\right\}$ **31.** $\{4\}$ **33.** $\left\{\dfrac{3}{2}\right\}$ **35.**   **37.**

39. 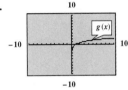   **41. (a)** 6   **(b)** 12   **(c)** 18   **(d)** $M(t) = 6 \log_4 (2t + 4)$ **43.** 8   **45.** 24   **49.** $(0, \infty);\quad (-\infty, \infty)$

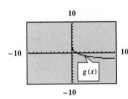

51. 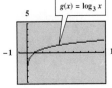   **53.** 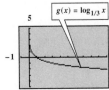   **55.** 4^9 **57.** $\dfrac{1}{5^{11}}$ **59.** $\dfrac{1}{9^6}$

SECTION 10.4 (PAGE 619)

◆ **CONNECTIONS**   **Page 619:**
$$\begin{aligned} \log_{10} 458.3 &\approx 2.661149857 \\ + \ \log_{10} 294.6 &\approx 2.469232743 \\ \hline &\approx 5.130382600 \end{aligned}$$
$10^{5.130382600} \approx 135,015.18$ A calculator gives $(458.3)(294.6) = 135,015.18$.

EXERCISES   **1.** sum   **3.** k **5.** $\log_7 4 - \log_7 5$ **7.** $\dfrac{1}{4} \log_2 8$ or $\dfrac{3}{4}$ **9.** $\log_4 3 + \dfrac{1}{2} \log_4 x - \log_4 y$

11. $\dfrac{1}{3} \log_3 4 - 2 \log_3 x - \log_3 y$ **13.** $\dfrac{1}{2} \log_3 x + \dfrac{1}{2} \log_3 y - \dfrac{1}{2} \log_3 5$ **15.** $\dfrac{1}{3} \log_2 x + \dfrac{1}{5} \log_2 y - 2 \log_2 r$ **19.** $\log_b xy$

21. $\log_a \dfrac{m^3}{n}$ **23.** $\log_a \dfrac{rt^3}{s}$ **25.** $\log_a \dfrac{125}{81}$ **27.** $\log_{10}(x^2 - 9)$ **29.** $\log_p \dfrac{x^3 y^{1/2}}{z^{3/2} a^3}$ **31.** 1.2552   **33.** 1.5562

35. .4771   **37.** true   **39.** false   **43.** $\log_2 8 - \log_2 4 = \log_2 \dfrac{8}{4} = \log_2 2 = 1$ **45.** 4   **46.** It is the exponent to which 3 must be raised in order to obtain 81.   **47.** 81   **48.** It is the exponent to which 2 must be raised in order to obtain 19.
49. 19   **50.** m **51.** $\log_{10} 10{,}000 = 4$ **53.** $\log_{10} .01 = -2$ **55.** $10^0 = 1$

SECTION 10.5 (PAGE 626)

◆ **CONNECTIONS**   **Page 623:** 2;   2.5; $2.\overline{6}$; $2.708\overline{3}$; $2.71\overline{6}$;   The difference is .0016151618. It approaches e fairly quickly.

EXERCISES   **1.** 10   **3.** 0 and 1   **5.** 19.2   **7.** 1.6335   **9.** 2.5164   **11.** -1.4868 **13.** 9.6776
15. 2.0592   **17.** -2.8896 **19.** 5.9613   **21.** 4.1506   **23.** 2.3026   **25. (a)** 2.552424846   **(b)** 1.552424846
(c) 0.552424846   **(d)** The whole number parts will vary but the decimal parts are the same.   **29. (a)**   **31. (a)** 56.0531
(b) 5.7466   **33. (a)** .0014   **(b)** .0580   **35.** 1.4315   **37.** 3.3030   **39.** Answers will vary. Suppose the name is Paul Bunyan, with $m = 4$ and $n = 6$.   **(a)** $\log_4 6$ is the exponent to which 4 must be raised in order to obtain 6.   **(b)** 1.29248125
(c) 6 (the value of n)   **41.** 11.6   **43.** 4.3   **45.** 4.0×10^{-8} **47.** 4.0×10^{-6} **49. (a)** 1.62 grams   **(b)** 1.18 grams
(c) .69 gram   **(d)** 2.00 grams   **51.** 2228 million (or approximately 2.2 billion)   **53.** 527 years   **55.** 3466 years
57. 789.3 thousand (or 789,300)   **59.** In 1989 (when $x = 10$), the number of births was about 730,000.

61. 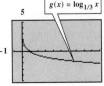   **63.** 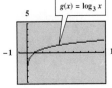   **65.** $\left\{-\dfrac{3}{5}\right\}$ **67.** $\{5\}$ **69.** $\{-3\}$
71. $\log(x + 2)(x - 3)$ or $\log(x^2 - x - 6)$

SECTION 10.6 (PAGE 635)

◆ **CONNECTIONS** **Page 630:** about 6.96 grams; about 4.84 grams; 5730 years **Page 634:** $1221.40; about 17.3 years; about 13.9 years; As r increases, doubling time decreases.

EXERCISES **1.** $\{3\}$ **3.** $\{6\}$; $\dfrac{\log 64}{\log 2} = 6$ **5.** $\{.827\}$ **7.** $\{.833\}$ **9.** $\{2.269\}$ **11.** $\{566.866\}$

13. $\{-18.892\}$ **15.** $\{43.301\}$ **19.** $\left\{\dfrac{2}{3}\right\}$ **21.** $\{-1 + \sqrt[3]{49}\}$ **25.** $\left\{\dfrac{1}{3}\right\}$ **27.** $\{2\}$ **29.** $\emptyset$ **31.** $\{8\}$ **33.** $\left\{\dfrac{4}{3}\right\}$

35. $\{8\}$ **37.** $2539.47 **39.** about 180 grams **41.** (a) $11,260.96 (b) $11,416.64 (c) $11,497.99 (d) $11,580.90
(e) $11,581.83 **43.** $137.41 **45.** 5.55 hours **47.** (a) 12.33 grams (b) 7.60 grams (c) about 5730 years

49. $\{.827\}$

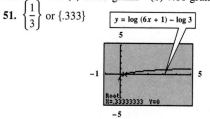

51. $\left\{\dfrac{1}{3}\right\}$ or $\{.333\}$

55. (a) 6 (b) $\dfrac{7}{2}$ (c) $\dfrac{8}{3}$ (d) $\dfrac{9}{4}$ **57.** (a) 0 (b) 2 (c) 6 (d) 12

CHAPTER 10 REVIEW EXERCISES (PAGE 641)

1. not one-to-one **3.** $f^{-1}(x) = \dfrac{x-7}{-3}$ or $\dfrac{7-x}{3}$ **5.** not one-to-one

7. **9.** **11.** **13.** $\{4\}$ **15.** $(0, 1)$ **17.** (a) 1 million barrels

(b) .50 million barrels (c) .17 million barrels (d) .03 million barrels (e) 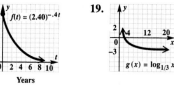 **19.**

21. $\left\{\dfrac{3}{2}\right\}$ **23.** $\{8\}$ **25.** $\{b \mid b > 0, b \neq 1\}$ **27.** a **29.** $\log_4 3 + 2\log_4 x$ **31.** $\log_b \dfrac{3x}{y^2}$ **33.** 1.4609

35. 4.8613 **37.** -1.3587 **39.** .9251 **41.** 1.4315 **43.** 8.4 **45.** (a) 500 grams (b) about 409 grams

47. $\{18.310\}$ **49.** $\{-6 + \sqrt[3]{25}\}$ **51.** $\left\{\dfrac{3}{298}\right\}$ **53.** $\{1\}$ **55.** $28,295.56 **57.** Plan A is better, since it would pay

$2.92 more. **59.** 67% **61.** $\{72\}$ **63.** $\left\{\dfrac{1}{9}\right\}$ **65.** $\{3\}$ **67.** $\left\{\dfrac{11}{3}\right\}$ **69.** $\dfrac{1}{4}, \dfrac{1}{2}, 1, 2, 4, 8$ **70.** $-2, -1, 0, 1, 2, 3$

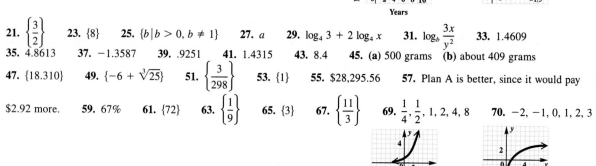

71. The roles of x and y are reversed. They are inverses. **72.** horizontal; vertical **73.** 5; 2; 3; 5
74. 4; 8 (or 8; 4) **75.** 3.700439718 (The number of displayed digits may vary.) **77.** 13 **78.** 13; The number in
Exercise 75 is the exponent to which 2 must be raised in order to obtain 13. **79.** 3.700439718 **80.** $\left\{-\dfrac{8}{5}\right\}$ **81.** $\{4\}$

CHAPTER 10 TEST (PAGE 645)

[10.1] 1. (a) not one-to-one **(b)** one-to-one   **2.** $f^{-1}(x) = x^3 - 7$ **3.**   **[10.2] 4.**

[10.3] 5.   **6.** Simply exchange the x and y values in the ordered pairs.   **[10.2] 7.** $\{-4\}$ **8.** $\left\{ -\dfrac{13}{3} \right\}$

9. (a) $5000   **(b)** $2973   **(c)** $1768   **(d)** 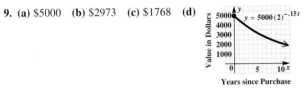   **[10.3] 10.** $\log_4 .0625 = -2$ **11.** $7^2 = 49$ **12.** $\{32\}$

13. $\left\{ \dfrac{1}{2} \right\}$ **14.** $\{2\}$ **15.** 5; 2; 5th; 32   **[10.4] 16.** $2 \log_3 x + \log_3 y$ **17.** $\dfrac{1}{2} \log_5 x - \log_5 y - \log_5 z$ **18.** $\log_b \dfrac{s^3}{t}$

19. $\log_b \dfrac{r^{1/4} s^2}{t^{2/3}}$ **[10.5] 20.** 1.3284   **21.** $-.8440$ **22. (a)** 2 and 3   **(b)** 2.1245   **[10.6] 23.** $\{3.9656\}$ **24.** $\{3\}$

25. (a) $2973   **(b)** after $15.48 \approx 15$ years

CUMULATIVE REVIEW (Chapters 1–10) (PAGE 646)

1. $-2, 0, 6, \dfrac{30}{3}$ (or 10)   **2.** $-\dfrac{9}{4}, -2, 0, .6, 6, \dfrac{30}{3}$ (or 10)   **3.** $-\sqrt{2}, \sqrt{11}$ **4.** all except $\sqrt{-8}$ **5.** 16   **6.** -27

7. -39 **8.** $\left\{ -\dfrac{2}{3} \right\}$ **9.** $[1, \infty)$ **10.** $\{-2, 7\}$ **11.** $\left[-\dfrac{16}{3}, \dfrac{16}{3} \right]$ **12.** $\left[\dfrac{7}{3}, 3 \right]$ **13.** $(-\infty, -3) \cup (2, \infty)$

14. $6p^2 + 7p - 3$ **15.** $16k^2 - 24k + 9$ **16.** $-5m^3 + 2m^2 - 7m + 4$ **17.** $2t^3 + 5t^2 - 3t + 4$ **18.** $x(8 + x^2)$

19. $(3y - 2)(8y + 3)$ **20.** $z(5z + 1)(z - 4)$ **21.** $(4a + 5b^2)(4a - 5b^2)$ **22.** $(2c + d)(4c^2 - 2cd + d^2)$

23. $(4r + 7q)^2$ **24.** $-\dfrac{1875 p^{13}}{8}$ **25.** $\dfrac{x + 5}{x + 4}$ **26.** $\dfrac{-3k - 19}{(k + 3)(k - 2)}$ **27.** $\dfrac{22 - p}{p(p - 4)(p + 2)}$ **28.** 6 pounds

29. $\dfrac{16}{25}$ **30.** $\dfrac{1}{6^5}$ **31.** $-27\sqrt{2}$ **32.** 41   **33.** $\left\{ \dfrac{1 + \sqrt{13}}{6}, \dfrac{1 - \sqrt{13}}{6} \right\}$ **34.** $(-\infty, -4) \cup (2, \infty)$

35. (a) 233.4 million   **(b)** 238.5 million   **(c)** 247 million   **(d)** 2030   **36.** $-5,587,500$ **37.** $3x - 4y = 19$

38.   **39.** $-4x + y \le 5$ **40.** $f(x) = \dfrac{1}{3}(x - 1)^2 + 2$ **41.** $\dfrac{x^2}{9} + \dfrac{y^2}{16} = 1$ **42.** $25x^2 - 16y^2 = 400$

43. $\{(4, 2)\}$ **44.** $\{(1, -1, 4)\}$ **45.** -1 **46.** -9 **47.**   **48.** $\{-1\}$ **49.**

50. (a) 25,000   **(b)** 30,500   **(c)** 37,300   **(d)** 68,000

CHAPTER 11   SEQUENCES AND SERIES

SECTION 11.1   (PAGE 655)

◆ **CONNECTIONS**   **Page 649: 1.** After the first term, each term is the sum of the two preceding terms.   **2.** We show only four generations in this sketch.

```
M   1
|
F   1
/\
F  M   2
/\ |
F M F   3
```

EXERCISES   1. linear   **3.** $4, \dfrac{5}{2}, 2, \dfrac{7}{4}, \dfrac{8}{5}$ **5.** 3, 9, 27, 81, 243   **7.** $1, \dfrac{1}{4}, \dfrac{1}{9}, \dfrac{1}{16}, \dfrac{1}{25}$ **9.** $-1, 1, -1, 1, -1$

11. -70 **13.** $\dfrac{49}{23}$ **15.** 171   **17.** $4n$ **19.** $\dfrac{1}{3^n}$ **21.** \$110, \$109, \$108, \$107, \$106, \$105; \$400

23. 15 meters, 30 meters, 45 meters, 60 meters;   150 meters   **25.** $3 + 6 + 11 = 20$ **27.** $\dfrac{1}{2} + \dfrac{1}{3} + \dfrac{1}{4} + \dfrac{1}{5} = \dfrac{77}{60}$

29. $-1 + 1 - 1 + 1 - 1 + 1 = 0$ **31.** $0 + 6 + 14 + 24 + 36 = 80$ **33.** $2x + 4x + 6x + 8x + 10x$

35. $x + 2x^2 + 3x^3 + 4x^4 + 5x^5$ **37.** $\displaystyle\sum_{i=1}^{5} (i + 2)$ **39.** $\displaystyle\sum_{i=1}^{5} \dfrac{1}{i + 1}$ **43.** $\dfrac{59}{7}$ **45.** 5

47. $\displaystyle\sum_{i=1}^{6} i^2 + \sum_{i=1}^{6} 3i + \sum_{i=1}^{6} 5$ **48.** $3\displaystyle\sum_{i=1}^{6} i$ **49.** $6 \cdot 5 = 30$ **50.** $\displaystyle\sum_{i=1}^{n} i = \dfrac{n(n + 1)}{2}$ **51.** $\displaystyle\sum_{i=1}^{n} i^2 = \dfrac{n(n + 1)(2n + 1)}{6}$

52. $91 + 3(21) + 6(5) = 184$ **53.** 572   **54.** -2620 **55.** $a = 6, d = 2$ **57.** 10

SECTION 11.2   (PAGE 661)

EXERCISES   3. $d = 3$ **5.** $d = -4$ **7.** not arithmetic   **9.** $d = \dfrac{1}{2}$ **11.** not arithmetic

13. $a_n = 5n - 3$ **15.** $a_n = \dfrac{9}{4} + \dfrac{3}{4}n$ **17.** $a_n = 3n - 6$ **19.** $m + b, 2m + b, 3m + b$ **20.** yes   **21.** m

22. $a_n = mn + b$ **23.** 76   **25.** 48   **27.** -1 **29.** 16   **31.** 6   **35.** 90   **37.** -90 **39.** 159   **41.** 136

43. $\dfrac{35}{2}$ **45.** 2,001,000   **47.** 2500   **49.** 6   **51.** \$4.65   **53.** 18   **55.** $\dfrac{1}{2}$

SECTION 11.3   (PAGE 669)

EXERCISES   3. $r = 2$ **5.** not geometric   **7.** $r = -3$ **9.** $r = -\dfrac{1}{2}$ **11.** $a_n = 5(2)^{n-1}$ **13.** $a_n = \dfrac{3^{n-1}}{9}$

15. $a_n = 10\left(-\dfrac{1}{5}\right)^{n-1}$ **17.** ab, ab^2, ab^3 **18.** yes; b **19.** $a_n = ab^n$ **21.** $2(5)^9$ **23.** $\dfrac{1}{2}\left(\dfrac{1}{3}\right)^{11}$ **25.** $2\left(\dfrac{1}{2}\right)^{24} = \dfrac{1}{2^{23}}$

27. $\dfrac{121}{243}$ **29.** -1.997 **31.** 2.662   **33.** -2.982 **35.** 9   **37.** $\dfrac{10{,}000}{11}$ **39.** $-\dfrac{9}{20}$ **41.** does not exist

43. $10\left(\dfrac{3}{5}\right)^4 \approx 1.3$ feet   **45.** $\left(\dfrac{2}{3}\right)^6 = \dfrac{64}{729}$ units   **47. (a)** $1.1(1.02)^5 \approx 1.21$ billion units   **(b)** about 35 years

49. $9x^2 + 12xy + 4y^2$ **51.** $a^3 - 3a^2b + 3ab^2 - b^3$

SECTION 11.4   (PAGE 676)

◆ **CONNECTIONS**   **Page 672: 1.** 21 and 34   **2.** 15, 21, 28, 36, 45   **3.** Answers will vary.

EXERCISES   1. The quantity raised to the nth power is a binomial.   **3.** $x^5 + 5x^4r + 10x^3r^2 + 10x^2r^3 + 5xr^4 + r^5$

5. $p^4 - 4p^3q + 6p^2q^2 - 4pq^3 + q^4$ **7.** $\dfrac{x^4}{16} - \dfrac{x^3y}{2} + \dfrac{3x^2y^2}{2} - 2xy^3 + y^4$ **9.** $x^8 + 4x^6 + 6x^4 + 4x^2 + 1$

11. $8p^6 - 12p^4q^2 + 6p^2q^4 - q^6$ **13.** $m^{20} - 20m^{19}n + 190m^{18}n^2 - 1140m^{17}n^3$

15. $2^{11}p^{11} + 33(2^{10})p^{10}q + 495(2^9)p^9q^2 + 4455(2^8)p^8q^3$ **17.** $x^{30} - 15x^{28}y^2 + 105x^{26}y^4 - 455x^{24}y^6$

19. $120(2^7)m^7n^3$ **21.** $\dfrac{7x^2y^6}{16}$ **23.** $36k^7$ **25.** $-160x^6y^3$ **27.** $4320\, x^9y^4$

CHAPTER 11 REVIEW EXERCISES (PAGE 679)

1. $-1, 1, 3, 5$ **3.** $1, 4, 9, 16$ **5.** $0, 3, 8, 15$ **7.** $2x + 3x^2 + 4x^3 + 5x^4 + 6x^5 + 7x^6$ **9.** 126 **11.** -4

13. arithmetic; $d = 4$ **15.** geometric; $r = -1$ **17.** geometric; $r = \dfrac{1}{2}$ **19.** arithmetic; $d = \dfrac{1}{2}$ **21.** 69

23. $a_n = 9 - 3n$ **25.** 22 **27.** 164 **29.** $a_n = -1(4)^{n-1} = -(4)^{n-1}$ **31.** $2(-3)^{10} = 118{,}098$ **33.** $\dfrac{341}{1024}$

35. $\dfrac{242}{81}$ **37.** does not exist   **39.** $x^8 + 12x^6y + 54x^4y^2 + 108x^2y^3 + 81y^4$

41. $m^{20} + 10m^{19}n + \dfrac{95m^{18}n^2}{2} + \dfrac{285m^{17}n^3}{2}$ **43.** $-18{,}400(3)^{22}k^3$ **45.** $-3(-2)^9 = 1536$; $2^{10} - 1 = 1023$

47. 38; 95 **49.** $a_n = 2(4)^{n-1}$ **51.** $a_n = 15 - 3n$ **53.** $\$78$ **55.** $50{,}000(.97)^6 \approx 42{,}000$

CHAPTER 11 TEST (PAGE 681)

[11.1] 1. $0, 2, 0, 2, 0$ **2.** $6, 12, 20, 30, 42$ **[11.2–11.3] 3.** $4, 6, 8, 10, 12$ **4.** $13, 9, 5, 1, -3$

5. $2, -6, 18, -54, 162$ **6.** $48, 24, 12, 6, 3$ **7.** 0 **8.** 29 **9.** $\dfrac{16}{9}$ **10.** $\dfrac{64}{3}$ or $-\dfrac{64}{3}$ **11.** 50

12. 75 **13.** $\dfrac{279}{2}$ **14.** 124 or 44 **15.** 70 **16.** 33 **17.** $125{,}250$ **18.** 42 **19.** $\dfrac{1}{3}$ **20.** does not exist

[11.4] 21. $\dfrac{14{,}080x^8y^4}{9}$ **22.** $81k^4 - 540k^3 + 1350k^2 - 1500k + 625$ **[11.3] 23.** $20(3^{11})$ **[11.2] 24.** $\$324$

25. It has a sum if $|r| < 1$.

CUMULATIVE REVIEW (Chapters 1–11) (PAGE 682)

1. $10, 0, \dfrac{45}{15}, -3$ **2.** $-\dfrac{8}{3}, 10, 0, \dfrac{45}{15}, .82, -3$ **3.** $\sqrt{13}, -\sqrt{3}$ **4.** all except $\sqrt{-7}$ **5.** 8 **6.** -35

7. -55 **8.** $\left\{\dfrac{1}{6}\right\}$ **9.** $[10, \infty)$ **10.** $\left\{-\dfrac{9}{2}, 6\right\}$ **11.** $\{9\}$ **12.** $\left(-3, \dfrac{3}{2}\right)$ **13.** $(-\infty, -3] \cup [8, \infty)$

14. $20p^2 - 2p - 6$ **15.** $9k^2 - 42k + 49$ **16.** $-5m^3 - 3m^2 + 3m + 8$ **17.** $2t^3 + 3t^2 - 4t + 2 + \dfrac{3}{3t - 2}$

18. $x(7 + x^2)$ **19.** $(7y - 4)(2y + 3)$ **20.** $z(3z + 4)(2z - 1)$ **21.** $(7a^2 + 3b)(7a^2 - 3b)$

22. $(c + 3d)(c^2 - 3cd + 9d^2)$ **23.** $(8r + 3q)^2$ **24.** $-\dfrac{27p^2}{10}$ **25.** $\dfrac{x + 7}{x - 2}$ **26.** $\dfrac{-2k - 23}{(k + 4)(k - 1)}$

27. $\dfrac{3p - 26}{p(p + 3)(p - 4)}$ **28.** $\varnothing$ **29.** 2 pounds   **30.** $\dfrac{9}{4}$ **31.** $\dfrac{1}{5^5}$ or $\dfrac{1}{3125}$ **32.** $10\sqrt{2}$ **33.** 73

34. $\left\{\dfrac{-5 + \sqrt{217}}{12}, \dfrac{-5 - \sqrt{217}}{12}\right\}$ **35.** $[-2, 3]$ **36. (a)** 2.98 million **(b)** 4.69 million **(c)** 6.4 million **(d)** 2000

37. $\dfrac{3}{4}$ **38.** $3x + y = 4$ **39.**   **40.**   **41.**   **42.**

43. $(x + 5)^2 + (y - 12)^2 = 81$ **44.** $\{(3, -5)\}$ **45.** $\{(2, 1, 4)\}$ **46.** -15 **47.** 7

48. $f^{-1}(x) = \dfrac{x - 5}{9}$ **49.**   **50.** $\left\{\dfrac{5}{2}\right\}$ **51.**   **52.** $\{2\}$ **53.** $-7, -2, 3, 8, 13$

54. 78 **55.** $\dfrac{75}{7}$ **56.** 30 **57.** $32a^5 - 80a^4 + 80a^3 - 40a^2 + 10a - 1$

Index

FORMULAS

FIGURE	FORMULAS	EXAMPLES
Square	Perimeter: $P = 4s$ Area: $A = s^2$	
Rectangle	Perimeter: $P = 2L + 2W$ Area: $A = LW$	
Triangle	Perimeter: $P = a + b + c$ Area: $A = \dfrac{1}{2}bh$	
Pythagorean Formula (for right triangles)	In a right triangle with legs a and b and hypotenuse c, $$c^2 = a^2 + b^2.$$	
Sum of the Angles of a Triangle	$A + B + C = 180°$	
Circle	Diameter: $d = 2r$ Circumference: $C = 2\pi r$ $C = \pi d$ Area: $A = \pi r^2$	
Parallelogram	Area: $A = bh$ Perimeter: $P = 2a + 2b$	
Trapezoid	Area: $A = \dfrac{1}{2}(B + b)h$ Perimeter: $P = a + b + c + B$	